Animal Lectins: Form, Function and Clinical Applications

G.S. Gupta

Animal Lectins: Form, Function and Clinical Applications

In Collaboration with Anita Gupta and Rajesh K. Gupta

Principal author
G.S. Gupta, Ph. D.
Panjab University
Chandigarh 160014
India

ISBN 978-3-7091-1064-5 ISBN 978-3-7091-1065-2 (eBook)
DOI 10.1007/978-3-7091-1065-2
Springer Wien Heidelberg New York Dordrecht London

Library of Congress Control Number: 2012945422

Printed on acid-free paper

Springer is part of Springer Science+Business Media (www.springer.com)

Foreword I

Lectins are typically carbohydrate-binding proteins that are widely distributed in Nature. With the growing interest in the field of glycobiology, the body of research related to animal lectins has grown at an explosive rate, particularly in the past 25 years. However, the Lectinology field is still relatively young, nascent, and evolving. "**Animal Lectins: Form, Function and Clinical Applications**" presents the most up-to-date analysis of these carbohydrate - binding, and potentially lifesaving proteins in two comprehensive volumes. Lectionology is an exciting area of research that has helped immensely in our understanding of host-pathogen interactions. Importantly, C-type lectins can act as pattern-recognition receptors (PRR) that sense invading pathogens.

The interactions between lectins and carbohydrates have been shown to be involved in diverse activities such as opsonization of microbes, cell adhesion and migration, cell activation and differentiation, and apoptosis. Developments in the area of lectin research has opened a new aspect in studying the immune system, and at the same time, provided new therapeutic routes for the treatment and prevention of diseases.

This present book on animal lectins discusses the biochemical and biophysical properties of animal lectins at length along with their functions in health and diseases. Importantly, the potential interrelationships between lectins of the innate immune system and latent viruses that reside within host cells, sometimes integrated into the genome have been beautifully highlighted. These interactions help to explain autoimmune diseases and shed light on the development of cancer diagnostics. The present book on animal lectins presents new insights into the biological roles of most animal lectins, including their role in prevention of infections through innate immunity. The contents of "**Animal Lectins: Form, Function and Clinical Applications**" provide functional explanation for the enormous diversity of glycan structures found on animal cells. There are still several other areas wherein lectins and their specificities are not well defined and the biological functions of the interactions remain elusive, thereby underscoring the need for further research. The book offers novel ideas for students of Immunology, Microbiology, as well as young researchers in the area of Biochemistry. I congratulate the authors in completing this truly enormous task.

Prof. N. K. Ganguly, MD, DSc (hc), FMed Sci (London),
FRC Path (London), FAMS, FNA, FASc, FNASc,
FTWAS (Italy), FIACS (Canada), FIMSA

President, JIPMER
Former Director General, ICMR
Former Director, PGIMER, Chandigarh

Pondicherry, India

Foreword II

Lectins are proteins that bind to soluble carbohydrates as well as to the functional groups of carbohydrate chains that are part of a glycoprotein or glycolipid found on the surfaces of cells. Lectins are known to be widespread in nature and play a role in interactions and communication between cells typically for recognition. Carbohydrates on the surface of one cell bind to the binding sites of lectins on the surface of another cell. For example, some bacteria use lectins to attach themselves to the cells of the host organism during infection. Binding results from numerous weak interactions which come together to form a strong attraction. A lectin usually contains two or more binding sites for carbohydrate units. In addition, the carbohydrate-binding specificity of a certain lectin is determined by the amino acid residues that bind to the carbohydrates. Most of the lectins are essentially nonenzymic in action and nonimmune in origin. They typically agglutinate certain animal cells and/or precipitate glycoconjugates. Plant lectins are highly resistant to breakdown from heat or digestion. They provide a defense for plants against bacteria, viruses, and other invaders, but can create problems for humans. In animals, lectins regulate the cell adhesion to glycoprotein synthesis, control protein levels in blood, and bind soluble extracellular and intracellular glycoproteins. Also, in the immune system, lectins recognize carbohydrates found specifically on pathogens or those that are not recognizable on host cells. Embryos are attached to the endometrium of the uterus through L-selectin. This activates a signal to allow for implantation. *E. coli* are able to reside in the gastrointestinal tract by lectins that recognize carbohydrates in the intestines. The influenza virus contains hemagglutinin, which recognizes sialic acid residues on the glycoproteins located on the surface of the host cell. This allows the virus to attach and gain entry into the host cell. Clinically, purified lectins can be used to identify glycolipids and glycoproteins on an individual's red blood cells for blood typing.

The journey of *Animal Lectins: Form, Function, and Clinical Applications*, essentially the Encyclopedia of Animal Lectins, starts with an introductory chapter on lectin families, in general, followed by specific animal lectin families, such as intracellular sugar-binding ER chaperones, calnexin and calreticulin; the P-type lectins working in the endocytic pathway, the lectins of ERAD pathway, and the complex mannose-binding ERGIC-53 protein and its orthologs circulating between the ER, ERGIC, and the Golgi apparatus; and the fairly small galectins that are synthesized in the cytosol but may be found at many locations. Chapters on R-type lectin families, pentraxins, siglecs, C-type lectins, regenerating gene family, tetranectin group of lectins, ficolins, F-type lectins, and chi-lectins have covered the up-to-date literature on the subject and contain illustrations of their structures, functions, and clinical applications. The emerging group of annexins as lectins has been given a place as a separate family.

The C-type lectins comprising 17 subfamilies such as collectins, selectins, NK cell lectin receptors, latest discoveries of lectins on dendritic cells and others form the backbone of Volume 2. The chapters are well written, although there is variability on how they are focused. Most of the chapters focus on lectin structures, functions, their ligands, and their medical relevance in terms of diagnosis and therapy. The journey ends with five reviews on clinical applications of lectins with a survey of literature on endogenous lectins as drug targets. This may reflect state of the art in each area or the interests of the author(s).

In post-genomic years, with the human genome sequence at hand, a complete overview of many human lectin genes has become available, as illustrated by C-type lectin domain 3D structures having as little as 30% amino acid sequence similarity. The major effort in the future will be on elaboration of similar studies on other families of lectins with carbohydrate specificities and in vivo lectin–ligand interactions. In this respect, further writings will bring the reader to the forefront of knowledge in the field. Thus, on thirst of learning about animal recognition systems, *Animal Lectins: Form, Function, and Clinical Applications* is an excellent reference book for those studying biochemistry, biotechnology, and biophysics with a specialization in the areas of immunology, lectinology, and glycobiology as well as for pharmacy students involved in drug discoveries through lectin–carbohydrate interactions. Both novice and advanced researchers in biomedical, analytical, and pharmaceutical fields need to understand animal lectins.

T.P. Singh, F.N.A., F.A.Sc., F.N.A.Sc., F.T.W.A.S
DBT-Distinguished Biotechnology Research Professor
Department of Biophysics,
All India Institute of Medical Sciences, New Delhi, India

Preface

Lectins are phylogenetically ancient proteins that have specific recognition and binding functions for complex carbohydrates of glycoconjugates, that is, of glycoproteins, proteoglycans/glycosaminoglycans, and glycolipids. They occur ubiquitously in nature and typically agglutinate certain animal cells and/or precipitate glycoconjugates without affecting their covalent linkages. Lectins mediate a variety of biological processes, such as cell–cell and host–pathogen interactions, serum–glycoprotein turnover, and innate immune responses. Although originally isolated from plant seeds, they are now known to be ubiquitously distributed in nature. The successful completion of several genome projects has made amino acid sequences of several lectins available. Their tertiary structures provide a good framework upon which all other data can be integrated, enabling the pursuit of the ultimate goal of understanding these molecules at the atomic level. With growing interest in the field of glycobiology, the function–structure relations of animal lectins have increased at an explosive rate, particularly in the last 20 years. Since lectins mediate important processes of adhesion and communication both inside and outside the cells in association with their ligands and associated co-receptor proteins, there is a need of a reference book that can describe emerging applications on the principles of structural biology of animal lectins at one point. No book has ever described in a coordinated fashion structures, functions, and clinical applications of 15 families of animal lectins presently known. Therefore, with the increasing information on animal lectins in biomedical research and their therapeutic applications, writing of a comprehensive document on animal lectins and associated proteins in the form of *Animal Lectins: Form, Function, and Clinical Applications* has been the main objective of the present work. The entire manuscript has been distributed into two volumes in order to produce an easily readable work with easy portability. Volume 1 comprises most of the superfamilies (Chaps. 1–21) of lectins, excluding C-type lectins or C-type-lectin-like domain. In volume 2, we have mainly focused on C-type lectins, which have been extensively studied in vertebrates with wider clinical applications (Chaps. 22–46).

Animal Lectins: Form, Function, and Clinical Applications reviews the current knowledge of animal lectins, their ligands, and associated proteins with a focus on their structures and functions, biochemistry and patho-biochemistry (protein defects as a result of disease), cell biology (exocytosis and endocytosis, apoptosis, cell adhesion, and malignant transformation), clinical applications, and their intervention for therapeutic purposes. The book emphasizes on the effector functions of animal lectins in innate immunity and provides reviews/chapters on extracellular animal lectins, such as C-type lectins, R-type lectins, siglecs, and galectins, and intracellular lectins, such as calnexin family (M-type, L-type, and P-type), recently discovered F-box lectins, ficolins, chitinase-like lectins, F-type lectins, and intelectins, mainly in vertebrates. The clinical significance of lectin–glycoconjugate interactions has been exemplified by inflammatory diseases, defects of immune defense, autoimmunity, infectious diseases, and tumorigenesis/metastasis, along with therapeutic perspectives of novel drugs that interfere with lectin–carbohydrate interactions.

Based on the information gathered on animal lectins in this book, a variety of medical and other applications are in the offing. Foremost among these is the lectin-replacement therapy for patients suffering from lectin deficiency defects. We have pointed out the advancements of such studies where such progress has been made. Other uses in different stages of development are antibacterial drugs; multivalent hydrophobic carbohydrates for anti-adhesion therapy of microbial diseases; highly effective inhibitors of the selectins for treatment of leucocyte-mediated pathogenic conditions, such as asthma, septic shock, stroke, and myocardial infarction; inhibitors of the galectins and other lectins involved in metastasis; and application of lectins for facile and improved disease diagnosis. Recent advances in the discovery of M6PR-homologous protein family (Chap. 5), lectins of ERAD pathway, F-box proteins and M-type lectins (Chap. 6), and mannose receptor–targeted drugs and vaccines (Chaps. 8 and 46) can form the basis of cutting-edge technology in drug delivery devices.

Based on the structures of animal lectins, classified into at least 15 superfamilies, C-type lectins and galectins are the classical major families. Galectins are known to be associated with carcinogenesis and metastasis. Galectin-3 is a pleiotropic carbohydrate-binding protein, involved in a variety of normal and pathological biological processes. Its carbohydrate-binding properties constitute the basis for cell–cell and cell–matrix interactions (Chap. 12) and cancer progression (Chap. 13). Studies lead to the recognition of galectin-3 as a diagnostic/prognostic marker for specific cancer types, such as thyroid and prostate. In interfering with galectin–carbohydrate interactions during tumor progression, a current challenge is the design of specific galectin inhibitors for therapeutic purposes. Anti-galectin agents can restrict the levels of migration of several types of cancer cells and should, therefore, be used in association with cytotoxic drugs to combat metastatic cancer (Chap. 13). The properties of siglecs that make them attractive for cell-targeted therapies have been reviewed in Chaps. 16, 17, and 46.

F-type domains are found in proteins from a range of organisms from bacteria to vertebrates, but exhibit patchy distribution across different phylogenetic taxa, suggesting that F-type lectin genes have been selectively lost even between closely related lineages, thus making it difficult to trace the ancestry of the F-type domain. The F-type domain has clearly gained functional value in fish, whereas the fate of F-type domains in higher vertebrates is not clear, rather it has become defunct. Two genes encoding three-domain F-type proteins are predicted in the genome of the opossum (*Monodelphis domestica*), an early-branching mammal. There is plenty of scope to discover F-type lectins in mammalian vertebrates (Chap. 20). Since the reports on the C-reactive protein (CRP) as a cardiovascular marker (Chap. 8), novel biomarkers in cardiovascular and other inflammatory diseases have emerged in recent years. The substantial knowledge on CRP is now being complemented by new markers such as YKL-40, a member of chi-lectins group of CTLD (Chap. 19). The YKL-40 (chitinase-3-like protein 1, or human cartilage glycoprotein-39) displays a typical fold of family 18 glycosyl hydrolases and is expressed and secreted by several types of solid tumors, including glioblastoma, colon cancer, breast cancer, and malignant melanoma. Chitinase-3-like protein 1 was recently introduced into clinical practice; yet its application is still restricted.

In volume 2, we have mainly focused on C-type lectins, which have been extensively studied in vertebrates. A C-type lectin is a type of carbohydrate-binding protein domain that requires calcium for binding interactions in general. Drickamer et al. classified C-type lectins into seven subgroups (I to VII) based on the order of the various protein domains in each protein. This classification was subsequently updated in 2002, leading to seven additional groups (VIII to XIV). A further three subgroups (XV to XVII) were added recently. The C-type lectins share structural homology in their high-affinity carbohydrate recognition domains (CRDs) and constitute a large and diverse group of extracellular proteins that have been extensively studied. Their activities have been implicated as indispensable players in carbohydrate recognition, suggesting their possible application in discrimination of various correlative microbes and developing biochemical tools. The C-type lectins, structurally characterized by a double loop composed of two highly conserved disulfide bridges located at the bases of the loops, are believed to mediate pathogen recognition and play important roles in the innate immunity of both vertebrates and invertebrates. A large number of these proteins have been characterized and more than 80 have been sequenced. Recent data on the primary sequences and 3D structures of C-type lectins have enabled us to analyze their molecular evolution. Statistical analysis of their cDNA sequences shows that C-type-lectin-like proteins, with some exceptions, have evolved in an accelerated manner to acquire their diverse functions.

The C-type lectin family includes the monocyte mannose receptor (MMR), mannose-binding lectin (MBL), lung surfactant proteins, ficolins, selectins, and others, which are active in immune functions and pathogen recognition (Chaps. 23–34 and 41–45). Several C-type lectins and lectin-like receptors have been characterized that are expressed abundantly on the surface of professional antigen-presenting cells (APCs). Dendritic cells (DCs) are equipped with varying sets of C-type lectin receptors that help them with the uptake of pathogens. Important examples are langerin, DC-SIGN, DC-SIGNR, DCAR, DCIR, dectins, DEC-205, and DLEC (Chaps. 34–36). DCs are key regulators in directing the immune responses and, therefore, are under extensive research for the induction of antitumor immunity. They scan their surroundings for recognition and uptake of pathogens. Intracellular routing of antigens through C-type lectins enhances loading and presentation of antigens through MHC class I and II, inducing antigen-specific $CD4^+$ and $CD8^+$ T-cell proliferation and skewing T-helper cells. These characteristics make C-type lectins interesting targets for DC-based immunotherapy. Extensive research has been performed on targeting specific tumor antigens to C-type lectins, using either antibodies or natural ligands such as glycan structures. In Chaps. 34–36, we have presented the current knowledge of DC receptors to exploit them for antitumor activity and drug targeting in the near future (Chap. 46).

The monocyte mannose receptor (MMR) or the mannose receptor (MR) (CD206) is a member of the Group VI C-type lectins along with ENDO180, DEC205, and the phospholipase A2 receptor. Expressed on a broad range of cell types, including tissue macrophages and various epithelial cells, the MMR is active in endocytosis and phagocytosis. It is also thought to be involved in innate immunity, though its exact role remains unclear. Structurally, MMR is a complex molecule, which has been reviewed as an R-type lectin in volume 1 (Chap. 15) and as a C-type lectin in volume 2 (Chap. 35). Further research is required to fully understand the function of MMR. In addition, the Reg family constitutes an interesting subset of the C-type lectin family. The Reg family members are small, secreted proteins, which have been implicated in a range of physiological processes such as acute phase reactants and survival/growth factors for insulin-producing pancreatic β-cells, neural cells, and epithelial cells of the digestive system (Chap. 39). The C-type lectin DC-SIGN is unique in the regulation of adhesion processes, such as DC trafficking and T-cell synapse formation, besides its well-studied function in antigen capture. In particular, the DC-SIGN and associated homologues contribute to the potency of DC to control immunity (Chaps. 36 and 46). There is always significant interest in the development of drug and antigen delivery systems via the oral route due to patient compliance and acceptability. The presence of DCs with knowledge of associated receptors in the gastrointestinal tract offers principles of methodology for the development of oral vaccines (Chap. 46).

The search of the database of NCBI revealed that the C-type lectins attract much more attention, which resulted in recent discoveries of novel groups of lectins (Groups XV–XVII; Chap. 40). The clinical applications of C-type lectins have been exemplified in Chaps. 42–46. Although a variety of lectins have enabled greater insight into the diversity and complexity of lectin repertoires in vertebrates in two volumes, the nature of the protein–carbohydrate interactions and the potential mechanisms of different functions for invertebrate lectins are under intense investigation. Future progress will elucidate the contribution of different lectin families and their cross talk with each other or with other molecules with respect to mounting protective immune responses in invertebrates and vertebrates.

MBL as a reconstitution therapy in genetically determined MBL deficiency has advanced significantly. Since the genetically determined MBL deficiency is very common and can be associated with increased susceptibility to a variety of infections, the potential benefits of MBL reconstitution therapy still need to be evaluated. In a phase I trial on MBL-deficient healthy adult volunteers, MBL did not show adverse clinical effects (Chap. 23). SP-A and SP-D have been recently categorized as "Secretory Pathogen Recognition Receptors." Treatment with a recombinant fragment of human SP-D consisting of a short collagen-like stalk (but not the entire collagen-like domain of native SP-D), neck, and CRD inhibited development of emphysema-like pathology in SP-D-deficient mice (Chaps. 24, 25, and 43). Autosomal dominant polycystic kidney disease (ADPKD) is a common inherited nephropathy, affecting over 1:1000 of the population worldwide. It is a systemic condition with frequent hepatic and cardiovascular manifestations in addition to the progressive development of fluid-filled renal cysts that eventually result in loss of renal function in the majority of affected individuals. The cysts that grow in the kidneys of the majority of ADPKD patients are the result of mutations within the genes *PKD1* and *PKD2* that code for polycystin-1 (PC-1) and PC-2, respectively (Chap. 45). The annexins or lipocortins are a multigene family of proteins that bind to acidic phospholipids and biological membranes. Some of the annexins bind to glycosaminoglycans (GAGs) in a Ca^{2+}-dependent manner and function as recognition elements for GAGs in extracellular space. The emerging groups of C-type lectins include layilin, tetranectin, and chondrolectin (Group VIII of CTLD) (Chap. 40) and CTLD-containing protein - CBCP in Group XVII. Fras1, QBRICK/Frem1, Frem2, and Frem3 form the family of Group XVII (Chap. 41).

There is plenty of scope to discover lectins in invertebrates and amphibians which offer novel biomaterials useful in therapeutics, with a hope that the list of native lectins as well as genetically modified derivatives will grow with time. Thus, understanding animal lectins and the associated network of proteins is of high academic value for those working in the field of protein chemistry and designing new drugs on the principle of protein–carbohydrate or protein–protein interactions. Refined information on the sites of interactions on glycoproteins in toto with lectins is the subject of future study. Efforts are being made to develop an integrated knowledge-based animal lectins database together with appropriate analytical tools. Thus, *Animal Lectins: Form, Function, and Clinical Applications*, the Encyclopedia of Vertebrate Lectins, is unique in its scope and differs from earlier publications on animal lectins. It is more than lectinology and is suitable to the students and researchers working in the areas of biochemistry, glycobiology, biotechnology, biophysics, microbiology and immunology, pharmaceutical chemistry, biomedicine, and animal sciences in general.

January 2012 **G.S. Gupta**

Acknowledgments

We are thankful to Professor R. C. Sobti, the Vice-Chancellor, and Professor M. L. Garg, Chairman, Department of Biophysics, Punjab University, Chandigarh, for providing the facilities from time to time to complete this project. Authors gratefully acknowledge the assistance of the following scientists for providing the reprints and the literature on the subject:

Dr. Borrego F, Rockville, USA; Dr. Clark DA, Hamilton, Canada; Dr. Dimaxi N, Genova, Italy; Dr. Doan LG, Golden, Colorado, USA; Dr. Dorner T, Berlin, Germany; Dr. Ezekowitz RAB, Boston, USA; Dr. Gabor DF, Vienna, Austria; Dr. Garg ML, Chandigarh, India; Dr. Girbes T, Valladolid, Spain; Dr. Goronzy JJ, Atlanta, USA; Dr. Hofer E, Vienna, Austria; Dr. Jepson MA, Bristol, UK; Dr. Kishore U, West London, UK; Dr. Le Bouteiller P, Toulouse, France; Dr. Markert U, Jena, Germany; Dr. Mincheva-Nilsson L, Umea, Sweden; Dr. Monsigny M, Orleans, France; Dr. Nan-Chi Chang, Taipei, Taiwan; Dr. Natarajan K, Bethesda, USA; Dr. Osborn HMI, Reading, UK; Dr. Palecanda A, Bozeman, USA; Dr. Petroff MG, Kansas City, USA; Dr. Piccinni MP, Florence, Italy; Dr. Radaev S, Rockville, USA; Dr. Rajbinder Kaur Virk, Chandigarh, India; Dr. Roos A, Leiden, the Netherlands; Dr. Sano H, Sapporo, Japan; Dr. Schon MP, Wurzburg, Germany; Dr. Sharon N, Rehovot, Israel; Dr. Shwu-Huey Liaw, Taipei, Taiwan; Dr. Singh TP, New Delhi, India; Dr. Soilleux EJ, Cambridge, UK; Dr. Steinhubl SR, Lexington, USA; Dr. Steinman RM, New York, USA; Dr. Zenclussen AC, Berlin, Germany; and Dr.Valladeau J, Lyon, France.

Anita Gupta acknowledges the technical help of Dr. Neerja Mittal and Ms. Sargam Preet from RBIEBT. G. S. Gupta is grateful to Mrs. Kishori Gupta for the moral support and patience at the time of need.

G.S. Gupta
Anita Gupta
Rajesh K. Gupta

Contents of Volume 1

Part 1 Introduction

1 Lectins: An Overview . . . 3
G.S. Gupta
1.1 Lectins: Characteristics and Diversity . . . 3
1.1.1 Characteristics . . . 3
1.1.2 Lectins from Plants . . . 4
1.1.3 Lectins in Microorganisms . . . 5
1.1.4 Animal Lectins . . . 5
1.2 The Animal Lectin Families . . . 5
1.2.1 Structural Classification of Lectins . . . 5
1.3 C-Type Lectins (CLEC) . . . 8
1.3.1 Identification of CLEC . . . 8
1.3.2 C-Type Lectin Like Domain (CTLD/CLRD) . . . 8
1.3.3 Classification of CLRD-Containing Proteins . . . 9
1.3.4 The CLRD Fold . . . 9
1.4 Disulfide Bonds in Lectins and Secondary Structure . . . 10
1.4.1 Disulfide Bond . . . 10
1.4.2 Pathway for Disulfide Bond Formation in the ER of Eukaryotic Cells . . . 10
1.4.3 Arrangement of Disulfide Bonds in CLRDs . . . 10
1.4.4 Functional Role of Disulfides in CLRD of Vertebrates . . . 11
1.4.5 Disulfide Bonds in Ca^{2+}-Independent Lectins . . . 12
1.5 Functions of Lectins . . . 12
1.5.1 Lectins in Immune System . . . 13
1.5.2 Lectins in Nervous Tissue . . . 15
1.6 The Sugar Code and the Lectins as Recepors in System Biology . . . 15
1.6.1 Host-Pathogen Interactions . . . 16
1.6.2 Altered Glycosylation in Cancer Cells . . . 17
1.6.3 Protein-Carbohydrate Interactions in Immune System . . . 18
1.6.4 Glycosylation and the Immune System . . . 19
1.6.5 Principles of Protein-Glycan Interactions . . . 19
1.7 Applications of Lectin Research and Future Perspectives . . . 21
1.7.1 Mannose Receptor-Targeted Vaccines . . . 21
References . . . 22

Part II Intracellular Lectins

2 **Lectins in Quality Control: Calnexin and Calreticulin** 29
G.S. Gupta
2.1 Chaperons 29
2.1.1 Calnexin 30
2.1.2 Calnexin Structure 30
2.1.3 Calnexin Binds High-Mannose-Type Oligosaccharides 31
2.1.4 Functions of Calnexin 32
2.1.5 Patho-Physiology of Calnexin Deficiency 35
2.2 Calreticulin 36
2.2.1 General Features 36
2.2.2 The Protein 37
2.2.3 Cellular Localization of Calreticulin 40
2.2.4 Functions of Calreticulin 41
2.2.5 Structure-Function Relationships in Calnexin and Calreticulin 48
2.2.6 Pathophysiological Implications of Calreticulin 48
2.2.7 Similarities and Differences Between Cnx and Crt 50
2.2.8 Calreticulin in Invertebrates 51
References 52

3 **P-Type Lectins: Cation-Dependent Mannose-6-Phosphate Receptor** 57
G.S. Gupta
3.1 The Biosynthetic/Secretory/Endosomal Pathways 57
3.1.1 Organization of Secretory Pathway 57
3.2 P-Type Lectin Family: The Mannose 6-Phosphate Receptors 62
3.2.1 Fibroblasts MPRs 63
3.2.2 MPRs in Liver 63
3.2.3 MPRs in CNS 63
3.2.4 CI-MPR in Bone Cells 64
3.2.5 Thyroid Follicle Cells 64
3.2.6 Testis and Sperm 64
3.2.7 MPRs During Embryogenesis 65
3.3 Cation-Dependent Mannose 6-Phosphate Receptor 65
3.3.1 CD-MPR- An Overview 65
3.3.2 Human CD-MPR 66
3.3.3 Mouse CD-MPR 66
3.4 Structural Insights 67
3.4.1 N-Glycosylation Sites in CD-MPR 67
3.4.2 3-D Structure of CD-MPR 67
3.4.3 Carbohydrate Binding Sites in MPRs 70
3.4.4 Similarities and Dis-similarities between two MPRs 70
3.5 Functional Mechanisms 71
3.5.1 Sorting of Cargo at TGN 71
3.5.2 TGN Exit Signal Uncovering Enzyme 72
3.5.3 Association of Clathrin-Coated Vesicles with Adaptor Proteins 72
3.5.4 Role of Di-leucine-based Motifs in Cytoplasmic Domains 73
3.5.5 Sorting Signals in Endosomes 74
3.5.6 Palmitoylation of CD-MPR is Required for Correct Trafficking 74
References 75

4 P-Type Lectins: Cation-Independent Mannose-6-Phosphate Reeptors 81
G.S. Gupta
4.1 Cation-Independent Mannose 6-Phosphate Receptor (CD222) 81
4.1.1 Glycoprotein Receptors for Insulin and Insulin-like Growth Factors 81
4.2 Characterization of CI-MPR/IGF2R 81
4.2.1 Primary Structures of Human CI-MPR and IGF2R Are Identical . . . 81
4.2.2 Mouse IGF2R/CI-MPR Gene 82
4.2.3 Bovine CI-MPR 82
4.2.4 CI-MPR in Other Species 83
4.3 Structure of CI-MPR 84
4.3.1 Domain Characteristics of IGF2R/CI-MPR (M6P/IGF2R) 84
4.3.2 Crystal Structure 85
4.4 Ligands of IGF2R/CI-MPR 86
4.4.1 Extracellular Ligands of IGF2R/CI-MPR 86
4.4.2 Binding Site for M6P in CI-MPR 86
4.4.3 The Non-M6P-Containing Class of Ligands 89
4.5 Complementary Functions of Two MPRS 90
4.5.1 Why Two MPRs 91
4.5.2 Cell Signaling Pathways 92
4.6 Functions of CI-MPR 92
4.7 Proteins Associated with Trafficking of CI-MPR 94
4.7.1 Adaptor Protein Complexes 94
4.7.2 Mammalian TGN Golgins 95
4.7.3 TIP47: A Cargo Selection Device for MPR Trafficking 95
4.7.4 Sorting Signals in GGA and MPRs at TGN 95
4.8 Retrieval of CI-MPR from Endosome-TO-GOLGI 97
4.8.1 Endosome-to-Golgi Retrieval of CIMPR Requires Retromer Complex 97
4.8.2 Retromer Complex and Sorting Nexins (SNX) 98
4.8.3 Small GTPases in Lysosome Biogenesis and Transport 100
4.8.4 Role for Dynamin in Late Endosome Dynamics and Trafficking of CI-MPR 100
4.9 CI-MPR/IGF2R System and Pathology 100
4.9.1 Deficiency of IGF2R/CI-MPR Induces Myocardial Hypertrophy . . . 100
4.9.2 MPRs in Neuromuscular Diseases 101
4.9.3 CI-MPR in Fanconi syndrome 101
4.9.4 Tumor Suppressive Effect of CI-MPR/IGF2R 101
References 102

5 Mannose-6-Phosphate Receptor Homologous Protein Family 109
G.S. Gupta
5.1 Recognition of High-Mannose Type N-Glycans in ERAD Pathways 109
5.2 Proteins Containing M6PRH Domains 110
5.3 GlcNAc-Phosphotransferase 110
5.4 α-Glucosidase II 112
5.4.1 Function of α-Glucosidase II 112
5.4.2 M6PRH Domain in GIIβ 113
5.4.3 Two Distinct Domains of β-Subunit of GII Interact with α-Subunit 113
5.4.4 Polycystic Liver Disease (PCLD) and β-Subunit of Glucosidase II 114

5.5 Osteosarcomas-9 (OS-9) 115
5.5.1 The Protein 115
5.5.2 Requirement of HRD1, SEL1L, and OS-9/XTP3-B for Disposal of ERAD-Substrates 115
5.5.3 OS-9 Recognizes Mannose-Trimmed N-Glycans 116
5.5.4 Dual Task for Xbp1-Responsive OS-9 Variants 116
5.5.5 Interactions of OS-9 116
5.6 YOS9 from *S. cerevisiae* 118
5.7 Erlectin/XTP3-B 119
5.8 *Drosophila* Lysosomal Enzyme Receptor Protein (LERP) 119
5.9 MRL1 119
References 120

6 Lectins of ERAD Pathway: F-Box Proteins and M-Type Lectins 123
G.S. Gupta
6.1 Intracellular Functions of N-Linked Glycans in Quality Control 123
6.2 The Degradation Pathway for Misfolded Glycoproteins 124
6.2.1 Endoplasmic Reticulum-Associated Degradation (ERAD) 124
6.2.2 Ubiquitin-Mediated Proteolysis 124
6.2.3 SCF Complex 125
6.2.4 F-Box Proteins: Recognition of Target Proteins by Protein-Protein Interactions 126
6.3 F-Box Proteins with a C-Terminal Sugar-Binding Domain (SBD) 127
6.3.1 Diversity in SCF Complex due to Lectin Activity of F-Box Proteins 127
6.3.2 Fbs Family 128
6.3.3 Fbs1 Equivalent Proteins 131
6.3.4 Ligands for F-Box Proteins 133
6.3.5 Evolution of F-Box Proteins 134
6.3.6 Localization of F-Box Proteins 134
6.3.7 Regulation of F-Box Proteins 134
6.4 α-Mannosidases and M-Type Lectins 135
6.4.1 α-Mannosidases 135
6.4.2 ER-associated Degradation-enhancing α-Mannosidase-like Proteins (EDEMs) 135
6.4.3 Functions of M-Type Lectins in ERAD 136
6.5 Derlin-1, -2 and -3 138
References 139

Part III L-Type Lectins

7 L-Type Lectins in ER-Golgi Intermediate Compartment 145
G.S. Gupta
7.1 L-Type Lectins 145
7.1.1 Lectins from Leguminous Plants 145
7.1.2 L-Type Lectins in Animals and Other Species 145
7.2 ER-Golgi Intermediate Compartment 146
7.3 Lectins of Secretory Pathway 146
7.4 ER-Golgi Intermediate Compartment Marker-53 (ERGIC-53) or LMAN1 . . 147
7.4.1 ERGIC-53 Is Mannose-Selective Human Homologue of Leguminous Lectins 147
7.4.2 Cells of Monocytic Lineage Express MR60: A Homologue of ERGIC-53 148

7.4.3 Rat Homologue of ERGIC53/MR60 . . . 149
7.4.4 Structure-Function Relations . . . 150
7.4.5 Functions of ERGIC-53 . . . 152
7.4.6 Mutations in ERGIC-53 *LMAN1* Gene and Deficiency of Coagulation Factors V and VIII lead to bleeding disorder . . . 154
7.5 Vesicular Integral Membrane Protein (VIP36) OR LMAN2 . . . 156
7.5.1 The Protein . . . 156
7.5.2 VIP36-SP-FP as Cargo Receptor . . . 157
7.5.3 Structure for Recognition of High Mannose Type Glycoproteins by VIP36 . . . 157
7.5.4 Emp47p of *S. cerevisiae:* A Homologue to VIP36 and ERGIC-53 . . . 158
7.6 VIP36-Like (VIPL) L-Type Lectin . . . 158
References . . . 159

8 Pentraxins: The L-Type Lectins and the C-Reactive Protein as a Cardiovascular Risk . . . 163
G.S. Gupta
8.1 Pentraxins and Related Proteins . . . 163
8.2 Short Pentraxins . . . 163
8.3 C-Reactive Protein . . . 164
8.3.1 General . . . 164
8.3.2 CRP Protein . . . 164
8.3.3 Structure of CRP . . . 165
8.3.4 Functions of CRP . . . 166
8.4 CRP: A Marker for Cardiovascular Risk . . . 168
8.4.1 CRP: A Marker for Inflammation and Infection . . . 168
8.4.2 CRP: A Marker for Cardiovascular Risk . . . 169
8.4.3 Role of Modified/Monomeric CRP . . . 170
8.5 Extra-Hepatic Sources of CRP . . . 171
8.6 Serum Amyloid P Component . . . 171
8.6.1 Genes Encoding SAP . . . 171
8.6.2 Characterization of SAP . . . 172
8.6.3 Interactions of SAP and CRP . . . 173
8.6.4 Functions of SAP . . . 174
8.6.5 SAP in Human Diseases . . . 175
8.6.6 SAP from *Limulus Polyphemus* . . . 176
8.7 Female Protein (FP) in Syrian Hamster . . . 176
8.7.1 Similarity of Female Protein to CRP and APC . . . 176
8.7.2 Structure of Female Protein . . . 177
8.7.3 Gene Structure and Expression of FP . . . 177
8.8 Long Pentraxins . . . 178
8.8.1 Long Pentraxins 1, -2, -3 . . . 178
8.9 Neuronal Pentraxins (Pentraxin-1 and -2) . . . 178
8.9.1 Functions of Pentraxin 1 and -2 . . . 178
8.10 Pentraxin 3 (PTX3) . . . 179
8.10.1 Characterization of PTX3 . . . 179
8.10.2 Cellular Sources of PTX3 . . . 180
8.10.3 Ligands . . . 180
8.10.4 Regulation of PTX3 . . . 181
8.10.5 Functions of PTX3 . . . 182
References . . . 183

Part IV Animal Galectins

9 Overview of Animal Galectins: Proto-Type Subfamily 191
Anita Gupta and G.S. Gupta
9.1 Galectins 191
9.2 Galectin Sub-Families 191
9.3 Galectin Ligands 192
9.4 Functions of Galectins 192
9.4.1 Functional Overlap/Divergence Among Galectins 193
9.4.2 Cell Homeostasis by Galectins 193
9.4.3 Immunological Functions 195
9.4.4 Signal Transduction by Galectins 196
9.4.5 Common Structural Features in Galectins 197
9.4.6 Galectin Subtypes in Tissue Distribution 198
9.5 Prototype Galectins (Mono-CRD Type) 199
9.5.1 Galectin-1 199
9.5.2 Galectin-2 199
9.5.3 Galectin-5 202
9.5.4 Galectin-7 202
9.5.5 Galectin-10 (Eosinophil Charcot-Leyden Crystal Protein) 203
9.5.6 Galectin-Related Inter-Fiber Protein (Grifin/Galectin-11) 204
9.5.7 Galectin-13 (Placental Protein -13) 205
9.5.8 Galectin 14 206
9.5.9 Galectin-15 206
9.6 Evolution of Galectins 207
9.6.1 Phylogenetic Analysis of Galectin Family 207
9.6.2 Galectins in Lower Vertebrates 207
References 208

10 Galectin-1: Forms and Functions 213
Anita Gupta
10.1 The Subcellular Distribution 213
10.2 Molecular Characteristics 213
10.2.1 Galectin-1 Gene 213
10.2.2 X-Ray Structure of Human Gal-1 214
10.2.3 Gal-1 from Toad (*Bufo arenarum Hensel*) Ovary 216
10.2.4 GRIFIN Homologue in Zebrafish (DrGRIFIN) 216
10.3 Regulation of *Gal-1* Gene 216
10.3.1 Gal-1 in IMP1 Deficient Mice 216
10.3.2 Blimp-1 Induces Galectin-1 Expression 217
10.3.3 Regulation by Retinoic Acid 217
10.3.4 Regulation by TGF-β, IL-12 and FosB Gene Products 218
10.3.5 Regulation by Metabolites/Drugs/Other Agents 218
10.4 Gal-1 in Cell Signaling 220
10.5 Ligands/Receptors of Gal-1 220
10.5.1 Each Galectin Recognizes Different Glycan Structures 220
10.5.2 Interactions of Galectin-1 221
10.6 Functions of Galectin-1 223
10.6.1 Role of Galectin-1 in Apoptosis 223
10.6.2 Gal-1 in Cell Growth and Differentiation 225
10.6.3 Gal-1 and Ras in Cell Transformation 227
10.6.4 Development of Nerve Structure 228
10.6.5 Skeletal Muscle Development 232

10.6.6 Gal-1 and the Immune System . . . 232
10.6.7 Role of Galetin-1 and Other Systems . . . 235
References . . . 236

11 Tandem-Repeat Type Galectins . . . 245
Anita Gupta
11.1 Galectin 4 . . . 245
11.1.1 Localization and Tissue Distribution . . . 245
11.1.2 Galectin-4 Isoforms . . . 246
11.1.3 Gal-4 from Rodents and Other Animals . . . 247
11.1.4 Ligands for Galectin-4 . . . 248
11.1.5 Functions of Galectin-4 . . . 249
11.1.6 Galectin-4 in Cancer . . . 251
11.2 Galectin-6 . . . 251
11.3 Galectin-8 . . . 251
11.3.1 Galectin-8 Characteristics . . . 251
11.3.2 Functions of Galectin-8 . . . 253
11.3.3 Clinical Relevance of Gal-8 . . . 253
11.3.4 Isoforms of Galectin-8 in Cancer . . . 253
11.4 Galectin-9 . . . 254
11.4.1 Characteristics . . . 254
11.4.2 Stimulation of Galectin-9 Expression by IFN-γ . . . 254
11.4.3 Crystal Structure of Galectin-9 . . . 255
11.4.4 Galectin-9 Recognizes *L. major* Poly-β-galactosyl Epitopes . . . 255
11.4.5 Functions of Galectin-9 . . . 255
11.4.6 Galectin-9 in Clinical Disorders . . . 257
11.4.7 Galectin-9 in Cancer . . . 258
11.5 Galectin-12 . . . 259
References . . . 260

12 Galectin-3: Forms, Functions, and Clinical Manifestations . . . 265
Anita Gupta
12.1 General Characteristics . . . 265
12.1.1 Galectin-3 Structure . . . 265
12.1.2 Galectin-3 Gene . . . 266
12.1.3 Tissue and Cellular Distribution . . . 266
12.2 Ligands for Galectin-3: Binding Interactions . . . 268
12.2.1 Extracellular Matrix and Membrane Proteins . . . 268
12.2.2 Intracellular Ligands . . . 268
12.2.3 Carbohydrate Binding . . . 269
12.2.4 Carbohydrate-Independent Binding . . . 271
12.3 Functions . . . 271
12.3.1 Galectin-3 is a Multifunctional Protein . . . 271
12.3.2 Role in Cell Adhesion . . . 272
12.3.3 Gal-3 at the Interface of Innate and Adaptive Immunity . . . 272
12.3.4 Regulation of T-Cell Functions . . . 274
12.3.5 Pro-apoptotic and Anti-apoptotic Effects . . . 274
12.3.6 Role in Inflammation . . . 276
12.3.7 Gal-3 in Wnt Signaling . . . 278
12.3.8 In Urinary System of Adult Mice . . . 278
12.3.9 Gal-3 in Reproductive Tissues . . . 278
12.3.10 Gal-3 on Chondrocytes . . . 278
12.3.11 Role of Gal-3 in Endothelial Cell Motility and Angiogenesis . . . 279
12.3.12 Role in CNS . . . 279

12.4 Clinical Manifestations of Gal-3 279
12.4.1 Advanced Glycation End Products (AGES) 279
12.4.2 GAL-3 and Protein Kinase C in Cholesteatoma 280
12.4.3 Gal-3 and Cardiac Dysfunction 280
12.4.4 Gal-3 and Obesity 281
12.4.5 Autoimmune Diseases 281
12.4.6 Myofibroblast Activation and Hepatic Fibrosis 282
12.5 Gal-3 as a Pattern Recognition Receptor 282
12.5.1 Gal-3 Binds to *Helicobacter pylori* 282
12.5.2 Recognition of *Candida albicans* by Macrophages Requires Gal-3 282
12.5.3 Gal-3 is Involved in Murine Intestinal Nematode and Schistosoma Infection 283
12.5.4 Up-Regulation of Gal-3 and Its Ligands by *Trypanosoma cruzi* Infection 283
12.6 Gal-3 as a Therapeutic Target 283
12.6.1 Gal-3: A Target for Anti-inflammatory/Anticancer Drugs 283
12.7 Xenopus-Cortical Granule Lectin: A Human Homolog of Gal-3 284
References 284

13 Galectin-3: A Cancer Marker with Therapeutic Applications 291
Anita Gupta
13.1 Galectin-3: A Prognostic Marker of Cancer 291
13.2 Discriminating Malignant Tumors from Benign Nodules of Thyroid 291
13.2.1 Large-Needle Aspiration Biopsy 292
13.2.2 Fine-Needle Aspiration Biopsy 292
13.2.3 Combination of Markers 293
13.2.4 Hashimoto's Thyroiditis 294
13.3 Breast Cancer 294
13.4 Tumors of Nervous System 295
13.4.1 Galectins and Gliomas 295
13.5 Diffuse Large B-Cell Lymphoma 296
13.6 Gal-3 in Melanomas 297
13.7 Head and Neck Carcinoma 298
13.8 Lung Cancer 298
13.9 Colon Neoplastic Lesions 299
13.10 Expression of Gal-3 in Other Tumors 300
13.11 Gal-3 in Metastasis 302
13.12 β1,6 N-acetylglucosaminyltransferase V in Carcinomas 302
13.13 Macrophage Binding Protein 303
13.14 Galectinomics 303
13.15 Mechanism of Malignant Progression by Galectin-3 304
13.16 Anti-Galectin Compounds as Anti-Cancer Drugs 305
References 306

Part V R-Type Animal Lectins

14 R-Type Lectin Families 313
Rajesh K. Gupta and G.S. Gupta
14.1 Ricinus Communis Lectins 313
14.1.1 Properties of Ricin 313
14.1.2 Other R-Type Plant Lectins 314
14.2 R-Type Lectins in Animals 315
14.3 Mannose Receptor Family 315

14.4 UDP-Galnac: Polypeptide α-N-Acetyl-galactosaminyltransferases 316
14.4.1 Characteristics of UDP-GalNAc: α-N-Acetylgalactosaminyltransferases 316
14.4.2 The Crystal Structure of Murine ppGalNAc-T-T1 318
14.4.3 Parasite ppGalNAc-Ts 319
14.4.4 Crystal Structure of CEL-III from *Cucumaria echinata* Complexed with GalNAc 320
14.5 Microbial R-Type Lectins 321
14.5.1 *S. olivaceoviridis* E-86 Xylanase: Sugar Binding Structure 321
14.5.2 The Mosquitocidal Toxin (MTX) from *Bacillus sphaericus* 321
14.6 R-Type Lectins in Butterflies 322
14.6.1 Pierisin-1 322
14.6.2 Pierisin-2, Pierisin-3 and -4 324
14.7 Discoidin Domain and Carbohydrate-Binding Module 324
14.7.1 The Discoidin Domain 324
14.7.2 Discoidins from *Dictyostelium discoideum* (DD) 325
14.7.3 Discoidin Domain Receptors (DDR1 and DDR2) 326
14.7.4 Earth Worm (EW)29 Lectin 327
References 327

15 Mannose Receptor Family: R-Type Lectins 331
Rajesh K. Gupta and G.S. Gupta
15.1 R-Type Lectins in Animals 331
15.2 Mannose Receptor Lectin Family 331
15.3 The Mannose Receptor (CD206) 332
15.3.1 Human Macrophage Mannose Receptor (MMR) 332
15.3.2 Structure-Function Relations 332
15.3.3 Cell and Tissue Distribution 334
15.3.4 Ligands 335
15.3.5 Functions of Mannose Receptor 336
15.3.6 Mouse Mannose Receptor 337
15.3.7 Interactions of MR with Branched Carbohydrates 339
15.3.8 Mannose Receptor-Targeted Drugs and Vaccines 339
15.4 Phospholipase A2-Receptors 339
15.4.1 The Muscle (M)-Type sPLA2 Receptors 340
15.4.2 Neuronal or N-Type PLA2 Receptor 342
15.5 DEC-205 (CD205) 342
15.5.1 Characterization 342
15.5.2 Functions of DEC-205 343
15.6 ENDO 180 (CD280)/uPARAP 343
15.6.1 Urokinase Receptor (uPAR)-Associated Protein 343
15.6.2 Interactions of Endo180 344
References 345

Part VI I-Type Lectins

16 I-Type Lectins: Sialoadhesin Family 351
G.S. Gupta
16.1 Sialic Acids 351
16.2 Sialic Acid-Binding Ig-Like Lectins (I-Type Lectins) 352
16.2.1 Two Subsets of Siglecs 352
16.2.2 Siglecs as Inhibitory Receptors 353
16.2.3 Binding Characteristics of Siglecs 353
16.2.4 Siglecs of Sialoadhesin Family 354

16.3 Sialoadhesin (Sn)/Siglec-1 (CD169) . . . 355
16.3.1 Characterization of Sialoadhesin/Siglec-1 . . . 355
16.3.2 Cellular Expression of Sialoadhesin . . . 355
16.3.3 Ligands for Sialoadhesin . . . 356
16.3.4 Sialoadhesin Structure . . . 357
16.3.5 Regulation of Sialoadhesin . . . 358
16.3.6 Functions of Sialoadhesin . . . 358
16.3.7 Lessons from Animal Experiments . . . 359
16.3.8 Interactions with Pathogens . . . 360
16.4 CD22 (Siglec-2) . . . 361
16.4.1 Characterization and Gene Organization . . . 361
16.4.2 Functional Characteristics . . . 361
16.4.3 Ligands of CD22 . . . 362
16.4.4 Regulation of CD22 . . . 364
16.4.5 Functions of CD22 . . . 365
16.4.6 Signaling Pathway of Human CD22 and Siglec-F in Murine . . . 365
16.4.7 CD22 as Target for Therapy . . . 367
16.5 Siglec-4 [Myelin-Associated Glycoprotein, (MAG)] . . . 367
16.5.1 MAG and Myelin Formation . . . 367
16.5.2 Characteristics of MAG . . . 368
16.5.3 MAG Isoforms . . . 368
16.5.4 Ligands of MAG: Glycan Specificity of MAG . . . 369
16.5.5 Functions of MAG . . . 370
16.5.6 MAG in Demyelinating Disorders . . . 371
16.5.7 Inhibitors of Regeneration of Myelin . . . 372
16.5.8 Axonal Regeneration by Overcoming Inhibitory Activity of MAG . . . 372
16.5.9 Fish Siglec-4 . . . 373
16.6 Siglec-15 . . . 373
References . . . 373

17 CD33 (Siglec 3) and CD33-Related Siglecs . . . 381
G.S. Gupta
17.1 Human CD33 (Siglec-3) . . . 381
17.1.1 Human CD33 (Siglec-3): A Myeloid-specific Inhibitotry Receptor . . . 381
17.2 CD33-Related Siglecs (CD33-rSiglecs) . . . 382
17.2.1 CD33-Related Siglecs Family . . . 382
17.2.2 CD33-rSiglec Structures . . . 382
17.2.3 Organization of CD33-rSiglec Genes on Chromosome 19q13.4 . . . 384
17.3 Siglec-5 (CD170) . . . 384
17.3.1 Characterization . . . 385
17.3.2 Siglec-5: An Inhibitory Receptor . . . 385
17.3.3 Siglec-5-Mediated Sialoglycan Recognition . . . 385
17.4 Siglec-6 . . . 386
17.4.1 Cloning and Gene Organization of Siglec-6 (OB-BP1) . . . 386
17.4.2 Siglec-6 (OB-BP1) and Reproductive Functions . . . 387
17.5 Siglec-7 (p75/AIRM1) . . . 387
17.5.1 Characterization . . . 387
17.5.2 Cytoplasmic Domain of Siglec-7 (p75/AIRM1) . . . 387
17.5.3 Crystallographic Analysis . . . 388
17.5.4 Interactions of Siglec-7 . . . 388
17.5.5 Functions of Siglec-7 . . . 389
17.6 Siglec-8 . . . 389
17.6.1 Characteristics and Cellular Specificity . . . 389
17.6.2 Ligands for Siglec-8 . . . 390

17.6.3 Functions in Apoptosis 390
17.6.4 Siglec-8 in Alzheimer's Disease 391
17.7 Siglec-9 391
17.7.1 Characterization and Phylogenetic Analysis 391
17.7.2 Functions of Siglec-9 391
17.8 Siglec-10, -11, -12, and -16 392
17.8.1 Siglec-10 392
17.8.2 Siglec-11 392
17.8.3 Siglec-12 393
17.8.4 Siglec-16 394
17.9 Mouse Siglecs 394
17.9.1 Evolution of Mouse and Human CD33-rSiglec Gene Clusters . . . 394
17.9.2 Mouse CD33/Siglec-3 394
17.9.3 Siglec-3-Related Siglecs in Mice 395
17.10 Glycoconjugate Binding Specificities of Siglecs 397
17.11 Functions of CD33-Related Siglecs 399
17.11.1 Endocytosis 399
17.11.2 Phagocytosis of Apoptotic Bodies 400
17.12 Siglecs as Targets for Immunotherapy 400
17.13 Molecular Diversity and Evolution of Siglec Family 401
References 402

Part VII Novel Super-Families of Lectins

18 Fibrinogen Type Lectins 409
Anita Gupta
18.1 Ficolins 409
18.1.1 Ficolins versus Collectins 409
18.1.2 Characterization of Ficolins 409
18.1.3 Ligands of Ficolins 411
18.1.4 X-ray Structures of M, L- and H-Ficolins 412
18.1.5 Functions of Ficolins 413
18.1.6 Pathophysiology of Ficolins 415
18.2 Tachylectins 415
18.2.1 Horseshoe Crab Tachylectins 415
18.2.2 X-ray Structure 416
References 417

19 Chi-Lectins: Forms, Functions and Clinical Applications 421
Rajesh K. Gupta and G.S. Gupta
19.1 Glycoside Hydrolase Family 18 Proteins in Mammals 421
19.1.1 Chitinases 421
19.2 Chitinase-Like Lectins: Chi-Lectins 421
19.3 YKL-40 [Chitinase 3-Like Protein 1 (CHI3L1)] 422
19.3.1 The Protein 422
19.3.2 Cell Distribution and Regulation 422
19.3.3 Ligands of YKL-40 423
19.3.4 The Crystal Structure of YKL-40 423
19.4 Human Cartilage 39-KDA Glycoprotein (or YKL-39)/(CHI3L2) 425
19.5 Ym1 and Ym2: Murine Proteins 426
19.5.1 The Protein 426
19.5.2 Crystal Structure of Ym1 427
19.5.3 Oviductin 427

19.6 Functions of CHI3L1 (YKL-40) 428
19.6.1 Role in Remodeling of Extracellular Matrix and Defense Mechanisms 428
19.6.2 Growth Stimulating Effect 428
19.7 Chi-Lectins As Markers of Pathogenesis 428
19.7.1 A Marker of Inflammation 428
19.7.2 CHI3L1 as Biomarker in Solid Tumors 431
19.7.3 Chitinase 3-Like-1 Protein (CHI3L1) or YKL-40 in Clinical Practice 432
19.8 Evolution of Mammalian Chitinases (-Like) of GH18 Family 433
References 434

20 Novel Groups of Fuco-Lectins and Intlectins 439
Rajesh K. Gupta and G.S. Gupta
20.1 F-type Lectins (Fuco-Lectins) 439
20.1.1 F-type Lectins in Mammalian Vertebrates 439
20.2 F-type Lectins in Fish and Amphibians 440
20.2.1 Anguilla Anguilla Agglutinin (AAA) 440
20.2.2 FBP from European Seabass 442
20.2.3 Other F-type Lectins in Fish 442
20.2.4 F-Type Lectins in Amphibians 443
20.3 F-type Lectins in Invertebrates 444
20.3.1 Tachylectin-4 444
20.3.2 F-type Lectins from *Drosophila melanogaster* 445
20.3.3 F -Type Lectins in Sea Urchin 445
20.3.4 Bindin in Invertebrate Sperm 445
20.4 F-type Lectins in Plants 445
20.5 F-type Lectins in Bacteria 446
20.6 Fuco-Lectins in Fungi 447
20.6.1 Fuco-Lectin from Aleuria Aurantia (AAL) 447
20.7 Intelectins 448
20.7.1 Intelectin-1 (Endothelial Lectin HL-1/Lactoferrin Receptor or *Xenopus* Oocyte Lectin) 448
20.7.2 Intelectin-2 (HL-2) and Intelectin-3 449
20.7.3 Intelectins in Fish 450
20.7.4 Eglectin (XL35) or Frog Oocyte Cortical Granule Lectins 450
References 451

21 Annexins (Lipocortins) 455
G.S. Gupta
21.1 Annexins 455
21.1.1 Characteristics of Annexins 455
21.1.2 Classification and Nomenclature 456
21.1.3 Annexins in Tissues 457
21.1.4 Functions of Annexins 458
21.2 Annexin Family Proteins and Lectin Activity 459
21.3 Annexin A2 (p36) 459
21.3.1 Annexin 2 Tetramer (A2t) 460
21.3.2 Crystal Analysis of Sugar-Annexin 2 Complex 461
21.3.3 Functions of Annexin A2 461
21.4 Annexin A4 (p33/41) 461
21.4.1 General Characteristics 461
21.4.2 Tissue Distribution 462
21.4.3 Characterization 463

21.4.4 Pathophysiology 463
21.4.5 Doublet p33/41 Protein 463
21.5 Annexin A5/Annexin V 464
21.5.1 Gene Encoding Human Annexin A5 464
21.5.2 Interactions of Annexin A5 464
21.5.3 Molecular Structure of Annexin A5 465
21.5.4 Annexin A5-Mediated Pathogenic Mechanisms 466
21.5.5 A Novel Assay for Apoptosis 466
21.5.6 Calcium-Induced Relocation of Annexins 4 and 5 in the Human Cells 467
21.6 Annexin A6 (Annexin VI) 467
21.6.1 Structure 467
21.6.2 Functions 467
References 468

Contents of Volume 2

Part VIII C-Type Lectins: Collectins

22 C-Type Lectins Family 473
Anita Gupta and G.S. Gupta
22.1 C-Type Lectins Family 473
22.1.1 The C-Type Lectins (CLEC) 473
22.1.2 C-Type Lectin Like Domain (CLRD/CTLD) 473
22.1.3 The CLRD Fold 474
22.1.4 Ligand Binding 477
22.1.5 C-Type Lectin-Like Domains in Model Organisms 477
22.2 Classification of CLRD/CTLD-Containing Proteins 477
22.3 Disulfide Bonds in Lectins and Secondary Structure 478
22.3.1 Arrangement of Disulfide Bonds in CTLDs 478
22.3.2 Functional Role of Disulfides in CTLD of Vertebrates 478
22.4 Collectins 480
22.4.1 Collectins: A Group of Collagenous Type of Lectins 480
References 480

23 Collectins: Mannan-Binding Protein as a Model Lectin 483
Anita Gupta
23.1 Collectins 483
23.1.1 N-Terminal Region 484
23.1.2 Collagenous Region 484
23.1.3 C-Type Lectin Domain 485
23.1.4 Comparative Genetics of Collagenous Lectins 486
23.1.5 Generalized Functions of Collectins 486
23.2 Human Mannan-Binding Protein 487
23.2.1 Characterization of Serum MBL 487
23.2.2 Gene Structure of MBP 489
23.2.3 Regulation of MBP Gene 490
23.2.4 Structure-Function Relations 491
23.2.5 Functions of MBL 492
23.3 MBP/MBL from Other Species 494
23.3.1 Rodents 494
23.3.2 Primates MBL 495
23.4 Similarity Between C1Q and Collectins/Defense Collagens 495
23.4.1 Structural Similarities 495
23.4.2 Functional Similarities 496
23.4.3 Receptors for Defense Collagens 496
References 496

24 Pulmonary SP-A: Forms and Functions 501
Anita Gupta and Rajesh K. Gupta
24.1 Pulmonary Surfactant Proteins 501
24.1.1 Pulmonary Surfactant 501
24.1.2 Pulmonary Surfactant Protein A 501
24.2 Structural Properties of SP-A 502
24.2.1 Human SP-A: Domain Structure 502
24.2.2 Structural Biology of Rat SP-A 504
24.2.3 3-D Structure of SP-A Trimer 505
24.2.4 Bovine SP-A 506
24.2.5 SP-A in Other Species 507
24.3 Cell Surface Receptors for SP-A 508
24.4 Interactions of SP-A 509
24.4.1 Protein-Protein Interactions 509
24.4.2 Protein-Carbohydrate Interactions 510
24.4.3 SP-A Binding with Lipids 510
24.5 Functions of SP-A 511
24.5.1 Surfactant Components in Surface Film Formation 511
24.5.2 Recognition and Clearance of Pathogens 512
24.5.3 SP-A: As a Component of Complement System in Lung 513
24.5.4 Modulation of Adaptive Immune Responses by SPs 513
24.5.5 SP-A Stimulates Chemotaxis of AMΦ and Neutrophils 515
24.5.6 SP-A Inhibits sPLA2 and Regulates Surfactant Phospholipid Break-Down 515
24.5.7 SP-A Helps in Increased Clearance of Alveolar DPPC 515
24.5.8 Protection of Type II Pneumocytes from Apoptosis 516
24.5.9 Anti-inflammatory Role of SP-A 516
24.6 Reactive Oxygen and Nitrogen-Induced Lung Injury 518
24.6.1 Decreased Ability of Nitrated SP-A to Aggregate Surfactant Lipids 518
24.7 Non-Pulmonary SP-A 518
24.7.1 SP-A in Epithelial Cells of Small and Large Intestine 518
24.7.2 SP-A in Female Genital Tract and During Pregnancy 519
References 519

25 Surfactant Protein-D 527
Rajesh K. Gupta and Anita Gupta
25.1 Pulmonary Surfactant Protein-D (SP-D) 527
25.1.1 Human Pulmonary SP-D 528
25.1.2 Rat Lung SP-D 529
25.1.3 Mouse SP-D 530
25.1.4 Bovine SP-D 531
25.1.5 Porcine Lung SP-D 531
25.2 Interactions of SP-D 531
25.2.1 Interactions with Carbohydrates 531
25.2.2 Binding with Nucleic Acids 533
25.2.3 Interactions with Lipids 533
25.3 Structure: Function Relations of Lung SP-D 533
25.3.1 Role of NH_2 Domain and Collagenous Region 533
25.3.2 Role of NH2-Terminal Cysteines in Collagen Helix Formation 534
25.3.3 D4 (CRD) Domain in Phospholipid Interaction 534

25.3.4 A Three Stranded α-Helical Bundle at Nucleation Site of Collagen Triple-Helix Formation . . . 535
25.3.5 Ligand Binding Amino Acids . . . 535
25.3.6 Ligand Binding and Immune Cell-Recognition . . . 535
25.4 Regulation of Sp-D by Various Factors . . . 536
25.4.1 Glucocorticoids . . . 536
25.4.2 1α,25-Dihydroxyvitamin D3 . . . 536
25.4.3 Growth Factors . . . 536
25.5 SP-D in Human Fetal and Newborns Lungs . . . 537
25.6 Non-Pulmonary SP-D . . . 537
25.6.1 Human Skin and Nasal Mucosa . . . 537
25.6.2 Digestive Tract, Mesentery, and Other Organs . . . 537
25.6.3 Male Reproductive Tract . . . 538
25.6.4 Female Genital Tract . . . 538
25.7 Functions of Lung SP-D . . . 539
25.7.1 Innate Immunity . . . 539
25.7.2 Effects on Alveolar Macrophages . . . 539
25.7.3 Functions of Neutrophils . . . 540
25.7.4 Protective Role in Allergy and Infection . . . 541
25.7.5 Adaptive Immune Responses . . . 543
25.7.6 Apoptosis . . . 544
25.7.7 Other Effects of SP-D . . . 544
25.8 Oxidative Stress and Hyperoxia . . . 544
References . . . 545

Part IX C-Type Lectins: Selectins

26 L-Selectin (CD62L) and Its Ligands . . . 553
G.S. Gupta
26.1 Cell Adhesion Molecules . . . 553
26.1.1 Selectins . . . 553
26.2 Leukocyte-Endothelial Cell Adhesion Molecule 1 (LECAM-1) (L-Selectin/CD62L or LAM-1) . . . 554
26.2.1 Leukocyte-Endothelial Cell Adhesion Molecule 1 in Humans . . . 554
26.2.2 Gene Structure of L-Selectin . . . 555
26.2.3 Murine PLN Homing Receptor/mLHR3 . . . 555
26.3 Functions of L-Selectin . . . 556
26.3.1 Lymphocyte Homing and Leukocyte Rolling and Migration . . . 556
26.3.2 Immune Responses . . . 557
26.4 L-Selectin: Carbohydrate Interactions . . . 559
26.4.1 Glycan-Dependent Leukocyte Adhesion . . . 559
26.5 Cell Surface Ligands for L-Selectin . . . 561
26.5.1 Subsets of Sialylated, Sulfated Mucins of Diverse Origins are Recognized by L-Selectin . . . 561
26.5.2 GlyCAM-1 . . . 561
26.5.3 CD34 . . . 562
26.5.4 Mucosal Addressin Cell Adhesion Molecule-1 (MAdCAM-1) . . . 563
26.5.5 PSGL-1 Binds L-Selectin . . . 564
26.6 L-Selectin IN Pathological States . . . 564
26.6.1 Gene Polymorphism in L-Selectin . . . 564
26.6.2 Antitumor Effects of L-Selectin . . . 565

26.6.3 Autoimmune Diseases . . . 566
26.6.4 CD62L in Other Conditions . . . 568
26.7 Oligonucleotide Antagonists IN Therapeutic Applications . . . 569
26.7.1 Monomeric and Multimeric Blockers of Selectins . . . 569
26.7.2 Synthetic and Semisynthetic Oligosaccharides . . . 569
References . . . 570

27 P-Selectin and Its Ligands . . . 575
G.S. Gupta
27.1 Platelet Adhesion and Activation . . . 575
27.2 P-Selectin (GMP-140, PADGEM, CD62): A Member of Selectin Adhesion Family . . . 575
27.2.1 Platelets and Vascular Endothelium Express P-Selectin . . . 575
27.2.2 Human Granule Membrane Protein 140 (GMP-140)/P-Selectin . . . 576
27.2.3 Murine P-Selectin . . . 576
27.2.4 P-Selectin Promoter . . . 577
27.2.5 Structure-Function Studies . . . 578
27.2.6 P-Selectin-Sialyl Lewis^x Binding Interactions . . . 580
27.2.7 Functions of P-Selectin . . . 580
27.3 P-Selectin Glycoprotein Ligand-1 (PSGL-1) . . . 583
27.3.1 PSGL-1: The Major Ligand for P-Selectin . . . 583
27.3.2 Genomic Organization . . . 585
27.3.3 Specificity of PSGL-1 as Ligand for P-Selectin . . . 585
27.3.4 Structural Polymorphism in PSGL-1 and CAD Risk . . . 585
27.3.5 Carbohydrate Structures on PSGL-1 . . . 586
27.3.6 Role of PSGL-1 . . . 587
27.4 Other Ligands of P-Selectin . . . 588
References . . . 590

28 E-Selectin (CD62E) and Associated Adhesion Molecules . . . 593
G.S. Gupta
28.1 E-Selectin (Endothelial Leukocyte Adhesion Molecule 1: ELAM-1) . . . 593
28.1.1 Endothelial Cells Express E-Selectin (ELAM-1/CD62E) . . . 593
28.2 E-Selectin Genomic DNA . . . 593
28.2.1 Human E-Selectin . . . 593
28.2.2 E-Selectin Gene in Other Species . . . 594
28.3 E-Selectin Gene Regulation . . . 594
28.3.1 Transcriptional Regulation of CAMs . . . 594
28.3.2 Induction of E-Selectin and Associated CAMs by TNF-α . . . 595
28.3.3 IL-1-Mediated Expression of CAMs . . . 596
28.4 Factors in the Regulation of CAMs . . . 596
28.4.1 NF-kB: A Dominant Regulator . . . 596
28.4.2 Cyclic AMP Inhibits NF-kB-Mediated Transcription . . . 597
28.4.3 Peroxisome Proliferator-Activated Receptors . . . 598
28.4.4 Endogenous Factors Regulating CAM Genes . . . 599
28.4.5 Tat Protein Activates Human Endothelial Cells . . . 599
28.4.6 Reactive Oxygen Species (ROS) . . . 599
28.4.7 Role of Hypoxia . . . 600
28.5 Binding Elements in E-Selectin Promoter . . . 600
28.5.1 Transcription Factors Stimulating *E-sel* Gene . . . 600
28.5.2 CRE/ATF Element or NF-ELAM1 . . . 601
28.5.3 HMG-I(Y) Mediates Binding of NF-kB Complex . . . 602
28.5.4 Phased-Bending of E-Selectin Promoter . . . 602

28.6 E-Selectin Ligands . . . 602
28.6.1 Carbohydrate Ligands (Lewis Antigens) . . . 602
28.6.2 E-Selectin Ligand-1 . . . 604
28.6.3 PSGL-1 and Relating Ligands . . . 605
28.6.4 Endoglycan, a Ligand for Vascular Selectins . . . 607
28.6.5 L-Selectin as E-Selectin Ligand . . . 607
28.7 Structural Properties of E-Selectin . . . 608
28.7.1 Soluble E-Selectin: An Asymmetric Monomer . . . 608
28.7.2 Complement Regulatory Domains . . . 608
28.7.3 Three-Dimensional Structure . . . 609
28.8 Functions of E-Selectin . . . 609
28.8.1 Functions in Cell Trafficking . . . 609
28.8.2 E-Selectin in Neutrophil Activation . . . 610
28.8.3 P- and E-Selectin in Differentiation of Hematopoietic Cells . . . 610
28.8.4 Transmembrane Signaling in Endothelial Cells . . . 611
References . . . 612

Part X C-Type Lectins: Lectin Receptors on NK Cells

29 KLRB Receptor Family and Human Early Activation Antigen (CD69) . . . 619
Rajesh K. Gupta and G.S. Gupta
29.1 Lectin Receptors on NK Cell . . . 619
29.1.1 NK Cell Receptors . . . 619
29.1.2 NKC Gene Locus . . . 620
29.2 The Ever-Expanding Ly49 Receptor Gene Family . . . 620
29.2.1 Activating and Inhibitory Receptors . . . 620
29.2.2 Crystal Analysis of CTLD of Ly49I and comparison with Ly29A, NKG2D and MBP-A . . . 622
29.3 NK Cell Receptor Protein 1 (NKR-P1) or KLRB1 . . . 623
29.3.1 *Ly49* and *Nkrp1 (Klrb1)* Recognition Systems . . . 623
29.3.2 NKRP1 . . . 623
29.4 Human NKR-P1A (CD161) . . . 626
29.4.1 Cellular Localization . . . 626
29.4.2 Transcriptional Regulation . . . 626
29.4.3 Ligands of CD161/ NKR-P1 . . . 627
29.4.4 Signaling Pathways . . . 628
29.4.5 Functions of NKR-P1 . . . 629
29.5 NKR-P1A in Clinical Disorders . . . 630
29.5.1 Autoimmune Reactions . . . 630
29.5.2 Other Diseases . . . 631
29.5.3 NKR-P1A Receptor (CD161) in Cancer Cells . . . 632
29.6 Human Early Activation Antigen (CD69) . . . 632
29.6.1 Organization of CD69 Gene . . . 633
29.6.2 Src-Dependent Syk Activation of CD69-Mediated Signaling . . . 633
29.6.3 Crystal Analysis of CD69 . . . 634
References . . . 634

30 NKG2 Subfamily C (KLRC) . . . 639
Rajesh K. Gupta and G.S. Gupta
30.1 NKG2 Subfamily C (KLRC) . . . 639
30.1.1 NKG2 Gene Family and Structural Organization . . . 639
30.1.2 Human NKG2-A, -B, and -C . . . 640
30.1.3 Murine NKG2A, -B, -C . . . 640

30.1.4 NKG2 Receptors in Monkey 641
30.1.5 NKG2 Receptors in Other Species 642
30.1.6 Inhibitory and Activatory Signals 642
30.2 CD94 (KLRD1) 642
30.2.1 Human CD94 in Multiple Transcripts 642
30.2.2 Mouse CD94 643
30.3 Cellular Sources of NKG2/CD94 643
30.4 The Crystal Analysis of CD94 644
30.4.1 An Intriguing Model for CD94/NKG2 Heterodimer 644
30.5 CD94/NKG2 Complex 646
30.5.1 CD94 and NKG2-A Form a Complex for NK Cells 646
30.6 Acquisition of NK Cell Receptors 646
30.7 Regulation of CD94/NKG2 647
30.7.1 Transcriptional Regulation of NK Cell Receptors 647
30.7.2 Regulation by Cytokines 647
30.8 Functions of CD94/NKG2 648
30.8.1 CD94/NKG2 in Innate and Adaptive Immunity 649
30.8.2 Modulation of Anti-Viral and Anti-Tumoral Responses of γ/δ T Cells 649
30.9 NK Cells in Female Reproductive Tract and Pregnancy 650
30.9.1 Decidual NK Cell Receptors 650
30.10 Signal Transduction by CD94/NKG2 651
30.10.1 Engagement of CD94/NKG2-A by HLA-E and Recruitment of Phosphatases 651
30.11 Ligands for CD94/NKG2 651
30.11.1 HLA-E as Ligand for CD94/NKG2A 651
30.11.2 CD94/NKG2-A Recognises HLA-G1 652
30.11.3 Non-Classical MHC-I Molecule Qa-1b as Ligand 653
30.12 Structure Analysis of CD94-NKG2 Complex 653
30.12.1 Crystal Analysis of NKG2A/CD94: HLA-E Complex 653
30.12.2 CD94-NKG2A Binding to HLA-E 654
30.13 Inhibitory Receptors in Viral Infection 656
30.14 Pathophysiological Role of CD94/NKG2 Complex 658
30.14.1 Polymorphism in NKG2 Genes 658
30.14.2 Phenotypes Associated with Leukemia 658
30.14.3 CD94/NKG2A on NK Cells in T Cell Lymphomas 659
30.14.4 CD94/NKG2 Subtypes on Lymphocytes in Melanoma Lesions 659
30.14.5 Cancers of Female Reproductive Tract 660
30.14.6 Disorders of Immune System 660
References 661

31 NKG2D Activating Receptor 667
Rajesh K. Gupta and G.S. Gupta
31.1 NKG2D Activating Receptor (CD314, Synonyms KLRK1) 667
31.2 Characteristics of NKG2D 667
31.2.1 The Protein 667
31.2.2 Orthologues to Human NKG2D 668
31.3 NKG2D Ligands 668
31.3.1 The Diversity of NKG2D Ligands 668
31.3.2 MHC Class I Chain Related (MIC) Proteins 669

31.3.3 Retinoic Acid Early (RAE) Transcripts . . . 669
31.3.4 Role of NKG2D Ligands . . . 671
31.3.5 Regulation of Ligands . . . 671
31.4 Crystal Structure of NKG2D . . . 672
31.4.1 Structures of NKG2D-Ligand Complexes . . . 672
31.5 DAP10/12 Adapter Proteins . . . 674
31.5.1 DAP10 . . . 674
31.5.2 KARAP (DAP12 or TYROBP) . . . 674
31.5.3 Characterization of DAP10 and DAP12 . . . 674
31.5.4 NKG2D Receptor Complex and Signaling . . . 675
31.6 Functions of NKG2D . . . 677
31.6.1 Engagement of NKG2D on γδ T Cells and Cytolytic Activity . . . 677
31.6.2 NKG2D: A Co-stimulatory Receptor for Naive $CD8^+$ T Cells . . . 678
31.6.3 NKG2D in Cytokine Production . . . 678
31.6.4 Heterogeneity of NK Cells in Umbilical Cord Blood . . . 679
31.7 Cytotoxic Effector Function and Tumor Immune Surveillance . . . 679
31.7.1 Anti-Tumor Activity . . . 679
31.7.2 Immune Evasion Mechanisms . . . 680
31.8 NKG2D in Immune Protection and Inflammatory Disorders . . . 681
31.8.1 NKG2D Response to HCMV . . . 681
31.8.2 HTLV-1-Associated Myelopathy . . . 682
31.8.3 Protection Against Bacteria . . . 682
31.8.4 Autoimmune Disorders . . . 682
31.9 Decidual/Placental NK Cell Receptors in Pregnancy . . . 683
31.10 Regulation of NKG2D Functions . . . 684
31.10.1 NKG2D Induction by Chronic Exposure to NKG2D Ligand . . . 684
31.10.2 Effects of Cytokines and Other Factors . . . 685
31.11 Role in Immunotherapy . . . 686
References . . . 686

32 KLRC4, KLRG1, and Natural Cytotoxicity Receptors . . . 693
Rajesh K. Gupta and G.S. Gupta
32.1 Killer Cell Lectin-Like Receptor F-1 (NKG2F/KLRC4/*CLEC5C*/NKp80) . . . 693
32.1.1 NKp80 or Killer Cell Lectin-Like Receptor Subfamily F-Member 1 (KLRF1) . . . 693
32.1.2 NKp80/KLRF1 Associates with DAP12 . . . 694
32.2 Activation-Induced C-Type Lectin (CLECSF2): Ligand for NKp80 . . . 694
32.3 KLRG1 (Rat MAFA/CLEC15A or Mouse 2F1-Ag) . . . 695
32.3.1 Mast Cell Function-Associated Antigen (MAFA) . . . 695
32.3.2 MAFA Is a Lectin . . . 695
32.3.3 Interactions of MAFA with FcεR . . . 695
32.3.4 MAFA Gene . . . 696
32.4 KLRG1 (OR 2F1-AG): A Mouse Homologue of MAFA . . . 696
32.4.1 KLRG1: A Mouse Homologue of MAFA . . . 696
32.4.2 NK Cell Maturation and Homeostasis Is Linked to KLRG1 Up-Regulation . . . 698
32.4.3 Cadherins as Ligands of KLRG1 . . . 699
32.5 MAFA-Like Receptor in Human NK Cell (MAFA-L) . . . 699
32.5.1 Characteristics and Biological Properties . . . 699
32.6 Killer Cell Lectin Like Receptor Subfamily E, Member 1 (KLRE1) . . . 699
32.7 KLRL1 from Human and Mouse DCs . . . 700

32.8 Natural Cytotoxicity Receptors 701
32.8.1 NK Cell Triggering: The Activating Receptors 701
32.8.2 Activating Receptors and Their Ligands 701
References 703

Part XI C-Type Lectins: Endocytic Receptors

33 Asialoglycoprotein Receptor and the Macrophage Galactose-Type Lectin . . . 709
Anita Gupta
33.1 Asialoglycoprotein Receptor: The First Animal Lectin Discovered 709
33.2 Rat Asialoglycoprotein Receptor 709
33.2.1 Characteristics 709
33.3 Human Asialoglycoprotein Receptor 710
33.3.1 Structural Characteristics 710
33.4 Ligand Binding Properties of ASGP-R 711
33.4.1 Interaction with Viruses 712
33.5 The Crystal Structure of H1-CRD 713
33.5.1 The Sugar Binding Site of H1-CRD 713
33.5.2 Sugar Binding to H1-CRD 714
33.6 Physiological Functions 715
33.6.1 Impact of ASGP-R Deficiency on the Development of Liver Injury 715
33.7 ASGP-R: A Marker for Autoimmune Hepatitis and Liver Damage 715
33.7.1 Autoimmune Hepatitis 715
33.7.2 Hepatocellular Carcinoma 716
33.7.3 Extra-Hepatic ASGP-R 716
33.8 ASGP-R: A Model Protein for Endocytosis 716
33.9 ASGP-R for Targeting Hepatocytes 717
33.10 Macrophage Galactose-Type Lectin (MGL) (CD301) 718
33.10.1 Human MGL (CD 301) 718
33.10.2 Murine MGL1/MGL2 (CD301) 718
33.10.3 Ligands of MGL 719
33.10.4 Functions of MGL 720
References 721

34 Dectin-1 Receptor Family 725
Rajesh K. Gupta and G.S. Gupta
34.1 Natural Killer Gene Complex (NKC) 725
34.2 β-Glucan Receptor (Dectin 1) (CLEC7A or CLECSF12) 725
34.2.1 Characterization of β-Glucan Receptor (Dectin 1) 725
34.2.2 Crystal Structure of Dectin-1 727
34.2.3 Interactions of Dectin-1 with Natural or Synthetic Glucans 727
34.2.4 Regulation of Dectin-1 727
34.2.5 Signaling Pathways by Dectin-1 728
34.2.6 Functions of Dectin-1 729
34.2.7 Genetic Polymorphism in Relation to Pathology 730
34.3 The C-Type Lectin-Like Protein-1 (CLEC-1) or CLEC-1A 731
34.3.1 *CLEC-1* Gene 731
34.4 CLEC-18 or the C-Type Lectin-Like Protein-2 (CLEC-2) or CLEC1B 731
34.4.1 Characterization 731
34.4.2 Ligands for CLEC-2 732
34.4.3 Crystal Structure of CLEC-2 733

34.4.4 Functions of CLEC-2 . . . 733
34.4.5 CLEC-2 Signaling . . . 734
34.5 CLEC9A (DNGR-1) . . . 734
34.6 Myeloid Inhibitory C-Type Lectin-Like Receptor (MICL/CLEC12A) . . . 735
34.7 CLEC12B or Macrophage Antigen H . . . 735
34.8 CLECSF7 (CD303) . . . 735
34.9 Lectin-Like Oxidized LDL Receptor (LOX-1) (CLEC8A) . . . 736
34.9.1 General Features of LOX-1 . . . 736
34.9.2 LOX-1 Gene in Human . . . 737
34.9.3 Ligands for LOX-1 . . . 738
34.9.4 Structural Analysis . . . 738
34.9.5 Functions of LOX-1 . . . 740
34.9.6 LOX-1 and Pathophysiology . . . 741
34.9.7 Macrophage Differentiation to Foam Cells . . . 742
References . . . 743

35 Dendritic Cell Lectin Receptors (Dectin-2 Receptors Family) . . . 749
Rajesh K. Gupta and G.S. Gupta
35.1 Dendritic Cells . . . 749
35.2 Dendritic Cell-Associated Lectins . . . 750
35.2.1 Type-I and Type-II Surface Lectins on DC . . . 750
35.3 Macrophage Mannose Receptor (CD206) on DC . . . 750
35.3.1 Expression and Characteristics . . . 750
35.3.2 Functions of Mannose Receptor in DC . . . 752
35.4 DEC-205 . . . 753
35.5 Langerin: A C-Type Lectin on Langerhans Cells . . . 753
35.5.1 Human Langerin (CD207) . . . 753
35.5.2 Mouse Langerin: Homology to Human Langerin . . . 754
35.5.3 Ligands of Langarin . . . 755
35.5.4 Functions of Langerin . . . 755
35.6 DC-SIGN and DC-SIGNR on DCs . . . 756
35.7 Dectin-2 Cluster in Natural Killer Gene Complex (NKC) . . . 757
35.7.1 Natural Killer Gene Complex . . . 757
35.7.2 Antigen Presenting Lectin-Like Receptor Complex (APLEC) . . . 759
35.8 Dectin-2 (CLECF4N) . . . 759
35.8.1 Characteristics . . . 759
35.8.2 Ligands of Dectin-2 . . . 760
35.9 The DC Immunoreceptor and DC-Immunoactivating Receptor . . . 762
35.9.1 Dendritic Cell Immunoreceptor (DCIR) (CLECsF6): Characterization . . . 762
35.9.2 Dendritic Cell Immunoactivating Receptor (DCAR) . . . 764
35.10 Macrophage-Inducible C-Type Lectin (Mincle) . . . 764
35.10.1 Recognition of Pathogens . . . 765
35.10.2 Recognition of Mycobacterial Glycolipid, Trehalose Dimycolate . . . 765
35.11 Blood Dendritic Cell Antigen-2 (BDCA-2) (CLEC-4C) . . . 765
35.11.1 BDCA-2: A Plasmacytoid DCs (PDCs)-Specific Lectin . . . 765
35.11.2 Characterization . . . 765
35.11.3 BDCA-2 Signals in PDC via a BCR-Like Signalosome . . . 766
35.11.4 Functions of BDCA-2 . . . 766
35.12 CLECSF8 . . . 767
35.13 Macrophage Galactose/N-acetylgalactosamine Lectin (MGL) . . . 767
References . . . 767

36 DC-SIGN Family of Receptors . . . 773
Rajesh K. Gupta and G.S. Gupta
36.1 DC-SIGN (CD209) Family of Receptors . . . 773
36.1.1 CD209 Family Genes in Sub-Human Primates . . . 773
36.2 DC-SIGN (CD209): An Adhesion Molecule on Dendritic Cells . . . 774
36.3 Ligands of DC-SIGN . . . 775
36.3.1 Carbohydrates as Ligands of DC-SIGN . . . 775
36.4 Structure of DC-SIGN . . . 775
36.4.1 Neck-Domains . . . 775
36.4.2 Crystal Structure of DC-SIGN (CD209) and DC-SIGNR (CD299) . . . 776
36.5 DC-SIGN versus DC-SIGN-RELATED RECEPTOR [DC-SIGNR or L-SIGN (CD 209)/LSEctin] (Refer Section 36.9) . . . 777
36.5.1 DC-SIGN Similarities with DC-SIGNR/L-SIGN/LSEctin . . . 777
36.5.2 Domain Organization of DC-SIGN and DC-SIGNR . . . 778
36.5.3 Differences Between DC-SIGN and DC-SIGNR/L-SIGN . . . 778
36.5.4 Recognition of Oligosaccharides by DC-SIGN and DC-SIGNR . . . 779
36.5.5 Extended Neck Regions of DC-SIGN and DC-SIGNR . . . 780
36.5.6 Signaling by DC-SIGN through Raf-1 . . . 781
36.6 Functions of DC-SIGN . . . 781
36.6.1 DC-SIGN Supports Immune Response . . . 781
36.6.2 DC-SIGN Recognizes Pathogens . . . 782
36.6.3 DC-SIGN as Receptor for Viruses . . . 783
36.6.4 HIV-1 gp120 and Other Viral Envelope Glycoproteins . . . 783
36.7 Subversion and Immune Escape Activities of DC-SIGN . . . 785
36.7.1 The entry and dissemination of viruses can be mediated by DC-SIGN . . . 785
36.7.2 DC-SIGN and Escape of Tumors . . . 785
36.7.3 *Mycobacterial* Carbohydrates as Ligands of DC-SIGN, L-SIGN and SIGNR1 . . . 786
36.7.4 Decreased Pathology of Human DC-SIGN Transgenic Mice During Mycobacterial Infection . . . 787
36.7.5 Genomic Polymorphism of DC-SIGN (CD209) and Consequences . . . 787
36.8 SIGNR1 (CD209b): The Murine Homologues of DC-SIGN . . . 789
36.8.1 Characterization . . . 789
36.8.2 Functions . . . 790
36.9 Liver and Lymph Node Sinusoidal Endothelial Cell C-Type Lectin (LSECtin) (or CLEC4G or L-SIGN or CD209L) . . . 791
36.9.1 Characterization and Localization . . . 791
36.9.2 Ligands of LSECtin . . . 792
36.9.3 Functions . . . 792
36.9.4 Role in Pathology . . . 793
References . . . 793

Part XII C-Type Lectins: Proteoglycans

37 Lectican Protein Family ... 801
G.S. Gupta
37.1 Proteoglycans ... 801
37.1.1 Nomenclature of PGs ... 801
37.1.2 Glycosaminoglycans ... 802
37.1.3 Chondroitin ... 803
37.1.4 Heparan Sulfate Proteoglycans (HSPG) ... 805
37.2 Proteoglycans in Tissues ... 805
37.3 Hyaluronan: Proteoglycan Binding Link Proteins ... 807
37.4 Lecticans (Hyalectans) ... 807
37.5 Cartilage Proteoglycan: Aggrecan ... 808
37.5.1 Skeletogenesis ... 808
37.5.2 Human Aggrecan ... 808
37.5.3 Rat Aggrecan ... 810
37.5.4 Mouse Aggrecan ... 812
37.5.5 Chick Aggrecan ... 812
37.6 Versican or Chondroitin Sulfate Proteoglycan Core Protein 2 ... 812
37.6.1 Expression ... 813
37.6.2 Structure ... 814
37.6.3 Interactions of Versican ... 814
37.6.4 Functions ... 815
37.7 Pathologies Associated with PGS ... 816
37.7.1 Proteoglycans Facilitate Lipid Accumulation in Arterial Wall ... 816
37.8 Diseases of Aggrecan Insufficiency ... 817
37.8.1 Aggrecanase-Mediated Cartilage Degradation ... 817
37.8.2 A Mutation in Aggrecan Gene Causes Spondyloepiphyseal Dysplasia ... 818
37.9 Expression of Proteoglycans in Carcinogenesis ... 818
37.10 Regulation of Proteoglycans ... 819
37.11 CD44: A Major Hyaluronan Receptor ... 819
37.11.1 CD44: A Hyaluronan Receptor ... 819
37.11.2 Hyaluronan Binding Sites in CD44 ... 819
37.11.3 CD44 in Cancer ... 820
References ... 820

38 Proteoglycans of the Central Nervous System ... 825
G.S. Gupta
38.1 Proteoglycans in Central Nervous System ... 825
38.1.1 Large Proteoglycans in Brain ... 825
38.1.2 Ligands of Lecticans ... 826
38.2 Neurocan ... 828
38.2.1 Cellular Sites of Synthesis ... 828
38.2.2 Characterization ... 828
38.2.3 Ligand Interactions ... 829
38.3 Brevican ... 830
38.3.1 Characterization ... 830
38.3.2 Murine Brevican Gene ... 830
38.4 Hyaluronan: Proteoglycan Binding Link Protein ... 831
38.4.1 Cartilage link protein 1 and Brain Link Proteins ... 831
38.4.2 Brain Enriched Hyaluronan Binding (BEHAB)/Brevican ... 832
38.5 Proteolytic Cleavage of Brevican ... 832
38.6 Proteoglycans in Sensory Organs ... 833

38.7 Functions of Proteoglycans in CNS 834
38.7.1 Functions of Chondroitin Sulphate Proteoglycans 834
38.7.2 Functions of Brevican and Neurocan in CNS 835
38.7.3 Neurocan in Embryonic Chick Brain 836
38.8 Chondroitin Sulfate Proteoglycans in CNS Injury Response 836
38.8.1 CS-PGs in CNS in Response to Injury 836
38.8.2 Glial Scar and CNS Repair 838
38.9 Other Proteoglycans in CNS 838
38.10 Regulation of Proteoglycans in CNS 839
38.10.1 Growth Factors and Cytokine Regulate CS-PGs by Astrocytes 839
38.10.2 Decorin Suppresses PGs Expression 839
References 840

Part XIII C-Type Lectins: Emerging Groups of C-Type Lectins

39 Regenerating (Reg) Gene Family 847
G.S. Gupta
39.1 The Regenerating Gene Family 847
39.2 Regenerating (*Reg*) Gene Products 847
39.3 Reg 1 848
39.3.1 Tissue Expression 848
39.3.2 Gastric Mucosal Cells 849
39.3.3 Ectopic Expression 849
39.3.4 Reg Protein 850
39.3.5 *REGIA (Reg Iα)* Gene in Human Pancreas 850
39.3.6 REGIB (Reg Iβ) Gene in Human Pancreas 852
39.4 Pancreatic Stone Protein/Lithostathine (PSP/LIT) 852
39.4.1 Characterization 852
39.4.2 3D-Structure 853
39.4.3 Secretory Forms of PSP 853
39.4.4 Functions of PSP/LIT 854
39.5 Pancreatic Thread Protein 855
39.5.1 Pancreatic Thread Proteins (PTPs) 855
39.5.2 Neuronal Thread Proteins 856
39.5.3 Pancreatic Proteins Form Fibrillar Structures upon Tryptic Activation 856
39.6 REG-II/REG-2/PAP-I 857
39.7 Reg-III 858
39.7.1 Murine *Reg-IIIα*, *Reg-IIIβ*, *Reg-III γ*, and *Reg-IIIδ* Genes 858
39.7.2 Human Reg-III or HIP/PAP 859
39.8 Rat PAPs 860
39.8.1 Three Forms of PAP in Rat 860
39.8.2 PAP-I/Reg-2 Protein (or HIP, p23) 861
39.8.3 PAP-II/PAP2 861
39.8.4 PAP-III 862
39.9 Functions of PAP 863
39.9.1 PAP: A Multifunctional Protein 863
39.9.2 PAP in Bacterial Aggregation 863
39.9.3 PAP-I: An Anti-inflammatory Cytokine 863
39.10 Panceatitis Associated Protein (PAP)/Hepatocarcinoma-Intestine Pancreas (HIP) 865
39.10.1 HIP Similarity to PAP 865
39.10.2 Characterization of HIP/PAP 866

39.10.3 PAP Ligands (PAP Interactions) . . . 867
39.10.4 Crystal Structue of human PAP (hPAP) . . . 867
39.10.5 Expression of PAP . . . 867
39.10.6 Similarity of PAP to Peptide 23 from Pituitary . . . 869
39.10.7 Serum PAP: An Indicator of Pancreatic Function . . . 870
39.11 Reg IV (RELP) . . . 870
39.11.1 Tissue Expression . . . 870
39.12 Islet Neogenesis-Associated Protein (INGAP) in Hamster . . . 871
39.12.1 Characterization . . . 871
39.12.2 Properties of Pentadecapeptide from INGAP . . . 872
References . . . 873

40 Emerging Groups of C-Type Lectins . . . 881
G.S. Gupta
40.1 Layilin Group of C-Type Lectins (Group VIII of CTLDS) . . . 881
40.1.1 Layilin: A Hyaluronan Receptor . . . 881
40.1.2 Interactions of Layilin . . . 881
40.1.3 Functions of Layilin . . . 881
40.1.4 Chondrolectin (CHODL) . . . 882
40.2 Tetranectin Group of Lectins (Group IX of CTLDs) . . . 882
40.2.1 Tetranectin . . . 883
40.2.2 Cell and Tissue Distribution . . . 884
40.2.3 Interactions of Tetranectin . . . 884
40.2.4 Crystal Structure of Tetranectin . . . 885
40.2.5 Functions of Tetranectin . . . 886
40.2.6 Association of TN with Diseases . . . 887
40.2.7 Cartilage-Derived C-Type Lectin (CLECSF1) . . . 888
40.2.8 Stem Cell Growth Factor (SCGF): Tetranectin Homologous Protein . . . 888
40.3 Attractin Group of CTLDs (Group XI) . . . 888
40.3.1 Secreted and Membrane Attractins . . . 888
40.3.2 Attractin has Dipeptidyl Peptidase IV Activity? . . . 889
40.3.3 Attractin-Like Protein . . . 889
40.3.4 Functions of Attractin . . . 890
40.3.5 Genetic and Phenotypic Studies of *Mahogany/Attractin* Gene . . . 892
40.3.6 Therapeutic Applications of ATRN/*Mahogany* Gene Products . . . 894
40.4 Eosinophil Major Basic Protein 1 (EMBP1) (Group XII of CTLD) . . . 894
40.4.1 Characterization of EMBP . . . 894
40.4.2 Functions of EMBP . . . 895
40.4.3 Crystal Structure of EMBP . . . 895
40.4.4 DEC-205-Associated C-Type Lectin (DCL)-1 . . . 895
40.5 Integral Membrane Protein, Deleted in Digeorge Syndrome (IDD) (Group XIII of CTLDs) . . . 896
40.5.1 DiGeorge Syndrome Critical Region 2 (DGCR2) . . . 896
References . . . 896

41 Family of CD93 and Recently Discovered Groups of CTLDs . . . 901
G.S. Gupta
41.1 Family of CD93 . . . 901
41.2 CD93 or C1q Receptor (C1qRp): A Receptor for Complement C1q . . . 901
41.2.1 CD93 Is Identical to C1qRp . . . 901
41.2.2 Characterization of CD93/C1qR$_P$ . . . 902
41.2.3 Two Types of Cell Surface C1q-Binding Proteins (C1qR) . . . 902
41.2.4 Tissue Expression and Regulation of CD93/C1qRp . . . 903

41.2.5 Functions 904
41.2.6 CD93 in Pathogenesis of SLE 904
41.3 Murine Homologue of Human C1qRp (AA4) 905
41.4 The gC1qR (p33, p32, C1qBP, TAP) 906
41.4.1 A Multi-Multifunctional Protein - Binding with C1q 906
41.4.2 Human gC1qR/p32 (C1qBP) Gene 907
41.4.3 Ligand Interactions 908
41.4.4 Role in Pathology 909
41.5 Thrombomodulin 910
41.5.1 Localization 910
41.5.2 Characteristics 910
41.5.3 Regulation of TM Activity 910
41.5.4 Structure-Function Relations - Binding to Thrombin 911
41.5.5 The Crystal Structure 913
41.5.6 Functions 914
41.5.7 Abnormalities Associated with Thrombomodulin Deficiency 916
41.6 Endosialin (Tumor Endothelial Marker-1 or CD248) 917
41.6.1 A Marker of Tumor Endothelium 917
41.6.2 Interactions with Ligands 918
41.6.3 Functions 919
41.7 Bimlec (DEC 205 Associated C-type lectin-1 or DCL-1) or (CD302) (Group XV of CTLD) 920
41.8 Proteins Containing SCP, EGF, EGF and CTLD (SEEC) (Group XVI of CTLD) 920
41.8.1 Nematocyst Outer Wall Antigen (NOWA) 921
41.8.2 The Cysteine-Rich Secretory Proteins (CRISP) Super-Family . . . 921
41.9 Calx-*b* and CTLD Containing Protein (CBCP) (Group XVII of CTLD) . . 922
41.9.1 CBCP/Frem1/QBRICK 922
41.9.2 Frem3 924
41.9.3 Zebrafish Orthologues of FRAS1, FREM1, or FREM2 924
References 924

Part XIV Clinical Significance of Animal Lectins

42 MBL Deficiency as Risk of Infection and Autoimmunity 933
Anita Gupta
42.1 Pathogen Recognition 933
42.1.1 MBL Characteristics 933
42.1.2 Pathogen Recognition and Role in Innate Immunity 933
42.2 MBL Deficiency as Risk of Infection and Autoimmunity 935
42.2.1 MBL Deficiency and Genotyping 935
42.2.2 MBL and Viral Infections 938
42.3 Autoimmune and Inflammatory Diseases 939
42.3.1 MBL Gene in Rheumatoid Arthritis 939
42.3.2 Systemic Lupus Erythematosus 940
42.3.3 Systemic Inflammatory Response Syndrome/Sepsis 941
42.3.4 MBL and Inflammatory Bowel Diseases 942
42.3.5 Rheumatic Heart Disease 943
42.3.6 MBL in Cardio-Vascular Complications 944
42.3.7 Other Inflammatory Disorders 945
42.4 Significance of MBL Gene in Transplantation 946
42.5 MBL in Tumorigenesis 947
42.5.1 Polymorphisms in the Promoter 947
42.5.2 MBL Binding with Tumor Cells 947

42.6 Complications Associated with Chemotherapy 947
42.6.1 Neutropenia 947
42.6.2 Animal Studies 948
42.7 MBL: A Reconstitution Therapy 949
References 949

43 Pulmonary Collectins in Diagnosis and Prevention of Lung Diseases 955
Anita Gupta
43.1 Pulmonary Surfactant 955
43.2 SP-A and SP-D in Interstitial Lung Disease 955
43.2.1 Pneumonitis 955
43.2.2 Interstitial Pneumonia (IP) 956
43.2.3 ILD Due to Inhaled Substances 957
43.2.4 Idiopathic Pulmonary Fibrosis 957
43.2.5 Cystic Fibrosis 958
43.2.6 Familial Interstitial Lung Disease 959
43.3 Connective Tissue Disorders 959
43.3.1 Systemic Sclerosis 959
43.3.2 Sarcoidosis 960
43.4 Pulmonary Alveolar Proteinosis 960
43.4.1 Idiopathic Pulmonary Alveolar Proteinosis 960
43.4.2 Structural Changes in SPs in PAP 961
43.5 Respiratory-Distress Syndrome and Acute Lung Injury 961
43.5.1 ARDS and Acute Lung Injury 961
43.5.2 Bronchopulmonary Dysplasia (BPD) 962
43.6 Chronic Obstructive Pulmonary Disease (COPD) 963
43.6.1 COPD as a Group of Diseases 963
43.6.2 Emphysema 964
43.6.3 Allergic Disorders 964
43.6.4 Interactions of SP-A and SP-D with Pathogens and Infectious Diseases 965
43.7 Pulmonary Tuberculosis 966
43.7.1 Enhanced Phagocytosis of *M. tuberculosis* by SP-A 966
43.7.2 SP-A Modulates Inflammatory Response in AΦs During Tuberculosis 967
43.7.3 Marker Alleles in *M. tuberculosis* 968
43.7.4 Interaction of SP-D with *M. tuberculosis* 968
43.7.5 Association of SPs with Diabetes 969
43.8 Expression of SPs in Lung Cancer 969
43.8.1 Non-Small-Cell Lung Carcinoma (NSCLC) 969
43.9 Other Inflammatory Disorders 971
43.9.1 Airway Inflammation in Children with Tracheostomy 971
43.9.2 Surfactant Proteins in Non-ILD Pulmonary Conditions 971
43.10 DNA Polymorphisms in SPs and Pulmonary Diseases 972
43.10.1 Association Between SP-A Gene Polymorphisms and RDS 972
43.10.2 SP-A and SP-B as Interactive Genetic Determinants of Neonatal RDS 973
43.10.3 RDS in Premature Infants 973
43.10.4 Gene Polymorphism in Patients of High-altitude Pulmonary Edema 973
43.10.5 SNPs in Pulmonary Diseases 974
43.10.6 Allergic Bronchopulmonary Aspergillosis and Chronic Cavitary Pulmonary Aspergillosis (CCPA) 974
43.10.7 Autoreactivity Against SP-A and Rheumatoid Arthritis 975

43.11 Inhibition of SP-A Function by Oxidation Intermediates of Nitrite 975
43.11.1 Protein Oxidation by Chronic Pulmonary Diseases 975
43.11.2 Oxidation Intermediates of Nitrite 976
43.11.3 BPD Treatment with Inhaled NO 976
43.12 Congenital Diaphragmatic Hernia 976
43.13 Protective Effects of SP-A and SP-D on Transplants 977
43.14 Therapeutic Effects of SP-A, SP-D and Their Chimeras 977
43.14.1 SP-A Effects on Inflammation of Mite-sensitized Mice 977
43.14.2 SP-D Increases Apoptosis in Eosinophils of Asthmatics 977
43.14.3 Targeting of Pathogens to Neutrophils Via Chimeric SP-D/Anti-CD89 Protein 978
43.14.4 Anti-IAV and Opsonic Activity of Multimerized Chimeras of rSP-D 978
43.15 Lessons from SP-A and SP-D Deficient Mice 979
References 980

44 Selectins and Associated Adhesion Proteins in Inflammatory disorders 991
G.S. Gupta
44.1 Inflammation 991
44.2 Cell Adhesion Molecules 991
44.2.1 Selectins 992
44.3 Atherothrombosis 992
44.3.1 Venous Thrombosis 992
44.3.2 Arterial Thrombosis 993
44.3.3 Thrombogenesis in Atrial Fibrillation 993
44.3.4 Atherosclerosis 993
44.3.5 Myocardial Infarction 996
44.3.6 Atherosclerotic Ischemic Stroke 997
44.3.7 Hypertension 998
44.3.8 Reperfusion Injury 998
44.4 CAMS in Allergic Inflammation 998
44.4.1 Dermal Disorders 998
44.4.2 Rhinitis and Nasal Polyposis 999
44.4.3 Lung Injury 999
44.4.4 Bronchial Asthma and Human Rhinovirus 999
44.5 Autoimmune Diseases 1000
44.5.1 Endothelial Dysfunction in Diabetes (Type 1 Diabetes) 1000
44.5.2 Rheumatic Diseases 1001
44.5.3 Rheumatoid Arthritis 1001
44.5.4 Other Autoimmune Disorders 1002
44.6 CAMs in System Related Disorders 1003
44.6.1 Gastric Diseases 1003
44.6.2 Liver Diseases 1004
44.6.3 Neuro/Muscular Disorders 1004
44.6.4 Acute Pancreatitis 1005
44.6.5 Renal Failure 1005
44.6.6 Other Inflammatory Disorders 1005
44.6.7 Inflammation in Hereditary Diseases 1006
44.7 Role of CAMs in Cancer 1006
44.7.1 Selectin Ligands in Cancer cells 1007
44.7.2 E-Selectin-Induced Angiogenesis 1007
44.7.3 E-Selectin in Cancer Cells 1007
44.7.4 Metastatic Spreading 1009
44.7.5 Survival Benefits of Heparin 1011

44.8 Adhesion Proteins in Transplantation 1012
44.9 Inflammation During Infection 1013
44.9.1 Microbial Pathogens 1013
44.9.2 Yeasts and Fungi 1014
44.9.3 Parasites and Amoeba 1014
44.10 Action of Drugs and Physical Factors on CAMS 1015
44.10.1 Inhibitors of Gene Transcription 1015
44.10.2 Anti-NF-kB Reagents 1015
44.10.3 Strategies to Combat Atherogenesis and Venous Thrombosis 1016
44.10.4 Anti-inflammatory Drugs 1017
References 1018

45 Polycystins and Autosomal Polycystic Kidney Disease 1027
G.S. Gupta
45.1 Polycystic Kidney Disease Genes 1027
45.1.1 Regulatory Elements in Promoter Regions 1028
45.2 Polycystins: The Products of PKD Genes 1028
45.2.1 Polycystins 1028
45.2.2 Polycystin-1 (TRPP1) with a C-type Lectin Domain 1028
45.2.3 Polycystin-2 (TRPP2) 1030
45.2.4 Interactions of Polycystins 1031
45.2.5 Tissue and Sub-Cellular Distribution of Polycystins 1031
45.3 Functions of Polycystin-1 and Polycystin-2 1032
45.3.1 Polycystin-1 and Polycystin-2 Function Together 1032
45.3.2 Cell-Cell and Cell-Matrix Adhesion 1032
45.3.3 Role in Ciliary Signaling 1033
45.3.4 Cilia and Cell Cycle 1034
45.3.5 Polycystins and Sperm Physiology 1034
45.4 Autosomal Dominant Polycystic Kidney Disease 1034
45.4.1 Mutations in *PKD1* and *PKD2* and Association of Polycystic Kidney Disease 1034
45.4.2 Proliferation and Branching Morphogenesis in Kidney Epithelial Cells 1035
References 1035

46 Endogenous Lectins as Drug Targets 1039
Rajesh K. Gupta and Anita Gupta
46.1 Targeting of Mannose-6-Phosphate Receptors and Applications in Human Diseases 1039
46.1.1 Lysosomal Storage Diseases 1039
46.1.2 Enzyme Replacement Therapy (ERT) 1040
46.1.3 M6PR-Mediated Transport Across Blood-Brain Barrier 1042
46.1.4 Other Approaches using CI-MPR as Target 1045
46.2 Cell Targeting Based on Mannan-Lectin Interactions 1045
46.2.1 Receptor-Mediated Uptake of Mannan-Coated Particles (Direct Targeting) 1046
46.2.2 Polymeric Glyco-Conjugates as Carriers 1046
46.2.3 Mannosylated Liposomes in Gene Delivery 1047
46.2.4 DC-Targeted Vaccines 1049
46.3 Asialoglycoprotein Receptor (ASGP-R) for Targeted Drug Delivery 1051
46.3.1 Targeting Hepatocytes 1051
46.4 Siglecs as Targets for Immunotherapy 1052

46.4.1 Anti-CD33-Antibody-Based Therapy of Human Leukemia 1052
46.4.2 CD22 Antibodies as Carrier of Drugs 1053
46.4.3 Immunogenic Peptides 1053
46.4.4 Blocking of CD33 Responses by SOCS3 1053
References 1053

About the Author 1059

Index 1061

Contributors

G.S. Gupta Former Professor and Chairman, Department of Biophysics, Punjab University, Chandigarh, India

Anita Gupta Assistant Professor, Department of Biomedical Engineering, Rayat and Bahra Institute of Engineering and Biotechnology, Kharar (Mohali), Punjab, India

Rajesh K. Gupta General Manager, Vaccine Production Division, Panacea Biotec, Lalru (Mohali), Punjab, India

Abbreviation

AA4	murine homologue of human C1qRp
AAA	*Anguilla Anguilla* Agglutinin
AAA	abdominal aortic aneurysm
AAL	aleuria aurantia lectin
ABPA	allergic bronchopulmonary aspergillosis
ACS	acute coronary syndrome
Ad/AD	Alzheimer's disease
ADAMTS	a disintegrin and metalloproteinase with thrombo-spondin motifs
ADPKD	autosomal dominant polycystic kidney disease
AF	atrial fibrillation
AFP	antifreeze polypeptide
AGE	advanced glycation end product
AICL	activation-induced C-type lectin
ALI	acute lung injury
AM	alveolar macrophage
AM	adrenomedullin
AML	acute myeloid leukemia
AMφ	alveolar macrophage
AP	adaptor protein
APC	antigen-presenting cell
APLEC	antigen presenting lectin-like receptor complex
APR	acute-phase response
ARDS	acute-RDS
ARDS	adult respiratory distress syndrome
ASGPR	asialoglycoprotein receptor
ATM	ataxia telangiectasia, mutated
ATRA	all-trans retinoic acid
ATRNL	attractin-like protein
BAL	bronchoalveolar lavage
BALF	bronchoalveolar lavage fluid
BBB	blood-brain barrier
BDCA-1	blood dendritic cell antigen 1
BDNF	brain derived neurotrophic factor
BEHAB	brain enriched hyaluronan binding
BG	Birbeck Granule
BiP	immunoglobulin binding protein
BOS	bronchiolitis obliterans syndrome
BPD	bronchopulmonary dysplasia
BRAL-1	brain link protein-1
BRAP	BRCA1-associated protein
C1q	complement C1q module
C1r	complement C1r module

C4S	chondroitin-4-sulfate
C6S	chondroitin-6-sulfate
CAMs	cell adhesion molecules
CARD9	caspase activating recruitment domain 9
CBM	carbohydrate binding module
CC16	clara cell 16
CCSP	clara cell specific protein
CCV	clathrin coated vesicles
CD	Celiac disease
CD	Crohn's disease
CDH	congenital diaphragmatic hernia
CD-MPR	cation dependent M6P receptor
CEA	carcinoembryonic antigen (CEA)
CEACAM1	carcinoembryonic antigen (CEA)-related cell adhesion molecule 1
CF	cystic fibrosis
CFTR	cystic fibrosis transmembrane conductance regulator
CGN	cis-Golgi network
CHODL	chondrolectin
CI-MPR	cation-independent mannose 6-phosphate receptor
CLECSF2	activation-induced C-type lectin
CLL-1	C-type lectin-like molecule-1
CLRD	C-type lectin domain
CNS	central nervous system
Cnx/CNX	calnexin
COAD	chronic obstructive airways disease
COPD	chronic obstructive pulmonary disease
COPI	coatomer protein complex I
COPII	coatomer protein complex II
CR	complement protein regulatory repeat/ complement receptor
CRC	colorectal cancer
CRD	carbohydrate recognition domain
CREB	cAMP responsive element binding protein
CRP	complement regulatory protein domain
CRP	C-Reactive Protein
Crt/CRT	calreticulin
CS	chondroitin sulfate
CS-GAGs	chondroitin sulfate glycosaminoglycans
CSPG	chondroitin sulfate proteoglycan
CSPG2	chondroitin sulfate proteoglycan 2 or PG-M or versican
CTLD	C-type lectin-like domain
DC	dendritic cell
DC-SIGN	dendritic cell-specific ICAM-3-grabbing nonintegrin
DDR	discoidin domain receptor
Dex	dexamethasone
dGal-1	dimeric galectin-1
DN	down's syndrome
DOPG	dioleylphosphatidylglycerol
DPPC	dipalmitoylphosphatidyl choline
DSPC	disaturated phosphatidyl choline
EAE	experimental autoimmune encephalomyelitis
EBM	Epstein-Barr virus
ECM	extracellular matrix
Ed	embryonic day

EDEM	ER degradation enhancing α-mannosidase-like protein
EE	early endosomes
EGF	epidermal growth factor
ELAM-1	endothelial-leukocyte adhesion molecule 1
LECAM2	leukocyte-endothelial cell adhesion molecule 2
EMBP	eosinophil major basic protein
EMSA	electrophoretic mobility shift assay
eNOS	endothelial cell NO synthase
ER	endoplasmic reticulum
ERAD	ER-associated degradation
ERK	extracellular signal-regulated kinase
ERT	enzyme replacement therapy
ESP	early secretoty pathway
FA5	blood coagulation factor 5
FA8	blood coagulation factor 8
FAK	focal adhesion kinase
FBG	fibrinogen-like (FBG)-domains
FDl	fibrinogen-like domain 1
FGF	fibroblast growth factor
FN	fibronectin
FN2	fibronectin type II module
FN3	fibronectin type III module
Fuc	fucose
FV	blood coagulation factor V
FVIII	blood coagulation factor VIII
GAGs	glycosaminoglycans
Gal	galactose
Gal-1/Gal-3	galectin-1/galectin-3
Gb3	globotriaosylceramide
Gb4	globotetraosylceramide
GGA	Golgi-localizing, gamma-adaptin ear homology domain, ARF binding
GlyCAM-1	glycosylation-dependent cell adhesion molecule 1
GM1	ganglioside 1
GM2	ganglioside 2
GM-CSF	granulocyte-macrophage colony-stimulating factor
GMP-140	granule membrane protein 140
Grifin	galectin-related inter-fiber protein
GVHD	graft versus host disease
HAPLN2	hyaluronan and proteoglycan link protein 2
HBV	hepatitis B virus
HCMV	human cytomegalo virus
HCV	hepatitis C virus
HEV	high endothelial venule
HHL	human hepatic lectin
HHL1	human hepatic lectin subunit-1
HHL2	human hepatic lectin subunit-2
HNSCC	head and neck squamous cell carcinoma
HSPG	heparan sulfate proteoglycan
HT	Hashimoto's thyroiditis
HUVEC	human umbilical vein endothelial cell
i.m	intramuscular
i.p	intraperitoneal
i.v.	intravascular

IBD	inflammatory bowel disease
ICAM	intercellular adhesion molecule
IDDM	insulin-dependent diabetes mellitus
IFN	interferon
Ig	immunoglobulin
IGF-1	insulin-like growth factor, type 1
IGF2/MPR	insulin-like growth factor 2 receptor (IGF2R)
IGF2R	insulin-like growth factor 2 receptor
IGF-II/CIMPR	insulin-like growth factor-II or cation-independent mannose 6-phosphate receptor
IgSF	immunoglobulin superfamily
IL	interleukin
ILT	immunoglobulin-like transcripts
INTL	intelectin
IPCD	interstitial pneumonia with collagen vascular diseases
IPF	idiopathic pulmonary fibrosis
iRNA/RNAi	RNA interference
ITAM	immunoreceptor tyrosine-based activatory motif
ITIM	immunoreceptor tyrosine-based inhibitory motif
KIR	killer cell Ig-like receptors
LAMP-1/Lamp1	lysosome-associated membrane protein 1
LDL	low density lipoprotein
Le^a	Lewis A
Le^b	Lewis B
Le^x	lewis X
LFA-2	leukocyte function antigen-2
LFA-3	leukocyte function antigen-3
LFR	lactoferrin (LF) receptor (R)
LL	dileucine
LN	lymph node
LOX-1	lectin-type oxidized LDL receptor 1
LPS	lipopolysaccharide
M-6-P	mannose-6-phosphate
M6PR/MPR	Mannose-6-phosphate receptor
MAdCAM-1	mucosal addressin cell adhesion molecule-1 (or addressin)
ManLAMs	mannose-capped lipoarabinomannan
MAPK	mitogen-activated protein kinase
MASP	MBL-associated serine protease
MBL	mannan-binding lectin
MBP	mannose binding protein
MCP-1	monocyte chemoattractant protein 1
MHC	major histocompatibility complex
MICA/MIC-A	MHC class I polypeptide-related sequence A
MICB/MIC-B	MHC class I polypeptide-related sequence B
MMP	matrix metalloproteinases
MMR	macrophage mannose receptor
MPS	mucopolysaccharidosis
MR/ManR	mannose receptor
NCAM	neural cell adhesion molecule
NG2	neuronglia antigen 2
NK cell	natural killer cell
NK	natural killer
NKC	natural killer gene complex

n-LDL	native LDL
NO	nitric oxide
NOD mice	non-obese diabetic mice
NPI	neuronal pentraxin I
NTP	neuronal thread protein
OLR1	oxidized low-density lipoprotein receptor 1
Ox-LDL	oxidized low density lipoprotein
PAMPs	pathogen-associated molecular patterns
PAP	pulmonary alveolar proteinosis
PB	peripheral blood
PC	polycystin
PD	Parkinson disease
pDC/PDC	plasmacytoid dendritic cell
PDI	protein disulfide isomerase
PECAM-1	platelet-endothelial cell adhesion molecule-1
PG	phosphatidylglycerol
PGN	peptidoglycan
PI3-kinase	phosphatidyl-inositol (PI)3-kinase
PKB	protein kinase B
PKC	protein kinase C
PLC	phospholipase C
PLN	peripheral lymph node
PNN	perineuronal nets
PPARγ	peroxisome proliferator-activated receptor γ
PRR	pattern-recognition receptor
PSA	prostate-specific antigen
PSGL-1	P-selectin glycoprotein ligand1
PTX	pentraxins
R	review
RAE-1/Rae-1	retinoic acid inducible gene-1
RB	retinoblastoma protein
RCC	renal cell carcinoma
RCMV	rat cytomegalovirus
RDS	respiratory distress syndrome
Reg	regenerating genes
RAET	retinoic acid early (RAE) transcript
rER	rough ER
RHL	rat hepatic lectin
RNAi	RNA interference
ROS	reactive oxygen species
RSV	respiratory syncytial virus/ respiratory syndrome virus
SAA	serum amyloid A
SAP	serum amyloid P component
SARS	severe acute respiratory syndrome
SC	scavenger receptor module
SCLC	small cell lung carcinoma
SCR	short consensus repeat
s-diLe^x	sulfated polysaccharide ligands
Siglec	sialic-acid-binding immunoglobulin-like lectin
Sjs	Sjogren syndrome
SH1/2	Src homology 1/2
SHP-1/2	Src homology 1/2 containing phosphatase
SLAM	signaling lymphocyte activated molecule

SLE	systemic lupus erythematosus
sLe^a/$s\text{-}Le^a$	Sialyl Lewis A
sLe^x/$s\text{-}Le^x$	Sialyl Lewis X
Sn	sialoadhesin
SNP	single-nucleotide polymorphism
snRNPs	small nuclear ribonucleoproteins
SOCS3	suppressor of cytokine signaling 3
SP-A	surfactant protein A
SP-D	surfactant protein D
ST	sialyl transferase
STAT	signal transducers and activators of transcription
STAT1	signal transducer and activator of transcription 1
T1DM	type 1 diabetes mellitus
T2DM	type 2 diabetes mellitus
TGF-β	transforming growth factorβ
TGN	trans-Golgi network
TIM	triosephosphate isomerase
TLR	toll-like receptor
TNF	tumor necrosis factor
Ub	ubiquitin
UDPG	uridine diphosphate (UDP)-glucose
UL16	unique long 16
ULBP	UL16 binding protein
uPAR	uPA receptor or urokinase receptor
uPARAP	urokinase receptor-associated protein
UPR	unfolded protein response
VCAM-1	vascular adhesion molecule-1
VEGF	vascular endothelial growth factor
VLA	very late antigen
VN	vitronectin
VSMC	vascular smooth muscle cell
vWF	von Willebrand factor
β2-m	β2-microglobulin

Part I

Introduction

Lectins: An Overview

1

G.S. Gupta

1.1 Lectins: Characteristics and Diversity

1.1.1 Characteristics

Glycan-recognizing proteins can be broadly classified into two groups: lectins [which typically contain an evolutionarily conserved carbohydrate-recognition domain (CRD)] and sulfated glycosaminoglycan (SGAG)-binding proteins (which appear to have evolved by convergent evolution). The term "lectin" is derived from the Latin word legere, meaning, among other things, "to select". Lectins are phylogenetically ancient and have specific recognition and binding functions for complex carbohydrates of glycoconjugates, i.e., glycoproteins, proteoglycans/glycosaminoglycans and glycolipids. Although lectins were first discovered more than 100 years ago in plants, they are now known to be present throughout nature and found in most of the organisms, ranging from viruses and bacteria to plants and animals. It is generally believed that the earliest description of a lectin was given by Peter Hermann Stillmark in his doctoral thesis in 1888 to the University of Dorpat. Stillmark isolated ricin, an extremely toxic hemagglutinin, from seeds of the castor plant (Ricinus communis) (Stillmark 1988). The animal hemagglutinins were noted quite early, almost in all invertebrates or lower vertebrates, but until the middle of the 1970s, only three of these (of eel, snail, and horseshoe crab) were isolated and characterized. The first of the animal lectins shown to be specific for a sugar (L-fucose) was from eel (Watkins and Morgan 1952). The isolation of the first mammalian lectin, the galactose-specific hepatic asialoglycoprotein receptor, was achieved by Gilbert Ashwell at the NIH and by Anatol G. Morell at the Albert Einstein Medical School (New York) in 1974 (Hudgin et al. 1974; Stockert et al. 1974). At the same time, Teichberg et al. 1975 isolated a β-galactose-specific lectin from electric eel that was designated galectins (Barondes et al. 1994), of which 15 members so far have been characterized. Since the beginning of 1980s, the number of purified animal lectins started to grow quickly, largely thanks to the advent of recombinant techniques.

Presently, lectins are known as a heterogeneous group of carbohydrate binding proteins of non-immune origin, which agglutinate cells and/or precipitate glycoconjugates without affecting their covalent linkages (Goldstein et al. 1980). This definition implied that each lectin molecule has two or more carbohydrate binding sites to allow cross-linking between cells and between sugar containing macromolecules. During last few decades, however, interest in lectins has been greatly intensified after realization that they act as mediators of cell recognition in biological system (Sharon 2006). Though, lectins are similar to antibodies in their ability to agglutinate red blood cells, they are not the product of immune system. Lectins not only distinguish between different monosaccharides, but also specifically bind to oligosaccharides, detecting subtle differences in complex carbohydrate structure. They are distinct also from carbohydrate modifying enzymes, as they do not carry out glycosidase or glycosyl transferase reactions. All foods of plant origin contain specific lectins (Goldstein et al. 1980) and other anti-nutritional factors. These lectins when consumed in raw form, both in food and feed, may have serious and deleterious effects (Liener 1986).

Lectins are capable of specific binding to oligosaccharide structures present on cell surfaces, the extracellular matrix, and secreted glycoproteins. They are involved in intra- and intercellular glycan routing using oligosaccharides as postal-code equivalents and acting as defense molecules homing in on foreign or aberrant glycosignatures, as crosslinking agent in biosignaling and as coordinator of transient or firm cell-cell/cell-matrix contacts. By delineating the driving forces toward complex formation, knowledge about the causes for specificity can be turned into design of custom-made high-affinity ligands for clinical application, e.g. in anti-adhesion therapy, drug targeting or diagnostic histopathology (Kaltner and Gabius 2001). Galectins and collectins (mannose-binding lectins and surfactant proteins) illustrate the ability of endogenous glycan-binding proteins to act as cytokines, chemokines or growth factors, and thereby modulating innate and adaptive immune responses under physiological or pathological conditions.

G.S. Gupta et al., *Animal Lectins: Form, Function and Clinical Applications*,
DOI 10.1007/978-3-7091-1065-2_1,

Understanding the pathophysiologic relevance of endogenous lectins in vivo will reveal novel targets for immunointervention during chronic infection, autoimmunity, transplantation and cancer (Toscano et al. 2007). Lectins by the ability to bind specific carbohydrate structures mediate cell-cell and cell-pathogen interactions. Since, the biological function of lectins is their ability to recognize and bind to specific carbohydrate structures involving cells and proteins, some viruses use lectins to attach themselves to the cells of the host organism during infection. Cyanobacterial and algal lectins have become prominent in recent years due to their unique biophysical traits, such as exhibiting novel protein folds and unusually high carbohydrate affinity, and ability to potently inhibit human immuno-deficiency virus (HIV-1) entry through high affinity carbohydrate-mediated interactions with the HIV envelope glycoprotein gp120. The antiviralcyanobacterial lectin *Microcystis viridis* lectin (MVL), which contains two high affinity oligomannose binding sites, is one such example. However, studies representing the demonstration of dual catalytic activity and carbohydrate recognition for discrete oligosaccharides at the same carbohydrate-binding site in a lectin have started emerging (Bourgeois et al. 2010; Shahzad-ul-Hussan et al. 2009). A large variety of lectins are expressed in the human organism and found in a variety of sizes and shapes, but can be grouped in families with similar structural features. Several reviews have focussed on current knowledge of human lectins with a focus on biochemistry and pathobiochemistry (principles of protein glycosylation and defects of glycosylation as a basis of disease) and cell biology (protein sorting, exocytosis and endocytosis, apoptosis, cell adhesion, cell differentiation, and malignant transformation). The clinical significance of lectin-glycoconjugate interactions is described by example of inflammatory diseases, defects of immune defense, autoimmunity, infectious diseases, and tumor invasion/metastasis. Moreover, therapeutic perspectives of novel drugs that interfere with lectin-carbohydrate interactions have appeared in literatue.

The combining sites of lectins are diverse, although they are similar in the same family. Specificities of lectins are determined by the exact shape of the binding sites and the nature of the amino acid residues to which the carbohydrate is linked. Small changes in the structure of the sites, such as the substitution of only one or two amino acids, may result in marked changes in specificity. The carbohydrate is linked to the protein mainly through hydrogen bonds, with added contributions from van der Waals contacts and hydrophobic interactions. Coordination with metal ions may occasionally play a role too. Surface lectins of viruses, bacteria and protozoa serve as a means of adhesion to host cells: a prerequisite for the initiation of infection. Blocking the adhesion by carbohydrates that mimic those to which the lectins bind prevents infection by these organisms. The way is thus open for the development of anti-adhesive therapy against microbial diseases.

1.1.2 Lectins from Plants

Although lectins were first discovered more than 100 years ago in plants, the first lectin to be purified, in 1960s, on a large scale was Concanavalin A (ConA), which is now the most-used lectin for characterization and purification of sugar-containing molecules and cellular structures. Plant lectins with potent biological activity occur in foods like cereals and vegetables. The legume lectins are classified under one family since they have high sequence similarities, very similar tertiary structure and biophysical properties. Furthermore, most of these legume lectins exist as oligomers in nature, and all of them require metal ions for their carbohydrate-binding activities. Foods with high concentrations of lectins may be harmful if consumed in excess in uncooked or improperly uncooked form. Adverse effects may include nutritional deficiencies, and immune (allergic) reactions. Possibly, most effects of lectins are due to gastrointestinal distress through interaction of the lectins with the gut epithelial cells. The toxicity of plant lectins can lead to diarrhoea, nausea, bloating, and vomiting. Soybean is the most important grain legume crop, the seeds of which contain high activity of soybean lectins (soybean agglutinin or SBA). The SBA is able to disrupt small intestinal metabolism and damage small intestinal villi via the ability of lectins to bind with brush border surfaces in the distal part of small intestine.

Many plant lectins possess anticancer properties in vitro, in vivo, and in human trial studies. The plant lectins represent a well-defined class of anti-HIV (microbicidal) drugs with a novel HIV drug resistance profile different from those of other existing anti-HIV drugs. The function of lectins in plants is still uncertain. It is hypothesized that they may serve as mediators of the symbiosis between nitrogen fixing microorganisms, primarily rhizobium, and leguminous plants, a process of immense importance in both the nitrogen cycle of terrestrial life and in agriculture (Sharon and Lis 2003; Shanmugham et al. 2006). The large concentration of lectins in plant seeds decreases with growth, and suggests a role in plant germination and perhaps in the seed's survival itself. Many legume seeds have been proven to contain high lectin activity, termed as hemagglutinating activity. Lectins from legume plants, such as PHA or Concanavalin A, have been widely used as model systems to understand the molecular basis of how proteins recognize carbohydrates, because they are relatively easy to obtain and have a wide variety of sugar specificities. The structures of many plant lectins have been deduced. For the most part, leguminous lectins assemble into a compact β-barrel configuration devoid of α helices and dominated by two antiparallel pleated sheets. Interestingly, the organization of the antiparallel β sheets and the overall tertiary structure of the leguminous plant lectins are very similar to that seen for animal galectins, despite the fact that leguminous lectins and galectins share no sequence homology and galectins do not require metals for activity. In leguminous lectins, the metal-binding sites are located on a single long loop.

Saccharide-binding specificity of ConA from seeds of the jack bean *Canavalia ensifomist* is shown toward the α–D-mannopyranoside or α–D-glucopyranoside ring with unmodified hydroxyl groups at the 3, 4 and 6 positions. ConA activates T cells in polyclonal activation and regulates Ca^{2+} entry in human neutrophils. It specifically recognizes the pentasaccharide core [β-GlcNAc-(1–2)-α-Man-(1–3)-(β-GlcNAc-(1–2)-α-Man-(1–6)]-Man) of N-linked oligosaccharides (Bouckaert et al. 1999). The crystal structures of several legume lectins have led to a detailed insight of the atomic interactions between carbohydrates and proteins. Several plant and animal Lecins are used as biochemical tool in affinity chromatography for purifying proteins. In general, proteins may be characterized with respect to glycoforms and carbohydrate structure by means of affinity chromatography, blotting, affinity electrophoresis and affinity immunoelectrophoreis with lectins. Legume lectins have both calcium and manganese binding sites. The calcium ion puts several conserved residues in positions that are critical for carbohydrate binding (Brinda et al. 2005).

1.1.3 Lectins in Microorganisms

Numerous bacterial strains and viruses produce surface lectins. Adhesion of microorganisms to host tissues is the pre-requisite for the initiation of majority of infectious diseases. Microorganisms have evolved along with lectins that interact with glycoproteins, proteoglycans, and glycolipids. Adhesion of infectious organism is mediated by lectins present on the surface of organim that binds to complementary carbohydrates on the surface of the host tissues. Until early 1980s, only bacteria specific for mannose were identified, namely type 1 fimbriated strains *E. coli*. Since then, *E. coli* strains with diverse specificities were discovered (Sharon 2006). These include urinary strains carrying P fimbriae that are specific for galabiose (Galα4Gal), and neural S fimbriated strains specific for NeuAc(α2–3) Galβ3GalNAc. In addition, bacteria with affinities for other sugars have been described, e.g., *Neisseria gonorrhoea*, a genital pathogen, which recognizes Nacetyllactosamine (Galβ4GlcNAc, LacNAc). *Helicobacter pylori*, the causative agent of peptic ulcer expresses a number of distinct binding specificities (Sharon 2006). Several of these lectins recognize NeuAc (α2–3) Galβ4Glc (Sia3′Lac) and its N-acetylglucosamine analog (Sia3′LacNAc) while others are specific for the Leb determinant Fucα2Galβ3(Fucα4) GlcNAc (R: Gupta et al. 2009). A broad range of proteins bind high-mannose carbohydrates found on the surface of the envelope protein gp120 of the HIV and thus interfere with the viral life cycle, providing a potential new way of controlling HIV infection. These proteins interact with the carbohydrate moieties in different ways (Bourgeois et al. 2010). Among bacterial lectins, best characterized are type 1 (mannose specific) fimbrial lectins of *E. coli* that consist almost exclusively of one class of subunit with a molecular mass of 17 kDa. The various bacterial surface lectins appear to function primarily in the initiation of infection by mediating bacterial adherence to epithelial cells, e.g. in the urinary and gastrointestinal tracts (Sharon 2006). Further studies of these systems may lead to a deeper understanding of the molecular basis of infectious diseases, and to new approaches for their prevention.

1.1.4 Animal Lectins

During last few decades, interest in animal lectins has been greatly intensified after realization that they act as mediators of cell recognition in biological systems. Lectin activities specific for different monosaccharides or glycans (fucose, galactose, mannose, N-acetylglucosamine, N-acetylgalactosamine, N-acetylneuraminic acid, fucose and heparin) have been identified. Most of them show a cellular specificity and developmental regulation. But some of them seem to be involved in signaling events both intracellularly (nuclear lectins) or at the cell surface by autocrine and paracrine mechanisms. As early as 1988, most animal lectins were thought to belong to one of two primary structural families, the C-type and S-type (presently known as galectins) lectins. However, it is now clear that animal lectin activity is found in association with an astonishing diversity of primary structures. At least 15 structural families are known to exist at present (Table 1.1), while many other lectins have structures apparently unique amongst carbohydrate-binding proteins, although some of those "orphans" belong to recognised protein families that are otherwise not associated with sugar recognition. Furthermore, many animal lectins also bind structures other than carbohydrates via protein–protein, protein–lipid or protein–nucleic acid interactions.

1.2 The Animal Lectin Families

1.2.1 Structural Classification of Lectins

Lectins occur in plants, animals, bacteria and viruses. Initially described for their carbohydrate-binding activity (Sharon and Lis 2001), they are now recognised as a more diverse group of proteins, some of which are involved in protein-protein, protein-lipid or protein-nucleic acid interactions (Kilpatrick 2002). The lectin superfamily as classified

Table 1.1 Summary of some lectin families

S. no	Lectin family	Typical saccharide ligands	Subcellular location	Functions
1	Calnexin and calreticulin	Glc_1Man_9	ER	Protein sorting and as molecular chaperones in ER (Ireland et al. 2006).
2	M-type lectins related to the α-mannosidases	Man_8	ER	Degradation of glycolproteins (Molinari et al. 2003).
3	L-type lectins (ERGIC-53 and VIP-36, Pentraxins)	Various	ER, ERGIC, Golgi	Protein sorting in ER (Arar et al. 1995; Bottazzi et al. 2006).
4	P-type lectins	Man6-phosphate, phosphomannosyl receptors	Secretory pathway	Protein sorting post-Golgi, glycoprotein trafficking, enzyme targeting (Dahms and Hancock 2002).
5	C-type lectins (collectins, selectins, lecticans, others Volume 2)	Mannosides, galactosides, sialic acids, others	Cell membrane, extracellular	Cell adhesion/selectins, glycoprotein clearance, innate immunity (Zelensky and Gready 2005).
6	S-type lectins (galectins)	β-Galactosides	Cytoplasm, extracellular	Glycan crosslinking in the extracellular matrix (Elola et al. 2007).
7	I-type lectins (siglec-1, -2, -3, others)	Sialic acid and other glycosaminoglycans	Cell membrane	Cell adhesion (Angata and Brinkman-Van der Linden 2002).
8	R-type lectins (macrophage mannose receptor; discoidins)	Various	Golgi, Cell membrane	Enzyme targeting, glycoprotein hormone turnover (Lord et al. 2004; Pluddemann et al. 2006).
9	F-box lectins (Fbs1, Fbs2, Fbg3, others)	$GlcNAc_2$	Cytoplasm	Degradation of misfolded glycoproteins (Ho et al. 2006).
10	Fibrinogen-type lectins (ficolins,tachy-lectins 5A and 5B)	GlcNAc, GalNAc	Cell membrane, extracellular	Innate immunity. (Matsushita and Fujita 2001; Mali et al. 2006).
11	Chi-lectins (CHI3L1, CHI3L4, Ym1, Ym2)	Chito-oligosaccharides	Extracellular	Collagen metabolism (Tharanathan and Kittur 2003).
12	F-type lectins (Eel aggutinins/ fucolectins)	Fuc-terminating oligosaccharides	Extracellular	Innate immunity (Odom and Vasta 2006).
13	Intelectins (Endothelial lectin HL-1; Eglectin)	Gal, galactofuranose, pentoses	Extracellular/ cell membrane	Innate immunity, fertilization, embryogenesis (Tsuji et al. 2001).
14	Annexins (Annexin IV, V, VI)	Glycosaminoglycans, heparin and heparan sulfate	Cell membrane	Cell-adhesion, formation of apical-secretory vesicles (Moss and Morgan 2004; Turnay et al. 2005).

by SCOP (structural classification of proteins) (Lo Conte et al. 2002) comprises 15 families that include the legume lectins, β-glucanases, endoglucanases, sialidases, galectins, pentraxins and calnexin/calreticulin, among others. All of them belong to all-β class and have ConA (concanavalin-A-like) jelly-roll fold that can be seen. The jelly-roll motif consists of three sets of anti-parallel β-sheets, as can also be seen from the legume lectins shown in Fig. 1.1. There is a six-stranded flat 'back' sheet, a curved seven-stranded 'front' sheet and a short five-member sheet at the 'top' of the molecule. The sheets are connected by several loops of various lengths. The legume lectins have high sequence similarities, very similar tertiary structure and biophysical properties and hence are classified under one family. Since last decade, new information has been acquired from structural analysis of proteins belonging to the legume lectin family. Studies indicate that legume lectins belong to an interesting family of proteins with very similar tertiary structures but varied quaternary structures. Presently legume lectins offer good models to study the role of primary structures in finding the modes of quaternary association. Most of the legume lectins exist as oligomers in nature, and all of them require metal ions for their carbohydrate-binding activities. The metal-binding and carbohydrate-binding sites in various legume lectins have overlapping structures. However, these lectins do differ widely in their carbohydrate specificities and in their quaternary associations. Other lectins in this superfamily viz, galectins, pentraxins, calnexin and calreticulin have less sequence similarity with the legume lectins. Despite this disimilarities, they do share the same jelly-roll tertiary structure and show different quaternary associations. The unique three-dimensional structure of both monomeric and oligomeric proteins is encoded in their sequence. The biological functions of proteins are dependent on their tertiary and quaternary structures, and hence it is important to understand the determinants of quaternary association in proteins. Although a large number of investigations have been carried out in this direction, the underlying principles of protein oligomerization are yet to be completely understood. Brinda et al. (2005) reviewed the results of a legume lectins and on animal lectins, namely galectins, pentraxins, calnexin, calreticulin and rhesus rotavirus Vp4

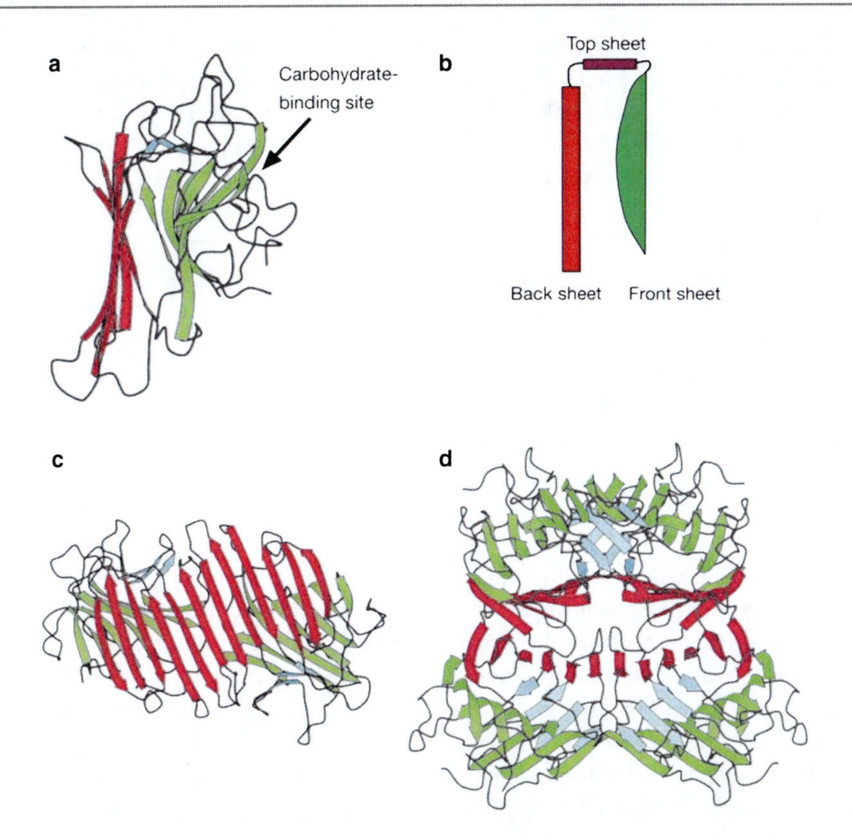

Fig. 1.1 Structure of concanavalin A (ConA). (**a**) The tertiary structure of the monomer is best described as a "jelly-roll fold." (**b**) This fold consists of a flat six-stranded antiparallel "back" β-sheet (*red*), a curved seven-stranded "front" β-sheet (*green*), and a five-stranded "top" sheet (*pink*) linked by loops of various lengths. (**c**) Dimerization of ConA involves antiparallel side-by-side alignment of the flat six-stranded "back" sheets, giving rise to the formation of a contiguous 12-stranded sheet (**d**) The tetramerization of ConA occurs by a back-to-back association of two dimers (Reprinted by permission from Srinivas et al. 2001© Elsevier)

sialic-acid-binding domain. The study has provided the signature sequence motifs for different kinds of quaternary association seen in lectins. The network association of lectin oligomers has enabled to detect the residues which make extensive interactions ('hubs') across the oligomeric interfaces that can be targetted for interface-destabilizing mutations. Brinda et al. (2005) illustrated the potential of such a representation in elucidating the structural determinants of protein-protein association in general and their significance to protein chemists and structural biologists in particular.

Animal lectins that recognize endogenous ligands appear to play a role in fertilization and development, and their function often involves cell-to-cell or cell-to-matrix interaction. Lectins that recognize exogenous ligands probably evolved for non-self discrimination, and they may be soluble or surface bound· The 15 families of animal lectins, defined in structural terms, participate in intra- and intercellular glycan routing using oligosaccharides as postal-code equivalents and acting as defense molecules homing in on foreign or aberrant glycosignatures, as crosslinking agent in biosignaling and as coordinator of transient or firm cell-cell/cell-matrix contacts. Ver tebrate lectins possess functionally diverse group of protein domains which can bind specific CRDs/oligosaccharide structures present on cell surfaces, the extracellular matrix, and secreted glycoproteins (Drickamer 1999). Of the eight well-established CRD groups, four contain lectins that are predominantly intracellular and four contain lectins that generally function outside the cell (Table 1.1). The intracellular lectins—calnexin family, M-type, L-type and P-type are located in luminal compartments of the secretory pathway and function in the trafficking, sorting and targeting of maturing glycoproteins. The extracellular lectins—C-type (collectins, selectins, mannose receptor, and others), R-type, siglecs, and S-type (galectins)—are either secreted into the extracellular matrix or body fluids, or localized to the plasma membrane, and mediate a range of functions including cell adhesion, cell signaling, glycoprotein clearance and pathogen recognition. Collagenous lectins such as mannan-binding protein (MBPs), ficolins, surfactant proteins (collectins), conglutinin, and related ruminant lectins are multimeric proteins with CRD aligned in a manner that facilitates binding to microbial surface polysaccharides. In soluble collectins such as MBL, the subunits possessing a single CRD associate to form a "bouquet"-like oligomer with all CRDs facing a potential target surface (Irache et al. 2008). However, in "cruciform" organization of conglutinin subunits, the single CRDs are arranged in a manner that can also cross-link multiple targets. Fibrinogen-type lectins include ficolins, tachylectins 5A and 5B, and Limax flavus (Spotted garden slug) agglutinin. These proteins have clear distinctions from one another, but they share a homologous fibrinogen-like domain used for carbohydrate binding. A less frequent organizational plan is the presence of tandem CRDs encoded within a polypeptide, such as observed in macrophage mannose receptor, the immulectins, and tandem-type galectins.

Recent findings point to the existence of additional new groups of animal lectins—F-box lectins, chitinase-like lectins (Chi-lectins), F-type lectins, intelectins—some of which have roles complementary to those of the well-established lectin families (Table 1.1). Some of the unclassified orphan lectins group includes: amphoterin, Cel-II, complement factor H, thrombospondin, sailic acid-binding lectins, adherence lectin, and cytokins (such as tumor-necrosis factor, and annexin and several interleukins). More detailed information on the structure, sugar-binding activity, biological function and evolution of proteins (Cambi and Figdor 2003, 2009) in each of the lectin families as well as annotated sequence alignments can be obtained in relevant chapters of this book.

1.3 C-Type Lectins (CLEC)

1.3.1 Identification of CLEC

A C-type lectin (CLEC) or a C-type lectin receptor (CLR) is a type of carbohydrate-binding protein domain known as a lectin. The C-type designation is from their requirement for calcium for binding. The C-type lectin superfamily is a large group of proteins which is characterized as having at least one carbohydrate recognition domain (CRD) (Drickamer 1999). The C-type lectin fold has been found in more than 1,000 proteins, and it represents a ligand-binding motif that is not necessarily restricted to binding sugars. Proteins that contain C-type lectin domains have a diverse range of functions including cell-cell adhesion, immune response to pathogens and apoptosis. However, many C-type lectins actually lack calcium- and carbohydrate-binding elements and thereby have been termed C-type lectin-like proteins. Therefore, CLR or CLEC broadly denotes proteins with a C-type lectin domain, regardless of their ability to bind sugars.

1.3.2 C-Type Lectin Like Domain (CTLD/CLRD)

Use of terms of "C-type lectin", "C-type lectin domain" (CLRD), "C-type lectin-like domain" (also abbreviated as CLRD), often used interchangeably and use of CRD in the literature, have been clarified by Zelensky and Gready (2005). With the large number of CLR sequences and structures now available, studies indicate that the implications of the CRD domain are broad and vary widely in function. In metazoans, most proteins with a CLR are not lectins. Moreover, proteins use the C-type lectin fold to bind other proteins, lipids, inorganic molecules (e.g., Ca_2CO_3), or even ice (e.g., the antifreeze glycoproteins). An increasing number of studies show that "atypical" C-type lectin-like proteins are involved in regulatory processes pertaining to various aspects of the immune system. Examples include the NK cell inhibitory receptor Ly49A, a C-type lectin-like protein, which is shown to complex with the MHC class I ligand (Correa and Raulet 1995), and the C-type lectin-like protein mast cell function-associated antigen which is involved in the inhibition of IgE-Fc γRI mediated degranulation of mast cell granules (Guthmann et al. 1995). Yet glycan binding by the C-type lectins is always Ca^{2+}-dependent because of specific amino acid residues that coordinate Ca^{2+} and bind the hydroxyl groups of sugars (Cummings and McEver 2009). To resolve the contradiction, a more general term C-type lectin like domain (CLRD) was introduced to distinguish a group of Ca^{2+}-independent carbohydrate-binding animal proteins from the Ca^{2+}-dependent C-type of animal lectins (CLR/CLEC). The usage of this term is however, somewhat ambiguous, as it is used both as a general name for the group of domains with sequence similarity to C-type lectin CRDs (regardless of the carbohydrate-binding properties), and as a name of the subset of such domains that do not bind carbohydrates, with the subset that does bind carbohydrates being called C-type CRDs. Also both 'C-type CRD' and 'C-type lectin domain' terms are still being used in relation to the C-type lectin homologues that do not bind carbohydrate, and the group of proteins containing the domain is still often called the 'C-type lectin family' or 'C-type lectins', although most of them are not in fact lectins. The abbreviation CRD is used in a more general meaning of 'carbohydrate-recognition domain', which encompasses domains from different lectin groups. Occasionally CRD is also used to designate the short amino-acid motifs (i.e. amino-acid domain) within CLRDs that direCLRy interact with Ca^{2+} and carbohydrate (Zelensky and Gready 2005). In this book authors use the term C-type lectin domain or C-type lectin-like domain (CLRD) interchangeably in its broadest definition to refer to protein domains that are homologous to the CRDs of the C-type lectins, or which have structure resembling the structure of the prototypic C-type lectin CRD or as used by different researchers in their work. More over, due to contradictions (Zelensky and Gready 2005) and uncertainties which may arise in future studies our sequence of chapters is not based on structure databases as in the SCOP; instead chapters are

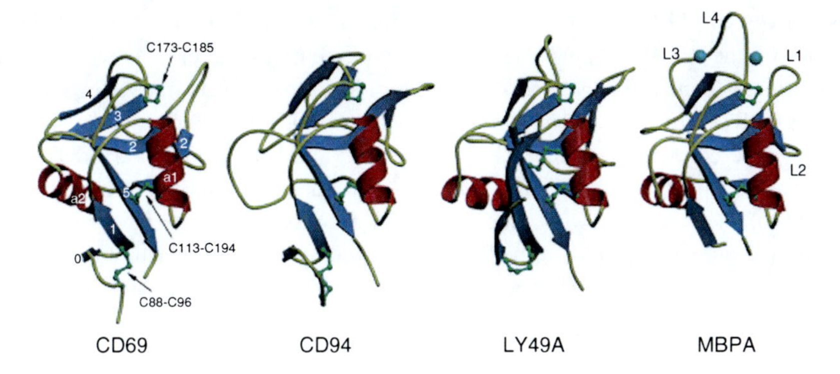

Fig. 1.2 CLRD Structures: Ribbon diagrams of the natural killer domains (NKDs) from human CD69 (Llera et al. 2001), a CTLD, divergent from true C-type lectins, human CD94 (PDB ID: 1B6E), mouse Ly49A (PDB ID: 1QO3), and the CRD of rat MBP-A (PDB ID: 1YTT) are shown in a common orientation obtained by pairwise superpositions. Ca^{2+} ions bound to MBP-A are shown as light blue spheres, whereas its loop regions without regular secondary structure, three of them involved in Ca^{2+} coordination, are labeled L1 to L4. Secondary structure elements in CD69 have been labeled following the numbering for MBP-A, prototype of the family. Therefore, the first β strand, which is absent in MBP-A and is characteristic of the long-form CTLDs, has been labeled as β0, whereas the strand that forms a β-hairpin with strand β2 has been named β2'. Differences in the orientation of helix α2, which is absent in CD94, are evident (Adapted from Llera et al. 2001)

linked more to cell functions or cell biology. Thus the C-type lectin fold is an evolutionarily ancient structure that is adaptable for many uses.

1.3.3 Classification of CLRD-Containing Proteins

The C-type lectin superfamily is a large group of proteins that are characterised by the presence of one or more CLRDs. The superfamily is divided into 17 groups based on their phylogeny and domain organisation (Drickamer and Fadden 2002; Zelensky and Gready 2005). Despite the presence of a highly conserved domain, C-type lectins are functionally diverse and have been implicated in various processes including cell adhesion, tissue integration and remodelling, platelet activation, complement activation, pathogen recognition, endocytosis, and phagocytosis.

C-type lectins can be divided into seventeen subgroups based on additional non-lectin domains and gene structure: The CLRD-containing proteins have been sorted based on two independent sets of criteria. The two approaches give essentially the same results, indicating that members of each group are derived from a common ancestor, which had already acquired the domain architecture that is characteristic of the group. Initially classified into seven subgroups (I to VII) based on the order of various protein domains in each protein: (I) hyalectans, (II) asialoglycoprotein receptors, (III) collectins, (IV) selectins, (V) NK cell receptors, (VI) Endocytic receptors, and (VII) simple (single domain) lectins (Drickamer 1993; McGreal et al. 2004). This classification was subsequently updated in 2002, leading to seven additional groups (VIII to XIV) (Drickamer and Fadden 2002). A further three subgroups were added (XV to XVII) by Zelensky and Gready (2005). The section of animal lectins genomics resource includes information on the structure and function of proteins in each group, as well as annotated sequence alignments, and a comprehensive database for all human and mouse CLRD-containing proteins (Chaps. 22–40).

1.3.4 The CLRD Fold

The CLRD fold has a double-loop structure (Fig. 1.2). The overall domain is a loop, with its N- and C-terminal β strands (β1, β5) coming close together to form an antiparallel β-sheet. The second loop, which is called the long loop region, lies within the domain; it enters and exits the core domain at the same location. Four cysteines (C1-C4), which are the most conserved CLRD residues, form disulfide bridges at the bases of the loops: C1 and C4 link β5 and α1 (the whole domain loop) and C2 and C3 link β3 and β5 (the long loop region). The rest of the chain forms two flanking α helices (α1 and α2) and the second ('top') β-sheet, formed by strands β2, β3 and β1. The long loop region is involved in Ca^{2+}-dependent carbohydrate binding, and in domain-swapping dimerization of some CLRDs (Fig. 1.2), which occurs via a unique mechanism (Mizuno et al. 1997; Feinberg et al. 2000; Mizuno et al. 2001; Hirotsu et al. 2001; Liu and Eisenberg 2002). For conserved positions involved in CLRD fold maintenance and their structural roles readers are referred elsewhere (Zelensky and Gready 2003). In addition to four conserved cysteines, one other sequence feature and the highly conserved 'WIGL' motif is located on the β2 strand and serves as a useful landmark for sequence analysis (Chap. 22).

1.3.4.1 Ca^{2+}-Binding Sites in CLRDs

Four Ca^{2+}-binding sites in the CLRD domain recur in CLRD structures from different groups. The site occupancy depends on the particular CLRD sequence and on the crystallization conditions; in different known structures zero, one, two or three sites are occupied. Sites 1, 2 and 3 are located in the upper lobe of the structure, while site 4 is involved in salt bridge formation between α2 and the β1/β5 sheet. Sites 1 and 2 were observed in the structure of rat MBP-A complexed with holmium, which was the first CLRD structure determined. Site 3 was first observed in the MBP-A complex with Ca^{2+} and oligomannose asparaginyl-oligosaccharide. It is located very close to site 1 and all the side chains coordinating Ca^{2+} in site 3 are involved in site 1 formation. As biochemical data indicate that MBP-A binds only two calcium atoms (Loeb and Drickamer 1988), Ca^{2+}-binding site 3 is considered a crystallographic artifact (Zelensky and Gready 2005).

1.4 Disulfide Bonds in Lectins and Secondary Structure

1.4.1 Disulfide Bond

Disulfide bond is a covalent bond, usually derived by the coupling of two thiol groups. The disulfide bond is a strong typical bond with dissociation energy being 60 kcal/mole. Disulfide bonds play an important role in the folding and stability of proteins, usually proteins which are secreted in to extracellular medium. Since most cellular compartments are reducing environments, disulfide bonds are generally unstable in the cytosol, with some exceptions. In proteins, disulfide bonds are formed between the thiol groups of cysteine amino acids. The methionine, cannot form disulfide bonds. A disulfide bond is typically denoted by hyphenating the abbreviations for cysteine, e.g., when referring to Ribonuclease A the "Cys26-Cys84 disulfide bond", or the "26–84 disulfide bond", or most simply as "C26-C84" where the disulfide bond is understood and does not need to be mentioned. The prototype of a protein disulfide bond is the two-amino-acid peptide, cystine, which is composed of two cysteine amino acids joined by a disulfide bond.The disulfide bond stabilizes the folded form of a protein in several ways: (1) It holds two portions of the protein together, biasing the protein towards the folded topology. Stated differently, the disulfide bond destabilizes the unfolded form of the protein by lowering its entropy. (2) The disulfide bond may form the nucleus of a hydrophobic core of the folded protein, i.e., local hydrophobic residues may condense around the disulfide bond and onto each other through hydrophobic interactions. (3) Related to #1 and #2, the disulfide bond link two segments of the protein chain, the disulfide bond increases the effective local concentration of protein residues and lowers the effective local concentration of water molecules. Since water molecules attack amide-amide hydrogen bonds and break up secondary structure, a disulfide bond stabilizes secondary structure in its vicinity. For example, researchers have identified several pairs of peptides that are unstructured in isolation, but adopt stable secondary and tertiary structure upon forming a disulfide bond between them.

1.4.2 Pathway for Disulfide Bond Formation in the ER of Eukaryotic Cells

The folding of many secretory proteins depends upon the formation of disulfide bonds. The formation of intra- and interchain disulfide bonds constitutes an integral part of the maturation of most secretory and membrane-bound proteins in the endoplasmic reticulum. Evidence indicates that members of the protein disulfide isomerase (PDI) superfamily are part of the machinery needed for proper oxidation and isomerization of disulfide bonds. Recent advances in genetics and cell biology have outlined a core pathway for disulfide bond formation in the ER of eukaryotic cells. In this pathway, oxidizing equivalents flow from the ER membrane protein Ero1p to secretory proteins via protein disulfide isomerase (PDI) (Frand et al. 2000; Ellgaard and Rudock 2005). Models based on in vitro studies predict that the formation of mixed disulfide bonds between oxidoreductase and substrate is intermediate in the generation of the native intrachain disulfide bond in the substrate polypeptide (Gruber et al. 2006). The ER-resident oxidoreductases PDI, together with the lectins calnexin and calreticulin, are central in glycoprotein folding in the endoplasmic reticulum of mammalian cells (Molinari and Helenius 1999). Studies suggest that the calnexin cycle has evolved with a specialized oxidoreductase to facilitate native disulfide formation in complex glycoproteins (Jessop et al. 2009).

1.4.3 Arrangement of Disulfide Bonds in CLRDs

Drickamer and Dodd (1999) summarized positions of six different disulfide bonds in CLRDs (Fig. 1.3). Chemical evidence for the presence of each of these bonds, except number 4, has been provided in at least one CLRD (Fuhlendorff et al. 1987; Usami et al. 1993). The positions of disulfide bonds designated 1, 2, and 3 have been demonstrated by X-ray crystallography as well (Weis et al. 1991; Nielsen et al. 1991), while homology modeling of CLRDs containing disulfide bonds 5 and 7 shows that they

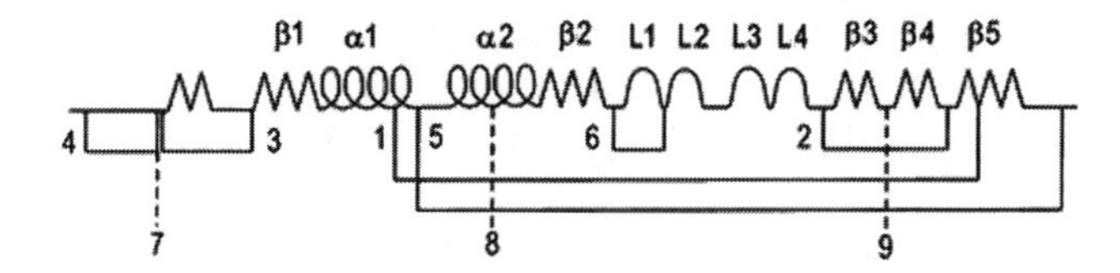

Fig. 1.3 Disulfide bonds in CLRDs. Secondary structure shared by most CLRDs is summarized, with coils representing α-helices, jagged lines denoting β-strands and loops shown as curved segments. The number of these elements corresponds to the secondary structure organisation of rat serum mannose-binding protein (Weis et al. 1991). Potential disulfide bonds within the CLRD are numbered 1 through 6 and cysteines that participate in interchain disulfide bonds are numbered 7 through 9 (Adapted by permission from Drickamer and Dodd 1999 © Oxford University Press)

could readily be accommodated into the C-type lectin fold. The patterns of cysteine residues in the CLRDs from *C.elegans* are consistent with the presence of disulfide bonds in each of the arrangements shown in Fig. 1.3 except for bond type 4. No additional pairs of cysteine residues within the CLRDs are consistently evident for any of the subgroups, indicating that the cysteine residues are mostly involved in disulfide bonds of the types already characterized in vertebrate homologues. CLRDs lacking one of a pair of cysteine residues almost invariably also lack the cysteine side chain to which the first residue would be linked. Like the CLRDs from other organisms, those from *C.elegans* each contain a subset of the possible disulfide bonds. CLRDs in a given subgroup generally show the same disulfide bonds, although a few domains contain extra unique pairs of cysteine residues that might form disulfides. CEL-I from the sea cucumber, Cucumaria echinata is composed of two identical subunits held by a single disulfide bond. A subunit of CEL-I is composed of 140 amino acid residues. Two intrachain (Cys3-Cys14 and Cys31-Cys135) and one interchain (Cys36) disulfide bonds were also identified from an analysis of the cystine-containing peptides obtained from the intact protein (Yamanishi et al. 2007). Thus, the similarity in disulfide bond structure in each subgroup reflects the overall similarity in sequence of the CLRDs (Drickamer and Dodd 1999).

1.4.4 Functional Role of Disulfides in CLRD of Vertebrates

Single cysteine residues appear in several of these groups at positions 7 and 8 as well as at a unique position 9. The turn between β-strands 3 and 4 is exposed on the surface of the domain, so it is expected that cysteine residues at position 9 would be accessible for formation of disulfide bonds. It is possible that such bonds could form with other cysteine residues within the same polypeptide but outside the CLRDs. However, no likely pairing partner is evident for any of these residues, suggesting that they are more likely to form interchain disulfide bonds. Homo- and hetero-dimer formation through cysteine residues at positions 7 and 8 has been particularly well documented in snake venom proteins containing CLRDs (Atoda et al. 1991; Usami et al. 1993). The botrocetin, which promotes platelet agglutination in the presence of von Willebrand factor, from venom of the snake Bothrops jararaca is a heterodimer composed of the α subunit and the β subunit held together by a disulfide bond. Seven disulfide bonds link half-cystine residues 2–13, 30–128, and 103–120 of the α subunit; 2–13, 30–121, and 98–113 of the β subunit; and 80 of the α subunit to 75 of the β subunit. In terms of amino acid sequence and disulfide bond location, two-chain botrocetin is homologous to echinoidin (a sea urchin lectin) and other C-type lectins (Usami et al. 1993). The disulfide bond pattern of Trimeresurus stejnegeri lectin (TSL), a CTLD, showed four intrachain disulfide bonds: Cys3-Cys14, Cys31-Cys131, Cys38-Cys133 and Cys106-Cys123, and two interchain linkages, Cys2-Cys2 and Cys86-Cys86 (Zeng et al. 2001). The antifreeze polypeptide (AFP) from the sea raven, Hemitripterus americanus, a member of the cystine-rich class of blood antifreeze proteins, contains 129 residues with 10 half-cystine residues and all 10 half-cystine residues appeared to be involved in disulfide bond formation. The disulfide bonds are linked at Cys7 to Cys18, Cys35 to Cys125, and Cys89 to Cys117. Similarities in covalent structure suggested that sea raven AFP, pancreatic stone protein, and several lectin-binding proteins may possess a common fold (Ng and Hew 1992).

Functional rat or human asialoglycoprotein receptors (ASGP-Rs), the galactose-specific C-type lectins, are hetero-oligomeric integral membrane glycoproteins. Rat ASGP-R contains three subunits, designated rat hepatic lectins (RHL) 1, 2, and 3; human ASGP-R contains two subunits, HHL1 and HHL2. Both receptors are covalently modified by fatty acylation (1996). Unfolded forms of the HHL2 subunit of the human ASGP-Rs are degraded in the ER, whereas folded forms of the protein can mature to the cell surface (Wikström and Lodish 1993). Deacylation ofASGP-Rs with hydroxylamine results in the spontaneous formation of dimers through reversible disulfide bonds, indicating that deacylation concomitantly generates free thiol groups. Results also show that Cys57 within the transmembrane domain of HHL1 is not normally palmitoylated. Thus, Cys35 in RHL1, Cys54 in RHL2 and RHL3, and Cys36 in HHL1 are fatty acylated, where as Cys57 in HHL1 and probably Cys56 in RHL1 are not palmitoylated (Zeng and Weigel 1996). Eosinophil granule major basic protein 2 (MBP2 or major basic protein homolog) is a paralog of major basic protein (MBP1) and, similar to MBP1, is cytotoxic and cytostimulatory in vitro. MBP2, a small protein of 13.4 kDa molecular weight, contains 10 cysteine residues. Mass spectrometry shows two cystine disulfide linkages (Cys20-Cys115 and Cys92-Cys107) and 6 cysteine residues with free sulfhydryl groups (Cys2, Cys23, Cys42, Cys43, Cys68, and Cys96). MBP2, similar to MBP1, has conserved motifs in common with C-type lectins. The disulfide bond locations are conserved among human MBP1, MBP2 and C-type lectins (Wagner et al. 2007; Swaminathan et al. 2001).

The biological functions of rat surfactant protein-A (SP-A), an oligomer composed of 18 polypeptide subunits are dependent on intact disulfide bonds. Reducible and collagenase-reversible covalent linkages of as many as six or more subunits in the molecule indicate the presence of at least two NH2-terminal interchain disulfide bonds. However, the reported primary structure of rat SP-A predicts that only Cys6 in this region is available for interchain disulfide formation. Direct evidence for a second disulfide bridge was obtained by analyses of a set of three mutant SP-As with telescoping deletions from the reported NH2-terminus. Two of the truncated recombinant proteins formed reducible dimers despite deletion of the domain containing Cys6. A novel post translational modification results in naturally occurring cysteinyl isoforms of rat SP-A which are essential for multimer formation (Elhalwagi et al. 1997). Pulmonary SP-D is assembled predominantly as dodecamers consisting of four homotrimeric subunits each. Association of these subunits is stabilized by interchain disulfide bonds involving two conserved amino-terminal cysteine residues (Cys15 and Cys20). Mutant recombinant rat SP-D lacking these residues (RrSP-Dser15/20) is secreted in cell culture as trimeric subunits rather than as dodecamers. Disulfide cross-linked SP-D oligomers are required for the regulation of surfactant phospholipid homeostasis and the prevention of emphysema and foamy macrophages in vivo (Zhang et al. 2001).

A knowledge-based approach and energy minimization suggest that three-dimensional structures of each of the lectin domains of P-selectin, E-selectin, and L-selectin contains 118 amino acids. The structures thus found for P-, L-, and E-selectin lectin domains share a common feature, i.e., they all contain two α-helices, and two antiparallel β-sheets of which one is formed by two strands (strands 1 and 5) and the other by three strands (strands 2, 3, and 4). Besides, they all possess two intact disulfide bonds formed by the pair of Cys19-Cys117, and the pair of Cys90-Cys109. A notable feature is the convergence-divergence duality of the 77–107 polypeptide in the three domains; i.e., part of the peptide is folded into a closely similar conformation, and part of it into a highly different one (Chou 1996).

1.4.5 Disulfide Bonds in Ca^{2+}-Independent Lectins

The lectin from Japanese frog (Rana japonica) eggs, which specifically agglutinates transformed cells, is a single-chain protein consisting of 111 residues, with a pyroglutamyl residue at the amino terminus. Four disulfide bonds link half-cystinyl residue 19–72, 34–82, 52–97, and 94–111. The sequence and the location of the disulfide bonds are highly homologous to those of bull frog (Rana catesbeiana) egg S-lectin. They are also homologous to human angiogenin, a tumor angiogenesis factor, and a family of pancreatic ribonucleases (Kamiya et al. 1990). The positions of three disulfide bonds of Selenocosmia huwena lectin-I (SHL-I) from the venom of the Chinese bird spider S. huwena could be assigned as Cys2-Cys14, Cys7-Cys19, Cys13-Cys26 (Li and Liang 1999). Galactose-specific lectin SEL 24 K from the egg of Chinook salmon Oncorhynchus tshawytscha contains seven disulfide bonds and one pair of free cysteines. After proteolysis, peptides containing one or two disulfide bonds were identified by reduction and mass spectral comparison (Yu et al. 2007).

ERGIC-53 is present exclusively as a hexameric complex in cells. However, the hexamers exist in two forms, one as a disulfide-linked, SDS-resistant complex, and the other as a SDS-sensitive complex made up of three disulfide-linked dimers that are likely to interact through the coiled-coil domains present in the luminal part of the protein (Neve et al. 2005). Recombinant human galectin-1 secretory protein exists as an oxidized form containing three intramolecular disulfide bonds (Cys2-Cys130, Cys16-Cys88 and Cys42-Cys60). Galectin-1 promotes axonal regeneration only in the oxidized form containing three intramolecular disulfide bonds, not in the reduced form which exhibits lectin activity (Inagaki et al. 2000). Ficolins form trimer-based multimers that are N-terminally linked by disulfide bonds (Ohashi and Erickson 2004; Hummelshoj et al. 2008). Retinoschisin or RS1 is a discoidin domain-containing protein encoded by the gene responsible for X-linked retinoschisis. The RS1 functions as a cell adhesion protein to maintain the cellular organization and synaptic structure of the retina. Cys residues involved in intramolecular and intermolecular disulfide bonds are essential for protein folding and subunit assembly (Kiedzierska et al. 2007; Wu and Molday 2003).

1.5 Functions of Lectins

Interactions between cells or between cell and substratum involve specific receptors and their ligands. Lectin-like receptors are involved in signal transduction in a great variety of ways; at the molecular level; they mimic in most of the cases the function of growth factor receptor either coupled to tyrosine kinase activity or to heterotrimeric G protein. They lead to a multiplicity of cellular events following their activation depending on factors such as cellular type, species and/or tissue. Nevertheless the potential of surface lectins as transducers is emphasized by the observation that in a few cases lectin-like receptors induce either novel signal transduction mechanism or new intracellular events with regards to what it has been observed as a consequence of growth factor receptor activation. This observation brings the idea that lectins may offer, as cell surface transducers, an alternative or additional signaling potential to cell. Lectins have no single function, and their relative abundance is not necessarily related to the importance of their function. Immune responses are mediated mainly by protein/protein interactions. In addition, protein/carbohydrate (sugar) interactions through specific lectin and chi-lectin are also involved in several immune and biological responses under not only the state of health but also inflammatory conditions. Interestingly, recent studies have identified unexpected roles of animal lectins and chi-lectin in intestinal inflammation. During fertilization, the sea urchin sperm acrosome reaction (AR), an ion channel-regulated event, is triggered by glycoproteins in egg jelly (EJ). During last two decades crystal structures of several lectins have been solved. These structures provide important insights into how these transmembrane-spanning receptors function (Rini and Lobsanov 1999).

1.5.1 Lectins in Immune System

The immune system consists of various types of cells and molecules that specifically interact with each other to initiate the host defense mechanism. Among cell surface receptors, the lectins are of peculiar interest because glycolipids, glycoproteins and proteoglycans have been shown to interact with lectins on the surface of animal cells. To initiate immune responses against infection, antigen presenting cells (APC) must recognize and react to microbes. Recognition is achieved by interaction of particular surface receptors on APC with corresponding surface molecules on infectious agents. The three types of professional APC are: (1) mature dendritic cells, found in lymphoid tissues and derived from immature tissue dendritic cells that interact with many distinct types of pathogens (Jenner et al. 2006), (2) macrophages, specialized to internalize extracellular pathogens, especially after they have been coated with antibody, and to present their antigens, and (3) B cells, which have antigen-specific receptors that enable them to internalize large amounts of specific antigen, process it, and present it naïve T cell for activation. Dendritic cells and Langerhans cells are specialized for the recognition of pathogens and have a pivotal role in the control of immunity. They are present in essentially every organ and tissue, where they operate at the interface of innate and acquired immunity. Recently, several C-type lectin and lectin-like receptors have been characterized that are expressed abundantly on the surface of these professional antigen-presenting cells. It is now becoming clear that lectin receptors not only serve as antigen receptors but also regulate the migration of dendritic cells and their interaction with lymphocytes. By contrast, pattern recognition receptors (PRR) recognize and interact with pathogens direCLRy. In addition to scavenger receptors and toll-like receptors, the PRR include the C-type lectin-like receptors (CLR) that bind carbohydrate moieties of many pathogens (Weis et al. 1998). The CLR include the following: (1) mannose receptors for mannose or its polymers (2) mannose-binding lectins for encapsulated group B or C meningococci, (3) DC-SIGN and structurally related receptors (DC-SIGNR) for mannose on HIV, *Leishmania*, and *Mycobacteria*, and (4) dectin-1 and dectin-2 for β-glucan on yeasts and fungi (Sato et al. 2006, 2009).

Collectins are a family of collagenous calcium-dependent defense lectins in animals. Their polypeptide chains consist of four regions: a cysteine-rich N-terminal domain, a collagen-like region, an alpha-helical coiled-coil neck domain and a C-terminal lectin or carbohydrate-recognition domain. These polypeptide chains form trimers that may assemble into larger oligomers. The best studied family members are the mannan-binding lectin, which is secreted into the blood by the liver, and the surfactant proteins A and D, which are secreted into the pulmonary alveolar and airway lining fluid. The collectins represent an important group of pattern recognition molecules, which bind to oligosaccharide structures and/or lipid moities on the surface of microorganisms. They bind preferentially to monosaccharide units of the mannose type, which present two vicinal hydroxyl groups in an equatorial position. High-affinity interactions between collectins and microorganisms depend, on the one hand, on the high density of the carbohydrate ligands on the microbial surface, and on the other, on the degree of oligomerization of the collectin. Apart from binding to microorganisms, the collectins can interact with receptors on host cells. Binding of collectins to microorganisms may facilitate microbial clearance through aggregation, complement activation, opsonization and activation of phagocytosis, and inhibition of microbial growth. In addition, the collectins can modulate inflammatory and allergic responses, affect apoptotic cell clearance and modulate the adaptive immune system (van de Wetering et al. 2004; Kerrigan and Brown 2009).

Lung surfactant proteins A and D bind essential carbohydrate and lipid antigens found on the surface of microorganisms via low affinity C-type lectin domains and regulate the host's response by binding to immune cell surface receptors. SP-A and SP-D contribute to host defense against respiratory viral infection. The most extensive body of evidence relates to influenza A viruses (IAV), and evidence from gene-deleted mice also indicate a role for surfactant collectins in defense against respiratory syncytial virus (RSV) and adenovirus. Despite extensive structural similarity, the two proteins show many functional differences and considerable divergence in their interactions with microbial surface components, surfactant lipids, and other ligands. Recent data have also highlighted their involvement in clearance of apoptotic cells, hypersensitivity and a number of lung diseases. The relative importance of antiviral versus anti-inflammatory effects of SP-A and SP-D in viral infections and the potential use of these collectins as therapeutics for viral infections are under investigation. Current research suggests that structural biology approaches will help to elucidate the molecular basis of pulmonary collectin-ligand recognition and facilitate development of new therapeutics based upon SP-A and SP-D (Palaniyar et al. 2002; Seaton et al. 2010; Waters et al. 2009).

Soluble mediators, including complement components and the MBL make an important contribution to innate immune protection and work along with epithelial barriers, cellular defenses such as phagocytosis, and pattern-recognition receptors that trigger pro-inflammatory signaling cascades. These four aspects of the innate immune system act in concert to protect from pathogen invasion. Mannan-binding lectin (MBL), L-ficolin, M-ficolin and H-ficolin are all complement activating soluble pattern recognition molecules with recognition domains linked to collagen-like regions. All four may form complexes with four structurally related proteins, the three MBL-associated serine proteases (MASPs), MASP-1, MASP-2, and MASP-3, and a smaller MBL-associated protein (MAp19). The four recognition molecules recognize patterns of carbohydrate or acetyl-group containing ligands. After binding to the relevant targets all four are able to activate the complement system. We thus have a system where four different and/or overlapping patterns of microbial origin or patterns of altered-self may be recognized, but in all cases the signaling molecules, the MASPs, are shared. The clinical impact of deficiencies of MBL and MASPs in humans have been reported (Thiel 2007; Ip et al. 2009). A similar lectin-based complement system, consisting of the lectin-protease complex and C3, is present in ascidians, our closest invertebrate relatives, and functions in an opsonic manner (Fujita et al. 2004).

Ficolins are soluble oligomeric proteins composed of trimeric collagen-like regions linked to fibrinogen-related domains (FReDs) that have the ability to sense molecular patterns on both pathogens and apoptotic cell surfaces and activate the complement system. The ficolins have acetyl-binding properties, which have been localized to different binding sites in the FReD-region (Thomsen et al. 2011). From a structural point of view, ficolins are assembled from basal trimeric subunits comprising a collagen-like triple helix and a globular domain composed of 3 fibrinogen-like domains. The globular domains are responsible for sensing danger signals whereas the collagen-like stalks provide a link with immune effectors. The structure and recognition properties of the **3** human ficolins have been studied in recent years by crystallographic analysis. The ligand binding sites have been identified in the 3 ficolins and their recognition mechanisms have been characterized at the atomic level (Garlatti et al. 2009).

The mammalian natural killer gene complex (NKC) contains several families of type II transmembrane C-type lectin-like receptors (CLRs) that are best known for their involvement in the detection of virally infected or transformed cells, through the recognition of endogenous (or self) proteinacious ligands. However, certain CLR families within the NKC, particularly those expressed by myeloid cells, recognize structurally diverse ligands and perform a variety of other immune and homoeostatic functions. One such family is the 'Dectin-1 cluster' of CLRs, which includes MICL, CLEC-2, CLEC12B, CLEC9A, CLEC-1, Dectin-1 and LOX-1. We reviewed each of these CLRs, exploring our current understanding of their ligands and functions and highlighting where they have provided new insights into the underlying mechanisms of immunity and homeostasis (Huysamen and Brown 2009).

The Dectin-2 family of C-type lectins includes Dectin-2, BDCA-2, DCIR, DCAR, Clecsf8 and Mincle whose genes are clustered in the telomeric region of the NK-gene cluster on mouse chromosome 6 and human chromosome 12. These type II receptors are expressed on myeloid and non-myeloid cells and contain a single extracellular carbohydrate recognition domain and have diverse functions in both immunity and homeostasis. DCIR is the only member of the family which contains a cytoplasmic signaling motif and has been shown to act as an inhibitory receptor, while BDCA-2, Dectin-2, DCAR and Mincle all associate with FcRγ chain to induce cellular activation, including phagocytosis and cytokine production. Dectin-2 and Mincle have been shown to act as pattern recognition receptors for fungi, while DCIR acts as an attachment factor for HIV. In addition to pathogen recognition, DCIR has been shown to be pivotal in preventing autoimmune disease by controlling DC proliferation, whereas Mincle recognizes a nuclear protein released by necrotic cells (Graham and Brown 2009). Binding of fungal PAMPs to PRRs triggers the activation of innate effector cells. Recent findings underscore the role of DCs in relaying PAMP information through their PRRs to stimulate the adaptive response. In agreement, deficiencies in certain PRRs strongly impair survival to Candida infections in mice and are associated with enhanced susceptibility to mucocutaneous fungal infections in humans. Understanding the complex signaling networks protecting the host against fungal pathogens is a challenging problem (Bourgeois et al. 2010).

Macrophage lectins contribute to host defence by a variety of mechanisms. The best characterised, mannose receptor (MR) and complement receptor three (CR3), are both able to mediate phagocytosis of pathogenic microbes and induce intracellular killing mechanisms. MR is a C-type lectin primarily expressed by macrophages and dendritic cells. Its three distinct extracellular binding sites recognise a wide range of both endogenous and exogenous ligands. The MR has been implicated in both homeostatic processes and pathogen recognition. However, the function of MR in host defence is not yet clearly understood as MR-deficient animals do not display enhanced susceptibility to pathogens bearing MR ligands. The regulation of the effector functions induced via MR is complex, and may involve both host and microbial factors (Gazi and Martinez-Pomares 2009; Linehan et al. 2000).

1.5.2 Lectins in Nervous Tissue

There is increasing evidence that lectins are widely distributed in mammalian tissues, including the nervous tissue. Based on histochemical techniques using neoglycoproteins, lectin activities specific for different monosaccharides or glycans have been identified (fucose, galactose, mannose, N-acetylglucosamine, N-acetylgalactosamine, N-acetylneuraminic acid and heparin). Most of them showed a cellular specificity and developmental regulation in the CNS. Several lectins isolated from the nervous tissue seem to play an essential role during ontogenetic processes, especially as far as cell adhesion and cell recognition mechanisms are concerned (axonal growth and fasciculation, neuron migration, synaptogenesis, myelination). But some of them seem to be involved in signaling events both intracellularly (nuclear lectins) or at the cell surface by autocrine and paracrine mechanisms (R: Zanetta 1998). Siglecs (sialic acid-binding Ig-like lectins) are mainly expressed in the immune system. Sn (sialoadhesin) (siglec-1), CD22 (siglec-2) and siglec-15 are well conserved, whereas the CD33-related siglecs are undergoing rapid evolution, as reflected in their repertoires among the different mammals studied. The CD33-related siglecs are both inhibitory and activating forms of receptors. Alzheimer's disease (AD) is a progressive neurodegenerative disease. Recent progress PRRs of monocytes and macrophages has revealed that the Siglec family of receptors is an important recognition receptor for sialylated glycoproteins and glycolipids. Recent studies have revealed that microglial cells contain only one type of Siglec receptors, Siglec-11, which mediates immunosuppressive signals and thus inhibits the function of other microglial pattern recognition receptors, such as TLRs, NLRs, and RAGE receptors. Recent studies clearly indicate that aggregating amyloid plaques are masked in AD by sialylated glycoproteins and gangliosides (Lopez and Schnaar 2009). This kind of immune evasion can prevent the microglial cleansing process of aggregating amyloid plaques in AD (Crocker and Redelinghuys 2008; Salminen and Kaarniranta 2009).

Endogenous Lectins in Tumors in CNS: The analysis of endogenous sugar receptors, as part of an intercellular information code system, may represent a way of studying the mechanism of tumor differentiation and its propagation. The carbohydrate part of cellular glycoconjugates -glycoproteins, glycolipids and proteoglycans—and specific endogenous sugar receptors, i.e. lectins, can establish a system of biological recognition based on protein-sugar interactions on the cellular and subcellular levels. Presence of sugar receptors has been noted in different types of meningiomas, glioblastomas, gangliocytomas, anaplastic and well-differentiated oligodendrogliomas and ependymomas as well as in neurinomas and neurofibromas of peripheral nerves. In comparison to well-differentiated ependymomas, the anaplastic form of this tumor shows generally a higher capacity to specifically bind the neoglycoproteins, containing α- or β-glucosides. Inverse intensity of glycohistochemical reaction is observed with galactose-6-phosphate-, galactose-β(1.3)-N-acetylglucosamine-N-acetyl-D-glucosamine- and mannose-(BSA-biotin), respectively, in anaplastic and differentiated oligodendrogliomas. Dedifferentiated-tumorous neurons, i.e. gangliocytomas, and distinct subtypes of meningiomas show an altered spectrum of endogenous sugar receptors in comparison to neurons of normal counterparts. Receptors for N-acetyl-D-galactosamine were present only in the anaplastic form, while glucuronic acid-specific receptors were found only in the meningotheliomatous meningiomas. Analysis of the spectrum of endogenous sugar receptors can serve to distinguish between different cell populations composing a given tumor, as shown in neurofibromas in the cases of Schwann cells and fibroblastoid cells stained with N-acetyl-D-glucosamine (Bardosi et al. 1991).

1.6 The Sugar Code and the Lectins as Recepors in System Biology

In order to understand intra- and intercellular processes, the parameters based on protein sequences are not sufficient to explain molecular events in the cellular processes such as cell adhesion or cell communication. Carbohydrates play a major role in such processes. Carbohydrates are often referred to as the third molecular chain of life, after DNA and proteins. Cabohydrate-protein interactions are responsible for important biological functions such as inter-cellular communication particularly in the immune system. Carbohydrates are uniquely suited for this role in molecular recognition, as they possess the capacity to generate an array of structurally diverse moieties from a relatively small number of monosaccharide units. This could be attributed to the fact, that unlike the components of nucleic acids, carbohydrates can link together in multiple, nonlinear ways because each building block has about four functional groups for linkage. They can even form branched chains. Hence, the number of possible polysaccharides is enormous. Since carbohydrates assume a large variety of configurations, many carbohydrate-binding proteins are being considered as targets for new medicines. The accurate *in silico* identification of carbohydrate-binding sites is a key issue in genome annotation and drug targeting. Different aspects of protein carbohydrate recognition have also been extensively studied.

Sugars in the form of monosaccharides, oligosaccharides, polysaccharides and glycoconjugates (glycoproteins, glycolipids) are vital components of pathogens and host cells, and are involved in cell signaling associated with modulation of inflammation in all integumental structures. Infact, sugars are the molecules most commonly involved in cell recognition and communication. In skin, they are essential to epidermal development and homeostasis. They play important roles in microbial adherence, colonization and biofilm formation, and in virulence. Thus, it shows that biological information is not only stored in protein sequences but also in the structure of the glycan part of the glycoconjugates. For example, in immune system, the spatially accessible carbohydrate structures that contribute to the cell's glycome are decoded by recognition systems in order to maintain the immune homeostasis of an organism. Hence the term 'Sugar code' has been introduced to bridge the gap between cell's communication and adhesion. Sugar code like genetic code yet to be defined in the form of monosaccharide alphabets, possesses more information and may have more storage capacity than other molecular units. As suggested (Gabius 2000) oligosaccharides surpass peptides by more than seven orders of magnitude in the theoretical ability to build isomers, when the total of conceivable hexamers is calculated.

1.6.1 Host-Pathogen Interactions

Study on protein-carbohydrate interactions is an exciting area of research with huge potential for development and exploration. This is particularly true for host-pathogen interactions that lead to infectious diseases; as the surfaces of cells and pathogens display complex carbohydrate structures and carbohydrate binding proteins on their surface. Most of the infectious diseases occur due to protein-carbohydrate interactions on the surface of host and disease causing pathogen (Malik and Ahmad 2010). As a result, considerable efforts have been directed toward understanding and mimicking the recognition processes and developing effective agents to develop carbohydrate-based therapeutics, targeting of drugs to specific disease cells via carbohydrate–lectin interactions; and carbohydrate based anti-thrombotic agents. Although, many human pathogens, including the influenza virus, possess surface proteins that complex with specific membrane-bound oligosaccharides on human cells (Sonnenburg et al. 2004; Takahashi et al. 2004; Varki et al. 2009), the cell surface glycans continually undergo structural variations during disease progression. Hence the variety of carbohydrate structures that occur on diseased cells gives rise to highly complex carbohydrate–lectin interactions and signaling processes. Emerging roles of carbohydrates and glycomimetics in anticancer drug design are being recognized. The information from these studies is useful for designing tailored compounds that inhibit a glycan-binding site.

Pathogens Are Recognized by Different Lectin Receptors: Innate immunity is the earliest response to invading microbes and acts to contain infection in the first minutes to hours of challenge. Unlike adaptive immunity that relies upon clonal expansion of cells that emerge days after antigenic challenge, the innate immune response is immediate. The glycocalyx is a glycan layer found on the surfaces of host cells as well as microorganisms and enveloped virus. Its thickness may easily exceed 50 nm. The glycocalyx does not only serve as a physical protective barrier but also contains various structurally different glycans, which provide cell- or microorganism-specific 'glycoinformation'. This information is decoded by host glycan-binding proteins, lectins. Pathogen recognition is central to the induction of adaptive immunity. Dendritic cells (DCs) express different pattern recognition receptors (PRRs), such as Toll-like receptors and C-type lectins that sense invading pathogens. Pathogens trigger a specific set of PRRs, leading to activation of intracellular signaling processes that shapes the adaptive immunity. It is becoming clear that cross talk between these signaling routes is crucial for pathogen-tailored immune responses. The DC-SIGN interacts with different mannose-expressing pathogens such as Mycobacterium tuberculosis and HIV-1. Mycobacterium tuberculosis (Mtb) is recognized by pattern recognition receptors on macrophages and DCs, thereby triggering phagocytosis, antigen presentation to T cells and cytokine secretion. The DC-SIGN has specificity for mannose-containing glycoconjugates and fucose-containing Lewis antigens. Mannosylated moieties of the mycobacterial cell wall were shown to bind to DC-SIGN on immature dendritic cells and macrophage subpopulations. This interaction reportedly impaired dendritic cell maturation, modulated cytokine secretion by phagocytes and DCs and was postulated to cause suppression of protective immunity to TB. However, experimental Mtb infections in mice transgenic for human DC-SIGN revealed that, instead of favoring immune evasion of mycobacteria, DC-SIGN may promote host protection by limiting tissue pathology. The dominant Mtb-derived ligands for DC-SIGN are presently uncertain, and a major role of DC-SIGN in the immune response to Mtb infection may lie in its capacity to maintain a balanced inflammatory state during chronic TB. For several pathogens that interact with DC-SIGN, including Mycobacterium tuberculosis and HIV-1,

Raf-1 activation leads to acetylation of NF-kB subunit p65, which induces specific gene transcription profiles. In addition, other DC-SIGN-ligands induce different signaling pathways downstream of Raf-1, indicating that DC-SIGN-signaling is tailored to the pathogen (Dunnen et al. 2009; Ehlers 2010; Gringhuis and Geijtenbeek 2010). Many of these C-type lectin receptors (CLRs), coupled to Syk kinase, signal via CARD9 leading to NF-κB activation, which in turn contributes to the induction of both innate and adaptive immunity. Dectin-1, Dectin-2 and Mincle are all CLRs that share this common signaling mechanism and have been shown to play key roles in antifungal immunity (Drummond et al. 2011). Thus, interaction of CLRs and TLRs with PAMPs initiates a cascade of events leading to production of reactive oxygen intermediates, cytokines and chemokines, and promotes inflammation. Thus sugars may provide valuable adjunctive anti-inflammatory and/or antimicrobial treatment (Geijtenbeek and Gringhuis 2009; Lloyd et al. 2007). Microbial PAMPs are recogniized by CLRs such as MBL, the tandem-repeat-type macrophage mannose receptor, DC-SIGN or dectin-1 of dendritic cells, certain TLRS or the TCR of NKT cells. Galectins, the key sensors reading the high-density sugar code, exert regulatory functions on activated T cells, among other activities. Autoimmune diseases are being associated with defined changes of glycosylation. This correlation deserves to be thoroughly studied on the levels of structural mimicry and dysregulation as well as effector molecules to devise innovative anti-inflammatory strategies (Buzás et al. 2006).

Pathogen recognition by dendritic cells (DCs) is central to the induction of adaptive immunity. PRRs on DCs interact with pathogens, leading to signaling events that dictate adaptive immune responses. It is becoming clear that C-type lectins are important PRRs that recognize carbohydrate structures. Triggering of C-type lectins induces signaling cascades that initiate or modulate specific cytokine responses and therefore tailor T cell polarization to the pathogens. Under steady state conditions DCs continuously sample antigens, leading to tolerance, whereas inflammatory conditions activate DCs, inducing immune activation. DCs express CLRs for antigen capture and presentation, whereas TLRs are involved in pathogen recognition and DC activation. Recent reports demonstrate that communication between TLRs and CLRs can affect the direction of immune responses. Several pathogens specifically target CLRs to subvert this communication to escape immune surveillance, either by inducing tolerance or skewing the protective immune responses (Geijtenbeek et al. 2004; van Kooyk et al. 2004). Langerhans cells (LC) are the first DCs to encounter pathogens entering the body via mucosa or skin. Equipped with PRRs, LCs are able to detect and respond to pathogens. Langarin is an important CLR, exclusively expressed by LC in humans. Langerin forms a protective barrier against HIV-1 infection by binding and degradation of this virus. In addition, antigens targeted to Langerin are presented to T cells to induce an adaptive immune response. How its functions and how Langerin polymorphisms influence the function of Langerhans cells, has been reviewed (van der Vlist and Geijtenbeek 2010). However, evidence is accumulating that many CLRs are also able to recognize endogenous 'self' ligands and that this recognition event often plays an important role in immune homeostasis. Endogenous ligands for human and mouse CLRs have been described. Special attention has been drawn to the signaling events initiated upon recognition of the self ligand and the regulation of glycosylation as a switch modulating CLR recognition, and finally in immune homeostasis (García-Vallejo and van Kooyk 2009).

Known biological function//targets of carbohydrates and combinatorial synthesis, structural analysis of natural//non-natural carbohydrates and further insights into the functional consequences of the sugar code's translation are thus expected to have notable repercussions for diagnostic and therapeutic procedures (Gabius 2011). To fill the need for expression analysis of glycogenes, Comelli et al. (2006) employed the Affymetrix technology to develop a focused and highly annotated glycogene-chip representing human and murine glycogenes, including glycosyltransferases, nucleotide sugar transporters, glycosidases, proteoglycans, and glycan-binding proteins. Comparison of gene expression profiles with MALDI-TOF profiling of N-linked oligosaccharides suggested that the α1-3 fucosyltransferase 9 is the enzyme responsible for terminal fucosylation in kidney and brain- a finding validated by analysis of fucosyltransferase 9 knockout mice. Two families of lectins, C-type lectins and Siglecs, are predominately expressed in the immune tissues, consistent with the emerging functions in both innate and acquired immunity.

1.6.2 Altered Glycosylation in Cancer Cells

Altered glycosylation is a universal feature of cancer cells, and certain glycan structures are well-known markers for tumor progression. Some glycan biosynthetic pathways are frequently altered in cancer cells. Reports suggest that the expression of branched and sialylated complex type N-oligosaccharides in human melanoma consistently increases in cells from metastatic sites, and support the view that carbohydrates are associated with the acquisition of the metastatic potential of tumor cells (Litynska et al. 2001; Ciolczyk-Wierzbicka et al. 2002). Glycosylation can be altered in various ways in malignancy. Classic reports of increased size of tumor cell–derived glycopeptides have been convincingly explained by an increase in β1–6 branching of N-glycans (Fig. 1.4), which results from enhanced expression of UDP-GlcNAc:N-glycan GlcNAc

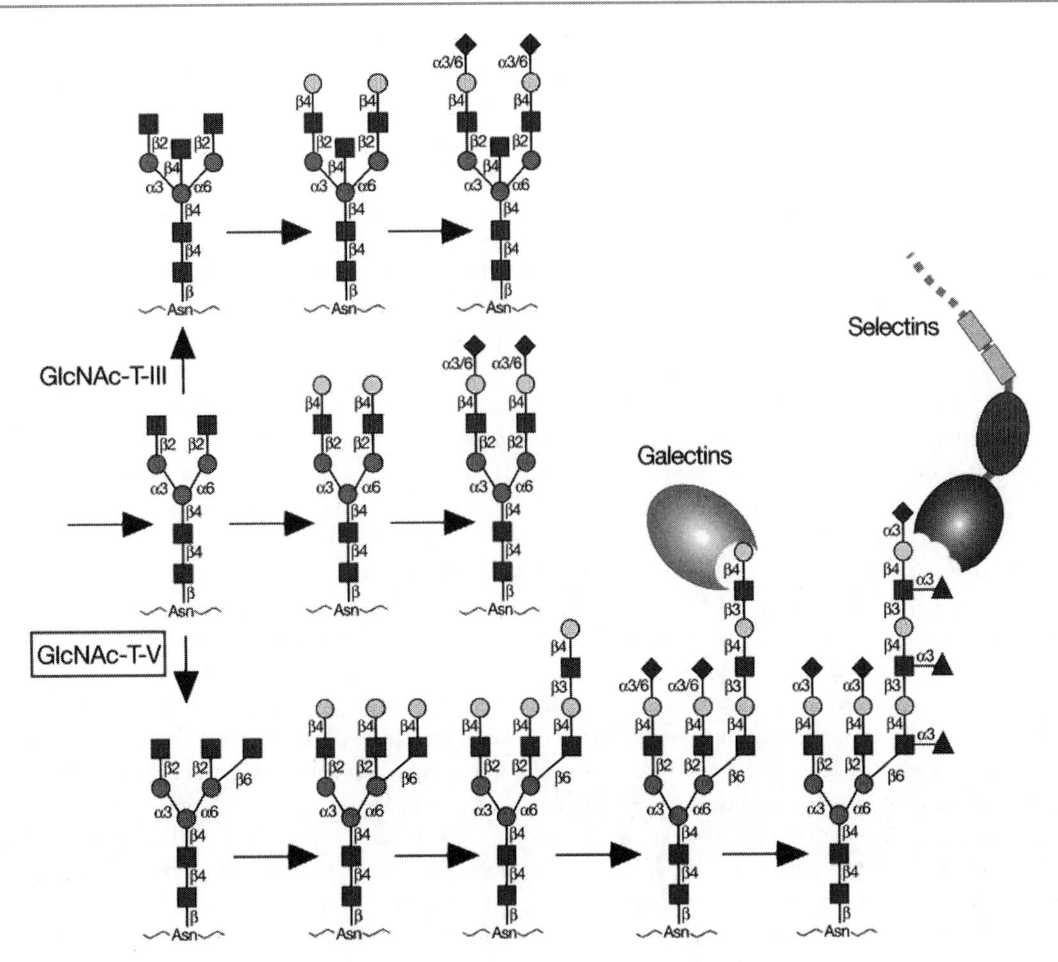

Fig.1.4 N-Glycan: Glycan covalently linked to an asparagine residue of a polypeptide chain in the consensus sequence: -Asn-X-Ser/Thr. Unless otherwise stated, the term N-glycan denotes the most common linkage region, Manβ1–4GlcNAcβ1–4GlcNAcβ1-N-Asn. The increased size of N-glycans that occurs upon transformation can be explained by an elevation in GlcNAc transferase-V (GNT-V) activity, which catalyzes the β1–6 branching of N-glycans. This, in turn, may lead to enhanced expression of poly-*N*-acetyllactosamines, which can also be sialylated and fucosylated. These structures are potentially recognized by galectins and selectins, respectively. The structural consequences of increased expression of GlcNAc transferase-III (GNT-III) are also shown (Adapted by permission from Varki et al. 2009 © The Consortium of Glycobiology Editors, La Jolla, California) ● Galactose (Gal); ● Mannose (Man); ■ N-Acetylglucosamine (GlcNac); ◆ N-Acetylneuraminic acid (Neu 5Ac); ▲ Fucose (Fuc)

transferase V (GlcNAcT-V). Reports indicate that GlcNAcT-V plays an important part in the biology of cancer. Enhanced expression of another glycosyltransferase affecting N-glycan structure UDP-GlcNAc:N-glycan GlcNAc transferase III (GlcNAcT-III), which catalyzes the addition of the bisecting GlcNAc branch, has been reported in certain tumors, such as rat hepatomas. Although GlcNAcT-III expression appears to affect the biology of tumors, the results are not as clear-cut and consistent as those seen with GlcNAcT-V (Varki et al. 2009).

1.6.3 Protein-Carbohydrate Interactions in Immune System

The immune system consists of various types of cells and molecules that specifically interact with each other to initiate the host defense mechanism. Recent studies have shown that carbohydrates and lectins play an essential role in mediating such interactions. Carbohydrates, due to their chemical nature, can potentially form structures that are more variable than proteins and nucleic acids. The interactions between lectins and carbohydrates have been shown to be involved in such activities as opsonization of microorganisms, cell adhesion and migration, cell activation and differentiation, and apoptosis. The number of lectins identified in the immune system is increasing at a rapid pace. The development in this area has opened a new aspect in studying the immune system, and at the same time, provided new therapeutic routes for the treatment and prevention of disease.

Protein-carbohydrate interactions are used for intercellular communication. Mammalian cells are known to bear a variety of glycoconjugates. The discovery of plant lectins as biochemical reagents gave a great impulse to modern glycobiology and in the biomedical research. Plant lectins have been fundamental in human immunological studies because some of them are mitogenic/activating to lymphocytes. The understanding the molecular basis of lectin-carbohydrate interactions and of the intracellular signaling evoked holds promise for the design of novel drugs for the treatment of infectious, inflammatory and malignant diseases. It may also be of help for the structural and functional investigation of glycoconjugates and their changes during physiological and pathological processes.

1.6.4 Glycosylation and the Immune System

1.6.4.1 Carbohydrate-Mediated Recognition in Immune System

Almost all key molecules involved in the innate and adaptive immune response are glycoproteins. In the cellular immune system, specific glycoforms participate in the folding, quality control, and assembly of peptide-loaded major MHC antigens and the T cell receptor complex. Recent evidence indicates that protein-glycan interactions play a critical role in different events associated with the physiology of T-cell responses including thymocyte maturation, T-cell activation, lymphocyte migration and T-cell apoptosis. Glycans decorating T-cell surface glycoproteins can modulate T-cell physiology by specifically interacting with endogenous lectins including selectins and galectins. Sometimes, the generation of peptide antigens from glycoproteins may require enzymatic removal of sugars before the protein can be cleaved. Oligosaccharides attached to glycoproteins at the junction between T cells and antigen-presenting cells help to orient binding faces, provide protease protection, and restrict nonspecific lateral protein-protein interactions. In the humoral immune system, all of the immunoglobulins and most of the complement components are glycosylated. Although a major function for sugars is to contribute to the stability of the proteins to which they are attached, specific glycoforms are involved in recognition events. Among carbohydrate-recognizing receptors of the innate immune system are the members of the C-type lectin family, which include the collectins and the selectins and which operate by ligating exogenous (microbial) or endogenous carbohydrates. In rheumatoid arthritis, an autoimmune disease, a galactosylated glycoforms of aggregated IgG may induce association with the mannose-binding lectin and contribute to the pathology.

1.6.4.2 The Impact of Differential Glycosylation on T Cell Responses

Protein-carbohydrate interactions may be controlled at different levels, including regulated expression of lectins during T-cell maturation and differentiation and the spatio-temporal regulation of glycosyltransferases and glycosidases, which create and modify sugar structures present in T-cell surface glycoproteins. Galactose (Gal)-containing structures are involved in both the innate and adaptive immune systems. Gal is a ligand for Gal/N-acetylgalactosmine (GalNAc) receptors and galectins. Galactose is part of the scaffold structure that synthesizes oligosaccharide ligands for selectins, siglecs and other lectins of the immune system. Gal residues are added to glycoproteins and glycolipids by members of a large family of galactosyltransferases. Specific galactosyltransferases have been shown to control cell adhesion and leukocyte functions. Antibodies need to be galactosylated for normal function, and under-galactosylated immunoglobulin (Ig) is associated with rheumatoid arthritis, while Gal is lacking in the IgA of patients with IgA nephropathy. Interactions involving Gal play important roles in host defenses; they can also result in serious pathophysiology. Galactosyltransferases represent potential targets for the control of cell growth and apoptosis, inflammation and infections (Brockhausen 2006). Galectins participate in a wide spectrum of immunological processes. These proteins regulate the development of pathogenic T-cell responses by influencing T-cell survival, activation and cytokine secretion. Administration of recombinant galectins or their genes into the cells modulate the development and severity of chronic inflammatory responses in experimental models of autoimmunity by triggering different immunoregulatory mechanisms. Thus galectins have potential use as novel anti-inflammatory agents or targets for immunosuppressive drugs, in autoimmune and chronic inflammatory disorders (Bianco et al. 2006; Daniels et al. 2002; Toscano et al. 2007).

1.6.5 Principles of Protein-Glycan Interactions

Emerging roles of carbohydrates and glycomimetics in drug design are being recognized. The information from these studies is useful for designing tailored compounds that inhibit a glycan-binding site. Understanding the principles of glycan–protein interactions led to the development of high-affinity inhibitors of carbohydrate-protein interactions such as for viral neuraminidase, which is involved in the removal of sialic acids from host glycoprotein receptors; these inhibitors have proven utility in reducing the duration and spread of influenza infection. However, knowledge of protein and carbohydrate structure alone does not ensure accurate prediction of function and biological activity. A well-known characteristic of protein–carbohydrate interactions is the low affinity of binding, usually in mM range, but the estimates of binding affinities are not accurate due

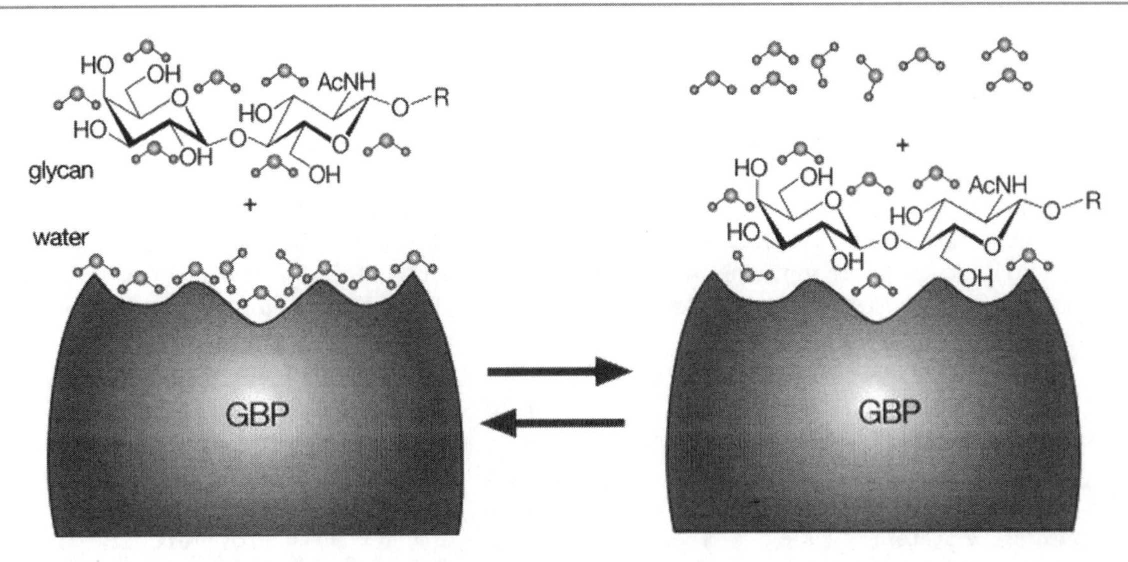

Fig. 1.5 Schematic diagram showing interaction between hydrated polyhydroxylated glycan and a hydrated protein-combining site of a GBP in water, resulting in the displacement of water (Reprinted with permission from Cummings and Esko, 2009 © The Consortium of Glycobiology Editors, La Jolla, California)

to several complications. As shown in Fig. 1.5, the overall process of binding typically involves the union of a hydrated polyhydroxylated glycan and a hydrated protein-combining site. If a surface on the glycan is complementary to the protein-combining site, water can be displaced and binding occurs. When the complex finally forms, it presents a new surface to the surrounding medium, which will also be hydrated. As solvation/desolvation energies are very large because of entropy from the disordering of water molecules, their contribution to binding cannot be reliably determined with existing models (Cummings and Esko 2009). Furthermore, glycans may undergo a conformational change upon binding, changing their internal energy and solvation. Schematic diagram of the binding of a glycan-binding protein (GBP) in water, which results in the displacement of water.bonding interactions in the combining site, the estimation of solvation energy changes is difficult. Another complicating issue concerns valency, which greatly increases the association of GBPs with their glycan ligands under biological conditions (Cummings and Esko 2009). The density of binding sites on the ligand can also affect the affinity of binding. Ligands for some GBPs may be glycoproteins that carry one or more multiantennary N-linked chains. Mucins present potentially hundreds of glycans, and their proximity can affect conformation and presentation of the ligands. Some polysaccharides, such as glycosaminoglycans, have multiple binding sites located along a single chain. In addition, nonglycan components, such as tyrosine sulfate, lipid, or peptide determinants, may also cooperate with glycans to provide relatively high affinity and specific interactions.

Glycan–protein complexes solved by X-ray crystallography and NMR spectroscopy revealed that the glycan-binding sites typically accommodate one to four sugar residues, although in some complexes, recognition extends over larger numbers of residues and may include aglycone components, such as the peptide or lipids to which the glycans are attached. Information on the three-dimensional structure of a glycan-protein complex can reveal the specificity of binding, changes in conformation that take place on binding, and the contribution of specific amino acids to the interaction. However, one would also like to determine the affinity of the interaction. Since the forces involved in the binding of a glycan to a protein are the same as for the binding of any ligand to its receptor (hydrogen bonding, electrostatic or charge interactions, van der Waals interactions, and dipole attraction), it is tempting to try to calculate their contribution to overall binding energy, which can be related to an affinity constant (K_a). Unfortunately, calculating the free energy of association is difficult for several reasons, including problems in defining the conformation of the unbound versus the bound glycan, changes in bound water within the glycan and the binding site, and conformational changes in the GBP upon binding.

The complete characterization of any binding interaction requires a quantification of the affinity, number of binding sites, and the thermodynamic parameters. Thermodynamic data, specifically enthalpy (ΔH) and entropy (ΔS), reveal the forces that drive complex formation and mechanism of action. Thermodynamics provide information on conformational changes, hydrogen bonding, hydrophobic interactions, and charge-charge interactions. These interactions are driven by a favourable enthalpy (heat is released), offset by the multiple contact points (hydrogen bonds, van der Waals' interactions and hydrophobic stacking) between the carbohydrate and the protein. Recent studies have shown that in the binding site, polar–polar interactions are actually stronger than in the corresponding protein–protein interactions. One reason for this is shorter hydrogen bond distances, partly because hydrogen bonds involve charged groups as opposed to the polar–polar interactions in proteins, which mainly involve non-charged groups. In addition, highly-organized hydrogen-binding atoms and a higher frequency of hydrogen bonds per unit area result in a closely spaced protein–carbohydrate interface (Holgersson et al. 2005).

The relatively high association of enthalpies of protein–carbohydrate (as compared to protein–protein) interactions has been ascribed to a higher degree of hydrogen bonds per unit area and shorter distances between the molecules involved in the hydrogen binding. The amphiphilic nature of carbohydrates results in different interactions with water molecules in the vicinity. The hydrophobic ring core represents a field of low dielectricity, resulting in densely packed water molecules. The hydroxyl groups, on the other hand, form hydrogen bonds with the closest water molecules, thus forcing these molecules to arrange themselves in a configuration that is energetically less favourable than the ones formed by bulk water molecules (Fig. 1.5). During protein–carbohydrate complexation, water molecules tend to escape to the bulk with a concomitant decrease or increase in energy depending on their pre-existing molecular interactions. After protein–carbohydrate binding has occurred, the complex formed will be resolvated and the water molecules rearrange themselves according to the new surface exposed (Holgersson et al. 2005). The affinity of most single glycan–protein interactions is generally low (mM–μM K_D values). In nature, many GBPs are oligomeric or may be membrane-associated proteins, which allows aggregation of the GBP in the plane of the membrane. Many of the glycan ligands for GBPs are also multivalent. In the case of cholera toxin, five B subunits present in the holotoxin interact with five molecules of GM1 normally present in the cell membrane. The interaction of multiple subunits with a multivalent display of GM1 raises the affinity of interaction by several orders of magnitude (K_D of ~40 pM).

1.7 Applications of Lectin Research and Future Perspectives

Lectins are used in wide variety of areas as in separation of glycoproteins and glycoconjugates, histochemistry of cells and tissues, cell differentiation, and in tracing cell surface pathways. These molecules are of great interest to immunologists mainly because of their ability to interact with lymphocytes and to induce blast cell transformation. Purified lectins are important in a clinical setting because they are used for blood typing . Some of the glycolipids and glycoproteins on an individual's red blood cells can be identified by lectins (such as lecin from Dolichos biflorus for A1 blood group, Ulex europaeus lectin for detection of H blood group antigen, and lecin from Vicia graminea in identification of N blood group antigen). The anterograde labeling method is used to trace the path of efferent axons with PHA-L, a lectin from the kidney bean (Carlson 2007). A lectin (BanLec) from bananas inhibits HIV-1 in vitro (Swanson et al. 2010). Lectins can be used in biochemical warfare. One example of the powerful biological attributes of lectins is the biochemical warfare agent ricin. Ricin from seeds of castor oil plant cleaves nucleobases from ribosomal RNA resulting in inhibition of protein synthesis and cell death (Lord et al. 1994). Growing insights into the functionality of lectin-carbohydrate interactions are identifying attractive new targets for drug design and drug targeting (Gupta et al. 2009). As glycan recognition is regulated by the structure of the sugar epitope and also by topological aspects of its presentation, a suitable arrangement of ligands in synthetic glycoclusters has the potential to enhance their avidity and selectivity. If adequately realized, such compounds might find medical applications (Lyra and Diniz 2007). Lectins of algal origin whose antiviral properties make them candidate agents for prevention of viral transmission through topical applications include cyanovirin-N, *Microcystis viridis* lectin, scytovirin, and griffithsin. Although all these proteins exhibit significant antiviral activity, their structures are unrelated and their mode of binding of carbohydrates differs significantly.

1.7.1 Mannose Receptor-Targeted Vaccines

Targeting antigens to endocytic receptors on professional APCs such as DCs represents an attractive strategy to enhance the efficacy of vaccines. Such APC-targeted vaccines have an exceptional ability to guide exogenous protein antigens into vesicles that efficiently process the antigen for MHC class I and class II presentation. The specific targeting of nanomedicines to mannose receptors (MR) and related C-type lectin receptors, highly expressed in cells of the immune system, perform a useful strategy for improving the efficacy of vaccines and chemotherapy (Gupta et al. 2009). The MR and related C-type lectin receptors are particularly designed to sample antigens (self and non-self), much like pattern recognition receptors, to integrate the innate with adaptive immune responses. Certainly, a better understanding of the mechanism associated with the induction of immune responses as a result of targeting antigens to the MR, will be important in exploiting MR-targeted vaccines not only for mounting immune defenses against cancer and infectious disease, but also for specific induction of tolerance in the treatment of autoimmune diseases (Ahlers and Belyakov 2009). The DC-SIGN whose expression is restricted to DCs has gained an increased attention because of its involvement in multiple aspects of immune function. Upon DC-SIGN engagement by mannose- or fucose-containing oligosaccharides, the latter leads to a tailored Toll-like receptor signaling, resulting in an altered DC-cytokine profile and skewing of Th1/Th2 responses (Svajger et al. 2010) (Refer Chap. 46).

References

Ahlers JD, Belyakov IM (2009) Strategies for recruiting and targeting dendritic cells for optimizing HIV vaccines. Trends Mol Med 15:263–274

Angata T, Brinkman-Van der Linden E (2002) I-type lectins. Biochim Biophys Acta 1572:294–316

Arar C, Carpentier V, Le Caer JP et al (1995) ERGIC-53, a membrane protein of the endoplasmic reticulum-Golgi intermediate compartment, is identical to MR60, an intracellular mannose-specific lectin of myelomonocytic cells. J Biol Chem 270:3551–3553

Atoda H, Hyuga M, Morita T (1991) The primary structure of coagulation factor IX/factor X-binding protein isolated from the venom of *Trimeresurus flavoviridis*: homology with asialoglycoprotein receptors, proteoglycan core protein, tetranectin and lymphocte Fce receptor for immunoglobulin E. J Biol Chem 266:14903–14911

Bardosi A, Brkovic D, Gabius HJ (1991) Localization of endogenous sugar-binding proteins (lectins) in tumors of the central and peripheral nervous system by biotinylated neoglycoproteins. Anticancer Res 11:1183–1187

Barondes SH, Castronovo V, Cooper DN et al (1994) Galectins: a family of animal beta-galactoside-binding lectins. Cell. 76: 597–598

Bianco GA, Toscano MA, Ilarregui JM, Rabinovich GA (2006) Impact of protein-glycan interactions in the regulation of autoimmunity and chronic inflammation. Autoimmun Rev 5:349–356

Blixt O, Head S, Mondala T et al (2004) Printed covalent glycan array for ligand profiling of diverse glycan binding proteins. Proc Natl Acad Sci U S A 101:17033–17037

Bottazzi B, Garlanda C, Salvatori G et al (2006) Pentraxins as a key component of innate immunity. Curr Opin Immunol 18:10–15

Bouckaert J, Hamelryck TW, Wyns L, Loris R (1999) The crystal structures of Man(α1-3)Man(α1-O)Me and Man(α1-6)Man(α1-O) Me in complex with concanavalin A. J Biol Chem 274: 29188–29195

Bourgeois C, Majer O, Frohner IE et al (2010) Fungal attacks on mammalian hosts: pathogen elimination requires sensing and tasting. Curr Opin Microbiol 13:401–408

Brinda KV, Surolia A, Vishveshwara S (2005) Insights into the quaternary association of proteins through structure graphs: a case study of lectins. Biochem J 39:1–15

Brockhausen I (2006) The role of galactosyltransferases in cell surface functions and in the immune system. Drug News Perspect 19: 401–407

Buzás EI, György B, Pásztói M et al (2006) Carbohydrate recognition systems in autoimmunity. Autoimmunity 39:691–704

Cambi A, Figdor CG (2003) Dual function of C-type lectin-like receptors in the immune system. Curr Opin Cell Biol 15: 539–546

Cambi A, Figdor C (2009) Necrosis: C-type lectins sense cell death. Curr Biol 19:R375–R377

Carlson NR (2007) Physiology of Behavior, 9th edn. Pearson Education, Inc, Boston, p 144

Chou KC (1996) Knowledge-based model building of the tertiary structures for lectin domains of the selectin family. J Protein Chem 15:161–168

Ciolczyk-Wierzbicka D, Gil D et al (2002) Carbohydrate moieties of N-cadherin from human melanoma cell lines. Acta Biochim Pol 49:991–997

Comelli EM, Head SR, Gilmartin T et al (2006) A focused microarray approach to functional glycomics: transcriptional regulation of the glycome. Glycobiology 16:117–131

Correa I, Raulet DH (1995) Binding of diverse peptides to MHC class I molecules inhibits target cell lysis by activated natural killer cells. Immunity 2:61–71

Crocker PR, Redelinghuys P (2008) Siglecs as positive and negative regulators of the immune system. Biochem Soc Trans 36(Pt 6): 1467–1471

Cummings RD, McEver RP (2009) C-type Lectins. In: Varki A, Cummings RD, Esko JD (eds) Essentials of Glycobiology, 2nd edn. Cold Spring Harbor, NY

Cummings RD, Esko JD et al (2009) Principles of Glycan Recognition. In: Varki A, Cummings RD, Esko JD (eds) Essentials of Glycobiology, 2nd edn. Cold Spring Harbor, NY

Dahms NM, Hancock MK (2002) P-type lectins. Biochim Biophys Acta 1572:317–340

Daniels MA, Hogquist KA, Jameson SC (2002) Sweet 'n' sour: the impact of differential glycosylation on T cell responses. Nat Immunol 3:903–910

Dodd RB, Drickamer K (2001) Lectin-like proteins in model organisms: implications for evolution of carbohydrate-binding activity. Glycobiology 11:71R–79R

Drickamer K (1988) Two distinct classes of carbohydrate-recognition domains in animal lectins. J Biol Chem 263:9557–9560

Drickamer K (1993) Evolution of Ca^{2+}-dependent animal lectins. Prog Nucleic Acid Res Mol Biol 45:207–232

Drickamer K (1999) C-type lectin-like domains. Curr Opin Struct Biol 9:585–590

Drickamer K, Dodd RB (1999) C-type lectin-like domains in *Caenorhabditis elegans*: predictions from the complete genome sequence. Glycobiology 9:1357–1367

Drickamer K, Fadden AJ (2002) Genomic analysis of C-type lectins. Biochem Soc Symp 69:59–72

Drummond RA, Saijo S, Iwakura Y et al (2011) The role of Syk/CARD9 coupled C-type lectins in antifungal immunity. Eur J Immunol 41: 276–281

Dunnen J, Gringhuis SI, Geijtenbeek TB (2009) Innate signaling by the C-type lectin DC-SIGN dictates immune responses. Cancer Immunol Immunother 58:1149–1157

Ehlers S (2010) DC-SIGN and mannosylated surface structures of Mycobacterium tuberculosis: a deceptive liaison. Eur J Cell Biol 89:95–101

Elhalwagi BM, Damodarasamy M, McCormack FX (1997) Alternate amino terminal processing of surfactant protein A results in cysteinyl isoforms required for multimer formation. Biochemistry 36:7018–7025

Ellgaard L, Ruddock LW (2005) The human protein disulfide isomerase family: substrate interactions and functional properties. EMBO Rep 6:28

Elola MT, Wolfenstein-Todel C, Troncoso MF et al (2007) Galectins: matricellular glycan-binding proteins linking cell adhesion, migration, and survival. Cell Mol Life Sci 64:1679–1700

Feinberg H, Park-Snyder S, Kolatkar AR et al (2000) Structure of a C-type carbohydrate recognition domain from the macrophage mannose receptor. J Biol Chem 275:21539–21548

Frand AR, Cuozzo JW, Kaiser CA (2000) Pathways for protein disulfide bond formation. Trends Cell Biol 10:203–210

Fuhlendorff J, Clemmensen I, Magnusson S (1987) Primary structure of tetranectin, a plasminogen kringle 4 binding plasma protein: homology with asialoglycoprotein receptors and cartilage proteoglycan core protein. Biochemistry 26:6757–6764

Fujita T, Matsushita M, Endo Y (2004) The lectin-complement pathway—its role in innate immunity and evolution. Immunol Rev 198:185–202

Gabius HJ (2000) Biological information transfer beyond the genetic code: the sugar code. Naturwissenschaften 87:108–121

Gabius HJ (2011) Glycobiomarkers by glycoproteomics and glycan profiling (glycomics): emergence of functionality. Biochem Soc Trans 39: 399–405

García-Vallejo JJ, van Kooyk Y (2009) Endogenous ligands for C-type lectin receptors: the true regulators of immune homeostasis. Immunol Rev 230:22–37

Garlatti V, Martin L, Lacroix M et al (2009) Structural insights into the recognition properties of human ficolins. J Innate Immun 2:17–23

Gazi U, Martinez-Pomares L (2009) Influence of the mannose receptor in host immune responses. Immunobiology 214:554–561

Geijtenbeek TB, Gringhuis SI (2009) Signaling through C-type lectin receptors: shaping immune responses. Nat Rev Immunol 9: 465–479

Geijtenbeek TB, van Vliet SJ, Engering A et al (2004) Self- and nonself-recognition by C-type lectins on dendritic cells. Annu Rev Immunol 22:33–54

Goldstein IJ, Hudges RC, Monsigny M et al (1980) What should be called lectin? Nature 285:66

Graham LM, Brown GD (2009) The Dectin-2 family of C-type lectins in immunity and homeostasis. Cytokine 48:148–155

Gringhuis SI, Geijtenbeek TB (2010) Carbohydrate signaling by C-type lectin DC-SIGN affects NF-kB activity. Methods Enzymol 480:151–164

Gruber CW, Čemažar M, Heras B et al (2006) Protein disulfide isomerase: the structure of oxidative folding. Trend Biochem Sci 31:455–464

Gupta A, Gupta R, Gupta GS (2009) Targeting cells for drug and gene delivery: emerging applications of mannans and mannan binding lectins. J Sci Indus Res 68:465–483

Guthmann MD, Tal M, Pecht I (1995) A new member of the C-type lectin family is a modulator of the mast cell secretory response. Int Arch Allergy Immunol 107:82–86

Hartley MR, Lord JM. (2004) Cytotoxic ribosome-inactivating lectins from plants. Biochim Biophys Acta. 1701:1–14

Hirotsu S, Mizuno H, Fukuda K et al (2001) Crystal structure of bitiscetin, a von Willebrand factor-dependent platelet aggregation inducer. Biochemistry 40(13592):7

Ho MS, Tsai PI, Chien CT (2006) F-box proteins: the key to protein degradation. J Biomed Sci 13:181–191

Holgersson J, Gustafsson A, Breimer ME (2005) Characteristics of protein–carbohydrate interactions as a basis for developing novel carbohydrate-based antirejection therapies. Immunol Cell Biol 83:694–708

Hudgin RL, Pricer WE Jr, Ashwell G, Stockert RJ, Morell AG (1974) The isolation and properties of a rabbit liver binding protein specific for asialoglycoproteins. J Biol Chem 249:5536–5543

Hummelshoj T, Fog LM, Madsen HO et al (2008) Comparative study of the human ficolins reveals unique features of ficolin-3 (Hakata antigen). Mole Immunol 45:1623–1632

Huysamen C, Brown GD (2009) The fungal pattern recognition receptor, Dectin-1, and the associated cluster of C-type lectin-like receptors. FEMS Microbiol Lett 290:121–128

Inagaki Y, Sohma Y, Horie H et al (2000) Oxidized galectin-1 promotes axonal regeneration in peripheral nerves but does not possess lectin properties. Eur J Biochem 267:2955–2964

Ip WK, Takahashi K, Ezekowitz RA, Stuart LM (2009) Mannose-binding lectin and innate immunity. Immunol Rev 230:9–21

Irache JM, Salman HH, Gamazo C, Espuelas S (2008) Mannose-targeted systems for the delivery of therapeutics. Expert Opin Drug Deliv 5:703–724

Ireland BS, Niggemann M, Williams DB (2006) In vitro assays of the functions of calnexin and calreticulin, lectin chaperones of the endoplasmic reticulum. Methods Mol Biol 347:331–342

Jenner J, Kerst G, Handgretinger R, Muller I (2006) Increased α2,6-sialylation of surface proteins on tolerogenic, immature dendritic cells and regulatory T cells. Exp Hematol 34:1212–1217

Jessop CE, Tavender TJ, Watkins RH et al (2009) Substrate specificity of the oxidoreductase ERp57 is determined primarily by its interaction with calnexin and calreticulin. J Biol Chem 284:2194–2202

Kaltner H, Gabius HJ (2001) Animal lectins: from initial description to elaborated structural and functional classification. Adv Exp Med Biol 491:79–91

Kamiya Y, Oyama F, Oyama R et al (1990) Amino acid sequence of a lectin from Japanese frog (Rana japonica) eggs. J Biochem 108: 139–143

Kerrigan AN, Brown GD (2009) C-type lectins and phagocytosis. Immunobiology 214:562–575

Kiedzierska A, Smietana K, Czepczynska H, Otlewski J (2007) Structural similarities and functional diversity of eukaryotic discoidin-like domains. Biochim Biophys Acta 1774:1069–1077

Kilpatrick DC (2002) Animal lectins: a historical introduction and overview. Biochim Biophys Acta 1572:187–197

Li F, Liang S (1999) Assignment of the three disulfide bonds of Selenocosmia huwena lectin-I from the venom of spider Selenocosmia huwena. Peptides 20:1027–1034

Liener IE (1986) Nutritional significance of lectins in the diet. In: Liener IE, Sharon N, Goldstein IJ (eds) The Lectins: Properties, Functions and applications in Biology and Medicine. Academic Press, New York, pp 527–547

Linehan SA, Martínez-Pomares L, Gordon S (2000) Macrophage lectins in host defence. Microbes Infect 2:279–288

Litynska A, Przybylo M et al (2001) Comparison of the lectin-binding pattern in different human melanoma cell lines. Melanoma Res 11:205–212

Liu Y, Eisenberg D (2002) 3D domain swapping: as domains continue to swap. Protein Sci 11:1285–1299

Llera AS, Viedma F, Sa'nchez-Madrid F, Tormo J (2001) Crystal structure of the C-type lectin-like domain from the human hematopoietic cell receptor CD69. J Biol Chem 276:7312–7319

Lloyd DH, Viac J et al (2007) Role of sugars in surface microbe-host interactions and immune reaction modulation. Vet Dermatol 18: 197–204

Lo Conte L, Brenner SE, Hubbard TJ et al (2002) SCOP database in 2002: refinements accommodate structural genomics. Nucleic Acids Res 30:264–267

Loeb JA, Drickamer K (1988) Conformational changes in the chicken receptor for endocytosis of glycoproteins. Modulation of ligand-binding activity by Ca^{2+} and pH. J Biol Chem 263:9752–9760

Lopez PH, Schnaar RL (2009) Gangliosides in cell recognition and membrane protein regulation. Curr Opin Struct Biol 19:549–557

Lord JM, Roberts LM, Robertus JD (1994) Ricin: structure, mode of action, and some current applications. FASEB J 8:201–208

Lyra PP, Diniz EM (2007) The importance of surfactant on the development of neonatal pulmonary diseases. Clinics (Sao Paulo) 62: 181–190

Mali B, Soza-Ried J, Frohme M, Frank U (2006) Structural but not functional conservation of an immune molecule: a tachylectin-like gene in Hydractinia. Dev Comp Immunol 30:275–281

Malik A, Ahmad S (2010) Sequence and structural features of carbohydrate binding in proteins and assessment of predictability using a neural network. BMC Struct Biol 7:1

Matsushita M, Fujita T (2001) Ficolins and the lectin complement pathway. Immunol Rev 180:78–85

McGreal EP, Martinez-Pomares L, Gordon S (2004) Divergent roles for C-type lectins expressed by cells of the innate immune system. Mol Immunol 41:1109–1121

Mizuno H, Fujimoto Z, Koizumi M et al (1997) Structure of coagulation factors IX/X-binding protein, a heterodimer of C-type lectin domains. Nat Struct Biol 4:438–441

Mizuno H, Fujimoto Z, Atoda H, Morita T (2001) Crystal structure of an anticoagulant protein in complex with the Gla domain of factor X. Proc Natl Acad Sci U S A 98:7230–7231

Molinari M, Helenius A (1999) Glycoproteins form mixed disulfides with oxidoreductases during folding in living cells. Nature 402 (6757):90–93

Molinari M, Calanca V et al (2003) Role of EDEM in the release of misfolded glycoproteins from the calnexin cycle. Science 299 (5611):1397–1400

Moss SE, Morgan RO (2004) The annexins. Genome Biol 5:219

Neve EP, Lahtinen U, Pettersson RF (2005) Oligomerization and interacellular localization of the glycoprotein receptor ERGIC-53 is independent of disulfide bonds. J Mol Biol 354:556–568

Ng NF, Hew CL (1992) Structure of an antifreeze polypeptide from the sea raven. Disulfide bonds and similarity to lectin-binding proteins. J Biol Chem 267:16069–16075

Nielsen BB, Kastrup JSHR, Holtet TL (1997) Crystal structure of tetranectin, a trimeric plasminogen-binding protein with an alpha-helical coiled coil. FEBS Lett 412:388–396

Odom EW, Vasta GR (2006) Characterization of a binary tandem domain f-type lectin from striped bass (*Morone saxatilis*). J Biol Chem 281:1698–13

Ohashi T, Erickson HP (2004) The disulfide bonding pattern in ficolin multimers. J Biol Chem 279:6534–6539

Palaniyar N, Nadesalingam J, Reid KB (2002) Pulmonary innate immune proteins and receptors that interact with gram-positive bacterial ligands. Immunobiology 205:575–591

Pluddemann A, Mukhopadhyay S, Gordon S (2006) The interaction of macrophage receptors with bacterial ligands. Expert Rev Mol Med 8:1–25

Rini JM, Lobsanov YD (1999) New animal lectin structures. Curr Opin Struct Biol 9: 578–584

Salminen A, Kaarniranta K (2009) Siglec receptors and hiding plaques in Alzheimer's disease. J Mol Med 87:697–701

Sato K, X-li Y, Yudate T et al (2006) Dectin-2 is a pattern recognition receptor for fungi that couples with the fc receptor γchain to induce innate immune responses. J Biol Chem 281:38854–38866

Sato S, St-Pierre C, Bhaumik P et al (2009) Galectins in innate immunity: dual functions of host soluble beta-galactoside-binding lectins as damage-associated molecular patterns (DAMPs) and as receptors for pathogen-associated molecular patterns (PAMPs). Immunol Rev 230:172–187

Seaton BA, Crouch EC, McCormack FX et al (2010) Structural determinants of pattern recognition by lung collectins. Innate Immun 16:143–150

Shahzad-ul-Hussan S, Cai M, Bewley CA (2009) Unprecedented glycosidase activity at a lectin carbohydrate-binding site exemplified by the cyanobacterial lectin MVL. J Am Chem Soc 131:16500–16507

Shanmugham LN, Castellani ML, Salini V et al (2006) Relevance of plant lectins in human cell biology and immunology. Riv Biol 99:227–247

Sharon N (2006) Carbohydrates as future anti-adhesion drugs for infectious diseases. Biochim Biophys Acta 1760:527–537

Sharon N (2008) Lectins: past, present and future. Biochem Soc Trans 36:1457–1460

Sharon N, Lis H (2001) The structural basis for carbohydrate recognition by lectins. Adv Exp Med Biol 491:1–16

Sharon N, Lis H (2003) Lectins, 2nd edn. Kluwer Academic Publishers, Dordrecht

Sonnenburg JL, Altheide TK, Varki A (2004) A uniquely human consequence of domain-specific functional adaptation in a sialic acid-binding receptor. Glycobiology 14:339–346

Srinivas VR, Reddy GB, Ahmad N et al (2001) Legume lectin family, the 'natural mutants of the quaternary state', provide insights into the relationship between protein stability and oligomerization. Biochim Biophys Acta 1527:102–111

Stillmark PH (1988) Über Ricin, ein giftiges Ferment aus den Samen von Ricinus comm. L. und einigen anderen Euphorbiaceen. Ph. D. thesis, University of Tartu

Stockert RJ, Morell AG, Scheinberg IH (1974) Mammalian hepatic lectin. Science 186:365–366

Svajger U, Anderluh M, Jeras M, Obermajer N (2010) C-type lectin DC-SIGN: an adhesion, signaling and antigen-uptake molecule that guides dendritic cells in immunity. Cell Signal 22:1397–1405

Swaminathan GJ, Weaver AJ, Loegering DA et al (2001) Crystal structure of the eosinophil major basic protein at 1.8 Å: an atypical lectin with a paradigm shift in specificity. J Biol Chem 23: 26197–26203

Swanson MD, Winter HC, Goldstein IJ (2010) A lectin isolated from Bananas is a potent inhibitor of HIV replication. J Biol Chem 285:8646–8655

Takahashi Y, Yajima A, Cisar JO, Konishi K (2004) Functional analysis of the *Streptococcus gordonii* DL1 sialic acid-binding adhesin and its essential role in bacterial binding to platelets. Infect Immun 72:3876–3882

Teichberg VI, Silman I, Beitsch DD, Resheff G (1975) A β-D-galactoside binding protein from electric organ tissue of electrophorus electricus. Proc Natl Acad Sci U S A 72:1383–1387

Tharanathan RN, Kittur FS (2003) Chitin: the undisputed biomolecule of great potential. Crit Rev Food Sci Nutr 43:61–87

Thiel S (2007) Complement activating soluble pattern recognition molecules with collagen-like regions, mannan-binding lectin, ficolins and associated proteins. Mol Immunol 44:3875–3888

Thomsen T, Schlosser A, Holmskov U, Sorensen GL (2011) Ficolins and FIBCD1: soluble and membrane bound pattern recognition molecules with acetyl group selectivity. Mol Immunol 48: 369–381

Toscano MA, Ilarregui JM, Bianco GA et al (2007) Dissecting the pathophysiologic role of endogenous lectins: glycan-binding proteins with cytokine-like activity? Cytokine Growth Factor Rev 18:57–71

Tsuji S, Uehori J, Matsumoto M et al (2001) Human intelectin is a novel soluble lectin that recognizes galactofuranose in carbohydrate chains of bacterial cell wall. J Biol Chem 276:23456–23463

Turnay J, Lecona E, Fernandez-Lizarbe S et al (2005) Structure-function relationship in annexin A13, the founder member of the vertebrate family of annexins. Biochem J 389:899–911

Usami Y, Fujimura Y, Suzuki M et al (1993) Primary structure of two-chain botrocetin, a von Willebrand factor modulator purified from the venom of *Bothrops jararaca*. Proc Natl Acad Sci U S A 90: 928–932

van de Wetering JK, van Golde LM, Batenburg JJ (2004) Collectins: players of the innate immune system. Eur J Biochem 271: 1229–1249

van der Vlist M, Geijtenbeek TB (2010) Langerin functions as an antiviral receptor on Langerhans cells. Immunol Cell Biol 88: 410–415

van Kooyk Y, Engering A et al (2004) Pathogens use carbohydrates to escape immunity induced by Dendritic cells. Curr Opin Immunol 16:488–493

Varki A, Kannag R, Toole BP (2009) Glycosylation Changes in Cancer. In: Varki A, Cummings RD, Esko JD et al (eds) Essentials of Glycobiology, 2nd edn. Cold Spring Harbor, NY

Wagner LA, Ohnuki LE et al (2007) Human eosinophil major basic protein 2: location of disulfide bonds and free sulfhydryl groups. Protein J 26:13–18

Waters P, Vaid M, Kishore U, Madan T (2009) Lung surfactant proteins A and D as pattern recognition proteins. Adv Exp Med Biol 653: 74–97

Watkins WM, Morgan WTJ (1952) Neutralization of the anti-H agglutinin in eel serum by simple sugars. Nature 169:825–826

Weis WI, Kahn R, Fourme R et al (1991) Structure of the calcium-dependent lectin domain from a rat mannose- binding protein determined by MAD phasing. Science 254:1608–1615

Weis WI, Taylor ME, Drickamer K (1998) The C-type lectin superfamily in the immune system. Immunol Rev 163:19–31

Wikström L, Lodish HF (1993) Unfolded H2b asialoglycoprotein receptor subunit polypeptides are selectively degraded within the endoplasmic reticulum. J Biol Chem 268:14412–14416

Wu WW, Molday RS (2003) Defective discoidin domain structure, subunit assembly, and endoplasmic reticulum processing of retinoschisin are primary mechanisms responsible for X-linked retinoschisis. J Biol Chem 278:28139–28146

Yamanishi T, Yamamoto Y, Hatakeyama T et al (2007) CEL-I, an invertebrate N-acetylgalactosamine-specific C-type lectin, induces TNF-alpha and G-CSF production by mouse macrophage cell line RAW264.7 cells. J Biochem 142:587–595

Yu H, Murata K, Hedrick JL et al (2007) The disulfide bond pattern of salmon egg lectin 24K from the Chinook salmon *Oncorhynchus tshawytscha*. Arch Biochem Biophys 463:1–11

Zanetta JP (1998) Structure and functions of lectins in the central and peripheral nervous system. Acta Anat (Basel) 161:180–195

Zelensky AN, Gready JE (2003) Comparative analysis of structural properties of the C-type-lectin-like domain (CLRD). Proteins 52:466–477

Zelensky AN, Gready JE (2005) The C-type lectin-like domain superfamily. FEBS J 272:6179–6217

Zeng FY, Weigel PH (1996) Fatty acylation of the rat and human asialoglycoprotein receptors. A conserved cytoplasmic cysteine residue is acylated in all receptor subunits. J Biol Chem 271: 32454–32460

Zeng R, Xu Q, Shao XX et al (2001) Determination of the disulfide bond pattern of a novel C-type lectin from snake venom by mass spectrometry. Rapid Commun Mass Spectrom 15:2213–2220

Zhang L, Ikegami M, Crouch EC et al (2001) Activity of pulmonary surfactant protein-D (SP-D) in vivo is dependent on oligomeric structure. J Biol Chem 276:19214–19219

Part II

Intracellular Lectins

Lectins in Quality Control: Calnexin and Calreticulin 2

G.S. Gupta

2.1 Chaperons

A long-standing enigma has been the role of *N*-linked glycans attached to many proteins in the endoplasmic reticulum (ER) and their co- and posttranslational remodelling along the secretory pathway. Evidence is accumulating that intracellular animal lectins play important roles in quality control and glycoprotein sorting along the secretory pathway. Calnexin and calreticulin in conjunction with associated chaperones promote correct folding and oligomerization of many glycoproteins in the ER. The discovery that one of these glycan modifications, mannose 6-phosphate, serves as a lysosomal targeting signal that is recognized by mannose 6-phosphate receptors has led to the notion that lectins may play more general roles in exocytotic protein trafficking (Chaps. 3–6). In present and subsequent chapters we discuss the role of intracellular lectins in quality control (QC) and their role in understanding the mechanisms underlying protein traffic in the secretory pathway (Chaps. 3–7) (Table 2.1).

In eukaryotic cells, the ER plays an essential role in the synthesis and maturation of a variety of important secretory and membrane proteins. The ER is also considered one of the most important and metabolically relevant sources of cellular Ca^{2+}. The ability of ER to control Ca^{2+} homoeostasis has profound effects on many cell functions. To achieve its function the ER and its lumen contain a characteristic set of resident proteins that are involved in every aspect of ER function. For glycoproteins, the ER possesses a dedicated maturation system, which assists folding and ensures the quality of final products before ER release. A chaperone is a protein that binds transiently to newly synthesized proteins and assists in newly synthesized proteins in folding (Ellgaard and Helenius 2003; Trombetta and Parodi 2003). Essential components of this system include the lectin chaperones calnexin (Cnx) and calreticulin (Crt) and their associated co-chaperone ERp57, a glycoprotein specific thiol-disulfide oxidoreductase. The significance of this system is underscored by the fact that Cnx and Crt interact with practically all glycoproteins investigated to date, and by the debilitating phenotypes revealed in knockout mice deficient in either gene. Compared to other important chaperone systems, such as the Hsp70s, Hsp90s and GroEL/GroES, the principles whereby this system works at the molecular level are relatively poorly understood. However, structural and biochemical data have provided important new insights into this chaperone system and present a solid basis for further mechanistic studies. Both Crt and Cnx act as lectins which recognize CRD and act as molecular chaperones. Both of them bind monoglucosylated proteins and associate with the thiol oxidoreductase ERp57, which is a protein disulfide isomerase (PDI)-like protein resident in ER and promotes disulfide formation/isomerization in glycoproteins. Calreticulin, together with Cnx and ERp57 comprise the so-called “calreticulin/calnexin cycle”, which is responsible for QC and folding in newly synthesized (glyco) proteins.

Crt/Cnx pathway is thought to monitor protein conformation through modification of N-linked glycans covalently attached to Asn residues through the activity of glycosyl transferases, glucosidases and chaperones. Folding enzymes such as PDI and ERp57 interact with polypeptides displaying non-native disulfide bonds, effectively reorganizing these covalent cross-links to their native pattern. ERp57, a thiol oxidoreductase that catalyzes disulfide formation in heavy chains of MHC class I molecules, also forms a mixed disulfide with tapasin within the class I peptide loading complex, stabilizing the complex and promoting efficient binding of peptides to class I molecules. ERp57 appears to play a structural rather than catalytic role within the peptide loading complex (Zhang et al. 2009). The PDI can help the formation of disulfide bonds that are the critical structure of protein secondary structure (Jun-Chao et al. 2006).

G.S. Gupta et al., *Animal Lectins: Form, Function and Clinical Applications*,
DOI 10.1007/978-3-7091-1065-2_2, © Springer-Verlag Wien 2012

Table 2.1 Lectins associated in quality control of proteins and the secretory pathway

Lectin	Localization	Sugar specificity	Major function
Calnexin	ER	Gluc + Man	Folding and degradation
Calreticulin	ER	Gluc + Man	Folding and degradation
F-Box Proteins	ER	Man	Degradation
EDEM	ER	Man	Degradation
CD-MPR[a]	LE	Man-6-P	Golgi-to-endosome transport
IGF-II/CI-MPR[b]	LE	Man-6-P	Golgi-to-endosome and plasma membrane-to - endosome transport
ERGIC-53	ERGIC	Man	ER-to-ERGIC transport
VIP36	cis-Golgi/ ERGIC	Man6–9	Retrieval?

[a]CD-MPR = Cation-dependent mannose 6-phosphate receptor
[b]IGF-II/CI-MPR = Insulin-like growth factor-II or cation-independent mannose 6-phosphate/receptor; ER = Endoplasmic reticulum; LE = late endosome

2.1.1 Calnexin

The Calnexin Protein

Calnexin is a 90-kDa integral membrane protein of the ER. It binds Ca^{2+} and functions as a chaperone in the transition of proteins from ER to outer cellular membrane. Calnexin has been cloned from placenta. A subdomain containing four internal repeats binds Ca^{2+} with highest affinity. This sequence is highly conserved when compared to calreticulin, and yeast and plant calnexin homologues. An adjacent subdomain, also highly conserved and containing four internal repeats, failed to bind Ca^{2+}. The carboxyl-terminal, cytosolic domain is highly charged and binds Ca^{2+} with moderate affinity. The calnexin amino-terminal domain (residues 1–253) also binds Ca^{2+}, in contrast to the amino-terminal domain of calreticulin, which is relatively less acidic. A subdomain containing four internal repeats binds Ca^{2+} with the highest affinity. This sequence is highly conserved when compared to Crt, and yeast and plant Cnx homologues. An adjacent subdomain, also highly conserved and containing four internal repeats, failed to bind Ca^{2+}. Cnx cDNA is highly conserved when compared to calreticulin, an Onchocerca surface antigen, and yeast and plant Cnx homologues. Comparison of mouse and rat calnexin sequences reveals very high conservation of sequence identity (93–98%), suggesting that calnexin performs important cellular functions. The gene for human calnexin is located on the distal end of the long arm of human chromosome 5 at 5q35 (Ellgaard et al. 1994; Tjoelker et al. 1994).

Calmegin: A Male Germ Cell Specific Homologue of Calnexin

During mammalian spermatogenesis, many specific molecules are expressed. A 93-kDa male meiotic germ cell-specific antigen (Meg 1) is exclusively expressed in germ cells from pachytene spermatocyte to spermatid stage (Watanabe et al. 1994). A cDNA from mouse testis showed this transcript of 2.3 kb in length and expressed only in testis and not in other somatic tissues or in ovary. The expression of the mRNA was first detected at pachytene spermatocyte stage of male germ cell development. The predicted protein consists of 611 amino acids, including a hydrophobic NH2 terminus characteristic of a signal peptide, two sets of internal repetitive sequences (four repeats of IPDPSAVKPEDWDD and GEWXPPMIPNPXYQ), and a hydrophilic COOH terminus. The deduced amino acid sequence showed 58% homology with dog Cnx and significant partial homology with Crt at repetitive sequence. The name calmegin was proposed for this antigen. Calmegin is a Ca^{2+}-binding protein that is specifically expressed in spermatogenesis (Watanabe et al. 1994). *Calmegin* gene contained GC-rich sequences and potential binding sites for AP2 and Sp1, but lacked TATA sequence. The CAT gene activity was detected exclusively in testes, indicating that the 330 bp calmegin 5′ sequence was sufficient for the testis-specific expression. The existence of testicular nuclear factors specifically bound to the putative promoter sequence was also demonstrated (Watanabe et al. 1995). The human homolog of mouse Calmegin showed 80% identity with the mouse calmegin and strong conservation of two sets of internal repetitive sequences (Ca^{2+} binding motif), and the hydrophilic COOH terminus, which corresponds to the putative ER retention motif. The transcript was 3 kb in length and was expressed exclusively in the testis. Human Calmegin gene was mapped to chromosome 4q28.3–q31.1 (c/r Gupta 2005). Calmegin functions as a chaperone for one or more sperm surface proteins that mediate the interactions between sperm and egg. The defective zona pellucida-adhesion phenotype of sperm from calmegin-deficient mice is reminiscent of certain cases of unexplained infertility in human males (Ikawa et al. 1997). These results suggest that spermatogenic cell endoplasmic reticulum has a unique calcium binding protein, calnexin-t, which appears to be a calnexin variant.

2.1.2 Calnexin Structure

The ER luminal segment of Cnx and its soluble paralog Crt share sequence similarity that is most pronounced in a central segment containing two proline-rich sequence motifs repeated in tandem (David et al. 1993). Motifs 1 and 2, which are repeated three times each in Crt and four times each in Cnx, have consensus sequences of I-DP(D/E) A-KPEDWD(D/E) and G-W-P-IN-P-Y, respectively. In addition, there are three segments of high sequence similarity, A, B, and C, with the last two flanking the repeat motifs (Fig. 2.1 and 2.2). The globular domain contains the oligosaccharide-binding site with amino acids that contact

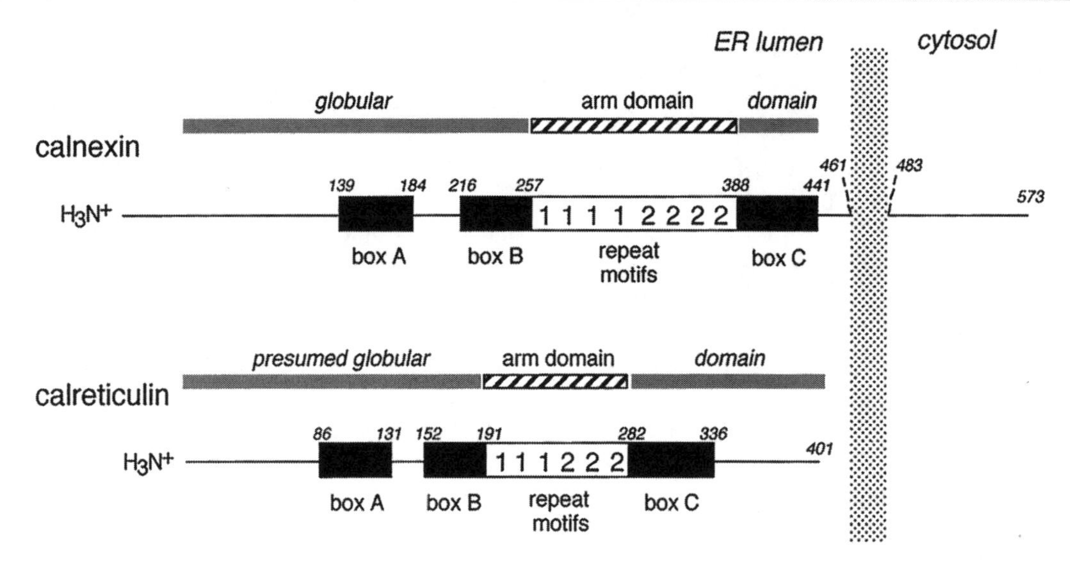

Fig. 2.1 Comparison of the linear sequences of CNX and CRT. Regions of sequence similarity are represented by *large rectangles*. The *white rectangles* correspond to segments with highest identity which are comprised of two sequence motifs repeated in tandem (indicated by the *numbers 1* and 2). *Black rectangles* represent segments that share substantial sequence identity, and they are termed *boxes A, B*, and *C* to facilitate discussion. Two domains identified in the x-ray crystal structure of CNX are shown: an arm domain that corresponds to the repeat motifs (*hatched bar*) and a globular domain (*gray bar*). An arm domain has also been identified in CRT (*hatched bar*) which corresponds to the repeat motifs. The structure of the remainder of the molecule has not been solved but is presumed to form a globular domain analogous to that of CNX (*gray bar*) (Reprinted by permission from Leach et al. 2002 © The American Society for Biochemistry and Molecular Biology)

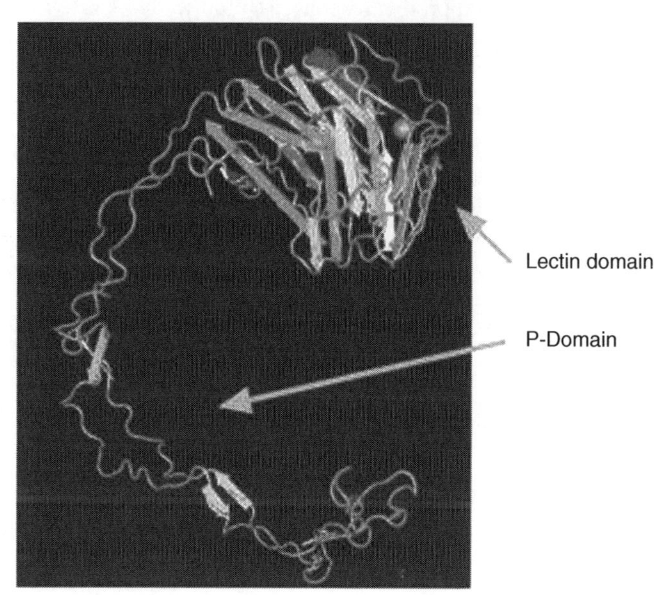

Fig. 2.2 Domain organization of calnaxin (Adapted with permission from Schrag et al. 2001 © Elsevier)

the terminal glucose residue. Each motif-1-repeat is paired with a motif-2-repeat on the opposite strand.

The lectin domain confers specific binding to glycoproteins bearing Asn-linked oligosaccharides of the form Glc1Man5–9GlcNAc2 (Leach et al. 2002; Ware et al. 1995). A 140-Å hairpin loop forms the arm domain, the tip of which includes the binding site for a thiol oxidoreductase, ERp57 (Leach et al. 2002; Pollock et al. 2004). In vitro studies have shown that the recruitment of a reduced glycoprotein to Cnx-ERp57 complex greatly enhances oxidative folding relative to ERp57 alone (Zapun et al. 1998). Both the globular and arm domains have been shown to bind Ca^{2+} but the crystal structure revealed only a single bound Ca^{2+} within the globular domain (Schrag et al. 2001). Although the complete structure of Crt is yet to be solved, ~39% overall sequence identity to Cnx, combined with conserved oligosaccharide binding specificity (Vassilakos et al. 1998), a shorter but similar arm domain structure as determined by NMR (Ellgaard et al. 2001a), and conserved ERp57 association properties (Frickel et al. 2002) all suggest a similar overall structure for the two chaperones. Ca^{2+} also binds to the globular and arm domains of Crt, which suggests that this ion is important for stabilizing both chaperones (Brockmeier and Williams 2006; Li et al. 2001).

2.1.3 Calnexin Binds High-Mannose-Type Oligosaccharides

In ER and in early secretory pathway, where repertoire of oligosaccharide structures is still small, the glycans play a pivotal role in protein folding, oligomerization, quality control, sorting, and transport. They are used as universal "tags" that allow specific lectins and modifying enzymes to establish order among the diversity of maturing glycoproteins (Yamashita et al. 1999). High-mannose-type oligosaccharides have been shown to play important roles in protein quality control (QC). Several intracellular proteins, such as lectins, chaperones and glycan-processing enzymes, are involved in this process. These include Cnx/Crt, UDP-glucose:glycoprotein glucosyltransferase (UGGT), cargo receptors (such as VIP36 and ERGIC-53), mannosidase-like proteins (e.g. EDEM and Htm1p) and ubiquitin ligase (Fbs). They are thought to recognize high-mannose-type glycans with subtly different structures, although the precise specificities are yet to be clarified. Calnexin binds mostly with N-glycosylated proteins. Soluble Cnx binds specifically with Glc1Man9GlcNAc2 oligosaccharide as an initial step in recognizing unfolded glycoprotein. Findings suggested that once complexes between calnexin and glycoproteins are formed, oligosaccharide binding does not contribute significantly to the overall interaction (Leach et al. 2002; Ware et al. 1995).

2.1.4 Functions of Calnexin

Cnx/Crt as Classical Chaperones of Protein Folding

The polypeptide-binding site on Cnx and Crt permits them to function as classical chaperones capable of recognizing non-native features of protein folding intermediates and suppressing their aggregation. This function was initially uncovered through in vitro experiments that demonstrated that both Cnx and Crt can suppress the aggregation not only of glycoproteins bearing monoglycosylated oligosaccharides but that of nonglycosylated proteins as well (Culina et al. 2004; Rizvi et al. 2004; Thammavongsa et al. 2005). Aggregation suppression ability was enhanced in presence of physiological ER Ca^{2+} concentrations as well as millimolar ATP, the latter causing an increased hydrophobic surface on chaperones (Brockmeier and Williams 2006). Several studies have validated the existence of functional polypeptide-based interactions between either Cnx or Crt and folding glycoproteins in living cells. ER molecular chaperones of the calreticulin/calnexin cycle have overlapping and complementary but not redundant functions. The absence of one chaperone can have devastating effects on the function of the others, compromising overall QC of the secretory pathway and activating unfolded protein response (UPR)-dependent pathways.

A number of proteins have been demonstrated to bind calnexin. The 90-kDa phosphoprotein (p90) of ER formed stable and transient complexes with other cellular proteins, and associated with heavy chain of MHC class I protein. A truncated version of integral membrane glycoprotein Glut 1 (GT_{155}) interacts with calnexin in vitro. Reports highlight the importance of the calnexin cycle in the functional maturation of the g-secretase complex (Hayashi et al. 2009b). γ-secretase is a membrane protein complex that catalyzes intramembrane proteolysis of a variety of substrates including the amyloid precursor protein of Alzheimer disease. Nicastrin (NCT), a single-pass membrane glycoprotein that harbors a large extracellular domain, is an essential component of the γ-secretase complex. Lectin deficient mutants of Cnx were shown to interact with heavy chains of MHC class I molecules in insect cells and to prevent their rapid degradation (Leach and Williams 2004). Similarly, Crt-deficient mutants were found not only to interact with a broad spectrum of newly synthesized proteins and dissociate with normal kinetics, but it was also able to complement all MHC class I biosynthetic defects associated with Crt deficiency (Ireland et al. 2008). Studies are consistent with a model wherein Cnx and Crt associate with folding glycoproteins through both lectin- and polypeptide-based interactions thereby increasing the avidity of association relative to either interaction alone (Stronge et al. 2001). Binding of Cnx or Crt serves to prevent premature release of folding intermediates from the ER and promotes proper folding by suppressing off-pathway aggregation and by providing a privileged environment in which associated ERp57 promotes thiol oxidation and isomerization reactions (Zapun et al. 1998). The Crt is essential for normal Cnx chaperone function. In the absence of Crt, Crt substrates are not "picked up" by Cnx but accumulate in ER lumen, resulting in the activation of unfolded protein response (UPR), which is activated to induce transcription of ER-localized molecular chaperones (Shen et al. 2004). Pancreatic ER kinase (PERK) and Ire1a, UPR-specific protein kinases, and eIF2a are also activated in the absence of Crt (Knee et al. 2003).

Cnx/Crt Cycle

Calnexin binds only those N-glycoproteins which possess GlcNAc2Man9Glc1 oligosaccharides. Oligosaccharides with three sequential glucose residues are added to asparagine residues of the nascent proteins in the ER. The monoglucosylated oligosaccharides that are recognized by Cnx result from trimming of two glucose residues by sequential action of two glucosidases (GLS), I and II. GLS II can also remove third and last glucose residue (Hebert et al. 1995). The ER also contains a uridine diphosphate (UDP)-glucose, glycoprotein transferase (UGGT), which can re-glucosylate chains that have been glucose-trimmed. If the glycoprotein is not properly folded, the UGGT will add the glucose residue back onto the oligosaccharide thus regenerating the glycoprotein's ability to bind to calnexin. The improperly-folded glycoprotein chain thus loiters in the ER, risking the encounter with α-mannosidase (MNS1), which eventually sentences the underperforming glycoprotein to degradation by removing its mannose residue. If the protein is correctly translated, the chance of it being correctly folded before it encounters MNS1 is high. ATP and calcium ions are two of the cofactors involved in substrate binding for calnexin. Together, UGGT and GLS II establish a cycle of de-glucosylation and re-glucosylation.

Importantly, UGGT discriminates between folded and unfolded proteins, adding back a glucose residue to unfolded proteins only. This results in "rebuilding" of the monoglucosylated oligosaccharide on unfolded substrates, enabling them to interact with Crt and/or Cnx again. This de-glucosylation/glucosylation cycle may be repeated several times before a newly synthesized glycoprotein is properly folded (Roth et al. 2003) (Fig. 2.3). The lectin-oligosaccharide interaction is regulated by the availability of the terminal glucose on Glc1Man5–9Glc-NAc2 oligosaccharides. If folding of glycoprotein does not occur promptly, the folding sensor UGGT recognizes non-native conformers and reglucosylates *N*-glycans, thereby allowing re-entry into the chaperone cycle (Caramelo et al. 2004; Ritter et al. 2005; Taylor et al. 2004). Terminally folding-defective glycoproteins are further processed by demannosylation that diverts them from Cnx/Crt cycle into ERAD disposal pathway (Helenius and Aebi 2004). The lectin-binding site of Cnx and Crt is localized to the Ca^{2+} binding P-domain of the protein and the bound Ca^{2+} is essential for the lectin-like function of these proteins. Moreover, while glycoproteins are bound to Cnx and Crt, the disulphide bonds of the substrates are rearranged by the PDI activity associated with ERp57 suggesting that Crt binding to carbohydrates may be a 'signal' to recruit other chaperones to assist in protein folding. It should be emphasized, however, that monoglucosylated high-mannose carbohydrates may not be a prerequisite for substrate binding to Crt. For example, castanospermine and 1-deoxynojirimycin, inhibitors of the glucosidase II, do not affect association between Crt and Factor VIII or between Crt and mucin. Calreticulin also binds directly to PDI, ERp57, perforin, the synthetic peptide KLGFFKR and the DNA-binding domain of steroid receptors, indicating that chaperone function of Crt may involve both protein–protein and protein–carbohydrate interactions (Michalak et al. 1999).

Site of Interaction/Substrate Specificity

Although lectin sites of Cnx and Crt have been well defined through structural and mutagenesis studies (Schrag et al. 2001; Kapoor et al. 2004; Thomson and Williams 2005), less is known about the location and substrate specificity of the polypeptide-binding sites. Deletion mutagenesis of rabbit Crt as well as Cnx suggested that their abilities to suppress the aggregation of nonglycosylated proteins reside primarily within their globular domains (Leach et al. 2002; Xu et al. 2004). Furthermore, in vitro binding experiments with nonglycosylated proteins such as citrate synthase and malate dehydrogenase have indicated that both chaperones interact preferentially with non-native conformers, suggesting that they act as folding sensors in addition to the role provided by UDP-glucose:glycoprotein glucosytransferase (Saito et al. 1999). To characterize the specificity of the polypeptide-binding site of Crt, Sandhu et al. (2007) and Duus et al. (2008) examined a panel of peptides for their binding to Crt using a competitive ELISA. Peptide binding required a minimum peptide length of five residues that were hydrophobic in nature. In another study, a hydrophobic Crt-binding peptide was shown to compete with the ability of Crt to suppress the thermally induced aggregation of a soluble MHC class I molecule (Rizvi et al. 2004). Collectively, these findings indicate the presence of a site on Cnx and Crt that recognizes non-native protein conformers and that, in the case of Crt, exhibits specificity for hydrophobic peptide segments.

Brockmeier et al. (2009) investigated the location and characteristics of the polypeptide-binding function of Cnx under physiological conditions of ER lumen. Using an assay in which soluble ER luminal domain of Cnx (S-Cnx) suppresses the aggregation of nonglycosylated firefly luciferase at 37 °C and 0.4 mM Ca^{2+} (Brockmeier and Williams 2006), Brockmeier et al. showed that this aggregation suppression function resides within the globular lectin domain but is enhanced by the presence of the full-length arm domain. Direct binding experiments revealed a single site of peptide binding in globular domain at a location distinct from the lectin site. The site in globular domain is responsible for aggregation suppression and is capable of binding hydrophobic peptides with μM affinity ($K_D = 0.9$ μM). Furthermore, binding studies with peptides and non-native proteins of increasing size revealed that the arm domain contributes to the aggregation suppression function of S-Cnx not through direct substrate binding but rather by sterically constraining large polypeptide chains.

Cnx as a PACS-2 Cargo Protein

Proteins of Cytosolc Sorting (Pacs): CK2-phosphorylatable acidic clusters are hallmark interacting sequences for proteins of cytosolc sorting (PACS) protein family, which includes PACS-1 and PACS-2. The interaction of these acidic motifs with PACS proteins mediates a variety of intracellular steps that include trafficking between *trans*-Golgi (TGN) network and endosomes, localization to mitochondria and retention in ER. Cargo proteins can usually interact with both PACS-1 and PACS-2. Changing the amount of Cnx on plasma membrane could affect cell surface properties and might have implications on phagocytosis or cell–cell interactions. Hence, the amount of Cnx on plasma membrane could depend on cell type or cellular homeostasis, and it might be the result of regulated intracellular retention (c/r Myhill et al. 2008). Myhill et al. (2008) identified Cnx as a PACS-2 cargo protein on ER. Cnx interacts with PACS-2 using its acidic CK2 motif. Results suggest that the phosphorylation state of the calnexin cytosolic domain and its interaction with PACS-2 sort this chaperone between domains of the ER and the plasma membrane.

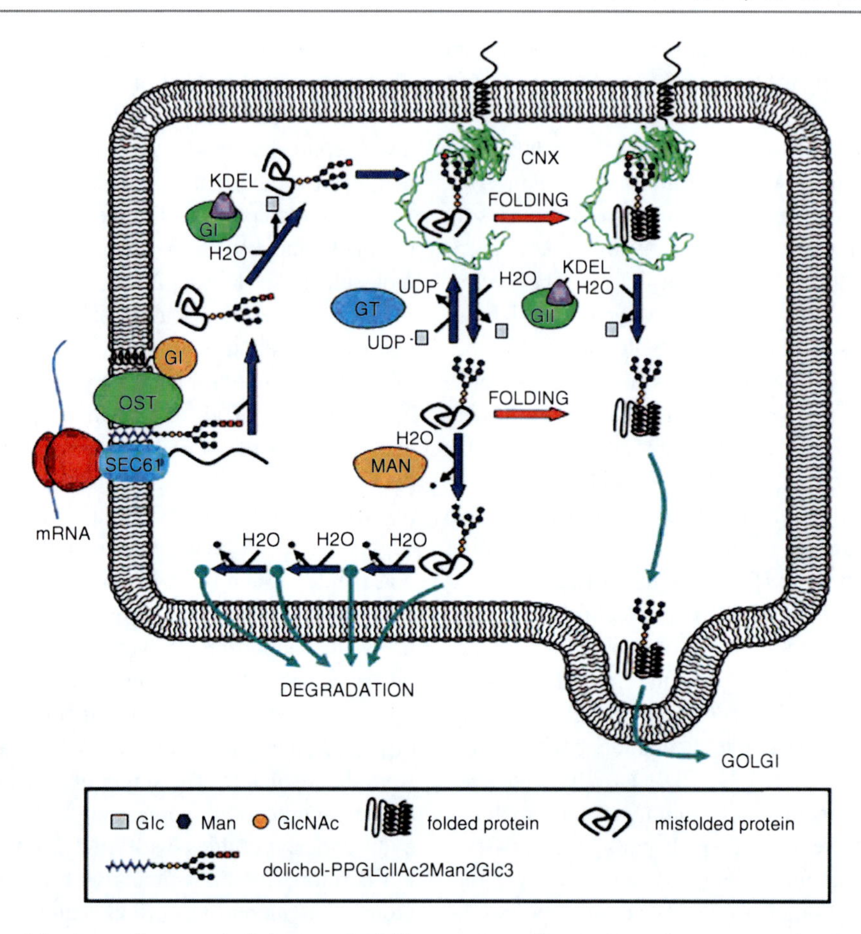

Fig. 2.3 Model proposed for the quality control of glycoprotein folding. Proteins entering the ER are *N*-glycosylated by the oligosaccharyltransferase (*OST*) as they emerge from the translocon. Two glucose units are removed by the sequential action of GI and GII to generate monoglucosylated species that are recognized by Cnx and/or Crt (only Cnx is shown), which is associated with ERp52. The complex between the lectins and folding intermediates/misfolded glycoproteins dissociates upon removal of the last glucose by GII and is reformed by GT activity. Once glycoproteins have acquired their native conformations, either free or complexed with the lectins, GII hydrolyzes the remaining glucose residue and releases the glycoproteins from the lectin anchors. These species are not recognized by GT and are transported to the Golgi. Glycoproteins remaining in misfolded conformations are retrotranslocated to the cytosol, where they are deglycosylated and degraded by the proteasome. One or more mannose residues may be removed during the whole folding process (Reprinted with permission from Caramelo and Parodi 2008 © American Society for Biochemistry and Molecular Biology)

In addition to folding intermediates, ribosomes and SERCA2b, Cnx also interacts with BAP31, an ER cargo receptor that mediates export of transmembrane proteins from the ER and shuttles them to the ER quality control compartment. Thus, Cnx can reach the plasma membrane and can also interact with numerous ER membrane proteins that are found on multiple domains of the ER such as the Mitochondria Associated Membrane (MAM). Although Cnx and other ER chaperones clearly localize to multiple cellular membranes, it is currently not understood whether the cell has mechanisms in place that control the distribution of chaperones between these various locations. Support for the hypothesis of a controlled distribution of ER proteins to specific membrane domains comes from pioneering studies on Cnx (Chevet et al. 1999; Roderick et al. 2000), which showed that PKC, ERK-1 and protein kinase CK2 (CK2) can phosphorylate the Cnx cytosolic domain. Phosphorylation by ERK-1 on serine 583 increases interaction of Cnx with ribosomes, but also interaction with SERCA2b. In addition, CK2 phosphorylation of serines 554 and 564 by CK2 synergizes with ERK-1 phosphorylation of serine 583 to promote interaction with ribosomes (Chevet et al. 1999). Hence, the Cnx phosphorylation state could lead to enrichment on the MAM and the rER, where these Cnx interactors are found. However, it is still unclear what happens to dephosphorylated v that has been demonstrated to exist in vivo (Myhill et al. 2008).

However, Calnexin deficiency is not embryonic lethal. This is amazing considering great structural and functional similarities between the Crt and Cnx chaperones. The $Cnx^{-/-}$ animals exhibit impaired motor function and die within the first 5 weeks of life. Studies with Crt and Cnx gene knockout mice indicate that these proteins are unable to compensate for the loss of each other, suggesting they have unique functions. The molecular chaperone function of calreticulin and calnexin may only partially explain phenotypes of $Cnx^{-/-}$ and $Crt^{-/-}$ mice (Denzel et al. 2002).

2.1.5 Patho-Physiology of Calnexin Deficiency

ER and Oxidative Stresses Are Common Mediators of Apoptosis: The field of the ER stress in mammalian cells has expanded rapidly during the past decade, contributing to understanding of the molecular pathways that allow cells to adapt to perturbations in ER homeostasis. One major mechanism is mediated by molecular ER chaperones which are critical not only for quality control of proteins processed in the ER, but also for regulation of ER signaling in response to ER stress. Proteome analysis of human diploid fibroblasts (HDF) showed that Cnx significantly decreased with aging. Oxidative stress-induced expression of Cnx also attenuated in old HDF compared to young cells (Choi and Kim 2004). Ni and Lee (2007) reviewed the properties and functions of Cnx, Crt, and their role in development and diseases. Many of the new insights are derived from constructed mouse models where the genes encoding the chaperones are genetically altered, providing invaluable tools for examining the physiological involvement of the ER chaperones in vivo. Wei et al. (2008) uncovered that chemical disruption of lysosomal homeostasis in normal cells causes ER stress, suggesting a cross-talk between the lysosomes and the ER. Most importantly, chemical chaperones that alleviate ER and oxidative stresses are also cytoprotective in all forms of LSDs studied. It was proposed that ER and oxidative stresses are common mediators of apoptosis in both neurodegenerative and non-neurodegenerative LSDs. Hepatic ER stress induced by burn injury was associated with compensatory upregulation Cnx and Crt, suggesting that ER calcium store depletion was the primary trigger for induction of ER stress response in mice. Thus, thermal injury causes long-term adaptive and deleterious hepatic function characterized by significant upregulation of ER stress response (Song et al. 2009). Pre-administration of α-tocopherol is protective against oxidative renal tubular damage and subsequent carcinogenesis by ferric nitrilotriacetate (Fe-NTA) in rats. In addition to scavenging effects, α-tocopherol showed significantly beneficial effects in renal protection. Results suggest that α-tocopherol modifies glycoprotein metabolism partially by conferring mild ER stress (Lee et al. 2006).

Calnexin in Apoptosis Induced by ER Stress in *S. pombe*: Stress conditions affecting the functions of the endoplasmic reticulum (ER) cause the accumulation of unfolded proteins. ER stress is counteracted by the unfolded-protein response (UPR). However, under prolonged stress the UPR initiates a proapoptotic response. Mounting evidence indicate that the Cnx is involved in apoptosis caused by ER stress. Overexpression of Cnx in *Schizosaccharomyces pombe* induces cell death with apoptosis markers. Guerin et al. (2008) argue for the conservation of the role of calnexin in apoptosis triggered by ER stress, and validate *S. pombe* as a model to elucidate the mechanisms of Cnx-mediated cell death. The ER is highly sensitive to stresses perturbing the cellular energy levels and ER lipid or glycolipid imbalances or changes in the redox state or Ca^{2+} concentration. Such stresses reduce the protein folding capacity of the ER, which results in the accumulation and aggregation of unfolded proteins, a condition referred to as ER stress. When the capacity of the ER to fold proteins properly is compromised or overwhelmed, a highly conserved UPR signal-transduction pathway is activated. Guerin et al. (2008) further showed that the apoptotic effect of calnexin is counteracted by overexpression of Hmg1/2p, the *S. pombe* homologue of the mammalian antiapoptotic protein HMGB1 (high-mobility group box-1 protein). Interestingly, the overexpression of mammalian Cnx also induced apoptosis in *S. pombe*, suggesting the functional conservation of the role of Cnx in apoptosis. Inositol starvation in *S. pombe* causes cell death with apoptotic features. Observations indicated that Cnx takes part in at least two apoptotic pathways in *S. pombe*, and suggested that the cleavage of Cnx has regulatory roles in apoptotic processes involving Cnx (Guerin et al. 2009).

Calnexin in Biogenesis of Cystic Fibrosis: Deletion of phenylalanine at position 508 (δ F508) in first nucleotide-binding fold of cystic fibrosis transmembrane conductance regulator (CFTR) is the most common mutation in patients with cystic fibrosis. Although retaining functional Cl^- channel activity, this mutant is recognized as abnormal by cellular "quality control" machinery and is retained within ER. This intracellular retention was restricted to the immature (or ER-associated) forms of the CFTR proteins. Study indicated that Cnx retains misfolded or incompletely assembled proteins in ER and thus is likely to contribute to the mislocalization of mutant CFTR (Pind et al. 1994).

Cnx in Rod Opsin Biogenesis: Misfolding mutations in rod opsin are a major cause of the inherited blindness retinitis pigmentosa. A report from *Drosophila* rhodopsin Rh1 suggests the requirement of Cnx for its maturation and correct localization to R1–6 rhabdomeres (Rosenbaum et al. 2006). However, unlike *Drosophila* Rh1, mammalian rod opsin biogenesis does not appear to have an absolute requirement for Cnx. Other chaperones are likely to be more important for mammalian rod opsin biogenesis and quality control (Kosmaoglou and Cheetham 2008). Furthermore, the over-expression of Cnx leads to an increased accumulation of misfolded P23H opsin but not the correctly folded protein.

Finally, the increased levels of Cnx in the presence of the pharmacological chaperone 11-cis-retinal increase the folding efficiency and result in an increase in correct folding of mutant rhodopsin. These results demonstrate that misfolded rather than correctly folded rhodopsin is a substrate for Cnx and that the interaction between Cnx and mutant- misfolded rhodopsin can be targeted to increase the yield of folded mutant protein (Noorwez et al. 2009). Thus, Cnx preferentially associates with misfolded mutant opsins during retinitis pigmentosa.

Congenital Disorders: Autosomal dominant polycystic liver disease (PCLD), a rare progressive disorder, is characterized by an increased liver volume due to many fluid-filled cysts of biliary origin. Disease causing mutations in PRKCSH or SEC63 are found in approximately 25% of the PCLD patients. Hepatocystin is directly involved in the protein folding process by regulating protein binding to Cnx/Crt in the ER. A separate group of genetic diseases affecting protein N-glycosylation in the ER is formed by the congenital disorders of glycosylation (CDG). In distinct subtypes of this autosomal recessive multisystem disease specific liver symptoms have been reported that overlap with PCLD. Recent research revealed novel insights in PCLD disease pathology such as the absence of hepatocystin from cyst epithelia indicating a two-hit model for PCLD (Janssen et al. 2010)

Peripheral Neuropathies: Schwann cell-derived peripheral myelin protein-22 (PMP-22) when mutated or over-expressed causes heritable neuropathies with a unexplained "gain-of-function" ER retention phenotype. Missense point mutations in Gas3/PMP22 are responsible for the peripheral neuropathies Charcot-Marie-Tooth 1A and Dejerine Sottas syndrome. These mutations induce protein misfolding with the consequent accumulation of proteins in ER and the formation of aggresomes. In Trembler-J (Tr-J) sciatic nerves, prolonged association of mutant PMP-22 with Cnx is found. In 293A cells overexpressing PMP-22(Tr-J), Cnx and PMP-22 colocalize in large intracellular structures. Similar intracellular myelin-like figures were also present in Schwann cells of sciatic nerves from homozygous Trembler-J mice with no detectable activation of stress response pathway as deduced from BiP and CHOP expression. Sequestration of Cnx in intracellular myelin-like figures may be relevant to the autosomal dominant Charcot-Marie-Tooth-related neuropathies (Dickson et al. 2002). During folding PMP22 associates with calnexin. Calnexin interacts with the misfolded transmembrane domains of PMP22. The emerging models indicate for a glycan-independent chaperone role for calnexin and for the mechanism of retention of misfolded membrane proteins in the endoplasmic reticulum (Fontanini et al. 2005)

2.2 Calreticulin

2.2.1 General Features

Calreticulin (Crt) was first identified as a Ca^{2+}-binding protein of the muscle sarcoplasmic reticulum in 1974. Calreticulin is a ubiquitous protein, found in a wide range of species and in all nucleated cell types, and has a variety of important biological functions. Since its discovery (Ostwald and MacLennan 1974) from rabbit skeletal muscle it has been cloned in several vertebrates and invertebrates, and also higher plants (Michalak et al. 1999). There is no Crt gene in yeast and prokaryotes whose genomes have been fully sequenced. The human gene for calreticulin contains nine exons and eight introns. The deduced amino acid sequence indicates that calreticulin has a 17 amino acid hydrophobic signal sequence at its N terminus and that mature calreticulin contains 400 amino acids. The structure of calreticulin has been well characterized. It has at least three structural and functional domains.

The genomic configuration of the mouse and human calreticulin gene is shown in Fig. 2.4. The protein is encoded by a single gene (McCauliffe et al. 1992; Waser et al. 1997), and only one species of 1.9 kb mRNA encoding calreticulin has been identified. There is no evidence for alternative splicing of the calreticulin mRNA. The Crt gene consists of nine exons and spans approx. 3.6 kb or 4.6 kb of human or mouse genomic DNA respectively (Fig. 2.4). Human and mouse genes have been localized to chromosomes 19 and 8 respectively (McCauliffe et al. 1992; Rooke et al. 1997). The exon–intron organization of human and mouse genes is almost identical. The nucleotide sequences of the mouse and human gene show greater than 70% identity, with the exception of introns 3 and 6, indicating a strong evolutionary conservation of the gene. In the mouse gene these introns are approximately twice the size of the corresponding introns in the human gene. The promoter of the mouse and human calreticulin genes contain several putative regulatory sites, including AP-1 and AP-2 sites, GC-rich areas, including an Sp1 site, an H4TF-1 site, and four CCAAT sequences (McCauliffe et al. 1992; Waser et al. 1997). AP-2 and H4TF-1 recognition sequences are typically found in genes that are active during cellular proliferation. There is no obvious nuclear factor of activated T-cell (NF-AT) and nuclear factor κB (NF-κB) sites in the calreticulin promoter. Several poly (G) sequences, including GGGNNGGG motifs, are also found in the promoter regions of calreticulin and other ER/sarcoplasmic reticulum (SR) luminal proteins, including glucose-regulated protein 78 (Grp78) and Grp94. These motifs may therefore play a role in regulation of the expression of luminal ER proteins and in ER stress-

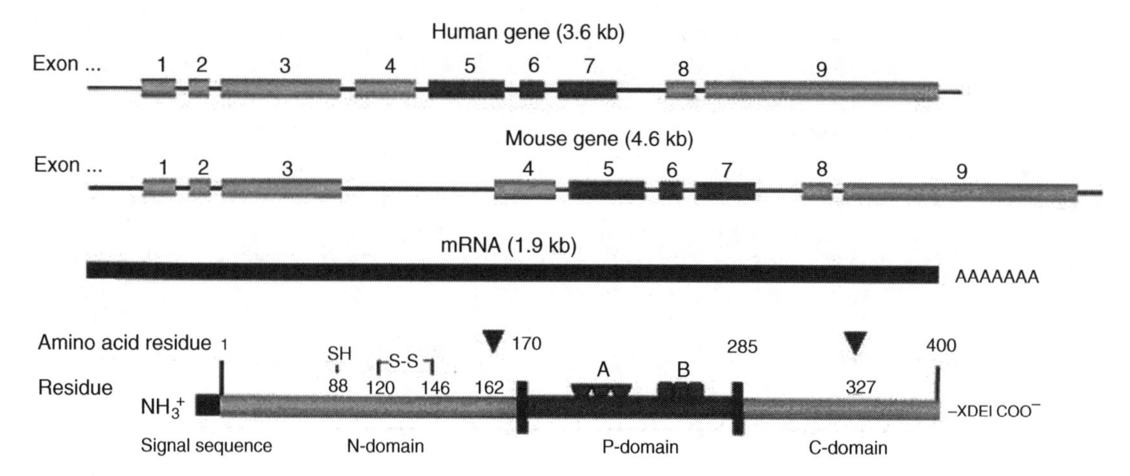

Fig. 2.4 The calreticulin gene: The Figure shows a schematic representation of the genomic configuration of domain structure of calreticulin protein. Structural predictions for calreticulin suggest that the protein has at least three structural and functional domains. Exons encoding the N domain (including the N-terminal signal sequence), and the C domain of calreticulin are in *grey* color while P is *black*. The N, P, and C domains are also presented in same color. The protein contains an N-terminal amino acid signal sequence (*black box*) and a C-terminal KDEL ER retrieval signal. The locations of 3 cysteine residues and the disulphide bridge in the N domain of calreticulin are indicated. The arrows indicate the location of potential glycosylation sites (residues 162 and 327). Repeats A (amino acid sequence PXXIXDPDAXKPEDWDE) and B (amino acid sequence GXWXPPXIXNPXYX) are also indicated (Reprinted with permission from Michalak et al. 1999; Biochem J. 344: 281–97 © The Biochemical Society)

dependent activation of the calreticulin gene (Michalak et al. 1999).

Depletion of Ca^{2+} stores induces severalfold activation of Crt promoter followed by increase in Crt mRNA and protein levels (Waser et al. 1997). Expression of Crt is also activated by bradykinin-dependent Ca^{2+} depletion of intracellular Ca^{2+} stores both in vitro and in vivo (Waser et al. 1997). The calreticulin promoter is activated by Zn^{2+} and heat shock. Expression of calreticulin is also induced by viral infection (Zhu 1996), by amino acid deprivation and in stimulated cytotoxic T-cells, further indicating that the calreticulin gene is activated by a variety of chemical and biological stresses. Since calreticulin has been implicated in a wide variety of cellular processes, the stress-dependent activation of the calreticulin gene may affect numerous biological and pathophysiological conditions (Michalak et al. 1999).

2.2.2 The Protein

Calreticulin is a 46-kDa protein with an N-terminal cleavable amino acid signal sequence and a C-terminal KDEL ER retrieval signal (Fig. 2.4). These specific amino acid sequences are responsible for targeting and retention of Crt in the ER lumen. Depending on species, Crt may have one or more potential N-linked glycosylation sites. The glycosylation pattern of the protein seems to be heterogeneous and does not appear to be a conserved property of the protein. The glycosylation of Crt is more common in plants than in animal cells (Navazio, et al. 1996). Heat shock may trigger glycosylation of calreticulin (Jethmalani, and Henle 1998); however, the functional consequence of this stress-induced glycosylation of the protein is presently not clear. Calreticulin has three cysteine residues, and all of them are located in the N-domain of the protein. Importantly, the location of these amino acid residues is conserved in calreticulin from higher plants to that in humans (Michalak et al. 1999). Two out of three cysteine residues found in the protein form a disulphide bridge (Cys^{120}–Cys^{146}) (Michalak et al. 1999; 2009), which may be important for proper folding of N-terminal region of Crt. ER localization of Crt is specified by two types of targeting signals, an N-terminal hydrophobic sequence that directs insertion into ER and a C-terminal KDEL sequence that is responsible for retention in the ER. Afshar et al. (2005) showed that Crt is fully inserted into ER, undergoes processing by signal peptidase, and subsequently undergoes retrotranslocation to the cytoplasm. C-terminal Ca^{2+} binding domain plays an important role in Crt retrotranslocation. Calreticulin is an ATP-binding protein but it does not contain detectable ATPase activity. Digestion of the protein with trypsin in presence of Mg^{2+} ATP protects the full-length protein. Results indicate that calreticulin may undergo frequent, ion-induced conformation changes, which may affect its function and its ability to interact with other proteins in lumen of ER (Corbett et al. 2000).

Calreticulin consists of various structural and functional domains. The N-domain of calreticulin, together with the central P-domain, is responsible for protein's chaperone function. Studies by Guo et al. (2003) and Michalak et al.

(2002a) with site-specific mutagenesis showed that mutation of a single His^{153} in calreticulin's N-domain destroys the protein's chaperone function. The P-domain of Crt (residues 181–290) contains a proline-rich region, forms an extended-arm structure, and interacts with other chaperones in the lumen of ER. The extended-arm structure is predicted to curve like that in Cnx that is similar to Crt, forming an opening that is likely to accommodate substrate binding, including the carbohydrate-binding site. As a molecular chaperone, Crt binds the monoglucosylated high mannose oligosaccharide (Glc1Man9GlcNAc2) and recognizes the terminal glucose and four internal mannoses in newly synthesized glycoproteins (Michalak et al. 2002b). Changes within ER, such as alterations in concentration of Ca^{2+}, Zn^{2+} or ATP, may affect the formation of these chaperone complexes and thus the ability of Crt to assist in protein folding (Trombetta and Parodi 2003).

Three Structural Domains

Similar to Cnx, Crt promotes the folding of proteins carrying N-linked glycans. Both proteins cooperate with an associated co-chaperone, the thiol-disulfide oxidoreductase ERp52. Three distinct structural domains have been identified in calreticulin: the amino-terminal, globular N-domain; the central P-domain; and the carboxyl-terminal C-domain (Michalak et al. 1999). NMR (Ellgaard et al. 2001), modeling (Michalak et al. 2002), and biochemical studies (Nakamura et al. 2001a) indicate that the globular N-domain and the "extended arm" P-domain of Crt may form a functional protein-folding unit (Michalak et al. 2002). This region of Crt contains a Zn^{2+} binding site and one disulfide bond, and it may also bind ATP (Baksh et al. 1995; Andrin et al. 2000; Corbett et al. 2000).

The exon–intron organization of the Crt gene suggests that the central P-domain of the protein may be encoded by exons 5, 6 and 7, whereas the first four exons and the last two exons may encode the N- and C-domain of the protein respectively (Fig. 2.4). Table 2.1 summarizes functional properties of these domains. The N-terminal part of the protein, encompassing the N- and P-domain of Crt, has the most conserved amino acid sequence (Michalak et al. 1999).

i). N-Domain

The N-terminal half of the molecule is predicted to be a highly folded globular structure containing eight anti-parallel β-strands connected by protein loops. The amino acid sequence of N-domain of Ctr is extremely conserved in all calreticulins. The N-domain binds Zn^{2+} (Baksh et al. 1995b, c; Andrin et al. 2000; Corbett et al. 2000) and it undergoes dramatic conformational changes (Michalak et al. 1999). Chemical modification of Crt has revealed that four histidines located in the N-domain of the protein (His^{25}, His^{82}, His^{128}, and His^{153}) are involved in Zn^{2+} binding (Baksh et al. 1995c). The Zn^{2+}-dependent conformational change in Crt affects its ability to bind to unfolded protein/glycoprotein substrates in vitro (Saito et al. 1999), suggesting that conformational changes in Crt may modify its chaperone function. ER of calreticulin-deficient cells with N-terminal histidine (His^{25}, His^{82}, His^{128}, and His^{153}) indicated that His^{153} chaperone function was impaired. Thus, mutation of a single amino acid residue in Crt has devastating consequences for its chaperone function, and may play a significant role in protein folding disorders (Guo et al. 2003). The N-domain interacts with the DNA-binding domain of the glucocorticoid receptor in vitro (Burns et al. 1994), with rubella virus RNA (Singh and Atreya 1994; Nakhasi et al. 1994), β-integrin and with protein disulphide-isomerase (PDI) and ER protein 57 (ERp57) (Baksh et al. 1995a, Corbett et al. 1999). Interaction of this region of Crt with PDI inhibits PDI chaperone function (Baksh et al. 1995a), but enhances ERp57 activity (Zapun et al. 1998). These protein–protein interactions are regulated by Ca^{2+} binding to the C-domain of Crt (Corbett et al. 1999). The N-domain of Crt also inhibits proliferation of endothelial cells and suppresses angiogenesis (Pike et al. 1998).

ii). P-Domain

The P-domain of Crt comprises a proline-rich sequence with three repeats of the amino acid sequence PXXIXDPDAXKPEDWDE (repeat A) followed by three repeats of the sequence GXWXPPXIXNPXYX (repeat B) (Fig. 2.5a). While the C-domain is responsible for the low-affinity and high-capacity Ca^{2+} binding, the P-domain of Crt binds Ca^{2+} with high affinity. The repeats may be essential for high-affinity Ca^{2+} binding of Crt (Baksh and Michalak 1991; Tjoelker et al. 1994) that interact with PDI (Corbett et al. 1999), NK2 homeobox 1 (Perrone et al. 1999), and perforin (Andrin et al. 1998; Fraser et al. 1998), a component of the cytotoxic T-cell granules. More importantly, repeats A and B are critical for the lectin-like chaperone activity of Crt (Vassilakos et al. 1998). The P-domain is one of the most interesting and unique regions of Crt because of its lectin-like activity and amino acid sequence similarities to other Ca^{2+}-binding chaperones, including Cnx (Bergeron et al. 1994), calmegin (Watanabe et al. 1994) and CALNUC, a Golgi Ca^{2+}-binding protein (Lin et al. 1998). However, the C-domain is less conserved than other domains of Crt (Michalak et al. 1999). Four amino acid residues (Glu^{239}, Asp^{241}, GLu^{243}, and Trp^{244}) at the tip of the 'extended arm' of P-domain are critical in chaperone function of Crt (Martin et al. 2006). Martin et al. (2006) focused studies on two cysteine residues (Cys^{88} and Cys^{120}), which form a disulfide bridge in N-terminal domain of Crt, a tryptophan residue located in CRD (Trp^{302}), and on certain residues located at the tip of "hairpin-like" P-domain of Crt (Glu^{238}, Glu^{239},

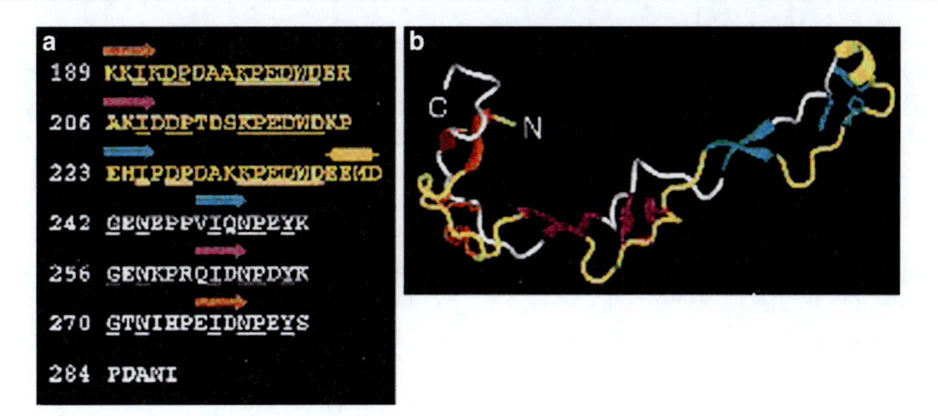

Fig. 2.5 (**a**). Alignment of the sequence repeats in rat Crt (189–288). The three 17-residue type 1 repeats are shown in yellow, and the three 14-residue type 2 repeats are white. The positions of the strands in three short antiparallel β-sheets are indicated by *orange*, *red*, and *blue arrows* above the sequence. The position of a helical turn, which includes a two-residue insert between the third and fourth repeat, is shown by a *yellow* cylinder above the sequence. Residues conserved in all three repeats of one type are underlined. (**b**). Bundles of the 20 energy-minimized conformers used to represent the NMR structure of Crt(189–288). (a) Superposition for best fit of the backbone atoms N, Ca, and C9 of the residues 219–258. (b) Superposition for best fit of the backbone atoms N, Ca, and C9 of the residues 189–209 and 262–284. In each drawing the polypeptide segments used for the superposition are colored *yellow*, and the remaining residues are *white*. (c) Cartoon of the conformer from *a* for which the *white* region is on the extreme *left*. The β-sheets and the helical turn on the extreme *right* are represented by ribbons and colored as in Fig. 4a. The same color code is used for the three associated hydrophobic clusters. The polypeptide segments that connect the β-strands are drawn as thin cylindrical rods, which are *yellow* for the type 1 repeats and white for the type 2 repeat (Reprinted with permission from Ellgaard et al. 2001a © National Academy of Sciences, USA)

Asp^{241}, Glu^{243}, and Trp^{244}). It was revealed that bradykinin-dependent Ca^{2+} release from ER was rescued by wild-type Crt and by Glu^{238}, Glu^{239}, Asp^{241}, and Glu^{243} mutants. Other amino acids mutants under study rescued the Crt-deficient phenotype only partially (~40%), or did not rescue it at all. Thus, amino acid residues Glu^{239}, Asp^{241}, Glu^{243}, and Trp^{244} at the hairpin tip of P-domain are critical in the formation of a complex between ERp57 and Crt. Although the Glu^{239}, Asp^{241}, and Glu^{243} mutants did not bind ERp57 efficiently, they fully restored bradykinin dependent Ca^{2+} release in $crt^{-/-}$ fibroblast cells.

The central, proline-rich P-domain of Crt, comprising residues 189–288, contains three copies of each of two repeat sequences (types 1 and 2), which are arranged in a characteristic '111222' pattern. The central proline-rich P-domain of Crt (189–288) shows an extended hairpin topology, with three short anti-parallel β-sheets, three small hydrophobic clusters, and one helical turn at the tip of the hairpin. The loop at the bottom of the hairpin consists of residues 227–247, and is closed by an anti-parallel β-sheet of residues 224–226 and 248–250. Two additional β-sheets contain residues 207–209 and 262–264, and 190–192 and 276–278. The 17-residue spacing of the β-strands in N-terminal part of hairpin and 14-residue spacing in the C-terminal part reflect the length of type 1 and type 2 sequence repeats. As a consequence of this topology the peptide segments separating the β-strands in the N-terminal part of the hairpin are likely to form bulges to accommodate the extra residues. Further, the residues 225–251 at the tip of Crt P-domain are involved in direct contacts with ERp52. The Crt P-domain fragment Crt(221–256) constitutes an autonomous folding unit, and has a structure highly similar to that of corresponding region in Crt(189–288). Of the 36 residues present in Crt(221–256), 32 form a well-structured core, making this fragment one of the smallest known natural sequences to form a stable non-helical fold in absence of disulfide bonds or tightly bound metal ions. Crt(221–256) comprises all the residues of intact P-domain that were shown to interact with ERp52.

The NMR structure of the rat calreticulin P-domain, comprising residues 189–288, Crt(189–288), shows a hairpin fold that involves the entire polypeptide chain, has the two chain ends in close spatial proximity, and does not fold back on itself. This globally extended structure is stabilized by three antiparallel β-sheets, with the β-strands comprising the residues 189–192 and 276–279, 206–209 and 262–265, and 223–226 and 248–251, respectively. The hairpin loop of residues 227–247 and the two connecting regions between the β-sheets contain a hydrophobic cluster, where each of the three clusters includes two highly conserved tryptophyl residues, one from each strand of the hairpin. The three β-sheets and the three hydrophobic clusters form a repeating pattern of interactions across the hairpin that reflects the periodicity of the amino acid sequence, which consists of three 17-residue repeats followed by three 14-residue repeats. Within the global hairpin fold there are two well-ordered subdomains comprising the residues 219–258, and 189–209 and 262–284, respectively. These are separated by a poorly ordered linker region, so that the relative orientation of the two subdomains cannot be precisely described. The structure for Crt(189–288) provides an additional basis for functional studies of the abundant endoplasmic reticulum chaperone calreticulin (Ellgaard et al. 2001).

iii). C-Domain

The C-terminal region of Crt (C-domain) is highly acidic and terminates with the KDEL ER retrieval sequence (Fig. 2.1). The C-domain of Crt is susceptible to proteolytic cleavage and that the N- and P-domains form a proteolytically stable tight association. The C-domain of Crt binds over 25 mol of Ca^{2+}/mol of protein (Baksh and Michalak 1991), binds to blood-clotting factors (Kuwabara et al. 1995) and inhibits injury-induced restenosis (Dai et al. 1997). Ca^{2+} sensitivity, confined to the C-terminal part of the protein (C-domain), suggests that the C-domain of Crt may play a role of Ca^{2+} 'sensor' in ER lumen. Ca^{2+} binding to C-domain of Crt plays a regulatory role in the control of Crt interaction with PDI, ERp57 and perhaps other chaperones (Corbett et al. 1999). A modified form of calreticulin lacking the C-terminal hexapeptide including KDEL ER retention sequon has been isolated. Such a truncation may point to a mechanism that allows escape of Crt from ER (Hojrup and Roepstorff 2001). A transcription-based reporter assay revealed an important role for C-domain in Crt retrotranslocation (Afshar et al. 2005). At present, no structural information is available for C-domain which is involved in Ca^{2+} storage in the lumen of ER (Nakamura et al. 2001b).

A study by Jin et al. (2009) shows that only one of the three forms of the ER folding helper Crt of the plant Arabidopsis thaliana interacts with a mutated form of BRI1, the plasma membrane leucine-rich-repeat kinase receptor for brassinosteroids, plant-specific hormones playing important roles in plant growth. Arabidopsis CRT1 and CRT2 are very similar and have homologs in nonplant organisms, but the BRI1-interacting CRT3 seems to be a plant-specific form, with orthologs in higher and lower plant species (Persson et al. 2003). Gene coexpression analysis indicates that CRT3 can be grouped with stress resistance genes, whereas CRT1 and CRT2 are coexpressed mainly with other folding helpers. The important observations by Jin et al. (2009) point to plant-specific functional divergence in CRT family.

2.2.3 Cellular Localization of Calreticulin

Numerous studies confirmed ER localization of the protein in many diverse species, including plants. Besides its main location in ER (Opas et al. 1996), Crt has been found to reside in the nuclear envelope, the spindle apparatus of the dividing cells (Denecke et al. 1995), the cell surface (Gardai et al. 2005), and the plasmodesmata (Laporte et al. 2003; Chen et al. 2005), indicating that Crt is essential for normal cell function. The protein has also been localized to the cytoplasmic granules of the cytotoxic T-cell (Andrin et al. 1998; Fraser et al. 1998; Dupuis et al. 1993), sperm acrosomes (Nakamura et al. 1993), tick saliva (Jaworski et al. 1996), the cell surface (Arosa et al. 1999; Basu and Srivastava 1999), and it may even be secreted in the bloodstream. However, Crt has also been found outside the ER, such as within the secretory granules of cytotoxic lymphocytes, the cell surface of melanoma cells and virus-infected fibroblasts, and the cytosol and nucleus of several cell types (Arosa et al. 1999). Given its lectin-like properties, Crt is considered to be an ER chaperone involved in the assembly and folding of nascent glycoproteins. Surprisingly, there is a considerable controversy concerning cellular localization of Crt (Michalak et al. 1999). It was proposed that this may be due to the direct interaction between Crt and the DNA-binding domain of steroid receptors (Burns et al. 1994; Dedhar et al. 1994) and the cytoplasmic tail of α-integrin (Dedhar 1994). For calreticulin to bind to these molecules, the protein would have to be present in the nucleus and/or cytosol. However, to date there have been no reports on the identification of calreticulin or calreticulin-like protein in the cytosol. Calreticulin-like immunoreactivity was detected in the nucleus of some cells (Opas et al. 1996; Dedhar et al. 1994), in squamous carcinoma cell nuclei in response to ionizing radiation (Ramsamooj et al. 1995) or in nucleus of dexamethasone-treated LM(TK^-) cells (Roderick et al. 1997). However studies on the biosynthesis of MHC class I molecules have never reported associations of ER chaperones with MHC class I molecules outside the ER and Golgi compartments (c/r Arosa et al. 1999). In human peripheral blood T lymphocytes calreticulin is expressed at the cell surface, where it is physically associated with a pool of unfolded MHC class I molecules. Michalak et al. (1996) indicated that Crt is not a nuclear resident protein, and identification of the protein in the nucleus (Opas et al. 1996) was likely an artifact of immunostaining. There is also accumulating evidence for diverse roles for Crt localized outside ER, including reports suggesting important roles for Crt localized to the outer cell surface of a variety of cell types, in the cytosol, and in the extracellular matrix (ECM). Moreover, the addition of exogenous Crt rescues numerous Crt -driven functions, such as adhesion, migration, phagocytosis, and immunoregulatory functions of Crt-null cells.

2.2.4 Functions of Calreticulin

Calreticulin Is a Multi-Process Molecule

Calreticulin (also referred as calregulin, CRP55, CaBP3, mobilferrin and calsequestrin-like protein) is a multifunctional protein involved in many biological processes that include the regulation of Ca^{2+} homeostasis (Michalak et al. 1999), intercellular or intracellular signaling, gene expression (Johnson et al. 2001), glycoprotein folding (Helenius and Aebi 2004), and nuclear transport (Holaska et al. 2001). Overexpression of Crt enhanced apoptosis in myocardiac H9c2 cells under conditions inductive to differentiation with retinoic acid (Kageyama et al. 2002) or under oxidative stress (Ihara et al. 2006). Moreover, Crt regulates p53 function to induce apoptosis by affecting the rate of degradation and nuclear localization of p53 (Mesaeli and Phillipson 2004). Michalak et al. (1999, 2009) focussed on calreticulin, as a ER luminal Ca^{2+}-binding chaperone implicated in playing a role in many cellular functions, including lectin-like chaperoning, Ca^{2+} storage and signaling, regulation of gene expression, cell adhesion and autoimmunity. Several excellent reviews have been published concerning the structure and function of Crt in animals (Bedard et al. 2005; Johnson et al. 2001; Michalak et al. 1999, 2009) and in plants (Jia et al. 2009)

Calreticulin binds Ca^{2+} in the lumen of ER with high capacity and also participates in the folding of newly synthesized proteins and glycoproteins. Hence, it is an ER luminal Ca^{2+}-buffering chaperone. The protein is involved in regulation of intracellular Ca^{2+} homoeostasis and ER Ca^{2+} capacity. The protein impacts on store-operated Ca^{2+} influx and influences Ca^{2+}-dependent transcriptional pathways during embryonic development. It is a component of the calreticulin/calnexin pathway (Ellgaard et al. 1999). Both Crt and Cnx proteins cooperate with an associated co-chaperone, the thiol-disulfide oxidoreductase ERp57, which catalyzes the formation of disulfide bonds in Cnx and Crt-bound glycoprotein substrates. Calreticulin has been implicated to participate in many (perhaps too many) cellular functions. This strongly exemplifies the central role that the ER plays in a variety of cellular functions. It is not surprising, therefore, that any changes in calreticulin expression and function have profound effects on many cellular functions. There is also accumulating evidence for diverse roles for Crt localized outside the ER, including roles for Crt localized to the outer cell surface of a variety of cell types, in the cytosol, and in the extracellular matrix. Furthermore, the addition of exogenous Crt rescues numerous Crt-driven functions, such as adhesion, migration, phagocytosis, and immunoregulatory functions of Crt-null cells. Nonetheless, it has become clear that Crt is a multicompartmental protein that regulates a wide array of cellular responses important in physiological and pathological processes, such as wound healing, the immune response, fibrosis, and cancer (Gold et al. 2010). Notwithstanding, there is a widespread agreement that calreticulin performs two major functions in the ER lumen: (1) chaperoning and (2) regulation of Ca^{2+} homoeostasis.

In a soluble form, and in association with the homologous membrane bound protein Cnx, Crt binds to glycoproteins, preventing aggregation and allowing the proteins to attain their correct folding conformation (Bedard et al. 2005). Calreticulin also plays an important role in maintenance of cellular calcium homeostasis (Michalak et al. 1992). By regulating the amount of free and bound calcium within the lumen of the endoplasmic reticulum, calreticulin affects many different cellular functions, including cell shape, adhesion and motility (Bedard et al. 2005). Other activities that influence the various roles that this multi-functional protein plays include its capacity to bind to zinc (Baksh et al. 1995b, c) and a number of hormone receptors (Burns et al. 1994; Dedhar et al. 1994).

Calreticulin as a Chaperone

Calreticulin forms part of the quality control systems of ER for newly synthesized proteins.similar to calnexin. Both chaperones participate in the 'quality-control' process during the synthesis of a variety of molecules, including ion channels, surface receptors, integrins, MHC class I molecules, and transporters. Signicant progress has been made in understanding how Crt/Cnx act jointly with other ER chaperones and assisting proteins, in correct folding of proteins in ER. Molecular chaperones prevent the aggregation of partially folded proteins, increase the yield of correctly folded proteins and assembly, and also increase the rate of correctly folded intermediates by recruiting other folding enzymes. Researches document that Crt functions as a lectin-like molecular chaperone for many proteins (c/r Michalak et al. 1999, 2009).

Correct folding of a protein is determined in large part by the sequence of the protein, but it is also assisted by interaction with enzymes and chaperones of the ER. Calreticulin, calnexin, and ERp57 are among the endoplasmic chaperones that interact with partially folded glycoproteins and determine if the proteins are to be released from ER to be expressed, or alternatively, if they are to be sent to proteosome for degradation (Bedard et al. 2005; Michalak et al. 2009). Calnexin and calreticulin share many substrates and may form a link of lectin-like chaperones handing over the glycoproteins from one to the other to ensure proper folding (Bass et al. 1998; Helenius and Aebi 2004; Van Leeuwen et al. 1996a, b; Wada et al. 1995). Proposed functions for calreticulin range from chaperoning in ER to antithrombotic effects at the cell surface, and from the regulation of Ca^{2+} signaling to the modulation of gene expression

Table 2.2 Putative functions of calreticulin domains[a]

(a) Structural features and function	P-domain	C-domain
Calreticulin CRT ensuring proper protein folding and preventing aggregation. Proceeded by the N-terminal signal sequence targeting the protein to the ER lumen; highly conserved amino acid sequence; potential phosphorylation site; potential glycosylation site (bovine proteins); putative autokinase activity; inhibits PDI activity; suppresses tumors; inhibits angiogenesis	The P-domain contains the lectin site and a high-affinity Ca-binding region and is proline-rich; the lectin site recognizes N-linked oligosaccharide processing intermediates of glycoproteins and prolonged interaction with misfolded proteins initiates rejection and subsequent direction to the proteasome for degradation. CRT also engages in direct protein–protein interactions. CRT amino acid sequence shows similarity to calnexin, calmegin and CANLUC.	The acidic carboxy-terminal C-domain contains the high-capacity, low-affinity Ca^{2+}, binding sequence and terminates in a KDEL (lysine, aspartic acid, glutamic acid, leucine) sequence for ER retrieval; Putative glycosylation site; antithrombotic activity; prevents restenosis Ca^{2+} 'sensor' of calreticulin–protein interactions. CRT was shown to exist outside the ER by retrotranslocation to the cytoplasm through C-domain (Afshar et al. 2005).
(b) Ion binding		
binds Zn^{2+}	High-affinity Ca^{2+}-binding site	High-capacity Ca^{2+}-binding site
(c) Molecules binding		
Binds to DNA-binding domain of steroid receptor; binds to α-subunit of integrin; binds rubella RNA; interacts with PDI and ERp57; weak interactions with perforin. CRT directly binds to the identical amino-acid sequence, GFFKR, in the alpha integrin cytoplasmic tails. The binding site in the N-terminus of CRT that binds heparin-binding domain I, to mediate TSP-1 signaling, has been localized to amino-acid residues 19–36	Binds to a set of ER proteins; strong interactions with PDI; strong interactions with perforin; lectin-like chaperone site. Cell surface CRT binds to the carbohydrate constituent (mannose) of the cell adhesion and basement membrane protein, laminin, important in cell migration through integrin binding	Binds a set of ER proteins; binds factor IX and factor X; binds to cell surface

[a]Michalak et al. (1999; Gold et al. (2006)

and cellular adhesion (Table.2.2). Two major functions of calreticulin in the ER lumen, ie, chaperoning and regulation of Ca^{2+} homeostasis, were intensively investigated and well characterized.

The assembly of MHC class I molecules is one of the most widely studied examples of protein folding in ER. It is also one of the most unusual cases of glycoprotein quality control involving ERp57 and chaperones Cnx/Crt. Pulse-chase experiments showed that Crt was associated with several proteins ER in T lymphocytes and suggested that it was expressed at the cell surface. The cell surface 46-kDa protein co-precipitated with Crt is unfolded MHC-I. Results show that after T cell activation, significant amounts of Crt are expressed on T cell surface, where they are found in physical association with a pool of β2-free MHC class I molecules (Arosa et al. 1999). Calreticulin promotes folding of HLA class I molecules to a state at which they spontaneously acquire peptide binding capacity. However, it does not induce or maintain a peptide-receptive state of class I-binding site, which is likely to be promoted by one or several other components of class I loading complexes (Culina et al. 2004).

MHC class I molecules consist of two non-covalently linked subunits, an integral membrane glycoprotein (β chain), a small soluble protein [β_2-microglobulin (β_2-m)], and a peptide of eight to eleven residues. The multi-step assembly of MHC class-I heavy chain with β_2-m and peptide is facilitated by these ER-resident proteins and further tailored by the involvement of a peptide transporter, aminopeptidases, and the chaperone-like molecule tapasin (Wearsch and Cresswell 2008). In mouse and human, β chains associate with calnexin soon after its synthesis via interactions with both immature glycans and with residues in the transmembrane domain of β chains. In human chain, β_2-m binding to β chain displaces Cnx, and the resulting β chain–β_2-m heterodimer binds Crt. MHC class I expression and transport to the cell surface are not changed in Cnx-deficient cells, suggesting that Cnx is not essential for MHC class I synthesis and transport. These findings indicate, albeit indirectly, that Crt can function in the absence of Cnx. Wearsch and Cresswell (2008) presented the roles of these general and class I-specific ER proteins in facilitating the optimal assembly of MHC class I molecules with high affinity peptides for antigen presentation. Bass et al. (1998) demonstrated a chaperone function for Cnx/Crt in human insulin receptor (HIR) folding in vivo and also provided evidence that folding efficiency and homo-dimerization are counter balanced.

ER Luminal Ca^{2+} and Calreticulin Function

Ca^{2+} is released from the ER and taken up to the ER lumen. Ca^{2+} storage capacity of ER lumen is enhanced by Ca^{2+}-binding chaperones. These include calreticulin, Grp94, immunoglobulin-heavy-chain-binding protein (BiP; Grp78), PDI, ERp72 and ER/calcistorin. Reduction of $[Ca^{2+}]_{ER}$ (ER Ca^{2+}-depletion conditions) leads to accumulation of misfolded proteins, activation of expression of ER chaperones and ER–nucleus and ER–plasma membrane 'signaling'. Ca^{2+} depletion inhibits ER–Golgi trafficking, blocks transport of molecules across the nuclear pore and affects chaperone function. Clearly, changes of the ER luminal $[Ca^{2+}]_{ER}$ have profound effects at multiple cellular sites, including the structure and function of the ER luminal Ca^{2+}-binding chaperones (Michalak et al. 1999).

Several studies of mammalian Crts have elucidated a number of key physiological functions, including the regulation of Ca^{2+} homeostasis and Ca^{2+}-dependent signal pathways (Michalak et al. 2002b; Gelebart et al. 2005), and integrin-dependent Ca^{2+} signaling at the extra-ER sites in mammalian cells (Coppolino et al. 1997; Krause and Michalak, 1997), and molecular chaperone activity in the folding of many proteins (Denecke et al. 1995; Williams, 2006). Ca^{2+} is a universal signaling molecule in the cell cytosol and can affects several processes in ER lumen, including modulation of chaperone–substrate and protein–protein interactions. For example, binding of carbohydrate to calreticulin and calnexin occurs at high $[Ca^{2+}]_{ER}$ and it is inhibited at low $[Ca^{2+}]_{ER}$ similar to Ca^{2+} depletion in stores. Several steps during chaprone action of both Cnx and Crt are regulated by Ca^{2+}. Studies suggest that C-domain of Crt plays a role of Ca^{2+} 'sensor' in the ER lumen. Association of Crt with Grp94, Grp78 and Cnx, maturation of thyroglobulin and apolipoprotein B may be regulated by Ca^{2+} binding.. The structure of Crt provides a unique feature enabling it to perform several functions in the ER lumen, while responding to continuous fluctuations of the free $[Ca^{2+}]_{ER}$. Calreticulin also affects Ca^{2+} homoeostasis (see below), and, in one case, the protein may even be taking advantage of its chaperone, lectin-like activity to modulate Ca^{2+} fluxes across the ER membrane (Leach et al. 2002; Michalak et al. 1999).

Calreticulin and Regulation of Ca^{2+} Homoeostasis: Calreticulin has two Ca^{2+}-binding sites: a high-affinity, low-capacity site ($K_d = 1\ \mu M$; $B_{max} = 1$ mol of Ca^{2+}/mol of protein) in the P-domain and a low-affinity high-capacity site ($K_d = 2$ mM; $B_{max} = 25$ mol of Ca^{2+}/mol of protein) in the C-domain. Overexpression of Crt in a variety of cells does not affect the cytoplasmic $[Ca^{2+}]$; however, it does result in an increased amount of intracellularly stored Ca^{2+}. Interestingly, Ca^{2+}-storage capacity of the ER is not changed in the calreticulin-deficient embryonic stem cells (ES) or mouse embryonic fibroblasts (MEF) (Michalak et al. 1999). It is reported that changes in intracellular Ca^{2+} homeostasis modulate the rate of apoptosis as in molecular chaperones and radiation-induced apoptosis of gliomas (Brondani et al. 2004). Using human glioma cell lines, overexpression of Crt modulated radiosensitivity of human glioblastoma cells by suppressing Akt/protein kinase B signaling for cell survival via alterations of cell Ca^{2+} homeostasis. The level of CRT was higher in neuroglioma H4 cells than in glioblastoma cells (U251MG and T98G), and was well correlated with the sensitivity to γ-irradiation (Okunaga et al. 2006).

Calreticulin Functions Outside ER

There is considerable evidence to indicate that Crt is found outside ER, although how the protein relocates from ER to outside of the ER remains unclear. Functions of Crt outside the ER include modulation of cell adhesion, integrin-dependent Ca^{2+} signaling, and steroid-sensitive gene expression as well as mRNA destabilization both in vitro and in vivo. One major controversy in Crt research field concerns the mechanisms involved in Crt-dependent modulation of functions outside ER (Yokoyama and Hirata 2005). Outside ER Crt is known to modulate nuclear-hormone receptor-mediated gene expression (Burns et al. 1994; Michalak et al. 1996), control of cell adhesion (Opas et al. 1996; Fadel et al. 1999; Fadel et al. 2001; Goicoechea et al. 2002), and integrin-dependent Ca^{2+} signaling in vitro and in vivo (Coppolino et al. 1997; Michalak et al. 1999). It is also involved in blood function and development (Kuwabara et al. 1995; Andrin et al. 1998; Mesaeli et al. 1999). In addition, Crt appears to play a role in the immune system (Guo et al. 2002) and apoptosis. For example, Crt-dependent shaping of Ca^{2+} signaling was found to be a critical contributor to the modulation of the T cell adaptive immune response (Porcellini et al. 2006). Surface Crt mediates muramyl dipeptide-induced RK13 cell apoptosis through activating the apoptotic pathway (Chen et al. 2005). One major controversy in the calreticulin field concerns the mechanisms involved in Crt-dependent modulation of functions outside ER.

During apoptosis, both CRT expression and the concentration of nitric oxide (NO) are increased. By using S-nitroso-l-cysteine-ethyl-ester, an intracellular NO donor and inhibitor of APLT, phosphotidylserine (PS) and CRT externalization occurred together in an S-nitrosothiol-dependent and caspase-independent manner. Furthermore, the CRT and PS are relocated as punctate clusters on the cell surface. Thus, CRT induced nitrosylation and its externalization with PS could explain how CRT acts as a bridging molecule during apoptotic cell clearance (Tarr et al. 2010b).

Calreticulin and Cell Adhesiveness: Calreticulin may be involved in integrin function and cell adhesion. Crt binds to KXFF($^{K}/_{R}$)R synthetic peptide, a region corresponding to conserved amino acid sequence found in C-terminal tail of β-subunit of integrin. It was suggested that Crt may bind to the C-terminal cytoplasmic tail of β-integrin and modulate its function (Coppolino et al. 1997; Leung-Hagesteijn et al. 1994). Differential adhesiveness correlates inversely with the expression level of mRNA and protein for the focal contact-associated cytoskeletal protein, vinculin, and the calreticulin. Furthermore, an inverse relationship exists between the level of calreticulin and the level of total cellular phosphotyrosine (Tyr(P)) such that the cells underexpressing calreticulin display a dramatic increase in the abundance of total cellular Tyr(P). This suggests that the effects of calreticulin on cell adhesiveness may involve modulation of the activities of protein-tyrosine kinases or phosphatases (Fadel et al. 1999). In addition, Crt can regulate cell adhesion indirectly from the ER lumen via modulation of gene expression of adhesion-related molecules such as vinculin and β-catenin (Opas et al. 1996; Fadel et al. 1999, 2001). It has also been shown that calreticulin associates transiently with the cytoplasmic domains of integrin α subunits during spreading and that this interaction can influence integrin-mediated cell adhesion to extracellular matrix (Coppolino et al. 1997; Leung-Hagesteijn et al. 1994; Goicoechea et al. 2002). Calreticulin serves as a cytosolic activator of integrin and a signal transducer between integrins and Ca^{2+} channels on the cell surface (Kwon et al. 2000).

Calreticulin can also modulate cell adhesion from the cell surface and mediate cell spreading on glycosylated laminin (Goicoechea et al. 2002) and thrombospondin-induced focal adhesion disassembly (Goicoechea et al. 2000; Pallero et al. 2008). Thrombospondin (TSP) is a member of a group of extracellular matrix proteins that exist in both soluble and extracellular matrix forms and regulates cellular adhesion (Goicoechea et al. 2002). When exposed to cells in its soluble form, thrombospondin has primarily anti-adhesive effects characterized by a reorganization of stress fibers and loss of focal adhesion plaques. A 19-amino acid sequence (aa 17–35) in the N-terminal heparin-binding domain of thrombospondin, referred to as the hep I peptide, has been shown to be sufficient for focal adhesion disassembly. Since, calreticulin can modulate cell adhesion from the cell surface and mediate thrombospondin-induced focal adhesion disassembly (Goicoechea et al. 2000), it was suggested that interactions between calreticulin and thrombospondin are Zn^{2+}- and Ca^{2+}-dependent and involve the RWIESKHKSDFGKFVLSS sequence in the N-terminal region of the N-domain of calreticulin (Goicoechea et al. 2002). TSP binding to Crt-LRP1 signals resistance to anoikis (Pallero et al. 2008).

Calreticulin Affects β-Catenin-Associated Pathways: It was shown that differential adhesiveness correlates inversely with the expression level of mRNA and protein for cytoskeletal protein, vinculin, and the Crt. Calreticulin has been shown to be important in cell adhesion. Furthermore, an inverse relationship existed between the level of Crt and the level of total cellular phosphotyrosine (Tyr(P)) such that cells underexpressing Crt display a dramatic increase in the abundance of total cellular Tyr(P). In either cell type, spatial distributions of Crt and Tyr(P) are complementary, with the former being confined to ER and the latter being found outside of it. Among proteins that are dephosphorylated in cells that over-express Crt is β-catenin, a structural component of cadherin-dependent adhesion complex and a part of the Wnt signaling pathway. To investigate the mechanisms behind Crt dependent modulation of cell adhesiveness, Fadel et al. (2001), using mouse L fibroblasts differentially expressing Crt, showed that stable over-expression of ER-targeted Crt correlates with an increased adhesiveness in transformed fibroblasts, such that their cohesion resembles that of epithelial cells in culture. Fadel et al. (2001) suggest that the changes in cell adhesiveness may be due to Crt-mediated effects on a signaling pathway from the ER, which impinges on Wnt signaling pathway via cadherin/catenin protein system and involves changes in the activity of protein-tyrosine kinases and/or phosphatases. Results suggest that calreticulin may play a role in a signaling pathway from ER, involving protein-tyrosine kinases and/or phosphatases This suggests that the effects of Crt on cell adhesiveness may involve modulation of the activities of protein-tyrosine kinases or phosphatases (Fadel et al. 1999, 2001). Protein phosphorylation/dephosphorylation of tyrosine is a major mechanism for regulation of cell adhesion. Although the mechanism(s) are still elusive, it is conceivable that the effects of Crt over-expression on cell adhesion may be due to Crt effects on a signaling pathway, which includes the vinculin/catenin–cadherin protein system and may involve changes in activity of tyrosine kinases and/or phosphatases. A direct implication of this for cell–substratum interactions is that calreticulin effects may target primarily focal-contact-mediated adhesion (Michalak et al. 1999).

Wound Repair: Crt was originally shown to be the biologically active component of a hyaluronic acid isolate from fetal sheep skin that accelerated wound healing in animal experimental models of cutaneous repair. Nanney et al. (2008) showed roles for exogenous Crt in both cutaneous wound healing and diverse processes associated with repair. Topical application of Crt to porcine excisional wounds enhanced the rate of wound re-epithelialization. The in vitro bioactivities provide mechanistic support for the positive biological effects of Crt observed on both the epidermis and dermis of wounds in vivo, underscoring a significant role for Crt in the repair of cutaneous wounds (Nanney et al. 2008).

Interaction of Calreticulin with C1q: C1q immobilized on a hydrophobic surface, exposed to heat-treatment or bound to Igs showed a strong, rapid and specific binding of Crt.

When both proteins were present in equal amounts in solution, no interaction could be demonstrated. Binding between C1q and Crt could be inhibited by serum amyloid P component. Results suggest Crt as a potential receptor for an altered conformation of C1q as occurs during binding to Igs. Thus, the chaperone and protein-scavenging function of Crt may extend from the ER to the topologically equivalent cell surface, where it may contribute to the elimination of immune complexes and apoptotic cells (Steinø et al. 2004).

Calreticulin in Signal Transduction: As indicated earlier, Crt is also found on cell surface of many cell types where it serves as a mediator of adhesion and as a regulator of the immune response. Calreticulin, present on the extracellular surface of mouse egg plasma membrane, is increased in perivitelline space after egg activation. The extracellular Crt appears to be secreted by vesicles in the egg cortex that are distinct from cortical granules. An anticalreticulin antibody binds to extracellular Crt on live eggs and inhibits sperm-egg binding but not fusion. In addition, engagement of cell surface Crt by incubation of mouse eggs in the presence of anticalreticulin antibodies results in alterations in the localization of cortical actin and the resumption of meiosis as indicated by alterations in chromatin configuration, decreases in cdc2/cyclin B1 and MAP kinase activities, and pronuclear formation. These events occur in absence of any observable alterations in intercellular calcium. These studies suggest that Crt functionally interacts with the egg cytoskeleton and can mediate transmembrane signaling linked to cell cycle resumption. Evidences suggest a role for Crt as a lectin that may be involved in signal transduction events during or after sperm-egg interactions at fertilization (Tutuncu et al. 2004).

Lessons from Calreticulin-Deficient Mice

The calreticulin-deficient mouse, created by the homologous-recombination, is embryonically lethal at 14.5–16.5 days *post coitus*. Calreticulin-deficient embryos most likely die from a lesion in cardiac development. In the adult, calreticulin is expressed mainly in non-muscle and smooth-muscle cells, and is only a minor component of the skeletal and cardiac muscle. However, the calreticulin gene is activated during cardiac development, concomitant with an elevated expression of the protein, which decreases sharply in the newborn heart. Therefore it was not surprising that calreticulin-deficient mice die from heart failure. Calreticulin may in general must play an important role in the formation of heart (Michalak et al. 2002a).

Calreticulin knockout ($Crt^{-/-}$) mice die at the embryonic stage due to impaired heart development, making it impossible to study mature $Crt^{-/-}$ CTLs and NK cells. To specifically investigate the role of Crt in CTL lytic function, Sipione et al. (2005) generated CTL lines from splenocytes derived from these mice and showed that in absence of Crt, CTL cytotoxicity is impaired. Sipione et al. (2005) suggest that Crt is dispensable for the cytolytic activity of granzymes and perforin, but it is required for efficient CTL-target cell interaction and for the formation of the death synapse. It was proposed that Crt may be involved in the mechanisms underlying target recognition by CTLs and/or formation of the death synapse. Calreticulin-deficient ES cells had impaired integrin-mediated adhesion, supporting the observation that changes in expression of calreticulin affect cell adhesion.

Impaired p53 Expression in Crt-Deficient Cells: The tumor suppressor protein, p53 is a transcription factor that not only activates expression of genes containing the p53 binding site but also can repress the expression of some genes lacking this binding site. Overexpression of wild-type p53 leads to apoptosis and cell cycle arrest. The level of Crt has been correlated with the rate of apoptosis. Crt-deficient cells ($crt^{-/-}$) demonstrated that Crt function is required for the stability and localization of the p53 protein. The observed changes in p53 in the $crt^{-/-}$ cells are due to the nuclear accumulation of Mdm2 (murine double minute gene). These results, lead us to conclude that Crt regulates p53 function by affecting its rate of degradation and nuclear localization (Mesaeli and Phillipson 2004).

Calreticulin and Steroid-Sensitive Gene Expression

Calreticulin is also found in nucleus, which suggests that it may play a role in transcriptional regulation. Calreticulin binds to the DNA-binding domain of steroid receptors and transcription factors containing the amino acid sequence KXFF($^{K}/_{R}$)R and prevents their interaction with DNA in vitro (Michalak et al. 1999). With the exception of the peroxisome-proliferator-activated receptor ('PPAR')–retinoid X heterodimers (Winrow et al. 1995), transcriptional activation by glucocorticoid, androgen, retinoic acid and vitamin D_3 receptors in vivo is modulated in cells overexpressing calreticulin (Burns et al. 1994; 1997; Dedhar et al. 1994; Desai et al. 1996; Michalak et al. 1996; St-Arnaud et al. 1995; Sela-Brown et al. 1998 Wheeler et al. 1995).

Outside ER, Crt appears to be affected by varios factors. The mRNA levels of Crt increased as a function of time after UV irradiation in transformed human keratinocytes (HaCaT cells) (Szegedi et al. 2001). A developmentally expressed cytosolic, trophoblast-specific, high Mr 57-kDa Ca-binding protein (CaBP) plays an important role in regulating and/or shuttling cytosolic Ca and represents the primary mechanism in fetal Ca homeostasis. The full-length cDNA of the mouse CaBP shows significant homology to Crt. In addition, the action of parathyroid hormone related protein (PTHrP) on placental trophoblast Ca transport is likely to involve the regulation of CaBP expression to handle the increasing Ca requirements of developing fetus (Hershberger and Tuan 1998).

Calreticulin binds to the synthetic peptide KLGFFKR, which is almost identical to an amino acid sequence in the DNA-binding domain of the superfamily of nuclear receptors. The amino terminus of calreticulin interacts with the DNA-binding domain of the glucocorticoid receptor and prevents the receptor from binding to its specific glucocorticoid response element. Calreticulin can inhibit the binding of androgen receptor to its hormone-responsive DNA element and can inhibit androgen receptor and retinoic acid receptor transcriptional activities in vivo, as well as retinoic acid-induced neuronal differentiation.

Crt and Glucocorticoid Receptor Pathways: Calreticulin play an important role in the regulation of glucocorticoid-sensitive pathway of expression of the hepatocytes specific genes during development (Burns et al. 1997). ER Crt but not cytosolic Crt is responsible for inhibition of glucocorticoid receptor-mediated gene expression. These effects are specific to Crt, since over-expression of ER lumenal proteins (BiP, ERp72, or calsequestrin) had no effect on glucocorticoid-sensitive gene expression. The N domain of Crt binds to the DNA binding domain of glucocorticoid receptor in vitro. However, the N + P domain of Crt, when synthesized without ER signal sequence, does not inhibit glucocorticoid receptor function in vivo Michalak et al. (1996) suggest that Crt and glucocorticoid receptor may not interact in vivo and that the Crt-dependent modulation of the glucocorticoid receptor function may therefore be due to a Crt-dependent signaling from the ER. Wnt signaling pathway is a multicomponent cascade involving interaction of several proteins and found to be important for development and function of various cells and tissues. There is increasing evidence that Wnt/beta-catenin pathway constitutes also one of the essential molecular mechanisms controlling the metabolic aspects of osteoblastic cells. However, in bone, glucocorticoids (GCs) have been reported to weaken Wnt signaling. Calreticulin, known to bind the DNA binding domain of glucocorticoid receptor (GR), was found to be involved in the GR-mediated down-regulation of Wnt signaling. Furthermore, GR and β-catenin were shown to exist in same immunocomplex, while interaction between Crt and beta-catenin was observed only in the presence of GR as a mediator molecule. In addition, the GR mutant lacking Crt binding ability impaired the complex formation between beta-catenin and Crt. Together with GR, β-catenin could thus be co-transported from the nucleus in a Crt-dependent way (Olkku and Mahonen A, 2009).

Overexpression of calreticulin in mouse L fibroblasts inhibits glucocorticoid-response-mediated transcriptional activation of a glucocorticoid-sensitive reporter gene and of the endogenous, glucocorticoid-sensitive gene encoding cytochrome P450. This indicates that calreticulin may be important in gene transcription, regulating the glucocorticoid receptor and perhaps other members of the super-family of nuclear receptors (Burns et al. 1994). Thus, calreticulin can act as an important modulator of the regulation of gene transcription by nuclear hormone receptors. These are surprising findings, since calreticulin is an ER-resident protein and steroid receptors are found in the cytoplasm or in the nucleus. What could be a physiological or pathophysiological relevance of calreticulin (ER)-dependent modulation of gene expression? Up-regulation of the calreticulin gene may correlate with increased resistance to steroids. For example, calreticulin is one of the androgen-sensitive genes in prostate cancer (Zhu et al. 1998; Zhu and Wang 1999; Wang et al. 1997). Steroid-dependent regulation of expression of calreticulin may affect differential sensitivity of patients to steroid therapies.

Over-expression of calreticulin and calsequestrin impairs cardiac function, leading to premature death. Calreticulin is vital for embryonic development, but also impairs glucocorticoid action. Glucocorticoid overexposure during late fetal life causes intra-uterine growth retardation and programmed hypertension in adulthood. In view of the known associations between cardiac calreticulin overexpression and impaired cardiac function, targeted up-regulation of calreticulin may contribute to the increased risk of adult heart disease introduced as a result of prenatal overexposure to glucocorticoids (Langdown et al. 2003; Michalak et al. 2002a). The neural cell adhesion molecule, N-CAM inhibits the proliferation of rat astrocytes both in vitro and in vivo. Exposure of astrocytes to N-CAM in vitro, the levels of mRNAs for glutamine synthetase and calreticulin increased while mRNA levels for N-CAM decreased. Glutamine synthetase and calreticulin are known to be involved in glucocorticoid receptor pathways. Inhibition of rat cortical astrocyte proliferation in culture by dexamethasone, corticosterone, and aldosterone suggests that astrocyte proliferation is in part regulated by alterations in glucocorticoid receptor pathways, which may involve Crt (Crossin et al. 1997).

Regulation by Androgens: Calreticulin is an intracellular protein in prostatic epithelial cells. Its expression in prostate is much higher than that in seminal vesicles, heart, brain, muscle, kidney, and liver. The expression of Crt in prostate is conserved evolutionarily. After castration, Crt mRNA and protein are down-regulated in the prostate and seminal vesicles and restored by androgen replacement. Because Crt is a major intracellular Ca^{2+}-binding protein with 1 high-affinity and 25 low-affinity Ca binding sites, observations suggest that Crt is a promising candidate that mediates androgen regulation of intracellular Ca^{2+} levels and/or signals in prostatic epithelial cells (Zhu et al. 1998). As expected, androgen protects androgen-sensitive LNCaP but not androgen-insensitive PC-3 cells from cytotoxic intracellular Ca^{2+} overload induced by Ca^{2+} ionophore A23182. Observations suggest that Crt mediates androgen regulation of the sensitivity to Ca^{2+} ionophore-induced apoptosis in LNCaP cells (Zhu and Wang, 1999; Meehan et al. 2004).

Estrogens: While estrogens are mitogenic in breast cancer cells, the presence of estrogen receptor α (ERα) clinically indicates a favorable prognosis in breast carcinoma. Calreticulin that could interact with amino acids 206–211 of ERα reversed hormone-independent ERα inhibition of invasion. However, since Crt alone also inhibited invasion, it was proposed that this protein probably prevents ERα interaction with another unidentified invasion-regulating factor. The inhibitor role of the unliganded ER was also suggested in three ERα-positive cell lines, where ERα content was inversely correlated with cell migration. It was concluded that ERα protects against cancer invasion in its unliganded form, probably by protein-protein interactions with the N-terminal zinc finger region, and after hormone binding by activation of specific gene transcription (Platet et al. 2000).

Neonatal treatment with diethylstilbestrol (DES) leads to disruption of spermatogenesis in adult animals after apparently normal testicular development during puberty indicating aberrant androgen action in DES-exposed adult hamsters. Analyses revealed that mRNA levels for AR-responsive genes calreticulin, SEC-23B, and ornithine decarboxylase were significantly decreased in DES-exposed animals and that neonatal DES exposure impairs the action of androgens on target organs in male hamsters (Karri et al. 2004).

Crt Enhances Transcriptional Activity of TTF-1: Calreticulin binds to thyroid transcription factor-1 (TTF-1), a homeodomain-containing protein implicated in the differentiation of lung and thyroid. The interaction between calreticulin and TTF-1 appears to have functional significance because it results in increased transcriptional stimulation of TTF-1-dependent promoters. Calreticulin binds to the TTF-1 homeodomain and promotes its folding, suggesting that the mechanism involved in stimulation of transcriptional activity is an increase of the steady-state concentration of active TTF-1 protein in the cell. It was also demonstrated that calreticulin mRNA levels in thyroid cells are under strict control by the thyroid-stimulating hormone, thus implicating calreticulin in the modulation of thyroid gene expression by thyroid-stimulating hormone (Perrone et al. 1999). The thyroid hormone receptor α1 (TRα) directly interacts with calreticulin, and point to the intriguing possibility that TRα follows a cooperative export pathway in which both calreticulin and CRM1 (Exportin) play a role in facilitating efficient translocation of TRα from the nucleus to cytoplasm (Grespin et al. 2008).

The increased level of calreticulin in activated T-cells suggests that the protein may be part of the Ca^{2+}-dependent transduction pathway(s) in stimulated T-cells, including activation of NF-AT. Therefore stress-dependent stimulation of the T-cells results in the activation of both the NF-AT and calreticulin pathways. Development of the heart as an organ must inflict a tremendous stress on cardiomyocytes. Therefore it is tempting to speculate that similar stress-induced signaling pathways are essential during cardiac development and the activation of the immune system.

Destabilization of 3′-Untranslated Region of mRNA by Crt

Angiotensin II is hypertrophic for cultured adult rat aortic vascular smooth muscle cells (VSMC), whereas platelet-derived growth factor and serum are hyperplastic. Hyperplastic and hypertrophic growth are accompanied by similar changes in protein expression, suggesting that both types of growth require up-regulation of the protein synthesis and folding machinery such as calreticulin and HSPs (Patton et al. 1995). Angiotensin II plays a central role in cardiovascular homeostasis and downregulates type 1 angiotensin II (AT_1) receptor, which appears to involve several mechanisms. The AT1 receptor plays a pivotal role in the pathogenesis of hypertension and atherosclerosis. AT1 receptor expression is regulated posttranscriptionally via destabilization of AT1 receptor mRNA by mRNA binding proteins. Nickening et al. discovered that mRNA binding protein, Crt, binds to the cognate sequence bases 2175–2195 within the 3′ untranslated region of the AT_1 receptor mRNA (Nickenig et al. 2002). Angiotensin II stimulation, which causes destabilization of AT_1 receptor mRNA, causes phosphorylation of Crt. This region comprises aAUUUUA hexamer and is considerably AU-rich. Phosphorylation of Crt is essential for binding of the AT1 receptor mRNA. Findings imply an important role of serine dephosphorylation and tyrosine phosphorylation on Crt mediated AT1 receptor mRNA stability in VSMC (Mueller et al. 2008).

Glucose transport in mammalian cells is mediated by a family of structurally related glycoproteins, the glucose transporters (GLUTs) (McGowan et al. 1995). Totary-Jain et al. report that calreticulin destabilized GLUT-1 mRNA expression in primary bovine aortic endothelial cells and smooth muscle cells under high glucose conditions

(Totary-Jain et al. 2005). They identified Crt as a specific destabilizing trans-acting factor that binds to a 10-nucleotide cis-acting element (CAE 2181–2190) in the 3′-untranslated region of GLUT-1 mRNA. CAE 2181–2190–Crt complex, which is formed in VSMCs and endothelial cells exposed to hyperglycemic conditions, renders GLUT-1 mRNA susceptible to degradation. RNA–protein interactions have been shown to influence many processes, including translation, RNA stability, mRNA transport and localization, splicing, and polyadenylation (Qi and Pekala, 1999, Yokoyama and Hirata 2005).

Calreticulin binds to antibodies in certain sera of systemic lupus and Sjogren patients that contain anti-Ro/SSA antibodies. Systemic lupus erythematosus is associated with increased autoantibody titers against Crt, but Crt is not a Ro/SS-A antigen. Earlier papers referred to Crt as an Ro/SS-A antigen, but this was later disproven. Increased autoantibody titer against human Crt is found in infants with complete congenital heart block of both the IgG and IgM classes.

2.2.5 Structure-Function Relationships in Calnexin and Calreticulin

The three-dimensional structure of the lumenal domain of the lectin-like chaperone calnexin determined to 2.9 A resolution reveals an extended 140 A arm inserted into a beta sandwich structure characteristic of legume lectins. The extended arm is curved, forming an opening, which likely accommodates specific substrates. The glucose-binding site of calnexin is located on the surface of the globular domain, facing the extended arm. The arm is composed of tandem repeats of two proline-rich sequence motifs which interact with one another in a head-to-tail fashion. Identification of the ligand binding site establishes calnexin as a monovalent lectin, providing insight into the mechanism by which the calnexin family of chaperones interacts with monoglucosylated glycoproteins. The globular domain of calnexin also contains a Ca2+ binding site and one disulfide bond. Another disulfide bond is located near the tip of the extended arm (Schrag et al. 2001). A model of calreticulin 3D structure predicts that the N-domain of calreticulin is globular and contains a glucose binding site and a disulfide bridge. The P-domain is also predicted to contain the unusual extended arm structure identified in calnexin. The globular N-domain together with the extended arm P-domain of calreticulin and calnexin may form a functional "protein-folding module". The C-terminal region of calreticulin, which is highly acidic, binds Ca2+ with high capacity and is involved in Ca^{2+} storage in the lumen of the ER in vivo (Nakamura et al. 2001b). Ca^{2+} binding to calreticulin and, consequently, changes in the ER Ca^{2+} storage capacity, affect the protein's chaperone function and thereby influence the "quality control" of the secretory pathway. Calreticulin interacts in a Ca^{2+}-dependent manner with other ER chaperones, modulating their function (Corbett et al. 2000).

Lectin-Deficient Crt Retains Full Functionality as a Chaperone: Calreticulin uses both a lectin site specific for $Glc_1Man_{5-9}GlcNAc_2$ oligosaccharides and a polypeptide binding site to interact with nascent glycoproteins. The latter mode of substrate recognition is controversial. To examine the relevance of polypeptide binding to protein folding in living cells, in Crt-deficient mutants, class I molecules exhibit inefficient loading of peptide ligands, reduced cell surface expression and aberrantly rapid export from ER. It suggested that Crt can use nonlectin-based modes of substrate interaction to effect its chaperone and quality control functions on class I molecules in living cells. Furthermore, lectin-deficient Crt bound to a similar spectrum of client proteins as wild-type Crt and dissociated with similar kinetics, suggesting that lectin-independent interactions are common place in cells that may be regulated during client protein maturation (Ireland et al. 2008).

2.2.6 Pathophysiological Implications of Calreticulin

Cardiovascular System

It is interesting that high glucose augments calreticulin expression in vascular smooth muscle cells and endothelial cells (Qi and Pekala 1999), although the mechanism and patho-physiological significance of these findings in the vascular cells has not yet been investigated. Dai et al. (1997) documented a profound inhibitory effect of intravenous administration of calreticulin on intimal hyperplasia in rat iliofemoral arteries after balloon injury in vivo. Because calreticulin can be found in extracellular locations including the blood, and it has been associated with regulation of immune responses, calreticulin has also been implicated in a number of pathological processes. The calreticulin gene knock-out study also indicates that the protein plays a role in the development of the heart (Gelebart et al. 2005; Michalak et al. 2008). It is shown that carticulin is upregulated in the heart during the middle stages of embryogenesis, whereas it is expressed at a low level after birth. Further studies are required to unravel the pathophysiological roles of calreticulin in the pathogenesis of various diseases.

Calreticulin and Cardiac Pathology: Calreticulin deficiency is embryonic lethal because it causes lesions during cardiac development (Mesaeli et al. 1999). However, overexpression of the protein in developing and postnatal heart leads to bradycardia, complete heart block and sudden death.

Ultrastructural evidence indicates that the deficiency associated with the absence of calreticulin in the heart may be due to a defect in the development of the contractile apparatus and/or a defect in development of the conductive system as well as a metabolic abnormality. Michalak et al. (2004) postulate that calreticulin and endoplasmic reticulum plays an important role in cardiac development and postnatal pathologies (Michalak et al. 2004).

Calreticulin-Deficient Mouse: Cells isolated from $Crt^{-/-}$ embryos have impaired agonist-induced Ca^{2+} release (Nakamura et al. 2001a), inhibited nuclear import of the transcription factors NF-ATc1, Mef2c and p53, modified sensitivity to apoptosis, compromised function of calnexin, and activated unfolded proteins response (UPR) indicating a major impact of calreticulin deficiency on ER and cellular functions. Remarkably, $Crt^{-/-}$ mice are rescued by expression of constitutively active calcineurin in the heart and exhibit severe postnatal pathology with death 7–35 days after birth (Guo et al. 2002). Calreticulin-deficient animals that have been rescued with cardiac expression of calcineurin go on to develop severe metabolic problems in cholesterol, lipid, and carbohydrate metabolism. The underlying cause of the metabolic aberrations in these mice is not understood but it indicates that many metabolic processes rely on ER function. Calreticulin expression is high in embryonic heart and declines sharply after birth, probably due to transcriptional control of the calreticulin gene. High expression of calreticulin in the heart of transgenic mice results in early postnatal death. Animals over-expressing calreticulin in the heart develop bradycardia associated with sinus node dysfunction, complete cardiac block, and death due to intractable heart failure. This indicates that calreticulin plays a role in the pathology of the heart's conductive system (Nakamura et al. 2001b).

Deletion of Crt gene leads to defects in the heart and the formation of omphaloceal. These defects could both be due to changes in the extracellular matrix composition. Matrix metalloproteinases (MMP)-2 and MMP-9 are two of the MMPs which are essential for cardiovascular remodeling and development. Wu et al. (2007) demonstrated that there is a significant decrease in the MMP-9 and increase in the MMP-2 activity and expression in $Crt^{-/-}$ deficient cells, and a significant increase in the expression of membrane type-1 matrix metalloproteinase (MT1-MMP).

Human Pregnancy and Pre-Eclampsia

Evidence indicates that pre-eclampsia involves widespread activation of maternal endothelial cells. Calreticulin has been shown to have both pro- and anti-inflammatory effects in vitro and in whole animals. In normal human pregnancy and in pre-eclampsia, there was a significant increase (5-fold) in calreticulin in plasma in term pregnant women compared with women who were not pregnant. Results indicate that calreticulin is increased in peripheral maternal blood early in pregnancy and remains elevated throughout normal gestation and that there is a further increase in calreticulin in pre-eclampsia (Gu et al. 2008). Calreticulin can be released into extracellular environment in some circumstances. For example, there is a tenfold increase in Crt in the blood of patients with systemic lupus erythematosus (Eggleton et al. 1997). The sources and roles of extracellular Crt are not clear. Nevertheless, evidence indicates that extracellular as well as intracellular Crt can also affect many cellular functions including adhesion, migration and proliferation (Bedard et al. 2005). In particular, its effects on vascular endothelial cells may be relevant to the normal pregnancy and pre-eclampsia.

Crt Interacts with HIV Envelope Protein

Calreticulin binds in vitro to a number of proteins isolated from ER. In cells expressing recombinant HIV envelope glycoprotein, gp160 bound transiently to calreticulin. The binding kinetics of calnexin and calreticulin to gp160 were very similar. Data suggested that most of the gp160 associated with calreticulin was also bound to calnexin but that only a portion of gp160 associated with calnexin was also bound to calreticulin (Otteken and Moss 1996).

Despite ER targeting and retention signals, calreticulin is also located within the nucleus where its presence increases due to its interaction with glucocorticoid receptors (Roderick et al. 1997). Therefore, Crt can inhibit steroid-regulated gene transcription by preventing receptor binding to DNA. Over-expression of Crt gene in B16 mouse melanoma cells resulted in a decrease in retinoic acid (RA)-stimulated reporter gene expression. Purified Crt inhibited the binding of endogenous RAR to a β-RA response element oligonucleotide, only if added prior to the addition of oligonucleotide. Cyclic AMP increased the expression of Crt. Cyclic AMP may act to antagonize RA action by both decreasing RAR expression and stimulating Crt levels (Desai et al. 1996).

Crt Is a Human Rheumatic Disease-Associated Autoantigen

Reports indicate that Crt is a human rheumatic disease-associated autoantigen. This protein shares an intimate relationship with Ro/SS-A autoantigen complex. Calcium ionophore, heat shock, and heavy metals such as zinc and cadmium are consistently found to increase Crt transcriptional activities in A431 cells (a human epidermoid squamous carcinoma cell line) under transient transfection conditions. Studies suggest that Crt is regulated at transcriptional level, and like some other LE-related autoantigens, Crt appears to function as a heat shock/stress-response gene (Nguyen et al. 1996). CRT was present at higher

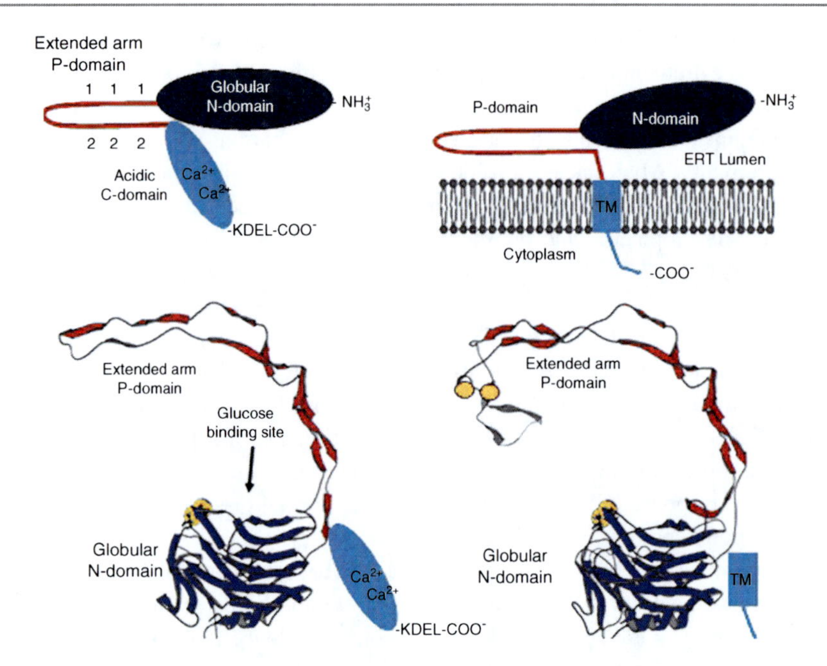

Fig 2.6 Structural models of calnexin and of calreticulin: *Left* panel shows schematic representation of calreticulin domains. *Right* panel shows a model of calnexin based on crystallographic studies of Cnx (1JHN). *Yellow* balls represent cysteines, which form an S-S bridge. Putative glucose-binding site is indicated (Adapted with permission from Schrag et al. 2001 © Elsevier and with permission from Michalak et al. 1999 Biochem J. 344: 281–97 © The Biochemical Society)

concentrations in the plasma and synovial fluid of RA patients. CRT had the capacity to bind directly to FasL, and inhibiting FasL- mediated apoptosis of Jurkat T cells, and might play a role in inhibiting apoptosis of inflammatory T cells in RA (Tarr et al. 2010a).

2.2.7 Similarities and Differences Between Cnx and Crt

Cnx performs the same service for soluble proteins as does calreticulin. Both proteins, Cnx and Crt, have the function of binding to oligosaccharides containing terminal glucose residues, thereby targeting them for degradation. Structural studies suggest that both proteins consist of a globular domain and an extended arm domain comprised of two sequence motifs repeated in tandem (Fig. 2.6). The primary lectin site of Cnx and Crt resides within the globular domain, but the results also point to a much weaker secondary site within the arm domain, which lacks specificity for monoglucosylated oligosaccharides. For both proteins, a site of interaction with ERp57 is centered on the arm domain, which retains ~50% of binding compared with full-length controls. This site is in addition to a Zn^{2+}-dependent site located within the globular domain of both proteins. Finally, calnexin and calreticulin suppress the aggregation of unfolded proteins via a polypeptide binding site located within their globular domains but require the arm domain for full chaperone function (Leach et al. 2002).

In normal cellular function, trimming of glucose residues off the core oligosaccharide added during N-linked glycosylation is a part of protein processing. If "overseer" enzymes note that residues are misfolded, proteins within the RER will re-add glucose residues so that other Calreticulin/Calnexin can bind to these proteins and prevent them from being exported from ER to Golgi. This leads these aberrantly folded proteins down a path whereby they are targeted for degradation.

Calnexin and Calreticulin at Mitochondrial Membranes: ER chaperones, particularly Ca^{2+}-binding chaperones (Cnx, Crt, and BiP), were also found to be compartmentalized at mitochondrial membranes (MM) (Hayashi and Su 2007; Myhill et al. 2008). Under physiological conditions, these chaperones serve as high-capacity Ca^{2+}-binding proteins at ER (Hendershot 2004). Calreticulin provides up to 45% of the Ca^{2+}-buffering capacity for a pool of the IP3-sensitive Ca^{2+} inside the ER (Bastianutto et al. 1995). The compart-mentalized chaperones at the MAM therefore serve as high-capacity Ca^{2+} pools in the ER. In addition, independent of its Ca^{2+}-buffering capacity in the ER, calreticulin inhibits IP3 receptor-mediated Ca^{2+} signaling by using its high-affinity-low-capacity Ca^{2+}-binding domain (Camacho and Lechleiter 1995). Further, calreticulin regulates the activity of Ca^{2+}-ATPase, providing dynamic control of ER Ca^{2+} homeostasis (Li and Camacho 2004). Calnexin can also regulate the activity of Ca^{2+}-ATPase via a direct protein-protein interaction (Roderick et al. 2000). In addition, the activity and action of calnexin and calreticulin are regulated by other chaperones or proteins most likely occurring at the MAM of the ER (Hayashi et al. 2009a).

Cnx/Crt in MHC Class I Assembly Pathway: MHC class I molecules are ligands for T-cell receptors of $CD8^+$ T cells and inhibitory receptors of natural killer cells. Assembly of the heavy chain, light chain, and peptide components of MHC class I molecules occurs in ER. The folding and assembly of class I molecules is assisted by molecular chaperones and folding catalysts that comprise the general ER quality control system which also monitors the integrity of the process, disposing of misfolded class I molecules through ER associated degradation (ERAD). Fu et al. (2009) showed that reduced class I expression in Crt deficient cells can be restored by direct delivery of peptides into the ER or by incubation at low temperature.

Crt deficient cells exhibited a TAP-deficient phenotype in terms of class I assembly, without loss of TAP expression or functionality. In the absence of Crt, ERp57 is up-regulated, which indicates that they collaborate with each other in class I antigen processing. Specific assembly factors and generic ER chaperones, collectively called the MHC class I peptide loading complex (PLC), are required for MHC class I assembly. Calreticulin has an important role within the PLC and induces MHC class I cell surface expression. Interactions with ERp57 and substrate glycans are important for the recruitment of calreticulin into the PLC and for its functional activities in MHC class I assembly. The glycan and ERp57 binding sites of calreticulin contribute directly or indirectly to complexes between calreticulin and the MHC class I assembly factor tapasin and are important for maintaining steady-state levels of both tapasin and MHC class I heavy chains. The generic polypeptide binding sites per se are insufficient for stable recruitment of calreticulin to PLC substrates in cells. However, such binding sites could contribute to substrate stabilization in a step that follows the glycan and ERp57-dependent recruitment of calreticulin to the PLC (Del Cid et al. 2010).

Association of HLA Class I Antigen Abnormalities with Disease Progression in Malignancies: MHC class I molecules are crucial in presenting antigenic peptide epitopes to cytotoxic T lymphocytes. Proper assembly of MHC class I molecules is dependent on several cofactors, e.g. chaperones Cnx and Crt residing in ER. Lectin deficient mutants of Cnx were shown to interact with heavy chains of major MHC class I molecules in insect cells and to prevent their rapid degradation. Similarly, lectin-deficient Crt was found not only to interact with a broad spectrum of newly synthesized proteins and dissociate with normal kinetics, but it was also able to complement all MHC class I biosynthetic defects associated with Crt deficiency. MHC class Ia downregulation has been repeatedly described on melanoma cells and is thought to be involved in failure of immune system to control tumor progression. Alterations in the expression of chaperones Cnx/Crt may have important implications for MHC class I assembly, peptide loading, and presentation on the tumor cell surface and thus may contribute to immune escape phenotype of tumor cells. Metastatic melanoma lesions exhibited significant downregulation of Cnx as compared to primary melanoma lesions. In contrast, Crt was expressed in melanoma cells of primary as well as of metastatic lesions. Data suggest that chaperone-downregulation, particularly Cnx-downregulation, may contribute to the metastatic phenotype of melanoma cells in vivo. Consistently, conserved chaperone expression in metastatic melanoma lesions may be a useful criterion for selection of patients for treatment with T cell-based immunotherapies (Dissemond et al. 2004). However, mutant human cells lacking Cnx, infected with recombinant vaccinia viruses encoding mouse MHC class I molecules, K^d, K^b, K^k, D^d, D^b, and L^d, indicated that Cnx is not required for the efficient assembly of MHC class I molecules with TAP-dependent or independent peptides (Prasad et al. 1998; Mehta et al. 2008). The IFN-γ inducible proteasome subunits LMP2 and LMP7, TAP1, TAP2, Cnx, Crt, ERp57, and tapasin are strongly expressed in the cytoplasm of normal prostate cells, whereas HLA class I heavy chain (HC) and β_2-microglobulin are expressed on cell surface. Most of antigen processing machinery (APM) components was downregulated in a substantial number of prostate cancers. Thus HLA class I APM component abnormalities are mainly due to regulatory mechanisms, play a role in the clinical course of prostate cancer and on the outcome of T cell-based immunotherapies (Dissemond et al. 2004; Seliger et al. 2010).

2.2.8 Calreticulin in Invertebrates

Crt is highly conserved in eukaryotic cells, which is indicated by sequence analysis on the deduced amino acids of the known Crt cDNA clones from several mammalian species and other organisms including nematode, fruit fly (Smith, 1992), marine snail (Kennedy et al. 1992), clawed frog (Treves et al. 1992), rainbow trout (Stephen et al. 2004), and Cotesia rubecula (Zhang et al. 2006). Kennedy et al. (1992) identified Crt in Aplysia where it was enriched in presynaptic varicosities. The steady-state level of Crt mRNA in Aplysia sensory neurons increases during the maintenance phase of long-term sensitization. This mRNA increase in expression late, some time after training, is consistent with the idea that long-term neuromodulatory changes underlying sensitization may depend on a cascade of gene expression in which the induction of early regulatory genes leads to the expression of late effector genes (Kennedy et al. 1992).

A human Ro/SS-A (Ro) autoantigen of 60-kDa, homologous to Crt and Aplysia "memory molecule" has a molecular mass, isoelectric point, and significant amino acid sequence similar to Aplysia californica snail neuronal protein 407 (McCauliffe et al. 1990). These homologies suggest that this Ro protein has a very basic cellular function(s) which may in part involve calcium binding (McCauliffe et al. 1990). The Ro autoantigens consist of at least four immunologically distinct proteins which are recognized by autoantibodies typically found in sera from patients with primary Sjogren's syndrome and in subsets of patients with lupus erythematosus. The mouse cDNA-encoded amino acid sequence was found to be 94% homologous to the human Ro sequence and is 100% homologous to murine calreticulin, a high affinity calcium-binding protein which resides in the endoplasmic and sarcoplasmic reticulum. The amino acid sequence of rabbit Crt is 92% homologous to both murine Crt and human Ro. Onchocerca volvulus and Drosophila melanogaster also have molecules that are highly homologous to human Ro.

References

Afshar N, Black BE, Paschal BM (2005) Retrotranslocation of the chaperone calreticulin from the endoplasmic reticulum lumen to the cytosol. Mol Cell Biol 25:8844–8853

Andrin C, Pinkoski MJ, Burns K et al (1998) Interaction between a Ca2 + -binding protein calreticulin and perforin, a component of the cytotoxic T-cell granules. Biochemistry 37:10386–10394

Andrin C, Corbett EF, Johnson S et al (2000) Expression and purification of mammalian calreticulin in *Pichia pastoris*. Protein Expr Purif 20:207–215

Arosa FA, de Jesus O, Porto G et al (1999) Calreticulin is expressed on the cell surface of activated human peripheral blood T lymphocytes in association with major histocompatibility complex class I molecules. J Biol Chem 274:16917–16922

Baksh S, Michalak M (1991) Expression of calreticulin in *Escherichia coli* and identification of its Ca^{2+} binding domains. J Biol Chem 266:21458–21465

Baksh S, Burns K, Andrin C et al (1995a) Interaction of calreticulin with protein disulfide isomerase. J Biol Chem 270:31338–31351

Baksh S, Spamer C, Oikawa K et al (1995b) Zn2+ binding to cardiac calsequestrin. Biochem Biophys Res Commun 209:310–315

Baksh S, Spamer C, Heilmann C et al (1995c) Identification of the Zn2+ binding region in calreticulin. FEBS Lett 376:53–57

Bass J, Chiu G, Argon Y et al (1998) Folding of insulin receptor monomers is facilitated by the molecular chaperones calnexin and calreticulin and impaired by rapid dimerization. J Cell Biol 141:637–646

Bastianutto C, Clementi E, Codazzi F et al (1995) Overexpression of calreticulin increases the Ca^{2+}capacity of rapidly exchanging Ca^{2+} stores and reveals aspects of their lumenal microenvironment and function. J Cell Biol 130:847–855

Basu S, Srivastava PK (1999) Calreticulin, a peptide-binding chaperone of the endoplasmic reticulum, elicits tumor- and peptide-specific immunity. J Exp Med 189:797–802

Bedard K, Szabo E, Michalak M, Opas M (2005) Cellular functions of endoplasmic reticulum chaperones calreticulin, calnexin, and ERp57. Int Rev Cytol 245:91–121

Bergeron JJ, Brenner MB, Thomas DY et al (1994) Calnexin: a membrane-bound chaperone of the endoplasmic reticulum. Trends Biochem Sci 19:124–128

Brockmeier A, Williams DB (2006) Potent lectin-independent chaperone function of calnexin under conditions prevalent within the lumen of the endoplasmic reticulum. Biochemistry 45:12906–12916

Brockmeier A, Brockmeier U, Williams DB (2009) Distinct contributions of the lectin and arm domains of calnexin to its molecular chaperone function. J Biol Chem 284:3433–3451

Brondani Da Rocha A, Regner A, Grivicich I et al (2004) Radioresistance is associated to increased Hsp70 content in human glioblastoma cell lines. Int J Oncol 25:777–785

Burns K, Opas M, Michalak M (1997) Calreticulin inhibits glucocorticoid- but not cAMP-sensitive expression of tyrosine aminotransferase gene in cultured McA-RH7777 hepatocytes. Mol Cell Biochem 171:37–43

Camacho P, Lechleiter JD (1995) Calreticulin inhibits repetitive intracellular Ca^{2+} wave. Cell 82:765–771

Caramelo JJ, Parodi AJ (2008) Getting in and out from calnexin/calreticulin cycles. J Biol Chem 283:10221–10225

Caramelo JJ, Castro OA, de Prat-Gay G et al (2004) The endoplasmic reticulum glucosyltransferase recognizes nearly native glycoprotein folding intermediates. J Biol Chem 279:46280–46285

Chen MH, Tian GW, Gafni Y, Citovsky V (2005) Effects of calreticulin on viral cell-to-cell movement. Plant Physiol 138:1866–1876

Chevet E, Wong HN, Gerber D et al (1999) Phosphorylation by CK2 and MAPK enhances calnexin association with ribosomes. EMBO J 18:3655–3666

Choi BH, Kim JS (2004) Age-related decline in expression of calnexin. Exp Mol Med 36:499–503

Coppolino MG, Woodside MJ, Demaurex N et al (1997) Calreticulin is essential for integrin-mediated calcium signaling and cell adhesion. Nature 386(6627):843–847

Corbett EF, Oikawa K, Francois P et al (1999) Ca^{2+} regulation of interactions between endoplasmic reticulum chaperones. J Biol Chem 274:6203–6211

Corbett EF, Michalak KM, Johnsoni KOS et al (2000) The conformation of calreticulin is influenced by the endoplasmic reticulum luminal environment. J Biol Chem 275:27177–27185

Crossin KL, Tai MH, Krushel LA et al (1997) Glucocorticoid receptor pathways are involved in the inhibition of astrocyte proliferation. Proc Natl Acad Sci U S A 94:2687–2692

Culina S, Lauvau G, Gubler B et al (2004) Calreticulin promotes folding of functional human leukocyte antigen class I molecules in vitro. J Biol Chem 279:54210–54215

Dai E, Stewart M, Ritchie B et al (1997) Calreticulin, a potential vascular regulatory protein, reduces intimal hyperplasia after arterial injury. Arterioscler Thromb Vasc Biol 17:2359–2368

David V, Hochstenbach F, Rajagopalan S et al (1993) Interaction with newly synthesized and retained proteins in the endoplasmic reticulum suggests a chaperone function for human integral membrane protein IP90 (calnexin). J Biol Chem 268:9585–9592

Dedhar S (1994) Novel functions for calreticulin: interaction with integrins and modulation of gene expression? Trends Biochem Sci 19:269–271

Dedhar S, Rennie PS, Shago M et al (1994) Inhibition of nuclear hormone receptor activity by calreticulin. Nature 367(6462):480–483

Del Cid N, Jeffery E, Rizvi SM et al (2010) Modes of calreticulin recruitment to the major histocompatibility complex class I assembly pathway. J Biol Chem 285:4520–4535

Denecke J, Carlsson LE, Vidal S et al (1995) The tobacco homolog of mammalian calreticulin is present in protein complexes *in vivo*. Plant Cell 7:391–406

Denzel A, Molinari M, Trigueros C et al (2002) Early postnatal death and motor disorders in mice congenitally deficient in calnexin expression. Mol Cell Biol 22:7398–7404

Desai D, Michalak M, Singh NK et al (1996) Inhibition of retinoic acid receptor function and retinoic acid-regulated gene expression in mouse melanoma cells by calreticulin. A potential pathway for cyclic AMP regulation of retinoid action. J Biol Chem 2711:5153–5159

Dickson KM, Bergeron JJ, Shames I et al (2002) Association of calnexin with mutant peripheral myelin protein-22 ex vivo: a basis for "gain-of-function" ER diseases. Proc Natl Acad Sci U S A 99:9852–9857

Dissemond J, Busch M, Kothen T et al (2004) Differential downregulation of endoplasmic reticulum-residing chaperones calnexin and calreticulin in human metastatic melanoma. Cancer Lett 203:225–231

Dupuis M, Schaerer E, Krause KH, Tschopp J (1993) The calcium-binding protein calreticulin is a major constituent of lytic granules in cytolytic T lymphocytes. J Exp Med 177:1–7

Duus K, Hansen PR, Houen G (2008) Interaction of calreticulin with amyloid beta peptide 1-47. Protein Pept Lett 15:103–107

Eggleton P, Reid K, Kishore U, Sontheimer RD (1997) Clinical relevance of calreticulin in systemic lupus erythematosus. Lupus 6:564–571

Ellgaard L, Frickel E-M (2003) Calnexin, calreticulin, and ERp57. Cell Biochem Biophys 39:223–247

Ellgaard L, Helenius A (2003) Quality control in the endoplasmic reticulum. Nat Rev Mol Cell Biol 4:181–191

Ellgaard L, Riek R, Braun D et al (1994) Human, mouse, and rat calnexin cDNA cloning: identification of potential calcium binding motifs and gene localization to human chromosome 5. Biochemistry 33:3229–3236

Ellgaard L, Molinari M, Helenius A (1999) Setting the standards: quality control in the secretory pathway. Science 286:1882–1888

Ellgaard L, Riek R, Herrmann T et al (2001a) NMR structure of the calreticulin P-domain. Proc Natl Acad Sci U S A 98:3133–3138

Ellgaard L, Riek R, Braun D et al (2001b) Three-dimensional structure topology of the calreticulin P-domain based on NMR assignment. FEBS Lett 488:69–73

Ellgaard L, Bettendorff P, Braun D, Herrmann T, Fiorito F, Jelesarov I, Güntert P, Helenius A, Wüthrich K (2002) NMR structures of 36 and 73-residue fragments of the calreticulin P-domain. J Mol Biol 322:773–784

Fadel MP, Dziak E, Lo CM et al (1999) Calreticulin affects focal contact-dependent but not close contact-dependent cell-substratum adhesion. J Biol Chem 274:15085–15094

Fadel MP, Szewczenko-Pawlikowski M, Leclerc P et al (2001) Calreticulin affects beta-catenin-associated pathways. J Biol Chem 276:27083–27089

Fontanini A, Chies R, Snapp EL et al (2005) Glycan-independent role of calnexin in the intracellular retention of Charcot-Marie-tooth 1A Gas3/PMP22 mutants. J Biol Chem 280:2378–2387

Fraser SA, Michalak M, Welch WH, Hudig D (1998) Calreticulin, a component of the endoplasmic reticulum and of cytotoxic lymphocyte granules, regulates perforin-mediated lysis in the hemolytic model system. Biochem Cell Biol 76:881–887

Frickel EM, Riek R, Jelesarov I et al (2002) TROSY-NMR reveals interaction between ERp57 and the tip of the calreticulin P-domain. Proc Natl Acad Sci U S A 99:1954–1959

Fu H, Liu C, Flutter B, Tao H, Gao B (2009) Calreticulin maintains the low threshold of peptide required for efficient antigen presentation. Mol Immunol 46:3198–3206

Gardai SJ, McPhillips KA, Frasch SC et al (2005) Cell-surface calreticulin initiates clearance of viable or apoptotic cells through trans-activation of LRP on the phagocyte. Cell 123: 321–334

Gelebart P, Opas M, Michalak M (2005) Calreticulin, a Ca^{2+}-binding chaperone of the endoplasmic reticulum. Int J Biochem Cell Biol 37:260–266

Goicoechea S, Orr AW, Pallero MA et al (2000) Thrombospondin mediates focal adhesion disassembly through interactions with cell surface calreticulin. J Biol Chem 275:36358–36368

Goicoechea S, Pallero MA, Eggleton P et al (2002) The anti-adhesive activity of thrombospondin is mediated by the n-terminal domain of cell surface calreticulin. J Biol Chem 277:37219–37228

Gold LI, Rahman M, Blechman KM et al (2006) Overview of the role for calreticulin in the enhancement of wound healing through multiple biological effects. J Invest Dermatol 11:57–65

Gold LI, Eggleton P, Sweetwyne MT et al (2010) Calreticulin: non-endoplamic reticulum functions in physiology and disease. FASEB J 24:665–683

Grespin ME, Bonamy GMC, Roggero VR et al (2008) Thyroid hormone receptor α1 follows a cooperative CRM1/calreticulin-mediated nuclear export pathway. J Biol Chem 283:25576–25588

Gu VY, Wong MH, Stevenson JL et al (2008) Calreticulin in human pregnancy and pre-eclampsia. Mol Hum Reprod 14:309–315

Guerin R, Arseneault G, Dumont S et al (2008) Calnexin is involved in apoptosis induced by endoplasmic reticulum stress in the fission yeast. Mol Biol Cell 19:4404–4420

Guérin R, Beauregard PB, Leroux A et al (2009) Calnexin regulates apoptosis induced by inositol starvation in fission yeast. PLoS One 4:e6251

Guo L, Nakamura K, Lynch J et al (2002) Cardiac-specific expression of calcineurin reverses embryonic lethality in calreticulin-deficient mouse. J Biol Chem 277:50776–50779

Guo L, Groenendyk J, Papp S et al (2003) Identification of an N-domain histidine essential for chaperone function in calreticulin. J Biol Chem 278:50645–50653

Gupta GS (2005) Quality control of germ cell proteins. In: Gupta GS (ed) Proteomics of spermatogenesis. Springer, New York, pp 749–776

Hayashi T, Su TP (2007) Sigma-1 receptor chaperones at the ER-mitochondrion interface regulate Ca^{2+} signaling and cell survival. Cell 131:596–10

Hayashi I, Takatori S, Urano Y et al (2009a) Single chain variable fragment against nicastrin inhibits the γ-secretase activity. J Biol Chem 284:27838–27847

Hayashi T, Rizzuto R, Hajnoczky G, Su T-P (2009b) MAM: more than just a housekeeper. Trends Cell Biol 19:81–88

Hebert DN, Foellmer B, Helenius A (1995) Glucose trimming and reglucosylation determine glycoprotein association with calnexin in the endoplasmic reticulum. Cell 81:425–433

Hendershot LM (2004) The ER function BiP is a master regulator of ER function. Mt Sinai J Med 71:289–297

Hershberger ME, Tuan RS (1998) Placental 57-kDa Ca^{2+}-binding protein: regulation of expression and function in trophoblast calcium transport. Dev Biol 199:80–92

Hojrup P, Roepstorff P, Houen G (2001) Human placental calreticulin - characterization of domain structure and post-translational modifications. Eur J Biochem 268:2558–2565

Holaska JM, Black BE, Love DC, Hanover JA, Leszyk J, Paschal BM (2001) Calreticulin is a receptor for nuclear export. J Cell Biol 152:127–140

Ihara Y, Goto US Y, Kondo T (2006) Role of calreticulin in the sensitivity of myocardiac H9c2 cells to oxidative stress caused by hydrogen peroxide. Am J Physiol Cell Physiol 290: C208–C221

Ikawa M, Wada I, Kominami K et al (1997) The putative chaperone calmegin is required for sperm fertility. Nature 387(6633):607–611

Ireland BS, Brockmeier U, Howe CM et al (2008) Lectin-deficient calreticulin retains full functionality as a chaperone for class I histocompatibility molecules. Mol Biol Cell 19:2413–2423

Janssen MJ, Waanders E, Woudenberg J, Lefeber DJ, Drenth JP (2010) Congenital disorders of glycosylation in hepatology: the example of polycystic liver disease. J Hepatol 52:432–440

Jethmalani SM, Henle KJ (1998) Calreticulin associates with stress proteins: implications for chaperone function during heat stress. J Cell Biochem 69:30–43

Jia XY, He LH, Jing RL, Li RZ (2009) Calreticulin: conserved protein and diverse functions in plants. Physiol Plant 136:127–138

Jin H, Hong Z, Su W, Li S (2009) A plant-specific calreticulin is a key retention factor for a defective brassinosteroid receptor in the endoplasmic reticulum. Proc Natl Acad Sci U S A 106:13612–13617

Johnson S, Michalak M, Opas M, Eggleton P (2001) The ins and outs of calreticulin: from the ER lumen to the extracellular space. Trends Cell Biol 11:122–129

Jun-Chao WU, Liang Z-Q et al (2006) Quality control system of the endoplasmic reticulum and related diseases. Acta Biochim Biophys Sin 38:219–226

Kageyama K, Ihara Y, Goto S et al (2002) Overexpression of calreticulin modulates protein kinase B/Akt signaling to promote apoptosis during cardiac differentiation of cardiomyoblast H9c2 cells. J Biol Chem 277:19255–19264

Kapoor M, Ellgaard L, Gopalakrishnapai J et al (2004) Mutational analysis provides molecular insight into the carbohydrate-binding region of calreticulin: pivotal roles of tyrosine-109 and aspartate-135 in carbohydrate recognition. Biochemistry 43:97–106

Karri S, Johnson H, Hendry WJ 3rd et al (2004) Neonatal exposure to diethylstilbestrol leads to impaired action of androgens in adult male hamsters. Reprod Toxicol 19:53–63

Kennedy TE, Kuhl D, Barzilai A et al (1992) Long-term sensitization training in Aplysia leads to an increase in calreticulin, a major presynaptic calcium-binding protein. Neuron 9:1013–1024

Knee R, Ahsan I, Mesaeli N et al (2003) Compromised calnexin function in calreticulin-deficient cells. Biochem Biophys Res Commun 304:661–666

Kosmaoglou M, Cheetham ME (2008) Calnexin is not essential for mammalian rod opsin biogenesis. Mol Vis 14:2466–2474

Krause K-H, Michalak M (1997) Calreticulin. Cell 88:439–443

Kuwabara K, Pinsky DJ, Schmidt AM et al (1995) Calreticulin, an antithrombotic agent which binds to vitamin K-dependent coagulation factors, stimulates endothelial nitric oxide production, and limits thrombosis in canine coronary arteries. J Biol Chem 270:8179–8187

Kwon MS, Park CS, Choi K et al (2000) Calreticulin couples calcium release and calcium influx in integrin-mediated calcium signaling. Mol Biol Cell 11:1433–1443

Langdown ML, Holness MJ, Sugden MC (2003) Effects of prenatal glucocorticoid exposure on cardiac calreticulin and calsequestrin protein expression during early development and in adulthood. Biochem J 371:61–69

Laporte C, Vetter G, Loudes AM et al (2003) Involvement of the secretory pathway and the cytoskeleton in intracellular targeting and tubule assembly of *Grapevine fanleaf virus* movement protein in tobacco BY-2 cells. Plant Cell 15:2058–2075

Leach MR, Williams DB (2004) Lectin-deficient calnexin is capable of binding class I histocompatibility molecules in vivo and preventing their degradation. J Biol Chem 279:9072–9079

Leach MR, Cohen-Doyle MF, Thomas DY et al (2002) Localization of the lectin, ERp57 binding, and polypeptide binding sites of calnexin and calreticulin. J Biol Chem 277:29686–29697

Lee WH, Akatsuka S, Shirase T et al (2006) Alpha-tocopherol induces calnexin in renal tubular cells: another protective mechanism against free radical-induced cellular damage. Arch Biochem Biophys 453:168–178

Leung-Hagesteijn CY, Milankov K, Michalak M et al (1994) Cell attachment to extracellular matrix substrates is inhibited upon down-regulation of expression of calreticulin, an intracellular integrin alpha-subunit-binding protein. J Cell Sci 107:589–600

Li Y, Camacho P (2004) Ca^{2+}-dependent redox modulation of SERCA 2b by ERp57. J Cell Biol 164:35–46

Li Z, Stafford WF, Bouvier M (2001) The metal ion binding properties of calreticulin modulate its conformational flexibility and thermal stability. Biochemistry 40:11193–11201

Lin P, Le-Niculescu H, Hofmeister R et al (1998) The mammalian calcium-binding protein, nucleobindin (CALNUC), is a Golgi resident protein. J Cell Biol 141:1515–1527

Martin V, Groenendyk J, Steiner SS et al (2006) Identification by mutational analysis of amino acid residues essential in the chaperone function of calreticulin. J Biol Chem 281:2338–2346

McCauliffe DP, Zappi E, Lieu TS et al (1990) A human Ro/SS-A autoantigen is the homologue of calreticulin and is highly homologous with onchocercal RAL-1 antigen and an aplysia "memory molecule". J Clin Invest 86:332–335

McCauliffe DP, Yang YS, Wilson J et al (1992) The 5'-flanking region of the human calreticulin gene shares homology with the human GRP78, GRP94, and protein disulfide isomerase promoters. J Biol Chem 267:2557–2562

McGowan KM, Long SD, Pekala PH (1995) Glucose transporter gene expression: regulation of transcription and mRNA stability. Pharmacol Ther 66:465–505

Meehan KL, Sadar MD (2004) Quantitative profiling of LNCaP prostate cancer cells using isotope-coded affinity tags and mass spectrometry. Proteomics 4:1116–1134

Mehta AM, Jordanova ES, Kenter GG et al (2008) Association of antigen processing machinery and HLA class I defects with clinicopathological outcome in cervical carcinoma. Cancer Immunol Immunother 57:197–206

Mesaeli N, Phillipson C (2004) Impaired p53 expression, function, and nuclear localization in calreticulin-deficient cells. Mol Biol Cell 15:1862–1870

Mesaeli N, Nakamura K, Zvaritch E et al (1999) Calreticulin is essential for cardiac development. J Cell Biol 144:857–868

Michalak M, Burns K, Andrin C et al (1996) Endoplasmic reticulum form of calreticulin modulates glucocorticoid-sensitive gene expression. J Biol Chem 271:29436–29437

Michalak M, Corbett EF, Mesaeli N et al (1999) Calreticulin: one protein, one gene, many functions. Biochem J 344:281–297

Michalak M, Lynch J, Groenendyk J et al (2002a) Calreticulin in cardiac development and pathology. Biochim Biophys Acta 1600:32–37

Michalak M, Robert Parker JM, Opas M (2002b) Ca^{2+} signaling and calcium binding chaperones of the endoplasmic reticulum. Cell Calcium 32:269–278

Michalak M, Guo L, Robertson M et al (2004) Calreticulin in the heart. Mol Cell Biochem 263:137–147

Michalak M, Groenendyk J, Szabo E et al (2009) Calreticulin, a multiprocess calcium-buffering chaperone of the endoplasmic reticulum. Biochem J 417:651–666

Mueller CF, Wassmann K, Berger A et al (2008) Differential phosphorylation of calreticulin affects AT1 receptor mRNA stability in VSMC. Biochem Biophys Res Commun 370:669–674

Myhill N, Lynes EM, Nanji JA et al (2008) The subcellular distribution of calnexin is mediated by PACS-7. Mol Biol Cell 19:2777–2788

Nakamura K, Robertson M, Liu G et al (2001a) Complete heart block and sudden death in mouse over-expressing calreticulin. J Clin Invest 107:1245–1253

Nakamura K, Zuppini A, Arnaudeau S et al (2001b) Functional specialization of calreticulin domains. J Cell Biol 154:961–972

Nakhasi HL, Singh NK, Pogue GP et al (1994) Identification and characterization of host factor interactions with cis-acting elements of rubella virus RNA. Arch Virol Suppl 9:255–267

Nanney LB, Woodrell CD, Greives MR et al (2008) Calreticulin enhances porcine wound repair by diverse biological effects. Am J Pathol 173:610–630

Navazio L, Baldan B, Mariani P et al (1996) Primary structure of the N-linked carbohydrate chains of calreticulin from spinach leaves. Glycoconj J 13:977–983

Nguyen TQ, Capra JD, Sontheimer RD (1996) Calreticulin is transcriptionally upregulated by heat shock, calcium and heavy metals. Mol Immunol 33:379–386

Ni M, Lee AS (2007) ER chaperones in mammalian development and human diseases. FEBS Lett 581:3641–3651

Nickenig G, Michaelsen F, Müller C et al (2002) Destabilization of AT1 receptor mRNA by calreticulin. Circ Res 90:53–58

Noorwez SM, Sama RR, Kaushal S (2009) Calnexin improves the folding efficiency of mutant rhodopsin in the presence of pharmacological chaperone 11-cis-retinal. J Biol Chem 284: 33333–33342

Okunaga T, Urata Y, Goto S, Matsuo T et al (2006) Calreticulin, a molecular chaperone in the endoplasmic reticulum, modulates radiosensitivity of human glioblastoma U251MG cells. Cancer Res 66:8662–8671

Olkku A, Mahonen A (2009) Calreticulin mediated glucocorticoid receptor export is involved in beta-catenin translocation and Wnt signaling inhibition in human osteoblastic cells. Bone 44:555–565

Opas M, Szewczenko-Pawlikowski M, Jass GH et al (1996) Calreticulin modulates cellular adhesiveness via regulation of expression of vinculin. J Cell Biol 135:1913–1923

Ostwald TJ, MacLennan DH (1974) Isolation of a high affinity calcium-binding protein from sarcoplasmic reticulum. J Biol Chem 249:974–979

Otteken A, Moss B (1996) Calreticulin interacts with newly synthesized human immunodeficiency virus type 1 envelope glycoprotein, suggesting a chaperone function similar to that of calnexin. J Biol Chem 271:97–103

Pallero MA, Elzie CA, Chen J et al (2008) Thrombospondin 1 binding to calreticulin-LRP1 signals resistance to anoikis. FASEB J 22:3968–3979

Patton WF, Erdjument-Bromage H, Marks AR et al (1995) Components of the protein synthesis and folding machinery are induced in vascular smooth muscle cells by hypertrophic and hyperplastic agents. Identification by comparative protein phenotyping and microsequencing. J Biol Chem 270:21404–21410

Perrone L, Tell G, Di Lauro R (1999) Calreticulin enhances the transcriptional activity of thyroid transcription factor-1 by binding to its homeodomain. J Biol Chem 274:4640–4645

Persson S et al (2003) Phylogenetic analyses and expression studies reveal two distinct groups of calreticulin isoforms in higher plants. Plant Physiol 133:1385–1396

Pike SE, Yao L, Jones KD et al (1998) Vasostatin, a calreticulin fragment, inhibits angiogenesis and suppresses tumor growth. J Exp Med 188:2349–2356

Pind S, Riordan JR, Williams DB (1994) Participation of the endoplasmic reticulum chaperone calnexin (p88, IP90) in the biogenesis of the cystic fibrosis transmembrane conductance regulator. J Biol Chem 269:12784–12788

Platet N, Cunat S, Chalbos D et al (2000) Unliganded and liganded estrogen receptors protect against cancer invasion via different mechanisms. Mol Endocrinol 14:999–1009

Pollock S, Kozlov G, Pelletier MF et al (2004) Specific interaction of ERp57 and calnexin determined by NMR spectroscopy and an ER two-hybrid system. EMBO J 23:1020–1029

Porcellini S, Traggiai E, Schenk U et al (2006) Regulation of peripheral T cell activation by calreticulin. J Exp Med 203:461–471

Prasad SA, Yewdell JW, Porgador A et al (1998) Calnexin expression does not enhance the generation of MHC class I-peptide complexes. Eur J Immunol 28:907–913

Qi C, Pekala PH (1999) Breakthroughs and views: the influence of mRNA stability on glucose transporter (GLUT1) gene expression. Biochem Biophys Res Commun 263:265–269

Ramsamooj P, Notario V, Dritschilo A (1995) Enhanced expression of calreticulin in the nucleus of radioresistant squamous carcinoma cells in response to ionizing radiation. Cancer Res 55:3016–3021

Ritter C, Quirin K, Kowarik M et al (2005) Minor folding defects trigger local modification of glycoproteins by the ER folding sensor GT. EMBO J 24:1730–1738

Rizvi SM, Mancino L, Thammavongsa V et al (2004) A polypeptide binding conformation of calreticulin is induced by heat shock, calcium depletion, or by deletion of the C-terminal acidic region. Mol Cell 15:913–923

Roderick HL, Campbell AK, Llewellyn DH (1997) Nuclear localisation of calreticulin in vivo is enhanced by its interaction with glucocorticoid receptors. FEBS Lett 405:181–185

Roderick HL, Lechleiter JD, Camacho P (2000) Cytosolic phosphorylation of calnexin controls intracellular Ca^{2+} oscillations via an interaction with SERCA2b. J Cell Biol 1491:235–248

Rooke K, Briquet-Laugier V, Xia YR et al (1997) Mapping of the gene for calreticulin (Calr) to mouse chromosome 8. Mamm Genome 8:870–871

Rosenbaum EE, Hardie RC, Colley NJ (2006) Calnexin is essential for rhodopsin maturation, Ca^{2+} regulation, and photoreceptor cell survival. Neuron 49:229–241

Roth J, Ziak M, Zuber C (2003) The role of glucosidase II and endomannosidase in glucose trimming of asparagine-linked oligosaccharides. Biochimie 85:287294

Saito Y, Ihara Y, Leach MR et al (1999) Calreticulin functions in vitro as a molecular chaperone for both glycosylated and non-glycosylated proteins. EMBO J 18:6718–6729

Sandhu N, Duus K, Jørgensen CS et al (2007) Peptide binding specificity of the chaperone calreticulin. Biochim Biophys Acta 1774:701–713

Schrag JD, Bergeron JJ, Li Y et al (2001) The structure of calnexin, an ER chaperone involved in quality control of protein folding. Mol Cell 8:633–644

Sela-Brown A, Russell J, Koszewski NJ et al (1998) Calreticulin inhibits vitamin D's action on the PTH gene in vitro and may prevent vitamin D's effect in vivo in hypocalcemic rats. Mol Endocrinol 12:1193–1200

Seliger B, Stoehr R, Handke D et al (2010) Association of HLA class I antigen abnormalities with disease progression and early recurrence in prostate cancer. Cancer Immunol Immunother 59:529–540

Shen X, Zhang K, Kaufman RJ (2004) The unfolded protein response—a stress signaling pathway of the endoplasmic reticulum. J Chem Neuroanat 28:7992

Singh NK, Atreya CD (1994) Nakhasi HL Identification of calreticulin as a rubella virus RNA binding protein. Proc Natl Acad Sci U S A 91:12770–12774

Sipione S, Ewen C, Shostak I et al (2005) Impaired cytolytic activity in calreticulin-deficient CTLs. J Immunol 174:3212–3219

Smith MJ (1992) Nucleotide sequence of a *Drosophila melanogaster* gene encoding a calreticulin homologue. DNA Seq J DNA Seq Mapp 3:247–250

Song J, Finnerty CC, Herndon DN et al (2009) Severe burn-induced endoplasmic reticulum stress and hepatic damage in mice. Mol Med 15:316–320

St-Arnaud R, Prud'homme J, Leung-Hagesteijn C, Dedhar S (1995) Constitutive expression of calreticulin in osteoblasts inhibits mineralization. J Cell Biol 131:1351–1359

Steinø A, Jørgensen CS, Laursen I, Houen G (2004) Interaction of C1q with the receptor calreticulin requires a conformational change in C1q. Scand J Immunol 59:485–495

Stephen K, Kazuhiro F, Dixon B (2004) Molecular cloning and characterization of calreticulin from rainbow trout (Oncorhynchus mykiss). Immunogenetics 55:717–723

Stronge VS, Saito Y, Ihara Y, Williams DB (2001) Relationship between calnexin and BiP in suppressing aggregation and promoting refolding of protein and glycoprotein substrates. J Biol Chem 276:39779–39787

Szegedi A, Irinyi B, Bessenyei B et al (2001) UVB light and 17-beta-estradiol have different effects on the mRNA expression of Ro/SSA and La/SSB autoantigens in HaCaT cells. Arch Dermatol Res 293:275–282

Tarr JM, Winyard PG, Ryan B et al (2010a) Extracellular calreticulin is present in the joints of rheumatoid arthritis patients and inhibits FasL (CD95L) mediated apoptosis of T cells. Arthritis Rheum 62:2919–2929

Tarr JM, Young PJ, Morse R et al (2010b) A mechanism of release of calreticulin from cells during apoptosis. J Mol Biol 401:799–812

Taylor SC, Ferguson AD, Bergeron JJ et al (2004) The ER protein folding sensor UDP-glucose glycoprotein-glucosyltransferase modifies substrates distant to local changes in glycoprotein conformation. Nat Struct Mol Biol 11:128–134

Thammavongsa V, Mancino L, Raghavan M (2005) Polypeptide substrate recognition by calnexin requires specific conformations of the calnexin protein. J Biol Chem 280:33497–33505

Thomson SP, Williams DB (2005) Delineation of the lectin site of the molecular chaperone calreticulin. Cell Stress Chaperones 10:242–251

Tjoelker LW, Seyfried CE, Eddy RL Jr et al (1994) Human, mouse, and rat calnexin cDNA cloning: identification of potential calcium binding motifs and gene localization to human chromosome 5. Biochemistry 33:3229–3236

Totary-Jain H, Naveh MT, Riahi Y et al (2005) Calreticulin destabilizes GLUT-1 mRNA in vascular endothelial and smooth muscle cells under high glucose conditions. Circ Res 97:1001–1008

Treves S, Zorzato F, Pozzan T (1992) Identification of calreticulin isoforms in the central nervous system. Biochem J 287:579–581

Trombetta ES, Parodi AJ (2003) Quality control and protein folding in the secretory pathway. Annu Rev Cell Dev Biol 19:649–676

Tutuncu L, Stein P, Ord TS et al (2004) Calreticulin on the mouse egg surface mediates transmembrane signaling linked to cell cycle resumption. Dev Biol 270:246–260

Van Leeuwen JE, Kearse KP (1996a) The related molecular chaperones calnexin and calreticulin differentially associate with nascent T cell antigen receptor proteins within the endoplasmic reticulum. J Biol Chem 271:25345–25349

van Leeuwen JE, Kearse KP (1996b) Calnexin associates exclusively with individual CD3 δ and T cell antigen receptor (TCR) α proteins containing incompletely trimmed glycans that are not assembled into multisubunit TCR complexes. J Biol Chem 271:9660–9665

Vassilakos A, Michalak M, Lehrman MA et al (1998) Oligosaccharide binding characteristics of the molecular chaperones calnexin and calreticulin. Biochemistry 37:3480–3490

Wada I, Imai S, Kai M et al (1995) Chaperone function of calreticulin when expressed in the endoplasmic reticulum as the membrane-anchored and soluble forms. J Biol Chem 270:20298–20304

Wang Z, Tufts R, Haleem R, Cai X (1997) Genes regulated by androgen in the rat ventral prostate. Proc Natl Acad Sci U S A 94:12999–13004

Ware FE, Vassilakos A, Peterson PA et al (1995) The molecular chaperone calnexin binds Glc1Man9GlcNAc2 oligosaccharide as an initial step in recognizing unfolded glycoproteins. J Biol Chem 270:4697–4704

Waser M, Mesaeli N, Spencer C, Michalak M (1997) Regulation of calreticulin gene expression by calcium. J Cell Biol 138:547–557

Watanabe D, Yamada K, Nishina Y et al (1994) Molecular cloning of a novel Ca^{2+}-binding protein (calmegin) specifically expressed during male meiotic germ cell development. J Biol Chem 269:7744–7749

Watanabe D, Okabe M, Hamajima N et al (1995) Characterization of the testis-specific gene 'calmegin' promoter sequence and its activity defined by transgenic mouse experiments. FEBS Lett 368:509–512

Wearsch PA, Cresswell P (2008) The quality control of MHC class I peptide loading. Curr Opin Cell Biol 20:624–631

Wei H, Kim SJ, Zhang Z et al (2008) ER and oxidative stresses are common mediators of apoptosis in both neurodegenerative and non-neurodegenerative lysosomal storage disorders and are alleviated by chemical chaperones. Hum Mol Genet 17:469–477

Wheeler DG, Horsford J, Michalak M et al (1995) Calreticulin inhibits vitamin D3 signal transduction. Nucleic Acids Res 23:3268–3274

Williams DB (2006) Beyond lectins: the calnexin/calreticulin chaperone system of the endoplasmic reticulum. J Cell Sci 119:615–623

Winrow CJ, Miyata KS, Marcus SL et al (1995) Calreticulin modulates the in vitro DNA binding but not the in vivo transcriptional activation by peroxisome proliferator-activated receptor/retinoid X receptor heterodimers. Mol Cell Endocrinol 111:175–179

Wu M, Massaeli H, Durston M et al (2007) Differential expression and activity of matrix metalloproteinase-2 and -9 in the calreticulin deficient cells. Matrix Biol 26:463–472

Xu X, Azakami H, Kato A (2004) P-domain and lectin site are involved in the chaperone function of *Saccharomyces cerevisiae* calnexin homologue. FEBS Lett 570:155–160

Yamashita K, Hara-Kuge S, Ohkura T (1999) Intracellular lectins associated with N-linked glycoprotein traffic. Biochim Biophys Acta 1473:147–160

Yokoyama M, Hirata K-i (2005) New function of calreticulin. Calreticulin-dependent mRNA destabilization. Circ Res 97:961–963

Zapun A, Darby NJ, Tessier DC et al (1998) Enhanced catalysis of ribonuclease B folding by the interaction of calnexin or calreticulin with ERp57. J Biol Chem 273:6009–6012

Zhang G, Schmidt O, Asgari S (2006) A calreticulin-like protein from endoparasitoid venom fluid is involved in host hemocyte inactivation. Dev Comp Immunol 30:756–764

Zhang Y, Kozlov G, Pocanschi CL et al (2009) ERp57 does not require interactions with calnexin and calreticulin to promote assembly of class I histocompatibility molecules, and it enhances peptide loading independently of its redox activity. J Biol Chem 284:10160–10173

Zhu J (1996) Ultraviolet B irradiation and cytomegalovirus infection synergize to induce the cell surface expression of 52-kD/Ro antigen. Clin Exp Immunol 103:47–53

Zhu N, Wang Z (1999) Calreticulin expression is associated with androgen regulation of the sensitivity to calcium ionophore-induced apoptosis in LNCaP prostate cancer cells. Cancer Res 59:1896–1902

Zhu N, Pewitt EB, Cai X et al (1998) Calreticulin: an intracellular Ca++-binding protein abundantly expressed and regulated by androgen in prostatic epithelial cells. Endocrinology 139:4337–4351

P-Type Lectins: Cation-Dependent Mannose-6-Phosphate Receptor

3

G.S. Gupta

In eukaryotic cells, post-translational modification of secreted proteins and intracellular protein transport between organelles are ubiquitous features. One of the most studied systems is the *N*-linked glycosylation pathway in the synthesis of secreted glycoproteins (Schrag et al. 2003). The *N*-linked glycoproteins are subjected to diverse modifications and are transported through ER and Golgi apparatus to their final destinations in- and outside the cell. Incorporation of cargo glycoproteins into transport vesicles is mediated by transmembrane cargo receptors, which have been identified as intracellular lectins. For example, mannose 6-phosphate receptors (Ghosh et al. 2003) function as a cargo receptor for lysosomal proteins in the *trans*-Golgi network, whereas ERGIC-53 (Zhang et al. 2003) and its yeast orthologs Emp46/47p (Sato and Nakano 2003) are transport lectins for glycoproteins that are transported out of ER.

3.1 The Biosynthetic/Secretory/Endosomal Pathways

In eucaryotic cells, proteins destined for secretion are first inserted into ER and then transported by a process of vesicle budding and fusion, through the Golgi complex and then to the cell surface. Various compartments that comprise this secretory pathway, despite being interconnected by vesicular traffic, differ in their lipid and protein composition. The maintenance of these differences requires that the incorporation of molecules into vesicles is a selective process, and that vesicles are directed to specific target membranes. Much effort has been directed in recent years in understanding these processes, and the ways in which they are integrated to produce organelles of characteristic size, morphology and composition.

3.1.1 Organization of Secretory Pathway

The early secretoty pathway (ESP) is defined as sequential compartments comprising the cisternal/tubular ER, pre-Golgi intermediates [also referred to as vesicular tubular clusters (VTCs) or ER-Golgi intermediates (ERGIC)] and the cis-Golgi network (CGN), the Golgi stacks and trans-Golgi network (TGN) as a final sorting station. The basic function of this pathway includes transport of proteins destined for secretion from ER to Golgi and further to the plasma membrane (PM) (Fig. 3.1). Synthesized material can also be recycled to ER from the Golgi. The compartments are connected with each other by vesicular traffic which mediates the transport of cargo between different organelles and also controls the composition and homeostasis of the structures (Rothman and Wieland 1996). The ESP transports biosynthetic material from ER to Golgi complex. Lipids, proteins and carbohydrates are modified and transported through Golgi to TGN in which they are sorted and packed into vesicles for further transport to various destinations. At TGN, specific sorting signals in the cargo molecules and the cellular sorting machineries are responsible for directing the cargo either to the PM, to regulated secretory granules, or to the endosomal/lysosomal system (Le Borgne and Hoflack 1998). The bulk flow secretory pathway operates in all cells and it leads to a continuous unregulated secretion or transport to the PM. Some specialized cells also possess a distinct regulated secretory pathway in which certain specific proteins are secreted to extracellular space in response to external signal(s). In general, the secretory pathway provides a framework by which proteins undergo a series of posttranslational modifications including proteolytic processing, folding and glycosylation (Storrie et al. 2000).

3.1.1.1 Endoplasmic Reticulum

The endoplasmic reticulum (ER) is a eukaryotic organelle that forms an interconnected network of tubules, vesicles, and cisternae within cells. The general structure of the ER is an extensive membrane network of cisternae (sac-like structures) held together by the cytoskeleton. The phospholipid membrane encloses a space, the cisternal space (or lumen), which is continuous with the perinuclear space but separate from

G.S. Gupta et al., *Animal Lectins: Form, Function and Clinical Applications*,
DOI 10.1007/978-3-7091-1065-2_3, © Springer-Verlag Wien 2012

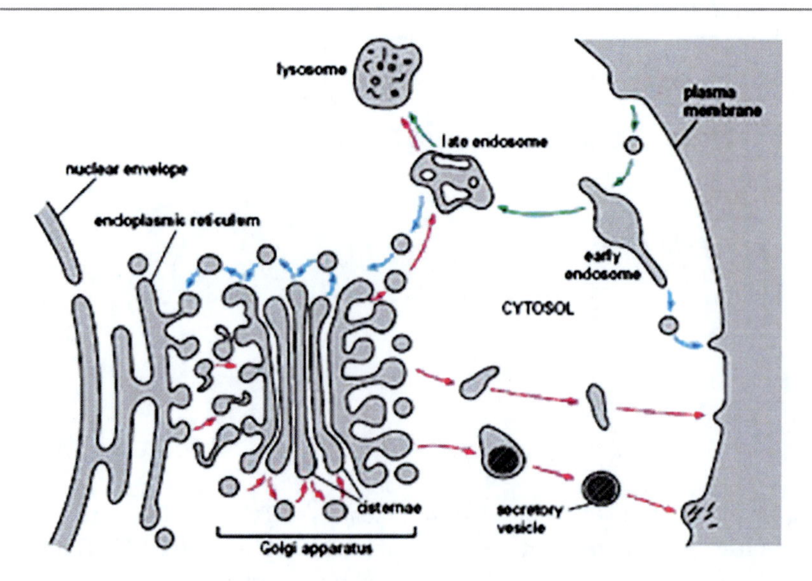

Fig. 3.1 A schematic representation of the different vesicle transport pathways originating from the *trans*-Golgi network (Reprinted by permission from from Le Borgne and Hoflack 1998 © Elsevier). *Arrows* indicate transport steps that are known to occur, but not discussed in detail in this article

the cytosol. The functions of the endoplasmic reticulum vary greatly depending on the exact type of endoplasmic reticulum and the type of cell in which it resides. The ER is the entry station for all proteins of the synthetic/secretory pathway and consists of nuclear envelope, rough ER (rER), smooth ER (sER), transitional ER (tER), and intermediate compartment (IC) (Marie et al. 2009; Lippincott-Schwartz et al. 2000). Rough endoplasmic reticula synthesize proteins, while smooth endoplasmic reticula synthesize lipids and steroids, metabolize carbohydrates and steroids, and regulate calcium concentration, drug metabolism, and attachment of receptors on cell membrane proteins. Different subcompartments of ER have characteristic biochemical and physiological properties and they serve specific subcellular functions. Structurally ER is seen as a three-dimensional, reticular network of continuous tubules and sheets creating the largest membranous organelle of the cell. Functionally ER is responsible for the synthesis and processing of secreted proteins, membrane proteins and organelle resident proteins. The ER is also part of a protein sorting pathway and seen as a compartment that participates in the assembly, sorting, and degradation of proteins as well as in the regulation of intracellular calcium concentration (Harter and Wieland 1996). It is, in essence, the transportation system of the eukaryotic cell. From ER, properly folded and assembled proteins are further transported via IC to the Golgi complex by specific carrier vesicles which bud on the ER and move to cis-Golgi membranes. Folded proteins may remain in the ER if it is their home compartment or else they are transported to the secretory pathway. COPII, a coat complex which forms a main structure of transport vesicles, is responsible for forward transport of cargo from the ER to Golgi complex. COPI vesicles in their turn carry cargo between Golgi and ER and in intra-Golgi transport (Barlowe et al. 1994; Letourneur et al. 1994; Schekman and Orci 1996). Sorting of synthesized proteins at the ER occurs by selective incorporation of secretory and membrane proteins into vesicles that bud from the ER (Pelham 1996; Wieland and Harter 1999). The majority of ER resident proteins are retained in the ER through a specific carboxy terminal retention motif. This motif is composed of four amino acids at the end of the protein sequence. The most common retention sequence is KDEL (*lys-asp-glu-leu*). However, variation on KDEL does occur and other sequences can also give rise to ER retention. There are three KDEL receptors in mammalian cells, and they have a very high degree of sequence identity. The functional differences between these receptors remain to be established. The lumenal soluble proteins of the ER carry a specific carboxyterminal KDEL signal which prevents the secretion of these proteins. Those ER-resident KDEL proteins which have escaped the ER, are recycled back from the Golgi to ER by COPI vesicles (Poussu 2001). Secretory proteins, mostly glycoproteins, are moved across the ER membrane. Proteins that are transported by the ER and from there throughout the cell are marked with an address tag called a signal sequence. The N-terminus (one end) of a polypeptide chain contains a few amino acids that work as an address tag, which are removed when the polypeptide reaches its destination. Proteins that are destined for places outside the ER are packed into transport vesicles and moved along the cytoskeleton toward their destination.

3.1.1.2 Golgi Complex

The Golgi is a stack of polarized tubular/saccular compartments with a defined cis- to trans content reflecting the presence of specialized processing enzymes that extensively modify newly synthesized proteins. The membrane-bound stacks are known as cisternae (singular: *cisterna*). Between four and eight stacks are usually present, although, in some protists as many as 60 have been observed. Each cisterna comprises a flattened membrane disk, and carries Golgi enzymes to help or to modify cargo proteins that travel through them. They are found in both plant and animal cells and the overall morphology of the ER-Golgi system can vary between cell types such as in budding to higher eukaryotes. The Golgi has earlier been viewed as a static station for the processing of secretory material, but now it seems that Golgi undergoes continuous remodeling. Traditionally, Golgi has been viewed as a series of stable compartments, named the *cis*-, medial- and *trans*-Golgi, as well as TGN (Glick 2000). The *trans*- face of the *trans*-Golgi network is the face from which vesicles leave the Golgi. These vesicles then proceed to later compartments such as the cell membrane, secretory vesicles or late endosomes. New *cisternae* form at the *cis*-Golgi network. The *cis*- and *trans*-Golgi networks are thought to be specialised *cisternae* leading in and out of the Golgi apparatus.

The Golgi apparatus is integral in modifying, sorting, and packaging macromolecules for cell secretion (exocytosis) or use within the cell. The Golgi apparatus plays an important role at the crossroads of the secretory pathway. It receives freshly synthesized proteins and lipids from the ER, modifies them, and then distributes cargo to various destinations. Proteins coming from ER to Golgi enter the organ on its tubulovesicular cis-face, travel across the stacks, and leave the Golgi on its trans-face. Cis-Golgi network not only receives the material from the ER but is also involved in sorting and recycling of lipids and proteins to the ER. On the way through the Golgi, newly synthesized glycoproteins are subjected to several posttranslational modifications such as ordered remodeling of their N-linked oligosaccharide side chains and biosynthesis of O-linked glycans. To effect such modifications, the Golgi complex is organized as polarized stacks of flattened cisternae enriched in transmembrane processing enzymes. To be able to send cargo even long distances through the cytoplasm, the Golgi complex is closely associated with the cytoskeleton. It is situated around the microtubule organizing center and is surrounded by actin cytoskeleton and actin-binding proteins (Holleran and Holzbaur 1998).

Enzymes within the cisternae are able to modify the proteins by addition of carbohydrates (glycosylation) and phosphates (phosphorylation). For example, the Golgi apparatus adds a mannose-6-phosphate label to proteins destined for lysosomes. The Golgi plays an important role in the synthesis of proteoglycans, which are molecules present in the extracellular matrix of animals. It is also a major site of carbohydrate synthesis. This includes the production of glycosaminoglycans (GAGs), long unbranched polysaccharides which the Golgi then attaches to a protein synthesised in the endoplasmic reticulum to form proteoglycans. Enzymes in the Golgi polymerize several of these GAGs via a xylose link onto the core protein. Another task of the Golgi involves the sulfation of certain molecules passing through its lumen via sulphotranferases that gain their sulphur molecule from a donor called PAPs. This process occurs on the GAGs of proteoglycans as well as on the core protein. The level of sulfation is very important to the proteoglycans' signalling abilities as well as giving the proteoglycan its overall negative charge.

3.1.1.3 Trans-Golgi Network

The trans-Golgi network is the site of the sorting and final exit of cargo from the Golgi. It refers to the trans-side of the Golgi and structurally it is seen as a sacculotubular network. The structure and the size of TGN varies remarkably from one cell type to another: in cells with a low number of secretory granules but with an extensive lysosomal system, TGN is massive, while secretory cells showing small or large secretory granules typically possess small TGN or even lack it (Clermont et al. 1995). Newly synthesized proteins traverse the Golgi stack until they reach the TGN. The trans-Golgi network sorts the proteins into several types of vesicles. Clathrin-coated vesicles carry certain proteins to lysosomes. At TGN, cargo molecules are sequestered into coated vesicles and directed to their correct destinations. For example, proteins with specific recognition signals are packed into clathrin-coated vescicles (CCVs) and transported to endosomal/lysosomal system in a selective pathway (Marks et al. 1997). Proteins carrying specific sorting signals are targeted and transported to plasma membrane through a so-called constitutive pathway (Pearse and Robinson 1990). In specialized cells producing large quantities of particular products in response to extracellular stimuli (e.g. hormonal or neural stimuli), there exists another secretion pathway leading to cell surface called the regulated secretory pathway (Traub and Kornfeld 1997). Other proteins are packaged into secretory vesicles for immediate delivery to the cell surface. Still other proteins are packaged into secretory granules, which undergo regulated secretion in response to specific signals. The sorting function of the Golgi apparatus allows the various organelles to grow while maintaining their distinct identities.

3.1.1.4 ER-Golgi Intermediate Compartment (ERGIC)

Protein traffic moving from ER to Golgi complex in mammalian cells passes through the tubulovesicular membrane clusters of the ERGIC, the marker of which is the lectin

ERGIC-53. Because the functional borders of the intermediate compartment (IC) are not well defined, the spatial map of the transport machineries operating between the ER and the Golgi apparatus remains incomplete (Fig. 3.1). However, studies showed that the IC consists of interconnected vacuolar and tubular parts with specific roles in pre-Golgi trafficking. The identification of ERGIC-53 has added to the complexity of the exocytic pathway of higher eukaryotic cells. Ffractional analysis of the ERGIC from Vero cells suggested that in the secretory pathway of Vero cells O-glycan initiation and sphingomyelin as well as glucosylceramide synthesis mainly occur beyond the ERGIC in the Golgi apparatus (Schweizer et al. 1994; Appenzeller-Herzog and Hauri 2006). Marie et al. (2009) provided novel insight into the compartmental organization of the secretory pathway and Golgi biogenesis, in addition to a direct functional connection between the IC and the endosomal system, which evidently contributes to unconventional transport of the cystic fibrosis transmembrane conductance regulator to the cell surface. The ERGIC defined by ERGIC-53 also participates in the maturation of (or is target for) several viruses such as corona virus, cytomegalovirus, flavivirus, poliovirus, Uukuniemi virus, and vaccinia virus. Understanding the targeting of viruses and viral proteins to the ERGIC could lead to development of general approaches for viral interference. Further analysis of the ERGIC-53 as a marker protein should provide novel results about the mechanisms controlling traffic in the secretory pathway (Chap. 7).

3.1.1.5 Protein Coats

Coatomer Protein Complex II (COPII) and COPI

Membrane traffic between the ER and the Golgi complex is regulated by two vesicular coat complexes, called coatomer protein complex II (COPII) and COPI. In addition, cells contain numerous clathrin/adaptor complexes—with each coat budding vesicles from a discrete subcellular location. COPII has been implicated in the selective packaging of anterograde cargo into coated transport vesicles budding from ER. In higher eukaryotes, transport from the ER is initiated by COPII mediated budding of vesicular carriers, but this is restricted to specialized, long-lived subdomains of the ER, the ER-exit sites (ERES). ERGIC clusters containing ERGIC-53 are close to but clearly distinct from ERES and delineate the subsequent stage in transport to the Golgi. The ERES are thought to generate transit vesicles and pleiomorphic tubular carriers through the activity of COPII coat machinery to yield pre-Golgi intermediates. In mammalian cells, these vesicles coalesce to form tubulo-vesicular transport complexes (TCs), which shuttle anterograde cargo from the ER to Golgi complex.

In contrast to COPII, COPI-coated vesicles are proposed to mediate recycling of proteins from the Golgi complex to the ER (David et al. 1999). The binding of COPI to COPII-coated TCs, however, has led to the proposal that COPI binds to TCs and specifically packages recycling proteins into retrograde vesicles for return to the ER. Observations, consistent with biochemical data, suggest a role for COPI within TCs *en route* to the Golgi complex. By sequestering retrograde cargo in the anterograde-directed TCs, COPI couples the sorting of ER recycling proteins to the transport of anterograde cargo (David et al. 1999; Rowe et al. 1996).

As observed, COPII-coated vesicles form on the ER to transport newly synthesized cargo to Golgi complex. Three proteins—Sec23/24, Sec13/31, and the ARF-family GTPase Sar1—are sufficient to bud ~60-nm COPII vesicles from native ER membranes and from synthetic liposomes. The COPII coat components coordinate to create a vesicle by locally generating membrane curvature and populating the incipient bud with the appropriate cargo (Lee and Miller 2007; Wiseman et al. 2007). COPII budding is initiated by the activation of Sar1 to its GTP-bound form, causing it to translocate to the membrane and embed an N-terminal α-1 helix in the bilayer (Fath et al. 2007).

Sequential Mode of Action for COPII and COPI

Exocytic transport from the ER to the Golgi complex has been visualized in living cells using a chimera of the temperature-sensitive glycoprotein of vesicular stomatitis virus and green fluorescent protein (ts-G-GFP[ct]). Upon shifting to permissive temperature, ts-G-GFP(ct) concentrates into COPII-positive structures close to ER, which then build up to form an intermediate compartment or transport complex, containing ERGIC-53 and the KDEL receptor, where COPII is replaced by COPI. These structures appear heterogenous and move in a microtubule-dependent manner toward the Golgi complex. These results suggest a sequential mode of COPII and COPI action and indicate that the transport complexes are ER-to-Golgi transport intermediates from which COPI may be involved in recycling material to the ER (Scales et al. 2000).

3.1.1.6 Clathrin Coated Vesicles (CCV)

Clathrin coats contain both clathrin and clathrin adaptor proteins. While clathrin acts as the scaffold, the clathrin adaptors bind to protein and lipid cargo. Specific cargos are recruited into clathrin-coated vesicles with the aid of CLASP proteins (clathrin-associated sorting proteins), such as ARH and Dab2. Clathrin-associated protein complexes are believed to interact with the cytoplasmic tails of membrane proteins, leading to their selection and concentration. At least 20 clathrin adaptors have been identified, which share a common design composed of a compact domain plus a long unstructured region that binds the clathrin beta-propeller. The two major types of clathrin adaptor complexes are: heterotetrameric adaptor protein

(AP) complexes, and the monomeric GGA (Golgi-localising, γ-adaptin ear domain homology, ARF-binding proteins) adaptors. Whereas clathrin heavy chain provides the structural backbone of the clathrin coat, it was suggested that clathrin light chains (CLCs) are not required for clathrin-mediated endocytosis but are critical for clathrin-mediated trafficking between the TGN and the endosomal system. In CLC-deficient mice CI-MPRs cluster near the TGN leading to a delay in processing of the lysosomal cathepsin D. In mammalian cells CLCs function in intracellular membrane trafficking by acting as recruitment proteins for huntingtin-interacting protein 1-related (HIP1R), enabling HIP1R to regulate actin assembly on clathrin-coated structures (Poupon et al. 2008).

3.1.1.7 Adaptor Protein Complexes

Adaptor protein (AP) complexes are found in coated vesicles and clathrin-coated pits. AP complexes connect cargo proteins and lipids to clathrin at vesicle budding sites, as well as binding accessory proteins that regulate coat assembly and disassembly. There are different AP complexes in mammals. AP1 is responsible for the transport of lysosomal hydrolases between the TGN and endosomes. AP2 associates with the plasma membrane and is responsible for endocytosis. The AP-1 and AP-2 complexes are the most abundant adaptors in CCVs, but clathrin-mediated trafficking can still occur in the absence of any detectable AP-1 or AP-2. AP3 is responsible for protein trafficking to lysosomes and other related organelles. AP4 is less well characterised. AP-4 is localized mainly in the Golgi complex, as well as on endosomes and transport vesicles. Mammary epithelial cells contain an unexpectedly high quantity of clathrin coated vesicles. Analysis of CCV adaptor composition showed that approximately 5–10% of total APs consist of AP-2 in mammary gland CCV whereas it represents approximately 70% of the total APs from bovine brain CCV. Relatively high quantities of furin and CI-MPR were detected in mammary CCV. AP-1 and the CI-MPR were localized in Golgi-associated vesicles and on the membrane of secretory vesicles. CCV in lactating mammary epithelial cells are involved in the transcytotic pathway, in sorting at the TGN and in the biogenesis of casein-containing secretory vesicles (Pauloin et al. 1999).

3.1.1.8 Lysosomes and Role of Mannose 6-Phosphate

The main function of lysosomes in the cell is the degradation of internalized material by lysosomal enzymes, the acid hydrolases. These enzymes are synthesized in the ER and during their maturation in the Golgi apparatus they acquire the mannose 6-phosphate (M6P) recognition marker. Most of these soluble hydrolases are transported to lysosomes through specific M6P receptors. The receptor-

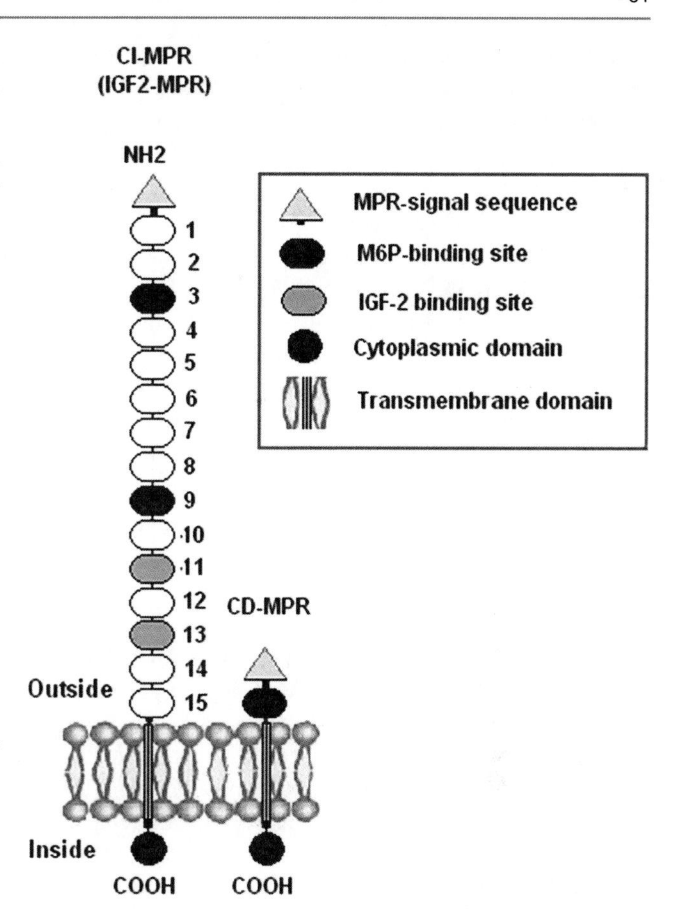

Fig. 3.2 Schematic diagram of the full-length MPRs. The MPRs are type I transmembrane glycoproteins that consist of an N-terminal (NH2) signal sequence, an extracytoplasmic region, a single transmembrane domain, and a C-terminal (COOH) cytoplasmic tail (Hancock et al. 2002)

enzyme complexes segregated in TGN are transferred to the lysosomal compartments *via* the secretory pathway through clathrin-coated vesicles.In mammalian cells, the targeting of newly synthesized hydrolases, as well as others ligands with M6P residues on their N-linked oligosaccharides, to the lysosomes depends on their recognition by two specific M6P receptors at the TGN: the cation-dependent M6P receptor (CD-MPR) and the cation-independent M6P receptor (CI-MPR), also called mannose 6-phosphate/insulin-like growth factor II receptor in eutherian mammals, in which it binds IGF-II (Kornfeld 1992; Varki and Kornfeld 2009) (Fig. 3.2). These two receptors (M6PRs) are considered sorting receptors because of their routing function. In the pre-lysosomal compartment, acidity induces release of enzymes from both MPR, which are then recycled into the Golgi apparatus. However, the CI-MPR, which participates to this cellular routing, is also anchored to the cell surface membrane and can internalize extracellular ligands (Kornfeld 1992; Munier-Lehmann et al. 1996; Dahms 1996). Thus, the pool of lysosomal acid hydrolases comes both from in situ synthesis, which involves the two

M6P receptors, and from endocytosis through the CI-MPR (Ni et al. 2006; Nykjaer et al. 1998). It is now well accepted that the TGN is the major site where proteins are sorted from the biosynthetic pathway for efficient delivery to endosomes. There, the MPRs and their bound ligands are segregated into nascent clathrin-coated vesicles, probably together with other transmembrane proteins destined to the endosomes/lysosomes. After budding and uncoating, these Golgi-derived vesicles fuse with endosomal compartments where the MPRs discharge their bound ligands. While the soluble lysosomal enzymes are directed toward the lysosomes, the MPRs recycle back to the TGN (Duncan and Kornfeld 1988) or to the cell surface where they are found in small amounts at steady state. At the plasma membrane, the MPRs undergo endocytosis via clathrin-coated pits like many other cell surface receptors (Fig. 3.1).

3.2 P-Type Lectin Family: The Mannose 6-Phosphate Receptors

The best-characterized function of two MPRs, CI-MPR and CD-MPR of 46 kDa and 300 kDa respectively, is their ability to direct the delivery of approximately 60 different newly synthesized soluble lysosomal enzymes bearing M6P on their N-linked oligosaccharides to lysosome. The CI-MPR is a multifunctional protein which binds at the cell surface to two distinct classes of ligands, the M6P bearing proteins and IGF-II. In addition to its intracellular role in lysosome biogenesis, the CI-MPR, but not the CD-MPR, participates in a number of other biological processes by interacting with various molecules at the cell surface. Though, the major function of CI-MPR is to bind and transport M6P-enzymes to lysosomes, but it can also modulate the activity of a variety of extracellular M6P-glycoproteins (i.e., latent TGFβ precursor, urokinase-type plasminogen activator receptor, Granzyme B, growth factors, Herpes virus). The synthesis and potential use of high affinity M6P analogues able to target this receptor have been described. Several M6P analogues with phosphonate, carboxylate or malonate groups display a higher affinity and a stronger stability in human serum than M6P itself. These derivatives can be used to favour the delivery of specific therapeutic compounds to lysosomes, notably in enzyme replacement therapies of lysosomal diseases or in neoplastic drug targeting (Gary-Bobo et al. 2007). In addition, their potential applications in preventing clinical disorders, which are associated with the activities of other M6P-proteins involved in wound healing, cell growth or viral infection, have been discussed.The list of extracellular ligands recognized by this multifunctional receptor has grown to include a diverse spectrum of M6P-containing proteins as well as several non-M6P-containing ligands. Structural studies have shown how these two receptors use related, but yet distinct, approaches in the recognition of phosphomannosyl residues (Dahms et al. 2008).

The two receptors, which share sequence similarities, constitute the P-type family of animal lectins. The CD-MPR (46 kDa) and the CI-MPR (300 kDa) are ubiquitously expressed throughout the animal kingdom and are distinguished from all other lectins by their ability to recognize phosphorylated mannose residues (Fig. 3.2). The P-type lectins play an essential role in the generation of functional lysosomes within the cells of higher eukaryotes by directing newly synthesized lysosomal enzymes bearing M6P signal to lysosomes. At the cell surface, the IGF2R/CI-MPR also binds to the nonglycosylated polypeptide hormone, IGF-II, targeting this potent mitogenic factor for degradation in lysosomes. The two MPRs have overlapping function in intracellular targeting of newly synthesized lysosomal proteins, but both are required for efficient targeting. Their main function is to transport lysosomal enzymes from TGN to the pre-lysosomal compartment. MPRs are conserved in the vertebrates from fish to mammals and show non-identical distribution among sub-cellular fractions in liver (Dahms and Hancock 2002; Messner et al. 1989; Messner 1993). Although much has been learned about the MPRs, it is unclear how these receptors interact with the highly diverse population of lysosomal enzymes. It is known that the terminal M6P is essential for receptor binding. Mannose receptor-enriched membranes of liver sinusoidal cells contain significant levels of CD-MPR, but not the CI-MPR. Both CD-MPRs and CI-MPRs bind their M6P-tagged cargo in the lumen of the Golgi apparatus in the cell. The CD-MPR shows greatly enhanced binding to M6P in the presence of divalent cations, such as manganese. M6P-containing proteins can be purified on immobilized MPRs. The sequences of zebrafish (*Danio rerio*) CD-MPR and CI-MPR (Nolan et al. 2006) indicate that targeting of lysosomal enzymes by MPRs is an ancient pathway in vertebrate cell biology. Yadavalli and Nadimpalli (2008) reported putative MPR receptors from starfish. Structural comparison of starfish receptor sequences with other vertebrate receptors gave structural homology with the vertebrate MPR-46 protein. The expressed protein in mpr$^{-/-}$ mouse embryonic fibroblast cells efficiently sorts lysosomal enzymes within the cells establishing a functional role for this protein (Yadavalli and Nadimpalli 2008). The insect cell line *Sf9* infected with a recombinant baculovirus containing the gene for human prorenin, cultured in presence of ^{3}H-mannose do not synthesize high-mannose-type oligosaccharides containing M6P, and consequently it appears unlikely that these cells utilize the MPR mediated pathway for targeting of lysosomal enzymes (Aeed and Elhammer 1994). Recently P-type CRD-like domains have been found in proteins with different architectures to the MPRs, and have been termed MPR homology (MRH) domains. Some of the MRH domains in non-MPR proteins are known to have sugar-binding activity, and glycan recognition may be a general function of the MRH domain.

3.2.1 Fibroblasts MPRs

The turnover of the phosphomannosyl receptor in fibroblasts is very slow, in contrast with its rate of internalization in endocytosis, and that its rate of degradation is not greatly altered by a variety of agents that affect lysosomal protein turnover and/or receptor-mediated endocytosis (Creek and Sly 1983). The MPR, on the surface of fibroblasts, accounts for the intracellular transport of newly synthesized enzymes to the lysosome. Fibroblasts MPRs internalized oligosaccharides of known specific activity bearing a single phosphate in monoester linkage with K_{uptake} of 3.2×10^{-7} M, whereas oligosaccha- rides bearing two phosphates in monoester linkage were internalized with a K_{uptake} of 3.9×10^{-8} M (Natowicz et al. 1983).

3.2.2 MPRs in Liver

During perinatal development, the CI-MPR expression decreases progressively from 18-day fetuses to adults, whereas the CD-MPR showed a transient decrease in newborn and at the fifth day after birth in rats. Both receptors localize to hepatocytes at all ages and, additionally, the CD-MPR was reactive in megakaryocytes at early stages. In adult rat liver, CIMPR is detected intensely in hepatocytes and weakly in sinusoidal Kupffer cells and interstitial cells in Glisson's capsule. A high level of expression of CI-MPR mRNAs in hepatocytes and of CD-MPR mRNA in Kupffer cells was detected by in situ hybridization. Differential changes during perinatal development and adults suggest that two MPRs play distinct roles during organ maturation (Romano et al. 2006; Waguri et al. 2001). It was found that the activity of glycosidases changes during development, reaching a peak at the tenth day after birth and correlated with the expression and binding properties of CD-MPR. It was suggested that lysosome maturation in rat liver occurs around tenth day after birth, and that the CD-MPR may participate in that event. In hepatocytes of MPR-deficient neonatal mice, lysosomal storage occurs when both MPRs are lacking, whereas deficiency of CI-MPR only has no effect on the ultrastructure of the lysosomal system (Schellens et al. 2003). A biochemical comparison between autophagosomes and amphisomes from rat liver showed that the amphisomes were enriched in early endosome markers [the asialoglycoprotein receptor and the early endosome-associated protein as well as in a late endosome marker (CI-MPR)]. Amphisomes would thus seem to be capable of receiving inputs both from early and late endosomes (Berg et al. 1998). In human HepG2 and BHK cells, the two receptors were identified at the same sites: the trans-Golgi reticulum (TGR), endosomes, electron-dense cytoplasmic vesicles, and the plasma membrane. It was suggested that the two MPRs exit TGR via same coated vesicles. However, on arrival in the endosomes CD-MPR is more rapidly than CI-MPR segregated into associated tubules and vesicles (ATV) which probably are destined to recycle MPRs to TGR (Klumperman et al. 1993).

3.2.3 MPRs in CNS

The two related but distinct MPRs have been localized in neurons of mouse CNS, with more intense labeling in the medial septal nucleus, the nucleus of the Broca's diagonal band, layers IV-VI of the cerebral neocortex, layers II-III of the entorhinal cortex, the habenular nucleus, the median eminence, several nuclei and structures of the brainstem, the Purkinje cell layer of the cerebellum, and in the ventral horn of the spinal cord. Although intense reactivities of both MPRs were observed in the same groups of neurons in the same regions, the spatial differences in immunoreactive intensity for CI-MPR were greater, particularly in the telencephalon such as the basal forebrain and cerebral cortex, than those for CD-MPR. While CD-MPR is ubiquitously necessary for the general function of neurons, the CI-MPR is selectively necessary for certain region- and neurotransmitter-specific functions of neurons (Konishi et al. 2005). In rat brain during perinatal development, the expression of CI-MPR decreases progressively from fetuses to adults, where as CD-MPR increases around the tenth day of birth, and maintained there after. This shows that the two receptors play a different role in rat brain during perinatal development; CD-MPR being mostly involved in lysosome maturation (Romano et al. 2005). Given the critical role of endosomal-lysosomal (EL) system in the clearance of abnormal proteins, it is likely that the increase in the CI-MPR and components of the EL system in surviving neurons after 192-IgG-saporin represents an adaptive mechanism to restore the metabolic/structural abnormalities induced by the loss of cholin-ergic neurons (Hawkes et al. 2006).

A lysosomal enzyme binding receptor protein from monkey brain shows protein kinase activity and undergoes phosphorylation on serine and tyrosine residues. The lysosomal enzyme fucosidase and M6P, which are ligands for the receptor, stimulated the activity of protein phosphatase associated with the receptor protein. A phosphorylation/dephosphorylation mechanism may be operative in the ligand binding and functions of the receptor (Panneerselvam and Balasubramanian 1993).

The prion encephalopathies are characterized by accumulation of the abnormal form PrPsc of a normal host gene product PrPc in the brain. In search of the mechanism and site of formation of PrPsc from PrPc in ME7 scrapie-infected mouse brain, it was found that proteinase K-resistant PrPsc is enriched in subcellular structures which contain CI-MPR, ubiquitin-protein conjugates, β-glucuronidase, and cathepsin B, termed late endosome-like organelles. The organelles may act as chambers for the conversion of PrPc into infectious PrPsc in murine model of scrapie (Arnold et al. 1995).

In neuroendocrine cells sorting of proteins from immature secretory granules (ISGs) occurs during maturation and is achieved by CCV containing AP-1. The MPRs are detected in ISGs of PC12 cells and more than 80% of the ISGs contained furin. Fifty percentage at most of the ISGs contained CI-MPR. Dittié et al. (1999) suggested the presence of two populations of ISGs: those that have both MPRs and furin, and those which contain only furin. It was shown that binding of adapter protein-1 (AP-1) requires casein kinase II phosphorylation of CI-MPRfusion protein, and in particular phosphorylation of Ser-2474.

The β-amyloid deposits in the brains of all patients of Alzheimer's disease (AD). Stephens and Austen (1996) defined major location of β-amyloid precursor protein fragments possessing the Asp-1 N-terminus of β-amyloid as the TGN or late endosome on the basis of colocalisation with a mAb to the CI-MPR. The co-localisation suggested that the p13 fragment and MPR are trafficked by alternative pathways from TGN. Hawkes and Kar (2004) delineated the role of the CI-MPR in the CNS, including its distribution, possible importance as well as its implications in neurodegenerative disorders such as AD.

3.2.4 CI-MPR in Bone Cells

The osteoclast is a polarized cell which secretes large amounts of newly synthesized lysosomal enzymes into an apical extracellular lacuna where bone resorption takes place. Osteoclast expresses large amounts of immunoreactive CI-MPR, despite the fact that most of the lysosomal enzymes it synthesizes are secreted. In osteoclast, M6P receptors are involved in the vectorial transport and targeting of newly synthesized lysosomal enzymes, presumably via a constitutive pathway, to the apical membrane where they are secreted into the bone-resorbing compartment. This mechanism could insure polarized secretion of lysosomal enzymes into the bone-resorbing lacuna (Baron 1989). The rapid inhibition of bone resorption by calcitonin involves the vesicular translocation of the apical membranes and the rapid arrest in the synthesis and secretion of lysosomal enzymes in osteoclasts (Baron et al. 1990). IGF2R/CI-MPR is present in rat calvarial osteoblasts. Osteoblasts bind IGF-II with high affinity (K_D ~ 2.0 nM). The osteoblastic Ca^{2+} response to IGF-II is caused by an intracellular release of Ca^{2+} which is mediated by the IGF-II/CI-MPR (Martinez et al. 1995). The phosphorylated monosaccharide, M6P stimulates alkaline phosphatase produced by osteoblasts. Glucose-6-phosphate and fructose-1-phosphate also stimulated osteoblast alkaline phosphatase production, but not to the same extent as M6P. Since, the stimulatory effect of M6P is similar to that of IGF-II, it supports similar mechanism for signal transduction for both IGF-II and M6P (Ishibe et al. 1991).

Secretory ameloblasts possesses strong immunoreactivity for MPR in the supranuclear Golgi region and in the cytoplasm between the Golgi region and the distal junctional complexes, where as cathepsin B immunoreactivity was mainly seen in the distal portion of Tomes' process, which was unreactive for MPR immunogenicity. Since MPR and lysosomal enzymes were also detected on the ruffled border of osteoclasts adjacent to alveolar bone, report provides strong evidence for a similarity between the maturation process in enamel, as mediated by ameloblasts, and bone resorption mediated by osteoclasts (Al Kawas et al. 1996).

3.2.5 Thyroid Follicle Cells

Thyroglobulin (Tg), the major secretory product of thyrocytes, is the macromolecular precursor of thyroid hormones. The Tg has been shown to be phosphorylated and to carry M6P signal in terminal position. In porcine thyroid follicle cells, the CI-MPR is primarily located in elements of the endocytic pathway such as coated pits and endosomes. This localization of the CI-MPR in thyrocytes differs from the receptor sites in other cell types by the rare occurrence of CI-MPR in cisternae of the Golgi complex. The CI-MPR in thyrocytes might be unable to bind and to convey Tg efficiently. The receptor is, however, a binding site for Tg at the apical plasma membrane and may, therefore, be involved in the binding of Tg and its transfer from the follicle lumen to lysosomes (Lemansky and Herzog 1992; Scheel and Herzog 1989). Using antibodies against Tg and CI-MPR, Kostrouch et al. (1991) suggested three types of endocytic structures: those slightly positive for MPR and ArS-A, those strongly positive for both markers, and those only positive for ArS-A. These compartments exhibited the properties of early endosomes (EE), late endosomes (LE), and lysosomes (L), respectively. The data indicate that internalized Tg molecules are transported to EE and then transferred from EE to LE.

3.2.6 Testis and Sperm

M6P receptors have been isolated from germ cells and Sertoli cells present in testes. Isolated mouse pachytene spermatocytes and round spermatids synthesize predominantly the 46 kDa CD-MPR and only low levels of the 270 kDa CI-MPR. In contrast, Sertoli cells synthesized substantial amounts of the CI-MPR, but little of CD-MPR. Like germ cells, Sertoli cells in primary culture endocytosed ^{125}I-M6P-bearing ligands at levels that were about 10% of the endocytic activity measured for 3T3 fibroblasts. This indicates that both spermatogenic and Sertoli cells have surface MPRs capable of mediating endocytosis (O'Brien et al. 1989, 1993). Tsuruta and O'Brien (1995) and Tsuruta et al. (2000) provided evidence that IGF-II/CI-MPR ligands secreted by Sertoli cells can modulate gene expression in spermatogenic cells and strongly suggest that they are important in the regulation of spermatogenesis. Moreno (2003) studied the dynamics of some components of the endosome/lysosome system, as a way to understand the complex membrane trafficking circuit established during spermatogenesis and suggested that the CI-MPR could be involved in membrane trafficking and/or acrosomal shaping during spermiogenesis.

A single CI-MPR transcript, approximately 10 kb in size, was present in mouse spermatogenic and Sertoli cells. Like the CI-MPR protein, its mRNA transcript was more abundant in Sertoli cells than in spermatogenic cells from adult testes. The CD-MPR was the predominant MPR synthesized by pachytene spermatocytes or round spermatids. Multiple CD-MPR transcripts were detected in these cells, including a 2.4-kb CD-MPR mRNA that was indistinguishable from CD-MPR transcripts in somatic tissues and Sertoli cells. Results suggested that alternate polyadenylation signals are used to produce multiple CD-MPR transcripts in spermatogenic cells (O'Brien et al. 1994). Low molecular weight M6P-receptors from bovine testis exhibited two isoforms with Mr of 45,000 (MPR-2A) and 41,000 (MPR-2B). These isoforms contain a common polypeptide core, but differ in their carbohydrate content (Li and Jourdian 1991). Two (pro)renin receptors have been characterized so far, the MPR and a specific receptor called (P)RR for (pro)renin receptor. Each receptor controls a different aspect of renin and prorenin metabolism. The MPR is a clearance receptor, whereas (P)RR mediates their cellular effects by activating intracellular signaling and up-regulating gene expression.

Belmonte et al. (1998) demonstratesd that α-mannosidase from rat epididymal fluid is a ligand for phosphomannosyl receptors on the sperm surface. Evidence is also presented that the CI-phosphomannosyl receptors are responsible for the interaction with alpha-mannosidase. These findings suggest a new role for extracellular transport mediated by the M6P receptor. Both MPRs undergo changes in distribution as spermatozoa passed from rete testis to cauda epididymis. CI-MPR was concentrated in the dorsal region of the head in rete testis sperm and that this labeling extended to the equatorial segment of epididymal spermatozoa. CD-MPR, however, changed from a dorsal distribution in rete testis, caput, and corpus to a double labeling on the dorsal and ventral regions in cauda spermatozoa; staining for either CI-MPR or CD-MPR increased from rete testis to epididymis. Changes in MPRs distribution may be related to a maturation process, which suggests new roles for the phosphomannosyl receptors (Belmonte et al. 2000). The targeted disruption of either MPR does not result in decreased acrosomal targeting efficiency (Chayko and Orgebin-Crist 2000).

3.2.7 MPRs During Embryogenesis

The MPR46 showed high expression at the sites of hemopoiesis and in the thymus while MPR300 was highly expressed in the cardiovascular system. Late in embryogenesis (day 17.5) a wide variety of tissues expressed the receptors, but still the expression pattern was almost non-overlapping. This unexpected spatially and temporally expression pattern points to specific functions of the two MPRs during mouse embryogenesis (Matzner et al. 1992).

3.3 Cation-Dependent Mannose 6-Phosphate Receptor

3.3.1 CD-MPR- An Overview

The cation-dependent mannose 6-phosphate receptor (or CD-MPR) is one of two proteins that bind M6P tags on acid hydrolase precursors in the Golgi apparatus that are destined for transport to the endosomal-lysosomal system. The CD-MPR recognizes the phosphomannosyl recognition marker of lysosomal enzymes. Homologues of CD-MPR are found in all eukaryotes. The CD-MPR is a type I transmembrane protein with a single transmembrane domain. The extracytoplasmic/lumenal M6P binding-domain consists of 157 amino acid residues. The bovine CD-MPR is composed of a 28-residue amino-terminal signal sequence, a 159-residue extracytoplasmic region, a 25-residue transmembrane region, and a 67-residue carboxyl-terminal cytoplasmic domain. The extracytoplasmic region of the CD-MPR contains 6 cysteine residues that are involved in the formation of three intramolecular disulfide bonds that play an essential role in the folding of the receptor (Wendland et al. 1991) (Fig. 3.2).

The CD-MPR from P388D1 macrophages lacks 215-kDa MPR. An identical protein was purified from bovine liver. The MPR binds efficiently to phosphomannosyl monoester-containing ligands in presence of MnCl2. The receptor contains both high mannose (or hybrid)- and complex-type oligosaccharide units on the basis of sensitivity to digestion with endo-beta-N-acetylglucosaminidase H and endo-β-N-acetylglucosaminidase F. The 46-kDa CD-MPR and the 215-kDa CI-MPR not only differ in their properties but are also immunologically distinct (Hoflack and Kornfeld 1985).

The receptor from human liver has a subunit molecular size of 43 kDa. It is rich in hydrophobic and charged amino acids and contains threonine at the N-terminus. The receptors from human and rat liver are antigenically related. Both are immunologically distinct from the CI-MPR of 215-kDa from human liver. The CD receptor exists in solution as a dimer or tetramer (Dahms and Hancock 2002). Modification of arginine and histidine residues reduced the binding of the receptor to immobilized ligands. Presence of M6P during modification of arginine residues protected the binding properties of the receptor, suggesting that arginine is a constituent of the M6P binding site of the receptor (Stein et al. 1987c). PC12 cells express CI-MPR, but not CD-MPR as much. The CD-MPR preferentially transports cathepsin B in PC12 cells, and cathepsins B and D participate in the regulation of PC12 cell apoptosis (Kanamori et al. 1998).

The cDNA clones encoding entire sequence of bovine 46-kDa CD-MPR, in *Xenopus laevis* oocytes results in a protein that binds specifically to phosphomannan-Sepharose and a deduced 279 amino acid sequence reveals a single polypeptide chain that contains a putative signal sequence and a transmembrane domain. The microsomal membranes containing the receptor and the location of the five potential N-linked glycosylation sites indicate that the receptor is a transmembrane protein with an extracytoplasmic amino terminus. This extracytoplasmic domain is homologous to the approximately 145 amino acid long repeating domains present in the 215-kDa CI-MPR (Dahms et al. 1987). The full-length cDNA for the goat CD-MPR46 protein was expressed in MPR deficient cells. It exhibits oligomeric nature as observed in the other species. The binding and sorting functions of the expressed protein to sort cathepsin D to lysosomes were similar to natural protein (Poupon et al. 2007). 46-kDa MPR mediates transport of endogenous but not endocytosis of exogenous lysosomal enzymes. Internalization of receptor antibodies indicated that the failure to mediate endocytosis of lysosomal enzymes is due to an inability of surface 46-kDa MPR to bind ligands rather than its exclusion from the plasma membrane or from internalization (Stein et al. 1987a, b).

3.3.2 Human CD-MPR

c-DNA clones for the human CD-MPR from a human placenta encoding the nucleotide sequence of the 2463-bp cDNA insert includes a 145-bp 5′ untranslated region, an ORF of 831 bp corresponding to 277 amino acids (Mr = 30,993), and a 1487-bp 3′ untranslated region. The deduced amino acid sequence is colinear with that determined by amino acid sequencing of the N-terminus peptide (41 residues) and nine tryptic peptides (93 additional residues). The receptor is synthesized as a precursor with a signal peptide of 20 amino acids. The hydrophobicity profile of the receptor indicates a single membrane-spanning domain, which separates an N-terminal region containing five potential N-glycosylation sites from a C-terminal region lacking N-glycosylation sites. Thus the N-terminal (Mr = 18,299) and C-terminal (Mr less than or equal to 7,648) segments of the mature receptor are assumed to be exposed to the extracytosolic and cytosolic sides of the membrane, respectively. The gene for the receptor is located on human Chromosome 12 *p13* (Ghosh et al. 2003).

The human MPR46 gene is distributed over 12 kb and divided into seven exons (110–1573 bp). All the intron/exon borders agree with the consensus sequences of splice junctions. Exon 1 codes for a 5′ untranslated sequence. The ATG initiation codon begins with the second nucleotide in exon 2. A signal sequence of 26 amino acid residues is followed by the extracytoplasmic (luminal) domain, which extends to exon 5. The transmembrane domain of the receptor spans exons 5 and 6 and the cytoplasmic domain is encoded by exons 6 and 7. The latter domain also codes for an extended 3′ untranslated sequence. The transcription-initiation site was defined by primer extension. The sequence upstream of the cap site has strong promoter activity and contains structural elements characteristic of promoters found in housekeeping genes. No correlation between the genomic organization and known protein domains of the MPR46 was apparent (Klier et al. 1991). Moreover, the sequence of about 150 amino acids within the luminal domain of MPR46, which is homologous to the 15 repeats that constitute the luminal domain of the MPR300, does not correlate with intron/exon borders. MPR46 and MPR300 have therefore diverged from a common ancestral gene before introduction of the present intron sequences.

3.3.3 Mouse CD-MPR

A cDNA clone for mouse MPR revealed a single open reading frame that codes for a protein of 278 residues. It shows an over-all amino-acid identity of 93% with the human receptor. Nine non-conservative amino-acid exchanges are found in the luminal domain, one non-conservative exchange of hydrophobic amino acids is in the transmembrane domain, while the cytoplasmic receptor tails are identical. All five potential N-glycosylation sites are conserved as well as amino acids that are important for ligand binding (Arg^{137} and His^{131}) and disulfide pairing (Cys^{32} and Cys^{78}, Cys^{132} and Cys^{167}, Cys^{145} and Cys^{179}). The absolute identity in the cytoplasmic MPR46 tail suggests the importance of this amino-acid sequence for the intracellular routing of the MPR46 (Köster et al. 1991). The 278-amino acid sequence deduced from the cDNA for the murine MPR46 shows 19 amino acid differences from that of the human MPR46, none of which are found in the 68-amino acid cytoplasmic tail. Binding of ligand to the murine MPR46 in permeabilized cells showed a pH optimum of 6.5, was completely inhibited by M6P, and was stimulated by divalent cations. Mn^{2+} was more effective than Ca^{2+} or Mg^{2+}. Endocytosis was demonstrated at pH 6.5 and was stimulated four- to sevenfold by Mn^{2+}. In its responsiveness to divalent cations and its preference for Mn^{2+}, the murine 46MPR resembled the bovine 46MPR more than the human 46MPR. It was no more efficient than the human 46MPR in correcting the sorting defect of IGF-IIR/MPR-deficient mouse L cells (Ma et al. 1991).

3.3.3.1 Pseudogene of Mouse CD-MPR

Ludwig et al. (1992) cloned the mouse CD-MPR gene and also a very unusual processed-type CD-MPR pseudogene. Both are present at one copy per haploid genome and map to chromosomes 6 and 3, respectively. Comparison of the complete 10-kb sequence of the functional gene with the cDNA indicates that it contains seven exons. Exon 1 encodes the 5′-untranslated region of the mRNA, the others (exons 2–7) encode the luminal, transmembrane, and cytoplasmic domains of the CD-MPR. Exon 7 also contains a 1.2-kb-long 3′-untranslated region of the mRNA. A unique transcription-initiation site was determined by primer extension of mouse liver mRNA. The promoter elements in the 5′ upstream region of this site resemble those contained in genes constitutively transcribed. However, Northern blot analysis demonstrates that the CD-MPR is variably expressed in adult mouse tissues and during mouse development. The pseudogene, which is flanked by direct repeats, is almost colinear with the cDNA indicating that it presumably arose by reverse transcription of an mRNA. However, the pseudogene differs from the cDNA. It contains at its 5′ end, an additional 340-nucleotide (nt) sequence homologous to the promoter region of the functional gene. This sequence exhibits some promoter activity in vitro. Furthermore, a 24-nt insertion interrupts the region homologous to the 5′-noncoding region of the cDNA. In the functional gene, this 24-nt sequence occurs between exon 1 and 2, where it is flanked by typical consensus sequences of exon/intron boundaries. Therefore, it may represent an additional exon of the functional gene. These two features of the pseudogene suggest that expression of the CD-MPR gene may be regulated by use of different promoters and/or alternative splicing.

3.4 Structural Insights

3.4.1 N-Glycosylation Sites in CD-MPR

The bovine CD-MPR contains five potential N-linked glycosylation sites, four of which are utilized. CD-MPR mutants lacking various potential glycosylation sites showed that the presence of a single oligosaccharide chain, particularly at position 87 significantly enhanced its M6P-binding ability when compared with non-glycosylated receptors. It was suggested that N-glycosylation of the bovine CD-MPR facilitates the folding of the nascent polypeptide chain into a conformation that is conducive for intracellular transport and ligand binding (Zhang and Dahms 1993). A soluble truncated form CD-MPR encoding only the extracytoplasmic region, Stop155, and a truncated glycosylation-deficient form of the CD-MPR, Asn81/Stop155, which has been modified to contain only one N-linked glycosylation site at position 81 instead of five, were purified from baculovirus-infected High Five insect cells. The extracellular region of the CD-MPR is sufficient for high-affinity binding and that oligosaccharides at positions 31, 57, and 87 do not influence ligand binding (Marron-Terada et al. 1998). The recombinant insect-produced CD-MPR existed as a dimer in the membrane. The cytoplasmic domains of the MPRs are sufficient to determine the steady-state distribution of the full-length proteins (Dahms and Hancock 2002; Mauxion et al. 1995).

Mammalian cell lysosomal enzymes or phosphorylated oligosaccharides derived from them are endocytosed by a MPR found on the surface of fibroblasts. Studies suggest that two residues of M6P in phosphomonoester linkage but not diester linkage (PDE) are essential for a high rate of uptake. The lysosomal enzymes of the slime mold *Dictyostelium discoideum* are also recognized by the MPR on these cells; however, none of the oligosaccharides from these enzymes contain two phosphomonoesters. Instead, most contain multiple sulfate esters and two residues of M6P in an unusual PDE linkage. Further study shows that nearly all of the α-mannosidase molecules contain the oligosaccharides required for uptake, and that each tetrameric, holoenzyme molecule has sufficient carbohydrate for an average of 10 Man8GlcNAc2 oligosaccharides. Results suggest that the interactions of multiple, weakly binding oligosaccharides, especially those with 2 PDE, are important for high rate of uptake of the slime mold enzymes. The conformation of the protein may be important in orienting the oligosaccharides in a favorable position for binding to MPR (Freeze 1985).

3.4.2 3-D Structure of CD-MPR

Roberts et al. (1998) reported the three-dimensional structure of a glycosylation-deficient, yet fully functional form of the extracytoplasmic domain of the bovine CD-MPR (residues 3–154) complexed with M6P at 1.8 A resolution. The extracytoplasmic domain of the CD-MPR crystallizes as a dimer, and each monomer folds into a nine-stranded flattened beta barrel, which bears a striking resemblance to avidin (Fig. 3.3). The distance of 40 A between the two ligand-binding sites of the dimer provides a structural basis for the observed differences in binding affinity exhibited by the CD-MPR toward various lysosomal enzymes.

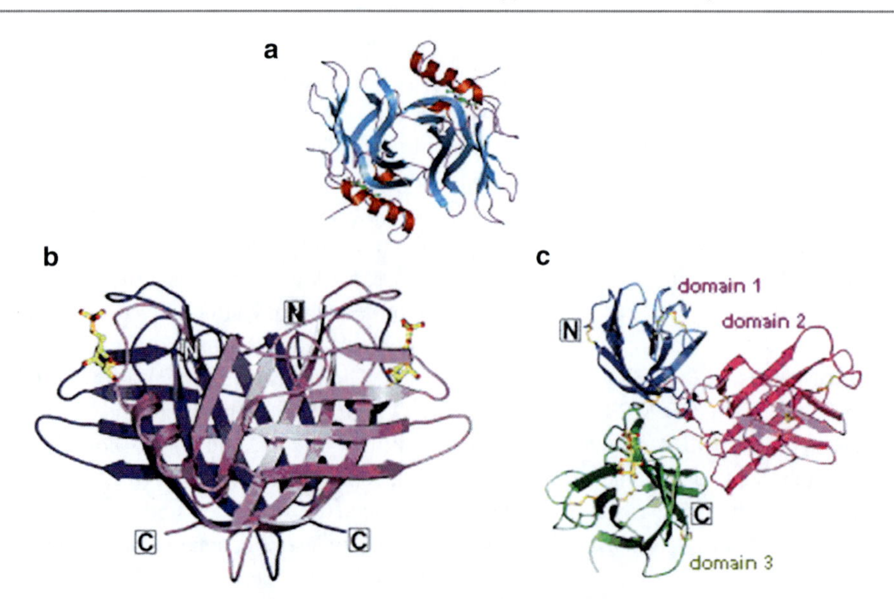

Fig. 3.3 Crystal structure of CD-MPR (46 kDa in humans) (**a**) (PDB ID: 1KEO). (**b**) Crystal structure of the extracytoplasmic region (residues 3–154) of the bovine CD-MPR in the presence of an oligosaccharide, pentamannosyl phosphate (PDB 1 C39). Note that only the terminal Man-6-P (*gold* ball-and-stick model) is shown for clarity. Both monomers (*light purple* and *dark purple*) of the CD-MPR dimer are shown in this ribbon diagram. The N-terminus (N) and C-terminus (C) are boxed. (**c**) Crystal structure of the N-terminal three domains (residues 7–432) of the bovine CI-MPR (PDB 1SZO). The N- and C-terminus of the protein encoding domain 1 (*blue*), domain 2 (*pink*), and domain 3 (*green*) are indicated. The location of Man-6-P (*gold* ball-and-stick model) is shown (Reprinted with permission from Dahms et al. 2008 © Oxford University Press)

Studies using synthetic oligosaccharides indicated that the binding site encompasses at least two sugars of the oligosaccharide. Olson et al. (1999b) reported the structure of the soluble extracytoplasmic domain of a glycosylation-deficient form of the bovine CD-MPR complexed to pentamannosyl phosphate. This construct consists of the amino-terminal 154 amino acids (excluding the signal sequence) with glutamine substituted for asparagine at positions 31, 57, 68, and 87. The binding site of the receptor encompasses the phosphate group plus three of the five mannose rings of pentamannosyl phosphate. Receptor specificity for mannose arises from protein contacts with the 2-hydroxyl on the terminal mannose ring adjacent to the phosphate group. Glycosidic linkage preference originates from the minimization of unfavorable interactions between the ligand and receptor.

Recent advances in the structural analyses of both CD-MPR and CI-MPR have revealed the structural basis for phosphomannosyl recognition by these receptors and provided insights into how the receptors load and unload their cargo. A surprising finding is that the CD-MPR is dynamic, with at least two stable quaternary states, the open (ligand-bound) and closed (ligand-free) conformations, similar to those of hemoglobin. Ligand binding stabilizes the open conformation; changes in the pH of the environment at the cell surface and in endosomal compartments weaken the ligand-receptor interaction and/or weaken the electrostatic interactions at the subunit interface, resulting in the closed conformation (Kim et al. 2009).

CD-MPR Adopts at Least Two Different Conformations

Crystallographic studies of CD-MPR have identified 11 amino acids within its carbohydrate binding pocket. Mutant receptors containing a single amino acid substitution toward a lysosomal enzyme showed that substitution of Gln^{66}, Arg^{111}, Glu^{133}, or Tyr^{143} results in a >800-fold decrease in affinity, suggesting that these four amino acids are essential for carbohydrate recognition by CD-MPR. Furthermore, Asp^{103} has been identified as the key residue which mediates the effects of divalent cations on the binding properties of the CD-MPR. The MPRs encounter a variety of conditions as they travel to various compartments where they bind and release their ligands. Key to their function is pH-dependence of ligand-protein interaction. Cells treated with reagents that raise the pH of endosomal/lysosomal compartments exhibit decreased sorting of lysosomal enzymes to lysosomes and a concomitant increase in the secretion of these enzymes into the medium (Imort et al. 1983). This observation implies that it is essential for MPRs to release their ligands in the acidic environment of endosomes in order to be able to recycle back to the TGN to retrieve additional lysosomal enzymes. To determine whether different pH conditions elicit conformational changes in the receptor that alters ligand binding affinities, CD-MPR structures were obtained under different conditions representing various environments encountered by the receptor: bound state at pH 6.5 and pH 7.4 and unbound state at pH 6.5 and pH 4.8 (Olson et al. 2002, 2008) (Fig. 3.4).

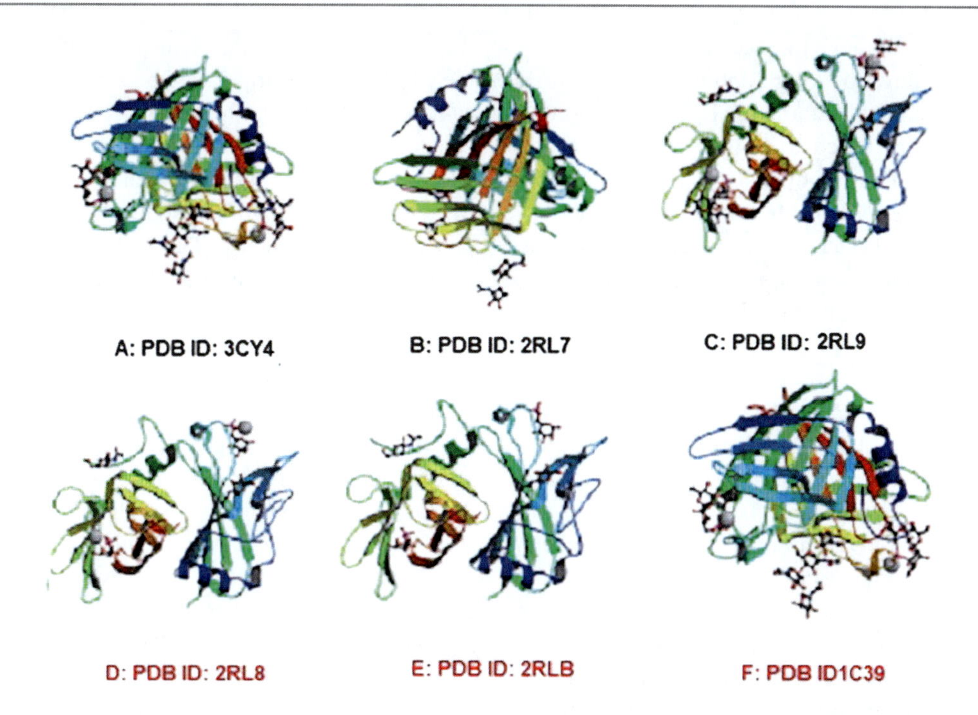

Fig. 3.4 **Crystal Structure of bovine CD-MPR.** (**a**) at pH 7.4 (Asymmetric Unit) (PDB ID: 3CY4 DOI:dx.doi.org); (**b**) at pH 4.8 (PDB ID: 2RL7); (**c**) at pH 6.5 bound to trimannoside (PDB ID: 2RL9 DOI:dx.doi.org); (**d**) at pH 6.5 bound to M6P (PDB ID: 2RL8 DOI:dx.doi.org); (**e**) at pH 6.5 bound to M6P in absence of Mn (PDB ID: 2RLB) DOI:dx.doi.org; (**f**) bound to pentamannosyl phosphate (PDB ID:1 C39)

These structures of CD-MPR were categorized into one of two conformations: an "open" conformation found in all structures containing ligand in the binding pocket and a "closed" conformation found in all structures missing bound carbohydrate (Dahms et al. 2008). Unlike what has been observed in other lectins, the structure of the ligand-free CD-MPR differs considerably from the ligand-bound form in that changes in both quaternary structure and positioning of loops involved in sugar binding, along with changes in the spacing of the two carbohydrate binding sites in the dimeric receptor (the Cα atoms of His^{105} located in Loop C are ~34 Å apart in the open conformation and ~26 Å apart in the closed conformation). Loop D (residues Glu^{134}–Cys^{141}) exhibits the most dramatic change in position, with Val^{138} displaying the largest displacement (Cα–Cα distance of 16 Å). The CD-MPR conformation differs dramatically from other lectins in an unbound state, where water molecules fill the shallow binding grooves of other most lectins in absence of bound sugar. Instead of essential side chain interactions being shifted from the carbohydrate hyroxyls to water, the pocket of CD-MPR undergoes restructuring: loop D swings into the binding pocket in the absence of ligand and provides contacts that hold essential residues in the proper orientation so that they are maintained in a "ready-state" to accept ligand. The two conformations also display a dramatic difference in their quaternary structure that can be described globally as a scissoring and twisting motion between the two subunits of the dimer. Results indicate that the CD-MPR is dynamic and must be able to transition between two conformations as it moves to different organelles, with changing environment of each (Dahms et al. 2008; Olson et al. 2008).

Based on these structures, distinct mechanisms for the dissociation of lysosomal enzymes at the cell surface and under the acidic conditions of the endosome were proposed for the CD-MPR (Olson et al. 2008). His^{105} is the only residue of the receptor in which a titratable side chain is involved in binding the phosphate group of M6P. Deprotonation of His^{105} and the phosphate moiety of M6P appear to be key elements in the release of ligand at the cell surface: loss of the electrostatic interaction between the uncharged His^{105} and M6P is predicted to facilitate dissociation of phosphorylated ligands at pH 7.4. In the acidic environment of the endosome, it is proposed that disruption, via protonation, of intermonomer electrostatic interactions that tie loop D of one monomer to the α-helix of the other monomer in the ligand bound conformation would "free" loop D to move into the binding pocket, resulting in the displacement of ligand. In addition, protonation of Glu^{133} that is located in the binding pocket is predicted to weaken its interaction with the 3- or 4-hydroxyl group of M6P and disrupt the electrostatic environment of the entire binding pocket, thereby enhancing the release of M6P. The repositioning of loop D into the binding pocket eliminates its intermonomer interaction with the N-terminal α-helix. This loss of intermonomer contact may trigger the reorientation of the two monomers as the receptor changes its quaternary structure, adopting a more closed conformation in the unbound state (Dahms et al. 2008; Olson et al. 2008). These results allowed to suggest that the receptor regulates its ligand binding upon changes in pH; the pK_a of Glu^{133} appeared to be responsible for ligand release in the acidic environment of the late endosomal compartment, and the pK_a values of the sugar phosphate and His^{105} were accountable for its inability to bind ligand at the cell surface where the pH was about 7.4.

Sequence comparison between CD-MPR and CI-MPR shows that they are related. In fact, the extracytoplasmic domain of CD-MPR is homologous to the approximately 145 amino acid long repeating domains present in the CI-MPR with sequence identity ranging from 14% to 28%. These studies allowed to conclude that these two receptors located on different chromosomes (12p13 and 6q26, respectively) have diverged from a common ancestral gene (Dahms et al. 1987; Klier et al. 1991). This receptor is a 46 kDa single polypeptide chain that contains a putative signal sequence and a transmembrane domain. The CD-MPR is a single membrane-spanning domain, which separates a N-terminal extracytoplasmic region with five potential Asn-linked glycosylation sites, from a C-terminal cytoplasmic region without Asn-glycosylation sites. Sequence analysis of the bovine CD-MPR revealed that it consists of a 28 amino-acid residue N-terminal signal sequence, a 159 amino acid residue luminal domain, a 25 amino acid residue transmembrane domain and a 67 amino acid residue C-terminal cytoplasmic domain. It is highly conserved from mouse to human (93% homology). The CD-MPR appears to be a homodimer at the membrane (Stein et al. 1987a), and either a dimer or a tetramer in solution (Dahms et al. 1987; Tong and Kornfeld 1989; Stein et al. 1987b).

3.4.3 Carbohydrate Binding Sites in MPRs

Soluble acid hydrolases constitute a group of over 60 heterogeneous enzymes that differ in size, oligomeric state, number of *N*-linked oligosaccharides, extent of phosphorylation, and the position of the M6P moiety and its linkage to the penultimate mannose residue in the oligosaccharide chain (Dahms et al. 2008). The two MPRs have been shown to display different affinities and capacities for transport of various acid hydrolases, and both receptors are necessary for the efficient sorting of all lysosomal enzymes to the lysosome as neither MPR can fully compensate for the other (Dahms et al. 2008). These studies indicate that the two MPRs recognize distinct but overlapping populations of acid hydrolases. A proteomic analysis of serum from mutant mice deficient in either the CD-MPR or CI-MPR revealed that several lysosomal proteins are preferentially sorted by the CD-MPR (e.g., tripeptidyl peptidase I) or CI-MPR (e.g., cathepsin D) (Qian et al. 2008). Amine-activated glycans, covalently printed on N-hydroxysuccinimide-activated glass slides, interrogated with different concentrations of rCD-MPR or soluble CI-MPR. Neither receptor bound to non-phosphorylated glycans. The CD-MPR bound weakly or undetectably to the phosphodiester derivatives, but strongly to the phosphomonoester-containing glycans with the exception of a single Man7GlcNAc2-R isomer that contained a single M6P residue. By contrast, the CI-MPR bound with high affinity to glycans containing either phosphomono- or -diesters although, like the CD-MPR, it differentially recognized isomers of phosphorylated Man7GlcNAc2-R. This differential recognition of phosphorylated glycans by the CI- and CD-MPRs has implications for understanding the biosynthesis and targeting of lysosomal hydrolases (Song et al. 2009). Future studies will shed light onto the functional significance of two distinct MPRs in a given cell type.

3.4.4 Similarities and Dis-similarities between two MPRs

The CD-MPR and CI-MPR share a number of similarities with respect to carbohydrate recognition. For example, both MPRs bind the M6P with essentially the same affinity ($7–8 \times 10^{-6}$ M). Mannose or glucose 6-phosphate interact poorly with the MPRs ($K_i = 1–5 \times 10^{-2}$ M) (Tong et al. 1989; Tong and Kornfeld 1989). Like mammalian MPRs, calotes MPR-300/IGF-IIR also binds IGF-II with high affinity (K_D ~12.02 nM). A number of synthetic analogs and those with the highest affinity to the CI-MPR were found to be isosteric to M6P. Several M6P analogues with phosphonate, carboxylate or malonate groups displayed a higher affinity and a stronger stability in human serum than M6P itself. These derivatives could be used to favour the delivery of specific therapeutic compounds to lysosomes, notably in enzyme replacement therapies of lysosomal diseases or in neoplastic drug targeting. Although analogues containing two negative charges were the best ligands, the presence of a phosphorous atom was not necessary for recognition (Gary-Bobo et al. 2007). In addition, linear mannose sequences which contain a terminal M6P linked α1, 2 to the penultimate mannose were shown to be the most potent inhibitors (Distler et al. 1991; Tomoda et al. 1991), suggesting that the MPRs bind an extended oligosaccharide structure which includes the M6P α1,2 Man sequence. The crystal structure of CD-MPR complexed with α1, 2-linked phosphorylated trimannoside has revealed the site of penultimate and prepenultimate mannose rings in the binding pocket and their hydrogen bond interactions with the receptor (Olson et al. 2008). Furthermore, multivalent interactions between the receptor and a lysosomal enzyme result in high affinity binding, of the order of 1–10 nM for both MPRs (Dahms et al. 2008; Watanabe et al. 1990). The two MPRs exhibit optimal ligand binding at ~pH 6.4 and no detectable binding below pH 5, which is in accordance with their function of releasing ligands in acidic environment of the endosome.

In contrast to these similarities, the two MPRs differ in their binding properties, which depend on pH, cations, and nature of phosphodiesters. The CI-MPR retains phosphomannosyl binding capabilities at neutral pH which corresponds well with the ability of this receptor to bind and internalize lysosomal enzymes at the cell surface. In contrast, the ligand binding ability of the CD-MPR is dramatically reduced at a pH > 6.4 (Tong et al. 1989; Tong and Kornfeld 1989) which is consistent with its decreased ability to bind and internalize lysosomal enzymes at the cell surface (Stein et al. 1987a, b). The inability to purify the CD-MPR by phosphomannosyl affinity chromatography performed in the absence of cations led to its designation as a "cation-dependent" receptor (Hoflack and Kornfeld 1985a, b). However, the presence of cations increases the binding affinity of the CD-MPR towards M6P (Tong and Kornfeld 1989) and lysosomal enzymes only fourfold (Sun et al. 2005) but has no effect on the binding affinity of CI-MPR. This finding differentiates the CD-MPR from C-type lectins which have an absolute requirement for calcium to carry out their sugar binding activities. Mutagenesis studies (Sun et al. 2005) indicated that a conserved aspartic acid residue at position 103 of the CD-MPR, which is not present in the CI-MPR, necessitates the presence of a divalent cation in the binding pocket to obtain high affinity ligand binding by functioning to neutralize the negative charge of Asp^{103} juxtaposed to the phosphate oxygen of M6P. The CI-MPR, unlike the CD-MPR, is able to recognize MP-GlcNAc phosphodiesters as well as lysosomal enzymes derived from *Dictyostelium discoideum* which contain mannose 6-sulfate residues and small methyl phosphodiesters, M6P-OCH_3, but not phosphomonoesters (c/r Dahms et al. 2008).

3.5 Functional Mechanisms

3.5.1 Sorting of Cargo at TGN

At TGN, sorting of cargo to different destinations is regulated by several mechanisms: (1) biochemically distinct coats can specify protein sorting,; (2) cytosol-oriented sorting signals of cargo proteins direct them to the appropriate export site; (3) TGN might be organized into discrete subdomains dedicated to assemble specific coat population (Traub and Kornfeld 1997). The constitutive pathway in polarized epithelial cells (e.g. MDCK cells) includes the apical- and basolateral routes. Sorting to the basolateral pathway is mediated by cytoplasmic sorting signals of the cargo molecules. They include tyrosine residues, the dileucine motif, or "adders" that contains neither dileucine- or tyrosine-motifs (Keller and Simons 1997). The machinery responsible for the basolateral sorting is currently unknown while some data implicate that it could be mediated by AP-1 and clathrin (Futter et al. 1998). However, unequivocal evidence showing that the TGN contains sorting mechanisms able to discriminate between proteins traveling to apical and basolateral surfaces has not been obtained. Sorting signals involved in the targeting to selective pathway, among others, include the M6P residues in lysosomal hydrolases and tyrosine- and di-leucine-based sorting determinants in membrane proteins which direct them into CCVs and further into the endosomal/lysosomal system (Kirchhausen et al. 1997; Marks et al. 1997; Rohn et al. 2000). The membrane proteins are e.g. the lysosomal associated membrane protein (LAMP) and the lysosomal integral membrane protein (LIMP). A conformation-dependent motif is suggested to destine the proteins for secretion for a regulated secretory pathway by Keller and Simons (1997).

At TGN, the sorting and transport of a group of soluble proteins via a selective pathway to lysosomes relies on the existence of M6P residues on their oligosaccharides (Le Borgne and Hoflack 1998a, b). They serve as recognition signals for MPRs. Sorting of MPRs and their bound ligands to their destinations is mediated preferentially by the interaction of tyrosine- and dileucine-based sorting signals present in their tails with the adaptor protein complex AP-1 and by transport in CCVs (Le Borgne et al.1996; Le Borgne and Hoflack 1998a). The pinching off of the CCVs from the membranes is effected by dynamin (Jones et al. 1998; Kasai et al. 1999).

Though the CD-MPR and the CI-MPR deliver soluble acid hydrolases to lysosome in higher eukaryotic cells by binding with high affinity to M6P residues, found on N-linked oligosaccharides of their ligands, for many other transmembrane proteins, the MPRs contain multiple molecular sorting signals in their cytoplasmic domains that mediate their intracellular traffic between distinct membrane-bound compartments (Lobel et al. 1989). A schematic representation of the different vesicle transport pathways originating from TGN is given (Fig. 3.1).

3.5.1.1 Interaction of Phosphorylated Oligosaccharides and Lysosomal Enzymes with CD-MPR and CI-MPR

Oligosaccharides with phosphomonoesters interact with the CD-MPR, and molecules with two phosphomonoesters showed the best binding. Lysosomal enzymes with several oligosaccharides containing only one phosphomonoester had a higher affinity for the receptor than did the isolated oligosaccharides, indicating the possible importance of multivalent interactions between weakly binding ligands and the receptor. The binding of phosphorylated lysosomal enzymes to the CD-MPR is markedly influenced by pH. At pH 6.3, almost all of the lysosomal enzymes bound to the receptor. Results indicated that at neutral pH the phosphorylated oligosaccharides on some lysosomal enzyme molecules are oriented in a manner which makes them inaccessible to the binding site of the CD-MPR. Since the same enzymes bind to the CI-MPR at neutral pH, at least a portion of the phosphomannosyl residues must be exposed. It appeared that small variations in the pH of the Golgi compartment where lysosomal enzymes bind to the receptors could potentially modulate the extent of binding to the two receptors (Hoflack et al. 1987).

Lysosomal enzymes bearing phosphomannosyl residues bind specifically to MPRs in the Golgi apparatus and the resulting receptor-ligand complex is transported to an acidic prelyosomal compartment where the low pH mediates the dissociation of the complex (Le Borgne and Hoflack 1998). The transport of proteins from the secretory to the endocytic pathway is mediated by carrier vesicles coated with the AP-1 Golgi assembly proteins and clathrin. The MPRs are segregated into these transport vesicles. Together with GTPase-ARF-1, these cargo proteins are essential components for the efficient translocation of cytosolic AP-1 onto membranes of the TGN, the first step of clathrin coat assembly. The transport of lysosomal enzymes to lysosomes requires two distinct determinants in the CD-MPR carboxyl-terminal domain, a casein kinase II phosphorylation site critical for the efficient interaction of AP-1 with its target membranes and the adjacent di-leucine motif which appears more important for a post AP-1 binding step in the CD-MPR cycling pathway.

3.5.2 TGN Exit Signal Uncovering Enzyme

According to Ghosh et al. (2003) dynamic fusion/fission between the late endosomal and lysosomal compartments results in selective delivery of the hydrolases to the lysosome. TIP47/Rab9 prevents the MPRs from reaching the lysosomes, in which they would otherwise be degraded. The return pathway from the early endosomal compartment to the Golgi is probably mediated by PACS-1-assisted packaging into AP1-containing CCVs, whereas that from the late endosomal (LE) compartments is mediated by TIP47 and Rab9. Some of the MPRs go to the cell surface either from early or late endosomes through the recycling endosome (RE), or from proximal TGN cisternae as a consequence of mis-sorting. The cell-surface receptors are internalized in AP2 CCVs and delivered back to the endosomes.

Nair et al. (2005) proposed that the human M6P uncovering enzyme participates in the uncovering of M6P recognition tag on lysosomal enzymes, a process that facilitates recognition of those enzymes by MPRs to ensure a delivery to lysosomes. Uncovering enzyme has been identified on TGN. The cytoplasmic tail of the uncovering enzyme does not possess any of the known canonical signal sequences for interaction with Golgi-associated γ-ear-containing adaptor proteins. The identification of a TGN exit signal in its cytoplasmic tail elucidates the trafficking pathway of uncovering enzyme, a crucial player in the process of lysosomal biogenesis (Nair et al. 2005). However, uncovered phosphates are not essential for optimal recognition by the phosphomannosyl receptor.

3.5.3 Association of Clathrin-Coated Vesicles with Adaptor Proteins

It is suggested that 46-kDa MPR contains multiple binding sites for clathrin adaptors (Honing et al. 1997). The Golgi-derived and plasma membrane-derived clathrin-coated vesicles can be distinguished by the nature of their underlying assembly proteins AP-1 and AP-2, two related heterotetrameric complexes (Robinson 1994). Localization studies are consistent with the notion that AP-1 is associated with TGN-derived vesicles, whereas AP-2 is found in plasma membrane-derived vesicles. In vitro studies have shown that the translocation of cytosolic AP-1 onto membranes requires ADP-ribosylation factor ARF-1 (Traub et al. 1993), a small GTPase also involved in coatomer binding and vesicular transport in early secretory pathway (Rothman 1994; Boman and Kahn 1995). The AP-1 complex interacts with sorting signals in the cytoplasmic tails of cargo molecules and targeted disruption of the mouse μ1A-adaptin gene causes embryonic lethality. Under normal conditions, MPRs are cargo molecules that exit the TGN via AP-1-clathrin-coated vesicles. But the steady-state distribution of MPR46 and MPR300 in μ1A-deficient cells is shifted to endosomes at the expense of TGN. Thus, MPR46 fails to recycle back from the endosome to the TGN, indicating that AP-1 is required for retrograde endosome to TGN transport of the receptor (Meyer et al. 2000).

Binding Sites in CI-MPR for AP-1

The trafficking of CI-MPR between the TGN and endosomes requires binding of sorting determinants in the cytoplasmic tail of the receptor to adaptor protein complex-1 (AP-1). A GST pull-down binding assay identified four binding motifs in the cytoplasmic tail of CI-MPR: a tyrosine-based motif ^{26}YSKV^{29}, an internal dileucine-based motif ^{39}ETEWLM^{44}, and two casein kinase 2 sites ^{84}DSEDE^{88} and ^{154}DDSDED^{159}. The YSKV motif mediated the strongest interaction with AP-1 and the two CK2 motifs bound AP-1 only when they were phosphorylated. The COOH-terminal dileucines were not required for interaction with AP-1 (Ghosh and Kornfeld 2004).

AP-3 Adaptor Complex Defines a Novel Endosomal Exit Site

The AP-3 adaptor complex has been implicated in the transport of lysosomal membrane proteins. The mammalian AP-3 adaptor-like complex mediates the intracellular transport of lysosomal membrane glycoproteins (Le Borgne et al. 1998b). Electron microscopy showed that AP-3 is associated with budding profiles evolving from a tubular endosomal compartment that also exhibits budding profiles positive for AP-1. AP-3 colocalizes with clathrin, but to a lesser extent than does AP-1. The AP-3- and AP-1-bearing tubular compartments contain low amounts of the CI-MPR and the lysosome-associated membrane proteins (LAMPs) 1 and 2. AP-3 defines a novel pathway by which lysosomal membrane proteins are transported from tubular sorting endosomes to lysosomes. In an attempt to find the site of action of AP-3 Chapuy et al. (2008) showed that sorting of TRP-1 and CD-MPR was AP-1 dependent, while budding of tyrosinase and LAMP-1 required AP-3. Depletion of clathrin inhibited sorting of all four cargo proteins, suggesting that AP-1 and AP-3 are involved in the formation of distinct types of CCVs, each of which is characterized by the incorporation of specific cargo membrane proteins. Harasaki et al. (2005) indicated that three proteins: CI-MPR, carboxypeptidase D (CPD) and low-density lipoprotein receptor-related protein 1 (LRP1) have AP-dependent sorting signals, which may help to explain the relative abundance of AP complexes in CCVs.

AP-4 as a Component of the Clathrin Coat Machinery

AP-4, a protein complex related to clathrin adaptors (Dell'Angelica et al. 1999) is localized mainly in the Golgi complex, as well as on endosomes and transport vesicles. Interestingly, AP-4 is localized with the clathrin coat machinery in the Golgi complex and in the endocytic pathway. Moreover, AP-4 is localized with the CI-MPR, but not with the transferrin receptor, LAMP-2 or invariant chain. The difference in morphology between CI-MPR/AP-4-positive vesicles and CI-MPR/AP-1-positive vesicles raises the possibility that AP-4 acts at a location different from that of AP-1 in the intracellular trafficking pathway of CI-MPR (Barois and Bakke 2005).

3.5.4 Role of Di-leucine-based Motifs in Cytoplasmic Domains

Ludwig et al. (1993) indicated that CD-MPR is required for efficient intracellular targeting of multiple lysosomal enzymes, although homozygous mice lacking CD-MPR suggested that other targeting mechanisms could partially compensate for the loss of CD-MPR in vivo. The cytoplasmic domain of the M6P/IGF2R has two signals for lysosomal enzyme sorting in the Golgi, a di-leucine-based motif (LLHV sequence) and the tyrosine-based endocytosis motif (YKYSKV sequence) (Johnson and Kornfeld 1992), whereas a di-leucine-based motif near the carboxyl terminus of the CD-MPR (HLLPM sequence) in cytoplasmic domain is essential for efficient targeting of newly synthesized lysosomal enzymes (Johnson and Kornfeld 1992). Several other transmembrane proteins destined to the lysosomes also contain di-leucine-based motifs in their cytoplasmic domains that are essential for their proper delivery to lysosomes (Sandoval and Bakke 1994). In the light of these different results, it has been proposed that di-leucine-based motifs mediate sorting of membrane proteins in the TGN. In both MPRs, the di-leucine motifs are flanked by casein kinase II phosphorylation sites that are phosphorylated in vivo (Méresse et al. 1990; Hemer et al. 1993). Such a post-translational modification occurs when the M6P/IGF2R exits from the TGN and represents a major, albeit transient, modification (Méresse and Hoflack 1993). Thus far, the functional importance of the phosphorylation sites in the M6P/IGF2R trafficking has remained controversial (Johnson and Kornfeld 1992; Chen et al. 1993). Mouse L cells deficient in the M6P/IGF-IIR were transfected with normal bovine CD-MPR cDNA or cDNAs containing mutations in the 67-amino acid cytoplasmic tail and assayed for their ability to target the lysosomal enzyme cathepsin D to lysosomes. Mutant receptors with the carboxyl-terminal His-Leu-Leu-Pro-Met67 residues deleted or replaced with alanines sorted cathepsin D below the base-line value (Johnson and Kornfeld 1992). Of the eight amino acids mutated in bovine CD-MPR, four (Gln^{66}, Arg^{111}, Glu^{133}, and Tyr^{143}) were found to be essential for ligand binding. In addition, mutation of the single histidine residue, His^{105}, within the binding site diminished the binding of the receptor to ligand, but did not eliminate the ability of the CD-MPR to release ligand under acidic conditions (Olson et al. 1999a).

A casein kinase II (CK-II) phosphorylation site in the cytoplasmic tail of CD-MPR determines the interaction of AP-1 Golgi assembly proteins with membranes. Mauxion et al. (1996) demonstrated that the casein kinase II phosphorylation site in the CD-MPR cytoplasmic domain determines the high affinity of AP-1 for membranes and that mutations introduced independently in the tyrosine-based or the di-leucine-based motifs are not sufficient to modify these interactions. MPR-negative fibroblasts have a low capacity of recruiting AP-1 which can be restored by re-expressing the MPRs in these cells. This property helped to identify the protein motif of the CD-MPR cytoplasmic domain that is essential for these interactions. It was found that the targeting of lysosomal enzymes requires the CD-PDR cytoplasmic domain that is different from tyrosine-based endocytosis motifs. The first is a casein kinase II phosphorylation site (ESEER) probably acts as a dominant determinant controlling CD-MPR sorting in the TGN. The second is the adjacent di-leucine motif (HLLPM), which, by itself, is not critical for AP-1 binding, but is absolutely required for a downstream sorting event (Mauxion et al. 1996).

Domain 5 of CD-MPR Preferentially Binds Phosphodiesters: Sequence alignment predicts that domain 5 contains four conserved residues (Gln, Arg, Glu, Tyr) which are essential for M6P binding by the CD-MPR and domains 1–3 and 9 of the CI-MPR. Surface plasmon resonance (SPR) analyses of constructs containing single amino acid substitutions showed that these conserved residues (Gln^{644}, Arg^{687}, Glu^{709}, Tyr^{714}) are critical for carbohydrate recognition by domain 5. Furthermore, the N-glycosylation site at position 711 of domain 5, which is predicted to be located near the binding pocket, has no influence on the carbohydrate binding affinity. Using endogenous ligands for the MPRs demonstrated that, unlike the CD-MPR or domain 9 of the CI-MPR, domain 5 exhibits a 14–18-fold higher affinity for MP-GlcNAc than M6P, implicating this region of the receptor in targeting phosphodiester-containing lysosomal enzymes to the lysosome (Chavez et al. 2007). Crystallographic studies have shown that at pH 6.5, the CD-MPR bound to M6P adopts a significantly different quaternary conformation than the CD-MPR in a ligand-unbound state, a feature unique among known lectin structures. Additional crystal structures of the available CD-MPR revealed the positional invariability of specific binding pocket residues which implicate intermonomer contact(s), as well as the protonation state of M6P, as regulators of pH-dependent carbohydrate binding (Olson et al. 2008).

Interaction of MPRs with GGA Proteins: The GGAs (Golgi-localizing, γ-adaptin ear homology domain, ARF-binding), the multidomain family of proteins have been implicated in protein trafficking between Golgi and endosomes. Evidence suggests that CI-MPR and CD-MPR bind specifically to VHS domains of GGAs through acidic cluster-dileucine motifs at the carboxyl ends of their cytoplasmic tails. However, the CD-MPR binds VHS domains more weakly than the CI-MPR. Alignment of C-terminal residues of two receptors revealed a number of non-conservative differences in the acidic cluster-dileucine motifs and the flanking residues. Studies indicate that GGAs participate in lysosomal enzyme sorting mediated by CD-MPR (Doray et al. 2002).

3.5.5 Sorting Signals in Endosomes

The endocytosis of cell surface proteins is mediated by tyrosine-based (Trowbridge et al. 1993) or di-leucine-based motifs (Sandoval and Bakke 1994). In case of CI-MPR/IGF2 receptor, its endocytosis requires a single YSKV sequence (Jadot et al. 1992), while that of CD-MPR requires two distinct motifs. The bovine CD-MPR cycles between TGN, endosomes and the plasma membrane. When the terminal 40 residues were deleted from the 67-amino acid cytoplasmic tail of the CD-MPR, the half-life of the receptor was drastically decreased and the mutant receptor was recovered in lysosomes; amino acids 34–39 being critical for avoidance of lysosomal degradation. Findings indicated that the cytoplasmic tail of the CD-MPR contains a signal that prevents the receptor from trafficking to lysosomes. The transmembrane domain of the CD-MPR also contributes to this function (Rohrer et al. 1995).

The 67-amino acid cytoplasmic tail of CD-MPR contains a signal(s) that prevents the receptor from entering lysosomes where it would be degraded. A receptor with a Trp^{19} –> Ala substitution in the cytoplasmic tail resulted into highly missorted to lysosomes whereas receptors with either Phe^{18} –> Ala or Phe^{13} –> Ala mutations were partially defective in avoiding transport to lysosomes. Results indicated that the di-aromatic motif (Phe^{18}-Trp^{19} with Trp^{19} as the key residue) in its cytoplasmic tail is required for the sorting of the receptor from late endosomes back to the Golgi apparatus. Because a diaromatic amino acid sequence is also present in the cytoplasmic tail of other receptors known to be internalized from the plasma membrane, this feature may be a general determinant for endosomal sorting (Schweizer et al. 1997). However, the CI-MPR lacks such a di-aromatic motif. Studies indicate that sorting of the CD-MPR in late endosomes requires a distinct di-aromatic motif with only limited possibilities for variations, in contrast to the CI-MPR, which seems to require a putative loop (Pro^{49}-Pro-Ala-Pro-Arg-Pro-Gly^{55}) along with additional hydrophobic residues in the cytoplasmic tail. This raises the possibility of two separate binding sites on Tip47 because both receptors require binding to Tip47 for endosomal sorting (Nair et al. 2003).

3.5.6 Palmitoylation of CD-MPR is Required for Correct Trafficking

Evasion of lysosomal degradation of the CD-MPR requires reversible palmitoylation of a cysteine residue in its cytoplasmic tail. Because palmitoylation is reversible and essential for correct trafficking, it presents a potential regulatory mechanism for the sorting signals within the cytoplasmic domain of the CD-MPR. The two cysteine residues (Cys^{30} and Cys^{34}) in the cytoplasmic tail of the CD-MPR are palmitoylated via thioesters and Cys^{34} residue influences the biologic function of the receptor. Mutation of Cys^{34} to Ala resulted in the gradual accumulation of the receptor in dense lysosomes and the total loss of cathepsin D sorting function in the Golgi. A Cys^{30} to Ala mutation had no biologic consequences, showing the importance of Cys^{34}. Mutation of amino acids 35–39 to alanines impaired palmitoylation of Cys^{30} and Cys^{34} and resulted in abnormal receptor trafficking to lysosomes and loss of cathepsin D sorting. The palmitoylation of Cys^{30} and Cys^{34} leads to anchoring of this region of the cytoplasmic tail to the lipid bilayer. Thus, anchoring via Cys^{34} is essential for the normal trafficking and lysosomal enzyme sorting function of the receptor (Schweizer et al. 1996). The palmitoylation of the CD-MPR occurs enzymatically by a membrane-bound palmitoyltransferase, which cycles between endosomes and the plasma membrane. The localization of the palmitoyltransferase indicates it as a regulator of the intracellular trafficking of the CD-MPR and also affects the sorting/activity of other receptors cycling through endosomes (Stöckli and Rohrer 2004).

References

Aeed PA, Elhammer AP (1994) Glycosylation of recombinant prorenin in insect cells: the insect cell line Sf9 does not express the mannose 6-phosphate recognition signal. Biochemistry 33:8793–8797

Al Kawas S, Amizuka N, Bergeron JJ et al (1996) Immunolocalization of the cation-independent mannose 6-phosphate receptor and cathepsin B in the enamel organ and alveolar bone of the rat incisor. Calcif Tissue Int 59:192–199

Appenzeller-Herzog C, Hauri HP (2006) The ER-Golgi intermediate compartment (ERGIC): in search of its identity and function. J Cell Sci 119(Pt 11):2173–2183

Arnold JE, Tipler C, Laszlo L et al (1995) The abnormal isoform of the prion protein accumulates in late-endosome-like organelles in scrapie-infected mouse brain. J Pathol 176:403–411

Barlowe C, Orci L, Yeung T et al (1994) COPII: a membrane coat formed by Sec proteins that drive vesicle budding from the endoplasmic reticulum. Cell 77:895–907

Barois N, Bakke O (2005) The adaptor protein AP-4 as a component of the clathrin coat machinery: a morphological study. Biochem J 385:503–510

Baron R (1989) Molecular mechanisms of bone resorption by the osteoclast. Anat Rec 224:317–324

Baron R, Neff L, Brown W et al (1990) Selective internalization of the apical plasma membrane and rapid redistribution of lysosomal enzymes and mannose 6-phosphate receptors during osteoclast inactivation by calcitonin. J Cell Sci 97:439–447

Belmonte SA, Challa A, Gutierrez LS, Bertini F, Sosa MA. (1998) Alpha-mannosidase from rat epididymal fluid is a ligand for phosphomannosyl receptors on the sperm surface. Int J Androl 21:277–282

Belmonte SA, Romano PS, Fornés WM, Sosa MA (2000) Changes in distribution of phosphomannosyl receptors during maturation of rat spermatozoa. Biol Reprod 63:1172–1178

Berg TO, Fengsrud M, Strømhaug PE et al (1998) Isolation and characterization of rat liver amphisomes. Evidence for fusion of autophagosomes with both early and late endosomes. J Biol Chem 273:21883–21892

Boman AL, Kahn RA (1995) Arf proteins: the membrane traffic police? Trends Biochem Sci 20:147–150

Chapuy B, Tikkanen R, Mühlhausen C et al (2008) AP-1 and AP-3 mediate sorting of melanosomal and lysosomal membrane proteins into distinct post-Golgi trafficking pathways. Traffic 9:1157–1172

Chavez CA, Bohnsack RN, Kudo M et al (2007) Domain 5 of the cation-independent mannose 6-phosphate receptor preferentially binds phosphodiesters (mannose 6-phosphate N-acetylglucosamine ester). Biochemistry 46:12604–12617

Chayko CA, Orgebin-Crist MC (2000) Targeted disruption of the cation-dependent or cation-independent mannose 6-phosphate receptor does not decrease the content of acid glycosidases in the acrosome. J Androl 21:944–953

Chen HJ, Remmler J, Delaney JC et al (1993) Mutational analysis of the cation-independent mannose 6-phosphate/insulin-like growth factor II receptor. A consensus casein kinase II site followed by 2 leucines near the carboxyl terminus is important for intracellular targeting of lysosomal enzymes. J Biol Chem 268:22338–22348

Clermont Y, Rambourg A, Hermo L (1995) Trans-Golgi network (TGN) of different cell types: three-dimensional structural characteristics and variability. Anat Rec 242:289–301

Creek KE, Sly WS (1983) Biosynthesis and turnover of the phosphomannosyl receptor in human fibroblasts. Biochem J 214:353–60

Dahms NM (1996) Insulin-like growth factor II/cation-independent mannose 6-phosphate receptor and lysosomal enzyme recognition. Biochem Soc Trans 24:136–141

Dahms NM, Hancock MK (2002) P-type lectins. Biochim Biophys Acta 1572:317–340

Dahms NM, Lobel P, Breitmeyer J et al (1987) 46 kD mannose 6-phosphate receptor: cloning, expression, and homology to the 215 kd mannose 6-phosphate receptor. Cell 50:181–192

Dahms NM, Olson LJ, Kim JJ (2008) Strategies for carbohydrate recognition by the mannose 6-phosphate receptors. Glycobiology 18:664–678

David TS, Scales SJ et al (1999) Segregation of COPI-rich and anterograde-cargo-rich domains in endoplasmic-reticulum-to-Golgi transport complexes. Curr Biol 9:821–S3

Dell'Angelica EC, Mullins C, Bonifacino JS (1999) AP-4, a novel protein complex related to clathrin adaptors. J Biol Chem 274:7278–7285

Distler JJ, Guo JF, Jourdian GW et al (1991) The binding specificity of high and low molecular weight phosphomannosyl receptors from bovine testes. Inhibition studies with chemically synthesized 6-*O*-phosphorylated oligomannosides. J Biol Chem 266:21687–21692

Dittié AS, Klumperman J, Tooze SA (1999) Differential distribution of mannose-6-phosphate receptors and furin in immature secretory granules. J Cell Sci 112:3955–3966

Doray B, Bruns K, Ghosh P et al (2002) Interaction of the cation-dependent mannose 6-phosphate receptor with GGA proteins. J Biol Chem 277:18477–18482

Duncan JR, Kornfeld S (1988) Intracellular movement of two mannose 6-phosphate receptors: return to the Golgi apparatus. J Cell Biol 106:617–628

Fath S, Mancias JD, Bi X, Goldberg J (2007) Structure and organization of coat proteins in the COPII cage. Cell 129:1325–1336

Freeze HH (1985) Interaction of dictyostelium discoideum lysosomal enzymes with the mammalian phosphomannosyl receptor. The importance of oligosaccharides which contain phosphodiesters. J Biol Chem 260:8857–8864

Futter CE, Gibson A, Allchin EH et al (1998) In polarized MDCK cells basolateral vesicles arise from clathrin γ-adaptin-coated domains on endosomal tubules. J Cell Biol 41:611–623

Gary-Bobo M, Nirdé P, Jeanjean A et al (2007) Mannose 6-phosphate receptor targeting and its applications in human diseases. Curr Med Chem 14:2945–2953

Ghosh P, Kornfeld S (2004) The cytoplasmic tail of the cation-independent mannose 6-phosphate receptor contains four binding sites for AP-1. Arch Biochem Biophys 426:225–230

Ghosh P, Dahms NH, Kornfeld S (2003) Mannose 6-phosphate receptors: new twists in the tale. Nature Rev Mole Cell Biol 4:202–212

Glick BS (2000) Organization of the Golgi apparatus. Curr Opin Cell Biol 12:450–456

Hancock MK, Haskins DJ, Sun G, Dahms NM (2002) Identification of residues essential for carbohydrate recognition by the insulin-like growth factor II/Mannose6-phosphate receptor. J Biol Chem 277:11255–11264

Harasaki K, Lubben NB, Harbour M et al (2005) Sorting of major cargo glycoproteins into clathrin-coated vesicles. Traffic 6:1014–1026

Harter C, Wieland F (1996) The secretory pathway: mechanisms of protein sorting and transport. Biochim Biophys Acta 1286:75–93

Hawkes C, Kar S (2004) The insulin-like growth factor-II/mannose-6-phosphate receptor: structure, distribution and function in the central nervous system. Brain Res Brain Res Rev 44:117–140

Hawkes C, Kabogo D, Amritraj A, Kar S (2006) Up-regulation of cation-independent mannose 6-phosphate receptor and endosomal-lysosomal markers in surviving neurons after 192-IgG-saporin administrations into the adult rat brain. Am J Pathol 169:1140–1154

Hemer F, Körner C, Braulke T (1993) Phosphorylation of the human 46-kDa mannose 6-phosphate receptor in the cytoplasmic domain at serine 56. J Bio Chem 268:17108–17113

Hoflack B, Kornfeld S (1985a) Lysosomal enzyme binding to mouse P388D1 macrophage membranes lacking the 215-kDa mannose 6-phosphate receptor: evidence for the existence of a second mannose 6-phosphate receptor. Proc Natl Acad Sci USA 82:4428–4432

Hoflack B, Kornfeld S (1985b) Purification and characterization of a cation-dependent mannose 6-phosphate receptor from murine P388D1 macrophages and bovine liver. J Biol Chem 260:12008–12014

Hoflack B, Fujimoto K, Kornfeld S (1987) The interaction of phosphorylated oligosaccharides and lysosomal enzymes with bovine liver cation-dependent mannose 6-phosphate receptor. J Biol Chem 262:123–129

Holleran EA, Holzbaur EL (1998) Speculating about spectrin: new insights into the Golgi-associated cytoskeleton. Trends Cell Biol 8:26–29

Honing S, Sosa M, Hille-Rehfeld A, von Figura K (1997) The 46-kDa mannose 6-phosphate receptor contains multiple binding sites for clathrin adaptors. J Biol Chem 272:19884–19890

Imort M, Zuhlsdorf M, Feige U et al (1983) Biosynthesis and transport of lysosomal enzymes in human monocytes and macrophages. Effects of ammonium chloride, zymosan and tunicamycin. Biochem J 214:671–678

Ishibe M, Rosier RN, Puzas JE (1991) Activation of osteoblast insulin-like growth factor-II/cation-independent mannose-6-phosphate receptors by specific phosphorylated sugars and antibodies induce insulin-like growth factor-II effects. Endocr Res 17:357–366

Jadot M, Canfield WM, Gregory W et al (1992) Characterization of the signal for rapid internalization of the bovine mannose 6-phosphate/insulin-like growth factor-II receptor. J Biol Chem 267:11069–11077

Johnson KF, Kornfeld S (1992) A His-Leu-Leu sequence near the carboxyl terminus of the cytoplasmic domain of the cation-dependent mannose 6-phosphate receptor is necessary for the lysosomal enzyme sorting function. J Biol Chem 267:17110–17115

Jones SM, Howell KE, Henley JR et al (1998) Role of dynamin in the formation of transport vesicles from the trans-Golgi network. Science 279:573–577

Kanamori S, Waguri S, Shibata M et al (1998) Overexpression of cation-dependent mannose 6-phosphate receptor prevents cell death induced by serum deprivation in PC12 cells. Biochem Biophys Res Commun 251:204–208

Kasai K, Shin HW, Shinotsuka C et al (1999) Dynamin II is involved in endocytosis but not in the formation of transport vesicles from the trans-Golgi network. J Biochem 125:780–789

Keller P, Simons K (1997) Post-Golgi biosynthetic trafficking. J Cell Sci 110:3001–3009

Kim JJ, Olson LJ, Dahms NM (2009) Carbohydrate recognition by the mannose-6-phosphate receptors. Curr Opin Struct Biol 19:534–542

Kirchhausen T, Bonifacino JS, Riezman H (1997) Linking cargo to vesicle formation: receptor tail interactions with coat proteins. Curr Opin Cell Biol 9:488–495

Klier HJ, von Figura K, Pohlmann R (1991) Isolation and analysis of the human 46-kDa mannose 6-phosphate receptor gene. Eur J Biochem 197:23–28

Klumperman J, Hille A, Veenendaal T et al (1993) Differences in the endosomal distributions of the two mannose 6-phosphate receptors. J Cell Biol 12:997–1010

Kölsch H, Ptok U, Majores M et al (2004) Putative association of polymorphism in the mannose 6-phosphate receptor gene with major depression and Alzheimer's disease. Psychiatr Genet 14:97–100

Konishi Y, Fushimi S, Shirabe T (2005) Immunohistochemical distribution of cation-dependent mannose 6-phosphate receptors in the mouse central nervous system: comparison with that of cation-independent mannose 6-phophate receptors. Neurosci Lett 378:7–12

Kornfeld S (1992) Structure and function of the mannose 6-phosphate/insulinlike growth factor II receptors. Annu Rev Biochem 61:307–330

Köster A, Nagel G, von Figura K et al (1991) Molecular cloning of the mouse 46-kDa mannose 6-phosphate receptor (MPR 46). Biol Chem Hoppe Seyler 372:297–300

Kostrouch Z, Munari-Silem Y, Rajas F et al (1991) Thyroglobulin internalized by thyrocytes passes through early and late endosomes. Endocrinology 129:2202–2211

Le Borgne R, Griffiths G, Hoflack B (1996) Mannose 6-phosphate receptors and ADP-ribosylation factors cooperate for high affinity interaction of the AP-1 Golgi assembly proteins with membranes. J Biol Chem 271:2162–2170

Le Borgne R, Hoflack B (1998a) Mechanisms of protein sorting and coat assembly: insights from the clathrin-coated vesicle pathway. Curr Opin Cell Biol 10:499–503

Le Borgne R, Hoflack B (1998b) Protein transport from the secretory to the endocytic pathway in mammalian cells. Biochim Biophys Acta 1404:195–209

Le Borgne R, Alconada A, Bauer U, Hoflack B (1998) The mammalian AP-3 adaptor-like complex mediates the intracellular transport of lysosomal membrane glycoproteins. J Biol Chem 273:29451–29461

Lee MCS, Miller EA (2007) Molecular mechanisms of COPII vesicle formation seminars in cell. Dev Biol 18:424–434

Lemansky P, Herzog V (1992) Endocytosis of thyroglobulin is not mediated by mannose-6-phosphate receptors in thyrocytes. Evidence for low-affinity-binding sites operating in the uptake of thyroglobulin. Eur J Biochem 209:111–119

Letourneur F, Gaynor EC, Hennecke S et al (1994) Coatomer is essential for retrieval of dilysine-tagged proteins to the endoplasmic reticulum. Cell 79:1199–1207

Lobel P, Fujimoto K, Ye RD, Griffiths G, Kornfeld S (1989) Mutations in the cytoplasmic domain of the 275 kd mannose 6-phosphate receptor differentially alter lysosomal enzyme sorting and endocytosis. Cell 57:787–796

Li MM, Jourdian GW (1991) Isolation and characterization of the two glycosylation isoforms of low molecular weight mannose

6-phosphate receptor from bovine testis. Effect of carbohydrate components on ligand binding. J Biol Chem 266:17621–17630

Lippincott-Schwartz J, Roberts TH, Hirschberg K (2000) Secretory protein trafficking and organelle dynamics in living cells. Annu Rev Cell Dev Biol 16:557–589

Ludwig T, Munier-Lehmann H, Bauer U et al (1994) Differential sorting of lysosomal enzymes in mannose 6-phosphate receptor-deficient fibroblasts. EMBO J 13:3430–3437

Ludwig T, Eggenschwiler J, Fisher P et al (1996) Mouse mutants lacking the type 2 IGF receptor (IGF2R) are rescued from perinatal lethality in Igf2 and Igf1r null backgrounds. Dev Biol 177:517–535

Ma ZM, Grubb JH, Sly WS (1991) Cloning, sequencing, and functional characterization of the murine 46-kDa mannose 6-phosphate receptor. J Biol Chem 266:10589–10595

Marie M, Dale HA, Sannerud R, Saraste J (2009) The function of the intermediate compartment in pre-Golgi trafficking involves its stable connection with the centrosome. Mol Biol Cell 20:4458–4470

Marks MS, Ohno H, Kirchhausen T, Bonifacino J (1997) Protein sorting by tyrosine-based signals: adapting to the Ys and wherefores. Trends Cell Biol 7:124–128

Marron-Terada PG, Bollinger KE, Dahms NM (1998) Characterization of truncated and glycosylation-deficient forms of the cation-dependent mannose 6-phosphate receptor expressed in baculovirus-infected insect cells. Biochemistry 37:17223–17229

Martinez DA, Zuscik MJ, Ishibe M et al (1995) Identification of functional insulin-like growth factor-II/mannose-6-phosphate receptors in isolated bone cells. J Cell Biochem 59:246–257

Matzner U, von Figura K, Pohlmann R (1992) Expression of the two mannose 6-phosphate receptors is spatially and temporally different during mouse embryogenesis. Development 114:965–972

Mauxion F, Schmidt A, Le Borgne R, Hoflack B (1995) Chimeric proteins containing the cytoplasmic domains of the mannose 6-phosphate receptors codistribute with the endogenous receptors. Eur J Cell Biol 66:119–126

Mauxion F, Le Borgne R, Munier-Lehmann H et al (1996) A casein kinase II phosphorylation site in the cytoplasmic domain of the cation-dependent mannose 6-phosphate receptor determines the high affinity interaction of the AP-1 Golgi assembly proteins with membranes. J Biol Chem 271:2171–2178

Méresse S, Hoflack B (1993) Phosphorylation of the cation-independent mannose 6-phosphate receptor is closely associated with its exit from the trans-Golgi network. J Cell Biol 120:67–75

Méresse S, Ludwig T, Frank R et al (1990) Phosphorylation of the cytoplasmic domain of the bovine cation-independent mannose 6-phosphate receptor. Serines 2421 and 2492 are the targets of a casein kinase II associated to the Golgi-derived HAI adaptor complex. J Biol Chem 265:18833–18842

Messner DJ (1993) The mannose receptor and the cation-dependent form of mannose 6-phosphate receptor have overlapping cellular and subcellular distributions in liver. Arch Biochem Biophys 306:391–401

Messner DJ, Griffiths G, Kornfeld S (1989) Isolation and characterization of membranes from bovine liver which are highly enriched in mannose 6-phosphate receptors. J Cell Biol 108:2149–2162

Meyer C, Zizioli D, Lausmann S, Eskelinen EL, Hamann J, Saftig P, von Figura K, Schu P (2000) mu1A-Adaptin-deficient mice: lethality, loss of AP-1 binding and rerouting of mannose 6-phosphate receptors. EMBO J 19:2193–2203

Moreno RD (2003) Differential expression of lysosomal associated membrane protein (LAMP-1) during mammalian spermiogenesis. Mol Reprod Dev 66:202–209

Munier-Lehmann H, Mauxion F, Hoflack B (1996) Function of the two mannose 6-phosphate receptors in lysosomal enzyme transport. Biochem Soc Trans 24:133–136

Nair P, Schaub BE, Rohrer J (2003) Characterization of the endosomal sorting signal of the cation-dependent mannose 6-phosphate receptor. J Biol Chem 278:24753–24758

Nair P, Schaub BE, Huang K et al (2005) Characterization of the TGN exit signal of the human mannose 6-phosphate uncovering enzyme. J Cell Sci 118:2949–2956

Natowicz M, Hallett DW, Frier C et al (1983) Recognition and receptor-mediated uptake of phosphorylated high mannose-type oligosaccharides by cultured human fibroblasts. J Cell Biol 96:915–919

Nolan CM, McCarthy K, Eivers E, Jirtle RL, Byrnes L (2006) Mannose 6-phosphate receptors in an ancient vertebrate, zebrafish. Dev Genes Evol 216:144–151

Ni X, Canuel M, Morales CR (2006) The sorting and trafficking of lysosomal proteins. Histol Histopathol 21:899–913

Nykjaer A, Christensen EI, Vorum H et al (1998) Mannose 6-phosphate/insulin-like growth factor-II receptor targets the urokinase receptor to lysosomes via a novel binding interaction. J Cell Biol 141:815–828

O'Brien DA, Gabel CA, Rockett DL, Eddy EM (1989) Receptor-mediated endocytosis and differential synthesis of mannose 6-phosphate receptors in isolated spermatogenic and sertoli cells. Endocrinology 125:2973–2984

O'Brien DA, Welch JE, Fulcher KD, Eddy EM (1994) Expression of mannose 6-phosphate receptor messenger ribonucleic acids in mouse spermatogenic and Sertoli cells. Biol Reprod 50:429–435

O'Brien DA, Gabel CA, Eddy EM (1993) Mouse Sertoli cells secrete mannose 6-phosphate containing glycoproteins that are endocytosed by spermatogenic cells. Biol Reprod 49:1055–1065

Olson LJ, Hancock MK, Dix D et al (1999a) Mutational analysis of the binding site residues of the bovine cation-dependent mannose 6-phosphate receptor. J Biol Chem 274:36905–36911

Olson LJ, Zhang J, Lee YC, Dahms NM, Kim JJ (1999b) Structural basis for recognition of phosphorylated high mannose oligosaccharides by the cation-dependent mannose 6-phosphate receptor. J Biol Chem 274:29889–29896

Olson LJ, Zhang J, Dahms NM, Kim JJ (2002) Twists and turns of the cation-dependent mannose 6-phosphate receptor. Ligand-bound versus ligand-free receptor. J Biol Chem 277:10156–10161

Olson LJ, Hindsgaul O, Dahms NM, Kim JJ (2008) Structural insights into the mechanism of pH-dependent ligand binding and release by the cation-dependent mannose 6-phosphate receptor. J Biol Chem 283:10124–10134

Panneerselvam K, Balasubramanian AS (1993) Stimulation by lysosomal enzymes and mannose-6-phosphate of a phosphoprotein phosphatase activity associated with the lysosomal enzyme binding receptor protein from monkey brain. Cell Signal 5:269–277

Pauloin A, Tooze SA, Michelutti I, Delpal S, Ollivier-Bousquet M (1999) The majority of clathrin coated vesicles from lactating rabbit mammary gland arises from the secretory pathway. J Cell Sci 112:4089–4100

Pearse BM, Robinson MS (1990) Clathrin, adaptors, and sorting. Annu Rev Cell Biol 6:151–171

Pelham RH (1996) The dynamic organisation of the secretory pathway. Cell Struct Funct 21:413–419

Poupon V, Girard M, Legendre-Guillemin V et al (2007) Biochemical and functional characterization of cation dependent (Mr 46,000) goat mannose 6-phosphate receptor. Glycoconj J 24:221–229

Poupon V, Girard M, Legendre-Guillemin V et al (2008) Clathrin light chains function in mannose phosphate receptor trafficking via regulation of actin assembly. Proc Natl Acad Sci USA 105:168–173

Poussu A (2001) Cloning and characterization of Vear, a novel Golgi-associated protein involved in vesicle trafficking. Dissertation Acta Universitatis Ouluensis, Medica, D 636

Qian M, Sleat DE, Zheng H et al (2008) Proteomics analysis of serum from mutant mice reveals lysosomal proteins selectively transported by each of the two mannose 6-phosphate receptors. Mol Cell Proteomics 7:58–70

Reaves BJ, Bright NA, Mullock BM, Luzio JP (1996) The effect of wortmannin on the localisation of lysosomal type I integral membrane glycoproteins suggests a role for phosphoinositide 3-kinase activity in regulating membrane traffic late in the endocytic pathway. J Cell Sci 109:749–762

Roberts DL, Weix DJ, Dahms NM, Kim JJ (1998) Molecular basis of lysosomal enzyme recognition: three-dimensional structure of the cation-dependent mannose 6-phosphate receptor. Cell 93:639–648

Robinson MS (1994) The role of clathrin, adaptors and dynamin in endocytosis. Curr Opin Cell Biol 6:538–544

Rohn WM, Rouille Y, Waguri S (2000) Hoflack B Bi-directional trafficking between the trans-Golgi network and the endosomal/lysosomal system. J Cell Sci 113:2093–2101

Rohrer J, Schweizer A, Johnson KF, Kornfeld S (1995) A determinant in the cytoplasmic tail of the cation-dependent mannose 6-phosphate receptor prevents trafficking to lysosomes. J Cell Biol 130:1297–1306

Romano PS, Carvelli L, López AC et al (2005) Developmental differences between cation-independent and cation-dependent mannose-6-phosphate receptors in rat brain at perinatal stages. Brain Res Dev Brain Res 158:23–30

Romano PS, Jofré G, Carvelli L et al (2006) Changes in phosphomannosyl ligands correlate with cation-dependent mannose-6-phosphate receptors in rat liver during perinatal development. Biochem Biophys Res Commun 344:605–611

Rothman JE (1994) Mechanisms of intracellular protein transport. Nature 372(6501):55–63

Rothman JE, Wieland FT (1996) Protein sorting by transport vesicles. Science 272:227–234

Rowe T, Aridor M, McCaffery JM et al (1996) COPII vesicles derived from mammalian endoplasmic reticulum microsomes recruit COPI. J Cell Biol 135:895–911

Sandoval IV, Bakke O (1994) Targeting of membrane proteins to endosomes and lysosomes. Trends Cell Biol 377:292–297

Sato K, Nakano A (2003) Oligomerization of a cargo receptor directs protein sorting into COPII-coated transport vesicles. Mol Biol Cell 14:3055–3063

Scales S, Gomez M, Kreis T (2000) Coat proteins regulating membrane traffic. Int Rev Cytol 195:67–144

Scheel G, Herzog V (1989) Mannose 6-phosphate receptor in porcine thyroid follicle cells. Localization and possible implications for the intracellular transport of thyroglobulin. Eur J Cell Biol 49:140–148

Schekman R, Orci L (1996) Coat proteins and vesicle budding. Science 271:1526–1533

Schellens JP, Saftig P, von Figura K, Everts V (2003) Deficiency of mannose 6-phosphate receptors and lysosomal storage: a morphometric analysis of hepatocytes of neonatal mice. Cell Biol Int 27:897–902

Schrag JD, Procopio DO, Cygler M et al (2003) Lectin control of protein folding and sorting in the secretory pathway. Trends Biochem Sci 28:49–57

Schweizer A, Clausen H, van Meer G, Hauri HP (1994) Localization of O-glycan initiation, sphingomyelin synthesis, and glucosylceramide synthesis in Vero cells with respect to the ER-Golgi intermediate compartment. J Biol Chem 269:4035–4047

Schweizer A, Kornfeld S, Rohrer J (1996) Cysteine34 of the cytoplasmic tail of the cation-dependent mannose 6-phosphate receptor is reversibly palmitoylated and required for normal trafficking and lysosomal enzyme sorting. J Cell Biol 132:577–584

Schweizer A, Kornfeld S, Rohrer J (1997) Proper sorting of the cation-dependent mannose 6-phosphate receptor in endosomes depends on a pair of aromatic amino acids in its cytoplasmic tail. Proc Natl Acad Sci USA 94:14471–14476

Song X, Lasanajak Y, Olson LJ et al (2009) Glycan microarray analysis of P-type lectins reveals distinct phosphomannose glycan recognition. J Biol Chem 284:35201–35214

Stein M, Braulke T, Krentler C, Hasilik A, von Figura K (1987a) 46-kDa mannose 6-phosphate-specific receptor: biosynthesis, processing, subcellular location and topology. Biol Chem Hoppe Seyler 368:937–947

Stephens DJ, Austen BM (1996) -amyloidogenic fragments of the Alzheimer's b -amyloid precursor protein accumulate in the *trans*-Golgi network and/or late endosome in 293 cells. JNeurosci Res 46:211–225

Stein M, Meyer HE, Hasilik A, von Figura K (1987b) 46-kDa mannose 6-phosphate-specific receptor: purification, subunit composition, chemical modification. Biol Chem Hoppe Seyler 368:927–936

Stein M, Zijderhand-Bleekemolen JE, Geuze H, Hasilik A, von Figura K (1987c) Mr 46,000 mannose 6-phosphate specific receptor: its role in targeting of lysosomal enzymes. EMBO J 6:2677–2681

Stöckli J, Rohrer J (2004) The palmitoyltransferase of the cation-dependent mannose 6-phosphate receptor cycles between the plasma membrane and endosomes. Mol Biol Cell 15:2617–2626

Storrie B, Pepperkok R, Nilsson T (2000) Breaking the COPI monopoly on Golgi recycling. Trends Cell Biol 10:385–391

Sun G, Zhao H, Kalyanaraman B, Dahms NM (2005) Identification of residues essential for carbohydrate recognition and cation dependence of the 46-kDa mannose 6-phosphate receptor. Glycobiology 15:1136–1149

Tomoda H, Ohsumi Y, Ichikawa Y et al (1991) Binding specificity of *D*-mannose 6-phosphate receptor of rabbit alveolar macrophages. Carbohydr Res 213:37–48

Tong PY, Kornfeld S (1989) Ligand interactions of the cation-dependent mannose 6-phosphate receptor. Comparison with the cation-independent mannose 6-phosphate receptor. J Biol Chem 264:7970–7975

Tong PY, Gregory W, Kornfeld S (1989) Ligand interactions of the cation-independent mannose 6-phosphate receptor. The stoichiometry of mannose 6-phosphate binding. J Biol Chem 264: 7962–7969

Traub LM, Kornfeld S (1997) The trans-Golgi network: a late secretory sorting station. Curr Opin Cell Biol 9:527–533

Traub LM, Ostrom JA, Kornfeld S (1993) Biochemical dissection of AP-1 recruitment onto Golgi membranes. J Cell Biol 123:561–573

Trowbridge IS, Collawn JF, Hopkins CR (1993) Signal-dependent membrane protein trafficking in the endocytic pathway. Annu Rev Cell Biol 9:129–161

Tsuruta JK, O'Brien DA (1995) Sertoli cell-spermatogenic cell interaction: the insulin-like growth factor-II/cation-independent mannose 6-phosphate receptor mediates changes in spermatogenic cell gene expression in mice. Biol Reprod 53:1454–1464

Tsuruta JK, Eddy EM, O'Brien DA (2000) Insulin-like growth factor-II/cation-independent mannose 6-phosphate receptor mediates paracrine interactions during spermatogonial development. Biol Reprod 63:1006–1013

Varki A, Kornfeld S (2009). P Type Lectins, In: Varki A, Cummings RD, Esko JD, Freeze HH, Stanley P, Bertozzi CR, Hart GW, Etzler ME, (eds) Essentials of glycobiology, vol 2. Laboratory Press, Cold Spring Harbor (NY)

Waguri S, Kohmura M, Kanamori S et al (2001) Different distribution patterns of the two mannose 6-phosphate receptors in rat liver. J Histochem Cytochem 49:1397–1405

Waguri S, Tomiyama Y, Ikeda H et al (2006) The luminal domain participates in the endosomal trafficking of the cation-independent mannose 6-phosphate receptor. Exp Cell Res 312:4090–4107

Watanabe H, Grubb JH, Sly WS (1990) The overexpressed human 46-kDa mannose 6-phosphate receptor mediates endocytosis and sorting of beta-glucuronidase. Proc Natl Acad Sci USA 87:8036–8040

Wendland M, von Figura K, Pohlmann R (1991) Mutational analysis of disulfide bridges in the Mr 46,000 mannose 6-phosphate receptor. Localization and role for ligand binding. J Biol Chem 266:7132–7136
Wieland FT, Harter C (1999) Mechanisms of vesicle formation: insights from the COP system. Curr Opin Cell Biol 11:440–446
Wiseman RL, Koulov A, Powers E et al (2007) Protein energetics in maturation of the early secretory pathway. Curr Opin Cell Biol 19:359–367
Yadavalli S, Nadimpalli SK (2008) Mannose-6-phosphate receptors (MPR 300 and 46) from the highly evolved invertebrate asterias Rubens (echinodermate): biochemical and functional characterization of MPR 46 protein. Glycoconj J 25:889–901
Zhang Y, Dahms NM (1993) Site-directed removal of N-glycosylation sites in the bovine cation-dependent mannose 6-phosphate receptor: effects on ligand binding, intracellular targetting and association with binding immunoglobulin protein. Biochem J 295:841–848
Zhang B, Cunningham MA, Nichols WC et al (2003) Bleeding due to disruption of a cargo-specific ER-to-Golgi transport complex. Nat Genet 34:220–225

P-Type Lectins: Cation-Independent Mannose-6-Phosphate Reeptors

4

G.S. Gupta

4.1 Cation-Independent Mannose 6-Phosphate Receptor (CD222)

4.1.1 Glycoprotein Receptors for Insulin and Insulin-like Growth Factors

Insulin and the insulin-like growth factors, IGF-I (IGF1) and –II (IGF2) are structurally related peptides that elicit a large number of similar biological effects in target cells. Three well-characterized receptor complexes bind one or more of these peptides with high affinity. Two of these receptors, denoted as type 1, are ligand-activated tyrosine kinases with similar heterotetrameric $\alpha 2\beta 2$ subunit structures which bind insulin or IGF-1, respectively, with highest affinity. Ligand-stimulated tyrosine autophosphorylation of these receptors further activates their intrinsic tyrosine kinase activities both in vitro and in intact cells. Rapid signal transduction follows such receptor autophosphorylation and tyrosine kinase activation, leading to increased serine phosphorylation of many cell proteins and decreased serine phosphorylation of several others. A third receptor in this group binds IGF-1 and -2, lacks kinase activity and is denoted as type II IGF receptor (IGF2R). The cell surface receptor for IGF2 also functions as a cation-independent M6PR. Therefore, cation-independent mannose 6-phosphate receptor (CI-MPR) is also referred as insulin-like growth factor 2 receptor (IGF2R) or IGF2/MPR. The CI-MPR/IGF2R is a single transmembrane domain glycoprotein that plays a major role in the trafficking of lysosomal enzymes from the trans-Golgi network (TGN) to the endosomal-lysosomal (EL) system. This CI-MPR/IGF2R has also a potential role in growth factor maturation and clearance, and mediates IGF2-activated signal transduction through a G-protein-coupled mechanism. The IGF2R/CI-MPR rapidly recycles between the cell surface membrane and intracellular membrane compartments, providing for the rapid uptake of both IGF2 and M6P-linked lysosomal enzymes. Insulin action markedly increases the proportion of receptors in the plasma membrane and the uptake of bound ligands. Embryonic development and normal growth require exquisite control of IGFs (Dahms et al. 2008; Gary-Bobo et al. 2007).

4.2 Characterization of CI-MPR/IGF2R

4.2.1 Primary Structures of Human CI-MPR and IGF2R Are Identical

The full-length cDNA of 9,104-nt for human CI-MPR contains 7,473 nt encoding a protein of 2,491 aa. The amino acid sequence includes a putative signal sequence of 40 aa, an extracytoplasmic domain consisting of 15 homologous repeat sequences of 134–167 aa, a transmembrane region of 23 aa, and a cytoplasmic domain of 164 aa. The predicted molecular size is greater than 270 kDa. Repeats 7–15 of the extracytoplasmic domain of the human receptor are highly homologous with the sequence of the bovine receptor (Lobel et al. 1987). The nucleotide sequence for the full-length cDNA and the deduced amino acid sequence for the CI-MPR has identity of 99.8% at nt level and 99.4% identity at aa level to the human IGF2R receptor from HepG2 hepatoma cells (Morgan et al. 1987). Kiess et al. (1988) also supported that the type II IGF receptor and the CI-MPR are the same protein, but the binding sites of IGF2 and M6P are distinct. The structural and biochemical features of the IGF2 receptor appeared to be identical to those of the CI-MPR (Kiess et al. 1988; Morgan et al. 1987; Oshima et al. 1988). Phenotypic changes in the transformed cell lines support the role of the CI-MPR in intracellular sorting and targeting of lysosomal enzymes (Oshima et al. 1988).

G.S. Gupta et al., *Animal Lectins: Form, Function and Clinical Applications*,
DOI 10.1007/978-3-7091-1065-2_4, © Springer-Verlag Wien 2012

4.2.2 Mouse IGF2R/CI-MPR Gene

The mouse IGF2R/CI-MPR gene is 93 kb long, comprising 48 exons, and codes for a predicted protein of 2,482 amino acids. The extracellular part of the receptor is encoded by exons 1–46, with each of 15 related repeating motifs being determined by parts of 3–5 exons. A single fibronectin type II-like element is found in exon 39. The transmembrane domain of MPR also is encoded by exon 46 and the cytoplasmic region by exons 46–48. The positions of exon-intron splice junctions are conserved between several of the repeats in IGF2R/CI-MPR and the homologous extracellular region of the gene for other known receptor, the CD-MPR (Szebenyi and Rotwein 1994).

Promoter Elements: The expression of CI-MPR is controlled by both epigenetic and tissue-specific factors. The 93-kb mouse gene has been characterized for its 48 exons. Transient transfection assays revealed that promoter gene of IGF2R/CI-MPR was orientation-specific and was maximal with a plasmid containing 266 bp of IGF2R/CI-MPR DNA. In vitro DNase I footprinting revealed an extended 54-bp footprint within the proximal promoter that contained two E-boxes and potential binding sites for transcription factors Sp1, NGF-IA, and related proteins. These results define a strong minimal IGF2R/CI-MPR promoter of no more than 266 bp and identify a 54-bp enhancer within this promoter fragment (Liu et al. 1995).

Imprinting of Mouse Igf2/Mpr or Igf2r Gene: *Igf2/Mpr* has been mapped to the mouse Tme locus and shown to be an imprinted gene, which further suggests a role in embryonic growth regulation. Igf2 is a general embryonic mitogen, and mice lacking *Igf2* are markedly reduced in size. *Igf2r* acts to fine tune the amount of growth factor, and embryos lacking this gene show overgrowth and die perinatally (Wang et al. 1994). Gametic imprinting is a developmental process that uses cis-acting epigenetic mechanisms to induce parental-specific expression in autosomal and X-linked genes. The biological function of imprinting in mammals is not fully understood. It has been proposed that CG rich sequences resembling CpG islands, which are associated with many imprinted genes and often subject to parental-specific methylation, could act as a common imprinting element. The link between imprinting and growth regulation is best exemplified by the *Igf2* and *Igf2r* genes. Both genes show parental-specific expression patterns in the embryo. The mouse *Igf2r* gene contains a CpG island known as region2, in the second intron. Region2 was proposed as the imprinting element of this gene because it inherited a methylation imprint from the female gamete that was maintained only on the maternal chromosome in diploid cells. Barlow and associates used yeast artificial chromosome transgenes carrying the complete Igf2r locus, to test if imprinting and parental-specific methylation of the mouse *Igf2r* gene is maintained when transferred to other chromosomal locations and to test whether imprinting is dependent on the intronic CpG island proposed as the imprinting element for this gene. Gametic imprints are epigenetic modifications which are imposed onto the gametic chromosomes and cause parental-specific differences in the expression of a small number of genes in the embryo. As a consequence, correct imposition of the imprints in the parental germlines is a prerequisite for successful development of mammals and any anomaly in the expression of imprinted genes is often accompanied by aberration of embryonic growth. The phenomenon of gametic imprinting is predicted to arise by an unusual regulation of the imposition and erasure of epigenetic modifications. One of events which are required for imprinting is that the imprint must be imposed in one of the gametes before fertilisation (Wutz and Barlow 1998; Ludwig et al. 1996).

Chromosomal Mapping of Gene for IGF2/CI-MPR: The gene for IGF2R that has been found to be identical to the CI-MPR has been mapped in the human and murine species. The genes are located in a region of other conserved syntenic genes on the long arm of human chromosome 6, region 6q25-q27, and mouse chromosome 17, region A-C. The CI-MPR/IGF2R locus in man is asyntenic with the genes encoding IGF2, the IGF-I receptor (IGF1R), and the CD-MPR (Laureys et al. 1988). The *Igf2r* gene in bovine is localized to BTA9 chromosome 9q27–28 (Friedl and Rottmann 1994).

4.2.3 Bovine CI-MPR

The 215-kDa CI-MPR from fetal calf liver has a carboxyl-terminal cytoplasmic domain of 163 amino acids that is rich in acidic residues, a 23-amino acid transmembrane segment, a 44-residue amino-terminal signal sequence, and an extracellular domain containing at least eight homologous repeats of approximately 145 amino acids. One of the repeats is similar to the type II repeat of fibronectin. Each repeat contains a highly conserved 13-amino acid unit bordered by cysteine residues that may be functionally important (Lobel et al. 1987). The large 2269-residue extracytoplasmic region is composed of 15 contiguous domains that display a similar size (147 residues) and distinctive pattern of 8 cysteine residues, and exhibit significant amino acid identity (14–38%) when compared to each other and to the CD-MPR, giving rise to the prediction that they have a similar

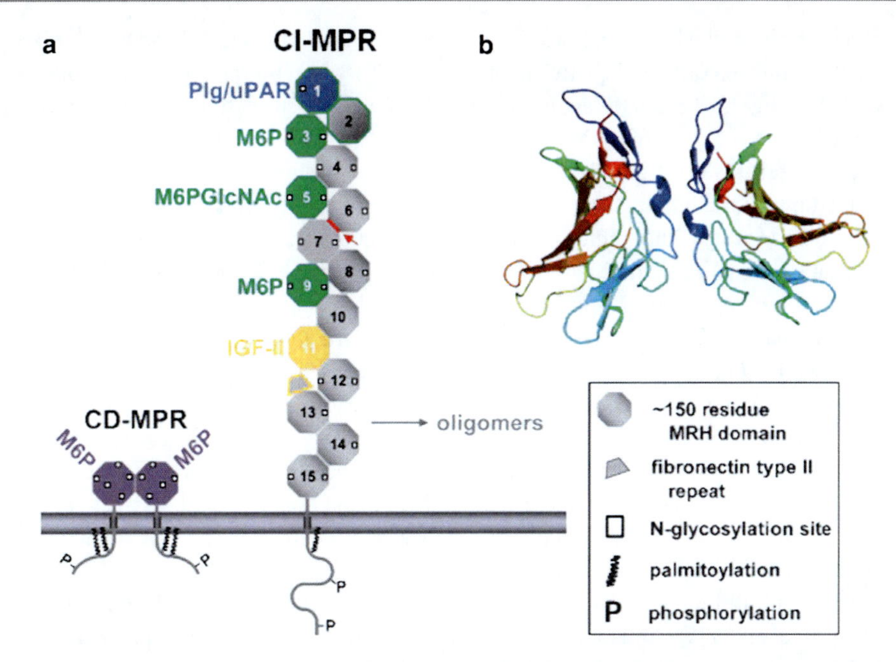

Fig. 4.1 (**a**) Domain organization of M6PRs: The 279-residue CD-MPR exists as a homodimer. The 2499-residue CI-MPR also undergoes oligomerization and most likely exists as a dimer. The M6P binding sites of the CD-MPR (*purple*) and CI-MPR (*green*) are indicated. Domains 1 and 2 are outlined in *green* since the presence of these two domains enhances the affinity of domain 3 for lysosomal enzymes by ~1,000-fold (Hancock et al. 2002b). The CD-MPR contains a single high affinity M6P binding site per polypeptide. In contrast, CI-MPR contains three carbohydrate recognition sites: two high affinity sites are localized to domains 1–3 and domain 9 and one low affinity site is contained within domain 5. The IGF2 (*gold*) and plasminogen (Plg)/uPAR (*blue*) binding sites are also indicated. The fibronectin type II repeat present in domain 13 is outlined in *yellow* since its presence increases the affinity of domain 11 for IGF2 by ~10-fold. The *red arrow* indicates the location of a proteolytically sensitive cleavage site between domains 6 and 7 (Westlund et al. 1991) (Adapted with permission from Dahms et al. 2008 © Oxford University Press). (**b**) The overall folding of CI-MPR is similar to that of CD-MPR, but unlike CD-MPR, CI-MPR is cation-independent. The CI-MPR (PDB ID: 1E6F) most likely exists as a dimmer (Dahms and Hancock 2002)

tertiary structure. This hypothesis has been confirmed, in part, by crystal structure determinations which show that the extracytoplasmic region of the CD-MPR and domains 1, 2, 3, 11, 12, 13, and 14 of the CI-MPR all exhibit the same fold (Olson et al. 1999a, 2004a, b; Brown et al. 2008) (see below). Except for domain 13 of the CI-MPR which has 12 cysteine residues due to a 43-residue insertion homologous to the type II repeat of fibronectin, each of the domains of the CI-MPR that has been crystallized contains eight cysteine residues that form four intramolecular disulfide bridges. The MPRs undergo several types of co- and posttranslational modifications, including N-glycosylation, palmitoylation, and phosphorylation (Fig. 4.1). Palmitoylation of the CD-MPR has been shown to prevent its degradation in lysosomes (Rohrer et al. 1995) while serine phosphorylation of both MPRs has been shown to influence their intracellular transport (Braulke 1999; Breuer et al. 1997). In addition, canonical sorting motifs (i.e., D/EXXLL and YXXφ) within the cytosolic regions of MPRs have been shown to be recognized by components of vesicular machinery that dictate the localization and intracellular trafficking of the receptors (Ghosh et al. 2003b; Bonifacino 2004). Alignment of the 15 domains and the extracytoplasmic domain of the CI-MPR shows that all have sequence similarities and suggests that all are homologous (Lobel et al. 1988). The bovine CI-MPR and the human IGF2R were shown to be 80% identical in their amino acid sequences as deduced from cDNA clones (Morgan et al. 1987). The CI-MPR (215 kDa) has been isolated from embryonic bovine tracheal cells and embryonic human skin fibroblasts. Results suggest the presence of amide-linked palmitic acid in the structure of the CI-MPR (Westcott and Rome 1988). Amino acid sequences deduced from rat c-DNA clones encoding the IGF2R closely resemble those of the bovine CI-MPR, suggesting they are identical structures. It is also shown that IGF2 receptors are adsorbed by immobilized pentamannosyl-6-phosphate and are specifically eluted with M6P. Results indicate that the type II IGF receptor contains cooperative, high-affinity binding sites for both IGF2 and M6P-containing proteins (MacDonald et al. 1988)

4.2.4 CI-MPR in Other Species

In contrast to the bovine, rat, and human CI-MPRs, which bind human IGF2 and IGF-I with nanomolar and micromolar affinities, respectively, the chicken receptor does not bind either radioligand at receptor concentrations as high as

1 μM. The bovine receptor binds chicken IGF2 with high affinity as compared to chicken receptor that binds this ligand with only low affinity (μM range). These data demonstrate that the chicken CI-MPR lacks the high affinity binding site for IGF2 (Canfield and Kornfeld 1989). Like chicken, frog CI-MPR also lacks high affinity IGF2-binding site.

Yang et al. (1991) demonstrated that chick embryo fibroblasts (CEFs) bind and internalize lysosomal enzymes in a M6P-inhibitable fashion, and possess a protein immunologically related to the mammalian IGF2/M6P receptor that binds lysosomal enzymes with M6P recognition markers but does not bind IGF2. The 8,767-bp full-length cDNA of chicken CI-MPR encodes a protein of 2,470 aa that includes a putative signal sequence, an extracytoplasmic domain consisting of 15 homologous repeat sequences, a 23-residue transmembrane sequence, and a 161-residue cytoplasmic domain. It shows 60% sequence identity with human and bovine CI-MPR homologs, and all but two of 122 cysteine residues are conserved (Zhou et al. 1995). American opossum (Didelphis virginiana) liver expressed both the CI-MPR and the CD-MPR. Both receptors contained Asn-linked oligosaccharides. In contrast to CD-MPRs isolated from other species, the opossum CD-MPR displayed heterogeneity with respect to the number of Asn-linked oligosaccharide chains it contains. The CI-MPR from opossum liver bound human recombinant IGF2. However, the opossum CI-MPR bound IGF2 with a lower affinity ($K_D = 14.5$ nM) than the bovine receptor ($K_D = 0.2$ nM) (Dahms et al. 1993a). The CI-MPR from the opossum binds bovine IGF2 with low affinity. The kangaroo CI-MPR has a lower affinity for IGF2 than its eutherian (placental mammal) counterparts. Furthermore, the kangaroo CI-MPR has a higher affinity for kangaroo IGF2 than for human IGF2. The cDNA sequence of the kangaroo CI-MPR indicates that there is considerable divergence in the area corresponding to the IGF2 binding site of the eutherian receptor

4.3 Structure of CI-MPR

4.3.1 Domain Characteristics of IGF2R/CI-MPR (M6P/IGF2R)

The IGF2R/CI-MPR is an ~300-kDa type I transmembrane glycoprotein that consists of an N-terminal signal sequence, a large extracytoplasmic (EC) region composed of 15 homologous repeating domains having ~145 amino acids each, a single transmembrane region, and a short C-terminal cytoplasmic domain (Fig. 4.1a), which lacks kinase activity (Ghosh et al. 2003a; Lobel et al. 1988; Oshima et al. 1988; Scott and Firth 2004). It appears to exist and function as a dimer (Fig. 4.1b). The ~46-kDa CD-MPR is a much smaller type I transmembrane glycoprotein that in some species requires divalent cations for optimal ligand binding (Tong et al. 1989a). Significantly, each of the 15 IGF2/MPR extracytoplasmic domains displays amino acid sequence identity (14–38%), similar size (~147 residues) (Ghosh et al. 2003a), and cysteine distribution to each other and to the single, EC domain of CD-MPR, giving rise to the suggestion that they exhibit similar disulfide bonding and tertiary structures (Lobel et al. 1988).

Domains 3, 5 and 9 of the IGF2R have been shown to contain M6P binding sites (Hancock et al. 2002a, b; Reddy et al. 2004) and the binding sites for other ligands, including the urokinase plasminogen activator receptor, plasminogen, TGFβ, and retinoic acid. A distinct IGF2 binding site is located on domain 11, and the additional presence of domain 13 is required for high affinity IGF2 binding equivalent to intact IGF2R (Dahms et al. 1994; Garmroudi and MacDonald 1994). Separate domains, expressed in *Pichia pastoris*, were tested for their ability to bind carbohydrate ligands and the carbohydrate modifications (mannose 6-sulfate and M6P methyl ester) found on *Dictyostelium discoideum* lysosomal enzymes. The carbohydrate binding sites of the IGF2/MPR were located to domains 3 and 9 of the extracytoplasmic region (Hancock et al. 2002b).

The crystal structures of one of the M6P binding sites (domains 1–3) (Olson et al. 2004a, b) and the IGF2 binding site (domains 11–14) (Brown et al. 2008) have been obtained. The structure of domains 1–3 is very compact and forms a triangular disk of 70 Å (each side) × 50 Å (thickness), with each corner of the triangle occupying one domain. In contrast, the relative orientations of domains 11–14 are very different. The structure of domains 11–14 is rather elongated (50 Å × 60 Å × 115 Å high) and resembles beads on a string, with each domain forming a bead. Combining these two modular structures, along with prediction that the remaining two A simplified representation of the entire extracellular portion (domains 1–15) of the CI-MPR is shown in Fig. 4.1. A similar model has been proposed by Brown et al. (2008). Consistent with the three-domain architecture for each M6P binding site is the presence of a proteolytically sensitive site between domains 6 and 7 (Westlund et al. 1991). The structure of the IGF2 binding site reveals that IGF2 binds in the same location as M6P, thus the same M6PR-H fold can function in protein–protein or protein–carbohydrate interactions. However, it is not clear how various ligand binding sites are oriented relative to each other and how conformation is influenced by pH or ligand binding. Studies are required to support the positioning of ligand binding sites on one face of the molecule as shown in model.

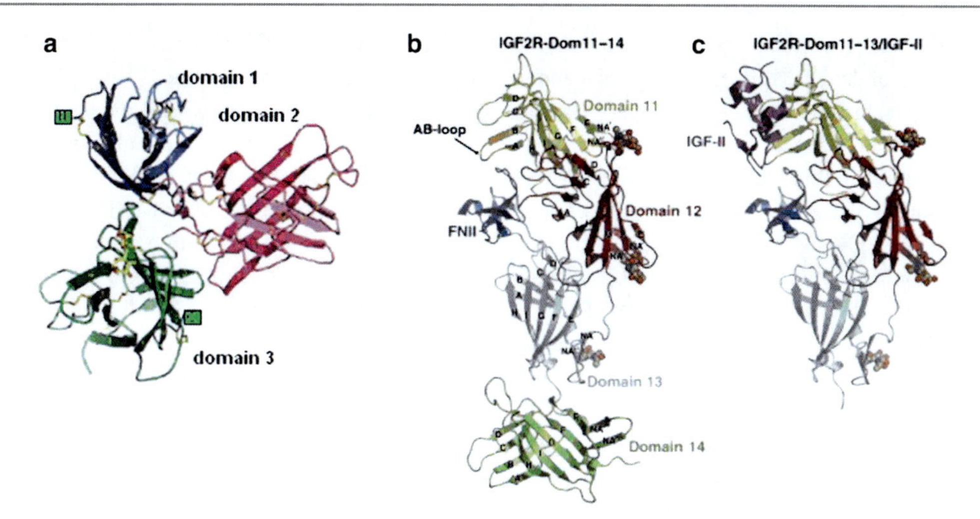

Fig. 4.2 Three-dimensional structure of domains 1–3 of bovine CI-MPR with M6P: (**a**) Stereo view of the M6P complex encompassing residues 7–432 of the bovine CI-MPR. The N and C terminus of protein encoding domain 1 (*blue*), domain 2 (*pink*), and domain 3 (*green*) are indicated, as are the disulfide bonds (*yellow*). β-Strands are labeled (see enlarged view), and the location of M6P (*gold ball-and-stick model*) is shown in green in domain 3 (see enlarged view) (PDB ID: 1SZ0). The CI-MPR contains multiple M6P binding sites that map to domains 3, 5, and 9 within its 15-domain extracytoplasmic region, functions as an efficient carrier of M6P-containing lysosomal enzymes (Olson et al. 2004a) (Adapted with permission from Olson et al. 2004a © American Society for Biochemistry and Molecular Biology). Figure (**b**) and (**c**) are representations of IGF2R fragment structures CI-MPR-Domain11–14 and the complex between CI-MPRDomain11–13 and IGF-II. Glycans are shown as spheres and strand labelling is given in panel B (Adapted by permission from Macmillan Publishers Ltd, EMBO J. Brown et al. © 2008)

4.3.2 Crystal Structure

Truncated forms of CD-MPR and CI-MPR: The CD-MPR has been crystallized under different pH conditions: in an unbound state at pH 6.5 and pH 4.8 (Olson et al. 2002, 2008), or at pH 7.4 in a bound state (Olson et al. 2008). The extracytoplasmic region of the CD-MPR has been studied with X-ray crystallography in complex form with a single sugar M6P (Roberts et al. 1998), or with various oligosaccharide structures (Olson et al. 1999b, 2008). Seven out of fifteen domains of the CI-MPR have also been crystallized: the N-terminal 432 residues (domains 1, 2, and 3), which house a high affinity M6P binding site, bound to either a mannose residue from a crystallographic neighbor or M6P (Olson et al. 2004a, b); the IGF2 binding site (domains 11–14) in the unbound state (Brown et al. 2002; Uson et al. 2003) or domains 11–13 in an IGF2 bound state (Brown et al. 2008). These studies provide a framework from which the mechanism of ligand binding by the MPRs can be inferred (Fig. 4.2).

The extracytoplasmic domain of CD-MPR has been crystallized as a dimer (Figs. 3.1 and 3.2; Chap. 3). Each polypeptide chain folds into an N-terminal α-helix followed by four anti-parallel β-strands, which together comprise the solvent-exposed front face. The dimer interface β-sheet (β5–β9), which accounts for ~20% of the surface area of the monomer, is composed of five anti-parallel β-strands, with strand 9 interjecting between strands 7 and 8. Subsequent determination of the structure of domains 1–3 (Fig. 4.2) and domains 11–14 of the CI-MPR showed that this overall topology is conserved with the exception of the N-terminal region: neither domains 1–3 nor 11–14 contain the α-helix; rather, this secondary structural element is replaced by two β-strands. The quaternary domain arrangement of CD-MPR is not conserved in the structure of domains 1–3 of the CI-MPR. The structure of the N-terminal region of the CI-MPR shows the three domains form a wedge with domains 1 and 2 oriented such that the four-stranded N-terminal sheet (β1–β4) of domain 1 and the five-stranded C-terminal sheet (β5–β9) of domain 2 form a continuous surface (Fig. 4.2b). In comparison to the CD-MPR in which extensive contacts exist between the two dimer interface β-sheets (β5–β9) (Fig. 3.1), the interaction between the three N-terminal domains of the CI-MPR is quite different and much less extensive; the contacts between the three domains are mediated mainly by residues within the linker regions and loops (Fig. 4.2). However, the contacts between domains 1, 2, and 3 are important for maintaining the integrity of the binding pocket housed within domain 3. The multiple interactions between residues of domains 1 and 2 with residues of loops C and D of domain 3 are likely to aid in the stabilization of the binding pocket and provide an explanation for the inability of a construct encoding domain 3 alone to generate a high affinity carbohydrate binding site (Hancock et al. 2002b).

A comparison of the sugar binding pocket of CD-MPR and domain 3 of CI-MPR reveals that residues which interact with the mannose ring (Gln, Arg, Glu, and Tyr) are located in a strikingly similar position in the base of the pocket and form the same contacts with the ligand. These four residues have been shown to be essential for M6P recognition by mutagenesis and are conserved in all species and in the other two M6P binding sites of the CI-MPR (i.e., domains 5 and 9). The presence of this signature motif for phosphomannosyl binding (Gln, Arg, Glu and Tyr) along with conserved cysteine residues led to predict that domain 5 of the CI-MPR contains a M6P binding site, a hypothesis which was confirmed by Reddy et al. (2004). Mutagenic studies of Gln, Arg, Glu, and Tyr in domains 5 (Chavez et al. 2007) and 9 (Hancock et al. 2002a) demonstrate their essential role in carbohydrate recognition by these binding sites. Thus, the strict requirement for a terminal mannose residue by both receptors is reflected in the similarities in that region of the binding pocket responsible for sugar recognition (Dahms et al. 2008).

In contrast, the phosphate recognition region of the binding site appears to have the most variability both in amino acid composition and in structure. In both receptors the lid is formed by residues joining β-strands 6 and 7 (loop C). This lid region is larger in the CD-MPR and the positioning of the disulfide anchors the loop in a more closed position which translates into a more sterically confined binding region. The conformationally constrained lid may account for the inability of CD-MPR to bind phosphodiesters. Shortening of both loops C and D effectively makes the binding pocket of domains 1–3 more open than that of the CD-MPR, allowing for this region of the CI-MPR to bind a larger repertoire of ligands, including phosphomonoesters, mannose 6-sulfate, and phosphodiesters. Thus, the diversity in ligand recognition by the two receptors appears to be accomplished by alterations in the receptor binding site architecture surrounding the phosphate moiety.

Brown et al. (2008) reported crystal structures of IGF2R domains 11-12, 11-12-13-14 and domains 11-12-13/IGF-II complex. A distinctive juxtaposition of these domains provides the IGF-II-binding unit, with domain 11 directly interacting with IGF-II and domain 13 modulating binding site flexibility. The complex shows that Phe19 and Leu53 of IGF-II lock into a hydrophobic pocket unique to domain 11 of mammalian IGF2Rs. Mutagenesis analyses confirmed this IGF-II 'binding-hotspot', revealing that IGF-binding proteins and IGF2R have converged on the same high-affinity site (Brown et al. 2008) (Fig 4.2b, c).

4.4 Ligands of IGF2R/CI-MPR

4.4.1 Extracellular Ligands of IGF2R/CI-MPR

The M6P/IGF2R has been shown to bind at least two classes of ligands, the M6P-containing and the non-M6P-containing polypeptide ligands, all of which bind to sites within the EC region (Braulke 1999). Newly synthesized M6P-containing ligands such as lysosomal acid hydrolases bind to the M6P/IGF2R in the *trans*-Golgi network through M6P residues on their *N*-linked oligosaccharides. Through binding of a large class of ligands, the M6P/IGF2R mediates several important cellular functions, such as the endocytosis and/or targeting of acid hydrolases to lysosomes (Kornfeld 1992), the proteolytic activation of latent TGF-β1 (Dennis and Rifkin 1991; Ghahary et al. 1999), mediation of the migration and angiogenesis induced by proliferin (Groskopf et al. 1997), and the internalization of granzyme B (Motyka et al. 2000).

Proliferin is a prolactin-related glycoprotein secreted by proliferating mouse cell lines and by mouse placenta. The proliferin secreted by cultured cell binds to CI-MPRs and may be a lysosomal protein targeted to lysosomes. The activity of MPRs in murine fetal and maternal liver and in placenta is regulated during pregnancy (Lee and Nathans 1988). The angiogenic activity of proliferin depends on its interaction with CI-MPR. This recognition involves at least in part the carbohydrate moiety of proliferin which competes with M6P (Lee and Nathans 1988). Tumor growth depends on its vascularisation and needs micro-vessels development to expand its cell population and to become invasive (Folkman 2002). The expression of proliferin seems to increase when a tumor becomes more invasive and exhibits a strong angiogenic activity.

M6P-containing ligands, such as latent form of TGF-β, placental hormone proliferin, and granzyme B bind at the cell surface (Ghosh et al. 2003a; Dahms and Hancock 2002). The IGF2/MPR binding to mannose 6-phosphorylated latent form of TGF-β at the cell surface results in the proteolytic activation of this critical growth factor that regulates the cellular differentiation and proliferation of many cell types (Dennis and Rifkin 1991; Godár et al. 1999). The activation of latent TGF-β to active TGF-β is a critical step in wound healing. Cell surface recognition of M6P-modified proliferin by the IGF2/MPR is required for proliferin-induced angiogenesis (Jackson and Linzer 1997.). The precursor form of the renin (prorenin), a key enzyme of the cardiac renin-angiotensin system, contains M6P residues that enable IGF2/MPR binding and internalization of prorenin, resulting in its subsequent proteolytic activation in endosomal compartments (Saris et al. 2001; Nguyen and Contrepas 2008).

4.4.2 Binding Site for M6P in CI-MPR

The interactions of the bovine CD-MPR with monovalent and divalent ligands have been studied. This receptor appears to be a homodimer or a tetramer. Each mole of receptor monomer bound 1.2 mol of the monovalent ligands, M6P and pentamannose phosphate with K_D values of $8 \times {}^{-6}$ M and 6×10^{-6} M, respectively and 0.5 mol of the divalent ligand, a high mannose oligosaccharide with two phosphomonoesters, with a K_D of 2×10^{-7} M. When Mn^{2+} was replaced by EDTA in the dialysis buffer, the K_D for pentamannose phosphate was 2.5×10^{-5} M (Tong et al. 1989b). The interactions of the bovine CI-MPR with a variety of phosphorylated ligands showed the K_D for M6P, pentamannose phosphate, bovine testes β-galactosidase, and a high mannose oligosaccharide with two phosphomonoesters as 7×10^{-6} M, 6×10^{-6} M, 2×10^{-8} M, and 2×10^{-9} M, and the mol of ligand bound/mol of receptor monomer as 2.17, 1.85, 0.9, and 1.0, respectively. Study indicates that the CI-MPR has two M6P-binding sites/ polypeptide chain (Tong et al. 1989a).

The two high affinity binding sites are not functionally equivalent with respect to ligand preference, having distinct K_D for the multivalent M6P-ligand β-glucuronidase (2.0 versus 4.3 nM for repeats 3 and 9, respectively) (Marron-Terada et al. 1998). The pH optimum for carbohydrate binding was also more acidic for repeat 9 than repeat 3 (pH 6.4 versus pH 6.9, respectively), and the two sites differed in their ability to recognize distinctive modifications found on *Dictyostelium discoideum* glycoproteins, such as mannose 6-sulfate and M6P methyl esters (Marron-Terada et al. 2000). Additionally, repeat 9 alone could fold into a high affinity ligand binding domain, whereas repeat 3 was dependent on residues in adjacent repeats 1 and/or 2 for optimal ligand binding (Hancock et al. 2002b). Although it exhibits significant sequence homology with repeats 3 and 9, as well as sharing four conserved residues key for M6P binding, repeat 5 had an ~300-fold lower affinity for M6P than repeat 9 or repeats 1–3, possibly due to the absence of two half-cystines that form a stabilizing disulfide bond in repeats 3 and 9 (Reddy et al. 2004).

The MPRs differ in number of M6P binding sites contained within their polypeptide chains. While CD-MPR contains one (Tong and Kornfeld 1989b), the CI-MPR contains three M6P binding sites (Tong et al. 1989b; Reddy et al. 2004). Expression of recombinant truncated forms of the CI-MPR revealed its three carbohydrate binding sites: two high affinity sites (K_i = ~10 μM for M6P) map to domains 1–3 and domain 9 (Hancock et al. 2002b) while domain 5 houses a low affinity (K_i = ~5 mM for M6P) binding site (Reddy et al. 2004). A comparison of the binding properties of the individual carbohydrate recognition sites demonstrated that domain 9 of the CI-MPR exhibits optimal binding at pH 6.4–6.5, similar to that of the CD-MPR. In contrast, the N-terminal M6P binding site (i.e., domains 1–3) has a significantly higher optimal binding pH of 6.9–7.0 (Marron-Terada et al. 2000). This observation may not only explain the relatively broad pH range of ligand binding by the CI-MPR but likely is a main contributor to the ability of the CI-MPR, as opposed to CD-MPR, to internalize exogenous ligands at slightly alkaline pH 7.4 present at cell surface. Domain 9 of the CI-MPR, like the CD-MPR, is highly specific for phosphomonoesters (Chavez et al. 2007). In contrast, the N-terminal carbohydrate recognition site of the CI-MPR is promiscuous in that, in addition to M6P, it efficiently binds mannose 6-sulfate and the M6P-OCH_3 phosphodiester with only a 20-fold or 10-fold, respectively, lower affinity than M6P (Marron-Terada et al. 2000). Surface plasmon resonance analyses demonstrated that unlike CD-MPR and domain 9 of CI-MPR, domain 5 of CI-MPR exhibits a 14–18-fold higher affinity for MP-GlcNAc than M6P and implicates this region of the receptor in targeting phosphodiester-containing lysosomal enzymes (i.e., those acid hydrolases that escaped the action of the uncovering enzyme in TGN) to the lysosome (Chavez et al. 2007). The presence of three distinct carbohydrate recognition sites in the CI-MPR likely accounts for the ability of the CI-MPR to recognize a greater diversity of ligands than the CD-MPR, both in vitro and in vivo (Dahms et al. 2008; Bohnsack et al. 2009).

Residues Corresponding to the M6P Binding: Residues corresponding to the M6P binding site of the CD-MPR were mutated to Ala, and Tyr^{1542}, Thr^{1570}, Phe^{1567}, and Ile^{1572} within this hydrophobic patch were shown to be critical for IGF2 binding. Ile^{1572} had been identified as essential for IGF2 binding by Garmroudi and MacDonald (1994). Interestingly, mutation of Glu^{1544} (which lies at the edge of this binding site) to Ala or Lys leads to an increase in binding affinity (Zaccheo et al. 2006). Although the binding site on domain 11 has been mapped, it was not clear whether IGF2 contacts domain 13.

Alanine mutagenesis of structurally determined IGF2R showed that two hydrophobic residues in the CD loop (F1567 and I1572) are essential for binding, with a further non-hydrophobic residue (T1570) that slows the dissociation rate. These findings have implications in the molecular architecture and evolution of the domain 11 IGF2-binding site, and the potential interactions with other domains of IGF2R (Zaccheo et al. 2006). Brown et al. (2008) reported crystal structures of IGF2R domains 11-12, 11-12-13-14 and domains 11-12-13/IGF2 complex. A distinctive juxtaposition of these domains provides the IGF2-binding unit, with domain 11 directly interacting with IGF2 and domain 13 modulates binding site flexibility. This complex showed that Phe^{19} and Leu^{53} of IGF2 lock into a hydrophobic pocket unique to domain 11 of mammalian IGF2Rs. Mutagenesis analyses confirmed this IGF2 'binding-hotspot', revealing that IGF-binding proteins and IGF2R have converged on the same high-affinity site (Brown et al. 2008).

Substitution of Arg^{435} in domain 3 of the amino-terminal binding site and Arg^{1334} in domain 9 of the second binding site results in a dramatic loss of ligand binding activity. Earlier Dahms et al. (1993b) and Marron-Terada et al. (2000) had identified a single arginine residue in each of the two IGF2/MPR M6P-binding sites, Arg^{435} in domain 3 and Arg^{1334} in domain 9, to be important for M6P binding by the IGF2/MPR. In comparison to CD-MPR, amino acid residues involved in carbohydrate recognition by the IGF2/MPR were studied by Hancock et al. (2002a). The three-dimensional structure of the EC region of the bovine CD-MPR complexed with M6P (Roberts et al. 1998) or pentamannosyl phosphate (Olson et al. 1999b) revealed the nature of carbohydrate recognition by the CD-MPR, and provided the framework from which to decipher the molecular basis of M6P recognition by the IGF2/MPR (Hancock et al. 2002a). Site-directed mutagenesis generated in soluble, truncated forms of the IGF2/MPR showed that residues of IGF2/MPR EC domains 3 and 9 predicted by a structure-based sequence alignment with the CD-MPR to be in the N- and C-terminal M6P-binding pockets of the IGF2/MPR, respectively, are essential for high affinity M6P binding. Pentamannosyl phosphate-agarose affinity chromatography revealed four amino acid residues (Gln-392, Ser-431, Glu-460, and Tyr-465) in domain 3 and four residues (Gln-1292, His-1329, Glu-1354, and Tyr-1360) in domain 9 as essential for carbohydrate recognition by the IGF2/MPR in addition to the previously determined Arg-435 (domain 3) and Arg-1334 (domain 9) residues. It is emphasized that the two IGF2/MPR M6P-binding sites utilize a mechanism similar to that of the CD-MPR for high affinity M6P binding and that the N- and C-terminal carbohydrate recognition domains of the IGF2/MPR are structurally similar to each other and to the CD-MPR.

A structure-based sequence alignment predicts that domain 5 contains four conserved key residues (Gln, Arg, Glu, and Tyr) identified as essential for carbohydrate recognition by CD-MPR and domains 3 and 9 of CI-MPR, but lacks two cysteine residues predicted to form a disulfide bond within the binding pocket. Reddy et al. (2004) demonstrated that the CI-MPR contains a third M6P recognition site that is located in domain 5 and that exhibits lower affinity than the carbohydrate-binding sites present in domains 1–3 and 9. While CI-MPR contains three distinct M6P binding sites located in domains 3, 5, and 9, only domain 5 exhibits a marked preference for phosphodiester-containing lysosomal enzymes. Although domain 5 contains only three of the four disulfide bonds found in the other seven domains whose structures have been determined to date, it adopts the same fold consisting of a flattened β-barrel. Structure determination of domain 5 bound to *N*-acetylglucosaminyl 6-phosphomethylmannoside, along with mutagenesis studies, revealed the residues involved in diester recognition, including Y679. These results show the mechanism by which the CI-MPR recognizes Man-P-GlcNAc-containing ligands and provides new avenues to investigate the role of phosphodiester-containing lysosomal enzymes in the biogenesis of lysosomes (Olson et al. 2010).

Recognition Sites of CI-MPR by X-ray Analysis: The crystal structure of the N-terminal 432 residues of the CI-MPR, encompassing domains 1–3, was solved in the presence of bound M6P. The structure reveals the unique architecture of this carbohydrate binding pocket and provides insight into the ability of this site to recognize a variety of mannose-containing sugars (Olson et al. 2004a). The three domains, which exhibit similar topology to each other and to the 46 kDa CD-MPR, assemble into a compact structure with the uPAR/plasminogen and the carbohydrate-binding sites situated on opposite faces of the molecule. Of the 15 IGF2R extracellular domains, domains 1–3 and 11 are known to have a conserved β-barrel structure similar to that of avidin and the CD-MPR, yet only domain 11 binds IGF2 with high specificity and affinity. The crystal structures for repeat 11 by Brown et al. (2002) and repeats 1–3 by Olson et al. (2004a, b) allowed these workers to propose different models for the overall structure of the EC domain of the IGF2R. The EC domain of the receptor shows considerable homology among repeats and the CD-MPR (16–38% identity) (Lobel et al. 1988). This high level of sequence identity accounts for structural similarities among domains, including conserved disulfide bond organization, random coil linker regions connecting the domains, and an overall core flattened β-barrel structure. The 1–3 triple-repeat crystal revealed a structure in which repeat 3 sits on top of repeats 1 and 2 (Olson et al. 2004b). Olson et al. (2004b) have proposed that the M6P/IGF2R forms distinct structural units for every three repeats of the EC region, producing five tri-repeat units that stack in a back-to-front manner. In this model, the IGF2-binding site is located on the opposite face of the structure relative to the M6P-binding sites.

Monomer Versus Oligomer: Traditionally thought to function as a monomer, the M6P/IGF2R is now considered to operate optimally in the membrane as an oligomer for high affinity M6P binding and efficient internalization of ligands (York et al. 1999; Byrd and MacDonald. 2000; Byrd et al. 2000). Intermolecular cross-linking of two M6P/IGF2R partners was shown to occur upon binding of the multivalent ligand, β-glucuronidase, resulting in increased rate of ligand internalization (York et al. 1999). The initial rate of internalization of β-glucuronidase was faster than for the monovalent ligand, IGF2, which showed that multivalent ligands enhance the rate of receptor movement, likely due to clustering of the M6P/IGF2R for improved interaction with the endocytic machinery in the formation of clathrin-coated pits (Le Borgne and Hoflack 1997, 1998; York et al. 1999). Further studies demonstrated that alignment of the M6P binding domains of monomeric partners of a receptor dimer is responsible for bivalent, high affinity binding, also supporting the importance of receptor oligomerization (Byrd and MacDonald 2000).

While studying the dimerization domain(s) of the M6P/IGF2R, Kreiling et al. (2005) hypothesized that one or more dimer interaction domains would be located at or near the ligand binding domains in the EC region of the receptor, and that these regions would contribute preferentially to receptor dimerization. It was concluded that a distinct dimerization domain for the M6P/IGF2R does not exist per se, but instead, interactions between monomeric receptor partners apparently occur all along the EC region of the receptor with special contribution made by repeat 12 (Kreiling et al. 2005).

Signal for Internalization of Bovine CI-MPR: The signal for the rapid internalization of bovine CI-MPR/IGF2R was localized to the inner half of the 163-amino acid cytoplasmic tail, including tyrosine 24 and tyrosine 26 (Lobel et al. 1989). Canfield et al. (1991) indicated that the sequence Tyr-Lys-Tyr-Ser-Lys-Val serves as the internalization signal for CI-MPR/IGF2 receptor. The crucial elements of this sequence are present in the cytoplasmic tails of a number of other membrane receptors and proteins known to undergo rapid internalization.

4.4.3 The Non-M6P-Containing Class of Ligands

The ability of the CI-MPR to interact with many different proteins and a lipophilic molecule is facilitated by the receptor's large (~2270 amino acids) extracytoplasmic region comprising 15 homologous domains, in which several distinct ligand-binding sites have been localized to individual domains. The CI-MPR interacts, in a M6P-independent fashion, on cell surface with IGF2 (Dahms et al. 1994), uPAR (Nykjaer et al. 1998), plasminogen (Godár et al. 1999), retinoic acid (Kang et al. 1997), serglycin (Lemansky et al. 2007), heparanase (Wood and Hulett 2008), latent TGFβ precursor, Granzyme B, growth factors, Herpes virus, and CD26 (Gary-Bobo et al. 2007). Limited studies on full-length (300 kDa) and truncated serum form of CI-MPR with retinoic acid indicated that the ~40 kDa C-terminal region of the receptor, which is absent in the serum form of the CI-MPR, is essential for retinoic acid binding (Kang et al. 1997). Among non-M6P-containing ligands, the interaction between CI-MPR and IGF2 has been studied extensively and as a result the CI-MPR is also referred to as the IGF2 receptor (IGF2-MPR). The IGF2 binding site has been located to domain 11 (Devi et al. 1998; Linnell et al. 2001). However, the nature of interaction between the CI-MPR and uPAR is unclear since serum form of CI-MPR binds uPAR in a M6P-dependent fashion whereas membrane-associated full-length CI-MPR binds uPAR in a M6P-independent manner (Kreiling et al. 2003). The uPAR and plasminogen binding sites have been localized to domain 1 (Leksa et al. 2002) (Fig. 4.1a). Although the interactions between the receptor and these ligands are protein–protein or protein–lipid mediated, the interaction between the CI-MPR and serglycin, a lysosomal soluble proteoglycan, appears to be carbohydrate-based via serglycin's chondroitin sulfate chains (Lemansky et al. 2007). Based on observation that domains 1–3, but not domain 9, of the CI-MPR can interact with Man-6-sulfate, it can be predicted that serglycin binds to the N-terminal region (i.e., domains 1–3) of the receptor (Marron-Terada et al. 2000).

IGF2 Binding: The underlying specificity of the interaction between IGF2 and IGF2R is not understood. IGF-I and IGF2 share >60% sequence identity, yet the IGF2R only binds IGF2 with high affinity. Residues Phe^{48}, Arg^{49}, and Ser^{50} have been implicated in IGF2R binding (Sakano et al. 1991). However, these residues are conserved between IGF-I and IGF2 and therefore would not drive the specificity of binding. Ala^{54} and Leu^{55} of IGF2 differ from the corresponding residues in IGF-I (Arg^{55}, Arg^{56}) and are important for IGF2R binding. Forbes et al. (2001) mutated residues Ala^{54} and Leu^{55} of IGF2 in the second A domain helix to arginine (found in the corresponding positions of IGF-I) and measured IGF2R binding. There is a 4- and 3.3-fold difference in dissociation constants for Ala^{54}R IGF2 and Leu^{55}R IGF2, respectively, and a 6.6-fold difference for Ala^{54}R Leu^{55}R IGF2 compared with IGF2. Hence, residues at positions 54 and 55 in IGF2 are important for and equally contribute to IGF2R binding. The IGF2-binding site has been mapped to repeat 11 of the EC region, with high affinity binding being conferred by residues contributed by the 13th repeat (Dahms et al. 1994; Garmroudi and MacDonald 1994; Devi et al. 1998).

The interaction of soluble forms of human CI-MPR/IGF2R with IGFs and mannosylated ligands was analyzed in real time. IGF2R proteins containing domains 1–15, 10–13, 11–13, or 11–12 were combined with rat CD4 domains 3 and 4. Results suggest that domain 13 acts as an enhancer of IGF2 affinity by slowing the rate of dissociation, but additional enhancement by domains other than 10–13 also occurs (Linnell et al. 2001). Structural analyses of repeat 11 identified the putative IGF2-binding site in a hydrophobic pocket at the end of a β-barrel structure (Brown et al. 2002).

In addition, IGF2 residues in the non-conserved C domain are important for IGF2R binding and play a role in conferring the specificity of binding (Roche et al. 2006). However, the effects of mutation on residue Ala^{54} or Leu^{55} or the C domain are insufficient to account for the complete lack of binding by IGF-I, suggesting that another determinant provides the IGF2 specificity. In addition to Thr^{16} as the major determinant on IGF2 responsible for the specificity of IGF2R binding, Delaine et al. (2007) revealed a hydrophobic patch critical for IGF2R binding, which encompasses residues Phe^{19} and Leu^{53} and includes Thr^{16} and Asp^{52}. Thr^{16} was identified as playing a major role in determining why IGF2, but not IGF-I, binds with high affinity to the IGF2R. In nut-shell, a binding surface on IGF2, important for binding to domain 11 of the IGF2R, is achieved predominantly through hydrophobic interactions (Delaine et al. 2007). The structure of the IGF2R·IGF2 complex has not been solved.

Urokinase-Type Plasminogen Activator Receptor and Plasminogen: The other members of the non-M6P-containing ligands are urokinase-type plasminogen activator receptor (uPAR) and plasminogen, whose binding sites have been mapped to a peptide region within EC repeat 1 (Godár et al. 1999; Leksa et al. 2002). The uPAR plays an important role on the cell surface in mediating extracellular degradative processes and formation of active TGF-β, and in nonproteolytic events such as cell adhesion, migration, and transmembrane signaling. The uPAR binds to CI-MPR with an affinity in low μM range, but not to CD-MPR. The binding is not perturbed by uPA and appears to involve domains DII + DIII of the uPAR protein moiety. The binding occurs at site(s) on the CI-MPR different from those engaged in binding of M6P epitopes or IGF2 (Godár et al. 1999; Leksa et al. 2002; Nykjaer et al. 1998). The uPAR-M6P/IGF2R interaction appears to be weak, of low affinity, and confined to a small subpopulation of uPAR molecules (Kreiling et al. 2003), which calls into question the physiological relevance of this interaction.

Interactions with Other Proteins: The CI-MPR/IGF2R has been observed to bind to soluble forms of glycosyl-phosphatidylinositol-linked molecules, one of mammalian origin (rat Thy-1) and two of protozoan origins. Of the two phosphate groups found on the soluble forms of the protozoan glycosyl-phosphatidylinositol-linked molecules: (1) the internal M6P diester (which forms a part of the ethanolamine bridge) and (2) the inositol-1,2 cyclic phosphate group (which arises after cleavage of the membrane associated form with phosphatidylinositol-specific PLC); only the former appears to be recognized by CI-MPR/IGF2R (Green et al. 1995). Human hDNase binds to immobilized CD-MPR, with the strongest binding exhibited by the protein bearing diphosphorylated oligosaccharides (Cacia et al. 1998). CI-MPR interacts with cubilin and megalin, the renal apical brush-border membrane (BBM) endocytic receptors (Yammani et al. 2002). The proliferin secreted by cultured cell binds to CI-MPRs and therefore may be a lysosomal protein targeted to lysosomes. The activity of MPRs in murine fetal and maternal liver and in placenta is regulated during pregnancy (Lee and Nathans 1988). No specific receptor capable of binding α-L-fucosidase independent of M6P was demonstrable, despite published results that support the existence of a M6P independent trafficking mechanism in lymphoid cells for this enzyme (Dicioccio and Miller 1992). Retinoic acid, a unique ligand for the M6P/IGF2R, binds the cytoplasmic region and is thought to function by altering intracellular trafficking of the M6P/IGF2R and its cargo (Kang et al. 1997).

Observations indicate the role for CI-MPR in the adherence and invasion of *L. monocytogenes* of mammalian cells, perhaps in combination with known mechanisms (Gasanov et al. 2006). Mordue and Sibley (1997) characterized the intracellular fate of Toxoplasma in bone marrow-derived macrophages following two modes of uptake: phagocytosis versus active invasion. It appeared that Toxoplasma evades endocytic processing due to an absence of host regulatory proteins including CI-MPR necessary to drive endocytic fusion. This divergence from normal maturation occurs during the formation of primary vacuole. Varicella-zoster virus (VZV) remains a public health issue around the globe. The CI-MPR is critical to both entry and egress of enveloped VZV and in pathogenesis of varicella (Hambleton 2005). Entry into cells and transmission between cells of herpes simplex virus can be facilitated by IGF2R binding of M6P-modified herpes simplex viral glycoprotein D (Brunetti et al. 1995, 1998; Chen et al. 2004)

4.5 Complementary Functions of Two MPRS

Lysosomal membrane proteins and soluble lysosomal proteins are transported from the trans-Golgi network (TGN) to endosomes and lysosomes via coated-vesicles, which bud from the donor compartment and are transported to and fuse with the proper acceptor compartment. The MPRs, while transporting lysosomal enzymes as cargo to endosomal compartments, are recognized by the Golgi-localized, γ-ear-containing, ARF-binding proteins (GGAs) family of clathrin adaptor proteins and accumulate in forming clathrin-coated vesicles (CCV) (Ghosh et al. 2003b; Ghosh and Kornfeld 2004). They are trafficked to the early endosome where, in the relatively low pH environment of the endosome, the MPRs release their cargo. The MPRs are recycled back to the Golgi by way of interaction with Golgi-localized, γ-ear-containing, GGAs and vesicles. The cargo proteins are then trafficked to the lysosome via late endosome independently of the MPRs. The proteins forming the vesicle-coat bind to the cytoplasmic domains of the cargo proteins and recruit additional proteins like clathrin to the site of vesicle formation. These proteins are hence called adaptor proteins (AP) or adaptor-protein complexes and their subunits are called adaptins. The family of heterotetrameric adaptor-protein complexes consists of AP-1, AP-2, AP-3 and AP-4 and all four are required for lysosome biogenesis. They are ubiquitously expressed in mammals and many of the adaptins also exist as tissue-specific isoforms encoded by different genes or generated by alternative splicing. Adaptor-protein complexes are compartment specific proteins and recruit their specific accessory proteins to the site of vesicle formation, which is believed to regulate vesicle budding and fission and vesicle transport (Fig. 3.4; Chap. 3) (Schu 2005). Endocytosed proteins destined for degradation in lysosomes are targeted mainly to early endosomes following uptake. Late endosomes are the major site for entry of newly synthesized lysosomal hydrolases via the CI-MPR into the degradative pathway. Immunoelectron microscopic data support a model in which early endosomes gradually mature into late endosomes (Stoorvogel et al. 1991). Runquist and Havel (1991) examined the distribution of CI-MPR in early and late endosomes and a receptor-recycling fraction isolated from livers of estradiol-treated rats.

4.5.1 Why Two MPRs

Two MPRs have been implicated in the M6P-dependent transport of lysosomal enzymes to lysosomes. It is not known why two different MPRs are present in most cell types. Ludwig et al. (1994), while studying targeting of lysosomal enzymes in mice that lack either or both MPRs, demonstrated that both receptors are required for efficient intracellular targeting of lysosomal enzymes. More importantly, comparison of phosphorylated proteins secreted by different cell types indicated that the two receptors may interact in vivo with different subgroups of hydrolases. This observation may provide a rational explanation for the existence of two distinct MPRs in mammalian cells (Ludwig et al. 1994). The fibroblast cell lines lacking either the CI-MPR or the CD-MPR, partially missort phosphorylated lysosomal enzymes and secrete, in a large part, different phosphorylated ligands. The analysis of the phosphorylated oligosaccharides showed that the ligands missorted in the absence of CD-MPR were slightly but significantly depleted in oligosaccharides with two M6P residues, when compared with those missorted in the absence of CI-MPR. While these results could explain some differences between the structure and the sorting function of the two MPRs, they strongly suggest the reason why cells express two different but related MPRs (Munier-Lehmann et al. 1996a, b).

Despite extensive investigation, the relative roles and specialized functions of each MPR in targeting of specific proteins remain questions of fundamental interest. One possibility is that most M6P-glycoproteins are transported by both MPRs, but there may be subsets that are preferentially transported by each. Proteomic approach of serum from mice lacking either of the MPRs revealed a number of proteins that appear specifically elevated in serum from each MPR-deficient mouse. M6P-glycoforms of cellular repressor of E1A-stimulated genes, tripeptidyl peptidase I, and heparanase were elevated in absence of the CD-MPR and M6P glycoforms of α-mannosidase B1, cathepsin D, and prosaposin were elevated in the absence of the CI-MPR. Qian et al. (2008) suggest that cellular targeting appears to be MPR-selective under physiological conditions. Mammalian CI-MPR also mediates endocytosis and clearance of IGF2. Mutant mice that lack the CD-MPR are viable. Mice that lack the CI-MPR accumulate high levels of IGF2 and usually die perinatally, whereas mice that lack both IGF2 and CI-MPR are viable. Thus, while lack of the CI-MPR appears to perturb lysosome function to a greater degree than lack of the CD-MPR, each MPR has distinct functions for the targeting of lysosomal enzymes in vivo (Sohar et al. 1998). Additionally the receptors mediate the secretion (CD-MPR) and the endocytosis (CI-MPR) of lysosomal enzymes. However, trafficking of acid hydrolases is only part of the story. Evidence is emerging that one of the receptors can regulate cell growth and motility, and that it functions as a tumor suppressor (Ghosh et al. 2003a).

Basolateral endocytic pathway converges with the autophagic pathway after the early endosome in pancreas. Tooze et al. (1990) identified three distinct classes of autophagic compartments, which were referred to as phagophores, Type I autophagic vacuoles, and Type II autophagic vacuoles. Phagophores, the earliest autophagic compartment, contained no CI-MPR and cathepsin D. Where as type I autophagic vacuoles contained only very low levels of cathepsin D and CI-MPR, type II autophagic vacuoles by contrast were enriched for lysosomal enzymes, and also enriched for CI-MPR. The lysosomal enzymes present in Type II autophagic vacuoles carry M6P monoester residues. Thus, type II autophagic vacuoles are a prelysosomal compartment in which the combined endocytic and autophagic pathways meet the delivery pathway of lysosomal enzymes. Studies in kidney on the intracellular polarity of the CI-MPR showed that the CI-MPR was present in basolateral early endosomes and in late endosomes but absent from apical early endosomes (Parton et al. 1989; Prydz et al. 1990).

Prelysosomal Endosome Compartment: Cells contain an intracellular compartment that serves as both the "prelysosomal" delivery site for newly synthesized lysosomal enzymes by the MPR and as a station along the endocytic pathway to lysosomes. The MPR structures are prelysosomes involved in lysosomal enzyme targeting in rat cardiac myocytes (Marjomäki et al. 1990). Cell fractionation studies indicated that prelysosomal endosome compartment (PLC) is the site of confluence of the endocytic and biosynthetic pathways to lysosomes (Park et al. 1991).

In NRK cells 90% of the labelling for the receptor was found in the PLC, with the rest distributed over the other compartments. The PLC is the first structure along the endocytic pathway that gives a significant reaction for acid phosphatase. However, the PLC is clearly distinct from the MPR-negative lysosomes, which are also acid phosphatase-positive (Griffiths et al. 1990b).

The cytoplasmic tail of CI-MPR plays an important role in the receptor trafficking, while the role of the luminal domain is controversial. It was noticed that the peripheral distribution of GFP, fused to the transmembrane and cytoplasmic domains of CI-MPR (G-CI-MPR-tail), was distinct from that of endogenous CI-MPR or of GFP fused to the full-length CI-MPR (G-CI-MPR-full). The CI-MPR luminal domain appeared to be required for tight interaction with endocytic compartments, and retention by them, and that there are additional transport steps, in which the binding to M6P-ligands is involved (Waguri et al. 2006).

Rate of Internalization of the IGF2R: The CI-MPR/IGF2R undergoes constitutive endocytosis, mediating the internalization of two unrelated classes of ligands: M6P-containing acid hydrolases and IGF2. To determine the role of ligand valency in CI-MPR/IGF2receptor-mediated endocytosis, the internalization rates of two ligands, β-glucuronidase (a homotetramer bearing multiple M6P moieties) and IGF2 were measured. The β-glucuronidase entered the cell three- to fourfold faster than IGF2. Purified IGF2R was present as a monomer, but its association with β-glucuronidase generated a complex composed of two receptors and one β-glucuronidase. Neither IGF2 nor the synthetic peptide induced receptor dimerization. It was indicated that intermolecular cross-linking of the IGF2R receptor occurs upon binding of a multivalent ligand, resulting in an increased rate of internalization (York et al. 1999).

M6P-Independent Pathways: The CI-MPR is known to play a role in endocytic uptake of granzyme B, and cells lacking MPR are considered poor targets for CTL that mediate allograft rejection or tumor immune surveillance. However, Trapani et al. (2003) suggest that the uptake of granzyme B into target cells is independent of MPR. Contrary to previous findings, mouse tumor allografts that lack MPR expression were rejected as rapidly as tumors that over-express MPR. Entry of granzyme B into target cells and its intracellular trafficking to induce target cell death in the presence of perforin are therefore not critically dependent on MPR or clathrin/dynamin-dependent endocytosis (Trapani et al. 2003). The β-glucocerebrosi- dase, the enzyme defective in Gaucher disease, is targeted to the lysosome independently of M6PR. Reports suggest a role for LIMP-2, a supporting protein for M6P as the M6P-independent trafficking receptor for β-glucocerebrosidase (Reczek et al. 2007).

4.5.2 Cell Signaling Pathways

The rat IGF2R develops transmembrane signaling by directly coupling to a G protein having a 40-kDa α subunit, Gi-2. By using vesicles reconstituted with the clonal human CI-MPR and G proteins, Murayama et al. (1990) indicated that the CI-MPR could stimulate guanosine 5′-O-(3-thiotriphosphate) (GTPγS) binding and GTPase activities of Gi proteins in response to IGF2. Results suggest that the human CI-MPR has two distinct signaling functions that positively or negatively regulate the activity of Gi-2 in response to the binding of IGF2 or M6P. The PKA type II seems to be associated with the membranes of precisely those subcellular compartments that are active in endocytosis and recycling of cell surface receptors. It appears to be related to the well-established role of cyclic AMP in signal transduction. It was proposed that activation of PKA in endocytic compartments may contribute to regulation (via phosphorylation) of the subcellular distribution of internalized surface receptors or their functional coupling to effector systems involved in signal transduction (Griffiths et al. 1990a).

4.6 Functions of CI-MPR

The mammalian IGF2R/CI-MPR (IGF2R) has multiple functions due to its ability to interact with a wide variety of ligands. (1) The IGF2R plays a major role in targeting M6P-labeled proteins via TGN and from the cellular membrane to lysosomes for degradation (Ghosh et al. 2003a, b). (2) It binds and internalizes IGF2, thereby maintaining correct levels of IGF2 locally and in the circulation. These functions are central in the processes of embryonic development and normal growth. (3) Circulating levels of the leukemia inhibitory factor are modulated via IGF2R-mediated endocytosis and targeting of this M6P-containing protein for degradation in the lysosomes (Blanchard et al. 1999). (4) Entry into cells and transmission between cells of herpes simplex virus can be facilitated by IGF2R binding of M6P-modified herpes simplex viral glycoprotein D (Brunetti et al. 1994; Brunetti et al. 1995, 1998). (5) The IGF2R also functions as a death receptor by mediating uptake of mannose 6-phosphorylated granzyme B, a serine proteinase that is essential for the rapid induction of target cell apoptosis by cytotoxic T cells (Motyka et al. 2000). (6) Further more, internalization of the M6P-containing T cell activation antigen, CD26, by the IGF2R plays an important role in CD26-mediated T cell costimulation (Ikushima et al. 2000). (7) The ability of the IGF2R to recognize with high affinity the M6P signal found on many functionally distinct ligands underscores the importance of the IGF2R and its involvement in a myriad of essential physiological pathways (Hancock et al. 2002b). (8) In addition, the IGF2R may also mediate transmembrane signal transduction in response to IGF2 binding under certain conditions. CI-MPR/IGF2R may also be functional in terms of supporting cell adhesion and proliferation of myeloma cells (Nishiura et al. 1996).

CI-MPR Blocks Apoptosis Induced by HSV 1 Mutants Lacking Glycoprotein D

The herpes simplex virus (HSV) glycoprotein D (gD) is essential for HSV entry into cells, it contains M6P on its Asn-linked oligosaccharides, and binds to CI-MPR. It was supposed that the interaction between gD and CI-MPR could play a role in some aspect of virus entry or egress (Brunetti et al.1994). The HSV1 mutants lacking the gene encoding glycoprotein D (gD) and the gD normally present in the envelope of the virus (gD$^{-/-}$ stocks) or mutants lacking the gD gene but containing trans-induced gD in their envelopes (gD$^{-/+}$) cause apoptosis in human SK-N-SH cells. The CI-MPR blocks apoptosis induced by HSV1 mutants lacking gD and is likely the target of antiapoptotic activity of glycoprotein. This conclusion is consistent with published reports that phosphorylated gD interacts with CI-MPR (Zhou and Roizman 2002). The domains of glycoprotein D required to block apoptosis induced by herpes simplex virus 1 are largely distinct from those involved in cell-cell fusion and binding to nectin1 (Zhou et al. 2003)

CI-MPR/IGF2R is a Death Receptor for Granzyme B During Cytotoxic T Cell-Induced Apoptosis

CTL and NK cells destroy target cells via directed exocytosis of lytic effector molecules such as perforin and granzymes. Whether the endocytic mechanism is nonspecific or is dependent on the CI-MPR is not certain since further experiments with Granzyme B (GzmB) indicated this was not essential. The GzmB, a serine protease of CTLs and NK cells, induces apoptosis by caspase activation after crossing the plasma membrane of target cells. The GzmB can enter cells by endocytosis and in a perforin-independent manner through CI-MPR/IGF2R. Inhibition of the GzmB-CI-MPR interaction prevented GzmB cell surface binding, uptake, and the induction of apoptosis. Significantly, expression of the CI-MPR was essential for cytotoxic T cell-mediated apoptosis of target cells in vitro and for the rejection of allogeneic cells in vivo (Motyka et al. 2000). Since granzyme B must access the substrates within the cell, the endocytosis through CI-MPR binding is a key step in the granule-mediated killing process (Veugelers et al. 2006).

Kun et al. (2008) investigated the effects of ω-3 polyunsaturated fatty acids on apoptosis and granzyme B, perforin, and CI-MPR expression of intestinal epithelial cells of chronic rejection after small intestinal transplantation. ω-3 polyunsaturated fatty acids can suppress the rejection to mucosal cells of allograft at the time of chronic rejection in small intestinal transplantation, which may be significant in increasing the surviving rate of allograft, delaying the chronic dysfunction, and prolonging the lifetime of both allograft and acceptor.

Loss of CI-MPR Expression Promotes the Accumulation of Lysobisphosphatidic Acid

The importance of lipid domains within endocytic organelles in the sorting and movement of integral membrane proteins has been highlighted. The role of the unusual phospholipid lysobisphosphatidic acid (LBPA), which appears to be involved in the trafficking of cholesterol and glycosphingolipids, and accumulates in a number of lysosomal storage disorders, has received attention. Disruption of LBPA function leads to mis-sorting of CI-MPRs. The converse is also true, since spontaneous loss of CI-MPRs from a rat fibroblast cell line led to the formation of aberrant late endocytic structures enriched in LBPA. Accumulation of LBPA was directly dependent upon the loss of the MPRs, and could be reversed by expression of bovine CI-MBP in the mutant cell line. Therefore, loss of such a protein may cause the accumulation of abarrant organelles. It appears that the formation of inclusion bodies in many lysosomal storage diseases is also due to an imbalance in membrane trafficking within the endocytic pathway (Reaves et al. 2000).

Hepatocyte is Direct Target for TGFβ Activation Via the IGF2R/CI-MPR

The activation of latent proTGFβ normally seen in cocultures of bovine aortic endothelial and bovine smooth muscle cells can be inhibited by coculturing the cells with either M6P or antibodies directed against CI-MPR. The binding to the CI-MPR/IFE2R appears to be a requirement for activation of proTGFβ (Dennis and Rifkin 1991). The CI-MPR, over-expressed in hepatocytes during liver regeneration, has been implicated in the maturation of latent proTGFβ. It appears that: (1) the induction of the CI-MPR gene during liver regeneration and hepatocyte culture occurs in mid G1 phase; and (2) the CI-MPR mediates latent proTGFβ activation and thus may act, by targeting TGFβ to hepatocytes, as a negative regulator of hepatocyte growth (Villevalois-Cam et al. 2003).

Regulation of MPRs During the Myogenic Development of C2 cells

The CD-MPR lacks an IGF2-binding site and participates only in the intracellular trafficking of lysosomal enzymes. During terminal differentiation of the myogenic C2 cell line there is an increase in cell surface expression of the IGF2/CI-MPR in parallel with a rise in secretion of IGF2. IGF2/CI-MPR mRNA increased by more than 10-fold during the initial 48 h of C2 muscle differentiation, similar to the rise in IGF2 mRNA. Comparable levels of both mRNAs are expressed in C2 myotubes and in primary cultures of fetal muscle. The differential regulation of the two M6P receptors during muscle differentiation suggests that they may serve distinct functions in development (Szebenyi and Rotwein 1991).

CI-MPR Binds Heparanase to Promote Extracellular Matrix Degradation

Heparanase, a β-D-endoglucuronidase, cleaves heparan sulfate, a structural component of ECM and vascular basement membrane (BM). The cleavage of heparan sulfate by heparanase-expressing cells facilitates degradation of the ECM/BM to promote cell invasion associated with inflammation, tumor metastasis, and angiogenesis. Heparanase has also been shown to act as a cell adhesion and/or signaling molecule upon interaction with cell surfaces. Three hundred-kDa CI-MPR is a cell surface receptor for heparanase. Furthermore, the tethering of heparanase to the surface of cells via CI-MPR was found to increase their capacity to degrade an ECM or a reconstituted BM. These results indicate a role for CI-MPR in the cell surface presentation of enzymatically active heparanase for the efficient passage of T cells into an inflammatory site and have implications for the use of this mechanism by other cell types to enhance cell invasion (Wood and Hulett 2008).

4.7 Proteins Associated with Trafficking of CI-MPR

4.7.1 Adaptor Protein Complexes

The AP-1 and AP-2 complexes are the most abundant adaptors in CCVs, but clathrin-mediated trafficking can still occur in the absence of any detectable AP-1 or AP-2. Results indicate that three proteins: CI-MPR, carboxypeptidase D (CPD) and low-density lipoprotein receptor-related protein 1 (LRP1) have AP-dependent sorting signals, which may help to explain the relative abundance of AP complexes in CCVs (Harasaki et al. 2005).

Four Binding Sites in CI-MPR for AP-1: The trafficking of CI-MPR between the TGN and endosomes requires binding of sorting determinants in the cytoplasmic tail of the receptor to adaptor protein complex-1 (AP-1). A GST pull-down binding assay identified four binding motifs in the cytoplasmic tailof CI-MPR: a tyrosine-based motif ^{26}YSKV^{29}, an internal dileucine-based motif ^{39}ETEWLM^{44}, and two casein kinase 2 sites ^{84}DSEDE^{88} and ^{154}DDSDED^{159}. The YSKV motif mediated the strongest interaction with AP-1 and the two CK2 motifs bound AP-1 only when they were phosphorylated. The COOH-terminal dileucines were not required for interaction with AP-1 (Ghosh and Kornfeld 2004).

AP-3 Adaptor Complex Defines a Novel Endosomal Exit Site: The mammalian AP-3 adaptor-like complex mediates the intracellular transport of lysosomal membrane glycoproteins (Le Borgne et al. 1998). Electron microscopy showed that AP-3 is associated with budding profiles evolving from a tubular endosomal compartment that also exhibits budding profiles positive for AP-1. AP-3 colocalizes with clathrin, but to a lesser extent than does AP-1. The AP-3- and AP-1-bearing tubular compartments contain low amounts of the CI-MPR and the lysosome-associated membrane proteins (LAMPs) 1 and 2. Chapuy et al. (2008) showed that sorting of TRP-1 and CD-MPR was AP-1 dependent, while budding of tyrosinase and Lamp-1 required AP-3. Depletion of clathrin inhibited sorting of all four cargo proteins, suggesting that AP-1 and AP-3 are involved in the formation of distinct types of CCVs, each of which is characterized by the incorporation of specific cargo membrane proteins.

AP-4 as a Component of the Clathrin Coat Machinery: AP-4 is localized mainly in the Golgi complex, as well as on endosomes and transport vesicles. Interestingly, AP-4 is localized with the clathrin coat machinery in the Golgi complex and in the endocytic pathway. Moreover, AP-4 is localized with the CI-MPR, but not with the transferrin receptor, LAMP-2 (lysosomal-associated membrane protein-2) or invariant chain. The difference in morphology between CI-MPR/AP-4-positive vesicles and CI-MPR/AP-1-positive vesicles raises the possibility that AP-4 acts at a location different from that of AP-1 in the intracellular trafficking pathway of CI-MPR (Barois and Bakke 2005).

Serines 2421 and 2492 Are the Targets of a CK II Associated to the Golgi-Derived HAI Adaptor Complex: A kinase activity of bovine brain CCV phosphorylates the bovine CI-MPR with high efficiency (Km ~ 50–100 nM). The kinase is part of the coat of Golgi-derived CCV. The kinase is associated to the 47-kDa subunit of the complex and exhibits properties similar to a casein kinase II. This posttranslational modification occurs on serines 2421 and 2492 of bovine CI-MPR precursor, residues which are located in typical casein-kinase II recognition sequences. The same serines are phosphorylated in vivo (Méresse et al. 1990). The phosphorylation of two serines in the CI-MPR cytoplasmic domain is associated with a single step of transport of its recycling pathways and occurs when this receptor is in the TGN and/or has left this compartment via clathrin-coated vesicles (Méresse and Hoflack 1993).

4.7.2 Mammalian TGN Golgins

Four mammalian golgins are specifically targeted to the TGN membranes via their C-terminal GRIP domains. The TGN golgins, p230/golgin-245 and golgin-97, are recruited via the GTPase Arl1, whereas the TGN golgin GCC185 is recruited independently of Arl1. Of the four TGN golgins, p230/golgin-245, golgin-97, GCC185, and GCC88, GCC88 defines a retrograde transport pathway from early endosomes (EE) to the TGN. Derby et al. (2007) suggest a dual role for the GCC185 golgin in the regulation of endosome-to-TGN membrane transport and in the organization of the Golgi apparatus.

The maintenance of MPRs that cycle between the TGN and endosomes requires the function of the mammalian Golgi-associated retrograde protein (GARP) complex. Depletion of any of the three GARP subunits, Vps52, Vps53, or Vps54, by RNAi impairs sorting of the precursor of the cathepsin D, to lysosomes and leads to its secretion into the culture medium. As a consequence, lysosomes become swollen, likely due to a buildup of undegraded materials. Results indicate that the mammalian GARP complex plays a general role in the delivery of retrograde cargo into the TGN. The Vps54 mutant protein in the Wobbler mouse strain being active in retrograde transport explains the viability of these mutant mice (Pérez-Victoria et al. 2008).

4.7.3 TIP47: A Cargo Selection Device for MPR Trafficking

MPRs transport newly synthesized lysosomal hydrolases from Golgi to prelysosomes and then return to the Golgi for another round of transport. A 47 kDa protein (TIP47 = tail-interacting protein of 47 kDa) that binds selectively to the cytoplasmic domains of CI-MPR and CD-MPR is present in cytosol and on endosomes and is required for MPR transport from endosomes to the TGN in vitro and in vivo. TIP47 recognizes a phenylalanine/tryptophan signal in the tail of the CD- MPR that is essential for its proper sorting within the endosomal pathway. Thus, TIP47 binds MPR cytoplasmic domains and facilitates their collection into transport vesicles destined for the Golgi (Díaz and Pfeffer 1998). Recombinant TIP47 binds more tightly to the CI-MPR ($K_D = 1$ μm) than to the CD-MPR ($K_D = 3$ μm). In addition, TIP47 fails to interact with the cytoplasmic domains of TGN38 and other proteins that are also transported from endosomes to the TGN. Thus, TIP47 recognizes a very select set of cargo molecules (Krise et al. 2000).

Recognition of the CI-MPR by TIP47: TIP47 interaction with the 163-residue CI-MPR cytoplasmic domain is conformation dependent and requires CI-MPR residues that are proximal to the membrane. CI-MPR cytoplasmic domain residues 1–47 are dispensable, whereas residues 48–74 are essential for high-affinity binding. However, residues 48–74 are not sufficient for high-affinity binding; residues 75–163 alone display weak affinity for TIP47, yet they contribute to the presentation of residues 48–74 in the intact protein. TIP47 binding is competed by the binding of the AP-2-clathrin adaptor at (and near) residues 24–29 but not by AP-1 binding at (and near) residues 160–161. Finally, TIP47 appears to recognize a putative loop generated by the sequence PPAPRPG and other hydrophobic residues in the membrane-proximal domain. These data provide structural basis for TIP47-CI-MPR association (Orsel et al. 2000). The MPRs, transported from endosomes to the TGN, require Rab9 and the cargo adaptor TIP4. The CI-MPRs are enriched in the Rab9 domain relative to the Rab7 domain. TIP47 is likely to be present in this domain because a TIP47 mutant disrupted endosome morphology and sequestered MPRs intracellularly.

4.7.4 Sorting Signals in GGA and MPRs at TGN

CI-MPR takes a complex route that involves multiple sorting steps in both early and late endosomes (Lin et al. 2004). Specific sorting signals direct transmembrane proteins to the compartments of the endosomal-lysosomal system. The Golgi-localizing, γ-adaptin ear homology domain, ADP ribosylation factor (Arf)-interacting (GGA) proteins constitute a family of clathrin adaptors that are mainly associated with the TGN and mediate the sorting of MPRs. MPRs and ADP-ribosylation factors cooperate for high affinity interaction of the AP-1 golgi assembly proteins with membranes (Le Borgne et al. 1996). Three mammalian GGAs, GGA1, 2, and 3 have been implicated in the sorting of MPR. This sorting is dependent on the interaction of the N-terminal of VHS (VPS-27, Hrs, and STAM) domain of the GGAs with acidic-cluster-dileucine (ACLL) signals in the cytosolic tails of the receptors msuch as the CI- and CD-MPRs. The GGA proteins have modular structures with an N-terminal VHS domain followed by a GAT (GGA and TOM1) domain, a connecting hinge segment, and a C-terminal GAE (γ-adaptin ear) domain.

Cytoplasmic Domain of the CI-MPR Contains Different Signals for Rapid Endocytosis: Mutant receptors with 40 and 89 residues deleted from the carboxyl terminus of the cytoplasmic tail functioned normally in endocytosis, but were partially impaired in sorting. Mutant receptors with larger deletions leaving only 7 and 20 residues of the cytoplasmic tail were defective in endocytosis and sorting.

A mutant receptor containing alanine instead of tyrosine residues at positions 24 and 26 was defective in endocytosis, and partially impaired in sorting. Receptors deficient in endocytosis accumulated at the cell surface. Results indicate that the cytoplasmic domain of the Cl-MPR contains different signals for rapid endocytosis and efficient lysosomal enzyme sorting (Lobel et al. 1989). The chimeric CI-MPRs containing bovine extracytoplasmic domain and the human or mouse trans-membrane and cytoplasmic domains function identically to the bovine receptor, thus demonstrating that sorting signals are conserved. Analysis of a series of truncation and alanine scanning mutants revealed that a consensus casein kinase II site followed by 2 leucines near the COOH terminus that has the sequence D^{10}DSDEDLL^{3} is important for receptor function in sorting of lysosomal enzymes (Chen et al. 1993).

Golgi-Localized, γ-ear-Containing, Arf-Binding Proteins (GGAs) Mediate Sorting of the MPRs at the TGN: The cytosolic tails of both receptors contain acidic-cluster-dileucine signals that direct sorting from the trans-Golgi network (TGN) to the endosomal-lysosomal system. The lysosomal trafficking of the M6P-receptor and sortilin require that the GGAs be recruited to Golgi membranes where they bind in the cytosolic tail of the receptors and recruit clathrin to form trafficking vesicles. GGA recruitment to membranes requires Arf1, a protein that cycle between a GDP-bound inactive state and GTP-bound active state. The guanine nucleotide exchange factors (GEFs) promote the formation of Arf-GTP, while the GTPase activating proteins induce hydrolysis of GTP to GDP. Evidences indicate that the GEF, GBF1, colocalizes with the GGAs and interacts with the GGAs. Depletion of GBF1 or expression of an inactive mutant prevents recruitment of the GGAs to Golgi membranes and results in the improper sorting of cargo. In short, GBF1 is required for GGA recruitment to Golgi membranes and plays a role in the proper processing and sorting of lysosomal cargo. The receptors and the GGAs leave the TGN on the same tubulo-vesicular carriers. A dominant-negative GGA mutant blocking the exit of the receptors from the TGN indicated that the GGAs mediate sorting of the M6P-receptors at the TGN (Lefrançois and McCormick 2007; Puertollano et al. 2001). Ghosh et al. (2003) indicated that the three mammalian GGAs cooperate to sort cargo and are required for maintenance of TGN structure.

Acidic Cluster/Dileucine Motif (ACLL) of MPRs in Lysosomal Enzyme Sorting: During endocytic trafficking of CI-MPR, a cluster of acidic amino acids followed by a dileucine (ACLL) motif in the cytoplasmic tail has been proposed to mediate receptor sorting from TGN to late endosomes. Mutations in this motif impair lysosomal enzyme sorting by preventing association of CI-MPR with coat proteins. Alanine cluster mutagenesis demonstrates that the major sorting determinant is a conserved casein kinase II site followed by a dileucine motif (D^{157}DSDEDLL^{164}). Small deletions or additions outside this region have severe to mild effects, indicating that context is important. Single residue mutagenesis indicates that cycles of serine phosphorylation/dephosphorylation are not obligatory for sorting. In addition, the two leucine residues and four of the five negatively charged residues can readily tolerate conservative substitutions. In contrast, aspartate 160 could not tolerate isoelectric or isosteric substitutions, implicating it as a critical component of the sorting signal. Thus, the ACLL motif of CI-MPR is critical for receptor sorting at early stages of intracellular transport following endocytosis (Chen et al. 1997; Tortorella et al. 2007).

The GGA1 binds to CD-MPR in the TGN and targets the receptor to clathrin-coated pits for transport from the TGN to endosomes. The motif of the CD-MPR that interacts with GGA1 was shown to be DXXLL. However, reports on increased affinity of cargo, when phosphorylated by CK2, to GGAs created interest on the effect of the CD-MPR CK2 site on binding to GGA1. It seems that Glu58 and Glu59 of the CK2 site are essential for high affinity GGA1 binding in vitro, whereas the phosphorylation of Ser57 of the CD-MPR has no influence on receptor binding to GGA1. In vivo interaction between GGA1 and CD-MPR was abolished when all residues, namely, Glu58, Glu59, Asp61, Leu64, and Leu65, involved in GGA1 binding were mutated. In contrast, AP-1 binding to CD-MPR required all the glutamates (namely, Glu55, Glu56, Glu58, and Glu59) that surround the phosphorylation site, but like GGA1 binding, was independent of the phosphorylation of Ser57 (Stöckli et al. 2004).

Autoinhibition of the Ligand-Binding Site of GGA1/3 VHS Domains by an Internal ACLL Motif: Isolated VHS domains bind specifically to ACLL motifs present in the cytoplasmic tails of the MPRs. Full-length cytoplasmic GGA1 and GGA3 but not GGA2 bind the CI-MPR very poorly because of autoinhibition. The inhibition depends on the phosphorylation of a serine located three residues upstream of the ACLL motif. Substitution of the GGA1 inhibitory sequence into the analogous location in GGA2, which lacks the ACLL motif, results in autoinhibition of the latter protein. Results indicate that the activity of GGA1 and GGA3 is regulated by cycles of phosphorylation/dephosphorylation (Doray et al. 2002). Crystallographic analyses demonstrated that the phosphoserine residue interacts electrostatically with two basic residues on the VHS domain of GGA3, thus providing an additional point of attachment of the ACLL signal to its recognition module (Kato et al. 2002).

PACS-1, GGA3 and CK2 Complex Regulates CI-MPR Trafficking: Cycling of CI-MPR between the TGN and early endosomes is mediated by GGA3, which directs TGN export, and PACS-1, which directs endosome-to-TGN retrieval. The acidic cluster on PACS-1, similar to acidic cluster motifs on cargo proteins, acts as an autoregulatory domain that controls PACS-1-directed sorting. Despite executing opposing sorting steps, GGA3 and PACS-1 bind to an overlapping CI-MPR trafficking motif and their sorting activity is controlled by the CK2 phosphorylation of their respective autoregulatory domains. Study suggests a CK2-controlled cascade regulating hydrolase trafficking and sorting of itinerant proteins in the TGN/endosomal system (Scott et al. 2003, 2006). The PACS-1 connects the clathrin adaptor AP-1 to acidic cluster sorting motifs contained in the cytoplasmic domain of cargo proteins such as furin and CI-MPR and in viral proteins such as HIV-1 Nef. The phosphorylation state of an autoregulatory domain controls PACS-1-directed protein traffic

VHS Domain of GGA2 Binds to Acidic Cluster-Dileucine Motif in the Cytoplasmic Tail of CI-MPR: Structural Basis for Sorting-Signal Recognition: Shiba et al. (2002) reported the X-ray structure of the GGA1 VHS domain alone, and in complex with the carboxy-terminal peptide of CI-MPRs containing an ACLL sequence. The VHS domain forms a super helix with eight α-helices, similar to the VHS domains of TOM1 and Hrs. Unidirectional movements of helices α6 and α8, and some of their side chains, create a set of electrostatic and hydrophobic interactions for correct recognition of the ACLL peptide. This recognition mechanism provides the basis for regulation of protein transport from the TGN to endosomes/lysosomes, which is shared by sortilin and low-density lipoprotein receptor-related protein (Shiba et al. 2002).

The VHS domain of GGA2 was shown to bind to the acidic cluster-dileucine motif in the cytoplasmic tail of CI-MPR by Zhu et al. (2001). Receptors with mutations in this motif were defective in lysosomal enzyme sorting. The hinge domain of GGA2 bound clathrin, suggesting that GGA2 could be a link between cargo molecules and clathrin-coated vesicle assembly. Thus, GGA2 binding to the CI-MPR is important for lysosomal enzyme targeting. Misra et al. (2002) reported the structures of the VHS domain of human GGA3 complexed with signals from both MPRs. The signals bind in an extended conformation to α-helices 6 and 8 of the VHS domain. The structures highlight an Asp residue separated by two residues from a dileucine sequence as critical recognition elements. The side chains of the Asp-X-X-Leu-Leu sequence interact with subsites consisting of one electropositive and two shallow hydrophobic pockets, respectively. The rigid spatial alignment of the three binding subsites leads to high specificity (Misra et al. 2002).

Interactions of GGA3 with Ubiquitin Sorting Machinery: Puertollano and Bonifacino (2004) demonstrated the existence of another population of GGAs that are associated with early endosomes. RNAi of GGA3 expression results in accumulation of the CI-MPR and internalized epidermal growth factor (EGF) within enlarged early endosomes. This impairs the degradation of internalized EGF, a process that is normally dependent on the sorting of ubiquitinated EGF receptors (EGFRs) to late endosomes. Protein interactions showed that the GGAs bind ubiquitin. The VHS and GAT domains of GGA3 were responsible for this binding, and with TSG101, a component of the ubiquitin-dependent sorting machinery. Thus, GGAs may have additional roles in sorting of ubiquitinated cargo.

4.8 Retrieval of CI-MPR from Endosome-TO-GOLGI

4.8.1 Endosome-to-Golgi Retrieval of CIMPR Requires Retromer Complex

After releasing the hydrolase precursors into the endosomal lumen, the unoccupied CI-MPR returns to the TGN for further rounds of sorting. The mammalian retromer complex participates in this retrieval pathway. Endosome-to-Golgi retrieval of the MPR is required for lysosome biogenesis. This pathway is poorly understood. Analyses in yeast identified a complex of proteins called "retromer" that is essential for endosome-to-Golgi retrieval of the carboxypeptidase Y receptor Vps10p. Retromer comprises five distinct proteins: Vps35p, 29p, 26p, 17p, and 5p, which are conserved in mammals. Mammalian retromer is localized to endosomes and comprises two distinct sub complexes: the vacuolar protein sorting Vps26 (vacuolar sorting protein 26), Vps29 and Vps35 proteins sub complex that binds cargo and the SNX-1/2 sub complex that tubulates endosomal membranes. The hVps35 subunit of retromer interacts with the cytosolic domain of the CI-MPR. This interaction occurs in an endosomal compartment, where most of the retromer is localized. In particular, retromer is associated with tubular-vesicular profiles that emanate from early endosomes or from intermediates in the maturation from early to late endosomes. Depletion of retromer by iRNA increases the lysosomal turnover of the CI-MPR, decreases cellular levels of lysosomal hydrolases, and causes swelling of lysosomes. These observations indicate that retromer prevents the delivery of the CI-MPR to lysosomes, probably by sequestration into endosome-derived tubules from where the CIMPR returns to the TGN (Arighi et al. 2004; Damen et al. 2006).

Cells lacking mammalian Vps26 fail to retrieve the CI-MPR, resulting in either rapid degradation of or mislocalization to the plasma membrane. Study supports the hypothesis that retromer performs a selective function in endosome-to-Golgi transport, mediating retrieval of the CI-MPR, but not furin (Seaman 2004). A conserved aromatic-containing sorting motif is critical for the endosome-to-TGN retrieval of the CI-MPR and for the interaction with retromer and the clathrin adaptor AP-1 (Seaman 2007).

4.8.2 Retromer Complex and Sorting Nexins (SNX)

Retromer Complex

In yeast, retromer is composed of Vps5p (the orthologue of SNX-1), Vps17p (a related SNX) and a cargo selective subcomplex composed of Vps26p, Vps29p and Vps35p. With the exception of Vps17p, mammalian orthologues of all yeast retromer components have been identified. For Vps17p, one potential mammalian orthologue is SNX-2. Sorting nexins (SNXs) are phox homology (PX) domain-containing proteins thought to regulate endosomal sorting of internalized receptors. SNX-1 and SNX-2 through their PX domain bind phosphatidylinositol 3-monophosphate [PtdIns(3)P] and phosphatidylinositol 3,5-bisphosphate [PtdIns(3,5)P(2)], and possess a Bin/Amphiphysin/Rvs domain that endows SNX1 with the ability to form dimers and to sense membrane curvature. However, in contrast to SNX-1, SNX-2 could not induce membrane tubulation in vitro or in vivo. Functionally, endogenous SNX-1/-2 co-localise on high curvature tubular elements of the 3-phosphoinositide-enriched early endosome, and that suppression of SNX-2 does not perturb the degradative sorting of receptors for epidermal growth factor or transferrin, nor the steady-state distribution of the CI-MPR. However, suppression of SNX-2 results in a subtle alteration in the kinetics of CI-MPR retrieval.

SNX1 Defines an Early Endosomal Recycling Exit for Sortilin and MPRs: In addition to MPRs, the multi-ligand receptor sortilin has also been implicated in transport of lysosomal hydrolases from the TGN to endosomes. The transport carriers involved have been identified recently. Sortilin of HepG2 cells predominantly localized to the TGN and endosomes. Sortilin and MPRs recycle to the TGN in SNX1-dependent carriers, which were named endosome-to-TGN transport carriers (ETCs). The ETCs emerge from early endosomes (EE), lack recycling plasma membrane proteins and exhibit unique structural features. Hence, ETCs are distinct from hitherto described EE-derived membranes involved in recycling. This study emphasizes an important role of EEs in recycling to the TGN (Mari et al. 2008).

SNX1 is associated with early endosomes, from where regulates the degradation of internalized epidermal growth factor (EGF) receptors through modulating endosomal-to-lysosomal sorting. Through coincidence detection, the BAR and PX domains efficiently target SNX1 to a microdomain of early endosome defined by high curvature and the presence of 3-phosphoinositides. In addition, the BAR domain endows SNX1 with an ability to tubulate membranes in-vitro and drive the tubulation of the endosomal compartment in-vivo. Using RNAi, SNX1 did not demonstrate a role in EGF or transferrin receptor sorting; rather it specifically perturbed endosome-to-TGN transport of CI-MPR (Carlton et al. 2004). Studies support an evolutionarily conserved function for SNX1 from yeast to mammals and provide functional insight into the molecular mechanisms underlying lipid-mediated protein targeting and tubular-based protein sorting. It appears that through coincidence detection SNX1 associates with a microdomain of the early endosome-characterized by high membrane curvature and the presence of 3-phosphoinositides-from where it regulates tubular-based endosome-to-TGN retrieval of the CI-MPR (Carlton et al. 2005). Bujny et al. (2007) suggest a role for SNX1 in the endosome-to-TGN transport of Shiga toxin and are indicative for a fundamental difference between endosomal sorting of Shiga and cholera toxins into endosome-to-TGN retrograde transport pathways.

Genetic Targeting of Retromer Components Leads to Embryonic Lethality: Mammalian retromer components sorting SNX-1 and SNX-2 result in embryonic lethality when simultaneously targeted for deletion in mice, whereas others have shown that Hβ58 (also known as mVps26), another retromer component, results in similar lethality when targeted for deletion. Genetic interaction of these mammalian retromer components in mice reveals a functional interaction between Hβ58, SNX-1, and SNX-2. SNX-2 plays a more critical role than SNX1 in retromer activity during embryonic development. This supports the existence of retromer complexes containing SNX-1 and SNX-2; SNX-2 as an important mediator of retromer biology. Moreover, mammalian retromer complexes containing SNX-1 and SNX-2 have an essential role in embryonic development that is independent of CI-MPR trafficking (Griffin et al. 2005). Observations indicate that the mammalian retromer complex assembles by sequential association of SNX1/2 and Vps26-Vps29-Vps35 subcomplexes on endosomal membranes and that SNX1 and SNX2 play interchangeable but essential roles in retromer structure and function (Rojas et al. 2007). However, Carlton et al. (2005) indicated that although SNX-2 may be a component of the retromer complex, its presence is not essential for the regulation of endosome-to-TGN retrieval of the CI-MPR (Carlton et al. 2005).

SNX5 and SNX6 in Retromer Regulate CI-MPR

In yeast, an additional sorting nexin-Vps17p-is a component of the membrane bound coat. Mammalian retromer may require a functional equivalent of Vps17p. Wassmer et al. (2007) identified two proteins, SNX5 and SNX6, that, when suppressed, induced a phenotype similar to that observed upon suppression of known retromer components. Whereas SNX5 and SNX6 colocalised with SNX1 on early endosomes, only SNX6 appeared to exist in a immune-complex with SNX1. Interestingly, suppression of SNX5 and/or SNX6 resulted in a significant loss of SNX1, an effect that seemed to result from post-translational regulation of the SNX1 level. Such data suggest that SNX1 and SNX6 exist in a stable, endosomally associated complex that is required for retromer-mediated retrieval of CI-MPR. SNX5 and SNX6 may therefore constitute functional equivalents of Vps17p in mammals. Proteomic profiles of endosomally enriched membranes from wild-type or retromer-deficient mouse cells showed Eps15 homology domain-containing protein-1 (EHD1) in endosomally enriched membrane fractions from retromer-deficient cells. EHD1 is localized to tubular and vesicular endosomes, partially colocalizes with retromer and is associated with retromer in vivo. The interaction between EHD1 and retromer and the requirement for EHD1 to stabilize SNX1-tubules establish EHD1 as a novel facilitating component of endosome-to-Golgi retrieval (Gokool et al. 2007).

DOCK180, a DOCK180-family guanine nucleotide exchange factor for small GTPases Rac1 and Cdc42, shares two conserved domains, called DOCK homology region (DHR)-1 and -2. In order to understand the function of DHR1, SNX-1, 2, 5, and 6 were searched as its binding partners which make up a multimeric protein complex mediating endosome-to-TGN retrograde transport of CI-MPR. Colocalization of DOCK180 with SNX5 at endosomes and the iRNA-mediated knockdowns of SNX5 and DOCK180, and the redistribution of CI-MPR from TGN to endosomes suggested that DOCK180 regulates CI-MPR trafficking via SNX5 and that this function was independent of its Rac1 (Hara et al. 2008).

OCRL1 Regulates Protein Trafficking Between Endosomes and TGN

Oculocerebrorenal syndrome of Lowe is caused by mutation of OCRL1, a phosphatidylinositol 4,5-bisphosphate 5-phosphatase localized at the Golgi apparatus. The OCRL1 is associated with clathrin-coated transport intermediates operating between TGN and endosomes. OCRL1 interacts directly with clathrin heavy chain and promotes clathrin assembly in vitro. Over-expression of OCRL1 results in redistribution of clathrin and the CI-MPR to enlarged endosomal structures that are defective in retrograde trafficking to the TGN. It seems that OCRL1 plays a role in clathrin-mediated trafficking of proteins from endosomes to the TGN and defects in this pathway might contribute to the Lowe syndrome phenotype (Choudhury et al. 2005).

Hrs and ESCRT Proteins Function at Different Stages in Endocytic Pathway: A ubiquitin-binding endosomal protein machinery is responsible for sorting endocytosed membrane proteins into intraluminal vesicles of multivesicular endosomes (MVEs) for subsequent degradation in lysosomes. The Hrs-STAM complex and endosomal sorting complex required for transport (ESCRT)-I, -II and -III are central components of this machinery. Raiborg et al. (2008) revealed that none of them was required for recycling of CI-MPRs from endosomes to TGN. It seemed that they function at distinct stages of endocytic pathway.

Cholesterol Requirement for CI-MPR Exit from Late Endosomes to Golgi

The regulation of endocytic traffic of receptors has central importance in the fine tuning of cell activities. Miwako et al. (2001) provided evidence that cholesterol is required for the exit of CI-MPR from the endosomal carrier vesicle/multivesicular bodies (ECV/MVBs) to the Golgi. A Chinese hamster ovary cell mutant, LEX2, exhibits arrested ECV/MVBs in which CI-MPR and lysosomal glycoprotein-B (lgp-B) are accumulated. Results suggest that cholesterol is required for ECV/MVB reorganization that drives the sorting/transport of materials destined for the Golgi out of the pathways towards lysosomes.

Cholesterol accumulated in late Endocytic Compartments Does Not Affect Distribution and Trafficking of CI-MPR: The treatment with 3β-[2-(diethylamino)ethoxy] androst-5-en-17-one (U18666A) is known to cause the accumulation of cholesterol in late endosomal/lysosomal compartments in BHK cells. The accumulation of cholesterol within late endosomes/lysosomes in Niemann-Pick type C (NPC) fibroblasts and U18666A-treated cells causes impairment of retrograde trafficking of CI-MPR/IGF2R from late endosomes to TGN, where as the accumulation of cholesterol in late endosomes/lysosomes did not affect the retrieval of CI-MPR from endosomes to the TGN. However, treatment of normal and NPC fibroblasts with chloroquine, which inhibits membrane traffic from early endosomes to the TGN, resulted in a redistribution of CI-MPR to EEA1 and internalized transferrin-positive, but LAMP-2-negative, early-recycling endosomes. It was proposed that in normal and NPC fibroblasts, CI-MPR is exclusively targeted from the TGN to early endosomes, from where it rapidly recycles back to the TGN without being delivered to late endosomes (Umeda et al. 2003). In HeLa cells, it was suggested that U18666A treatment primarily suppresses the CI-MPR transport pathways to late endosomes and from transferrin-containing endosomes, both of which may be dependent on cholesterol function (Tomiyama et al. 2004).

4.8.3 Small GTPases in Lysosome Biogenesis and Transport

Abundant information is available about the subcellular distribution and function of some of the endocytosis-specific Rabs (e.g. Rab5, Rab4, Rab7, Rab15, Rab22). Soon after endocytosis, internalized material is sorted along different pathways in a process that requires the coordinated activity of several Rab proteins. The molecular machinery behind lysosome biogenesis and the maintenance of the perinuclear aggregate of late endocytic structures is not well understood. A likely candidate for being part of this machinery is the small GTPase Rab7. Rab GTPases belong to the Ras GTPase superfamily and are key regulators of membrane traffic. Rab7 also appears to play a fundamental role in controlling late endocytic membrane traffic (Vitelli et al. 1997). In contrast to the wild-type protein, a rab7 mutant with a reduced GTPase activity is in part associated with lysosomal membranes. Results implicated rab7 as a GTPase functioning on terminal endocytic structures in mammalian cells (Méresse et al. 1995). Bucci et al. (2000) demonstrated that Rab7, controlling aggregation and fusion of late endocytic structures/lysosomes, is essential for maintenance of the perinuclear lysosome compartment. Stable BHK cell lines inducibly expressing wild-type or dominant negative mutant forms of the rab7 GTPase were used to study the role of a rab7-regulated pathway in lysosome biogenesis. Expression of mutant rab7N125I protein induced a dramatic redistribution of CI-MPR from its normal perinuclear localization to large peripheral endosomes.

Rab22a associates with early and late endosomes. Conversely, overexpression of Rab22aQ64L which strongly affects the morphology of endosomes, did not inhibit bulk endocytosis. Over-expression of Rab22a hampers the transport between endosomes and the Golgi apparatus. Moreover, these cells accumulated the CI-MPR in endosomes. Observations indicate that Rab22a can affect the trafficking from endosomes to the Golgi apparatus probably by promoting fusion among endosomes and impairing the proper segregation of membrane domains required for targeting to the TGN (Mesa et al. 2001, 2005).

4.8.4 Role for Dynamin in Late Endosome Dynamics and Trafficking of CI-MPR

It is well established that dynamin is involved in clathrin-dependent endocytosis and the endogenous dynamin plays an important role in the molecular machinery behind the recycling of the CI-MPR from endosomes to the TGN. Nicoziani et al. (2000) proposed that dynamin is required for the final scission of vesicles budding from endosome tubules.

Granzyme B (GzmB) endocytosis is facilitated by dynamin in many endocytic pathways. Uptake of and killing by purified granzyme B occurred by both dynamin-dependent and -independent mechanisms. It was proposed that under physiological conditions serglycin-bound granzyme B is critically endocytosed by a MPR, and receptor binding is enhanced by cell surface heparan sulfate (Veugelers et al. 2006). The CI-MPR participates in lysosomal and granular targeting of serglycin and basic proteins such as lysozyme associated with the proteoglycan in hematopoietic cells (Lemansky et al. 2007). However, the role of the CI-MPR has been refuted and that membrane receptors for GzmB on target cells are not crucial for killer cell-mediated apoptosis (Kurschus et al. 2005).

4.9 CI-MPR/IGF2R System and Pathology

4.9.1 Deficiency of IGF2R/CI-MPR Induces Myocardial Hypertrophy

Mice lacking the IGF2R have increased levels of circulating IGF2, are born larger, and die soon after birth due to cardiac hyperplasia. This phenotype is overcome by concomitant deletion of the IGF2 gene (Ludwig et al. 1996). The IGF2R function in extra-cellular matrix remodeling is associated with TGF-β activation and plasmin in the proteolytic cleavage caused by the interaction between latent TGF-β and uPAR. IGF2 and IGF2R have also been correlated with the progression of hypertrophy remodeling following abdominal aorta ligation. IGF2R is expressed in myocardial infarction scars. While exploring the function of IGF2R and IGF2 in myocardial extra-cellular matrix remodeling, it was suggested that after binding with IGF2, IGF2R may trigger intracellular signaling cascades involved in the progression of cardiac hypertrophy. Using $Leu^{27}IGF2$, an analog of IGF2 which interacts selectively with the IGF2R, the binding of $Leu^{27}IGF2$ to IGF2R resulted into an increase in the phosphorylation of PKCα and calcium/calmodulin-dependent protein kinase II (CaMKII) in a $G_{\alpha q}$-dependent manner. By the inhibition of PKCα/CaMKII activity, the IGF2 and $Leu^{27}IGF2$-induced cell hypertrophy and up-regulation of ANP and BNP were significantly suppressed (Chu et al. 2008). It was suggested that IGF2R signaling inhibition may have potential use in the development of therapies preventing heart fibrosis progression and a new insight into the effects of IGF2R and its downstream signaling in cardiac hypertrophy (Chang et al. 2008; Chu et al. 2008).

Transport of lysosomal enzymes is mediated by CD-MPR and a CI-MPR. To address the consequences of abnormalities of cellular morphology and function on CI-MPR subcellular localization, fibroblasts from Pompe disease patients with different genotypes and phenotypes have been studied. In these cells, which showed abnormalities of cellular morphology, CI-MPR is mislocalized and its availability at the plasma membrane is reduced. These abnormalities in CI-MPR distribution result in a less efficient uptake of rhGAA by Pompe disease fibroblasts. CI-MPR-mediated endocytosis of rhGAA is an important pathway by which the enzyme is delivered to the affected lysosomes of Pompe muscle cells (Chap. 46). In hepatocytes of MPR-deficient neonatal mice lysosomal storage occurs when both MPRs are lacking, whereas deficiency of CI-MPR only has no effect on the ultrastructure of the lysosomal system. Some structural features have been shown to be crucial for the binding of M6P to CI-MPR.

4.9.2 MPRs in Neuromuscular Diseases

M6PRs in heart muscle function in the endocytosis and transport of lysosomal enzymes in cardiomyocytes. In association, the activity of β-N-acetyl-glucosaminidase significantly increases in the muscles of patients with myopathies (polymyositis and muscular dystrophies) but not in those with neurogenic muscle atrophies (amyotrophic lateral sclerosis, polyneuropathy or other neurogenic muscle disease). The content of M6PRs correlated with the muscular activity of β-N-acetylglucosaminidase, muscle atrophy index, and serum creatine kinase activity (Salminen et al. 1988).

Alzheimer's Disease (AD): The expression of the CD-MPR is increased in Alzheimer's disease. Endosomal and lysosomal changes are invariant features of neurons in AD. These changes include increased levels of lysosomal hydrolases in early endosomes and increased expression of CD-MPR, which is partially localized to early endosomes. The redistribution of lysosomal hydrolases to early endocytic compartments mediated by increased expression of CD-MPR may represent a potentially pathogenic mechanism for accelerating amyloid beta (Aβ) generation in sporadic AD, where the mechanism of amyloidogenesis is unknown (Mathews et al. 2002). However, despite CD-MPR gene being located next to a region on chromosome 12 linked to AD, Kölsch et al. (2004) could not find an association of C/T polymorphism in CD-MPR with AD.

CI-MPR Trafficking in Batten disease: The neuronal ceroid lipofuscinoses (NCLs, Batten disease) are a group of inherited childhood-onset neurodegenerative disorders characterized by the lysosomal accumulation of undigested material within cells. Cause of dysfunction may be associated with trafficking of CI-MPR, which delivers the digestive enzymes to lysosomes. A common form of NCL is caused by mutations in CLN3, a multipass transmembrane protein of unknown function. The ablation of CLN3 causes accumulation of CI-MPR in the TGN, reflecting a 50% reduction in exit. This CI-MPR trafficking defect is accompanied by a fall in maturation and cellular activity of lysosomal cathepsins. CLN3 is therefore essential for trafficking along the route needed for delivery of lysosomal enzymes, and its loss thereby contributes to and may explain the lysosomal dysfunction underlying Batten disease (Metcalf et al. 2008).

4.9.3 CI-MPR in Fanconi syndrome

In renal Fanconi syndrome (FS), Norden et al. (2008) suggest that the underlying gene defects in FS may disrupt normal membrane trafficking of CI-MPR, leading to mistrafficking of lysosomal enzymes via a default pathway from the Golgi to the apical surface of proximal tubule cells rather than to lysosomes. Lysosomal enzymes are then secreted into the tubular fluid and excreted in the urine. This contrasts with the widely held view that cell necrosis is the cause of lysosomal enzymuria in renal disease. Moreover, cathepsin D in FS urine is M6P-tagged.

4.9.4 Tumor Suppressive Effect of CI-MPR/IGF2R

The M6P/IGF2R gene is considered a "candidate" tumor suppressor gene. The phenotypic consequences of loss of M6P/IGF2R through somatic mutation are potentially very complex since M6P/IGF2R has a number of roles in cellular physiology. Loss of function mutations in M6P/IGF2R gene could contribute to multi-step carcinogenesis (Hébert 2006; Kreiling et al. 2003). Mutation of M6P/IGF2R causes both diminished growth suppression and augmented growth stimulation. The M6P/IGF2R is a negative regulator of cell growth. The genetic alterations in hepatocarcinomas and a few breast cancers suggest that this receptor behaves as a tumor suppressor. Mutational and functional evidences are consistent with CI-MPR/IGF2R being a tumor suppressor in human colon, liver, lung, breast, and ovarian cancers.

Decreased levels of functional M6P/IGF2R directly contribute to the process of carcinogenesis and loss of IGF2R activity in many cancers has been associated with increased tumor cell growth and tumor progression. This can arise from the loss of heterozygosity or mutation of the receptor (De Souza et al. 1995; Devi et al. 1998; Killian et al. 2001), leading to an increase in bioavailable IGF2, which then acts via the type 1 IGF receptor (IGF-1R) to promote cancer growth (Alexia et al. 2004). As cancer cell proliferation can be abrogated by blocking mRNA or protein products of these genes, tumors with extensive involvement of the IGF2 pathway are candidates for the therapeutics strategies aimed at interference with this pathway.

Studies on Cancer Cells: Loss or mutation of the M6P/IGF2R has been found in breast cancer. Human breast cancer cells, MCF-7 cells with the adenovirus carrying a ribozyme targeting the CI-MPR/IGF2R mRNA dramatically reduced the level of transcripts and the functional activity of IGF2R and exhibited a higher growth rate and a lower apoptotic index than control cells. The decreased expression of IGF2R enhanced IGF2-induced proliferation and reduced cell susceptibility to TNF-induced apoptosis (Chen et al. 2002). Tumor suppressive effect of IGF2R has been displayed by 66cl4, a mouse mammary tumor cell line deficient in the receptor (Li and Sahagian 2004). Furthermore, patients with heterozygosity (LOH) are much more likely to have elevated plasma TGFβ, suggesting an inability to normally process this cytokine (Kong et al. 2001). Ionizing radiation induces the rapid expression of M6P/IGF2R in a dose-dependent manner in MCF7 cancer cells. This increase is mediated, at least in part, by a stabilization of M6P/IGF2R transcripts by radiation in both estrogen receptor (ER) positive (MCF7 and T47D) and ER negative (MDA-MB-231) breast cancer cell lines (Iwamoto and Barber 2007).

In opposing studies, MDA-MB-231 breast cancer cells stably transfected with M6P/IGF2R cDNA exhibited not only a greatly reduced ability to form tumors but also a markedly reduced growth rate in nude mice. In vitro, increased M6P/IGF2R expression resulted in twofold reduced uptake of IGF2 and was associated with reduced cellular invasiveness and motility (Lee et al. 2003). A frequent loss of heterozygosity (LOH) in the 6q27-qter region in ovarian carcinomas confirmed the role of *IGF2R* gene in ovarian carcinomas and breast- and ovarian-cancer cell lines. The 2491 amino-acid sequence of M6P/IGF2R was perfectly conserved in 9 out of 10 samples, including MCF7 and MDA-MB231 cells and five ovarian carcinomas with LOH. The only amino-acid change (Thr → Ala) was in BG1 ovarian-cancer cells, and was due to an A → G substitution on one allele at nucleotide 2561. Rey et al. (2000) proposed that, in breast and ovarian cancers, the frequent loss of one allele, associated with over-expression of some of its ligands, might be sufficient to saturate the receptor protein, displace the ligands to other sites, and consequently facilitate tumor progression.

Members of the IGF family are involved in the pathogenesis of gastric cancer, probably by autocrine/paracrine stimulation of cell growth. Such tumors might be excellent candidates for therapeutic strategies aimed at interference with this pathway (Pavelić et al. 2007). Decreased IGF2R expression could partly account for the increased growth of lymph node carcinoma of the prostate (LNCaP). It appears that the IGF2- and M6P-binding functions of the IGF2R have opposing activities, with respect to growth of prostate cancer cells (Schaffer et al. 2003). Pavelić et al. (2007) suggested that IGF1, IGF2 and their receptors were involved in the progression of endometrial adenocarcinomas. M6P/IGF2R is expressed in human lung-cancer cells (Bredin et al. 1999). Harper et al. (2006) showed rescue of the *Igf2*-dependent intestinal and adenoma phenotype. This evidence shows the functional potency of allelic dosage of an epigenetically regulated gene in cancer and supports the application of an IGF2 ligand-specific therapeutic intervention in colorectal cancer. The MPRs are expressed in a polarized fashion in human adenocarcinoma cell line Caco-2 and the IGF2R/CI-MPR present on apical membranes, unlike the IGF-2R/CI-MPR expressed on the basolateral surface, is not functional in endocytosing lysosomal enzymes (Dahms et al. 1996). Reduced supply of IGF2 is detrimental to tumor growth, and this suggests that gain of function of IGF2 is a molecular target for human cancer therapy (Chap. 46).

References

Alexia C, Fallot G, Lasfer M et al (2004) An evaluation of the role of insulin-like growth factors (IGF) and of type-I IGF receptor signalling in hepatocarcinogenesis and in the resistance of hepatocarcinoma cells against drug-induced apoptosis. Biochem Pharmacol 68:1003–1015

Arighi CN, Hartnell LM, Aguilar RC, Haft CR, Bonifacino JS (2004) Role of the mammalian retromer in sorting of the cation-independent mannose 6-phosphate receptor. J Cell Biol 165:123–133

Barois N, Bakke O (2005) The adaptor protein AP-4 as a component of the clathrin coat machinery: a morphological study. Biochem J 385:503–510, 17067–17074

Blanchard F, Duplomb L, Raher S et al (1999) Mannose 6-phosphate/insulin-like growth factor ii receptor mediates internalization and degradation of leukemia inhibitory factor but not signal transduction. J Biol Chem 274:24685–24693

Bohnsack RN, Song X, Olson LJ et al (2009) Cation-independent mannose 6-phosphate receptor: a composite of distinct phosphomannosyl binding sites. J Biol Chem 284:35215–35226

Bonifacino JS (2004) The GGA proteins: adaptors on the move. Nat Rev Mol Cell Biol 5:23–32

Brady RO (2006) Enzyme replacement for lysosomal diseases. Annu Rev Med 57:283–296

Braulke T (1999) Type-2 IGF receptor: a multi-ligand binding protein. Horm Metab Res 3:242–246

Bredin CG, Liu Z, Hauzenberger D et al (1999) Growth-factor-dependent migration of human lung-cancer cells. Int J Cancer 82:338–345

Breuer P, Korner C, Boker C et al (1997) Serine phosphorylation site of the 46-kDa mannose 6-phosphate receptor is required for transport to the plasma membrane in Madin–Darby canine kidney and mouse fibroblast cells. Mol Biol Cell 8:567–576

Brown J, Esnouf RM, Jones MA et al (2002) Structure of a functional IGF2R fragment determined from the anomalous scattering of sulfur. EMBO J 21:1054–1062

Brown J, Delaine C, Zaccheo OJ et al (2008) Structure and functional analysis of the IGF2/IGF2R interaction. EMBO J 27:265–276

Brunetti CR, Burke RL, Hoflack B et al (1995) Role of mannose-6-phosphate receptors in herpes simplex virus entry into cells and cell-to-cell transmission. J Virol 69:3517–3528

Brunetti CR, Burke RL, Kornfeld S, Gregory W, Masiarz FR, Dingwell KS, Johnson DC (1994) Herpes simplex virus glycoprotein D acquires mannose 6-phosphate residues and binds to mannose 6-phosphate receptors. J Biol Chem 269:17067–17074

Brunetti CR, Dingwell KS, Wale C et al (1998) Herpes simplex virus gD and virions accumulate in endosomes by mannose 6-phosphate-dependent and -independent mechanisms. J Virol 72:3330–3339

Bucci C, Thomsen P, Nicoziani P, McCarthy J, van Deurs B (2000) Rab7: a key to lysosome biogenesis. Mol Biol Cell 11:467–480

Bujny MV, Popoff V, Johannes L, Cullen PJ (2007) The retromer component sorting nexin-1 is required for efficient retrograde transport of Shiga toxin from early endosome to the trans Golgi network. J Cell Sci 120:2010–2021

Byrd JC, MacDonald RG (2000) Mechanisms for high affinity mannose 6-phosphate ligand binding to the insulin-like growth factor II/mannose 6-phosphate receptor. J Biol Chem 275:18638–18646

Cacia J, Quan CP, Pai R, Frenz J (1998) Human DNase I contains mannose 6-phosphate and binds the cation-independent mannose 6-phosphate receptor. Biochemistry 37:15154–15161

Canfield WM, Kornfeld S (1989) The chicken liver cation-independent mannose 6-phosphate receptor lacks the high affinity binding site for insulin-like growth factor II. J Biol Chem 264:7100–7103

Canfield WM, Johnson KF, Ye RD et al (1991) Localization of the signal for rapid internalization of the bovine cation-independent mannose 6-phosphate/insulin-like growth factor-II receptor to amino acids 24-29 of the cytoplasmic tail. J Biol Chem 266:5682–5688

Carlton J, Bujny M, Peter BJ et al (2004) Sorting nexin-1 mediates tubular endosome-to-TGN transport through coincidence sensing of high- curvature membranes and 3-phosphoinositides. Curr Biol 14:1791–1800

Carlton JG, Bujny MV, Peter BJ et al (2005) Sorting nexin-2 is associated with tubular elements of the early endosome, but is not essential for retromer-mediated endosome-to-TGN transport. J Cell Sci 118:4527–4539

Chang MH, Kuo WW, Chen RJ et al (2008) IGF2/mannose 6-phosphate receptor activation induces metalloproteinase-9 matrix activity and increases plasminogen activator expression in H9c2 cardiomyoblast cells. J Mol Endocrinol 41:65–74

Chapuy B, Tikkanen R, Mühlhausen C et al (2008) AP-1 and AP-3 mediate sorting of melanosomal and lysosomal membrane proteins into distinct post-golgi trafficking pathways. Traffic 9:1157–1172

Chavez CA, Bohnsack RN, Kudo M et al (2007) Domain 5 of the cation-independent mannose 6-phosphate receptor preferentially binds phosphodiesters (mannose 6-phosphate *N*-acetylglucosamine ester). Biochemistry 46:12604–12617

Chen HJ, Remmler J, Delaney JC, Messner DJ, Lobel P (1993) Mutational analysis of the cation-independent mannose 6-phosphate/insulin-like growth factor II receptor. A consensus casein kinase II site followed by 2 leucines near the carboxyl terminus is important for intracellular targeting of lysosomal enzymes. J Biol Chem 268:22338–22346

Chen HJ, Yuan J, Lobel P (1997) Systematic mutational analysis of the cation-independent mannose 6-phosphate/insulin-like growth factor II receptor cytoplasmic domain. An acidic cluster containing a key aspartate is important for function in lysosomal enzyme sorting. J Biol Chem 272:7003–7012

Chen Z, Ge Y, Landman N et al (2002) Decreased expression of the mannose 6-phosphate/insulin-like growth factor-II receptor promotes growth of human breast cancer cells. BMC Cancer 2:18

Chen JJ, Zhu Z, Gershon AA et al (2004) Mannose 6-phosphate receptor dependence of varicella zoster virus infection in vitro and in the epidermis during varicella and zoster. Cell 119:915–926

Choudhury R, Diao A, Zhang F et al (2005) Lowe syndrome protein OCRL1 interacts with clathrin and regulates protein trafficking between endosomes and the trans-Golgi network. Mol Biol Cell 16:3467–3479

Chu CH, Tzang BS, Chen LM et al (2008) IGF2/mannose-6-phosphate receptor signaling induced cell hypertrophy and atrial natriuretic peptide/BNP expression via Gαq interaction and protein kinase C-α/CaMKII activation in H9c2 cardiomyoblast cells. J Endocrinol 197:381–390

Dahms NM (1996) Insulin-like growth factor II/cation-independent mannose 6-phosphate receptor and lysosomal enzyme recognition. Biochem Soc Trans 24:136–141

Dahms NM, Hancock MK (2002) P-type lectins. Biochim Biophys Acta 1572:317–340

Dahms NM, Lobel P, Breitmeyer J et al (1987) 46 kD mannose 6-phosphate receptor: cloning, expression, and homology to the 215 kd mannose 6-phosphate receptor. Cell 50:181–192

Dahms NM, Brzycki-Wessell MA, Ramanujam KS, Seetharam B (1993a) Characterization of mannose 6-phosphate receptors (MPRs) from opossum liver: opossum cation-independent MPR binds insulin-like growth factor-II. Endocrinology 133:440–446

Dahms NM, Rose PA, Molkentin JD et al (1993b) The bovine mannose 6-phosphate/insulin-like growth factor II receptor. The role of arginine residues in mannose 6-phosphate binding. J Biol Chem 268:5457–5463

Dahms NM, Wick DA, Brzycki-Wessell MA (1994) The bovine mannose 6-phosphate/insulin-like growth factor II receptor. Localization of the insulin-like growth factor II binding site to domains 5-11. J Biol Chem 269:3802–3809

Dahms NM, Seetharam B, Wick DA (1996) Expression of insulin-like growth factor (IGF)-I receptors, IGF2/cation-independent mannose 6-phosphate receptors (CI-MPRs), and cation-dependent MPRs in polarized human intestinal Caco-2 cells. Biochim Biophys Acta 1279:84–92

Dahms NM, Olson LJ, Kim JJ (2008) Strategies for carbohydrate recognition by the mannose 6-phosphate receptors. Glycobiology 18:664–678

Damen E, Krieger E, Nielsen JE et al (2006) The human Vps29 retromer component is a metallo-phosphoesterase for a cation-independent mannose 6-phosphate receptor substrate peptide. Biochem J 398:399–409

De Souza AT, Hankins GR, Washington MK, Orton TC, Jirtle RL (1995) M6P/IGF2R gene is mutated in human hepatocellular carcinomas with loss of heterozygosity. Nat Genet 11:447–449

Delaine C, Alvino CL, McNeil KA et al (2007) A novel binding site for the human insulin-like growth factor-II (IGF2)/mannose 6-phosphate receptor on IGF2. J Biol Chem 282:18886–18894

Dennis PA, Rifkin DB (1991) Cellular activation of latent transforming growth factor β requires binding to the cation-independent mannose 6-phosphate/insulin-like growth factor type II receptor. Proc Natl Acad Sci USA 88:580–584

Derby MC, Lieu ZZ, Brown D et al (2007) The trans-Golgi network golgin, GCC185, is required for endosome-to-Golgi transport and maintenance of Golgi structure. Traffic 8:758–773

Devi GR, Byrd JC, Slentz DH et al (1998) An insulin-like growth factor ii (IGF2) affinity-enhancing domain localized within extracytoplasmic repeat 13 of the IGF2/mannose 6-phosphate receptor. Mol Endocrinol 12:166–172

Díaz E, Pfeffer SR (1998) TIP47: a cargo selection device for mannose 6-phosphate receptor trafficking. Cell 93:433–449

Dicioccio RA, Miller AL (1992) Binding receptors for α-L-fucosidase in human B-lymphoid cell lines. Glycoconj J 9:56–62

Doray B, Bruns K, Ghosh P, Kornfeld SA (2002) Autoinhibition of the ligand-binding site of GGA1/3 VHS domains by an internal acidic cluster-dileucine motif. Proc Natl Acad Sci USA 99:8072–8077

Folkman J (2002) Role of angiogenesis in tumor growth and metastasis. Semin Oncol 29(Suppl 16):15–8

Forbes BE, McNeil KA, Scott CD et al (2001) Contribution of residues A54 and L55 of the human insulin-like growth factor-II (IGF2) A domain to Type 2 IGF receptor binding specificity. Growth Factors 19:163–173

Friedl R, Rottmann O (1994) Assignment of the cation independent mannose 6-phosphate/insulin-like growth factor II receptor to bovine chromosome 9q27-28 by fluorescent in situ hybridization. Anim Genet 25:191–193

Garmroudi F, MacDonald RG (1994) Localization of the insulin-like growth factor II (IGF2) binding/cross-linking site of the IGF2/ mannose 6-phosphate receptor to extracellular repeats 10-11. J Biol Chem 269:26944–26952

Gary-Bobo M, Nirdé P, Jeanjean A et al (2007) Mannose 6-phosphate receptor targeting and its applications in human diseases. Curr Med Chem 14:2945–2953

Gasanov U, Koina C, Beagley KW, Aitken RJ, Hansbro PM (2006) Identification of the insulin-like growth factor II receptor as a novel receptor for binding and invasion by Listeria monocytogenes. Infect Immun 74:566–577

Ghahary A, Tredget EE, Mi L, Yang L (1999) Cellular response to latent TGF-beta1 is facilitated by insulin-like growth factor-II/mannose-6-phosphate receptors on MS-9 cells. Exp Cell Res 251:111–120

Ghosh P, Kornfeld S (2004) The GGA proteins: key players in protein sorting at the trans-Golgi network. Eur J Cell Biol 83:257–262

Ghosh P, Dahms NM, Kornfeld S (2003a) Mannose 6-phosphate receptors: new twists in the tale. Nat Rev Mol Cell Biol 4:202–212

Ghosh P, Griffith J, Geuze HJ, Kornfeld S (2003b) Mammalian GGAs act together to sort mannose 6-phosphate receptors. J Cell Biol 163:755–766

Godár S, Horejsi V, Weidle UH et al (1999) M6P/IGF2-receptor complexes urokinase receptor and plasminogen for activation of transforming growth factor-β1. Eur J Immunol 29:1004–1013

Gokool S, Tattersall D, Seaman MN (2007) EHD1 interacts with retromer to stabilize SNX1 tubules and facilitate endosome-to-Golgi retrieval. Traffic 8:1873–1886

Green PJ, Ferguson MA, Robinson PJ, Feizi T (1995) The cation-independent mannose-6-phosphate receptor binds to soluble GPI-linked proteins via mannose-6-phosphate. FEBS Lett 360:34–38

Griffin CT, Trejo J, Magnuson T (2005) Genetic evidence for a mammalian retromer complex containing sorting nexins 1 and 2. Proc Natl Acad Sci USA 102:15173–15177

Griffiths G, Hollinshead R, Hemmings BA, Nigg EA (1990a) Ultrastructural localization of the regulatory (RII) subunit of cyclic AMP-dependent protein kinase to subcellular compartments active in endocytosis and recycling of membrane receptors. J Cell Sci 96:691–703

Griffiths G, Matteoni R, Back R, Hoflack B (1990b) Characterization of the cation-independent mannose 6-phosphate receptor-enriched prelysosomal compartment in NRK cells. J Cell Sci 95:441–461

Groskopf JC, Syu LJ, Saltiel AR, Linzer DI (1997) Proliferin induces endothelial cell chemotaxis through a G protein-coupled, mitogen-activated protein kinase-dependent pathway. Endocrinology 138:2835–2840

Hambleton S (2005) Chickenpox. Curr Opin Infect Dis 18:235–40

Hancock MK, Haskins DJ, Sun G, Dahms NM (2002a) Identification of residues essential for carbohydrate recognition by the insulin-like growth factor ii/mannose 6-phosphate receptor. J Biol Chem 277:11255–11264

Hancock MK, Yammani RD, Dahms NM (2002b) Localization of the carbohydrate recognition sites of the insulin-like growth factor II/ mannose 6-phosphate receptor to domains 3 and 9 of the extracytoplasmic region. J Biol Chem 277:47205–47212

Hara S, Kiyokawa E, Iemura S et al (2008) The DHR1 domain of DOCK180 binds to SNX5 and regulates cation-independent mannose 6-phosphate receptor transport. Mol Biol Cell 19:3823–3835

Harasaki K, Lubben NB, Harbour M et al (2005) Sorting of major cargo glycoproteins into clathrin-coated vesicles. Traffic 6:1014–1026

Harper J, Burns JL, Foulstone EJ et al (2006) Soluble IGF2 receptor rescues Apc(Min/+) intestinal adenoma progression induced by Igf2 loss of imprinting. Cancer Res 66:1940–1948

Hébert E (2006) Mannose-6-phosphate/insulin-like growth factor II receptor expression and tumor development. Biosci Rep 26:7–17

Ikushima H, Munakata Y, Ishii T et al (2000) Internalization of CD26 by mannose 6-phosphate/insulin-like growth factor II receptor contributes to T cell activation. Proc Natl Acad Sci USA 97:8439–8444

Iwamoto KS, Barber CL (2007) Radiation-induced posttranscriptional control of M6P/IGF2r expression in breast cancer cell lines. Mol Carcinog 46:497–502

Jackson D, Linzer DI (1997) Proliferin transport and binding in the mouse fetus. Endocrinology 138:149–155

Kang JX, Li Y, Leaf A (1997) Mannose-6-phosphate/insulin-like growth factor-II receptor is a receptor for retinoic acid. Proc Natl Acad Sci USA 94:13671–13676

Kato Y, Misra S, Puertollano R et al (2002) Phosphoregulation of sorting signal-VHS domain interactions by a direct electrostatic mechanism. Nat Struct Biol 9:532–536

Kiess W, Blickenstaff GD, Sklar MM et al (1988) Biochemical evidence that the type II insulin-like growth factor receptor is identical to the cation-independent mannose 6-phosphate receptor. J Biol Chem 263:9339–9344

Killian JK, Oka Y, Jang HS, Fu X, Waterland RA, Sohda T, Sakaguchi S, Jirtle RL (2001) Mannose 6-phosphate/insulin-like growth factor 2 receptor (M6P/IGF2R) variants in American and Japanese populations. Hum Mutat 18:25–31

Kölsch H, Ptok U, Majores M et al (2004) Putative association of polymorphism in the mannose 6-phosphate receptor gene with major depression and Alzheimer's disease. Psychiatr Genet 14:97–100

Kong FM, Anscher MS, Sporn TA et al (2001) Loss of heterozygosity at the mannose 6-phosphate insulin-like growth factor 2 receptor (M6P/IGF2R) locus predisposes patients to radiation-induced lung injury. Int J Radiat Oncol Biol Phys 49:35–41

Kornfeld S (1992) Structure and function of the mannose 6-phosphate/ insulinlike growth factor II receptors. Annu Rev Biochem 61:307–330

Kreiling JL, Byrd JC, Deisz RJ et al (2003) Binding of urokinase-type plasminogen activator receptor (uPAR) to the mannose 6-phosphate/insulin-like growth factor II receptor: contrasting interactions of full-length and soluble forms of uPAR. J Biol Chem 278:20628–20637

Kreiling JL, Byrd JC, Macdonald RG (2005) Domain interactions of the mannose 6-phosphate/insulin-like growth factor II receptor. J Biol Chem 280:21067–21077

Krise JP, Sincock PM, Orsel JG, Pfeffer SR (2000) Quantitative analysis of TIP47-receptor cytoplasmic domain interactions: implications for endosome-to-trans Golgi network trafficking. J Biol Chem 275:25188–25193

Kun Z, Haiyun Z, Meng W et al (2008) Dietary ω-3 polyunsaturated fatty acids can inhibit expression of granzyme B, perforin, and cation-independent mannose 6-phosphate/insulin-like growth factor receptor in rat model of small bowel transplant chronic rejection. JPEN J Parenter Enteral Nutr 32:12–17

Kurschus FC, Bruno R, Fellows E et al (2005) Membrane receptors are not required to deliver granzyme B during killer cell attack. Blood 105:2049–2058

Laureys G, Barton DE, Ullrich A, Francke U (1988) Chromosomal mapping of the gene for the type II insulin-like growth factor receptor/cation-independent mannose 6-phosphate receptor in man and mouse. Genomics 3:224–229

Le Borgne R, Hoflack B (1997) Mannose 6-phosphate receptors regulate the formation of clathrin-coated vesicles in the TGN. J Cell Biol 137:335–345

Le Borgne R, Hoflack B (1998) Protein transport from the secretory to the endocytic pathway in mammalian cells. Biochim Biophys Acta 1404:195–209

Le Borgne R, Griffiths G, Hoflack B (1996) Mannose 6-phosphate receptors and ADP-ribosylation factors cooperate for high affinity interaction of the AP-1 Golgi assembly proteins with membranes. J Biol Chem 271:2162–2170

Le Borgne R, Alconada A, Bauer U, Hoflack B (1998) The mammalian AP-3 adaptor-like complex mediates the intracellular transport of lysosomal membrane glycoproteins. J Biol Chem 273:29451–29461

Lee SJ, Nathans D (1988) Proliferin secreted by cultured cells binds to mannose 6-phosphate receptors. J Biol Chem 263:3521–3527

Lee JS, Weiss J, Martin JL, Scott CD (2003) Increased expression of the mannose 6-phosphate/insulin-like growth factor-II receptor in breast cancer cells alters tumorigenic properties in vitro and in vivo. Int J Cancer 107:564–570

Lefrançois S, McCormick PJ (2007) The Arf GEF GBF1 is required for GGA recruitment to Golgi membranes. Traffic 8:1440–1451

Leksa V, Godár S, Cebecauer M et al (2002) The N terminus of mannose 6-phosphate/insulin-like growth factor 2 receptor in regulation of fibrinolysis and cell migration. J Biol Chem 277:40575–40582

Lemansky P, Fester I, Smolenova E et al (2007) The cation-independent mannose 6-phosphate receptor is involved in lysosomal delivery of serglycin. J Leukoc Biol 81:1149–1158

Li J, Sahagian GG (2004) Demonstration of tumor suppression by mannose 6-phosphate/insulin-like growth factor 2 receptor. Oncogene 23:9359–9368

Lin SX, Mallet WG, Huang AY et al (2004) Endocytosed cation-independent mannose 6-phosphate receptor traffics via the endocytic recycling compartment en route to the trans-Golgi network and a subpopulation of late endosomes. Mol Biol Cell 15:721–733

Linnell J, Groeger G, Hassan AB (2001) Real time kinetics of insulin-like growth factor II (IGF2) interaction with the IGF2/mannose 6-phosphate receptor: the effects of domain 13 and pH. J Biol Chem 276:23986–23991

Liu Z, Mittanck DW, Kim S, Rotwein P (1995) Control of insulin-like growth factor-II/mannose 6-phosphate receptor gene transcription by proximal promoter elements. Mol Endocrinol 9:1477–1487

Lobel P, Dahms NM, Breitmeyer J et al (1987) Cloning of the bovine 215-kDa cation-independent mannose 6-phosphate receptor. Proc Natl Acad Sci USA 84:2233–2237

Lobel P, Dahms NM, Kornfeld S (1988) Cloning and sequence analysis of the cation-independent mannose 6-phosphate receptor. J Biol Chem 263:2563–2570

Lobel P, Fujimoto K, Ye RD, Griffiths G, Kornfeld S (1989) Mutations in the cytoplasmic domain of the 275 kD mannose 6-phosphate receptor differentially alter lysosomal enzyme sorting and endocytosis. Cell 57:787–796

Ludwig T, Munier-Lehmann H, Bauer U et al (1994) Differential sorting of lysosomal enzymes in mannose 6-phosphate receptor-deficient fibroblasts. EMBO J 13:3430–3437

Ludwig T, Eggenschwiler J, Fisher P et al (1996) Mouse mutants lacking the type 2 IGF receptor (IGF2R) are rescued from perinatal lethality in Igf2 and Igf1r null backgrounds. Dev Biol 177:517–535

MacDonald RG, Pfeffer SR, Coussens L et al (1988) A single receptor binds both insulin-like growth factor II and mannose-6-phosphate. Science 239(4844):1134–1137

Mari M, Bujny MV, Zeuschner D, Geuze HJ et al (2008) SNX1 defines an early endosomal recycling exit for sortilin and mannose 6-phosphate receptors. Traffic 9:380–393

Marjomäki VS, Huovila AP, Surkka MA, Jokinen I, Salminen A (1990) Lysosomal trafficking in rat cardiac myocytes. J Histochem Cytochem 38:1155–1164

Marron-Terada PG, Brzycki-Wessell MA, Dahms NM (1998) The Two mannose 6-phosphate binding sites of the insulin-like growth factor-II/mannose 6-phosphate receptor display different ligand binding properties. J Bio Chem 273:22358–22366

Marron-Terada PG, Hancock MK, Haskins DJ, Dahms NM (2000) Recognition of *Dictyostelium discoideum* lysosomal enzymes is conferred by the amino-terminal carbohydrate binding site of the insulin-like growth factor II/mannose 6-phosphate receptor. Biochemistry 39:2243–2253

Mathews PM, Guerra CB, Jiang Y et al (2002) Alzheimer's disease-related overexpression of the cation-dependent mannose 6-phosphate receptor increases Aβ secretion: role for altered lysosomal hydrolase distribution in β-amyloidogenesis. J Biol Chem 277:5299–5307

Méresse S, Hoflack B (1993) Phosphorylation of the cation-independent mannose 6-phosphate receptor is closely associated with its exit from the trans-Golgi network. J Cell Biol 120:67–75

Méresse S, Ludwig T, Frank R et al (1990) Phosphorylation of the cytoplasmic domain of the bovine cation-independent mannose 6-phosphate receptor. Serines 2421 and 2492 are the targets of a casein kinase II associated to the Golgi-derived HAI adaptor complex. J Biol Chem 265:18833–18842

Méresse S, Gorvel JP, Chavrier P (1995) The rab7 GTPase resides on a vesicular compartment connected to lysosomes. J Cell Sci 108:3349–3358

Mesa R, Salomón C, Roggero M et al (2001) Rab22a affects the morphology and function of the endocytic pathway. J Cell Sci 114:4041–4049

Mesa R, Magadán J, Barbieri A et al (2005) Overexpression of Rab22a hampers the transport between endosomes and the Golgi apparatus. Exp Cell Res 304:339–353

Metcalf DJ, Calvi AA, Seaman MNJ et al (2008) Loss of the Batten disease gene CLN3 prevents exit from the TGN of the mannose 6-phosphate receptor. Traffic 9:1905–1914

Misra S, Puertollano R, Kato Y et al (2002) Structural basis for acidic-cluster-dileucine sorting-signal recognition by VHS domains. Nature 415(6874):933–937

Miwako I, Yamamoto A, Kitamura T, Nagayama K, Ohashi M (2001) Cholesterol requirement for cation-independent mannose 6-phosphate receptor exit from multivesicular late endosomes to the Golgi. J Cell Sci 114:1765–1776

Mordue DG, Sibley LD (1997) Intracellular fate of vacuoles containing Toxoplasma gondii is determined at the time of formation and depends on the mechanism of entry. J Immunol 159:4452–4459

Morgan DO, Edman JC, Standring DN et al (1987) Insulin-like growth factor II receptor as a multifunctional binding protein. Nature 329:301–307

Motyka B, Korbutt G, Pinkoski MJ et al (2000) Mannose 6-phosphate/insulin-like growth factor II receptor is a death receptor for granzyme B during cytotoxic T cell-induced apoptosis. Cell 103:491–500

Munier-Lehmann H, Mauxion F, Bauer U et al (1996a) Re-expression of the mannose 6-phosphate receptors in receptor-deficient fibroblasts. Complementary function of the two mannose 6-phosphate receptors in lysosomal enzyme targeting. J Biol Chem 271:15166–15174

Munier-Lehmann H, Mauxion F, Hoflack B (1996b) Function of the two mannose 6-phosphate receptors in lysosomal enzyme transport. Biochem Soc Trans 24:133–136

Murayama Y, Okamoto T, Ogata E et al (1990) Distinctive regulation of the functional linkage between the human cation-independent mannose 6-phosphate receptor and GTP-binding proteins by insulin-like growth factor II and mannose 6-phosphate. J Biol Chem 265:17456–17462

Nguyen G, Contrepas A (2008) The (pro)renin receptors. J Mol Med 86:643–646

Nicoziani P, Vilhardt F, Llorente A et al (2000) Role for dynamin in late endosome dynamics and trafficking of the cation-independent mannose 6-phosphate receptor. Mol Biol Cell 11:481–495

Nishiura T, Karasuno T, Yoshida H et al (1996) Functional role of cation-independent mannose 6-phosphate/insulin-like growth factor II receptor in cell adhesion and proliferation of a human myeloma cell line OPM-2. Blood 88:3546–3554

Norden AG, Gardner SC, Van't Hoff W et al (2008) Lysosomal enzymuria is a feature of hereditary Fanconi syndrome and is related to elevated CI-mannose-6-P-receptor excretion. Nephrol Dial Transplant 23:2795–2803

Nykjaer A, Christensen EI, Vorum H et al (1998) Mannose 6-phosphate/insulin-like growth factor-II receptor targets the urokinase receptor to lysosomes via a novel binding interaction. J Cell Biol 141:815–828

Olson LJ, Hancock MK, Dix D, Kim J-JP, Dahms NM (1999a) Mutational analysis of the binding site residues of the bovine cation-dependent mannose 6-phosphate receptor. J Biol Chem 274:36905–36911

Olson LJ, Zhang J, Lee YC, Dahms NM, Kim J-JP (1999b) Structural basis for recognition of phosphorylated high mannose oligosaccharides by the cation-dependent mannose 6-phosphate receptor. J Biol Chem 274:29889–29896

Olson LJ, Zhang J, Dahms NM et al (2002) Twists and turns of the CD-MPR: ligand-bound versus ligand-free receptor. J Biol Chem 277:10156–10161

Olson LJ, Dahms NM, Kim JJ (2004a) The N-terminal carbohydrate recognition site of the cation-independent mannose 6-phosphate receptor. J Biol Chem 279:34000–34009

Olson LJ, Yammani RD, Dahms NM, Kim JJ (2004b) Structure of uPAR, plasminogen, and sugar-binding sites of the 300 kDa mannose 6-phosphate receptor. EMBO J 23:2019–2028

Olson LJ, Hindsgaul O, Dahms NM, Kim JJ (2008) Structural insights into the mechanism of pH-dependent ligand binding and release by the cation-dependent mannose 6-phosphate receptor. J Biol Chem 283:10124–10134

Olson LJ, Peterson FC, Castonguay A et al (2010) Structural basis for recognition of phosphodiestercontaining lysosomal enzymes by the cationindependent mannose 6-phosphate receptor. Proc Natl Acad Sci USA 107:12493–12498

Orsel JG, Sincock PM, Krise JP et al (2000) Recognition of the 300-kDa mannose 6-phosphate receptor cytoplasmic domain by 47-kDa tail-interacting protein. Proc Natl Acad Sci USA 97:9047–9051

Oshima A, Nolan CM, Kyle JW et al (1988) The human cation-independent mannose 6-phosphate receptor. Cloning and sequence of the full-length cDNA and expression of functional receptor in COS cells. J Biol Chem 263:2553–2562

Park JE, Lopez JM, Cluett EB, Brown WJ (1991) Identification of a membrane glycoprotein found primarily in the prelysosomal endosome compartment. J Cell Biol 112:245–255

Parton RG, Prydz K, Bomsel M, Simons K, Griffiths G (1989) Meeting of the apical and basolateral endocytic pathways of the Madin-Darby canine kidney cell in late endosomes. J Cell Biol 109:3259–3272

Pavelić K, Kolak T, Kapitanović S et al (2003) Gastric cancer: the role of insulin-like growth factor 2 (IGF 2) and its receptors (IGF 1R and M6-P/IGF 2R). J Pathol 201:430–438

Pérez-Victoria FJ, Mardones GA, Bonifacino JS (2008) Requirement of the human GARP complex for mannose 6-phosphate-receptor-dependent sorting of cathepsin D to Lysosomes. Mol Biol Cell 19:2350–2362

Prydz K, Brändli AW, Bomsel M, Simons K (1990) Surface distribution of the mannose 6-phosphate receptors in epithelial Madin-Darby canine kidney cells. J Biol Chem 265:12629–12635

Puertollano R, Bonifacino JS (2004) Interactions of GGA3 with the ubiquitin sorting machinery. Nat Cell Biol 6:244–251

Puertollano R, Aguilar RC, Gorshkova I et al (2001) Sorting of mannose 6-phosphate receptors mediated by the GGAs. Science 292 (5522):1712–1716

Qian M, Sleat DE, Zheng H, Moore D, Lobel P (2008) Proteomics analysis of serum from mutant mice reveals lysosomal proteins selectively transported by each of the two mannose 6-phosphate receptors. Mol Cell Proteomics 7:58–70

Raiborg C, Malerød L, Pedersen NM, Stenmark H (2008) Differential functions of Hrs and ESCRT proteins in endocytic membrane trafficking. Exp Cell Res 314:801–813

Reaves BJ, Row PE, Bright NA, Luzio JP, Davidson HW (2000) Loss of cation-independent mannose 6-phosphate receptor expression promotes the accumulation of lysobisphosphatidic acid in multilamellar bodies. J Cell Sci 113:4099–4108

Reczek D, Schwake M, Schröder J, Hughes H, Blanz J, Jin X, Brondyk W, Van Patten S, Edmunds T, Saftig P (2007) LIMP-2 is a receptor for lysosomal mannose-6-phosphate-independent targeting of β-glucocerebrosidase. Cell 131:770–783

Reddy ST, Chai W, Childs RA, Page JD, Feizi T, Dahms NM (2004) Identification of a low affinity mannose 6-phosphate-binding site in domain 5 of the cation-independent mannose 6-phosphate receptor. J Biol Chem 279:38658–38667

Rey JM, Theillet C, Brouillet JP, Rochefort H (2000) Stable amino-acid sequence of the mannose-6-phosphate/insulin-like growth-factor-II receptor in ovarian carcinomas with loss of heterozygosity and in breast-cancer cell lines. Int J Cancer 85:466–473

Roberts DL, Weix DJ, Dahms NM, Kim J-JP (1998) Molecular basis of lysosomal enzyme recognition: three-dimensional structure of the cation-dependent mannose 6-phosphate receptor. Cell 93:639–648

Roche P, Brown J, Denley A, Forbes BE, Wallace JC, Jones EY, Esnouf RM (2006) Computational model for the IGF2/IGF2r complex that is predictive of mutational and surface plasmon resonance data. Proteins 64:758–768

Rohrer J, Schweizer A, Johnson KF, Kornfeld S (1995) A determinant in the cytoplasmic tail of the cation-dependent mannose 6-phosphate receptor prevents trafficking to lysosomes. J Cell Biol 130:1297–1306

Rojas R, Kametaka S, Haft CR, Bonifacino JS (2007) Interchangeable but essential functions of SNX1 and SNX2 in the association of retromer with endosomes and the trafficking of mannose 6-phosphate receptors. Mol Cell Biol 27:1112–1124

Runquist EA, Havel RJ (1991) Acid hydrolases in early and late endosome fractions from rat liver. J Biol Chem 266:22557–22563

Sakano K, Enjoh T, Numata F et al (1991) The design, expression, and characterization of human insulin-like growth factor II (IGF2) mutants specific for either the IGF2/cation-independent mannose 6-phosphate receptor or IGF-I receptor. J Biol Chem 266:20626–20635

Salminen A, Marjomäki V, Tolonen U, Myllylä VV (1988) Phosphomannosyl receptors of lysosomal enzymes of skeletal muscle in neuromuscular diseases. Acta Neurol Scand 77:461–467

Saris JJ, Derkx FH, De Bruin RJ et al (2001) High-affinity prorenin binding to cardiac man-6-P/IGF2 receptors precedes proteolytic activation to renin. Am J Physiol Heart Circ Physiol 280: H1706–H1715

Schaffer BS, Lin MF, Byrd JC et al (2003) Opposing roles for the insulin-like growth factor (IGF)-II and mannose 6-phosphate (Man-6-P) binding activities of the IGF2/Man-6-P receptor in the growth of prostate cancer cells. Endocrinology 144:955–966

Schu P (2005) Adaptor Proteins in Lysosomal Biogenesis. In: Saftig P (ed) Lysosomes. Spriger, Boston

Scott CD, Firth SM (2004) The role of the M6P/IGF2 receptor in cancer: tumor suppression or garbage disposal? Horm Metab Res 36:261–271

Scott GK, Gu F, Crump CM et al (2003) The phosphorylation state of an autoregulatory domain controls PACS-1-directed protein traffic. EMBO J 22:6234–6244

Scott GK, Fei H, Thomas L et al (2006) A PACS-1, GGA3 and CK2 complex regulates CI-MPR trafficking. EMBO J 25:4423–4435

Seaman MN (2004) Cargo-selective endosomal sorting for retrieval to the Golgi requires retromer. J Cell Biol 165:111–122

Seaman MN (2007) Identification of a novel conserved sorting motif required for retromer-mediated endosome-to-TGN retrieval. J Cell Sci 120:2378–2389

Shiba T, Takatsu H, Nogi T et al (2002) Structural basis for recognition of acidic-cluster dileucine sequence by GGA1. Nature 415 (6874):937–941

Sivaramakrishna Y, Amancha PK, Siva Kumar N (2009) Reptilian MPR 300 is also the IGF2R: cloning, sequencing and functional characterization of the IGF2 binding domain. Int J Biol Macromol 44:435–440

Sohar I, Sleat D, Gong Liu C et al (1998) Mouse mutants lacking the cation-independent mannose 6-phosphate/insulin-like growth factor II receptor are impaired in lysosomal enzyme transport: comparison of cation-independent and cation-dependent mannose 6-phosphate receptor-deficient mice. Biochem J 330:903–908

Stöckli J, Höning S, Rohrer J (2004) The acidic cluster of the CK2 site of the cation-dependent mannose 6-phosphate receptor (CD-MPR) but not its phosphorylation is required for GGA1 and AP-1 binding. J Biol Chem 279:23542–23549

Stoorvogel W, Strous GJ, Geuze HJ et al (1991) Late endosomes derive from early endosomes by maturation. Cell 65:417–427

Szebenyi G, Rotwein P (1991) Differential regulation of mannose 6-phosphate receptors and their ligands during the myogenic development of C2 cells. J Biol Chem 266:5534–5539

Szebenyi G, Rotwein P (1994) The mouse insulin-like growth factor II/cation-independent mannose 6-phosphate (IGF2/MPR) receptor gene: molecular cloning and genomic organization. Genomics 19:120–129

Tomiyama Y, Waguri S, Kanamori S et al (2004) Early-phase redistribution of the cation-independent mannose 6-phosphate receptor by U18666A treatment in HeLa cells. Cell Tissue Res 317:253–264

Tong PY, Kornfeld S (1989) Ligand interactions of the cation-dependent mannose 6-phosphate receptor. Comparison with the cation-independent mannose 6-phosphate receptor. J Biol Chem 264:7970–7975

Tong PY, Gregory W, Kornfeld S (1989) Ligand interactions of the cation-independent mannose 6-phosphate receptor. The stoichiometry of mannose 6-phosphate binding. J Biol Chem 264:7962–7969

Tooze J, Hollinshead M, Ludwig T et al (1990) In exocrine pancreas, the basolateral endocytic pathway converges with the autophagic pathway immediately after the early endosome. J Cell Biol 111:329–345

Tortorella LL, Schapiro FB, Maxfield FR (2007) Role of an acidic cluster/dileucine motif in cation-independent mannose 6-phosphate receptor traffic. Traffic 8:402–413

Trapani JA, Sutton VR, Thia KY et al (2003) A clathrin/dynamin- and mannose-6-phosphate receptor-independent pathway for granzyme B-induced cell death. J Cell Biol 160:223–233

Umeda A, Fujita H, Kuronita T et al (2003) Distribution and trafficking of MPR300 is normal in cells with cholesterol accumulated in late endocytic compartments: evidence for early endosome-to-TGN trafficking of MPR300. J Lipid Res 44:1821–1832

Uson I, Schmidt B, von Bulow R et al (2003) Locating the anomalous scatterer substructures in halide and sulfur phasing. Acta Crystallogr D Biol Crystallogr 59:57–66

Veugelers K, Motyka B, Goping IS et al (2006) Granule-mediated killing by granzyme B and perforin requires a mannose 6-phosphate receptor and is augmented by cell surface heparan sulfate. Mol Biol Cell 17:623–633

Villevalois-Cam L, Rescan C, Gilot D et al (2003) The hepatocyte is a direct target for transforming-growth factor β activation via the insulin-like growth factor II/mannose 6-phosphate receptor. J Hepatol 38:156–163

Vitelli R, Santillo M, Lattero D et al (1997) Role of the small GTPase Rab7 in the late endocytic pathway. J Biol Chem 272:4391–4397

Wang ZQ, Fung MR, Barlow DP, Wagner EF (1994) Regulation of embryonic growth and lysosomal targeting by the imprinted Igf2/Mpr gene. Nature 372(6505):464–467

Waguri S, Tomiyama Y, Ikeda H, Hida T, Sakai N, Taniike M, Ebisu S, Uchiyama Y (2006) The luminal domain participates in the endosomal trafficking of the cation-independent mannose 6-phosphate receptor. Exp Cell Res 312:4090–4107

Wassmer T, Attar N, Bujny MV et al (2007) A loss-of-function screen reveals SNX5 and SNX6 as potential components of the mammalian retromer. J Cell Sci 120:45–54

Westcott KR, Rome LH (1988) Cation-independent mannose 6-phosphate receptor contains covalently bound fatty acid. J Cell Biochem 38:23–33

Westlund B, Dahms NM, Kornfeld S (1991) The bovine mannose 6-phosphate/insulin-like growth factor II receptor. Localization of mannose 6-phosphate binding sites to domains 1–3 and 7–11 of the extracytoplasmic region. J Biol Chem 266:23233–23239

Wood RJ, Hulett MD (2008) Cell surface-expressed cation-independent mannose 6-phosphate receptor (CD222) binds enzymatically active heparanase independently of mannose 6-phosphate to promote extracellular matrix degradation. J Biol Chem 283:4165–4176

Wutz A, Barlow DP (1998) Imprinting of the mouse Igf2r gene depends on an intronic CpG island. Mol Cell Endocrinol 140:9–14

Yammani RR, Sharma M, Seetharam S et al (2002) Loss of albumin and megalin binding to renal cubilin in rats results in albuminuria after total body irradiation. Am J Physiol Regul Integr Comp Physiol 283:R339–R346

Yang YW, Robbins AR, Nissley SP et al (1991) The chick embryo fibroblast cation-independent mannose 6-phosphate receptor is functional and immunologically related to the mammalian insulin-like growth factor-II (IGF2)/man 6-P receptor but does not bind IGF2. Endocrinology 128:1177–1189

York SJ, Arneson LS, Gregory WT et al (1999) The rate of internalization of the mannose 6-phosphate/insulin-like growth factor II receptor is enhanced by multivalent ligand binding. J Biol Chem 274:1164–1171

Zaccheo OJ, Prince SN, Miller DM et al (2006) Kinetics of insulin-like growth factor II (IGF2) interaction with domain 11 of the human IGF2/mannose 6-phosphate receptor: function of CD and AB loop solvent-exposed residues. J Mol Biol 359:403–421

Zhou G, Roizman B (2002) Cation-independent mannose 6-phosphate receptor blocks apoptosis induced by herpes simplex virus 1 mutants lacking glycoprotein D and is likely the target of antiapoptotic activity of the glycoprotein. J Virol 76:6197–6204

Zhou M, Ma Z, Sly WS (1995) Cloning and expression of the cDNA of chicken cation-independent mannose-6-phosphate receptor. Proc Natl Acad Sci USA 92:9762–9766

Zhou G, Avitabile E, Campadelli-Fiume G et al (2003) The domains of glycoprotein D required to block apoptosis induced by herpes simplex virus 1 are largely distinct from those involved in cell-cell fusion and binding to nectin1. J Virol 77:3759–3767

Zhu Y, Doray B, Poussu A et al (2001) Binding of GGA2 to the lysosomal enzyme sorting motif of the mannose 6-phosphate receptor. Science 292(5522):1716–1718

Mannose-6-Phosphate Receptor Homologous Protein Family

5

G.S. Gupta

5.1 Recognition of High-Mannose Type N-Glycans in ERAD Pathways

Protein quality control in the endoplasmic reticulum (ER) is an elaborate process conserved from yeast to mammals, ensuring that only newly synthesized proteins with correct conformations in the ER are sorted further into the secretory pathway. The ER discriminates between native and nonnative protein conformations, selectively transporting properly folded proteins to their final destinations through the secretory pathway, or alternatively, retrotranslocating misfolded proteins to the cytosol to be degraded by proteasomes. In the quality control process, high-mannose type N-glycans play important roles in protein-folding events. Proteins that fail to achieve proper folding or proper assembly are degraded in a process known as ER-associated degradation (ERAD). The ERAD pathway comprises multiple steps including substrate recognition and targeting to the retro-translocation machinery, retrotranslocation from the ER into the cytosol, and proteasomal degradation through ubiquitination. The quality-control system also surveys the ER lumen for terminally misfolded proteins. Polypeptides singled out by this system are ultimately degraded by the cytosolic ubiquitin-proteasome pathway. Key components of both the ER quality-control system and the ERAD machinery have been identified, but a connection between the two systems has remained elusive (Yoshida and Tanaka 2010). Recent studies have documented the important roles of sugar-recognition (lectin-type) molecules for trimmed high-mannose type N-glycans and glycosidases in the ERAD pathways in both ER and cytosol. Since the ER is distributed throughout the cytosol, studies suggest that the cytosolic face of the ER membrane serves as a "platform" for degradation of misfolded cytosolic proteins (Metzger et al. 2008). The fundamental system that monitors glycoprotein folding in the ER and the unique roles of the sugar-recognizing ubiquitin ligase and peptide:N-glycanase (PNGase) in the cytosolic ERAD pathway has been reviewed (Yoshida and Tanaka 2010).

Glycosylation of asparagine residues in Asn-x-Ser/Thr motifs is a common covalent modification of proteins in the lumen of the ER. Protein glycosylation itself and processing of the glycan transferred play a key role in the folding and conformation discrimination of glycoproteins within the ER. By substantially contributing to the overall hydrophilicity of the polypeptide, pre-assembled core glycans inhibit possible aggregation caused by the inevitable exposure of hydrophobic patches on the as yet unstructured chains. Thereafter, N-glycans are modified by ER-resident enzymes glucosidase I (GI), glucosidase II (GII), UDP-glucose:glycoprotein glucosyltransferase (UGT) and mannosidase(s) and become functional appendices that determine the fate of the associated polypeptide. The glycan ($Glc_3Man_9GlcNAc_2$) transferred to Asn residues is first deglucosylated by glucosidase I, a type II membrane protein with a lumenal hydrolytic domain, which removes the outermost Glc of the glycan (Fig. 5.1). The $Glc_2Man_9GlcNAc_2$ (G2M9) thus produced is then deglucosylated by glucosidase II (GII) that successively generates $Glc_1Man_9GlcNAc_2$ (G1M9) and $Man_9GlcNAc_2$ (M9) upon cleavage of Glcα1,3Glc (cleavage 1) and Glcα1,3Man (cleavage 2) bonds. Both GII-mediated cleavages play a determining role in the quality control of glycoprotein folding in the ER. Monoglucosylated glycan-bearing glycoproteins may interact with calnexin (CNX) and/or calreticulin (CRT), two ER-resident lectin chaperones that enhance folding efficiency by preventing aggregation and facilitating correct disulfide bond formation through their interaction with ERp57, a protein disulfide isomerase (Chap. 2). Furthermore, the interaction of folding intermediates and misfolded glycoproteins and the lectin-chaperones prevent their exit from the ER to the Golgi. The second GII-mediated cleavage that generates M9 abolishes the glycoprotein-lectin-chaperone interaction, thus allowing glycoproteins to pursue their transit through the secretory

G.S. Gupta et al., *Animal Lectins: Form, Function and Clinical Applications*,
DOI 10.1007/978-3-7091-1065-2_5, © Springer-Verlag Wien 2012

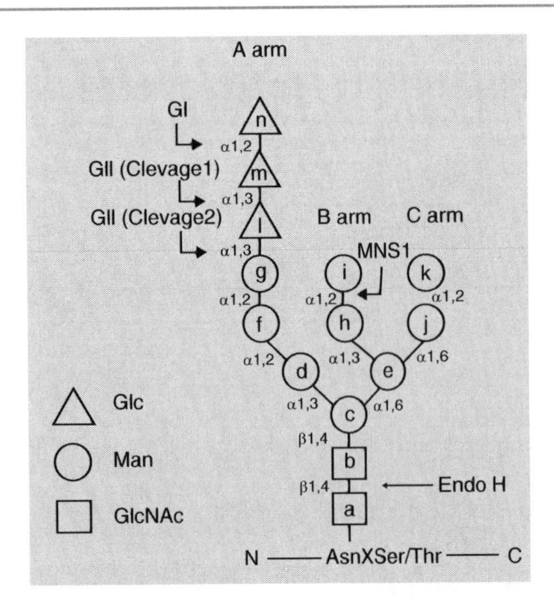

Fig. 5.1 Glycan structures: The structure depicted is that of the glycan transferred to Asn residues in *N*-glycosylation. Lettering (*a*, *b*, *c* ...) follows the order of addition of monosaccharides in the synthesis of the dolichol-P-P derivatives. Glucosidase I removes residue n and GII, residues *l* and *m*. GT re-adds residue *l*. ER mannosidase cleaves residue *i*. G2M9, G1M9, and M9 comprise residues *a–m*, *a–l*, and *a–k*, respectively, except when the glycans had been previously treated with Endo H. Because this enzyme cleaves the bond between residues *a* and *b*, those glycans are devoid of residue *a*. Jack bean α-mannosidase removes residues *h–k* and partially also residue *e* from the glucosylated glycans. Therefore, G2M4-5 comprise residues *b–d*, *f*, *g*, *l*, and *m* and partially also residue *e*. G1M4-5 is similar to G2M4-5 but lacks residue *m*. *Arm A* comprises residues *d*, *f*, *g*, and *l–n*; *arm B* comprises residues *h* and *i*; and *arm C* comprises residues *j* and *k* (Stigliano et al. 2009).

pathway. However, if not yet properly folded, glycoproteins may be reglucosylated by the uridine 5′-diphosphate (UDP)-Glc:glycoprotein glucosyltransferase (GT), an enzyme that specifically glucosylates nonnative conformers and regenerates monoglucosylated glycans. These, in turn, interact again with the lectin chaperones. Cycles of reglucosylation and deglucosylation catalyzed by the opposing activity of GT and GII continue until species acquire their proper tertiary structure (Trombetta and Parodi 2003).

Recent studies have improved our understanding of how the removal of terminal glucose residues from N-glycans allows newly synthesized proteins to access the calnexin chaperone system; how substrate retention in this specialized chaperone system is regulated by de-/re-glucosylation cycles catalyzed by GII and UGT1; and how acceleration of N-glycan dismantling upon induction of EDEM variants promotes ERAD under conditions of ER stress. In particular, characterization of cells lacking certain ER chaperones has revealed important new information on the mechanisms regulating protein folding and quality control. Tight regulation of N-glycan modifications is crucial to maintain protein quality control, to ensure the synthesis of functional polypeptides and to avoid constipation of the ER with folding-defective (Ruddock and Molinari 2006)

The mannose-6-phosphate (M6P) based system for targeting lysosomal hydrolases to lysosomes is conserved in mammals, birds, amphibians, and crustaceans but is absent in the unicellular protozoa *Trypanosoma* (Huete-Perez et al. 1999) and *Leishmania* (Clayton et al. 1995) (Chaps. 3, 4). *D. discoideum* and *Acanthamoeba castellani* both exhibit GlcNAc-phosphotransferase activity and can transfer GlcNAc-1-PO_4 to mannose residues (Lang et al. 1986). However, MPRs have not been identified in these species. The CD-MPR (Nadimpalli and von Figura 2002) and CI-MPR (Lakshmi et al. 1999; Nadimpalli et al. 2004) have also been reported in the invertebrate mollusc *Unio*. These studies demonstrate that numerous species in animal kingdom express *bone fide* MPRs that are capable of binding phosphomannosyl residues (Figs. 5.2, 5.3).

5.2 Proteins Containing M6PRH Domains

In recent years, several proteins (Mrl1, LERP, GlcNAc-1-phosphotransferase γ-subunit, ER glucosidase II β-subunit, OS-9, and XTP3-B/Erlectin) have been identified that contain mannose-6-phosphate receptor homology (MRH or M6PRH) domains (Dodd and Drickamer 2001; Munro 2001; Cruciat et al. 2006) and implicated in N-glycan recognition but none has been shown to bind M6P. Of these, only LERP has been shown to function in the delivery of lysosomal enzymes to the lysosome. In humans, none of them has transmembrane region (Fig. 5.2).

5.3 GlcNAc-Phosphotransferase

Lysosomal enzymes are targeted to the lysosome through binding to M6PRs because their glycans are modified with M6P. This modification is catalyzed by UDP-N-acetylglucosamine: lysosomal enzyme N-acetylglucosamine-1-phosphotransferase (GlcNAc-phosphotransferase), which transfers GlcNAc 1-phosphate from UDP-GlcNAc to mannose in high mannose-type glycans. The second step is removal of GlcNAc by *N*-acetylglucosamine-1-phosphodiester α-GlcNAcase (UCE) to expose mannose 6-phosphate. Bovine GlcNAc-phosphotransferase has been characterized as a multisubunit enzyme with the subunit structure $\alpha_2\beta_2\gamma_2$. The cDNA for the human γ-subunit was cloned and its gene has been localized to chromosome 16p. The α- and β-subunits have enzyme activity, whereas the γ-subunit (GNPTG), which contains the MRH domain, lacks catalytic activity and is believed to participate in substrate binding of the enzyme (Lee et al. 2007). However, the sugar-binding property of GNPTG is currently unknown. The α- and β -subunits have been cloned by Kudo et al. (2005). While the α- and β-subunits of human GlcNAc-phosphotransferase are encoded by a single cDNA as a

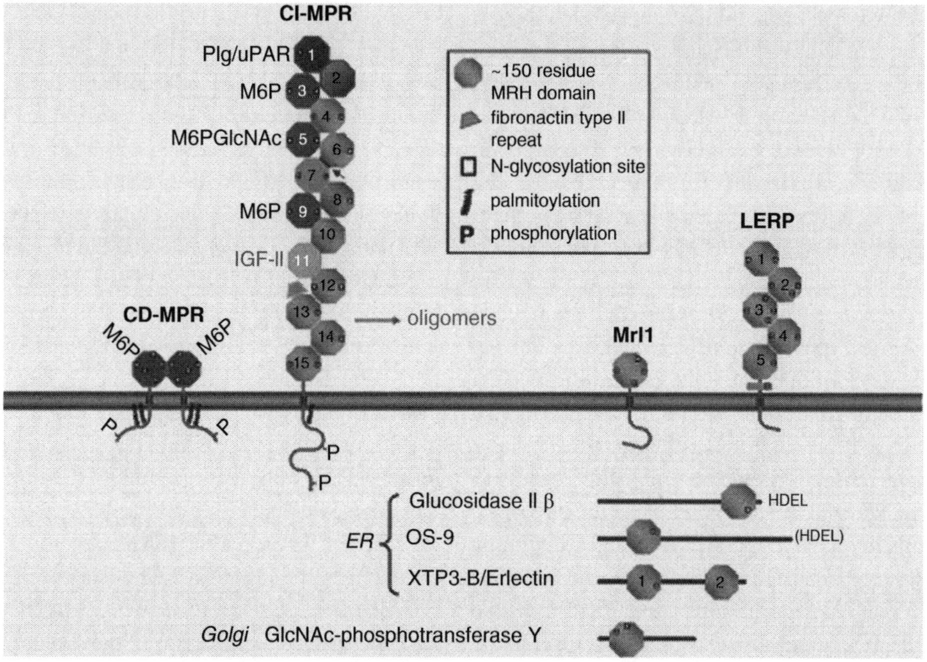

Fig. 5.2 Schematic diagram of M6PR-H domain containing proteins. The MPRs, 381-residue Mrl1, and 886-residue LERP are type I transmembrane glycoproteins. The 279-residue CD-MPR exists as a homodimer. The 2499-residue CI-MPR also undergoes oligomerization and most likely exists as a dimer. The M6P binding sites of the CD-MPR and CI-MPR are indicated. Domains 1 and 2 are outlined since the presence of these two domains enhances the affinity of domain 3 for lysosomal enzymes by ~1,000-fold (Hancock et al. 2002). The CD-MPR contains a single high affinity M6P binding site per polypeptide. In contrast, CI-MPR contains three carbohydrate recognition sites: two high affinity sites are localized to domains 1–3 and domain 9 and one low affinity site is contained within domain 5. The IGF-II and plasminogen (Plg)/uPAR binding sites are also indicated. The fibronectin type II repeat present in domain 13 increases the affinity of domain 11 for IGF-II by ~10-fold. The *arrow* indicates the location of a proteolytically sensitive cleavage site between domains 6 and 7 (Westlund et al. 1991). The 528-residue glucosidase II β subunit, 667-residue OS-9, and 483-residue XTP3-B/Erlectin are soluble resident ER proteins. Glucosidase II β subunit and the yeast ortholog of OS-9 contain a C-terminal ER retention signal (HDEL). The 305-residue N-acetylglucosamine-1-phosphotransferase (GlcNAc-phosphotransferase) γ-subunit, in complex with the catalytic α/β subunits, is enriched in *cis*-Golgi cisternae (Adapted by permission from Dahms et al. 2008 © Oxford University Press)

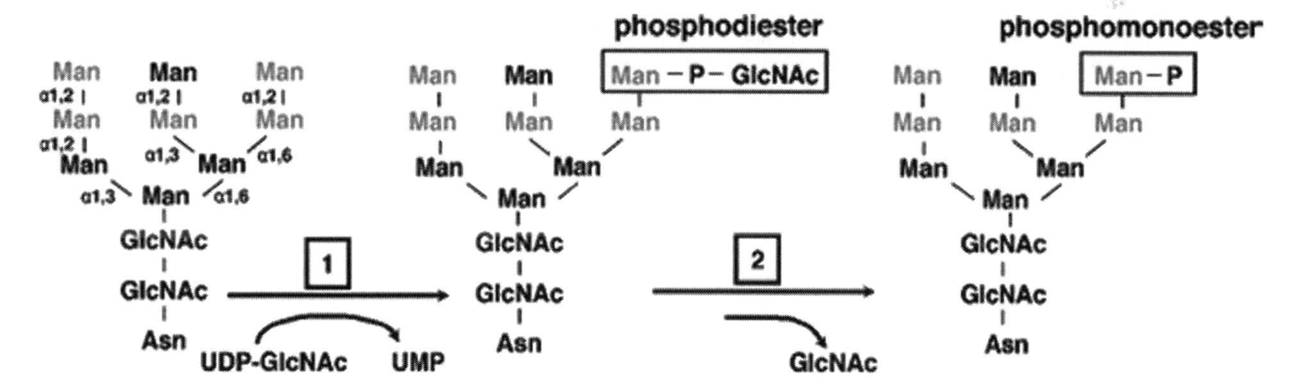

Fig. 5.3 Generation of the Man-6-P tag on *N*-linked oligosaccharides. Phosphorylation of mannose residues on *N*-linked oligosaccharides occurs in two steps. First, the GlcNAc-phosphotransferase transfers GlcNAc-1-phosphate from UDP-GlcNAc to the C-6 hydroxyl group of mannose to form the Man-P-GlcNAc phosphodiester intermediate. Second, the uncovering enzyme removes the GlcNAc moiety in the TGN, revealing the Man-6-P phosphomonoester. The five potential sites of phosphorylation are indicated (*gray*) (Reprinted by permission from Dahms et al. 2008 © Oxford University Press)

precursor and appear to be generated by a proteolytic cleavage at the Lys928-Asp929 bond, the γ-subunit is encoded by a second gene. The hydropathy plots of the deduced amino acid sequences revealed that the α- and β-subunits contain amino and carboxyl terminus transmembrane domains, respectively, whereas the γ-subunit does not. The mechanism responsible for the processing of the α/β-subunits precursor into the mature α- and β-subunits is not yet known, but likely to be the result of protease(s) in the ER or *cis*-Golgi, where GlcNAc-phosphotransferase resides. Recombinant soluble GlcNAc-phosphotransferase exhibited specific activity and substrate preference similar to the wild type bovine GlcNAcphosphotransferase and was able to phosphorylate acid α-glucosidase in vitro (Kudo and Canfield 2006). The comparative domain organization of six human MRH-containing proteins has been presented in Fig. 5.2.

The M6PR-H domain is the only recognized domain in the γ-subunit of GlcNAc phosphotransferase. This enzyme catalyzes the first step in the biosynthesis of M6P sorting signals in the *cis*-Golgi (Fig. 5.3) and the M6PR-H domain may be involved in the recognition of substrate glycans. GlcNAc phoshotransferase is present only in organisms which also possess M6PRs, and raises the interesting possibility that the proteins involved in the synthesis and recognition of M6P evolved together from a common ancestor. The M6PRs and the GlcNAc-phosphotransferase, both essential components of the lysosomal enzyme targeting machinery, contain M6PR-H domains (Munro 2001). Lee et al. (2007), however, have shown that transgenic mice deficient in the γ-subunit of GlcNAc-phosphotransferase retain substantial activity toward acid hydrolases and that γ-subunit is not essential for substrate recognition. The α/β-subunits, in addition to their catalytic function, have some ability to recognize acid hydrolases as specific substrates. However, Lee et al. (2007) suggest that this specific recognition is somehow enhanced by the presence of γ-subunit. Clearly, detailed analysis of the putative carbohydrate binding properties of these M6PR-H-containing proteins is lacking. Alignment studies predict that M6PR-H domains containing three out of four conserved residues (Arg, Glu, Tyr) of M6PRs interact with 2-, 3-, and 4-hydroxyl groups of the mannose ring of M6P (Olson et al. 2004). It is likely that M6PR-H-containing proteins bind specifically to high mannose-type oligosaccharides, a concept in agreement with their proposed functions in ER -Golgi compartments (c/r Dahms et al. 2008; Dahms and Hancock 2002).

5.4 α-Glucosidase II

5.4.1 Function of α-Glucosidase II

The α-Glucosidase II (GII) is a soluble ER-resident heterodimer composed of two tightly but noncovalently bound α and β chains (GIIα and GIIβ) (Trombetta et al. 2001). GIIα is a 95–110 kDa protein that contains the consensus sequence (G/F)(L/I/V/M)WXDMNE) of the active site of family 31 glycosyl hydrolases and lacks the ER retention signal – XDEL at its C-terminus (Trombetta et al. 1996; D'Alessio et al. 1999). This subunit has a single active site but it has been proposed to have different kinetics for the first and second cleavages (Alonso et al. 1993), although recent work suggests that the differential trimming rates of both Glc units is not operative at the high protein concentrations occurring within the ER lumen (Totani et al. 2008).

Glucosidase II is responsible for the sequential removal of the two innermost glucose residues from the glycan G_3M_9 transferred to Asn residues in proteins. GII participates in the calnexin/calreticulin cycle; it removes the single glucose unit added to folding intermediates and misfolded glycoproteins by the UDP-Glc:glycoprotein glucosyltransferase. Trimming of the first α1,3-linked glucose from $Glc_2Man_9GlcNAc_2$ (G2M9) is important for a glycoprotein to interact with calnexin/calreticulin (CNX/CRT), and cleavage of the innermost glucose from $Glc_1Man_9GlcNAc_2$ (G1M9) sets glycoproteins free from the CNX/CRT cycle and allows them to proceed to the Golgi apparatus.

Using a genetic approach, the heterodimeric nature of GII was demonstrated in the fission yeast *Schizosaccharomyces pombe* and the microsomes from ΔGIIα and ΔGIIβ mutant cells were devoid of GII activity when using $Glc_1Man_9GlcNAc_2$ (G1M9) as substrate in the assays. Nevertheless, whereas *N*-glycans formed in intact ΔGIIα cells were identified as $Glc_2Man_9GlcNAc_2$ (G2M9), ΔGIIβ cells formed, in addition, small amounts of G1M9 (D'Alessio et al. 1999). It was suggested then that this last compound was formed either by GIIα transiently present in the ER in its way to secretion and/or by low amounts of GIIα that folded successfully in the absence of GIIβ. Moreover, ΔGIIβ cells presented the unfolding protein response as the BiP gene was induced in these mutant cells, thus showing that the subunit plays a key role in the efficient folding of glycoproteins. *S. pombe* is an organism that displays a glycoprotein folding quality control mechanism similar to that occurring in mammalian cells and it expresses an active GT (Fernández et al. 1994; D'Alessio et al. 1999). Furthermore, *S. pombe* GIIβ presents a high homology to its mammalian counterpart, including the presence of a consensus ER retention/retrieval sequence at its C terminus. Stigliano et al. (2009) reported that in the absence of GIIβ, the catalytic subunit GIIα of the fission yeast *S. pombe* folds to an active conformation able to hydrolyze *p*-nitrophenyl α-D-glucopyranoside. However, the heterodimer is required to efficiently deglucosylate the physiological substrates $Glc_2Man_9GlcNAc_2$ (G_2M_9) and $Glc_1Man_9GlcNAc_2$ (G_1M_9) (Fig. 5.1). The interaction of the mannose 6-phosphate receptor homologous domain present in GIIβ and mannoses in the B and/or C arms of the glycans mediates glycan hydrolysis enhancement. Evidence indicates that also in mammalian cells GIIβ modulates G_2M_9 and G_1M_9 trimming (Stigliano et al. 2009).

The roles of GIIβ are controversial and have been object of growing interest in the last years, as autosomal dominant polycystic liver disease may develop in individuals carrying mutations in GIIβ gene (Drenth et al. 2005; Davila et al. 2004) and GIIβ is induced in differentiating neuritic rat progenitor cells and in response to the glial cell-derived neurotrophic factor (Hoffrogge et al. 2007). GIIβ is a 50- to 60-kDa subunit that has been suggested to be responsible for GIIα maturation to an active conformation in mammals (Pelletier et al. 2000; Treml et al. 2000) as well as for its presence in the ER as GIIβ displays an ER retention/retrieval consensus sequence at its C-terminus (Trombetta et al. 1996; D'Alessio et al. 1999). In contrast, the GIIβ subunit of the budding yeast *S. cerevisiae* does not display a consensus ER retention/retrieval sequence. Furthermore, GIIα was also retained in the ER of GIIβ null mutants and G_1M_9 was the *N*-glycan formed in these last cells, indicating that GIIβ is required for the second but not for the first cleavage (Wilkinson et al. 2006). It is to be noted that this microorganism lacks a classical CNX/CRT cycle because it does not express an active GT (Fernández et al. 1994).

5.4.2 M6PRH Domain in GIIβ

The β subunit of ER GII contains an M6PR-H domain preceded by two regions of coiled-coil and two complement-like repeats. A study has suggested a possible involvement of the M6PR-H domain of GIIβ (GIIβ-M6PR-H) in the glucose trimming process via its putative sugar-binding activity. Human GIIβ-M6PR-H binds to high-mannose-type glycans most strongly to the α1,2-linked mannobiose structure. In absence of GIIβ, the catalytic subunit GIIα of fission yeast *S. pombe* folds to an active conformation able to hydrolyze p-nitrophenyl α-d-glucopyranoside. Deletion of the homologue of the noncatalytic β-subunit in *S. pombe* drastically reduces glucosidase II activity. However, the heterodimer is required to efficiently deglucosylate the physiological substrates G_2M_9 and G_1M_9. The interaction of the M6PR-H domain present in GIIβ and mannoses in the B and/or C arms of the glycans mediates glycan hydrolysis enhancement. Results demonstrate the capacity of the GIIβ-M6PR-H to bind high-mannose-type glycans and its importance in efficient glucose trimming of N-glycans (Hu et al. 2009; Stigliano et al. 2009). As in GlcNAc phosphotransferase, the M6PR-H domain may serve to recognize the substrate glycan. Through its C-terminal HDEL signal, the β-subunit of GIIβ may retain the complete α1β1 complex in the ER (Trombetta et al. 2001).

5.4.3 Two Distinct Domains of β-Subunit of GII Interact with α-Subunit

The β-subunit, which contains a C-terminal His-Asp-Glu-Leu (HDEL) motif, may function to link the catalytic subunit to the KDEL receptor as a retrieval mechanism. DNA sequencing has revealed that both subunits are encoded by gene products that undergo alternative splicing in T lymphocytes. The catalytic α-subunit possesses two variably expressed segments, box A-1, consisting of 22 amino acids located proximal to the amino-terminus, and box A-2, composed of 9 amino acids situated between the amino-terminus and the putative catalytic site in the central region of the molecule. Box B1, a variably expressed seven amino acid segment in the β-subunit of glucosidase II, is located immediately downstream of an acidic stretch near the carboxyl-terminus. Screening of reverse transcribed RNA by PCR confirms the variable inclusion of each of these segments in transcripts obtained from a panel of T-lymphocyte cell lines. Thus, distinct isoforms of glucosidase II exist that may perform specialized functions (Arendt et al. 1999).

The human cDNA sequences of the α-subunit predicted a soluble protein (104 kDa) devoid of known signals for residence in the ER. It showed homology with several other glucosidases but not with glucosidase I. The α- subunit was the functional catalytic subunit of glucosidase II. The sequence of the β subunit (58 kDa) showed no sequence homology with other known proteins. It encoded a soluble protein rich in glutamic and aspartic acid with a putative ER retention signal (HDEL) at the C terminus. This suggested that the β subunit is responsible for the ER localization of the enzyme (Trombetta et al. 1996). The two subunits form a defined complex, composed of one catalytic subunit and one accessory subunit $\alpha_1\beta_1$ with a molecular mass of 161 kDa. The complex had an s value of 6.3 S, indicative of a highly nonglobular shape. The β subunit could be proteolytically removed from $\alpha_1\beta_1$ complex without affecting catalysis, demonstrating that it is not required for glucosidase II activity in vitro. It was suggested that the catalytic core of glucosidase II resides in a globular domain of the α-subunit, which can function independently of β subunit, while the complete α- and β subunits assemble in a defined heterodimeric complex with a highly extended conformation, which may favor interaction with other proteins in ER. Through its C-terminal HDEL signal, the β subunit may retain the complete $\alpha_1\beta_1$ complex in the ER (Trombetta et al. 2001).

Arendt and Ostergaard (2000) mapped the regions of the mouse β-subunit protein responsible for mediating the association with the α-subunit and identified two non-overlapping interaction domains (ID1 and ID2) within the β-subunit. ID1 encompasses 118 amino acids at the N-terminus of the mature polypeptide, spanning the cysteine-rich element in this region. ID2, located near the C-terminus, is contained within amino acids 273–400, a region occupied in part by a stretch of acidic residues. Based on experimental facts it was postulated that the catalytic subunit of glucosidase II binds synergistically to ID1 and ID2, explaining the high associative stability of the enzyme complex (Arendt and Ostergaard 2000). *S. cerevisiae* gene *GTB1* encodes, a polypeptide with 21% sequence similarity to the β-subunit of human glucosidase II. The Gtb1 protein was shown to be a soluble glycoprotein (96–102 kDa) localized to the ER lumen where it was present in a complex together with the yeast α-subunit homologue Gls2p. Surprisingly, Δgtb1 mutant cells were specifically defective in the processing of monoglucosylated glycans. Thus, although Gls2p is sufficient for cleavage of the penultimate glucose residue, Gtb1p is essential for cleavage of the final glucose. Data demonstrate that Gtb1p is required for normal glycoprotein biogenesis and reveal that the final two glucose-trimming steps in N-glycan processing are mechanistically distinct (Wilkinson et al. 2006).

Watanabe et al. (2009) reported an in vivo enzymatic analysis using gene disruptants lacking either the G-II α- or β-subunit in the filamentous fungus *Aspergillus oryzae*. The fraction lacking the β-subunit retained hydrolytic activity toward p-nitrophenyl α-D-glucopyranoside but was inactive toward both $Glc_2Man_9GlcNAc_2$ and $Glc_1Man_9GlcNAc_2$. When the fraction containing the β-subunit was added to the one including the α-subunit, the glucosidase activity was restored. These results suggested that the β-subunit confers the substrate specificity toward di- and monoglucosylated glycans on the glucose-trimming activity of the α-subunit (Watanabe et al. 2009). In order to further dissect these activities Quinn et al. (2009) mutagenized a number of conserved residues across the protein. Both the M6PRH and G2B domains of Gtb1p contribute to the Glc2 trimming event but the M6PR-H domain is essential for Glc1 trimming (Quinn et al. 2009).

5.4.4 Polycystic Liver Disease (PCLD) and β-Subunit of Glucosidase II

Autosomal-dominant polycystic liver disease (PCLD) is a rare disorder that is characterized by the progressive development of fluid-filled biliary epithelial cysts spread throughout the liver parenchyma and characterized by an increased liver volume due to many (>20) fluid-filled cysts of biliary origin. Positional cloning has identified two genes that are mutated in patients with polycystic liver disease, *PRKCSH and SEC63*, which encode the hepatocystin (β-subunit of glucosidase II) and Sec63, respectively. Both proteins are components of the molecular machinery involved in the translocation, folding and quality control of newly synthesized glycoproteins in the ER (Drenth et al. 2003). Most mutations are truncating and probably lead to a complete loss of the corresponding proteins and the defective processing of a key regulator of biliary cell growth. The finding that PCLD is caused by proteins involved in oligosaccharide processing was unexpected and implicates a new avenue for research into neocystogenesis, and might ultimately result in the identification of novel therapeutic drugs (Davila et al. 2004; Drenth et al. 2005) (see Chap. 45).

Hepatocystin is a protein kinase C substrate, a component of a cytosolic signal transduction complex, a receptor for advanced glycation end products, a vacuolar protein, and the β subunit of ER glucosidase II. The exact localization and cellular function of hepatocystin remain unclear. Normal hepatocystin localizes to the ER, where it assembles with the glucosidase II α-subunit. The 1338–2A—>G truncating mutation in hepatocystin observed in some polycystic liver disease patients produces a protein that is not retained in the ER but is secreted into the medium. This mutant protein fails to assemble with the glucosidase II α-subunit. As a consequence, mutant hepatocystin is undetectable in liver cysts. In addition, levels of normal hepatocystin and of the glucosidase II α-subunit are substantially reduced in liver and Epstein-Barr virus-immortalized B lymphoblasts from patients with polycystic liver disease. These findings are consistent with a role of hepatocystin in carbohydrate processing and quality control of newly synthesized glycoproteins in the ER. Therefore, altered ER processing of some key regulator of cell proliferation may underlie polycystic liver disease (Drenth et al. 2004).

Disease causing mutations in *PRKCSH* or *SEC63* are found in approximately 25% of the PCLD patients. Both gene products function in the ER, however, the molecular mechanism behind cyst formation remains to be elucidated. As part of the translocon complex, SEC63 plays a role in protein import into the ER and is implicated in the export of unfolded proteins to the cytoplasm during ERAD. *PRKCSH* codes for the β-subunit of glucosidase II (hepatocystin), which cleaves two glucose residues of $Glc_3Man_9GlcNAc_2$ N-glycans on proteins. Hepatocystin is thereby directly involved in the protein folding process by regulating protein binding to calnexin/calreticulin in the ER. A separate group of genetic diseases affecting protein N-glycosylation in the ER is formed by the congenital disorders of glycosylation (CDG). In distinct subtypes of this autosomal recessive multisystem disease specific liver symptoms have been reported that overlap with PCLD. Recent research revealed novel insights in PCLD disease pathology such as the absence of hepatocystin from cyst epithelia indicating a two-hit model for PCLD cystogenesis. This opens the way to speculate about a recessive mechanism for PCLD pathophysiology and shared molecular pathways between CDG and PCLD (Janssen et al. 2010).

PCLD is often asymptomatic, but if symptoms arise, they are usually due to the mass effect of cysts. The phenotype is more severe in females and correlates with the number of pregnancies or estrogen use. The gene for PCLD has been assigned to chromosome 19p13.2–13.1. Mutations found in *PRKCSH* introduce stopcodons in the m-RNA, resulting in premature termination of translation to protein. The protein, designated as hepatocystin, is predicted to be localised in the ER (Jansen et al. 2009).

The PCLD is a distinct clinical and genetic entity that can occur independently from autosomal dominant polycystic kidney disease (PCKD) (Chap. 45). In contrast to PKD1, PKD2, and PKHD1, normal PRKCSH encodes a human protein termed "protein kinase C substrate 80 K-H" or "noncatalytic β-subunit of glucosidase II also designated as hepatocystin (Jansen et al. 2009)," which is highly conserved, expressed in all tissues, and contains a leader sequence, an LDLα domain, two EF-hand domains, and a conserved C-terminal HDEL sequence. Its function may be dependent on calcium binding, and its putative actions include the regulation of N-glycosylation of proteins and signal transduction via fibroblast growth-factor receptor. In light of the focal nature of liver cysts in PCLD, the apparent loss-of-function mutations in PRKCSH, and the two-hit mechanism operational in dominant polycystic kidney disease, PCLD may also occur by a two-hit mechanism (Li et al. 2003).

5.5 Osteosarcomas-9 (OS-9)

5.5.1 The Protein

OS-9 is a *N*-glycosylated protein expressed in two splice variants in the ER lumen. Transcription of both OS-9 variants is enhanced upon activation of Ire1/Xbp1 pathway in cells exposed to acute ER stress. OS-9 variants do not associate with folding-competent proteins, but form non-covalent complexes with misfolded ones. OS-9 association prevents secretion from the ER of misfolded NHK conformers and facilitates NHK disposal. OS-9 variants play a crucial role in maintaining the tightness of retention-based ER quality control and in promoting disposal of misfolded proteins from the mammalian ER. The OS-9 gene maps to a region (q13–15) of chromosome 12 that is highly amplified in human osteosarcomas and encodes a protein of unknown function. Su et al. (1996) identified two genes (OS-9 and OS-4) within 12q13–15, a region frequently amplified in human cancers. The full-length OS-9 cDNA sequence consists of 2,785 bp with an ORF of 667 amino-acids. The predicted polypeptide was water soluble and acidic. The OS-9 gene encoded a 2.8-kb mRNA transcribed, ubiquitously expressed in human tissues and revealed significant similarities with two ORFs from genomic sequences in *C. elegans* and *S. cerevisiae*. The ORF region comprises a functional domain present in a novel evolutionarily conserved gene family defined by OS-9. The gene spanned ~30.4 kbp and had 15 exons. The 1,010 bp sequence of the 5′ upstream region contains binding-sequence motifs TATA and CCAAT for general transcription. Primer extension analysis revealed two putative transcription start sites (Kimura et al. 1997).

Three isoforms of OS-9 cDNA were found in a myeloid leukemia HL-60 cDNA library. Isoform 1 consisted of 2,700 bp from which a 667 amino acid sequence was found to be identical with that of OS-9 cDNA from osteosarcoma cells. Isoform 2 cDNA lacked a 165 nt sequence in the coding region. Isoform 3 cDNA had an additional 45 bp deletion in coding region. Isoforms 2 and 3 encode 612 and 597 amino acid polypeptides, respectively. These three isoforms were found to be splice variants. Isoform 2 mRNA expressed predominantly in myeloid leukemia HL-60 cells, osteosarcoma OsA-CL cells and rhabdomyosarcoma Rh30 cells and in various tumor cell lines of sarcoma cells, carcinoma cells and myeloid leukemia cells, but 3–4 times higher expression in OsA-CL cells and Rh30 cells containing a homogeneously staining region of 12q13-15 (Kimura et al. 1998).

OS-9 is conserved in eukaryotes and functions together with M-type lectins in ERAD pathway used to get rid of irreversibly misfolded glycoproteins. OS-9 appears to identify glycoprotein targets through recognition of both N-linked glycans and other determinants.

5.5.2 Requirement of HRD1, SEL1L, and OS-9/XTP3-B for Disposal of ERAD-Substrates

In the ER, lectins and processing enzymes are involved in quality control of newly synthesized proteins for productive folding as well as in ERAD of misfolded proteins. Misfolded glycoproteins are translocated from the ER into the cytoplasm for proteasome-mediated degradation. While Htm1/EDEM protein has been proposed to act as a "degradation lectin" for ERAD of misfolded glycoproteins, OS-9 and XTP3-B/Erlectin are ER-resident glycoproteins that bind to ERAD substrates and, through the SEL1L adaptor, to the ER-membrane-embedded ubiquitin ligase HRD1. OS-9 protein is thought to participate in ERAD. Both proteins contain conserved mannose 6-phosphate receptor homology (MRH) domains, which are required for interaction with SEL1L, but not with substrate. OS-9 associates with ER chaperone GRP94 which, together with HRD1 and SEL1L, is required for the degradation of an ERAD substrate, mutant α1-antitrypsin. It is suggested that XTP3-B and OS-9 are components of distinct, partially redundant, quality control surveillance pathways that coordinate protein folding with membrane dislocation and ubiquitin conjugation in mammalian cells (Christianson et al. 2008; Tamura et al. 2008; Bernasconi et al. 2010). Surprisingly, however, OS-9 is not required for ubiquitination or degradation of this nonglycosylated ERAD substrate. In a model, OS-9 recognises terminally misfolded proteins via polypeptide-based rather than glycan-based signals, but is only required for transferring those bearing N-glycans to the ubiquitination machinery (Alcock and Swanton 2009).

5.5.3 OS-9 Recognizes Mannose-Trimmed N-Glycans

Mannose trimming from the N-glycans plays an important role in targeting of misfolded glycoproteins for ERAD. It was demonstrated that the recombinant hOS-9 M6P-R homology domain specifically binds N-glycans lacking the terminal mannose from the C branch in vitro. The ability of hOS-9 to enhance glycoprotein ERAD depended on the N-glycan structures on NHK. Hosokawa et al. (2009) proposed a model for mannose trimming and the requirement for hOS-9 lectin activity in glycoprotein ERAD in which N-glycans lacking the terminal mannose from C branch are recognized by hOS-9 and targeted for degradation (Hosokawa et al. 2009) (Fig. 5.4). Results suggest that trimming of the outermost α1,2-linked mannose on the C-arm is a critical process for misfolded proteins to enter ERAD (Hosokawa et al. 2009; Mikami et al. 2010; Riemer et al. 2009).

5.5.4 Dual Task for Xbp1-Responsive OS-9 Variants

Alternative splicing products of OS-9 gene, OS-9.1 and OS-9.2, are ubiquitously expressed in human tissues and are amplified in tumors. They are transcriptionally induced upon activation of the Ire1/Xbp1 ER-stress pathway. OS-9 variants do not associate with folding-competent proteins. Rather, they selectively bind folding-defective ones thereby inhibiting transport of non-native conformers through the secretory pathway. The intralumenal level of OS-9.1 and OS-9.2 inversely correlates with the fraction of a folding-defective glycoprotein, the Null(hong kong) (NHK) variant of α1-antitrypsin that escapes retention-based ER quality control. OS-9 up-regulation does not affect NHK disposal, but reduction of the intralumenal level of OS-9.1 and OS-9.2 substantially delays disposal of this model substrate. OS-9.1 and OS-9.2 also associate transiently with non-glycosylated folding-defective proteins, but association is unproductive. Finally, OS-9 activity does not require an intact mannose 6-P homology domain. Thus, OS-9.1 and OS-9.2 play a dual role in mammalian ER quality control: first as crucial retention factors for misfolded conformers, and second as promoters of protein disposal from the ER lumen (Bernasconi et al. 2008). Vourvouhaki et al. (2007) suggested that OS-9 promotes cell viability and confers resistance to apoptosis, potentially implicating OS-9 in the survival of cancer cells.

5.5.5 Interactions of OS-9

N-copine is a two C2 domain protein that shows Ca^{2+}-dependent phospholipid binding and membrane association. OS-9 is capable of interacting with N-copine. The second C2 domain of N-copine binds with the carboxy-terminal region of OS-9 in Ca^{2+}-dependent manner. N-copine and OS-9 are co-expressed in the same regions of human brain (Nakayama et al. 1999). OS-9 interacts with the membrane proteinase meprin β found in brush border membranes of kidney and small intestine. This cytoplasmic region is indispensable for the maturation of meprin β, which included an ER-to-Golgi translocation. OS-9 associates with meprin β only transiently, coinciding with ER-to-Golgi transport of meprin β. The OS-9-binding site in the cytoplasmic domain of meprin β overlaps the region essential for this transport. The alternatively spliced forms of rat and mouse OS-9 have been characterized, but that only the non-spliced form of OS-9 binds to meprin β. It indicated that OS-9 may be involved in the ER-to-Golgi transport of meprin β (Litovchick et al. 2002). While the presence of a single *N*-linked oligosaccharide in the middle of the M6PR-H domain of OS-9 would inhibit the ability of the M6PR-H domain to function in stabilizing a multiprotein complex (Bernasconi et al. 2008), arguments of Dahms et al. (2008) suggest additional studies to be performed to test the hypothesis that the presence of an *N*-glycan chain inhibits the lectin activity of OS-9 and/or XTP3-B/Erlectin.

The presence of structural lesions in the luminal, transmembrane, or cytosolic domains determines the classification of misfolded polypeptides as ERAD-L, -M, or -C substrates and results in selection of distinct degradation pathways. It was shown that disposal of soluble (nontransmembrane) polypeptides with luminal lesions (ERAD-L_S substrates) is strictly dependent on the E3 ubiquitin ligase HRD1, the associated cargo receptor SEL1L, and two interchangeable ERAD lectins, OS-9 and XTP3-B. These

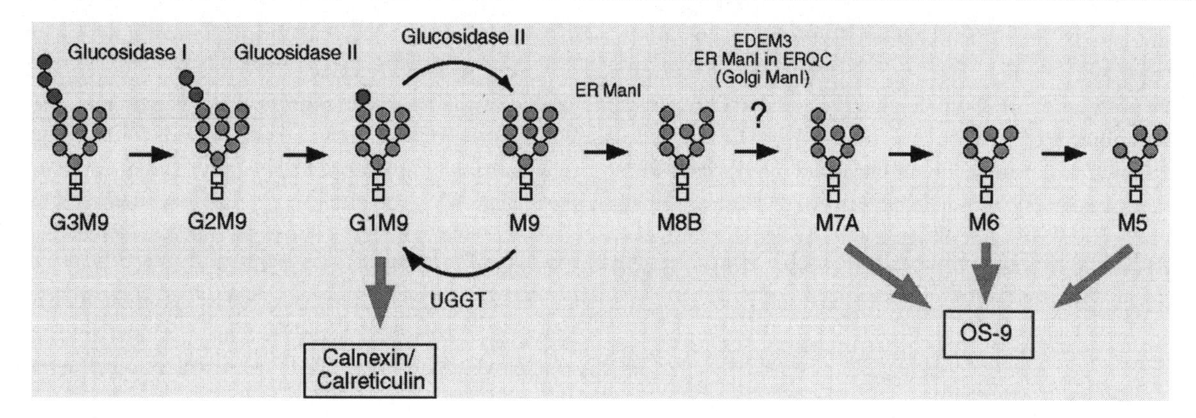

Fig. 5.4 *N*-glycan processing and recognition in mammals. Calnexin and calreticulin bind to monoglucosylated *N*-glycans and assist in the folding of glycoproteins. These lectins dissociate from the glycoprotein upon removal of the terminal glucose on monoglucosylated *N*-glycans by glucosidase II. In mammals, UDP-glucose:glycoprotein glucosyltransferase (*UGGT*), which is absent from the yeast genome, adds a glucose back onto the *N*-glycan if the glycoprotein has still not acquired the native conformation, thus forming the calnexin/calreticulin or monoglucose cycle. ER ManI generates M8B, followed by further mannose trimming to produce M7A ($Man_7GlcNAc_2$ isomer A), M6 ($Man6GlcNAc_2$) and M5 ($Man5GlcNAc_2$) *N*-glycans, which are recognized by OS-9. The candidate α1,2-mannosidases responsible for this process include EDEM3, ER ManI concentrated in the ER quality control compartment (*ERQC*), and Golgi α1,2-mannosidases for the ERAD substrates that may recycle between the ER and Golgi prior to degradation (Adapted by permission from Hosokawa et al. 2010 © Oxford University Press)

ERAD factors become dispensable for degradation of the same polypeptides when membrane tethered (ERAD-L_M substrates). Studies reveal that, in contrast to budding yeast, tethering of mammalian ERAD-L substrates to the membrane changes selection of the degradation pathway (Bernasconi et al. 2010). Several *N*-glycans containing terminal ∝1,6-linked mannose in the Man∝1,6(Man∝1,3) Man∝1,6(Man∝1,3)Man structure are good ligands for OS-9^{M6PRH}, having K_a values of approximately 10^4 M^{-1} and that trimming of either an ∝1,6-linked mannose from the C-arm or an ∝1,3-linked mannose from the B-arm abrogated binding to OS-9^{M6PRH}. An immunoprecipitation experiment demonstrated that the ∝1-antitrypsin variant $null^{Hong\ Kong}$, but not wild-type ∝1-antitrypsin, selectively interacted with OS-9 in the cells in a sugar-dependent manner. These results suggest that trimming of the outermost ∝1,2-linked mannose on the C-arm is a critical process for misfolded proteins to enter ERAD (Mikami et al. 2010).

An Auxiliary Protein for TRPV4 Maturation: Transient receptor potential (TRP) proteins constitute a family of cation-permeable channels that are formed by homo- or heteromeric assembly of four subunits. However, the mechanisms governing TRP channel assembly and biogenesis in general, remain largely unknown. OS-9 interacts with the cytosolic N-terminal tail of TRPV4. The underlying mechanisms revealed that OS-9 preferably binds TRPV4 monomers and other ER- variants of TRPV4 and attenuates their polyubiquitination. Thus, OS-9 regulates the secretory transport of TRPV4 and appears to protect TRPV4 subunits from the precocious ubiquitination and ER-associated degradation. Data suggest that OS-9 functions as an auxiliary protein for TRPV4 maturation (Wang et al. 2007).

OS9 Interacts with DC-STAMP: Dendritic cell-specific transmembrane protein (DC-STAMP) has been identified in a cDNA library of human monocyte-derived DCs. DC-STAMP is a multimembrane spanning protein that has been implicated in skewing haematopoietic differentiation of bone marrow cells towards the myeloid lineage, and in cell fusion during osteoclastogenesis and giant cell formation. (Jansen et al. 2009) reported that amplified OS9 physically interacts with DC-STAMP, and that both proteins co-localize in the ER in various cell lines, including immature DC. The TLR-induced maturation of DC leads to the translocation of DC-STAMP from the ER to the Golgi while OS9 localization is unaffected. Collectively, results indicate that OS9 is critically involved in the modulation of ER-to-Golgi transport of DC-STAMP in response to TLR triggering, suggesting a novel role for OS9 in myeloid differentiation and cell fusion (Jansen et al. 2009, 2010).

Interaction of OS9 with HIF-1: Hypoxia-inducible factor 1 (HIF-1) functions as a master regulator of oxygen homeostasis in metazoan. HIF-1 mediates changes in gene transcription in response to changes in cellular oxygenation. OS-9 interacts with both HIF-1α and HIF-1α prolyl hydroxylases. OS-9 gain-of-function promotes HIF-1α hydroxylation, VHL binding, proteasomal degradation of HIF-1α, and inhibition of HIF-1-mediated transcription. The loss of function of OS-9 by iRNA increases HIF-1α protein levels, HIF-1-mediated transcription, and vascular endothelial growth factor

(VEGF) mRNA expression under nonhypoxic conditions. These data indicate that OS-9 is an essential component of a multiprotein complex that regulates HIF-1α levels in an O_2-dependent manner (Baek et al. 2005). HIF-1α is believed to promote tumor growth and metastasis, and many efforts have been made to develop new anticancer agents based on HIF-1α inhibition. YC-1 is a widely used HIF-1α inhibitor both in vitro and in vivo, and is being developed as a novel class of anticancer drug (Kim et al. 2006).

5.6 YOS9 from *S. cerevisiae*

Due to easy manipulating of yeast genome, many aspects and components of ERAD have been discovered in *S. cerevisiae* (Brodsky 2007; Molinari et al. 2007). Studies on the involvement of mammalian ortholog OS-9 in ERAD have been hampered by data showing that OS-9 is a cytosolic protein (Litovchick et al. 2002). Studies were followed by a series of publications in which experimental design and interpretation of the data were based on the assumption that OS-9 is a cytosolic protein (Baek et al. 2005; Flashman et al. 2005; Wang et al. 2007). A homolog, Yos9, from *S. cerevisiae* is a membrane-associated glycoprotein that localizes to ER. Reports revealed that Yos9p is required for disposal of substrates with luminal folding defects, whereas it is dispensable for disposal of proteins with defects in the transmembrane and cytosolic domains (Carvalho et al. 2006; Denic et al. 2006; Szathmary et al. 2005). In yeast, the key pathway of ER quality control consists of a two-lectin receptor system made of Yos9p and Htm1/Mnl1p that recognizes N-linked glycan signals embedded in substrates. Szathmaryet al. (2005) provided genetic and biochemical evidence that Yos9 protein in *S. cerevisiae* is essential for efficient degradation of mutant glycoproteins. Yos9 is a member of the OS-9 protein family, which is conserved among eukaryotes and shows similarities with M6PRs. The amino acids conserved among OS-9 family members and M6PRs are essential for Yos9 protein function. Immunoprecipitation studies showed that Yos9 specifically associates with misfolded carboxypeptidase Y (CPY*), an ERAD substrate, but only when it carried $Man_8GlcNAc_2$ or $Man_5GlcNAc_2$ N-glycans. Experiments further suggested that Yos9 acts in the same pathway as Htm1/EDEM. Yos9 protein is important for glycoprotein degradation and may act via its M6PRH domain as a degradation lectin-like protein in the glycoprotein degradation pathway. Yos9p forms a complex with substrates and has a sugar binding pocket that is essential for its ERAD function. Nonetheless, substrate recognition persists even when the sugar binding site is mutated or CPY* is unglycosylated (Bhamidipati et al. 2005). However, Yos9p binds directly to substrates to discriminate misfolded from folded proteins. Substrates displaying cytosolic determinants can be degraded independently of this system. Studies suggest that mechanistically divergent systems collaborate to guard against passage and accumulation of misfolded proteins in secretory pathway. These and other considerations suggest that Yos9p plays a critical role in the bipartite recognition of terminally misfolded glycoproteins (Bhamidipati et al. 2005; Kim et al. 2005; Cormier et al. 2005).

YOS9 gene interacts with genes involved in ER-Golgi transport, particularly SEC34, whose temperature-sensitive mutant is rescued by *YOS9* over-expression. Interestingly, Yos9 appears to play a direct role in the transport of glycosylphosphatidylinositol (GPI)-anchored proteins to the Golgi apparatus. Yos9 binds directly to Gas1 and Mkc7 and accelerates Gas1 transport and processing in cells over-expressing. As Yos9 is not a component of the Emp24 complex, it may act as a novel escort factor for GPI-anchored proteins in ER-Golgi transport in yeast and possibly in mammals (Friedmann et al. 2002; Vigneron et al. 2002). Gauss et al. (2006a) reported an association between Yos9p and Hrd3p, a component of the ubiquitin-proteasome system that links these pathways. These workers identified designated regions in the luminal domain of Hrd3p that interact with Yos9p and the ubiquitin ligase Hrd1p. Binding of misfolded proteins occurs through Hrd3p, suggesting that Hrd3p recognizes proteins that deviate from their native conformation, whereas Yos9p ensures that only terminally misfolded polypeptides are degraded. Key components of ERAD are ER membrane-bound ubiquitin ligases. These ligases associate with the cytoplasmic AAA-ATPase Cdc48p/p97, which is thought to support the release of malfolded proteins from the ER. Gauss et al. (2006b) characterized a yeast protein complex containing ubiquitin ligase Hrd1p and the ER membrane proteins Hrd3p and Der1p. Hrd3p binds mis-folded proteins in ER lumen enabling their delivery to downstream components. Therefore, it was proposed that Hrd3p acts as a substrate recruitment factor for the Hrd1p ligase complex. Hrd3p function is also required for the association of Cdc48p with Hrd1p. Moreover, results demonstrated that recruitment of Cdc48p depends on substrate processing by Hrd1p ligase complex. Thus, Hrd1p ligase complex unites substrate selection in ER lumen and polyubiquitination in the cytoplasm and links these processes to the release of ER proteins via Cdc48p complex (Gauss et al. 2006b). Thus, Yos9p associates with the membrane-embedded ubiquitin ligase complex, Hrd1p-Hrd3p, and provides a proofreading mechanism for ERAD. The mammalian homologues of Yos9p, OS-9 and XTP3-B are also ER resident proteins that associate with the HRD1-SEL1L ubiquitin ligase complex and are important for the regulation of ERAD. Recent studies have also identified the N-glycan species with which both yeast Yos9p and mammalian OS-9 associate as M7A, a $Man_7GlcNAc_2$ isomer that lacks the α1,2-linked terminal mannose from both the B and C branches. M7A has been known since then as a degradation signal in both yeast and mammals (Hosokawa et al. 2010). Taken together, a clear picture has not yet emerged concerning the role of M6PR-H domains in Yos9, OS-9, and XTP3-B/Erlectin.

5.7 Erlectin/XTP3-B

Erlectin, also referred to as XTP3-B, is a luminal ER-resident protein first characterized in *Xenopus*. In *Xenopus* XTP3-B has been shown to be involved in the regulation of glycoprotein trafficking and to be essential during development. It is present in vertebrate and invertebrate animals. XTP3-B is a soluble ER-resident protein that contains two M6PR-H domains in its sequence. XTP3-B interacts with a membrane-associated ubiquitin ligase complex, and is thought to participate in ERAD. Kremen1 and 2 (Krm1/2) are co-receptors for Dickkopf1 (Dkk1), an antagonist of Wnt/β-catenin signaling, and play a role in head induction during early *Xenopus* development. Erlectin (XTP3-B) containing M6PR-H- or PRKCSH- domains is implicated in N-glycan binding. Like other members of M6PR-H family, Erlectin is essential for Krm2 binding, and this interaction is abolished by Krm2 deglycosylation. Results indicate that Erlectin functions in N-glycan recognition in the endoplasmic reticulum, suggesting that it may regulate glycoprotein traffic (Cruciat et al. 2006).

Human XTP3-B (hXTP3-B) has two transcriptional variants, and both isoforms retard ERAD of the human α1-antitrypsin variant null Hong Kong (NHK), a terminally misfolded glycoprotein. The hXTP3-B long isoform strongly inhibited ERAD of NHK-QQQ, which lacks all of the N-glycosylation sites of NHK, but the short transcriptional variant of hXTP3-B had no effect. Examination revealed that the hXTP3-B long isoform associates with the HRD1-SEL1L membrane-anchored ubiquitin ligase complex and BiP and forms a 27 s ER quality control scaffold complex. The hXTP3-B short isoform, however, is excluded from scaffold formation. Another MRH domain-containing ER lectin, hOS-9, is incorporated into this large complex, but gp78, another mammalian homolog of the yeast ubiquitin ligase Hrd1p, is not (Hosokawa et al. 2008). XTP3-B specifically binds to AT (NHK) via a C-terminal M6PR-H domain in a glycan-dependent manner (Yamaguchi et al. 2010). These results indicate that the large ER quality control scaffold complex, containing ER lectins, a chaperone, and a ubiquitin ligase, provides a platform for the recognition and sorting of misfolded glycoproteins as well as nonglycosylated proteins prior to retrotranslocation into the cytoplasm for degradation (Hosokawa et al. 2008).

5.8 *Drosophila* Lysosomal Enzyme Receptor Protein (LERP)

A type I transmembrane protein termed lysosomal enzyme receptor protein (LERP) with partial homology to the mammalian M6PR-300 encoded by gene CG31072 was identified in *Drosophila* (Dennes et al. 2005). LERP contains 5 lumenal repeats that share homology to the 15 lumenal repeats found in all M6PR-300. Four of the repeats display the P-lectin type pattern of conserved cysteine residues. However, the arginine residues identified to be essential for M6P binding are not conserved. The LERP cytoplasmic domain shows highly conserved interactions with *Drosophila* and mammalian GGA adaptors known to mediate Golgi-endosome traffic of M6PRs and other transmembrane cargo. Moreover, LERP rescues missorting of soluble lysosomal enzymes in MPR-deficient cells, giving strong evidence for a function that is equivalent to the mammalian counterpart. However, unlike the mammalian M6PRs, LERP did not bind to the multimeric M6P ligand phosphomannan. LERP plays important role in biogenesis of *Drosophila* lysosomes; the GGA functions in the receptor-mediated lysosomal transport system in the fruit fly (Dennes et al. 2005). Authors showed that LERP, which is able to interact with *Drosophila* and mammalian Golgi-localized, Gamma-ear-containing, ADP-ribosylation factor-binding (GGA) adaptors that have been shown to sort M6PRs in clathrin-coated vesicles at the TGN (Bonifacino 2004; Ghosh and Kornfeld 2004), mediates lysosomal enzyme targeting and rescues the missorting of lysosomal enzymes that occurs in MPR-deficient mammalian cells. The protein-sorting machinery in fly cells is well conserved relative to that in mammals, enabling the use of fly cells to dissect CCV biogenesis and clathrin-dependent protein trafficking at the TGN of higher eukaryotes (Kametaka et al. 2010).

5.9 MRL1

In yeast, the mechanism by which soluble hydrolases, such as carboxypeptidase Y and proteinase A, reach the vacuole (functional equivalent of the lysosome) is very similar to that found in mammalian cells except that the vacuolar sorting signal is not carbohydrate-based, but rather resides in the propeptide region of hydrolase and is recognized by a receptor, Vps10, that cycles between the Golgi and endosomal compartments (Ni et al. 2006). The yeast genome contains an ORF that encodes a membrane protein that is distantly related to mammalian M6PRs. The protein encoded by this gene (which was termed MRL1) cycles through late endosome (Whyte and Munro 2001). MRL1 is a type I membrane glycoprotein that contains a single M6PR-H domain like CD-MPR and co-localizes with Vps10 in *S. cerevisiae*. The vacuolar hydrolases of yeast *S. cerevisiae* do not contain 6-phosphate monoesters on their N-glycans, but contain the product of VPS10 gene as receptor for hydrolases. MRL1 also cycles between the Golgi and late endosome, but is unrelated to vertebrate M6PRs, and recognizes a specific amino acid sequence of carboxypeptidase Y. The delivery of carboxypeptidase Y or proteinase A was not affected in *S. cerevisiae* strains lacking MRL1. Moreover, there is a strong synergistic effect on the maturation of proteinases A and B when both MRL1 and VPS10 are deleted, which suggests that MRL1 may serve as a sorting receptor in the delivery of vacuolar hydrolases.

References

Alcock F, Swanton E (2009) Mammalian OS-9 is upregulated in response to endoplasmic reticulum stress and facilitates ubiquitination of misfolded glycoproteins. J Mol Biol 385:1032–1042

Alonso JM, Santa-Cecilia A, Calvo P (1993) Effect of bromoconduritol on glucosidase II from rat liver. A new kinetic model for the binding and hydrolysis of the substrate. Eur J Biochem 215:37–42

Arendt CW, Ostergaard HL (2000) Two distinct domains of the β-subunit of glucosidase II interact with the catalytic α–subunit. Glycobiology 10:487–492

Arendt CW, Dawicki W, Ostergaard HL (1999) Alternative splicing of transcripts encoding the α– and β-subunits of mouse glucosidase II in T lymphocytes. Glycobiology 9:277–283

Baek JH, Mahon PC, Oh J et al (2005) OS-9 interacts with hypoxia-inducible factor 1alpha and prolyl hydroxylases to promote oxygen-dependent degradation of HIF-1alpha. Mol Cell 17:503–512

Bernasconi R, Pertel T, Luban J et al (2008) A dual task for the Xbp1-responsive OS-9 variants in the mammalian endoplasmic reticulum: inhibiting secretion of misfolded protein conformers and enhancing their disposal. J Biol Chem 283:16446–16454

Bernasconi R, Galli C, Calanca V et al (2010) Stringent requirement for HRD1, SEL1L, and OS-9/XTP3-B for disposal of ERAD-LS substrates. J Cell Biol 188:223–235

Bhamidipati A, Denic V, Quan EM, Weissman JS (2005) Exploration of the topological requirements of ERAD identifies Yos9p as a lectin sensor of misfolded glycoproteins in the ER lumen. Mol Cell 19:741–751

Bonifacino JS (2004) The GGA proteins: adaptors on the move. Nat Rev Mol Cell Biol 5:23–32

Brodsky JL (2007) The protective and destructive roles played by molecular chaperones during ERAD (endoplasmic-reticulum-associated degradation). Biochem J 404:353–363

Carvalho P, Goder V, Rapoport TA (2006) Distinct ubiquitin-ligase complexes define convergent pathways for the degradation of ER proteins. Cell 126:361–373

Christianson JC, Shaler TA, Tyler RE, Kopito RR (2008) OS-9 and GRP94 deliver mutant alpha1-antitrypsin to the Hrd1-SEL1L ubiquitin ligase complex for ERAD. Nat Cell Biol 10:272–282

Clayton C, Hausler T, Blattner J (1995) Protein trafficking in kinetoplastid protozoa. Microbiol Rev 59:325–344

Cormier JH, Pearse BR, Hebert DN (2005) Yos9p: a sweet-toothed bouncer of the secretory pathway. Mol Cell 19:717–726

Cruciat CM, Hassler C, Niehrs C (2006) The M6PRH protein Erlectin is a member of the endoplasmic reticulum synexpression group and functions in N-glycan recognition. J Biol Chem 281:12986–12993

D'Alessio C, Fernández F, Trombetta ES, Parodi AJ (1999) Genetic evidence for the heterodimeric structure of glucosidase II. The effect of disrupting the subunit-encoding genes on glycoprotein folding. J Biol Chem 274:25899–25905

Dahms NM, Hancock MK (2002) P-type lectins. Biochim Biophys Acta 1572:317–340

Dahms NM, Olson LJ, Kim JJ (2008) Strategies for carbohydrate recognition by the mannose 6-phosphate receptors. Glycobiology 18:664–678

Davila S, Furu L, Gharavi AG et al (2004) Mutations in SEC63 cause autosomal dominant polycystic liver disease. Nat Genet 36:575–577

Denic V, Quan EM, Weissman JS (2006) A luminal surveillance complex that selects misfolded glycoproteins for ER-associated degradation. Cell 126:349–359

Dennes A, Cromme C, Suresh K et al (2005) The novel *Drosophila* lysosomal enzyme receptor protein mediates lysosomal sorting in mammalian cells and binds mammalian and *Drosophila* GGA adaptors. J Biol Chem 280:12849–12857

Dodd RB, Drickamer K (2001) Lectin-like proteins in model organisms: implications for evolution of carbohydrate-binding activity. Glycobiology 11:71R–79R

Drenth JP, Morsche RH, Smink R et al (2003) Germline mutations in PRKCSH are associated with autosomal dominant polycystic liver disease. Nat Genet 33:345–347

Drenth JP, Martina JA, Te Morsche RH et al (2004) Molecular characterization of hepatocystin, the protein that is defective in autosomal dominant polycystic liver disease. Gastroenterology 126:1819–1827

Drenth JP, Martina JA, van de Kerkhof R, Bonifacino JS, Jansen JB (2005) Polycystic liver disease is a disorder of cotranslational protein processing. Trends Mol Med 11:37–42

Fernández FS, Trombetta SE, Hellman U, Parodi AJ (1994) Purification to homogeneity of UDP-glucose:glycoprotein glucosyltransferase from *Schizosaccharomyces pombe* and apparent absence of the enzyme from *Saccharomyces cerevisiae*. J Biol Chem 269:30701–30706

Friedmann E, Salzberg Y, Weinberger A, Shaltiel S, Gerst JE (2002) YOS9, the putative yeast homolog of a gene amplified in osteosarcomas, is involved in the endoplasmic reticulum (ER)-Golgi transport of GPI-anchored proteins. J Biol Chem 277: 35274–35281

Flashman E, McDonough MA, Schofield CJ (2005) OS-9: another piece in the HIF complex story. Mol Cell 17:472–473

Gauss R, Jarosch E, Sommer T, Hirsch C (2006a) A complex of Yos9p and the HRD ligase integrates endoplasmic reticulum quality control into the degradation machinery. Nat Cell Biol 8:849–854

Gauss R, Sommer T, Jarosch E (2006b) The Hrd1p ligase complex forms a linchpin between ER-lumenal substrate selection and Cdc48p recruitment. EMBO J 25:1827–1835

Ghosh P, Kornfeld S (2004) The GGA proteins: key players in protein sorting at the trans-Golgi network. Eur J Cell Biol 83:257–262

Hancock MK, Haskins DJ, Sun G, Dahms NM (2002) Identification of residues essential for carbohydrate recognition by the insulin-like growth factor ii/mannose 6-phosphate receptor. J Biol Chem 277:11255–11264

Hoffrogge R et al (2007) 2-DE profiling of GDNF overexpression-related proteome changes in differentiating ST14A rat progenitor cells. Proteomics 7:33–46

Hosokawa N, Wada I, Nagasawa K et al (2008) Human XTP3-B forms an endoplasmic reticulum quality control scaffold with the HRD1-SEL1L ubiquitin ligase complex and BiP. J Biol Chem 283:20914–20924

Hosokawa N, Kamiya Y, Kamiya D et al (2009) Human OS-9, a lectin required for glycoprotein endoplasmic reticulum-associated degradation, recognizes mannose-trimmed N-glycans. J Biol Chem 284:17061–17068

Hosokawa N, Kamiya Y, Kato K (2010) The role of M6PRH domain-containing lectins in ERAD. Glycobiology 20:651–660

Hu D, Kamiya Y, Totani K et al (2009) Sugar-binding activity of the M6PRH domain in the ER alpha-glucosidase II beta subunit is important for efficient glucose trimming. Glycobiology 19:1127–1135

Huete-Perez JA, Engel JC, Brinen LS et al (1999) Protease trafficking in two primitive eukaryotes is mediated by a prodomain protein motif. J Biol Chem 274:16249–16256

Jansen BJ, Eleveld-Trancikova D, Sanecka A et al (2009) OS9 interacts with DC-STAMP and modulates its intracellular localization in response to TLR ligation. Mol Immunol 46:505–515

Janssen MJ, Waanders E, Woudenberg J, Lefeber DJ, Drenth JP (2010) Congenital disorders of glycosylation in hepatology: the example of polycystic liver disease. J Hepatol 52:432–440

Kametaka S, Sawada N, Bonifacino JS, Waguri S (2010) Functional characterization of protein-sorting machineries at the trans-Golgi network in Drosophila melanogaster. J Cell Sci 123:460–471

Kim W, Spear ED, Ng DT (2005) Yos9p detects and targets misfolded glycoproteins for ER-associated degradation. Mol Cell 19:753–764

Kim HL, Yeo EJ, Chun YS et al (2006) A domain responsible for HIF-1alpha degradation by YC-1, a novel anticancer agent. Int J Oncol 29:255–260

Kimura Y, Nakazawa M, Tsuchiya N et al (1997) Genomic organization of the OS-9 gene amplified in human sarcomas. J Biochem 122:1190–1195

Kimura Y, Nakazawa M, Yamada M (1998) Cloning and characterization of three isoforms of OS-9 cDNA and expression of the OS-9 gene in various human tumor cell lines. J Biochem 123:876–882

Kudo M, Canfield WM (2006) Structural requirements for efficient processing and activation of recombinant. Human UDP-N-acetylglucosamine:lysosomal-enzyme-N-acetylglucosamine-1-phosphotransfe rase. J Biol Chem 28(281):11761–11768

Kudo M, Bao M, D'Souza A et al (2005) The α- and β-subunits of the human UDP-N-acetylglucosamine:lysosomal enzyme N-acetylglucosamine-1-phosphotransferase [corrected] are encoded by a single cDNA. J Biol Chem 280:36141–36149

Lakshmi YU, Radha Y, Hille-Rehfeld A et al (1999) Identification of the putative mannose 6-phosphate receptor protein (MPR 300) in the invertebrate unio. Biosci Rep 19:403–409

Lang L, Couso R, Kornfeld S (1986) Glycoprotein phosphorylation in simple eucaryotic organisms. Identification of UDP-GlcNAc: glycoprotein *N*-acetylglucosamine-1-phosphotransferase activity and analysis of substrate specificity. J Biol Chem 261:6320–6325

Lee WS, Payne BJ, Gelfman CM et al (2007) Murine UDP-GlcNAc: lysosomal enzyme N-acetylglucosamine-1-phospho transferase lacking the gamma-subunit retains substantial activity toward acid hydrolases. J Biol Chem 282:27198–27203

Li A, Davila S, Furu L et al (2003) Mutations in PRKCSH cause isolated autosomal dominant polycystic liver disease. Am J Hum Genet 72:691–703

Litovchick L, Friedmann E, Shaltiel S (2002) A selective interaction between OS-9 and the carboxyl-terminal tail of meprin β. J Biol Chem 277:34413–34423

Metzger MB, Maurer MJ, Dancy BM, Michaelis S (2008) Degradation of a cytosolic protein requires ER-associated degradation machinery. J Biol Chem 283:32302–32316

Mikami K, Yamaguchi D, Tateno H et al (2010) The sugar-binding ability of human OS-9 and its involvement in ER-associated degradation. Glycobiology 20:310–321

Molinari M (2007) N-glycan structure dictates extension of protein folding or onset of disposal. Nat Chem Biol 3:313–320

Munro S (2001) The M6PRH domain suggests a shared ancestry for the mannose 6-phosphate receptors and other *N*-glycan-recognising proteins. Curr Biol 11:R499–R501

Nadimpalli SK, von Figura K (2002) Identification of the putative mannose 6-phosphate receptor (MPR 46) protein in the invertebrate mollusc. Biosci Rep 22:513–521

Nadimpalli SK, Padmanabhan N, Koduru S (2004) Biochemical and immunological characterization of a glycosylated alpha-fucosidase from the invertebrate Unio: interaction of the enzyme with its in vivo binding partners. Protein Expr Purif 37:279–287

Nakayama T, Yaoi T, Kuwajima G et al (1999) Ca^{2+}-dependent interaction of N-copine, a member of the two C2 domain protein family, with OS-9, the product of a gene frequently amplified in osteosarcoma. FEBS Lett 453:77–80

Ni X, Canuel M, Morales CR (2006) The sorting and trafficking of lysosomal proteins. Histol Histopathol 21:899–913

Olson LJ, Dahms NM, Kim JJ (2004) The N-terminal carbohydrate recognition site of the cation-independent mannose 6-phosphate receptor. J Biol Chem 279:34000–34009

Pelletier MF, Marcil A, Sevigny G et al (2000) The heterodimeric structure of glucosidase II is required for its activity, solubility, and localization in vivo. Glycobiology 10:815–827

Quinn RP, Mahoney SJ, Wilkinson BM, Thornton DJ, Stirling CJ (2009) A novel role for Gtb1p in glucose trimming of N-linked glycans. Glycobiology 19:1408–1416

Riemer J, Appenzeller-Herzog C, Johansson L et al (2009) A luminal flavoprotein in endoplasmic reticulum-associated degradation. Proc Natl Acad Sci USA 106:14831–14836

Ruddock LW, Molinari M (2006) N-glycan processing in ER quality control. J Cell Sci 119:4373–4380

Stigliano ID, Caramelo JJ, Labriola CA et al (2009) Glucosidase II beta subunit modulates N-glycan trimming in fission yeasts and mammals. Mol Biol Cell 20:3974–3984

Szathmary R, Bielmann R, Nita-Lazar M et al (2005) Yos9 protein is essential for degradation of misfolded glycoproteins and may function as lectin in ERAD. Mol Cell 19:765–775

Tamura T, Cormier JH, Hebert DN (2008) Sweet bays of ERAD. Trends Biochem Sci 33:298–300

Totani K, Ihara Y, Matsuo I, Ito Y (2008) Effects of macromolecular crowding on glycoprotein processing enzymes. J Am Chem Soc 130:2101–2107

Treml K, Meimaroglou D, Hentges A, Bause E (2000) The α– and β-subunits are required for expression of catalytic activity in the hetero-dimeric glucosidase II complex from human liver. Glycobiology 10:493–502

Trombetta ES, Parodi AJ (2003) Quality control and protein folding in the secretory pathway. Annu Rev Cell Dev Biol 19:649–676

Trombetta ES, Simons JF, Helenius A (1996) ER glucosidase II is composed of a catalytic subunit, conserved from yeast to mammals, and a tightly bound noncatalytic HDEL-containing subunit. J Biol Chem 271:27509–27516

Trombetta ES, Fleming KG, Helenius A (2001) Quaternary and domain structure of glycoprotein processing glucosidase II. Biochemistry 40:10717–10722

Vigneron N, Ooms A, Morel S et al (2002) Identification of a new peptide recognized by autologous cytolytic T lymphocytes on a human melanoma. Cancer Immun 2:9

Vourvouhaki E, Carvalho C, Aguiar P (2007) Model for Osteosarcoma-9 as a potent factor in cell survival and resistance to apoptosis. Phys Rev E Stat Nonlin Soft Matter Phys 76:011926

Wang Y, Fu X, Gaiser S et al (2007) OS-9 regulates the transit and polyubiquitination of TRPV4 in the endoplasmic reticulum. J Biol Chem 282:36561–36570

Watanabe T, Totani K, Matsuo I et al (2009) Genetic analysis of glucosidase II β-subunit in trimming of high-mannose-type glycans. Glycobiology 19:834–840

Westlund B, Dahms NM, Kornfeld S (1991) The bovine mannose 6-phosphate/insulin-like growth factor II receptor. Localization of mannose 6-phosphate binding sites to domains 1–3 and 7–11 of the extracytoplasmic region. J Biol Chem 266:23233–23239

Whyte JR, Munro S (2001) A yeast homolog of the mammalian mannose 6-phosphate receptors contributes to the sorting of vacuolar hydrolases. Curr Biol 11:1074–1078

Wilkinson BM, Purswani J, Stirling CJ (2006) Yeast GTB1 encodes a subunit of glucosidase II required for glycoprotein processing in the ER. J Biol Chem 281:6325–6333

Yamaguchi D, Hu D, Matsumoto N, Yamamoto K (2010) Human XTP3-B binds to alpha1-antitrypsin variant null(Hong Kong) via the C-terminal M6PRH domain in a glycan-dependent manner. Glycobiology 20:348–355

Yoshida Y, Tanaka K (2010) Lectin-like ERAD players in ER and cytosol. Biochim Biophys Acta 1800:172–180

Lectins of ERAD Pathway: F-Box Proteins and M-Type Lectins

6

G.S. Gupta

6.1 Intracellular Functions of N-Linked Glycans in Quality Control

Glycoprotein folding and degradation in endoplasmic reticulum (ER) is mediated by ER quality control system. Quality control in ER ensures that only properly folded proteins are retained in the cell through mechanisms that recognize and discard misfolded or unassembled proteins in a process called endoplasmic reticulum-associated degradation (ERAD). The ERAD pathway comprises multiple steps including substrate recognition and targeting to the retro-translocation machinery, retrotranslocation from the ER into the cytosol, and proteasomal degradation through ubiquitination. Roles of lectin-type molecules for trimmed high-mannose type N-glycans and glycosidases in the ERAD pathways in both ER and cytosol have been documented in recent years. Yoshida and Tanaka (2010) reviewed the fundamental system that monitors glycoprotein folding in the ER and the unique roles of sugar-recognizing ubiquitin ligase and peptide:N-glycanase (PNGase) in the cytosolic ERAD pathway. Mannose trimming plays an important role by forming specific *N*-glycans that permit the recognition and sorting of terminally misfolded conformers for ERAD. The EDEM (ER degradation enhancing α-mannosidase-like protein) subgroup of proteins belonging to Class I α1,2-mannosidase family (glycosylhydrolase family 47) has been shown to enhance ERAD. However, the mechanisms of substrate recognition and sorting to the ERAD pathway are poorly defined.

Post-translational modification of proteins regulates many cellular processes. For example, the attachment of *N*-linked glycans to nascent proteins in ER facilitates proper folding, whereas retention of high mannose glycans on misfolded glycoproteins serves as a signal for retrotranslocation and ubiquitin-mediated proteasomal degradation. N-glycosylation of proteins in ER has a central role in protein quality control. In ER and in early secretory pathway, where the repertoire of oligosaccharide structures is still small, the glycans play a pivotal role in protein folding, oligomerization, quality control, sorting, and transport. The glycans not only promote folding directly by stabilizing polypeptide structures but also indirectly by serving as recognition "tags" that allow glycoproteins to interact with a variety of lectins, glycosidases, and glycosyltranferases. Some of these (such as glucosidases I and II, calnexin, and calreticulin) have a central role in folding and retention, while others (such as α-mannosidases and EDEM) target unsalvageable glycoproteins for ER-associated degradation. Each residue in the core oligosaccharide and each step in the modification program have significance for the fate of newly synthesized glycoproteins. In the Golgi complex, the glycans acquire more complex structures and a new set of functions. The division of synthesis and processing between the ER and the Golgi complex represents an evolutionary adaptation that allows efficient exploitation of the potential of oligosaccharides (Helenius and Aebi 2004).

Most secretory and membrane proteins present outside cells are glycosylated. The oligosaccharides attached covalently to asparagine (N-linked) or serine/threonine (O-linked) of mature proteins have highly diverse structures and are implicated in many extracellular processes (1). This diversity of oligosaccharides is provided by hundreds of glycosyltransferases in Golgi complex. On the other hand, the N-linked oligosaccharides attached to newly synthesized polypeptides in ER are homogenous, relatively simple high-mannose oligosaccharides. High-mannose oligosaccharides are deeply involved in promoting protein folding and quality control, and in the trafficking and sorting of glycoproteins in that they are recognized by specific lectin-type molecules (2). Of these intracellular events, information on the role of high mannose oligosaccharides as tags for the quality control of proteins has been accumulated. The ER through a quality control mechanism discriminates correctly folded proteins from incorrectly folded or incompletely assembled

G.S. Gupta et al., *Animal Lectins: Form, Function and Clinical Applications*,
DOI 10.1007/978-3-7091-1065-2_6, © Springer-Verlag Wien 2012

non-functional proteins. (3) Protein folding and oligomer formation are assisted by ER chaperone proteins, and only correctly folded proteins are transported out of ER to Golgi complex. Any misfolded proteins are retained in the ER, and they are retro-translocated into the cytosol when misfolding persists, where they are degraded by the proteasome through ubiquitylation. This mechanism is known as ER-associated degradation (ERAD). (4) Recent studies indicate that various ER-resident high mannose oligosaccharide recognition molecules are involved in the several processes of the quality control system, such as individual steps of ER-retention, selection, and targeting of ERAD substrates. In this process (ERAD) the role of group of lectins has started to be appreciated. As an illustration, the F-box protein-1 (Fbs1) that recognizes sugar chains, specifically binds to the high-mannose oligosaccharides uniquely at the innermost position (5). Unlike other intracellular oligosaccharide recognition molecules, Fbs1 is present in the cytosol, where glycoproteins rarely exist. Although N-linked glycans are attached in the lumen of the ER or Golgi and are present on the outside of cells or inside vesicular compartments, it has been found that glycosylated molecules become detectable in the cytosol when cytosolic peptide: N-glycanase (PNGase) is knocked down by siRNA, suggesting that glycoproteins are deglycosylated after relocation from the ER into the cytosol (6). Fbs1 forms the SCF-type E3 ubiquitin ligase complex, and SCF-Fbs1 is probably responsible for the ubiquitylation of N-linked glycoproteins through the ERAD pathway. N-glycan serves as a signal for degradation by the Skp1-Cullin1-Fbx2-Roc1 (SCF-Fbx2) ubiquitin ligase complex. Like Fbs1, F-box protein Fbx2 binds specifically to proteins attached to N-linked high-mannose oligosaccharides and subsequently contributes to ubiquitination of N-glycosylated proteins. Reports indicate that SCF-Fbx2 ubiquitinates N-glycosylated proteins that are translocated from the ER to the cytosol by the quality control mechanism (Yoshida et al. 2002; Tai 2006).

6.2 The Degradation Pathway for Misfolded Glycoproteins

Eukaryotic cells have two main systems for intracellular protein degradation. Autophagy is an intracellular bulk degradation process and less selective. On the other hand, ubiquitin-proteasome system is responsible for most of the selective protein degradation for determination of life span of each intracellular protein. As stated above, in ubiquitin-proteasome system, a series of enzymes, E3 ubiquitin-ligases, play the most important role in the selection of target proteins.

6.2.1 Endoplasmic Reticulum-Associated Degradation (ERAD)

The quality control system includes the calnexin-calreticulin cycle, a unique chaperone system that recognizes $Glc_1Man_{9-6}GlcNAc_2$ and assists refolding of misfolded or unfolded proteins. When the improperly folded or incompletely assembled proteins fail to restore their functional states, they are degraded by ERAD system, which involves a retrograde transfer of proteins from the ER to the cytosol followed by degradation by 26S proteasome, a 2 MDa multicatalytic protease complex (Tsai et al. 2002). Precisely how they are selected for ERAD remains unclear, but what is clear is that the trimming of N-glycans plays a key role in the selection process. Studies have demonstrated that $Man_8GlcNAc_2$ structures serve as part of the signal needed for ERAD and that a lectin for $Man_8GlcNAc_2$ in ER accelerates the turnover rate of the misfolded glycoprotein (Hosokawa et al. 2001; Jakob et al. 2001; Nakatsukasa et al. 2001). It has been reported that many E3s are involved in the ERAD pathway, such as ER-embedded Hrd1 and Doa10, which have overlapping functions in yeast, and gp78, CHIP and Parkin, which ubiquitylate ER membrane proteins such as cystic fibrosis transmembrane conductance regulator (CFTR) and the Pael receptor in mammals (c/r Yoshida et al. 2003). In addition, the ERAD-linked E3 family, SCF^{Fbx2} participates in ERAD for selective elimination of glycoproteins (Yoshida et al. 2002). Misfolding or misassembly might be the general feature of all substrates; however, Fbx2 is expressed mainly in neuronal cells in the adult brain. Winston et al. (1999) and Ilyin et al. (2002) reported that several F-box proteins, including Fbx2, contain a conserved motif F-box-associated (FBA) domain or G-domain (sharing similarity with bacterial protein ApaG) in their C termini. Thus, SCF contains three core subunits, and a number of less critical components:

6.2.2 Ubiquitin-Mediated Proteolysis

Ubiquitin (Ub)-mediated proteolysis has a regulatory function in many diverse cellular processes. The ubiquitylation of a specific protein is carried out by the sequential activities of three enzymes, an activating enzyme (E1), an ubiquitin-conjugating enzyme (E2), and an ubiquitin-ligase (E3) (Fig. 6.3a). In ubiquitin pathway, E3 is responsible for the selection of target proteins. In ubiquitin ligase system E1 activates ubiquitin, E2 transfers ubiquitin on to target proteins, and E3 selects proteins for ubiquitination. E3s are believed to exist as molecules with a large diversity, presumably in more than hundreds of species, which are classified into many subfamilies. Five types of E3s have been identified that differ according to their subunit organization and or mechanism of Ub transfer. Ubiquitin-ligases E3s, identified in the ERAD pathway, include Hrd1 (Bays et al. 2001) and Doa10 (Swanson et al. 2001) in yeast, and gp78 (Fang et al. 2001), CHIP (Meacham et al. 2001) and Parkin (Imai et al. 2001) in mammals. In addition, a member of the ERAD-linked E3 family, SCF^{Fbs} which participates in ERAD for selective elimination of glycoproteins (Yoshida et al. 2002, 2003) has been identified. Whereas Hrd1, Doa10 and gp78 are localized in the ER, SCF^{Fbs} complexes are localized in the cytosol similar to CHIP and Parkin.

6.2.3 SCF Complex

One of the best characterized E3 families is the Skp1-Cullin1-F-box protein-Roc1 (SCF) complex (Deshaies 1999). In yeast (*Saccharomyces cerevisiae*), E3 type SCF complex is composed of Cullin1/Cdc53, Skp1, Roc1/Rbx1, and one member of family of F-box proteins, which are involved in trapping target proteins (Deshaies 1999; Ho et al. 2006). The cullin1, Rbx1, and Skp1 subunits appear to form the core ligase activity, with Rbx1 recruiting the E2 bearing an activated Ub (Hershko and Ciechanover 1998; Skowyra et al. 1999). In SCF complex, F-box protein is a variable component among 100 different members. F-box proteins in human are thought to allow the specific ubiquitylation of a wide range of substrates. In many instances, the F-box proteins serve as receptors for substrates that have various covalent modifications (phosporylation seems to be the predominant signal).

- Skp1—Bridging protein, forms part of the horseshoe-shaped complex, along with cullin (cul1). Skp1 is essential in the recognition and binding of the F-box (Figs. 6.1, 6.2a).
- Cullin (CUL1) forms the major structural scaffold of the SCF complex, linking the skp1 domain with the Rbx1 domain.
- Rbx1—Rbx1 contains a small zinc-binding domain called the RING Finger, to which the E2-ubiquitin conjugate binds, allowing the transferral of the ubiquitin to a lysine residue on the target protein.
- F-Box protein - As shown in Fig. 6.2b, Fbs1 has an N-terminal P domain whose function is not determined, an F-box domain that is required for Skp1 binding, and a substrate-binding domain that recognizes N-glycans. F box motif consisting of ~50 amino acid residues and a C-terminal region that interacts with the substrate and, thereby, the function of F-box protein is to trap target proteins. Fbs1 is named as the *F-box* protein that recognizes sugar chains (Fig. 6.2c). Thus, F-box proteins perform the crucial role of delivering appropriate targets to the complex. These proteins all containing a N-terminal motif (F-box) interact with the rest of the SCF

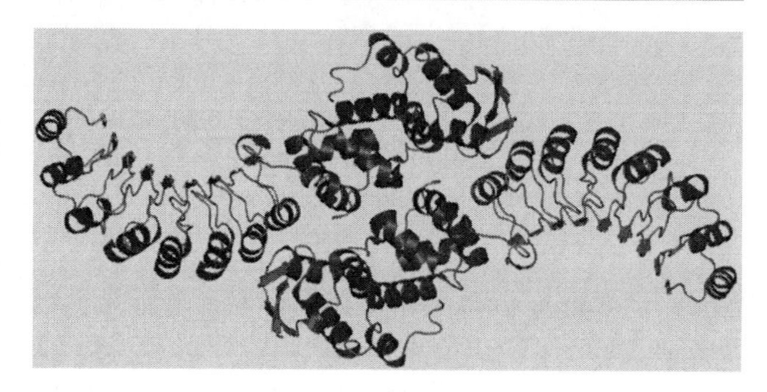

Fig. 6.1 Crystal structure of the human F-box protein Skp2 bound to Skp1 or (Skp1-Skp2) as horseshoe-shaped complex (Schulman et al. 2000) (PDB ID: 1FS2). Skp1 recruits the F-box protein through a bipartite interface involving both the F-box and the substrate-recognition domain. The structure raises the possibility that different Skp1 family members evolved to function with different subsets of F-box proteins, and suggests that the F-box protein may not only recruit substrate, but may also position it optimally for the ubiquitination reaction

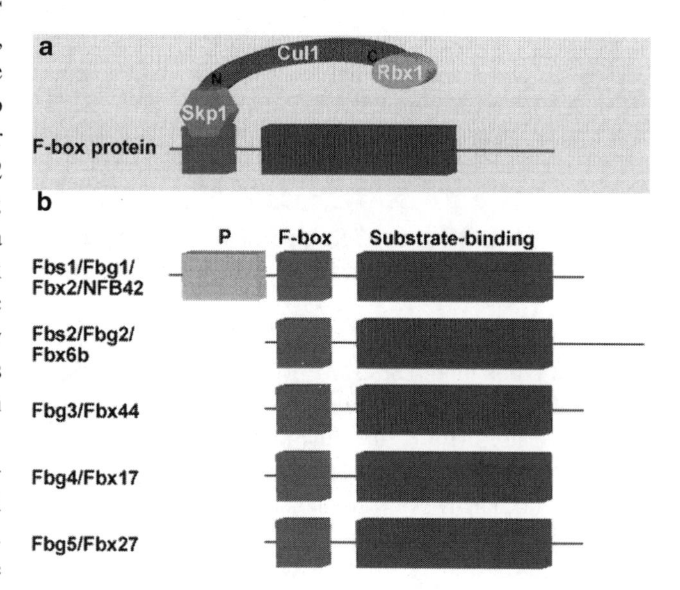

Fig. 6.2 The Fbs Family: (a) The SCF complex; (b) the domain structure of the Fbs family. In ubiquitin-proteasome system, E3 ubiquitin-ligases play important role in the selection of target proteins. E3s exist in hundreds of species. The largest known family of E3s is the SCF complex consisting of Skp1, Cul1, Rbx1/Roc1, and F-box proteins. In SCF complex, only F-box proteins are variable components, and at least 100 different members of F-box proteins in human are thought to allow for the specific ubiquitylation of a wide range of substrates. Fbs1 has an N-terminal pest (*P*) domain whose function is not determined, an F-box domain that is required for Skp1 binding, and a substrate-binding domain that recognizes *N*-glycans. Fbs1 might bind to denatured glycoproteins which have exposed inner chitobiose in *N*-glycans. Improperly folded proteins in the ER are retro-grade translocated into the cytosol where they are destroyed by the ubiquitin-proteasome system through ERAD. It has been reported that Fbs1 has at least four F-box proteins that show high homology in the substrate-binding domain. Among these proteins, Fbs1/Fgb1/Fbx2/NFB42, Fbs2/Fbg2/Fbx6b, Fbg3/Fbx44, Fbg4/Fbx17, and Fbg5/Fbx27, only Fbs2 and Fbs1 have the ability to interact with *N*-glycans, but the *N*-glycan binding activities of other proteins are not known (Adapted by permission from Yoshida 2007 © Japan Society for Bioscience, Biotechnology, and Agrochemistry)

complex by binding to the Skp subunit. The C-terminal portions of F-box proteins typically contain a variable protein-interaction domain that binds the target and thus confers specificity to the SCF complex presumably participate in substrate recognition.

Regulated protein degradation is a key recurring theme in multiple aspects of cell-cycle regulation. Progression of eukaryotic cell cycle is regulated through synthesis-degradation and phosphorylation-dephosohorylation of cell cycle regulating proteins. Two ubiquitin ligases are crucial in cell cycle. The anaphase-promoting complex or cyclosome (APC/C) controls metaphase-anaphase transition with its activator Cdc20. The APC/C is a ubiquitin-protein ligase required for the completion of mitosis in all eukaryotes. Mechanistic studies reveal how this remarkable enzyme combines specificity in substrate binding with flexibility in Ub transfer, thereby allowing the modification of multiple lysines on the substrate as well as specific lysines on ubiquitin itself (Matyskiela et al. 2009; Vodermaier 2004; Nakayama et al. 2005). Its activity is required for sister chromatids separation. APC/C with another activator, Cdh1, is also active in G1 phase and controls levels of mitosis regulating proteins

6.2.4 F-Box Proteins: Recognition of Target Proteins by Protein-Protein Interactions

F box proteins typically have a bipartite structure with an N terminal F box motif consisting of ~50 amino acid residues (Fig. 6.2c) and a C-terminal region that interacts with the substrate and, thereby, the function of F-box protein is to trap target proteins (Winston et al. 1999; Kipreos and Pagano 2000). However, it remains unknown how E3s accurately recognize target proteins. Studies suggest that phosphorylation of target proteins is a prerequisite for their recognition by SCF complexes (Hershko and Ciechanover 1998; Deshaies 1999; Kipreos and Pagano 2000). In addition, proline hydroxylation of transcription factor hypoxia-induced factor 1α (HIF1α) serves as a signal for ubiquitylation by the SCF-like Cullin2-based VBC ubiquitin-ligase (Ivan et al. 2001; Jaakkola et al. 2001). The F-box component of the SCF machineries is responsible for recognizing different substrates for ubiquitination. Interaction with components of the SCF complex is mediated through the F-box motif of the F-box protein while it associates with phosphorylated substrates through its second protein-protein interaction motif such as Trp-Asp (WD) repeats or leucine-rich repeats (LRRs). By targeting diverse substrates, F-box proteins exert controls over stability of proteins and regulate the mechanisms for a wide-range of cellular processes. F-box proteins play important function in various cellular settings such as tissue development, cell proliferation, and cell death as demonstrated in the modeling organism *Drosophila* (Ho et al. 2006). On the other hand, Fbx2 forms an SCF^{Fbx2} ubiquitin ligase complex that targets sugar chains in N-linked glycoproteins for ubiquitylation (Yoshida et al. 2002). Thus, Fbx2 is an example of F-box proteins that have evolved to recognize protein modifications other than phosphorylation and hydroxylation.

Winston et al. (1999) reported the identification of a family of 33 novel mammalian F-box proteins. The large number of these proteins in mammals suggests that the SCF system controls a correspondingly large number of regulatory pathways in vertebrates. Four of these proteins contain a novel conserved motif, the F-box-associated (FBA) domain, which may represent a new protein-protein interaction motif. The identification of these genes will help uncover pathways controlled by ubiquitin-mediated proteolysis in mammals. In *S. cerevisiae*, the F-box proteins Cdc4, Grr1 and Met^{30} target cyclin-dependent kinase inhibitors, G1 cyclins and transcriptional regulators for ubiquitination and reviewed in (Kipreos and Pagano 2000). Presently more than 700 F-box proteins that act as recognition modules to specifically target their dedicated substrates for ubiquitylation are known. Some F-box proteins function as phytohormone or light receptors, which directly perceive signals and facilitate specific target-protein degradation to regulate downstream pathways. If this new connection between ligand-regulated proteolysis and signaling proves to be more extensive, an entirely new way of understanding the control of signal transduction is in the offing (Somers and Fujiwara 2009).

The F-box protein family is the largest protein superfamily known, the total number of members ranging from approx. 20 in yeasts to several hundreds in higher eukaryotes. F-box proteins exhibit a typical bipartite structure. The N-terminal conserved F-box domain of 40 amino acid residues is required for direct interaction with Skp1 in the SCF complex (Feldman et al. 1997). With its C-terminal domain, the F-box protein recruits the target protein into the SCF complex, mostly by protein–protein interactions. Depending on the substrate-specific motifs at the C-terminus, F-box proteins are classified into three groups (Jin et al. 2004). The two classes of binding domains are WD40 and leucine-rich repeats, and hence they are named the Fbw (or FBXW) and Fbl (or FBXL) families respectively. The third class of F-box proteins is the Fbx (or FBXO) family, which does not contain any WD40 repeats or leucine-rich repeats. The Fbx (or FBXO) comprises of all F-box proteins with other C-terminal domains such as Kelch domains, Armadillo and tetratricopeptide repeats, zinc fingers and proline-rich domains.

Recently, a small subfamily of the Fbx family that comprises F-box proteins with a C-terminal SBD (sugar binding domain) was identified in mammals. Biochemical and structural studies demonstrated that these sugar-binding F-box proteins (or Fbs proteins) specifically interact with N-linked glycans (mostly with Man3−9GlcNAc2 structures) present on target glycoproteins and are assumed to play a key role in glycoprotein homoeostasis (Jin et al. 2004; Winston et al. 1999; Ilyin et al. 2002). Most substrates are targeted to SCF E3 ligases after they are covalently modified. While phosphorylation targets numerous substrates to the SCF complex, high-mannose oligosaccharide modification is required for substrate binding by Fbs1 in the SCF complex. Fbs1 belongs to a subfamily consisting of at least five homologous (FBXO2, FBXO6, FBXO17, FBXO27 and FBXO44) F-box proteins that contain a conserved F-box-associated (FBA) motif at their C-termini (Ilyin et al. 2002). All five proteins can co-precipitate with components of the SCF complex, and have been proposed to target misfolded glycoproteins for degradation by the Ub/proteasome system. Among these, mammalian FBXO2 and FBXO6 (also called Fbs1 and Fbs2 respectively) recognize high-mannose oligosaccharides to form SCF-type ubiquitin ligases.

In yeast and animals, a number of F-box proteins are present, easily classified by the nature of this interaction domain. In yeast, for example, the F-box protein Grr1p contains C-terminal leucine-rich repeats (LRRs) that recruit the phosphorylated forms of the cyclins Cln1p and Cln2p to the SCFGrr1 complex, whereas the Cdc4 F-box protein contains WD-40 repeats that recruit the phosphorylated form of the cyclin-dependent kinase inhibitor Sic1 to the SCFCdc4 complex. Studies in *Arabidopsis thaliana* indicate that F-box proteins and SCF complexes play critical roles in various aspects of plant growth and development. ASK1, one of the 19 Skp proteins in *Arabidopsis*, is involved in male gametogenesis and floral organ identity. Gagne et al. (2002) identified 694 potential F-box genes in *Arabidopsis*, making this gene superfamily one of the largest currently known in plants. Most of the encoded proteins contain interaction domains C-terminal to the F-box that presumably participate in substrate recognition.

6.3 F-Box Proteins with a C-Terminal Sugar-Binding Domain (SBD)

F-box protein is present in ubiquitin-ligase E3 and is responsible for substrate selection. Each F-box protein contains a conserved F-box domain, which interacts with other subunits in E3, and a substrate recognition domain, which determines the target specificity of the E3 complex. Many F-box proteins play a determining role in the substrate specificity of this degradation pathway. In most cases, selective recognition of the target proteins relies on protein–protein interactions mediated by the C-terminal domain of the F-box proteins. In mammals, the occurrence of F-box proteins with a C-terminal SBD (sugar-binding domain) that specifically interacts with high-mannose N-glycans on target glycoproteins has been documented. The identification and characterization of sugar-binding F-box proteins demonstrated that F-box proteins do not exclusively use protein–protein interactions but also protein–carbohydrate interactions in Ub/proteasome pathway. F-box protein (Such as Cdc4) contributes to the specificity of SCF by aggregating to target proteins independently of the complex and then binding to the Skp1 component, thus allowing the protein to be brought into proximity with the functional E2 protein. The F-box is also essential in regulating SCF activity during the course of the cell-cycle. SCF levels are thought to remain constant throughout the cell-cycle. Instead, F-box affinity for protein substrates is regulated through cdk/cyclin mediated phosphorylation of target proteins (Mizushima et al. 2007).

6.3.1 Diversity in SCF Complex due to Lectin Activity of F-Box Proteins

N-glycosylation acts as a targeting signal to eliminate intracellular glycoproteins by Fbx2-dependent ubiquitylation and subsequent proteasomal degradation. N-glycosylation of the proteins occurs when newly synthesized proteins enter the ER through translocation channel “translocon.” N-glycans play an important role in glycoprotein transport and sorting (Ellgaard et al. 1999; Helenius and Aebi 2004), in particular at the initial step of secretion that occurs in ER compartment (Ellgaard et al. 1999; Helenius and Aebi 2004). N-linked glycoproteins are subjected to “quality control” in which aberrant proteins are distinguished from properly folded proteins and retained in ER (Helenius and Aebi 2004).

Ilyin et al. (2000) characterized 10 mammalian F-box proteins. Five of them (FBL3 to FBL7) share structural similarities with Skp2 and contain C-terminal leucine-rich repeats. The other five proteins have different putative protein-protein interaction motifs. Specifically, FBS and FBWD4 proteins contain Sec7 and WD40-repeat domains, respectively. The C-terminal region of FBA shares similarity with bacterial protein ApaG while FBG2 shows homology with the F-box protein NFB42. The marked differences in - F-box gene expression in human tissues suggest their distinct role in ubiquitin-dependent protein degradation. The cDNAs encoding FBG3, FBG4 and FBG5 display similarity with NFB42 (FBX2) and FBG2 (FBX6) proteins.

All five proteins are characterized by an approximately 180-amino-acid (aa) conserved C-terminal domain and thus constitute a third subfamily of mammalian F-box proteins. Genomic organization of the five FBG genes revealed that all of them consist of six exons and five introns. FBG1, FBG2 and FBG3 genes are located in tandem on chromosome 1p36, and FBG4 and FBG5 are mapped to chromosome 19q13. FBG genes are expressed in a limited number of human tissues including kidney, liver, brain and muscle tissues. Specifically, FBG2 mRNA was expressed in foetal liver, decreased after birth and re-accumulated in adult liver (Ilyin et al. 2002). Glenn et al. (2008) examined the substrate specificity of the only family of ubiquitin ligase subunits thought to target glycoproteins through their attached glycans. Five proteins comprising this FBA family (FBXO2, FBXO6, FBXO17, FBXO27, and FBXO44) contain a conserved G domain that mediates substrate binding. Glenn et al. (2008) showed that each family member has differing specificity for glycosylated substrates and suggested that the F-box proteins in the FBA family bind high mannose and sulfated glycoproteins, with one FBA protein, FBX044, failing to bind any glycans on the tested arrays. Differences in substrate recognition, SCF complex formation, and tissue distribution suggested that FBA proteins play diverse roles in glycoprotein quality control.

6.3.2 Fbs Family

6.3.2.1 Fbs1 and Fbs2 in ERAD Pathway

Studies suggest that Fbx6b/FBG2 binds several glycoproteins, though other F-box proteins failed to bind any of the glycoproteins tested. Based on these functional studies, Fbx2/FBG1 and Fbx6b/FBG2 were named as Fbs1 (F-box protein that recognizes sugar chains 1) and Fbs2, respectively. Fbs2 is widely distributed in a variety of mouse tissues, differing from the restricted expression of Fbs1 in adult brain and testis. Furthermore, a dominant negative Fbs2 mutant suppressed degradation of a typical ERAD substrate, the T cell receptor α subunit (TCRα). The human genome contains about 70 genes for F-box proteins, and at least five homologous F-box proteins containing a conserved motif in their C-termini are thought to recognize sugar chain of N-linked glycoproteins. Among these, Fbs1 and Fbs2 are involved in ERAD pathway and have the ability to bind to proteins modified with high-mannose oligosaccharides, whose modification occurs in the ER. Yoshida (2007) reviewed the in vivo function of Fbs1 and homologous proteins, intracellular oligosaccharide recognition molecules involved in the quality control system. Screening for proteins bound to various glycoproteins led to the identification of Fbs1 from mouse brain extracts (Yoshida et al. 2002). Similar to the ubiquitously expressed Fbs2, Fbs1 recognizes N-glycans at the innermost position as a signal for unfolded glycoproteins, probably in the ERAD pathway. Majority of Fbs1 is present as Fbs1-Skp1 heterodimers or Fbs1 monomers but not SCF(Fbs1) complex in situ. In vitro, Fbs1 prevented the aggregation of the glycoprotein through N-terminal unique sequence of Fbs1. Results suggest that Fbs1 assists clearance of aberrant glycoproteins in neuronal cells by suppressing aggregates formation, independent of ubiquitin ligase activity, and thus functions as a unique chaperone for those proteins (Yoshida et al. 2007) (Fig. 6.3).

Fbs1 can form an SCF complex specific to N-linked glycoproteins and can bind to proteins modified with high-mannose oligosaccharides, which occur in ER. N-linked glycoproteins are normally not accessible to the ubiquitylation machinery in the cytosol, because they reside within the lumen of the ER and other compartments of the secretory pathway. One of the identified Fbs1 substrates is pre-integrin β1 modified with high mannose oligosaccharides, and their physical association was detected in the cytosol in the presence of the proteasome inhibitor (Yoshida et al. 2002). Moreover, over-expression of the dominant-negative form Fbs1 lacking the F-box domain essential to the formation of the SCF complex led to inhibition of the degradation of ERAD substrates, suggesting that SCF(Fbs1) is involved in the ERAD pathway (Fig. 6.3).

6.3.2.2 X-Ray Crystallographic Structure of Fbs1

X-ray crystallographic and nuclear magnetic resonance (NMR) studies of the substrate-binding domain of Fbs1 have revealed that Fbs1 interacts with the inner chitobiose in *N*-glycans of glycoproteins by a small hydrophobic pocket located at the top of the β-sandwich. The intramolecular interactions of the innermost chitobiose and the polypeptide moiety generally hamper the intermolecular binding of macromolecules such as proteins. Therefore, Fbs1 might bind to denatured glycoproteins which have exposed inner chitobiose in N-glycans. Improperly folded proteins in the ER are retro-grade translocated into the cytosol where they are destroyed by the ubiquitin-proteasome system, a disposal system, called ER-association degradation (ERAD). The reason why Fbs1 interacts with internal chitobiose of N-glycans is because Fbs1 recognizes the innermost position of N-glycans as the signal for unfolded glycoproteins in the ERAD pathway. Hypothetically, it is suggested that although Fbs1 cannot access the inner chitobiose protected by polypeptide moieties in folded glycoproteins, Fbs1 can bind to exposed inner chitobiose by denaturation (Mizushima et al. 2004). The CRD in the murine F-box protein Fbs1 (Fbx2) bound to GlcNAc disaccharide is

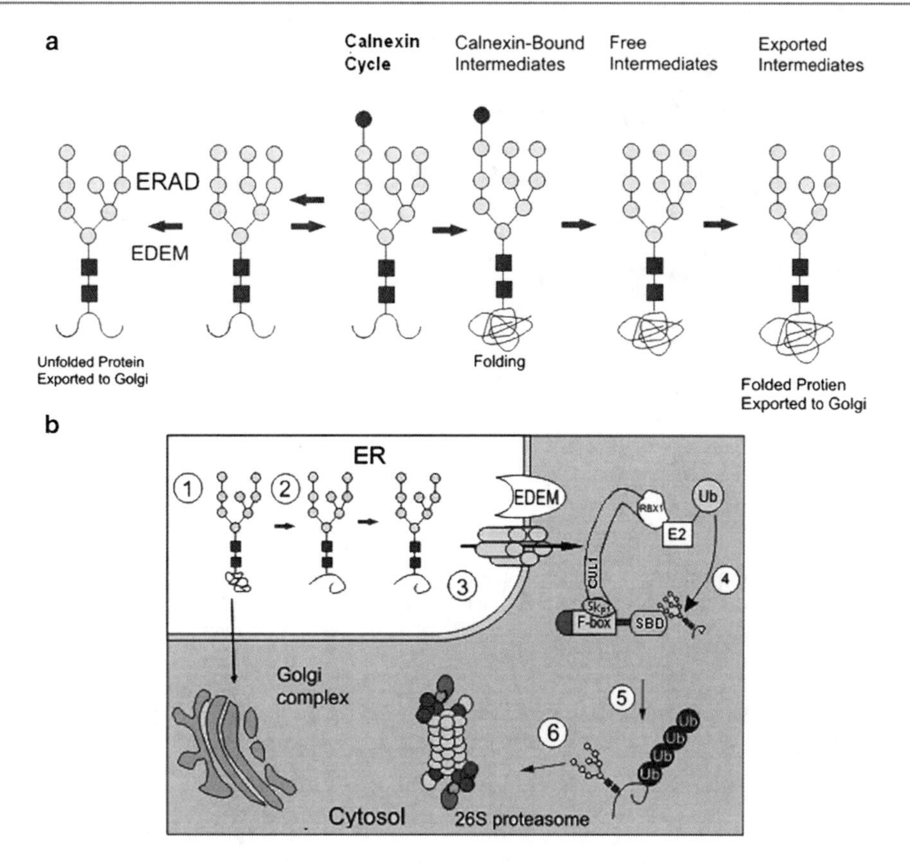

Fig. 6.3 (a) **N-linked glycan structures in ERAD.** (b) **Protein–carbohydrate interactions in the ERAD pathway:** Proper folding of newly synthesized glycoproteins occurs in the ER. Correctly folded glycoproteins (with $Man_9GlcNAc_2$ glycan) can exit the ER via vesicles to enter the secretory pathway *1*. Misfolded or unassembled glycoproteins (with $Man_8GlcNAc_2$ glycan) *2* are recognized by an ER-residing lectin, EDEM *3*, and retro-translocated from the ER into the cytosol where they are assembled in an active SCF complex through binding of their N-glycan with the SBD of an Fbs protein *4*. After conjugation with Ub *5*, the target proteins are recognized and degraded by the 26S proteasome *6*. *Circles* represent mannose and *squares* represent GlcNAc

shown in Fig. 6.4. Homologous domains present in F-box proteins from a range of animals may also function as CRDs.

6.3.2.3 Fbs1 and Fbs2 Homologous Proteins

Other glycoproteins modified with high-mannose oligosaccharides in the brain (Murai-Takebe et al. 2004; Kato et al. 2005) are also degraded in the ERAD pathway, mediated by SCFFbs1. SHP substrate-1 (SHPS-1), a transmembrane glycoprotein that regulates cytoskeletal reorganization and cell–cell communication, is particularly abundant in CNS and is a physiological substrate for SCF(NFB42) E3 ubiquitin ligase. Ectopic expression of Fbs1 resulted in elimination of misfolded SHPS-1 molecules from the ER and led to substantial up-regulation of SHPS-1 expression to the cell surface (Murai-Takebe 2004). Fbs1 mediated neuronal activity-dependent degradation of NR1, one of the NMDA subunits (Kato et al. 2005). Since the sugar moieties on the cell-surface NR1 are high-mannose, it appears possible that the mechanism of Fbs1 regulation of NMDA receptors is due not only to the ERAD pathway but also to retrograde transfer of endocytosed protein. It appears likely that Fbs1 assists in the clearance of aberrant glycoproteins in neuronal cells by suppressing aggregate formation, independent of ubiquitin ligase activity. Hence, Fbs1 has an additional function as a unique chaperone for those proteins.

Besides Fbs1, at least four F-box proteins that show high homology in SBD have been reported (Winston et al. 1999; Ilyin et al. 2002). These F-box proteins are composed of a highly homologous F-box domain, a substrate-binding domain, and a low homologous short linker sequence between F-box and SBD (Fig. 6.5). Although the homologies between Fbs2 and Fbg3 and between Fbg4 and Fbg5 in the short linker sequence are high, the linker sequence of Fbs1 shows no homology to any of them. It has been reported that the inefficient SCF complex formation of Fbs1 and the restricted presence of SCF^{Fbs1} bound to the ER membrane are due to the short linker sequence. The Fbs1, Fbs2, and Fbg3 genes are localized on human chromosome1p36.11–36.23 and on mouse chromosome 4E2, and the Fbg4 and Fbg5 genes are on human 19q13.2 and mouse 7B1. In the NCBI database Entrez gene, Fbs1, Fbs2, and Fbg5 encode single proteins, but Fbg3 and Fbx17 encode two proteins. Although Fbx17

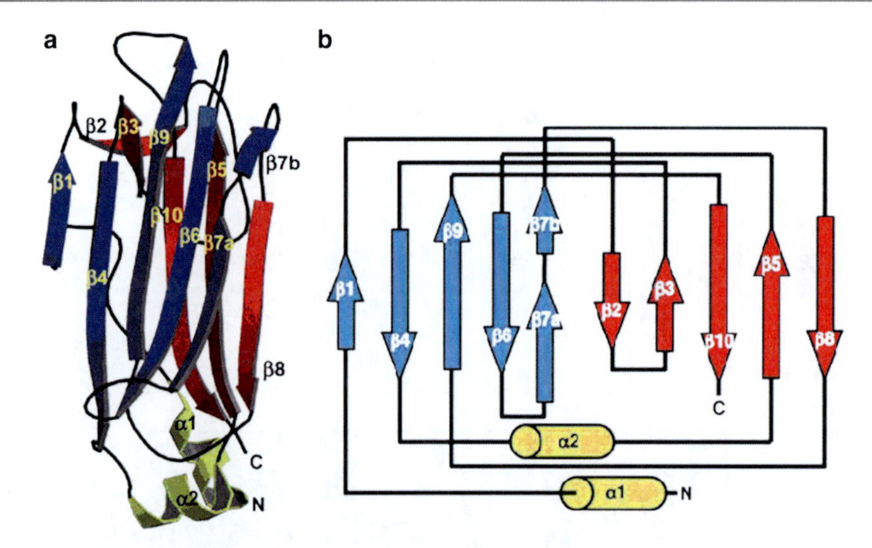

Fig. 6.4 Tertiary structure of SBD in Fbs1. (**a**) Overall structure of SBD of Fbs1 shown as a ribbon diagram (PDB ID: 1UMH; Mizushima et al. 2004). The SBD is composed of a ten-stranded antiparallel β-sandwich. Strands β1, β4, β6, β7 and β9 form one β-sheet, whereas other β-sheet consists of strands β2, β3, β5, β8 and β10. These two sheets are referred to as S1 and S2 sheets. β-strands belonging to S1 and S2 are *blue* and *red*, respectively. Loops and helices are *black* and *yellow*, respectively. (**b**) A topology diagram of SBD. The α-helices are *yellow cylinders* labeled α1 and α2. The β-strands are *arrows* labeled β1–β10. The left and right forms of β-strands correspond to S1 and S2, respectively, as in (**a**). *N* and *C* indicate N and C termini, respectively (Adapted by permission from Mizushima et al. 2004 © Macmillan Publishers Ltd)

encodes two proteins whose differences lie only in their N-terminal 9 amino acids by the different use of the first exon, Fbg3 encodes proteins non-homologous in their SBDs (isoform1 and isoform2).

The Fbg3 isoform1 protein is translated without skipping exons and shows high identity with Fbs2, as shown in Fig. 6.5. On the other hand, Fbg3 isoform2 is synthesized by skipping the fourth exon, and the resulting protein contains the 255 amino-acid region, which is different from C-terminal half of isoform1 (after residue-123). The expression of Fbs1, Fbg4, and Fbg5 is tissue specific, but that of Fbs2 and Fbg3 is relatively ubiquitous (Ilyin et al. 2002; Yoshida et al. 2003). Fbs1 is strongly expressed in neural cells in the adult brain and weakly in the testis. Fbg4 expression is predominant in the kidney and weak in the liver. Fbg5 is detected in the brain, heart, kidney, and liver.

Among five homologous F-box proteins (Ilyin et al. 2002), the FBG3 protein exhibits 75% identity with Fbs2, and the identity of Fbs1 and Fbs2 is similar to that of Fbs1 and FBG3. Interestingly, *Fbs1, FBG3*, and *Fbs2* genes are located in tandem on chromosome, and the expression of FBG3 is observed ubiquitously but strongly in the brain and testis (Yoshida et al. 2003). It is anticipated that FBG3 can recognize high mannose *N*-glycans because of their high homology. However, there was no sugar-binding activity for FBG3 despite several assay systems were used.

Despite the high homology between Fbs2 and Fbg3, no sugar-binding activity of Fbg3 has been detected. Of the Fbs family members, Fbs1 has the ability to bind strongly to the high-mannose oligosaccharides-modified proteins, while Fbs2 and Fbg5 bind them more weakly (Yoshida et al. 2002, 2003; Nelson et al. 2006), but Yoshida (2007) did not detect any sugar-binding activity for Fbg3 and Fbg4. X-ray crystallographic study of the substrate-binding domain of Fbs1 revealed that Fbs1 interacts with the inner-most chitobiose of glycoproteins by Phe177, Tyr279, Try280, and Lys281 on the loop structures in mouse Fbs1 and in human Fbs1, Phe173, Tyr278, Trp279, and Lys280, as shown in Fig. 6.5 (Mizushima et al. 2004) Among these residues, the Phe177Ala, Tyr2798Ala, or Trp280Ala mutants lost the ability to interact with glycoproteins. The alignment of human Fbs1 and its homologs shows that these residues, except for Lys280, are conserved in Fbs1, Fbs2, Fbg3, and Fbg5. It is anticipated that Fbg4 does not interact with the chitobiose moiety, because the residue corresponding to Tyr278 of human Fbs1 is serine in Fbg4, but these four residues are conserved in Fbs2 and Fbg3. Further analysis of the structure of the substrate-binding domain of Fbs2 and Fbg3 is needed to elucidate the mechanism by which Fbg3 fails to bind to chitobiose.

Although Fbs1, Fbs2, and Fbg5 have similar abilities to bind to glycoproteins modified with high-mannose oligosaccharide, they have different potencies of binding to glycoproteins. To gain insight into the differences in their mechanisms for oligosaccharide recognition related to their cellular functions, further structural and biological analysis is required. Moreover, it would be intriguing to identify the substrates for Fbg3 and Fbg4, which are not capable of

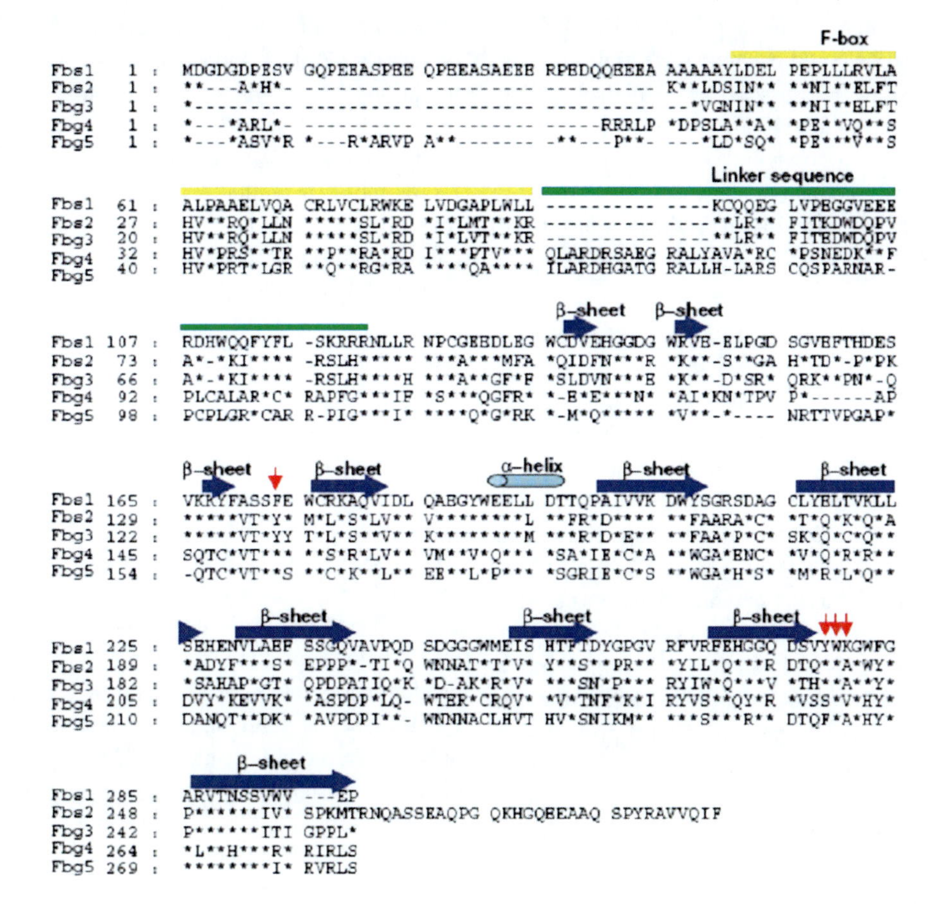

Fig. 6.5 Alignment of Human Fbs Family Proteins. Identical residues with Fbs1 are indicated by *asterisks*. F-box domain and the linker sequence are indicated by *yellow* and *green bars* respectively. Secondary structure elements of SBD in Fbs1 are shown above the amino acid sequence. Chitobiose binding residues are marked by *red arrows* (Adapted by permission from Yoshida 2007 © Japan Society for Bioscience, Biotechnology, and Agrochemistry)

binding to glycoproteins. Such analysis is clearly necessary for a better understanding of the overall functions of Fbs family proteins in vivo. There is little known at present about the organization of the genes encoding F-box proteins, with most studies focusing on the protein products (Kipreos and Pagano 2000).

How the SCF(Fbs1 and -2) complexes interact with unfolded glycoproteins? The SCF(Fbs1) complex was found associated with p97/VCP AAA ATPase and bound to integrin-β1, one of the SCF(Fbs1) substrates, in the cytosol in a manner dependent on p97 ATPase activity. Both Fbs1 and Fbs2 proteins interacted with denatured glycoproteins, which were modified with not only high-mannose but also complex-type oligosaccharides, more efficiently than native proteins. Given that Fbs proteins interact with innermost chitobiose in N-glycans, it was proposed that Fbs proteins distinguish native from unfolded glycoproteins by sensing the exposed chitobiose structure (Yoshida et al. 2005). Yamaguchi et al. (2007) suggested that Fbs1 captures malfolded glycoproteins, protecting them from the attack of PNGase, during the chaperoning or ubiquitinating operation in the cytosol (Yamaguchi et al. 2007).

6.3.3 Fbs1 Equivalent Proteins

It has been found that Fbs1 has at least four F-box proteins that show high homology in the substrate-binding domain. Among these proteins, Fbs1/Fgb1/Fbx2/NFB42, Fbs2/Fbg2/Fbx6b, Fbg3/Fbx44, Fbg4/Fbx17, and Fbg5/Fbx27, only Fbs2 and Fbs1 have the ability to interact with N-glycans, but the N-glycan binding activities of other proteins have not been reported. Fbs2 is widely expressed in various tissues in contrast to the limited expression of Fbs1 in neural cells in adult brain (Fbs1 was first named NFB42, meaning neural F Box 42 kDa). As ERAD might be the general feature of all cells, Fbs2 is thought to be a general ERAD E3 in mammals. Although Fbs1 and Fbs2 recognize high-mannose N-glycans, Fbs1 seems to possess 10^{2-3} higher affinity for oligosaccharides (Yoshida 2003; Yoshida et al. 2002). Thus, Fbs1 is equivalent to neural F-box protein of 42 kDa (NFB42)), Fbx2, (Winston et al. 1999; Cenciarelli et al. 1999), F-box only protein 2 (FBXO2), Fbg1 (Ilyin et al. 2000).

6.3.3.1 NFB42 Mediates Nuclear Export of UL9 Protein After Viral Infection

Neural F-box 42-kDa protein: The neural F-box 42-kDa protein (NFB42) is a component of the SCF(NFB42) E3 ubiquitin ligase that is expressed in all major areas of brain; it is not detected in nonneuronal tissues. Similar to other F-box proteins, NFB42 interacts with Skp1 through its F-box domain (Winston et al. 1999; Cenciarelli et al. 1999). NFB42 is highly enriched in the nervous system, as a binding partner for the herpes simplex virus 1 (HSV-1) UL9 protein. Co-expression of NFB42 and UL9 in human embryonic kidney (293T) cells led to a significant decrease in the level of UL9 protein. The interaction of UL9 protein with NFB42 results in its polyubiquitination and subsequent degradation by the 26S proteasome (Eom and Lehman 2003). The HSV-1 infection promotes the shuttling of NFB42 between the cytosol and the nucleus in both 293T cells and primary hippocampal neurons, permitting NFB42 to bind to the phosphorylated UL9 protein, which is localized in the nucleus. Because the intranuclear localization of the UL9 protein, along with other viral and cellular factors, is an essential step in viral DNA replication, degradation of the UL9 protein in neurons by means of nuclear export through its specific interaction with NFB42 may prevent active replication and promote neuronal latency of HSV-1 (Eom et al. 2004). Although NFB42 is present in the cytosol whereas the UL9 protein is located predominantly in the nucleus, HSV-1 infection promotes nuclear import of NFB42 and NFB42 mediates the nuclear export of the UL9 protein, leading to cytosolic degradation of UL9 protein (Eom et al. 2004)

6.3.3.2 OCP1 and OCP2

Role of selective cochlear degeneration in mice deficient in Fbx2 with specificity for high-mannose glycoproteins has been studied. Originally described as a brain-enriched protein (Yoshida et al. 2002), Fbx2 is also highly expressed in the organ of Corti (the sensory organ of cochlea) and protein, named as organ of Corti protein 1 (OCP1) (Thalmann et al. 1997). OCP1 and OCP2, the most abundant proteins in cochlea, are subunits of an SCF E3 ubiquitin ligase. OCP1 co-localizes exactly with OCP2 in the epithelial gap junction region of the guinea pig organ of Corti (OC), which is rich in OCP1. The full-length OCP1 cDNA—1,180 nt in length—includes a 67 nt 5′ leader sequence, 300 codons (including initiation and termination signals), and a 216 nt 3′ untranslated region. The cDNA encodes a protein having a predicted molecular weight of 33.7 kDa and harbors an F-box motif spanning residues 52–91, consistent with a role for OCP1 and OCP2 in the proteasome-mediated degradation of select OC proteins. OCP1 is equivalent to F-box proteins: Fbs1, Fbx2, or NFB42 - known to bind N-glycosylated proteins and believed to function in the retrieval and recycling of misfolded proteins. The OCP1 displays extensive homology to the rat brain NFB42. But clustered sequence non-identities indicate that the two proteins are transcribed from distinct genes. Located on chromosome 1p35, the inferred translation product exhibits 94% identity with the guinea pig OCP1 coding sequence (Henzl et al. 2001). Although transcribed from a distinct gene, OCP2 is identical to Skp1. The high concentrations of OCP1 and OCP2 in the cochlea suggest that the OCP1-OCP2 heterodimer may serve an additional function independent of its role in a canonical SCF complex. OCP1 and OCP2 associate tightly at room temperature. However, DSC data for the complex suggest that they denature independently, consistent with the highly exothermic enthalpy of complex formation (Tan and Henzl 2009).

Fbs1-deficient mice indicated that the Fbs1-Skp1 dimer is essential to inner ear homeostasis (Nelson et al. 2007). The orthologs of Fbs1 and Skp1 are abundantly present in guinea pig organ of Corti. They are called OCP1 and OCP2 respectively (Henzl et al. 2001). Mice with targeted deletion of Fbs1 develop age-related hearing loss with cochlear degeneration. The inner ear gap junction protein, connexin 26, which interacts with OCP1 (Henzl et al. 2004) increases in $Fbs1^{-/-}$ mouse cochlea. Furthermore, loss of Fbs1 leads to decrease in cochlear Skp1. On the other hand, the Skp1 level remains unchanged and no defects are seen in $Fbs1^{-/-}$ mouse brain (Nelson et al. 2007).

6.3.3.3 F-Box Proteins with a C-Terminal Domain is Homologous to Tobacco Lectin

During the last decade it was shown that plants synthesize minute amounts of carbohydrate-binding proteins upon exposure to stress situations like drought, high salt, hormone treatment, pathogen attack or insect herbivory. In contrast to the 'classical' plant lectins, which are typically found in storage vacuoles or in the extracellular compartment this new class of lectins is located in the cytoplasm and the nucleus in animals. Based on these observations the concept was developed that lectin-mediated protein-carbohydrate interactions in the cytoplasm and the nucleus play an important role in the stress physiology of the plant cell. Hitherto, six families of nucleocytoplasmic lectins have been identified. The putative sugar-binding F-box proteins have been identified in plants. Genome analyses in Arabidopsis and rice revealed the presence of F-box proteins with a C-terminal lectin-related domain homologous with Nictaba, a jasmonate-inducible lectin from tobacco that was shown to interact with the core structure of high-mannose and complex N-glycans. Owing to the high similarity in structure and specificity between Nictaba and the SBD of the mammalian Fbs proteins, a similar role for the plant F-box proteins with a Nictaba domain in nucleocytoplasmic protein degradation in plant cells is suggested (Lannoo and Van Damme 2010; Lannoo et al. 2008).

6.3.4 Ligands for F-Box Proteins

The F-box proteins can be classified into five major families, which can be further organized into multiple subfamilies. Sequence diversity within the subfamilies suggests that many F-box proteins have distinct functions and or substrates. Representatives of all of the major families interact in yeast two-hybrid experiments with members of the *Arabidopsis* Skp family supporting their classification as F-box proteins. Reports show that *Arabidopsis* has exploited the SCF complex and the ubiquitin_26S proteasome pathway as a major route for cellular regulation and that a diverse array of SCF targets is likely present in plants.

6.3.4.1 Binding to High Mannose Oligosaccharides

Fbs1 (equivalent to Fbx2 or NFB42) binds specifically to proteins attached with high mannose oligosaccharides and contributes to elimination of *N*-glycoproteins in cytosol (Yoshida et al. 2002). Fbs2 is another F-box protein that recognizes *N*-glycan and forms an SCF^{Fbs2} ubiquitin ligase complex that targets sugar chains in *N*-glycoproteins for ubiquitylation and plays a role in ERAD pathway (Ilyin et al. 2002; Yoshida et al. 2003). Pull-down analysis revealed that $Man_{3–9}GlcNAc_2$ glycans were required for efficient Fbs2 binding, whereas modifications of mannose residues by other sugars or deletion of inner GlcNAc reduced Fbs2 binding. Fbs2 interacted with *N*-glycans of T-cell receptor α-subunit (TCRα), a typical substrate of ERAD pathway, and the mutant Fbs2ΔF, which lacks the F-box domain essential for forming the SCF complex.

X-ray crystallographic and NMR studies of the SBD of Fbs1 have confirmed that Fbs1 interacts with the innermost chitobiose (GlcNAc-GlcNAc) in N-glycans of glycoproteins by way of a small hydrophobic pocket located at the top of the β sandwich (Mizushima et al. 2004). In general, the internal chitobiose of N-glycans in many native glycoproteins is not accessible by means of macromolecules. Fbs1 interacted with denatured glycoproteins more efficiently than native proteins did, indicating that the innermost position of N-linked oligosaccharides becomes exposed upon protein denaturation and is used as a signal of unfolded glycoproteins to be recognized by Fbs1 (Yoshida et al. 2005)

6.3.4.2 Chitobiose Is More Accessible for SBD

Both Fbs1 and Fbs2 are cytosolic proteins that preferentially bind to free N-glycans with a high-mannose oligosaccharide ($Man_{3–9}$) attached to a core chitobiose (GlcNAc2) and to high-mannose, N-linked glycoproteins. Fbs1 and Fbs2 do not bind to Man_3, Man_4 or Man_5 structures lacking $GlcNAc_2$, demonstrating the importance of the chitobiose core. In general, the core $GlcNAc_2$ of N-glycans present on correctly folded (native) proteins is not accessible to lectins, since most lectins bind to the non-reducing terminal sugar groups of carbohydrates. However, in unfolded (denatured) N-linked glycoproteins, the chitobiose core of glycan is more accessible for the SBD of Fbs1 and Fbs2. Yoshida et al.(2002, 2003) confirmed that SCF complexes with either Fbs1 or Fbs2 recognize denatured ERAD substrates in the cytosol and act as E3 Ub ligases in the ERAD pathway (Fig. 6.3). This implies that mammalian cells possess a ubiquitination system that uses protein–carbohydrate interactions instead of protein–protein interactions for a specific proteasome mediated degradation of glycoproteins. The crystal structures of sugar-binding domain (SBD) of Fbs1 alone and in complex with chitobiose has been solved. The SBD is composed of a ten-stranded antiparallel β-sandwich. The structure of the SBD-chitobiose complex includes hydrogen bonds between Fbs1 and chitobiose and insertion of the methyl group of chitobiose into a small hydrophobic pocket of Fbs1. Moreover, the amino acid residues adjoining the chitobiose-binding site interact with the outer branches of the carbohydrate moiety. Considering that the innermost chitobiose moieties in N-glycans are usually involved in intramolecular interactions with the polypeptide moieties, it was proposed that Fbs1 interacts with the chitobiose in unfolded N-glycoprotein, pointing the protein moiety toward E2 for ubiquitination (Mizushima et al. 2004).

The substrate recognition domain in Fbs1 has an ellipsoid shape formed by a sandwich of five-stranded, antiparallel β-sheets. One end of the ellipsoid is capped by two alpha-helices and is also the location of the domain N- and C-termini. The sugar binding site is located at the opposite end of the domain and is formed by the loops connecting β-strands 3–4 and 9–10. The binding site is specific for the N-acetylglucosamine (GlcNAc) disaccharide chitobiose. One GlcNAc residue stacks against a tryptophan aromatic ring, while the other inserts a methyl group into a hydrophobic pocket. Both residues make hydrogen bonds to the protein (Fig. 6.4). Specificity for chitobiose enables Fbs1 to bind to misfolded glycoproteins that have been translocated back into the cytoplasm: a GlcNAc disaccharide is at the core of all N-linked oligosaccharides added to proteins in the ER. In folded proteins, these GlcNAc residues interact with the polypeptide and are shielded, but in unfolded proteins they may become exposed and act as a novel sugar-based degradation signal. In contrast, outer residues in N-linked glycans appear to make only minor interactions with Fbs1.

6.3.5 Evolution of F-Box Proteins

F-box proteins contain a wide range of secondary motifs including zinc fingers, cyclin domains, leucine zippers, ring fingers, tetratricopeptide (TPR) repeats, and proline-rich regions. The diversity of associated protein domains suggests that F-box motifs have been transferred into existing proteins multiple times during eukaryotic evolution. Evolutionary constraints are higher for certain classes of F-box proteins: all of the human FBXW or FBXL proteins have counterparts in *C. elegans* with most also conserved in yeast, but only about half of the human FBXO class of proteins is conserved in nematodes or yeast. An interesting observation is the huge number of F-box proteins in *C. elegans*. The F-box motif is the fourth most common protein domain in *C. elegans*, with their number dwarfing the F-boxes found in other species by a factor often. Over half of the predicted *C. elegans* F-box proteins (135) are found with another motif known as DUF38 (domain of unknown function 38) or FTH (FOG-2 homology). The FTH/DUF38 domain is found mostly in nematodes, with none in humans or yeast. A second domain, PfamB-45, is found in another 56 *C. elegans* F-box proteins. Both of these cases suggest the expansion of single progenitor genes within nematodes (Kipreos and Pagano 2000).

6.3.6 Localization of F-Box Proteins

There have been a limited number of studies analyzing the subcellular localization of F-box proteins, and in all but a couple of cases analysis was performed with over-expressed tagged proteins (Winston et al. 1999). Some F-box proteins were found to be distributed both in the cytoplasm and in the nucleus. The identical localization of wild-type and mutant F-box proteins demonstrates that the presence of the F-box and the F-box-dependent binding to Skp1 does not determine the subcellular localization of these proteins. While the expression of mRNAs encoding some F-box proteins has been found in all tissues tested, others are clearly tissue-specific. Because of the large number of F-box proteins, this information is too complex to be summarized here (Kipreos and Pagano 2000).

6.3.7 Regulation of F-Box Proteins

F-box proteins are regulated by several mechanisms and at different levels: such as synthesis, degradation, and association with SCF components. The three yeast F-box proteins Cdc4, Grr1, and Met30 are intrinsically unstable proteins whose levels do not oscillate during cell cycle. It appears that they are subjected to ubiquitin-proteasome mediated degradation by an autocatalytic mechanism. Whereas the degradation of Cdc4 and Grr1 is dependent on their abilities to bind Skp1 through their F-boxes (Galan and Peter 1999), Met30 seems to be ubiquitinated in a cullin-dependent manner but in an F-box-independent manner (Rouillon et al. 2000).

Mammalian Skp2 is degraded by the ubiquitin-proteasome pathway but its expression is mostly regulated at a transcriptional level (Zhang et al. 1995). The expression of both Skp2 mRNA (Zhang et al. 1995) and Skp2 protein (Carrano et al. 1999) are cell-cycle-regulated, peaking in S phase and declining as cells progress through M phase. In contrast, the expression of other subunits of the SCF-Skp2 ligase complex (Cull, Skp1 and Roc1), as well as its ubiquitin-conjugating enzyme (Ubc3) does not fluctuate through cell cycle. Thus, although the ubiquitination of Skp2 substrates is regulated by their own phosphorylation, which allows their recognition by Skp2, a second level of control is ensured by the cell-cycle oscillations in Skp2 levels. The only characterized post-translational modification of an F-box protein is phosphorylation of Skp2 on Ser76 by the cyclin A-cdk2 complex (Yam et al. 1999), but the significance of this modification is not well known.

Enforced expression of β-catenin induces the expression of the F-box protein β-TrCP (Spiegelman et al. 2000). Although β-catenin can act as a transcriptional regulator, induction of β-TrCP by β-catenin is due to a stabilization of β-TrCP mRNA. As β-catenin is an SCFβ-Trcp substrate, stimulation of β-TrCP expression by β-catenin results in an accelerated degradation of β-catenin itself, suggesting that a negative feedback loop may control the β-catenin pathway. Finally, the association of Grr1 with Skp1 is regulated by glucose levels (Li and Johnston 1997). Grr1 is required to transduce the glucose signal to transcriptional regulatory proteins. When glucose levels are high, the post-translational association of Grr1 with Skp1 is markedly increased; this effect is dependent on the carboxy-terminal region of Grr1 (Kipreos and Pagano 2000; Frescas and Pagano 2008).

6.4 α-Mannosidases and M-Type Lectins

6.4.1 α-Mannosidases

α-Mannosidases in eukaryotic cells participates in both glycan biosynthetic reactions and glycan catabolism. Two broad families of enzymes have been identified that cleave terminal mannose linkages from Asn-linked oligosaccharides (Moremen 2000), including the Class 1 mannosidases (CAZy GH family 47 (Henrissat and Bairoch 1996)) of the early secretory pathway involved in the processing of N-glycans and quality control and the Class 2 mannosidases (CAZy family GH38 (Henrissat and Bairoch 1996) involved in glycoprotein biosynthesis or catabolism. Within the Class 1 family of α-mannosidases, three subfamilies of enzymes have been identified (Moremen 2000). The ER α1,2-mannosidase I (ERManI) subfamily acts to cleave a single residue from Asn-linked glycans in ER. The Golgi α-mannosidase I (GolgiManI) subfamily has at least three members in mammalian systems (Lal et al. 1998; Tremblay and Herscovics 2000) that help in glycan maturation in the Golgi complex to form the $Man_5GlcNAc_2$ processing intermediate. The third subfamily of GH47 proteins comprises the ER degradation, enhancing α-mannosidase-like proteins (EDEM proteins) (Helenius and Aebi 2004; Hirao et al. 2006; Mast et al. 2005; Mast and Moremen 2006). These proteins have been proposed to accelerate the degradation of misfolded proteins in the lumen of the ER by a lectin function that leads to retrotranslocation to the cytosol and proteasomal degradation. Recent studies have also indicated that ERManI acts as a timer for initiation of glycoprotein degradation via the ubiquitin-proteasome pathway (Hosokawa et al. 2003; Wu et al. 2003). Gong et al. (2005) discussed methods for analysis of the GH47 α-mannosidases, including expression, purification, activity assays, generation of point mutants, and binding studies by surface plasmon resonance. ERManI inhibitor kifunensine and down-regulation of EDEM suppressed the degradation of Y611H mutant channel proteins (Gong et al. 2005). ERAD of the misfolded genetic variant-null Hong Kong (NHK) α1-antitrypsin is enhanced by overexpression of ER processing α1,2-mannosidase (ERManI) in HEK 293 cells, indicating the importance of ERManI in glycoprotein quality control. In addition, EDEM, the enzymatically inactive mannosidase homolog, interacts with misfolded α1-antitrypsin and accelerates its effect showing a combined effect of ER ManI and EDEM on ERAD of misfolded α1-antitrypsin degradation (Hosokawa et al. 2001). It was also shown that misfolded α1-antitrypsin NHK contains labeled $Glc_1Man_9GlcNAc$ and $Man_{5\text{-}9}GlcNAc$ released by endo-β-N-acetylglucosaminidase.

Humans have four α-Mannosidases: ER mannosidase and Golgi mannosidases IA, IB and IC (gene names Man1B1, Man1A1, Man1A2 and Man1C1 respectively). ER mannosidase mediates the first mannose trimming reaction in the processing of N-linked glycans, reducing a $Man_9GlcNAc_2$ oligosaccharide to $Man_8GlcNAc_2$ by removal of a specific terminal mannose residue. ER mannosidase is conserved in eukaryotes; the yeast protein is known as Mns1. The trio of *cis*-Golgi mannosidases performs further specific trimming reactions on the $Man_8GlcNAc_2$ oligosaccharide, reducing the glycan to a $Man_5GlcNAc_2$ core that may be elaborated upon by a number of glycosyltransferases in the medial and *trans*-Golgi. The three Golgi mannosidases are conserved in vertebrates only and often have just a single homologue in invertebrate organisms. High-resolution crystal structures of the human and yeast ER α-1,2-mannosidases have been determined, the former complexed with the inhibitors 1-deoxymannojirimycin and kifunensine, both of which bind in its active site in the unusual $^1C_4$ conformation. Data revealed the roles of potential catalytic acid and base residues and the identification of a novel $^3S_1$ sugar conformation for the bound substrate analog. The co-crystal structure, in combination with $^1C_4$ conformation of a previously identified co-complex with the glycone mimic, 1-deoxymannojirimycin, indicated that glycoside bond cleavage proceeds through a least motion conformational twist of a properly predisposed substrate in the −1 subsite. A novel $^3H_4$ conformation is proposed as the exploded transition state (Karaveg et al. 2005).

6.4.2 ER-associated Degradation-enhancing α-Mannosidase-like Proteins (EDEMs)

The M-type lectins are members of glycosylhydrolase family 47 protein structural group. They are closely related to α-mannosidases of ER and *cis*-Golgi and function alongside these proteins in the handling of N-linked glycoproteins. These lectins lack key catalytic residues, as well as a key disulphide bond thought to be essential for enzymatic activity. As a result they bind to high mannose glycans attached to glycoproteins in ER lumen, but have no catalytic function (Hosokawa et al. 2001). Like other intracellular lectin families, the M-type lectin family is modest in size. Mammals have three M-type lectins, EDEM1, EDEM2 and EDEM3 (ER-associated degradation-enhancing α-mannosidase-like proteins), and these are generally conserved in metazoa, although EDEM1 is missing in *Drosophila* for example. Yeast has a single M-type lectin, Mnl1 that is related to EDEM1. Different M-type lectins are found in plants. M-type lectins are type II transmembrane proteins with very short cytoplasmic tails (although in some cell types the expressed protein may lack transmembrane domain). In common with the ligand/substrate binding domains of glycoside hydrolase family 47 proteins, such as human ER mannosidase, the M-type CRD is a barrel-like structure with both α-helices and β-sheets. Some of the M-type lectins have C-terminal extensions after the M-type CRD that are not related to known protein folds (Fig. 6.3). The ERAD system is upregulated under stress conditions, when a greater proportion of proteins remain unfolded or misfolded, and serves to clear the secretory pathway of 'waste' proteins and thus prevent clogging. EDEM1 was the first M-type lectin established as having a role in recognizing unfolded or misfolded proteins in ERAD (Hosokawa et al. 2001).

More recently, EDEMs 2 and 3 have also been shown to accelerate ERAD, but EDEM3 is unique in retaining mannosidase activity, suggesting it enhances ERAD by a different mechanism to EDEMs 1 and 2. Endogenous EDEM1 exists mainly as a soluble glycoprotein. High-resolution analysis showed that endogenous EDEM1 is sequestered in buds that form along cisternae of the rough ER at regions outside of the transitional ER. They give rise to approximately 150-nm vesicles scattered throughout the cytoplasm that are lacking a recognizable COPII coat. Some of the EDEM1 vesicles also contain Derlin-2 and the misfolded Hong Kong variant of α-1-antitrypsin, a substrate for EDEM1 and ERAD. Experiments demonstrate the existence of a vesicle budding transport pathway out of the rough ER that does not involve the canonical transitional ER exit sites and therefore represents a passageway to remove potentially harmful misfolded luminal glycoproteins from ER (Zuber et al. 2007). EDEM may function as an acceptor of terminally misfolded glycoproteins. To test this hypothesis, Hosokawa et al. (2006) constructed several genetically manipulated cell lines and suggested that EDEM may function as a molecular chaperone. To examine this possibility, researchers found that the accumulation of covalent NHK dimers was selectively prevented by the over-expression of EDEM, and therefore EDEM may maintain the retrotranslocation competence of NHK by inhibiting aggregation so that unstable misfolded proteins can be accommodated by the dislocon for ERAD.

The structure of ER mannosidase with a bound substrate analogue suggests that the ligand binding site in M-type CRDs is a deep cleft at one end of the barrel. This unusually deep binding site for a lectin allows selective interaction with high mannose glycans rather than just terminal residues (Fig. 6.6).

6.4.3 Functions of M-Type Lectins in ERAD

The folding of glycoproteins in ER lumen is assisted by chaperone proteins including the calnexin family lectins (Chap. 2). Proteins which have failed to fold correctly are re-glucosylated to signal that re-binding to calnexin is necessary. Trimming of mannose from a $Man_9GlcNAc_2$ glycan on an unfolded glycoprotein is the signal that the glycoprotein should be exported from ER to cytoplasm for degradation, and concordantly inhibits further participation of the glycoprotein in the calnexin cycle. Despite several dedicated chaperones and folding factors that ensure efficient maturation, protein folding remains error-prone and mutations in the polypeptide sequence may significantly reduce folding-efficiency. Folding-incompetent proteins carrying N-glycans are extracted from futile folding cycles in the calnexin chaperone system upon intervention of EDEM1, EDEM2 and EDEM3, three ER-stress-induced members of the glycosyl hydrolase 47 family (Olivari and Molinari 2007). The plant toxin ricin is transported retrogradely from the cell surface to the ER from where the enzymatically active part is retrotranslocated to the cytosol, presumably by the same mechanism as used by misfolded proteins. The EDEM is responsible for directing aberrant proteins for ERAD (Fig. 6.3). EDEM is involved in ricin retrotranslocation. from the ER to cytosol (Slominska-Wojewodzka et al. 2006).

All glycoproteins - folded or unfolded - bear a $Man_8GlcNAc_2$ glycan following ER mannosidase processing, so ERAD receptors are required to distinguish between folded and unfolded glycoproteins bearing this structure, in order to trigger degradation of the unfolded ones only. If the protein is permanently misfolded, the mannose residue in the middle branch of the oligosaccharide is removed by ERManI. This leads to recognition by EDEM, which probably targets glycoproteins for ERAD. The EDEM proteins are believed to fulfil this role, probably alongwith other proteins such as the P-type lectin OS-9. EDEMs in turn hand over bound glycoproteins to the translocon machinery for retrotranslocation into the cytoplasm, where the glycoproteins are recognized by proteins including F-box lectins and targeted for proteasomal degradation. That the protein generating the $Man_8GlcNAc_2$ degradation signal (ER mannosidase) and the proteins recognizing the signal (EDEMs) are structurally related points to concurrent evolution of both aspects of the ERAD system in a very early eukaryotic ancestor.

EDEM has been shown to interact with calnexin, but not with calreticulin, through its transmembrane region. Both binding of substrates to calnexin and their release from calnexin were required for ERAD to occur. Over-expression of EDEM accelerated ERAD by promoting the release of terminally misfolded proteins from calnexin. Thus, EDEM appeared to function in the ERAD pathway by accepting substrates from calnexin (Oda et al. 2003). Overexpression of ERManI greatly increases the formation of $Man_8GlcNAc$, induced the formation of Glc1Man8GlcNAc and increased trimming to $Man_{5-7}GlcNAc$. A model whereby the misfolded glycoprotein interacts with ERManI and with EDEM, before being recognized by downstream ERAD

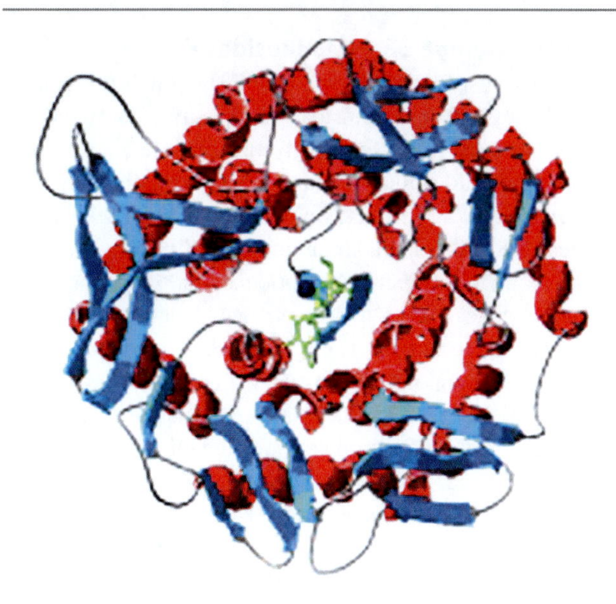

Fig. 6.6 Human ER class I α1,2-mannosidase structure showing catalytic domain of human ER mannosidase with bound substrate analogue. The thiodisaccharide substrate analogue is shown in *green* and the Ca^{2+} ion in *dark blue* (PDB: ID: 1X9D; Karaveg et al. 2005).

components, suggests that the carbohydrate recognition determinant triggering ERAD may not be restricted to $Man_8GlcNAc2$ isomer B as suggested earlier (Hosokawa et al. 2003). It was further demonstrated that overexpression of Golgi α1,2-mannosidase IA, IB, and IC also accelerates ERAD of terminally misfolded human α1-antitrypsin variant, and mannose trimming from the N-glycans on NHK in 293 cells (Hosokawa et al. 2007).

EDEM extracts misfolded glycoproteins, but not glycoproteins undergoing productive folding, from the calnexin cycle. EDEM overexpression resulted in faster release of folding-incompetent proteins from the calnexin cycle and earlier onset of degradation, whereas EDEM down-regulation prolonged folding attempts and delayed ERAD. Up-regulation of EDEM during ER stress may promote cell recovery by clearing the calnexin cycle and by accelerating ERAD of terminally misfolded polypeptides (Molinari et al. 2003). Non-maintenance of efficient glycoprotein folding in cells with defective ERAD caused by lack of adaptation of the intralumenal level of EDEM may increase in the ER cargo load. Eriksson et al. (2004) indicated that up-regulation of EDEM to strengthen the ERAD machinery (but not up-regulation of calnexin to reinforce the folding machinery) was instrumental in maintaining folding efficiency and secretory capacity and thus underscore the important role for degradation machinery in maintaining a functional folding environment in ER.

The efficiency of ERAD, which orchestrates the clearance of structurally aberrant proteins under basal conditions, is boosted by the unfolded protein response (UPR) as one of several means to relieve ER stress. However, the underlying mechanism that links the two systems in higher eukaryotes has remained elusive. Results of transient expression, RNAi-mediated knockdown and functional studies demonstrate that the transcriptional elevation of EDEM1 boosts the efficiency of glycoprotein ERAD through the formation of a complex that suppresses the proteolytic downregulation of ER mannosidase I (ERManI). A model is proposed in which ERManI, by functioning as a downstream effector target of EDEM1, represents a checkpoint activation paradigm by which the mammalian UPR coordinates the boosting of ERAD (Termine et al. 2009). EDEM1 is a novel chaperone of rod opsin and its expression promoted the degradation of P23H rod opsin and decreased its aggregation. Thus, EDEM1 can be used to promote correct folding, as well as enhanced degradation, of mutant proteins in ER to combat protein-misfolding disease (Kosmaoglou et al. 2009).

6.4.3.1 EDEM1

EDEM-1 is a crucial regulator of ERAD. Under normal growth conditions, the intralumenal level of EDEM1 must be low to prevent premature interruption of ongoing folding programs. In unstressed cells, EDEM1 is segregated from the bulk ER into LC3-I-coated vesicles and is rapidly degraded. The rapid turnover of EDEM1 is regulated by a novel mechanism that shows similarities but is clearly distinct from macroautophagy (Cali et al. 2009). Cells with defective EDEM1 turnover contain unphysiologically high levels of EDEM1, show enhanced ERAD activity and are characterized by impaired capacity to efficiently complete maturation of model glycopolypeptides. EDEM1 specifically binds nonnative proteins in a glycan-independent manner. Inhibition of mannosidase activity with kifunensine or disruption of the EDEM1 mannosidase-like domain by mutation had no effect on EDEM1 substrate binding but diminished its association with the ER membrane adaptor protein SEL1L. These results support a model whereby EDEM1 binds nonnative proteins and uses its mannosidase-like domain to target aberrant proteins to the ER membrane dislocation and ubiquitination complex containing SEL1L (Cormier et al. 2009).

Endogenous over-expression of EDEM1 not only resulted in inappropriate occurrence throughout the ER but also caused cytotoxic effects. Proteasome inhibitors had no effect on the clearance of endogenous EDEM1 in non-starved cells. However, EDEM1 could be detected in purified autophagosomes. Furthermore, influencing the lysosome-autophagy by vinblastine or pepstatin A/E64d and inhibiting autophagosome formation by 3-methyladenine or siRNA knockdown stabilized EDEM1. It was demonstrated that endogenous EDEM1 in cells not stressed by the expression of a transgenic misfolded protein reaches the cytosol and is degraded by basal autophagy (Le Fourn et al. 2009). Overexpression of EDEM1 produces $Glc_1Man_8GlcNAc_2$ isomer C on terminally misfolded null Hong Kong 1-antitrypsin (NHK) in vivo. It was suggested that EDEM1 activity trims mannose from the C branch of *N*-glycans in vivo (Hosokawa et al. 2010; Ushioda et al. 2008).

6.4.3.2 EDEM2

In humans there are a total of three EDEM homologs. One of the EDEM homologs from Homo sapiens was termed EDEM2 (C20orf31). Over-expression of EDEM2 accelerates the degradation of misfolded α1-antitrypsin, indicating that the protein is involved in ERAD (Mast et al. 2005). The transcriptional up-regulation of EDEM2 depends on the ER stress-activated transcription factor Xbp1 and selectively accelerates ERAD of terminally misfolded glycoproteins by facilitating their extraction from the calnexin cycle (Olivari et al. 2005).

6.4.3.3 EDEM3

Mouse EDEM3, a soluble homolog of EDEM consists of 931 amino acids and has all the signature motifs of Class I α-mannosidases in its N-terminal domain and a protease-associated motif in its C-terminal region. EDEM3 accelerates glycoprotein ERAD in transfected HEK293 cells, as shown by increased degradation of misfolded α1-antitrypsin variant (null (Hong Kong) and of TCRα. Overexpression of EDEM3 also greatly stimulates mannose trimming not only from misfolded α1-AT null (Hong Kong) but also from total glycoproteins, in contrast to EDEM, which has no apparent α1,2-mannosidase activity. Results suggest that EDEM3 has α1,2-mannosidase activity in vivo, suggesting that the mechanism whereby EDEM3 accelerates glycoprotein ERAD is different from that of EDEM (Hirao et al. 2006). Overexpression of EDEM3 enhances glycoprotein ERAD with a concomitant increase in mannose-trimming activity in vivo.

6.4.3.4 HTM1

In Saccharomyces cerevisiae, Jakob et al. (2001) identified a gene coding for a non-essential protein that is homologous to mannosidase I (HTM1), which is required for degradation of glycoproteins. Deletion of the HTM1 gene does not affect oligosaccharide trimming, yet, deletion of HTM1 did reduce the rate of degradation of the mutant glycoproteins but not of mutant non-glycoprotein. This mannosidase homolog is a lectin that recognizes $Man_8GlcNAc_2$ oligosaccharides that serve as signals in the degradation pathway. Clerc et al. (2009) defined the function of the Htm1 protein as an α1,2-specific exomannosidase that generates the $Man_7GlcNAc_2$ oligosaccharide with a terminal α1,6-linked mannosyl residue on degradation substrates. This oligosaccharide signal is decoded by ER-localized lectin Yos9p that in conjunction with Hrd3p triggers the ubiquitin-proteasome-dependent hydrolysis of these glycoproteins. The Htm1p exomannosidase activity requires processing of the N-glycan by glucosidase I, glucosidase II, and mannosidase I, resulting in a sequential order of specific N-glycan structures that reflect the folding status of the glycoprotein.

6.4.3.5 *S. pombe* ER α-Mannosidase

It has been postulated that creation of $Man_8GlcNAc_2$ isomer B (M_8B) by ER α-mannosidase I constitutes a signal for driving irreparably misfolded glycoproteins to proteasomal degradation. ER α-mannosidase I is present in extremely feeble form in *Schizosaccharomyces pombe*. The enzyme yielded M8B on degradation of $Man_9GlcNAc_2$ and was inhibited by kifunensin. Disruption of the α-mannosidase encoding gene almost totally prevented degradation of a misfolded glycoprotein. The enzyme, behaving as a lectin binding polymannose glycans of varied structures, would belong together with its enzymatically inactive homologue Htm1p/Mnl1p/EDEM, to a transport chain responsible for delivering irreparably misfolded glycoproteins to proteasomes. Kifunensin and 1-deoxymannojirimycin, being mannose homologues, would behave as inhibitors of ER mannosidase or/and Htm1p/Mnl1p/EDEM putative lectin properties (Movsichoff et al. 2005).

6.5 Derlin-1, -2 and -3

Derlin-1, a member of a family of proteins that bears homology to yeast Der1p, is a factor that is required for the human cytomegalovirus US11-mediated dislocation of class I MHC heavy chains from the ER membrane to the cytosol. Derlin-1 acts in concert with the AAA ATPase p97 to remove dislocation substrate proteins from the ER membrane. Mammalian genomes encode two additional, related proteins (Derlin-2 and Derlin-3). The similarity of the mammalian Derlin-2 and Derlin-3 proteins to yeast Der1p suggested that Derlins also play a role in ER protein degradation. Derlin-1 and Derlin-2 are ER-resident proteins that participates in the degradation of proteins from the ER. Furthermore, Derlin-2 forms a robust multiprotein complex with the p97 AAA ATPase as well as the mammalian orthologs of the yeast Hrd1p/Hrd3p ubiquitin-ligase complex (Lilley and Ploegh 2005).

Derlin-2 and -3 showed weak homology to Der1p, a transmembrane protein involved in yeast ERAD. Both Derlin-2 and -3 are up-regulated by unfolded protein response (UPR), and at least Derlin-2 is a target of the IRE1 branch of the response, which is known to up-regulate EDEM and EDEM2, receptor-like molecules for misfolded glycoprotein. Over-expression of Derlin-2 or -3 accelerated degradation of misfolded glycoprotein, whereas their knock-down blocked degradation. Derlin-2 and -3 are associated with EDEM and p97, a cytosolic ATPase responsible for extraction of ERAD substrates. Report indicates that Derlin-2 and -3 provide missing link between EDEM and p97 in the process of degrading misfolded glycoproteins (Lilley et al. 2006; Oda et al. 2006; Dixit et al. 2008).

References

F-Box Lectins

Bays NW, Gardner RG, Seelig LP et al (2001) Hrd1p/Der3p is a membrane-anchored ubiquitin ligase required for ER-associated degradation. Nat Cell Biol 3:24–29

Carrano AC, Eytan E, Hershko A et al (1999) SKP2 is required for ubiquitin-mediated degradation of the CDK inhibitor p27. Nat Cell Biol 1:193–199

Cenciarelli C, Chiaur DS, Guardavaccaro D et al (1999) Identification of a family of human F-box proteins. Curr Biol 9:1177–1179

Deshaies RJ (1999) SCF and Cullin/Ring H2-based ubiquitin ligases. Annu Rev Cell Dev Biol 15:435–467

Ellgaard L, Molinari M, Helenius A (1999) Setting the standards: quality control in the secretory pathway. Science 286:1882–1888

Eom CY, Lehman IR (2003) Replication-initiator protein (UL9) of the herpes simplex virus 1 binds NFB42 and is degraded via the ubiquitin-proteasome pathway. Proc Natl Acad Sci USA 100:9803–9807

Eom CY, Heo WD, Craske ML et al (2004) The neural F-box protein NFB42 mediates the nuclear export of the herpes simplex virus type 1 replication initiator protein (UL9 protein) after viral infection. Proc Natl Acad Sci USA 101:4036–4040

Fang S, Ferrone M, Yang C et al (2001) The tumour autocrine motility factor receptor, gp78, is a ubiquitin protein ligase implicated in degradation from the endoplasmic reticulum. Proc Natl Acad Sci USA 98:14422–14427

Feldman RM, Correll CC, Kaplan KB et al (1997) A complex of Cdc4p, Skp1p, and Cdc53p/cullin catalyzes ubiquitination of the phosphorylated CDK inhibitor Sic1p. Cell 91:221–230

Frescas D, Pagano M (2008) Deregulated proteolysis by the F-box proteins SKP2 and β-TrCP: tipping the scales of cancer. Nat Rev Cancer 8(441):438–449

Gagne JM, Brian P, Downes BP et al (2002) The F-box subunit of the SCF E3 complex is encoded by a diverse superfamily of genes in *Arabidopsis*. Proc Natl Acad Sci USA 99:11519–11524

Galan JM, Peter M (1999) Ubiquitin-dependent degradation of multiple F-box proteins by an autocatalytic mechanism. Proc Natl Acad Sci USA 96:9124–9129

Glenn KA, Nelson RF, Wen HM et al (2008) Diversity in tissue expression, substrate binding, and SCF complex formation for a lectin family of ubiquitin ligases. J Biol Chem 283:12717–12729

Helenius A, Aebi M (2004) Roles of N-linked glycans in the endoplasmic reticulum. Annu Rev Biochem 73:1019–1049

Henrissat B, Bairoch A (1996) Updating the sequence-based classification of glycosyl hydrolases. Biochem J 316:695–696

Henzl MT, O'Neal J, Killick R et al (2001) OCP1, an F-box protein, co-localizes with OCP2/SKP1 in the cochlear epithelial gap junction region. Hear Res 157:100–111

Henzl MT, Thalmann I, Larson JD et al (2004) The cochlear F-box protein OCP1 associates with OCP2 and connexin 26. Hear Res 191:101–109

Hershko A, Ciechanover A (1998) The ubiquitin system. Annu Rev Biochem 67:425–479

Ho MS, Tsai PI, Chien CT (2006) F-box proteins: the key to protein degradation. J Biomed Sci 13:181–191

Hosokawa N, Wada I, Hasegawa K et al (2001) A novel ER alpha-mannosidase-like protein accelerates ER-associated degradation. EMBO Rep 2:415–422

Ilyin GP, Rialland M, Pigeon C, Guguen-Guillouzo C (2000) cDNA cloning and expression analysis of new members of the mammalian F-box protein family. Genomics 67:40–47

Ilyin GP, Serandour AL, Pigeon C, Rialland M, Glaise D, Guguen-Guillouzo C (2002) A new subfamily of structurally related human F-box proteins. Gene 296:11–20

Imai Y, Soda M, Inoue H et al (2001) An unfolded putative transmembrane polypeptide, which can lead to endoplasmic reticulum stress, is a substrate of Parkin. Cell 105:891–902

Ivan M, Kondo K, Yang H et al (2001) HIFalpha targeted for VHL-mediated destruction by proline hydroxylation: implications for O_2 sensing. Science 292(5516):464–468

Jaakkola P, Mole DR, Tian YM et al (2001) Targeting of HIF-alpha to the von Hippel-Lindau ubiquitylation complex by O_2-regulated prolyl hydroxylation. Science 292(5516):468–472

Jakob CA, Bodmer D, Spirig U et al (2001) Htm1p, a mannosidase-like protein, is involved in glycoprotein degradation in yeast. EMBO Rep 2:423–430

Jin J, Cardozo T, Lovering RC et al (2004) Systematic analysis and nomenclature of mammalian F-box proteins. Genes Dev 18:2573–2580

Kato A, Rouach N, Nicoll RA, Bredt DS (2005) Activity-dependent NMDA receptor degradation mediated by retrotranslocation and ubiquitination. Proc Natl Acad Sci USA 102:5600–5605

Kipreos ET, Pagano M (2000) The F-box protein family. Genome Biol 1:3002.1–3002.7

Lal A, Pang P, Kalelkar S et al (1998) Substrate specificities of recombinant murine Golgi alpha1, 2-mannosidases IA and IB and comparison with endoplasmic reticulum and Golgi processing alpha1,2-mannosidases. Glycobiology 8:981–995

Lannoo N, Van Damme EJ (2010) Nucleocytoplasmic plant lectins. Biochim Biophys Acta 1800:190–201

Lannoo N, Peumans WJ, Van Damme EJ (2008) Do F-box proteins with a C-terminal domain homologous with the tobacco lectin play a role in protein degradation in plants? Biochem Soc Trans 36:843–847

Li FN, Johnston M (1997) Grr1 of Saccharomyces cerevisiae is connected to the ubiquitin proteolysis machinery through Skp1: coupling glucose sensing to gene expression and the cell cycle. EMBO J 16:5629–5638

Matyskiela ME, Rodrigo-Brenni MC, Morgan DO (2009) Mechanisms of ubiquitin transfer by the anaphase-promoting complex. J Biol Chem 8:92

Meacham GC, Patterson C, Zhang W et al (2001) The Hsc70 co-chaperone CHIP targets immature CFTR for proteasomal degradation. Nat Cell Biol 3:100–105

Mizushima T, Hirao T, Yoshida Y et al (2004) Structural basis ofsugar-recognizing ubiquitin ligase. Nat Struct Mol Biol 11:365–370

Mizushima T, Yoshida Y, Kumanomidou T et al (2007) Structural basis for the selection of glycosylated substrates by SCF(Fbs1) ubiquitin ligase. Proc Natl Acad Sci USA 104:5777–5781

Moremen KW (2000) α-mannosidases in asparagine-linked oligosaccharide processing and catabolism. In: Ernst B, Hart G, Sinay P (eds) Oligosaccharides in chemistry and biology: a comprehensive handbook, Part 1, vol 2. Wiley, Weinheim, pp 81–117

Movsichoff F, Castro OA, Parodi AJ (2005) Characterization of *Schizosaccharomyces pombe* ER alpha-mannosidase: a reevaluation of the role of the enzyme on ER-associated degradation. Mol Biol Cell 16:4714–4724

Murai-Takebe R, Noguchi T, Ogura T et al (2004) Ubiquitination-mediated regulation of biosynthesis of the adhesion receptor SHPS-1 in response to endoplasmic reticulum stress. J Biol Chem 279:11616–11625

Nakatsukasa K, Nishikawa S, Hosokawa N et al (2001) Mnl1p, an alpha -mannosidase-like protein in yeast Saccharomyces cerevisiae, is required for endoplasmic reticulum-associated degradation of glycoproteins. J Biol Chem 276:8635–8638

Nakayama KI, Nakayama K (2005) Regulation of the cell cycle by SCF-type ubiquitin ligases. Semin Cell Dev Biol 16:323

Nelson RF, Glenn KA, Miller VM et al (2006) A novel route for F-box protein-mediated ubiquitination links CHIP to glycoprotein quality control. J Biol Chem 281:20242–20251

Nelson RF, Glenn KA, Zhang Y et al (2007) Selective cochlear degeneration in mice lacking the F-box protein, Fbx2, a glycoprotein-specific ubiquitin ligase subunit. J Neurosci 27:5163–5171

Rouillon A, Barbey R, Patton EE et al (2000) Feedback-regulated degradation of the transcriptional activator Met4 is triggered by the SCF(Met30) complex. EMBO J 19:282–294

Schulman BA, Carrano AC, Jeffrey PD et al (2000) Insights into SCF ubiquitin ligases from the structure of the Skp1-Skp2 complex. Nature 408(6810):381–386

Skowyra D, Craig KL, Tyers M et al (1997) F-box proteins are receptors that recruit phosphorylated substrates to the SCF ubiquitin-ligase complex. Cell 91:209–219

Somers DE, Fujiwara S (2009) Thinking outside the F-box: novel ligands for novel receptors. Trends Plant Sci 14:206–213

Spiegelman VS, Slaga TJ, Pagano M et al (2000) Wnt/β-catenin signaling induces the expression and activity of βTrCP ubiquitin ligase receptor. Mol Cell 5:877–882

Swanson R, Locher M, Hochstrasser M (2001) A conserved ubiquitin ligase of the nuclear envelope/endoplasmic reticulum that functions in both ER-associated and Matα2 repressor degradation. Genes Dev 15:2660–2674

Tai T (2006) Identification of N-glycan-binding proteins for E3 ubiquitin ligases. Methods Enzymol 415:20–30

Tan A, Henzl MT (2009) Conformational stabilities of guinea pig OCP1 and OCP2. Biophys Chem 144:108–118

Thalmann R, Henzl MT, Thalmann I (1997) Specific proteins of the organ of Corti. Acta Otolaryngol 117:265–268

Tremblay LO, Herscovics A (2000) Characterization of a cDNA encoding a novel human golgi α1,2-mannosidase (IC) involved in N-glycan biosynthesis. J Biol Chem 275:31655–31660

Tsai B, Ye Y, Rapoport TA (2002) Retro-translocation of proteins from the endoplasmic reticulum into the cytosol. Nat Rev Mol Cell Biol 3:246–255

Vodermaier HC (2004) APC/C and SCF: controlling each other and the cell cycle. Curr Biol 14:787

Winston JT, Koepp DM, Zhu C et al (1999) A family of mammalian F-box proteins. Curr Biol 9:1180–1182

Wu Y, Swulius MT, Moremen KW, Sifers RN (2003) Elucidation of the molecular logic by which misfolded alpha 1-antitrypsin is preferentially selected for degradation. Proc Natl Acad Sci USA 100:8229–8234

Yam CH, Ng RW, Siu WY et al (1999) Regulation of cyclin A-Cdk2 by SCF component Skp1 and F-box protein Skp2. Mol Cell Biol 19:635–645

Yamaguchi Y, Hirao T, Sakata E et al (2007) Fbs1 protects the malfolded glycoproteins from the attack of peptide:N-glycanase. Biochem Biophys Res Commun 362:712–716

Yoshida Y (2003) A novel role for *N*-glycans in the ERAD system. J Biochem (Tokyo) 134:183–190

Yoshida Y (2007) F-box proteins that contain sugar-binding domains. Biosci Biotechnol Biochem 71:2623–2631

Yoshida Y, Tanaka K (2010) Lectin-like ERAD players in ER and cytosol. Biochim Biophys Acta 1800:172–180

Yoshida Y, Chiba T, Tokunaga F et al (2002) E3 ubiquitin ligase that recognizes sugar chains. Nature 418(6896):438–442

Yoshida Y, Tokunaga F, Chiba T et al (2003) Fbs2 is a new member of the E3 ubiquitin ligase family that recognizes sugar chains. J Biol Chem 278:43877–43884

Yoshida Y, Adachi E, Fukiya K et al (2005) Glycoprotein-specific ubiquitin ligases recognize N-glycans in unfolded substrates. EMBO Rep 6:239–244

Yoshida Y, Murakami A, Iwai K, Tanaka K (2007) A neural-specific F-box protein Fbs1 functions as a chaperone suppressing glycoprotein aggregation. J Biol Chem 282:7137–7144

Zhang H, Kobayashi R, Galaktionov K, Beach D (1995) p19-Skp1 and p45-Skp2 are essential elements of the cyclin A-CDK2 S phase kinase. Cell 82:915–925

M-Type Lectins

Calì T, Galli C, Olivari S et al (2008) Segregation and rapid turnover of EDEM1 by an autophagy-like mechanism modulates standard ERAD and folding activities. Biochem Biophys Res Commun 371:405–410

Clerc S, Hirsch C, Oggier DM et al (2009) Htm1 protein generates the N-glycan signal for glycoprotein degradation in the endoplasmic reticulum. J Cell Biol 184:159–172

Cormier JH, Tamura T, Sunryd JC, Hebert DN (2009) EDEM1 recognition and delivery of misfolded proteins to the SEL1L-containing ERAD complex. Mol Cell 34:627–633

Dixit G, Mikoryak C, Hayslett T et al (2008) Cholera toxin up-regulates endoplasmic reticulum proteins that correlate with sensitivity to the toxin. Exp Biol Med (Maywood) 233:163–175

Eriksson KK, Vago R, Calanca V et al (2004) EDEM contributes to maintenance of protein folding efficiency and secretory capacity. J Biol Chem 279:44600–44605

Gong Q, Keeney DR, Molinari M, Zhou Z (2005) Degradation of trafficking-defective long QT syndrome type II mutant channels by the ubiquitin-proteasome pathway. J Biol Chem 280:19419–19425

Hirao K, Natsuka Y, Tamura T et al (2006) EDEM3, a soluble EDEM homolog, enhances glycoprotein endoplasmic reticulum-associated degradation and mannose trimming. J Biol Chem 281:9650–9658

Hosokawa N, Tremblay LO, You Z et al (2003) Enhancement of endoplasmic reticulum (ER) degradation of misfolded Null Hong Kong α1-antitrypsin by human ER mannosidase I. J Biol Chem 278:26287–26294

Hosokawa N, Wada I, Natsuka Y, Nagata K (2006) EDEM accelerates ERAD by preventing aberrant dimer formation of misfolded alpha1-antitrypsin. Genes Cells 11:465–476

Hosokawa N, You Z, Tremblay LO et al (2007) Stimulation of ERAD of misfolded null Hong Kong α1-antitrypsin by Golgi α1,2-mannosidases. Biochem Biophys Res Commun 362:626–632

Hosokawa N, Tremblay LO, Sleno B et al (2010) EDEM1 accelerates the trimming of α1,2-linked mannose on the C branch of *N*-glycans. Glycobiology 20:567–575

Karaveg K, Siriwardena A, Tempel W et al (2005) Mechanism of class 1 (glycosylhydrolase family 47) {α}-mannosidases involved in N-glycan processing and endoplasmic reticulum quality control. J Biol Chem 280:16197–16207

Kosmaoglou M, Kanuga N, Aguilà M et al (2009) A dual role for EDEM1 in the processing of rod opsin. J Cell Sci 122:4465–4472

Le Fourn V, Gaplovska-Kysela K, Guhl B et al (2009) Basal autophagy is involved in the degradation of the ERAD component EDEM1. Cell Mol Life Sci 66:1434–1445

Lilley BN, Ploegh HL (2005) Multiprotein complexes that link dislocation, ubiquitination, and extraction of misfolded proteins from the endoplasmic reticulum membrane. Proc Natl Acad Sci USA 102:14296–14301

Lilley BN, Gilbert JM, Ploegh HL, Benjamin TL (2006) Murine polyomavirus requires the endoplasmic reticulum protein Derlin-2 to initiate infection. J Virol 80:8739–8744

Mast SW, Moremen KW (2006) Family 47 α-mannosidases in N-glycan processing. Methods Enzymol 415:31–46

Mast SW, Diekman K, Karaveg K et al (2005) Human EDEM2, a novel homolog of family 47 glycosidases, is involved in ER-associated degradation of glycoproteins. Glycobiology 15:421–436

Molinari M, Calanca V, Galli C et al (2003) Role of EDEM in the release of misfolded glycoproteins from the calnexin cycle. Science 299(5611):1397–1400

Oda Y, Hosokawa N, Wada I et al (2003) EDEM as an acceptor of terminally misfolded glycoproteins released from calnexin. Science 299(5611):1394–1397

Oda Y, Okada T, Yoshida H et al (2006) Derlin-2 and Derlin-3 are regulated by the mammalian unfolded protein response and are required for ER-associated degradation. J Cell Biol 172:383–393

Olivari S, Molinari M (2007) Glycoprotein folding and the role of EDEM1, EDEM2 and EDEM3 in degradation of folding-defective glycoproteins. FEBS Lett 581:3658–3664

Olivari S, Galli C, Alanen H et al (2005) A novel stress-induced EDEM variant regulating endoplasmic reticulum-associated glycoprotein degradation. J Biol Chem 280:2424–2428

Slominska-Wojewodzka M, Gregers TF, Wälchli S, Sandvig K (2006) EDEM is involved in retrotranslocation of ricin from the endoplasmic reticulum to the cytosol. Mol Biol Cell 17:1664–1675

Termine DJ, Moremen KW, Sifers RN (2009) The mammalian UPR boosts glycoprotein ERAD by suppressing the proteolytic downregulation of ER mannosidase I. J Cell Sci 122:976–984

Ushioda R, Hoseki J, Araki K et al (2008) ERdj5 is required as a disulfide reductase for degradation of misfolded proteins in the ER. Science 321(5888):569–572

Zuber C, Cormier JH, Guhl B et al (2007) EDEM1 reveals a quality control vesicular transport pathway out of the endoplasmic reticulum not involving the COPII exit sites. Proc Natl Acad Sci USA 104:4407–4412

Part III

L-Type Lectins

L-Type Lectins in ER-Golgi Intermediate Compartment

7

G.S. Gupta

7.1 L-Type Lectins

7.1.1 Lectins from Leguminous Plants

L-type lectins possess a luminal carbohydrate recognition domain (CRD) that binds to high-mannose-type oligosaccharides in a Ca^{2+}-dependent manner. The L-type CRD is named after the lectins found in abundance in the seeds of leguminous plants, such as concanavalin A from jack beans. The history of L-type lectins is as old as discovery of plant lectins from seeds of leguminous plants in nineteenth century. The structural motifs of L-type lectins are now known to be present in a variety of glycan-binding proteins from other eukaryotic organisms. The domain is present in plant, fungal, and animal proteins, but plant and animal L-type lectins have divergent sequences and different molecular properties. While plant lectins are secreted-soluble proteins and found at high level in specialised tissues, animal L-type lectins are (often membrane-bound) luminal proteins that are found at low levels in many different cell types. This observation suggests that animal L-type lectins have different functions. The crystal structures of some of the legume seed lectins show structural similarities among these lectins and to some other lectins, including the galectins and a variety of other lectins. Therefore, the term "L-lectins" has been designated as a classification for all lectins with this legume seed lectin-like structure. The L-type lectin-like domain has an overall globular shape composed of a β-sandwich of two major twisted antiparallel β-sheets. The β-sandwich comprises a major concave β-sheet and a minor convex β-sheet, in a variation of the jelly roll fold (Velloso et al. 2002, 2003; Satoh et al. 2006, 2007).

7.1.2 L-Type Lectins in Animals and Other Species

L-type lectins in animal cells are involved in protein sorting in luminal compartments of animal cells. In humans and other mammals there are four L-type lectins: ERGIC-53, ERGIC-53 like (ERGL), vesicular integral membrane protein-36 (VIP36), and VIP36 like (VIPL). ERGIC-53 is found only in mammals and VIP36 is restricted to vertebrates, but ERGIC-53 and VIPL are also found in invertebrates. A protein similar to ERGIC-53 is present in the slime mold *Dictyostelium dyscoideum*, a very simple eukaryote. L-type lectins have different intracellular distributions and dynamics in the ER-Golgi system of secretory pathway and interact with N-glycans of glycoproteins in a Ca^{2+}-dependent manner, suggesting a role in glycoprotein sorting and trafficking. The VIP36 is an intracellular animal lectin that acts as a putative cargo receptor, which recycles between the Golgi and the ER (Hauri et al. 2002).

Proteins more distantly related to ERGIC-53 and VIP36 are present in yeast and other fungi and in protozoa. Emp46p and Emp47p are L-type lectins from *S. cerevisiae* which cycle between the ER and the Golgi to facilitate the exit of N-linked glycoproteins from the ER. Unlike ERGIC-53, binding of high mannose glycans does not require a Ca^{2+} ion. Emp46p binds a K^+ ion, which is essential for glycoprotein transport, at a different location to that of Ca^{2+} ion in ERGIC-53, and Emp47p does not bind any metal ions. The differences in metal binding are evident in the primary structure of the proteins. Members of the galectin family of lectins may be considered as members of the L-type lectin family and are the subject of a separate chapter. The pentraxins are a superfamily of plasma proteins that are involved in innate immunity in invertebrates and vertebrates. They contain L-type lectin folds and require Ca^{2+} ions for ligand binding (Chap. 8). Other carbohydrate-binding proteins that may fit into this category are discussed in this chapter.

VP4 is a monomeric sialic acid–binding domain with an L-type lectin fold. This domain is required for infectivity of most animal rotaviruses. Sialidases or neuraminidases are a superfamily of N-acylneuraminate-releasing (sialic-acid-releasing) exoglycosidases found mainly in higher eukaryotes

G.S. Gupta et al., *Animal Lectins: Form, Function and Clinical Applications*,
DOI 10.1007/978-3-7091-1065-2_7, © Springer-Verlag Wien 2012

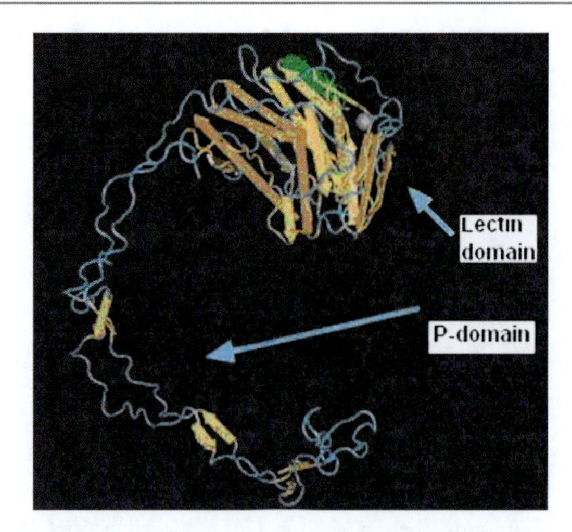

Fig. 7.1 Domain organization of calnaxin (Adapted with permission from Schrag et al. 2001 © Elsevier)

and in some, mostly pathogenic, viruses, bacteria and protozoans and contain L-type lectin domain. Several bacterial toxins such as exotoxin A (ETA) from *Pseudomonas aeruginosa* and Leech intramolecular trans-Sialidase and Vibrio cholerae Sialidase are known to possess CRD as observed in many lectins. Calnexin (Cnx) and calreticulin (Crt) are homologous chaperones that mediate quality control of proteins in the ER (see Chap. 2). Both Crt and Cnx are Ca^{2+}-binding proteins, and their carbohydrate-binding activity is sensitive to changes in Ca^{2+} concentration. Cnx is a type I membrane protein with its carboxy-terminal end in the cytoplasm. The lumenal portion of the protein is divided into three domains: a Ca^{2+}-binding domain (which is adjacent to the transmembrane domain), a proline-rich long hairpin loop called the P domain, and the amino-terminal L-type lectin domain. The Crt has a similar structure, but it is missing the cytoplasmic and transmembrane regions; it is retained in the ER through its KDEL-retrieval signal at the carboxyl terminus (Fig. 7.1)

7.2 ER-Golgi Intermediate Compartment

Protein traffic moving from ER to Golgi complex in mammalian cells passes through the tubulovesicular membrane clusters of the ER-Golgi intermediate compartment (ERGIC), the marker of which is the lectin ERGIC-53. Because the functional borders of the intermediate compartment (IC/ERGIC) are not well defined, the spatial map of the transport machineries operating between the ER and the Golgi apparatus remains incomplete. However, studies showed that the ERGIC consists of interconnected vacuolar and tubular parts with specific roles in pre-Golgi trafficking. The identification of ERGIC-53 has added to the complexity of the exocytic pathway of higher eukaryotic cells. A subcellular fractionation procedure for the isolation of the ERGIC from Vero cells provided a means to study more precisely the compartmentalization of the various enzymic functions along the early secretory pathway. The results suggested that in the secretory pathway of Vero cells O-glycan initiation and sphingomyelin as well as glucosylceramide synthesis mainly occur beyond the ERGIC in the Golgi apparatus (Schweizer et al. 1994). The dynamic nature and functional role of the ERGIC have been debated for quite some time. In the most popular current view, the ERGIC clusters are mobile transport complexes that deliver secretory cargo from ER-exit sites to the Golgi. Recent live-cell imaging data revealing the formation of anterograde carriers from stationary ERGIC-53-positive membranes, however, suggest a stable compartment model in which ER-derived cargo is first shuttled from ER-exit sites to stationary ERGIC clusters in a COPII-dependent step and subsequently to the Golgi in a second vesicular transport step. This model can better accommodate previous morphological and functional data on ER-to-Golgi traffic. Such a stationary ERGIC would be a major site of anterograde and retrograde sorting that is controlled by coat proteins, Rab and Arf GTPases, as well as tethering complexes, SNAREs and cytoskeletal networks. The ERGIC also contributes to the concentration, folding, and quality control of newly synthesized proteins (Appenzeller-Herzog and Hauri 2006). Marie et al. (2009)provided novel insight into the compartmental organization of the secretory pathway and Golgi biogenesis, in addition to a direct functional connection between the intermediate compartment and the endosomal system, which evidently contributes to unconventional transport of the cystic fibrosis transmembrane conductance regulator to the cell surface. The ERGIC defined by ERGIC-53 also participates in the maturation of (or is target for) several viruses such as corona virus, cytomegalovirus, flavivirus, poliovirus, Uukuniemi virus, and vaccinia virus. Further analysis of the function of ERGIC-53, and the use of ERGIC-53 as a marker protein, should provide novel results about the mechanisms controlling traffic in the secretory pathway.

7.3 Lectins of Secretory Pathway

The most prominent cycling lectin is mannose-binding type I membrane protein ERGIC-53 (ERGIC protein of 53 kDa), a marker for ERGIC, which functions as a cargo receptor to facilitate export of an increasing number of glycoproteins with different characteristics from the ER. The ERGIC-53 is a homo-hexameric transmembrane lectin localized to the ERGIC that exhibits mannose-selective properties in vitro. Two ERGIC-53-related proteins, VIP36 (vesicular integral

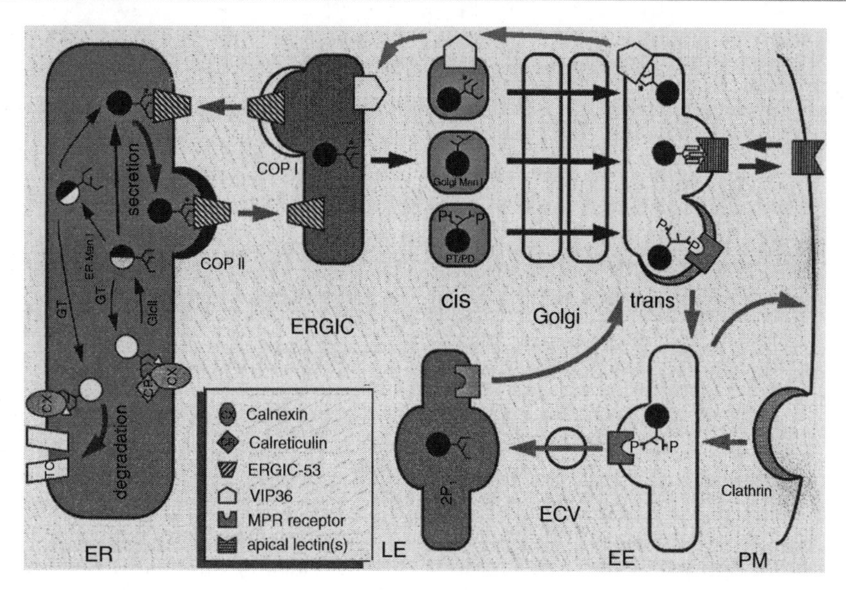

Fig. 7.2 Lectin-mediated glycoprotein transport in the secretory pathway. The secretory pathway is composed of membrane compartments specialized in protein folding, modification, transport, and sorting. Several lines of evidence indicate that glycan moieties are essential for folding, sorting and targeting of glycoproteins through the secretory pathway to various cellular compartments. Numerous transient protein-protein interactions guide the transport-competent proteins through the secretory pathway (Lee et al. 2004). Crystallographic and NMR studies of proteins located in ER, Golgi complex and ERGIC have illuminated their roles in glycoprotein folding and secretion. Calnexin and calreticulin, both ER-resident proteins with lectin domains are crucial for their function as chaperones (Chap. 2). After synthesis and removal of the two outermost glucose residues of their *N*-glycans many glycoproteins bind to calnexin and/or calreticulin which recognize monoglucosylated *N*-glycans. Subsequently the glycoproteins are trimmed by glucosidase II (*GlcII*). Thel incompletely folded protein (marked in *white*) is reglucosylated by ER enzyme UDP-glucose:glycoprotein glucosyltransferase (*GT*) which redirects them to another cycle of quality control. After prolonged time in the ER, ER mannosidase I (*ER Man I*) removes one mannose residue of the middle branch of the *N*-glycan. Incompletely folded, and thus reglucosylated, Man_8 glycoproteins are targeted for calnexin-dependent retrotranslocation to the cytosol and subsequent degradation by the proteasome. By contrast, correctly folded proteins (marked in *black*) are no longer recognized by GT after deglucosylation by GlcII and are transport-competent. They may or may not undergo some additional trimming by mannosidase I and II before leaving the ER. Some of these Man_{7-9} glycan-bearing proteins (*) now bind to the lectin ERGIC-53 which recruits them to COPII buds and thereby facilitates transport to the ERGIC. Dissociation of ERGIC-53 and its glycoprotein ligand occurs in the ERGIC and free ERGIC-53 recycles to the ER via COPI vesicles. In the *cis*-Golgi, glycoproteins are either trimmed to Man_5 prior to reglucosylation by Golgi glycosyltransferases, or tagged with the lysosomal signal Man-6-P by sequential action of phosphotransferase (*PT*) and phosphodiesterase (*PD*). Some glycoproteins escape *cis*-Golgi trimming but may be recognized by VIP36 in the *trans*-Golgi and recycled to the *cis*-Golgi for another trimming attempt. Proteins carrying Man-6-P residues are recognized by MPRs in the *trans*-Golgi and sorted to endosomes via clathrin-coated vesicles. Secreted Man-6-P-bearing glycoproteins can also be internalized from the plasma membrane by the large MPR. *N*-Glycans also serve as signals for Golgi exit and apical targeting in epithelial cells. These processes may also be mediated by lectins that are, however, unknown. *ECV* endosomal carrier vesicles, *EE* early endosome, *LE* late endosome, *TC* translocation channel (Reprinted with permission from Hauri et al. 2000a © Elsevier)

membrane protein 36) and a ERGIC-53-like protein (ERGL) are also found in early secretory pathway. The homologous lectin VIP36 may operate in quality control of glycosylation in the Golgi. In addition to well-understood role of mannose 6-phosphate receptors in lysosomal protein sorting, the VIP36 functions as a sorting receptor by recognizing high-mannose type glycans containing α1—>2Man residues for transport from Golgi to cell surface in polarized epithelial cells (Hauri et al. 2000a, b). The ERGL may act as a regulator of ERGIC-53. Exit from the Golgi of lysosomal hydrolases to endosomes requires mannose 6-phosphate receptors and exit to the apical plasma membrane may also involve traffic lectins. Analysis of the cycling of ERGIC-53 uncovered a complex interplay of trafficking signals and revealed novel cytoplasmic ER-export motifs that interact with COP-II coat proteins. These motifs are common to type I and polytopic membrane proteins including presenilin 1 and presenilin 7. The results support the notion that protein export from the ER is selective (Yamashita et al. 1999; Hauri et al. 2000b, 2002; Yerushalmi et al. 2001) (Fig. 7.2).

7.4 ER-Golgi Intermediate Compartment Marker-53 (ERGIC-53) or LMAN1

7.4.1 ERGIC-53 Is Mannose-Selective Human Homologue of Leguminous Lectins

Secretory proteins are cotranslationally translocated into lumen of ER, where they interact with ER-resident chaperones such as calnexin, and/or calreticulin. Only secretory proteins that fold correctly are transported through the Golgi apparatus to their final destinations. Several proteins are known to be transported by specific receptors. Such receptors may include membrane proteins ERGIC-53, VIP36, the p24 family, and Erv29p, which cycle between ER and the Golgi apparatus (Nichols et al. 1998; Appenzeller et al. 1999; Muniz et al. 2000; Belden and Barlowe 2001). Protein ERGIC-53, in humans, is encoded by the *LMAN1* gene. The protein encoded by this gene is a type I integral membrane protein localized in the intermediate region between the endoplasmic reticulum and the Golgi, presumably recycling between the two compartments. Also named LMAN1, the protein is a mannose-specific lectin and is a member of L-type lectin family.

ERGIC-53 bears homology to leguminous lectins and binds to mannose (Itin et al. 1996). It is, therefore, proposed to recognize high mannose-type oligosaccharides attached to proteins and to transport these glycoproteins from the ER to Golgi apparatus. The ERGIC-53, a member of a putative new class of animal lectins, is associated with the secretory pathway. It is type I transmembrane lectin, which facilitates the efficient export of a subset of secretory glycoproteins from ER. Indeed, the lack of ERGIC-53 impairs the secretion of procathepsin C and blood coagulation factors V (FV) and VIII (FVIII) glycoproteins (Nichols et al. 1998; Vollenweider et al. 1998). Chemical cross-linking studies revealed that ERGIC-53 interacts with procathepsin Z in a mannose- and calcium-dependent manner (Appenzeller et al. 1999; Appenzeller-Herzog et al. 2005). However, ERGIC-53 and its mutant, which is unable to bind to mannose, both coimmunoprecipitate with FVIII, and treatment with tunicamycin does not reduce the interaction between ERGIC-53 and FVIII (Cunningham et al. 2003), which indicates that protein–protein interactions also contribute to this interaction. It is therefore possible that ERGIC-53 also acts as a molecular chaperone in addition to transporting glycoproteins.

Although ERGIC-53 has selectivity for mannose yet it has a low affinity for glucose and GlcNAc, but not for galactose. Since leguminous family of lectin proteins possesses a highly conserved invariant asparagine essential for carbohydrate binding, the corresponding mutation in ERGIC-53 as well as a mutation affecting a second site in the putative CRD abolished mannose-column binding and co-staining with mannosylated neoglycoprotein. Based on its monosaccharide specificity, domain organization, and recycling properties, it was proposed that ERGIC-53 functions as a sorting receptor for glyco-proteins in the early secretory pathway (Itin et al. 1996). ERGIC-53 is present as reduction-sensitive homo-oligomers, i.e. as a balanced mixture of disulfide-linked hexamers and dimers, with the two cysteine residues located close to the transmembrane domain playing a crucial role in oligomerization. It is present exclusively as a hexameric complex in cells. Beyond its interest as a transport receptor, ERGIC-53 is an attractive probe for studying numerous aspects of protein trafficking in the secretory pathway, including traffic routes, mechanisms of anterograde and retrograde traffic, retention of proteins in the ER, and the function of the ERGIC.

7.4.2 Cells of Monocytic Lineage Express MR60: A Homologue of ERGIC-53

Most mammalian macrophages express mannose receptor (membrane lectin of 175 kDa) allowing endocytosis of their ligands, but cells of monocytic lineage (HL60, U937, monocyte) lack this receptor. However, after permeabilization, promyelocytic, promonocytic cells and monocytes bind D-mannose-terminated neoglycoproteins. The intracellular mannose binding protein from the human promyelocytic cell line HL60 is a 60 kDa protein (MR60). Under similar conditions, mouse macrophages express a 175 kDa mannose receptor but not the 60 kDa receptor (Pimpaneau et al. 1991). Under non-reducing conditions, MR60 migrates as a 120 kDa protein. MR60 does not contain any N-glycan moiety that could be cleaved by N-glycanase. MR60 induces a sugar selective aggregation of beads coated with α-D-mannosyl residues while beads bearing α-D-glucosyl residues are not.

Independently, a promyelocytic protein (MR60) was purified by mannose-column chromatography, and a cDNA was isolated that matched MR60 peptide sequences. This cDNA was identical to that of ERGIC-53 and homologies with the animal lectin family of the galectins were noticed. Not all peptide sequences of MR60, however, were found in ERGIC-53, raising the possibility that another protein associated with ERGIC-53 might possess lectin activity. This lectin is a type I transmembrane protein which includes a luminal N-terminal domain, a transmembrane domain, and a short C-terminal cytosolic domain (Fig. 7.3). The sequence of MR60 is identical (except for one amino acid) to that of ERGIC-53 (Arar et al. 1995; Schindler et al. 1993). The sequence of MR60/ERGIC-53 from human cells and that of the rat homologue p58 revealed 89% identity at the amino acid level (Lahtinen et al. 1996). Homologous proteins of MR60/ERGIC-53 have been characterized in *Caenorrhabditis elegans* and *Xenopus*. MR60 is not present at the cell surface and is structurally and immunochemically distinct from 175 kDa mannose receptor of mature macrophages. MR60/ERGIC-53 shares significant homologies with VIP-36 and with leguminous plant lectins (Fiedler and Simons 1994).

The recombinant protein binds mannosides and is oligomeric, up to the hexameric form. Two truncated proteins

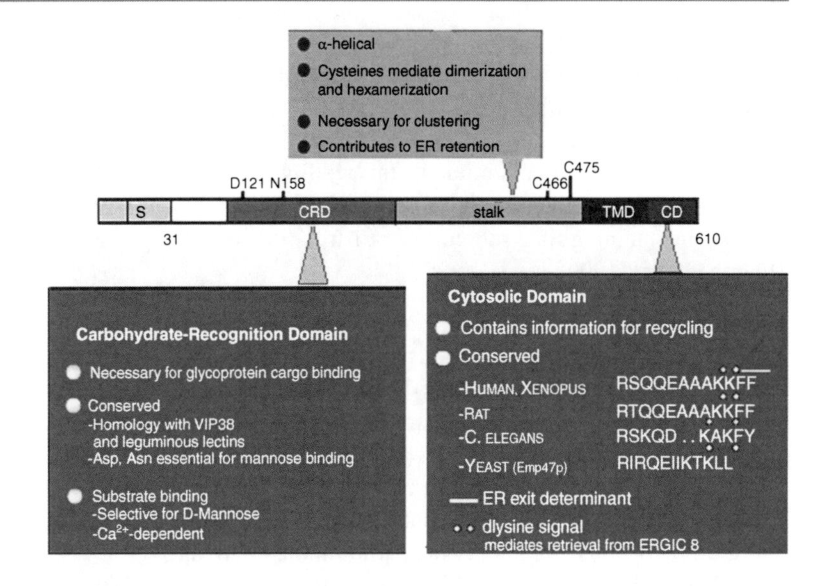

Fig. 7.3 Schematic diagram of ERGIC-53 showing functional domains. *CD* cytosolic domain, *S* signal sequence, *TMD* transmembrane domain. Some functionally important amino acid sequences (Adapted from Hauri et al. 2000b)

showed that the luminal moiety of MR60/ERGIC-53 contains a device, which allows both its oligomeric pattern and its sugar binding capability (Carriere et al. 1999a, b). ERGIC-53 is present as reduction-sensitive homo-oligomers, i.e. as a balanced mixture of disulfide-linked hexamers and dimers, with two cysteine residues located close to transmembrane domain playing a crucial role in oligomerization. It is present exclusively as a hexameric complex in cells. However, the hexamers exist in two forms, one as a disulfide-linked, Triton X-100, perfluoro-octanic acid, and SDS-resistant complex, and the other as a non-covalent, Triton X-100, perfluoro-octanoic acid-resistant, but SDS-sensitive, complex made up of three disulfide-linked dimers which are likely to interact through the coiled-coil domains present in the luminal part of the protein (Neve et al. 2005).

The cDNA sequence of the monkey homologue of ERGIC-53 revealed a sequence of 2,422 nt with 96.2% identity to the human ERGIC-53 cDNA and 87% and 67% identity to the rat and amphibian cDNA, respectively. The translated CV1 ERGIC-53 protein is 96.47% identical to the human ERGIC-53, 87% identical to rat p58 and 66.98% to *Xenopus laevis* protein. ERGIC-53 is expressed as a major transcript of about 5.5 kb in either monkey CV1 or in human CaCo7. A shorter transcript of 7.3 kb was detected in both cell lines and in mRNAs derived from human pancreas and placenta (Sarnataro et al. 1999).

7.4.3 Rat Homologue of ERGIC53/MR60

7.4.3.1 The 58-kDa Protein in Rat

The 58-kDa (p58), the type I homodimeric and hexameric microsomal membrane protein, has been characterized and localized to tubulo-vesicular elements at the ER-Golgi interface and the cis-Golgi cisternae in pancreatic acinar cells in rat (Lahtinen et al. 1992). The rat cDNA encodes a 517-amino acid protein having a putative signal sequence, a transmembrane domain close to the C terminus and a short cytoplasmic tail. The C-terminal tail contains a double-lysine motif (KKFF), known to mediate retrieval of proteins from the Golgi back to the endoplasmic reticulum. The rat p58 sequence is 89% identical with those of ERGIC-53 and MR60 and a strong homology with the frog sequence, indicating a high evolutionary conservation. Over-expression of c-Myc-tagged p58 resulted in accumulation of the protein both in the ER and in an apparently enlarged Golgi complex, as well as its leakage to the plasma membrane. The C-terminal tail of p58 located in the ER and transport intermediates is hidden, but becomes exposed when the protein reaches the Golgi complex (Lahtinen et al. 1996).

Shortly after synthesis, p58 forms dimers and hexamers, after which they are in equilibrium. The mature p58 contains four cysteine residues in the lumenal domain which are capable of forming disulphide bonds. The membrane-proximal half of the lumenal domain consists of four predicted alpha-helical domains, one heavily charged and three amphipathic in nature, all candidates for electrostatic or coiled-coil interactions. Using single-stranded mutagenesis, the cysteine residues were individually changed to alanine and the contribution of each of the alpha-helical domains was probed by internal deletions. The N-terminal cysteine to alanine mutants C198A and C238A and the double mutant, C198/238A oligomerized like the wild-type protein. The two membrane-proximal cysteines were found to be necessary for the oligomerization of p58. Mutants lacking one of the membrane proximal cysteines, either C473A or C482A, were unable to form hexamers, while dimers were formed normally. The C473/482A double mutant formed only monomers. Deletion of any of the individual alpha-helical domains had no effect on oligomerization. The dimeric and hexameric forms bound equally well to D-mannose. The dimeric and monomeric mutants displayed a cellular distribution similar to the wild-type protein, indicating that the oligomerization status played a minimal role in maintaining the sub-cellular distribution of p58 (Lahtinen et al. 1999).

7.4.3.2 Crystal Structure

The structures of many of L-type lectins have been thoroughly characterized, and many are employed in a wide range of biomedical and analytical procedures. All soluble L-type lectins found to date are multimeric proteins, although all do not have the same quaternary structure. Thus, these lectins are multivalent with more than one glycan-binding site per lectin molecule. The same multivalent principle applies to the membrane bound L-type lectins because the presence of two or more molecules on a membrane surface essentially presents a multivalent situation. The L-type CRD is a β-sandwich structure with a concave sheet of seven β-strands and a convex sheet of five. The ligand binding site is in a negatively-charged cleft of conserved residues. Most L-type CRDs require metal ions for ligand binding: in concanavalin A, a transition metal is bound at one site and Ca^{2+} at the second site. Despite divergent sequences and function, it seems likely that the L-type CRDs have retained similar mechanisms of sugar binding. Certain key residues in four loop regions that contribute to the binding sites in the plant proteins are conserved in all of the animal and plant L-type CRDs. The L-type lectins are distinguished from other lectins primarily on the basis of tertiary structure. In general, either the entire lectin monomer or the CRDs of the more complex lectins are composed of antiparallel β-sheets connected by short loops and β-bends, and they are usually devoid of any α-helical structure. These sheets form a dome-like structure related to the 'jelly-roll fold,' and it is often called a 'lectin fold'. The carbohydrate-binding site is generally localized toward the apex of this dome (Srinivas et al. 2001).

Velloso et al. (2002) determined the crystal structure of CRD of p58, the rat homologue of human ERGIC-53, to 1.46 Å resolution. The CRD of rat p58 was over-expressed in insect cells and *E. coli,* purified and crystallized using Li_2SO_4 as a precipitant. The crystals belong to space group I222, with unit-cell parameters a = 49.6, b = 86.1, c = 128.1 Å, and contain one molecule per asymmetric unit. The fold and ligand binding site are most similar to those of leguminous lectins. The structure also resembles that of the CRD of the ER folding chaperone calnexin and the neurexins, a family of non-lectin proteins expressed on neurons. The CRD comprises one concave and one convex β-sheet packed into a β-sandwich. The ligand binding site resides in a negatively charged cleft formed by conserved residues. A large surface patch of conserved residues with a putative role in protein-protein interactions and oligomerization lies on the opposite side of the ligand binding site (Fig. 7.4a). Knowledge of the structure of p58/ERGIC-53 provides a starting model for understanding receptor-mediated glycoprotein sorting between the ER and the Golgi (Velloso et al. 2002).

7.4.4 Structure-Function Relations

7.4.4.1 The Recycling of ERGIC-53 in the Early Secretory Pathway

COPII proteins are necessary to generate secretory vesicles at ER. Investigation on the targeting of ERGIC-53 by site-directed mutagenesis revealed that its lumenal and transmembrane domains together confer ER retention (Schokman and Orci 1996). In addition the cytoplasmic domain is required for exit from the ER indicating that ERGIC-53 carries an ER-exit determinant. Two phenylalanine residues at the C terminus are essential for ER-exit. Thus, ERGIC-53 contains determinants for ER retention as well as anterograde transport which, in conjunction with a dilysine ER retrieval signal, control the continuous recycling of ERGIC-53 in the early secretory pathway. In vitro binding studies revealed a specific phenylalanine-dependent interaction between an ERGIC-53 cytosolic tail peptide and the COPII coat component Sec23p. Results suggest that the ER-exit of ERGIC-53 is mediated by direct interaction of its cytosolic tail with the Sec23pSec24p complex of COPII and that protein sorting at the level of the ER occurs by a mechanism similar to receptor-mediated endocytosis or Golgi to ER retrograde transport (Kappeler et al. 1997).

Nufer et al. (2003a) suggested that the ER export determinant of ERGIC-53, which cycles in the early secretory pathway, requires a phenylalanine motif at the C-terminus, known to mediate COPII interaction, which is assisted by a glutamine in the cytoplasmic domain. Disulfide bond-stabilized oligomerization is also required. Efficient hexamerization depends on the presence of a polar and two aromatic residues in the transmembrane domain (TMD). ER export is also influenced by TMD length, 21 amino acids being most efficient. Results suggest an ER-export mechanism in which transmembrane and luminal determinants mediate oligomerization required for efficient recruitment of ERGIC-53 into budding vesicles via the C-terminal COPII-binding phenylalanine motif (Nufer et al. 2003a).

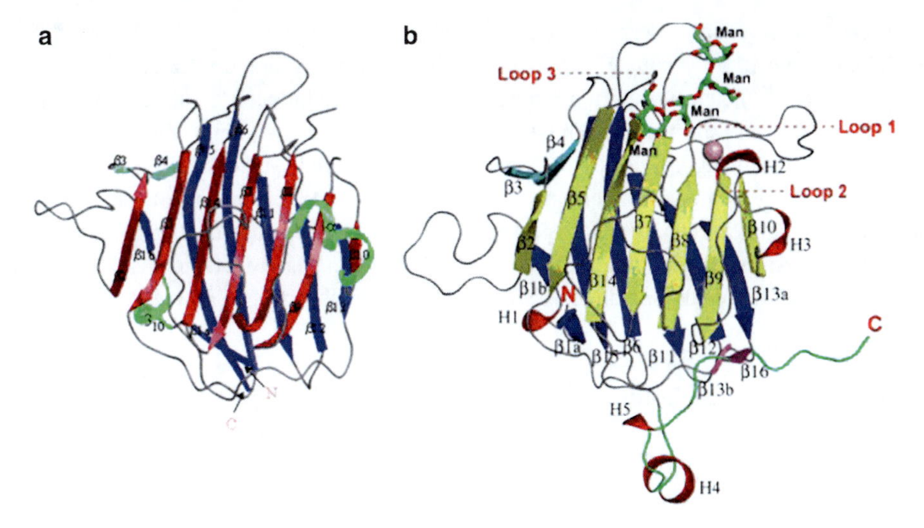

Fig. 7.4 (a) **Overall structure of Rat homologue (p58) of ERGIC53/MR60.** Ribbon diagram of p58 monomer shown perpendicular to the β-sheets. Secondary structure elements are labeled. Positions of the N and C termini are indicated. β-Strands belonging to the concave and convex β-sheets are shown in *red* and *dark blue*, respectively. Strands that do not take part in the β-sheets and are variable when compared with leguminous lectin structures are shown in *light blue*. Loops and helices are shown in *gray* and *green*, respectively (PDB ID: 1GV9 (Adapted with permission from Velloso et al. 2003 © Elsevier). (**b**) **Overall structure of exoplasmic/luminaldomain of VIP36.** Ribbon model of VIP36 (Man2-bound form) (PDB ID: 2DUR) is shown. The secondary structures are highlighted (β-strands belonging to concave β-sheets, *yellow*; β-strands belonging to convex β-sheets, *blue*; β-strands belonging to β-hairpin, *cyan*; β-strands belonging to short β-sheet formed between the stalk domain and one of the loops of the CRD, *magenta*; helices, *red*), and the loops of the CRD and stalk domain are colored *gray* and *green*, respectively. The bound Ca^{2+} is shown as a *pink sphere*. The bound oligomannoses are superimposed from the VIP36 complex structures with Man-α-1,2-Man and Man-α-1,2-Man-α-1,3-Man and are shown as a *green stick* model. The reducing-end mannose residue in the Man2-bound form is omitted because its position is almost the same as that of the $Man_3GlcNAc$-bound form. Positions of Loops 1, 2, and 3, which are bound to the oligomannose, are indicated (Adapted by permission from Satoh et al. 2007 © The American Society for Biochemistry and Molecular Biology)

7.4.4.2 Cytoplasmic ER-Retrieval Signal in ERGIC-53

Although ERGIC-53 is not a resident protein of the rough ER its cDNA sequence carries a double lysine ER retention motif at cytoplasmically exposed COOH terminus. Cell surface ERGIC-53 is efficiently endocytosed by a mechanism that is disturbed when the two critical lysines of the ER retention motif are replaced by serine residues. Results suggested a similarity between pre-Golgi retention by the double lysine motif and lysine based endocytosis (Kappeler et al. 1994). Although ERGIC-53 contains a cytoplasmic ER-retrieval signal, KKFF, over-expressed ERGIC-53 is transported to the cell surface and rapidly endocytosed. The ERGIC-53 endocytosis signal KKFF and like the ER-retrieval signal require a C-terminal position. In fact, the minimal consensus sequence determined by substitutional mutagenesis was also related to the ER-retrieval consensus (K-K-X-X). Evidence shows that internalization of VIP36, a protein that cycles between plasma membrane and Golgi, is mediated by a signal at its C-terminus that matches the internalization consensus sequence. The relatedness of the two signals suggests that coatomer-mediated retrieval of proteins may be mechanistically more related to clathrin-dependent sorting than previously anticipated (Itin et al. 1995a). Further dissection of the cytoplasmic domain revealed a COOH-terminal di-lysine ER-retrieval signal, KKFF, and an RSQQE targeting determinant adjacent to the transmembrane domain. Surprisingly, the two COOH-terminal phenylalanine residues influence the targeting. They reduce the ER-retrieval capacity of the di-lysine signal and modulate the RSQQE determinant (Itin et al. 1995b).

KKAA Retrieval Signal in Yeast: Studies on ERGIC-53 KKAA as a signal revealed a mechanism for static retention of mammalian proteins in the ER. This mechanism is conserved in yeast. Making use of a genetic assay, this signal was shown to induce COPI-dependent ER retrieval. ER retention of KKAA-tagged proteins was impaired in yeast mutants affected in COPI subunits. Furthermore, post-ER carbohydrate modifications detected on reporter proteins indicated that KKAA-tagged proteins recycle continuously within early compartments of the secretory pathway. Therefore in yeast, the KKAA signal might only function as a classical dilysine ER retrieval signal (Dogic et al. 2001).

A Single C-Terminal Valine Is Export Signal: The ERGIC-53 carries a C-terminal diphenylalanine motif that is required for efficient ER export. Replacement of C-terminal diphenylalanine motif by a single C-terminal valine accelerates transport of inefficiently exported reporter constructs and hence operates as an export signal. The valine signal is position dependent. Results suggest that cytoplasmic C-terminal amino-acid motifs, either alone or in conjunction with other transport determinants, accelerate ER export of numerous type I and probably polytopic membrane proteins by mediating interaction with COPII coat components (Nufer et al. 2002).

7.4.4.3 Two Distinct Pathways for Golgi-To-ER Transport

The cytosolic COP-I interacts with cytoplasmic 'retrieval' signals present in membrane proteins that cycle between ER and the Golgi complex, and is required for both anterograde and retrograde transport in the secretory pathway. The role of COP-I in Golgi-to-ER transport of distinct marker proteins has been described. For example, anti-COP-I antibodies inhibit retrieval of ERGIC-53 and of the KDEL receptor from the Golgi to the ER. Transport to the ER of protein toxins, which contain a sequence that is recognized by the KDEL receptor, is also inhibited. Results indicated the existence of at least two distinct pathways for Golgi-to-ER transport, one COP-I dependent and the other COP-I independent. The COP-I-independent pathway is specifically regulated by Rab6 and is used by Golgi glycosylation enzymes and Shiga toxin/Shiga-like toxin-1 (Girod et al. 1999).

7.4.4.4 Site of Transport Arrest of Mitotic Cells

Using ERGIC-53/p58 and plasma membrane protein CD8, Farmaki et al. (1999) established the site of transport arrest between ER and Golgi stack of mitotic animal cells. Recycled ERGIC 53/p58 and newly synthesised CD8 accumulate in ER cisternae but not in COPII-coated export structures or more distal sites. During mitosis the tubulo-vesicular ER-related export sites were depleted of the COPII component Sec13p, which indicated that COPII budding structures are the target for mitotic inhibition. Findings established that the site of ER-Golgi transport arrest of mitotic cells is COPII budding structures (Farmaki et al. 1999).

7.4.5 Functions of ERGIC-53

7.4.5.1 ERGIC-53 as a Cargo Receptor

Soluble secretory proteins are transported from the ER to ERGIC in vesicles coated with COP-II coat proteins. The sorting of secretory cargo into these vesicles is thought to involve transmembrane cargo-receptor proteins. According to Appenzeller et al. (1999) a cathepsin-Z-related glycoprotein binds to ERGIC-53. Binding that occurs in the ER, is carbohydrate- and calcium- dependent and is affected by untrimmed glucose residues. Binding does not, however, require oligomerization of ERGIC-53, although oligomerization is required for exit of ERGIC-53 from the ER. Dissociation of ERGIC-53 occurs in the ERGIC and is delayed if ERGIC-53 is mislocalized to the ER. These results indicate that ERGIC-53 may function as a receptor facilitating ER-to-ERGIC transport of soluble glycoprotein cargo (Appenzeller et al. 1999).

7.4.5.2 ERGIC-53 in Traffic and in the Secretory Pathway

Functional deficiency in ERGIC-53 leads to a selective defect in secretion of glycoproteins in cultured cells and to hemophilia in humans. Beyond its interest as a transport receptor, ERGIC-53 is an attractive probe for studying numerous aspects of protein trafficking in the secretory pathway (Hauri et al. 2000b). Studies suggest that ERGIC is a dynamic membrane system composed of a constant average number of clusters and that the major recycling pathway of ERGIC-53 bypasses the Golgi apparatus (Klumperman et al. 1998). To investigate if ERGIC-53 is involved in glycoprotein secretion, a mutant form of this protein was generated that is incapable of leaving the ER. If expressed in HeLa cells in a tetracycline-inducible manner, this mutant accumulated in the ER and retained the endogenous ERGIC-53 in this compartment, thus preventing its recycling. It was suggested that recycling of ERGIC-53 is required for efficient intracellular transport of a small subset of glycoproteins, but it does not appear to be essential for the majority of glycoproteins (Vollenweider et al. 1998).

pH-Induced Conversion of ERGIC-53 Triggers Glycoprotein Release: Binding of cargo to ERGIC-53 in the ER requires Ca^{2+}. Cargo release occurs in the ERGIC, but the signals required for the cargo-receptor interaction are largely unknown. Though the efficient binding of ERGIC-53 to immobilized mannose occurs at pH 7.4, the pH of the ER, binding fails at slightly lower pH. pH sensitivity of the lectin was more prominent when Ca^{2+} concentrations were low. A conserved histidine in the center of the CRD was required for lectin activity suggesting it may serve as a molecular pH/Ca^{2+} sensor. Acidification of cells inhibited the association of ERGIC-53 with the known cargo cathepsin Z-related protein (Appenzeller-Herzog et al. 2004).

7.4.5.3 ERGIC-53 as a Marker of ER-Cargo Exit Site

A two-step reconstitution system for the generation of ER cargo exit sites from starting ER-derived low density microsomes (LDMs) has been described. The first step involves the hydrolysis of Mg^{2+}ATP and Mg^{2+}GTP, leading to the formation of a transitional ER (tER) with the soluble cargo albumin, transferrin, and the ER-to-Golgi recycling membrane enriched proteins α2p24 and p58/ERGIC-53. Upon further incubation (step two) with cytosol and mixed nucleotides, interconnecting smooth ER tubules within tER transform into vesicular tubular clusters (VTCs). The cytosolic domain of α2p24 and cytosolic COPI coatomer affect VTC formation. This was observed from the effect of Abs to the C-terminal tail of α2p24, but not of Abs to the C-terminal tail of calnexin on this reconstitution. Therefore, the p24 family member, α2p24, and its cytosolic coat ligand, COPI coatomer play a role in the de novo formation of VTCs and the generation of ER cargo exit sites (Lavoie et al. 1999).

Stephens and Pepperkok (2002) examined the ER-to-Golgi transport of procollagen, which, when assembled in the lumen of the ER, is thought to be physically too large to fit in classically described 60–80 nm COPI- and COPII-coated transport vesicles. Using ERGIC-53 as a marker, data indicated the existence of an early COPI-dependent, pre-Golgi cargo sorting step in mammalian cells (Stephens and Pepperkok 2002).

Carbohydrate- and Conformation-Dependent Cargo Capture for ER-Exit: The targeting motif in ERGIC-53 is composed of a high-mannose type oligosaccharide intimately associated with a surface-exposed peptide β-hairpin loop. The motif accounts for ERGIC-53-assisted ER-export of the lyososomal enzyme pro-cathepsin Z. The second oligosaccharide chain of pro-cathepsin Z exhibits no binding activity for ERGIC-53, illustrating the selective lectin properties of ERGIC-53. Evidences suggest that the conformation-based motif is only present in fully folded pro-cathepsin Z and that its recognition by ERGIC-53 reflects a quality control mechanism that acts complementary to the primary folding machinery in the ER. A similar oligosaccharide/β-hairpin loop structure is present in cathepsin C, another cargo of ERGIC-53, suggesting the general nature of this ER-exit signal. Perhaps, the molecular mechanism underlying reversible lectin/cargo interaction involves the ERGIC as the earliest low pH site of the secretory pathway. Possibly, this is a report on an ER-exit signal in soluble cargo in conjunction with its decoding by a transport receptor (Appenzeller-Herzog et al. 2004, 2005).

Regeneration of Golgi Stacks at Peripheral ER Exit Sites: After microtubule depolymerization, Golgi membrane components are found to redistribute to a distinct number of peripheral sites that are not randomly distributed, but correspond to sites of protein exit from the ER. Whereas Golgi enzymes redistributed gradually over hours to these peripheral sites, ERGIC-53 redistributed rapidly (within minutes) to these sites after first moving through the ER. It was proposed that a slow but constitutive flux of Golgi resident proteins through the same ER/Golgi cycling pathways as ERGIC-53 underlies Golgi dispersal upon microtubule depolymerization. Both ERGIC-53 and Golgi proteins would accumulate at peripheral ER exit sites due to failure of membranes at these sites to cluster into the centrosomal region. Regeneration of Golgi stacks at these peripheral sites would re-establish secretory flow from the ER into the Golgi complex and result in Golgi dispersal (Cole et al. 1996).

Heat Shock Affects Translation and Recycling Pathway of ERGIC-53: ERGIC-53 accumulates during heat shock response. However, at variance with unfolded protein response, which results in enhanced transcription of ERGIC-53 mRNA, heat shock leads only to enhanced translation of ERGIC-53 mRNA. In addition, the half-life of the protein does not change during heat shock. Therefore, distinct pathways of the cell stress modulate the ERGIC-53 protein level. Heat shock also affects the recycling pathway of ERGIC-53. The protein rapidly redistributes in a more peripheral area of the cell, in a vesicular compartment that has a low sedimentation density in comparison to the compartment that contains majority of ERGIC-53 at 37°C. Moreover, the anterograde transport of two unrelated reporter proteins is not affected. Interestingly, MCFD2, which interacts with ERGIC-53 to form a complex required for the ER-to-Golgi transport of specific proteins, is regulated similar to ERGIC-53 in response to cell stress. These results support the view that ERGIC-53 alone, or in association with MCFD2, plays important role during cellular response to various stress conditions (Spatuzza et al. 2004).

ER stress induces transcription of ERGIC-53. The ERGIC-53 promoter contains a single cis-acting element that mediates induction of gene by thapsigargin and other ER stress-causing agents. This ER stress response element is highly conserved in mammalian ERGIC-53 genes. The ER stress response element contains a 5′-end CCAAT sequence that constitutively binds NFY/CBF and, 9 nt away, a 3′-end region (5′-CCCTGTTGGCCATC-3′) that is equally important for ER stress-mediated induction of the gene. This sequence is the binding site for endogenous YY1 at the 5′-CCCTGTTGG-3′ part and for undefined factors at the CCATC 3′-end. ATF6 alpha-YY1, but not XBP1, interacted with the ERGIC-53 regulatory region and activated ERGIC-53 ER stress response element-dependent transcription (Renna et al. 2007).

7.4.5.4 ERGIC-53 (LMAN1) and MCFD2 Form a Cargo Receptor Complex

Interaction of ERGIC-53 and MCFD2 with Factor VIII: The ERGIC is the site of segregation of secretory proteins for anterograde transport, via packaging into COPII-coated transport vesicles (Schekman and Orci 1996). Correctly folded proteins destined for secretion are packaged in the ER into COPII-coated vesicles, which subsequently fuse to form the ERGIC. Multiple coagulation factor deficiency 2 (MCFD2) is a soluble EF-hand-containing protein that is retained in ER through its interaction with ERGIC-53. Exit of soluble secretory proteins from ER can occur by receptor-mediated export as exemplified by blood coagulation factors V and VIII. Their efficient secretion requires ERGIC-53 and its soluble luminal interaction partner MCFD2, which form a cargo receptor complex in early secretory pathway. ERGIC-53 also interacts with two lysosomal glycoproteins cathepsin Z and cathepsin C. In absence of ERGIC-53, MCFD2 was secreted, whereas knocking down MCFD2 had no effect on the localization of ERGIC-53. Endogenous LMAN1 and MCFD2 are present primarily in complex with each other with a 1:1 stoichiometry, although MCFD2 is not required for oligomerization of ERGIC-53. Coimmunoprecipitation of ERGIC-53 and FVIII from transfected HeLa and COS-1 cells and results of the crystal structure of CRD of ERGIC-53 demonstrated an interaction between ERGIC-53 and FVIII in vivo, mediated via high mannose-containing asparagine-linked oligosaccharides, which were densely situated within B domain of FVIII, as well as protein-protein interactions (Cunningham et al. 2003). Perhaps ERGIC-53 and MCFD2 form a cargo receptor complex and that the primary sorting signals residing in the B domain direct the binding of factor VIII to ERGIC-53 1-MCFD2 through calcium-dependent protein-protein interactions. MCFD2 may function to specifically recruit factor V and factor VIII to sites of transport vesicle budding within the ER lumen (Zhang et al. 2003, 2005). These findings suggest that MCFD2-ERGIC-53 complex forms a specific cargo receptor for the ER-to-Golgi transport of selected proteins. However, MCFD2 is dispensable for the binding of cathepsin Z and cathepsin C to ERGIC-53, since ERGIC-53 can bind cargo glycoproteins in an MCFD2-independent fashion and that MCFD2 is a recruitment factor for blood coagulation factors V and VIII (Nyfeler et al. 2006).

7.4.5.5 ERGIC-53 in Quality Control

Activating Transcription Factors 6, a Regulator of Mammalian UPR: Newly synthesized secretory and transmembrane proteins are folded and assembled in ER where an efficient quality control system operates so that only correctly folded molecules are allowed to move along the secretory pathway. The productive folding process in ER has been thought to be supported by the unfolded protein response (UPR). The accumulation of unfolded proteins in the ER triggers a signaling response. In yeast the UPR affects several hundred genes that encode ER chaperones and proteins operating at later stages of secretion. In mammalian cells the UPR appears to be more limited to chaperones of the ER and genes assumed to be important after cell recovery from ER stress that are not important for secretion.

The mRNA of ERGIC-53 and its related protein VIP36 is induced by inducers of ER stress, tunicamycin and thapsigargin. The rate of synthesis of the ERGIC-53 protein is also induced by these agents. The response was due to the UPR since it was also triggered by castanospermine, a specific inducer of UPR, and inhibited by genistein. Thapsigargin-induced upregulation of ERGIC-53 could be fully accounted for by the activating transcription factors 6 (ATF6) pathway of UPR. It has been suggested that in mammalian cells the UPR also affects traffic from and beyond the ER (Nyfeler et al. 2003). However, a dilemma has emerged; activation of ATF6, the key regulator of mammalian UPR, requires intracellular transport from the ER to the Golgi apparatus. This suggests that unfolded proteins might be leaked from the ER together with ATF6 in response to ER stress, exhibiting proteotoxicity in the secretory pathway. It has been found that ATF6 and correctly folded proteins are transported to the Golgi apparatus via the same route and by the same mechanism under conditions of ER stress, whereas unfolded proteins are retained in the ER. Thus, the activation of UPR is compatible with the quality control in the ER and the ER possesses a remarkable ability to select proteins to be transported in mammalian cells in marked contrast to yeast cells, which actively utilize intracellular traffic to deal with unfolded proteins accumulated in the ER (Nadanaka et al. 2004).

ERGIC-53 in the Formation of Russell Bodies: Owing to the impossibility of reaching the Golgi for secretion or the cytosol for degradation, mutant Ig-micro chains that lack the first constant domain (micro-δCH1) accumulate as detergent-insoluble aggregates in dilated ER cisternae, called Russell bodies. The pathological role(s) of similar structures in ER storage diseases remains obscure. In cells containing smooth Russell bodies, ERGIC-53 co-localizes with micro-δCH1 aggregates in a Ca^{2+}-dependent fashion. Studies suggest that interaction with light chains or ERGIC-53 seed micro-δCH1condensation in different stations of the early secretory pathway (Mattioli et al. 2006).

7.4.6 Mutations in ERGIC-53 *LMAN1* Gene and Deficiency of Coagulation Factors V and VIII lead to bleeding disorder

Mutations in the ERGIC-53 gene are associated with a coagulation defect. Using positional cloning, the gene was identified as the disease gene leading to combined factor V-factor (FV) and VIII (FVIII) deficiency, a rare, autosomal recessive disorder in which both coagulation factors V and VIII are diminished. ERGIC-53 was mapped to a YAC and BAC contig containing the critical region for the combined factors V and VIII deficiency gene. DNA sequence analysis identified two different mutations, accounting for all affected individuals in nine families studied. Findings indicated that ERGIC-53 may function as a molecular chaperone for the transport from ER to Golgi of a specific subset of secreted proteins, including coagulation factors V and VIII (Nichols et al. 1998). The crystal structure of CRD of ERGIC-53 complements the biochemical and functional characterization of the protein, confirming that a lectin domain is essential for this protein in sorting and transfer of glycoproteins from the ER to the Golgi complex. The lectin domains of calnexin and ERGIC-53 are structurally similar, although there is little primary sequence similarity. By contrast, sequence similarity between ERGIC-53 and VIP36, a Golgi-resident protein, leaves little doubt that a similar lectin domain is central to the transport and/or sorting functions of VIP36. The theme emerging from these studies is that carbohydrate recognition and modification are central to mediation of glycoprotein folding and secretion (Schrag et al. 2003).

Studies suggest that mutations in genes *LMAN1* and *MCFD2* are responsible for FVFVIII action. The binding of ERGIC-53 to sugar is enhanced by its interaction with MCFD2, and defects in this interaction in FV/FVIII patients may be the cause for reduced secretion of factors V and VIII (Kawasaki et al. 2008). Though clinically indistinguishable, MCFD2 mutations generally exhibit lower levels of FV and FVIII than LMAN1 mutations. The LMAN1 is a mannose-specific lectin which cycles between ER and ER-Golgi intermediate compartment. MCFD2 is an EF-hand domain protein that forms a calcium-dependent heteromeric complex with LMAN1 in cells. Missense mutations in the EF-hand domains of MCFD2 abolish the interaction with LMAN1. The LMAN1-MCFD2 complex may serve as a cargo receptor for the ER-to-Golgi transport of FV and FVIII, and perhaps a number of other glycoproteins. The B domain of FVIII may be important in mediating its interaction with the LMAN1-MCFD2 complex (Zhang 2009). Loss of functional ERGIC-53 leads to a selective defect in secretion of glycoproteins in cultured cells. Studies on the effect of defective ER to Golgi cycling by ERGIC-53 on the secretion of factors V and VIII showed that efficient trafficking of factors V and VIII requires a functional ERGIC-53 cycling pathway and that this trafficking is dependent on post-translational modification of a specific cluster of asparagine (N)-linked oligosaccharides to a fully glucose-trimmed, mannose9 structure (Moussalli et al. 1999).

7.4.6.1 DNA Polymorphism in ERGIC-53/*LMAN1* and *MCFD2* Genes

Mutations in a candidate gene of, *ERGIC-53/LMAN1*, were found to be associated with the coagulation defect in human population. Single-strand conformation and sequence analysis of the *ERGIC-53* gene in families of different ethnic origins identified 13 distinct mutations accounting for 52 of 70 mutant alleles. These were 3 splice site mutations, 6 insertions and deletions resulting in translational frame shifts, 3 nonsense codons, and elimination of the translation initiation codon. These mutations predict the synthesis of either a truncated protein product or no protein at all. This study revealed that FVFVIII shows extensive allelic heterogeneity and all ERGIC-53 mutations resulting in FV/FVIII are "null." Approximately 26% of the mutations have not been identified, suggesting that lesions in regulatory elements or severe abnormalities within the introns may be responsible for the disease in these individuals. In two such families, ERGIC-53 protein was detectable at normal levels in patients' lymphocytes, raising the further possibility of defects at other genetic loci (Neerman-Arbez et al. 1999; Nyfeler et al. 2005).

Nineteen additional families were analyzed by direct sequence analysis of the entire coding region and the intron/exon junctions. Seven novel mutations were identified in ten families, with one additional family found to harbor one of the two previously described mutations. All of the identified mutations were predicted to result in complete absence of functional ERGIC-53 protein. In 8 of 19 families, no mutation was identified. Thus a significant subset of combined factors V and VIII deficiency is due to mutation in one or more additional genes (Nichols et al. 1999). Two mutations in ERGIC-53 gene have been observed in Jews: a guanine (G) insertion in exon 1 among Middle Eastern Jewish families, and a thymidine (T) to cytosine (C) transition in intron 9 at a donor splice site among Tunisian families. All affected Tunisian families belonged to an ancient Jewish community. Screening this community for the intron 9T—>C transition, among 233 apparently unrelated individuals five heterozygotes were detected, predicting an allele frequency of 0.0107, while among 259 North African Jews none was found to carry the mutation. The G insertion in exon 1 was found in one of 245 Iraqi Jews, predicting an allele frequency of 0.0022, but in none of 180 Iranian Jews examined. In view of the relatively low frequency of the mutations in the respective populations it seems reasonable to advocate carrier detection and prenatal diagnosis only in affected families (Segal et al. 2004).

Three Indian families with FV/FVIII were analyzed for the presence of mutations in their LMAN1 and MCFD2 genes. One of the three families showed the presence of a G to A substitution in exon 2 of the MCFD2 gene, whereas another family showed a nonsense mutation, i.e., G to T substitution, in exon 2 of the LMAN1 gene, the latter being a novel mutation not previously reported. The third family did not show mutations in either of the two genes, suggesting that a significant subset of FV/FVIII cases may be due to additional genes resulting in a similar phenotype (Mohanty et al. 2005). Immunoprecipitation and Western blot analysis detected a low level of LMAN1-MCFD2 complex in lymphoblasts derived from patients with missense mutations in LMAN1 (C475R) or MCFD2 (I136T), suggesting that complete loss of the complex may not be required for clinically significant reduction in FV and FVIII (Zhang et al. 2006).

7.4.6.2 *LMAN1* Expression in MSI-H Tumorigenesis

Roeckel et al. (2009) analyzed mutation frequencies of genes of glycosylation machinery in microsatellite unstable (MSI-H) tumors, focusing on frameshift mutations in coding MNRs (cMNRs). Among 28 candidate genes, LMAN1/ERGIC53 showed high mutation frequency in MSI-H colorectal cancer cell lines (52%; 12 of 23), carcinomas (45%; 72 of 161), and adenomas (40%; 8 of 20). Analysis of LMAN1-mutated carcinomas and adenomas revealed regional loss of LMAN1 expression due to biallelic LMAN1 cMNR frameshift mutations. In LMAN1-deficient colorectal cancer cell lines, secretion of α-1-antitrypsin (A1AT), an inhibitor of angiogenesis and tumor growth, was significantly impaired but could be restored upon LMAN1 re-expression. Results suggest that LMAN1 mutational inactivation is a frequent and early event potentially contributing to MSI-H tumorigenesis.

7.5 Vesicular Integral Membrane Protein (VIP36) OR LMAN2

7.5.1 The Protein

The vesicular integral membrane lectin (VIP36), also called - Lectin, mannose binding 2 (LMAN2) belongs to a family of lectins, conserved from yeast to mammals, trafficking in secretory pathway and closely related to lectin ERGIC-53 that acts as a cargo receptor, facilitating ER to Golgi transport of certain glycoproteins. VIP36 was originally identified as a component of apical post-Golgi vesicles in virally infected, polarized Madin-Darby canine kidney (MDCK) cells (Fiedler et al. 1994). VIP36 was shown to localize not only to the early secretory pathway but also to the plasma membrane of MDCK and Vero cells. The VIP36 recognizes high-mannose type glycans containing α1—>2 Man residues and α-amino substituted asparagine. The binding of VIP36 to high-mannose type glycans was independent of Ca^{2+} at optimum pH 6.0 at 37°C. The association constant between Man7-9GlcNAc2 in porcine thyroglobulin and immobilized VIP36 was $7.1 \times 10^8\ M^{-1}$. This shows that VIP36 functions as an intracellular lectin recognizing glycoproteins which possess high-mannose type glycans, (Manα1—>2)$_{2-4}$Man$_5$ GlcNAc$_2$ (Hara-Kuge et al. 1999).

Although VIP36 interacts with glycoproteins carrying high mannose-type oligosaccharides, further analysis using the frontal affinity chromatography (FAC) and the sugar-binding properties of a rCRD of VIP36 (VIP36-CRD) have shown that glucosylation and trimming of D1 mannosyl branch (Fig. 7.5) interfere with the binding of VIP36-CRD. VIP36-CRD exhibits an optimal pH value of ~6.5. Examining the specificity and optimal pH of sugar -VIP36 interaction and its subcellular localization, along with organellar pH, it was suggested that VIP36 binds glycoproteins that retain the intact D1 mannosyl branch in cis-Golgi network and recycles to ER where, due to higher pH, it releases its cargos, thereby contributing to quality control of glycoproteins.

In the plasma membrane, VIP36 exhibited an apical-predominant distribution, the apical/basolateral ratio being approximately 7. Localization of over-expressed VIP36 to plasma membrane, endosomes, and Golgi structures, together with evidence for lectin activity lead to the hypothesis that it functions to segregate apical cargo into distinct vesicles within the *trans*-Golgi network by binding specific *N*-glycans. Results indicated that VIP36 was involved in the transport and sorting of glycoproteins carrying high mannose-type glycan(s) (Fiedler and Simons 1994, 1995, 1996; Hara-Kuge et al. 2002). Punctate cytoplasmic structures co-localize with coatomer and ERGIC-53, labeling ER-Golgi intermediate membrane structures. Cycling of VIP36 is suggested by colocalisation with anterograde cargo trapped in pre-Golgi structures and modification of its N-linked carbohydrate by glycosylation enzymes of medial Golgi cisternae. Furthermore, after brefeldin A treatment VIP36 is segregated from resident Golgi proteins and codistributes with ER-Golgi recycling proteins (Fullekrug et al. 1997, 1999).

VIP36 shares significant homology with leguminous lectins as well as with ERGIC-53. Its ability to recognize high-mannose type glycans (Hara-Kuge et al. 1999; Kamiya et al. 2005, 2008) and its broad localization from ER to *cis*-Golgi (Fullekrug et al. 1999; Shimada et al. 2003a, 2003b) indicates that VIP36 also functions as a cargo receptor that facilitates the transport of various glycoproteins. Proteins that interact with VIP36 during quality control of secretory proteins have been identified. An 80 kDa immunoglobulin-binding protein (BiP), a major protein of Hsp70 chaperone family, binds VIP36. The interaction between VIP36 and BiP is not due to chaperone-substrate complex. The interaction depends on divalent cations but not on ATP. These observations suggest a new role for VIP36 in the quality control of secretory proteins (Nawa et al. 2007; Kamiya et al. 2008). It was speculated that VIP36 binds to sugar residues of glycosphingolipids and/or glycosylphosphatidyl-inositol anchors and might provide a link between the extracellular/luminal face of glycolipid rafts and the cytoplasmic protein segregation machinery (Fiedler et al. 1994).

VIP36 is highly expressed in salivary glands, especially the parotid gland, which secretes α-amylase in large quantities. Endogenous VIP36 is localized in trans-Golgi network, on immature granules, and on mature secretory granules in acinar cells and co-localized with amylase. VIP36 is involved in the post-Golgi secretory pathway, suggesting that VIP36 plays a role in trafficking and sorting of secretory and/or membrane proteins during granule formation. EM demonstrated that VIP36 was primarily localized to secretory vesicles in glandula parotis of the rat, where α-amylase also resided. It was suggested that VIP36 is involved in the secretion of α-amylase in the rat parotid gland (Hara-Kuge et al. 2004; Shimada et al. 2003a). In GH3 cells endogenous VIP36 is localized mainly in 70–100-nm-diameter uncoated transport vesicles between the exit site on ER and the neighboring cis-Golgi cisterna. The thyrotrophin-releasing hormone (TRH) stimulation and treatment with actin filament-perturbing agents, cytochalasin D or B or latrunculin-B, caused marked aggregation of the VIP36-positive vesicles and the appearance of a VIP36-positive clustering structure located near cis-Golgi cisterna. The size of this structure, which comprised conspicuous clusters of VIP36, depended on the TRH concentration. Furthermore, VIP36 colocalized with filamentous actin in the paranuclear Golgi area and its vicinity. It suggests that actin filaments are involved in glycoprotein transport between the ER and cis-Golgi cisterna by using the lectin VIP36 (Shimada et al. 2003b).

7.5.2 VIP36-SP-FP as Cargo Receptor

To investigate the trafficking of transmembrane lectin VIP36 and its relation to cargo-containing transport carriers (TCs), Dahm et al. (2001) analyzed a C-terminal fluorescent-protein (FP) fusion, VIP36-SP-FP. At moderate levels of expression, VIP36-SP-FP is localized to the ER, Golgi apparatus, and intermediate transport structures, and colocalized with epitope-tagged VIP36. VIP36-SP-FP recycles in the early secretory pathway, exhibiting trafficking representative of a class of transmembrane cargo receptors, including the closely related lectin ERGIC-53. The VIP36-SP-FP trafficking structures comprised tubules and globular elements, which translocated in a salutatory manner. Simultaneous visualization of anterograde secretory cargo and VIP36-SP-FP indicated that the globular structures were pre-Golgi carriers, and that VIP36-SP-FP segregated from cargo within the Golgi and was not included in post-Golgi TCs (Dahm et al. 2001).

7.5.3 Structure for Recognition of High Mannose Type Glycoproteins by VIP36

It has been shown that ERGIC-53 interacts with glycoproteins carrying high mannose type glycan by endo-β-*N* acetylglucosaminidase H treatment (Appenzeller 1999; Moussalli 1999) and binds glycoproteins in a Ca^{2+} and pH-dependent manner (Appenzeller-Herzog et al. 2004, 2005). VIP36 has high affinity for high mannose type glycans containing Man-α-1,2-Man residues in Man7–9(GlcNAc)2-Asn peptides (Hara-Kuge et al. 1999). Kamiya et al. (2005) reported carbohydrate binding properties of VIP36 by frontal affinity chromatography and suggested Ca^{2+} dependence of carbohydrate binding and the specificity for D1 arm, Man-α-1,2-Man-α- 1,2-Man residues, of high mannose type glycans (corresponding to Man(D1)-Man (C)-Man(4) (Fig. 7.5).

The exoplasmic/luminal domain of VIP36 as well as the luminal domain of ERGIC-53 and Emp46/47p share homology with L (leguminous)-type lectins and are thus identified as CRDs. It has been shown that ERGIC-53 interacts with glycoproteins carrying high mannose type glycan by endo-β-*N*-acetylgluco- saminidase H treatment (Appenzeller et al. 1999; Moussalli, et al. 1999; Appenzeller-Herzog et al. 2005) and binds glycoproteins in a Ca^{2+}- and pH-dependent manner (Appenzeller-Herzog et al. 2004). It has been found that VIP36 has high affinity for high mannose type glycans containing Man-α-1,2- Man residues in Man7–9(GlcNAc)2-Asn peptides (Hara-Kuge et al. 1999). Kamiya et al. (2005) reported the carbohydrate binding properties of VIP36 by frontal affinity chromatography. This study suggested Ca^{2+} dependence of carbohydrate binding and the specificity for D1 arm, Man-α-1,2-Man-α-1,2-Man residues, of high mannose type glycans (corresponding to Man(D1)-Man(C)-Man (4); (Fig. 7.5). In addition, using a VIP36 binds glycoproteins carrying high mannose type glycans (Kawasaki et al. 2007). These observations suggested that VIP36 is involved in the transport of glycoproteins via high mannose type glycans.

Crystal structures of the CRD of rat ERGIC-53 in the absence and presence of Ca^{2+} have been determined, confirming its structural similarity to the L-type lectins (Velloso et al. 2002, 2003). In these reports, it was shown that the putative ligand-binding site of ERGIC-53 is similar to the mannose-binding site of the L-type lectins. The crystal structures of the CRD of Ca^{2+}-independent K^+-bound Emp46p and the metalfree form of Emp47p have also been reported (Satoh et al. 2006). Satoh et al. (2007) determined structures of transport lectin in complex with high mannose type glycans. and determined crystal structures of the exoplasmic/luminal domain of VIP36 alone and in complex

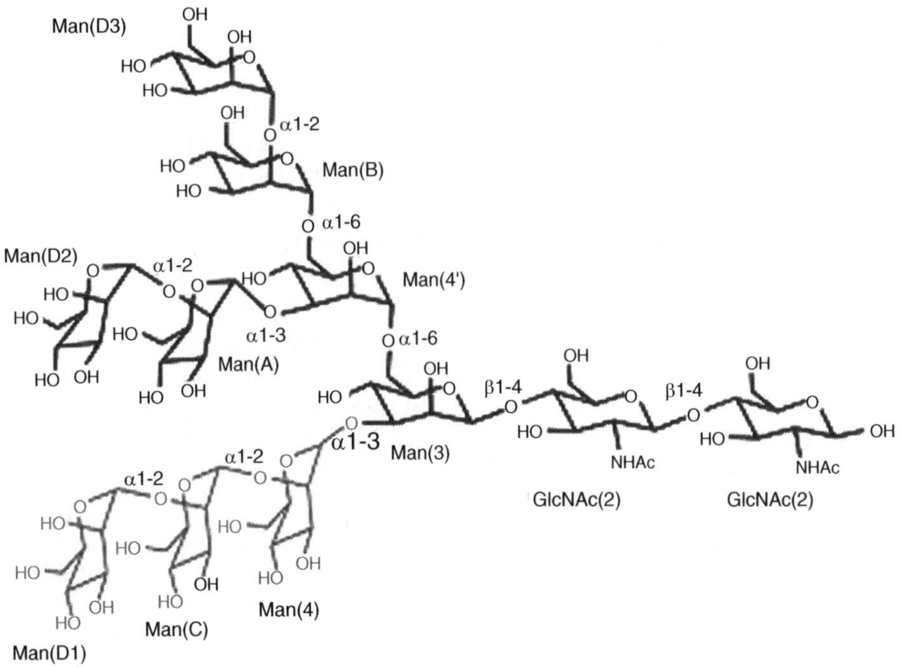

Fig. 7.5 Chemical structures of $Man_9(GlcNAc)_7$. The individual carbohydrate residues of $Man_9(GlcNAc)_2$ are labeled. The D1 arm of $Man_9(GlcNAc)_2$ is colored in *light grey*

with Ca^{2+} and mannose, Man-α-1,2-Man (termed Man2, which corresponds to Man(D1)-Man(C), Man(C)-Man(4), Man(D2)-Man(A), or Man(D3)-Man(B) of Man9(GlcNAc) 2; and Man-α-1,2- Man-α-1,3-Man-β-1,4-GlcNAc (termed Man3GlcNAc, which corresponds to Man(C)-Man(4)-Man (3)-GlcNAc(2) (Fig. 7.5).

Satoh et al. (2007) reported the crystal structure of VIP36 exoplasmic/luminal domain comprising a CRD and a stalk domain and in complexed form with Ca^{2+} and mannosyl ligands. The CRD is composed of a 17-stranded antiparallel β-sandwich and binds one Ca^{2+} adjoining the carbohydrate-binding site. The structure reveals that a coordinated Ca^{2+} ion orients the side chains of Asp131, Asn166, and His190 for carbohydrate binding. This result explains the Ca^{2+}-dependent carbohydrate binding of this protein. The Man-α-1,2-Man-α-1,2-Man, which corresponds to the D1 arm of high mannose type glycan, is recognized by eight residues through extensive hydrogen bonds. The complex structures reveal the structural basis for high mannose type glycoprotein recognition by VIP36 in a Ca^{2+}-dependent and D1 arm-specific manner (Fig. 7.4b).

7.5.4 Emp47p of *S. cerevisiae:* A Homologue to VIP36 and ERGIC-53

Whereas mannose 6-phosphate receptor functions as a cargo receptor for lysosomal proteins in the *trans*-Golgi network, ERGIC-53 (Hauri et al. 2000a, b; Zhang et al. 2003) and its yeast orthologs Emp46/47p (Sato and Nakano 2003; Schroder et al. 1995) are transport lectins for glycoproteins that are transported out of ER. The *S. cerevisiae EMP47* gene encodes a type-I transmembrane protein with sequence homology to ERGIC-53 and VIP36. The 12-amino acid COOH-terminal cytoplasmic tail of Emp47p ends in the sequence KTKLL, which agrees with the consensus for di-lysine-based signals in ER. Despite the presence of this motif, Emp47p is known as a Golgi protein at steady-state. The di-lysine motif of Emp47p was functional when transplanted onto Ste2p, a plasma membrane protein, conferring ER localization. Emp47p cycles between the Golgi apparatus and the ER and requires a di-lysine motif for its α-COP-independent, steady state localization in Golgi (Schroder et al. 1995).

7.6 VIP36-Like (VIPL) L-Type Lectin

The profiles of human L-type lectin-like membrane proteins ERGIC-53, ERGL, and VIP36 and optimal alignment of entire CRD of these proteins revealed numerous orthologous and homologous L-type lectin-like proteins in animals, protozoans, and yeast, as well as the sequence of a novel family member related to VIP36, named VIPL for VIP36-like. VIPL has 43% similarity to ERGIC-53 and 68% similarity to VIP36. Its orthologues are broadly distributed in many eukaryotes from human to fission yeast, whereas VIP36 is restricted to higher organisms. This phylogenetic finding suggests that VIP36 evolved from VIPL by gene duplication. VIPL is also predicted to have type I transmembrane topology with a putative CRD homologous to L-type lectins (Neve et al. 2003; Nufer et al. 2003b). Although the structural similarity between VIP36 and VIPL is very high, their characteristics are quite different. VIP36 and VIPL have the same cytoplasmic ER exit motif (KRFY) on their C-termini, but VIPL has an additional arginine at the fifth position from the C-terminus, creating the ER localization motif RKR. As a result, VIPL is a resident of ER (Neve et al. 2003; Nufer et al. 2003b), unlike ERGIC-53 and VIP36, which cycles in the secretory pathway. The sugar-binding activity of VIPL is also thought to be different from those of ERGIC-53 and VIP36; for example, HA-tagged VIPL does not bind to immobilized Man or BSA conjugated to Glc, Man, or GlcNAc as ERGIC-53 does (Nufer et al. 2003b). Moreover, VIPL does not compete with ERGIC-53 for binding to immobilized Man (Nufer et al. 2003b). Based on these observations, VIPL is thought to be an L-type lectin without sugar-binding activity. On the other hand, another group reported that knockdown of VIPL mRNA using short interfering RNA in HeLa cells significantly slowed down the secretion of two glycoproteins (35 and 250 kDa) into the medium (Neve et al. 2003), suggesting that VIPL does function in the secretion of glycoproteins, possibly by serving as an ER export receptor, similarly to ERGIC-53 (c/r Yamaguchi et al. 2007).

Although VIPL is structurally similar to VIP36, VIPL was thought not to be a lectin, because its sugar-binding activity had not been detected in several experiments. Yamaguchi et al. (2007) examined the sugar-binding activity and specificity of recombinant soluble VIPL (sVIPL) using cells with modified cell surface carbohydrates as ligands. Competition experiments with several high-mannose-type *N*-glycans indicated that VIPL recognizes the Man1–2Man1–2Man sequence and that the glucosylation of the outer mannose residue of this portion decreased the binding. Although the biochemical characteristics of VIPL are similar to those of VIP36, the sugar-binding activity of VIPL was stronger at neutral pH, corresponding to the pH in the lumen of the ER, than under acidic conditions. Unlike VIP36 and ERGIC-53 that are predominantly associated with ER membranes and cycle in the early secretory pathway, VIPL is a non-cycling resident protein of the ER. The results suggest that VIPL may function as a regulator of ERGIC-53 (Nufer et al. 2003a).

VIPL has been conserved through evolution from zebra fish to man. The 7.4-kb VIPL mRNA was widely expressed to varying levels in different tissues. The 32-kDa VIPL protein was detected in various cell lines. VIPL localized primarily to the ER and partly to the Golgi complex. Like VIP36, the cytoplasmic tail of VIPL terminates in the sequence KRFY, a motif characteristic for proteins recycling between the ER and ERGIC/cis-Golgi. Mutating the retrograde transport signal KR to AA resulted in transport of VIPL to the cell surface. Knockdown of VIPL mRNA using siRNA slowed down the secretion of two glycoproteins (M_r 35 and 250 kDa) to the medium, suggesting that VIPL may also function as an ER export receptor (Neve et al. 2003).

References

Appenzeller C, Andersson H, Kappeler F, Hauri H-P (1999) The lectin ERGIC-53 is a cargo transport receptor for glycoproteins. Nat Cell Biol 1:330–334

Appenzeller-Herzog C, Hauri HP (2006) The ER-Golgi intermediate compartment (ERGIC): in search of its identity and function. J Cell Sci 119(Pt 11):2173–2183

Appenzeller-Herzog C, Nyfeler B, Burkhard P et al (2005) Carbohydrate- and conformation-dependent cargo capture for ER-exit. Mol Biol Cell 16:1258–1267

Appenzeller-Herzog C, Roche AC, Nufer O, Hauri HP (2004) pH-induced conversion of the transport lectin ERGIC-53 triggers glycoprotein release. J Biol Chem 279:12943–12950

Arar C, Carpentier V, Le Caer JP et al (1995) ERGIC-53, a membrane protein of the endoplasmic reticulum-Golgi intermediate compartment, is identical to MR60, an intracellular mannose-specific lectin of myelomonocytic cells. J Biol Chem 270:3551–3553

Belden WJ, Barlowe C (2001) Role of Erv29p in collecting soluble secretory proteins into ER-derived transport vesicles. Science 294:1528–1531

Carriere V, Landemarre L, Altemayer V, Motta G, Monsigny M, Roche AC (1999a) Intradermal DNA immunization: antisera specific for the membrane lectin MR60/ERGIC-53. Biosci Rep 19:559–569

Carriere V, Piller V, Legrand A et al (1999b) The sugar binding activity of MR60, a mannose-specific shuttling lectin, requires a dimeric state. Glycobiology 9:995–1002

Cole NB, Sciaky N, Marotta A et al (1996) Golgi dispersal during microtubule disruption: regeneration of Golgi stacks at peripheral endoplasmic reticulum exit sites. Mol Biol Cell 7:631–650

Cunningham MA, Pipe SW, Zhang B et al (2003) LMAN1 is a molecular chaperone for the secretion of coagulation factor VIII. J Thromb Haemost 1:2360–2367

Dahm T, White J, Grill S, Fullekrug J, Stelzer EH (2001) Quantitative ER<–>Golgi transport kinetics and protein separation upon Golgi exit revealed by vesicular integral membrane protein 36 dynamics in live cells. Mol Biol Cell 12:1481–1498

Dogic D, Dubois A, de Chassey B, Lefkir Y, Letourneur F (2001) ERGIC-53 KKAA signal mediates endoplasmic reticulum retrieval in yeast. Eur J Cell Biol 80:151–155

Farmaki T, Ponnambalam S, Prescott AR et al (1999) Forward and retrograde trafficking in mitotic animal cells. ER-Golgi transport arrest restricts protein export from the ER into COPII-coated structures. J Cell Sci 112:589–600

Fiedler K, Parton RG, Kellner R et al (1994) VIP36, a novel component of glycolipid rafts and exocytic carrier vesicles in epithelial cells. EMBO J 13:1729–1740

Fiedler K, Simons K (1994) A putative novel class of animal lectins in the secretory pathway homologous to leguminous lectins. Cell 77:625–626

Fiedler K, Simons K (1995) The role of N-glycans in the secretory pathway. Cell 81:309–312

Fiedler K, Simons K (1996) Characterization of VIP36, an animal lectin homologous to leguminous lectins. J Cell Sci 109:271–276

Fullekrug J, Scheiffele P, Simons K (1999) VIP36 localisation to the early secretory pathway. J Cell Sci 112:2813–2821

Fullekrug J, Sonnichsen B, Schafer U et al (1997) Characterization of brefeldin A induced vesicular structures containing cycling proteins of the intermediate compartment/cis-Golgi network. FEBS Lett 404:75–81

Girod A, Storrie B, Simpson JC et al (1999) Evidence for a COP-I-independent transport route from the Golgi complex to the endoplasmic reticulum. Nat Cell Biol 1:423–430

Hara-Kuge S, Ohkura T, Ideo H et al (2002) Involvement of VIP36 in intracellular transport and secretion of glycoproteins in polarized Madin-Darby canine kidney (MDCK) cells. J Biol Chem 277: 16332–16339

Hara-Kuge S, Ohkura T, Seko A, Yamashita K (1999) Vesicular-integral membrane protein, VIP36, recognizes high-mannose type glycans containing alpha1–>2 mannosyl residues in MDCK cells. Glycobiology 9:833–839

Hara-Kuge S, Seko A, Shimada O et al (2004) The binding of VIP36 and alpha-amylase in the secretory vesicles via high-mannose type glycans. Glycobiology 14:739–744

Hauri H, Appenzeller C, Kuhn F, Nufer O (2000a) Lectins and traffic in the secretory pathway. FEBS Lett 476:32–37

Hauri HP, Kappeler F, Andersson H, Appenzeller C (2000b) ERGIC-53 and traffic in the secretory pathway. J Cell Sci 113:587–596

Hauri HP, Nufer O, Breuza L et al (2002) Lectins and protein traffic early in the secretory pathway. Biochem Soc Symp 69:73–82

Itin C, Kappeler F, Linstedt AD, Hauri HP (1995a) A novel endocytosis signal related to the KKXX ER-retrieval signal. EMBO J 14:2250–2256

Itin C, Roche AC, Monsigny M, Hauri HP (1996) ERGIC-53 is a functional mannose-selective and calcium-dependent human homologue of leguminous lectins. Mol Biol Cell 7:483–493

Itin C, Schindler R, Hauri HP (1995b) Targeting of protein ERGIC-53 to the ER/ERGIC/cis-Golgi recycling pathway. J Cell Biol 131: 57–67

Kamiya Y, Kamiya D, Yamamoto K et al (2008) Molecular basis of sugar recognition by the human L-type lectins ERGIC-53, VIPL, and VIP36. J Biol Chem 283:1857–1861

Kamiya Y, Yamaguchi Y, Takahashi N et al (2005) Sugar-binding properties of VIP36, an intracellular animal lectin operating as a cargo receptor. J Biol Chem 280:37178–37182

Kappeler F, Itin C, Schindler R, Hauri HP (1994) A dual role for COOH-terminal lysine residues in pre-Golgi retention and endocytosis of ERGIC-53. J Biol Chem 269:6279–6281

Kappeler F, Klopfenstein DR, Foguet M et al (1997) The recycling of ERGIC-53 in the early secretory pathway. ERGIC-53 carries a cytosolic endoplasmic reticulum-exit determinant interacting with COPII. J Biol Chem 272:31801–31808

Kawasaki N, Ichikawa Y, Matsuo I et al (2008) The sugar-binding ability of ERGIC-53 is enhanced by its interaction with MCFD2. Blood 111:1972–1979

Kawasaki N, Matsuo I, Totani K et al (2007) Detection of weak sugar binding activity of VIP36 using VIP36-streptavidin complex and membrane-based sugar chains. J Biochem 141:221–229

Klumperman J, Schweizer A, Clausen H et al (1998) The recycling pathway of protein ERGIC-53 and dynamics of the ER-Golgi intermediate compartment. J Cell Sci 111(Pt 22):3411–3425

Lahtinen U, Dahllof B, Saraste J (1992) Characterization of a 58 kDa cis-Golgi protein in pancreatic exocrine cells. J Cell Sci 103:321–333

Lahtinen U, Hellman U, Wernstedt C et al (1996) Molecular cloning and expression of a 58-kDa cis-Golgi and intermediate compartment protein. J Biol Chem 271:4031–4037

Lahtinen U, Svensson K, Pettersson RF (1999) Mapping of structural determinants for the oligomerization of p58, a lectin-like protein of the intermediate compartment and cis-Golgi. Eur J Biochem 260 (2):392–397

Lavoie C, Paiement J, Dominguez M et al (1999) Roles for α2p24 and COPI in endoplasmic reticulum cargo exit site formation. J Cell Biol 146:285–299

Lee MC, Miller EA, Goldberg J et al (2004) Bidirectional protein transport between the ER and Golgi. Annu Rev Cell Dev Biol 20:87–123

Marie M, Dale HA, Sannerud R, Saraste J (2009) The function of the intermediate compartment in pre-Golgi trafficking involves its stable connection with the centrosome. Mol Biol Cell 20:4458–4470

Mattioli L, Anelli T, Fagioli C et al (2006) ER storage diseases: a role for ERGIC-53 in controlling the formation and shape of Russell bodies. J Cell Sci 119:2532–2547

Mohanty D, Ghosh K, Shetty S et al (2005) Mutations in the MCFD2 gene and a novel mutation in the LMAN1 gene in Indian families with combined deficiency of factor V and VIII. Am J Hematol 79:262–266

Moussalli M, Pipe SW, Hauri HP, Nichols WC, Ginsburg D, Kaufman RJ (1999) Mannose-dependent ERGIC-53-mediated ER to Golgi trafficking of coagulation factors V and VIII. J Biol Chem 274:32539–32542

Muniz M, Nuoffer C, Hauri HP, Riezman H (2000) The Emp24 complex recruits a specific cargo molecule into endoplasmic reticulum-derived vesicles. J Cell Biol 148:925–930

Nadanaka S, Yoshida H, Kano F et al (2004) Activation of mammalian unfolded protein response is compatible with the quality control system operating in the endoplasmic reticulum. Mol Biol Cell 15:2537–2548

Nawa D, Shimada O, Kawasaki N et al (2007) Stable interaction of the cargo receptor VIP36 with molecular chaperone BiP. Glycobiology 17:913–921

Neerman-Arbez M, Johnson KM, Morris MA et al (1999) Molecular analysis of the ERGIC-53 gene in 35 families with combined factor V-factor VIII deficiency. Blood 93:2253–2260

Neve EP, Lahtinen U, Pettersson RF (2005) Oligomerization and interacellular localization of the glycoprotein receptor ERGIC-53 is independent of disulfide bonds. J Mol Biol 354:556–568

Neve EP, Svensson K, Fuxe J, Pettersson RF (2003) VIPL, a VIP36-like membrane protein with a putative function in the export of glycoproteins from the endoplasmic reticulum. Exp Cell Res 288: 70–83

Nichols WC, Seligsohn U, Zivelin A et al (1998) Mutations in the ER-Golgi intermediate compartment protein ERGIC-53 cause combined deficiency of coagulation factors V and VIII. Cell 93:61–70

Nichols WC, Terry VH, Wheatley MA et al (1999) ERGIC-53 gene structure and mutation analysis in 19 combined factors V and VIII deficiency families. Blood 93:2261–2266

Nufer O, Guldbrandsen S, Degen M et al (2002) Role of cytoplasmic C-terminal amino acids of membrane proteins in ER export. J Cell Sci 115(Pt 3):619–628

Nufer O, Kappeler F, Guldbrandsen S, Hauri HP (2003a) ER export of ERGIC-53 is controlled by cooperation of targeting determinants in all three of its domains. J Cell Sci 116(Pt 21):4429–4440

Nufer O, Mitrovic S, Hauri HP (2003b) Profile-based data base scanning for animal L-type lectins and characterization of VIPL, a novel VIP36-like endoplasmic reticulum protein. J Biol Chem 278:15886–15896

Nyfeler B, Michnick SW, Hauri HP (2005) Capturing protein interactions in the secretory pathway of living cells. Proc Natl Acad Sci USA 102:6350–6355

Nyfeler B, Nufer O, Matsui T et al (2003) The cargo receptor ERGIC-53 is a target of the unfolded protein response. Biochem Biophys Res Commun 304:599–604

Nyfeler B, Zhang B, Ginsburg D et al (2006) Cargo selectivity of the ERGIC-53/MCFD2 transport receptor complex. Traffic 7:1473–1481

Pimpaneau V, Midoux P, Monsigny M, Roche AC (1991) Characterization and isolation of an intracellular D-mannose-specific receptor from human promyelocytic HL60 cells. Carbohydr Res 213:95–108

Renna M, Caporaso MG, Bonatti S et al (2007) Regulation of ERGIC-53 gene transcription in response to endoplasmic reticulum stress. J Biol Chem 282:22499–22512

Roeckel N, Woerner SM, Kloor M et al (2009) High frequency of LMAN1 abnormalities in colorectal tumors with microsatellite instability. Cancer Res 69:292–299

Sarnataro S, Caporaso MG, Bonatti S et al (1999) Sequence and expression of the monkey homologue of the ER-golgi intermediate compartment lectin, ERGIC-53. Biochim Biophys Acta 1447:334–340

Sato K, Nakano A (2003) Oligomerization of a cargo receptor directs protein sorting into COPII-coated transport vesicles. Mol Biol Cell 14:3055–3063

Satoh T, Cowieson NP, Hakamata W et al (2007) Structural basis for recognition of high mannose type glycoproteins by mammalian transport lectin VIP36. J Biol Chem 282:28246–28255

Satoh T, Sato K, Kanoh A et al (2006) Structures of the carbohydrate recognition domain of Ca^{2+}-independent cargo receptors Emp46p and Emp47p. J Biol Chem 281:10410–10419

Schekman R, Orci L (1996) Coat proteins and vesicle budding. Science 271:1526–1533

Schindler R, Itin C, Zerial M et al (1993) ERGIC-53, a membrane protein of the ER-Golgi intermediate compartment, carries an ER retention motif. Eur J Cell Biol 61:1–9

Schrag JD, Bergeron JJ, Li Y et al (2001) The structure of calnexin, an ER chaperone involved in quality control of protein folding. Mol Cell 8:633–644

Schrag JD, Procopio DO, Cygler M, Thomas DY, Bergeron JJ (2003) Lectin control of protein folding and sorting in the secretory pathway. Trends Biochem Sci 28:49–57

Schroder S, Schimmoller F, Singer-Kruger B, Riezman H (1995) The Golgi-localization of yeast Emp47p depends on its di-lysine motif but is not affected by the ret1-1 mutation in alpha-COP. J Cell Biol 131:895–912

Schweizer A, Clausen H, van Meer G, Hauri HP (1994) Localization of O-glycan initiation, sphingomyelin synthesis, and glucosylceramide synthesis in Vero cells with respect to the ER-Golgi intermediate compartment. J Biol Chem 269:4035–4047

Segal A, Zivelin A, Rosenberg N et al (2004) A mutation in LMAN1 (ERGIC-53) causing combined factor V and factor VIII deficiency is prevalent in Jews originating from the island of Djerba in Tunisia. Blood Coagul Fibrinolysis 15:99–102

Shimada O, Hara-Kuge S, Yamashita K et al (2003a) Clusters of VIP-36-positive vesicles between endoplasmic reticulum and Golgi apparatus in GH3 cells. Cell Struct Funct 28:155–163

Shimada O, Hara-Kuge S, Yamashita K et al (2003b) Localization of VIP36 in the post-Golgi secretory pathway also of rat parotid acinar cells. J Histochem Cytochem 51:1057–1063

Spatuzza C, Renna M, Faraonio R et al (2004) Heat shock induces preferential translation of ERGIC-53 and affects its recycling pathway. J Biol Chem 279:42535–42544

Srinivas VR, Reddy GB, Ahmad N et al (2001) Legume lectin family, the 'natural mutants of the quaternary state', provide insights into the relationship between protein stability and oligomerization. Biochim Biophys Acta 1527:102–111

Stephens DJ, Pepperkok R (2002) Imaging of procollagen transport reveals COPI-dependent cargo sorting during ER-to-Golgi transport in mammalian cells. J Cell Sci 115:1149–1160

Velloso LM, Svensson K, Pettersson RF et al (2003) The crystal structure of the carbohydrate-recognition domain of the glycoprotein sorting receptor p58/ERGIC-53 reveals an unpredicted metal-binding site and conformational changes associated with calcium ion binding. J Mol Biol 334:845–851

Velloso LM, Svensson K, Schneider G et al (2002) Crystal structure of the carbohydrate recognition domain of p58/ERGIC-53, a protein involved in glycoprotein export from the endoplasmic reticulum. J Biol Chem 277:15979–15984

Vollenweider F, Kappeler F, Itin C, Hauri HP (1998) Mistargeting of the lectin ERGIC-53 to the endoplasmic reticulum of HeLa cells impairs the secretion of a lysosomal enzyme. J Cell Biol 142:377–389

Yamashita K, Hara-Kuge S, Ohkura T (1999) Intracellular lectins associated with N-linked glycoprotein traffic. Biochim Biophys Acta 1473:147–160

Yerushalmi N, Keppler-Hafkemeyer A, Vasmatzis G et al (2001) ERGL, a novel gene related to ERGIC-53 that is highly expressed in normal and neoplastic prostate and several other tissues. Gene 265:55–60

Yamaguchi D, Kawasaki N, Matsuo I et al. (2007) VIPL has sugar-binding activity specific for high-mannose-type *N*-glycans, and glucosylation of the α1,2 mannotriosyl branch blocks its binding. Glycobiology 17:1061–1069

Zhang B (2009) Recent developments in the understanding of the combined deficiency of FV and FVIII. Br J Haematol 145:15–23

Zhang B, Cunningham MA, Nichols WC et al (2003) Bleeding due to disruption of a cargo-specific ER-to-Golgi transport complex. Nat Genet 34:220–225

Zhang B, Kaufman RJ, Ginsburg D (2005) LMAN1 and MCFD2 form a cargo receptor complex and interact with coagulation factor VIII in the early secretory pathway. J Biol Chem 280:25881–25886

Zhang B, McGee B, Yamaoka JS, Tuddenham EG, Ginsburg D et al (2006) Combined deficiency of factor V and factor VIII is due to mutations in either LMAN1 or MCFD2. Blood 107:1903–1907

Pentraxins: The L-Type Lectins and the C-Reactive Protein as a Cardiovascular Risk

8

G.S. Gupta

8.1 Pentraxins and Related Proteins

The pentraxins (PTX) are a superfamily of plasma proteins that are involved in innate immunity in invertebrates and vertebrates. They contain L-type lectin folds and require Ca^{2+} ions for ligand binding. Three of the principal members of the pentraxin family are serum proteins: namely, C-reactive protein (CRP), serum amyloid P component protein (SAP), and female protein (FP). PTX3 (or TSG-14) protein is a cytokine-induced protein that is homologous to CRPs and SAPs, but its exact function has not yet been determined. Beckmann et al. (1998) identified the superfamily of protein modules comprising the pentraxin families. Beckmann et al. (1998) predicted a jellyroll fold for all members of this superfamily. Pentraxins are made up of five noncovalently bound identical subunits that are arranged in a annular pentameric disc in shape. Proteins with this type of configuration are known as *pentraxins*. Based on the primary structure of the subunit, the pentraxins are divided into two groups: short pentraxins and long pentraxins. C-reactive protein (CRP), the first innate immunity receptor and serum amyloid P-component (SAP) are the two short pentraxins. Soluble pentraxins act as pattern recognition receptors with a dual role: protection against extracellular microbes and autoimmunity. The prototype protein of the long pentraxin group is pentraxin 3 (PTX3). The "long pentraxins" are an emerging family of genes that have conserved in their carboxy-terminal halves a pentraxin domain homologous to the prototypical acute phase protein pentraxins (CRP and SAP component) and acquired novel amino-terminal domains. Long pentraxins, including the prototype PTX3, are expressed in a variety of tissues and cells and in particular by innate immunity cells and most notably by dendritic cells and macrophages, in response to Toll-like receptor (TLR) engagement and in response to proinflammatory signals. PTX3 interacts with several ligands, including growth factors, extracellular matrix components and selected pathogens, playing a role in complement activation and facilitating pathogen recognition by phagocytes. Some long pentraxins are expressed in the brain and some are involved in neuronal plasticity and degeneration. PTX3 acts as a functional ancestor of antibodies, recognizing microbes, activating complement, and facilitating pathogen recognition by phagocytes, hence playing a non-redundant role in resistance against selected pathogens. In addition, PTX3 is essential in female fertility because it acts as a nodal point for the assembly of the cumulus oophorus hyaluronan-rich extracellular matrix. Thus, the prototypic long pentraxin PTX3 is a multifunctional soluble pattern recognition receptor at the crossroads between innate immunity, inflammation, matrix deposition, and female fertility. Unlike the classical pentraxins, the PTX3 is expressed after exposure to the inflammatory cytokines to IL-1β and TNF-α, but not to IL-6, in various cell types. PTX3 has been shown to be produced in response to microbial infections, and highly elevated levels have been reported in patients with sepsis (al-Ramadi et al. 2004; Bottazzi et al. 2006; Garlanda et al. 2005; Mantovani et al. 2006).

8.2 Short Pentraxins

C-reactive protein (CRP) and serum amyloid P component (SAP) are short pentraxins which have been conserved through evolution. In humans both have pentameric structures and both play complex roles in the immune response, CRP being the classical acute-phase (AP) reactant produced in response to tissue damage and inflammation. Serum amyloid A (SAA) and the pentraxins CRP and SAP are major acute-phase proteins: their serum levels can rise by 1,000-fold, indicating that they play a critical role in defense and/or the restoration of homeostasis. The name "pentraxin" relates to the radial symmetry of five monomers forming a ring approximately 95 Å across and 35 Å deep (Fig 8.1–8.3). SAP exhibits multispecific calcium-dependent binding to oligosaccharides with terminal N-acetyl-galactosamine, mannose and glucuronic acid.

G.S. Gupta et al., *Animal Lectins: Form, Function and Clinical Applications*,
DOI 10.1007/978-3-7091-1065-2_8, © Springer-Verlag Wien 2012

Pentraxins have been conserved in evolution and share sequence homology, similar subunit assembly and the capacity for calcium-dependent ligand binding. The sequence homology and gene organization indicate that CRP and SAP arose from a gene duplication of an ancestral pentraxin gene. SAP is an integral component of all amyloid deposits. Inflammation induces dramatic changes in the biosynthetic profile of the liver, leading to increased serum concentrations of positive acute-phase proteins and decreased concentrations of negative acute-phase proteins. Hepatocytes are the single major site of pentraxin clearance and catabolism in vivo. This is consistent with the observation that SAP that has localized to amyloid deposits persists there and is not degraded (Hutchinson et al. 1994).

8.3 C-Reactive Protein

8.3.1 General

C-reactive protein was originally discovered in 1930 as a substance in the serum of patients with acute inflammation that reacted with the C polysaccharide of *pneumococcus* (Tillett and Francis 1930). It was named as C-reactive protein for its ability to precipitate the 'C' polysaccharide extracted from the pneumococcal cell wall (Black et al. 2004). Initially it was thought that CRP might be a pathogenic secretion, as it was elevated in people with a variety of illnesses, including cancer (Pepys and Hirschfield 2003). However, with the discovery of hepatic synthesis and secretion of CRP closed that debate. Nuclear localization of CRP with a high nuclear to cytoplasmic ratio is consistent with its active nuclear transport. Nuclear localization of SAP, a related protein, with a homologous nuclear localization signal was also identified. Because CRP has been known to inhibit RNA transcription and enhance chromatin degradation it was proposed that CRP may play a unique role in injured cells to alter processing of damaged nuclei (Du Clos 1996).

CRP is a glycoprotein and may have sugars—sialic acid, glucose, galactose and mannose—attached to it. In different disease states, one or two amino-acids get lopped off CRP. It retains its activity, but these losses open it up to glycosylation. Different diseases, which raise CRP add sugars to it in different patterns. The patterns of CRP are different across diseases, but similar among patients that had the same disease. Patients with lupus, leukemia, tuberculosis, leishmaniasis, Cushing's syndrome, and bone cancer have been reported in 2003. Although CRP increased the rate at which a particular parasite could invade blood cells, Tanusree et al. (2003) showed that different CRPs had very different potencies in this regard. It was speculated that subtyping CRP may give us more insight into heart attack mechanisms. The study offered circumstantial evidence that proves that the differing glycation is part of CRPs mode-of-action. CRPs, purified from several samples in different pathological conditions demonstrated that human CRP is glycosylated differently in some pathological conditions (Das et al. 2003). In rat serum, CRP is complexed with lipoprotein and interacts with apolipoprotein E. In contrast, human CRP showed no evidence of an interaction with rat serum or with the affinity-purified proteins. This selectivity coincided with the ability of these pentraxins to bind to O-phosphorylethanolamine (O-PE) with high affinity. The sedimentation properties of serum lipoproteins suggested an interaction with rat CRP (Schwalbe et al. 1995).

8.3.2 CRP Protein

CRP is a 224-residue protein with a monomer molar mass of 25,106 Da. Native CRP is somewhat different, as it has 10 subunits making two pentameric discs, with an overall molecular mass of 251,060 Da. Nucleotide sequencing of the coding regions of both cDNA and genomic DNA of CRP revealed an additional 19 amino acid peptide. The CRP gene contains a single 278 base pair intron within the codon specifying the third residue of mature CRP. The intron contains a repetitive sequence (GT)15 G(GT)3 which is similar to structures capable of adopting the Z-DNA form. A comparison of CRP coding and amino acid sequences with those of serum amyloid P component revealed striking overall homology which was not uniform: a region of limited conservation is bounded by two highly conserved regions (Woo et al. 1985).

The *CRP* gene is located on chromosome 1 (1q21-q23) and contains only one intron, which separates the region encoding the signal peptide from that encoding the mature protein. Induction of CRP in hepatocytes is principally regulated at the transcriptional level by cytokine IL-6, an effect which can be enhanced by IL-1β (Kushner et al. 1995). Both IL-6 and IL-1β control expression of many acute phase protein genes through activation of the transcription factors STAT3, C/EBP proteins, and Rel proteins (NF-kB). The unique regulation of each acute phase gene is due to cytokine-induced specific interactions of these and other transcription factors on their promoters. Thus, while C/EBP family members C/EBPβ and C/EBPδ are essential for induction of CRP, the NF-κB is essential for serum amyloid A genes. Interactions among these factors that result in enhanced stable DNA binding of C/EBP proteins result in maximum induction of the gene (Agrawal et al. 2003; Black et al. 2004).

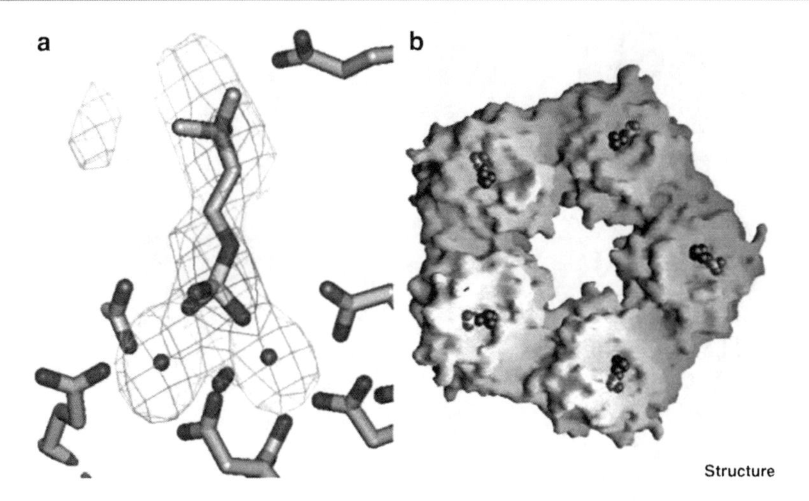

Fig. 8.1 Crystal structure of C-reactive protein complexed with phosphocholine. Ribbon diagram of the x-ray crystal structure of CRP-phosphocholine complex (PDB ID: 1B09). (**a**) Difference Fourier map at 2.5 Å resolution, contoured at 2σ, shows the positions of the two calcium ions (*spheres*) and a molecule of phosphocholine. (**b**) GRASP representation of CRP illustrating the positions of the five bound molecules of phosphocholine (*gray* and *black*) (Adapted by permission from Thompson et al. 1999 © Elsevier)

8.3.3 Structure of CRP

The CRP exists as a pentamer in the presence or absence of Ca^{2+}. Pentamers of CRP were shown to form mixed decamers in Ca^{2+}-free buffer; however, in the presence of Ca^{2+}, this interaction was not observed. Furthermore, no exchange of monomeric subunits was observed between the SAP and CRP oligomers, suggesting a remarkable stability of the individual pentameric complexes (Aquilina and Robinson 2003). The CRP structure contains a remarkable crystal contact, where the calcium binding loop including Glu^{147} from one protomer (Type II) coordinates into the calcium site of a protomer (Type I) in a symmetry related pentamer, revealing the mode of binding of the principal ligand PC and providing information concerning conformational changes associated with calcium binding. The Glu^{147}-Phe^{146} dipeptide from this loosely associated 140–150 loop mimics phosphate-choline (PC) binding in the accepting Type I protomer. The movement of the loop also results in the loss of calcium in the donating Type II protomer where large concerted movements of the structure, involving 43–48, 67–72 and 85–91 are seen. A striking structural cleft on the pentameric face opposite to the PC binding site, suggests an important functional role, perhaps in complement activation. There are significant conformational differences from SAP, both at the tertiary and molecular levels. In the region of the two-residue insertion with respect to SAP, human CRP shows large concerted movements of three loops due to both the insertion and sequence differences. These movements are reflected in the assembly of five protomers to form the pentamer, with a 15° rotation of each protomer, about the conformational axis, with respect to SAP. The calcium coordination differs from that in SAP due to the substitution $Asp^{147}Glu$ (Shrive et al. 1996).

CRP consists of five identical, noncovalently associated ~23-kDa protomers arranged symmetrically around a central pore. Each protomer has been found to be folded into two antiparallel β-sheets with a flattened jellyroll topology similar to lectins such as concanavalin A (Shrive et al. 1996; Thompson et al. 1999). Each protomer has a recognition face with a phosphocholine binding site consisting of two coordinated calcium ions adjacent to a hydrophobic pocket. The co-crystal structure of CRP with phosphocholine (Fig. 8.1) suggests that Phe^{66} and Glu^{81} are the two key residues mediating the binding of phosphocholine to CRP (Thompson et al. 1999). Phe^{66} provides hydrophobic interactions with the methyl groups of phosphocholine whereas Glu^{81} is found on opposite end of the pocket where it interacts with positively charged choline nitrogen. The importance of both residues has been confirmed by mutagenesis studies (Agrawal et al. 2002; Black et al. 2003, 2004).

The opposite face of the pentamer is the effector face, where complement C1q binds and Fcγ receptors are presumed to bind. A cleft extends from the center of the protomer to the central pore of the pentamer, and several residues along the boundaries of this cleft have been shown to be critical for the binding of CRP to C1q, including Asp^{112} and Tyr^{175} (Agrawal and Volanakis 1994; Agrawal et al. 2001). Gaboriaud et al. (2003) proposed a model for C1q binding to CRP. The model suggests that top of the predominantly positively charged C1q head interacts with the

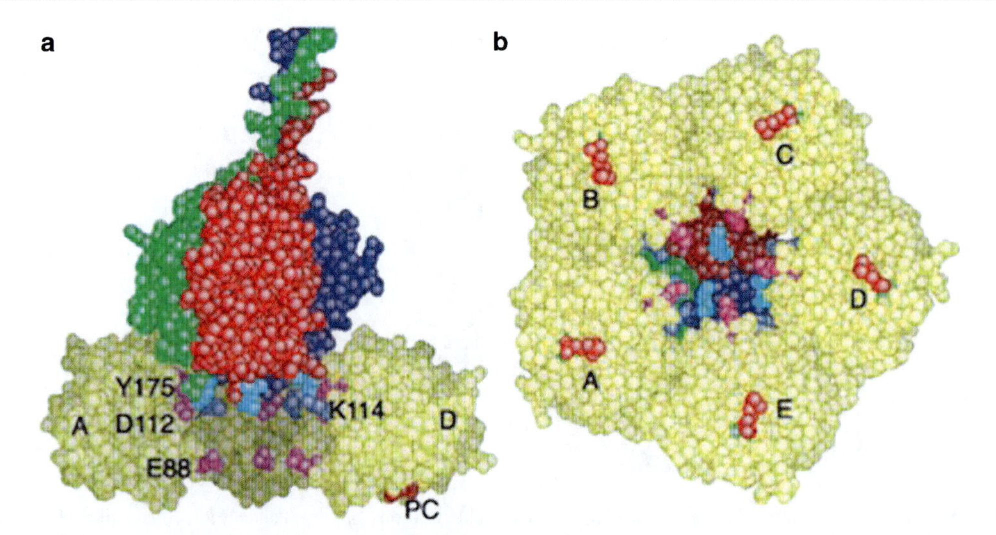

Fig. 8.2 Model of the interaction of CRP with C1q from Gaboriaud et al. (2003). (**a**) Side view. Subunits B and C of CRP have been omitted for clarity. (**b**) Perpendicular bottom view. Modules A, B, and C of the C1q subunit are shown in *blue*, *green*, and *red*, respectively. The lysines at the *top* of the C1q head (Ala^{173}, Ala^{200}, Ala^{201}, Cys^{170}) and Tyr^{B175} are in *light blue*. A–E designate the CRP protomers as described by Shrive et al. (1996). The phosphocholine (PC) ligand is in *red*, and the nearby Ca^{2+} ion is in *green*. Color coding for CRP mutations is as follows. Mutations impairing complement activation (Glu^{88}, Asp^{112}, Tyr^{175}) are *magenta*, and mutations enhancing complement activation (Lys^{114}) are *blue* (Adapted by permission from Gaboriaud et al. (2003) © The American Society for Biochemistry and Molecular Biology)

predominantly negatively charged central pore of the CRP pentamer; the globular head of C1q spans the central pore of CRP and interacts with two of the five protomers of the pentamer (Fig. 8.2). The steric requirements for CRP interaction with C1q imply that optimal C1q binding is accompanied by conformational changes in CRP structure (Gaboriaud et al. 2003). These conformational changes appear to differ depending on the ligand to which CRP is bound (c/r Black et al. 2004).

8.3.4 Functions of CRP

8.3.4.1 Innate Immunity

CRP is a member of the class of acute-phase reactants as its levels rise dramatically during inflammatory processes occurring in the body. This increment is due to a rise in the plasma concentration of IL-6, which is produced predominantly by macrophages (Pepys and Hirschfield 2003) as well as adipocytes (Lau et al. 2005). Its rapid synthesis after infection suggests it contributes to host defense. Unlike the cytokines, IL-6, IL-1β, and TNF-α that elicit CRP production from the liver (Mortensen 2001), CRP has a plasma half-life of 19 h and is quite stable in vitro (Aziz et al. 2003). Synthesis of the CRP is stimulated in response to many pathogens including gram-positive (Mold et al. 1981) and gram-negative pathogens, fungi, and malarial parasites (Volanakis 2001; Szalai 2002). It is thought to bind to phosphocholine, thus initiating recognition and phagocytosis of damaged cells (Pepys and Hirschfield 2003).

Volanakis and Kaplan (1971) identified the specific ligand for CRP in the pneumococcal C polysaccharide as phosphocholine, part of the techoic acid of the pneumococcal cell wall. CRP binding to phosphocholine is thought to assist in complement binding to foreign and damaged cells and to enhance phagocytosis by macrophages, which express a receptor for CRP (Tharia et al. 2002). C-reactive protein not only opsonizes a bacterium, but can also activate the complement cascade by binding to C1q, the first component of the classical pathway of complement activation (Gaboriaud et al. 2003). By binding to specific ligands of the pathogen's cell wall, CRP provides a means of defense against the invading pathogen (Mold et al. 1999). Unlike the activation of complement by immunoglobulin, complement activation initiated by CRP is limited to C1–C4 by the complement-control protein, factor H (Du Clos 2002; Giannakis et al. 2001, 2003). Therefore, CRP promotes phagocytosis of particles without generating a strong inflammatory response (Du Clos 2002). As a result, the strong inflammatory responses typically associated with C5a and the C5–C9 are limited.

In addition to interacting with various ligands and activating the classical complement pathway, CRP can stimulate phagocytosis, and bind to Ig receptors (FcγR). There is evidence that CRP can interact with the Ig receptors FcγRI and FcγRII as well, eliciting a response from phagocytic cells. The ability to recognize pathogens with subsequent recruitment and activation of complement, as well as effects on phagocytic cells, constitute important components of the first line of host defense. Although phosphocholine was the first defined ligand for CRP, a number of other ligands have since been identified. Phosphocholine is found in a number of bacterial species and is a constituent of sphingomyelin and phosphatidylcholine in eukaryotic membranes. However, the head groups of these phospholipids are inaccessible to CRP in normal cells, so that CRP can bind to these molecules only in damaged and apoptotic cells (Gershov et al. 2000; Chang et al. 2002).

8.3.4.2 Clearance of Apoptotic Cells

Besides phosphocholine, CRP can bind to a variety of other ligands, including phosphoethanolamine, chromatin, histones, fibronectin, small nuclear ribonucleoproteins, laminin, and polycations (Black et al. 2003; Szalai et al. 1999). Due to interaction of CRP with phosphorylcholine in calcium-dependent fashion human CRP is classified as a phosphocholine (PC)-binding protein, whereas SAP is identified as a polysaccharide-binding protein (Christner and Mortensen 1994a). Mouse SAP showed binding interactions and specificity similar to human SAP. Female protein (FP) from hamster and rat CRP showed a hybrid specificity and bound to both phosphoryl- ethanolamine and phosphorylcholine. All of the proteins that bound phosphorylethanolamine also associated with human C4b-binding protein (C4BP) suggesting more functions for pentraxins (Schwalbe et al. 1992). Gershov et al. (2000) assessed binding of apoptotic lymphocytes with CRP and the effect of binding on innate immunity. As expected, CRP bound to apoptotic cells and augmented the classical pathway of complement activation but protected the cells from assembly of the terminal complement components. Furthermore, CRP enhanced opsonization and phagocytosis of apoptotic cells by macrophages associated with the expression of the antiinflammatory cytokine TGF-β. The antiinflammatory effects of CRP required C1q and factor H and were not effective once cells had become necrotic. Thus, CRP and the classical complement components act in concert to promote noninflammatory clearance of apoptotic cells and may help to explain how deficiencies of the classical pathway and certain pentraxins lead to impaired handling of apoptotic cells and increased necrosis with the likelihood of immune response to self (Gershov et al. 2000).

8.3.4.3 Pro- and Anti-inflammatory Activity

CRP has pleiotropic effects and shows both "pro-inflammatory" and "anti-inflammatory" activities. CRP has been shown to induce the expression of IL-1 receptor antagonist (Tilg et al. 1993) and increase release of the anti-inflammatory cytokine IL-10 (Mold et al. 2002a, b; Szalai et al. 2002b) while repressing synthesis of interferon-γ (Szalai et al. 2002b). Expression of IL-10 is induced by CRP's binding to Fcγ receptors on macrophages (Ogden and Elkon 2005). However, many other functions that can be regarded as pro-inflammatory are recognized. For example, CRP activates complement and enhances phagocytosis. CRP up-regulates the expression of adhesion molecules in endothelial cells, inhibits endothelial nitric-oxide synthase expression in aortic endothelial cells (Venugopal et al. 2002), stimulates IL-8 release from several cell types, increases plasminogen activator inhibitor-1 expression and activity, and increases the release of IL-1, IL-6, IL-18, and tumor necrosis factor-α (Ballou and Lozanski 1992). The CRP also exhibits a distinct anti-inflammatory activity indicated by its protective effects against endotoxic shock, allergic encephalitis, inflammatory alveolitis, nephrotoxic nephritis, and systemic lupus erythematosus (SLE) (Black et al. 2004; Rodriguez et al. 2005; Szalai et al. 2000a). This activity is believed to be mediated, at least in part, by the immunosuppressive cytokine IL-10. In addition, CRP appears to play a very important role in preventing autoimmunity (Du Clos and Mold 2004; Russell et al. 2004) by targeting apoptotic and necrotic cells removal (Du Clos 1996; Black et al. 2004); this suggests for at least this autoimmune disease, a failure of the normal CRP clearance mechanisms.

Increased levels of CRP are associated with endothelial dysfunction. The glycocalyx decorates the luminal surface and affords critical protection of the endothelium. C-reactive protein dose-dependently increased HA release in vitro and in vivo. There was a significant positive correlation between HA release and monocyte–endothelial cell adhesion, plasminogen activator inhibitor-1, and ICAM-1 release and a negative correlation with endothelial nitric oxide synthase activity (Devaraj et al. 2009b). Although some of these in vitro properties are consistent with the net in vivo effects of CRP observed in mice and described below, it is likely that the function of CRP is context-dependent and that it can either enhance or dampen inflammatory responses depending on the circumstance.

CRP Receptor: Functional effects of CRP on phagocytic cells have been recognized for many years. Recently the receptors for CRP have been identified as the already known receptors for IgG, FcγRI and FcγRII. Two general classes of FcγRs include stimulatory receptors and inhibitory receptors. The stimulatory receptors are characterized by a cytoplasmic immunoreceptor tyrosine-based activation motif (ITAM) sequence. The inhibitory receptor is characterized by the presence of an immunoreceptor tyrosine-based inhibition motif (ITIM) sequence. Biological responses triggered by ITAM-containing FcγRs include phagocytosis, respiratory bursts, and secretion of cytokines. ITIM-containing FcγRs, when found co-aggregated with ITAM-containing FcγRs, negatively regulate ITAM-mediated activity. In both humans and mice, CRP binds to ITAM- and ITIM-containing receptors, which include FcγRI and FcγRII. Phagocytosis of CRP-opsonized particles and apoptotic cells has been shown to proceed through FcγRI in the mouse (Mold et al. 2001, 2002). CRP has also been shown to induce signaling through human FcγRIIa, an ITAM-containing receptor, in granulocytes (Chi et al. 2002). The enhancement of phagocytosis by CRP is likely due to its interactions with FcγRs. However, some investigators have voiced doubts about CRP binding to FcγRs (Hundt et al. 2001).

In Vivo Effects: In contrast to humans, plasma levels of mouse CRP rarely exceed 2 μg/ml following inflammatory stimuli. The murine CRP response represents an evolutionary oddity, a natural knockdown that has been exploited in a variety of studies utilizing exogenous or transgenic CRP to study the effects of CRP in vivo. The ability of CRP to protect mice against bacterial infection by various species has been well established. These species include *S. pneumoniae* (Szalai et al. 1995) and *Haemophilus influenza* (Weiser et al. 1998; Lysenko et al. 2000), which have phosphocholine-rich surfaces, and *Salmonella enterica* serovar Typhimurium, which has no known surface phosphocholine, although its cell membrane is known to be rich in phosphoethanolamine (Szalai et al. 2000). Protection appears to be mediated through CRP binding to phosphocholine or phosphoethanolamine, followed by activation of classical complement pathway. CRP protection of mice infected with *S. pneumoniae* has been shown to require an intact complement system (Mold et al. 2002b) but does not require interaction with FcγRs (Mold et al. 2002a; Szalai et al. 2002b). CRP has been also shown to play a protective role in a variety of inflammatory conditions, including protecting mice from lethal challenge with bacterial LPS and various mediators of inflammation (Xia and Samols 1997). In addition, CRP has been found to delay the onset and development of experimental allergic encephalomyelitis, an aseptic animal model of multiple sclerosis (Szalai et al. 2002a). In a murine model of chemotactic factor-induced alveolitis, CRP has also been shown to inhibit the influx of neutrophils and protein into the lungs (Heuertz et al. 1994; Ahmed et al. 1996). Taken together, these experiments suggest that the net effect of CRP in mice is anti-inflammatory.

It is of interest that CRP may exert an ameliorative effect upon murine models of SLE in which CRP levels are often unexpectedly low (ter Borg et al. 1990). Two reports (Szalai et al. 2003; Du Clos 1996) and mouse models implicate polymorphism in human CRP gene resulting in a lower basal level of CRP with an increased risk of developing SLE (Russell et al. 2004). These findings suggest the possibility that decreased amounts of CRP may contribute to the pathogenesis of SLE. Since CRP is important in clearance of the cellular debris of necrotic and apoptotic cells by binding to damaged cell membranes and nuclear material, decreased clearance of such material might enhance development of autoantibodies to them.

8.4 CRP: A Marker for Cardiovascular Risk

8.4.1 CRP: A Marker for Inflammation and Infection

C-reactive protein is well known to rheumatologists. Levels of CRP in the blood serve as a reliable marker of disease activity in rheumatoid arthritis and various vasculitides. In systemic lupus erythematosus (SLE), the expression of CRP does not appear to correlate with disease activity, and levels generally remain low despite active disease. CRP is also known to infectious disease specialists; it serves as a marker of infection and participates in host defense (Du Clos 2003). Cardiologists have become intensely interested in CRP, and it has been widely examined as a risk factor for cardiovascular disease. These properties suggest that CRP is a sensitive marker for acute and chronic inflammation of diverse causes and various infection (Du Clos 2003). However, since many things can cause elevated CRP, this is not a very specific prognostic indicator (Lloyd-Jones et al. 2006). A high-sensitivity CRP test measures low levels of CRP and a level above 2.4 mg/l has been associated with a doubled risk of a coronary event compared to levels below 1 mg/l (Pepys and Hirschfield 2003). C-reactive protein is not normally found in the blood of healthy people. It appears after an injury, infection, or inflammation and disappears when the injury heals or the infection or inflammation disappears. CRP rises up to 50,000-fold in acute inflammation, such as infection. It rises above normal limits within 6 h, and peaks at 48 h. Serum amyloid A is a related acute-phase marker that responds rapidly in similar circumstances. Viral infections tend to give a lower CRP level than bacterial infection. CRP is used mainly as a marker of inflammation. Apart from liver failure, there are few known factors that interfere with CRP production (Pepys and Hirschfield 2003). Consumption of red meat is associated with increased colon cancer risk. This association might be due to the heme content of red meat. In rat, PCR confirmed the strong heme-induced down-regulation of mucosal pentraxin gene (*Mptx*) (Van Der Meer-Van Kraaij et al. 2003). The role of inflammation in cancer is now well known. Blood samples of persons with colon cancer have an average CRP concentration of 2.69 mg/l. Persons without colon cancer average 1.97 mg/l (Erlinger et al. 2004). These findings concur with previous studies that indicate that anti-inflammatory drugs could lower colon cancer risk (Baron et al. 2003).

8.4.2 CRP: A Marker for Cardiovascular Risk

Patients with prolonged elevated levels of CRP are at an increased risk for heart disease, stroke, hypertension (high blood pressure), diabetes, and metabolic syndrome (insulin resistance, a precursor of type 2 diabetes). The amount of CRP produced by the body varies from person to person, and this difference is affected by lifestyle. Higher CRP levels tend to be found in individuals who smoke, have high blood pressure, are overweight and do not exercise, whereas lean, athletic individuals tend to have lower CRP levels. The research shows that too much inflammation can sometimes have adverse effects on the blood vessels which transport oxygen and nutrients throughout the body. Atherosclerosis, which involves the formation of fatty deposits or plaques in the inner walls of the arteries, is considered in many ways an inflammatory disorder of the blood vessels, similar to arthritis, which is considered an inflammatory disorder of the bones and joints. Inflammation affects the atherosclerotic phase of heart disease and can cause plaques to rupture, which produces a clot and interfere with blood flow, causing a heart attack or stroke. CRP has been localized to monocytes and tissue macrophages, which are present in the necrotic core of lesions prone to plaque rupture. Leukocyte-derived myeloperoxidase (MPO), primarily hosted in human polymorphonuclear cells (PMNs), has also been shown to be present in human atherosclerotic lesions. CRP stimulates MPO release both in vitro and in vivo, providing further cogent data for the proinflammatory effect of CRP. These results might further support the role of CRP in patients of acute coronary syndrome (Singh et al. 2009).

Inflammation is pivotal in all phases of atherosclerosis. Among the numerous inflammatory biomarkers, the largest amount of published data supports a role for CRP as a robust and independent risk marker in the prediction of primary and secondary adverse cardiovascular events. In addition to being a risk marker, there is much evidence indicating that CRP may indeed participate in atherogenesis. This correlation applies even to apparently healthy men and women who have normal cholesterol levels. Recent research suggests that patients with elevated basal levels of CRP are at an increased risk of diabetes, hypertension and cardiovascular disease (Dehghan et al. 2007; Pradhan et al. 2001). CRP level can be used by physicians as part of the assessment of a patient's risk for heart disease since it can be easily measured in blood. Studies provided evidence that the CRP level in human blood is an important and highly accurate predictor of future heart disease. CRP is an indicator of inflammation in the walls of arteries. Studies show that reducing the inflammation by lowering CRP levels with a class of drugs known as statins significantly lowers the rate of heart attacks and coronary-artery disease in people with acute heart disease. In fact, studies indicated that CRP levels may be as important—if not more important—in predicting and preventing heart disease as cholesterol levels are. Devaraj et al. (2009b) suggested that CRP is clearly a risk marker for cardiovascular disease and is recommended for use in primary prevention. In addition, CRP appears also to contribute to atherogenesis. A study of over 700 nurses showed that those in the highest quartile of trans fat consumption had blood levels of CRP that were 73% higher than those in the lowest quartile (Lopez-Garcia et al. 2005). Although CRP may only be a moderate risk factor for cardiovascular disease (Danesh et al. 2004), Reykjavik Study found to have some problems for this type of analysis related to the characteristics of the population, and an extremely long follow-up time, which may have attenuated the association between CRP and future outcomes (Verma et al. 2003). Others have shown that CRP can exacerbate ischemic necrosis in a complement-dependent fashion and that CRP inhibition can be a safe and effective therapy for myocardial and cerebral infarcts; so far, this has been demonstrated only in animal models (Pepys et al. 2006).

The JUPITER trial was conducted to determine if patients with elevated high-sensitivity CRP levels but without hyperlipidemia might benefit from statin therapy. Statins were selected because they have been proven to reduce levels of CRP. The trial found that patients taking rosuvastatin with elevated high-sensitivity CRP levels experienced a decrease in the incidence of major cardiovascular events (Armani and Becker 2005; Ridker et al. 2008). The trial specifically found, "the number of patients who would need to be treated with rosuvastatin for 2 years to prevent the occurrence of one primary end point is 95, and the number needed to treat for 4 years is 8." In other words, after 4 years of treatment, out of every 31 patients, one cardiovascular event would be prevented.

To clarify whether CRP is a bystander or active participant in atherogenesis, a study compared people with various genetic CRP variants. Those with a high CRP due to genetic variation had no increased risk of cardiovascular disease compared to those with a normal or low CRP (Zacho et al. 2008). However, the clinical utility of CRP measurement in cardiovascular risk prediction is still not well defined (Yeh 2005). Furthermore, there is an intense debate on whether CRP is merely a marker of inflammation or a direct participant. The finding by Dr Janos Filep provides additional insights into the current CRP debate (Khreiss et al. 2005).

Evidence in support of the possibility that CRP itself plays a role in the pathogenesis of atherosclerosis has been elucidated (Devaraj et al. 2009; Jialal et al. 2004). Examples include the finding that CRP binds the phosphocholine of oxidized LDL (Chang et al. 2002), up-regulates the expression of adhesion molecules in endothelial cells, increases LDL uptake into macrophages (Zwaka et al. 2001), inhibits endothelial nitric-oxide synthase expression in aortic endothelial cells (Venugopal et al. 2002), and increases plasminogen activator inhibitor-1 expression and activity. A recent study utilizing a mouse strain expressing transgenic CRP and deficient in apolipoprotein E reported a modest acceleration in aortic atherosclerosis in male animals expressing high levels of CRP (Paul et al. 2004). Another report demonstrated increased arterial occlusion in transgenic mice expressing CRP in a model of vascular injury (Danenberg et al. 2003). With the recognition that inflammation plays a role in cardiovascular disease (CVD) and precedes myocardial infarction, numerous reports have emerged with plausible explanations for an association between and CVD (Pepys and Hirschfield 2003) and for the characterization of high sensitivity serum C-reactive protein (hs-CRP) as an independent predictor of future cardiovascular events (Verma 2004). Despite these suggestive findings, a role for CRP in the pathogenesis of atherosclerosis is far from established.

CRP is strongly associated with obesity, and weight loss has been shown to decrease CRP in nine of ten studies in which it has been evaluated (Dietrich and Jialal 2005). Race and gender also strongly influence serum hs-CRP concentration (Khera et al. 2005). In Dallas County, characterized as a typical multiethnic U.S. urban population, the median hs-CRP level is 30% higher in blacks than in whites, and almost twice as high in women as men (Khera et al. 2005). A new study demonstrated that cardiorespiratory fitness level, hormone replacement therapy use, and high-density lipoprotein cholesterol accounted for the gender difference in hs-CRP (Huffman et al. 2006). It is also increasingly evident that genetic factors, including apoE genotype (Marz et al. 2004) and polymorphisms in the hs-CRP gene (Russell et al. 2004; Szalai et al. 2002a; Suk et al. 2005), regulate basal hs-CRP concentrations. The substantial variability in hs-CRP concentrations in people of different ethnic origins, led Anand et al. to conclude that uniform hs-CRP cut-points were not appropriate for defining vascular risk across diverse populations (Anand et al. 2004).

8.4.2.1 Lowering of CRP by Drugs

Persons with moderate or high levels of CRP can often reduce the levels with lifestyle changes, including quitting smoking, engaging in regular exercise, taking in healthy nutrition, taking a multivitamin daily, replacing saturated fats such as butter with monounsaturated fats (particularly olive oil), increasing intake of Omega-3 fatty acids, losing weight if overweight, and increasing fiber intake. Statins, usually used to reduce high levels of low density lipoproteins, can also reduce CRP levels. These drugs include: *lovastatin* (Mevacor), simvastatin (Zocor), rosuvastatin (Crestor), pravastatin (Pravacol) and *atorvastatin* (Lipitor). Other drugs that lower CRP levels include the anti-cholesterol drug *ezetimibe* (Zetia) and the diabetes medication *rosiglitazone* (Avandia). Not all physicians are convinced the two studies published in 2005 are accurate, noting that both studies were funded by pharmaceutical companies (Pfizer and Bristol-Meyer Squibb) that make statin drugs used to reduce CRP levels. Dietary nitrite prevents hypercholesterolemic microvascular inflammation and reverses endothelial dysfunction (Stokes et al. 2009).

8.4.3 Role of Modified/Monomeric CRP

Denatured and aggregated forms of CRP (neo-CRP or modified CRP) are pro-inflammatory in a number of experimental systems, although the existence of this material in vivo has not been unequivocally established (Shields 1993). Khreiss et al. (2005) found that at local sites of deposition, small amounts of modified CRP may be generated with a set of properties distinct from those of the native protein. It has recently been reported that modified CRP increased the release of the inflammatory mediators monocyte chemoattractant protein-1 and IL-8 and up-regulated the expression of ICAM-1 in endothelial cells. In this model, modified CRP was shown to be a much more potent inducer than native CRP.

The capacity of human CRP to activate/regulate complement may be an important characteristic that links CRP and inflammation with atherosclerosis. Emerging advances suggest that in addition to classical pentameric CRP, a conformationally distinct isoform of CRP, termed modified or monomeric CRP (mCRP), may also play an active role in atherosclerosis. Monomeric CRP activates endothelial cells via interaction with lipid raft microdomains (Ji et al. 2009). The capacity of mCRP to interact with and activate the complement cascade was studied by Ji et al. (2006). mCRP binds avidly to C1q, and this binding occurred primarily through collagen-like region of C1q. Fluid phase mCRP inhibited the activation of complement cascade via engaging C1q from binding with other complement activators. In contrast, when immobilized or bound to oxidized or enzymatically modified LDL, mCRP could activate classical complement pathway. Low-level generation of sC5b-9 indicated that the activation largely bypassed the terminal sequence of complement, which appears to involve recruitment of Factor H. Therefore, mCRP can both inhibit and activate the classical complement pathway by binding C1q, depending on whether it is in fluid phase or surface-bound state (Ji et al. 2006, 2007, 2009). The CRP isoforms differ in their effects on thrombus growth (Molins et al. 2008).

Eisenhardt et al. (2009) suggested dissociation of pentameric to monomeric CRP on activated platelets. Since platlet CRP (pCRP) is found neither in healthy nor in diseased vessels, Eisenhardt et al. (2009) suggested that dissociation of pCRP to mCRP involves activated platelets, which play a central role in cardiovascular events. Activated platelets mediate the dissociation of pCRP to mCRP via lysophosphatidylcholine, which is present on activated but not resting platelets. The dissociation of pCRP to mCRP can also be mediated by apoptotic monocytic THP-1 and Jurkat T cells. This mechanism provides a potential link between circulating pCRP and localized platelet-mediated inflammatory and proatherogenic effects (Filep 2009).

Emerging evidence indicates that calcium-dependent binding of pCRP to membranes, including liposomes and cell membranes, results into a rapid but partial structural change, producing molecules that express CRP subunit antigenicity but with retained native pentameric conformation. This hybrid molecule is herein termed $mCRP_m$. The formation of $mCRP_m$ was associated with significantly enhanced complement fixation. The $mCRP_m$ can further detach from membrane to form the well-recognized mCRP isoform converted in solution ($mCRP_s$) and exert potent stimulatory effects on endothelial cells. The membrane-induced pCRP dissociation not only provides a physiologically relevant scenario for mCRP formation but may represent an important mechanism for regulating CRP function (Ji et al. 2009).

8.5 Extra-Hepatic Sources of CRP

The CRP is traditionally thought to be produced by the liver in response to inflammatory cytokines. However, extrahepatic synthesis of CRP has also been reported in neurons, atherosclerotic plaques, monocytes, and lymphocytes (Jialal et al. 2004). Extra-hepatic sources of CRP production point to a more systemic generation of CRP in our body (Yeh 2005). The mechanisms regulating synthesis at these sites are unknown, and it is unlikely that they substantially influence plasma levels of CRP. Studies have shown that both epithelial cells of the respiratory tract and renal epithelium can produce CRP under certain circumstances (Jabs et al. 2003). Moreover, neuronal cells also synthesise acute phase reactants involved in the pathogenesis of neurodegenerative diseases (Yasojima et al. 2000). These new sources provided only tenuous link to atherosclerosis. CRP has been shown to colocalize with the terminal complement complex in atherosclerotic plaques (Torzewski et al. 2000; Yasojima et al. 2001). Human coronary artery smooth muscle cells, but not human umbilical vein endothelial cells, can synthesize CRP after stimulation by inflammatory cytokines. This locally produced CRP may directly participate in the pathogenesis of atherosclerosis. Moreover, human adipocytes can produce CRP after stimulation by inflammatory cytokines and by a specific adipokine, resistin (Calabro et al. 2005).

8.6 Serum Amyloid P Component

In humans serum Amyloid P component (SAP) is a constitutive serum protein that is synthesized by hepatocytes. The SAP is the identical serum form of amyloid P component, a 25 kDa pentameric protein first identified as the pentagonal constituent of in vivo pathological deposits called amyloid (Cathcart et al. 1967). Amyloid P component makes up 14% of the dry mass of amyloid deposits and is thought to be an important contributor to the pathogenesis of a related group of diseases called the Amyloidoses (Botto et al. 1997). These conditions are characterised by the ordered aggregation of normal globular proteins and peptides into insoluble fibres which disrupt tissue architecture and are associated with cell death. Amyloid P component is thought to decorate and stabilise aggregates by preventing proteolytic cleavage and hence inhibiting fibril removal via the normal protein scavenging mechanisms (Tennent et al. 1995).

The SAP, is a component of all amyloid plaques, is also a normal component of a number of basement membranes including the glomerular basement membrane. The association and distribution of SAP within the glomerular basement membrane are altered or completely disrupted in a number of nephritides (e.g. Alport's Syndrome, type II membranoproliferative glomerulonephritis, and membranous glomerulo- nephritis). SAP binds to human laminin and merosin as well as mouse and rat laminins. The K_D of this interaction is 2.74×10^{-7} M at SAP/laminin molar ratio of 1:7.1. The binding of SAP to laminin is inhibited by both SAP and its analog, CRP, as well as phosphatidylethanolamine. Binding of SAP to extracellular matrix components such as type IV collagen, proteoglycans, and fibronectin in concert with above observation suggests that SAP determines the properties of those basement membranes with which it is associated (Zahedi 1997).

8.6.1 Genes Encoding SAP

8.6.1.1 The Human Gene

Both SAP and CRP are evolutionary conserved in all vertebrates and also found in distant invertebrates such as the horseshoe crab (Limulus polyphemus) (Pepys et al. 1997). The serum amyloid P component gene (*APCS*) is human gene for SAP (Floyd-Smith et al. 1986; Ohnishi et al. 1986). It is encoded by a single copy gene on chromosome 1. C-DNA clones corresponding to the human SAP mRNA have been analyzed. The nt sequences of the cDNA and the corresponding regions of the genomic SAP DNA which were identical, revealed that after coding for a signal peptide of 19 amino acids and the first two amino acids of the mature SAP protein, there is one small intron of 115-bp, followed by a nt sequence coding for the remaining 202 amino acid residues. The SAP gene has an ATATAAA sequence 29-bp upstream from the cap site, but there is no CAAT box-like sequence. A possible polyadenylation signal sequence, ATTAAA, was found to be located 28-bp upstream from the polyadenylation site. A comparison of the genomic SAP DNA sequence with that of human CRP revealed a striking overall homology which was not uniform: several highly conserved regions were bounded by non-homologous regions. This comparison supports for the hypothesis that SAP and CRP are products of a gene duplication event. The genes encoding CRP and SAP are located on the proximal long arm of human chromosome 1, more precisely between bands q12 and q23 by to human metaphase chromosomes (Floyd-Smith et al. 1986; Ohnishi et al. 1986).

8.6.1.2 Mouse SAP Gene

The mouse SAP genomic clone contains the entire SAP gene and specifies a primary transcript of 1,065 nt residues. This comprises a first exon of 206 nt residues containing the mRNA 5′-untranslated region and sequence encoding the pre-SAP leader peptide and the first two amino acid residues of mature SAP separated by a single 110-base intron from a 749-nt-residue second exon containing sequence encoding the bulk of the mature SAP and specifying the mRNA 3′-untranslated region. The overall organization is similar to that of the human SAP gene, and the coding region and intron sequences are highly conserved. The 5′-region of the mouse SAP gene contains modified CAAT and TATA promoter elements preceded by a putative hepatocyte-nuclear-factor-1-recognition site; these structures are in a region that is highly homologous to the corresponding region of the human SAP gene (Whitehead and Rits 1989).

8.6.1.3 Guinea Pig SAP and CRP

CRP and SAP have been isolated from bovine and guinea pig sera (Rubio et al. 1993). In guinea pig, neither SAP nor the CRP is the major acute phase reactant. Both genes have organizations typical of the pentraxin genes in other species. However, some differences were observed in the regions that potentially determine the capacity of the pentraxin gene to be induced during acute inflammation. Nucleotide substitutions in coding regions have occurred at similar rates in the two pentraxin genes. Nonsynonymous substitution rates indicated that SAP and CRP are subject to similar, relatively low levels of constraint. An estimate of the phylogenetic relationship among the pentraxin genes suggested that SAP and CRP arose as the result of a gene duplication event that occurred very early in mammalian evolution, but subsequent to the divergence of the reptilian ancestors of the mammalian and avian lineages. This raised doubts about the identity of proteins from fish, which have been characterized as CRP and SAP (Rubio et al. 1993).

8.6.2 Characterization of SAP

Fourier transform infrared spectroscopy provided estimations of about 50% β-sheet, 12% α-helix, 24% β-turn, and 14% unordered structure for CRP and about 54% β-sheet, 12% α-helix, 25% β-turn, and 9% unordered structure for SAP. With both proteins significant calcium-dependent changes were observed in conformation-sensitive amide I regions assigned to each type of structure. The CRP spectrum was also affected by Mg^{2+}, but the changes differed from those induced by Ca^{2+}. The SAP spectrum was not affected by Mg. Phosphorylcholine (PC) in the presence of Ca^{2+} also affected the spectrum of CRP but not the spectrum of SAP. This study provided a comparison of the secondary structures of human CRP and SAP and hamster female protein (Dong et al. 1992). It indicated that three pentraxins have similar secondary structure compositions and calcium-dependent conformational changes, but differ significantly in their responses to phosphorylcholine and Mg^{2+} (Dong et al. 1994).

8.6.2.1 Structure

SAP is characterised by calcium dependent ligand binding and distinctive flattened β-jellyroll structure similar to that of the legume lectins (Emsley et al. 1994). Human SAP has 51% sequence homology with CRP and is a more distant relative to the "long" pentraxins such as PTX3 (a cytokine modulated molecule) and several neuronal pentraxins. The oligomeric state of human SAP in the absence and presence of known ligands has been investigated using ionization MS. At pH 8.0, in the absence of Ca^{2+}, SAP was shown to consist of pentameric and decameric forms. In the presence of physiological levels of Ca^{2+}, SAP exists primarily as a pentamer, reflecting its in vivo state. dAMP was shown not only to promote decamerization, but also to lead to decamer stacking involving up to 30 monomers. A mechanism for this finding was proposed by Aquilina and Robinson (2003). Furthermore, no exchange of monomeric subunits was observed between the SAP and CRP oligomers, suggesting a remarkable stability of the individual pentameric complexes (Aquilina and Robinson 2003). The three-dimensional structure of pentameric human SAP component at high resolution reveals that the tertiary fold is remarkably similar to that of the legume lectins (Fig. 8.3). Carboxylate and phosphate compounds bind directly to two Ca^{2+}; interactions with a carboxyethylidene ring are mediated by Asn^{59} and Gin^{148} ligands of the Ca^{2+}. X-ray results indicate the probable modes of binding of the biologically important ligands,

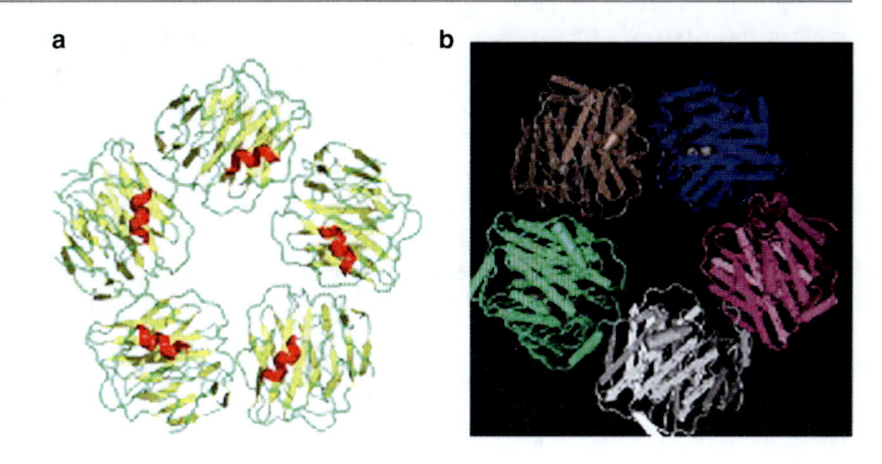

Fig. 8.3 (**a**) Cartoon model of SAP showing helices in *red*, sheets in *yellow* and coils in *green* (**b**) Crystal structure of a decameric complex of human serum Amyloid P component with bound dAMP (PDB ID: 1LGN; Hohenester et al. 1997)

DNA and amyloid fibrils (Emsley et al. 1994). Hohenester et al. (1997) revealed the crystal structure of the SAP-dAMP complex as a decamer in which all interactions between SAP pentamers were mediated by the ligand. The two calcium ions of SAP were bridged by the dAMP phosphate group and five hydrogen bonds are formed between the protein and the ligand, including specific interactions made by the adenine base. The SAP-dAMP decamer is stabilized mainly by base-stacking of adjacent ligand molecules and possibly by electrostatic interactions involving the dAMP phosphate groups; decamerization buries 1,000 A2 (2.6%) of the pentamer solvent-accessible surface.

8.6.3 Interactions of SAP and CRP

8.6.3.1 CRP and SAP Interact with Nuclear Antigens

In the past several years it has been demonstrated that both of these pentraxins interact with nuclear antigens including chromatin and small nuclear ribonucleoproteins (snRNPs). Both CRP and SAP have nuclear transport signals which facilitate their entry into the nuclei of intact cells. Furthermore, these pentraxins have been shown to affect the clearance of nuclear antigens in vivo. It is now believed that one of the major functions of the pentraxins could be to interact with the nuclear antigens released from apoptotic or necrotic cells. This interaction could mitigate against deposition of these antigens in tissue and autoimmune reactivity (Du Clos 1996).

C-reactive protein (CRP) binds to several nuclear Ag, including chromatin, histones, and small nuclear ribonucleoproteins. Binding to sites of tissue inflammation and the nuclei of inflammatory cells has been demonstrated in vivo. A significant similarity has been noticed between CRP and nucleoplasmin, a molecule with nuclear localization activity. Binding of CRP and SAP to chromatin may be involved in the solubilization and clearance of nuclear material. CRP binding to chromatin is mediated by histones. These interactions demonstrated under relatively physiological conditions, with native pentraxins unseparated from serum and with nuclear constituents in situ, are likely to be of functional importance in vivo (Pepys et al. 1994; Shephard et al. 1991).

The CRP and SAP specifically bind to each other only when the CRP is in an immobilized form bound to one of its ligands or to an antibody. CRP did not bind to immobilized SAP. The binding of SAP to immobilized CRP was Ca^{2+}-dependent with sufficient affinity in presence of serum or purified serum proteins. SAP bound preferentially to a synthetic peptide corresponding to the Ca^{2+}-binding region of CRP (Swanson et al. 1992; Christner and Mortensen 1994b).

In man CRP and SAP activate complement through the classical pathway and participate in opsonization of particulate antigens and bacteria. The CRP is an activator of the classical complement pathway. Cleavage of chromatin is C-dependent in the presence of CRP and serum. Oligomers of SAP have been found to bind C1q and consume total C and C4, indicating that SAP can activate complement as well. SAP differs from CRP in being able to bind to DNA. SAP binds to histones H1 and H2A as well as to chromatin. In contrast to CRP, SAP binding to chromatin did not require H1. SAP partially inhibited CRP binding to chromatin and to H1. However, neither pentraxin inhibited binding of the other to H2A. Binding of either CRP or SAP to H2A activated complement in SAP-depleted serum leading to the deposition of C4 and C3. C activation required C1q and produced C4d indicating that it occurred through the classical pathway. These findings demonstrated that both pentraxins can activate the classical C pathway after ligand binding (Hicks et al. 1992).

8.6.3.2 Regulation of Pentraxin Function by Lactic Acid

Carboxylated compounds, especially lactic acid, were capable of dissociating pentraxins from several macromolecular binding sites. Lactate dissociated the hSAP-membrane complex and prevented hSAP self-association. The only interaction that was not dissociated by 10 mM lactate was the hSAP-heparin complex. This suggested that the carboxyl group plus a hydrogen-bonding site on the hydrocarbon chain was important, but a charged amino group was not a contributor to function when the anion was provided by a carboxyl group. Other pentraxins also interacted with lactic acid, but with lower affinities. Importantly, lactic acid was capable of dissociating rat CRP from lipoproteins in rat serum. Human CRP bound very weakly to lactate, so that lactate probably is not a significant regulator of this pentraxin (Evans and Nelsestuen 1995).

8.6.4 Functions of SAP

8.6.4.1 Control of Chromatin Degradation

SAP shows specific calcium-dependent binding to DNA and chromatin in physiological conditions. The avid binding of SAP displaces H1-type histones and thereby solubilizes native long chromatin, which is otherwise profoundly insoluble at the physiological ionic strength of extracellular fluids. Furthermore, SAP binds in vivo both to apoptotic cells, the surface blebs of which bear chromatin fragments, and to nuclear debris released by necrosis. SAP may therefore participate in handling of chromatin exposed by cell death. Mice with targeted deletion of the SAP gene spontaneously develop antinuclear autoimmunity and severe glomerulonephritis, a phenotype resembling human SLE, a serious autoimmune disease. The $SAP^{-/-}$ mice also have enhanced anti-DNA responses to immunization with extrinsic chromatin; and the degradation of long chromatin is retarded in the presence of SAP both in vitro and in vivo. Therefore, SAP has an important physiological role, inhibiting the formation of pathogenic autoantibodies against chromatin and DNA, probably by binding to chromatin and regulating its degradation (Bickerstaff et al. 1997). However, the SAP-deficient mice displayed no obvious phenotypic abnormalities. Though, Soma et al. (2001) reaffirmed that the SAP-deficient mice had high titers of anti-nuclear antibody but did not develop severe glomerulonephritis. On the other hand, as reported that SAP bound to gram-negative bacteria via LPS prevented LPS-mediated activation of a classical complement pathway. Thus, contrary to documented data, SAP-deficient mice do not develop serious autoimmune disease. It was suggested that SAP has a critical role in LPS toxicity (Soma et al. 2001).

8.6.4.2 Amyloid Deposition is Delayed in Mice with Targeted Deletion of SAP Gene

The tissue amyloid deposits that characterize systemic amyloidosis, Alzheimer's disease and the transmissible spongiform encephalopathies always contain SAP bound to the amyloid fibrils. It was proposed that this normal plasma protein may contribute to amyloidogenesis by stabilizing the deposits. Systemic amyloidosis in a transgenic mouse model for an autosomal dominant disease, familial amyloidotic polyneuropathy (FAP) suggested that SAP is not important for the initiation and progression of amyloid deposition (Tashiro et al. 1991). On the contrary, the induction of reactive amyloidosis is retarded in mice with targeted deletion of the SAP gene (Botto et al. 1997). A precise role for SAP in amyloid deposition in vivo is not known. Maeda (2003) indicated that lack of SAP in AA amyloid deposits does not enhance regression of the deposits in vivo and that the dissociation of bound SAP from AA amyloid deposits would not accelerate regression of the deposits in vivo.

8.6.4.3 SAP Forms a Stable Complex with Human C5b6

In serum, SAP binds tightly to C5b6, which is formed by activating C7-depleted human serum with zymosan. The C5b6-SAP complex did not dissociate in the presence of EDTA, which distinguishes SAP-C5b6 binding from SAP's usual Ca^{2+}-dependent binding to other molecules. Purified SAP was able to bind to preformed C5b6, which was isolated from purified components. Functionally, the C5b6-SAP could bind C7, and the resulting C5b67-SAP complex had only moderately lower specific hemolytic activity than that of C5b67. In addition, hemolytically inactive C5b67-SAP, like hemolytically inactive C5b67, was chemotactically active for neutrophils, while isolated SAP had no effect on cell mobility. It is likely that the addition of SAP to terminal complement complexes may affect the fate of these complexes (Barbashov et al. 1997). C4b-binding protein (C4bp) is the main regulatory protein of complement system and regulates C3 convertase activity in classical way of complement activation. The major regulatory function of C4bp is related to its interaction with activated form of C4b. C4bp may also interact with SAP that inhibits complement-regulatory functions of C4bp.

8.6.4.4 Uptake of Apoptotic Cells Through FcγRI and/or FcγRIII

SAP and CRP are opsonins that react with nuclear autoantigens targeted in systemic autoimmunity. CRP and SAP bind to apoptotic and necrotic cells, which are potential sources of these autoantigens. Although SAP binds to DNA and chromatin and affects clearance of these autoantigens, FcγRI and FcγRIII are receptors for SAP in the mouse.

CRP as an opsonin binds to FcγR. The use of FcγR by the pentraxins links innate and adaptive immunity and may have important consequences for processing, presentation, and clearance of the self-Ags to which these proteins bind (Mold et al. 2001). In human neutrophils (PMN) and the Jurkat T-cell line SAP treatment of apoptotic human PMN increased ingestion by autologous macrophages. Mold et al. (2002) suggested that pentraxins promote uptake of apoptotic cells through FcγRI and/or FcγRIII. Ingestion through these receptors is expected to alter the pattern of cytokine production and antigen presentation in response to apoptotic cells.

8.6.4.5 Binding of Pentraxins and IgM to Newly Exposed Epitopes on Late Apoptotic Cells

A random distribution of phospholipids among the inner and outer leaflet of the cell membrane occurs during apoptosis and is known as membrane flip-flop. Flip-flopped cells have binding sites for various plasma proteins, such as IgM and the pentraxins CRP and SAP. Except for SAP which also bound to early apoptotic cells, pentraxins and IgM preferentially bound to late apoptotic cells. It revealed that CRP, SAP, and part of the IgM molecule bind to phospholipid head groups exposed on apoptotic cells. This shared specificity as well as their shared capability to activate complement suggest that IgM and the pentraxins CRP and SAP exert similar functions in the removal of apoptotic cells (Ciurana and Hack 2006).

8.6.4.6 Pentraxins as Receptors for Microbial Pathogens

Attempt to identify LPS-binding proteins from the hemolymph of the horseshoe crab led to the isolation and identification of CRP as the predominant LPS-recognition protein during *Pseudomonas* infection. Investigation of CRP response to *Pseudomonas aeruginosa* unveiled a robust innate immune system in the horseshoe crab, which displays rapid suppression at a dosage of 10^6 CFU of bacteria in the first hour of infection and effected complete clearance of the pathogen by 3 days. Results provide the importance of CRP as a conserved molecule for pathogen recognition (Ng et al. 2004). However, pentraxins do not participate significantly in normal human serum promastigote C3 opsonization. It was indicated that successful *Leishmania* infection in man must immediately follow promastigote transmission (Dominguez et al. 2002).

SAP can bind to influenza A virus and inhibit agglutination of erythrocytes mediated by the virus subtypes H1N1, H2N2 and H3N2. SAP also inhibits the production of haemagglutinin (HA) and the cytopathogenic effect of influenza A virus in MDCK cells. Of several monosaccharides tested only D-mannose interfered with SAP's inhibition of both HA and infectivity. The glycosaminoglycans heparan sulfate and heparin, which bind SAP, reduced SAPs binding to the virus. The results indicate that the inhibition by SAP is due to steric effects when SAP binds to terminal mannose on oligosaccharides of the HA trimer (Andersen et al. 1997).

8.6.5 SAP in Human Diseases

8.6.5.1 Amyloid P Component from Patient of Amyloidosis

Amyloid P component (AP) makes up 14% of the dry mass of amyloid deposits and is thought to be an important contributor to the pathogenesis of a related group of diseases called the Amyloidoses (Botto et al. 1997). Amyloid P component extracted from spleen of a patient with primary idiopathic amyloidosis is a glycoprotein composed of a pair of noncovalently bound pentameric discs with a subunit size of 23–25 kDa. The precursor of AP is the SAP, which has an identical amino acid sequence to that of amyloidosis patient. It shared 52% homology with the amended sequence of human CAP, and 68% homology with the Syrian hamster female protein, whose response is modulated by sex steroids (Prelli et al. 1985).

The mRNAs and proteins of both CRP and AP are concentrated in pyramidal neurons and are upregulated in affected areas of AD brain. Results suggested a more direct role for serum AP and CRP in the pathogenesis of AD. Controlling pentraxin production at the tissue level may be important in reducing inflammatory damage in AD (Duong et al. 1998; Yasojima et al. 2000). Formerly thought to be made primarily if not solely in liver, recent work has shown that AP and CRP are made not only in the brain but in other tissues such as heart and arteries. Their synthesis is markedly upregulated in affected brain regions in AD. Since they are known to activate the complement cascade in an antibody-independent fashion and chronic activation can cause destruction of host tissue, these pentraxins may be important initiators of an autodestructive process. As such, they may be prime targets for therapeutic intervention (McGeer et al. 2001; McGeer and McGeer 2004).

AP is thought to decorate and stabilise aggregates by preventing proteolytic cleavage and hence inhibiting fibril removal via the normal protein scavenging mechanisms (Tennent et al. 1995). This association is utilised in the routine clinical diagnostic technique of SAP scintigraphy whereby radio-labelled protein is injected into patients to locate areas of amyloid deposition (Hawkins and Pepys 1995). The SAP-amyloid association has also been identified as a possible drug target for anti-amyloid therapy, with the development first stage clinical trials of a compound called CPHPC (R-1-[6-[R-2-carboxy-pyrrolidin-1-yl]-6-oxohexanoyl] pyrrolidine-2-carboxylic acid), a small molecule able to strip AP from deposits by reducing levels of circulating SAP (Pepys et al. 2002).

8.6.5.2 SAP in Development of Atherosclerosis

SAP is present in human atherosclerotic lesions. SAP from the intima was indistinguishable from plasma or purified SAP with respect to immunological character and molecular weight. Studies suggest a role for SAP in atherogenesis and encourage efforts to determine more precisely the physiological contributions of the pentraxin family to the development of atherosclerosis (Li et al. 1995). It seems that SAP and CRP may represent different facets of inflammation. The association of SAP with CVD in older adults further supports the role of innate immunity in atherosclerosis (Jenny et al. 2007).

8.6.6 SAP from *Limulus Polyphemus*

The structure of SAP from horseshoe crab *Limulus polyphemus* was determined by molecular replacement. Contrary to popular opinion, both CRP and SAP are present in *Limulus* haemolymph. The two independent protomers in the asymmetric unit are related by a ncs two fold bisecting the tetragonal two fold axes, producing a physiological molecule of 16 protomers. In contrast to the known mammalian pentraxin structures, all of which are pentameric, *Limulus* SAP consists of 16 protomers, each with the pentraxin fold, arranged in a novel doubly stacked cyclic octameric structure. The calcium site and the pentraxin helix, which are situated on opposite faces of the homo-oligomeric pentamer in the mammalian pentraxins, are situated on the external and internal edges of the octameric ring in *Limulus* SAP. Interprotomer interactions throughout the molecular assembly are non-covalent and pairs of protomers, one from each octameric ring, form continuous, twisted beta-sheet structures across the hexadecamer. The protomers of the doubly-stacked octameric *Limulus* SAP retain the cyclic oligomeric nature of the pentraxins but aggregate in a distinctly different manner when compared to the known mammalian pentraxins, displaying a 75° rotational shift with respect to the cyclic axis. Alignment of the *Limulus* and human SAP structures with that of the three known polymorphic *Limulus* CRP sequences suggests that the majority of the 10% sequence difference between them is localised in the vicinity of the pentraxin ligand-binding site, giving rise to differing, but overlapping ligand binding specificities (Shrive et al. 1999).

8.7 Female Protein (FP) in Syrian Hamster

8.7.1 Similarity of Female Protein to CRP and APC

Female protein (FP) is a pentraxin of Syrian hamster which is a homologue of CRP and amyloid P component (AP). The pentraxin in the Syrian hamster is unique because it is preferentially expressed in the female at high constitutive levels and accordingly called female protein (FP) or FP (SAP) due to its close homology with human SAP. The high levels of FP in female serum (100-fold greater than male serum) suggested its role in hamster pregnancy, one of the shortest of any eutherian mammal. Serum FP concentration in pregnant Syrian hamsters shows marked decrease (>80%) at term with the nadir at parturition with subsequent increase. A similar down-regulation of FP was found in the normal female Syrian hamster after injury (acute phase response), so in both cases the assumed beneficial effects were achieved with less, rather than more pentraxin, a paradoxical pentraxin response. The fall in serum FP concentration could represent a response to protect the fetus from the high and potentially toxic level of FP normally found in the female that is harmful because of its association with amyloidosis. An FP that is 97.5% identical to Syrian hamster FP is found in the Turkish hamster (Mesocricetus brandti), although serum levels in females are much lower, and amyloid is very rare (Coe et al. 1999).

The serum of Armenian hamster (*Cricetulus migratorius*) contains a protein homologous to Syrian (golden) hamster FP. Whereas serum concentration of FP in Syrian hamsters (SFP) is many fold greater (200–300-fold) in females vs. males, Armenian hamster FP (AFP) is only moderately elevated (approximately three fold) in female vs. males and only for the fall-winter months of the year. In the Armenian hamster testosterone administration to females or castration of males has no effect on AFP serum levels, whereas in Syrian hamster these treatments change SFP serum concentration to that characteristic of the opposite sex. Some sex steroid control of hepatic AFP synthesis is evident, however, as serum levels decrease after exogenous estrogen treatment. In contrast to Syrian hamster FP, normal levels of AFP are dependent on an intact hypophysial-pituitary axis and also are influenced by the season of the year. As an acute-phase protein, AFP responds in a typical fashion, with increasing serum levels detected in both sexes in contrast to the divergent sex-limited response in Syrian hamsters. Although AFP and SFP are similar structurally, morphologically, and antigenetically and share common binding specificities, the regulation of FP synthesis in Armenian hamster is very different from that found in Syrian hamster (Coe and Ross 1990).

Functionally, FP has been shown to be similar to CRP, although FP has more homology at the amino terminus with AP. As an amyloid component, FP appears to be functionally similar to human AP. However, FP synthesis is under sex steroid control and the unique sex-limited expression of this pentraxin was associated with an equally novel propensity for deposition of amyloid in female hamsters under normal or experimental conditions. Thus, a high serum level of FP, as found in normal females or diethylstilbestrol-treated males, was associated with enhanced amyloidosis (Coe and Ross 1985). Female protein (FP) was found to be similar to CRP and SAP in the following ways: (a) hamster FP complexes with phosphorylcholine (PC) in a Ca^{2+} -dependent; (b) electron microscopy of FP indicated a pentameric structure similar in size and appearance to other pentraxins; (c) the parent molecule of FP (150 kDa mol wt) was composed of five noncovalantly assembled subunits of 30 kDa mol wt; and (d) the amino acid analysis and terminal NH2 sequence of FP clearly showed homology with SAP-CRP. Although FP evolved from an ancestral gene common to SAP and CRP, and shares functional, morphological and structural properties with these acute-phase proteins, the biological homology of FP appears quite diverse as this protein is a prominent serum constituent (1 2 mg/ml) of normal female hamsters and under hormonal control (testosterone suppression) in males (Coe et al. 1981). FP is unusual in that it is apparently the only pentraxin produced in hamsters, it is under hormonal control and it shares binding characteristics with both CRP and SAP. The response to inflammation is divergent; FP levels decrease in females and increase in males during an acute phase response (Dong et al. 1992).

Like CRP and SAP, FP shares the ability to bind to chromatin and histones. Similar to CRP, FP bound to histones H1 and H2A, and chromatin. FP shared with SAP the ability to bind to DNA. However, FP binding was inhibited by PC for all ligands, whereas SAP binding was not. FP and SAP also failed to compete with each other for binding to DNA. By cross-inhibition FP bound much less well to PC than CRP, but was a very effective inhibitor of CRP binding to H2A. These studies demonstrate that chromatin and histone binding are conserved among pentraxins (Saunero-Nava et al. 1992). In rats FP was closely related to hamster FP with respect to hormonal regulation and APP nature as well.

8.7.2 Structure of Female Protein

Syrian hamster female protein (SFP), a serum oligomer composed of five identical subunits, was reassociated in vitro from monomer subunits. The reconstituted pentamer was genuine by morphologic, antigenic, and structural criteria. Another female protein (FP), a homologue from Armenian hamsters (AFP), also reassociated into a pentamer after dissociation with 5 M guanidine hydrochloride. These two FP's hybridized when a mixture of them was dissociated and then reassociated. The in vitro dissociation-reassociation of female protein described herein may reflect an in vivo dissociation-reassociation which is functionally important and a common metabolic feature within this family of proteins (Coe and Ross 1987).

Hamster FP is composed of glycosylated subunits of 25,655 MW containing a single intrachain disulphide bridge. In the presence of EDTA the subunits are non-covalently associated as pentamers of mass approximately 128,000 MW, and in the presence of calcium they aggregate further, probably to form decamers. This pentamer-decamer transition at physiological ionic strength has not been described in other pentraxins. It also resembles human CRP in binding only weakly to agarose, to human AA amyloid fibrils in vitro, and to mouse AA amyloid deposits in vivo. It thus differs markedly from human and mouse SAP but it is nevertheless deposited in hamster AA amyloid in vivo and clearly is the hamster counterpart of SAP in other species. These results illustrate the subtle diversity among members of the otherwise conserved pentraxin family of vertebrate plasma proteins (Tennent et al. 1993).

FT infrared spectroscopy indicated that FP is composed of 50% β-sheet, 11% α-helix, 29% β-turn, and 10% random structures. Two putative calcium-binding sites were proposed for FP (residues 93–109 and 150–168) as well as other members of the pentraxin family on the basis of the theoretical secondary structure predictions and the similarity in sequence between the pentraxins and EF-hand calcium-binding proteins. The changes in protein conformation detected upon binding of calcium and PC provide a mechanism for the effects of these ligands on physiologically important properties of the protein, e.g., activation of complement and association with amyloids (Dong et al. 1992).

8.7.3 Gene Structure and Expression of FP

Hamster FP is a unique pentraxin because pretranslational expression of this gene is modulated by mediators of inflammation and sex steroids. The FP gene encodes a 211 amino acid residue mature polypeptide as well as a 22-residue signal peptide. The intron/exon organization is similar to that of other pentraxins, but additional transcripts are generated from alternate polyadenylation sites in the 3′ region. Circulating levels of FP and the corresponding hepatic transcript levels are augmented by estrogen, while testosterone, dexamethasone, and progesterone cause a decrease in these levels. In addition the cytokines interleukin-1, -6, and tumor necrosis factor mediate a decrease in hepatic FP transcript levels in female hamsters but did not cause a significant elevation of FP mRNA in livers of male hamsters. The differences in expression of the FP gene between male and female hamsters and between unstimulated male hamsters and male hamsters stimulated with an injection of lipopolysaccharide are due, at least in part, to alterations in transcription (Rudnick and Dowton 1993a, b).

8.8 Long Pentraxins

8.8.1 Long Pentraxins 1, -2, -3

The earliest described pentraxins, CRP and SAP, are cytokine-inducible acute phase proteins implicated in innate immunity whose concentrations in the blood increase dramatically upon infection or trauma. The highly conserved family of pentraxins was thought to consist solely of approximately 25 kDa proteins. Later, several distinct larger proteins were identified in which only the C-terminal halves showed characteristic features of the pentraxin family. One of these described "long" pentraxins (TSG-14/PTX3) is inducible by TNF-α or IL-1 and is produced during the acute phase response. Other newly identified long pentraxins are constitutively expressed proteins associated with sperm-egg fusion (apexin/p50), may function at the neuronal synapse (neuronal pentraxin I, NPI), or may serve yet other, unknown functions (NPII and XL-PXN1). Evidence obtained by molecular modeling and by direct physicochemical analysis suggests that TSG-14 protein retains some characteristic structural features of the pentraxins, including the formation of pentameric complexes (Goodman et al. 1996).

8.9 Neuronal Pentraxins (Pentraxin-1 and -2)

Murine or rat neuronal pentraxin 1 (NP1) and human neuronal pentraxin 2 (NPTX2) are expressed in CNS. NP1, NP2, and neuronal pentraxin receptor (NPR) are members of a new family of proteins identified through interaction with a presynaptic snake venom toxin taipoxin. Neuronal pentraxins represent a novel neuronal uptake pathway that may function during synapse formation and remodeling. NP1 and NP2 are secreted, exist as higher order multimers (probably pentamers), and interact with taipoxin and taipoxin-associated calcium-binding protein 49 (TCBP49). NPR is expressed on the cell membrane and does not bind taipoxin or TCBP49 by itself, but it can form heteropentamers with NP1 and NP2 that can be released from cell membranes. Heteromultimerization of pentraxins and release of a pentraxin complex by proteolysis directly effect the localization and function of neuronal pentraxins in neuronal uptake or synapse formation and remodeling (Kirkpatrick et al. 2000).

8.9.1 Functions of Pentraxin 1 and -2

8.9.1.1 Expression of Narp in Orexin Neurons

Evidences suggest that orexin (also known as hypocretin) neurons play a central role in the pathophysiology of narcolepsy, though targeted deletion of orexin does not fully mimic the functional deficits induced by selective ablation of these neurons. Reti et al. (2002) demonstrated that orexin neurons displayed robust expression of neuronal activity-regulated pentraxin (Narp), a secreted neuronal pentraxin, implicated in regulating clustering of α-amino-3-hydroxy-5-methylisoxazole-4-propionate (AMPA) receptors. Furthermore, it was found that hypothalamic melanin-concentrating hormone (MCH) neurons, which form a peptidergic pathway thought to oppose the effects of the orexin system, express another neuronal pentraxin, NP1. Thus, these pathways utilize neuronal pentraxins, in addition to neuropeptides, as synaptic signaling molecules (Reti et al. 2002).

Glycogen Synthase Kinase 3 Mediates NP 1 Expression and Cell Death: Expression of NP1 is part of the apoptotic cell death program activated in mature cerebellar granule neurons when potassium concentrations drop below depolarizing levels. NP1 is involved in both synaptogenesis and synaptic remodeling. Both activation of the phosphatidylinositol 3-kinase/Akt (PI-3-K/AKT) pathway by insulin-like growth factor I and blockage of the stress activated JNK offer transitory neuroprotection from the cell death evoked by nondepolarizing concentrations of potassium. Impairing the activity of glycogen synthase kinase 3 (GSK3) completely blocks NP1 over-expression induced by potassium depletion and provides transient protection against cell death. Results showed that both the JNK and GSK3 signaling pathways are the main routes by which potassium deprivation activates apoptotic cell death, and that NP1 overexpression is regulated by GSK3 activity independently of the PI-3-K/AKT or JNK pathway (Enguita et al. 2005).

Requirement of NP1 in Hypoxic Neuronal Injury: Neonatal hypoxic-ischemic brain injury is a major cause of neurological disability and mortality. Neuronal pentraxin 1 (NP1), a member of subfamily of "long pentraxins," plays a role in the HI injury cascade. After hypoxia-ischemia (HI) there was an elevated neuronal expression of NP1 in the ipsilateral cerebral cortex and in the hippocampal CA1 and CA3 regions. Results suggest that NP1 induction mediates hypoxic-ischemic injury probably by interacting with and modulating AMPA glutamate receptor subunit (GluR1) and potentially other excitatory glutamate receptors (Hossain et al. 2004).

Neuronal Pentraxins Mediate Synaptic Refinement in Developing Visual System
Neuronal pentraxins (NPs) have been hypothesized to be involved in activity-dependent synaptic plasticity. NP1/2 knock-out mice exhibited defects in the segregation of eye-specific retinal ganglion cell (RGC) projections to the dorsal lateral geniculate nucleus, a process that involves activity-dependent synapse formation and elimination. Retinas from mice lacking NP1 and NP2 had cholinergically driven waves of activity that occurred at a frequency similar to that of wild-type mice, but several other parameters of retinal activity were altered. Studies indicate that NPs are necessary for early synaptic refinements in the mammalian retina and dorsal lateral geniculate nucleus (Bjartmar et al. 2006).

8.9.1.2 Neuronal Pentraxin Receptor (NPR) in Small Cell Lung Cancer

A global gene analysis showed that the neuronal pentraxin receptor (NPR) is highly and relatively specifically expressed in Small cell lung cancer (SCLC), consistent with the neuroendocrine features of this cancer. Normally, NPR is exclusively expressed in neurons, where it associates with the homologous proteins NP1 and NP2 in complexes capable of binding the snake venom neurotoxin taipoxin. The receptor in SCLC is surface associated. Microarray signals for NP1 and NP2mRNA was detected in a subset of SCLC-cell lines Furthermore, NP1 protein was detected in a few SCLC-cell lines. A number of SCLC-cell lines showed marked sensitivity to taipoxin (IC50: 3–130 nM) (Poulsen et al. 2005).

8.10 Pentraxin 3 (PTX3)

8.10.1 Characterization of PTX3

The long pentraxin 3 (PTX3) is a member of the pentraxin superfamily, a family of proteins highly conserved during evolution and characterized by a multimeric, usually pentameric structure (Bottazzi et al. 2006; Garlanda et al. 2005). PTX3 shares similarities with the classical, short pentraxins but differs for the presence of an unrelated long N-terminal domain coupled to the carboxy-terminal pentraxin domain, and differ, with respect to short pentraxins, in their gene organization, chromosomal localization, cellular source, and in their stimuli-inducing and ligand-recognition ability. PTX3 consists of a C terminal

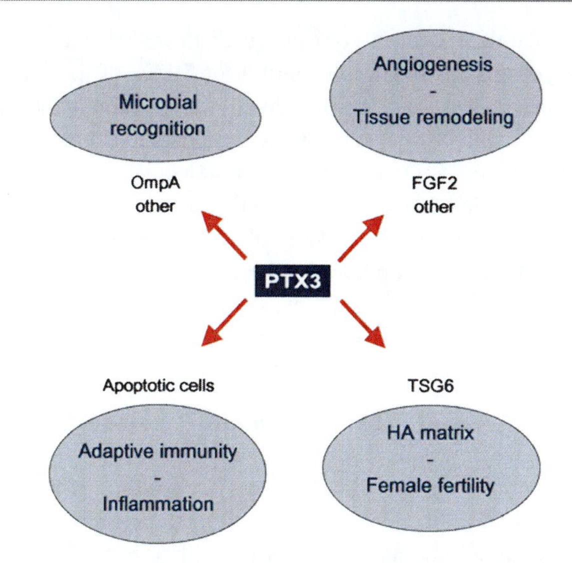

Fig. 8.4 **PTX3: a soluble, multifunctional protein**. Through interaction with its numerous ligands (fibroblast growth factor 2 (FGF2)) (Bottazzi et al. 2006), the ECM protein TNF-α-induced protein 6 (TSG-6) (Salustri et al. 2004), and the outer membrane protein A from *Klebsiella pneumoniae* (KpOmpA), PTX3 plays complex functions in vivo, recognizing microbes, tuning inflammation, and editing self-nonself discrimination. In addition, PTX3 participates in tissue remodeling and plays an essential role in female fertility by acting as a nodal point in the ECM architecture

203-amino acid pentraxin-like domain coupled with an N-terminal 178-amino acid unrelated portion. PTX3 is induced by primary proinflammatory signals in various cell types, most prominently macrophages and endothelial cells. PTX3 is produced by a variety of cells and tissues, most notably dendritic cells and macrophages, upon Toll-like receptor (TLR) engagement and inflammatory cytokines. PTX3 recognizes microbial moieties, opsonizes fungi and selected bacteria, activates the classic pathway of complement cascade, and participates in the formation of ECM (Fig. 8.4) and apoptotic cell clearance. In addition, PTX3 is involved and female fertility. Unlike short pentraxins, PTX3 primary sequence and regulation are highly conserved in man and mouse. Thus, gene targeting identified PTX3 as multifunctional soluble pattern recognition receptors acting as a nonredundant component of the humoral arm of innate immunity and involved in tuning inflammation, matrix deposition, and female fertility (Deban et al. 2010).

8.10.1.1 Multimer Formation by PTX3

The structure and ligand binding properties of human PTX3 were compared with the CRP and SAP component. Sequencing of CHO cell-expressed PTX3 revealed that the mature secreted protein starts at residue 18 (Glu). Lectin binding and treatment with N-glycosidase F showed that PTX3 is N-glycosylated, sugars accounting for 5 kDa of the monomer mass (45 kDa). The protein consists predominantly of β-sheets with a minor α-helical component. While in gel filtration the PTX3 is eluted with a molecular mass of congruent with 900 kDa, gel electrophoresis revealed that PTX3 forms multimers predominantly of 440 kDa apparent molecular mass, corresponding to decamers, and that disulfide bonds are required for multimer formation. PTX3 does not have coordinated Ca^{2+}.

8.10.1.2 LG/LNS Domains: Multiple Functions

The three-dimensional structures of LG/LNS domains from neurexin, the laminin α2 chain and sex hormone-binding globulin reveal a close structural relationship to the carbohydrate-binding pentraxins and other lectins. However, these LG/LNS domains appear to have a preferential ligand-interaction site distinct from the carbohydrate-binding sites found in lectins, and this interaction site accommodates not only sugars but also steroids and proteins. In fact, the LG/LNS domain interaction site has features reminiscent of the antigen-combining sites in immunoglobulins. The LG/LNS domain presents an interesting case in which the fold has remained conserved but the functional sites have evolved; consequently, making predictions of structure-function relationships on the basis of the lectin fold alone is difficult (Rudenko et al. 2001).

8.10.1.3 Human PTX3 Gene

A genomic fragment of 1,371 nt from human PTX3 gene was characterized as a promoter on TNFα and IL-1β exposure in transfected 8,387 human fibroblasts. The minimal promoter contains one NF-kB element which is necessary for induction and able to bind p50 homodimers and p65 heterodimers but not c-Rel. Mutants in this site lose the ability to bind NF-kB proteins and to respond to TNFα and IL-1β in functional assays. Sp1- and AP-1 binding sites lying in proximity to the NF-kB site do not seem to play a major role for cytokine responsiveness. The cotransfection experiments with expression vectors validate that the natural promoter contains a functional NF-kB site (Basile et al. 1997).

The mouse *ptx3* gene is organized into three exons on chromosome 3: the first (43 aa) and second exon (175 aa) code for the signal peptide and for a protein portion with no high similarity to known sequences the third (203 aa) for a domain related to classical pentraxins, which contains the "pentraxin family signature." Analysis of the N terminal portion predicts a predominantly α-helical structure, while the pentraxin domain of ptx3 is accommodated comfortably in the tertiary structure fold of SAP (Introna et al. 1996).

8.10.2 Cellular Sources of PTX3

Cells producing PTX3 include DC, mononuclear phagocytes, fibroblasts, endothelial cells, smooth muscle cells, adipocytes, synovial cells, and chondrocytes (Abderrahim-Ferkoune et al. 2003; Bottazzi et al. 2006; Doni et al. 2003; Jeannin et al. 2005). PTX3 is also produced by cells of epithelial origin, such as renal and alveolar epithelial cells (c/r Bottazzi et al. 2006). DCs are major producers of PTX3 in vitro. While production is restricted to DCs of myelomonocytic lineage, plasmacytoid DC do not produce PTX3 (Doni et al. 2003; Baruah et al. 2006a, b). In brain, PTX3 expresses in glial cells, in the white matter (corpus callosum, fimbria) and meningeal pia mater as well as in dentate gyrus hilus and granule cells (Polentarutti et al. 2000). Scleroderma fibroblasts, unlike normal fibroblasts, constitutively expressed high levels of PTX3 in the absence of deliberate stimulation. The constitutive expression of PTX3 in SSc fibroblasts was not modified by anti-TNF-α antibodies or IL-1 receptor antagonist. In contrast, IFN-γ and TGF-β inhibited the constitutive but not the stimulated expression of PTX3 in SSc fibroblasts (Luchetti et al. 2004). IL-1 and TNF-α strongly stimulate the expression of PTX3 by human proximal renal tubular epithelial cells (PTECs). In addition, activation of PTECs with IL-17 and CD40L, respectively, but not with IL-6 or IL-4, resulted in strongly increased production of PTX3, whereas granulocyte-macrophage colony-stimulating factor (GM-CSF) inhibited IL-1-induced PTX3 production. Results suggest that local expression of PTX3 may play a role in the innate immune response and inflammatory reactions in the kidney (Nauta et al. 2005).

Human monocyte-derived DC produced copious amounts of PTX3 in response to microbial ligands engaging different members of the Toll-like receptor (TLR) family (TLR1 through TLR6), whereas engagement of the mannose receptor had no substantial effect. DCs were better producers of PTX3 than monocytes and macrophages. In contrast, plasmacytoid DC exposed to influenza virus or to CpG oligodeoxynucleotides engaging TLR9, did not produce PTX3. PTX3-expressing DCs were present in inflammatory lymph nodes from HIV-infected patients. Thus, DCs of myelomonocytic origin are a major source of PTX3, which facilitates pathogen recognition and subsequent activation of innate and adaptive immunity (Doni et al. 2003).

8.10.3 Ligands

Unlike the classical pentraxins CRP and SAP, PTX3 did not bind phosphoethanolamine, phosphocholine, or high pyruvate agarose. PTX3 in solution, bound to immobilized C1q, but not C1s. Binding of PTX3 to C1q is specific with a K_D

7.4×10^{-8} M. The Chinese hamster ovary cell-expressed pentraxin domain bound C1q when multimerized. Thus, on the basis of computer modeling, the long pentraxin PTX3 forms multimers, which differ from those formed by classical pentraxins in terms of protomer composition and requirement for disulfide bonds, and does not recognize CRP/SAP ligands (Bottazzi et al. 1997; Nauta et al. 2003). In addition, PTX3 binds with high affinity to a number of different ligands, the fibroblast growth factor 2 (FGF2) (Bottazzi et al. 2006), the ECM protein TNF-α-induced protein 6 (TSG-6) (Salustri et al. 2004), and the outer membrane protein A from *Klebsiella pneumoniae* (KpOmpA) (Jeannin et al. 2005). Moreover, a specific interaction with apoptotic cells has been described by Rovere et al. (2000). Interaction of PTX3 with its ligands is fundamental for some of the physiological functions attributed to PTX3 (Fig. 8.4).

8.10.4 Regulation of PTX3

Regulation: PTX3 is produced and released by a variety of cell types upon exposure to primary inflammatory signals, such as TNF-α and IL-1ß, TLR ligands, and microbial moieties, such as LPS, lipoarabinomannans, and outer membrane proteins. Normal and transformed fibroblasts, undifferentiated and differentiated myoblasts, normal endothelial cells, and mononuclear phagocytes express *mptx3* mRNA and release the protein in vitro on exposure to IL-1β and TNFα. The *mptx3* was induced by bacterial LPS in vivo in a variety of organs and, strongly in the vascular endothelium of skeletal muscle and heart. Thus, *mptx3* shows a distinct pattern of in vivo expression indicative of a significant role in cardiovascular and inflammatory pathology. Human peripheral blood mononuclear cells exposed to LPS or IL-1β expressed significant levels of PTX3 mRNA. TNF-α was a less-effective inducer of PTX3, whereas IL-6, monocyte chemotactic protein-1, MC-SF, GM-CSF, and IFN-γ were inactive. Thus, PTX3, unlike the classical pentraxins CRP and SAP, is expressed and released by cells of the monocyte-macrophage lineage exposed to inflammatory signals (Bottazzi et al. 2006; Garlanda et al. 2005; Jeannin et al. 2005). IFN-γ, which generally has a synergistic effect with LPS (Ehrt et al. 2001), inhibits LPS induction of PTX3 in DC as well as in monocytes and endothelial cells (Doni et al. 2006; Polentarutti et al. 2000). Conversely, IL-10 is a modest and inconsistent inducer of PTX3 in DC and monocytes (Bottazzi et al. 2006) but significantly amplifies the response to LPS, TLR ligands, and IL-1ß (Doni et al. 2006). PTX3 superinduction, in response to IL-10, is likely to play a role in matrix deposition, tissue repair, and remodeling. Moreover, data suggest that besides stimulation of the humoral arm of adaptive immunity, IL-10 also stimulates the humoral arm of innate immunity (Moore et al. 2001). Cell-specific regulation of *ptx3* by glucocorticoid hormones in hematopoietic and nonhematopoietic cells has been reported by Doni et al. (2008).

Mycobacterial lipoarabinomannan (LAM) induced expression of PTX3 mRNA in human peripheral blood mononuclear cells. The non-mannose-capped version of lipoarabinomannan (AraLAM) was considerably more potent than the mannose-capped version ManLAM or the simpler version phosphatidylinositol mannoside. Whole mycobacteria (Mycobacterium bovis BCG) strongly induced PTX3 expression. LAM-induced PTX3 expression was associated with the production of immunoreactive PTX3. IL-10 and IL-13 did not inhibit the induction of PTX3 by LAM. In contrast, IFNγ inhibited LAM-induced PTX3 expression. Thus, in addition to IL-1, TNF, and LPS, mycobacterial cell wall components also induce expression and production of long pentraxin PTX3 (Vouret-Craviari et al. 1997).

Inducible Expression of PTX3 in CNS: PTX3 is expressed in brain and PTX3 mRNA is induced in the mouse brain by LPS. In contrast NP1 is constitutively expressed in the murine CNS and is not modulated by LPS administration. IL-1β was also a potent inducer of PTX3 expression in the CNS, whereas TNF-α was substantially less effective and IL-6 induced a barely detectable signal. Central administration of LPS and IL-1 induced PTX3 also in the periphery (heart), whereas the reverse did not occur. Expression of PTX3 was also observed in the brain of mice infected with *Candida albicans (C. albicans)* or *Cryptococcus neoformans.* In situ hybridization revealed that i.c.v. injection of LPS induced a strong PTX3 expression in presumptive glial cells, in the white matter (corpus callosum, fimbria) and meningeal pia mater as well as in dentate gyrus hilus and granule cells (Polentarutti et al. 2000).

Elevation of Pentraxins in Haemodialysis Patients: Elevation of CRP in patients with renal failure and its association with cardiovascular disease is well described. PTX3 levels are markedly elevated in haemodialysis (HD) patients. The increase in PTX3 production in whole blood after HD indicates that the HD procedure itself contributes to elevated PTX3 levels in HD patients. The association between PTX3 and cardiovascular morbidity suggests a possible connection of PTX3 with atherosclerosis and cardiovascular disease in HD patients (Boehme et al. 2007).

8.10.5 Functions of PTX3

8.10.5.1 Innate Immunity

Studies so far suggest that PTX3 plays complex, nonredundant functions in vivo, ranging from innate resistance against selected pathogens to the assembly of a hyaluronic acid (HA)-rich extracellular matrix (ECM) and female fertility (Bottazzi et al. 2006; Garlanda et al. 2005). Two main features that characterize *ptx3*-deficient mice are that they are sterile and more susceptible to infections to selected pathogens (Garlanda et al. 2002; Salustri et al. 2004; Varani et al. 2002). The *ptx3*-deficient mice are more susceptible to infection with selected fungal and bacterial microorganisms, such as *A. fumigatus, P. aeruginosa*, and *S. typhymurium*, while they are resistant to infection with *Listeria monocytogens* and *S. aureus* (Garlanda et al. 2005). Susceptibility of *ptx3*-null mice was associated with defective recognition of conidia by alveolar macrophages and DCs, as well as inappropriate induction of an adaptive type 2 response (Garlanda et al. 2002). Binding of PTX3 has been demonstrated for conidia of *A. fumigatus* as well as for *P. aeruginosa* and *S. typhymurium*, while PTX3 does not recognize *L. monocytogens*. Moreover, macrophages from *ptx3*-transgenic mice have an improved phagocytic activity toward zymosan and *Paracoccidioides brasiliensis* (Diniz et al. 2004). These findings provide evidences for a role of PTX3 as an opsonin and imply the existence of a receptor for this molecule.

Pathogen recognition is a common feature among the members of the pentraxin family. PTX3 does not bind LPS as well as lipoteichoic acid, N-acetylmuramyl-L-alanyl-D-isoglutamine, exotoxin A, and enterotoxin A and B, but it binds with high affinity the recombinant KpOmpA, which binds and activates macrophages and DC in a TLR2-dependent way and activates genetic induction of PTX3. In turn, PTX3 binds KpOmpA and plays a crucial role in the amplification of the inflammatory response to this microbial protein, as demonstrated in *ptx3*-deficient mice (Jeannin et al. 2000, 2005) (Fig. 8.4).

Pentraxin 3 in Inflammatory and Immune Responses: Clinical significance of different levels of pentaxins, their role in the induction or protection from autoimmunity, and the presence of specific autoantibodies against them in the different autoimmune diseases have been reviewed (Kravitz et al. 2005; Kravitz and Shoenfeld 2006; Kunes 2005) (Fig. 8.4). In humans, PTX3 blood levels are barely detectable in normal conditions and increase rapidly during a range of inflammatory and infectious conditions. PTX3 levels are elevated in critically ill patients, reflecting the severity of disease (Muller et al. 2001). In patients with acute myocardial infarction (AMI), PTX3 is a prognostic marker of death (Latini et al. 2004). CRP levels increase in AMI, possibly as a consequence of a systemic response to myocardial injury. Emerging studies on rapidity of increase of PTX3 in blood compared with CRP in humans, together with a lack of correlation between CRP and PTX3 reflect a role for this pentraxin in the pathogenesis of damage (Napoleone et al. 2004). Patients with active vasculitis, dengue virus infections or active pulmonary tuberculosis have significantly higher plasma levels of PTX3 than patients with quiescent disease (Fazzini et al. 2001; Mairuhu et al. 2005; Azzurri et al. 2005). Increased levels of blood PTX3 have been observed in patients with pulmonary infection and acute lung injury (He et al. 2007) and in systemic sclerosis (Iwata et al. 2009).

Role of PTX3 in inflammatory conditions was evaluated in *Ptx3*-transgenic and -deficient mice. *Ptx3* over-expression increases resistance to LPS toxicity and *ptx3*-transgenic mice show an exacerbated, inflammatory response and reduced survival rate following intestinal ischemia reperfusion injury (Dias et al. 2001). Moreover, *ptx3*-deficient mice had more widespread and severe IL-1-induced neuronal damage. In these mice, PTX3 confers resistance to neurodegeneration, and rescuing neurons from otherwise irreversible damage (Ravizza et al. 2001). The PTX3 is crucial for tissue inflammation after intestinal ischemia and reperfusion in mice (Souza et al. 2009).

Expression of PTX3 in RA

The expression of PTX3 was studied in dissociated rheumatoid arthritis (RA) and osteoarthritis (OA) type B synoviocytes, cultured in the presence and in the absence of inflammatory cytokines. OA synoviocytes were induced to express high levels of PTX3 mRNA by TNF-α, but not by other cytokines including IL-1β and IL-6. RA synoviocytes, unlike OA synoviocytes, constitutively expressed high levels of PTX3 in the absence of deliberate stimulation. In contrast, IFNγ and TGF-β inhibited PTX3 constitutive expression in RA synoviocytes. The joint fluid from RA patients contained higher levels of immunoreactive PTX3 than controls and the synovial tissue contained endothelial cells and synoviocytes positive for PTX3. Thus, PTX3 may play a role in inflammatory circuits of RA (Luchetti et al. 2000).

8.10.5.2 PTX3 Bound Apoptotic Cells Are Regulated by APCs

The PTX3 is produced in tissues under the proinflammatory signals, such as LPS, IL-1β, and TNF-α, which also promote maturation of DCs. Cell death commonly occurs during inflammatory reactions. It was shown that PTX3 specifically binds to dying cells in dose dependent and saturable manner. Recognition was restricted to extranuclear membrane domains and to a chronological window after UV irradiation or after CD95 cross-linking-induced or spontaneous cell death in vitro. PTX3 bound to necrotic cells to a lesser extent. Studies suggest that PTX3 sequesters cell remnants from antigen-presenting cells, possibly contributing to preventing the onset of autoimmune reactions in inflamed tissues (Rovere et al. 2000).

PTX3 Limits C1q-Mediated Complement Activation and Phagocytosis of Apoptotic Cells by DCs: The PTX3 and C1q are innate opsonins involved in the disposal of dying cells by phagocytes. C1q increases the phagocytosis of apoptotic cells by DC and the release of interleukin-12 in the presence of TLR4 ligands and apoptotic cells; PTX3 inhibits both events. Results suggest that the coordinated induction by primary, proinflammatory signals of C1q and PTX3 and their reciprocal regulation during inflammation influences the clearance of apoptotic cells by APCs and possibly plays a role in immune homeostasis (Baruah et al. 2006a). PTX3 levels have been shown to parallel disease activity in small-vessel vasculitis, which is often characterized by leukocytoclasia, a persistence of leukocyte remnants in the vessel wall. Therefore, PTX3 can play a role in the development of leukocytoclasia by affecting the clearance of apoptotic PMNs, thereby inducing their accumulation in the vessel wall (van Rossum et al. 2004).

PTX3 is Recruited at the Synapse Between Dying and Dendritic Cells: The PTX3 is specifically recruited at both sides of the phagocytic synapse between DCs and dying cells and remains stably bound to the apoptotic membranes. Apoptotic cells per se influence the production of PTX3 by maturing DCs. When both microbial stimuli and dying cells are present, PTX3 behaves as a flexible adaptor of DC function, regulating the maturation program and the secretion of soluble factors. Moreover a key event associated with autoimmunity (i.e., the cross-presentation of epitopes expressed by apoptotic cells to T cells) abates in the presence of PTX3, as evaluated using self, viral, and tumor-associated model antigens (vinculin, NS3, and MelanA/MART1). In contrast, PTX3 did not influence the presentation of exogenous soluble antigens, an event required for immunity against extracellular pathogens. These data suggest that PTX3 acts as a third-party agent between microbial stimuli and dying cells, contributing to limit tissue damage under inflammatory conditions and the activation of autoreactive T cells (Baruah et al. 2006b).

8.10.5.3 Role in Female Fertility

Ptx3 deficiency results in a severe defect in female fertility (Garlanda et al. 2002; Salustri et al. 2004; Varani et al. 2002). Infertility of *ptx3*-deficient mice is due to an abnormal cumulus oophorus characterized by an unstable ECM in which cumulus cells are dispersed uniformly instead of radiating out from a central oocyte (Salustri et al. 2004). However, oocytes from *ptx3*-deficient mice can be fertilized in vitro, suggesting that oocyte develops normally in the absence of PTX3 and that lack of in vivo is a result of the defective cumulus expansion. Cumulus cells express *ptx3* mRNA, and no or barely detectable expression was observed in peripheral granulosa cells (Salustri et al. 2004; Zhang et al. 2005). PTX3 has been identified as a potential marker of oocyte competence expressed in bovine cumulus cells which were matured with follicle-stimulating hormone and/or phorbol myristate acetate in vitro (Assidi et al. 2008). Human cumulus cells also express *PTX3* (Zhang et al. 2005), and PTX3 is present in human cumulus matrix, suggesting some role for PTX3 in human female fertility. Results show a higher abundance of *PTX3* mRNA in cumulus cells from fertilized oocytes compared with cumulus cells from unfertilized oocytes, indicating in *PTX3* a possible marker for oocyte quality (Zhang et al. 2005), although this is not the case for PTX3 protein (Paffoni et al. 2006). In addition, pre-eclampsia is associated to elevated levels of PTX3 (Cetini et al. 2006). Thus, it is important to assess the potential of PTX3 in human fertility and pregnancy.

References

Abderrahim-Ferkoune A, Bezy O, Chiellini C et al (2003) Characterization of the long pentraxin PTX3 as a TNFα-induced secreted protein of adipose cells. J Lipid Res 44:994–1000

Agrawal A, Volanakis JE (1994) Probing the C1q-binding site on human C-reactive protein by site- directed mutagenesis. J Immunol 152:5404–5410

Agrawal A, Shrive AK, Greenhough TJ, Volanakis JE (2001) Topology and structure of the C1q-binding site on C-reactive protein. J Immunol 166:3998–4004

Agrawal A, Simpson MJ, Black S et al (2002) A C-reactive protein mutant that does not bind to phosphocholine and pneumococcal C-polysaccharide. J Immunol 169:3217–3222

Agrawal A, Samols D, Kushner I (2003) Transcription factor c-Rel enhances C-reactive protein expression by facilitating the binding of C/EBPbeta to the promoter. Mol Immunol 40:373–380

Ahmed N, Thorley R, Xia D et al (1996) ransgenic mice expressing rabbit C-reactive protein exhibit diminished chemotactic factor-induced alveolitis. Am J Respir Crit Care Med 153:1141–1147

Al-Ramadi BK, Ellis M, Pasqualini F, Mantovani A (2004) Selective induction of pentraxin 3, a soluble innate immune pattern recognition receptor, in infectious episodes in patients with haematological malignancy. Clin Immunol 112:221–224

Anand SS, Razak F, Yi Q et al (2004) C-reactive protein as a screening test for cardiovascular risk in a multiethnic population. Arterioscler Thromb Vasc Biol 24:1509–1515

Andersen O, Vilsgaard Ravn K, Juul Sorensen I et al (1997) Serum amyloid P component binds to influenza A virus haemagglutinin and inhibits the virus infection in vitro. Scand J Immunol 46:331–337

Aquilina JA, Robinson CV (2003) Investigating interactions of the pentraxins serum amyloid P component and C-reactive protein by mass spectrometry. Biochem J 375:323–328

Armani A, Becker RC (2005) The biology, utilization, and attenuation of C-reactive protein in cardiovascular disease: part II. Am Heart J 149:977–983

Assidi M, Dufort I, Ali A et al (2008) Identification of Potential Markers of Oocyte Competence Expressed in Bovine Cumulus Cells Matured with Follicle-Stimulating Hormone and/or Phorbol Myristate Acetate In Vitro. Biol Reprod 79:209–222

Aziz N, Fahey JL, Detels R, Butch AW (2003) Analytical performance of a highly sensitive C-reactive protein-based immunoassay and the effects of laboratory variables on levels of protein in blood. Clin Diagn Lab Immunol 10:652–657

Azzurri A, Sow OY, Amedei A et al (2005) IFN-γ-inducible protein 10 and pentraxin 3 plasma levels are tools for monitoring inflammation and disease activity in Mycobacterium tuberculosis infection. Microbes Infect 7:1–8

Ballou SP, Lozanski G (1992) Induction of inflammatory cytokine release from cultured human monocytes by C-reactive protein. Cytokine 4:361–368

Barbashov SF, Wang C, Nicholson-Weller A (1997) Serum amyloid P component forms a stable complex with human C5b6. J Immunol 158:3830–3835

Baron JA, Cole BF, Sandler RS et al (2003) A randomized trial of aspirin to prevent colorectal adenomas. N Engl J Med 348:891–899

Baruah P, Dumitriu IE, Peri G, Russo V, Mantovani A, Manfredi AA, Rovere-Querini P (2006a) The tissue pentraxin PTX3 limits C1q-mediated complement activation and phagocytosis of apoptotic cells by dendritic cells. J Leukoc Biol 80:87–95

Baruah P, Propato A, Dumitriu IE et al (2006b) The pattern recognition receptor PTX3 is recruited at the synapse between dying and dendritic cells, and edits the cross-presentation of self, viral, and tumor antigens. Blood 107:151–158

Basile A, Sica A, d'Aniello E et al (1997) Characterization of the promoter for the human long pentraxin PTX3. Role of NF-kB in tumor necrosis factor-α and interleukin-1β regulation. J Biol Chem 272:8172–8178

Beckmann G, Hanke J, Bork P, Reich JG (1998) Merging extracellular domains: fold prediction for laminin G-like and amino-terminal thrombospondin-like modules based on homology to pentraxins. J Mol Biol 275:725–730

Bickerstaff MC, Botto M, Hutchinson WL et al (1999) Serum amyloid P component controls chromatin degradation and prevents antinuclear autoimmunity. Nat Med 5:694–697

Bijl M, Horst G, Bijzet J et al (2003) Serum amyloid P component binds to late apoptotic cells and mediates their uptake by monocyte-derived macrophages. Arthritis Rheum 48:248–254

Bisoendial RJ, Kastelein JJ, Levels JH et al (2005) Activation of inflammation and coagulation after infusion of C-reactive protein in humans. Circ Res 96:714–716

Bjartmar L, Huberman AD, Ullian EM et al (2006) Neuronal pentraxins mediate synaptic refinement in the developing visual system. J Neurosci 26:6269–6281

Black S, Agrawal A, Samols D (2003) The phosphocholine and the polycation-binding sites on rabbit C-reactive protein are structurally and functionally distinct. Mol Immunol 39:1045–1054

Black S, Kushner I, Samols D (2004) C-reactive Protein. J Biol Chem 279:48487–48490

Boehme M, Kaehne F, Kuehne A et al (2007) Pentraxin 3 is elevated in haemodialysis patients and is associated with cardiovascular disease. Nephrol Dial Transplant 22:2224–2229

Bottazzi B, Vouret-Craviari V, Bastone A et al (1997) Multimer formation and ligand recognition by the long pentraxin PTX3. Similarities and differences with the short pentraxins C-reactive protein and serum amyloid P component. J Biol Chem 272:32817–32823

Bottazzi B, Bastone A, Doni A et al (2006) The long pentraxin PTX3 as a link among innate immunity, inflammation, and female fertility. J Leukoc Biol 79:909–912

Botto M, Hawkins PN, Bickerstaff MC et al (1997) Amyloid deposition is delayed in mice with targeted deletion of the serum amyloid P component gene. Nat Med 3:855–859

Calabro P, Chang D, Willerson JT, Yeh ET (2005) Release of C-reactive protein in response to inflammatory cytokines by human adipocytes. J Am Coll Cardiol 46:1112–1113

Cathcart ES, Shirahama T, Cohen AS (1967) Isolation and identification of a plasma component of amyloid. Biochim Biophys Acta 147:392–393

Cetini I, Cozzi V, Pasqualini F et al (2006) Elevated maternal levels of the long pentraxin PTX3 in preeclampsia and intrauterine growth restriction. Am J Obstet Gynecol 194:1347–1353

Chang MK, Binder CJ, Torzewski M, Witztum JL (2002) C-reactive protein binds to both oxidized LDL and apoptotic cells through recognition of a common ligand: phosphorylcholine of oxidized phospholipids. Proc Natl Acad Sci U S A 99:13043–13048

Chi M, Tridandapani S, Zhong W et al (2002) C-reactive protein induces signaling through Fc γ RIIa on HL-60 granulocytes. J Immunol 168:1413–1418

Christner RB, Mortensen RF (1994a) Binding of human serum amyloid P-component to phosphocholine. Arch Biochem Biophys 314:337–343

Christner RB, Mortensen RF (1994b) Specificity of the binding interaction between human serum amyloid P-component and immobilized human C-reactive protein. J Biol Chem 269:9760–9766

Ciurana CL, Hack CE (2006) Competitive binding of pentraxins and IgM to newly exposed epitopes on late apoptotic cells. Cell Immunol 239:14–21

Coe JE, Ross MJ (1985) Hamster female protein, a sex-limited pentraxin, is a constituent of Syrian hamster amyloid. J Clin Invest 76:66–74

Coe JE, Ross MJ (1987) Hamster female protein, a pentameric oligomer capable of reassociation and hybrid formation. Biochemistry 26:704–710

Coe JE, Ross MJ (1990) Armenian hamster female protein: a pentraxin under complex regulation. Am J Physiol 259:R341–R349

Coe JE, Margossian SS, Slayter HS, Sogn JA (1981) Hamster female protein. A new pentraxin structurally and functionally similar to C-reactive protein and amyloid P component. J Exp Med 153:977–991

Coe JE, Vomachka AJ, Ross MJ (1999) Effect of hamster pregnancy on female protein, a homolog of serum amyloid P component. Proc Soc Exp Biol Med 221:369–375

Danenberg HD, Szalai AJ, Swaminathan RV et al (2003) Increased thrombosis after arterial injury in human C-reactive protein-transgenic mice. Circulation 108:512–515

Danesh J, Wheeler JG, Hirschfield GM et al (2004) C-Reactive protein and other circulating markers of inflammation in the prediction of coronary heart disease. N Engl J Med 350:1387–1397

Das T, Sen AK, Kempf T et al (2003) Induction of glycosylation in human C-reactive protein under different pathological conditions. Biochem J 373:345–355

Deban L, Russo CR, Sironi M et al (2010) Regulation of leukocyte recruitment by the long pentraxin PTX3. Nat Immunol 11:328–334

Dehghan A, Kardys I, de Maat MP et al (2007) Genetic variation, C-reactive protein levels, and incidence of diabetes. Diabetes 56:872–878

Devaraj S, Singh U, Jialal I (2009a) The evolving role of C-reactive protein in atherothrombosis. Clin Chem 55:229–238

Devaraj S, Yun J-M, Adamson G et al (2009b) C-reactive protein impairs the endothelial glycocalyx resulting in endothelial dysfunction. Cardiovasu Res 84:479–484

Dias AA, Goodman AR, Dos Santos JL et al (2001) TSG-14 transgenic mice have improved survival to endotoxemia and to CLP-induced sepsis. J Leukoc Biol 69:928–936

Dietrich M, Jialal I (2005) The effect of weight loss on a stable biomarker of inflammation, C-reactive protein. Nutr Rev 63:22–28

Diniz SN, Nomizo R, Cisalpino PS et al (2004) PTX3 function as an opsonin for the dectin-1-dependent internalization of zymosan by macrophages. J Leukoc Biol 75:649–656

Dominguez M, Moreno I, Lopez-Trascasa M, Torano A (2002) Complement interaction with trypanosomatid promastigotes in normal human serum. J Exp Med 195:451–459

Dong A, Caughey B, Caughey WS et al (1992) Secondary structure of the pentraxin female protein in water determined by infrared spectroscopy: effects of calcium and phosphorylcholine. Biochemistry 31:9364–9370

Dong A, Caughey WS, Du Clos TW (1994) Effects of calcium, magnesium, and phosphorylcholine on secondary structures of human C-reactive protein and serum amyloid P component observed by infrared spectroscopy. J Biol Chem 269:6424–6430

Doni A, Peri G, Chieppa M et al (2003) Production of the soluble pattern recognition receptor PTX3 by myeloid, but not plasmacytoid, dendritic cells. Eur J Immunol 33:2886–2893

Doni A, Mosca M, Bottazzi B et al (2006) Regulation of PTX3, a key component of the humoral innate immunity, in human dendritic cells: stimulation by IL-10 and inhibition by IFNγ. J Leukoc Biol 79:797–802

Doni A, Mantovani G, Porta C et al (2008) Cell-specific Regulation of PTX3 by Glucocorticoid Hormones in Hematopoietic and Nonhematopoietic Cells. J Biol Chem 283:29983–29992

Du Clos TW (1996) The interaction of C-reactive protein and serum amyloid P component with nuclea antigens. Mol Biol Rep 23:253–260

Du Clos T (2002) C-reactive protein and the immune response. Sci Med 8:108–117

Du Clos TW (2003) C-reactive protein as a regulator of autoimmunity and inflammation. Arthritis Rheum 48:1475–1477

Du Clos TW, Mold C (2004) C-reactive protein: an activator of innate immunity and a modulator of adaptive immunity. Immunol Res 30:261–277

Duong T, Acton PJ, Johnson RA (1998) The in vitro neuronal toxicity of pentraxins associated with Alzheimer's disease brain lesions. Brain Res 813:303–312

Ehrt S, Schnappinger D, Bekiranov S et al (2001) Reprogramming of the macrophage transcriptome in response to interferon-γ and Mycobacterium tuberculosis: signaling roles of nitric oxide synthase-2 and phagocyte oxidase. J Exp Med 194:1123–1140

Eisenhardt SU, Habersberger J, Murphy A et al (2009) Dissociation of pentameric to monomeric c-reactive protein on activated platelets localizes inflammation to atherosclerotic plaques. Circ Res 105:128–137

Emsley J, White HE, O'Hara BP et al (1994) Structure of pentameric human serum amyloid P component. Nature 367(6461):338–345

Enguita M, DeGregorio-Rocasolano N, Abad A et al (2005) Glycogen synthase kinase 3 activity mediates neuronal pentraxin 1 expression and cell death induced by potassium deprivation in cerebellar granule cells. Mol Pharmacol 67:1237–1246

Erlinger TP, Platz EA, Rifai N et al (2004) C-reactive protein and the risk of incident colorectal cancer. J Am Med Assoc 291:585–590

Evans TC Jr, Nelsestuen GL (1995) Dissociation of serum amyloid P from C4b-binding protein and other sites by lactic acid: potential role of lactic acid in the regulation of pentraxin function. Biochemistry 34:10440–10447

Fazzini F, Peri G, Doni A et al (2001) PTX3 in small-vessel vasculitides: an independent indicator of disease activity produced at sites of inflammation. Arthritis Rheum 4:42841–42850, 30

Filep JG (2009) Platelets affect the structure and function of C-reactive protein. Circ Res 105:109–111

Floyd-Smith G, Whitehead AS, Colten HR, Francke U (1986) The human C-reactive protein gene (CRP) and serum amyloid P component gene (APCS) are located on the proximal long arm of chromosome 1. Immunogenetics 24:171–176

Gaboriaud C, Juanhuix J, Gruez A et al (2003) The crystal structure of the globular head of complement protein C1q provides a basis for its versatile recognition properties. J Biol Chem 278:46974–46982

Garlanda C, Hirsch E, Bozza S et al (2002) Non-redundant role of the long pentraxin PTX3 in anti-fungal innate immune response. Nature 420(6912):182–186

Garlanda C, Bottazzi B, Bastone A, Mantovani A (2005) Pentraxins at the crossroads between innate immunity, inflammation, matrix deposition, and female fertility. Annu Rev Immunol 23:337–366

Garlanda C, Bottazzi B, Salvatori G et al (2006) Pentraxins in innate immunity and inflammation. Novartis Found Symp 279:80–86

Gershov D, Kim S, Brot N, Elkon KB (2000) C-Reactive protein binds to apoptotic cells, protects the cells from assembly of the terminal complement components, and sustains an antiinflammatory innate immune response: implications for systemic autoimmunity. J Exp Med 192:1353–1364

Giannakis E, Male DA, Ormsby RJ et al (2001) Multiple ligand binding sites on domain seven of human complement factor H. Intern Immunopharmacol 1:433–443

Giannakis E, Jokiranta TS, Male DA et al (2003) A common site within factor H SCR 7 responsible for binding heparin, C-reactive protein and streptococcal M protein. Eur J Immunol 33:962–969

Goodman AR, Cardozo T, Abagyan R et al (1996) Long pentraxins: an emerging group of proteins with diverse functions. Cytokine Growth Factor Rev 7:191–202

Hawkins PN, Pepys MB (1995) Imaging amyloidosis with radiolabelled SAP. Eur J Nucl Med 22:595–599

He X, Han B, Liu M (2007) Long pentraxin 3 in pulmonary infection and acute lung injury. Am J Physiol Lung Cell Mol Physiol 292: L1039–L1049

Heuertz R, Xia D, Samols D, Webster R (1994) Inhibition of C5a des Arg-induced neutrophil alveolitis in transgenic mice expressing C-reactive protein. Am J Physiol 266:L649–L654

Hicks PS, Saunero-Nava L, Du Clos TW, Mold C (1992) Serum amyloid P component binds to histones and activates the classical complement pathway. J Immunol 149:3689–3694

Hohenester E, Hutchinson WL, Pepys MB, Wood SP (1997) Crystal structure of a decameric complex of human serum amyloid P component with bound dAMP. J Mol Biol 269:570–578

Hossain MA, Russell JC, O'Brien R, Laterra J (2004) Neuronal pentraxin 1: a novel mediator of hypoxic-ischemic injury in neonatal brain. J Neurosci 24:4187–4196

Huffman K, Samsa G, Slentz C et al (2006) Response of high-sensitivity C-reactive protein to exercise training in an at-risk population. Am Heart J 152:793–800

Hundt M, Zielinska-Skowronek M, Schmidt RE (2001) Lack of specific receptors for C-reactive protein on white blood cells. Eur J Immunol 31:3475–3483

Hutchinson WL, Noble GE, Hawkins PN, Pepys MB (1994) The pentraxins, C-reactive protein and serum amyloid P component, are cleared and catabolized by hepatocytes in vivo. J Clin Invest 94:1390–1396

Introna M, Alles VV, Castellano M et al (1996) Cloning of mouse ptx3, a new member of the pentraxin gene family expressed at extrahepatic sites. Blood 87:1862–1872

Iwata Y, Yoshizaki A, Ogawa F et al (2009) Increased serum pentraxin 3 in patients with systemic sclerosis. J Rheumatol 36:976–983

Jabs WJ, Logering BA, Gerke P et al (2003) The kidney as a second site of human C-reactive protein formation in vivo. Eur J Immunol 33:152–161

Jaillon S, Peri G, Delneste Y et al (2007) The humoral pattern recognition receptor PTX3 is stored in neutrophil granules and localizes in extracellular traps. J Exp Med 204:793–804

Jeannin P, Renno T, Goetsch L et al (2000) OmpA targets dendritic cells, induces their maturation and delivers antigen into the MHC class I presentation pathway. Nat Immunol 1:502–509

Jeannin P, Bottazzi B, Sironi M et al (2005) Complexity and complementarity of outer membrane protein A recognition by cellular and humoral innate immunity receptors. Immunity 2:2551–2560, 12

Jenny NS, Arnold AM, Kuller LH, Tracy RP, Psaty BM (2007) Serum amyloid P and cardiovascular disease in older men and women: results from the Cardiovascular Health Study. Arterioscler Thromb Vasc Biol 27:352–358

Ji S-R, Wu Y, Potempa LA et al (2006) Effect of modified c-reactive protein on complement activation. a possible complement regulatory role of modified or monomeric c-reactive protein in atherosclerotic lesions. Arterioscler Thromb Vasc Biol 26:935

Ji S-R, Wu Y, Zhu L et al (2007) Cell membranes and liposomes dissociate C-reactive protein (CRP) to form a new, biologically active structural intermediate. $mCRP_m$. FASEB J 21:284–294

Ji S-R, Ma L, Bai C-J et al (2009) Monomeric C-reactive protein activates endothelial cells via interaction with lipid raft microdomains. FASEB J 23:1806–1816

Jialal I, Devaraj S, Venugopal SK (2004) C-reactive protein: risk marker or mediator in atherothrombosis? Hypertension 44:6–11

Khera A, McGuire DK, Murphy SA et al (2005) Race and gender differences in C-reactive protein levels. J Am Coll Cardiol 46:464–469

Khreiss T, Jozsef L, Potempa LA, Filep JG (2004) Conformational rearrangement in C-reactive protein is required for proinflammatory actions on human endothelial cells. Circulation 109:2016–2022

Khreiss T, Jozsef L, Potempa LA, Filep JG (2005) Loss of pentameric symmetry in C-reactive protein induces interleukin-8 secretion through peroxynitrite signaling in human neutrophils. Circ Res 97:690–697

Kirkpatrick LL, Matzuk MM, Dodds DC et al (2000) Biochemical interactions of the neuronal pentraxins. Neuronal pentraxin (NP) receptor binds to taipoxin and taipoxin-associated calcium-binding protein 49 via NP1 and NP2. J Biol Chem 275:17786–17792

Kravitz MS, Shoenfeld Y (2006) Autoimmunity to protective molecules: is it the perpetuum mobile (vicious cycle) of autoimmune rheumatic diseases? Nat Clin Pract Rheumatol 2:481–490

Kravitz MS, Pitashny M, Shoenfeld Y (2005) Protective molecules–C-reactive protein (CRP), serum amyloid P (SAP), pentraxin3 (PTX3), mannose-binding lectin (MBL), and apolipoprotein A1 (Apo A1), and their autoantibodies: prevalence and clinical significance in autoimmunity. J Clin Immunol 25:582–591

Kunes P (2005) The role of pentraxin 3 in the inflammatory and immune response. Cas Lek Cesk 144:377–382 [Abstract]

Kushner I, Jiang SL, Zhang D, Lozanski G, Samols D (1995) Do post-transcriptional mechanisms participate in induction of C-reactive protein and serum amyloid A by IL-6 and IL-1? Ann N Y Acad Sci 762:102–107

Latini R, Maggioni AP, Peri G et al (2004) Prognostic significance of the long pentraxin PTX3 in acute myocardial infarction. Circulation 110:2349–2354

Lau DC, Dhillon B, Yan H et al (2005) Adipokines: molecular links between obesity and atheroslcerosis. Am J Physiol Heart Circ Physiol 288:H2031–H2041

Li XA, Hatanaka K, Ishibashi-Ueda H et al (1995) Characterization of serum amyloid P component from human aortic atherosclerotic lesions. Arterioscler Thromb Vasc Biol 15:252–257

Lloyd-Jones DM, Liu K, Tian L, Greenland P (2006) Narrative review: assessment of C-reactive protein in risk prediction for cardiovascular disease. Ann Intern Med 145:35–42

Lopez-Garcia E, Schulze MB, Meigs JB et al (2005) Consumption of trans- fatty acids is related to plasma biomarkers of inflammation and endothelial dysfunction. J Nutr 135:562–566

Luchetti MM, Piccinini G, Mantovani A et al (2000) Expression and production of the long pentraxin PTX3 in rheumatoid arthritis (RA). Clin Exp Immunol 119:196–202

Luchetti MM, Sambo P, Majlingova P et al (2004) Scleroderma fibroblasts constitutively express the long pentraxin PTX3. Clin Exp Rheumatol 22(3 Suppl 33):S66–S72

Lysenko E, Richards JC, Cox AD et al (2000) The position of phosphorylcholine on the lipopolysaccharide of Haemophilus influenzae affects binding and sensitivity to C-reactive protein-mediated killing. Mol Microbiol 35:234–245

Maeda S (2003) Use of genetically altered mice to study the role of serum amyloid P component in amyloid deposition. Amyloid 10 (Suppl 1):17–20

Mairuhu AT, Peri G, Setiati TE et al (2005) Elevated plasma levels of the long pentraxin, pentraxin 3, in severe dengue virus infections. J Med Virol 6:547–552

Mantovani A, Garlanda C, Bottazzi B et al (2006) The long pentraxin PTX3 in vascular pathology. Vascul Pharmacol 45:326–330

Marz W, Scharnagl H, Hoffmann MM et al (2004) The apolipoprotein E polymorphism is associated with circulating C-reactive protein (the Ludwigshafen risk and cardiovascular health study). Eur Heart J 25:2109–2119

McGeer PL, McGeer EG (2004) Inflammation and the degenerative diseases of aging. Ann N Y Acad Sci 1035:104–116

McGeer EG, Yasojima K, Schwab C, McGeer PL (2001) The pentraxins: possible role in Alzheimer's disease and other innate inflammatory diseases. Neurobiol Aging 22:843–848

Mold C, Nakayama S, Holzer TJ et al (1981) C-reactive protein is protective against Streptococcus pneumoniae infection in mice. J Exp Med 154:1703–1708

Mold C, Gewurz H, Du Clos TW (1999) Regulation of complement activation by C-reactive protein. Immunopharmacology 42:23–30

Mold C, Gresham HD, Du Clos TW (2001) Serum amyloid P component and C-reactive protein mediate phagocytosis through murine FcγRs. J Immunol 166:1200–1205

Mold C, Baca R, Du Clos TW (2002a) Serum amyloid P component and C-reactive protein opsonize apoptotic cells for phagocytosis through Fcγ receptors. J Autoimmun 19:147–154

Mold C, Rodic-Polic B, Du Clos TW (2002b) Protection from *Streptococcus pneumoniae* infection by C-reactive protein and natural antibody requires complement but not Fc γ receptors. J Immunol 168:6375–6381

Mold C, Rodriguez W, Rodic-Polic B, Du Clos TW (2002c) C-reactive protein mediates protection from lipopolysaccharide through interactions with Fcγ R. J Immunol 169:7019–7025

Molins B, Pena E, Vilahur G et al (2008) C-Reactive Protein Isoforms Differ in Their Effects on Thrombus Growth. Arterioscler Thromb Vasc Biol 28:2239–2246

Moore KW, de Waal MR, Coffman RL et al (2001) Interleukin-10 and the interleukin-10 receptor. Annu Rev Immunol 1:9683–9765

Mortensen RF (2001) C-reactive protein, inflammation, and innate immunity. Immunol Res 24:163–176

Muller B, Peri G, Doni A et al (2001) Circulating levels of the long pentraxin PTX3 correlate with severity of infection in critically ill patients. Crit Care Med 29:1404–1407

Napoleone E, di Santo A, Peri G et al (2004) The long pentraxin PTX3 up-regulates tissue factor in activated monocytes: another link between inflammation and clotting activation. J Leukoc Biol 76:203–209

Nauta AJ, Bottazzi B, Mantovani A et al (2003) Biochemical and functional characterization of the interaction between pentraxin 3 and C1q. Eur J Immunol 33:465–473

Nauta AJ, de Haij S, Bottazzi B et al (2005) Human renal epithelial cells produce the long pentraxin PTX3. Kidney Int 67:543–553

Ng PM, Jin Z, Tan SS, Ho B, Ding JL (2004) C-reactive protein: a predominant LPS-binding acute phase protein responsive to *Pseudomonas* infection. J Endotoxin Res 10:163–174

Ogden CA, Elkon KB (2005) Single-dose therapy for lupus nephritis: C-reactive protein, nature's own dual scavenger and immunosuppressant. Arthritis Rheum 52:378–381

Ohnishi S, Maeda S, Shimada K, Arao T (1986) Isolation and characterization of the complete complementary and genomic DNA sequences of human serum amyloid P component. J Biochem 100:849–858

Paffoni A, Ragni G, Doni A et al (2006) Follicular fluid levels of the long pentraxin PTX3. J Soc Gynecol Investig 13:226–231

Paul A, Ko KW, Li L, Yechoor V et al (2004) C-reactive protein accelerates the progression of atherosclerosis in apolipoprotein E-deficient mice. Circulation 109:647–655

Pearson TA, Mensah GA, Alexander RW et al (2003) Markers of inflammation and cardiovascular disease: application to clinical and public health practice: a statement for healthcare professionals from the Centers for Disease Control and Prevention and the American Heart Association. Circulation 107:499–511

Pepys MB (2005) CRP or not CRP? That is the question. Arterioscler Thromb Vasc Biol 25:1091–1094

Pepys MB, Hirschfield GM (2003) C-reactive protein: a critical update. J Clin Invest 111:1805–1812

Pepys MB, Booth SE, Tennent GA et al (1994) Binding of pentraxins to different nuclear structures: C-reactive protein binds to small nuclear ribonucleoprotein particles, serum amyloid P component binds to chromatin and nucleoli. Clin Exp Immunol 97:152–157

Pepys MB, Booth DR, Hutchinson WL et al (1997) Amyloid P component. A critical review. Amyloid: Int J Exp Clin Invest 4:274–295

Pepys MB, Herbert J, Hutchinson WL et al (2002) Targeted pharmacological depletion of serum amyloid P component for treatment of human amyloidosis. Nature 417(6886):254–259

Pepys MB, Hirshfield GM, Tennent GA et al (2003) C-reactive protein and interleukin-6 in vascular disease: culprits or passive bystanders? J Hypertens 21:1787–1803

Pepys MB, Hirschfield GM, Tennent GA et al (2006) Targeting C-reactive protein for the treatment of cardiovascular disease. Nature 440:1217–1221

Polentarutti N, Bottazzi B, Di Santo E et al (2000) Inducible expression of the long pentraxin PTX3 in the central nervous system. J Neuroimmunol 106:87–94

Poulsen TT, Pedersen N, Perin MS et al (2005) Specific sensitivity of small cell lung cancer cell lines to the snake venom toxin taipoxin. Lung Cancer 50:329–337

Pradhan AD, Manson JE, Rifai N et al (2001) C-reactive protein, interleukin 6, and risk of developing type 2 diabetes mellitus. JAMA 286:327–334

Prelli F, Pras M, Frangione B (1985) The primary structure of human tissue amyloid P component from a patient with primary idiopathic amyloidosis. J Biol Chem 260:12895–12898

Ravizza T, Moneta D, Bottazzi B et al (2001) Dynamic induction of the long pentraxin PTX3 in the CNS after limbic seizures: evidence for a protective role in seizure-induced neurodegeneration. Neuroscience 105:43–53

Reti IM, Reddy R, Worley PF, Baraban JM (2002) Selective expression of Narp, a secreted neuronal pentraxin, in orexin neurons. J Neurochem 82:1561–1565

Ridker PM (2004) High-sensitivity C-reactive protein, inflammation, and cardiovascular risk: from concept to clinical practice to clinical benefit. Am Heart J 148:S19–S26

Ridker PM, Danielson E et al (2008) Rosuvastatin to prevent vascular events in men and women with elevated c-reactive protein. N Engl J Med 359:2195–2207

Rodriguez W, Mold C, Kataranovski M et al (2005) Reversal of ongoing proteinuria in autoimmune mice by treatment with C-reactive protein. Arthritis Rheum 52:642–650

Rovere P, Peri G, Fazzini F et al (2000) The long pentraxin PTX3 binds to apoptotic cells and regulates their clearance by antigen-presenting dendritic cells. Blood 96:4300–4306

Rubio N, Sharp PM, Rits M et al (1993) Structure, expression, and evolution of guinea pig serum amyloid P component and C-reactive protein. J Biochem (Tokyo) 113:277–284

Rudenko G, Hohenester E, Muller YA (2001) LG/LNS domains: multiple functions – one business end? Trends Biochem Sci 26:363–368

Rudnick CM, Dowton SB (1993a) Serum amyloid P (female protein) of the Syrian hamster. Gene structure and expression. J Biol Chem 268:21760–21769

Rudnick CM, Dowton SB (1993b) Serum amyloid-P component of the Armenian hamster: gene structure and comparison with structure and expression of the SAP gene from Syrian hamster. Scand J Immunol 38:445–450

Russell AI, Cunninghame Graham DS et al (2004) Polymorphism at the C-reactive protein locus influences gene expression and predisposes to systemic lupus erythematosus. Hum Mol Genet 13:137–147

Salustri A, Garlanda C, Hirsch E et al (2004) PTX3 plays a key role in the organization of the cumulus oophorus extracellular matrix and in in vivo fertilization. Development 131:1577–1586

Saunero-Nava L, Coe JE, Mold C, Du Clos TW (1992) Hamster female protein binding to chromatin, histones and DNA. Mol Immunol 29:837–845

Schwalbe RA, Dahlback B, Coe JE et al (1992) Pentraxin family of proteins interact specifically with phosphorylcholine and/or phosphorylethanolamine. Biochemistry 31:4907–4915

Schwalbe RA, Coe JE, Nelsestuen GL (1995) Association of rat C-reactive protein and other pentraxins with rat lipoproteins containing apolipoproteins E and A1. Biochemistry 34: 10432–10439

Shields MJ (1993) A hypothesis resolving the apparently disparate activities of native and altered forms of human C-reactive protein. Immunol Res 12:37–47

Shrive AK, Cheetham GM, Holden D et al (1996) Three dimensional structure of human C-reactive protein. Nat Struct Biol 3: 346–354

Shrive AK, Metcalfe AM, Cartwright JR, Greenhough TJ (1999) C-reactive protein and SAP-like pentraxin are both present in Limulus polyphemus haemolymph: crystal structure of Limulus SAP. J Mol Biol 290:997–1008

Singh U, Devaraj S, Jialal I (2009) C-Reactive protein stimulates myeloperoxidase release from polymorphonuclear cells and monocytes: implications for acute coronary syndromes. Clin Chem 55:361–364

Soma M, Tamaoki T, Kawano H et al (2001) Mice lacking serum amyloid P component do not necessarily develop severe autoimmune disease. Biochem Biophys Res Commun 286:200–205

Souza DG, Amaral FA, Fagundes CT et al (2009) The long pentraxin PTX3 is crucial for tissue inflammation after intestinal ischemia and reperfusion in mice. Am J Pathol 174:1309–1318

Stokes KY, Dugas TR, Tang Y et al (2009) Dietary nitrite prevents hypercholesterolemic microvascular inflammation and reverses endothelial dysfunction. Am J Physiol Heart Circ Physiol 296: H1281–H1288

Suk HJ, Ridker PM, Cook NR, Zee RY (2005) Relation of polymorphism within the C-reactive protein gene and plasma CRP levels. Atherosclerosis 178:139–145

Swanson SJ, Christner RB, Mortensen RF (1992) Human serum amyloid P-component (SAP) selectively binds to immobilized or bound forms of C-reactive protein (CRP). Biochim Biophys Acta 1160:309–316

Szalai AJ (2002) The antimicrobial activity of C-reactive protein. Microbes Infect 4:201–205

Szalai AJ, Briles DE, Volanakis JE (1995) Human C-reactive protein is protective against fatal Streptococcus pneumoniae infection in transgenic mice. J Immunol 155:2557–2563

Szalai AJ, Agrawal A, Greenhough TJ et al (1999) C-reactive protein: structural biology and host defense function. Clin Chem Lab Med 37:265–270

Szalai AJ, Van Ginkel FW, Wang Y et al (2000a) Complement-dependent acute-phase expression of C-reactive protein and serum amyloid P-component. J Immunol 165:1030–1035

Szalai AJ, VanCott JL, McGhee JR et al (2000b) Human C-reactive protein is protective against fatal Salmonella enterica serovar typhimurium infection in transgenic mice. Infect Immun 68:5652–5656

Szalai AJ, McCrory MA, Cooper GS et al (2002a) Association between baseline levels of C-reactive protein (CRP) and a dinucleotide repeat polymorphism in the intron of the CRP gene. Genes Immun 3:14–19

Szalai AJ, Nataf S, Hu XZ, Barnum SR (2002b) Experimental allergic encephalomyelitis is inhibited in transgenic mice expressing human C-reactive protein. J Immunol 168:5792–5797

Szalai AJ, Weaver CT, McCrory MA et al (2003) Delayed lupus onset in (NZB x NZW)F1 mice expressing a human C-reactive protein transgene. Arthritis Rheum 48:1602–1611

Tanusree DAS et al (2003) Induction of glycosylation in human C-reactive protein under different pathological conditions. Biochem J 373:345–355

Tashiro F, Yi S, Wakasugi S et al (1991) Role of serum amyloid P component for systemic amyloidosis in transgenic mice carrying human mutant transthyretin gene. Gerontology 37(Suppl 1):56–62

Tennent GA, Baltz ML, Osborn GD et al (1993) Studies of the structure and binding properties of hamster female protein. Immunology 80:645–651

Tennent GA, Lovat LB, Pepys MB (1995) Serum amyloid P component prevents proteolysis of the amyloid fibrils of Alzheimer disease and systemic amyloidosis. Proc Natl Acad Sci U S A 92:4299–4303

ter Borg EJ, Horst G, Limburg PC et al (1990) C-reactive protein levels during disease exacerbations and infections in systemic lupus erythematosus: a prospective longitudinal study. J Rheumatol 17:1642–1648

Tharia HA, Shrive AK, Mills JD et al (2002) Complete cDNA sequence of SAP-like pentraxin from Limulus polyphemus: implications for pentraxin evolution. J Mol Biol 316:583–597

Thompson D, Pepys MB, Wood SP (1999) The physiological structure of human C-reactive protein and its complex with phosphocholine. Structure 7:169–177

Tilg H, Vannier E, Vachino G et al (1993) Antiinflammatory properties of hepatic acute phase proteins: preferential induction of interleukin I receptor antagonist over interleukin Ip synthesis by human peripheral blood mononuclear cells. J Exp Med 178:1629–1636

Tillett WS, Francis T Jr (1930) Serological reactions in pneumonia with a nonprotein somatic fraction of pneumococcus. J Exp Med 52:561–585

Torzewski M, Rist C, Mortensen RF (2000) et al C-reactive protein in the arterial intima: role of C-reactive protein receptor-dependent monocyte recruitment in atherogenesis. Arterioscler Thromb Vasc Biol 20:2094–2099

Van Der Meer-Van Kraaij C, Van Lieshout EM et al (2003) Mucosal pentraxin (Mptx), a novel rat gene 10-fold down-regulated in colon by dietary heme. FASEB J 17:1277–1285

van Rossum AP, Fazzini F, Limburg PC et al (2004) The prototypic tissue pentraxin PTX3, in contrast to the short pentraxin serum amyloid P, inhibits phagocytosis of late apoptotic neutrophils by macrophages. Arthritis Rheum 50:2667–2674

Varani S, Elvin JA, Yan C et al (2002) Knockout of pentraxin 3, a downstream target of growth differentiation factor-9, causes female subfertility. Mol Endocrinol 16:1154–1167

Venugopal SK, Devaraj S, Yuhanna I et al (2002) Demonstration that C-reactive protein decreases eNOS expression and bioactivity in human aortic endothelial cells. Circulation 106:1439–1441

Verma S (2004) C-reactive protein incites atherosclerosis. Can J Cardiol 20:29B–31B

Verma S, Yeh ET (2003) C-reactive protein and atherothrombosis-beyond a biomarker: an actual partaker of lesion formation. Am J Physiol Regul Integr Comp Physiol 285:R1253–R1256

Verma S, Szmitko PE, Ridker PM (2005) C-reactive protein comes of age. Nat Clin Pract 2:29–36

Volanakis JE, Kaplan MH (1971) Specificity of C-reactive protein for choline phosphate residues of pneumococcal C-polysaccharide. Proc Soc Exp Biol Med 136:612–614

Volanakis JE (2001) Human C-reactive protein: expression, structure, and function. Mol Immunol 38:189–197

Vouret-Craviari V, Matteucci C, Peri G et al (1997) Expression of a long pentraxin, PTX3, by monocytes exposed to the mycobacterial cell wall component lipoarabinomannan. Infect Immun 65: 1345–1350

Weiser JN, Pan N, McGowan KL et al (1998) Phosphorylcholine on the lipopolysaccharide of Haemophilus influenzae contributes to persistence in the respiratory tract and sensitivity to serum killing mediated by C-reactive protein. J Exp Med 187:631–640

Whitehead AS, Rits M (1989) Characterization of the gene encoding mouse serum amyloid P component. Comparison with genes encoding other pentraxins. Biochem J 263:25–31

Woo P, Korenberg JR, Whitehead AS (1985) Characterization of genomic and complementary DNA sequence of human C-reactive protein, and comparison with the complementary DNA sequence of serum amyloid P component. J Biol Chem 260: 13384–13388

Xia D, Samols D (1997) Transgenic mice expressing rabbit C-reactive protein are resistant to endotoxemia. Proc Natl Acad Sci U S A 94:2575–2580

Yasojima K, Schwab C, McGeer EG, McGeer PL (2000) Human neurons generate C-reactive protein and amyloid P: upregulation in Alzheimer's disease. Brain Res 887:80–89

Yasojima K, Schwab C, McGeer EG, McGeer PL (2001) Generation of C-reactive protein and complement components in atherosclerotic plaques. Am J Pathol 158:1039–1051

Yeh ETH (2005) A New Perspective on the Biology of C-Reactive Protein. Circ Res 97:609–611

Zacho J, Tybjaerg-Hansen A, Jensen JS et al (2008) Genetically elevated C-reactive protein and ischemic vascular disease. N Engl J Med 359:1897–1908

Zahedi K (1997) Characterization of the binding of serum amyloid P to laminin. J Biol Chem 272:2143–2148

Zhang X, Jafari N, Barnes RB et al (2005) Studies of gene expression in human cumulus cells indicate pentraxin 3 as a possible marker for oocyte quality. Fertil Steril 83(Suppl 1):1169–1179

Zwaka TP, Hombach V, Torzewski J (2001) C-reactive protein-mediated low density lipoprotein uptake by macrophages: implications for atherosclerosis. Circulation 103:1194–1197

Part IV

Animal Galectins

9 Overview of Animal Galectins: Proto-Type Subfamily

Anita Gupta and G.S. Gupta

9.1 Galectins

Lectins recognize and bind carbohydrates covalently linked to proteins and lipids on the cell surface and within the extracellular matrix, and they mediate many cellular functions ranging from cell adhesion to pathogen recognition. Phylogenetically conserved family of Galectins was defined in 1994 as a shared consensus of amino-acid-sequences of about 130 amino acids and the CRD responsible for ß-galactoside binding (Barondes et al. 1994a, b). The galectin (Gal) CRDs bind small β-galactosides/poly-N-acetyllactosamine-enriched glycoconjugates. But the overall binding affinity for more complex glycoconjugates varies substantially. To date, 15 members of the mammalian galectin family have been identified. Some, such as galectin-1, are isolated as dimers and have a single CRD in each monomer, whereas others, such as galectin-4, are isolated as monomers and have two CRDs in a single polypeptide chain. While CRDs of all galectins share an affinity for minimum saccharide ligand N-acetyllactosamine—a common disaccharide found on many cellular glycoproteins—individual galectins can also recognize different modifications to this minimum saccharide ligand and so demonstrate the fine specificity of certain galectins for tissue- or developmentally-specific ligands (Ahmad et al. 2004a). Location studies of galectins have established that these proteins can segregate into multiple cell compartments in function of the status of the cells in question (Danguy et al. 2002; Liu and Rabinovich 2005). Although galectins as a whole do not have the signal sequence required for protein secretion through usual secretory pathway, some galectins are secreted and are found in the extracellular space. While the intracellular activity of galectin-1 is mainly independent on its lectin activity, its extracellular activity is mainly dependent on it. The functions and distribution of Gal-1 and Gal-3 are well characterized.

9.2 Galectin Sub-Families

Galectins are localized to the cytoplasm, nucleus, and the extracellular environment (Liu et al. 2002). Secretion of galectins occurs via a nonclassical secretion pathway that requires association of galectins with glycosylated counter-receptors (Seelenmeyer et al. 2005). Nuclear expression of galectins has been studied in cultured stromal cells of human bone marrow and human/porcine keratinocytes. Binding studies revealed positivity for galectin-1, where as galectins-3, -5, and -7 were not reactive with nuclear sites in bone marrow stromal cells and keratinocytes. Presence of binding sites in nucleus adds a new aspect to the functional analysis of these lectins.

Thus far, 15 mammalian galectins have been identified and sequenced (Rabinovich and Gruppi 2005; Elola et al. 2005a, b). Hirabayshi and Kasai (1993) proposed designating galectin subfamilies as: Proto-, tandem-repeat, and Chimera-types based on their domain organization. Galectin-1, -2, -5, -7, -10, -11, -13, -14, and -15, which belong to the mono-CRD type and are found in many cells and tissues, have molecular weights of 14–18 kDa with a single galactoside-binding domain/CRD with a short N-terminal sequence. But the tandem-repeat type galectins (galectin-4, -6, -8, -9, and -12) are composed of two non-identical CRDs joined by a short linker peptide sequence. The single chimera-type galectin (galectin-3) has one CRD and an extended N-terminal tail containing several repeats of proline-tyrosine-glycine-rich motif with a collagenous domain. Galectin-3 with an apparent molecular weight between 26 and 30 kDa is abundant in activated macrophages and epithelial cells. Many of these β-galactoside-binding proteins have also received other names, according to their function, localization, or biochemical features, i.e., galectin-1 (L-14, BHL, or galaptin), galectin-3 [Mac-2, L-29, CBP-35, or εBP for "IgE-binding protein"], galectin-9 (ecalectin), galectin-10 (Charcot-Leyden crystal

G.S. Gupta et al., *Animal Lectins: Form, Function and Clinical Applications*,
DOI 10.1007/978-3-7091-1065-2_9, © Springer-Verlag Wien 2012

eosinophil protein), galectin-11 (GRIFIN for galectin-related interfiber protein) and galectin-13 (PP13). The vertebrate galectin CRDs are always encoded by three exons with two subtypes and are defined by the exon–intron structure (F4-CRD and F3 CRD). The F4-CRD-linker-F3-CRD gene structure is shared among all vertebrate bi-CRD galectins, one *Ciona intestinalis* galectin (Houzelstein et al. 2004), and the *Stronglocentrotus purpuratus* galectin. The chordate galectins share a common ancestor, and bi-CRD galectins are derived from an ancestral tandem-duplication of mono-CRD galectin either before or in early chordate evolution (Houzelstein et al. 2004).

Analysis of GenBank databases has led to the identification of more galectin-like proteins in mammals, invertebrates, plants, and microorganisms, confirming that these carbohydrate-binding proteins are highly conserved throughout evolution. The structures of several galectin CRDs have been reported, and all exhibit a β-sandwich fold containing two antiparallel β-sheets (Liao et al. 1994; Lobsanov et al. 1993; Seetharaman, et al. 1998; Leonidas et al. 1998). However, their quaternary structures differ. Galectin-1 and -2 form non-covalently associated homodimers through extended β-sheet interactions (Loris 2002). The association state of galectin-3 is regulated by its N-terminal domain, and it can exist in monomeric or oligomeric forms (Birdsall et al. 2001). Finally, because the tandem-repeat type galectins possess two different CRDs, they may adopt more complex assembly states. There are a variety of potential glycoconjugate targets for galectins in mammalian cells, but the molecular mechanisms of carbohydrate recognition remain unclear.

9.3 Galectin Ligands

Galectins have been shown to be involved in multiple biological functions such as cell-matrix and cell-cell interactions, cell proliferation, cell differentiation, cellular transformation, or apoptosis mainly through their binding properties to specific ligands. Galactose is a ligand for galectins. Galactose is part of scaffold structure that synthesizes oligosaccharide ligands for selectins, siglecs and other lectins of the immune system. Galactose residues are added to glycoproteins and glycolipids by members of a large family of galactosyltransferases. The expression of many of these enzymes is regulated by the action of cytokines, and becomes altered in various disease states. Antibodies need to be galactosylated for normal function, and under-galactosylated immunoglobulin (Ig) is associated with rheumatoid arthritis. Thus galectins have potential use as novel anti-inflammatory agents or targets for immunosuppressive drugs, in autoimmune and chronic inflammatory disorders (Bianco et al. 2006). Since galactose containing structures are involved in both the innate and adaptive immune systems, galectins play important roles in host-pathogen interactions, and in tumorigenesis (Rabinovich and Gruppi 2005). Nakahara and Raz (2006) discussed the role of galectins in signal transduction, dividing these proteins into extracellular and intracellular galectins. Galectin-1 is an important link between the sugar signal and the intracellular response. The growth-regulatory interaction between GM1 on human neuroblastoma cells and an endogenous lectin provides an example for glycan functionality.

9.4 Functions of Galectins

Galectins can be found in the extracellular space, in the cell membrane, in the cytoplasm, and even in the nucleus. They can bind to proteins via a carbohydrate interaction or via direct protein-protein interactions. Most galectins have multiple functions and they can mediate (1) receptor cross-linking or lattice formation, (2) cell-extracellular matrix interactions, (3) cell-to-cell interactions, (4) intracellular signaling, and (5) posttranscriptional splicing (Thijssen et al. 2007) (Fig. 9.1). Galectins can regulate inflammation, cell proliferation, the cell cycle, transcription processes, and cell death (Rabinovich et al. 2002a, b). Galectins lack a traditional signal sequence, and several are secreted by an unorthodox mechanism to exert their extracellular functions (Mehul and Hughes 1997). Some galectins are differentially regulated during pre- or post-natal development. Structural studies have shown that the Gal-specific proteins encompass a diverse range of primary and tertiary structures. The binding sites for galactose also seem to vary in different protein-galactose complexes. Sugar-encoded information of glyco-conjugates is translated into cellular responses by endogenous lectins. This remarkable versatility warrants close scrutiny of their emerging network. Cell localization studies have established that these proteins can segregate into multiple intracellular compartments, and the preference for segregation is dependent on the status of the cell. Localization, therefore, likely corresponds to compartmental function. Toscano et al. (2007) reviewed immunoregulatory roles of galectins (particularly galectin-1) and collectins to illustrate the ability of endogenous glycan-binding proteins to act as cytokines, chemokines or growth factors, and thereby modulating innate and adaptive immune responses under physiological or pathological conditions. Understanding the pathophysiologic relevance of endogenous galectins in vivo would reveal novel targets for immunointervention during chronic infection, autoimmunity, transplantation and cancer (Ilarregui et al. 2005; Toscano et al. 2007) (Table 9.1).

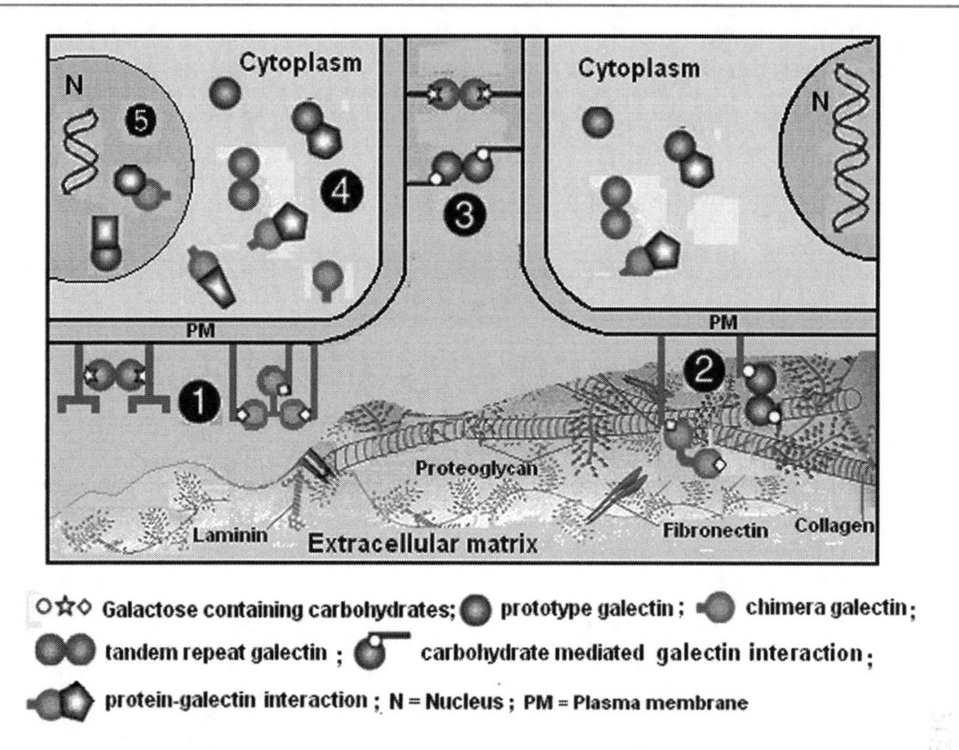

Fig. 9.1 Cellular location and functions of galectins. Galectins can be found in the extracellular space, in the cell membrane, in the cytoplasm, and in the nucleus. Functional interactions of galectins with cell-surface glycoconjugates and extracellular glycoconjugates can lead to cell adhesion and cell signaling. Interactions of galectins with intracellular ligands may also contribute to the regulation of intracellular pathways (Cummings and Liu 2009). They can bind to proteins via a carbohydrate interaction or via direct protein-protein interactions. Most galectins can mediate (1) receptor cross-linking or lattice formation, (2) cell-extracellular matrix interactions, (3) cell-to-cell interactions, (4) intracellular signaling, and (5) posttranscriptional splicing (Thijssen et al. 2007)

9.4.1 Functional Overlap/Divergence Among Galectins

To define the extent of functional overlap/divergence among galectins, a comparative profiling has been made in mouse as a model organism, combining sequence analysis, expression patterns and structural features in homodimeric galectins-1, -2 and -7. Close relationship was apparent at the level of global gene organization. RT-PCR mapping from an array of 17 organs revealed significant differences, separating rather ubiquitous gene expression of galectin-1 from the more restricted individual patterns of galectins-2 and -7. Nuclear presence was seen in case of galectin-1. In addition to nonidentical expression profiles the mapping of CRDs of galectins-1 and -7 by homology modelling and docking of naturally occurring complex tetra- and pentasaccharides disclosed a series of sequence deviations which may underlie disparate affinities for cell surface glycans/glycomimetic peptides (Lohr et al. 2007). A distinct expression profile of Galectin-3 was determined in various murine organs when set into relation to homodimeric galectins-1 and -7. Lohr et al. (2008) demonstrated cell-type specificity and cycle-associated regulation for galectin-3 with increased presence in atretic preantral follicles and in late stages of luteolysis. Staining patterns for galectin-8 were studied in hypo-pharyngeal and laryngeal tumor progression and relate d these parameters to galectins 1, 3 and 7 in the quest to explore the galectin network. Marked upregulation in galectin-8 staining intensity and immunopositive area in malignancy versus dysplasia was seen in hypopharyngeal cancer, in laryngeal cancer for labeling index and high-grade dysplasia/carcinoma. No correlation to recurrence was delineated. Data revealed a divergence within the galectin-1, -3, -7 and -8 network during tumor progression (Cludts et al. 2009).

9.4.2 Cell Homeostasis by Galectins

Galectin are known to participate in cellular homeostasis by modulating cell growth, controlling cell cycle progression, and inducing or inhibiting apoptosis. Both intracellular and extracellular activities of galectins have been described, with the former typically independent of lectin activity, and the latter mediated by lectin activity. Galectin-1 and -3 are recognized as activators and inducers of cell homeostasis in extracellular capacities. By using recombinant proteins, a number of galectins have been shown to interact with

Table 9.1 Schematic representation, biochemical and functional properties of different members of Galectin family (Adapted by permission from Rabinovich et al. (2002c) © Society for Leukocyte Biology)

Galectins/structure	Localisation	Biochemical and functional properties
Galectin-1	Abundant in most Organs: muscle, heart, liver, prostate, lymph nodes, spleen, thymus, placenta, testis, retina, macrophages, B-cells, T-cells, tumors	Non-covalent homodimer
		Induces apoptosis of activated T cells and immature thymocytes
		Induces a polarized Th2 immune response
		Modulates cell-cell and cell-matrix interactions
		Inhibits acute inflammation:blocks arachidonic acid release, mast cell degranulation and neutrophil extravasation
		Suppresses chronic inflammation and autoimmunity
Galectin-2	Stomach epithelial cells	Non-covalent homodimers
		Expressed at minor levels in tumor cells
Galectin-3	Mainly in tumor cells, macrophages, epithelial cells, fibroblasts, activated T cells	Non-lectin domain linked to CRD
		Anti apoptotic and proinfammatory functions
		Modulates cell adhesion and migration
		Induces chemotaxis of monocytes
		Potentiates pro-inflammatory (IL-1) cytokine secretion
		Induces nitric oxide induced apoptosis and ancilkis
		Down regulates IL-5 gene transcription
Galectin-4	GIT	Compose of two distinct CRD in a single polypeptide chain
		Expressed at sites of tumor cell adhesion
Galectin-5	Erythrocytes	Prototype galecin: monomer
		No function assigned
Galectin-6	GIT	Composed of two distinct CRD in a single polypeptide chain
		Closely linked to galectin—4
Galectin-7	Skin	Prototype galecin: monomer
		Used as a marker of stratified epithelium
		Increases susceptibility of keratinocytes to UVB-induced apoptosis
Galectin-8	Liver, kidney, Cardiac Muscle, Prostate, Brain	Compose of two distinct CRD in a single polypeptide chain
		Modulates integrin interactions with extracellular matrix
Galectin-9	Thymus, T-cells, Kidney, Hodgkin's lymphoma	Composed of two distinct CRD in a single polypeptide chain
		Induces eosinophil chemotaxis
		Induces apoptosis of murine thymocytes

(continued)

Table 9.1 (continued)

Galectins/structure	Localisation	Biochemical and functional properties
Galectin-10	Eosinophils and basophils	Prototype galecin: monomer Mainly expressed by eosinophils, formally called 'Charcot-Leyden Crystal Protein'
Galectin-11	Lens	Also called 'Grifin' May represent a new lens crystalline Lacks affinity for β–galactoside sugars
Galectin-12	Adipocytes	Composed of two distinct CRD in a single polypeptide chain Induces apoptosis and cell cycle arrest
Galectin-13	Human placenta	Similar to proto type 'galectins'? Also called 'pp-13'
Galectin-14	Eosinophil (Sheep)	Plays a role in eosinophil function and allergic inflammation
Galectin-15	Uterus (Sheep)	Intracellular gal-15 regulates cell survival, differentiation function Extracellular secreted gal-15 regulates implants and placentation

cell-surface and extracellular matrix glycoconjugates through lectin-carbohydrate interactions. Through this action, they can affect a variety of cellular processes, and the most extensively documented function is induction of apoptosis. Evidences suggest that some of these functions involve binding to cytoplasmic and nuclear proteins, through protein-protein interactions, and modulation of intracellular signaling pathways. Galectin-1, -7, -8, -9 and -12 are characterized as promoters or inducers of apoptosis, while galectin-3 is demonstrated as an inhibitor of apoptosis intracellularly. Thus, some galectins are pro-apoptotic, whereas others are anti-apoptotic; while some galectins induce apoptosis by binding to cell surface glycoproteins, others regulate apoptosis through interactions with intracellular proteins. Involvement of galectin-1, -2, -3, -7, -8, -9, and -12 in apoptosis has been reviewed (Hsu et al. 2006).

9.4.3 Immunological Functions

Ggalectins have attracted the attention of immunologists as novel regulators of host–pathogen interaction (Almkvist and Karlsson 2004; Rabinovich and Gruppi 2005). Galectins participate in a wide spectrum of immunological processes. These proteins regulate the development of pathogenic T-cell responses by influencing T-cell survival, activation and cytokine secretion. Administration of recombinant galectins or their genes into the cells modulate the development and severity of chronic inflammatory responses in experimental models of autoimmunity by triggering different immunoregulatory mechanisms. Galectins are differentially expressed by various immune cells and their expression levels appear to be dependent on cell differentiation and activation. They can interact with cell-surface and extracellular matrix glycoconjugates (glycoproteins and glycolipids), through lectin-carbohydrate interactions. Current research indicates that galectins play important roles in the immune response through regulating the homeostasis and functions of the immune cells. Galectins have important roles in cancer, where they contribute to neoplastic transformation, tumor cell survival, angiogenesis and tumor metastasis. They can modulate the immune and inflammatory responses and might have a key role helping tumors to escape immune surveillance (Liu and Rabinovich 2005; Liu 2005). Galectins operate at different levels of innate and adaptive immune responses, by modulating cell survival and cell activation or by influencing the Th1/Th2 cytokine balance. The influence of galectins in immunological processes relevant to microbial infection has been explored.

Galectins are novel anti-inflammatory agents or targets for immunosuppressive drugs, and have impact in the development and/or resolution of chronic inflammatory disorders, autoimmunity, and cancer. Inflammation involves the sequential activation of signaling pathways leading to the production of both pro-inflammatory and anti-inflammatory mediators (Ilarregui et al. 2005; Rabinovich and Gruppi 2005). The role of galectins in the initiation, amplification and resolution of the inflammatory response has been reviewed by Rubinstein et al. (2004) who examined the influence of each member of this family in regulating cell adhesion, migration, chemotaxis, antigen presentation, immune cell activation and apoptosis.

Galectin-1 and galectin-3 are upregulated in gastric epithelial cells infected with *Helicobacter pylori*, which suggests that galectins might contribute to bacterial invasion. In addition, galectin-3 has the ability to bind to Gram-negative (G^-) bacteria through the recognition of different bacterial LPSs (Mey et al. 1996) and induce the death of Candida species expressing specific ß-1,2-linked mannans (Kohatsu et al. 2006). The elevation of galectin-9 is involved in the inflammatory response of periodontal ligament cells exposed to *Porphylomonas gingivalis* LPS in vitro and in vivo (Kasamatsu et al. 2005) and can recognize the *Leishmania major*-specific polygalactostosyl epitope (Pelletier et al. 2003).

Role in Chemotaxis: The involvement of galectins in chemotaxis has been reported for galectin-3 which stimulates both monocyte and macrophage chemotaxis (Sano et al. 2000; Papaspyridonos et al. 2008) and galectin- 9 (Hirashima 1999; Matsumoto et al. 1998) which is a selective chemoattractant for eosinophils. Malik et al. (2009) demonstrated a role for galectin-1 in monocyte chemotaxis which differed from galectin-3 in that macrophages were nonresponsive. Study revealed that galectin-1-induced migration of monocytes was mediated by a pertussis toxin (PTX)-sensitive pathway and by the P44/42 MAP kinase pathway.

9.4.4 Signal Transduction by Galectins

9.4.4.1 Galectin-Glycan Lattices Regulate Cell-Surface Glycoprotein Organization and Signaling

The formation of multivalent complexes of soluble galectins with glycoprotein receptors on the plasma membrane helps to organize glycoprotein assemblies on the surface of the cell. In some cell types, this formation of galectin-glycan lattices or scaffolds is critical for organizing plasma membrane domains, such as lipid rafts, or for targeted delivery of glycoproteins to the apical or basolateral surface. Galectin-glycan lattice formation is also involved in regulating the signaling threshold of some cell-surface glycoproteins, including T-cell receptors (TCRs) and growth factor receptors. Finally, galectin-glycan lattices can determine receptor residency time by inhibiting endocytosis of glycoprotein receptors from the cell surface, thus modulating duration of signaling from the cell surface (Garner and Baum 2008). As an example, T-cell activation requires clustering of a threshold number of TCRs at the site of antigen presentation. A deficiency in β1,6 N-acetylglucosaminyltransferase V (Mgat5), an enzyme in the N-glycosylation pathway, lowers T-cell activation thresholds by directly enhancing TCR clustering. Dysregulation of Mgat5 in humans may increase susceptibility to autoimmune diseases, such as multiple sclerosis (Demetriou et al. 2001). Galectin-mediated ligation of glycoproteins on T-cell activation markers induces an increase in the cytosolic calcium concentration $(Ca^{2+})_i$ originating from a transient $(Ca^{2+})_i$ release of internal stores as well as a sustained influx across the plasma membrane (Walzel et al. 1996; Walzel et al. 2002). In transiently transfected Jurkat T-lymphocytes, galectins differentially stimulate the expression of reporter gene constructs, which were activated by the nuclear factor of activated T-cells (NFAT) or the transcription factor, activator protein 1 (AP-1), respectively. Electrophoretic mobility shift assays (EMSAs) provided evidence for gal-1-stimulated increase in the binding of nuclear extracts to a synthetic oligonucleotide with an AP-1 consensus sequence (Walzel et al. 2002). Cha et al. (2008) demonstrated that a galectin-1 lattice is responsible for retaining the renal epithelial Ca^{2+} channel TRPV5 at the plasma membrane, providing evidence for galectin lattice-mediated regulation of ion balance via regulation of channel residency at the cell surface (Cha et al. 2008).

9.4.4.2 Galectins, Integrins and Cell Migration

The interaction of galectins with integrins modulates cell migration as well as other processes. Galectin-1 interacts

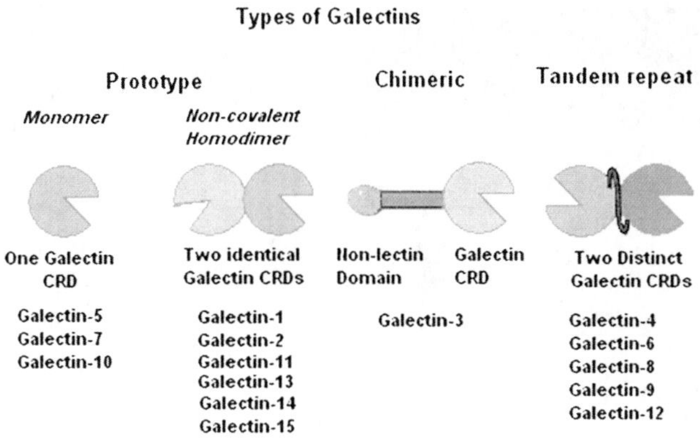

Types of Galectins

Prototype
Monomer
Non-covalent Homodimer
Chimeric
Tandem repeat
One Galectin CRD
Two identical Galectin CRDs
Non-lectin Domain
Galectin CRD
Two Distinct Galectin CRDs
Galectin-5
Galectin-7
Galectin-10
Galectin-1
Galectin-2
Galectin-11
Galectin-13
Galectin-14
Galectin-15
Galectin-3
Galectin-4
Galectin-6
Galectin-8
Galectin-9
Galectin-12

Fig. 9.2 Schematic representation of overall structures of galectins as assembled proteins

with the β1 integrin subunit inducing the phosphorylation of FAK, which modulates cell migration (Moiseeva et al. 2003). Binding of Gal-1 to integrin is involved in cell adhesion (Moiseeva et al. 1999). Moreover, Gal-1 also regulates the expression of the protein ADAM-15 that is involved in integrin-mediated adhesion (Camby et al. 2005). Gal-1 also induces growth inhibition via its interaction with α5β1 (Fischer et al. 2005). This interaction results in the inhibition of the Ras–MEK–ERK pathway and the consecutive transactivation of Sp1, which induces p27 transcription (Fischer et al. 2005). In addition, Gal-1 is involved in the PKCε/vimentin controlled trafficking of integrin β1, a process that is important for cell migration (Fortin et al. 2010). However, it is not known with which molecule(s) Gal-1 is interacting, or in which intra- or extracellular location this interaction is taking place in order to initiate this signaling. Finally, Gal-1 is also involved in cell motility via the Gal-1-induced expression of RhoA and the alteration of the polimerization of the actin cytoskeleton (Camby et al. 2002).

Galectin-3 regulates cell adhesion via binding to α1β1 (Ochieng et al. 1998). Gal-3 also forms a complex with α3β1 and the proteoglycan NG2 (Fukushi et al. 2004). This interaction regulates endothelial cell motility and angiogenesis. In addition, Gal-3 has been shown to regulate the expression of integrin α6β1 and actin cytoskeleton organization (Debray et al. 2004). It is not known with which molecule (s) Gal-3 is interacting to initiate this signaling. Gal-8 interacts with several integrins including α1β1, α3β1, α5β1 and α6β1. These interactions participate in cell adhesion and apoptosis (Hadari et al. 2000).

9.4.5 Common Structural Features in Galectins

Lectins are generally organized as oligomers of noncovalently bound subunits, each displaying a carbohydrate recognition domain (CRD) that binds to the sugar ligand, usually a nonreducing terminal monosaccharide or oligosaccharide. Although lectin-ligand interactions are relatively weak as compared with other immune recognition molecules, high avidity for the target is achieved by association of peptide subunits into oligomeric structures in which multiple CRDs interact with ligand simultaneously. The overall structures of galectins are shown schematically in Fig. 9.2. Galectin-1 and -2 are homodimers composed of subunits of about 130 amino acids. Each subunit folds as one compact globular domain as shown in Fig. 9.3. Galectin-3 and -4 include one or two such domains, as well as others. The shared domain has been referred to as carbohydrate-binding domain. The sequence of each carbohydrate-binding domain has been shown to be mainly encoded by 3 exons (Barondes et al. 1994). Most of the residues that are conserved among galectins are found in the sequence encoded by one of these threee axons. This sequence includes four contiguous β-strands and intervening loops in the structure of galectin-l (Liao et al. 1994) and galectin-2- (Lobsanov et al. 1993) and contains all residues that interact directly with a carbohydrate ligand (Barondes et al. 1994).

The structures of proto-type galectins have been resolved revealing a common jellyroll topology, typical of legume lectins (Liao et al. 1994; Vasta et al. 2004b) (Composed of two distinct CRD in a single polypeptide chain). In galectin dimer, each subunit is composed of an 11-strand antiparallel ß-sandwich that contains a CRD, with the N- and C-termini located at the dimer interface. The structure of the bovine (*Bos taurus*) gal1-LacNAc (N-acetyllactosamine) complex revealed that the amino acids H^{44}, N^{46}, R^{48}, N^{61}, E^{71}, and R^{73} are directly involved in hydrogen bonding with 4 and 6 hydroxyls of galactose unit, and 3 hydroxyl of N-acetylglucosamine group of LacNAc. W^{69} provides a strong hydrophobic interaction with galactose ring of LacNAc (Liao et al. 1994). Thus, galectin-ligand complex has provided a detailed description of lectin-carbohydrate-interactions. The importance of some of these residues for carbohydrate binding activity is also supported by site-directed mutagenesis. Deletion of sequences

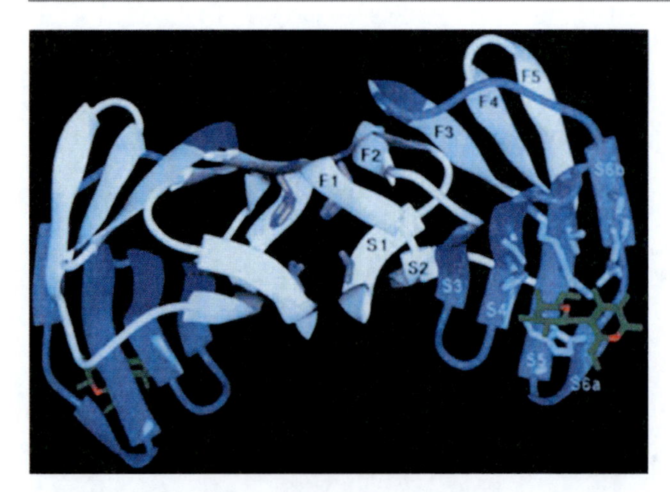

Fig. 9.3 X-ray crystal structure of human galectin-2. The figure shows a dimer of two globular carbohydrate-bindin domains (*blue* and *white ribbon* diagram) with bound lactose (*green stick representation* with *red* ring oxygens). The part of each domain (subunit) encoded by exon III is shown in *blue* and the other parts in *white*. The dimer interface is in the middle and is highlighted by *purple coloring* of the major contributing residues (Val and Ile). Each subunit consists of a sandwich of two β-sheets of 6 and 5 strands. In the subunit to the right the strands of each sheet are labeled Sl-5, S6a, and S6b (strand 6 is interrupted by a β-bulge), and Fl-5. The amino acid side chains interacting with the carbohydrate are displayed as *light blue stick* figures. In the right subunit the carbohydrate-binding site is facing toward the viewer, whereas in the left subunit the carbohydrate-binding site is facing away from the viewer (Reprinted by permission from Lobsanov et al. (1993) © American Society for Biochemistry and Molecular Biology)

encoded by the other two exons that encode the carbohydrate-binding domain also impairs activity (Abbott and Feizi 1991). The amino acid identity in the carbohydrate-binding domains among different known galectins from one mammalian species ranges from about 20% to 40%.

9.4.5.1 Binding Sites in Galactose-Specific Proteins

With the assumption that common recognition principles must exist for common substrate recognition, a study was undertaken to identify and characterize any unique galactose-binding site by analyzing the three-dimensional structures of 18 protein-galactose complexes. These proteins belong to seven nonhomologous families; thus, there is no sequence or structural similarity across the families. Within each family, the binding site residues and their relative distances were well conserved, but there were no similarities across families. Sujatha and Balaji (2004) furnished a potential galactose-binding site signature, evaluated by incorporation into the program COTRAN to search for potential galactose-binding sites in proteins that share the same fold as the known galactose-binding proteins. The deduced galactose-binding site signature is strongly validated and can be used to search for galactose-binding sites in proteins. PROSITE-type signature sequences have also been inferred for galectin and C-type animal lectin-like fold families of Gal-binding proteins (Sujatha and Balaji 2004).

9.4.5.2 Role of Aromatic Residue in Galactose-Binding Sites

The presence of an aromatic residue (Trp, Phe, Tyr) facing the nonpolar face of galactose is a common feature of galactose-specific lectins. The interactions such as those between the C-H groups of galactose and the pi-electron cloud of aromatic residues were characterized as weak hydrogen bonds between soft acids and soft bases, largely governed by dispersive and charge transfer interactions. An analysis of the binding sites of several galactose-specific lectins revealed that the spatial position-orientation of galactose relative to the binding site aromatic residue varies substantially. The strength of the C-H. . .pi interactions in galactose-aromatic residue complexes was comparable to that of a hydrogen bond. The study showed that the aromatic residue is important for discriminating galactose from glucose, in addition to its contribution to binding energy (Sujatha et al. 2005).

9.4.5.3 Galectins Bind Multivalent Glycoprotein Asialofetuin

Dam et al. (2002) showed that the binding of multivalent carbohydrates to the Man/Glc-specific lectins ConA and Dioclea grandiflora lectin (DGL) involve negative binding cooperativity that was due to the carbohydrate ligands and not the proteins. The negative cooperativity was associated with the decreasing functional valence of carbohydrates upon progressive binding of their epitopes. Negative cooperativity was also shown in binding of asialofetuin (ASF), a glycoprotein that possesses nine LacNAc epitopes, to galectin-1, -2, -3, -4, -5, and -7, and truncated, monomer versions of galectin-3 and -5. Study indicated that the galectins bind with fractional, high affinities to multivalent glycoproteins such as ASF, independent of the quaternary structures of the galectins. The report has important implications for the binding of galectins to multivalent carbohydrate receptors (Dam et al. 2005).

9.4.6 Galectin Subtypes in Tissue Distribution

9.4.6.1 Galectin Subtypes in Mouse Digestive Tract

Mucosal epithelium showed region/cell-specific localization of each galectin subtype. Gastric mucous cells exhibited intense immunoreactions for galectin-2 and galectin-4/6 with a limited localization of galectin-3 at the surface of the gastric mucosa. Epithelial cells in the small intestine showed characteristic localizations of galectin-2 and galectin-4/6 in the cytoplasm of goblet cells and the baso-lateral membrane of enterocytes in association with maturation, respectively. Galectin-3 expressed only at the villus tips was concentrated at the myosin-rich terminal web of fully matured enterocytes. Epithelial cells of large intestine contained intense reactions for galectin-3 and galectin-4/6 but not for galectin-2. The stratified squamous epithelium of the forestomach was immunoreactive for galectin-3 and galectin-7. Outside the epithelium, Only galectin-1 was localized in the connective tissue, smooth muscles, and neuronal cell bodies. The subtype-specific localization of galectin suggests its important roles in host-pathogen interaction and epithelial homeostasis such as membrane polarization and trafficking in gut (Nio-Kobayashi et al. 2009).

9.4.6.2 Multiple Galectins in Primary Olfactory Pathway

Primary olfactory neurons project axons from the olfactory neuroepithelium lining the nasal cavity to the olfactory bulb in the brain. These axons grow within large mixed bundles in the olfactory nerve and then sort out into homotypic fascicles in the nerve fiber layer of the olfactory bulb before terminating in topographically fixed glomeruli. The expression of the lactoseries binding galectins in the primary olfactory system and their interactions with carbohydrates on the cell surface have been implicated in axon sorting within the nerve fiber layer. In particular, galectin-3 is expressed by ensheathing cells surrounding nerve fascicles in the submucosa and nerve fiber layer, where it may mediate cross-linking of axons. Galectin-4, -7, and -8 are expressed by the primary olfactory axons as they grow from the nasal cavity to the olfactory bulb. A putative role for NOC-7 and NOC-8 in axon fasciculation and the expression of multiple galectins in the developing olfactory nerve suggest that these molecules may be involved in the formation of this pathway, particularly in the sorting of axons as they converge towards their target (Storan et al. 2004).

9.5 Prototype Galectins (Mono-CRD Type)

Galectin-1, -2, -5, -7, -10, -11, -13, -14, and -15 belong to the mono-CRD type and are found in many cells and tissues, have molecular weights of 14–18 kDa with a single galactoside-binding domain/CRD with a short N-terminal sequence. But the tandem-repeat type galectins (galectin-4, -6, -8, -9, and -12) are composed of two non-identical galactoside-binding domain/CRDs joined by a short linker peptide sequence

9.5.1 Galectin-1

Galectin-1 is differentially expressed by various normal and pathological tissues and appears to be functionally polyvalent, with a wide range of biological activity. The intracellular and extracellular activity of Gal-1 has been described (Chap. 10). Evidence points to Gal-1 and its ligands as one of the master regulators of such immune responses as T-cell homeostasis and survival, T-cell immune disorders, inflammation and allergies as well as host-pathogen interactions. Gal-1 expression or over-expression in tumors and/or the tissue surrounding them must be considered as a sign of the malignant tumor progression that is often related to the long-range dissemination of tumor cells (metastasis), to their dissemination into the surrounding normal tissue, and to tumor immune-escape. Gal-1 in its oxidized form plays a number of important roles in the regeneration of the central nervous system after injury. The targeted over-expression (or delivery) of Gal-1 should be considered as a method of choice for the treatment of some kinds of inflammation-related diseases, neurodegenerative pathologies and muscular dystrophies. A homodimeric (prototype) galectin-1 effectively suppressed symptoms of autoimmune disease in two models at cellular level delineating this lectin as inhibitory (apoptotic) effector for activated T cells (Perillo et al. 1995) (see Chap. 13). Gal-1 has solidly proven its potency to keep inflammatory and autoimmune responses in check in various systems (Gabius 2001; Rabinovich et al. 2002b). Its selection for distinct glycoprotein ligands, including CD3, CD7, CD43, and CD45, and its cross-linking ability appear to be crucial for induction of T cell apoptosis (Nguyen et al. 2001; Rabinovich et al. 2002a).

9.5.2 Galectin-2

Galectin-2, encoded by *LGALS2* gene, is a soluble β-galactoside binding protein. It is found as a homodimer and can bind to lymphotoxin-α. Galectin-2 is structurally related to galectin-1, but has a distinct expression profile primarily confined to gastrointestinal tract (Gitt and Barondes 1991; Gitt et al. 1992; Oka et al. 1999). In fact, Gal-2 and -7 are prototype proteins that are expressed by human tumor cells with cell type specificity (Kopitz et al. 2003). Structurally, Gal-2shares 43% amino acid sequence identity with Gal-1 (Sturm et al. 2004). Gal-2 can bind to T cells and trigger apoptosis. Prominent differences in the proximal promoter regions between galectins-2 and -1 concern Sp1-, hepatocyte NF-3, and T cell-specific factor-1 binding sites. These sequence elements are positioned equally in respective regions for human and rat galectins-2. Galectin-2 binds to T cells in a β-galactoside-specific manner. In contrast to galectin-1, glycoproteins CD3 and CD7 are not ligands, while β1 integrin (or a closely associated glycoprotein) accounts for a substantial extent of cell surface binding. In general, positivity of galectin-2 was predominantly epithelial without restriction of staining to certain tissue types. Staining was not limited to the cytoplasm but also included nuclear sites (Saal et al. 2005). A single nucleotide polymorphism in an intron of this gene can alter the transcriptional level of the protein, with a resultant increased risk of myocardial infarction (Ozaki and Tanaka 2005; Tanaka and Ozaki 2006).

The brush border of pig small intestine is a local hotspot for galectins. Galectins 3–4, intelectin, and lectin-like anti-glycosyl antibodies have been localized at this boundary. Galectins offer a maximal protection of the brush border against exposure to bile, pancreatic enzymes and pathogens (Thomsen et al. 2009). A lactose-sensitive 14-kDa galectin-2 is enriched in microvillar detergent resistant fraction. Its release from closed, right-side-out microvillar membrane vesicles shows that at least some of the galectin-2 resides at the lumenal surface of brush border, indicating that it plays a role in the organization/stabilization of lipid raft domains.

9.5.2.1 Galectin-2: Inducer of T Cell Apoptosis

The carbohydrate-dependent binding of galectin-2 induces apoptosis in activated T cells. Enhanced cytochrome C release, disruption of the mitochondrial membrane potential, and an increase of the Bax/Bcl-2 ratio by opposite regulation of expression of both proteins add to the evidence that the intrinsic apoptotic pathway is triggered. Notably, galectins-1 and -7 reduce cyclin B1 expression, defining functional differences between the structurally closely related galectins. Cytokine secretion of activated T cells was significantly shifted to the Th2 profile. Thus, galectin-2 acts as proapoptotic effector for activated T cells, suggesting a therapeutic perspective. However, the effects on regulators of cell cycle progression are markedly different between structurally closely related galectins. A single nucleotide polymorphism in an intron of this gene can alter the transcriptional level of the protein, with a resultant increased risk of myocardial infarction (Sturm et al. 2004).

Galectin-2 Induces Apoptosis of Lamina Propria T Lymphocytes: Paclik et al. (2008a) studied the therapeutic effect of galectin-2 in experimental colitis. Galectin-2-constitutively expresses mainly in epithelial compartment of mouse intestine and binds to lamina propria mononuclear cells. During colitis, galectin-2 expression was reduced, but could be restored to normal levels by immunosuppressive treatment. Galectin-2 treatment induced apoptosis of mucosal T cells and thus ameliorated acute and chronic dextran-sodium-sulfate-induced colitis and in a T-helper-cell driven model of antigen-specific transfer colitis. Furthermore, the pro-inflammatory cytokine release was inhibited by galectin-2 treatment. This study provides evidence that galectin-2 induces apoptosis in vivo and ameliorates acute and chronic murine colitis. Furthermore, galectin-2 had no significant toxicity over a broad dose range, suggesting that it may serve as a new therapeutic agent in the treatment of inflammatory bowel disease.

Gal-2 Suppresses Contact Allergy by Inducing Apoptosis in Activated CD8$^+$ T Cells: Galectin-2 regulates cell-mediated inflammatory bowel disease and colitis in mice. Loser et al. (2009) sensitized groups of naive mice to the contact allergen 2,4-dinitro-1-fluorobenzene and systemically treated with galectin-2 to analyze the effects of galectin-2 on contact allergy. Galectin-2 is expressed in murine skin and is up-regulated upon cutaneous inflammation. Interestingly, treatment of mice with galectin-2 significantly reduced the contact allergy response. This effect was long-lasting since rechallenge of galectin-2-treated mice after a 14-day interval still resulted in a decreased ear swelling. Further, galectin-2 induced a reduction of MHC class I-restricted immune responses in the treated animals, which was mediated by the induction of apoptosis specifically in activated CD8$^+$ T cells. Additionally, the galectin-2-binding protein CD29 is up-regulated on the surface of activated CD8$^+$ T cells compared with naive CD8$^+$ T cells or CD4$^+$ T cells, suggesting that increased galectin-2/CD29 signaling might be responsible for proapoptotic effects of galectin-2 on activated CD8$^+$ T cells. Data indicate that galectin-2 may represent a novel therapeutic alternative for the treatment of CD8-mediated inflammatory disorders such as contact allergy (Loser et al. 2009).

9.5.2.2 SNP in *LGALS2* and Lymphotoxin-α as Risk Factor in Cardiovascular Disease

The genotypes for 296 polymorphisms of 202 candidate genes revealed that *LGALS2* is one of the susceptibility loci for atherothrombotic cerebral infarction among Japanese individuals with metabolic syndrome. Genotypes for these polymorphisms may prove informative for the prediction of genetic risk for atherothrombotic cerebral infarction among such individuals (Yamada et al. 2008). Genetic and clinical studies implicate Lymphotoxin-α (LTA), and its binding and regulatory partner galectin-2 as risk factors in the pathogenesis of cardiovascular diseases including myocardial infarction (MI), aortic aneurysm, and cerebral infarction. The LTA gene variability is also associated with an increased level of C-reactive protein, an inflammatory marker (Naoum et al. 2006). The LTA has multiple functions in regulating the immune system and may contribute to inflammatory processes leading to CHD (coronary heart disease) with susceptibility to MI. (Ozaki et al. 2004). After genotyping approximately 65,000 SNPs in 1,000 patients, Tanaka and Ozaki (2006) found two SNPs located within LTA gene showing significant association with MI. These SNPs seemed to be involved in inflammation by both qualitatively and quantitatively modifying the function of LTA protein, thereby conferring a risk of MI. Ozaki and Tanaka (2005) identified functional SNPs within the LTA located on chromosome 6p21 that conferred susceptibility to myocardial infarction (Tanaka and Ozaki 2006; Topol et al. 2006; Ozaki and Tanaka 2005). The *LGALS2* gene variant is significantly associated with a decreased risk of CHD in women. In addition, the *LGALS2* polymorphism was directly associated with CRP (C-reactive protein) levels in two studies. The LTA gene polymorphisms were directly associated with levels of sTNFRs (soluble tumor necrosis factor receptors) and VCAM-1 in both women and men with CHD (Asselbergs et al. 2007).

The 252GG homozygote variant of LTA is considered as a susceptibility factor for arteriosclerosis and cardiovascular diseases. By contrast, galectin-2-encoding gene *LGALS2* 3279TT homozygote variant has been shown to exert protection against MI by reducing the transcriptional level of galectin-2, thereby leading to a reduced extracellular secretion of LTA. The combination of *LGALS2* 3279TT and LTA 252GG homozygote was significantly less frequent in ischemic stroke group (1.56%) than in controls. This finding suggests a gene-gene interaction (Szolnoki et al. 2009). However, Freilinger et al. (2009) reported that genetic variation in lymphotoxin-α cascade (LTA, LGALS2, and PSMA6) is not a major risk factor. The gene encoding LTA is associated with insulin resistance. Using the *LGALS2* genotype, which affects LTA secretion but is located on another chromosome than the HLA gene cluster or TNF, Christensen et al. (2006) examined the relationship between the LTA pathway and traits of the metabolic syndrome in the British Women for the physically unlinked *LGALS2*. Studies invite attention for further study of *LGALS2* and the LTA pathway for their influence on glucose-insulin regulation (Christensen et al. 2006).

For populations of other genetic background, the relevance of these polymorphisms in the pathogenesis of MI remains controversial. A case–control study on British MI patients and controls showed 98% of power to detect a significant association at OR of 1.57, and 80% power to detect an association with an OR of 1.35 (recessive model). Despite this, there was no significant association of allele frequency with risk of MI. Stratification for age, gender and other cardiovascular risk factors also failed to reveal an association of this polymorphism with MI (Mangino et al. 2007). Kimura et al. (2007) studied SNP of LTA (LTA 252A > G in LTA intron 1) and that of *LGALS2* (*LGALS2* 3279C>T in *LGALS2* intron 1) in Japanese and Korean populations. Although significant associations with MI were not observed in either population, LTA 252GG was significantly associated with severity of the disease for both the Japanese and Korean populations. On the other hand, the polymorphism of *LGALS2* was not associated with the severity of coronary atherosclerosis. These observations showed that, while the LTA 252GG genotype might modify the development of coronary atherosclerosis, the relation of LTA and *LGALS2* to MI itself remained much less certain (Kimura et al. 2007). Moreover, no association could be found in two MI populations of European descent for any of the examined SNPs in the LTA genomic region and *LGALS2* gene. These variants are unlikely to play a significant role in populations of European origin (Sedlacek et al. 2007). Although hypertension, a very prevalent entity in rheumatoid arthritis (RA), is one of the greatest risk factors for MI, the possible association of *LGALS2* 3279C/T and hypertension was genotyped by Panoulas et al. (2009) among 386 RA patients, 272 hypertensives and 114 normotensives. Diastolic blood pressure was significantly lower in TT compared to CC homozygotes even when adjusted for multiple confounders (Panoulas et al. 2009). SNP of genes for LTA and galectin-2 have been implicated as genetic risk factors for RA compared with non-RA controls, thus explaining some of the increased CVD in rheumatoid arthritis, and may be more frequent among patients with RA with prevalent CVD compared with patients with RA without CVD. The LT-A 252GG genotype occurs more frequently among patients with RA than the general population. In RA, this genotype appears to associate with increased likelihood of suffering an myocardial infarction (Panoulas et al. 2008).

BRAP (BRCA1-associated protein) is another galectin-2-binding protein. Association of SNPs in BRAP is risk factor with MI as found in Japanese cohort and in additional Japanese and Taiwanese cohorts. BRAP expression was observed in smooth muscle cells (SMCs) and macrophages in human atherosclerotic lesions. BRAP knockdown by siRNA using cultured coronary endothelial cells suppressed activation of NF-kB, a central mediator of inflammation (Ozaki et al. 2009).

In addition to *LGALS2*, polymorphic sites have been identified within 11-kb region containing the gene encoding galectin-1 (*LGALS1*). The map includes 14 SNPs and two genetic variations of other types detected in a Japanese population sample. Five of the 14 SNPs were not among those deposited in the dbSNP database in NCBI and appeared to be novel. Investigation of haplotype structure within the *LGALS1* locus revealed five common haplotypes covering more than 95% of the test population. One, or a pair, of the SNPs described in this study might serve as a "tag" for detecting associations between complex diseases and genes in this local segment of chromosome 22q13.1 (Iida et al. 2005).

9.5.3 Galectin-5

The monomeric galectin-5 from extracts of rat lung has been localized to erythrocytes. The deduced amino acid sequence of the cDNA predicts a protein with a M_r of 16,199, with no evidence of a signal peptide. The deduced sequence is identical to the sequences of seven proteolytic peptides derived from purified lectin. Peptide analysis indicated that the N-terminal methionine is cleaved and that serine 2 was acetylated. Galectin-5 is a weak agglutinin of rat erythrocytes, despite its monomeric structure. The gene encoding galectin-5 (*LGALS5*) has been mapped in mouse to chromosome 11, approximately 50 centimorgans from the centromere and 1.8 ± 1.8 centimorgans from the polymorphic marker D11Mit34n, a region syntenic with human chromosome 17q11 (Gitt et al. 1995).

Galectin-5, although mainly cytosolic, is also present on the cell surface of rat reticulocytes and erythrocytes. In reticulocytes, it resides in endosomal compartment. Galectin-5 translocates from cytosol into the endosome lumen, leading to its secretion in association with exosomes. Galectin-5 bound onto the vesicle surface may function in sorting galactose-bearing glycoconjugates. Data imply galectin-5 functionality in the exosomal sorting pathway during rat reticulocyte maturation (Barres et al. 2010). Galectin 1 and galectin 3 are first expressed in the trophectoderm cells of the implanting embryo and have been implicated in the process of implantation. However, presence of galectin 5 in the blastocyst at the time of implantation could be important in the process (Colnot et al. 1998).

9.5.3.1 Interaction with Saccharides and Glycoproteins

Interaction of cell-surface glycans with galectins triggers a wide variety of responses. Among 45 natural glycans tested for binding, galectin-5 reacted best with glycoproteins (gps) presenting a high density of Galβ1-3/4GlcNAc (I/II) and multiantennary N-glycans with II termini. Wu et al. (2006) suggested that (1) Galβ1-3/4GlcNAc and other Galβ1-related oligosaccharides with α1-3 extensions are essential for binding, their polyvalent form in cellular glycoconjugates being a key recognition force for galectin-5; (2) the combining site of galectin-5 appears to be of a shallow-groove type sufficiently large to accommodate a substituted β-galactoside, especially with α-anomeric extension at the non-reducing end (e.g., human blood group B-active II and B-active IIβ1-3 L); (3) the preference within β-anomeric positioning is Galβ1-4 $\geq$ Galβ1-3 $>$ Galβ1-6; and (4) hydrophobic interactions in the vicinity of core galactose unit can enhance binding. Systematic comparison of ligand selection in this family of adhesion/growth-regulatory effectors offers potential for medical applications (Wu et al. 2006). Lensch et al. (2006) studied galectins-5 and -9. After ascertaining species specificity of occurrence of galectin-5, constituted by a short section of rat galectin-9′s N-terminal part and its C-terminal carbohydrate recognition domain, the results on galectin-5 relative to galectin-9 intimated distinct functions especially in erythropoiesis and implied currently unknown mechanisms to compensate its absence from the galectin network in other mammals.

9.5.4 Galectin-7

Galectin-7 is a β-galactoside-binding animal lectin specifically expressed in stratified epithelia. It is expressed in all surface epithelium, glandular epithelium, and connective tissue in human nasal polyps. It contributes to different events associated with the differentiation and development of pluristratified epithelia. It is also associated with epithelial cell migration, which plays a crucial role in the re-epithelialization process of corneal or epidermal wounds. Galectin-7 is a proapoptotic protein, and the ectopic expression of galectin-7 in HeLa cells renders the cells more sensitive to a variety of apoptotic stimuli. In nasal polyposis model for the study of inflammatory processes, Galectin-7 expression coincides with the degree of epithelial stratification, and is subject to upregulation in connective tissue in response to treatment with budesonide. Budesonide modulates galectin-7 expression differently in surface epithelia of polyps from allergic and non-allergic patients (Delbrouck et al. 2005). Expression of galectin-7 is inducible by p53 and is down-regulated in squamous cell carcinomas. Galectin-7 has a suppressive effect on tumor growth, suggesting that galectin-7 gene transfer or other means of specifically inducing galectin-7 expression may be a new approach for management of cancers (Ueda et al. 2004). Studies are consistent with a role for galectin-7 in the regulation of cell growth through a pro-apoptotic effect.

Galectin-7 corresponds to IEF (isoelectric focusing) 17 (12,700 Da; pI, 7.6) in the human keratinocyte protein data base, and is strikingly down-regulated in SV40 transformed keratinocytes (K14). The protein encoded by the galectin-7 clone co-migrated with IEF 17 in 2D gel electrophoresis and bound lactose. The galectin-7 gene was mapped to chromosome 19 (Madsen et al. 1995). The 14-kDa lectin of pI7 predicted by 136-amino-acid ORF is specifically expressed in keratinocytes. It is expressed at all stages of epidermal differentiation. It is moderately repressed by retinoic acid, a behavior contrasting with those of other keratinocyte markers sensitive to this agent, which, either basal, are induced, or suprabasal, are repressed (Magnaldo et al. 1995). In human, rat and mouse epithelia, Galectin-7 was found to be expressed in interfollicular epidermis and in the outer root sheath of the hair follicle, but not in the hair matrix, nor in the sebaceous glands. It was present in esophagus and oral epithelia, cornea, Hassal's corpuscles of the thymus, but not in simple and transitional epithelia. Galectin-7 can thus be considered as a marker of all subtypes of keratinocytes (Magnaldo et al. 1998). Significant amounts of galectin-7 was detected in trachea and ovaries and localized in pseudostratified epithelium of the trachea and stromal epithelium of ovaries. It suggests that galectin-7 protein might be produced primarily in stratified epithelia, but also in some wet epithelia, and plays a unique role in cell-mucus contact, or the growth of ovarian follicles (Sato et al. 2002). Galectin-7 has been cloned from human, rat and mouse. In the adult, galectin-7 is expressed in all cell layers of epidermis and of other stratified epithelia such as the cornea and the lining of the oesophagus. Galectin-7 mRNA was identified in mouse embryos starting from E13.5, in bilayered ectoderm, and stronger expression was found in areas of embryonic epidermis where stratification was more advanced. In contrast, no expression of galectin-7 was found in epithelia derived from endoderm, such as lining of the intestine, kidney and lung. Thus, galectin-7 is expressed in all stratified epithelia examined so far, and that the onset of its expression coincides with the first visible signs of stratification (Timmons et al. 1999).

Galectin-7 is also associated with epithelial cell migration, which plays a crucial role in the re-epithelialization process of corneal or epidermal wounds. In addition, recent evidence indicates that galectin-7, a product of p53-induced gene 1 (designated as PIG1), is a regulator of apoptosis through JNK activation and mitochondrial cytochrome c release. The increase of galectin-7 is parallel to P53 stabilization. Defects in apoptosis constitute one of the major hallmarks of human cancers, and galectin-7 can act as either a positive or a negative regulatory factor in tumor development, depending on the histological type of the tumor (Saussez and Kiss 2006). UVB irradiation of skin is associated with sunburn/apoptotic keratinocytes. The galectin-7 mRNA and protein are increased rapidly after UVB irradiation of epidermal keratinocytes and involved in UV-induced apoptosis (Bernerd et al. 1999). Galectin-7 acts on a common point in apoptosis signaling pathways. It is a pro-apoptotic protein that functions intracellularly upstream of JNK activation and cytochrome C release (Kuwabara et al. 2002). Galectin-7 was found to be highly inducible by tumor suppressor protein p53 in a colon carcinoma cell line, DLD-1 and its gene is an early transcriptional target of p53 (Polyak et al. 1997).

Galectin-7 has a potential to reduce cell proliferation after carbohydrate-dependent surface binding in neuroblastoma cells (Kopitz et al. 2003). Suppressive effect of galectin-7 on tumor growth suggests that galectin-7 gene transfer or other means of specifically inducing galectin-7 expression may be a new approach for management of cancers (Ueda et al. 2004). Galectin-3 and galectin-7 are potential tumor markers for differentiating thyroid carcinoma from its benign counter part (see Chap. 13). They are supposed to be p53-related regulators in cell growth and apoptosis, being either anti-apoptotic or pro-apoptotic. The immunochemical localisation of Galectin-3 is a useful marker in conjunction with routine haematoylin and eosin staining in differentiating benign from malignant thyroid lesions, while there is no significant adjunct diagnostic value in galectin-7 for thyroid malignancy (Than et al. 2008b).

Leonidas et al. (1998) reported the crystal structure of human galectin-7 in free form and in presence of galactose, galactosamine, lactose, and N-acetyl-lactosamine at high resolution. The galectin-7 structure shows a fold similar to that of prototype galectins -1 and -2, but has greater similarity to galectin-10. Even though the carbohydrate-binding residues are conserved, there are significant changes in this pocket due to shortening of a loop structure. The monomeric human galectin-7 exists as a dimer in the crystals, but adopts a packing arrangement considerably different from that of Gal-1 and Gal-2, which has implications for carbohydrate recognition. N-acetyllactosamine thioureas are good inhibitors of galectin-7 and 9 N (K_D = 23 and 47 μM respectively for 3-pyridylmethyl-thiourea derivative) and represented more than an order of magnitude affinity enhancement over parent natural N-acetyllactosamine (Salameh et al. 2006).

9.5.5 Galectin-10 (Eosinophil Charcot-Leyden Crystal Protein)

The Charcot-Leyden crystal (CLC) protein is a major autocrystallizing constituent of human eosinophils and basophils, comprising ~10% of total cellular protein in these granulocytes. Identification of the distinctive hexagonal bipyramidal crystals of CLC protein in body fluids and secretions has long been considered a hallmark of eosinophil-associated allergic inflam-mation. Although CLC protein possesses lysophospholipase activity, its role(s) in eosinophil or basophil function or associated inflammatory responses has remained speculative. Across mammalian species, human galectin/CLC and ovine galectin-14 are unique in their expression in eosinophils and their release into lung and gastrointestinal tissues following allergen or parasite challenge. The X-ray crystal structure of the CLC protein is very similar to the structure of the galectins, including a partially conserved galectin CRD. Structural studies on the carbohydrate binding properties of CLC protein demonstrate no affinity for β-galactosides but binds mannose in a manner different from those of other related galectins that have been shown to bind lactosamine. The partial conservation of residues involved in carbohydrate binding led to significant changes in the topology and chemical nature of the CRD. Carbohydrate recognition by the CLC protein in vivo suggests its functional role in the process of inflammation (Swaminathan et al. 1999). Galectin-10 is a novel marker for evaluating celiac disease tissue damage and eosinophils as a possible target for therapeutic approaches (De Re et al. 2009). In sensitive patients, aspirin is associated with nasal and bronchial inflammation, eliciting local symptoms. Galectin-10 mRNA is overexpressed in aspirin-induced asthma, suggesting a novel candidate gene and a potentially innovative pathway for mucosal inflammation in aspirin intolerance (Devouassoux et al. 2008).

$CD4^+CD25^+Foxp3^+$ regulatory T cells ($CD25^+$ T_{reg} cells) direct maintenance of immunological self-tolerance by active suppression of autoaggressive T-cell populations. Galectin-10 is constitutively expressed in human $CD25^+$ T_{reg} cells, while they are nearly absent in resting and activated $CD4^+$ T cells. As expressed in human T lymphocytes, galectin-10 is essential for the functional properties of $CD25^+$ T_{reg} cells (Kubach et al. 2007). The CLC protein possesses weak lysophospholipase activity of eosinophil, but it shows no sequence similarities to any known lysophospholipases. In contrast, CLC protein has moderate sequence similarity, conserved genomic organization, and near structural identity to members of the galectin superfamily, and designated galectin-10. Ackerman et al. (2002) reassessed its enzymatic activity in peripheral blood eosinophils and an eosinophil myelocyte cell line (AML14.3D10). The affinity-purified CLC protein lacked significant lysophospholipase activity. X-ray crystallographic structures of CLC protein in complex with the inhibitors showed that p-chloromercuribenzenesulfonate bound CLC protein via disulfide bonds with Cys-29 and with Cys-57 near CRD, whereas N-ethylmaleimide bound to the galectin-10 CRD via ring stacking interactions with Trp-72, in a manner highly analogous to mannose binding to this CRD. Results clearly showed that CLC protein is not one of the eosinophil's lysophospholipases but it does interact with inhibitors of this lipolytic activity (Ackerman et al. 2002).

Transcriptional Regulation

Analysis of the minimal promoter revealed nine consensus-binding sites for transcription factors, including several that are also found in the minimal promoters of galectins -1, -2, and -3. The decrease in gal-10 promoter activity after disruption of either the GC box (−44 to −50) or the Oct site (−255 to −261) suggests that these sites, along with GATA and EoTF sites, are necessary for full promoter activity. Transcription factors Sp1 and Oct1 bind to the consensus GC box and the Oct site, respectively. Similar to gal-1, gal-10 expression is induced by butyric acid, an effect that is lost upon ablation of GC box. Additionally, AML3 binds to consensus AML site and YY1 binding to the Inr sequence, both elements functioning as silencers in the gal-10 promoter (Dyer and Rosenberg 2001).

Crystal Structure

The crystal structure of the CLC protein at 1.8 A resolution shows overall structural similarity to that of galectins -1 and -2. The CLC protein possesses a CRD comprising most, but not all, of the carbohydrate-binding residues that are conserved among the galectins. The protein exhibits specific (albeit weak) carbohydrate-binding activity for simple saccharides including N-acetyl-D-glucosamine and lactose. Despite CLC protein having no significant sequence or structural similarities to other lysophospholipase catalytic triad has also been identified within the CLC structure, making it a unique dual-function polypeptide. These structural findings suggest a potential intracellular and/or extracellular role(s) for the galectin-associated activities of CLC protein in eosinophil and basophil function in allergic diseases and inflammation (Leonidas et al. 1995).

9.5.6 Galectin-Related Inter-Fiber Protein (Grifin/Galectin-11)

The ability of the vertebrate lens to focus light on retina derives from a number of properties including the expression at high levels of a selection of soluble proteins referred to as the crystallins. Soluble lens protein, though related to the family of galectins, does not bind beta-galactoside sugars and has atypical sequences at normally conserved regions of the carbohydrate-binding domain. Like some galectin family members, it can form a stable dimer. It is expressed only in the lens and is located at the interface between lens fiber cells despite the apparent lack of any membrane-targeting motifs. This protein is designated GRIFIN (galectin-related inter-fiber protein) to reflect its exclusion from the galectin family given the lack of affinity for beta-galactosides. Although the abundance, solubility, and lens-specific expression of GRIFIN would argue that it represents a new crystallin, its location at the fiber cell interface might suggest that its primary function is executed at the membrane. Adipose tissue plays an active role in the development of obesity. Among 29 proteins, differentially expressed between lean and diet-induced obese rats, changes in grifin were associated with obesity (Barceló-Batllori et al. 2005).

In the past decade, several galectin-like proteins such as the lens crystallin protein GRIFIN (Galectin related interfiber protein) and the galectin-related protein GRP (previously HSPC159; hematopoietic stem cell precursor) have been identified (Cooper 2002; Zhou et al. 2006). cDNA sequences have been obtained for guinea pig lens GRIFIN (Simpanya et al. 2008). Although GRIFIN and GRP have close similarity to galectin sequences, due to lack of carbohydrate-binding activity they are not considered to be members of galectin family, but putative products of evolutionary co-option. GRIFIN, described in the rat as a novel crystallin, is expressed at the interface between lens fiber cells, and was proposed to have evolved from a galectin precursor that has lost its capacity to bind lactose. Crystallins are water-soluble structural proteins that account for transparency of the vertebrate lens. They constitute a heterogeneous family, composed of four major groups (α–δ) and several minor groups, with roles as both molecular chaperones and structural proteins (Piatigorsky 1989). The binding interaction between α-crystallin and GRIFIN was enhanced in the presence of ATP. Binding data supported the hypothesis that GRIFIN is a binding partner of α-crystallin in lens (Barton et al. 2009).

The zebrafish (*Danio rerio*) has been established as a useful animal model for studies of early development. Thus, it may constitute the model of choice for gaining further insight into the biological roles of galectins (Vasta et al. 2004a). Ahmed and Vasta (2004c) characterized the zebrafish galectin repertoire as follows: three gal1-like proteins (Drgal1-L1, Drgal1-L2, Drgal1-L3), one chimera type galectin (Drgal3), and two gal9-like proteins (Drgal9-L1 and Drgal9-L2) (Ahmed and Vasta 2004). In silico analysis of the nearly completed genome sequence has enabled the identification of additional members, such as gal4-, gal5-, and gal8-like proteins. Ahmed et al. (2008) identified a homologue of the GRIFIN in zebrafish (Danio rerio) (designated DrGRIFIN), which like the mammalian equivalent is expressed in the lens, particularly in fiber cells of 2 dpf (days post fertilization) embryos. It is weakly expressed in the embryos as early as 21 hpf (hour post fertilization) but strongly at all later stages tested (30 hpf and 3, 4, 6, and 7 dpf). In adult zebrafish tissues, however, DrGRIFIN is also expressed in oocytes, brain, and intestine. Unlike the mammalian homologue, DrGRIFIN contains all amino acids critical for binding to carbohydrate ligands and its activity and confirmed by purification of recombinant DrGRIFIN by affinity chromatography on a lactosyl-Sepharose column. Therefore, DrGRIFIN is a bona fide galectin family member that in addition to its carbohydrate-binding properties, may also function as a crystallin.

9.5.7 Galectin-13 (Placental Protein -13)

Placental protein 13 (PP13) is a galectin expressed by syncytiotrophoblast. PP13 from human term placenta shows structural and functional homology to members of the galectin family. It effectively agglutinates erythrocytes. Similar to human eosinophil Charcot-Leyden crystal protein but not other galectins, galectin-13 possesses weak lysophospholipase activity. N-acetyl-lactosamine, mannose and N-acetyl-glucosamine residues, expressed in human placenta have the strongest binding affinity to galectin-13/PP13. The protein is a homodimer of 16 kDa subunits linked together by disulphide bonds, a process differing from earlier known galectins. The syncytiotrophoblasts show its expression in perinuclear region, while strong labelling of syncytiotrophoblasts' brush border membrane confirmed its externalization activity at the cell surface. PP13/galectin-13 may have special haemostatic and immunobiological functions at the common feto-maternal blood-spaces or developmental role in the placenta (Than et al. 2004). PP-13 has been sequenced and expressed in *E. coli*. The primary structure of PP13 is highly homologous to several members of β-galactoside-binding S-type lectin family. By multiple sequence alignment and structure-based secondary structure prediction, the secondary structure of PP13 was identical with 'proto-type' galectins consisting of a five- and a six-stranded β-sheet, joined by two α-helices. Highly conserved CRD was also present in PP13. Of eight consensus residues in CRD, four identical and three conservatively substituted were shared by PP13. By docking simulations PP13 possessed sugar-binding activity with highest affinity to N-acetyllactosamine and lactose typical of most galectins. All ligands were docked into putative CRD of PP13. Based on several lines of evidence, PP13 was designated as galectin-13 (Visegrády et al. 2001).

PP-13 levels slowly increase during pregnancy. Parallel to its decreased placental expression, an augmented membrane shedding of PP13 contributes to the increased third trimester maternal serum PP13 in women with preterm pre-eclampsia and HELLP syndrome (Nicolaides et al. 2006; Than et al. 2008a). Galectin-13 transcripts have been identified in several normal and malignant tissues. Its possible role in promoting apoptosis has been suggested in U-937 human macrophage cell line. Galectin-13 over-expression facilitated paclitaxel-induced cell death and nuclear translocation of apoptosis-inducing factor (AIF) and endonuclease-G without inducing mitochondrial cytochrome-c release or caspase-3 activation. Galectin-13 promoted apoptosis presumably by activating Ask-1 kinase-JNK and p38-MAPK pro-apoptotic pathways and by suppressing the PI-3 K-Akt and ERK1/2 cytoprotective pathways (Boronkai et al. 2009).

9.5.8 Galectin 14

A full-length cDNA encodes a lectin with subunit M_r 14.6. Like its relative, called L-14-I, the L-14-II exists as a homodimer in solution. The two related human lectins have 43% amino acid sequence identity. The genomic DNA encoding L-14-II (LGALS2) contains four exons with similar intron placement to L-14-I (LGALS1); but the genomic upstream region, which contains several sequences characteristic of regulatory elements, differs significantly from L-14-I (Gitt et al. 1992). The x-ray crystal structure of human dimeric S-Lac lectin, L-14-II, in complex with lactose revealed a twofold symmetry. The twofold symmetric dimer is made up of two extended anti-parallel β-sheets, which associate in a β-sandwich motif. Remarkably, the L-14-II monomer shares not only the same topology, but a very similar β-sheet structure with that of leguminous plant lectins, suggesting a conserved structure-function relationship. Carbohydrate binding by L-14-II was found to involve protein residues that are very highly conserved among all S-Lac lectins. These residues map to a single DNA exon, suggesting a carbohydrate binding cassette common to all S-Lac lectins (Lobsanov et al. 1993).

9.5.8.1 Functional Characterization

A galectin cDNA (galectin-14) from ovine eosinophil-rich leukocytes encodes a prototype galectin that contains one putative CRD and exhibits most identity to galectin-9/ ecalectin, a potent eosinophil chemoattractant. The sugar binding properties of recombinant protein were confirmed by a hemagglutination assay and lactose inhibition. The mRNA and protein of galectin-14 are expressed at high levels in eosinophil-rich cell populations. The protein localizes to the cytoplasmic, but not the granular, compartment of eosinophils. Galectin-14 was not detected in neutrophils, macrophages, or lymphocytes. But it is released from eosinophils into the lumen of the lungs after challenge with house dust mite allergen. The restricted expression of this galectin to eosinophils and its release into the lumen of the lung in a sheep asthma model indicates that it may play an important role in eosinophil function and allergic inflammation (Dunphy et al. 2002).

Galectin-14 secretion from peripheral blood eosinophils can be induced by same stimuli that induce eosinophil degranulation. Recombinant galectin-14 can bind in vitro to eosinophils, neutrophils and activated lymphocytes. Glycan array screening indicated that galectin-14 recognizes terminal N-acetyllactosamine residues which can be modified with α1-2-fucosylation and, uniquely for a galectin, prefers α2- over α2-sialylation. Galectin-14 showed greatest affinity for lacto-N-neotetraose, an immunomodulatory oligosaccharide expressed by helminths. Galectin-14 binds specifically to laminin in vitro, and to mucus and mucus producing cells on lung and intestinal tissue sections. In vivo, galectin-14 is abundantly present in mucus scrapings collected from either lungs or gastrointestinal tract following allergen or parasite challenge, respectively. In vivo secretion of eosinophil galectins may be specifically induced at epithelial surfaces after recruitment of eosinophils by allergic stimuli, and that eosinophil galectins may be involved in promoting adhesion and changing mucus properties during parasite infection and allergies (Young et al. 2009).

9.5.8.2 Chicken Galectins

Two avian galectins have been detected in chicken liver (chicken galectin-16 CG-16) and intestine (chicken galectin-14; CG-14) with different developmental regulation. Epithelial cells of the mesonephric proximal tubules of kidney were immunoreactive for CG-14 from day 5 onwards. For CG-16 a rather similar pattern of staining was seen, additional positivity in early glomerular podocytes being notable. Results demonstrate quantitative differences in the developmental regulation of the two avian galectins with obvious similarities in the cell-type pattern but with a separate intracellular localisation profile (Stierstorfer et al. 2000). Kaltner et al. (2008) reported the cloning and expression of a third prototype CG. It has deceptively similar electrophoretic mobility compared with recombinant C-14, the protein first isolated from embryonic skin, and turned out to be identical with the intestinal protein. Hydrodynamic properties unusual for a homodimeric galectin and characteristic traits in the proximal promoter region set it apart from the two already known CGs. Their structural vicinity to galectin-1 prompts their classification as CG-1A (CG-16)/CG-1B (CG-14), whereas sequence similarity to mammalian galectin-2 gives reason to refer to the intestinal protein as CG-2. Overall results reveal a network of three prototype galectins in chicken (Kaltner et al. 2008).

To examine how sequence changes affect carbohydrate specificity, the two closely related proto-type chicken galectins CG-14 and CG-16 were selected as models and tested for binding of 56 free saccharides and 34 well-defined glycoproteins. The two galectins share preference for the II (Galβ1-4GlcNAc) versus I (Galβ1-3GlcNAc) version of β-galactosides. A pronounced difference was found owing to the reactivity of CG-14 with histo-blood group ABH active oligosaccharides and A/B active glycoproteins. This study identifies activity differences and provides information on their relation to structural divergence (Wu et al. 2007).

9.5.9 Galectin-15

Secretions of the uterus support survival and growth of the conceptus (embryo/fetus and associated membranes) during pregnancy. Galectin-15, also known as OVGAL11 and a member of the galectin family of secreted β-galactoside lectins containing a conserved CRD and a separate putative integrin binding domain, is present in the uterus of sheep. In endometria of cyclic and pregnant sheep, galectin-15 mRNA was expressed specifically in the endometrial luminal epithelium but not in the conceptus. The intracellular role of galectin-15 is to regulate cell survival, differentiation and function, while the extracellular role of secreted galectin-15 is to regulate implantation and placentation by functioning as a heterophilic cell adhesion molecule between the conceptus trophectoderm and endometrial LE (Gray et al. 2005). In the uterine lumen, secreted galectin-15 protein increased between days 14 and 16 of pregnancy. Galectin-15 protein was functional in binding lactose and mannose sugars and immunologically identical to the unnamed Mr 14 kDa protein from the ovine uterus that forms crystalline inclusion bodies in endometrial epithelia and trophectoderm (Tr) conceptus. Based on the functional studies of other galectins, galectin-15 is hypothesized to function extracellularly to regulate Tr migration and adhesion to the endometrial epithelium and intracellularly to regulate Tr cell survival, growth, and differentiation (Gray et al. 2005; Satterfield et al. 2006).

9.6 Evolution of Galectins

During past two decades, substantial progress has been made in elucidation of structural diversity of lectin repertoires of invertebrates, protochordates and ectothermic vertebrates, providing particularly valuable information on those groups that constitute the invertebrate/vertebrate 'boundary'. Although representatives of lectin families typical of mammals, such as C-type lectins, galectins and pentraxins, have been described in these taxa, the detailed study of selected model species has yielded either novel variants of structures described for mammalian lectin representatives or novel lectin families with unique sequence motifs, multidomain arrangements and a new structural fold. Along with high structural diversity of lectin repertoires in these taxa, a wide spectrum of biological roles is starting to emerge, underscoring the value of invertebrate and lower vertebrate models for gaining insight into structural, functional and evolutionary aspects of lectins (Vasta et al. 2004a).

9.6.1 Phylogenetic Analysis of Galectin Family

Galectins are widely distributed from higher vertebrates to lower invertebrates, including mammals, amphibians, fish, birds, nematodes, and sponges, having even been found in the mushroom *Coprinus cinereus*. Although the identification of many galectin relatives in widely divergent organisms (including *Arabidopsis, Drosophila, Caenorhabditis, Danio, Xenopus*, and human) has added significantly to the size and complexity of this intriguing protein family, several common themes arise, which suggest promising new research targets (Cooper 2002). Galectins have been found in the genome of the insects *Drosophila* and *Anopheles*, of viruses and of the plant *Arabadopsis*. To elucidate the evolutionary history of galectin-like proteins in chordates, Houzelstein et al. (2004) exploited three independent lines of evidence: (1) location of galectin encoding genes (*LGALS*) in the human genome; (2) exon-intron organization of galectin encoding genes; and (3) sequence comparison of CRDs of chordate galectins. Results of Houzelstein et al. (2004) suggest that a duplication of a mono-CRD galectin gene gave rise to an original bi-CRD galectin gene, before or early in chordate evolution. The N-terminal and C-terminal CRDs of this original galectin subsequently diverged into two different subtypes, defined by exon-intron structure (F4-CRD and F3-CRD). It was inferred that all vertebrate mono-CRD galectins known belong to either the F3- or F4- subtype. A sequence of duplication and divergence events of the different galectins in chordates was proposed (Houzelstein et al. 2004). Therefore, the galectin gene family must have evolved from the start of multicellular organisms, at least a million of years ago. In shrimp, investigations of the penaeidins, which are constitutively expressed peptides, have highlighted the importance of hemocytes and hematopoiesis as major elements of the immune response, providing both local and systemic reactions. The activation of hematopoiesis must be regarded as a regulatory way for the expression and distribution of constitutively expressed immune effectors (Bachère et al. 2004).

9.6.2 Galectins in Lower Vertebrates

First reptilian galectin was isolated from the skin of the lizard *Podarcis hispanica* (Solis et al. 2000). Up to five lactose-binding proteins were isolated from the lizard *Podarcis hispanica* on asialofetuin-Sepharose. The main component, abundantly expressed in skin, and under native conditions behaved as a monomer with a molecular mass of 14.5 kDa and an isoelectric pH 6.3. Based on sequence homology of the 58 *N*-terminal amino acid residues with galectins, and on its demonstrated galactoside-binding activity, this lectin was named LG-14 (from *Lizard Galectin* and 14 kDa). LG-14 falls into and strengthens the still thinly populated category of monomeric prototype galectins.

The crystal structures of congerin I and -II, the galectins from conger eel, have been determined. The congerin I revealed a fold evolution via strand swap; however, the structure of congerin II resembles other prototype galectins. A comparison of the two congerin genes with that of several other galectins suggests acceralated evolution of both congerin genes following gene duplication. The presence of a Mes (2-[N-morpholino]ethanesulfonic acid) molecule near the carbohydrate-binding site in the crystal structure points to the possibility of an additional binding site in congerin II. The binding site consists of a group of residues that had been replaced following gene duplication suggesting that the binding site was built under selective pressure. Congerin II may be a protein specialized for biological defense with an affinity for target carbohydrates on parasites' cell surface (Shirai et al. 2002).

The bi-CRD *Branchiostoma belcheri tsingtauense* galectin (BbtGal)-L together and its alternatively spliced mono-CRD isoform BbtGal-S from amphioxus intestine are encoded by a 9488-bp gene with eight exons and seven introns. BbtGal is a member of the galectin family. Phylogenetic analysis suggested that the BbtGal gene was the primitive form of chordate galectin family. BbtGal-L mRNA was mainly expressed in the immunity-related organs, such as hepatic diverticulum, intestine, and gill, but BbtGal-S was ubiquitously expressed in all tissues (Yu et al. 2007).

References

Abbott WM, Feizi T (1991) Soluble 14-kDa β-galactoside-specific bovine lectin. Evidence from mutagenesis and proteolysis that almost the complete polypeptide chain is necessary for integrity of the carbohydrate recognition domain. J Biol Chem 266:5552–5557

Ackerman SJ, Liu L, Kwatia MA et al (2002) Charcot-Leyden crystal protein (galectin-10) is not a dual function galectin with lysophospholipase activity but binds a lysophospholipase inhibitor in a novel structural fashion. J Biol Chem 277:14859–14868

Ahmad N, Gabius HJ, Andre S et al (2004a) Galectin-3 precipitates as a pentamer with synthetic multivalent carbohydrates and forms heterogeneous cross-linked complexes. J Biol Chem 279:10841–10847

Ahmad N, Gabius HJ, Sabesan S et al (2004b) Thermodynamic binding studies of bivalent oligosaccharides to galectin-1, galectin-3, and the carbohydrate recognition domain of galectin-3. Glycobiology 14:817–825

Ahmed H, Vasta GR (2008) Unlike mammalian GRIFIN, the zebrafish homologue (DrGRIFIN) represents a functional carbohydrate-binding galectin. Biochem Biophys Res Commun 371:350–355

Ahmed H, Pohl J et al (1996) The primary structure and carbohydrate specificity of a β-galactosyl-binding lectin from toad (*Bufo arenarum Hensel*) ovary reveal closer similarities to the mammalian galectin-1 than to the galectin from the clawed frog *Xenopus laevis*. J Biol Chem 271:33083–33094

Ahmed H, Du SJ, O'Leary N, Vasta GR (2004) Biochemical and molecular characterization of galectins from zebrafish (Danio rerio): notochord-specific expression of a prototype galectin during early embryogenesis. Glycobiology 14:219–232

Almkvist J, Karlsson A (2004) Galectins as inflammatory mediators. Glycoconj J 19:575–581

Asselbergs FW, Pai JK, Rexrode KM et al (2007) Effects of lymphotoxin-alpha gene and galectin-2 gene polymorphisms on inflammatory biomarkers, cellular adhesion molecules and risk of coronary heart disease. Clin Sci (Lond) 112:291–298

Bachère E, Gueguen Y, Gonzalez M et al (2004) Insights into the antimicrobial defense of marine invertebrates: the penaeid shrimps and the oyster Crassostrea gigas. Immunol Rev 198:149–168

Barceló-Batllori S, Corominola H, Claret M et al (2005) Target identification of the novel antiobesity agent tungstate in adipose tissue from obese rats. Proteomics 5:4927–4935 See more articles cited in this paragraph

Barondes SH, Cooper DNW, Gitt MA et al (1994a) Galectins: structure and function of a large family of animal lectins. J Biol Chem 269: 20807–20810

Barondes SH, Castronovo V, Cooper DN et al (1994b) Galectins: a family of animal beta-galactoside-binding lectins. Cell 76:597–598

Barres C, Blanc L, Bette-Bobillo P et al (2010) Galectin-5 is bound onto the surface of rat reticulocyte exosomes and modulates vesicle uptake by macrophages. Blood 115:696–705

Barton KA, Hsu CD, Petrash JM (2009) Interactions between small heat shock protein alpha-crystallin and galectin-related interfiber protein (GRIFIN) in the ocular lens. Biochemistry 48:3956–3966

Bernerd F, Sarasin A, Magnaldo T (1999) Galectin-7 overexpression is associated with the apoptotic process in UVB-induced sunburn keratinocytes. Proc Natl Acad Sci U S A 96:11329–11334

Bianco GA, Toscano MA, Ilarregui JM, Rabinovich GA (2006) Impact of protein-glycan interactions in the regulation of autoimmunity and chronic inflammation. Autoimmun Rev 5:349–356

Birdsall B, Feeney J, Burdett ID et al (2001) NMR solution studies of hamster galectin-3 and electron microscopic visualization of surface-adsorbed complexes: evidence for interactions between the N- and C-terminal domains. Biochemistry 40:4859–4866

Boronkai A, Bellyei S, Szigeti A et al (2009) Potentiation of paclitaxel-induced apoptosis by galectin-13 overexpression via activation of Ask-1-p38-MAP kinase and JNK/SAPK pathways and suppression of Akt and ERK1/2 activation in U-937 human macrophage cells. Eur J Cell Biol 88:753–763

Camby I, Belot N, Lefranc F et al (2002) Galectin-1 modulates human glioblastoma cell migration into the brain through modifications to the actin cytoskeleton and levels of expression of small GTPases. J Neuropathol Exp Neurol 61:585–596

Camby I, Decaestecker C, Lefranc F et al (2005) Galectin-1 knocking down in human U87 glioblastoma cells alters their gene expression pattern. Biochem Biophys Res Commun 335:27–35

Cha SK, Ortega B, Kurosu H et al (2008) Removal of sialic acid involving Klotho causes cell-surface retention of TRPV5 channel via binding to galectin-1. Proc Natl Acad Sci U S A 105: 9805–9810

Christensen MB, Lawlor DA, Gaunt TR et al (2006) Genotype of galectin 2 (LGALS2) is associated with insulin-glucose profile in the British Women's Heart and Health Study. Diabetologia 49:673–677

Cludts S, Decaestecker C, Mahillon V et al (2009) Galectin-8 up-regulation during hypopharyngeal and Laryngeal tumor progression and comparison with galectin-1, -3 and -7. Anticancer Res 29: 4933–4940

Colnot C, Fowlis D, Ripoche MA et al (1998) Embryonic implantation in galectin 1/galectin 3 double mutant mice. Dev Dyn 211:306–313

Cooper DN (2002) Galectinomics: finding themes in complexity. Biochim Biophys Acta 1572:209–231

Cummings RD, Liu FT (2009) Chapter 33; Galectins. In: Essentials of glycobiology. Cold Spring Harbor Laboratory Press, Cold Spring Harbor

Dam TK, Roy R, Pagé D, Brewer CF (2002) Negative cooperativity associated with binding of multivalent carbohydrates to lectins. Thermodynamic analysis of the "multivalency effect". Biochemistry 41:1351–1358

Dam TK, Gabius HJ, Andre S et al (2005) Galectins bind to the multivalent glycoprotein asialofetuin with enhanced affinities and a gradient of decreasing binding constants. Biochemistry 44: 12564–12571

Danguy A, Camby I, Kiss R (2002) Galectins and cancer. Biochim Biophys Acta 1572:285–293

De Re V, Simula MP, Cannizzaro R et al (2009) Galectin-10, eosinophils, and celiac disease. Ann N Y Acad Sci 1173:357–364

Debray C, Vereecken P et al (2004) Multifaceted role of galectin-3 on human glioblastoma cell motility. Biochem Biophys Res Commun 325:1393–1398

Delbrouck C, Souchay C, Kaltner H et al (2005) Regulation of expression of galectin-7 in human nasal polyps by budesonide. B-ENT 1:137–144

Demetriou M, Granovsky M, Quaggin S, Dennis JW (2001) Negative regulation of T-cell activation and autoimmunity by Mgat5 N-glycosylation. Nature 409(6821):733–739

Devouassoux G, Pachot A et al (2008) Galectin-10 mRNA is overexpressed in peripheral blood of aspirin-induced asthma. Allergy 63:125–131

Dunphy JL, Barcham GJ, Bischof RJ et al (2002) Isolation and characterization of a novel eosinophil-specific galectin released into the lungs in response to allergen challenge. J Biol Chem 277:14916–14924

Dyer KD, Rosenberg HF (2001) Transcriptional regulation of galectin-10 (eosinophil Charcot-Leyden crystal protein): a GC box (−44 to −50) controls butyric acid induction of gene expression. Life Sci 69:201–212

Elola MT, Chiesa ME, Alberti AF et al (2005a) Galectin-1 receptors in different cell types. J Biomed Sci 12:13–29

Elola MT, Chiesa ME, Fink NE (2005b) Activation of oxidative burst and degranulation of porcine neutrophils by a homologous spleen galectin-1 compared to N-formyl-L-methionyl-L-leucyl-L-phenylalanine and phorbol 12-myristate 13-acetate. Comp Biochem Physiol B Biochem Mol Biol 141:23–31

Fischer C, Sanchez-Ruderisch H, Welzel M et al (2005) Galectin-1 interacts with the {alpha}5{beta}1 fibronectin receptor to restrict carcinoma cell growth via induction of p21 and p27. J Biol Chem 280:37266–37277

Fortin S, Le Mercier M, Camby I et al (2010) Galectin-1 is implicated in the protein kinase C epsilon/vimentin-controlled trafficking of integrin-beta1 in glioblastoma cells. Brain Pathol 20:39–49

Freilinger T, Bevan S, Ripke S et al (2009) Genetic variation in the lymphotoxin-alpha pathway and the risk of ischemic stroke in European populations. Stroke 40:970–972

Fukushi J, Makagiansar IT, Stallcup WB (2004) NG2 proteoglycan promotes endothelial cell motility and angiogenesis via engagement of galectin-3 and alpha3beta1 integrin. Mol Biol Cell 15:3580–3590

Gabius H-J (2001) Probing the cons and pros of lectin-induced immunomodulation: case studies for the mistletoe lectin and galectin-1. Biochimie 83:659

Garner OB, Baum LG (2008) Galectin-glycan lattices regulate cell-surface glycoprotein organization and signalling. Biochem Soc Trans 36:1472–1477

Gerton GL, Hedrick JL (1986) The vitelline envelope to fertilization envelope conversion in eggs of *Xenopus laevis*. Dev Biol 116:1–7

Gitt MA, Barondes SH (1991) Genomic sequence and organization of two members of a human lectin gene family. Biochemistry 30:82–89

Gitt MA, Massa SM, Leffler H, Barondes SH (1992) Isolation and expression of a gene encoding L-14-II, a new human soluble lactose-binding lectin. J Biol Chem 267:10601–10606

Gitt MA, Wiser MF, Leffler H et al (1995) Sequence and mapping of galectin-5, a beta-galactoside-binding lectin, found in rat erythrocytes. J Biol Chem 270:5032–5038

Gray CA, Dunlap KA, Burghardt RC, Spencer TE (2005) Galectin-15 in ovine uteroplacental tissues. Reproduction 130:231–240

Hadari YR, Arbel-Goren R, Levy Y et al (2000) Galectin-8 binding to integrins inhibits cell adhesion and induces apoptosis. J Cell Sci 113:2385–2397

Hirabayshi J, Kasai K (1993) The family of metazoan metal-independent -galactoside-binding lectins: structure, function and molecular evolution. Glycobiology 3:297–304

Hirashima M (1999) Ecalectin as a T cell-derived eosinophil chemoattractant. Int Arch Allergy Immunol 120(Suppl 1):7–10

Hokama A, Mizoguchi E, Sugimoto K et al (2004) Induced reactivity of intestinal CD4(+) T cells with an epithelial cell lectin, galectin-4, contributes to exacerbation of intestinal inflammation. Immunity 20:681–693

Houzelstein D, Goncalves IR, Fadden AJ et al (2004) Phylogenetic analysis of the vertebrate galectin family. Mol Biol Evol 21: 1177–1187

Hsu DK, Yang RY, Liu FT (2006) Galectins in apoptosis. Methods Enzymol 417:256–273

Iida A, Ozaki K, Tanaka T et al (2005) Fine-scale SNP map of an 11-kb genomic region at 22q13.1 containing the galectin-1 gene. J Hum Genet 50:42–45

Ilarregui JM, Bianco GA, Toscano MA et al (2005) The coming of age of galectins as immunomodulatory agents: impact of these carbohydrate binding proteins in T cell physiology and chronic inflammatory disorders. Ann Rheum Dis 64(Suppl 4):96–103

Kaltner H, Solis D, Kopitz J et al (2008) Prototype chicken galectins revisited: characterization of a third protein with distinctive hydrodynamic behaviour and expression pattern in organs of adult animals. Biochem J 409:591–599

Kasamatsu A, Uzawa K, Shimada K et al (2005) Elevation of galectin-9 as an inflammatory response in the periodontal ligament cells exposed to Porphylomonas gingivalis lipopolysaccharide in vitro and in vivo. Int J Biochem Cell Biol 37:397–408

Kemp JM, Robinson NA, Meeusen EN, Piedrafita DM (2009) The relationship between the rapid rejection of Haemonchus contortus larvae with cells and mediators in abomasal tissues in immune sheep. Int J Parasitol 39:1589–1594

Kimura A, Takahashi M, Choi BY et al (2007) Lack of association between LTA and LGALS2 polymorphisms and myocardial infarction in Japanese and Korean populations. Tissue Antigens 69:265–269

Kohatsu L, Hsu DK et al (2006) Galectin-3 induces death of Candida species expressing specific β-1,2-linked mannans. J Immunol 177: 4718–4726

Kopitz J, André S, von Reitzenstein C et al (2003) Homodimeric galectin-7 (p53-induced gene 1) is a negative growth regulator for human neuroblastoma cells. Oncogene 22:6277–6288

Kubach J, Lutter P, Bopp T et al (2007) Human CD4+CD25+ regulatory T cells: proteome analysis identifies galectin-10 as a novel marker essential for their anergy and suppressive function. Blood 110:1550–1558

Kuwabara I, Kuwabara Y, Yang RY et al (2002) Galectin-7 (PIG1) exhibits pro-apoptotic function through JNK activation and mitochondrial cytochrome c release. J Biol Chem 277:3487–3497

Lensch M, Lohr M, Russwurm R et al (2006) Unique sequence and expression profiles of rat galectins-5 and -9 as a result of species-specific gene divergence. Int J Biochem Cell Biol 38: 1741–1758

Leonidas DD, Elbert BL, Zhou Z et al (1995) Crystal structure of human Charcot-Leyden crystal protein, an eosinophil lysophospholipase, identifies it as a new member of the carbohydrate-binding family of galectins. Structure 3:1379–1393

Leonidas DD, Vatzaki EH, Vorum H et al (1998) Structural basis for the recognition of carbohydrates by human galectin-7. Biochemistry 37:13930–13940

Liao DI, Kapadia G, Ahmed H et al (1994) Structure of S-lectin, a developmentally regulated vertebrate β-galactoside-binding protein. Proc Natl Acad Sci U S A 91:1428–1432

Liu FT (2005) Regulatory roles of galectins in the immune response. Int Arch Allergy Immunol 136:385–400

Liu FT, Rabinovich GA (2005) Galectins as modulators of tumour progression. Nat Rev Cancer 5:29–41

Liu FT, Patterson RJ, Wang JL (2002) Intracellular functions of galectins. Biochim Biophys Acta 1572:263–273

Lobsanov YD, Gitt MA, Leffler H et al (1993) X-ray crystal structure of the human dimeric S-Lac lectin, L-14-II, in complex with lactose at 2.9-A resolution. J Biol Chem 268:27034–27038
Lohr M, Lensch M, Andre S et al (2007) Murine homodimeric adhesion/growth-regulatory galectins-1, -2 and -7: comparative profiling of gene/promoter sequences by database mining, of expression by RT PCR/immunohistochemistry and of contact sites for carbohydrate ligands by computational chemistry. Folia Biol (Praha) 53:109–128
Lohr M, Kaltner H, Lensch M et al (2008) Cell-type-specific expression of murine multifunctional galectin-3 and its association with follicular atresia/luteolysis in contrast to pro-apoptotic galectins-1 and -7. Histochem Cell Biol 130:567–581
Loris R (2002) Principles of structures of animal and plant lectins. Biochim Biophys Acta 1572:198–208
Loser K, Sturm A et al (2009) Galectin-2 suppresses contact allergy by inducing apoptosis in activated $CD8^+$ T cells. J Immunol 182: 5419–5429
Madsen P, Rasmussen HH, Flint T et al (1995) Cloning, expression, and chromosome mapping of human galectin-7. J Biol Chem 270:5823–5829
Magnaldo T, Bernerd F, Darmon M (1995) Galectin-7, a human 14-kDa S-lectin, specifically expressed in keratinocytes and sensitive to retinoic acid. Dev Biol 168:259–271
Magnaldo T, Fowlis D, Darmon M (1998) Galectin-7, A marker of all types of stratified epithelia. Differentiation 63:159–168
Malik RK, Ghurye RR, Lawrence-Watt DJ, Stewart HJ (2009) Galectin-1 stimulates monocyte chemotaxis via the p44/42 MAP kinase pathway and a pertussis toxin-sensitive pathway. Glycobiology 19:1402–1407
Mangino M, Braund P, Singh R et al (2007) LGALS2 functional variant rs7291467 is not associated with susceptibility to myocardial infarction in Caucasians. Atherosclerosis 194:112–115
Matsumoto R, Matsumoto H, Seki M et al (1998) Human ecalectin, a variant of human galectin-9 is a novel eosinophil chemoattractant produced by T lymphocytes. J Biol Chem 273:16976–16984
Mehul B, Hughes RC (1997) Plasma membrane targetting, vesicular budding and release of galectin 3 from the cytoplasm of mammalian cells during secretion. J Cell Sci 110:1169–1178
Mey A, Leffler H, Hmama Z et al (1996) The animal lectin galectin-3 interacts with bacterial lipopolysaccharides via two independent sites. J Immunol 156:1572–1577
Moiseeva EP, Spring EL, Baron JH et al (1999) Galectin 1 modulates attachment, spreading and migration of cultured vascular smooth muscle cells via interactions with cellular receptors and components of extracellular matrix. J Vasc Res 36:47–58
Moiseeva EP, Williams B, Goodall AH et al (2003) Galectin-1 interacts with beta-1 subunit of integrin. Biochem Biophys Res Commun 310:1010–1016
Nakahara S, Raz A (2006) On the role of galectins in signal transduction. Methods Enzymol 417:273–289
Naoum JJ, Chai H et al (2006) Lymphotoxin-α and cardiovascular disease: clinical association and pathogenic mechanisms. Med Sci Monit 12:RA121–RA124
Nguyen JT, Evans DP, Galvan M et al (2001) CD45 modulates galectin-1-induced T cell death: regulation by expression of core 2 O-glycans. J Immunol 167:5697–5707
Nicolaides KH, Bindra R, Turan OM et al (2006) A novel approach to first-trimester screening for early pre-eclampsia combining serum PP-13 and Doppler ultrasound. Ultrasound Obstet Gynecol 27:13–17
Nio J, Takahashi-Iwanaga H, Morimatsu M et al (2006) Immunohistochemical and in situ hybridization analysis of galectin-3, a β-galactoside binding lectin, in the urinary system of adult mice. Histochem Cell Biol 126:45–56
Nio-Kobayashi J, Takahashi-Iwanaga H, Iwanaga T (2009) Immunohistochemical localization of six galectin subtypes in the mouse digestive tract. J Histochem Cytochem 57:41–50
Ochieng J, Leite-Browning ML, Warfield P (1998) Regulation of cellular adhesion to extracellular matrix proteins by galectin-3. Biochem Biophys Res Commun 246:788–791
Oka T, Murakami S, Arata Y et al (1999) Identification and cloning of rat galectin-2: expression is predominantly in epithelial cells of the stomach. Arch Biochem Biophys 361:195
Ozaki K, Tanaka T (2005) Genome-wide association study to identify SNPs conferring risk of myocardial infarction and their functional analyses. Cell Mol Life Sci 62:1804–1813
Ozaki K, Inoue K, Sato H, Iida A et al (2004) Functional variation in LGALS2 confers risk of myocardial infarction and regulates lymphotoxin-alpha secretion in vitro. Nature 429(6987):72–75
Ozaki K, Sato H, Inoue K et al (2009) SNPs in BRAP associated with risk of myocardial infarction in Asian populations. Nat Genet 41:329–333
Paclik D, Berndt U, Guzy C et al (2008) Galectin-2 induces apoptosis of lamina propria T lymphocytes and ameliorates acute and chronic experimental colitis in mice. J Mol Med 86:1395–1406
Panoulas VF, Nikas SN, Smith JP et al (2008) Lymphotoxin 252A > G polymorphism is common and associates with myocardial infarction in patients with rheumatoid arthritis. Ann Rheum Dis 67:1550–1556
Panoulas VF, Douglas KM, Smith JP et al (2009) Galectin-2 (LGALS2) 3279 C/T polymorphism may be independently associated with diastolic blood pressure in patients with rheumatoid arthritis. Clin Exp Hypertens 31:93–104
Papaspyridonos M, McNeill E, de Bono JP et al (2008) Galectin-3 is an amplifier of inflammation in atherosclerotic plaque progression through macrophage activation and monocyte chemoattraction. Arterioscler Thromb Vasc Biol 28:433–440
Pelletier I, Hashidate T, Urashima T et al (2003) Specific recognition of *Leishmania major* poly-β-galactosyl epitopes by galectin-9: possible implication of galectin-9 in interaction between *L. major* and host cells. J Biol Chem 278:22223–22230
Perillo NL, Pace KE, Seilhamer J, Baum LG (1995) Apoptosis of T cells mediated by galectin-1. Nature 378:736–739
Piatigorsky J (1989) Lens crystallins and their genes: diversity and tissue-specific expression. FASEB J 3:1933–1940
Polyak K, Xia Y, Zweier JL et al (1997) A model for p53-induced apoptosis. Nature 389:300–305
Purkrábková T, Smetana K Jr, Dvořánková B et al (2003) New aspects of galectin functionality in nuclei of cultured bone marrow stromal and epidermal cells: biotinylated galectins as tool to detect specific binding sites. Biol Cell 95:535–545
Quill TA, Hedrick JL (1996) The fertilization layer mediated block to polyspermy in *Xenopus laevis*: Isolation of the cortical granule lectin ligand. Arch Biochem Biophys 333:326–332
Rabinovich GA (2005) Galectin-1 as a potential cancer target. Br J Cancer 92:1188–1192
Rabinovich GA, Gruppi A (2005) Galectins as immunoregulators during infectious processes: from microbial invasion to the resolution of the disease. Parasite Immunol 27:103–114
Rabinovich GA, Ramhorst RE, Rubinstein N et al (2002a) Induction of allogenic T-cell hyporesponsiveness by galectin-1-mediated apoptotic and non-apoptotic mechanisms. Cell Death Differ 9:661–670
Rabinovich GA, Rubinstein N et al (2002b) Role of galectins in inflammatory and immunomodulatory processes. Biochim Biophys Acta 1572:274
Rabinovich GA, Rubinstein N, Fainboim L (2002c) Unlocking the secrets of galectins: a challenge at the frontier of glyco-immunology. J Leuk Biol 71:741–752
Rubinstein N, Ilarregui JM, Toscano MA, Rabinovich GA (2004) The role of galectins in the initiation, amplification and resolution of the inflammatory response. Tissue Antigens 64:1–12
Rumilla KM, Erickson LA, Erickson AK, Lloyd RV (2006) Galectin-4 expression in carcinoid tumors. Endocr Pathol 17:243–249, Fall

Saal I, Nagy N, Lensch M et al (2005) Human galectin-2: expression profiling by RT-PCR/immunohistochemistry and its introduction as a histochemical tool for ligand localization. Histol Histopathol 20:1191–1208

Salameh BA, Sundin A et al (2006) Thioureido N-acetyllactosamine derivatives as potent galectin-7 and 9 N inhibitors. Bioorg Med Chem 14:1215–1220

Sano H, Hsu DK, Yu L et al (2000) Human galectin-3 is a novel chemoattractant for monocytes and macrophages. J Immunol 165:2156–2164

Sato M, Nishi N, Shoji H et al (2002) Quantification of galectin-7 and its localization in adult mouse tissues. J Biochem 131:255–260

Satterfield MC, Bazer FW, Spencer TE (2006) Progesterone regulation of preimplantation conceptus growth and galectin 15 (LGALS15) in the ovine uterus. Biol Reprod 75:289–296

Saussez S, Kiss R (2006) Galectin-7. Cell Mol Life Sci 63:686–697

Sedlacek K, Neureuther K, Mueller JC et al (2007) Lymphotoxin-alpha and galectin-2 SNPs are not associated with myocardial infarction in two different German populations. J Mol Med 85: 997–1004

Seelenmeyer C, Wegehingel S, Tews I et al (2005) Cell surface counter receptors are essential components of the unconventional export machinery of galectin-1. J Cell Biol 171:373–381

Seetharaman J, Kanigsberg A, Slaaby R et al (1998) X-ray crystal structure of the human galectin-3 carbohydrate recognition domain at 2.1-A resolution. J Biol Chem 273:13047–13052

Shirai T, Matsui Y, Shionyu-Mitsuyama C et al (2002) Crystal structure of a conger eel galectin (congerin II) at 1.45A resolution: implication for the accelerated evolution of a new ligand-binding site following gene duplication. J Mol Biol 321:879–889

Simpanya MF, Wistow G et al (2008) Expressed sequence tag analysis of guinea pig (Cavia porcellus) eye tissues for NEIBank. Mol Vis 14:2413–2427

Solis D, Lopez-Lucendo MIF, Leon S, Varela J, Diaz-Maurino T (2000) Description of a monomeric prototype galectin from the lizard *Podarcis hispanica*. Glycobiology 10:1325–1331

Stierstorfer B, Kaltner H, Neumüller C et al (2000) Temporal and spatial regulation of expression of two galectins during kidney development of the chicken. Histochem J 32:325–336

Storan MJ, Magnaldo T, Biol-N'Garagba MC et al (2004) Expression and putative role of lactoseries carbohydrates present on NCAM in the rat primary olfactory pathway. J Comp Neurol 475:289–302

Sturm A, Lensch M et al (2004) Human galectin-2: novel inducer of T cell apoptosis with distinct profile of caspase activation. J Immunol 173:3825–3837

Sujatha MS, Balaji PV (2004) Identification of common structural features of binding sites in galactose-specific proteins. Proteins 55:44–65

Sujatha MS, Sasidhar YU, Balaji PV (2005) Insights into the role of the aromatic residue in galactose-binding sites: MP2/6-311 G++** study on galactose- and glucose-aromatic residue analogue complexes. Biochemistry 44:8554–8562

Swaminathan GJ, Leonidas DD, Savage MP et al (1999) Selective recognition of mannose by the human eosinophil Charcot-Leyden crystal protein (galectin-10): a crystallographic study at 1.8 A resolution. Biochemistry 38:13837–13843

Szolnoki Z, Maasz A, Magyari L et al (2009) Galectin-2 3279TT variant protects against the lymphotoxin-alpha 252GG genotype associated ischaemic stroke. Clin Neurol Neurosurg 111:227–230

Tanaka T, Ozaki K (2006) Inflammation as a risk factor for myocardial infarction. J Hum Genet 51:595–604

Than NG, Pick E, Bellyei S et al (2004) Functional analyses of placental protein 13/galectin-13. Eur J Biochem 271:1065–1078

Than NG, Abdul Rahman O, Magenheim R et al (2008a) Placental protein 13 (galectin-13) has decreased placental expression but increased shedding and maternal serum concentrations in patients presenting with preterm pre-eclampsia and HELLP syndrome. Virchows Arch 453:387–400

Than TH, Swethadri GK, Wong J et al (2008b) Expression of Galectin-3 and Galectin-7 in thyroid malignancy as potential diagnostic indicators. Singapore Med J 49:333–338

Thijssen VLJL, Poirier F, Baum LG et al (2007) Galectins in the tumor endothelium: opportunities for combined cancer therapy. Blood 110:2819–2827

Thomsen MK, Hansen GH, Danielsen EM (2009) Galectin-2 at the enterocyte brush border of the small intestine. Mol Membr Biol 26:347–355

Timmons PM, Colnot C, Cail I et al (1999) Expression of galectin-7 during epithelial development coincides with the onset of stratification. Int J Dev Biol 43:229–235

Topol EJ, Smith J, Plow EF et al (2006) Genetic susceptibility to myocardial infarction and coronary artery disease. Hum Mol Genet 15(Spec No 2):R117–R123

Toscano MA, Ilarregui JM, Bianco GA et al (2007) Dissecting the pathophysiologic role of endogenous lectins: glycan-binding proteins with cytokine-like activity? Cytokine Growth Factor Rev 18:57–71

Ueda S, Kuwabara I, Liu FT (2004) Suppression of tumor growth by galectin-7 gene transfer. Cancer Res 64:5672–5676

Vasta GR, Ahmed H, Du S, Henrikson D (2004a) Galectins in teleost fish: Zebrafish (*Danio rerio*) as a model species to address their biological roles in development and innate immunity. Glycoconj J 21:503–521

Vasta GR, Ahmed H, Odom EW (2004b) Structural and functional diversity of lectin repertoires in invertebrates, protochordates and ectothermic vertebrates. Curr Opin Struct Biol 14:617–630

Visegrády B, Than NG, Kilár F et al (2001) Homology modelling and molecular dynamics studies of human placental tissue protein 13 (galectin-13). Protein Eng 14:875–880

Walzel H, Blach M, Hirabayashi J et al (2002) Galectin-induced activation of the transcription factors NFAT and AP-1 in human Jurkat T-lymphocytes. Cell Signal 14:861–868

Wu AM, Singh T, Wu JH et al (2006) Interaction profile of galectin-5 with free saccharides and mammalian glycoproteins: probing its fine specificity and the effect of naturally clustered ligand presentation. Glycobiology 16:524–537

Wu AM, Singh T, Liu JH et al (2007) Activity-structure correlations in divergent lectin evolution: fine specificity of chicken galectin CG-14 and computational analysis of flexible ligand docking for CG-14 and the closely related CG-16. Glycobiology 17:165–184

Yamada Y, Kato K, Oguri M et al (2008) Association of genetic variants with atherothrombotic cerebral infarction in Japanese individuals with metabolic syndrome. Int J Mol Med 21: 801–808

Young AR, Barcham GJ et al (2009) Functional characterization of an eosinophil-specific galectin, ovine galectin-14. Glycoconj J 26:423–432

Yu Y, Yuan S, Yu Y et al (2007) Molecular and biochemical characterization of galectin from amphioxus: primitive galectin of chordates participated in the infection processes. Glycobiology 17:774–783

Zhou D, Sun J, Zhao W et al (2006) Expression, purification, crystallization and preliminary X-ray characterization of the GRP carbohydrate-recognition domain from Homo sapiens. Acta Crystallogr Sect F Struct Biol Cryst Commun 62:474–476

Galectin-1: Forms and Functions

10

Anita Gupta

Galectin-1 (Gal-1) is an animal lectin ranging from *Caenorhabditis elegans* to humans, which is defined by the affinity for β-galactosides and by significant sequence similarity in carbohydrate-binding site. Gal-1 is differentially expressed by various normal and pathological tissues and appears to be functionally polyvalent, with a wide range of biological activity. Evidences indicate that Gal-1 and its ligands are important regulators of immune responses such as T-cell homeostasis and survival, inflammation and allergies as well as host–pathogen interactions. Overexpression of Gal-1 in tumors is considered as a sign of the malignant tumor progression that is often related to the long-range dissemination of tumoral cells (metastasis), to their dissemination into the surrounding normal tissue, and to tumor immune-escape. Gal-1 in its oxidized form plays a number of important roles in the regeneration of the central nervous system after injury. The targeted overexpression of Gal-1 should be considered as a method of choice for the treatment of some kinds of inflammation-related diseases, neurodegenerative pathologies and muscular dystrophies. In contrast, the targeted inhibition of Gal-1 expression is what should be developed for therapeutic applications against cancer progression. Gal-1 is thus a promising molecular target for the development of new and original therapeutic tools.

10.1 The Subcellular Distribution

Galectin-1 shows both intracellular and extracellular functions. Hence it shows characteristics of typical cytoplasmic protein as well as an acetylated N-terminus and a lack of glycosylations (Clerch et al. 1988). It is present in cell nuclei and cytosol and also translocates to the intracellular side of cell membranes. Though Gal-1 lacks a secretion signal sequence and does not pass through endoplasmic reticulum/Golgi pathway (Hughes 1999), it is secreted and found on the extracellular side of all cell membranes as well as in the extracellular matrices of various normal and neoplastic tissues (van den Brule et al. 2003; Camby et al. 2006; von Wolff et al. 2005). Evidence suggests that this protein is secreted like FGF-2 via inside-out transportation involving direct translocation across the plasma membrane of mammalian cells (Nickel 2005). The ß-galactoside-binding site may constitute the primary targeting motif for galectin export machinery using ß-galactoside-containing surface molecules as export receptors for intracellular Gal-1 (Seelenmeyer et al. 2005). Nuclear presence of galectins suggests a role of these endogenous lectins in the regulation of transcription, pre-mRNA splicing and transport processes. Stromal cells of human bone marrow and human/porcine keratinocytes revealed positive reaction for galectin-1, whereas galectins-3, -5, and -7 were not reactive with nuclear sites under identical conditions (Purkrábková et al. 2003).

10.2 Molecular Characteristics

10.2.1 Galectin-1 Gene

Gal-1 is encoded by *LSGALS1* gene located on chromosome 22q12. Although it is not yet clear whether the 15 galectins identified so far have functions in common, a striking common feature among all galectins is their strong modulation of expression during development, differentiation and under different physiological or pathological conditions. This suggests that the expression of different galectins is finely tuned and possibly coordinated. Among galectin genes, a small region spanning over initial transcription start site (−63/+45) is sufficient in the promoter region of Gal-1 gene for its transcriptional activity in mice (Chiariotti et al. 2004). Both an upstream and a downstream position-dependent cis-element are necessary for efficient transcriptional activity; an additional start-up site has been mapped at position-31; and an Sp1-binding site (−57/−48) and a consensus initiator (Inr) element (which

G.S. Gupta et al., *Animal Lectins: Form, Function and Clinical Applications*,
DOI 10.1007/978-3-7091-1065-2_10, © Springer-Verlag Wien 2012

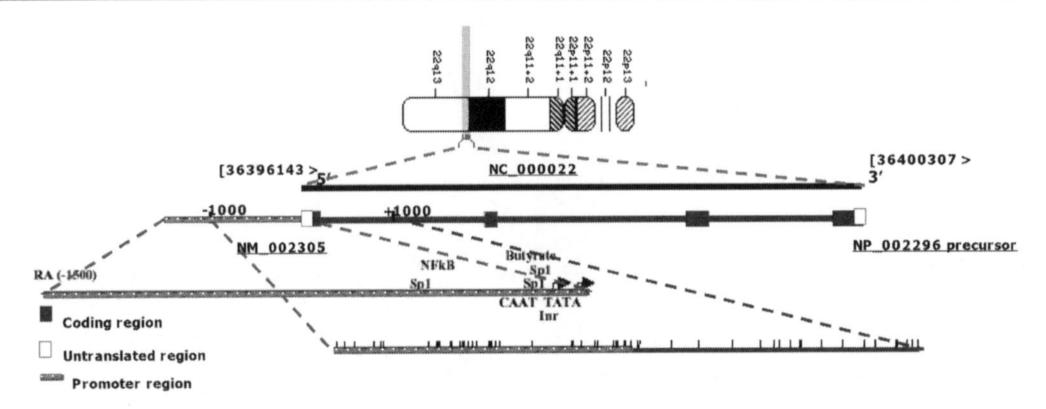

Fig. 10.1 Map of the Gal-1 gene on the human chromosome 22q12. The sequences were retrieved using the MapViewer program and the Entrez genome database on the NCBI website and were analyzed using the GeneWorks program produced by IntelliGenetics Inc. The *curved arrows* indicate the initial transcription start-up sites (Chiariotti et al. 2004). The 0.6 kb transcript results from the splicing of four exons (the *boxes* indicate four coding regions) and encodes for a protein of 135 aminoacids (Reprinted by permission from Camby et al. 2006 © Oxford University Press)

partially overlaps a non-canonical TATA box) direct RNA initiation (Chiariotti et al. 2004). The upstream transcripts contribute to more than half of the Gal-1 mRNA population (Fig. 10.1). The 5′-end of this transcript is extremely GC-rich and may fold into a stable hairpin structure which can influence translation (Chiariotti et al. 2004). The approximate position of the other putative and/or characterized regulatory elements is indicated in Fig. 10.1 and relates to the CAAT box, to NF-kB binding site and to the sodium butyrate and retinoic acid response sequences. Full length cDNAs coding for a 14-kDa was isolated from HL-60 cells and human placenta.

The cDNA clones for lectins had identical sequences with short 5′- and 3′-untranslated regions and coded for 135-amino acid protein which lacks a hydrophobic signal peptide sequence. Biochemical data showed that, despite the presence of a possible N-linked glycosylation site, the protein was not found glycosylated. The 14-kDa lectin is encoded by a single gene. The lectin expressed in *E. coli* was biologically active. Another full-length cDNA for this protein from a HepG2 cDNA library encodes a protein with subunit molecular weight of 14,650. The *E. coli* expressed product bound to a lactose affinity column and specifically eluted with lactose. Like L-14-I, the new lectin, L-14-II, exists as a homodimer in solution. The two related human lectins have 43% amino acid sequence identity. The genomic DNA encoding L-14-II (*LGALS2*) contains four exons with similar intron placement to L-14-I (*LGALS1*); but the genomic upstream region, which contains several sequences characteristic of regulatory elements, differs significantly from L-14-I.(Gitt et al. 1992) Gal-1 occurs as a monomer as well as a non-covalent homodimer consisting of subunits of one CRD (dGal-1, 29.5 kDa) (Barondes et al. 1994; Cho and Cummings 1995). Each form is associated with different biological activities. Gal-1 with a subunit molecular mass of 14.5 kDa contains six cysteine residues per subunit. Non-covalent homodimer of 29.5-kDa is widely expressed in many tissues. The cysteine residues should be in a free state in order to maintain a molecular structure that is capable of showing lectin activity. The 0.6 kb transcript is the result of splicing of four exons encoding a protein with 135 amino acids (Fig. 10.1).

10.2.2 X-Ray Structure of Human Gal-1

The overall folding of Gal-1 involves a β sandwich consisting of two anti-parallel β-sheets (Fig. 10.2). This jelly-roll topology of the CRD constitutes typical folding patterns of galectins. Human Gal-1 exists as a dimer in solution. The integrity of this dimer is maintained principally by interactions between the monomers at the interface and through the well-conserved hydrophobic core, a factor which explains the observed stability of dimer in molecular terms (Lopez-Lucendo et al. 2004). Single-site mutations introduced at some distance from the CRD can affect the lectin fold and influence sugar binding. Both the substitutions introduced in C2S and R111H mutants altered the presentation of loop, harbouring Asp123 in common "jelly-roll" fold. The orientation of side-chain was inverted 180° and positions of two key residues in sugar-binding site of R111H mutant were shifted, i.e. His52 and Trp68. The decrease in ligand affinity in both mutants and a significant increase in the entropic penalty were found to outweigh a slight enhancement of enthalpic contribution. The position of SH-groups in galectin appeared to considerably restrict the potential to form intra-molecular disulphide bridges and was assumed to be the reason for unstable lectin activity in absence of reducing agent. However, this offers no obvious explanation for the improved stability of the C2S mutant

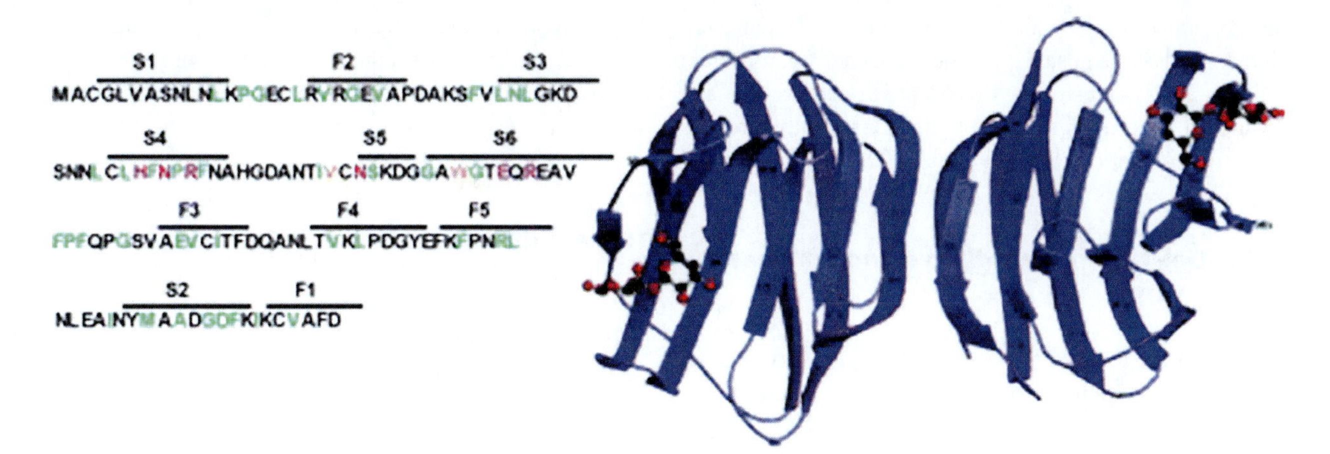

Fig. 10.2 *Right Panel*: **Ribbon diagram of the homodimeric human galectin-1.** The β-strands in the five-stranded (F1–F5) and six-stranded (S1–S6a/S6b) β-sheets are indicated by the letter-number code. The overall folding of Gal-1 involves a ß sandwich consisting of two anti-parallel ß-sheets of five (F1–F5) and six (S1–S6a/b) strands respectively. The N and the C termini of each monomer are positioned at the dimer interface and the CRDs are located at the far ends of the same face, a configuration which constitutes a long, negatively charged cleft in the cavity. The 3D ribbon diagram of the homodimeric human Gal-1 was designed with MOLSCRIPT by Lopez-Lucendo et al. (2004). *Left Panel*: **Sequence of human galectin-1**. Residue numbers of Leg1_Human are indicated. The positions of the β-strands for Leg1_Human are indicated by *horizontal arrows*. The *green* amino acid symbols illustrate highly conserved residues. The key residues of the CRD, which is known to interact directly with bound carbohydrate by means of hydrogen bonds, are *colored pink*, while those interacting with carbohydrates via van der Waals interactions are *orange* (Cooper 2002) and include His^{44}, Asn^{46}, Arg^{48}, Val^{59}, Asn^{61}, Trp^{68}, Glu^{71}, and Arg^{73}. A lactose (Galß1–4Glc) is illustrated in the CRD (Reprinted by permission from Lopez-Lucendo et al 2004 © Elsevier)

under oxidative conditions. The noted long-range effects in single-site mutants are relevant for the functional divergence of closely related galectins and in more general terms, the functionality definition of distinct amino acids (Lopez-Lucendo et al. 2004). The main characteristics of dGal-1 is that it spontaneously dissociates at low concentrations (K_D ~7 μM) into a monomeric form that is still able to bind to carbohydrates (Cho and Cummings 1995), but with a lower affinity (Leppanen et al. 2005).

10.2.2.1 Solution Structures of Galectins

Ligand presence and a shift to an aprotic solvent typical for bioaffinity chromatography might alter the shape of homodimeric human lectin in solution. Using small angle neutron and synchrotron x-ray scattering, the radius of gyration of human galectin-1 decreased from 19.1 ± 0.1 Å in absence of ligand to 18.2 ± 0.1 Å after ligand interaction. In presence of aprotic solvent dimethyl sulfoxide, which did not impair binding capacity, Gal-1 formed dimers of a dimer, yielding tetramers with a cylindrical shape. Surprisingly, no dissociation into subunits occurred. NMR monitoring was in accord with these data. In contrast, an agglutinin from mistletoe sharing galactose specificity showed a reduced radius of gyration from ~62 Å in water to 48.7 Å in dimethyl sulfoxide. Evidently, the solvent caused opposite responses in two tested galactoside-binding lectins with different folding patterns. Thus, ligand presence and an aprotic solvent significantly affect the shape of galectin-1 in solution (He et al. 2003).

10.2.2.2 Cross-Linking Properties of Galectins

Evidence suggests that biological activities of galectins are related to their multivalent binding properties since most galectins possess two carbohydrate recognition domains and are therefore bivalent (Chap. 9). For example, Gal-1, which is dimeric, binds and cross-links specific glycoprotein counter-receptors on the surface of human T-cells leading to apoptosis. Different galectin-1 counter-receptors associated with specific phosphatase or kinase activities form separate clusters on the surface of the cells as a result of the lectin binding to the carbohydrate chains of the respective glycoproteins. Importantly, monovalent Gal-1 is inactive in this system. This indicates that the separation and organization of signaling molecules that result from Gal-1 binding is involved in the apoptotic signal. The separation of specific glycoprotein receptors induced by Gal-1 binding was modeled on the basis of molecular and structural studies of the binding of lectins to multivalent carbohydrates resulting in the formation of specific two- and three-dimensional cross-linked lattices. Brewer (2002) reviewed the binding and cross-linking properties of Gal-1 and other lectins as a model for the biological signal transduction properties of the galectin family of animal lectins. Bovine heart Gal-1 forms stable dimers. In contrast, recombinant murine Gal-3, as well as its proteolytical derived C-terminal domain is predominantly monomeric (Morris et al. 2004).

Gal-1 regulates leukocyte turnover by inducing cell surface exposure of phosphatidylserine (PS), a ligand that targets cells for phagocytic removal, in absence of apoptosis.

Gal-1 monomer-dimer equilibrium appears to modulate Gal-1-induced PS exposure. Monomer-dimer equilibrium regulates Gal-1 sensitivity to oxidative inactivation and provides a mechanism whereby ligand partially protects Gal-1 from oxidation (Stowell et al. 2009).

10.2.3 Gal-1 from Toad (*Bufo arenarum Hensel*) Ovary

A galectin from toad (*Bufo arenarum Hensel*) ovary consists of identical single-chain polypeptide subunits composed of 134 amino acids (calculated mass, 14,797 Da), and its N-terminal residue, alanine, is N-acetylated. In comparison to sequences of known galectins, the *B. arenarum galectin* exhibits highest identity (48% for whole molecule and 77% for CRD) with the bovine spleen Gal-1, but surprisingly less identity (38% for whole molecule and 47% for CRD) with a Gal from *Xenopus laevis* skin (Marschal et al. 1992). Unlike the *X. laevis* galectin, the binding activity of the *B. arenarum* galectin for N-acetyllactosamine, the human blood group A tetrasaccharide and Gaβ1,3GalNAc relative to lactose, was in agreement with that observed for Gal-1 subgroup and those galectins having "conserved" (type I) CRDs (Ahmed and Vasta 1994). Moreover, the toad galectin shares three of the six cysteine residues that are conserved in all mammalian Gals-1, but not in the galectins from *X. laevis*, fish, and invertebrates described so far (Hirabayashi and Kasai 1993). Furthermore, galectins with conserved (type I) CRDs, represented by the *B. arenarum* ovary galectin, and those with "variable" (type II) CRDs, represented by *X. laevis* 16-kDa galectin, clearly constitute distinct subgroups in the extant amphibian taxa and may have diverged early in the evolution of chordate lineages (Ahmed et al. 1996).

Crystal structures of *B. arenarum* Gal-1 in complex with two related carbohydrates, LacNAc and TDG (TDG, thiodigalactoside (ß-D-galactopyranosyl-1-thio-ß-D-galactopyrano-side) showed that the topologically equivalent hydroxyl groups in two disaccharides have identical patterns of interaction with Gal-1. Groups that are not equivalent between two sugars present in second moiety of disaccharide, interact differently with the protein, but use the same number and quality of interactions. Structures showed additional protein-carbohydrate interactions not present in other reported lectin-lactose complexes. These contacts provide an explanation for the enhanced affinity of Gal-1 for TDG and LacNAc relative to lactose. Comparison of *B. arenarum* with other Gal-1 structures shows that among different galectins there are significant changes in accessible surface area buried upon dimer formation, providing a rationale for the variations observed in free-energies of dimerization. The structure of *B. arenarum* Gal-1 has a large cleft with a strong negative potential that connects two binding sites at the surface of the protein. Such a striking characteristic suggests that this cleft is probably involved in interactions of the galectin with other intra or extra-cellular proteins (Bianchet et al. 2000).

10.2.4 GRIFIN Homologue in Zebrafish (DrGRIFIN)

Galectin-like proteins such as rat lens crystalline protein GRIFIN (Galectin-related inter fiber protein) and the galectin-related protein GRP (previously HSPC159; hematopoietic stem cell precursor) lack carbohydrate-binding activity. Their inclusion in the galectin family has been debated as they are considered products of evolutionary co-option. A homologue of GRIFIN is present in zebrafish (Danio rerio) (designated DrGRIFIN), which like the mammalian equivalent is expressed in the lens, particularly in the fiber cells, as revealed in dpf (days post fertilization) embryos. As evidenced, it is weakly expressed in the embryos as early as 21 hpf (hour post fertilization) but strongly at all later stages tested (30 hpf and 3, 4, 6, and 7 dpf). In adult zebrafish tissues, however, DrGRIFIN is also expressed in oocytes, brain, and intestine. Unlike the mammalian homologue, DrGRIFIN contains all amino acids critical for binding to carbohydrate ligands and its activity was confirmed as the recombinant DrGRIFIN could be purified to homogeneity by affinity chromatography on a lactosyl-Sepharose column. Therefore, DrGRIFIN is a bona fide galectin family member that in addition to its carbohydrate-binding properties may also function as a crystalline (Ahmed and Vasta 2008).

10.3 Regulation of *Gal-1* Gene

10.3.1 Gal-1 in IMP1 Deficient Mice

Various physicochemical agents are known to modulate the expression of Gal-1 (Table 10.1). The methylation status of Gal-1 promoter is also a very important mechanism that controls the expression of the gene (Chiariotti et al. 2004). Downregulation of *Igf2* translation and the postnatal intestine showed reduced expression of transcripts encoding ECM components, such as galectin- 1, lumican, tenascin-C, procollagen transcripts, and the Hsp47 procollagen chaperone. It demonstrates that Insulin-like growth factor II mRNA-binding protein 1(IMP1) is essential for normal growth and development and may facilitate intestinal morphogenesis via regulation of ECM formation (Hansen et al. 2004).

Table 10.1 The regulation of Gal-1 expression[a]

Agent/factor	Tissues/cells	Effects	References
Budesonide (glucocorticoid)	Human nasal polyps	Increase	Delbrouck et al. 2002; Goldstone and Lavin 1991
Progesterone	Mouse uterine tissue	Increase	Choe et al 1997
Estrogen	Mouse uterine tissue/Rat1a embryo cells	Increase	Choe et al 1997; Tahara et al 2003
TSH	Rat thyroid	Increase	Chiariotti et al 1994a
Retinoic acid (all-trans)	1. Embryonal carcinomas, mouse myoblastic cells	Increase	Lu et al 2000; Choufani et al 2001
	2. Transformed rat neural cells	Decrease	Chiariotti et al. 1994a
TGFβ1	Metastatic mammary adenocarcinoma (LM3) cells, murine lung adenocarcinoma LP07 cells, and the human breast adenocarci noma (MCF-7) cells	Increase	Daroqui et al 2007
IL-12	Peripheral blood mononuclear cells and cord blood $CD4^+$ cells	Decrease	Filén et al. 2006
δFosB	Quiescent neuronal precursor cells	Increase	Kurushima et al 2005; Miura et al 2005
5-Azacytidine	Human hepatic carcinomas, osteosarcoma cells; human T leukemia cells; rat liver and thyroid cells	Increase	Kondoh et al 2003; Chiariotti et al. 2004; Poirier et al 2001a
Sodium butyrate	Human colon carcinomas; HNSCCs, globelet cells; prostate cancers; mouse embryonal carcinomas	Increase	Ellerhorst et al 1999a; Gaudier et al 2004; Gillenwater et al 1998; Lu and Lotan 1999; Ohannesian et al 1994
Cyclophosphamide	Rat lymphomas	Decrease	Rabinovich et al. 2002b
Benzodiazepine	Jurkat T lymphoma cells	Decrease	Rochard et al. 2004
Valproic acid	Mouse embryo	Increase	Kultima et al 2004
Oxidized low-density lipoprotein	Human endothelial cells	Increase	Baum et al 1995; Perillo et al 1995

[a]Modified after permission from Camby et al. 2006 © Oxford University Press

10.3.2 Blimp-1 Induces Galectin-1 Expression

B lymphocyte-induced maturation protein-1 (Blimp-1), a master regulator for plasma cell differentiation, was necessary and sufficient to induce Gal-1 expression. Blimp-1 is a transcriptional repressor (Keller and Maniatis 1992). Microarray study revealed that Gal-1 is up-regulated upon ectopic expression of Blimp-1 in mature human B cell lines (Shaffer et al. 2002). The ectopic expression of Gal-1 in mature B cells increased Ig μ-chain transcript levels as well as the overall level of Ig production. The effect of Gal-1 on promoting Ig production appears to be mediated through an extracellular receptor(s) and to depend on the binding of β-galactosides before terminal differentiation of B cells (Tsai et al. 2008). The jck murine model, which results from a double point mutation in *nek8* gene, suggests that *Nek8* mutation causes over-expression of Gal-1, sorcin, and vimentin and accumulation of major urinary protein in renal cysts of jck mice in PKD proteome relative to wild type (Valkova et al. 2005).

Regulation by Ovarian Steroids: Gal-1 mRNA is abundantly expressed in mouse reproductive organs such as uterus and ovary. Uterine expression of Gal-1 mRNA is specifically regulated during embryo implantation. Its expression increased on fifth day post coitum (dpc 5) when embryos hatched into endometrial epithelial cells. In absence of embryos, however, Gal-1 in mouse uterus decreased on dpc 5. In delayed implantation mice, Gal-1 mRNA level was augmented by termination of delay of implantation. Ovarian steroids progesterone and estrogen differentially regulated Gal-1 mRNA level in uterine tissues (Choe et al. 1997).

Induction by TSH in Thyroid Cells: Expression of Gal-1 is low in normal and very high in transformed thyroid cells. Gal-1 gene expression is transiently induced after TSH stimulation of normal quiescent FRTL-5 rat thyroid cells. Permanent activation of Gal-1 gene expression was obtained in same cells infected with a wild-type and a temperature sensitive mutant of Kirsten murine sarcoma virus, both at the permissive and non permissive temperature for transformation (Chiariotti et al. 1994a).

10.3.3 Regulation by Retinoic Acid

Growth stimulation and induction of cell differentiation are accompanied by strong modulation of Gal-1 gene expression. Gal-1 mRNA is undetectable in rat brain but abundantly present in rat oligodendrocytes precursors transformed by polyoma middle T oncogene. Retinoic acid (RA) treatment of these transformed cells leads to acquisition of a differentiated phenotype accompanied by a 30-fold decrease of Gal-1 mRNA. Removal of RA restored both the transformed undifferentiated phenotype and high Gal-1 expression (Chiariotti et al. 1994a). A RA responsiveness region was found within sequence from −1,578 to −1,448 upstream of transcription start site (+1). In contrast, high constitutive Gal-1 expression in C2C12 cells appeared to be mediated by a sequence within the promoter region from −62 to +1, which contains a Sp1 consensus sequence. EMSA indicated that transcription factor SP1 bound to Gal-1 Sp1 site; and mutagenesis of this Sp1 site abolished both the binding of nuclear proteins to the mutated Sp1 site and the high constitutive expression of Gal-1 gene.

RARβ Receptors and Expression of Galectin-1, -3 and -8 in Human Cholesteatomas: In human cholesteatomas, there is predominant expression of RARβ and Gal-1. Furthermore, the level of RARβ expression correlated highly with the level of galectin-8 expression, which also correlated with the level of RARα and RARγ. In addition, this parameter also correlated with the level of Gal-1 and Gal-3 expression. Simon et al. (2001) suggest that cholesteatomas may originate in an undifferentiated population of keratinocytes, and that a relation may exist between retinoid activity and galectins.

Macrophage migration inhibitory factor (MIF) expression is significantly higher in the epithelium and vessels of connective tissues of recurrent as opposed to non-recurrent cholesteatoma and significantly correlated to RARβ expression in non-infected cholesteatomas and to MMP-3 and anti-apoptotic Gal-3 in infected cholesteatomas. Together with galectin-3, MIF could cooperate to form an anti-apoptotic feedback loop (Choufani et al. 2001).

10.3.4 Regulation by TGF-β, IL-12 and FosB Gene Products

Daroqui et al. (2007) suggested a potential cross talk between TGF-β and Gal-1 in highly metastatic mammary adenocarcinoma (LM3) cells, murine lung adenocarcinoma LP07 cells, and the human breast adenocarcinoma MCF-7 cell lines. TGF-β1-mediated upregulation of Gal-1 expression was specifically mediated by TGFβRI and TGFβRII. The presence of three putative binding sites for Smad4 and Smad3 transcription factors in Gal-1 gene was consistent with the ability of TGF-β1 to trigger a Smad-dependent signaling pathway in these cells. Thus, TGF-β1 may control Gal-1 expression, suggesting that distinct mechanisms might cooperate in tilting the balance toward an immunosuppressive environment at the tumor site. IL-12 enhances the generation of Th1 lymphocytes and inhibits the production of Th2 subset. IL-12 regulated proteins were studied in microsomal fraction of Th cells. Gal-1 and CD7 are known to interact with each other, and regulate immunity through influencing apoptosis and cytokine production. Filén et al. (2006) indicated that IL-12 down-regulates the expression of both Gal-1 and CD7 in the microsomal fraction of peripheral blood mononuclear cells and cord blood $CD4^+$ cells. The down-regulation of these proteins is likely to have a role in specific Th cell selection and cytokine environment creation.

In rat 3Y1 embryo cell lines expressing FosB and δFosB as fusion proteins (ER-FosB, ER-δFosB) with ligand-binding domain of human estrogen receptor (ER) (Nishioka et al. 2002), the binding of estrogen to the fusion proteins resulted in their nuclear translocation. Among several proteins whose expression was affected after estrogen administration, one of the proteins was galectin-1 (Tahara et al. 2003).

10.3.5 Regulation by Metabolites/Drugs/Other Agents

The expression of Gal-1 can be induced in cultured hepatoma-derived cells by treatment with 5-AzaCytidine (AzaC), a DNA demethylating drug. Interestingly, Gal-1 and AzaC individually have been shown to affect similar cell processes in cancer cells, including differentiation, and growth inhibition (Chiariotti et al. 1994b; c/r Camby et al. 2006). The sensitivity to Gal-1 is associated with repression of endogenous Gal-1 gene whereas non-sensitive cells express high levels of Gal-1. Repression of Gal-1 gene in sensitive cells is associated with hyper-methylation of promoter region. Transient treatment of non-expressing cells with demethylating agent AzaC led to irreversible demethylation and subsequent reactivation of galectin-1 gene (Salvatore et al. 2000). The expression of Gal-1, accompanied by its secretion and its binding to cell surface receptors, could be involved in AzaC observed effects in hematopoietic cells where Gal-1 modulates differentiation or apoptosis.

Poirier et al. (2001a) investigated the effect of AzaC and Gal-1 on human lymphoid B cells phenotype. Treatment of lymphoid B cells with AzaC resulted in: (1) a decrease in cell growth with an arrest of cell cycle at Go/G1 phase, (2) phenotypic changes consistent with a differentiated phenotype, and (3) the expression of p16, a tumor-suppressor gene whose expression was dependent of its promoter demethylation, and of Gal-1. The targeting of Gal-1 to the plasma membrane followed its cytosolic expression. Recombinant Gal-1 added to BL36 cells displayed growth inhibition and phenotypic changes consistent with a commitment toward differentiation. It seems that AzaC-induced Gal-1 expression and consequent binding of Gal-1 on its cell membrane receptor may, in part, be involved in AzaC-induced plasmacytic differentiation. Poirier et al. (2001a) proposed that the released Gal-1 was immediately recruited to modulate cell activity. Gal-1 may do this by interacting with and modulating cell receptors via its carbohydrate recognition domains because the Gal-1-receptor interaction is abrogated by thiodigalactoside (Fouillit et al. 2000). While Gal-1 binds to T and B lymphoblastoid cells (Fouillit et al. 2000; Baum et al. 1995), studies have demonstrated that galectins are immunosuppressive in animal models of autoimmune diseases (Rabinovich et al. 1999). Whereas the full role of Gal-1 in modulating immune function is not yet understood, the increase in Gal-1 expression by AzaC in BL cells suggests that Gal-1 may play a role in the behaviour of normal leukocytes and of tumor cells.

The underlying mechanism, triggered by demethylating stimulus probably involves the stimulation of a signaling cascade that regulates cell proliferation and viability. Demethylating stimulus may modify a pathway activated by the membrane-anchored protein-tyrosine phosphatase CD45. Engagement of CD45 is known to regulate Src tyrosine kinases phosphorylation, phospholipase Cγ regulation, inositol phosphate production, diacylglycerol production, PKC activation, and calcium mobilisation (c/r Poirier et al. 2001a). Increased synthesis and secretion of Gal-1 by the cell could account for part of the phenotypic alterations detected in AzaC treated cells. Gal-1-induced dimerisation and/or segregation might inhibit the catalytic site in CD45, thereby blocking tyrosine phosphates activity. Because Gal-1 binding to cell surface receptors results in tyrosine phosphorylation (Fouillit et al. 2000; Vespa et al. 1999), it may allow a kinase-dependent signal to be transduced. Several studies have linked Gal-1 expression with growth inhibition and cell death (Goldstone and Lavin 1991; Delbrouck et al. 2002; Perillo et al. 1995). However, the reports that some growth inhibitory agents did not induce Gal-1 expression indicated that Gal-1 expression is not dependent on the cell's growth state in general, though it may be involved in growth suppression (Gillenwater et al. 1998). Moreover, it is likely that Gal-1 acts in a manner to regulate specific signal transduction processes that is determined by the cell type and by the state of cell differentiation, since, exogenous rGal-1 added to BL cells inhibited cell growth. Besides, Gal-1 as well as AzaC induced an expression of the cell surface plasma cell antigen, CD138, a phenotypic marker that identifies cells with plasmacytic differentiation (Kopper and Sebestyen 2000). This is consistent with the hypothesis that AzaC and Gal-1 share similar signals for differentiation, however, since there was a significant difference in the expression of CD19 and CD23 after AzaC or Gal-1 treatments it is likely that some pathways are specifically modified by AzaC. The mechanisms involved in these different pathways, important in clinical therapy, remain to be elucidated in the future (Poirier et al. 2001b).

Cyclophosphamide Modulates Gal-1 Expression: Gal-1 has been shown to contribute to tumor cell evasion of immune responses by modulating survival and differentiation of effector T cells. A single low dose of cyclophosphamide has an antimetastatic effect on lymphoma (L-TACB)-bearing rats by modulating the host immune response. Galectin-1 has potent immunomodulatory properties by regulating cell-matrix interactions and T-cell apoptosis. A single low dose of cyclophosphamide modulates the expression of Gal-1 and Bcl-2 by tumors, which could in turn influence the apoptotic threshold of spleen mononuclear cells. Conversely, Gal-1 expression was significantly reduced in spleen cells and lymph node metastasis through period of study. Cyclophosphamide treatment restored the basal levels of Gal-1 expression in primary tumors and spleens. Results suggest that, in addition to other well-known functions of cyclophosphamide, Cy may also modulate Gal-1 expression and function during tumor growth and metastasis with critical implications for tumor-immune escape and immunotherapy (Rabinovich et al. 2002b; ZacarÃas Fluck et al. 2007).

Effects of Butyrate and Valproic Acid: Colonic mucin glycosylation can be modified by butyrate. In HT29-Cl.16E cells, the most striking effect of butyrate was on galectin-1 gene, which increased 8- to 18-fold, with a central and apical intracellular localization. Butyrate effects have possible link with mucins expressed by HT29-Cl.16E cells (Gaudier et al. 2004). The antiepileptic drug valproic acid (VPA) is a potent inducer of neural tube defects (NTDs) in human and mouse embryos. Some VPA-responsive genes have been associated with NTDs or VPA effects, whereas others provide putative VPA targets, associated with processes relevant to neural tube formation such as, galectin-1, fatty acid synthase (Fasn), annexins A5 (Kultima et al. 2004). The ErbB protein family or epidermal growth factor receptor (EGFR), a family of four structurally related receptor tyrosine kinases is associated with genes involved in cell-matrix interactions including galectin 1 and galectin 3. These data represent profiles of transcriptional changes associated with ErbB2-related pathways in the breast, and identify potentially useful targets for prognosis and therapy (Mackay et al. 2003).

Global gene expression profiles were analyzed in European wild boar naturally infected with *Mycobacterium bovis*. While some proteins overexpressed in infected animals, lower expression was observed for galectin-1, complement component C1qB, certain HLA class I and MHC class II antigens and Ig chains in infected animals (Galindo et al. 2009).

Effect of Hypoxia: The expression of Gal-1 has been shown to be regulated by hypoxia. Ectopically expressed hypoxia-inducible factor (HIF) 1α protein, an oxygen-sensitive subunit of HIF-1 that is a master factor for cellular response to hypoxia, significantly increases galectin-1 expression at both m-RNA and protein levels in colorectal cancer (CRC) cell lines. The knockdown of Gal-1 by its specific shRNA can significantly reduce hypoxia-induced invasion and migration of CRC cell line, proposing that Gal-1 mediates the HIF-1-induced migration and invasion of CRC cells during hypoxia. The CCAAT/enhancer binding protein-α (C/EBPα), a critical transcriptional factor for hematopoietic cell differentiation, can directly activate galectin-1 through binding to the −48 to −42 bp region of its promoter. Moreover, knockdown or chemical inhibition of galectin-1 partially blocks the differentiation induced by HIF-1α or C/EBPα, which can be rescued by recombinant galectin-1. These observations give new insights on the mechanisms for Gal-1 expression regulation and HIF-1α- and C/EBPα-induced leukemic cell differentiation (Zhao et al. 2011).

10.4 Gal-1 in Cell Signaling

Gal-1 is present both inside and outside cells, and has both intracellular and extracellular functions. The extracellular functions require the carbohydrate-binding properties of dGal-1 while the intracellular ones are associated with carbohydrate-independent interactions between Gal-1 and other proteins. The Gal-1 induced growth inhibition requires functional interactions with α5β1 integrin (Fischer et al. 2005). The antiproliferative effects result from inhibition of Ras-MEK-ERK pathway and consecutive transcriptional induction of p27: two Sp1-binding sites in p27 promoter are crucial to Gal-1 responsiveness (Fischer et al. 2005). The inhibition of the Ras-MEK-ERK cascade by Gal-1 increases Sp1 transactivation, with DNA binding relating to the reduced threonine-phosphorylation of Sp1 (Fischer et al. 2005). Furthermore, Gal-1 induces p21 transcription and selectively increases p27 protein stability. The Gal-1-mediated accumulation of p27 and p21 inhibits Cdk2 activity and, ultimately, results in G1 cell cycle arrest and growth inhibition (Fischer et al. 2005). The Gal-1-induced increase in cell motility involves Gal-1-induced increase in rhoA expression and the alteration of polymerization of actin cytoskeleton (Camby et al. 2002). Gal-1 is recruited from cytosol to cell membrane by H-Ras-GTP in a lactose-independent manner with the resulting stabilization of H-Ras-GTP, the clustering of H-RAS-GTP and Gal-1 in non-raft microdomains (Prior et al. 2003), the subsequent binding to Raf-1 (but not to PI3Kinase) and the activation of ERK signaling pathway (Elad-Sfadia et al. 2002). Nuclear Gal-1 interacts with Gemin4 and is co-immunoprecipitated with the nuclear SMN complexes involved in the splicing pathway (Vyakarnam et al. 1997)

10.5 Ligands/Receptors of Gal-1

10.5.1 Each Galectin Recognizes Different Glycan Structures

Human galectins have divergent roles and each galectin recognizes different glycan receptors (Camby et al. 2006; Ilarregui et al. 2005) although most of the members of galectin family bind weakly to simple disaccharide lactose (Galβ1-4Glc). While Gal-1 inhibits mast cell degranulation (Rubinstein et al. 2004b), Gal-3 induces degranulation in mast cells independently of IgE-mediated antigen stimulation (Frigeri et al. 1993). Gal-1 blocks leukocyte chemotaxis (La et al. 2003), whereas Gal-3 has opposite effect, inducing leukocyte chemotaxis (Sano et al. 2000) and release of preformed IL-8 from neutrophils (Jeng et al. 1994), which further augments chemotaxis of leukocytes. In addition, Gal-1 inhibits acute inflammatory responses through various mechanisms, including suppression of phospholipase A_2-induced edema (Rabinovich et al. 2000) and inhibition of neutrophil extravasation (La et al. 2003). Gal-3 enhances the extravasation of neutrophils, and Gal-3 null mice also exhibit attenuated leukocyte infiltration following challenge (Hsu et al. 2000). Interestingly, patients with reduced Gal-2 expression were found to have reduced risk for myocardial infarction, suggesting that Gal-2 may also have pro-inflammatory roles (Ozaki et al. 2004). Furthermore, Gal-1, Gal-2, and Gal-3 have all been reported to signal T cells through different receptors (Hahn et al. 2004; Sturm et al. 2004; Stillman et al. 2006). These types of studies suggest that Gal-1, Gal-2, and Gal-3 recognize distinct receptors on leukocytes.

Functional studies indicated that Gal-1, Gal-2, and Gal-3 recognized distinct receptors on leukocytes although different galectins could also recognize related receptors. For example, Gal-3 attenuates Gal-1 inhibition of growth in neuroblastoma cells at the receptor level, and both Gal-1 and Gal-3 induce superoxide production in human neutrophils (Almkvist et al. 2001). Gal-1 and Gal-2 both induce surface exposure of phosphatidylserine (PS) in activated human neutrophils in the absence of apoptosis through a Ca^{2+}-dependent pathway (Karmakar et al. 2005). In this way, galectins likely exhibit unique versatility in a wide range of biological functions (Ilarregui et al. 2005). Although some differences have been reported in glycan recognition by these galectins (Hirabayashi et al. 2002; Brewer 2004), there are many questions remaining about their glycan recognition and subsequent effects on galectin binding. It has been suggested that differences in biological effects of Gal-1 and Gal-3 result from differences in tertiary structure, rather than ligand binding properties (Perillo et al. 1995; Ahmad et al. 2004), because Gal-3 was thought to behave primarily as a monomer (Massa et al. 1993). However, Gal-3 can form homo-oligomeric structures (Nieminen et al. 2007), which supports the likelihood that the major differences in biological functions by these lectins are because of differences in glycan recognition.

10.5.2 Interactions of Galectin-1

Although Gal-1 was first described to bind ß-galactosides, it is now clear that besides its interaction with ß-galactosides it is also engaged in protein–protein interactions (Camby et al. 2006). Infact, lectin activity of Gal-1 is observed when it is extracellular, while the protein–protein interactions of Gal-1 participate in its intracellular functions. As a result, Gal-1 has large number of binding partners.

10.5.2.1 Protein–Carbohydrate Interactions

The lectin activity of Gal-1 relates to its carbohydrate-binding site (Fig. 10.2). Thermodynamic calculations indicate that sugar binding to Gal-1 is enthalpically driven, and suggest the concept that weak interactions constitute the main forces in stability of complex formation (Lopez-Lucendo et al. 2004). The dissociation constant of dGal-1 with various glycoproteins is ~5 μM (Symons et al. 2000). Galectin-1 preferentially recognizes multiple Galβ1,4GlcNAc units, which may be presented on the branches of *N*- or *O*-linked glycans on cell surface glycoproteins. Although dGal-1 binds preferentially to glycoconjugates containing disaccharide N-acetyllactosamine (Gal-ß1–3/4 GlcNAc/LacNAcII), its binding to individual lactosamine units is characterized by relatively low affinity (K_D ~50 μM). However, the arrangement of lactosamine disaccharides in multiantennary repeating chains increases the binding avidity (K_D ~4 μM) (Ahmad et al. 2004). In contrast, there is no increase in avidity when recognition unit is repeated in a string (poly-N-lactosamine) (Ahmad et al. 2004). In polysaccharides, dGal-1 does not bind glycans that lack a terminal non-reducing unmodified N-acetyllactosamine (Stowell et al. 2004). In search of molecular mechanism of Gal-1 secretion, a Gal-1 export mutant identified 26 single amino acid changes that cause a defect of both export and binding to counter receptors. It was indicated that β-galactoside binding site represents the primary targeting motif of galectins defining a galectin export machinery that makes use of β-galactoside-containing surface molecules as export receptors for intracellular galectin-1 (Seelenmeyer et al. 2005). Isothermal titration calorimetry measurements determined affinities of five galactose-containing ligands for Gal-1. Although the galactose moiety of each oligosaccharide is necessary for binding, it is not sufficient by itself. The nature of both the reducing sugar in the disaccharide and the interglycosidic linkage play essential roles in binding to human Gal-1 (Meynier et al. 2009).

The specificity of Gal-1 for glycans depends not only on the structure of glycans but also on the mode of their presentation viz., solid phase or in solution. In solution-based assays (Blixt et al. 2004; Hirabayashi et al. 2002; Leppanen et al. 2005), Gal-1 binds glycans with a single *N*-acetyllactosamine (LacNAc) unit (Galβ1-4GlcNAc) equivalently to those with poly-*N*-acetyllactosamine [poly(LacNAc)] sequences (Galβ1-4GlcNAc)$_n$; however, the dimeric form of Gal-1 showed a significant preference for poly(LacNAc)-containing glycans in solid phase assays. Human recombinant homodimer (dGal-1) binds with high affinity (K_D ~ 2–4 μM) to immobilized extended glycans containing terminal N-acetyllactosamine (LacNAc; Galβ1-4GlcNAc) sequences on poly-N-acetyllactosamine [poly(LacNAc)/(3Galβ1-4GlcNAcβ1-)$_n$] sequences, complex-type biantennary N-glycans, or novel chitin-derived glycans modified to contain terminal LacNAc. Although terminal Gal residues are important for dGal-1 recognition, dGal-1 bound similarly to α3-sialylated and α2-fucosylated terminal LacNAc, but not to α6-sialylated and α3-fucosylated terminal LacNAc. Results suggested that dGal-1 functions as a dimer to recognize LacNAc units on extended poly(LacNAc) on cell surfaces (Leppanen et al. 2005). Gal-1 also exists in an oxidized form, which lacks lectin activity (Outenreath and Jones 1992).

Significantly, Gal-1 does not recognize internal Lac-NAc units within poly(LacNAc) (Leppanen et al. 2005; Stowell et al. 2004), suggesting that this preference likely reflects favorable poly(LacNAc) conformational constraints of terminal LacNAc unit that are enhanced by immobilization. But Gal-1 recognized poly-(LacNAc)-containing glycans on leukocyte surfaces with a similar affinity as observed for immobilized poly(LacNAc) glycans (Leppanen et al. 2005). Likewise, frontal affinity chromatography or isothermal calorimetry, in which glycans are free in solution, also did not reveal significant difference in carbohydrate recognition (Hirabayashi et al. 2002; Ahmad et al. 2004). This suggests that galectin-glycan interactions may be tested in context of immobilized glycan presentation. Stowell et al. (2008a, b) and others evaluated Gal-1, Gal-2, and Gal-3 interactions using immobilized glycans in a glycan microarray format that included hundreds of structurally diverse glycans, and compared the results to binding determinants toward promyelocytic HL60 cells, which respond to signals by these galectins resulting in exposure of surface PS (Stowell et al. 2007; Fukumori et al. 2003) and tested binding to human erythrocytes. These galectins exhibited differences in glycan binding and demonstrated that each of these galectins mechanistically differed in their binding to glycans on the microarrays and that these differences were reflected in the determinants required for cell binding and signaling. The specific glycan recognition by each galectin underscores the basis for differences in their biological activities. Gal-1 binds to a number of ECM components in ß-galactoside-dependent manner in following order: laminin > cellular fibronectin > thrombospondin > plasma fibronectin > vitronectin > osteopontin (Moiseeva et al. 2000, 2003). Laminin and cellular fibronectin are glycoproteins which are highly N-glycosylated with bi- and tetra-antennary poly-N-lactosamines (Carsons et al. 1987).

Gal-1 is a receptor for the angiogenesis inhibitor anginex, and is crucial for tumor angiogenesis. Gal-1 is overexpressed in endothelial cells of different human tumors. The role of gal-1 in tumor angiogenesis is demonstrated in zebrafish model and in gal-1-null mice, in which tumor growth is markedly impaired because of insufficient tumor angiogenesis. Thus, gal-1 regulates tumor angiogenesis and is a target for angiostatic cancer therapy (Thijssen et al. 2006).

N-Glycans on CD45 Negatively Regulate Galectin-1: Although terminal galactose residues are important for dGal-1 recognition, dGal-1 binds similarly to α3-sialylated and α2-fucosylated terminal N-acetyllactosamine, but not to α6-sialylated or α3-fucosylated terminal N-acetyllactosamine (Amano et al. 2003; Leppanen et al. 2005). The addition of sialic acid to T cell surface glycoproteins influences essential T cell functions such as selection in the thymus and homing in the peripheral circulation. Addition of α2,6-linked sialic acid to Galβ1,4GlcNAc sequence, the preferred ligand for galectin-1, inhibits recognition of this saccharide ligand by Gal-1. SAα2,6Gal sequences, created by ST6Gal I enzyme (sialyltransferase), are present on medullary thymocytes resistant to Gal-1-induced death but not on galectin-1-susceptible cortical thymocytes. Because the ST6Gal I preferentially utilizes N-glycans as acceptor substrates, it was shown that N-glycans are essential for galectin-1-induced T cell death. Expression of ST6Gal I specifically resulted in increased sialylation of N-glycans on CD45 which is a T cell receptor for galectin-1. ST6Gal I expression abrogated the reduction in CD45 tyrosine phosphatase activity that results from galectin-1 binding. Sialylation of CD45 by the ST6Gal I also prevented galectin-1-induced clustering of CD45 on the T cell surface, an initial step in galectin-1 cell death. Thus, regulation of glycoprotein sialylation may control susceptibility to cell death at specific points during T cell development and peripheral activation (Amano et al. 2003) (see Camby et al. 2006).

Galectin-1 Interacts with Subunit of Integrin: The activity of integrin adhesion receptors is essential for normal cellular function and survival. N-glycosylations of ß-integrins regulate ß1 integrin functions by modulating their heterodimerization with α chains and ligand-binding activity (Gu and Taniguchi 2004). By direct binding to ß1 integrins (without cross-linking them) dGal-1 increases the amounts of partly activated ß1 integrins, but does not induce dimerization with α subunits (Moiseeva et al. 2003). In case of vascular smooth muscle cells this interaction of Gal-1 with α1ß1 integrin has been reported both as transiently phosphorylating the focal adhesion kinase (FAK) and as modulating attachment of cells and their spreading and migration on laminin, but not on cellular fibronectin (Moiseeva et al. 2003). Gal-1 is secreted during skeletal muscle differentiation and accumulates with laminin in the basement membrane surrounding each myofiber (Gu and Taniguchi 2004). The coincidence of Gal-1 secretion with the onset of myoblast differentiation and fusion and the transition in myoblast adhesion and mobility on laminin are regulated by the interaction of Gal-1 with laminin and the α7ß1 integrin. Fischer et al. (2005) have shown that Gal-1-induced growth inhibition requires functional interactions with α5ß1 integrin. Gal-1 from mouse macrophages has been found to specifically associate with other integrins such as α_Mß2 (the CR3) (Avni et al. 1998).

IgA1 as Premier Serum Glycoprotein Recognized by Human Galectin-1: Endogenous glycoproteins co-purified with human heart galectin-1 (HHL) are excellent ligands for HHL. These glycoproteins are rich in T antigen (Galβ1→3 GalNAc-) of O-linked oligosaccharides. It seems that galectin-1 plays important role in anchoring of microbial and cancer cells known to be rich in T antigen, in high serum IgA1 turn over and in tissue sequestering of IgA1 immune complexes especially after their microbial desialylation in IgA nephropathy and other immune complex-mediated disorders (Sangeetha and Appukuttan 2005).

CD2, CD3, CD7 CD43, CD45, CA125 as Gal-1 Receptors: Elola et al. (2005a) reviewed galectin-1 receptors, and some of the mechanisms by which this lectin affects different cell types. Several Gal-1 receptors are discussed such as CD45, CD7, CD43, CD2, CD3, CD4, CD107, CEA, actin, extracellular matrix proteins such as laminin and fibronectin, glycosaminoglycans, integrins, a β-lactosamine glycolipid, GM1 ganglioside, polypeptide HBGp82, glycoprotein 90 K/MAC-2BP, CA125 ovarian cancer antigen, and pre-B cell receptor (*R:* Elola et al. 2005a). A number of T-cell glycoproteins from MOLT-4 and Jurkat human T cells have been shown to be specific receptors for mammalian Gal-1 binding (Pace et al. 1999; Walzel et al. 1999; Symons et al. 2000; Fajka-Boja et al. 2002; Seelenmeyer et al. 2003). Primary structure of CA125 is a giant mucin-like glycoprotein present on cell surface of ovarian tumor. CA125 is a counter receptor for Gal-1, as both soluble and membrane-associated fragments of CA125 from HeLa cell lysates bind specifically to human Gal-1 with high efficiency. This interaction depends on β-galactose-terminated, O-linked oligosaccharide chains of CA125, and has preference for Gal-1 versus Gal-3. Results suggest that CA125 might be a factor involved in the regulation of Gal-1 export to the cell surface (Seelenmeyer et al. 2003). CD45 can positively and negatively regulate Gal-1-induced T cell death, depending on the glycosylation status of the cells. The CD45 inhibitory effect involved the phosphatase domain. Oligosaccharide-mediated clustering of CD45 facilitated galectin-1-induced cell death (Nguyen et al. 2001).

Gal-1 is a major receptor for the carbohydrate portion of the ganglioside GM1 exposed on the surface of human neuroblastoma cells (Kopitz et al. 1998; Andre et al. 2004). Cell confluence increases the surface presentation of dGal-1. Under these circumstances Gal-1 acts as a negative growth regulator of neuroblastoma cells, though without being pro-apoptotic (Kopitz et al. 1998, 2001).

10.5.2.2 Protein–Protein Interactions

The proteins identified so far that interact in a carbohydrate-independent manner with Gal-1 are not structurally related to each other and do not seem to share any common domains or motifs. The galectin sites that are involved in protein–protein interactions have not been fully established. Gemin4, a cytoplasmic and also found in the nucleus of cells, is a member of the survival of motor neuron protein (SMN) complex and the miRNP particle (microRNA [ribonucleoprotein [RNP]). Nuclear Gal-1 interacts with Gemin4 and is co-immunoprecipitated with nuclear SMN complexes involved in splicing pathway (Vyakarnam et al. 1997). The interactions of Gal-1 with Gemin4 and nuclear SMN complex (Park et al. 2001) are loci of action in spliceosome assembly.

10.6 Functions of Galectin-1

10.6.1 Role of Galectin-1 in Apoptosis

A growing body of evidence indicates that Gal-1 functions as a homeostatic agent by modulating innate and adaptative immune responses. A high level of expression is found in lymphatic organs, which feature high rates of apoptosis. Furthermore, Gal-1 can initiate T cell apoptosis and potetiates apoptosis in the epithelial tumor cell lines (MCF7 and BeWo). It induces the inhibition of cell growth and cell-cycle arrest and promotes apoptosis of activated, but not resting immune cells (He and Baum 2004; Rabinovich et al. 2002b). The suggested-resting T cells are sensitized to CD95/Fas-mediated cell death by Gal-1 (Matarrese et al. 2005). Furthermore, the Gal-1 expressed by thymic epithelial cells promotes the apoptosis of immature cortical thymocytes in vitro (Perillo et al. 1997), suggesting a potential role for this protein in the processes of positive and/or negative selection within thymic microenvironment. Gal-1 also suppresses the secretion of the pro-inflammatory cytokine interleukin-2 (IL-2) and favors the secretion of the anti-inflammatory cytokine IL-10 (van der Leij et al. 2004). Battig et al. (2004) have shown that the dimeric form of Gal-1 is a dramatically more potent inducer of apoptosis in T cells than wild-type Gal-1. Evidence indicates that the amount of Gal-1 secreted by different cell types in the ECM is sufficient to kill T cells (Perillo et al. 1995; He and Baum 2004).

A number of T-cell glycoproteins from human MOLT-4 and Jurkat T cells have been shown to be specific receptors for mammalian Gal-1 and are involved in Gal-1-mediated T-cell death: CD45, CD43, CD7 (c/r Pace et al. 1999; Walzel et al. 1999; Symons et al. 2000; Fajka-Boja et al. 2002). However, although the deletion mutants of the glycoproteins confirm their importance in the apoptotic response to Gal-, the role of CD45 in T-cell apoptosis mediated by Gal-1 remains controversial since Gal-1 induces apoptosis in CD45-deficient T cells (Walzel et al. 1999; Fajka-Boja et al. 2002). Earl et al. (2010) showed that Gal-1 signaling through CD45, which carries both N- and O-glycans, is regulated by CD45 isoform expression, core 2 O-glycan formation and the balance of N-glycan sialylation. Regulation of Gal-1 T cell death by O-glycans is mediated through CD45 phosphatase activity. As for CD7, it seems that only specific spliced isoforms or glycoforms of CD45 may be important in signaling Gal-1-induced cell death (Nguyen et al. 2001; Xu and Weiss 2002; Amano et al. 2003; See Camby et al. 2006).

Gal-1 interferes with the Fas-associated apoptosis cascade in T-cell lines Jurkat and MOLT-4. Gal-1 and an Apo-1 mAb induced DNA-fragmentation in Jurkat T-cells whereas MOLT-4 cells were resistant. Gal-1 stimulated DNA-fragmentation could be inhibited by caspase-8 inhibitor II (Z-IETD-FMK) and a neutralizing Fas mAb. Fas could be identified as a target for Gal-1 recognition. Gal-1 stimulates the activation and proteolytic processing of procaspase-8 and downstream procaspase-3 in Jurkat-T cells. Inhibition of Gal-1 induced procaspase-8 activation by a neutralizing Fas mAb strongly suggests that Gal-1 recognition of Fas is associated with caspase-8 activation.

This report provides experimental evidence for targeting of Gal-1 to glycotopes on Fas and the subsequent activation of the apoptotic death-receptor pathway (Brandt et al. 2008).

Galectin-Induced Activation of NFAT and AP-1 in Human Jurkat T-Lymphocytes: Galectin-mediated ligation of glycoepitopes on T-cell activation markers induces an increase in the cytosolic $(Ca^{2+})_i$ originating from a transient Ca^{2+} release of internal stores as well as a sustained influx across the plasma membrane. The signal transduction events that lead to galectin-induced cell death in activated T cells involve several intracellular mediators including the induction of specific transcription factors (i.e., NFAT, AP-1), the activation of Lck/ZAP-70/MAPK signaling pathway, the modulation of Bcl-2 protein production, the depolarization of the mitochondrial membrane potential and cytochrome C release, the activation of caspases and the participation of the ceramide pathway (Rabinovich et al. 2000; Walzel et al. 2000; Hahn et al. 2004; Ion et al. 2005; Matarrese et al. 2005). Electrophoretic mobility shift assays (EMSAs) provided evidence for Gal-1-stimulated increase in the binding of nuclear extracts to a synthetic oligonucleotide with an AP-1 consensus sequence (Walzel et al. 2002). However, the Gal-1-induced apoptosis in human T leukemia MOLT-4 cells deficient in Fas-induced cell death is not dependent on the activation of caspase-3 or on cytochrome c release—two hallmarks of apoptosis—but involves the rapid nuclear translocation of EndoG from mitochondria (Hahn et al. 2004), implying that Gal-1-induced cell death might also relate to one of the other types of cell death (Broker et al. 2005). In addition, in vivo studies on experimental autoimmunity models have revealed the ability of Gal-1 to skew the balance toward a T2-type cytokine response by reducing the levels of IFNγ, TNFα, IL-2, and IL-12 and increasing the level of IL-5 secretion (Santucci et al. 2000, 2003). Galectin-1 tunes TCR binding and signal transduction to regulate CD8 burst size (Liu et al. 2009a).

Induction of Apoptosis in Galectin-1-Stimulated Jurkat T Cells: Galectin-1 induces death of diverse cell types including lymphocytes and tumor cells. Gal-1 sensitizes human resting T cells to Fas (CD95)/caspase-8-mediated cell death. The treatment of Jurkat E6.1 cells with N-glycan processing inhibitors (1-deoxymannojirimycin and swainsonine) strongly reduced the cell binding of Gal-1-biotin, conjugate binding to cell lysate glycoproteins, and to CD3 immunoprecipitates as well as the binding of CD2 and CD3 to immobilized Gal-1. Study provided evidence that Gal-1 triggers through binding to N-linked glycans a Ca^{2+}-sensitive apoptotic pathway (Walzel et al. 2006) and that the JNK/c-Jun/AP-1 pathway plays a key role for T-cell death regulation in response to gal-1 stimulation (Brandt et al. 2010).

Galectin-1 is highly expressed in thymus and induces apoptosis of specific thymocyte subsets and activated T cells. Galectin-1 binds to N- and O- glycans on several glycoprotein receptors, including CD7, CD43, and CD45. Galectin-1 kills immature thymocytes and activated peripheral T cells by binding to glycans on T cell glycoproteins including CD7, CD45, and CD43. Role for CD43 in galectin-1-induced death and the effects of O-glycan modification on galectin-1 binding to CD43 has been demonstrated. It appeared that CD43 bearing either core 1 or core 2 O-glycans can positively regulate T cell susceptibility to galectin-1, identifying a function for CD43 in controlling cell death. Studies also demonstrated that different T cell glycoproteins on the same cell have distinct requirements for glycan modifications that allow recognition and cross-linking by galectin-1 (Hernandez et al. 2006).

Galectin-1-Induced Apoptosis of Immature Thymocytes is Regulated by NF-kB: CD7, one of the galectin-1 receptors, has crucial roles in Gal-mediated apoptosis of activated T-cells and T-lymphoma progression in peripheral tissues. CD7 promoter activity was increased by NF-kB and this activity was synergistic when Sp1 was co-expressed in immature T-cell line L7. Furthermore, the regulation of CD7 gene expression through NF-kB activation induced by TCR/CD28 might have significant implications for T-cell homeostasis (Koh et al. 2008). However, while understanding regulatory pathway in functional mature T-cells it was revealed that CD7 expression was downregulated by Twist2 in Jurkat cells, a human T-cell lymphoma cell line, and in EL4 cells, a murine T-cell lymphoma cell line. Furthermore, ectopic expression of Twist2 in Jurkat cells reduced galectin-1-induced apoptosis. Based on these results, it was concluded that upregulation of Twist2 increases the resistance to galectin-1-mediated-apoptosis, which may have significant implications for the progression of some T-cells into tumors such as Sezary cells (Koh et al. 2009).

Cell Death via Mitochondrial Pathway: A high level of expression is found in lymphatic organs, which feature high rates of apoptosis. However, the galectin-1 T cell death pathway is distinct from other death pathways, including those initiated by Fas and corticosteroids. Galectin-1-induced cell death proceeds via a caspase-independent pathway that involves a unique pattern of mitochondrial events, and different galectin family members can coordinately regulate susceptibility to cell death (Hahn et al. 2004). Furthermore, Gal-1 triggers an apoptotic program involving an increase of mitochondrial membrane potential and participation of the ceramide pathway. In addition, Gal-1 induces mitochondrial coalescence, budding, and fission accompanied by an increase and/or redistribution of fission-associated molecules h-Fis and DRP-1. Importantly, these changes are detected in both resting and activated human T cells, suggesting that Gal-1-induced cell death might become an excellent model to analyze the morphogenetic changes of mitochondria during the execution of cell death. This suggests the association among Gal-1, Fas/Fas ligand-mediated cell death, and the mitochondrial pathway, providing a rational basis for the immunoregulatory properties of Gal-1 in experimental models of chronic inflammation and cancer (Matarrese et al. 2005).

Galectin-1 initiates the acid sphingomyelinase mediated release of ceramide, which is critical in further steps. Elevation of ceramide level coincides with exposure of phosphatidylserine on the outer cell membrane. The downstream events such as decrease of Bcl-2 protein, depolarization of the mitochondria and activation of the caspase 9 and caspase 3 depend on production of ceramide. All downstream steps, including production of ceramide, require the generation of membrane rafts and the presence of two tyrosine kinases, p56 (lck) and ZAP70. Acid sphingomyelinase mediated release of ceramide is essential to trigger the mitochondrial pathway of apoptosis by galectin-1. Based on these findings a model of the mechanism of galectin-1 triggered cell has been suggested (Ion et al. 2006).

Galβ1-4GlcNAc and Galβ1-3GalNAc Epitopes on BeWo Cells Have Regulatory Effects on Cell Proliferation: Galectin-1 shows apoptotic potential in the epithelial tumor cell lines (MCF7 and BeWo) in vitro, only with additional stress stimuli (Wiest et al. 2005). Gal-1 inhibited BeWo cell proliferation in a concentration-dependent manner. The lectin decreased cellular hCG and progesterone production as well as hCGβ gene transcription by BeWo cells. Gal-1 mediated inhibition of cellular progesterone production was reduced in presence of a Thomsen-Friedenreich (TF)-polyacrylamide conjugate. Therefore, ligation of Galβ1-4GlcNAc and Galβ1-3GalNAc epitopes on BeWo cells may have regulatory effects on cell proliferation and hCG and progesterone production (Jeschke et al. 2004, 2006).

Dimeric Form of Gal-1 is a Powerful Inducer of Apoptosis: Since the affinity of the monomers for each other is rather low, the in vivo efficacy of galectin-1 is limited because the equilibrium is shifted towards the inactive monomeric form at lower concentrations. A covalently linked form of the dimer based on the galectin-1 crystal structure is a potent inducer of apoptosis in murine thymocytes as well as murine mature T cells at concentrations 10-fold lower than wild-type galectin-1. This structurally optimized form of galectin-1 may be a powerful tool to treat chronic inflammatory diseases (Battig et al. 2004). Moreover, recombinant human galectin-1 shows biphasic effect on the growth and death of early hematopoietic cells (Vas et al. 2005).

Surface Exposure of Phosphatidylserine in Activated Human Neutrophils: Cell turn-over depends on the surface exposure of phosphatidylserine (PS) in apoptotic cells, leading to their phagocytic recognition and removal. Galectins induce PS exposure in a carbohydrate-dependent fashion in activated, but not resting, human neutrophils and in several leukocyte cell lines. Apoptotic cells redistribute PS to the cell surface by both Ca^{2+} -dependent and -independent mechanisms. Binding of dGal-1 to glycoconjugates on N-formyl-Met-Leu-Phe (fMLP)-activated neutrophils exposes PS and facilitates neutrophil phagocytosis by macrophages, yet it does not initiate apoptosis. It appeared that dGal-1 initiated Ca^{2+} fluxes are required to redistribute PS to the surface of activated neutrophils. Results suggest that transient Ca^{2+} fluxes contribute to a sustained redistribution of PS on neutrophils activated with fMLP and dGal-1 (Karmakar et al. 2005). Hence surface PS exposure is not always associated with apoptosis in activated neutrophils. The exposure of PS in cell lines treated with galectins is sustained and does not affect cell viability. Unexpectedly, galectins-1, -2, and -4 bind well to activated T lymphocytes, but do not induce either PS exposure or apoptosis, indicating that galectin's effects are cell specific. These results suggest immunoregulatory contribution of galectins in regulating leukocyte turnover independently of apoptosis (Stowell et al. 2007).

10.6.2 Gal-1 in Cell Growth and Differentiation

Regulation of Cell Growth: While extracellular Gal-1 has no effect on the growth rates of naïve T cells (Endharti et al. 2005) or of astrocytic (Camby et al. 2002) or colon (Hittelet et al. 2002) tumor cell lines, Gal-1 is mitogenic for various types of normal or pathological murine and human cells, such as murine Thy-1-negative spleen or lymph node cells (Symons et al. 2000), mammalian vascular cells (Moiseeva et al. 2000), and hepatic stellate cells (Maeda et al. 2003). Gal-1 inhibits the growth of other cell types such as neuroblastoma (Kopitz et al. 2001) and stromal bone marrow cells (Andersen et al. 2003). Interestingly, it seems that depending on the dose involved, Gal-1 causes the biphasic modulation of cell growth. While high doses (~1 μM) of recombinant Gal-1 inhibit cell proliferation independently of Gal-1 sugar-binding activity, low doses (~1 nM) of Gal-1 are mitogenic and are susceptible to inhibition by lactose (Adams et al. 1996; Vas et al. 2005). While the knock-down of Gal-1 expression in murine melanomas (Rubinstein et al. 2004a) and human glioma cells (Camby et al. 2006) does not affect their growth rate in vitro, it does decrease it in 9 L rat gliosarcomas (Yamaoka et al. 2000). Furthermore, Gal-1 can also regulate cell cycle progression in human mammary tumor cells (Wells et al. 1997). The paradoxical biphasic effects of Gal-1 on cell growth are highly dependent on cell type and cell activation state, and might also be influenced by the relative distribution of monomeric versus dimeric, or intracellular versus extracellular forms. However, Gal-1 signaling in leukocytes requires expression of complex-type N-glycans (Karmakar et al. 2008)

Regulation of Cell Motility: While cell migration is the net result of adhesion, motility, and invasion (Lefranc et al. 2005), Gal-1 modifies each of these three steps in cell migration-related processes. Gal-1 increases adhesion of various normal and cancer cells to ECM via cross-linking of integrins exposed on cell surfaces with carbohydrate moieties of ECM components such as laminin and fibronectin (Ellerhorst et al. 1999b; Moiseeva et al. 1999; van den Brule et al. 2003). In addition, Gal-1 can also mediate homotypical cell interaction, so favoring the aggregation of human melanoma cells (Tinari et al. 2001) and heterotypical cell interactions such as the interaction between cancer and endothelial cells, which, in its turn, favors the dispersion of tumor cells (Clausse et al. 1999; Glinsky et al. 2000).

Gal-1 causes the increased motility of glioma cells and the reorganization of the actin cytoskeleton associated with an increased expression of RhoA, a protein that modulates actin polymerization and depolymerization (Camby et al. 2002). Conversely, the knock-down of Gal-1 expression in glioma cells reduces motility and adhesiveness (Camby et al. 2002, 2005). Oxidized Gal-1 stimulates the migration of Schwann cells from both proximal and the distal stumps of transected nerves and promotes axonal regeneration after peripheral nerve injury (Fukaya et al. 2003). In colon carcinomas a Gal-1-enriched ECM decreases colon carcinoma cell motility (Hittelet et al. 2002). Using a proteomic approach based on the comparison of highly and poorly invasive mammary carcinoma cell lines, Harvey et al. (2001) identified the membrane expression of Gal-1 as a signature of cell invasiveness.

Regulation of Chemotaxis: Gal-1 blocks leukocyte chemotaxis (La et al. 2003), whereas Gal-3 has the opposite effect, inducing leukocyte chemotaxis (Sano et al. 2000) and the release of pre-formed IL-8 from neutrophils (Jeng et al. 1994), which further augments chemotaxis of leukocytes (Baggiolini et al. 1992). Galectin-1 stimulates monocyte migration but is not chemotactic for macrophages. Galectin-1-induced monocyte chemotaxis is blocked by lactose and inhibited by an anti-galectin-1 antibody. Furthermore, galectin-1-mediated monocyte migration is significantly inhibited by MEK inhibitors suggesting that MAP kinase pathways are involved in galectin-1. Migration involves G-protein in galectin-1-induced chemotaxis. A role for galectin-1 in monocyte chemotaxis differs from galectin-3 in that macrophages are nonresponsive. Furthermore, observations suggest that galectin-1 may be involved in chemoattraction at sites of inflammation in vivo and may contribute to disease processes such as atherosclerosis (Malik et al. 2009).

Differentiation of the Myogenic Lineage: During the course of myoblast differentiation intracellular Gal-1 is externalized as myoblasts fused into myotubes (Cooper and Barondes 1990). The role of Gal-1 in the case of myoblast fusion may be explained by the fact that the adherence of the myoblast to the extracellular component laminin is disrupted in the presence of Gal-1 via the selective modulation by Gal-1 of the interaction between the $\alpha 7\beta 1$ integrin and fibronectin and laminin (Gu et al. 1994). Although the exact role of Gal-1 in myogenesis remains to be seen, this galectin has been shown to induce non-committed myogenic cells in the dermis to express myogenic markers. It increases the terminal differentiation of committed myogenic cells and has a role to play in the development and regenerative ability of muscles (Cooper and Barondes 1990; Harrison and Wilson 1992; Goldring et al. 2002). Gal-1 may thus be regarded as a potentially important tool in the treatment of cases of human muscular dystrophy (Goldring et al. 2002).

Differentiation of Hematopoietic Lineage: Mesenchymal cells give rise to the stromal marrow environment that supports hematopoiesis. These cells constitute a wide range of differentiation potentials (e.g., adipocytes, osteoblasts, chondrocytes, lymphocytes, erythrocytes, macrophages) as well as a complex relationship with hematopoietic and endothelial cells. Numerous reports have suggested that Gal-1 may be a key element in the course of hematopoietic cell differentiation (Andersen et al. 2003; Silva et al. 2003; Wang et al. 2004; Vas et al. 2005). The K562 human leukemia cell line expresses Gal-1 in the cytosol, but upon treatment with erythropoietin these cells develop an erythroid phenotype that leads to the externalization of cytosolic Gal-1 (Lutomski et al. 1997). Similarly, Gal-1 is externalized during adipocyte differentiation (Wang et al. 2004) and is able to modulate osteoblastic differentiation (Andersen et al. 2003) as well as the proliferation and death of hematopoietic stem and progenitor cells (Vas et al. 2005).

Gal-1 During Lineage Commitment of $CD4^+CD8^+$ Thymocyte Cell: During linage commitment of $CD4^+CD8^+$ thymocytes, $CD4^+CD8^+$ show skewed differentiation of into $CD4^-CD8^+$ thymocytes in presence of 2,3,7,8-tetrachlorodibenzo-p-dioxin (TCDD). DPK cells, a $CD4^+CD8^+$ thymic lymphoma cell line can differentiate into $CD4^+CD8^-$ thymocytes after antigen stimulation. Among the 10 up-regulated proteins, S100A4, S100A6, and galectin-1 were highly up-regulated and dramatically increased after antigen stimulation, similar to transcription factors intimately associated with lineage commitment. In thymus S100A4, S1006, and galectin-1 were most prominently expressed in $CD4^+CD8^+$ thymocytes, but not at all in $CD4^-CD8^+$ and $CD4^-CD8^-$ thymocytes. In spleen, expression of S100A4, S1006, and galectin-1 was greater in CD4 than in CD8 splenocytes. On addition of TCDD to antigen-stimulated DPK cells, antigen-induced up-regulation of S100A4, S1006, and galectin-1 was remarkably reduced, probably accounting for the skewed differentiation of $CD4^+CD8^+$ into $CD4^-CD8^+$ thymocytes induced by TCDD (Jeon et al. 2009).

In the thymus, during T-cell differentiation, the expression of peripheral benzodiazepine receptor (PBR) modulates. Rochard et al. (2004) studied role of PBR in Jurkat cells, which are immature T lymphocytes. These cells are PBR negative and were stably transfected to achieve PBR levels similar to that in mature T cells. A majority of modulated genes encode proteins playing direct or indirect roles during lymphocyte maturation process. In particular, PBR expression induced several differentiation markers, or key regulating elements. By contrast, some regulators of TCR signaling were reduced. PBR expression also affected the expression of critical apoptosis regulators: the proapoptotic lipocortin I, galectin-1, and galectin-9 were reduced while the antiapoptotic Bcl-2 was induced. Results supported the hypothesis that PBR controls T-cell maturation and suggested mechanisms through which PBR may regulate thymocyte-positive selection (Rochard et al. 2004).

Menstrual Cycle, Early Gestation and Embryogenesis: Gal-1 expression has been reported in male and female gonads (Timmons et al. 2002; Dettin et al. 2003). In testis, Gal-1 was identified in Leydig cell. Exogenously added Gal-1 has an inhibitory effect both on the steroidogenic activity of Leydig cells in testicles (Martinez et al. 2004) and on the granulosa cells in ovary (Jeschke et al. 2004; Walzel et al. 2004). Gal-1 influences rat Leydig cells morphology, decrease in cell viability in culture. Testosterone production was reduced after addition of Gal-1, reaching a minimum of 26% after 24 h. Study indicated that the reduction in viability and in steroidogenesis was caused by apoptosis induced by Gal-1. Besides, addition of Gal-1 caused Leydig cell detachment. The CRD is involved in inducing apoptosis. These findings indicated that Gal-1 and laminin-1 interactions form the molecular basis of Leydig cell function and survival. Galectin-1 plays a role in biphasic growth of Leydig tumor cells (Biron et al. 2006).

In the uterus Gal-1 expression is restricted to the endometrium and varies during the menstrual cycle and the early phases of gestation (von Wolff et al. 2005). The expression of Gal-1 increases significantly in late secretory phase endometrium and in decidual tissue (von Wolff et al. 2005), and shows a specific pattern of expression in trophoblastic tissue (Maquoi et al. 1997; Vicovac et al. 1998). During first trimester of embryogenesis Gal-1 is expressed in connective tissue, in smooth and striated muscles, and in some epithelia such as skin, gonads, thyroid gland, and kidneys (Hughes 2004; von Wolff et al. 2005).

10.6.3 Gal-1 and Ras in Cell Transformation

H-Ras–Gal-1 interactions establish an essential link between Gal-1 and Ras associated with cell transformation and human malignancies that can be exploited to selectively target oncogenic Ras proteins. In fact, H-Ras-GTP recruits Gal-1 from the cytosol to the cell membrane with the resulting stabilization of H-Ras-GTP, the clustering of H-Ras-GTP and Gal-1 in non-raft microdomains (Prior et al. 2003), the subsequent binding to Raf-1 (but not to PI3Kinase), the activation of the ERK signaling pathway and, finally, increased cell transformation (Elad-Sfadia et al. 2002). So, in addition to increasing and prolonging H-Ras activation, the Gal-1-H-Ras complex renders the activated molecule selective toward Raf-1, but not toward PI3K (Ashery et al. 2006). Fischer et al. (2005) have observed that the antiproliferative potential of Gal-1 in a number of carcinoma cell lines requires functional interaction with the $\alpha5\beta1$ integrin. Antiproliferative effects result from the inhibition of the Ras-MEK-ERK pathway and the consecutive transcriptional induction of p27, whose promoter contains two Sp1-binding sites crucial for Gal-1 responsiveness (Fischer et al. 2005). The inhibition of the Ras-MEK-ERK cascade by Gal-1 increases Sp1 transactivation and DNA binding due to the reduced threonine phosphorylation of Sp1. In addition, Gal-1 induces p21 transcription and selectively increases p27 protein stability, while the Gal-1-mediated accumulation of p27 and p21 inhibits cyclin-dependent kinase 2 activity, a process which ultimately results in G1 cell cycle arrest and growth inhibition (Fischer et al. 2005).

Gal-1 interacts in a lactose-independent manner with H-Ras-guanosine triphosphate (H-Ras-GTP) through its farnesyl cystein carboxymethylester and so strengthens its membrane association (Paz et al. 2001). Farnesylthiosalicylic acid (FTS) disrupted H-Ras(12V)-galectin-1 interactions. Overexpression of Gal-1 increased membrane-associated Ras, Ras-GTP, and active ERK resulting in cell transformation, which was blocked by dominant negative Ras. H-Ras(12V)-galectin-1 interactions establish an essential link between two proteins associated with cell transformation and human malignancies that can be exploited to selectively target oncogenic Ras proteins. Rotblat et al. (2004) identified a hydrophobic pocket in Gal-1, analogous to Cdc42 geranylgeranyl-binding cavity in RhoGDI, having homologous isoprenoid-binding residues, including critical L11, whose RhoGDI L77 homologue changes on Cdc42 binding. By substituting L11A, Rotblat et al. (2004) obtained a dominant interfering Gal-1 that possessed normal carbohydrate-binding capacity but inhibited H-Ras GTP-loading and ESR-kinase activation, dislodged H-Ras (G12V) from cell membrane, and attenuated H-Ras(G12V) fibroblast transformation and PC12-cell neurite outgrowth. Thus, independent of carbohydrate binding, Gal-1 cooperates with Ras, whereas galectin-1inhibits its activity (Rotblat et al. 2004). The binding of Gal-1 to Ras is an interesting and potentially significant function of Gal-1. These functions have been discussed in following sections.

10.6.4 Development of Nerve Structure

Galectin-1 seems to have a variety of biological functions, which vary according to time and the site at which biological function is taking place. In addition, these functions could vary according to the structure of Gal-1 by which a particular biological function is taking place. Disulfide bond formation alters the structure of Gal-1 that confers the ability to promote axonal regeneration. However, structural analysis revealed that the axonal regeneration-promoting factor exists as an oxidized form of Gal-1, containing three intramolecular disulfide bonds. The Gal-1 subunits are not covalently linked but the monomers are in a dynamic equilibrium with dimeric form. Since the affinity of monomers for each other is rather low (in range of 10^{-5} M), the in vivo efficacy of Gal-1 is limited because the equilibrium is shifted towards the inactive monomeric form at lower concentrations.

10.6.4.1 Galectin-1 in Sensory and Motor Neurons

Galectin-1 is widely distributed in central and peripheral nervous systems of rodents during their development. Galectin-1 is important for the embryonic development of primary sensory neurons and their synaptic connections in spinal cord. Galectin-1 is colocalized c-Ret mRNA in small DRG neurons. About 20% of the DRG neurons showed intense galectin-1-reactivity (IR). On the other hand, only 6.8% displayed TrkA mRNA positive signals. Galectin-1-IR was increased in the dorsal horn at 1 to 2 weeks after axotomy. The endogenous galectin-1 may potentiate neuropathic pain after the peripheral nerve injury at least partly by increasing SPR in the dorsal horn (Imbe et al. 2003).

Gal-1 plays a number of important roles in the formation of neural network of olfactory bulb of mice (Puche et al. 1996). Gal-1 homozygous null mutant (Gal-1$^{-/-}$) mice are viable and can grow into adults without any phenotypical abnormalities except for a deficiency in olfactory network (Tenne-Brown et al. 1998) and a reduced thermal sensitivity (McGraw et al. 2005). In these mice the neuronal subpopulation in olfactory bulb, which normally expresses Gal-1, does not reach appropriate targets in olfactory glomeruli. During its development into adulthood, a rat's sensory neurons from the dorsal root ganglion express Gal-1, as do some spinal motor neurons. The initial expression in sensory neurons begins as they finish their final mitotic division and begin their growth toward their targets in the dorsal horn of spinal cord. When Gal-1-expressing neurons reach their targets, Gal-1 expression remains high, albeit at lower levels (Hynes et al. 1990; Sango et al. 2004). In addition to neurons, Gal-1 mRNAs are also detected in non-neuronal cells such as the pia mater, the choroid plexus, and the pineal gland as well as in reactive astrocytic and Schwann cells (Akazawa et al. 2004; Sango et al. 2004; Egnaczyk et al. 2003; Gaudet et al. 2005; c/r Camby et al. 2006; Gaudet et al. 2005).

Gal-1 has been shown to promote axonal regeneration through the activation of macrophages. Also, Gal-1 may act within the injured neuron to enhance re-growth: the injury-induced regulation of Gal-1 in numerous types of peripherally- and centrally-projecting neurons correlates positively with the regenerative potential of their axons. Kopitz et al. (2004) used two chicken proto-type galectins, i.e., monomeric CG-14 and dimeric CG-16, with very similar carbohydrate affinities, and rat hippocampal neurons in culture to assess the involvement of carbohydrate-protein interaction in axonal growth and directionality, neurite sprouting and axon regenerative capacity after section. In view of the concept of sugar code, Kopitz et al. (2004) indicated that biological effects triggered by glycan binding engaging an endogenous lectin can be modulated by carbohydrate affinity and/or by other factors like differential cross-linking capacity.

Galectin-1 regulates neural progenitor cell proliferation in two neurogenic regions: the subventricular zone (SVZ) of the lateral ventricle and the subgranular zone of the hippocampal dentate gyrus. The temporal profile of endogenous Gal-1 expression and the effects of human recombinant Gal-1 on neurogenesis and neurological functions in an experimental focal ischemic model suggest that Gal-1 is one of the principal regulators of adult SVZ neurogenesis through its carbohydrate-binding ability and provide evidence that Gal-1 protein has a role in improvement of sensorimotor function after stroke (Ishibashi et al. 2007).

Neural Stem Cells: In the subventricular zone of the adult mammalian forebrain, neural stem cells (NSCs) reside and proliferate to generate young neurons. Sakaguchi et al. (2006) identified galectin-1, which is expressed in a subset of slowly dividing subventricular zone astrocytes, which include NSCs. Based on results from intraventricular infusion experiments and phenotypic analyses of knockout mice, it was demonstrated that galectin-1 is an endogenous factor that promotes the proliferation of NSCs in the adult brain. Further experiments revealed that mouse adult NSCs as well as OP9 cells express galectin-1. The molecular mechanism by which galectin-1 enhances proliferation of NSCs is unknown. It is tempting, however, to suggest that galectin-1 binds to the carbohydrate chains of signaling molecules and modulates signal transduction as do HSPGs and/or cystatin C.

Plachta et al. (2007) indicated that galectin-1 actively participates in the elimination of neuronal processes after lesion, and that engineered embryonic stem cells (ESC) are a useful tool for studying relevant aspects of neuronal degeneration that have been hitherto difficult to analyze (Plachta et al. 2007). Akama et al. (2008) identified seven proteins with increased expression and one protein with decreased expression from ESC to neural stem cells (NSC), and eight proteins with decreased expression from NSC to neurons. Laminin-binding protein, galectin 1, increased in NSC, and Gal-1 and a cell adhesion receptor, laminin receptor (RSSA), decreased in neurons. The mRNA of Gal-1 was also up-regulated in NS cells and down-regulated in neurons, implying an important role of Gal-1 in regulating the differentiation. The differentially expressed proteins provide insight into the molecular basis of neurogenesis from ES cells to NSC and neurons.

Proangiogenic and Promigratory Effects of Gal-1 in Hs683 Cells: Galectin 1 is involved in Hs683 oligodendroglioma chemoresistance, neoangiogenesis, and migration. Down-regulating Gal-1 expression in Hs683 cells through targeted siRNA provokes a marked decrease in expression of brain-expressed X-linked gene: BEX2. Decreasing BEX2 expression in Hs683 cells increases the survival of Hs683 orthotopic xenograft-bearing mice. Furthermore, the decrease in BEX2 expression impairs vasculogenic mimicry channel formation in vitro and angiogenesis in vivo, and modulates glioma cell adhesion and invasive features through modification of several genes, reported to play a role in cancer cell migration, including MAP2, plexin C1, SWAP70, and integrin β_6 (Le Mercier et al. 2009).

Altered Primary Afferent Anatomy and Reduced Thermal Sensitivity in Gal-1$^{-/-}$ Mice: The transmission of nociceptive information occurs along non-myelinated, or thinly myelinated, primary afferent axons. During neuronal development and following injury, trophic factors and their respective receptors regulate their survival and repair. Reports show that Gal-1, which is expressed by nociceptive primary afferent neurons during development and into adulthood, is involved in axonal path finding and regeneration. McGraw et al. (2005) characterized anatomical differences in dorsal root ganglia (DRG) of Gal-1 homozygous null mutant mice (Gal-1$^{-/-}$), as well as behavioural differences in tests of nociception. While there was no difference in the total number of axons in the dorsal root of Gal-1$^{-/-}$ mice, there were an increased number of myelinated axons, suggesting that in the absence of Gal-1, neurons that are normally destined to become IB4-binding instead become NF200-expressing. In addition, mice lacking Gal-1 had a decreased sensitivity to noxious thermal stimuli. The exogenous application of rGal-1 has been shown to promote the rate of peripheral nerve regeneration. Galectin-1 null mutant mice showed an attenuated rate of functional recovery of whisking movement after a facial nerve. Thus, Gal-1 is involved in nociceptive neuronal development and that the lack of this protein results in anatomical and functional deficits in adulthood (McGraw et al. 2004a; 2005).

Mouse Gal-1 Inhibits Toxicity of Glutamate by Modifying NR1 NMDA Receptor Expression: Neuronal death in neurodegenerative disease is caused by the neurotransmitter glutamate. This cell death can arise from either excess levels of glutamate due to decreased astrocyte clearance or due to increased susceptibility. However, galectin-1 is a potential neuroprotective factor secreted by astrocytes and protects mouse and rat cerebellar neurons from the toxic effects of glutamate. The effect is mediated through increased expression of NMDA receptor NR1 and increased proportion of the NR1a subunit subtype. Galectin-1 also decreased the expression of PKC associated with increased resistance to glutamate toxicity. Results suggested that the astrocytic galectin-1 can protect neurons against the effects of excito-toxicity as seen in stroke and ischemic injury (Lekishvili et al. 2006).

δFosB Together with Galectin-1 Mediate Neuroprotection and Neurogenesis in Response to Brain Insult: Jun and Fos family proteins are components of an AP-1 (activator protein-1) complex, and are known to regulate the transcription of various genes involved in cell proliferation, differentiation and apoptosis. δFosB, one of the AP-1 subunits, triggers one round of proliferation in quiescent rat embryo cell lines, followed by a different cell fate such as morphological alteration or delayed cell death. Transient forebrain ischemia causes selective induction of δFosB subunit, in cells within the ventricle wall or those in the dentate gyrus in the rat brain prior to neurogenesis, followed by induction of nestin (a marker for neuronal precursor cells) or galectin-1. Different lines of approaches suggested that δFosB can promote the proliferation of quiescent neuronal precursor cells, thus enhancing neurogenesis after transient forebrain ischemia (Kurushima et al. 2005). Miura et al. (2005) demonstrated that the expression of galectin-1 is required for the proliferative activation of quiescent rat1A cells by δFosB, thus indicating that galectin-1 is one of functional targets of δFosB in rat3Y1 cell line. This is supported by the facts that the expression of δFosB is highly inducible in the adult brain in response to various insults such as ischemic reperfusion injury, seizure induced by electric stimulation or cocaine administration. On the other hand, galectin-1 has also been shown to be involved in the regeneration of damaged axons in the peripheral nerve, as well as in neurite outgrowth or synaptic connectivity in the olfactory system during development. Miura et al. (2005) proposed that δFosB together with galectin-1 mediate neuroprotection and neurogenesis in response to brain damage (**R:** Miura et al. 2005)

Miura et al. (2004) indicated that N-terminally processed form of galectin-1, galectin-1β (Gal-1β), a natural monomeric form of galectin-1 lacking its six amino-terminal residues promotes axonal regeneration but not cell death. Expression of Gal-1β was induced by δFosB. The properties and biological functions of Gal-1β have been compared with the full-length form of galectin-1 (Gal-1α). The rmGal-1α exists as a monomer under oxidized conditions and forms a dimer under reduced conditions, while the rmGal-1β exists as a monomer regardless of redox conditions. The affinity of rmGal-1β to β-lactose was two-fold lower than that of rmGal-1α under reduced conditions. In contrast, both rmGal-1α and rmGal-1β exhibited an equivalent capacity to promote axonal regeneration from the dorsal root ganglion explants. Results suggest that the biochemical properties of rmGal-1β determine its biological functions.

Galectin-1 Induces Expression of BDNF in Astrocytes: Astrocytes, in CNS, are considered to act in cooperation with neurons and other glial cells and to participate in the development and maintenance of functions of the CNS. Immature astrocytes possess a polygonal shape and have no processes, and continue to proliferate, while mature astrocytes have stellate cell morphology, and proliferate slowly. Stellate astrocytes, which immediately appear at the site of brain lesions caused by ischemia or other brain injuries, are known to produce several neurotrophic factors to protect neurons from delayed post-lesion death. Brain-derived neurotrophic factor (BDNF), a neuroprotective polypeptide, is considered to be responsible for neuron proliferation, differentiation, and survival. BDNF is known to promote neuronal survival, guide axonal path finding, and participate in activity-dependent synaptic plasticity during development. An agent that enhances production of BDNF is expected to be useful for the treatment of neurodegenerative diseases. In this context, galectin-1 induces astrocyte differentiation and strongly inhibits astrocyte proliferation, and then the differentiated astrocytes greatly enhance the production of BDNF. The effect of galectin-1 is astrocyte-specific and does not have any effect on neurons. Prevention of neuronal loss during CNS injuries is important to maintain brain function. Induction of neuroprotective factors in astrocytes by an endogenous mammalian lectin may be a new mechanism for preventing neuronal loss after brain injury, and may be useful for the treatment of neurodegenerative disorders (**R:** Endo 2005; Sasaki et al. 2004).

10.6.4.2 Oxidized Gal-1 in Promotion of Axonal Regeneration

Although many factors have been implicated in the regenerative response of peripheral axons to nerve injury, the signals that prompt neurons to extend processes in peripheral nerves after axotomy are not well-understood. The oxidized galectin-1 shows no lectin activity and exists as a monomer in a physiological solution. Oxidized galectin-1 has been shown to promote axonal regeneration from transected-nerve sites in an in vitro dorsal root ganglion (DRG) explant model as well as in in vivo peripheral nerve axotomy model. It advances the restoration of nerve function after peripheral nerve injury. Oxidized galectin-1 likely acts as an autocrine or paracrine factor to promote axonal regeneration, functioning more like a cytokine than as a lectin (Review: Kadoya and Horie 2005; Kadoya et al. 2005). In an in vitro peripheral nerve regeneration model of adult rats, the exogenous oxidized recombinant human galectin-1 (rGal-1/Ox) increased the number and diameter of regenerating myelinated fibers. At a similarly low concentration, rGal-1/Ox also was effective in enhancing axonal regeneration in vivo. Since Gal-1 is expressed in the regenerating sciatic nerves as well as in both sensory and motor neurons, results indicated that Gal-1, which is secreted into the extra-cellular space, is subsequently oxidized and then may regulate initial repair after axotomy. It was proposed that axonal regeneration occurs in axotomized peripheral nerves as a result of cytosolic reduced Gal-1 being released from Schwann cells and injured axons, which then becomes oxidized in the extracellular space. Gal-1/Ox in the extracellular space stimulates macrophages to secrete a factor that promotes axonal growth and Schwann cell migration, thus enhancing peripheral nerve regeneration and functional recovery (**R:** Horie et al. 2005).

Oxidized Galectin-1 as Therapy of ALS: Amyotrophic lateral sclerosis (ALS) is a fatal neurodegenerative disease that affects almost selectively motor neurons in the CNS. Most ALS patients die within 5 years of onset. One of the neuropathological features of ALS is an "axonal spheroid," a large swelling of a motor axon within the anterior horn of the spinal cord; this abnormal structure seems to be related to the pathogenesis of motor neuron degeneration in ALS. In 2001, acumulation of galectin-1 was observed in ALS spheroids in close association with neurofilaments. A marked depletion of galectin-1 is another abnormality frequently observed in the skin of ALS patients. Findings, therefore, suggest that galectin-1 may be involved in the pathogenesis of ALS. Oxidized galectin-1 may have a beneficial effect on the pathophysiology of ALS since administered oxidized galectin-1 to transgenic mice with H46R mutant SOD1, an ALS model mouse, improved the motor activity, delayed the onset of symptoms, and prolonged the survival of the galectin-1-treated mice. Furthermore, the number of remaining motor neurons in the spinal cord was more preserved in the galectin-1-treated mice than in the non-treated mice. Administration of rhGAL-1/ox to the mice delayed the onset of their disease and prolonged the life of the mice and the duration of their illness. Motor function was improved in rhGAL-1/ox-treated mice. Studies suggested that rhGAL-1/ox administration could be a new therapeutic strategy for ALS (Chang-Hong et al. 2005; Kadoya et al. 2005; Kato et al. 2005).

Macrophage Activation Enables Cat Retinal Ganglion Cells to Regenerate Injured Axons into the Mature Optic Nerve: Retinal ganglion cells (RGCs) in mature mammals are generally unable to regenerate injured axons into the optic nerve. However, an intravitreal injection of either of two macrophage activators, oxidized galectin-1 or zymosan, strongly enhanced the regeneration of transected RGC axons beyond an optic nerve crush site in adult cats. Data indicated that RGCs of adult cats are capable of reverting to an active growth state and at least partially overcoming an inhibitory CNS environment as a result of intravitreal macrophage activation (Okada et al. 2005).

Gal-1 in its oxidized form promotes neurite outgrowth (Outenreath and Jones 1992) and enhances axonal regeneration in peripheral (Inagaki et al. 2000; Fukaya et al. 2003; Kadoya et al. 2005) and central (McGraw et al. 2004a; Rubinstein et al. 2004b; McGraw et al. 2005) nerves even at relatively low concentrations (picoM range) (Horie and Kadoya 2000). The marked axonal regeneration-promoting activity of oxidized Gal-1 is likely to be paracrine. Indeed, Gal-1 is expressed in dorsal root ganglion neurons and motor neurons, with immunoreactivity restricted to the neuronal cell bodies, the axons, and the Schwann cells of adult rodents (Hynes et al. 1990; Fukaya et al. 2003; Horie and Kadoya 2000). After axonal injury, cytosolic reduced Gal-1 is likely to be externalized from growing axons and reactive Schwann cells to an extracellular space where some of the molecules may be converted into an oxidized form and may enhance axonal regeneration (Horie and Kadoya 2000; McGraw et al. 2004a; Rubinstein et al. 2004b; Miura et al. 2004; Sango et al. 2004). Miura et al. (2004) have identified a novel, naturally occurring, N-terminally processed form of Gal-1 that lacks the six amino-terminal residues of full length Gal-1. This isoform of Gal-1, which is monomeric under both reducing and oxidizing conditions, promotes axonal regeneration (Miura et al. 2004). Since oxidized Gal-1-induced neurite outgrowth is not observed on isolated neurons, the secreted Gal-1 probably influences the non-neuronal cells surrounding the axons, including the Schwann cells (Fukaya et al. 2003), and in so doing recruits macrophages, fibroblasts, and perineuronal cells (Horie et al. 2004). In this respect, macrophages are potential candidates since they secrete an axonal regeneration-promoting factor when stimulated by oxidized Gal-1 (Horie et al. 2004). A preclinical study using rats with surgically transected sciatic nerves has shown that the administration by an osmotic pump of oxidized Gal-1 at the site of surgery restores nerve function (Kadoya et al. 2005). In contrast to the effect of oxidized Gal-1 on axonal regeneration as suggested, the effects of Gal-1 on astrocyte differentiation and BDNF production depend on carbohydrate-binding activity and are astrocyte-specific since no effects on neurons have been observed (Sasaki et al. 2004). Thus, Gal-1 may thus be considered as a means for the prevention of neuronal loss in cases of injury to CNS (Egnaczyk et al. 2003).

Galectin-1 in Rat Nervous System: Proteomic approach identified two members of galectin family, namely galectin-1 and galectin-3 in primary rat cerebellar astrocytes (Yang et al. 2006). In rat nervous system, galectin-1 mRNA is predominantly observed in the cell bodies of neurons such as oculomotor nucleus (III), trochlear nucleus (IV), trigeminal motor nucleus (V), abducens nucleus (VI), facial nucleus (VII), hypoglossal nucleus (XII), red nucleus, and locus ceruleus. Neurons in pineal gland and dorsal root ganglia expressed galectin-1 mRNA. In the adult rats, the axotomy of facial nerve induced transient up-regulation of galectin-1 mRNA around 6 h after axotomy. These results indicated that galectin-1 may play roles in the early event of nerve injury and regeneration through the transient change of its expression level (Akazawa et al. 2004). The differential expression pattern of Gal-1 following peripheral axotomy and dorsal rhizotomy suggested that endogenous Gal-1 might be important to the regenerative response of injured axons (McGraw et al. 2005).

Galectin-1 from Bovine Brain: A β-galactoside-specific soluble 14-kD lectin from sheep brain was isolated, sequenced, and compared with similar galectins from other species. The isolated galectin shares all the absolutely preserved and critical residues of the mammalian galectin-1 subfamily. The isolated sheep brain galectin (SBG) showed more than 90% amino acid sequence (92%) and carbohydrate recognition domain identity (96%) with human brain galectin-1. Conformational changes were found induced by interaction of the protein with its specific disaccharide and oxidizing agent (hydrogen peroxide). Upon oxidation a drastic change in the environment of aromatic residues and conformation of the galectin was observed with the loss of hemagglutination activity, while no significant change was observed upon addition of D-lactose (Gal(β1-4)Glc) in the far-UV and near-UV spectra, suggesting no significant modification in the secondary as well as tertiary structures of sheep brain galectin (Shahwan et al. 2004). The galectin-1 from buffalo brain is a dimeric protein of 24.5 kDa with subunit mass of 13.8. The most potent inhibitor of the galectin activity was lactose, giving complete inhibition of hemagglutination at 0.8 mM. Galectin showed higher specificity towards human blood group A (Shamsul et al. 2007).

10.6.5 Skeletal Muscle Development

Regulation of Myotube Growth in Regenerating Skeletal Muscles: Galectin-1 is highly expressed in skeletal muscle and has been implicated in skeletal muscle development and in adult muscle regeneration, but also in the degeneration of neuronal processes and/or in peripheral nerve regeneration. Galectin-1 is involved in muscle stem cell behavior and in tissue regeneration after muscle injury in adult mice. The Gal-1 mRNA is expressed in the extrasynaptic and perisynaptic regions of the muscle, and its immunoreactivity can be detected in both regions (Svensson and Tågerud 2009). Muscle satellite cells play a critical role in skeletal muscle regeneration. In intact adult muscles, galectin-1 was associated with basement membranes of myofibers. Kami and Senba (2005) suggest that Gal-1 is a novel factor that promotes both myoblast fusion and axonal growth following muscle injury, and consequently, regulates myotube growth in regenerating skeletal muscles (**R:** Kami and Senba 2005; Cerri et al. 2008). Galectin-1 is an early marker of myogenesis as the transcripts and protein are initially confined to the somites, starting from day 9.0 of embryogenesis. By comparing the spatio-temporal distribution of Galectin-1 transcripts in control and Myf5 null mutant embryos, it was established that it acts downstream of Myf5. However, early myogenesis did not seem to be affected in Galectin-1 null mutant embryos indicating that, unlike in adult, Galectin-1 does not play a role in muscle fate acquisition during development (Shoji et al. 2009).

Knockdown experiments in zebrafish embryos targeted to the 5′-UTR sequence of DrGal-1-L2 resulted in a phenotype with a bent tail and disorganized muscle fibers. However, DrGal-1-L1 knockdown embryos showed no similar morphological defects, indicating that the observed effects are sequence-specific, and not due to the toxicity of the morpholino-modified oligos. Further, ectopic expression of native DrGal-1-L2 specifically rescued the phenotype. These results suggest that galectins produced by notochord play a key role in somatic cell differentiation and development (Ahmed et al. 2009b).

Incorporation of Vitronectin and Chondroitin Sulfate B in ECM of SMC: Gal-1 binds to a number of ECM components in ß-galactoside-dependent manner in following order: laminin > cellularfibronectin > thrombospondin > plasma fibronectin > vitronectin > osteopontin (Moiseeva et al. 2000, 2003). Laminin and cellular fibronectin are glycoproteins which are highly N-glycosylated with bi- and tetra-antennary poly-N-lactosamines (Carsons et al. 1987). Gal-1 is also involved in ECM assembly and remodeling: it inhibits the incorporation of vitronectin and chondroitin sulphate B into ECM of vascular smooth muscle cells (Moiseeva et al. 2003); the interaction seems to depend on vitronectin conformation since it preferentially recognizes unfolded vitronectin multimers rather than inactive folded monomers (Moiseeva et al. 2003). Some of the major cell surface-binding partners of Gal-1 have been reviewed (Camby et al. 2006).

10.6.6 Gal-1 and the Immune System

10.6.6.1 Galectin-1 Modulates Immune System in a Number of Ways

Galectin-1 produced by thymic epithelial cells causes apoptosis in human thymocytes (Perillo et al. 1997). In peripheral blood, galectin-1 causes apoptosis of activated T cells, but it supports the survival of naive T cells (Vespa et al. 1999; Perillo et al. 1995; Endharti et al. 2005). Galectin-1 has also been proposed to shift the T cell polarization reaction from Th1 to Th2 by triggering apoptosis in Th1 cells (Rabinovich et al. 2004; Toscano et al. 2007). Galectin-1 also promotes surface exposure of phosphatidylserine (PS) without accompanying apoptosis in human T cell lines (Stowell et al. 2007). Galectins in general and Gal-1 in particular, are known to be deeply involved in the initiation, amplification, and resolution of inflammatory responses (Almkvist and Karlsson 2004).

Galectin-1 has pleiotropic immunomodulatory functions, including regulation of lymphocyte survival and cytokine secretion in autoimmune, transplant disease, and parasitic infection models. Galectins are differentially expressed by various immune cells and their expression levels appear to be dependent on cell differentiation and activation. Galectin-1 inhibits clonal expansion and induces apoptosis of antigen-primed T lymphocytes and suppresses the development of T-cell-mediated autoimmune diseases in vivo. Since galectin-1 is expressed in activated but not resting T cells, it has been hypothesized that Gal-1-induced apoptosis may constitute an autocrine suicide mechanism to eliminate activated T cells leading to the termination of an effector immune response. Gal-1 plays a key role in human platelet activation and function. It binds to human platelets in a carbohydrate-dependent manner and synergizes with ADP or thrombin to induce platelet aggregation and ATP release. Furthermore, Gal-1 induces F-actin polymerization, up-regulation of P-selectin, and GPIIIa expression; promotes shedding of microvesicles and triggers conformational changes in GPIIb/IIIa (Pacienza et al. 2008).

Innate Immune Functions for Galectin-1: Inhibition of Cell Fusion by Nipah Virus Envelope Glycoproteins: Nipah virus (NiV) is an emerging pathogen that causes severe febrile encephalitis. The primary targets of NiV are endothelial cells. NiV envelope-mediated cell-cell fusion is blocked by Gal-1. This inhibition is specific to the Paramyxoviridae family because Gal-1 did not inhibit fusion triggered by envelope glycoproteins of other viruses. The physiologic dimeric form of Gal-1 was required for fusion inhibition since a monomeric Gal-1 mutant had no inhibitory effect on cell fusion. Gal-1 binds to specific N-glycans on NiV glycoproteins and aberrantly oligomerizes NiV. Gal-1also increases pro-inflammatory cytokines such as IL-6, known to be protective in the setting of other viral diseases such as Ebola infections. Thus, Gal-1 may have direct antiviral effects and may also augment the innate immune response against this emerging pathogen (Levroney et al. 2005).

Oxidative Burst and Degranulation of Porcine Neutrophils by Galectin-1: Local galectin-1 concentrations under physiological conditions might reach suitable levels for pig PMN stimulation, and might be a natural inducer of O_2^- formation or degranulation. Porcine galectins might produce enhanced responses in vivo when they stimulate neutrophils in combination with some other stimuli (Elola et al. 2005b). Galectin-1 tunes TCR binding and signal transduction to regulate CD8 burst size. T cell burst is regulated by duration of TCR engagement and balanced control of Ag-induced activation, expansion, and apoptosis. Galectin-1-deficient CD8 T cells undergo greater cell division in response to TCR stimulation, with fewer dividing cells undergoing apoptosis. TCR-induced ERK signaling was sustained in activated galectin-1-deficient CD8 T cells and antagonized by recombinant galectin-1, indicating galectin-1 modulates TCR feed-forward/feedback loops involved in signal discrimination and procession. Therefore, galectin-1, inducibly expressed by activated CD8 T cells, functions as an autocrine negative regulator of peripheral CD8 T cell TCR binding, signal transduction, and burst size. Together with recent findings demonstrating that Gal-1 promotes binding of agonist tetramers to the TCR of OT-1 thymocytes, these studies identify galectin-1 as a tuner of TCR binding, signaling, and functional fate determination that can differentially specify outcome, depending on the developmental and activation stage of the T cell (Liu et al. 2009b).

10.6.6.2 Activity of Galectin-1 at the Crossroad of Innate and Adaptive Immunity

Gal-1 can act as a link between innate and adaptive immunity by modulating the physiology of neutrophils, monocytes, and dendritic cells. The naturally occurring population of dedicated regulatory T cells that co-express $CD4^+$ and $CD25^+$ plays a key role in the maintenance of peripheral T-cell tolerance, though their mechanism of action has remained obscure. Galectin-1 shows immunoregulatory activity in vivo in experimental models of autoimmunity and cancer. Galectin-1 affects T cell fate and regulates monocyte and macrophage physiology. Treatment with galectin-1 in vitro differentially regulates constitutive and inducible FcγRI expression on human monocytes and FcγRI-dependent phagocytosis. In addition, galectin-1 inhibits IFN-γ-induced MHC class II expression and MHC-II-dependent Ag presentation in a dose-dependent manner. These regulatory effects were also evident in mouse macrophages following treatment with r-galectin-1 and in galectin-1-deficient mice. In these functions galectin-1 does not appear to affect survival of human monocytes, but rather influences FcγRI- and MHC-II-dependent functions through active mechanisms involving modulation of an ERK1/2-dependent pathway (Barrionuevo et al. 2007).

Gal-1-Matured Human Monocyte-Derived DCs: Galectin-1 induces a phenotypic and functional maturation in human monocyte-derived DCs (MDDCs) similar to but distinct from the activity of the exogenous pathogen stimuli, LPS. Immature human MDDCs exposed to galectin-1 up-regulated cell surface markers characteristic of DC maturation (CD40, CD83, CD86, and HLA-DR), secreted high levels of IL-6 and TNF- α, stimulated T cell proliferation, and showed reduced endocytic capacity, similar to LPS-matured MDDCs. However, unlike LPS-matured DCs, galectin-1-treated MDDCs did not produce the Th1-polarizing cytokine IL-12. In addition to modulating many of the same DC maturation genes as LPS, galectin-1 also uniquely up-regulated a significant subset of genes related to cell migration through the ECM. Findings suggested that galectin-1 is an endogenous activator of human MDDCs that up-regulates a significant subset of genes distinct from those regulated by an exogenous stimulus (LPS). The unique effect of galectin-1 on increase in DC migration through ECM suggests that galectin-1 is an important component in initiating an immune response (Fulcher et al. 2006).

Tolerogenic Signals Delivered by DCs to T Cells Through a Gal-1-Driven Immunoregulatory Circuit: Gal-1-DC represent a novel tool to control differentially the afferent and efferent arms of the T cell response (Perone et al. 2006b). Galectin-1 can endow DCs with tolerogenic potential. After exposure to Gal-1, DCs acquired an IL-27--dependent regulatory function, promoted IL-10-mediated T cell tolerance and suppressed autoimmune neuroinflammation. Consistent with its regulatory function, Gal-1 had its highest expression on DCs exposed to tolerogenic stimuli and was most abundant from the peak through the resolution of autoimmune pathology. DCs lacking Gal-1 had greater immunogenic potential and an impaired ability to halt inflammatory disease. These studies identify a tolerogenic circuit linking Gal-1 signaling, IL-27-producing DCs and IL-10-secreting T cells, which has broad therapeutic implications in immunopathology (Ilarregui et al. 2009).

Regulation of BCR Signaling Through an Association OCA-B with Galectin-1: OCA-B is a B cell-specific transcriptional co-activator for OCT factors during the activation of immunoglobulin genes. In addition, OCA-B is crucial for B cell activation and germinal center formation. Yu et al. (2006) identified galectin-1, and related galectins as a OCA-B-interacting protein. The galectin-1 binding domain in OCA-B is localized to the N terminus of OCA-B. In B cells lacking OCA-B expression, increased galectin-1 expression, secretion, and cell surface association are observed. Galectin-1 is shown to negatively regulate B cell proliferation and tyrosine phosphorylation upon BCR stimulation. Results raised the possibility that OCA-B may regulate BCR signaling through an association with galectin-1.

Galectin-1-Dependent pre-B Cell Receptor Relocalization and Activation: Interactions between B cell progenitors and bone marrow stromal cells are essential for normal B cell differentiation. It has been noticed that an immune developmental synapse is formed between human pre-B and stromal cells in vitro, leading to the initiation of signal transduction from the pre-BCR. This process relies on the direct interaction between the pre-BCR and the stromal cell-derived Gal-1 and is dependent on Gal-1 anchoring to cell surface glycosylated counter-receptors, present on stromal and pre-B cells (Gauthier et al. 2002, Vas et al. 2005). Rossi et al. (2006) identified α4β1 (VLA-4), α5β1 (VLA-5), and α4β7 integrins as major Gal-1-glycosylated counter-receptors involved in synapse formation. Pre-B cell integrins and their stromal cell ligands (ADAM15/fibronectin), together with the pre-BCR and Gal-1, form a homogeneous lattice at the contact area between pre-B and stromal cells. Results suggest that during pre-B/stromal cell synapse formation, relocalization of pre-B cell integrins mediated by their stromal cell ligands drives pre-BCR clustering and activation in a Gal-1-dependent manner.

In late-stage B cell activation and maturation, soluble galectin-1 produced by activated B cells resulting from *Trypanosoma cruzi* infection in mice causes T cell apoptosis and affects IFN- production (Zuniga et al. 2001). Galectin-1 regulates apoptotic pathways in human naive and IgM^+ memory B cells through altering balances in Bcl-2 family proteins (Tabrizi et al. 2009). Intracellular galectin-1 may associate with B cell-specific Oct-1-associated coactivator, OCA-B, to negatively regulate BCR signaling (Yu et al. 2006).

Galectin-1: A Key Effector of Regulation Mediated by $CD4^+CD25^+$ T Cells: Galectin-1 is over-expressed in regulatory T cells, and its expression is increased after activation. Recent evidence indicates that Gal-1 contributes to the immunosuppressive activity of $CD4^+$ $CD25^+$ $FOXP3^+$ regulatory T cells. *In vivo*, direct administration of Gal-1 suppresses chronic inflammation in experimental models of autoimmunity by skewing the balance of the immune response toward a T_H2 cytokine profile. Analysis of the mechanistic basis of this anti-inflammatory effect revealed that T_H1 and T_H17-differentiated cells share a common glycan motif, which can be specifically targeted by Gal-1, providing a novel link between differential glycosylation, susceptibility to cell death, and the regulation of the inflammatory response. Most importantly, blockade of Gal-1 binding significantly reduced the inhibitory effects of human and mouse $CD4^+CD25^+$T cells. Reduced regulatory activity was observed in $CD4^+CD25^+$ T cells obtained from Gal-1-homozygous null mutant mice. These results suggested that Gal-1 is the key effector of the regulation mediated by these cells (Garin et al. 2007). In addition, expression of Gal-1 at sites of tumor growth and metastasis can influence tumor progression by regulating cell-cell and cell-matrix interactions, tumor cell invasiveness, and angiogenesis. Furthermore, Gal-1 can also function as a soluble mediator employed by tumor cells to evade the immune response (c/r Pacienza et al. 2008).

Galectin-1 Co-clusters CD43/CD45 on DCs and Induces Cell Activation: Galectin-1 activates human monocyte-derived dendritic cells (MDDCs) and triggers a specific genetic program that up-regulates DC migration through the extracellular matrix, an integral property of mucosal DCs. Fulcher et al. (2009) identified the Gal-1 receptors on MDDCs and immediate downstream effectors of Gal-1-induced MDDC activation and migration. Galectin-1 binding to surface CD43 and CD45 on MDDCs induced an unusual unipolar co-clustering of these receptors and activates calcium flux that is abrogated by lactose. Syk and protein kinase C tyrosine kinases are effectors of the DC activation by Gal-1. Galectin-1, but not lipopolysaccharide, stimulated Syk phosphorylation and recruitment of phosphorylated Syk to the CD43 and CD45 co-cluster on MDDCs. Inhibitors of Syk and protein kinase C signaling abrogated galectin-1-induced DC activation as monitored by interleukin-6 production; and MMP-1, -10, and -12 gene up-regulation; and enhanced migration through the extracellular matrix. The latter two are specific features of galectin-1-activated DCs. Interestingly, galectin-1 can also prime DCs to respond more quickly to low dose lipopolysaccharide stimulation (Fulcher et al. 2009).

Cross-Linking of GM1 Ganglioside by Gal-1 Mediates Regulatory T Cell Activity: Several autoimmune disorders are suppressed in animal models by treatment with GM1 cross-linking units of certain toxins such as B subunit of cholera toxin (CtxB). GM1 being a binding partner for Gal-1, which is known to ameliorate symptoms in certain animal models of autoimmune disorders, such as murine experimental autoimmune encephalomyelitis (EAE) and further highlighted the role of GM1 in demonstrating enhanced susceptibility to EAE. Results indicate GM1 in murine $CD4^+$ and $CD8^+$ effector T (T_{eff}) cells to be primary target of Gal-1 expressed by T_{reg} cells, the resulting co-cross-linking and TRPC5 channel activation contributing to the mechanism of autoimmune suppression (Wang et al. 2009).

Galectin-1 Affects the Cross-Liking of HIV-1 Infection: The HIV-1 infection is initiated by the stable attachment of the virion to the target cell surface. Although this process relies primarily upon interaction between virus-encoded gp120 and cell surface CD4, a number of other interactions may influence binding of HIV-1 to host cells. For example, galectin-1 acts as a soluble adhesion molecule by facilitating attachment of HIV-1 to the cell surface. Experiments using fusion inhibitor T-20 confirmed that galectin-1 is primarily affecting HIV-1 attachment. Therefore, it was proposed that galectin-1, which is released in an exocrine fashion at HIV-1 replication sites, can cross-link HIV-1 and target cells and promote a firmer adhesion of the virus to the cell surface, thereby augmenting the efficiency of the infection process. Overall, findings suggest that galectin-1 might affect the pathogenesis of HIV-1 infection (Ouellet et al. 2005). Galectin-1 is known to interact for example with ganglioside GM1 and also the hydrophobic tail of oncogenic H-Ras. Observations indicate the potential of galectin-1 to affect membrane properties beyond the immediate interaction with cell surface epitopes (Gupta et al. 2006).

10.6.7 Role of Galetin-1 and Other Systems

Galectin-1 in Keratinocytes

Multipotent stem cells are localized in the bulge region of the outer root sheath of hair follicles, while stem cells giving rise to interfollicular epidermis reside in its basal. Galectin-1 reactivity is present in squamous epithelial cells (Chovanec et al. 2004; Klima et al. 2005). Since keratin 19 and nuclear reactivity to galectin-1 are potential markers of epidermal stem cells, Dvorankova et al. (2005) detected the expression of these markers in adult cells migrating from the hair follicle, where cells expressing keratin 19 are located in the bulge region. Observations indicated the transient expression of keratin 19 and nuclear galectin-1 binding sites in originally negative interfollicular epidermal cells induced by adhesion. Studies on expression of the lysosome-associated membrane protein 1 (Lamp-1) and expression of the epidermal galectins-1, -3 and -7, in human keratinocytes indicated that the up-regulated Lamp-1 expression at confluence could be involved in keratinocyte differentiation, but apparently not through interaction with galectin-3 (Sarafian et al. 2006).

Reproductive Tissues

Gal-1 has been detected in pig granulosa cell lysates. The lectin stimulated the proliferation of granulose cells from pig ovaries in culture and inhibited the FSH-stimulated progesterone synthesis of granulosa cells. It appeared to interfere with the receptor-dependent mechanism of FSH-stimulated progesterone production. It was suggested that Gal-1 exerts its inhibitory effect on steroidogenic activity of granulosa cells by interfering with the hormone-receptor interaction resulting in decreased responses to FSH stimulation (Walzel et al. 2004). In human endometrium, Gal-1 was localized mainly in stromal cells, whereas Gal-3 was predominantly found in epithelial cells. Cycle-dependent expression of Gal-1 in stromal cells and Gal-3 in epithelial cells suggested these lectins to be involved in the regulation of different endometrial cellular functions (von Wolff et al. 2005). A possible implication of galectins-1 and -3 has been indicated in the invasiveness of the transformed trophoblastic cell (Bozic et al. 2004).

Galectin-1 in Fetomaternal Tolerance

Phylogenetic footprinting and shadowing unveiled conserved cis motifs, including an estrogen responsive element in the 5′ promoter of *LGALS1* that could account for sex steroid regulation of *LGALS1* expression, suggesting a role of Gal-1 in immune-endocrine cross-talk and emergence of hormonal and redox regulation of Gal-1 in placental mammals at maternal-fetal immune tolerance. Galectin-1 is expressed in immune privileged sites and is implicated in establishing maternal-fetal immune tolerance, which is essential for successful pregnancy in eutherian mammals. A successful pregnancy requires synchronized adaptation of maternal immune-endocrine mechanisms to the fetus. Gal-1 has a pivotal role in conferring fetomaternal tolerance. Consistently with a marked decrease in Gal-1 expression during failing pregnancies, Gal-1-deficient mice showed higher rates of fetal loss compared to wild-type mice in allogeneic matings, whereas fetal survival was unaffected in syngeneic matings. Thus, Gal-1 is a pivotal regulator of fetomaternal tolerance that has potential therapeutic implications in threatened pregnancies (Blois et al. 2007;

Ilarregui et al. 2009). Gal-1 may be involved in the regulation of the inflammatory responses to chorioamniotic infection (Than et al. 2008a, b). Gal-1, Gal-1 ligand, Thomsen-Friedenreich (TF) (Galβ1-3GalNAc-) and Gal-3 are expressed and up-regulated on the membrane of extravillous trophoblast (EVT) in preeclamptic placentas. In addition, the expression of Gal-1 is significantly up-regulated in decidual tissue of preeclamptic placentas and villous trophoblast tissue of HELLP placentas (Than et al. 2008b). Jeschke et al. (2007) speculated that expression of both galectins and TF on the membrane of preeclamptic EVT and up-regulation of Gal-1 in preeclamptic decidual cells may at least in part compensate for the apoptotic effects of maternal immune cells.

Interaction Between Chondrocytes and a Lactose-Modified Chitosan

Gal-1 plays an important role in enhancing cell adhesion to extracellular matrix and inducing cell proliferation. Chitosan is a derivative of chitin extracted from lobsters, crabs and shrimps' exoskeletons. Although chitosan membranes show no cytotoxicity, some cell types (e.g. 3 T3 cells) fail to attach and proliferate on their surface. Over-expression of Gal-1 does not enhance 3 T3 cell proliferation on chitosan membranes. However, coating the chitosan membrane with recombinant Gal-1 proteins significantly expedites 3 T3 cells proliferation. Findings support a role for altered levels of protein phosphorylation in Gal-1-mediated cell attachment and proliferation on chitosan membranes (Chang et al. 2004). The Chitlac glycopolymer has been shown to promote pig chondrocyte aggregation and to induce extracellular matrix production. Recombinant Galectin-1 interacts in a dose-dependent manner with Chitlac. Expression level of galectin-1 gene was significantly higher in chondrocytes cultivated on Chitlac. Data indicated the role of Galectin-1 as a bridging agent between Chitlac and chondrocyte aggregates (Marcon et al. 2005).

References

Adams L, Scott GK, Weinberg CS (1996) Biphasic modulation of cell growth by recombinant human galectin-1. Biochim Biophys Acta 1312:137–144

Ahmad N, Gabius HJ, Sabesan S et al (2004) Thermodynamic binding studies of bivalent oligosaccharides to galectin-1, galectin-3, and the carbohydrate recognition domain of galectin-3. Glycobiology 14:817–825

Ahmed H, Pohl J, Fink NE et al (1996) The primary structure and carbohydrate specificity of a -galactosyl-binding lectin from toad (*Bufo arenarum Hensel*) ovary reveal closer similarities to the mammalian galectin-1 than to the galectin from the clawed frog *Xenopus laevis*. J Biol Chem 271:33083–33094

Ahmed H, Vasta GR (1994) Galectins: conservation of functionally and structurally relevant amino acid residues defines two types of carbohydrate recognition domains. Glycobiology 4:545–548

Ahmed H, Vasta GR (2008) Unlike mammalian GRIFIN, the zebrafish homologue (DrGRIFIN) represents a functional carbohydrate-binding galectin. Biochem Biophys Res Commun 371:350–355

Ahmed H, Du SJ, Vasta GR (2009) Knockdown of a galectin-1-like protein in zebrafish (*Danio rerio*) causes defects in skeletal muscle development. Glycoconj J 26:277–283

Akama K, Tatsuno R, Otsu M, Inoue N et al (2008) Proteomic identification of differentially expressed genes in mouse neural stem cells and neurons differentiated from embryonic stem cells in vitro. Biochim Biophys Acta 1784:773–782

Akazawa C, Nakamura Y, Sango K et al (2004) Distribution of the galectin-1 mRNA in the rat nervous system: its transient upregulation in rat facial motor neurons after facial nerve axotomy. Neuroscience 125:171–178

Almkvist J, Karlsson A (2004) Galectins as inflammatory mediators. Glycoconj J 19:575–581

Almkvist J, Faldt J, Dahlgren C et al (2001) Lipopolysaccharide-induced gelatinase granule mobilization primes neutrophils for activation by galectin-3 and formylmethionyl-Leu-Phe. Infect Immun 69:832–837

Amano M, Galvan M, He J, Baum LG (2003) The ST6Gal I sialyltransferase selectively modifies N-glycans on CD45 to negatively regulate galectin-1-induced CD45 clustering, phosphatase modulation, and T cell death. J Biol Chem 278:7469–7475

Andersen H, Jensen ON, Moiseeva EP, Eriksen EF (2003) A proteome study of secreted prostatic factors affecting osteoblastic activity: galectin-1 is involved in differentiation of human bone marrow stromal cells. J Bone Miner Res 18:195–203

Andre S, Kaltner H, Furuike T et al (2004) Persubstituted cyclodextrin-based glycoclusters as inhibitors of protein-carbohydrate recognition using purified plant and mammalian lectins and wild-type and lectin-gene-transfected tumor cells as targets. Bioconjug Chem 15:87–98

Ashery U, Yizhar O, Rotblat B et al (2006) Spatiotemporal organization of Ras signaling: rasosomes and the galectin switch. Cell Mol Neurobiol 26:471–495

Avni O, Pur Z, Yefenof E, Baniyash M (1998) Complement receptor, 3 of macrophages is associated with galectin-1-like protein. J Immunol 160:6151–6158

Baggiolini M, Clark-Lewis I (1992) Interleukin-8, a chemotactic and inflammatory cytokine. FEBS Lett 307:97–101

Barondes SH, Castronovo V, Cooper DN et al (1994) Galectins: a family of animal β-galactoside-binding lectins. Cell 76:597–598

Barrionuevo P, Beigier-Bompadre M, Ilarregui JM et al (2007) A novel function for galectin-1 at the crossroad of innate and adaptive immunity: galectin-1 regulates monocyte/macrophage physiology through a nonapoptotic ERK-dependent pathway. J Immunol 178:436–445

Battig P, Saudan P, Gunde T, Bachmann MF (2004) Enhanced apoptotic activity of a structurally optimized form of galectin-1. Mol Immunol 41:9–18

Baum LG, Seilhamer JJ, Pang M et al (1995) Synthesis of an endogeneous lectin, galectin-1, by human endothelial cells is up-regulated by endothelial cell activation. Glycoconj J 12:63–68

Berberat PO, Friess H, Wang L et al (2001) Comparative analysis of galectins in primary tumors and tumor metastasis in human pancreatic cancer. J Histochem Cytochem 49:539–549

Bianchet MA, Ahmed H, Vasta GR, Amzel LM (2000) Soluble β-galactosyl-binding lectin (galectin) from toad ovary: crystallographic studies of two protein-sugar complexes. Proteins 40:378–388

Biron VA, Iglesias MM, Troncoso MF, Besio-Moreno M, Patrignani ZJ, Pignataro OP, Wolfenstein-Todel C (2006) Galectin-1: biphasic growth regulation of Leydig tumor cells. Glycobiology 16:810–821

Blixt O, Head S et al (2004) Printed covalent glycan array for ligand profiling of diverse glycan binding proteins. Proc Natl Acad Sci USA 101:17033–17038

Blois SM, Ilarregui JM, Tometten M et al (2007) A pivotal role for galectin-1 in fetomaternal tolerance. Nat Med 13:1450–1457

Bochner BS, Alvarez RA, Mehta P et al (2005) Glycan array screening reveals a candidate ligand for Siglec-8. J Biol Chem 280:4307–4312

Bouffard DY, Momparler LF, Momparler RL (1994) Enhancement of the antileukemic activity of 5-aza-2′-deoxycytidine by cyclopentenyl cytosine in HL-60 leukemic cells. Anticancer Drugs 5:223–228

Bozic M, Petronijevic M et al (2004) Galectin-1 and galectin-3 in the trophoblast of the gestational trophoblastic disease. Placenta 25:797–802

Brandt B, Büchse T, Abou-Eladab EF et al (2008) Galectin-1 induced activation of the apoptotic death-receptor pathway in human Jurkat T lymphocytes. Histochem Cell Biol 129:599–609

Brandt B, Abou-Eladab EF et al (2010) Role of the JNK/c-Jun/AP-1 signaling pathway in galectin-1-induced T-cell death. Cell Death Dis 1:e23

Brewer FC (2002) Binding and cross-linking properties of galectins. Biochim Biophys Acta 1572:255–262

Brewer CF (2004) Thermodynamic binding studies of galectin-1,-3 and -7. Glycoconj J 19:459–465

Broker LE, Kruyt FA, Giaccone G (2005) Cell death independent of caspases: a review. Clin Cancer Res 11:3155–3162

Calame KL, Lin KI et al (2003) Regulatory mechanisms that determine the development and function of plasma cells. Annu Rev Immunol 21:205–230

Camby I, Belot N, Rorive S et al (2001) Galectins are differentially expressed in supratentorial pilocytic astrocytomas, astrocytomas, anaplastic astrocytomas and glioblastomas, and significantly modulate tumor astrocyte migration. Brain Pathol 11:12–26

Camby I, Belot N, Lefranc F et al (2002) Galectin-1 modulates human glioblastoma cell migration into the brain through modifications to the actin cytoskeleton and levels of expression of small GTPases. J Neuropathol Exp Neurol 61:585–596

Camby I, Decaestecker C, Lefranc F et al (2005) Galectin-1 knocking down in human U87 glioblastoma cells alters their gene expression pattern. Biochem Biophys Res Commun 335:27–35

Camby I, Mercier ML, Lefranc F, Kiss R (2006) Galectin-1: a small protein with major functions. Glycobiology 16:137R–157R

Carsons S, Lavietes BB, Slomiany A et al (1987) Carbohydrate heterogeneity of fibronectins. Synovial fluid fibronectin resembles the form secreted by cultured synoviocytes but differs from the plasma form. J Clin Invest 80:1342–1349

Cerri DG, Rodrigues LC, Stowell SR et al (2008) Degeneration of dystrophic or injured skeletal muscles induces high expression of galectin-1. Glycobiology 18:842–850

Chadli A, LeCaer JP, Bladier D et al (1997) Purification and characterization of a human brain galectin-1 ligand. J Neurochem 68:1640–1647

Chang-Hong R, Wada M, Koyama S et al (2005) Neuroprotective effect of oxidized galectin-1 in a transgenic mouse model of amyotrophic lateral sclerosis. Exp Neurol 194:203–211

Chang YY, Chen SJ, Liang HC et al (2004) The effect of galectin 1 on 3 T3 cell proliferation on chitosan membranes. Biomaterials 25:3603–3611

Chiariotti L, Benvenuto G, Salvatore P et al (1994a) Expression of the soluble lectin L-14 gene is induced by TSH in thyroid cells and suppressed by retinoic acid in transformed neural cells. Biochem Biophys Res Commun 199:540–546

Chiariotti L, Benvenuto G, Zarrilli R et al (1994b) Activation of the galectin-1 (L-14-I) gene from nonexpressing differentiated cells by fusion with undifferentiated and tumorigenic cells. Cell Growth Differ 5:769–775

Chiariotti L, Berlingieri MT, Battaglia C et al (1995) Expression of galectin-1 in normal human thyroid gland and in differentiated and poorly differentiated thyroid tumors. Int J Cancer 64:171–175

Chiariotti L, Salvatore P, Frunzio R, Bruni CB (2004) Galectin genes: regulation of expression. Glycoconj J 19:441–449

Cho M, Cummings RD (1995) Galectin-1, a β-galactoside-binding lectin in Chinese hamster ovary cells. I. Physical and chemical characterization. J Biol Chem 270:5198–5206

Choe YS, Shim C, Choi D et al (1997) Expression of galectin-1 mRNA in the mouse uterus is under the control of ovarian steroids during blastocyst implantation. Mol Reprod Dev 48:261–266

Choufani G, Nagy N, Saussez S et al (1999) The levels of expression of galectin-1, galectin-3, and the Thomsen-Friedenreich antigen and their binding sites decrease as clinical aggressiveness increases in head and neck cancers. Cancer 86:2353–2363

Choufani G, Ghanooni R, Decaestecker C et al (2001) Detection of macrophage migration inhibitory factor (MIF) in human cholesteatomas and functional implications of correlations to recurrence status and to expression of matrix metalloproteinases-3/9, retinoic acid receptor-β, and anti-apoptotic galectin-3. Laryngoscope 111:1656–1662

Chovanec M, Smetana K Jr, Dvorankova B et al (2004) Decrease of nuclear reactivity to growth-regulatory galectin-1 in senescent human keratinocytes and detection of non-uniform staining profile alterations upon prolonged culture for galectin-1 and -3. Anat Histol Embryol 33:348–354

Chung CD, Patel VP, Moran M, Lewis LA, Miceli MC (2000) Galectin-1 induces partial TCR zeta-chain phosphorylation and antagonizes processive TCR signal transduction. J Immunol 165:3722–3729

Cindolo L, Benvenuto G et al (1999) Galectin-1 and galectin-3 expression in human bladder transitional-cell carcinomas. Int J Cancer 84:39–43

Clausse N, van den Brule F, Waltregny D et al (1999) Galectin-1 expression in prostate tumor-associated capillary endothelial cells is increased by prostate carcinoma cells and modulates heterotypic cell-cell adhesion. Angiogenesis 3:317–325

Clerch LB, Whitney P, Hass M et al (1988) Sequence of a full-length cDNA for rat lung β-galactoside-binding protein: primary and secondary structure of the lectin. Biochemistry 27:692–29

Cooper DN (2002) Galectinomics: finding themes in complexity. Biochim Biophys Acta 1572:209–231

Cooper DN, Barondes SH (1990) Evidence for export of a muscle lectin from cytosol to extracellular matrix and for a novel secretory mechanism. J Cell Biol 110:1681–1691

D'Haene N, Maris C, Sandras F et al (2005) The differential expression of galectin-1 and galectin-3 in normal lymphoid tissue and non-Hodgkins and Hodgkins lymphomas. Int J Immunopathol Pharmacol 18:431–443

Daroqui CM, Ilarregui JM, Rubinstein N et al (2007) Regulation of galectin-1 expression by transforming growth factor β1 in metastatic mammary adenocarcinoma cells: implications for tumor-immune escape. Cancer Immunol Immunother 56:491–499

Delbrouck C, Doyen I, Belot N et al (2002) Galectin-1 is overexpressed in nasal polyps under budesonide and inhibits eosinophil migration. Lab Invest 82:147–158

Dettin L, Rubinstein N, Aoki A et al (2003) Regulated expression and ultrastructural localization of galectin-1, a proapoptotic β-galactoside-binding lectin, during spermatogenesis in rat testis. Biol Reprod 68:51–59

Dvorankova B, Smetana K Jr, Chovanec M et al (2005) Transient expression of keratin 19 is induced in originally negative interfollicular epidermal cells by adhesion of suspended cells. Int J Mol Med 16:525–531

Earl LA, Bi S, Baum LG (2010) N- and O-glycans modulate galectin-1 binding, CD45 signaling and T cell death. J Biol Chem 285:2232–2244

Egnaczyk GF, Pomonis JD, Schmidt JA et al (2003) Proteomic analysis of the reactive phenotype of astrocytes following endothelin-1 exposure. Proteomics 3:689–698

Elad-Sfadia G, Haklai R, Ballan E et al (2002) Galectin-1 augments Ras activation and diverts Ras signals to Raf-1 at the expense of phosphoinositide 3-kinase. J Biol Chem 277:37169–37175

Ellerhorst J, Nguyen T, Cooper DN et al (1999a) Induction of differentiation and apoptosis in the prostate cancer cell line LNCaP by sodium butyrate and galectin-1. Int J Oncol 14:225–232

Ellerhorst J, Nguyen T, Cooper DN et al (1999b) Differential expression of endogenous galectin-1 and galectin-3 in human prostate cancer cell lines and effects of overexpressing galectin-1 on cell phenotype. Int J Oncol 14:217–224

Elola MT, Chiesa ME, Alberti AF, Mordoh J, Fink NE (2005a) Galectin-1 receptors in different cell types. J Biomed Sci 12:13–29

Elola MT, Chiesa ME et al (2005b) Activation of oxidative burst and degranulation of porcine neutrophils by a homologous spleen galectin-1 compared to N-formyl-L-methionyl-L-leucyl-L-phenylalanine and phorbol 12-myristate 13-acetate. Comp Biochem Physiol B Biochem Mol Biol 141:23–31

Endharti AT, Zhou YW, Nakashima I, Suzuki H (2005) Galectin-1 supports survival of naive T cells without promoting cell proliferation. Eur J Immunol 35:86–97

Endo T (2005) Glycans and glycan-binding proteins in brain: galectin-1-induced expression of neurotrophic factors in astrocytes. Curr Drug Targets 6:427–436

Fajka-Boja R, Szemes M, Ion G et al (2002) Receptor tyrosine phosphatase, CD45 binds galectin-1 but does not mediate its apoptotic signal in T cell lines. Immunol Lett 82:149–154

Filén J-J, Nyman TA, Korhonen J et al (2006) Characterization of microsomal fraction proteome in human lymphoblasts reveals the down-regulation of galectin-1 by interleukin-12. Proteomics 5:4719–4732

Fischer C, Sanchez-Ruderisch H, Welzel M et al (2005) Galectin-1 interacts with the α5β1 fibronectin receptor to restrict carcinoma cell growth via induction of p21 and p27. J Biol Chem 280:37266–37277

Fitzner B, Walzel H, Sparmann G et al (2005) Galectin-1 is an inductor of pancreatic stellate cell activation. Cell Signal 17:1240–1247

Fouillit M, Joubert-Caron R et al (2000) Regulation of CD45-induced signaling by galectin-1 in Burkitt lymphoma B cells. Glycobiology 10:413–419

Frigeri LG, Zuberi RI, Lui FT (1993) Epsilon BP, a β-galactoside-binding animal lectin, recognizes IgE receptor (Fc ε RI) and activates mast cells. Biochemistry 32:7644–7649

Fukaya K, Hasegawa M, Mashitani T et al (2003) Oxidized galectin-1 stimulates the migration of Schwann cells from both proximal and distal stumps of transected nerves and promotes axonal regeneration after peripheral nerve injury. J Neuropathol Exp Neurol 62:162–172

Fukumori T, Takenaka Y, Yoshii T et al (2003) CD29 and CD7 mediate galectin-3-induced type ii t-cell apoptosis. Cancer Res 63:8302–8311

Fulcher JA, Hashimi ST, Levroney EL et al (2006) Galectin-1-matured human monocyte-derived dendritic cells have enhanced migration through extracellular matrix. J Immunol 177:216–226

Fulcher A, Chang MH, Wang S et al (2009) Galectin-1 co-clusters CD43/CD45 on dendritic cells and induces cell activation and migration through Syk and protein kinase c signaling. J Biol Chem 284:26860–26870

Gabius HJ, Andre S, Gunsenhauser I et al (2002) Association of galectin-1- but not galectin-3-dependent parameters with proliferation activity in human neuroblastomas and small cell lung carcinomas. Anticancer Res 22:405–410

Galindo RC, Ayoubi P, Naranjo V et al (2009) Gene expression profiles of European wild boar naturally infected with *Mycobacterium bovis*. Vet Immunol Immunopathol 129:119–125

Garin MI, Chu CC, Golshayan D, Cernuda-Morollon E, Wait R, Lechler RI (2007) Galectin-1: a key effector of regulation mediated by CD4 + CD25+ T cells. Blood 109:2058–2065

Gaudet AD, Steeves JD, Tetzlaff W, Ramer MS (2005) Expression and functions of galectin-1 in sensory and motoneurons. Curr Drug Targets 6:419–425

Gaudier E, Forestier L, Gouyer V et al (2004) Butyrate regulation of glycosylation-related gene expression: evidence for galectin-1 upregulation in human intestinal epithelial goblet cells. Biochem Biophys Res Commun 325:1044–1051

Gauthier L, Rossi B, Roux F, Termine E, Schiff C (2002) Galectin-1 is a stromal cell ligand of the pre-B cell receptor (BCR) implicated in synapse formation between pre-B and stromal cells and in pre-BCR triggering. Proc Natl Acad Sci USA 99:13014–13019

Gillenwater A, Xu XC, Estrov Y et al (1998) Modulation of galectin-1 content in human head and neck squamous carcinoma cells by sodium butyrate. Int J Cancer 75:217–224

Gitt MA, Massa SM, Leffler H, Barondes SH (1992) Isolation and expression of a gene encoding L-14-II, a new human soluble lactose-binding lectin. J Biol Chem 267:10601–10606

Glinsky VV, Huflejt ME, Glinsky GV et al (2000) Effects of Thomsen-Friedenreich antigen-specific peptide P-30 on β-galactoside-mediated homotypic aggregation and adhesion to the endothelium of MDA-MB-435 human breast carcinoma cells. Cancer Res 60:2584–2588

Goldring K, Jones GE, Thiagarajah R, Watt DJ (2002) The effect of galectin-1 on the differentiation of fibroblasts and myoblasts in vitro. J Cell Sci 115:355–366

Goldstone SD, Lavin MF (1991) Isolation of a cDNA clone, encoding a human β-galactoside binding protein, overexpressed during glucocorticoid-induced cell death. Biochem Biophys Res Commun 178:746–750

Grutzmann R, Pilarsky C, Ammerpohl O et al (2004) Gene expression profiling of microdissected pancreatic ductal carcinomas using high-density DNA microarrays. Neoplasia 6:611–622

Gu J, Taniguchi N (2004) Regulation of integrin functions by N-glycans. Glycoconj J 21:9–15

Gu M, Wang W, Song WK et al (1994) Selective modulation of the interaction of 71 integrin with fibronectin and laminin by L-14 lectin during skeletal muscle differentiation. J Cell Sci 107:175–181

Gunnersen JM, Spirkoska V, Smith PE et al (2000) Growth and migration markers of rat C6 glioma cells identified by serial analysis of gene expression. Glia 32:146–154

Gupta RK, Pande AH, Gulla KC et al (2006) Carbohydrate-induced modulation of cell membrane. VIII. Agglutination with mammalian lectin galectin-1 increases osmofragility and membrane fluidity of trypsinized erythrocytes. FEBS Lett 580:1691–1695

Hahn HP, Pang M, He J, Hernandez JD et al (2004) Galectin-1 induces nuclear translocation of endonuclease G in caspase- and cytochrome c-independent T cell death. Cell Death Differ 11:1277–1286

Hansen TV, Hammer NA, Nielsen J et al (2004) Dwarfism and impaired gut development in insulin-like growth factor II mRNA-binding protein 1-deficient mice. Mol Cell Biol 24:4448–4464

Harrison FL, Wilson TJ (1992) The 14 kDa β-galactoside binding lectin in myoblast and myotube cultures: localization by confocal microscopy. J Cell Sci 101:635–646

Harvey S, Zhang Y, Landry F, Miller C, Smith JW (2001) Insights into a plasma membrane signature. Physiol Genomics 5:129–136

He J, Baum LG (2004) Presentation of galectin-1 by extracellular matrix triggers T cell death. J Biol Chem 279:4705–4712

He L, Andre S, Siebert HC et al (2003) Detection of ligand- and solvent-induced shape alterations of cell-growth-regulatory human lectin galectin-1 in solution by small angle neutron and X-ray scattering. Biophys J 85:511–524

Hernandez JD, Nguyen JT et al (2006) Galectin-1 binds different CD43 glycoforms to cluster CD43 and regulate T cell death. J Immunol 177:5328–5336

Hirabayashi J, Kasai K-I (1993) The family of metazoan metal-independent ß-galactoside-binding lectins: structure, function and molecular evolution. Glycobiology 3:297–304

Hirabayashi J, Hashidate T, Arata Y et al (2002) Oligosaccharide specificity of galectins: a search by frontal affinity chromatography. Biochim Biophys Acta 1572:232–254

Hittelet A, Legendre H, Nagy N et al (2002) Upregulation of galectins-1 and -3 in human colon cancer and their role in regulating cell migration. Int J Cancer 103:370–379

Horie H, Kadoya T (2000) Identification of oxidized galectin-1 as an initial repair regulatory factor after axotomy in peripheral nerves. Neurosci Res 38:131–137

Horie H, Kadoya T, Hikawa N et al (2004) Oxidized galectin-1 stimulates macrophages to promote axonal regeneration in peripheral nerves after axotomy. J Neurosci 24:1873–1880

Horie H, Kadoya T, Sango K et al (2005) Oxidized galectin-1 is an essential factor for peripheral nerve regeneration. Curr Drug Targets 6:385–394

Horiguchi N, Arimoto K, Mizutani A et al (2003) Galectin-1 induces cell adhesion to the extracellular matrix and apoptosis of non-adherent human colon cancer Colo201 cells. J Biochem (Tokyo) 134:869–874

Hsu DK, Yang RY, Pan Z et al (2000) Targeted disruption of the galectin-3 gene results in attenuated peritoneal inflammatory responses. Am J Pathol 156:1073–1083

Hughes RC (1999) Secretion of the galectin family of mammalian carbohydrate-binding proteins. Biochim Biophys Acta 1473:172–185

Hughes RC (2004) Galectins in kidney development. Glycoconj J 19:621–629

Hutvagner G, Zamore PD (2002) A microRNA in a multiple-turnover RNAi enzyme complex. Science 297:2056–2060

Hynes MA, Gitt M, Barondes SH et al (1990) Selective expression of an endogenous lactose-binding lectin gene in subsets of central and peripheral neurons. J Neurosci 10:1004–1013

Ilarregui JM, Bianco GA, Toscano MA et al (2005) The coming of age of galectins as immunomodulatory agents: impact of these carbohydrate binding proteins in T cell physiology and chronic inflammatory disorders. Ann Rheum Dis 64(Suppl 4):96–103

Ilarregui JM, Croci DO, Bianco GA et al (2009) Tolerogenic signals delivered by dendritic cells to T cells through a galectin-1-driven immunoregulatory circuit involving interleukin 27 and interleukin 10. Nat Immunol 10:981–991

Imbe H, Okamoto K, Kadoya T et al (2003) Galectin-1 is involved in the potentiation of neuropathic pain in the dorsal horn. Brain Res 993:72–83

Inagaki Y, Sohma Y, Horie H et al (2000) Oxidized galectin-1 promotes axonal regeneration in peripheral nerves but does not possess lectin properties. Eur J Biochem 267:2955–2964

Ion G, Fajka-Boja R, Toth GK et al (2005) Role of p56lck and ZAP70-mediated tyrosine phosphorylation in galectin-1-induced cell death. Cell Death Differ 12:1145–1147

Ion G, Fajka-Boja R, Kovacs F et al (2006) Acid sphingomyelinase mediated release of ceramide is essential to trigger the mitochondrial pathway of apoptosis by galectin-1. Cell Signal 18:1887–1896

Ishibashi S, Kuroiwa T, Sakaguchi M et al (2007) Galectin-1 regulates neurogenesis in the subventricular zone and promotes functional recovery after stroke. Exp Neurol 207:302–313

Jeng KC, Frigeri LG, Liu FT (1994) An endogenous lectin, galectin-3 (epsilon BP/Mac-2), potentiates IL-1 production by human monocytes. Immunol Lett 42:113–116

Jeon CH, Kim HL, Park JH (2009) Induction of S100A4, S100A6, and galectin-1 during the lineage commitment of CD4 + CD8+ thymocyte cell line is suppressed by 2,3,7,8-tetrachlorodibenzo-p-dioxin. Toxicol Lett 187:157–163

Jeschke U, Reimer T, Bergemann C et al (2004) Binding of galectin-1 (Gal-1) on trophoblast cells and inhibition of hormone production of trophoblast tumor cells in vitro by Gal-1. Histochem Cell Biol 121:501–508

Jeschke U, Karsten U, Wiest I et al (2006) Binding of galectin-1 (Gal-1) to the Thomsen-Friedenreich (TF) antigen on trophoblast cells and inhibition of proliferation of trophoblast tumor cells in vitro by Gal-1 or an anti-TF antibody. Histochem Cell Biol 126:437–444

Jeschke U, Mayr D, Schiessl B et al (2007) Expression of galectin-1, -3 (Gal-1, Gal-3) and the Thomsen-Friedenreich (TF) antigen in normal, IUGR, preeclamptic and HELLP placentas. Placenta 28:1165–1173

Jeschke U, Walzel H, Mylonas I et al (2009) The human endometrium expresses the glycoprotein mucin-1 and shows positive correlation for thomsen-friedenreich epitope expression and galectin-1 binding. J Histochem Cytochem 57:871–881

Joubert R, Caron M, Avellana-Adalid V et al (1992) Human brain lectin: a soluble lectin that binds actin. J Neurochem 58:200–203

Kadoya T, Oyanagi K, Kawakami E et al (2005) Oxidized galectin-1 advances the functional recovery after peripheral nerve injury. Neurosci Lett 380:284–288

Kami K, Senba E (2005) Galectin-1 is a novel factor that regulates myotube growth in regenerating skeletal muscles. Curr Drug Targets 6:395–405

Karmakar S, Cummings RD, McEver RP (2005) Contributions of Ca2+ to galectin-1-induced exposure of phosphatidylserine on activated neutrophils. J Biol Chem 280:28623–28631

Karmakar S, Stowell SR, Cummings RD, McEver RP (2008) Galectin-1 signaling in leukocytes requires expression of complex-type N-glycans. Glycobiology 18:770–778

Kato T, Ren CH, Wada M et al (2005) Galectin-1 as a potential therapeutic agent for amyotrophic lateral sclerosis. Curr Drug Targets 6:407–418

Keller AD, Maniatis T (1992) Only two of the five zinc fingers of the eukaryotic transcriptional repressor PRDI-BF1 are required for sequence-specific DNA binding. Mol Cell Biol 12:1940–1949

Klima J, Smetana K Jr et al (2005) Comparative phenotypic characterization of keratinocytes originating from hair follicles. J Mol Histol 36:89–96

Koh HS, Lee C, Lee KS et al (2008) CD7 expression and galectin-1-induced apoptosis of immature thymocytes are directly regulated by NF-kB upon T-cell activation. Biochem Biophys Res Commun 370:149–153

Koh HS, Lee C, Lee KS et al (2009) Twist2 regulates CD7 expression and galectin-1-induced apoptosis in mature T-cells. Mol Cells 28(6):553–558

Kondoh N, Hada A, Ryo A et al (2003) Activation of galectin-1 gene in human hepatocellular carcinoma involves methylation-sensitive complex formations at the transcriptional upstream and downstream elements. Int J Oncol 23:1575–1583

Kopitz J, von Reitzenstein C, Burchert M et al (1998) Galectin-1 is a major receptor for ganglioside GM1, a product of the growth-controlling activity of a cell surface ganglioside sialidase, on human neuroblastoma cells in culture. J Biol Chem 273:11205–11211

Kopitz J, von Reitzenstein C, Andre S et al (2001) Negative regulation of neuroblastoma cell growth by carbohydrate-dependent surface binding of galectin-1 and functional divergence from galectin-3. J Biol Chem 276:35917–35923

Kopitz J, Russwurm R, Kaltner H et al (2004) Hippocampal neurons and recombinant galectins as tools for systematic carbohydrate structure-function studies in neuronal differentiation. Brain Res Dev Brain Res 153:189–196

Kopper L, Sebestyen A (2000) Syndecans and the lymphoid system. Leuk Lymphoma 38:271–281

Kultima K, Nystrom AM, Scholz B et al (2004) Valproic acid teratogenicity: a toxicogenomics approach. Environ Health Perspect 112:1225–1235

Kurushima H, Ohno M, Miura T et al (2005) Selective induction of DeltaFosB in the brain after transient forebrain ischemia accompanied by an increased expression of galectin-1, and the implication of DeltaFosB and galectin-1 in neuroprotection and neurogenesis. Cell Death Differ 12:1078–1096

La M, Cao TV, Cerchiaro G et al (2003) A novel biological activity for galectin-1: inhibition of leukocyte–endothelial cell interactions in experimental inflammation. Am J Pathol 163:1505–1515

Lefranc F, Tatjana M, Christine D et al (2005) Monitoring the expression profiles of integrins and adhesion/growth-regulatory galectins in adamantinomatous craniopharyngiomas: their ability to regulate tumor adhesiveness to surrounding tissue and their contribution to prognosis. Neurosurgery 56:763–776

Le Mercier M, Fortin S, Mathieu V et al (2009) Galectin 1 proangiogenic and promigratory effects in the Hs683 oligodendroglioma model are partly mediated through the control of BEX2 expression. Neoplasia 11:485–496

Lekishvili T, Hesketh S, Brazier MW, Brown DR (2006) Mouse galectin-1 inhibits the toxicity of glutamate by modifying NR1 NMDA receptor expression. Eur J Neurosci 24:3017–3025

Leppanen A, Stowell S, Blixt O, Cummings RD (2005) Dimeric galectin-1 binds with high affinity to α2,3-sialylated and non-sialylated terminal N-acetyllactosamine units on surface-bound extended glycans. J Biol Chem 280:5549–5562

Levroney EL, Aguilar HC, Fulcher JA et al (2005) Novel innate immune functions for galectin-1: galectin-1 inhibits cell fusion by Nipah virus envelope glycoproteins and augments dendritic cell secretion of proinflammatory cytokines. J Immunol 175: 413–420

Lin KI, Tunyaplin C, Calame K (2003) Transcriptional regulatory cascades controlling plasma cell differentiation. Immunol Rev 194:19–28

Liu F-T (2000) Galectins: a new family of regulators of inflammation. Clin Immunol 97:79–88

Liu FT, Rabinovich GA (2005) Galectins as modulators of tumour progression. Nat Rev Cancer 5:29–41

Liu FT, Patterson RJ, Wang JL (2002) Intracellular functions of galectins. Biochim Biophys Acta 1572:263–273

Liu SD, Tomassian T, Bruhn KW, Miller JF, Poirier F, Miceli MC (2009) Galectin-1 tunes TCR binding and signal transduction to regulate CD8 burst size. J Immunol 182:5283–5295

Lopez-Lucendo MF, Solis D et al (2004) Growth-regulatory human galectin-1: crystallographic characterisation of the structural changes induced by single-site mutations and their impact on the thermodynamics of ligand binding. J Mol Biol 343:957–970

Lu Y, Lotan R (1999) Transcriptional regulation by butyrate of mouse galectin-1 gene in embryonal carcinoma cells. Biochim Biophys Acta 1444:85–91

Lu Y, Lotan D, Lotan R (2000) Differential regulation of constitutive and retinoic acid-induced galectin-1 gene transcription in murine embryonal carcinoma and myoblastic cells. Biochim Biophys Acta 1491:13–19

Lutomski D, Fouillit M, Bourin P et al (1997) Externalization and binding of galectin-1 on cell surface of K562 cells upon erythroid differentiation. Glycobiology 7:1193–1199

Mackay A, Jones C, Dexter T et al (2003) cDNA microarray analysis of genes associated with ERBB2 (HER2/neu) overexpression in human mammary luminal epithelial cells. Oncogene 22:2680–2688

Maeda N, Kawada N, Seki S et al (2003) Stimulation of proliferation of rat hepatic stellate cells by galectin-1 and galectin-3 through different intracellular signaling pathways. J Biol Chem 278:18938–18944

Mahanthappa NK, Cooper DN, Barondes SH, Schwarting GA (1994) Rat olfactory neurons can utilize the endogenous lectin, L-14, in a novel adhesion mechanism. Development 120:1373–1384

Malik RKG, Ghurye RR, Lawrence-Watt DJ, Stewart HJS (2009) Galectin-1 stimulates monocyte chemotaxis via the p44/42 MAP kinase pathway and a pertussis toxin-sensitive pathway. Glycobiology 19:1402–1407

Maquoi E, van den Brule FA, Castronovo V, Foidart JM (1997) Changes in the distribution pattern of galectin-1 and galectin-3 in human placenta correlates with the differentiation pathways of trophoblasts. Placenta 18:433–439

Marcon P, Marsich E, Vetere A et al (2005) The role of Galectin-1 in the interaction between chondrocytes and a lactose-modified chitosan. Biomaterials 26:4975–4984

Marschal P, Herrmann J et al (1992) Sequence and specificity of a soluble lactose-binding lectin from *Xenopus laevis* skin. J Biol Chem 267:12942–12949

Martinez VG, Pellizzari EH et al (2004) Galectin-1, a cell adhesion modulator, induces apoptosis of rat Leydig cells in vitro. Glycobiology 14:127–137

Massa SM, Cooper DN et al (1993) L-29, an endogenous lectin, binds to glycoconjugate ligands with positive cooperativity. Biochemistry 32:260–267

Matarrese P, Tinari A, Mormone E et al (2005) Galectin-1 sensitizes resting human T lymphocytes to Fas (CD95)-mediated cell death via mitochondrial hyperpolarization, budding, and fission. J Biol Chem 280:6969–6985

McGraw J, McPhail LT, Oschipok LW et al (2004a) Galectin-1 in regenerating motoneurons. Eur J Neurosci 20:2872–2880

McGraw J, Oschipok LW, Liu J et al (2004b) Galectin-1 expression correlates with the regenerative potential of rubrospinal and spinal motoneurons. Neuroscience 128:713–719

McGraw J, Gaudet AD et al (2005) Altered primary afferent anatomy and reduced thermal sensitivity in mice lacking galectin-1. Pain 114:7–18

Meynier C, Feracci M, Espeli M et al (2009) NMR and MD investigations of human galectin-1/oligosaccharide complexes. Biophys J 97:3168–3177

Miura T, Takahashi M, Horie H et al (2004) Galectin-1β, a natural monomeric form of galectin-1 lacking its six amino-terminal residues promotes axonal regeneration but not cell death. Cell Death Differ 11:1076–1083

Miura T, Ohnishi Y et al (2005) Regulation of the neuronal fate by DeltaFosB and its downstream target, galectin-1. Curr Drug Targets 6:437–444

Moiseeva EP, Javed Q, Spring EL, de Bono DP (2000) Galectin 1 is involved in vascular smooth muscle cell proliferation. Cardiovasc Res 45:493–502

Moiseeva EP, Spring EL, Baron JH et al (1999) Galectin 1 modulates attachment, spreading and migration of cultured vascular smooth muscle cells via interactions with cellular receptors and components of extracellular matrix. J Vasc Res 36:47–58

Moiseeva EP, Williams B, Goodall AH et al (2003) Galectin-1 interacts with β-1 subunit of integrin. Biochem Biophys Res Commun 310:1010–1016

Morris S, Ahmad N, Andre S et al (2004) Quaternary solution structures of galectins-1, -3, and -7. Glycobiology 14:293–300

Nagy N, Legendre H, Engels O et al (2003) Refined prognostic evaluation in colon carcinoma using immunohistochemical galectin fingerprinting. Cancer 97:1849–1858

Nguyen JT, Evans DP, Galvan M et al (2001) CD45 modulates galectin-1-induced T cell death: regulation by expression of core 2 O-glycans. J Immunol 167:5697–5707

Nickel W (2005) Unconventional secretory routes: direct protein export across the plasma membrane of mammalian cells. Traffic 6:607–614

Nieminen J, Kuno A, Hirabayashi J, Sato S (2007) Visualization of galectin-3 oligomerization on the surface of neutrophils and endothelial cells using fluorescence resonance energy transfer. J Biol Chem 282:1374–1383

Nishioka T, Sakumi K, Miura T et al (2002) FosB gene products trigger cell proliferation and morphological alteration with an increased expression of a novel processed form of galectin-1 in the rat 3Y1embryo cell line. J Biochem 131:653–661

Ohannesian DW, Lotan D, Lotan R (1994) Concomitant increases in galectin-1 and its glycoconjugate ligands (carcinoembryonic antigen, lamp-1, and lamp-2) in cultured human colon carcinoma cells by sodium butyrate. Cancer Res 54:5992–6000

Okada T, Ichikawa M, Tokita Y et al (2005) Intravitreal macrophage activation enables cat retinal ganglion cells to regenerate injured axons into the mature optic nerve. Exp Neurol 196:153–163

Ouellet M, Mercier S, Pelletier I et al (2005) Galectin-1 acts as a soluble host factor that promotes HIV-1 infectivity through stabilization of virus attachment to host cells. J Immunol 174:4120–4126

Outenreath RL, Jones AL (1992) Influence of an endogenous lectin substrate on cultured dorsal root ganglion cells. J Neurocytol 21:788–795

Ozaki K, Inoue K, Sato H et al (2004) Functional variation in LGALS2 confers risk of myocardial infarction and regulates lymphotoxin-alpha secretion in vitro. Nature 429:72–75

Ozeki Y, Matsui T, Yamamoto Y et al (1995) Tissue fibronectin is an endogenous ligand for galectin-1. Glycobiology 5:255–261

Pace KE, Lee C, Stewart PL, Baum LG (1999) Restricted receptor segregation into membrane microdomains occurs on human T cells during apoptosis induced by galectin-1. J Immunol 163:3801–3811

Pace KE, Hahn HP, Pang M et al (2000) CD7 delivers a pro-apoptotic signal during galectin-1-induced T cell death. J Immunol 165:2331–2334

Pacienza N, Pozner RG et al (2008) The immunoregulatory glycan-binding protein galectin-1 triggers human platelet activation. FASEB J 22:1113–1123

Park JW, Voss PG, Grabski S et al (2001) Association of galectin-1 and galectin-3 with Gemin4 in complexes containing the SMN protein. Nucleic Acids Res 29:3595–3602

Paz A, Haklai R, Elad-Sfadia G, Ballan E, Kloog Y (2001) Galectin-1 binds oncogenic H-Ras to mediate Ras membrane anchorage and cell transformation. Oncogene 20:7486–7493

Perillo NL, Pace KE, Seilhamer J, Baum LG (1995) Apoptosis of T cells mediated by galectin-1. Nature 378:736–739

Perillo NL, Uittenbogaart CH, Nguyen JT et al (1997) Galectin-1, an endogenous lectin produced by thymic epithelial cells, induces apoptosis of human thymocytes. J Exp Med 185:1851–1858

Perillo NL, Marcus ME, Baum LG (1998) Galectins: versatile modulators of cell adhesion, cell proliferation, and cell death. J Mol Med 76:402–412

Perone MJ, Larregina AT, Shufesky WJ et al (2006) Transgenic galectin-1 induces maturation of dendritic cells that elicit contrasting responses in naive and activated T cells. J Immunol 176:7207–7220

Plachta N, Annaheim C, Bissière S et al (2007) Identification of a lectin causing the degeneration of neuronal processes using engineered embryonic stem cells. Nat Neurosci 10:712–719

Poirier F, Bourin P, Bladier D et al (2001a) Effect of 5-azacytidine and galectin-1 on growth and differentiation of the human B lymphoma cell line Bl36. Cancer Cell Int 1:2

Poirier F, Pontet M, Labas V et al (2001b) Two-dimensional database of a Burkitt lymphoma cell line (DG 75) proteins: protein pattern changes following treatment with 5′-azycytidine. Electrophoresis 22:1867–1877

Prior IA, Muncke C, Parton RG et al (2003) Direct visualization of Ras proteins in spatially distinct cell surface microdomains. J Cell Biol 160:165–170

Puche AC, Poirier F, Hair M et al (1996) Role of galectin-1 in the developing mouse olfactory system. Dev Biol 179:274–287

Purkrábková T, Smetana K Jr, Dvoránková B et al (2003) New aspects of galectin functionality in nuclei of cultured bone marrow stromal and epidermal cells: biotinylated galectins as tool to detect specific binding sites. Biol Cell 95:535–545

Rabinovich GA (2005) Galectin-1 as a potential cancer target. Br J Cancer 92:1188–1192

Rabinovich GA, Gruppi A (2005) Galectins as immunoregulators during infectious processes: from microbial invasion to the resolution of the disease. Parasite Immunol 27:103–114

Rabinovich GA, Ariel A, Hershkoviz R et al (1999) Specific inhibition of T-cell adhesion to extracellular matrix and proinflammatory cytokine secretion by human recombinant galectin-1. Immunology 97:100–106

Rabinovich GA, Alonso CR, Sotomayor CE et al (2000) Molecular mechanisms implicated in galectin-1-induced apoptosis: activation of the AP-1 transcription factor and downregulation of Bcl-2. Cell Death Differ 7:747–753

Rabinovich GA, Ramhorst RE, Rubinstein N et al (2002a) Induction of allogenic T-cell hyporesponsiveness by galectin-1-mediated apoptotic and non-apoptotic mechanisms. Cell Death Differ 9:661–670

Rabinovich GA, Rubinstein N, Matar P et al (2002b) The antimetastatic effect of a single low dose of cyclophosphamide involves modulation of galectin-1 and Bcl-2 expression. Cancer Immunol Immunother 50:597–603

Rabinovich GA, Sotomayor CE, Riera CM, Bianco I, Correa SG (2000) Evidence of a role for galectin-1 in acute inflammation. Eur J Immunol 30:1331–1339

Rabinovich GA, Toscano MA, Ilarregui JM, Rubinstein N (2004) Shedding light on the immunomodulatory properties of galectins: novel regulators of innate and adaptive immune responses. Glycoconj J 19:565–573

Rochard P, Galiegue S, Tinel N et al (2004) Expression of the peripheral benzodiazepine receptor triggers thymocyte differentiation. Gene Expr 12:13–27

Rorive S, Belot N, Decaestecker C et al (2001) Galectin-1 is highly expressed in human gliomas with relevance for modulation of invasion of tumor astrocytes into the brain parenchyma. Glia 33:241–255

Rossi B, Espeli M, Schiff C, Gauthier L (2006) Clustering of pre-B cell integrins induces galectin-1-dependent pre-B cell receptor relocalization and activation. J Immunol 177:796–803

Rotblat B, Niv H, Andre S et al (2004) Galectin-1(L11A) predicted from a computed galectin-1 farnesyl-binding pocket selectively inhibits Ras-GTP. Cancer Res 64:3112–3118

Rubinstein N, Alvarez M, Zwirner NW et al (2004a) Targeted inhibition of galectin-1 gene expression in tumor cells results in heightened T cell-mediated rejection; a potential mechanism of tumor-immune privilege. Cancer Cell 5:241–251

Rubinstein N, Ilarregui JM, Toscano MA et al (2004b) The role of galectins in the initiation, amplification and resolution of the inflammatory response. Tissue Antigens 64:1–12

Sakaguchi M, Shingo T, Shimazaki T et al (2006) A carbohydrate-binding protein, galectin-1, promotes proliferation of adult neural stem cells. Proc Natl Acad Sci USA 103:7112–7117

Salvatore P, Benvenuto G, Pero R et al (2000) Galectin-1 gene expression and methylation state in human T leukemia cell lines. Int J Oncol 17:1015–1018

Sangeetha SR, Appukuttan PS (2005) IgA1 is the premier serum glycoprotein recognized by human galectin-1 since T antigen (Galβ1–>3GalNAc-) is far superior to non-repeating N-acetyl lactosamine as ligand. Int J Biol Macromol 35:269–276

Sango K, Tokashiki A, Ajiki K et al (2004) Synthesis, localization and externalization of galectin-1 in mature dorsal root ganglion neurons and Schwann cells. Eur J Neurosci 19:55–64

Sano H, Hsu DK, Yu L et al (2000) Human galectin-3 is a novel chemoattractant for monocytes and macrophages. J Immunol 165:2156–2164

Santucci L, Fiorucci S, Cammilleri F et al (2000) Galectin-1 exerts immunomodulatory and protective effects on concanavalin A-induced hepatitis in mice. Hepatology 31:399–406

Santucci L, Fiorucci S, Rubinstein N et al (2003) Galectin-1 suppresses experimental colitis in mice. Gastroenterology 124:1381–1394

Sarafian V, Jans R, Poumay Y (2006) Expression of lysosome-associated membrane protein 1 (lamp-1) and galectins in humankeratinocytes is regulated by differentiation. Arch Dermatol Res 298:73–81

Sasaki T, Hirabayashi J, Manya H, Kasai K, Endo T (2004) Galectin-1 induces astrocyte differentiation, which leads to production of brain-derived neurotrophic factor. Glycobiology 14:357–363

Saussez S, Nonclercq D, Laurent G et al (2005) Toward functional glycomics by localization of tissue lectins: immunohistochemical galectin fingerprinting during diethylstilbestrol-induced kidney tumorigenesis in male Syrian hamster. Histochem Cell Biol 123:29–41

Saussez S, Cucu DR, Deacestecker C et al (2006) Galectin-7 (p53-induced gene-1): a new prognostic predictor of recurrence and survival in stage IV hypopharyngeal cancer. Ann Surg Oncol 13:999–1009

Schwarz G Jr, Remmelink M, Decaestecker C et al (1999) Galectin fingerprinting in tumor diagnosis. Differential expression of galectin-3 and galectin-3 binding sites, but not galectin-1, in benign vs malignant uterine smooth muscle tumors. Am J Clin Pathol 111:623–631

Seelenmeyer C, Wegehingel S, Lechner J, Nickel W (2003) The cancer antigen CA125 represents a novel counter receptor for galectin-1. J Cell Sci 116:1305–1318

Seelenmeyer C, Wegehingel S, Tews I et al (2005) Cell surface counter receptors are essential components of the unconventional export machinery of galectin-1. J Cell Biol 171:373–381

Shaffer AL, Lin KI, Kuo TC et al (2002) Blimp-1 orchestrates plasma cell differentiation by extinguishing the mature B cell gene expression program. Immunity 17:51–62

Shahwan M, Al-Qirim MT et al (2004) Physicochemical properties and amino acid sequence of sheep brain galectin-1. Biochemistry (Mosc) 69:506–512

Shamsul Ola M, Tabish M, Khan FH, Banu N (2007) Purification and some properties of galectin-1 derived from water buffalo (*Bubalus bubalis*) brain. Cell Biol Int 31:578–585

Shapiro-Shelef M, Lin KI, McHeyzer-Williams LJ et al (2003) Blimp-1 is required for the formation of immunoglobulin secreting plasma cells and pre-plasma memory B cells. Immunity 19:607–620

Shen J, Person MD, Zhu J et al (2004) Protein expression profiles in pancreatic adenocarcinoma compared with normal pancreatic tissue and tissue affected by pancreatitis as detected by two-dimensional gel electrophoresis and mass spectrometry. Cancer Res 64:9018–9026

Shimonishi T, Miyazaki K, Kono N et al (2001) Expression of endogenous galectin-1 and galectin-3 in intrahepatic cholangiocarcinoma. Hum Pathol 32:302–310

Shoji H, Deltour L, Nakamura T et al (2009) Expression pattern and role of galectin-1 during early mouse myogenesis. Dev Growth Differ 51:607–615

Silva WA Jr, Covas DT, Panepucci RA et al (2003) The profile of gene expression of human marrow mesenchymal stem cells. Stem Cells 21:661–669

Simon P, Decaestecker C, Choufani G et al (2001) The levels of retinoid RARβ receptors correlate with galectin-1, -3 and -8 expression in human cholesteatomas. Hear Res 156:1–9

Sorme P, Kahl-Knutsson B et al (2003) Design and synthesis of galectin inhibitors. Methods Enzymol 363:157–169

Stillman BN, Hsu DK, Pang M et al (2006) Galectin-3 and galectin-1 bind distinct cell surface glycoprotein receptors to induce T cell death. J Immunol 176:778–789

Stowell SR, Dias-Baruffi M, Penttila L et al (2004) Human galectin-1 recognition of poly-N-acetyllactosamine and chimeric polysaccharides. Glycobiology 14:157–167

Stowell SR, Karmakar S, Stowell CJ et al (2007) Human galectin-1, -2, and -4 induce surface exposure of phosphatidylserine in activated human neutrophils but not in activated T cells. Blood 109:219–227

Stowell SR, Arthur CM, Mehta P et al (2008a) Galectin-1, -2, and -3 exhibit differential recognition of sialylated glycans and blood group antigens. J Biol Chem 283:10109–10123

Stowell SR, Qian Y, Karmakar S et al (2008b) Differential roles of galectin-1 and galectin-3 in regulating leukocyte viability and cytokine secretion. J Immunol 180:3091–3102

Stowell SR, Cho M, Feasley CL et al (2009) Ligand reduces galectin-1 sensitivity to oxidative inactivation by enhancing dimer formation. J Biol Chem 284:4989–4999

Sturm A, Lensch M, André S et al (2004) Human galectin-2: novel inducer of T cell apoptosis with distinct profile of caspase activation. J Immunol 173:3825–3837

Svensson A, Tågerud S (2009) Galectin-1 expression in innervated and denervated skeletal muscle. Cell Mol Biol Lett 14:128–138

Symons A, Cooper DN, Barclay AN (2000) Characterization of the interaction between galectin-1 and lymphocyte glycoproteins CD45 and Thy-1. Glycobiology 10:559–563

Szoke T, Kayser K et al (2005) Prognostic significance of endogenous adhesion/growth-regulatory lectins in lung cancer. Oncology 69:167–174

Tabrizi SJ, Niiro H, Masui M et al (2009) T cell leukemia/lymphoma 1 and galectin-1 regulate survival/cell death pathways in human naive and IgM^+ memory B cells through altering balances in Bcl-2 family proteins. J Immunol 182:1490–1499

Tahara K, Tsuchimoto D, Tominaga Y et al (2003) DeltaFosB, but not FosB, induces delayed apoptosis independent of cell proliferation in the Rat1a embryo cell line. Cell Death Differ 10:496–507

Tenne-Brown J, Puche AC, Key B (1998) Expression of galectin-1 in the mouse olfactory system. Int J Dev Biol 42:791–799

Than NG, Kim SS, Abbas A et al (2008a) Chorioamnionitis and increased galectin-1 expression in PPROM – an anti-inflammatory response in the fetal membranes? Am J Reprod Immunol 60:298–311

Than NG, Romero R, Erez O et al (2008b) Emergence of hormonal and redox regulation of galectin-1 in placental mammals: implication in maternal-fetal immune tolerance. Proc Natl Acad Sci USA 105:15819–15824

Thijssen VL, Postel R, Brandwijk RJ et al (2006) Galectin-1 is essential in tumor angiogenesis and is a target for antiangiogenesis therapy. Proc Natl Acad Sci USA 103:15975–15980

Timmons PM, Rigby PW, Poirier F (2002) The murine seminiferous epithelial cycle is pre-figured in the Sertoli cells of the embryonic testis. Development 129:635–647

Tinari N, Kuwabara I, Huflejt ME et al (2001) Glycoprotein, 90K/MAC-2BP interacts with galectin-1 and mediates galectin-1-induced cell aggregation. Int J Cancer 91:167–172

Toscano MA, Bianco GA, Ilarregui JM et al (2007) Differential glycosylation of TH1, TH2 and TH-17 effector cells selectively regulates susceptibility to cell death. Nat Immunol 8:825–834

Tsai C-M, Chiu Y-K, Hsu T-L et al (2008) Galectin-1 promotes immunoglobulin production during plasma cell differentiation. J Immunol 181:4570–4579

Tsutsumi T, Suzuki T, Moriya K et al (2003) Hepatitis C virus core protein activates ERK and p38 MAPK in cooperation with ethanol in transgenic mice. Hepatology 38:820–828

Valkova N, Yunis R, Mak SK et al (2005) Nek8 mutation causes overexpression of galectin-1, sorcin, and vimentin and accumulation of the major urinary protein in renal cysts of jck mice. Mol Cell Proteomics 4:1009–1018

van den Brule F, Califice S, Garnier F et al (2003) Galectin-1 accumulation in the ovary carcinoma peritumoral stroma is induced by ovary carcinoma cells and affects both cancer cell proliferation and adhesion to laminin-1 and fibronectin. Lab Invest 83:377–386

van der Leij J, van den Berg A, Blokzijl T et al (2004) Dimeric galectin-1 induces IL-10 production in T-lymphocytes: an important tool in the regulation of the immune response. J Pathol 204:511–518

Vas V, Fajka-Boja R, Ion G et al (2005) Biphasic effect of recombinant galectin-1 on the growth and death of early hematopoietic cells. Stem Cells 23:279–287

Vasta GR, Ahmed H, Odom EW (2004) Structural and functional diversity of lectin repertoires in invertebrates, protochordates and ectothermic vertebrates. Curr Opin Struct Biol 14:617–630

Vespa GN, Lewis LA, Kozak KR et al (1999) Galectin-1 specifically modulates TCR signals to enhance TCR apoptosis but inhibit IL-2 production and proliferation. J Immunol 162:799–806

Vicovac L, Jankovic M, Cuperlovic M (1998) Galectin-1 and -3 in cells of the first trimester placental bed. Hum Reprod 13:730–735

von Wolff M, Wang X, Gabius HJ, Strowitzki T (2005) Galectin fingerprinting in human endometrium and decidua during the menstrual cycle and in early gestation. Mol Hum Reprod 11:189–194

Vyakarnam A, Dagher SF, Wang JL et al (1997) Evidence for a role for galectin-1 in pre-mRNA splicing. Mol Cell Biol 17:4730–4737

Walzel H, Schulz U et al (1999) Galectin-1, a natural ligand for the receptor-type protein tyrosine phosphatase CD45. Immunol Lett 67:193–202

Walzel H, Blach M et al (2000) Involvement of CD2 and CD3 in galectin-1 induced signaling in human Jurkat T-cells. Glycobiology 10:131–140

Walzel H, Blach M, Hirabayashi J et al (2002) Galectin-induced activation of the transcription factors NFAT and AP-1 in human Jurkat T-lymphocytes. Cell Signal 14:861–868

Walzel H, Brock J, Pohland R et al (2004) Effects of galectin-1 on regulation of progesterone production in granulose cells from pig ovaries in vitro. Glycobiology 14:871–881

Walzel H, Fahmi AA, Eldesouky MA et al (2006) Effects of N-glycan processing inhibitors on signaling events and induction of apoptosis in galectin-1-stimulated Jurkat T lymphocytes. Glycobiology 16:1262–1271

Wang P, Mariman E, Keijer J et al (2004) Profiling of the secreted proteins during 3 T3-L1 adipocyte differentiation leads to the identification of novel adipokines. Cell Mol Life Sci 61:2405–2417

Wang J, Lu ZH, Gabius HJ et al (2009) Cross-linking of GM1 ganglioside by galectin-1 mediates regulatory T cell activity involving TRPC5 channel activation: possible role in suppressing experimental autoimmune encephalomyelitis. J Immunol 182:4036–4045

Wasano K, Hirakawa Y (1997) Recombinant galectin-1 recognizes mucin and epithelial cell surface glycocalyces of gastrointestinal tract. J Histochem Cytochem 45:275–283

Wells V, Davies D, Mallucci L (1997) Cell cycle arrest and induction of apoptosis by β galactoside binding protein (β GBP) in human mammary cancer cells. A potential new approach to cancer control. Eur J Cancer 35:978–983

Wiest I, Seliger C et al (2005) Induction of apoptosis in human breast cancer and trophoblast tumor cells by galectin-1. Anticancer Res 25:1575–1580

Woynarowska B, Skrincosky DM et al (1994) Inhibition of lectin-mediated ovarian tumor cell adhesion by sugar analogs. J Biol Chem 269:22797–22803

Xu Z, Weiss A (2002) Negative regulation of CD45 by differential homodimerization of the alternatively spliced isoforms. Nat Immunol 3:764–771

Yamaoka K, Mishima K, Nagashima Y et al (2000) Expression of galectin-1 mRNA correlates with the malignant potential of human gliomas and expression of antisense galectin-1 inhibits the growth of 9 glioma cells. J Neurosci Res 59:722–730

Yang JW, Kang SU et al (2006) Mass spectrometrical analysis of galectin proteins in primary rat cerebellar astrocytes. Neurochem Res 31:945–955

Yu X, Siegel R, Roeder RG (2006) Interaction of the B cell-specific transcriptional coactivator OCA-B and galectin-1 and a possible role in regulating BCR-mediated B cell proliferation. J Biol Chem 281:15505–15516

ZacarÃas Fluck MF, Rico MJ et al (2007) Low-dose cyclophosphamide modulates galectin-1 expression and function in an experimental rat lymphoma model. Cancer Immunol Immunother 56: 237–248

Zhao XY, Zhao KW, Jiang Y et al (2011) Synergistic induction of galectin-1 by C/EBPα and HIF-1α and its role in differentiation of acute myeloid leukemic cells. J Biol Chem 286(42):36808–36819, Aug 31

Zou W (2005) Immunosuppressive networks in the tumour environment and their therapeutic relevance. Nat Rev Cancer 5: 263–274

Zuniga E, Rabinovich GA, Iglesias MM, Gruppi A (2001) Regulated expression of galectin-1 during B-cell activation and implications for T-cell apoptosis. J Leukoc Biol 70:73–79

Tandem-Repeat Type Galectins

11

Anita Gupta

Hirabayashi and Kasai (1993) proposed designating galectin subfamilies as proto-, tandem-repeat, and chimera- types based on their domain organization. The prototype galectins (galectin-1, -2, -5, -7, -10, -11, -13, -14, and -15) consist of a single CRD with a short N-terminal sequence and have been discussed in Chap. 9, The tandem-repeat type galectins (galectin-4, -6, -8, -9, and -12) are composed of two non-identical CRDs joined by a "hinge" region (short linker peptide sequence) and form a sub-family of galectins (see Fig. 9.2). The single chimera-type galectin (galectin-3) has one CRD and an extended N-terminal tail containing several repeats of proline-tyrosine-glycine-rich motif and will be discussed in Chap. 12. Alternative splicing leads to the formation of distinct splice variants (isoforms) of galectin-8 and galectin-9 with tandem-repeat-type structures. Galectin-8- consists of several isoforms, each made of two domains of approximately 140 amino-acids, both having a CRD. These domains are joined by a 'link peptide' of variable length. The isoforms share identical CRDs and differ only in linker region.

11.1 Galectin 4

11.1.1 Localization and Tissue Distribution

Galectin-4 belongs to a subfamily of galectins composed of two CRD within the same peptide chain. The two domains have all conserved galectin signature amino acids, but their overall sequences are only ~40% identical. Both domains bind lactose with a similar affinity as other galectins, but their respective preferences for other disaccharides, and larger saccharides, are distinctly different. Thus galectin-4 has a property of a natural cross-linker, but in a modified sense each domain prefers a different subset of ligands. Similarly to other galectins, galectin-4 is synthesized as a cytosolic protein, but can be externalized. During development and in adult normal tissues, galectin-4 is expressed only in alimentary tract, from the tongue to the large intestine. It is often found in relatively insoluble complexes, as a component of either adherens junctions or lipid rafts in the microvillus membrane, and it has been proposed to stabilize these structures. Strong expression of galectin-4 can be induced, however, in cancers from other tissues including breast and liver. Within a collection of human epithelial cancer cell lines, galectin-4 is over-expressed and soluble in those forming highly differentiated polarized monolayers, but absent in less differentiated ones. In cultured cells, intracellular galectin-4 may promote resistance to nutrient starvation, whereas—as an extracellular protein—it can mediate cell adhesion. Because of its distinct induction in breast and other cancers, it may be a valuable diagnostic marker and target for the development of inhibitory carbohydrate-based drugs (Huflejt and Leffler 2004). EM examination of galectin 4 confirmed highly elevated levels of the protein in endocrine, parietal, and chief cells in weaned rats than in suckling rats. Galectin 4 was strongly localized in weaned rats than in suckling rats in the nuclei of all cell types and in secretory granules in endocrine and chief cells, suggesting that galectin 4 is implicated in nuclear events and perhaps in secretory processes (Niepceron et al. 2004).

Galectin-4 has been found to be localized in the epithelium of the alimentary tract, including oral mucosa, esophagus, and intestinal mucosa (Chiu et al. 1994; Wooters et al. 2005a). In gastric mucosa, galectin 4 was present at lower levels in suckling than in weaned rats, but mRNA levels did not change significantly during postnatal development. It was more strongly localized in the nuclei of all cell types and in or over secretory granules in endocrine and chief cells of weaned rats compared to suckling rats, suggesting that galectin 4 is implicated in nuclear events and perhaps in secretory processes (Niepceron et al. 2004). Chiu et al (1994) found that galectin-4 in porcine oral and upper esophageal epithelium is water-insoluble as a component of adherens junction complexes. Danielsen and Deurs (1997) reported that galectin-4 forms detergent-insoluble complexes with apically sorting brush-border enzymes in

G.S. Gupta et al., *Animal Lectins: Form, Function and Clinical Applications*,
DOI 10.1007/978-3-7091-1065-2_11, © Springer-Verlag Wien 2012

porcine small intestine, when the isolation was performed at low temperature. On the contrary, Huflejt et al. (1997) showed the localization of galectin-4 on the basal membrane of human colon adenocarcinoma T84 cells in a confluent and polarized condition, whereas galectin-3 tends to be concentrated in granular inclusions mostly localized on the apical side. They also showed that galectin-4 was concentrated at the leading edges of lamellipodia in semiconfluent T84 cells. These results suggest that galectin-4 is located in a different area of cells dependently on tissues and culture conditions and that it plays a role in cell adhesion or cell migration.

Mouse galectin-4 is expressed in small intestine, colon, liver, kidney, spleen, heart and P19X1 cells in BW-5147 and 3T3 cell lines. Galectin-4 is expressed in spermatozoons and oocytes and its expression during early mouse embryogenesis appears in eight-cell embryos and remains in later stages. Mature epithelial cells at the villous tips stained the most intensely with the mAb, with progressively less intense staining observed along the sides of villi and into the crypts. Galectin-4 was also associated with nuclei in villous tip cells, indicating that some galectin-4 may migrate to the nucleus during terminal maturation of these cells. In intestinal crypts, a specific subset of cells, which may be enteroendocrine cells, expressed galectin-4 at a relatively high level (Wooters et al. 2005a). Lactose-binding proteins with molecular masses of 14-, 17-, 18-, 28-, and 34-kDa are present in extracts from porcine small intestinal mucosa. Thirty four-kilodalton protein, galectin-4 was the most abundant of these proteins. RT-PCR identified two galectin-4 isoforms that differed in the length of their linker region. The larger isoform, galectin-4.1, is nine amino acids longer in its linker region than the smaller isoform, galectin-4.2. Based on nucleotide sequence similarities, the two isoforms are likely splice variants of galectin-4 pre-mRNA and not products of separate genes like murine galectins-4 and -6 (Wooters et al. 2005b).

Nio et al. (2005) detected signals for five galectin subtypes (galectin-2, -3, -4/6, and -7) exclusively in the epithelia of gut. In the glandular stomach, galectin-2 and -4/6 were predominantly expressed from gastric pits to neck of gastric glands. The small intestine exhibited intense, maturation-associated expressions of galectin-2, -3, and -4/6 mRNAs. Galectin-2 was intensely expressed from crypts to the base of villi, whereas transcripts of galectin-3 gathered at villous tips. Signals for galectin-4/6 were most intense at the lower half of villi. Galectin-2 was also expressed in goblet cells of the small intestine but not in those of the large intestine. In the large intestine, galectin-4/6 predominated, and the upper half of crypts simultaneously contained transcripts of galectin-3. Stratified epithelium from the lip to fore-stomach and anus intensely expressed galectin-7 with weak expressions of galectin-3. To assess intestinal lipid rafts functions through the characterization of their protein markers, Nguyen et al. (2006) isolated lipid rafts of rat mucosa. Membrane preparations were enriched in cholesterol, ganglioside GM1, and N aminopeptidase (NAP) known as intestinal lipid rafts markers. Together results indicated that some digestive enzymes, trafficking and signaling proteins may be functionally distributed in the intestine lipid rafts. Mucosal epithelium in mouse showed region/cell-specific localization of each galectin subtype. Gastric mucous cells exhibited intense immunoreactions for galectin-2 and galectin-4/6 with a limited localization of galectin-3 at the surface of gastric mucosa. Epithelial cells in small intestine showed characteristic localizations of galectin-2 and galectin-4/6 in the cytoplasm of goblet cells and the basolateral membrane of enterocytes in association with maturation, respectively. Epithelial cells of large intestine contained intense reactions for galectin-3 and galectin-4/6 but not for galectin-2. The stratified squamous epithelium of the forestomach was immunoreactive for galectin-3 and galectin-7. Outside the epithelium, Only galectin-1 was localized in the connective tissue, smooth muscles, and neuronal cell bodies. The subtype-specific localization of galectin suggests its important roles in host-pathogen interaction and epithelial homeostasis such as membrane polarization and trafficking in the gut (Nio-Kobayashi et al. 2009).

11.1.2 Galectin-4 Isoforms

Lactose-binding proteins with molecular masses of 14-, 17-, 18-, 28-, and 34-kDa were identified in extracts from porcine small intestinal mucosa. The most abundant of these proteins, was identified as porcine galectin-4. Galectin-4 is a member of the tandem-repeat subfamily of monomer divalent galectins. Galectin-4 (LGALS4) is encoded in humans by *LGALS4* gene. Galectin-4 was initially discovered as a soluble 17-kDa lectin in rat intestinal extracts by Leffler et al. (1989); later the cDNA cloning of galectin-4 revealed that it is a 36-kDa protein (Oda et al. 1993). The 323-amino acid galectin-4 protein contains two homologous, ~150-amino acid carbohydrate recognition domains and all amino acids typically conserved in galectins. Wooters et al. (2005b) identified two galectin-4 isoforms that differed in the length of their linker region. The larger isoform, galectin-4.1, is nine amino acids longer in its linker region than the smaller isoform, galectin-4.2. Based on nucleotide sequence similarities, the two isoforms are likely splice variants of galectin-4 pre-mRNA and not products of separate genes like murine galectins-4 and -6. The crystal of N-terminal domain, CRD1, of mGal-4 in complex with lactose belongs to tetragonal space group $P42_12$ with unit-cell parameters $a = 91.1$, $b = 91.16$, $c = 57.10$ A The initial electron-density maps of X-ray diffraction indicated the presence of one lactose molecule. The crystal structure of CRD1of mouse Gal-4 in complex

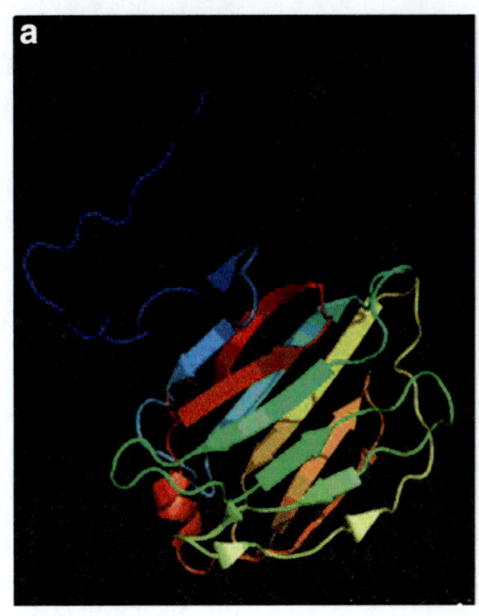

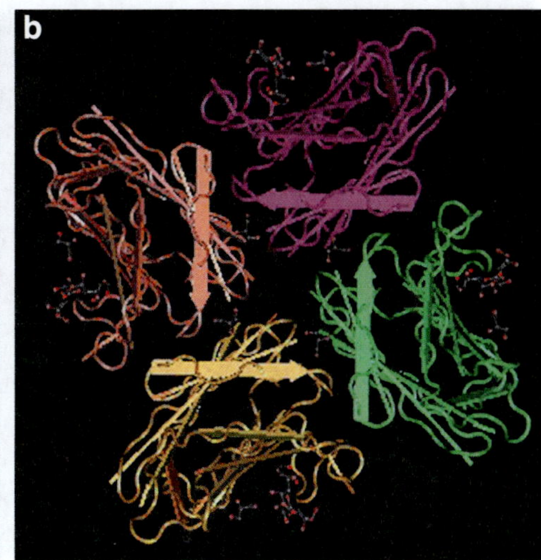

Fig. 11.1 (**a**) Solution structure of the C-terminal domain of gal-binding lectin domain of human galectin-4 by NMR (PDB ID:1X50). The 323-amino acid LGALS4 protein contains two homologous, ~150-amino acid CRDs (Tomizawa et al. PDB ID: 1X50). (**b**) Structure of N-terminal CRD1of mouse Gal-4 in complex with lactose (PDB: ID 3I8T) (Krejčiříková et al. 2011)

with lactose was reported recently (Fig. 11.1b). Two lactose-binding sites were identified: a high-affinity site with a K_{D1} value of 600 ± 70 μM and a low-affinity site with K_{D2} = 28 ± 10 mM (Krejčiříková et al. 2008, 2011). Solution Structure of the C-terminal domain of Galactose-binding lectin domain of human Gal-4 by NMR (PDB ID:1X50 is shown in Fig. 11.1a.

Multiple members of galectin family are expressed in primary olfactory system. In particular, galectin-3 is expressed by ensheathing cells surrounding nerve fascicles in the submucosa and nerve fiber layer, where it may mediate cross-linking of axons. Galectin-4, -7, and -8 are expressed by the primary olfactory axons as they grow from the nasal cavity to the olfactory bulb. A putative role for nitric oxide donor compound-7 (NOC-7) and NOC-8 in axon fasciculation and the expression of multiple galectins in the developing olfactory nerve suggest that these molecules may be involved in the formation of this pathway, particularly in the sorting of axons as they converge towards their target (Storan et al. 2004). Cell-specific surface carbohydrates and carbohydrate-binding proteins within embryonic nervous system have raised the possibility that carbohydrate recognition contribute to the interactions of developing neurons. A cDNA from rat brain encodes a lectin, RL-14.5, which is homologous and identical in primary sequence to a lectin present in non-neural tissue. High levels of RL-14.5 mRNA are present in primary sensory neurons and motor neurons in the spinal cord and brain stem. The selective expression of oligosaccharide ligands for RL-14.5 on the same neurons are consistent with the idea that carbohydrate-mediated interactions contribute to the development of this subset of mammalian neurons (Hynes et al. 1990).

11.1.3 Gal-4 from Rodents and Other Animals

Gal-4 from Rat Intestine: Of the multiple soluble lactose-binding lectins in rat intestine, the major one, tentatively designated RI-H, was isolated as a polypeptide of M_r 17,000 Da. Surprisingly the cDNA encodes a protein of M_r 36,000, and this protein contains two homologous but distinct domains each with sequence elements that are conserved among all S-Lac lectins. The C-terminal domain, designated domain II, corresponds to lectin with M_r of 17,000 previously isolated from intestinal extracts and shown to have lactose binding activity. The new lectin, which was designated L-36, is highly expressed in full-length form in rat small and large intestine and stomach but was not detected in other tissues including lung, liver, kidney, and spleen. Each domain has approximately 35% sequence identity with the other domain and with the carbohydrate-binding domain of L-29, another S-Lac lectin, but only about 15% identity with other known S-Lac Lectins (Oda et al. 1993). Tardy et al. (1995) detected five β-galactoside binding lectins of 14–20 kDa in rat small intestinal mucosa and purified the prominent proteins of 17 and 19 kDa. Direct N-terminal sequencing of 17 kDa protein and intrachain sequencing of 19 kDa protein produced sequences which are part of N-terminal domain of L-36/galectin-4. The 17 and 19 kDa lectins were related to 36 kDa protein in human undifferentiated HT29 cells (Tardy et al. 1995).

Adherens Junction Lectin: A pig junction protein of M_r 37,000 was found in oral epithelium but not in epidermis, limited to suprabasal cells, and colocalizing with adherens junction proteins. Secondary structure predictions indicated that the 37% identical 16 to 17-kDa N- and C-terminal domains from β-sheet-rich barrels are linked by a compact proline-rich segment. The protein is 72% identical in amino acid sequence and shares symmetrical two-domain structure with L-36, a lectin from rat intestine, indicating that the 37-kDa protein is the porcine form of L-36. The expressed protein binds a glycoprotein of 120 kDa from pig tongue epithelium and also inhibited by lactose. The lectin may be involved in the assembly of adherens junctions (Chiu et al. 1994).

Rabbit Bladder Galectin-4: The cDNA encoding rabbit bladder galectin-4 has been cloned and sequenced. The deduced 328 amino acid sequence predicts a multidomain structure consisting of an N-terminal peptide (19 residues) and two CRD (130 residues each) connected by a linker region (49 residues). Comparison of rabbit galectin-4 with related proteins reveals that two peptide motifs, M-A-F/Y-V-P-A-P-G-Y-Q-P-T-Y-N-P-T-L-P-Y in the N terminus and A-F-H-F-N-P-R-F-D-G-W-D-K-V-V-F in the first CRD are highly conserved in human, pig, rat, and mouse galectin-4 as well as in mouse galectin-6. The two peptide motifs were proposed as the signature sequences to identify new members of galectin-4 subfamily (Jiang et al. 1999). Comparison of rabbit galectin-4 sequence with those of human, pig, rat, and mouse revealed two invariant peptide motifs that are proposed as signature sequences for identifying related galectins (Bhavanandan et al. 2001).

Endogenous lactose binding lectin in retina has a subunit molecular weight of 16 kDa and a pI about 4.5 (Beyer et al. 1980) or C-16 (Sakakura et al. 1990) form of chicken endogenous soluble lactose-binding lectins. Retinal lectin might have a functional role during terminal differentiation of retinal cells (Castagna and Landa 1994). Two types of lectins have been isolated from extracts of axolotl (Ambystoma mexicanum) larvae: A thiol-independent lectin of subunit of 15 kDa and a thiol-dependent lectin of subunit of 18 kDa (Allen et al. 1992). A 16-kDa lactose-binding lectin comprises 5% or more of soluble protein in *Xenopus laevis* skin. This lectin is localized in the cytoplasm of granular gland cells and released onto skin surface by holocrine secretion in response to stress. Comparison of peptide sequences revealed expression of at least two isolectins, which differ in sequence at only two or three amino acids (Marschal et al. 1992).

11.1.4 Ligands for Galectin-4

Interaction with Linear Tetrasaccharides: The combining sites of mammalian galectins have overlapping carbohydrate specificities. Wu et al. (2002) analyzed CRD-I near the N-terminus of recombinant rat galectin-4 and suggested that among 35 glycans tested for lectin binding, galectin-4 reacted best with human blood group ABH precursor glycoproteins, and asialo porcine salivary glycoproteins, which contain high densities of blood group Ii determinants Galβ1-3GalNAc (the mucin-type sugar sequence on the human erythrocyte membrane) and/or GalNAcα1-Ser/Thr (Tn), whereas this lectin domain reacted weakly or not at all with most sialylated glycoproteins. Among oligosaccharides tested, Galβ1-3GlcNAcβ1-3Galβ1-4Glc was the best. Galectin-4 has a preference for the β-anomer of Gal at the non-reducing ends of oligosaccharides with a Galβ1-3 linkage, over Galβ1-4 and Galβ1-6. The fraction of Tn glycopeptide from asialo ovine submandibular glycoprotein was 8.3 times more active than Galβ1-3GlcNAc. The overall carbohydrate specificity of Galectin-4 can be defined as Galβ1-3GlcNAcβ1-3Galβ1-4Glc (lacto- N-tetraose) > Galβ1-4GlcNAcβ1-3Galβ1-4Glc (lacto- N -neo-tetraose) and Tn clusters > Galβ1-4Glc and GalNAcβ1-3Gal > Galβ1-3GalNAc > Galβ1-3GlcNAc > Galβ1-4GlcNAc > GalNAc > Gal. The definition of this binding profile provides the basis to detect differential binding properties relative to the other galectins with ensuing implications for functional analysis (Wu et al. 2002).

The core aspects of the carbohydrate specificity of domain-I of recombinant tandem-repeat-type galectin-4 from rat gastrointestinal tract (G4-N), especially its potent interaction with the linear tetrasaccharide Galβ1-3GlcNAcβ1-3Galβ1-4Glc (Iβ1-3L) have been defined. Analysis indicated that a high-density of polyvalent Galβ1-3/4GlcNAc (I/II), Galβ1-3GalNAc (T) and/or GalNAcα1-Ser/Thr (Tn) strongly favors G4-N/glycoform binding. These glycans were up to 2.3×10^6, 1.4×10^6, 8.8×10^5, and 1.4×10^5 more active than Gal, GalNAc, monomeric I/II and T, respectively. The distinct binding features of G4-N established the important concept of affinity enhancement by high density polyvalencies of glycotopes (vs. multi-antennary I/II) and by introduction of an ABH key sugar to Galβ1-terminated core glycotopes (Wu et al. 2004).

Binding to Glycosphingolipids: Galectin-4 specifically binds to an $SO3^-$ —> 3Galβ1 —> 3GalNAc pyranoside with high affinity (Ideo et al. 2002). Galectin-4 binds to glycosphingolipids carrying 3-O-sulfated Gal residues, such as SB1a, SM3, SM4s, SB2, SM2a, and GM1, but not to glycosphingolipids with 3-O-sialylated Gal, such as sLc4Cer, snLc4Cer, GM3, GM2, and GM4. Galectin-4 was colocalized with SB1a, GM1, and carcinoembryonic antigen (CEA) in the patches on the cell surface of human colon adenocarcinoma cells. It appeared that SB1a and CEA in the patches on the cell surface of human colon adenocarcinoma cells could be biologically important ligands for galectin-4 (Ideo et al. 2005).

Interaction with Sulfatides: 1-benzyl-2-acetamido-2-deoxy-α-D-galactopyranoside (GalNAc α-O-bn), an inhibitor of glycosylation, perturbs apical biosynthetic trafficking in polarized HT-29 cells suggesting the involvement of a lectin-based mechanism. Galectin-4 is one of the major components of detergent-resistant membranes (DRMs) isolated from HT-29 5M12 cells and also found in post-Golgi carrier vesicles. In galectin-4-depleted HT-29 5M12 cells apical membrane markers accumulated intracellularly. Galectin-4 depletion altered also the DRM association characteristics of apical proteins. Sulfatides with long chain-hydroxylated fatty acids, which were also enriched in DRMs, were identified as high-affinity ligands for galectin-4. It was proposed that interaction between galectin-4 and sulfatides plays a functional role in the clustering of lipid rafts for apical delivery (Delacour et al. 2005). Galectin-4 also binds to cholesterol 3-sulfate, which has no β-galactoside moiety. This characteristic of galectin-4 is unique within the galectin family. The site-directed mutated galectin-4-R45A suggested that Arg-45 of galectin-4 is indispensable for cholesterol 3-sulfate recognition. Results suggested that not only sulfated glycosphingolipids but also cholesterol 3-sulfate are endogenous ligands for galectin-4 in vivo (Ideo et al. 2007).

Murine Galectin-4: Murine (m) mGalectin-4 binds to α-GalNAc and α-Gal A and B type structures with or without fucose. While the CRD2 domain had a high specificity and affinity for A type-2 α-GalNAc structures, the CRD1 domain had a broader specificity in correlation to the total binding profile. Study suggested that CRD2 might be the dominant binding domain of mouse galectin-4. On mouse cryosections, all three forms (CRD1, CRD2 and mGalectin-4) bound to alveolar macrophages, macrophages of red pulp of the spleen and proximal tubuli of the kidney. However, mGalectin-4, but not CRD forms, bound to the suprabasal layer of squamous epithelium of the tongue, suggesting that the link region also plays an important role in ligand recognition (Markova et al. 2006).

11.1.5 Functions of Galectin-4

In absorptive cells, such as the small intestinal enterocyte and the kidney proximal tubule cell, the apical cell membrane is formed as a brush border, composed of regular, dense arrays of microvilli. Hydrolytic ectoenzymes make up the bulk of the microvillar membrane proteins, endowing the brush border with a huge digestive capacity. Several of the major enzymes are localized in lipid rafts, which are organized in a unique fashion, for enterocytes in particular. Glycolipids, rather than cholesterol, together with the divalent lectin galectin-4, define these rafts, which are stable and probably quite large. The architecture of these rafts supports a digestive/absorptive strategy for nutrient assimilation, but also serves as a portal for a large number of pathogens (Danielsen and Hansen 2006). Wasano and Hirakawa (1999) suggested that colorectal Gal-4 may be involved in crosslinking the lateral cell membranes of the surface-lining epithelial cells, thereby reinforcing epithelial integrity against mechanical stress exerted by the bowel lumen.

Role of Galectin-4 as Lipid Raft Stabilizer: The pig small intestinal brush border is a glycoprotein- and glycolipid-rich membrane that functions as a digestive/absorptive surface for dietary nutrients as well as a permeability barrier for pathogens. Galectins 3–4, intelectin, and lectin-like anti-glycosyl antibodies have been localized at brush border of pig small intestine. Together with the membrane glycolipids these lectins form stable lipid raft microdomains that also harbour several major digestive microvillar enzymes. Galectin-2 is enriched in microvillar detergent resistant fraction. Results suggest that galectins offer a maximal protection of the brush border against exposure to bile, pancreatic enzymes and pathogens (Hansen et al. 2005; Thomsen et al. 2009). Lipid rafts (glycosphingolipid/cholesterol-enriched membrane microdomains) have been isolated from many cell types. Superrafts are enriched in galectin-4 and harbor glycosylphosphatidylinositol-linked alkaline phosphatase and transmembrane aminopeptidase N. In microvillar membrane, galectin-4 functions as a core raft stabilizer/organizer for other, more loosely raft-associated proteins. The glycolipid rafts are stabilized by galectin-4 that cross-links galactosyl (and other carbohydrate) residues present on membrane lipids and several brush border proteins, including some of the major hydrolases. These supramolecular complexes are further stabilized by intelectin (Chap. 20) that also functions as an intestinal lactoferrin receptor. As a result, brush border hydrolases, otherwise sensitive to pancreatic proteinases, are protected from untimely release into the gut lumen. Finally, anti-glycosyl antibodies, synthesized by plasma cells locally in the gut, are deposited on the brush border glycolipid rafts, protecting epithelium from lumenal pathogens that exploit lipid rafts as portals for entry to the organism (Braccia et al. 2003; Danielsen and Hansen 2006).

Stechly et al. (2009) presented evidence of a lipid raft-based galectin-4-dependent mechanism of apical delivery of glycoproteins in these cells. First, galectin-4 recruits the apical glycoproteins in detergent-resistant membranes (DRMs) because these glycoproteins were depleted in DRMs isolated from galectin-4-knockdown (KD) HT-29 5M12 cells. DRM-associated glycoproteins were identified as ligands for galectin-4. Structural analysis showed that DRMs were markedly enriched in a series of complex

N-glycans in comparison to detergent-soluble membranes. Second, in galectin-4-KD cells, the apical glycoproteins still exit the Golgi but accumulated inside the cells, showing that their recruitment within lipid rafts and their apical trafficking required the delivery of galectin-4 at a post-Golgi level. This lectin that is synthesized on free cytoplasmic ribosomes is externalized from HT-29 cells mostly in the apical medium and follows an apical endocytic-recycling pathway that is required for the apical biosynthetic pathway. Together, data show that the pattern of N-glycosylation of glycoproteins serves as a recognition signal for endocytosed galectin-4, which drives the raft-dependent apical pathway of glycoproteins in enterocyte-like HT-29 cells.

Galectin-4 in Controlling of Intestinal Inflammation: Galectin-4 is selectively expressed and secreted by intestinal epithelial cells and binds potently to activated peripheral and mucosal lamina propria T-cells at CD3 epitope. The carbohydrate-dependent binding of galectin-4 at CD3 epitope is fully functional and inhibited T cell activation, cycling and expansion. Galectin-4 induced apoptosis of activated peripheral and mucosal lamina propria T cells via calpain-, but not caspase-dependent, pathways. Further, galectin-4 blockade by antisense oligonucleotides reduced TNF-α inhibitor induced T cell death and reduced pro-inflammatory cytokine secretion including IL-17 by T cells. In a model of experimental colitis, galectin-4 ameliorated mucosal inflammation, induced apoptosis of mucosal T-cells and decreased the secretion of pro-inflammatory cytokines. Thus, galectin-4 plays a unique role in the intestine and suggests a novel role of this protein in controlling intestinal inflammation by a selective induction of T cell apoptosis and cell cycle restriction. Thus, galectin-4 is a novel anti-inflammatory agent that could be therapeutically effective in diseases with a disturbed T cell expansion and apoptosis such as inflammatory bowel disease (Paclik et al. 2008b).

Galectins in Inflammatory Bowel Disease: Inflammatory bowel disease (IBD) is an immune-mediated intestinal inflammatory condition that is associated with an increase in autoantibodies, which bind to epithelial cells. The epithelial galectin-4 specifically stimulates IL-6 production by $CD4^+$ T cells. The galectin-4-mediated production of IL-6 is MHC class II independent and induced by PKCθ-associated pathway through the immunological synapse. The galectin-4-mediated stimulation of $CD4^+$ T cells is shown to exacerbate chronic colitis and delays the recovery from acute intestinal injury. These studies suggested the presence of an immunogenic, endogenous lectin in the intestine and indicated the biological role of lectin/$CD4^+$ T cell interactions under inflammatory conditions (Hokama et al. 2004).

Recent studies have identified immunoregulatory roles of galectins in intestinal inflammatory disorders. Unexpected roles of galectin-1 and galectin-4 and chi-lectin (chitinase 3-like-1) have been reported in intestinal inflammation. Galectin-1 and -2 contribute to the suppression of intestinal inflammation by the induction of effector T cell apoptosis. In contrast, galectin-4 is involved in the exacerbation of this inflammation by specifically stimulating intestinal $CD4^+$ T cells to produce IL-6 (Mizoguchi and Mizoguchi 2007). Hokama et al. (2008) reviewed how different members of galectins provide inhibitory or stimulatory signals to control intestinal immune response under intestinal inflammation. Inflammatory bowel diseases are characterized by various degrees of mucosal surface damage and subsequent impairment of intestinal barrier function. Resealing of the epithelial barrier requires intestinal cell migration and proliferation. Galectins are increasingly recognized as novel regulators of inflammation. Gal-2 and Gal-4 bind to epithelial cells at the E-cadherin/β-catenin complex. Both galectins significantly enhanced intestinal epithelial cell restitution in vitro. This enhancement of epithelial cell restitution was TGF-β-independent. In contrast, Gal-1 decreased epithelial cell migration TGF-β dependently. Gal-2 and Gal-4 were found to increase cyclin B1 expression and consequently cell cycle progression, while Gal-1 inhibited cell cycling. Studies on the influence of Gal-2 and Gal-4 on epithelial cell apoptosis showed no induction of apoptosis, whereas Gal-1 significantly induced apoptosis of epithelial cells caspase-independently. Thus, these galectins play a significant role in intestinal wound-healing processes and might exert beneficial effects in diseases characterized by epithelial barrier disruption like IBDs.

Galectin expression is related to the genetic background of control animals. In acute and chronic experimental colitis produced in C57BL/6 and BALB/c mice with acute dextran sodium sulphate colitis and in 129 Sv/Ev IL-10 knock-out ($IL\text{-}10^{-/-}$) mice, inflammation was associated with chronic colitis in $IL\text{-}10^{-/-}$ mice with increased galectin-4 expression. In contrast with two other models, no galectin-1 and -3 alterations were observed in $IL\text{-}10^{-/-}$ mice. Acute colitis in C57BL/6 and BALB/c mice showed increased galectin-3 expression in the lamina propria and the crypt epithelium, together with a decreased nuclear expression. These results suggest an involvement of galectins in the development and perpetuation of colonic inflammation and illustrate that the choice of the mouse strain for studying galectins might influence the outcome of the experiments (Mathieu et al. 2008).

Interaction of Galectin-4 with p27 in the yeast two-hybrid assay suggested that galectin-4 is involved in p27-mediated activation of myelin basic protein gene, possibly through modulation of the glycosylation status of transcription factor Sp1 (Wei et al. 2007).

11.1.6 Galectin-4 in Cancer

Galectin-4 Expression in Carcinoid Tumors: Galectin-4 is localized in the enterochromaffin cells of the porcine and murine small intestine. A differential distribution of galectin-4 is observed in carcinoid tumors in different locations of the GI tract and the lungs. In primary and metastatic human ileal carcinoid tumors as well as in carcinoid tumors of the stomach, lung, and rectum, galectin-4 was most highly expressed in the ileal carcinoids and the levels of expression tended to be higher in primary ileal carcinoids compared to the metastatic tumors. Pancreatic neuroendocrine tumors were negative for Gal-1, Gal-3, and Gal-4. Gastric carcinoids also expressed Gal-4, but very few pulmonary or rectal carcinoids were positive for Gal-4. Lower levels of Gal-1 and Gal-3 expression were present in ileal carcinoids compared to primary pulmonary and rectal tumors (Rumilla et al. 2006).

Integrins and Galectins in Adamantinomatous Craniopharyngiomas: Lefranc et al. (2005) suggested that at least part of the adhesiveness of craniopharyngiomas to the surrounding tissue, such as optical chiasms and pituitary stalks, could be explained by the interactions between α2 β1 integrin expressed by craniopharyngiomas and collagens on the one hand, and vitronectin expressed by the surrounding tissue on the other. In addition, the levels of galectin-4 contribute significant information toward the delay in recurrence independently of surgical status (Lefranc et al. 2005).

Sinonasal adenocarcinomas are uncommon tumors which develop in ethmoid sinus after exposure to wood dust. Within microarray study in sinonasal adenocarcinoma Tripodi et al. (2009) identified LGALS4 and CLU, that were significantly differentially expressed in tumors compared to normal tissue. A further evaluation is necessary to evaluate the possibility of using them as diagnostic markers (Tripodi et al. 2009).

Gal-4 in Colorectal and Other Cancers: The Gal-4 mRNA is under expressed in colorectal cancer and restricted to small intestine, colon and rectum. In patients of colon, decreased galectin-4 mRNA expression may be an early event in colon carcinogenesis (Rechreche et al. 1997). Among five cell lines derived from colon carcinoma, only two (HT29 and LS174T) expressed galectin-4 mRNA (Rechreche et al. 1997). Huflejt et al. (1997) found two prominent proteins in human colon adenocarcinoma T84 cell line. Cloning of 36-kDa protein was human homolog of galectin-4, which was localized in the epithelial cells of rat and porcine alimentary tract. The other, a 29-kDa protein, was galectin-3, found in a number of different cell types including human intestinal epithelium. Despite the marked similarities in the CRDs of these two galectins, their cellular distribution patterns were strikingly different. In subconfluent T84 cells, each galectin is distributed to specific domains of lamellipodia, with galectin-4 concentrated in the leading edge and galectin-3 more proximally. The localization of galectin-4 suggests a role in cell adhesion which is also supported by the ability of immobilized recombinant galectin-4 to stimulate adhesion of T84 cells. Furthermore, galectin-4 is up-regulated in human hepatocellular carcinoma (Kondoh et al. 1999) and metastatic gastric cancer cells (Hippo et al. 2001). Galectin 4 (*LGALS4*) is highly and specifically expressed in Mucinous epithelial ovarian cancers (MOC), but expressed at lower levels in benign mucinous cysts and borderline (atypical proliferative) tumors, supporting a malignant progression model of MOC. *LGALS4* may have application as an early and differential diagnostic marker of MOC (Heinzelmann-Schwarz et al. 2006).

11.2 Galectin-6

In the course of studying mouse colon mRNA for galectin-4, Gitt et al. (1998a, b) detected a related mRNA of Galectin-6, which has two CRDs in a single peptide chain. The *Lgals4* gene encoding galectin-4 is distinct from *Lgals6*. The coding sequence of galectin-6 is specified by eight exons. The upstream region contains two putative promoters. Both *Lgals6* and the closely related *Lgals4* are clustered together about 3.2 centimorgans proximal to *apoE* gene on mouse chromosome 7. The syntenic human region is 19q13.1–13.3. The galectin-6 lacks a 24-amino acid stretch in the link region between the two CRDs that is present in galectin-4. Otherwise, these two galectins have 83% amino acid identity. Expression of both galectin-4 and galectin-6 is confined to the epithelial cells of the embryonic and adult GI tract. Expression of mouse galectin-4 and galectin-6 indicates that both are expressed in the small intestine, colon, liver, kidney, spleen and heart and P19X1 cells while only galectin-4 is expressed in BW-5147 and 3T3 cell lines. In situ hybridization confirmed the presence of galectin-4/-6 transcripts in the liver and small intestine.

11.3 Galectin-8

11.3.1 Galectin-8 Characteristics

Galectin-8 and galectin-9, which each consists of two carbohydrate recognition domains (CRDs) joined by a linker peptide, belong to the tandem-repeat-type subclass of the galectin family. Alternative splicing leads to the formation of distinct splice variants (isoforms) of galectin-8 and galectin-9 with tandem-repeat-type structures. Galectin-8 consists of several isoforms, each made of two domains of approximately 140 amino-acids, both having a CRD. These domains are joined by a 'link peptide' of variable length. The isoforms share identical CRDs and differ only in linker region. Galectin-8 was discovered in prostate cancer cells and has been studied extensively in the last few years (Bidon et al. 2008). It is widely expressed in tumoral tissues and seems to be involved in integrin-like cell interactions. The human galectin-8 gene covers 33 kbp of genomic DNA. It is localized on chromosome 1 (1q42.11) and contains 11 exons. The gene produces by alternative splicing 14 different transcripts, altogether encoding 6 proteins. Galectin-8, like other galectins, is a secreted protein. Upon secretion galectin-8 acts as a physiological modulator of cell adhesion. Studies showed that the *LGALS8* gene encodes for almost seven mRNAs by alternative splicing pathways and various polyadenylation sites. These mRNAs could encode for six isoforms of galectin-8, of which three belong to the tandem-repeat galectin group (with two carbohydrate binding domains) and the three others to the prototype group (one carbohydrate binding domain). All these isoforms seem to be differentially expressed in various tumoral cells. This untypical galectin-8 subfamily seems to have a complex expression regulation that could be involved in cancer phenomena (Bidon et al. 2008; Bidon-Wagner and Pennec 2004).

A 35 kDa galectin was cloned from rat liver. Deduced amino acid sequence of galectin-8 contains two domains with conserved motifs that are implicated in carbohydrate binding of galectins. This protein was named galectin-8. In vitro, translation products of galectin-8 are biologically active and possess sugar binding and hemagglutination activity. The expected size (34 kDa) that binds to lactosyl-Sepharose is present in rat. Overall, galectin-8 is structurally related (34% identity) to galectin-4. Nonetheless, galectin-4 is confined to intestine and stomach where as galectin-8 is expressed in liver, kidney, cardiac muscle, lung, and brain. Unlike galectin-4, but similar to galectins-1 and -2, galectin-8 contains four Cys residues. The link peptide of galectin-8 is unique and bears no similarity to any known protein. The N-terminal carbohydrate-binding region of galectin-8 contains a unique WG-E-I motif instead of consensus WG-E-R/K motif implicated as playing an essential role in sugar-binding of all galectins (Hadari et al. 1995).

Linker Peptide in Functioning of Galectin-8 and Galectin-9 Galectin-8 has two covalently linked carbohydrate recognition domains (CRDs). The two CRDs are joined by a linker peptide. Ligation of integrins by galectin-8 induces a distinct cytoskeletal organization, associated with activation of the ERK and phosphatidylinositol 3-kinase signaling cascades. These properties of galectin-8 were mediated by the concerted action of its two CRDs and involved both protein-sugar and protein-protein interactions. Accordingly, the isolated N- or C-CRD domains of galectin-8 or galectin-8 mutated at selected residues implicated in sugar binding showed reduced sugar binding, resulting in severe impairment in the capacity of these mutants to promote the adhesive, spreading, and signaling functions of galectin-8. Deletion of the linker region similarly impaired the biological effects of galectin-8. These results provided evidence that cooperative interactions between the two CRDs and the "hinge" domain are required for proper functioning of galectin-8 (Levy et al. 2006; Sato et al. 2002).

Gal-8 and Gal-9 with Longest Linker Peptide are Susceptible to Thrombin Cleavage: The isoforms of galectin-8 and galectin-9 with the longest linker peptide, i.e. galectin-8L and galectin-9L (G8L and G9L), are highly susceptible to thrombin cleavage, whereas the predominant isoforms, galectin-8M and galectin-9M (G8M and G9M), and other members of human galectin family were resistant to thrombin. Amino acid sequence analysis of proteolytic fragments and site-directed mutagenesis showed that the thrombin cleavage sites (-IAPRT- and -PRPRG- for G8L and G9L, respectively) resided within the linker peptides. Although intact G8L stimulated neutrophil adhesion to substrate more efficiently than G8M, the activity of G8L but not of G8M decreased on thrombin digestion. Similarly, thrombin treatment almost completely abolished eosinophil chemoattractant (ECA) activity of G9L. Studies suggest that G8L and G9L play unique roles in relation to coagulation and inflammation (Nishi et al. 2006). Mutant forms lacking the entire linker peptide were highly stable against proteolysis and retained their biological activities. These mutant proteins need to be evaluated for the therapeutic potential (Nishi et al. 2005).

N-Glycans as Major Ligands: Binding of the galectins to the different CHO glycosylation mutants revealed that complex N-glycans are the major ligands for each galectin except the N-terminal CRD of galectins-8, and involve some fine differences in glycan recognition. Interestingly, increased binding of galectin-1 at 4°C correlated with increased propidium iodide (PI) uptake, whereas galectin-3 or -8 binding did not induce permeability to PI (Patnaik et al. 2006). Among galectins (N- and C-terminal domains) Galectin-4C and 8N were found to prefer the d-talopyranose configuration to the natural ligand d-galactopyranose configuration. Derivatization at talose O2 and/or O3 provided selective inhibitors for these two galectins (Oberg et al. 2008).

Gal-8 binding with β1 Integrins on Jurkat T Cells: Carcamo et al. (2006) studied the effects of immobilized Gal-8 on Jurkat T cells. Immobilized Gal-8 bound integrins α1β1, α3β1 and α5β1 but not α2β1 and α4β1 and adhered to these cells with similar kinetics to immobilized fibronectin (FN). The α5β1 was the main mediator of cell adhesion to galectin-8. Gal-8, but not FN, induced extensive cell spreading frequently leading to a polarized phenotype characterized by an asymmetric lamellipodial protrusion. Gal-8-induced Rac-1 activation and binding to α1 and α5 integrins are not known in any other cellular system. Strikingly, Gal-8 was also a strong stimulus on Jurkat cells in triggering ERK1/2 activation. Human galectin-8 induces reversible adhesion of peripheral blood neutrophils but not eosinophils, to a plastic surface in a lactose-sensitive manner and modulates neutrophil function related to transendothelial migration and microbial killing (Nishi et al. 2003). Galectin-8 induces neutrophil-adhesion through binding to integrin αM. Galectin-8, as well as tandem-repeat galectin-9, and several multivalent plant lectins, induced Jurkat T-cell adhesion to a culture plate, whereas single-CRD galectins-1 and -3 did not. Galectin-8 also induced the adhesion of peripheral blood leucocytes to human umbilical vein endothelial cells. It appeared that the di- or multivalent structure of galectin-8 is essential for the induction of cell adhesion and that this ability exhibits broad specificity for leucocytes (Yamamoto et al. 2008).

11.3.2 Functions of Galectin-8

Modulator of Cell Adhesion: When immobilized, it functions as a matrix protein equipotent to fibronectin in promoting cell adhesion by ligation and clustering of a selective subset of cell surface integrin receptors. Complex formation between galectin-8 and integrins involves sugar-protein interactions and triggers integrin-mediated signaling cascades such as Tyr phosphorylation of FAK and paxillin. In contrast, when present in excess as a soluble ligand, galectin-8 (like fibronectin) forms a complex with integrins that negatively regulates cell adhesion. Such a mechanism allows local signals emitted by secreted galectin-8 to specify territories available for cell adhesion and migration. Due to its dual effects on the adhesive properties of cells and its association with fibronectin, galectin-8 might be considered as a novel type of a matricellular protein (Zick et al. 2004).

Galectin-8 as a Modulator of Cell Growth: Galectin-8 functions as an extracellular matrix protein that forms high affinity interactions with integrins. Soluble galectin-8 inhibits cell cycle progression and induces growth arrest. These effects were not related to interference with cell adhesion but were attributed to an increase in the cellular content of the cyclin-dependent kinase inhibitor p21. The increase in p21 levels was preceded by an increase in JNK and protein kinase B (PKB) activities. This process involves activation of JNK, which enhances the synthesis of p21, along with the activation of PKB, which inhibits p21 degradation. These effects, due to protein-sugar interactions, were induced when galectin-8 was present as a soluble ligand or when it was overexpressed in cells (Arbel-Goren et al. 2005). Results indicated galectin-8 as a modulator of cell growth through up-regulation of p21.

11.3.3 Clinical Relevance of Gal-8

Galectin-8 Auto-Antibodies in SLE Patients: Patients with systemic lupus erythematosus (SLE) produce Gal-8-autoantibodies that impede both its binding to integrins and cell adhesion. These function-blocking autoantibodies reported for a galectin implied that Gal-8 constitutes an extracellular stimulus for T cells, able to bind specific β1 integrins and trigger signaling pathways conducive to cell spreading. Gal-8 could modulate a wide range of T cell-driven immune processes that eventually become altered in autoimmune disorders. During a search for antigens recognized by antibodies produced by a patient with SLE, Pardo et al. (2006) found reactivity against Gal-8, for which autoantibodies were not previously described. Higher frequency of autoantibodies against galectin-8 in patients with SLE suggests a pathogenic role. Further studies are needed to determine their clinical relevance (Pardo et al. 2006).

Galectin-8 in Primary Open Angle Glaucoma: Primary open angle glaucoma (POAG), a major blindness-causing disease, is characterized by elevated intraocular pressure due to an insufficient outflow of aqueous humor. The trabecular meshwork (TM) lining the aqueous outflow pathway modulates the aqueous outflow facility. TM cell adhesion, cell-matrix interactions, and factors that influence Rho signaling in TM cells are thought to play a pivotal role in the regulation of aqueous outflow. Galectin-8 modulates the adhesion and cytoskeletal arrangement of TM cells and that it does so through binding to β1 integrins and inducing Rho signaling. Thus studies that Gal8 modulates TM cell adhesion and spreading, at least in part, by interacting with α2-3-sialylated glycans on β1 integrins (Diskin et al. 2009).

11.3.4 Isoforms of Galectin-8 in Cancer

Studies showed that galectin-8 is widely expressed in tumor tissues as well as in normal tissues. The level of galectin-8 expression may correlate with the malignancy of human colon cancers and the degree of differentiation of lung squamous cell carcinomas and neuro-endocrine tumors. The differences in galectin-8 expression levels between normal and tumor tissues have been used as a guide for the selection of strategies for the prevention and treatment of lung squamous cell carcinoma. Experiments suggest the potential of galectin-8 in understanding of, and possibly prevent, the process of neoplastic transformation (Bidon-Wagner and Pennec 2004). Galectin-8 levels of expression positively correlate with certain human neoplasms, prostate cancer being the best example studied thus far. The overexpressed lectin might give these neoplasms some growth and metastasis related advantages due to its ability to modulate cell adhesion and cellular growth. Hence, galectin-8 may modulate cell-matrix interactions and regulate cellular functions in a variety of physiological and pathological conditions (Zick et al. 2004).

11.4 Galectin-9

11.4.1 Characteristics

Galectin-9, a tandem-repeat type galectin, is a 40-kDa protein consisting of 353 amino acids. The sequence identity between the N- and C-terminal CRDs is 35%. The C-terminal CRD (CCRD) is highly homologous to rat galectin-5 CRD with an amino acid sequence identity of 70%, but the N-terminal CRD (NCRD) is only moderately homologous with known galectins. Among these, the galectin-9 NCRD shows the highest sequence identity (40%) with the galectin-3 CRD. Galectin-9 was first cloned from tumor cells from Hodgkin disease, a condition characterized by blood and tissue eosinophilia. Moreover, the recombinant galectin-9 causes thymocyte apoptosis in mouse cells, suggesting a possible role for galectin-9 in negative selection during T-cell development (Wada and Kanwar 1997; Wada et al. 1997). Interestingly, galectin-9 was shown to be related to a novel eosinophil chemoattractant produced by T lymphocytes, previously designated "ecalectin" (Matsumoto et al. 1998). Mutation studies showed that both the NCRD and CCRD of galectin-9 were required for the eosinophil chemoattraction activity (Matsushita et al. 2000). Additionally, galectin-9 interacts with Tim-3, which is specifically expressed on the surface of T helper type 1 (Th1) cell, through recognition of Tim-3 carbohydrates, and the Tim3-galectin9 pathway induces cell death in Th1cells. This suggests that galectin-9 plays a role in down-regulating the effector Th1 responses (Zhu et al. 2005). Galectin-9 interacts with carbohydrate(s) covalently attached to the surface of Tim-3, but the molecular and structural basis for this recognition is unknown. In vitro analyses showed that galectin-9 has a high affinity for a variety of oligosaccharides containing β-galactosides (Hirabayashi et al. 2002), and the NCRD and CCRD of galectin-9 have different oligosaccharide-binding affinities. The biological activities of galectin-9 may be related to the ligand binding specificity of each CRD and the multivalent binding conferred by two CRDs. To date, the structures of many CRDs from fungi to human have been solved, but there is no structural information about the structure of tandem-repeat type galectin CRDs. Such information should greatly clarify the mechanism of carbohydrate recognition by the CRDs and the multivalent properties that lead to multiple functions for a single protein.

A 36-kDa lectin, galectin-9 from mouse embryonic kidney, has the characteristic conserved sequence motif of galectins. Galectin-9 from liver and thymus, as well as recombinant galectin-9 exhibited specific binding for lactosyl group. It had two distinct N- and C-terminal CRDs connected by a link peptide, with no homology to any other protein. Galectin-9 had an alternate splicing isoform, exclusively expressed in small intestine with a 31-amino acid insertion between the N-terminal domain and link peptide. Sequence analysis revealed that C-terminal CRD of mouse galectin-9 showed an extensive similarity to that of monomeric rat galectin-5, which was demonstrated in mouse. Sequence comparison of rat galectin-5 and rat galectin-9 cDNA did not reveal identical nucleotide sequences in the overlapping C-terminal carbohydrate-binding domain, indicating that galectin-9 is not an alternative splicing isoform of galectin-5. However, galectin-9 had a sequence identical with that of its intestinal isoform in the overlapping regions in both species. Genomic analyses indicated the presence of a novel gene encoding galectin-9 in both mice and rats. In contrast to galectin-5, which is mainly expressed in erythrocytes, galectin-9 was found to be widely distributed, i.e. in liver, small intestine, thymus > kidney, spleen, lung, cardiac and skeletal muscle > reticulocyte, brain (Wada and Kanwar 1997).

11.4.2 Stimulation of Galectin-9 Expression by IFN-γ

Galectin-9 was detected in membrane and cytosolic fractions of human umbilical vein endothelial cells (HUVECs) after stimulation with IFN-γ, which also enhanced the adhesion of human eosinophilic leukemia-1 cells to endothelial monolayers. The polyinosinic-polycytidylic acid (poly IC), a double-stranded RNA (dsRNA), induces the expression of galectin-9 in HUVECs and involves TLR3, PI3K, and IFN regulatory factor 3 (IRF3) (Imaizumi et al. 2007). The expression of galectin-9 by endothelium from patients with inflammatory diseases indicates that IFN-γ-induced production of galectin-9 plays an important role in immune responses by regulating interactions between vascular wall and eosinophils (Imaizumi et al. 2002). IFN-γ up-regulated the expression of galectin-9 in primary human dermal fibroblasts and surface expression of galectin-9 on human lung fibroblast cell line, HFL-1. It suggests that IFN-γ-induced galectin-9 expression in fibroblasts mediates eosinophil adhesion to the cells, suggesting a crucial role of galectin-9 in IFN-γ-stimulated fibroblasts at the inflammatory sites (Asakura et al. 2002).

11.4.3 Crystal Structure of Galectin-9

Nagae et al. (2006) reported the crystal structures of mouse galectin-9 N-terminal CRD (NCRD) in free and complexed form with four ligands. X-ray structure of galectin-9 NCRD showed it to be composed of six-stranded (S1–S6) and five stranded (F1–F5) β-sheets, which together form a β-sandwich arrangement (Fig. 11.2a). Based on this structure, the galectin-9 NCRD is not buried in membranes. The carbohydrate binding site is formed by the S4, S5, and S6 β-strands, and the carbohydrate recognition mechanism is similar to those of other galectins. All structures formed same dimmer (Fig. 11.2b), which was quite different from the canonical two-fold symmetric dimer seen for galectin-1 and -2. Thus, the β-galactoside recognition mechanism in the galectin-9 NCRD is highly conserved among other galectins. In the apo form structure, water molecules mimiced the ligand hydrogen-bond network. The galectin-9 NCRD could bind both N-acetyllactosamine (Galβ1-4GlcNAc) and T-antigen (Galβ1-3GalNAc) with the proper location of Arg-64. Moreover, the structure of the N-acetyllactosamine dimer (Galβ1-β1-3Galβ1-4GlcNAc) complex showed a unique binding mode of galectin-9. Surface plasmon resonance assay showed that the galectin-9 NCRD forms a homophilic dimer not only in the crystal but also in solution (Nagae et al. 2006).

Dimer formation—In asymmetric unit of apo form1, two molecules, referred as chains A and B, are related by a two-fold non-crystallographic axis perpendicular to the β-sheets. They form a continuous 12-stranded antiparallel β-sheet through interactions between the β strands of chain-A S6 and chain-B S6 (Fig. 11.2b). On the dimer interface, the main chain oxygen and nitrogen atoms of Arg-86 form hydrogen bonds with the corresponding main-chain atoms of Arg-86 of the other monomer (Fig. 11.2c). The N and C termini of each monomer are positioned at the opposite side of the dimer interface (Fig. 11.2a) (Nagae et al. 2006).

Nagae et al. (2008, 2009) also reported the crystal structures of human galectin-9 NCRD in presence of lactose and Forssman pentasaccharide. Human galectin-9 NCRD exists as a monomer in crystals, despite a high sequence identity to the mouse homologue. Comparative frontal affinity chromatography analysis of the mouse and human galectin-9 NCRDs revealed different carbohydrate binding specificities, with disparate affinities for complex glycoconjugates. Human galectin-9 NCRD exhibited a high affinity for Forssman pentasaccharide; the association constant for which was 100-fold more than for mouse galectin-9 NCRD. The combination of structural data with mutational studies demonstrated that non-conserved amino acid residues on the concave surface were important for determination of target specificities. The human galectin-9 NCRD exhibited greater inhibition of cell proliferation than the mouse NCRD. The biochemical and structural differences have been reported between highly homologous proteins from different species (Nagae et al. 2008).

11.4.4 Galectin-9 Recognizes *L. major* Poly-β-galactosyl Epitopes

The glycan epitopes on Leishmania parasites are involved in the pathogenesis of Leishmaniasis. Established species-specific glycan structures is the poly-β-galactosyl epitope (Galβ1-3)n found on *L. major*, which can develop cutaneous infection with strong inflammatory responses. The host galectin-3 can distinguish *L. major* from other species through its binding to poly-β-galactosyl epitope, proposing a role for galectin-3 as an immunomodulator that could influence the *L. major*-specific immune responses in Leishmaniasis. In addition, Galectin-9 can also recognize *L. major* by binding to *L. major*-specific polygalactosyl epitope. The galectin-9 affinity for polygalactose was enhanced in proportion to the number of Galβ1-3 units present. Although both Galectin-3 and Galectin-9 have comparable affinities toward the poly-galactosyl epitopes, Pelletier et al. (2003) suggested that only galectin-9 can promote the interaction between *L. major* and macrophages in the development of leishmaniasis in the host. Compared with galectin-9 C-terminal CRD, the N-terminal CRD shows striking affinities for complex glycoconjugates such as Forssman pentasaccharide and polymerized *N*-acetyllactosamine (Hirabayashi et al. 2002; Walser et al. 2004). The specific interactions of galectin-9 NCRD with the carbohydrates is thought to be the clue for understanding the physiological mechanism of galectin-9. Nagae et al. (2006) reported the crystal structures of mouse galectin-9 N-terminal CRD in absence and in presence of carbohydrate ligands. These structures suggest a potential mechanism for the specificity of carbohydrate recognition and binding.

11.4.5 Functions of Galectin-9

Galectin-9 as a Ligand for T-Cell Immunoglobulin Mucin-3: T cell immunoglobulin and mucin domain

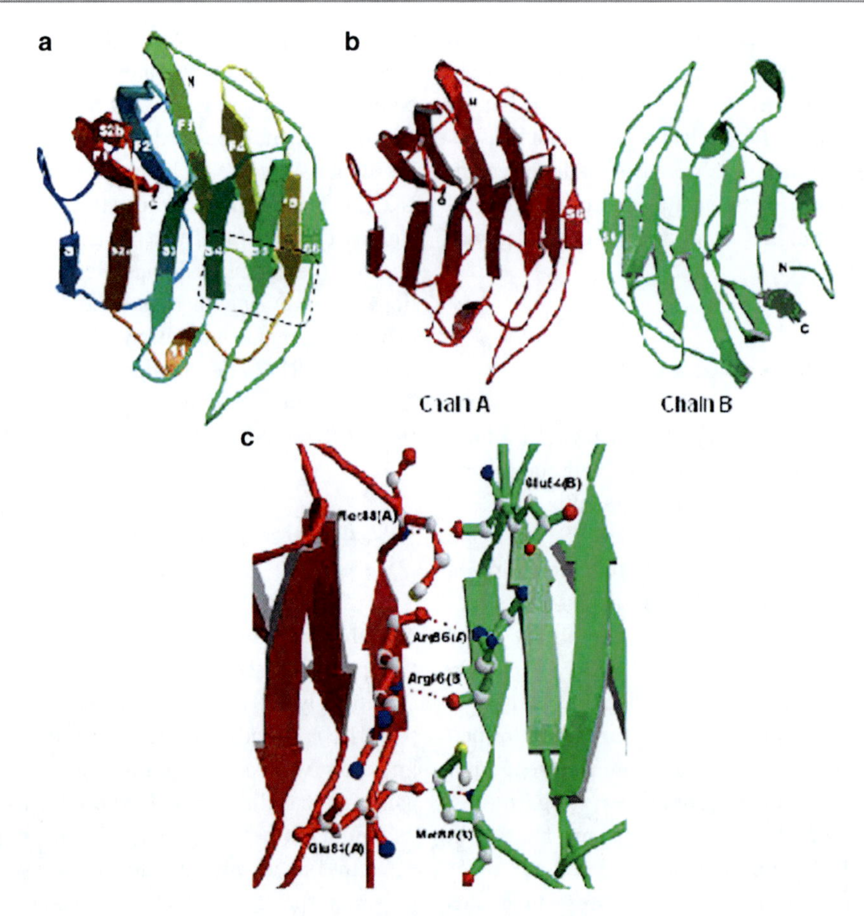

Fig. 11.2 Crystal structure of the mouse galectin-9 *N*-terminal CRD (NCRD). (**a**) Ribbon diagram of monomeric structure of apo form 1 of galectin-9 NCRD is shown. The five-stranded (*F1–F5*) and six-stranded (S1–S6) β-sheets and one short helix (H1) are indicated by the *number code*. The carbohydrate binding site is shown by a *dotted box*. (**b**) Dimeric structure of galectin-9 NCRD is shown. Two monomers in an asymmetric unit in apo form1 crystal are shown for chain-A (in *red*) and chain-B (*green*). (**c**) Close up view of the dimer interface. The amino acid residues involved in the dimer formation are shown in *ball-and-stick* model. The carbon, oxygen, nitrogen, and sulfur atoms are shown in *white*, *red*, *blue*, and *yellow spheres*, respectively. Hydrogen bonds are depicted by *dotted lines* (Adapted by permission from Nagae et al. 2006 © American Society for Biochemistry and Molecular Biology)

(TIM)-3 is a member of the TIM family of proteins (T-cell immunoglobulin mucin) involved in the regulation of $CD4^+$ T-cells. TIM-3 is a Th1-specific type 1 membrane protein that regulates Th1 proliferation and the development of tolerance. TIM-3 is a molecule expressed on terminally differentiated murine Th1 cells but not on Th2 cells. TIM-3 seems to regulate Th1 responses and the induction of peripheral tolerance. Zhu et al. (2005) showed that galectin-9 is the TIM-3 ligand. Galectin-9-induced intracellular calcium flux, aggregation and death of Th1 cells were TIM-3-dependent in vitro, and administration of galectin-9 in vivo resulted in selective loss of IFN-γ-producing cells and suppression of Th1 autoimmunity. These data suggested that the TIM-3-galectin-9 pathway may have evolved to ensure effective termination of effector Th1 cells (Zhu et al. 2005). Stimulation of TIM-3 by its ligand galectin-9 results in increased phosphorylation of Y265, suggesting that this tyrosine residue plays an important role in downstream signalling events regulating T-cell fate. Given the role of TIM proteins in autoimmunity and cancer, the conserved SH2 binding domain surrounding Y265 could represent a possible target site for pharmacological intervention (van de Weyer et al. 2006). The immunoregulatory TIM-3 in TGF-β-stimulated mast cells and melanoma cells may support the survival of this tumor type (Wiener et al. 2007).

Identification of galectin-9 as a ligand for TIM-3 has now firmly established the TIM-3-galectin-9 pathway as an important regulator of Th1 immunity and tolerance induction. TIM-3 is similarly expressed on human Th1 cells and not on Th2 cells, which suggests that TIM-3 might also contribute to Th1 regulation in humans. Binding of galectin-9 to the extracellular domain of TIM-3 resulted in apoptosis of Th1 cells, though the intracellular pathways involved in the regulatory function of TIM-3 remained unknown. In addition, genetic data associate the TIM locus and specific TIM-3 polymorphisms with various immune-mediated diseases. Most importantly, recent data suggest a novel paradigm in which dysregulation of the TIM-3-galectin-9 pathway could underlie chronic autoimmune disease states, such as multiple sclerosis (Anderson and Anderson 2006). Unlike TIM-1, which is expressed in renal epithelia and cancer, TIM-3 has been described in neuronal besides T-cells. TIM-3 has been found in malignant and non-malignant epithelial tissues. Lensch et al. (2006) indicated distinct functions for galectin-5 and -9 especially in erythropoiesis and unknown mechanisms to compensate its absence from galectin network in other mammals.

Maturation of Human Monocyte-Derived Dendritic Cells: Dai et al. (2005) assessed the role of galectin-9 in DC maturation. Culture of immature DCs with exogenous Gal-9 markedly increased the surface expression of CD40, CD54, CD80, CD83, CD86, and HLA-DR in a dose-dependent manner, although Gal-9 had no or little effect on differentiation of human monocytes into immature DCs. Gal-9-treated DCs secreted IL-12 but not IL-10, and elicited the production of Th1 cytokines (IFN-γ and IL-2) but not that of the Th2 cytokines (IL-4 and IL-5) by allogeneic CD4+ T cells. Gal-9 induced phosphorylation of the MAPK p38 and ERK1/2 in DCs, and an inhibitor of p38 signaling, but not inhibitors of signaling by either ERK1/2 or PI3K, blocked Gal-9-induced up-regulation of costimulatory molecule expression and IL-12 production. These findings suggest that Gal-9 plays a role not only in innate immunity but also in acquired immunity by inducing DC maturation and promoting Th1 immune responses (Dai et al. 2005).

Endometrial Epithelial Marker: Galectin-9 mRNA is expressed in the human endometrium, specifically in the human endometrial epithelial cells but not in stromal or immune cells. It is expressed at very low concentrations during the proliferative phase and the early-secretory phase and shows a sharp and significant increase in the mid- and late-secretory phases, the window of implantation, as well as in the decidua. Galectin-9 protein is also exclusively increased in human endometrial epithelial cells during the mid- and late-secretory phases and in the decidua, however, not in endometrial stromal cells or decidualized cells in vivo or in vitro. A regulation in vitro by estradiol, progesterone, epidermal growth factor, and IFN-γ could not be detected. Based on the functional studies of other galectins, galectin-9 was suggested as a novel endometrial marker for the mid- and late-secretory and decidual phases (Popovici et al. 2005).

Gal-9 and Gal-1 Determine Distinct T Cell Death Pathways: Galectin-9 and galectin-1 both kill thymocytes, peripheral T cells, and T cell lines; however, galectin-9 and galectin-1 require different glycan ligands and glycoprotein receptors to trigger T cell death. The two galectins also utilize different intracellular death pathways, as galectin-9, but not galectin-1, T cell death was blocked by intracellular Bcl-2, whereas galectin-1, but not galectin-9, T cell death was blocked by intracellular galectin-3. To define structural features responsible for distinct activities of the tandem repeat galectin-9 and dimeric galectin-1, Shuguang et al. (2008) created a series of bivalent constructs with galectin- 9 and galectin-1 CRDs connected by different peptide linkers and suggested that the N-terminal CRD and linker peptide contributed to the potency of these constructs. However, the C-terminal CRD was the primary determinant of receptor recognition, death pathway signaling, and target cell susceptibility. Thus, CRD specificity, presentation, and valency make distinct contributions to the specific effects of different galectins in initiating T cell death (Shuguang et al. 2008).

11.4.6 Galectin-9 in Clinical Disorders

Galectin-9 is known as a versatile immunomodulator that is involved in various aspects of immune regulations. It induces various biological reactions such as chemotaxis of eosinophils and apoptosis of T cells. A 36-kDa protein with characteristics of S-type lectins was detected in Hodgkin's disease. While the N-terminal lectin domain showed merely moderate homologies with known galectins, the C-terminal lectin domain was highly homologous to rat galectin-5 with an amino acid sequence identity of 70%. In accordance, the galactoside binding protein was designated as galectin-9 (Türeci et al. 1997). Drug-induced liver injury, often caused by an allergic mechanism, is occasionally accompanied by eosinophilic infiltration. High expression of Gal-9 is suggested to be a specific finding of drug-induced liver injury. However, tissue eosinophilia in drug-induced liver injury cannot be explained by the augmentation of Gal-9 expression (Takahashi et al. 2006).

The mechanisms of leucocyte traffic across the vascular endothelium induced by dsRNA involved the expression of galectin-9 as one of key molecules in the regulation of the interaction between vascular wall and white blood cells. In HUVEC in culture upregulation of galectin-9 by poly-IC in the vascular endothelium may explain part of the mechanism for leucocyte traffic through the vascular wall (Ishikawa et al. 2004).

Involvement of Gal-9 in Dermal Eosinophilia of Th1-Polarized Skin Inflammation: Skin eosinophilia is a common feature of allergic skin diseases, but it is not known how epidermal and dermal eosinophil infiltration is controlled. It appears that among eosinophil-specific chemoattractants in dermal fibroblasts and epidermal keratinocytes, eotaxin-3 contributes to dermal and epidermal eosinophil infiltration in Th2-polarized skin inflammation in which IL-4 is produced. In contrast, IFN-γ-dominated inflammation appears to mediate eosinophil extravasation into the dermis and eosinophil adhesion to dermal fibroblasts via galectin-9 in association with decreased chemoattractant activity of epidermal galectin-9. A mechanism of dermal eosinophilia in IFN-γ-mediated skin inflammation reflects concerted chemoattractant production involving dermal and/or epidermal eosinophilia during changes in the local cytokine profile (Igawa et al. 2006). Yamamoto et al. (2007) investigated the role of Gal-9 in asthma model in guinea pigs. In guinea pig, Gal-9 exists in three isoforms that differ only in the length of their linker peptides. Guinea pig Gal-9 was chemoattractant for eosinophils and promoted induction of apoptosis in sensitized but not non-sensitized T lymphocytes. Results provide evidence that Gal-9 is not involved in airway hypersensitivity, but is partly involved in prolonged eosinophil accumulation in the lung. Galectin-9 is expressed in human melanoma cells. Among these, MM-BP proliferated with colony formation, but MM-RU failed. High galectin-9 expression was inversely correlated with the progression of this disease, suggesting that high galectin-9 expression in primary melanoma links to better prognosis (Kageshita et al. 2002).

Galectin-9 as Eosinophil Chemoattractant in Nasal Polyps: Ecalectin (galectin-9) is an eosinophil chemoattractant from T lymphocytes. Ecalectin, is produced in the mucosa of nasal polyps and, seems to play an important role in the accumulation and activation of eosinophils in nasal polyps, regardless of the presence or absence of atopic predisposition. The role of ecalectin in nasal polyp tissues associated with various nasal and paranasal diseases and in the pathogenesis of eosinophilia in patients having chronic sinusitis with nasal polyposis has been clarified by the presence of ecalectin-positive cells in the subepithelial layer, where many EG2-positive cells were present. This was substantiated by the presence of many ecalectin mRNA-positive cells in nasal polyps with an accumulation of EG2-positive cells (Iino et al. 2006).

Gal-9 is a High Affinity IgE-Binding Lectin with Anti-allergic Effect: *Gal* attenuated asthmatic reaction in guinea pigs and suppressed passive-cutaneous anaphylaxis in mice and stabilizing effect on mast cells. In vitro studies of mast cell demonstrated that Gal-9 suppressed degranulation from the cells stimulated by IgE plus antigen and that the inhibitory effect was completely abrogated in presence of lactose thus indicating that lectin activity of Gal-9 is critical. It was found that Gal-9 strongly and specifically bound IgE, which is a heavily glycosylated Ig, and that the interaction prevented IgE-antigen complex formation, clarifying the mode of action of the anti-degranulation effect. The fact that immunological stimuli of MC/9 cells augmented Gal-9 secretion from cells implies that Gal-9 is an autocrine regulator of mast cell function to suppress excessive degranulation. Findings suggest a beneficial role of Gal-9 for the treatment of allergic disorders including asthma (Niki et al. 2009).

Periodontal Ligament Cells (PDL): Considerable evidence suggests that periodontal disease not only is caused by bacterial infection but also is associated with host susceptibility. Lipopolysaccharide (LPS) extracted from *Porphylomonas gingivalis (P. gingivalis* LPS) enhances the expression level of galectin-9 mRNA and protein in a time-dependent manner together with interleukin-8. In addition, strong immunoreaction for galectin-9 was detected in the PDL consisting of the periodontal pocket of a patient with severe periodontal disease. Furthermore, significant up-regulation of galectin-9 mRNA expression was detected in the mRNA from PDLs of patients with periodontal disease when compared with healthy donors. These results suggest that galectin-9 expression is associated with inflammatory reactions in the PDL (Kasamatsu et al. 2005).

The mechanisms of leucocyte traffic across the vascular endothelium induced by dsRNA involved the expression of galectin-9 as one of key molecules in the regulation of the interaction between vascular wall and white blood cells. In HUVEC in culture upregulation of galectin-9 by poly IC in the vascular endothelium may explain part of the mechanism for leucocyte traffic through the vascular wall (Ishikawa et al. 2004).

11.4.7 Galectin-9 in Cancer

In Management of T-Cell Leukemia: Galectin-9 is known as an apoptosis inducer of activated T lymphocytes. Caspase-dependent and-independent death pathways exist in Jurkat cells, and the main pathway might vary with the

T-cell type (Lu et al. 2007). ATL is a fatal malignancy of T lymphocytes caused by human T-cell leukemia virus type I (HTLV-I) and remains incurable. The protease-resistant galectin-9 by modification of its linker peptide, hG9NC (null), prevents cell growth of HTLV-I-infected T-cell lines and primary ATL cells. The suppression of cell growth was inhibited by lactose, but not by sucrose, indicating that β-galactoside binding is essential for hG9NC(null)-induced cell growth suppression. The hG9NC(null) induced cell cycle arrest by reducing the expression of cyclin D1, cyclin D2, cyclin B1, Cdk1, Cdk4, Cdk6, Cdc25C and c-Myc, and apoptosis by reducing the expression of XIAP, c-IAP2 and survivin. Hence, hG9NC(null) can be a suitable agent for the management of ATL(Okudaira et al. 2007).

Antimetastatic Potential in Breast Cancer: Galectin-9 induces aggregation of certain cell types. Irie et al. (2005), who assessed the contribution of galectin-9 to the aggregation of breast cancer cells as well as the relation between galectin-9 expression in tumor tissue and distant metastasis in patients with breast cancer, showed that MCF-7 subclones with a high level of galectin-9 expression formed tight clusters during proliferation in vitro, whereas a subclone (K10) with the lowest level of galectin-9 expression did not. However, K10 cells stably transfected with a galectin-9 expression vector aggregated in culture and in nude mice. Galectin-9 is a possible prognostic factor with antimetastatic potential in breast cancer (Irie et al. 2005). Yamauchi et al. (2006) suggested a relationship between galectin-9 expression in tumor tissue and distant metastasis exists in breast cancer. Tumors in 42 of the 84 patients were galectin-9-positive, and tumors in 19 of the 21 patients with distant metastasis were galectin-9-negative (Yamauchi et al. 2006).

Oral Squamous Cell Carcinomas: Oral squamous cell carcinomas (OSCCs) are the most common neoplasms of the head and neck. Galectin-9 is correlated with cellular adhesion and aggregation in melanoma cells. Galectin-9 mRNA and protein were commonly down-regulated in OSCC cell lines (Ca9-22, HSC-2, and HSC-3) and normal oral keratinocytes (NOKs). An adhesion assay revealed an increased cellular adherence ratio in overexpressed galectin-9 cells compared with nontransfected cells. Results supported that galectin-9 is correlated with oral cancer cell-matrix interactions and may therefore play an important role in the metastasis of OSCCs (Kasamatsu et al. 2005).

Nasopharyngeal Carcinomas: Nasopharyngeal carcinomas (NPC) are etiologically related to the Epstein-Barr virus (EBV), and malignant NPC cells have consistent although heterogeneous expression of the EBV latent membrane protein 1 (LMP1). LMP1 trafficking and signaling require its incorporation into membrane rafts. Conversely, raft environment is likely to modulate LMP1 activity. Galectin 9 was identified as a novel LMP1 partner. Galectin 9 is abundant in NPC biopsies as well as in LCLs, whereas it is absent in Burkitt lymphoma cells (Pioche-Durieu et al. 2005). Study provides the proof of concept that NPC cells can release HLA class-II positive exosomes containing galectin 9 and/or LMP1. It confirms that the EBV-encoded LMP1 has intrinsic T-cell inhibitory activity (Keryer-Bibens et al. 2006).

Galectin-9 Isoforms Influence Adhesion Between CCCs and HUVEC In Vitro: Galectin-9 plays multiple roles in a variety of cellular functions, including cell adhesion, aggregation, and apoptosis. Galectin-9 has three isoforms (named galectin-9L, galectin-9M, and galectin-9S), which differ in their functions. Transient expression of galectin-9L decreased E-selectin levels, while transient expression of galectin-9M or galectin-9S increased E-selectin levels in LoVo cells, which do not express endogenous galectin-9. Over-expression of three galectin-9 isoforms led to increased attachment of LoVo cells to extracellular matrix proteins respectively, while over-expression of galectin-9M or galectin-9S increased the adhesion of LoVo cells to HUVEC in vitro. Findings indicate that different isoforms of galectin-9 exhibit distinct biological functions (Zhang et al. 2009).

11.5 Galectin-12

Galectin-12 consists of conserved CRDs and is preferentially expressed in peripheral blood leukocytes and adipocytes. It is a major regulator of adipose tissue development and has been shown to be a predominantly adipocyte-expressed protein. Galectin-12 from human adipose tissue contained two potential CRDs with the second CRD being less conserved compared with other galectins (Hotta et al. 2001). In vitro translated galectin-12 bound to a lactosyl-agarose column far less efficiently than galectin-8. Galectin-12 mRNA was predominantly expressed in adipose tissue of human and mouse and in differentiated 3T3-L1 adipocytes. Caloric restriction and treatment of obese animals with troglitazone increased galectin-12 mRNA levels and decreased the average size of the cells in adipose tissue. The induction of galectin-12 expression by the thiazolidinedione, troglitazone, was paralleled by an increase in the number of apoptotic cells in adipose tissue. Galectin-12 was localized in the nucleus of adipocytes, and transfection with galectin-12 cDNA induced apoptosis of COS-1 cells. Hotta et al. (2001) suggested that galectin-12, an adipose-expressed galectin-like molecule, may participate in the apoptosis of adipocytes.

Galectin-12 is induced by cell cycle block at G_1 phase and causes G_1 arrest when overexpressed (Yang et al. 2001). The galectin-12 gene is expressed in mouse preadipocytes and is up-regulated when preadipocytes undergo cell cycle arrest, concomitant with acquisition of the competence to undergo differentiation in response to adipogenic hormone stimulation. Down-regulation of endogenous galectin-12 expression by RNA interference greatly reduced the expression of the adipogenic transcription factors CCAAT/enhancer-binding protein-β and -α and peroxisome proliferator-activated receptor-γ and severely suppressed adipocyte differentiation as a result of defective adipogenic signaling. It was suggested that galectin-12 is required for signal transduction that conveys hormone stimulation to the induction of adipogenic factors essential for adipocyte differentiation (Yang et al. 2004). Treatment of 3T3-L1 cells with isoproterenol, insulin, TNFα, and dexamethasone reduced galectin-12 gene expression between 47% and 85%; the inhibitory effect of isoproterenol could be reversed by pretreatment with the β-adrenergic antagonist propranolol and mimicked by stimulation of G(S)-proteins with cholera toxin or by activation of adenylyl cyclase with forskolin and dibutyryl-cAMP. Findings imply a role for galectin-12 in the pathogenesis of insulin resistance (Fasshauer et al. 2002).

The transcription factors CCAAT enhancer-binding protein α, β, and δ, and PPAR-γ are known to be crucial to the differentiation of adipocytes and are expressed in sebaceous gland cells. Galectin-12, resistin, SREBP-1, and stearoyl-CoA desaturase mRNAs are expressed in SZ95 sebocytes. Evidence suggests that pathways of differentiation in adipocytes and sebocytes could be similar and therefore further understanding of sebaceous gland differentiation and lipogenesis and potential therapies for sebaceous gland disorders may be obtained from our knowledge of adipocyte differentiation (Harrison et al. 2007).

References

Allen HJ, Ahmed H, Sharma A (1992) Isolation of lactose-binding lectins from axolotl (Ambystoma mexicanum). Comp Biochem Physiol B 103:313–315

Anderson AC, Anderson DE (2006) TIM-3 in autoimmunity. Curr Opin Immunol 18:665–669

Arbel-Goren R, Levy Y, Ronen D, Zick Y (2005) Cyclin-dependent kinase inhibitors and JNK act as molecular switches, regulating the choice between growth arrest and apoptosis induced by galectin-8. J Biol Chem 280:19105–19114

Asakura H, Kashio Y, Nakamura K et al (2002) Selective eosinophil adhesion to fibroblast via IFN-γ-induced galectin-9. J Immunol 169:5912–5918

Beyer EC, Zweig SE, Barondes SH (1980) Two lactose binding lectins from chicken tissues. Purified lectin from intestine is different from those in liver and muscle. J Biol Chem 255:4236–4239

Bhavanandan VP, Puch S et al (2001) Galectins and other endogenous carbohydrate-binding proteins of animal bladder. Adv Exp Med Biol 491:95–108

Bidon N, Brichory F, Bourguet P, Le Pennec JP et al (2008) Galectin-8: a complex sub-family of galectins (review). Int J Mol Med 8:245–250

Bidon-Wagner N, Le Pennec JP (2004) Human galectin-8 isoforms and cancer. Glycoconj J 19:557–563

Braccia A, Villani M, Immerdal L et al (2003) Microvillar membrane microdomains exist at physiological temperature. Role of galectin-4 as lipid raft stabilizer revealed by "superrafts". J Biol Chem 278:15679–15684

Carcamo C, Pardo E, Oyanadel C et al (2006) Galectin-8 binds specific β1 integrins and induces polarized spreading highlighted by asymmetric lamellipodia in Jurkat T cells. Exp Cell Res 312:374–386

Castagna LF, Landa CA (1994) Isolation and characterization of a soluble lactose-binding lectin from postnatal chicken retina. J Neurosci Res 37:750–758

Chiu ML, Parry DA, Feldman SR, Klapper DG, O'Keefe EJ (1994) An adherens junction protein is a member of the family of lactose-binding lectins. J Biol Chem 269:31770–31776

Dai SY, Nakagawa R, Itoh A et al (2005) Galectin-9 induces maturation of human monocyte-derived dendritic cells. J Immunol 175:2974–2981

Danielsen EM, van Deurs B (1997). Galectin-4 and small intestinal brush border enzymes form clusters. Mol Biol Cell. 8: 2241–2251

Danielsen EM, Hansen GH (2006) Lipid raft organization and function in brush borders of epithelial cells. Mol Membr Biol 23:71–79

Delacour D, Gouyer V et al (2005) Galectin-4 and sulfatides in apical membrane trafficking in enterocyte-like cells. J Cell Biol 169:491–501

Diskin S, Cao Z, Leffler H, Panjwani N (2009) The role of integrin glycosylation in galectin-8-mediated trabecular meshwork cell adhesion and spreading. Glycobiology 19:29–37

Fasshauer M, Klein J, Lossner U, Paschke R (2002) Negative regulation of adipose-expressed galectin-12 by isoproterenol, tumor necrosis factor α, insulin and dexamethasone. Eur J Endocrinol 147:553–559

Gitt MA, Colnot C, Poirier F et al (1998a) Galectin-4 and galectin-6 are two closely related lectins expressed in mouse gastrointestinal tract. J Biol Chem 273:2954–2960

Gitt MA, Xia Y-R, Atchison RE et al (1998b) Sequence, structure, and chromosomal mapping of the mouse *Lgals6* gene, encoding galectin-6. J Biol Chem 273:2961–2970

Hadari YR, Paz K, Dekel R, Mestrovic T, Accili D, Zick Y (1995) Galectin-8. A new rat lectin, related to galectin-4. J Biol Chem 270:3447–3453

Hansen GH, Pedersen ED, Immerdal L et al (2005) Anti-glycosyl antibodies in lipid rafts of the enterocyte brush border: a possible host defense against pathogens. Am J Physiol Gastrointest Liver Physiol 289:G1100–G1107

Harrison WJ, Bull JJ, Seltmann H et al (2007) Expression of lipogenic factors galectin-12, resistin, SREBP-1, and SCD in human sebaceous glands and cultured sebocytes. J Invest Dermatol 127:1309–1317

Heinzelmann-Schwarz VA, Gardiner-Garden M, Henshall SM et al (2006) A distinct molecular profile associated with mucinous epithelial ovarian cancer. Br J Cancer 94:904–913

Hippo Y, Yoshiro M, Ishii M et al (2001) Differential gene expression profiles of scirrhous gastric cancer cells with high metastatic potential to peritoneum or lymph nodes. Cancer Res 61:889–895

Hirabayashi J, Kasai K (1993) The family of metazoan metal-independent β-galactoside-binding lectins: structure, function and molecular evolution. Glycobiology 3:297–304

Hirabayashi J, Hashidate T, Arata Y et al (2002) Oligosaccharide specificity of galectins: a search by frontal affinity chromatography. Biochim Biophys Acta 1572:232–254

Hokama A, Mizoguchi E, Sugimoto K et al (2004) Induced reactivity of intestinal CD4(+) T cells with an epithelial cell lectin, galectin-4, contributes to exacerbation of intestinal inflammation. Immunity 20:681–693

Hokama A, Mizoguchi E, Mizoguchi A (2008) Roles of galectins in inflammatory bowel disease. World J Gastroenterol 14:5133–5137

Hotta K, Funahashi T, Matsukawa Y et al (2001) Galectin-12, an adipose-expressed galectin-like molecule possessing apoptosis-inducing activity. J Biol Chem 276:34089–34097

Huflejt ME, Jordan ET, Gitt MA et al (1997) Striking different localization of galectin-3 and galectin-4 in human colon adenocarcinoma T84 cells. J Biol Chem 272:14294–14303

Huflejt ME, Leffler H (2004) Galectin-4 in normal tissues and cancer. Glycoconj J 20: 247–255

Hynes MA, Gitt M, Barondes SH et al (1990) Selective expression of an endogenous lactose-binding lectin gene in subsets of central and peripheral neurons. J Neurosci 10:1004–1013

Ideo H, Seko A et al (2002) High-affinity binding of recombinant human galectin-4 to SO_3^-3Galß13GalNAc pyranoside. Glycobiology 12:199–208

Ideo H, Seko A, Yamashita K (2005) Galectin-4 binds to sulfated glycosphingolipids and carcinoembryonic antigen in patches on the cell surface of human colon adenocarcinoma cells. J Biol Chem 280:4730–4737

Ideo H, Seko A, Yamashita K (2007) Recognition mechanism of galectin-4 for cholesterol 3-sulfate. J Biol Chem 282:21081–21089

Igawa K, Satoh T, Hirashima M et al (2006) Regulatory mechanisms of galectin-9 and eotaxin-3 synthesis in epidermal keratinocytes: possible involvement of galectin-9 in dermal eosinophilia of Th1-polarized skin inflammation. Allergy 61:1385–1391

Iino Y, Miyazawa T, Kakizaki K et al (2006) Expression of ecalectin, a novel eosinophil chemoattractant, in nasal polyps. Acta Otolaryngol 126:43–50

Imaizumi T, Kumagai M et al (2002) Interferon-γ stimulates the expression of galectin-9 in cultured human endothelial cells. J Leukoc Biol 72:486–491

Imaizumi T, Yoshida H, Nishi N et al (2007) Double-stranded RNA induces galectin-9 in vascular endothelial cells: involvement of TLR3, PI3K, and IRF3 pathway. Glycobiology 17:12C–15C

Irie A, Yamauchi A, Kontani K et al (2005) Galectin-9 as a prognostic factor with antimetastatic potential in breast cancer. Clin Cancer Res 11:2962–2968

Ishikawa A, Imaizumi T, Yoshida H et al (2004) Double-stranded RNA enhances the expression of galectin-9 in vascular endothelial cells. Immunol Cell Biol 82:410–414

Jiang W, Puch S, Guo X, Bhavanandan VP (1999) Signature sequences for the galectin-4 subfamily. IUBMB Life 48:601–605

Kageshita T, Kashio Y, Yamauchi A et al (2002) Possible role of galectin-9 in cell aggregation and apoptosis of human melanoma cell lines and its clinical significance. Int J Cancer 99:809–816

Kasamatsu A, Uzawa K, Nakashima D et al (2005a) Galectin-9 as a regulator of cellular adhesion in human oral squamous cell carcinoma cell lines. Int J Mol Med 16:269–273

Kasamatsu A, Uzawa K, Shimada K et al (2005b) Elevation of galectin-9 as an inflammatory response in the periodontal ligament cells exposed to porphylomonas gingivalis lipopolysaccharide in vitro and in vivo. Int J Biochem Cell Biol 37:397–408

Keryer-Bibens C, Pioche-Durieu C, Villemant C et al (2006) Exosomes released by EBV-infected nasopharyngeal carcinoma cells convey the viral latent membrane protein 1 and the immunomodulatory protein galectin 9. BMC Cancer 6:283

Kondoh N, Wakatuski T, Ryo A et al (1999) Identification and characterization of genes associated with human hepatocellular carcinogenesis. Cancer Res 59:4990–4996

Krejčiříková V, Fábry M, Marková V et al (2008) Crystallization and preliminary x-ray diffraction analysis of mouse galectin-4 N-terminal carbohydrate recognition domain in complex with lactose. Acta Crystallogr Sect F Struct Biol Cryst Commun 64:665–667

Krejčiříková V, Pachl P, Fábry M (2011) Structure of the mouse galectin-4 N-terminal carbohydrate-recognition domain reveals the mechanism of oligosaccharide recognition. Acta Crystallogr D Biol Crystallogr 67:204–211

Lefranc L, Tatjana M, Christine D et al (2005) Monitoring the expression profiles of integrins and adhesion/growth-regulatory galectins in adamantinomatous craniopharyngiomas: their ability to regulate tumor adhesiveness to surrounding tissue and their contribution to prognosis. Neurosurgery 56:763–776

Leffler H, Masiarz FR, Barondes SH (1989) Soluble lactose-binding vertebrate lectins: a growing family. Biochemistry 28:9222–9229

Lensch M, Lohr M, Russwurm R et al (2006) Unique sequence and expression profiles of rat galectins-5 and -9 as a result of species-specific gene divergence. Int J Biochem Cell Biol 38: 1741–1758

Levy Y, Auslender S, Eisenstein M et al (2006) It depends on the hinge: a structure-functional analysis of galectin-8, a tandem-repeat type lectin. Glycobiology 16:463–476

Lu LH, Nakagawa R, Kashio Y, Ito A et al (2007) Characterization of galectin-9-induced death of Jurkat T cells. J Biochem (Tokyo) 141:157–172

Markova V, Smetana K Jr, Jenikova G et al (2006) Role of the carbohydrate recognition domains of mouse galectin-4 in oligosaccharide binding and epitope recognition and expression of galectin-4 and galectin-6 in mouse cells and tissues. Int J Mol Med 18:65–76

Marschal P, Herrmann J et al (1992) Sequence and specificity of a soluble lactose-binding lectin from *Xenopus laevis* skin. J Biol Chem 267:12942–12949

Mathieu A, Nagy N, Decaestecker C et al (2008) Expression of galectins-1, -3 and -4 varies with strain and type of experimental colitis in mice. Int J Exp Pathol 89:438–446

Matsumoto R, Matsumoto H, Seki M et al (1998) Human ecalectin, a variant of human galectin-9, is a novel eosinophil chemoattractant produced by T lymphocytes. J Biol Chem 273:16976–16984

Matsushita N, Nishi N, Seki M et al (2000) Requirement of divalent galactoside-binding activity of ecalectin/galectin-9 for eosinophil chemoattraction. J Biol Chem 275:8355–8360

Mizoguchi E, Mizoguchi A (2007) Is the sugar always sweet in intestinal inflammation? Immunol Res 37:47–60

Nagae M, Nishi N, Murata T et al (2006) Crystal structure of the galectin-9 N-terminal carbohydrate recognition domain from Mus musculus reveals the basic mechanism of carbohydrate recognition. J Biol Chem 281:35884–35893

Nagae M, Nishi N, Nakamura-Tsuruta S et al (2008) Structural analysis of the human galectin-9 N-terminal carbohydrate recognition domain reveals unexpected properties that differ from the mouse orthologue. J Mol Biol 375:119–135

Nagae M, Nishi N, Murata T et al (2009) Structural analysis of the recognition mechanism of poly-N-acetyllactosamine by the human galectin-9 N-terminal carbohydrate recognition domain. Glycobiology 19:112–117

Nguyen HT, Amine AB, Lafitte D et al (2006) Proteomic characterization of lipid rafts markers from the rat intestinal brush border. Biochem Biophys Res Commun 342:236–244

Niepceron E, Simian-Lerme F, Louisot P, Biol-N'garagba MC (2004) Expression and localization of galectin 4 in rat stomach during postnatal development. Int J Biochem Cell Biol 36:909–919

Niki T, Tsutsui S, Hirose S et al (2009) Galectin-9 is a high affinity IgE-binding lectin with anti-allergic effect by blocking IgE-antigen complex formation. J Biol Chem 284:32344–32352

Nio J, Kon Y, Iwanaga T (2005) Differential cellular expression of galectin family mRNAs in the epithelial cells of the mouse digestive tract. J Histochem Cytochem 53:1323–1334

Nio-Kobayashi J, Takahashi-Iwanaga H, Iwanaga T (2009) Immunohistochemical localization of six galectin subtypes in the mouse digestive tract. J Histochem Cytochem 57:41–50

Nishi N, Shoji H, Seki M et al (2003) Galectin-8 modulates neutrophil function via interaction with integrin αM. Glycobiology 13:755–763

Nishi N, Itoh A, Fujiyama A, Yoshida N et al (2005) Development of highly stable galectins: truncation of the linker peptide confers protease-resistance on tandem-repeat type galectins. FEBS Lett 579:2058–2064

Nishi N, Itoh A, Shoji H et al (2006) Galectin-8 and galectin-9 are novel substrates for thrombin. Glycobiology 16:15C–20C

Oberg CT, Blanchard H, Leffler H, Nilsson UJ (2008) Protein subtype-targeting through ligand epimerization: talose-selectivity of galectin-4 and galectin-8. Bioorg Med Chem Lett 18:3691–3694

Oda Y, Hermann J, Gitt MA et al (1993) Soluble lactose-binding lectin from rat intestine with two different carbohydrate-binding domains in the same peptide chain. J Biol Chem 268: 5929–5939

Okudaira T, Hirashima M, Ishikawa C et al (2007) A modified version of galectin-9 suppresses cell growth and induces apoptosis of human T-cell leukemia virus type I-infected T-cell lines. Int J Cancer 120:2251–2261

Paclik D, Danese S, Berndt U, Wiedenmann B, Dignass A, Sturm A (2008a) Galectin-4 controls intestinal inflammation by selective regulation of peripheral and mucosal T cell apoptosis and cell cycle. PLoS One 3:e2629

Paclik D, Lohse K, Wiedenmann B et al (2008b) Galectin-2 and -4, but not galectin-1, promote intestinal epithelial wound healing in vitro through a TGF-β-independent mechanism. Inflamm Bowel Dis 14:1366–1372

Pardo E, Carcamo C, Massardo L, Mezzano V, Jacobelli S, Gonzalez A, Soza A (2006) Antibodies against galectin-8 in patients with systemic lupus erythematosus. Rev Med Chil 134:159–166 (Article in Spanish)

Patnaik SK, Potvin B, Carlsson S et al (2006) Complex N-glycans are the major ligands for galectin-1, -3, and -8 on Chinese hamster ovary cells. Glycobiology 16:305–317

Pelletier I, Hashidate T, Urashima T et al (2003) Specific recognition of *Leishmania major* poly-β-galactosyl epitopes by galectin-9: possible implication of galectin-9 in interaction between *L. major* and host cells. J Biol Chem 278:22223–22230

Pioche-Durieu C, Keryer C, Souquere S et al (2005) In nasopharyngeal carcinoma cells, Epstein-Barr virus LMP1 interacts with galectin 9 in membrane raft elements resistant to simvastatin. J Virol 79:13326–13337

Popovici RM, Krause MS, Germeyer A et al (2005) Galectin-9: a new endometrial epithelial marker for the mid- and late-secretory and decidual phases in humans. J Clin Endocrinol Metab 90:6170–6176

Rechreche H, Mallo GV, Montalto G, Dagorn J-C, Iovanna JL (1997) Cloning and expression of the mRNA of human galectin-4, an S-type lectin down-regulated in colorectal cancer. Eur J Biochem 248:225–230

Rumilla KM, Erickson LA, Erickson AK, Lloyd RV (2006) Galectin-4 expression in carcinoid tumors. Endocr Pathol 17:243–249, Fall

Sakakura Y, Hirabayashi J, Oda Y, Ohyama Y, Kasai K (1990) Structure of chicken 16-kDa β-galactoside-binding lectin. Complete amino acid sequence, cloning of cDNA, and production of recombinant lectin. J Biol Chem 265:21573–21579

Sato M, Nishi N, Shoji H et al (2002) Functional analysis of the carbohydrate recognition domains and a linker peptide of galectin-9 as to eosinophil chemoattractant activity. Glycobiology 12:191–197

Shuguang Bi, Lesley A, Linsey E et al (2008) Structural features of galectin-9 and galectin-1 that determine distinct T cell death pathways. J Biol Chem 283:12248–12258

Stechly L, Morelle W, Dessein AF et al (2009) Galectin-4-regulated delivery of glycoproteins to the brush border membrane of enterocyte-like cells. Traffic 10:438–450

Storan MJ, Magnaldo T, Biol-N'Garagba MC, Zick Y, Key B (2004) Expression and putative role of lactoseries carbohydrates present on NCAM in the rat primary olfactory pathway. J Comp Neurol 475:289–302

Takahashi Y, Fukusato T, Kobayashi Y et al (2006) High expression of eosinophil chemoattractant ecalectin/galectin-9 in drug-induced liver injury. Liver Int 26:106–115

Tardy F, Deviller P, Louisot P, Martin A (1995) Purification and characterization of the N-terminal domain of galectin-4 from rat small intestine. FEBS Lett 359:169–172

Thomsen MK, Hansen GH, Danielsen EM (2009) Galectin-2 at the enterocyte brush border of the small intestine. Mol Membr Biol 26:347–355

Tomizawa T, Kigawa T, Saito K, Koshiba S, Inoue M, Yokoyama S. Solution structure of the C-terminal gal-bind lectin domain from human galectin-4 (to be published)

Tripodi D, Quemener S, Renaudin K et al (2009) Gene expression profiling in sinonasal adenocarcinoma. BMC Med Genomics 2:65

Türeci O, Schmitt H, Fadle N et al (1997) Molecular definition of a novel human galectin which is immunogenic in patients with Hodgkin's disease. J Biol Chem 272:6416–6422

van de Weyer PS, Muehlfeit M et al (2006) A highly conserved tyrosine of Tim-3 is phosphorylated upon stimulation by its ligand galectin-9. Biochem Biophys Res Commun 351:571–576

Wada J, Kanwar YS (1997) Identification and characterization of galectin-9, a novel β-galactoside-binding mammalian lectin. J Biol Chem 272:6078–6086

Wada J, Ota K, Kumar A et al (1997) Developmental regulation, expression, and apoptotic potential of galectin-9, a β-galactoside binding lectin. J Clin Invest 99:2452–2461

Walser PJ, Haebel PW, Künzler M et al (2004) Structure and functional analysis of the fungal galectin CGL2. Structure 12:689–702

Wasano K, Hirakawa Y (1999) Two domains of rat galectin-4 bind to distinct structures of the intercellular borders of colorectal epithelia. J Histochem Cytochem 47:75–82

Wei Q, Eviatar-Ribak T et al (2007) Galectin-4 is involved in p27-mediated activation of the myelin basic protein promoter. J Neurochem 101:1214–1223

Wiener Z, Kohalmi B, Pocza P et al (2007) TIM-3 is expressed in melanoma cells and is upregulated in TGF-β stimulated mast cells. J Invest Dermatol 127:906–914

Wooters MA, Hildreth MB, Nelson EA, Erickson AK (2005a) Immunohistochemical characterization of the distribution of galectin-4 in porcine small intestine. J Histochem Cytochem 53:197–205

Wooters MA, Ropp SL, Erickson AK (2005b) Identification of galectin-4 isoforms in porcine small intestine. Biochimie 87:143–149

Wu AM, Wu JH, Tsai MS et al (2002) Fine specificity of domain-I of recombinant tandem-repeat-type galectin-4 from rat gastrointestinal tract (G4-N). Biochem J 367:653–664

Wu AM, Wu JH, Liu JH et al (2004) Effects of polyvalency of glycotopes and natural modifications of human blood group ABH/Lewis sugars at the Galβ1-terminated core saccharides on the binding of domain-I of recombinant tandem-repeat-type galectin-4 from rat gastrointestinal tract (G4-N). Biochimie 86:317–326

Yamamoto H, Kashio Y et al (2007) Involvement of galectin-9 in guinea pig allergic airway inflammation. Int Arch Allergy Immunol 143(Suppl 1):95–105

Yamamoto H, Nishi N, Shoji H et al (2008) Induction of cell adhesion by galectin-8 and its target molecules in Jurkat T-cells. J Biochem 143:311–324

Yamauchi A, Kontani K, Kihara M et al (2006) Galectin-9, a novel prognostic factor with antimetastatic potential in breast cancer. Breast J 12(5 Suppl 2):S196–S200

Yang RY, Hsu DK, Yu L et al (2001) Cell cycle regulation by galectin-12, a new member of the galectin superfamily. J Biol Chem 276:20252–20260

Yang RY, Hsu DK, Yu L et al (2004) Galectin-12 is required for adipogenic signaling and adipocyte differentiation. J Biol Chem 279:29761–29766

Zhang F, Zheng MH, Qu Y et al (2009) Galectin-9 isoforms influence the adhesion between colon carcinoma LoVo cells and human umbilical vein endothelial cells in vitro by regulating the expression of E-selectin in LoVo cells. Zhonghua Zhong Liu Za Zhi 31:95–98

Zhu C, Anderson AC, Schubart A et al (2005) The Tim-3 ligand galectin-9 negatively regulates T helper type 1 immunity. Nat Immunol 6:1245–1252

Zick Y, Eisenstein M, Goren RA et al (2004) Role of galectin-8 as a modulator of cell adhesion and cell growth. Glycoconj J 19:517–526

12 Galectin-3: Forms, Functions, and Clinical Manifestations

Anita Gupta

12.1 General Characteristics

Galectins-3 (Gal-3) (formerly known as CBP35, Mac2, L-29, L-34, IgEBP, and LBP) is widely spread among different types of cells and tissues, found intracellularly in nucleus and cytoplasm or secreted via non-classical pathway out of cell, thus being found on the cell surface or in the extracellular space. Through specific interactions with a variety of intra- and extracellular proteins Gal-3 affects numerous biological processes and seems to be involved in different physiological and pathophysiological conditions, such as development, immune reactions, and neoplastic transformation and metastasis. Most members of the galectin family including Gal-1 possess apoptotic activities, whereas Gal-3 possesses anti-apoptotic activity as well. Information on structural, biochemical and intriguing functional properties of Gal-3 have been reviewed extensively (Dumic et al. 2006).

12.1.1 Galectin-3 Structure

Galectin-3 is structurally unique member of galectin family. It is a 31-kDa chimeric galectin characterized by a single C-terminal carbohydrate recognition domain (CRD) for carbohydrate binding and an N-terminal aggregating domain that interacts with a noncarbohydrate ligand and allows the formation of oligomers. Gal-3 has an extra long and flexible N-terminal domain consisting of 100–150 amino acid residues, according to species of origin, made up of repetitive sequence of nine amino acid residues rich in proline, glycine, tyrosine and glutamine and lacking charged or large side-chain hydrophobic residues (Hughes et al. 1994; Birdsall et al. 2001) (Fig. 12.1a). The N-terminal domain functions to cross-link both carbohydrate and noncarbohydrate ligands (Hughes 1994; Birdsall et al. 2001; Brewer et al. 2002) and contains sites for phosphorylation (Ser^{6}, Ser^{12}) (Huflejt et al. 1993; Mazurek et al. 2000) and other determinants important for the secretion of the lectin by a novel, nonclassical mechanism (Menon and Hughes 1999). The CRD consists of about 135 amino acid residues and displays an identical topology and very similar three-dimensional structure to that reported for CRD of homodimeric Gal-1 and -2 (Liao et al. 1994; Lobsanov et al. 1993) with which it shares 20–25% sequence identity. Like Gal-1 and -2, it is arranged in 12β strands (F1-F5 and S1-S6a/6b) (Seetharaman et al. 1998).

Of multiple soluble lactose-binding lectins in rat intestine, the major one, tentatively designated RI-H, was isolated as a polypeptide of M_r 17,000. Surprisingly the cDNA encodes a protein of M_r 36,000, and this protein contains two homologous but distinct domains each with sequence elements that are conserved among all S-Lac lectins. The C-terminal domain, designated domain II, corresponds to lectin with M_r of 17,000 isolated from intestinal extracts and shown to have lactose binding activity. The new lectin, which was designated L-36, is highly expressed in full-length form in rat small and large intestine and stomach but was not detected in other tissues including lung, liver, kidney, and spleen. Each domain has approximately 35% sequence identity with the other domain and with the carbohydrate-binding domain of L-29, another S-Lac lectin, but only about 15% identity with other known S-Lac lectins (Oda et al. 1993).

Sugar Binding Subsites in Galectins: A carbohydrate-binding subsite has been suggested in hamster Gal-3 involving Arg^{139}, Glu^{230}, and Ser^{232} for NeuNAc-[α]2,3-; Arg^{139} and Glu^{160} for fucose-[α]1,2-; and Arg^{139} and Ile^{141} for GalNAc-[α]1,3- substituents on the primary galactose. Each of these positions is variable within the whole galectin family. Two of these residues, Arg^{139} and Ser^{232}, were probed for their importance in this putative subsite. Mutagenesis studies indicated that residue 139 adopts main-chain dihedral angles characteristic of an isolated bridge structural feature, while residue 232 is C-terminal residue of [β]-strand-11, and is

G.S. Gupta et al., *Animal Lectins: Form, Function and Clinical Applications*,
DOI 10.1007/978-3-7091-1065-2_12, © Springer-Verlag Wien 2012

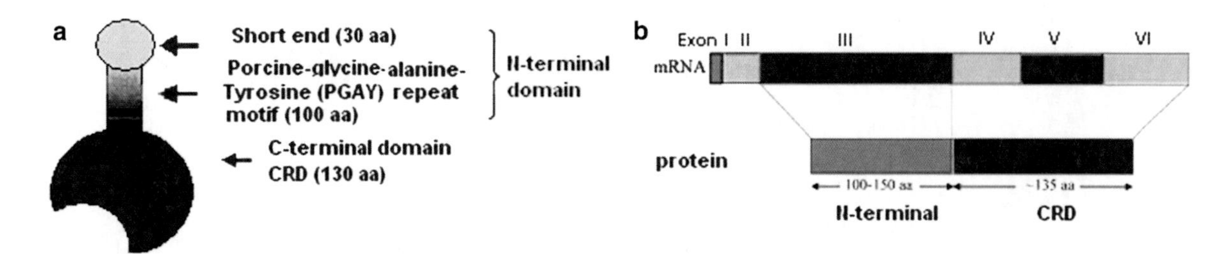

Fig. 12.1 The structure (**a**) and gene organization (**b**) of galectin-3

followed immediately by an inverse [γ]-turn. Arg[139] was suggested to recognize the extended sialylated ligand (Henrick et al. 1998).

The murine Gal-3 is a monomer in absence and presence of N-acetyllactosamine (LacNAc), a monovalent sugar. However, Gal-3 precipitates as a pentamer with a series of divalent pentasaccharides with terminal LacNAc residues. Although the majority of Gal-3 in solution is a monomer, a rapid equilibrium exists between the monomer and a small percentage of pentamer. The latter, in turn, precipitates with divalent oligosaccharides, resulting in rapid conversion of monomer to pentamer by mass action equilibria. It forms heterogenous, disorganized cross-linking complexes with the multivalent carbohydrates. This contrasts with Gal-1 and many plant lectins (Ahmad et al. 2004a).

12.1.2 Galectin-3 Gene

The structure of Gal-3 gene is consistent with the multi-domain organization of the protein. The gene for Gal-3 is composed of six exons and five introns (human locus 14q21-22). Exon I encodes the major part of 5′untranslated sequence mRNA. Exon II contains the remaining part of 5′untranslated sequence, the protein translation initiation site and the first six amino acids including the initial methionine. Galectin-3 is coded by a single gene in the human genome. The gene is composed of six exons and five introns, spanning a total of approximately 17 kb. Based on primer extension and ribonuclease protection analyses, there are two transcription initiation sites located 52 and 50 nt upstream of the exon I-intron 1 border, and defined as +1a and +1b, respectively. The translation start site is in exon II. The ribonucleoprotein-like N-terminal domain, containing the proline-glycine-alanine-tyrosine (PGAY) repeat motif, is found entirely within exon III. Genomic fragments encompassing −836 to +141 nt (relative to +1a) have significant promoter activity. Because galectin-3 is an immediate-early gene whose expression is dependent on the proliferative state of the cell, results provide the basis for determining the molecular mechanisms of transcriptional regulation in neoplasia or cellular senescence. Exons IV, V and VI code the C-terminal half of the protein. The carbohydrate recognition sequence is found entirely within exon V (Fig. 12.1b) (Hughes et al. 1994, 1997; Kadrofske et al. 1998).

Murine Galectin-3 (Mac-2) Gene: The murine Mac-2 gene is composed of six exons dispersed over 10.5 kb. S1 nuclease mapping showed multiple transcription initiation sites, clustered within a 30-bp region. Sequence analysis revealed that a consensus initiator sequence is located in this area which lacks a TATA motif. The untranslated first exon contains an alternative splice donor site, confirming the existence of two cDNA species with the potential to encode proteins differing at their NH2 termini. In vitro expression and translocation experiments demonstrated that both the alternatively spliced variants of Mac-2 encode proteins which lack a functional signal peptide. Most of the Mac-2 protein is present in the cytosol. Results support the view that Mac-2 is exported from the cell by an unusual mechanism which does not depend on the presence of a signal peptide (Rosenberg et al. 1993).

12.1.3 Tissue and Cellular Distribution

Gal-3 has been detected in many immunocompetent/inflammatory cells, the epithelium of gastrointestinal and respiratory tracts, the kidneys and some sensory neurons (Hughes et al. 1997, 1999). In immunocompetent/inflammatory cells, Gal-3 is expressed in monocytes, dendritic cells (DCs), macrophages, eosinophils, mast cells, NK cells, and activated T and B cells (Chen et al. 2005; Dumic et al. 2006; Sato and Nieminen 2004). Moreover, Gal-3 displays pathological expression in many tumors, e.g., human pancreas, colon and thyroid carcinomas (see Chap. 13). It is mainly found in the cytoplasm, although, depending on cell type and proliferative state, a significant amount of this lectin can also be detected in the nucleus, on cell surface or in extracellular environment. In mouse fibroblasts, the nuclear versus cytoplasmic distribution of Gal-3 depends on the proliferation state of analysed cells. In quiescent cultures of fibroblasts, phosphorylated derivative of Gal-3 (pI ~8.2) was predominantly cytoplasmic; however, proliferating cultures of same cells showed intense nuclear staining for nonphosphorylated native polypeptide (pI ~8.7). The intracellular location of Gal-3 is connected with its role in the regulation of nuclear pre-mRNA splicing and protection against apoptosis. On the other hand, its extracellular location on the cell surface and in the extracellular milieu indicates its participation in cell-cell and cell-matrix adhesion. A distinct expression profile of Gal-3 was determined in various murine organs when set into relation to homodimeric galectins-1 and -7. Lohr et al. (2008) demonstrated cell-type specificity and cycle-associated regulation for Gal-3 with increased presence in atretic preantral follicles and in late stages of luteolysis. Wang et al. (2004) reviewed lectins, which have been localized in the cytoplasm and nucleus of the cells. Gal-1 and Gal-3 were identified as pre-mRNA splicing factors in the nucleus, in conjunction with their interacting ligand, Gemin4. Gal-1 and Gal-3 are directly associated with splicing complexes throughout the splicing pathway in a mutually exclusive manner and they bind a common splicing partner through weak protein-protein interactions (Wang et al. 2006).

Transport of Gal-3 Between the Nucleus and Cytoplasm: Gal-3 shuttles between nucleus and the cytoplasm. Gal-3 compartmentalization in nucleus versus cytoplasm affects the malignant phenotype of various cancers. The mechanism by which Gal-3 translocates into nucleus remains debatable. In mouse fibroblasts, the Gal-3 (residues 1–263) fusion protein was localized predominantly in the nucleus. Mutants of this construct, containing truncations of the Gal-3 polypeptide from amino terminus suggested that the amino-terminal half was dispensable for nuclear import. Mutants of the same construct, containing truncations from the carboxyl terminus, showed loss of nuclear localization. Site-directed mutagenesis of the sequence ITLT (residues 253–256) suggested that nuclear import was dependent on the IXLT type of nuclear localization sequence, first shown in the *Drosophila* protein Dsh (dishevelled). In Gal-3 polypeptide, the activity of this nuclear localization sequence is modulated by a neighboring leucine-rich nuclear export signal (Davidson et al. 2006). It appeared that the export of Gal-3 from the nucleus may be mediated by the CRM1 receptor. The nuclear export signal fitting the consensus sequence recognized by CRM1 can be found between residues 240 and 255 of the murine Gal-3 sequence. Results indicated that residues 240–255 of the Gal-3 polypeptide contain a leucine-rich nuclear export signal that overlaps with the region (residues 252–258) identified as important for nuclear localization (Li et al. 2006).

Nakahara et al. (2006b) suggested that Gal-3 nuclear translocation is governed by dual pathways, whereas the cytoplasmic/nuclear distribution may be regulated by multiple processes, including cytoplasmic anchorage, nuclear retention, and or nuclear export. Gal-3 can be imported into the nucleus through at least two pathways; via passive diffusion and/or active transport (Nakahara et al. 2006a). The process mediated by the active nuclear transport of Gal-3 involves a nuclear localization signal (NLS)-like motif in its protein sequence, ^{223}HRVKKL^{228}, that resembles p53 and c-Myc NLSs [^{378}SRHKKL^{383}, ^{322}AKRVKL^{327}], respectively. It was suggested that Gal-3, in part, is translocated to the nucleus via the importin-α/β route and that Arg-224 residue of human Gal-3 is essential for its active nuclear translocation and molecular stability (Nakahara et al. 2006a).

Secretion: Gal-3 is secreted/externalized from cells through a nonclassical and a less understood transport pathway called ectocytosis, which is independent of classical secretory pathway through ER and Golgi system (Dumic et al. 2006; Henderson et al. 2006; Krześlak and Lipińska 2004; Nickel 2003; Rabinovich et al. 2002). Gal-3 is synthesized on free ribosomes in cytoplasm and lacks any signal sequence for translocation into ER (Hughes et al. 1999). As an extracellular protein, it interacts with glycoproteins within extracellular matrix to form a glycoprotein lattice or act as a soluble ligand to crosslink with carbohydrates of surface proteins by N-terminal oligomerization, thus evoking signal transduction and cell functions. Although it does not traverse endoplasmic reticulum/Golgi network, there is abundant evidence for Gal-3 also having an extracellular location. The N-terminus of Gal-3 has been proposed to contain targeting information for nonclassical secretion (Menon and Hughes 1999). A hamster Gal-3 CRD fragment lacking N-terminal domains, when expressed in transfected Cos cells, is not secreted, whereas addition of N-terminal segment to a normal cytosolic protein such as CAT-fusion protein efficiently exported it from transfected Cos cells. A short segment of Gal-3 N-terminal sequence comprising residues 89–96 (Tyr-Pro-Ser-Ala-Pro-Gly-Ala-Tyr) plays a critical role in Gal-3 secretion. However, this sequence is not sufficient on its own to cause direct secretion of CAT fusion protein, indicating that it is operative in association with a large part of N-terminal sequence of Gal-3 (Menon and Hughes, 1999). Studies indicate that the first step in Gal-3 secretion is its accumulation at cytoplasmic side of the plasma membrane. The acylation of Lck was an essential requirement for the retention and functioning of these proteins at the cytoplasmic side of plasma membranes (Zlatkine et al. 1997). In next step of secretion, the evaginating membrane domain(s) is/are pinched of and extracellular vesicles in which Gal-3 is protected against proteolysis is released. Isolated vesicles were more stable, and indicated that rapid breakdown of vesicles requires some factor(s) released by cells. Hughes (1999) suggested that a PLA2 may catalyse the hydrolysis of an sn-2 fatty acyl bond of phospholipids and liberate free fatty acids and lysophospholipids (Krześlak and Lipińska 2004).

12.2 Ligands for Galectin-3: Binding Interactions

12.2.1 Extracellular Matrix and Membrane Proteins

Potential ligands for Gal-3 are the lysosomal associated membrane proteins 1 and 2 (Lamp 1 and 2) (Sarafian et al. 1998). Lamps are lysosomal membrane proteins, and are rarely found on the plasma membrane of normal cells. But Lamp 1 and 2 increase on cell surface of tumor cells, especially in highly metastatic ones, where they are major carriers for poly-N acetyllactosamines. Recombinant Gal-3 binds to Lamp-expressing metastasing melanoma cells, which indicates that Lamps could be ligands for cell adhesion molecules and participate in the process of tumor invasion and metastasis. MP20, the lens membrane integral protein and a member of the tetraspanin superfamily, also interacts with Gal-3 (Gonen et al. 2001; Serafian et al. 1998). It is not known exactly what role the MP20/Gal-3 complex could play in the lens. It is conceivable that Gal-3 plays an essential role in modulating the ability of MP20 to form adhesive junctions at this critical stage of development (Gonen et al. 2001). Because of its affinity for polylactosamine glycans, Gal-3 binds to glycosylated extracellular matrix components, including laminin, fibronectin, tenascin and Mac-2 binding protein (Woo et al. 1990; Rosenberg et al. 1991; Koths et al. 1993; Sato and Hughes 1992; Probstmeier et al. 1995; Ochieng and Warfield 1995) (Table 12.1). Some cell- surface adhesion molecules, for instance integrins, are also ligands for Gal-3. Evidence suggests that galectins, through binding to extracellular domains of one or both subunits of an integrin, may modulate integrin activation, and affect the binding with extracellular ligands (Hughes 2001). A major ligand for Gal-3 on mouse macrophage is α-subunit of the integrin αMβ2, also known as CD11b/18 (Hughes 2001; Dong and Hughes 1997). Moreover, Gal-3 also interacts with integrin α1β1 via its CRD domain in a lactose dependent manner (Ochieng et al. 1998). Gal-3 also seems to be an endogenous cross-linker of CD98 antigen, leading to the activation of integrin mediated adhesion (Hughes 2001). While gal-1 specifically recognizes CD2, CD4, CD7, CD43, and CD45, Gal-3 binds to IgE, Fc receptors, CD66, and CD98 (Pace et al. 1999; Walzel et al. 2000; Hughes 2001). The two CRDs of Gal-4 have different preferences for carbohydrate ligands, suggesting that bivalent galectins may cross-link different ligands (Oda et al. 1993). Recognition of unique glycan ligands probably allows different galectins to exert distinct biological effects in various tissues.

12.2.2 Intracellular Ligands

All above listed ligands for Gal-3 are extracellular matrix or membrane proteins. However, Gal-3 is also known to have an intracellular location and to interact with several proteins inside the cell, i.e., cytokeratins (Goletz et al. 1997), CBP70 (Sève et al. 1993), Chrp (Menon et al. 2000), Gemin4 (Park et al. 2001), Alix/AIP-1 (Liu et al. 2002), and Bcl-2 (Akahani et al. 1997). Nuclear Gal-3 interacts with Gemin-4, a component of a complex containing ~15 polypeptides, including SMN (survival of motor neuron) protein, Gemin-2, Gemin-3, some of the Sm core proteins of snRNPs and others. The identification of Gemin-4 as an interacting partner of Gal-3 provides evidence that Gal-3 can play a role in splicesome assembly in vivo (Liu et al. 2002). The precise mechanism by which Gal-3 and other cytosolic proteins that lack signal peptides are secreted is yet to be elucidated. It is worth noting that almost all of the mentioned intracellular ligands interact with Gal-3 via protein-protein rather than lectin-glycoconjugate interactions. However, Goletz et al. (1997) have shown that cytokeratins of some human cells carry a posttranslational modification, a glycan with a terminal α linked GalNAc, which are recognized in vitro by Gal-3. A carbohydrate binding protein of 70 kDa (CBP70), which is a nuclear and cytoplasmic lectin glycosylated by the addition of N- and O-linked oligosaccharide chains (Rousseau et al. 2000), interacts with Gal-3 via a protein-protein interaction mediated by the addition of lactose, probably resulting in modification of the Gal-3 conformational structure. Menon et al. (2000) reported a cytoplasmic cysteine-histidine rich protein – called Chrp that assists intracellular trafficking of Gal-3. Chrp recognizes the CRD domain of Gal-3 and not the N-terminal repeat sequence (Bawumia et al. 2003). The functional significance of Chrp-Gal-3 interaction remains to be elucidated.

Another Gal-3 binding protein is a human homologue of ALG-2 linked protein X (Alix) or ALG-2 interacting protein-1 (AIP-1) (Liu et al. 2002). This protein interacts with ALG-2, a calcium binding protein necessary for cell death induced by different stimuli. AIP-1 cooperates with ALG-2 in executing the calcium dependent requirements along the cell death pathway (Vito et al. 1999). Vlassara et al. (1995) suggested that Gal-3 is one of the AGE receptors, which might be present in macrophages, astrocytes and umbilical vein endothelial cells (Pricci et al. 2000). Results

Table 12.1 Ligands for galectin-3: Binding interactions (Iacobini et al. 2003)

	Ligand	Source/cell	Reference
Extracellular matrix proteins	IgE	Eosinophils and neutrophils	Liu 2005
	Laminin	EHS, macrophage, neutrophil, placenta	Woo et al. 1990; Ochieng and Warfield 1995; Kuwabara et al. 1996
	Fibronectin	Foetal	Sato and Hughes 1992
	Tenascin	Brain	Probstmeier et al. 1995
	M2BP	Brain	Rosenberg et al. 1991; Koths et al. 1993
	Collagen I/IV	Madin-Darby canine kidney cells	Friedrichs et al. 2008
	gp90kDa	Colorectal cancer	Iacovazzi et al. 2010
	Elastin	Breast carcinoma	Ochieng et al. 1999
	AGE	Ubiquitous	Vlassara et al. 1995; Pricci et al. 2000; Zhu et al. 2001
Membrane proteins	IgE receptor	Mast cell (and basophil)	Pricci et al. 2000
	Integrins:		
	αM/β2(CD11b/18)	Macrophage	Dong and Hughes 1997
	α1/β1	Adenocarcinoma	Ochieng et al. 1998
	N-CAM	Mouse brain	Probstmeier et al. 1995
	L1	Mouse brain	Probstmeier et al. 1995
	MAG	Mouse brain	Probstmeier et al. 1995
	LAMP-1,2	Ubiquitous	Dong and Hughes 1997; Sarafian et al. 1998; Ohannesian et al. 1995
	MP20	Rat lens	Gonen et al. 2001
	CD98	Human T lymphoma Jurkat cells	Yang et al. 1996
	Mucin	Colon cancer	Bresalier et al. 1998
	CD66	Human neutrophils	Feuk-Lagerstedt et al. 1999
	Bacterial LPS	*Klebsiella pneumoniae*	Mey et al. 1996
Intracellular proteins	Cytokeratins	HeLa, MCF-7	Goletz, et al. 1997
	Chrp	Murine 3T3 fibroblasts	Menon, et al. 2000; Bawumia et al. 2003
	CBP70	HL60	Sève, et al. 1993
	Alix/AIP-1	Human T lymphoma Jurkat cells	Liu et al. 2002
	Bcl-2	Human T lymphoma Jurkat cells	Yang et al. 1996
	β-Catenin	Breast cancer cells; colon cancer cells	Shimura et al. 2004
	Gemin-4	HeLa	Park et al. 2001
	CEA and other glycoconjugates	Colon carcinoma cells	Ohannesian et al. 1995
	98 kDa (Mac-2-BP) and 70 kDa species	Human melanoma cells	Inohara and Raz 1994

showed that AGE binds to Gal-3 CHO cells, followed by endocytosis and subsequent lysosomal degradation. Further studies revealed that acetylated and oxidized low density lipoproteins also undergo receptor-mediated endocytosis by these cells. GGal-3 is likely to play an important role in the formation of atherosclerotic lesions in vivo by the modification of the endocytic uptake of AGE and by modified low density lipoproteins (Zhu et al. 2001; Krześlak and Lipińska 2004).

It is found that Gal-3 can interact directly with membrane lipids in solid phase binding assays and has its own capacity to traverse the lipid bilayer. The interaction of Gal-3 with the plasma membrane may involve cholesterol-rich membrane domains where Gal-3 can be concentrated and form multimers or interact covalently with other proteins (Lukyanov et al. 2005).

12.2.3 Carbohydrate Binding

Gal-3 has an affinity for lactose (Lac) and N-acetyllactosamine (LacNAc) and acts as a receptor for ligands containing poly-N-acetyllactosamine sequences, which consist of many disaccharide units: Gal α1,4GlcNAc binds to each other by α1,3 linkage. The binding site of Lac/LacNAc is located to β-strands S4-S6a/S6b. In recent years, it has been demonstrated that Gal-3 binds to multimeric LN (LacNAc) and LDN (LacdiNAc) forming GalNAcß1-4GlcNAc motifs (van den Berg et al. 2004) as well as to other carbohydrate structures on glycoproteins and glycolipids (including LPS) from many pathogens such as (myco)bacteria, protozoan parasites, and yeast (Ochieng et al. 2004). Gal-3 appears to have an increased affinity for complex oligosaccharides (Hughes 1999; Hirabayashi et al. 2002). Gal-1, -2, and -3 exhibit differential

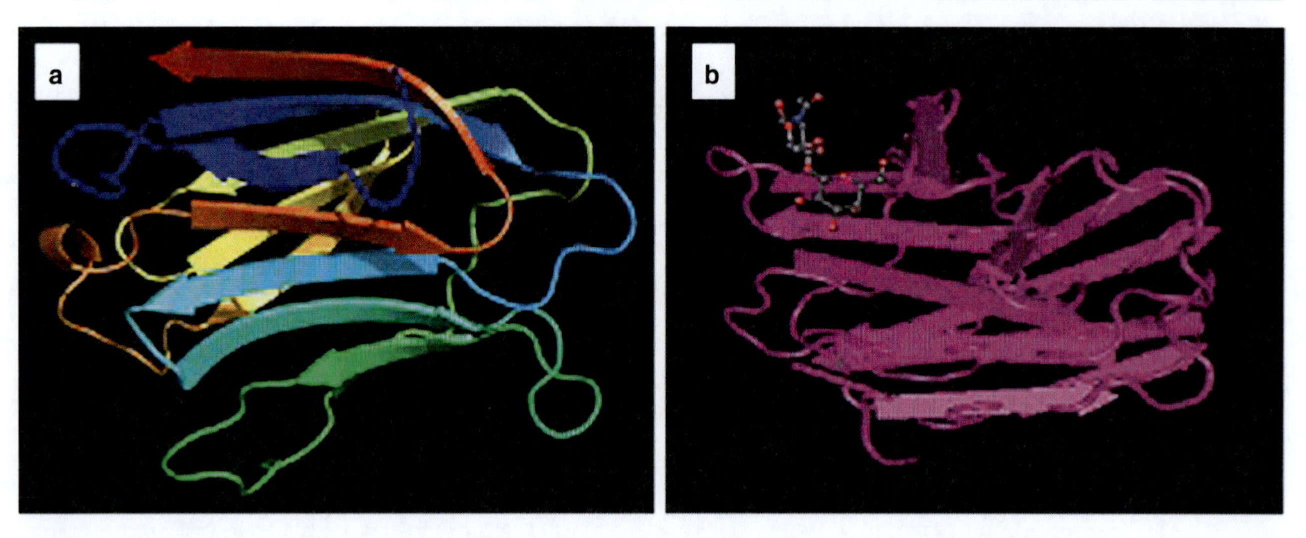

Fig. 12.2 Structure of human galectin-3 CRD (**a**) (Seetharaman et al. 1998) at 2.1 Å (PDB ID: 1A3K) and (**b**) bound to lactose at 1.35 Å (PDB ID: 2NN8) (Collins et al. 2007)

recognition of sialylated glycans and blood group antigens (Stowell et al. 2008). Complex N-glycans are the major ligands for Gal-1, -3, and -8 on Chinese hamster ovary cells (Patnaik et al. 2006). Extension at the nonreducing end of disaccharide units with NeuNAcα2, 3 or with GalNAcα1,3 and Fuc α1,2 substituents greatly enhances affinity for Gal-3 (Hughes 1999, 2001). Structural and mutagenic studies helped in identification of contact residues in Gal-3 CRD responsible for recognition of complex carbohydrates (Seetharaman et al. 1998). Studies revealed that Arg-144 in human Gal-3 is well positioned to interact with GlcNAc moiety or other saccharide residues linked to O-3 of terminal galactose (Seetharaman et al. 1998) (Figs. 12.2, 12.3). The C-4 hydroxyl group of galactose moiety (Gal) plays a central role in binding, probably accepting hydrogen bonds from the highly conserved residues His-158 and Arg-162, and at the same time donating hydrogen bonds to Asn-160 and a water molecule (W1) (Seetharaman et al. 1998). The galactose C-6 hydroxyl group also displays this cooperative hydrogen bonding pattern, interacting with Glu-184, Asn-174 and W3.

In case of N-acetylglucosamine (GlcNAc), only its C-3 hydroxyl group makes direct hydrogen bonds with Glu-184 and Arg-162 of protein. The only other contacts involving GlcNAc moiety are mediated through its N-acetyl group; the amide proton is hydrogen bonded through W2 to Glu-165 and the methyl group makes a van der Waals contact with guanidine head group of Arg-186. The van der Waals interaction and the strength of hydrogen bond involving two position of Glc/GlcNAc moiety represent the only significant differences between Lac and LacNAc complexes, and presumably account for the ~5-fold higher binding affinity for N-acetyllactosamine over lactose shown by human Gal-3 (Seetharaman et al. 1998) (Fig. 12.3). It seems that the N-terminal non-CRD domain is responsible for enhanced affinity of Gal-3 for extended structures of basic recognition units such as Lac or LacNAc (Hirabayashi et al. 2002). Affinity chromatography revealed that intact Gal-3 has 3.8 times higher affinity for oligosaccharides terminating in fucose or sialic acid residues than its deletion product in which N-terminal domain was removed by collagenase digestion.

Binding of Bivalent Oligosaccharides to Gal-3: Binding properties of recombinant murine Gal-3 to synthetic analogs containing two LacNAc residues separated by a varying number of methylene groups, as well as biantennary analogs possessing two LacNAc residues and binding of multivalent carbohydrates to C-terminal CRD domain of Gal-3 showed that each bivalent analog is bound by both LacNAc residues to two galectins. However, Gal-1 showed a lack of enhanced affinity for bivalent straight chain and branched-chain analogs, although Gal-3 showed enhanced affinity for only lacto-N-hexaose, a natural branched chain carbohydrate (Ahmad et al. 2004b).

Human Monocytes Recognize Porcine Endothelium via Interaction of Galectin 3: Gal-3 is mainly expressed in human monocytes, not lymphocytes and plays a key role in CD13-mediated homotypic aggregation of key inflammatory monocytic cells (Mina-Osorio et al. 2007). Monocytes, recruited to xenografts, play an important role in delayed xenograft rejection and have the ability to bind to major xenoantigen [Gal-α(1,3)Gal-β(1,4) GlcNAc-R]. Evidences show that Gal-3 is the receptor that recognizes β-galactosides (Gal-β(1-3/4)GlcNAc) and plays diverse roles in many physiological and pathological events. Human monocyte binding is strikingly

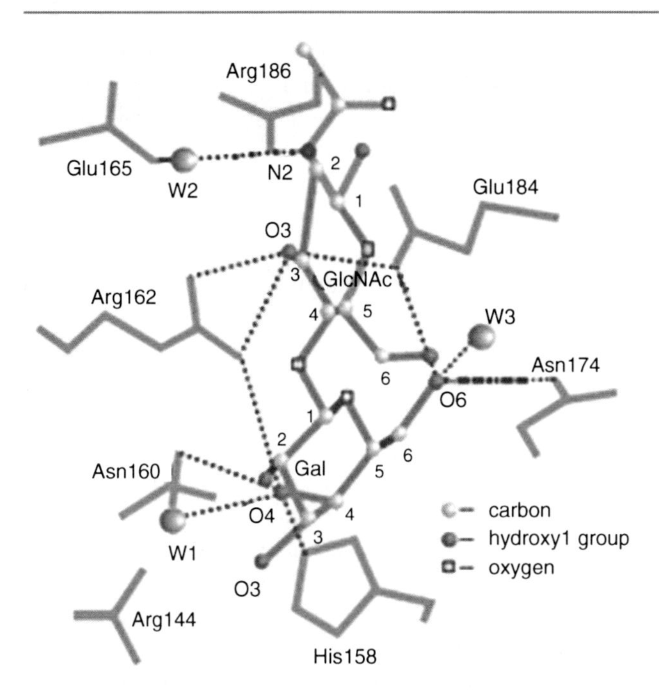

Fig. 12.3 The interaction of human Gal-3 with the N-acetyllactosamine moiety (for details, see text). The water molecules are labeled W1–W3. Potential hydrogen bonds are shown as *dotted lines*. The positions of the carbon atoms and the main hydroxyl groups are numbered (Adapted by permission from Seetharaman et al. 1998 © The American Society for Biochemistry and Molecular Biology)

increased on porcine aortic endothelial cells (PAEC), which express high levels of [Gal-α(1,3)Gal-β(1,4) GlcNAc-R], compared with human aortic endothelial cells. Human monocytes from healthy donors and purified Gal-3 bind to Gal-α(1,3)Gal-β(1,4)GlcNAc-R at variable intensities. Soluble Gal-3 binds preferentially to PAEC vs human aortic endothelial cells, and this binding can be inhibited by lactose, indicating dependence on CRD of Gal-3. Thus, Gal-3 expressed in human monocytes is a receptor for the major xenoantigen [Gal-α(1,3)Gal-β(1,4) GlcNAc-R], expressed on porcine endothelial cells (Jin et al. 2006).

12.2.4 Carbohydrate-Independent Binding

In addition to carbohydrate-dependent extracellular functions, Gal-3 participates in carbohydrate-independent intracellular signaling pathways, including apoptosis, via protein-protein interactions, some of which engage the carbohydrate-binding groove. When ligands bind within this site, conformational rearrangements are induced. The information on unliganded Gal-3 is therefore valuable for structure-based drug design. Removal of cocrystallized lactose from human Gal-3 CRD was achieved via crystal soaking, but took weeks despite low affinity (Collins et al. 2007).

The composition of mycolic acids (MA), the major constituents of *Mycobacterium tuberculosis* (Mtb) cell envelope, and other cell wall-associated lipids contribute to the virulence of a given strain. Gal-3, present mainly in cytoplasm of inflammatory cells and also on cell surface, can recognize mycobacterial MA. The MA can inhibit the lectin self-association but not its carbohydrate-binding abilities and can selectively interfere in the interaction of lectin with its receptors on temperature-sensitive DC line. This suggested that Gal-3 could be involved in the recognition of trafficking mycolic acids and can participate in their interaction with host cells (Barboni et al. 2005).

12.3 Functions

12.3.1 Galectin-3 is a Multifunctional Protein

Gal-3 is a multifunctional protein implicated in a variety of biological functions, including tumor cell adhesion, proliferation, differentiation, angiogenesis, cancer progression and metastasis. Multiple functions of Gal-3 depend on its location in the cell. Gal-3 is the only family member that is composed of a glycine/proline rich N-terminal repeated sequence and a C-terminal carbohydrate-binding domain. Although galectins lack a classical signal sequence, they are present in extracellular fluid and on the cell surface and are also located inside the cells (Hughes 1999; Liu et al. 2002). Extracellular Gal-3 can bind to cell surface through glycosylated proteins and thereby trigger or modulate cellular responses such as mediator release or apoptosis. Intracellular Gal-3 has been reported to inhibit apoptosis, regulation of tumor proliferation and angiogenesis, regulation of cell cycle, and participate in the nuclear splicing of pre-mRNA. Studies implicate Gal-3 in both innate and adaptive immune responses, where it participates in the activation or differentiation of immune cells. It is the only chimera-type member of galectin family of endogenous lectins, which share specificity with β-galactosides and have a jelly-roll-like folding pattern. Intracellular Gal-3 exhibits the activity to suppress drug induced apoptosis and anoikis (apoptosis induced by the loss of cell anchorage) that contribute to cell survival. Resistance to apoptosis is essential for cancer cell survival and plays a role in tumor progression. Conversely, it was shown that tumor cells secreted Gal-3 induces T-cells' apoptosis, thus playing a role in the immune escape mechanism during tumor progression through induction of apoptosis of cancer-infiltrating T-cells. Nakahara et al. (2005) summarized evidences on the role of Gal-3 as an anti-apoptotic and/or pro-apoptotic factor in various cell types and discussed the understanding of the molecular mechanisms of Gal-3 in apoptosis (Nakahara et al. 2005).

Gal-3 has been implicated in many facets of inflammatory response including neutrophil adhesion and activation (Kuwabara and Liu 1996), chemoattraction of monocytes/macrophages (Sano et al. 2000), and activation of mast cells and lymphocytes. Secreted Gal-3 can cross-link surface glycoproteins and activate pathways involved in several innate immune responses such as the oxidative burst in neutrophils (Janeway and Medzhitov 2002; Nieminen et al. 2005; Liu and Robinovich 2005) and degranulation in mast cells (Frigeri et al. 1993). Gal-3 also contributes to chemotaxis by mediating cell-cell and cell-substratum adhesion (Dumic et al. 2006). Mice lacking Gal-3 develop a dominant Th1 phenotype and exhibit abnormalities in several inflammatory disease models including asthma and diabetes, suggesting that Gal-3 may be involved in the regulation of inflammatory and Th1 responses (Pugliese et al. 2001; Zuberi et al. 2004; Bernardes et al. 2006; Nomoto et al. 2006). Gal-3 regulates peritoneal B1-cell differentiation into plasma cells (Oliveira et al. 2009).

12.3.2 Role in Cell Adhesion

Through its ability to bind polylactosamine structures [such as $(\text{Gal}\beta1\text{-}4\text{GlcNAc})_n$] on endogenous ligands (Hirabayashi et al. 2002), Gal-3 intervenes in many cellular processes in vitro. Despite the specific binding of galectin to glycoconjugates, there are still controversies as to whether galectins facilitate or inhibit cell adhesion. Gal-3 was shown to reduce the adhesion and spreading of baby hamster kidney epithelial cells on laminin 1 coated wells (Sato and Hughes 1992; Sato et al 1993). On the other hand, purified Gal-3 was shown to promote the adhesion of human neutrophils to laminin (Kuwabara and Liu 1996). This binary action of Gal-3 may be related to its concentration as well as to the expression level and glycosylation of its cell surface and matrix ligands (Hughes 2001). In particular, it favors cell-cell and cell-matrix glycoprotein adhesion (Ochieng et al. 1998), exerts chemotactic effects (Sano et al. 2000), controls cell proliferation (Demetriou et al. 2001; Yang et al. 1996), and promotes phagocytosis by macrophages (Sano et al. 2003; van den Berg et al. 2004).

12.3.3 Gal-3 at the Interface of Innate and Adaptive Immunity

Chen et al. (2005) reviewed the role of Gal-3 in the immune system and discussed the possible underlying mechanisms. Bernardes et al. (2006) studied Gal-3-deficient (*gal3*$^{-/-}$) mice after their response to *Toxoplasma gondii* infection, which is characterized by inflammation in affected organs, Th-1-polarized immune response, and accumulation of cysts in the CNS. In wild (*gal3*$^{+/+}$) orally infected mice, Gal-3 was highly expressed in the leukocytes infiltrating the intestines, liver, lungs, and brain. Compared with *gal3*$^{+/+}$, infected *gal3*$^{-/-}$ mice developed reduced inflammatory response in all of these organs but the lungs. Brain of *gal3*$^{-/-}$ mice displayed a significantly reduced number of infiltrating monocytes/macrophages and $CD8^+$ cells and a higher parasite burden. It shows that Gal-3 exerts an important role in innate immunity, including not only a pro-inflammatory effect but also a regulatory role on dendritic cells, capable of interfering in the adaptive immune response.

Gal-3 exerts cytokine-like regulatory actions in rat and mouse brain-resident immune cells. Both the expression of Gal-3 and its secretion into the extracellular compartment were significantly enhanced in glia under IFN-γ-stimulated, inflamed conditions. After exposure to Gal-3, glial cells produced high levels of proinflammatory mediators and exhibited activated properties. Notably, within minutes after exposure to Gal-3, JAK2 and STAT1, STAT3, and STAT5 showed considerable enhancement of tyrosine phosphorylation; thereafter, downstream events of STAT signaling were also significantly enhanced. Thus, Gal-3 acts as an endogenous danger signal under pathological conditions in the brain, providing a potential explanation for the molecular basis of Gal-3-associated pathological events (Jeon et al. 2010).

Gal-3 as an Opsonin: Galectins recognizes saccharide ligands on a variety of microbial pathogens, including viruses, bacteria, and parasites. It has been suggested that Gal-3 may serve as a pathogen pattern recognition receptor to visualize PAMPs from bacteria (Mandrell et al. 1994; Gupta et al. 1997), parasites (Pelletier and Sato 2002; van den Berg et al. 2004), and fungi (Kohatsu et al. 2006; Fradin et al. 2000). Gal-3, a galectin expressed by macrophages, dendritic cells, and epithelial cells, binds bacterial and parasitic pathogens including *Leishmania major*, *Trypanosoma cruzi*, and *Neisseria gonorrhoeae*. Gal-3 bound only to *Candida albicans* species that bear α-1,2-linked oligomannans on the cell surface, but did not bind *Saccharomyces cerevisiae* that lacks α-1,2-linked oligomannans. Binding induced death of *Candida* species containing specific α-1,2-linked oligomannosides by directly interaction (Kohatsu et al. 2006).

Gal-3 has been demonstrated to have antimicrobial activity toward the pathogenic fungus *C. albicans* (Kohatsu et al. 2006). After pneumococcal infection of lungs, Gal-3 accumulates in the alveolar space, and this correlates with the onset of neutrophil extravasation. However, although neutrophils were actively recruited into *E. coli* pneumonia-infected lungs, there was no increase in Gal-3 expression. Furthermore Gal-3 was released by alveolar macrophages on incubation with *S. pneumoniae* membrane fraction (Sato et al. 2002). In addition, LPS expressed on *E. coli* has been shown to down-regulate Gal-3 expression (Sato et al. 2002). The antimicrobial activity may also be relevant for other pathogens, thus revealing an interesting therapeutic use of this galectin. However, Gal-3 plays a role in leukocyte recruitment in a murine model of lung infection by *S. pneumoniae* (Nieminen et al. 2008). Gal-3$^{-/-}$ mice develop more severe pneumonia after infection with *S. pneumoniae*, as demonstrated by increased septicemia and lung damage compared to WT mice. Neutrophil recruitment to the alveolar space was reduced in Gal-3$^{-/-}$ mice; however, myeloperoxidase activity in lung homogenates was not reduced in these mice compared to WT. This would suggest that neutrophils accumulate in the interstitial lung tissue during pneumonia in Gal-3$^{-/-}$ mice but are hindered from transmigrating into the alveolar space in the absence of Gal-3 (Farnworth et al. 2008).

Although mouse neutrophils express very low levels of endogenous Gal-3 (Sato et al. 2002), they can be activated by extracellular Gal-3, which is up-regulated in the surrounding tissue environment after infection. However, in Gal-3-null mice this Gal-3 up-regulation does not occur, resulting in reduced neutrophil recruitment into the alveolar spaces, activation, and phagocytosis. Gal-3 expression is up-regulated after pneumococcal infection, and macrophages are capable of secreting large amounts of Gal-3 (Sato et al. 2002). Longevity of human neutrophils is increased after incubation with exogenous Gal-3, thus delaying spontaneous apoptosis in vitro. Furthermore, Gal-3$^{-/-}$ macrophages displayed reduced phagocytosis of apoptotic human neutrophils compared to WT. The resultant accumulation of apoptotic neutrophils in the lungs of Gal-3$^{-/-}$ mice after infection would cause considerable damage to lung tissue, thus allowing bacteria to traverse the lung epithelia and enter the blood stream resulting in septicemia.

During *S. pneumoniae* infection the Gal-3$^{-/-}$ mouse mounts a greater Th1 response as demonstrated by increased IL-6 and TNF-α cytokine levels compared to WT mice or Gal-3$^{-/-}$ mice treated with recombinant Gal-3. This increased Th1 response may contribute to lung damage and subsequent septicemia. Gal-3$^{-/-}$ macrophages demonstrated a deficit in their ability to adopt an anti-inflammatory alternative (M2) phenotype (MacKinnon et al. 2008). It was proposed that, in addition to reduced neutrophil activation and apoptotic neutrophil clearance by macrophages in the Gal-3$^{-/-}$ mice, these mice were less able to dampen down the excessive inflammation and destructive potential of pneumococcal infection.

Gal-3 can directly activate both human and mouse neutrophils and potentiate the effect of fMLP. Whole blood neutrophils can be directly activated by concentrations of Gal-3 lower than those that directly activate isolated neutrophils. This suggests that in whole blood there may be factors present in the serum (eg, GM-CSF) functioning to prime circulating neutrophils. The Gram-positive *Streptococcus pneumoniae* is the leading cause of community-acquired pneumonia worldwide, resulting in high mortality. In vivo studies show that Gal-3$^{-/-}$ mice develop more severe pneumonia after infection with *S. pneumoniae*, as demonstrated by increased bacteremia and lung damage compared to wild-type mice and that Gal-3 reduces the severity of pneumococcal pneumonia in part by augmenting neutrophil function. Specifically, (1) Gal-3 directly acts as a neutrophil-activating agent and potentiates the effect of fMLP, (2) exogenous Gal-3 augments neutrophil phagocytosis of bacteria and delays neutrophil apoptosis, (3) phagocytosis of apoptotic neutrophils by Gal-3$^{-/-}$ macrophages is less efficient compared to wild type, and (4) Gal-3 demonstrates bacteriostatic properties against *S. pneumoniae* in vitro. Furthermore, recombinant Gal-3 in vivo protects Gal-3-deficient mice from developing severe pneumonia. Studies demonstrate that Gal-3 is a key molecule in the host defense against pneumococcal infection. Strategies designed to augment Gal-3 expression in the lung may result in the development of novel treatments for pneumococcal pneumonia (Farnworth et al. 2008).

Modulation of Neutrophil Activation: Exogenously added Gal-3 increased the uptake of apoptotic neutrophils by monocyte-derived macrophages (MDM). The effect was lactose-inhibitable and required Gal-3 affinity for N-acetyllactosamine, a saccharide typically found on cell surface glycoproteins. Perhaps, Gal-3 functions as a bridging molecule between phagocyte and apoptotic prey, acting as an opsonin. Results imply that the increased levels of Gal-3 often found at inflammatory sites could potently affect apoptosis (Karlsson et al. 2009).

Human neutrophils are activated by Gal-3, provided cells are primed by in vivo extravasation or by in vitro preactivation with, for example, LPS. Removal of terminal sialic acid can change neutrophil functionality and responsiveness due to exposure of underlying glycoconjugate receptors or change in surface charge. Such alteration of cell surface carbohydrate composition can alter the responsiveness of the cells to Gal-3. Earlier studies had shown that priming of the neutrophil response to Gal-3 with LPS was paralleled by degranulation of intracellular vesicles and granules and upregulation of potential Gal-3 receptors (Almkvist et al. 2004).

While investigating the effects of Gal-3 on central effector functions of human neutrophils, Fernandez et al. (2005) supported the notion that Gal-3 and soluble fibrinogen are two physiological mediators present at inflammatory sites that activate different components of the MAPK pathway and could be acting in concert to modulate the functionality and life span of neutrophils. Whereas those activities are likely to be associated with ligand cross-linking by this lectin, Gal-3, unlike other members of the galectin family, exists as a monomer. Consequently, it was proposed that oligomerization of the N-terminal domains of Gal-3, after ligand binding by the C-terminal domain, is responsible for this cross-linking. The oligomerization status of Gal-3 could, thus, control the majority of its extracellular activities. Data of Nieminen et al. (2007) suggested that Gal-3 lattices are robust and could be involved in the restriction of receptor clustering. However, little is known about the actual mode of action through which Gal-3 exerts its function.

Annexin 1 (ANXA1), Gal-1 and Gal-3 regulate leukocyte migration. The expression of these proteins was studied in human neutrophils and endothelial cells (ECs) during a transmigration process induced by IL-8. ANXA1 and Gal-3 changed in their content and localization when neutrophils adhered to endothelia, suggesting a process of sensitive-balance between two endogenous anti- and pro-inflammatory mediators (Gil et al. 2006). Chronic morphine treatment in an *S. pneumoniae* infection model suppresses NF-kB gene transcription in lung resident cells, which, in turn, modulates the transcriptional regulation of MIP-2 and inflammatory cytokines. The decreased synthesis of MIP-2 and inflammatory cytokines coupled with the decreased release of Gal-3 result in reduced migration of neutrophils to the site of infection, thereby increasing susceptibility to *S. pneumoniae* infection after morphine treatment (Wang et al. 2005).

12.3.4 Regulation of T-Cell Functions

The rapid expansion of the field of galectin research has positioned Gal-3 as a key regulator of T-cell functions (Hsu et al. 2009). Gal-3 is absent in resting $CD4^+$ and $CD8^+$ T cells but is inducible by various stimuli. These include viral transactivating factors, T-cell receptor (TCR) ligation, and calcium ionophores. In addition, Gal-3 is constitutively expressed in human regulatory T cells and $CD4^+$ memory T cells. It exerts extracellular functions because of its lectin activity and recognition of cell surface and extracellular matrix glycans. Formation of lattices can result in restriction of receptor mobility and cause attenuation of receptor functions. Because of Gal-3 presence in intracellular locations, several functions have been described for Gal-3 inside T cells. These include inhibition of apoptosis, promotion of cell growth, and regulation of TCR signal transduction. Gal-3 takes part in control of T-cell and monocyte survival and activation. For instance, extracellular Gal-3, by associating with *N*-glycans on TCR has been shown to down-modulate TCR responsiveness and to regulate the production of Th1 and Th2 cytokines by differentiated T cells (Demetriou et al. 2001; Joo et al. 2001; Morgan et al. 2004). On the other hand, intracellular Gal-3 can positively or negatively impact intracellular signaling pathways by regulating the activities of various kinases, including protein kinases C, mitogen-activated protein kinase, and phosphatidylinositol 3-kinase (Chen et al. 2005). Studies of cell surface glycosylation have led to convergence of glycobiology and galectin biology and provided new clues on how Gal-3 may participate in the regulation of cell surface receptor activities (Hsu et al. 2009).

Ocklenburg et al. (2006) reported the identification of the ubiquitin-like gene, diubiquitin (UBD) as a downstream element of FOXP3 in human activated regulatory $CD4^+CD25^{hi}$ T cells (T_{reg}). Ocklenburg et al. (2006) suggested that UBD contributes to the anergic phenotype of human regulatory T cells and acts downstream in FOXP3 induced regulatory signaling pathways, including regulation of *LGALS3* expression. High level of *LGALS3* expression represents a FOXP3-signature of human antigen-stimulated $CD4^+CD25^{hi}$ derived regulatory T cells (Ocklenburg et al. 2006).

12.3.5 Pro-apoptotic and Anti-apoptotic Effects

Extracellular Gal-3 induces T-cell apoptosis (Fukumori et al. 2003) whereas intracellular Gal-3 results in an inhibition of apoptosis (Yang et al. 1996). Furthermore, peritoneal macrophages taken from Gal-$3^{-/-}$ mice are more prone to undergo apoptosis than wild-type (WT) macrophages (Hsu et al. 2000). Clearance of apoptotic neutrophils by macrophages is a key step in the resolution of inflammation. Without this step, apoptotic neutrophils will undergo secondary necrosis resulting in the release of damaging toxic products. Removal of potentially toxic apoptotic neutrophils results in the release of anti-inflammatory and preparative cytokines such as TGF-β1. These clearance and resolution phases help to limit the degree of tissue injury. Intracellular Gal-3 suppresses drug induced apoptosis and anoikis (apoptosis induced by the loss of cell anchorage) that contribute to cell survival. Resistance to apoptosis is essential for cancer cell survival and plays a role in tumor progression. Conversely, tumor cells' secreted Gal-3 induces T-cells' apoptosis, thus playing a role in immune escape mechanism during tumor progression through induction of apoptosis of cancer-infiltrating T-cells. Cells with over-expresed Gal-3 display increased resistance to the apoptotic stimuli induced by the anti-Fas antibody, staurosporine, TNF, radiation and nitric oxide (Akahani et al. 1997; Matarrese et al. 2000; Moon et al. 2001; Yang et al. 1996). Gal-3 has significant sequence similarity with Bcl-2 protein, a suppressor of apoptosis. The four amino acid motif, Asn-Trp-Gly-Arg (NWGR) of lectin is a highly conserved within BH1 domain of the Bcl-2 family proteins and is crucial for Bcl-2 protein function in the inhibition of programmed cell death (Akahani et al. 1997; Yang et al. 1996). An amino acid substitution of Gly to Ala at position 182 in this motif of Gal-3 prevents its anti-apoptotic activity (Akahani et al. 1997). Yang et al. (1996) demonstrated that Gal-3 can interact with Bcl-2 in a lactose-inhibitable manner, which is surprising since Bcl-2 is not a glycoprotein. Yang et al. (1996) suggested that the Asn-Trp-Gly-Arg motif is present within CRD of Gal-3, and is closely involved in interaction with Bcl-2. Lactose binding to Gal-3 may induce a conformational change in the critical region of this protein, which prevents its interaction with Bcl-2. Though the mechanism by which Gal-3 regulates apoptosis induced by different agents remains to be elucidated, it is possible that Gal-3 can replace or mimic Bcl-2 protein. Bcl-2 is located on the outer membranes of mitochondria and regulates apoptosis by blocking the release of cytochrome C from the mitochondria (Nagata 2000; Zörnig et al. 2001). Moon et al. (2001) showed that Gal-3 inhibition of nitrogen free radical-mediated apoptosis in breast carcinoma BT549 cells involved the protection of mitochondrial integrity, the inhibition of cytochrome C release and the activation of caspase. Thus, Gal-3 appears to be a mitochondrial-associated apoptotic regulator in addition to Bcl-2 (Matarrese et al. 2000; Moon et al. 2001). Studies have demonstrated that Gal-3 translocates into the mitochondrial membrane following a variety of apoptotic stimuli (Yu et al. 2002). Moreover, Gal-3 prevents mitochondrial damage and cytochrome C release. Such a location of Gal-3 is regulated by Gal-3 interacting proteins such as a synexin, a 51 kDa member of annexin family of proteins, which can bind to phospholipid membranes. Gal-3 regulates the time course of the apoptotic process in pancreatic acinar cells (Gebhardt et al. 2004).

Anti-apoptotic Function of Gal-3 and Cell Cycle Arrest: Gal-3 can protect against apoptosis induced by loss of cell anchorage (anoikis). Although BT549 cells (human breast epithelial cells) undergo anoikis, Gal-3-overexpressing BT549 cells respond to the loss of cell adhesion by inducing G1 arrest without detectable cell death. Gal-3-mediated G1 arrest involves down-regulation of G1-S cyclin levels (cyclin E and cyclin A) and up-regulation of their inhibitory protein levels (p21(WAF1/CIP1) and p27KIP1). After the loss of cell anchorage, Rb protein becomes hypophosphorylated in Gal-3-overexpressing cells. Interestingly, Gal-3 induces cyclin D1 expression (an early G1 cyclin) and its associated kinase activity in the absence of cell anchorage. Kim et al. (1999) proposed that Gal-3 inhibition of anoikis involves cell cycle arrest at an anoikis-insensitive point (late G1) through modulation of gene expression and activities of cell cycle regulators. The study suggests that Gal-3 may be a critical determinant for anchorage-independent cell survival of disseminating cancer cells in the circulation during metastasis. Studies have revealed that Gal-3 suppresses apoptosis and anoikis that contribute to cell survival during metastatic cascades. Human Gal-3 undergoes post-translational signaling modification of Ser-6 phosphorylation that acts as an "on/off" switch for its sugar-binding activity. Nakahara et al. (2005) summarized evidences on the role of Gal-3 as an anti-apoptotic and/or pro-apoptotic factor in various cell types and discussed the role of Gal-3 in apoptosis.

Apart from cell cycle arrest at late G1 in response to loss of cell adhesion, Gal-3 influences G2/M arrest of BT 549 cells following genistein treatment (Lin et al. 2000). Lin et al. (2000) showed that genistein effectively induces apoptosis without detectable cell cycle arrest in BT 549, which does not express Gal-3 at a detectable level. It is also likely that nuclear Gal-3 may directly modulate gene expression through regulation of transcription and/or mRNA splicing (Lin et al. 2002). Lin et al. (2002) showed that Gal-3 induces cyclin D1 promoter activity in BT 549 cells through multiple cis-elements, including SP1 and CREB binding sites. The Gal-3 induction of cyclin D1 promoter activity may result from stabilization of nuclear protein-DNA complex formation at the CRE site of cyclin D1 promoter.

Studies suggest that Gal-3 phosphorylation is required for its anti-apoptotic activity and anti-anoikis activity (Yoshii et al. 2002). Human Gal-3 is phosphorylated at Ser 6 by casein kinase I. Ser 6 phosphorylation of human Gal-3 significantly reduces its binding to ligands, e.g. laminin and asialomucin, while dephosphorylation fully restores the sugar binding activity (Mazurek et al. 2000). Yoshii et al. (2002) demonstrated that Gal-3 phosphorylation can also influence its other biological activities. Ser 6 mutation resulted in a relative decline in the level of Gal-3's ability to protect cells against *cis*platin-induced cell death and poly(ADP-ribose)polymerase from degradation compared with WT Gal-3.

Gal-3 and Gal-1 Bind Distinct Cell Surface Receptors to Induce T Cell Apoptosis: Extracellular Gal-1 directly induces death of T cells and thymocytes, while intracellular Gal-3 blocks T cell death. In contrast to the antiapoptotic function of intracellular Gal-3, extracellular Gal-3 directly induces death of human thymocytes and T cells. However, events in Gal-3- and Gal-1-induced cell death differ in a number of ways. Thymocyte subsets demonstrate different susceptibility to the two galectins: whereas Gal-1 kills double-negative and double-positive human thymocytes with equal efficiency, Gal-3 preferentially kills double-negative thymocytes. Gal-3 binds to a complement of T cell surface glycoprotein receptors distinct from that recognized by Gal-1. Of these glycoprotein receptors, CD45 and CD71, but not CD29 and CD43, appear to be involved in Gal-3-induced T cell death. In addition, CD7 that is required for Gal-1-induced death is not required for death triggered by Gal-3. Following Gal-3 binding, CD45 remains uniformly distributed on the cell surface, in contrast to the CD45 clustering induced by Gal-1. Thus, extracellular Gal-3 and Gal-1 induce death of T cells through distinct cell surface events. However, as Gal-3 and Gal-1 cell death are neither additive nor synergistic, the two death pathways may converge inside the cell (Stillman et al. 2006).

Gal-3 as a CD95-Binding Partner in Selecting Apoptotic Signaling Pathways: Studies on CD95 (APO-1/Fas), a member of death receptor family, have revealed that it is involved in two primary CD95 apoptotic signaling pathways, one regulated by the large amount of active caspase-8 (type I) formed at the death-inducing signaling complex and the other by the apoptogenic activity of mitochondria (type II). It is still unclear which pathway has to be activated in response to an apoptotic insult. Fukumori et al. (2004) demonstrated that the Gal-3, which contains the four amino acid-anti-death-motif (NWGR) conserved in the BH1 domain of the Bcl-2 member proteins, is expressed only in type I cells. In addition, Gal-3 is complexed with CD95 in vivo identifying Gal-3 as a novel CD95-binding partner that determines which of the CD95 apoptotic signaling pathways the cell will select (Fukumori et al. 2004).

Repression of Gal-3 by HIPK2-Activated p53: A New Apoptotic Pathway: It has been demonstrated that p53-induced apoptosis is associated with transcriptional repression of Gal-3. Phosphorylation of p53 at Ser46 is important for transcription of proapoptotic genes; induction of apoptosis and the homeodomain-interacting protein kinase 2 (HIPK2) is specifically involved in these functions. Cecchinelli et al. (2006) showed that HIPK2 cooperates with p53 in Gal-3 repression and this cooperation required HIPK2 kinase activity. Gene-specific RNA interference demonstrated that HIPK2 is essential for repression of Gal-3 upon induction of p53-dependent apoptosis. Furthermore, the expression of a nonrepressible Gal-3 prevents HIPK2- and p53-induced apoptosis. These results revealed a new apoptotic pathway induced by HIPK2-activated p53 requiring repression of the antiapoptotic factor Gal-3.

Nucling Mediates Apoptosis by Inhibiting Expression of Gal-3: Nucling is involved in cytochrome c/Apaf 1/caspase-9 apoptosome induction following pro-apoptotic stress. It is able to interact with Gal-3, which participates in apoptotic cell death. Nucling was found to down-regulate the expression level of Gal-3 mRNA/protein. Nucling-deficient cells, in which Gal-3 expression is up-regulated, appeared to be resistant to some forms of pro-apoptotic stress as compared with wild-type cells. In addition, the preputial gland from Nucling-deficient mice expressed a significant level of Gal-3 and exhibited a high incidence of inflammatory lesions, indicating that Nucling plays a crucial role in the homoeostasis of this gland by interacting with the Gal-3 and regulating the expression of Gal-3. Up regulation of Gal-3 was also observed in the heart, kidney, lung, testis and ovary of the Nucling-deficient mice. Nucling was shown to interfere with NF-kB activation via the nuclear translocation process of NF-kB/p65, thus inhibiting the expression of Gal-3. It was proposed that Nucling mediates apoptosis by interacting and inhibiting expression of Gal-3 (Liu et al. 2004).

Gal-3 Gene Encodes a Mitochondrial Protein that Promotes Cytochrome C Release: A Gal-3 internal gene (*Galig*) as an internal gene transcribed from the second intron of the human Gal-3 gene is implicated in cell growth, cell differentiation and cancer development. *Galig* expression causes morphological alterations in human cells, such as cell shrinkage, cytoplasm vacuolization, nuclei condensation, and ultimately cell death. *Galig* encodes a mitochondrial-targeted protein named mitogaligin. Structure studies revealed that the mitochondrial mitogaligin relies on an internal sequence that is required for the release of cytochrome c and cell death upon cell transfection. Moreover, incubation of isolated mitochondria with peptides derived from mitogaligin induces cytochrome c release. Thus, *Galig* is a cell death gene that encodes mitogaligin, a protein promoting cytochrome c release upon direct interaction with the mitochondria (Duneau et al. 2005).

12.3.6 Role in Inflammation

The role of Gal-3 in promotion and control of inflammation and in the regulation of the immune response has been evaluated in different models. Gal-3 has been proposed to be a powerful proinflammatory signal in vitro (Hsu et al. 2000) and in Gal-3-deficient mice (Zuberi et al. 2004). For instance, the lack of Gal-3 protects mice against asthma reactions, an effect associated with an enhanced Th1 response in challenged animals (Zuberi et al. 2004). On the other hand, administration of Gal-3 (by means of gene therapy) inhibits asthmatic reactions (Lopez et al. 2006). These studies reveal opposite effects of the endogenous and exogenous Gal-3 and suggest that, according to its location (extracellular versus intracellular), Gal-3 differentially controls immune/inflammatory cell survival, migration, and cytokine release. Along with its role in inflammation and immune responses in noninfectious conditions, Gal-3 can also sense certain microorganisms through multimeric LacNAc and LacdiNAc forming GalNAcß1-4GlcNAc motifs (van den Berg et al. 2004) as well as to other carbohydrate structures on glycoproteins and glycolipids (including LPS) from many pathogens such as (myco)bacteria, protozoan parasites, and yeast (Ochieng et al. 2004). Through this activity, Gal-3 participates in the phagocytosis of some microorganisms by macrophages and triggers host responses to pathogens, at least in vitro (Jouault et al. 2006; Sano et al. 2003). Whether it also contributes to signaling events in accessory cells during in vivo infection is unclear. Likewise, its role in signaling pathways triggered by Toll-like receptors (TLRs), which represent key sensors in innate cells (Janeway and Medzhitov 2002), is still elusive. The in vivo role of Gal-3 during infection has not been extensively studied. It appears to clear late *Mycobacterium tuberculosis* infection (Beatty et al. 2002), to contribute to neutrophil recruitment at the site of *Streptococcal pneumonia* infection (Sato et al. 2002), and to exert an important role in innate immunity during *Toxoplasma gondii* infection (Bernardes et al. 2006).

Gal-3 is a Negative Regulator of LPS-Mediated Inflammation: LPS, a major pathogen-associated molecular pattern (PAMP) from the outer membrane of Gram-negative bacteria, is a potent immune activator closely associated with many infectious and inflammatory diseases. LPS consists of hydrophobic lipid A, the 0-polysaccharide chain, and a core oligosaccharide that may be recognized differently by the host immune system. The visualization of LPS requires the TLR4 complex and triggers MyD88-dependent and independent signaling pathways. This leads to the activation of NF-κB and kinases, including MAPK, ERK1/2, p38, and JNK, which subsequently turn on the expression of many inflammatory genes including NADPH oxidase and inducible NO synthase. Since LPS is a powerful immune activator and may be fatal, the response to LPS must be tightly regulated to maintain the immune response at an appropriate level (Takeda et al. 2003; Miller et al. 2005; Liew et al. 2005; Li et al. 2008). Gal-3 interacts with LPS via both N′ and C′ terminals (Mey et al. 1996).

Li et al. (2008) demonstrated that Gal-3 is a negative regulator for LPS function and negatively regulates TCR-mediated CD4+ T-cell activation at the immunological synapse (Chen et al. 2009). Macrophages spontaneously express Gal-3, which specifically binds to LPS. Macrophages from Gal-3-deficient mice have elevated inflammatory cytokine production in response to LPS and lipid A compared with wild-type cells. This is accompanied by an increased phosphorylation of JNK, p38, ERK, and NF-κBp65. The increased inflammatory cytokine production by Gal-3 knockout cells could be normalized by recombinant Gal-3 protein. In vivo, mice lacking Gal-3 excessively produced inflammatory cytokines and NO and were more susceptible to LPS shock. On the other hand, such mice were more resistant to *Salmonella* infection due to the skewing of a Th1 response and increasing the levels of NO and hydrogen peroxide. Thus, Gal-3 is a natural negative regulator for LPS function, which protects against endotoxic shock but may be detrimental by helping in early *Salmonella* infection.

Modulation of T Cell Activity in the Inflamed Intestinal Epithelium: Gal-3, although mostly described as proinflammatory, can also act as an immunomodulator by inducing apoptosis in T cells. In inflamed intestinal mucosa Gal-3 is expressed at comparable levels as in controls and inflammatory bowel disease (IBD) patients in remission. In the normal mucosa, Gal-3 protein was mainly observed in differentiated enterocytes, preferentially at the basolateral side. However, Gal-3 was significantly downregulated in inflamed biopsies from IBD patients. Study suggested that down-regulation of epithelial Gal-3 in the inflamed mucosa reflects a normal immunological consequence, whereas under noninflammatory conditions, its constitutive expression may help to prevent inappropriate immune responses against commensal bacteria or food compounds. Hence, Gal-3 may be a useful parameter for manipulating disease activity (Muller et al. 2006).

Gal-3 in intestine binds to specific *C. albicans* glycans and is involved in inflammation. Inflammation strongly promoted *C. albicans* colonization. Conversely, *C. albicans* augmented inflammation induced by dextran sulfate sodium (DSS). The absence of Gal-3 reduced DSS inflammation and abolished the response of TLR-2 and TNF-α to *C. albicans* colonization. DSS-induced colitis provides a model for establishing *C. albicans* colonization in mice. This model reveals that *C. albicans* augments inflammation and confirms the role of Gal-3 in both inflammation and the control of host responses to *C. albicans* (Jawhara et al. 2008).

Gal-3 Regulates JNK Gene in Mast Cells: Gal-3 plays an important role in mast cell biology. To determine the role of Gal-3 in the function of mast cells, Chen et al. (2006) studied bone marrow-derived mast cells (BMMC) from wild-type ($gal3^{+/+}$) and Gal-3-deficient ($gal3^{-/-}$) mice. Results indicated that there is a defect in the response of mast cells in $gal3^{-/-}$ mice. Unexpectedly, $gal3^{-/-}$ BMMC showed a dramatically low level of JNK1 protein compared with $gal3^{+/+}$ BMMC, which was probably responsible for the lower IL-4 production. The decreased JNK1 level in $gal3^{-/-}$ BMMC was accompanied by a lower JNK1 mRNA level, suggesting that Gal-3 regulates the transcription of the JNK gene or processing of its RNA (Chen et al. 2006).

12.3.7 Gal-3 in Wnt Signaling

Gal-3, a pleiotrophic protein, is an important regulator of tumor metastasis, which, like, β-catenin shuttles between the nucleus and the cytosol in a phosphorylation-dependent manner. Reports show that β-catenin stimulation of cyclin D1 and c-myc expression is Gal-3 dependent. Gal-3 binds to β-catenin/Tcf complex, colocalizes with β-catenin in the nucleus, and induces the transcriptional activity of Tcf-4. The β-catenin- Gal-3-binding sequences which are in the NH2 and COOH termini of the proteins encompassing amino acid residues 1–131 and 143–250 respectively have been identified. Report indicated that Gal-3 is a binding partner for β-catenin involved in the regulation of Wnt/β-catenin signaling pathway (Shimura et al. 2004). The human Gal-3 sequence revealed a structural similarity to β-catenin as it also contains the consensus sequence (S92XXXS96) for glycogen synthase kinase-3β (GSK-3β) phosphorylation and can serve as its substrate. In addition, Axin, a regulator protein of Wnt that complexes with β-catenin, also binds Gal-3 using the same sequence motif identified by a deletion mutant analysis. Shimura et al. (2005) gave credence to the suggestion that Gal-3 is a key regulator in the Wnt/β-catenin signaling pathway and highlighted the functional similarities between Gal-3 and β-catenin.

12.3.8 In Urinary System of Adult Mice

Gal-3 is expressed in murine urinary system, starting from kidney to distal end of urethra; renal cortex expresses Gal-3 more intensely than medulla. An EM study demonstrated diffuse cytoplasmic localization of Gal-3 in principal cells of the collecting ducts and in the bladder epithelial cells. Urethral Gal-3 expression at pars spongiosa decreased in intensity near external urethral orifice, where predominant subtype of galectin was substituted by galectin-7. Observations indicated that adult urinary system shows intense and selective expression of Gal-3 in epithelia of the uretic bud- and cloaca-derivatives (Nio et al. 2006). Gal-3 modulates collecting duct growth/differentiation in vitro, and is expressed in human autosomal recessive polycystic kidney disease in cyst epithelia, almost all of which arise from collecting ducts. Moreover, exogenous Gal-3 restricts growth of cysts generated by Madin-Darby canine kidney collecting duct-derived cells in three-dimensional culture in collagen. The results suggest that Gal-3 may act as a natural brake on cystogenesis in cpk mice, perhaps via ciliary roles (Chiu et al. 2006). Bichara et al. (2006) Showed that Gal $3^{-/-}$ mice have mild renal chloride loss, which causes chronic ECF volume contraction and reduced blood pressure levels.

The adaptation of cortical collecting duct (CCD) to metabolic acidosis requires the polymerization and deposition in the extracellular matrix of the protein hensin. $HCO3^-$-secreting β-intercalated cells remove apical Cl^- : $HCO3^-$ exchangers and may reverse functional polarity to secrete protons. In intercalated cells in culture, Gal-3 facilitated hensin polymerization thereby causing their differentiation into the H^+-secreting cell phenotype. It appeared that Gal-3 may play important roles in the CCD, including mediating the adaptation of β-intercalated cells during metabolic acidosis (Schwaderer et al. 2006).

12.3.9 Gal-3 in Reproductive Tissues

Gal-3 expression has been identified in human, rat and porcine testes where it is under hormonal control. Gal-3 is present in Sertoli cells and appears to be absent in human and in rat germ cells. Expression of 31 kDa Gal-3 in cultured porcine Sertoli cells was under positive control of FSH as well as of cytokines EGF and TNF-α. Gal-3 expression in Sertoli cells is also under the control of mature germ cells since an increased expression was observed in adult rat testes depleted in spermatocytes or spermatids (Deschildre et al. 2007). In human testes, Gal-3 is specifically expressed in mature Sertoli cells and Leydig cells, and is absent from fetal and pre-pubertal testes, suggesting a hormone-dependence of this gene (Devouassoux-Shisheboran et al. 2006). Presence of Gal-3 in the connective tissues in the male reproductive organs suggests its role in extracellular matrix (Kim et al. 2006). Gal-1 and Gal-3 are predominantly expressed in the mouse ovary, where they mediate progesterone production and metabolism in luteal cells via different mechanisms (Nio and Iwanaga 2007). In human ovaries, Gal-3 is absent from granulosa cells, as well as from granulosa cell and Sertoli-Leydig cell tumors, and is not a useful marker in Sertoli-Leydig cell tumors (Devouassoux-Shisheboran et al. 2006). Conceptus tissue expresses potential receptors for endometrial HT1 antigen. Carbohydrate-lectin interactions may facilitate attachment of the apical surfaces of uterine epithelial cells and trophectoderm during the early stages of placentation (Woldesenbet et al. 2004).

12.3.10 Gal-3 on Chondrocytes

Endochondral bone formation is characterized by the progressive replacement of cartilage anlagen by bone at the growth plate with a tight balance between the rates of chondrocyte proliferation, differentiation, and cell death. Gal-3 is detected in mature and early hypertrophic chondrocytes. Notochord-specific expression of a prototype galectin has been characterized during early embryogenesis in zebrafish (Danio rerio) (Ahmed et al. 2004). Gal-3 may play a part in osteoarthritis (OA), possibly related to the interaction of chondrocytes and the cartilage matrix (Guevremont et al. 2004). Presence of Gal-3 at the surface of chondrocytes shows a strong correlation with integrin-β1. Deficiency of matrix metalloproteinase-9 (MMP-9) leads to an accumulation of late hypertrophic chondrocytes. It was found that Gal-3, the in vitro substrate of MMP-9, accumulates in the late hypertrophic chondrocytes and their surrounding ECM in the expanded hypertrophic cartilage zone. These results indicated that extracellular Gal-3 could be an endogenous substrate of MMP-9 that acts downstream to regulate hypertrophic chondrocyte death and osteoclast recruitment during endochondral bone formation. Thus, the disruption of growth plate homeostasis in *Mmp*-9 null mice links Gal-3 and MMP-9 in the regulation of the clearance of late chondrocytes through regulation of their terminal differentiation (Ortega et al. 2005; Li et al. 2009). Because of its ubiquitous expression, Gal-3 cannot be used as a marker of notochordal cells in the postnatal rat disc (Oguz et al. 2007).

12.3.11 Role of Gal-3 in Endothelial Cell Motility and Angiogenesis

During cross-talk between pericytes and endothelial cells (EC), binding of soluble NG2 proteoglycan to the EC surface induces cell motility and multicellular network formation in vitro and stimulated corneal angiogenesis in vivo. The process seems to involve both Gal-3 and α3β1 integrin in the EC response to NG2 and the formation of a complex on cell surface involving NG2, Gal-3, and α3β1. It appears that Gal-3-dependent oligomerization may potentiate NG2-mediated activation of α3β1. In conjunction with earlier studies, this study suggested that pericyte-derived NG2 is an important factor in promoting EC migration and morphogenesis during the early stages of neovascularization (Fukushi et al. 2004).

In two different models of corneal wound healing, re-epithelialization of wounds was significantly slower in Gal-3-deficient (gal3$^{-/-}$) mice compared with wild-type (gal3$^{+/+}$) mice. Exogenous Gal-3 accelerated re-epithelialization of wounds in gal3$^{+/+}$ mice but, surprisingly, not in the gal3$^{-/-}$ mice. In corresponding experiments, recombinant Gal-1 did not stimulate the corneal epithelial wound closure rate. The extent of acceleration of re-epithelialization of wounds with both Gal-3 and galectin-7 was greater than that observed in most of the published studies using growth factors. These findings have broad implications for developing novel therapeutic strategies for treating nonhealing wounds (Cao et al. 2002a, b).

12.3.12 Role in CNS

Microglia is the major cell type expressing Gal-3 in CNS. Ablation of Gal-3 did not affect PrP(Sc)-deposition and development of gliosis. However, Gal-3$^{-/-}$-mice showed prolonged survival times upon intracerebral and peripheral scrapie infections. Gal-3 plays a detrimental role in prion infections of the CNS, and the endo-/lysosomal dysfunction in combination with reduced autophagy may contribute to disease development (Mok et al. 2007). Schwann cells are considered to be closely involved in the success of peripheral nerve regeneration. The absence of Gal-3 allowed faster regeneration, which may be associated with increased growth of Schwann cells and expression of beta-catenin. This would favor neuron survival, followed by faster myelination, culminating in a better morphological and functional outcome (Narciso et al. 2009). Depletion of Gal-3 from MDCK cells results in missorting of non-raft-dependent apical membrane proteins to the basolateral cell pole. This suggests a direct role of Gal-3 in apical sorting as a sorting receptor (Delacour et al. 2006).

12.4 Clinical Manifestations of Gal-3

12.4.1 Advanced Glycation End Products (AGES)

The accumulation of irreversible advanced glycation end products (AGEs) on long-lived proteins, and the interaction of AGEs with cellular receptors such as AGE-R3/Gal-3 and receptors for AGE (RAGE) are considered the key events in the development of long-term complications of diabetes mellitus, Alzheimer's disease, uremia and ageing.

The expression and sub-cellular distribution of Gal-3, as well as its possible modulation by AGEs, has been investigated in MC3T3E1 mouse calvaria-derived osteoblasts and in UMR 106 rat osteosarcoma cells, both cell lines express 30 kDa (monomeric) Gal-3. Dimeric (70 kDa) Gal-3 was also observed in the UMR106 cells. Exposure to AGEs-BSA increased the cellular content of 30 kDa Gal-3 and decreased Gal-3 in culture media. These results confirmed the expression of Gal-3 in osteoblastic cells. Osteoblastic exposure to AGEs alters the expression of Gal-3, which may have significant consequences on osteoblast metabolism and thus on bone turnover (Mercer et al. 2004).

Corneal endothelial cell loss is a change that occurs with age. Interaction between AGE and its receptors is implicated in the corneal endothelial cell loss with age. AGEs accumulate with ageing and may have a significant impact on age related dysfunction of the retinal pigment epithelium (RPE). Expression of RAGE and Gal-3 was detected in bovine corneal endothelial cells. Gal-3 was important in the internalization of AGE. In contrast, RAGE was important in the generation of reactive oxygen species and induction of apoptosis. Based on these data, the interaction of AGE in aqueous humor and AGE receptors expressed on the corneal endothelial cells was speculated to have a role in the corneal endothelial cell loss with age (Kaji 2005). Stitt et al. (2005) reported a significant suppression of angiogenesis by the retinal microvasculature during diabetes and implicated AGEs and AGE-receptor interactions in its causation (Stitt et al. 2005). The role of the Gal-3 receptor component was examined by transfection and overexpression using the D407 cell line. Primary cultures of bovine RPE cells and also a human cell line showed a pathological response to AGE exposure, an effect which appears to be modulated by the Gal-3 component of the receptor complex (McFarlane et al. 2005).

Mice lacking Gal-3/AGE-receptor 3 develop accelerated diabetic glomerulopathy. Gal-3 ablation is associated with increased susceptibility to diabetes- and AGE-induced glomerulopathy, thus indicating a protective role of Gal-3 as an AGE receptor (Iacobini et al. 2004, 2005). Gal-3 in chronic kidney disease induced by unilateral ureteral obstruction (UUO) showed significantly increased expression compared with basal levels. The degree of renal damage was more extensive in Gal-3-deficient mice at days 14 and 21. Therefore, Gal-3 not only protects renal tubules from chronic injury by limiting apoptosis but it may lead to enhanced matrix remodeling and fibrosis attenuation (Okamura et al. 2010)

12.4.2 GAL-3 and Protein Kinase C in Cholesteatoma

Cholesteatoma is a benign disease characterized by the presence of an unrestrained growth and the accumulation of keratin in the middle ear cavity. Ghanooni et al. (2006) studied the expression of protein kinase C-α, -δ, -γ, -η, and -ζ in epithelial tissues of human cholesteatomas and their correlations with distributions of p53, Gal-3, retinoic acid receptor-β and macrophage migration inhibitory factor (MIF). The patterns of PKC-α, and -δ expression, but not of PKC-γ, -η and -ζ correlated significantly and positively with Gal-3 expression. In addition, the correlation levels between the expression of PKC-α, and -δ and that of Gal-3 varied depending on the infection and recurrence status. Thus, modifications occurring at the level of keratinocyte differentiation in human cholesteatomas involve the activation of PKC-α, -δ, -γ, -η and -ζ (Ghanooni et al. 2006).

12.4.3 Gal-3 and Cardiac Dysfunction

Inflammatory mechanisms have been proposed to be important in heart failure (HF), and cytokines have been implicated to enhance the progression of HF. In a comprehensive study, Gal-3 has emerged as the most robustly overexpressed gene in failing *versus* functionally compensated hearts from homozygous transgenic TGRmRen2-27 (Ren-2) rats. Myocardial biopsies obtained at an early stage of hypertrophy before apparent HF showed that expression of Gal-3 was increased specifically in the rats that later rapidly developed HF. Gal-3 colocalized with activated myocardial macrophages. Gal-3-binding sites were found in cardiac fibroblasts and the extracellular matrix. Sharma et al. (2004) showed that an early increase in Gal-3 expression identifies failure-prone hypertrophied hearts. Gal-3, a macrophage-derived mediator, induces cardiac fibroblast proliferation, collagen deposition, and ventricular dysfunction. This implies that HF therapy needs to antagonize multiple inflammatory mediators, including Gal-3. Nonetheless, the amino-terminal pro-brain natriuretic peptide (NT-proBNP) is considered superior to either apelin or Gal-3 for diagnosis of acute HF, although Gal-3 levels were significantly higher in subjects with HF compared with those without. However, recent data showed potential utility of Gal-3 as a useful marker for evaluation of patients with suspected or proven acute HF, whereas apelin measurement was not useful for these indications. Moreover, the combination of Gal-3 with NT-proBNP was the best predictor for prognosis in subjects with acute HF (van Kimmenade et al. 2006).

In the ApoE-deficient mouse model of atherosclerosis Nachtigal et al. (2008) showed an age-related increase in the incidence of aorta atheromatous plaques and periaortic vascular channels in ApoE-deficient mice. By contrast ApoE/Gal-3 double-knockout mice did not show an increase in pathological changes with age. The effect of Gal-3 deficiency on atherogenesis decrease could be related to its function in macrophage chemotaxis, angiogenesis, lipid loading, and inflammation. Atherosclerosis and renal disease are related conditions, sharing several risk factors. This includes hyperlipidaemia, which may result in enhanced lipoprotein accumulation and chemical modification, particularly oxidation, with formation of advanced lipoxidation end products (ALEs). Mice knocked out for $Gal3^{-/-}$ have shown that Gal-3 exerts a significant role in the uptake and effective removal of modified lipoproteins with diversion of these products from RAGE-dependent pro-inflammatory pathways associated with downregulation of RAGE expression. Iacobini et al. (2009a, b) suggest a unique protective role for Gal-3 in the uptake and effective removal of modified lipoproteins, with concurrent down-regulation of proinflammatory pathways responsible for atherosclerosis initiation and progression.

12.4.4 Gal-3 and Obesity

In streptozotocin induced diabetic C57BL/6 Gal-$3^{+/+}$ and Gal-$3^{-/-}$ mice, Gal-3 is involved in immune mediated β cell damage and is required for diabetogenesis in this mouse model (Mensah-Brown et al. 2009). Gal-3 might be associated with the pathophysiology of obesity in obesity-prone C57BL/6 J mice (Kiwaki et al. 2007). Adipocytes synthesize Gal-3 whose deficiency protects from inflammation associated with metabolic diseases. Circulating Gal-3 was elevated in type 2 diabetes (T2D) and obesity compared with normal-weight individuals. In T2D patients, Gal-3 was increased in serum of patients with elevated C-reactive protein and negatively correlated with glycated hemoglobin. It was suggested that systemic Gal-3 is elevated in obesity and negatively correlates with glycated hemoglobin in T2D patients, pointing to a modifying function of Gal-3 in human metabolic diseases (Weigert et al. 2010). There remains a need for robust mouse models of diabetic nephropathy (DN) that mimic key features of advanced human DN. The recently developed mouse strain BTBR with the ob/ob leptin-deficiency mutation develops severe type 2 diabetes, hypercholesterolemia, elevated triglycerides, and insulin resistance. The abnormalities closely resemble advanced human DN more rapidly than most other murine models, making this strain particularly attractive for testing therapeutic interventions (Hudkins et al. 2010).

12.4.5 Autoimmune Diseases

Gal-3 in Induction of Type 1 Diabetes: Pro-apoptotic cytokines have been associated with the pathogenesis of Type 1 diabetes (T1D). Rat islets identified Gal-3 as the most up-regulated protein. Combined proteome-transcriptome-genome and functional analyses identified gal-3 as a candidate gene/protein in T1D susceptibility that may prove valuable in future intervention/prevention strategies (Karlsen et al. 2006). Macrophages are potent immune regulators in the development of autoimmune diabetes and are important effector cells during diabetogenesis. The role of macrophages in autoimmune diabetes has been described with particular emphasis on the role of Gal-3 and T1/ST2, an IL-1 receptor-like protein, both of which play significant roles in the immunomodulatory functions of macrophages in BALB/c mice. It appeared that functional capacity of macrophages influences their participation in Th1-mediated autoimmunity and the development of autoimmune diabetogenesis (Mensah-Brown et al. 2006).

Gal-3 in Rheumatoid Arthritis: Galectin 3 is present in the inflamed synovium in patients with rheumatoid arthritis suggesting that the protein is associated with the pathogenesis of this disease. In murine model, Gal-3 plays a pathogenic role in the development and progression of antigen-induced arthritis (AIA) and the disease severity is accompanied by alterations of antigen-specific IgG levels, systemic levels of TNFα and IL-6, and frequency of IL-17-producing T cells. Rheumatoid arthritis (RA) is a chronic debilitating autoimmune disease that results in joint destruction and subsequent loss of function. Gal-3 is expressed in synovial tissue of patients with RA, particularly at sites of joint destruction. Gal-3 is induced either by proinflammatory cytokines or by adhesion to cartilage components. Results suggest four times more RA synovial fibroblasts (SF) than osteoarthritis SF adhered to COMP (cartilage oligomeric matrix protein) coated plates. The adhesion of RA SF to COMP was found to increase the intracellular level of Gal-3. In contrast, intracellular Gal-3 decreased after exposure to TNF-α. It was concluded that the increase of Gal-3 occurs after adhesion to COMP, and the αVβ3 receptor (CD51/CD61) has a pivotal role in this process (Neidhart et al. 2005). To better understand the pathogenesis of RA, Shou et al. (2006) profiled the rat model of collagen-induced arthritis (CIA) to discover and characterize blood biomarkers for RA. Genes known to be involved in autoimmune response and arthritis, such as those encoding Gal-3, Versican, and Socs3, were identified and validated. Analysis confirmed that Gal-3 was secreted over time in plasma as well as in supernatant of cultured tissue synoviocytes of the arthritic rats, which is consistent with disease progression (Shou et al. 2006; Forsman et al. 2011). Gal-3 also plays an important disease-exacerbating role in EAE through its multifunctional roles in preventing cell apoptosis and increasing IL-17 and IFN-γ synthesis, but decreasing IL-10 production (Jiang et al. 2009).

Gal-3 in Psoriatic Skin: Gal-3 is critical for development of allergic inflammatory response in a mouse model of atopic dermatitis (Saegusa et al. 2009). Contrary to normal epidermis, the psoriatic epithelium does not express Gal-3 and glycoligands for Gal-1. Strong expression of Gal-3/Gal-3-reactive glycoligands in capillaries of psoriatic dermis represents one of the most important findings demonstrating the activation of endothelium in the course of the disease. The altered galectin expression and binding pattern in psoriatic skin indicates the modified process of keratinocyte maturation in hyperactivated psoriatic epithelium. The enhanced expression of Gal-3/Gal-3-reactive glycoligands in dermal capillaries of psoriatic skin can be important for rearrangement of the capillary network and migration of inflammatory cells to psoriatic skin (Lacina et al. 2006). Keratinocytes undergo apoptosis in a variety of physiological and pathological conditions. Gal-3 mRNA was transiently upregulated in ultraviolet-B (UVB)-irradiated wild-type keratinocytes. Gal-$3^{-/-}$ keratinocytes were significantly more sensitive to apoptosis induced by UVB as well as various other stimuli, both in vitro and in vivo, than wild-type cells. It suggests that endogenous Gal-3 is an anti-apoptotic molecule in keratinocytes functioning by suppressing ERK activation and enhancing AKT activation and may play a role in the development of apoptosis-related skin diseases (Saegusa et al. 2008).

12.4.6 Myofibroblast Activation and Hepatic Fibrosis

Gal-3 expression is up-regulated in human fibrotic liver disease and is temporarily and spatially related to the induction and resolution of experimental hepatic fibrosis. Disruption of the Gal-3 gene blocks myofibroblast activation and procollagen (I) expression in vitro and in vivo, markedly attenuating liver fibrosis. The reduction in hepatic fibrosis was observed in the gal$3^{-/-}$ mouse despite equivalent liver injury and inflammation, and similar tissue expression of TGF-β. TGF-β failed to transactivate gal$3^{-/-}$ hepatic stellate cells, in contrast with WT (gal$3^{+/+}$) hepatic stellate cells. It indicates that Gal-3 is required for TGF-β mediated myofibroblast activation and matrix production. Further, in vivo siRNA knockdown of Gal-3 inhibited myofibroblast activation after hepatic injury and may therefore provide an alternative therapeutic approach to the prevention and treatment of liver fibrosis (Henderson et al. 2006).

Gal-3 may play an important role in inflammatory responses. Non-alcoholic fatty liver disease (NAFLD) is increasingly recognized as a liver condition that may progress to end-stage liver disease. Based on the known functions of Gal-3, it was hypothesized that Gal-3 might play a role in the development of NAFLD. Study in Gal-3 knockout gal$3^{-/-}$ mice suggested that the absence of Gal-3 can cause clinicopathological features in male mice similar to those of NAFLD (Nomoto et al. 2006).

Colonic lamina propria fibroblasts (CLPFs) play an important role in the pathogenesis of fibrosis and strictures in Crohn's disease. Soluble Gal-3 was identified as a strong activator of CLPFs produced by CEC. Gal-3 induced NF-kB activation and IL-8 secretion in these cells may be a target for future therapeutic approaches to reduce or avoid stricture formation (Lippert et al. 2007).

12.5 Gal-3 as a Pattern Recognition Receptor

The galectin family of lectins recognizes saccharide ligands on a variety of microbial pathogens, including viruses, bacteria, yeasts, and parasites. Gal-3, expressed by macrophages, dendritic cells, and epithelial cells, binds bacterial and parasitic pathogens including *Leishmania major*, *Trypanosoma cruzi*, and *Neisseria gonorrhoeae*. Galectins may have direct effects on microbial viability. Kohatsu et al. (2006) showed that Gal-3 binds only to those *C. albicans* species that bear β-1,2-linked oligomannans on the cell surface, but did not bind *S. cerevisiae* that lacks β-1,2-linked oligomannans. Surprisingly, binding directly induced death of *Candida* species containing specific β-1,2-linked oligomannosides. Thus, Gal-3 can act as a pattern recognition receptor that recognizes a unique pathogen-specific oligosaccharide sequence (Kohatsu et al. 2006).

12.5.1 Gal-3 Binds to *Helicobacter pylori*

Helicobacter pylori causes gastritis and some infections result in peptic ulceration, gastric adenocarcinoma or gastric lymphoma. The role for the lipopolysaccharide O-antigen side-chain in this process has been identified. Evidence indicates that the receptor recognized by the O-antigen side-chain is Gal-3. Expression of Gal-3, is upregulated by gastric epithelial cells following adhesion of *H. pylori*, suggesting that in addition to colonization this protein also plays a role in the host response to infection. Upregulation of Gal-3 is inhibited by treating gastric epithelial cells with MAPK inhibitors and does not occur in cells infected with either *H. pylori* cagE or cagA isogenic mutants. This implies that *H. pylori*-mediated expression of Gal-3 is dependent on delivery of CagA into the host cell cytosol and the subsequent stimulation of MAPK signalling. A further consequence of *H. pylori* adhesion is that it elicits a rapid release of Gal-3 from infected cells (Fowler et al. 2006).

12.5.2 Recognition of *Candida albicans* by Macrophages Requires Gal-3

β-1,2-linked oligomannoside residues are associated with mannan and a glycolipid, the phospholipomannan, at the *Candida albicans* cell wall surface. β-1,2-linked oligomannoside residues act as adhesins for macrophages and stimulate these cells to undergo cytokine production. The macrophage receptor involved in the recognition of *C. albicans* β-1,2-oligomannoside repeatedly led to detection of a 32-kDa macrophage protein. The purified peptides from the 32-kDa tryptic digest showed complete homology to Gal-3, an endogenous lectin which is expressed in a wide variety of cell types with which *C. albicans* interacts as a saprophyte or a parasite (Fradin et al. 2000). Glycans present in both *C. albicans* and *S. cerevisiae* cell walls have been shown to act as ligands for different receptors leading to different stimulating pathways, some of which need receptor co-involvement. However, among these ligand-receptor couples, none has been shown to discriminate the pathogenic yeast *C. albicans*. Jouault et al. (2006) explored the role of Gal-3, which binds *C. albicans* β-1,2 mannosides and suggested that macrophages differently sense *C. albicans* and *S. cerevisiae* through a mechanism involving TLR2 and Gal-3, which probably associate for binding of ligands expressing β-1,2 mannosides specific to the *C. albicans* cell wall surface.

12.5.3 Gal-3 is Involved in Murine Intestinal Nematode and Schistosoma Infection

Conflicting studies have addressed the role of endogenous Gal-3 during *Schistosoma* infection (Bickle and Helmby 2007; Oliveira et al. 2007). Gal-3 recognizes the GalNAcβ1-4GlcNAc (LDN) epitope present on many helminth antigens, including those of the schistosome eggs. However, Gal-3 is not a critical component in the development of Th2 responses during helminth infection in vivo, nor it is essential for schistosome egg granuloma formation (Bickle and Helmby 2007). The N-terminal lectin domain (Nh) of the tandem repeat-type nematode galectin LEC-1 has a lower affinity for sugars than the C-terminal lectin domain. LEC-1 forms a complex with N-acetyllactosamine-containing glycoproteins. The formation of a crosslinked product with the Q38C mutant suggested the low-affinity interaction of Nh with the glycoprotein (Arata et al. 2006). van den Berg et al. demonstrated that LDN motifs, which are expressed by *Schistosoma mansoni* eggs (Khoo and Dell 2001; Srivatsan et al. 1992), bind Gal-3 and suggested that this interaction may play an important role in Th2-mediated inflammatory response that occurs in the liver (van den Berg et al. 2004). Oliveira et al. (2007) and Breuilh et al. (2007) showed that, relative to wild-type (WT) mice, *S. mansoni*-infected Gal-3-deficient mice develop reduced granuloma formation and have a dramatically decreased number of total lymphocytes in spleen. Although Gal-3 deficiency in DCs does not impact their differentiation and maturation processes, it greatly influences the strength (but not the nature) of adaptive immune response that they trigger, suggesting that Gal-3 deficiency in some other cell types may be important during murine schistosomiasis. As a whole, Gal-3 is a modulator of immune/inflammatory responses during helminthic infection and that Gal-3 expression in DCs is pivotal to control the magnitude of T-lymphocyte priming (Breuilh et al. 2007).

12.5.4 Up-Regulation of Gal-3 and Its Ligands by *Trypanosoma cruzi* Infection

Human Gal-3 binds to the surface of *Trypanosoma cruzi trypomastigotes* and human coronary artery smooth muscle (CASM) cells. CASM cells express Gal-3 on their surface and secrete it. Exogenous Gal-3 increased the binding of *T. cruzi* to CASM cells. Trypanosome binding to CASM cells was enhanced when either *T. cruzi* or CASM cells were preincubated with Gal-3. Thus, host Gal-3 expression is required for *T. cruzi* adhesion to human cells and exogenous Gal-3 enhances this process, leading to parasite entry (Kleshchenko et al. 2004). The expression of Gal-3 and its ligands on splenic DCs (sDCs) from *T. cruzi* infected mice are markedly up-regulated and adhesiveness is increased with Gal-3-coated substratum. Results documented that a parasitic infection can modulate both in vivo and in vitro the expression of Gal-3 and of ligands for this lectin in DCs with functional consequences on their capacities of adhesion and migration (Vray et al. 2004). During acute infection with *T. cruzi*, the thymus undergoes intense atrophy followed by a premature escape of $CD4^{+}CD8^{+}$ immature cortical thymocytes. Studies provide evidence of a role for Gal-3 in the regulation of thymus physiology and in identifying a potential mechanism based on protein-glycan interactions in thymic atrophy associated with acute *T. cruzi* infection (Silva-Monteiro et al. 2007).

12.6 Gal-3 as a Therapeutic Target

12.6.1 Gal-3: A Target for Anti-inflammatory/Anticancer Drugs

Dabelic et al. (2006) investigated the effects of non-steroidal anti-inflammatory drugs (aspirin and indomethacin) and glucocorticoids (hydrocortisone and dexamethasone) on macrophage Gal-3, which in general acts as a strong pro-inflammatory signal. All immunomodulatory drugs in clinically relevant doses affect both the gene (LGALS3) and protein expression level of Gal-3. Study revealed Gal-3 as a new target molecule of immunomodulatory drugs, thus suggesting an additional pathway of their action on immune response (Dabelic et al. 2006). Targeting efficacy of anticancer drug (doxorubicin) was improved when the drug was conjugated to N-(2-hydroxypropyl) methacrylamide (HPMA)-based copolymers with bio-recognizable groups, such as simple carbohydrates. HPMA-based copolymers are efficient carriers for anticancer drugs because of their good biocompatibility. Thus, the binding of the glycoside-bearing HPMA copolymer-DOX conjugates to the cells was mediated not only by Gal-3, but HPMA Copolymer conjugates bearing multivalent galactoside residues can improve their cytotoxicity (David et al. 2004).

Galectins-3 is implicated in asthma. In a murine model of ovalbumin (OVA)-induced asthma, (1) peribronchial inflammatory cells express large amounts of Gal-3; (2) bronchoalveolar lavage fluid from OVA-challenged mice contained higher levels of Gal-3 compared to control mice; and (3) macrophages in bronchoalveolar lavage fluid were the major cell type that contained Gal-3. Further investigations revealed that OVA-sensitized gal3$^{-/-}$ mice developed fewer eosinophils and lower goblet cell metaplasia, after airway OVA challenge compared to similarly treated gal3$^{+/+}$ mice. In addition, the OVA-sensitized gal3$^{-/-}$ mice developed significantly less airway hyperresponsiveness and developed a lower Th2 response, but a higher Th1 response, suggesting that Gal-3 regulates the Th1/Th2 response after airway OVA challenge. Thus, Gal-3 might play an important role in the pathogenesis of asthma and inhibitors of this lectin could prove useful for treatment of this disease (Zuberi et al. 2004). The inhibitory effects of Gal-3 on eosinophilic inflammation in guinea pig asthma models provided evidence for an eosinophil recruitment from bone marrow to circulation blood to lung in asthmatic response, in which overexpression of IL-5, Eotaxin, and CCR-3 could be involved. Gal-3, a selective inhibitor of IL-5 mRNA transcription, might potentially suppress eosinophilc inflammation and be a compromising specific anti-asthma reagent (Li et al. 2005). The role of galectins in chronic obstructive pulmonary disease (COPD), characterized by epithelial changes and neutrophil infiltration remains unknown. Studies support the hypothesis that distal airways represent an important site for detecting changes in COPD: viz., in patients with severe disease increased Gal-3 expression and neutrophil accumulation in the small airway epithelium, correlating with epithelial proliferation and airway obstruction (Pilette et al. 2007). In chronic asthmatic mice, treatment with Gal-3 gene led to an improvement in the eosinophil count and the normalization of hyperresponsiveness to methacholine. Concomitantly, this treatment resulted in an improvement in mucus secretion and sub-epithelial fibrosis in the chronically asthmatic mice, with a measured reduction in lung collagen, a prominent feature of airway remodeling (Lopez et al. 2006).

Studies have revealed that Gal-3 demonstrates anti-apoptotic effects which contribute to cell survival in several types of cancer cells. Intracellular Gal-3 in particular, which contains the NWGR anti-death motif of the Bcl-2 family, inhibits cell apoptosis induced by chemotherapeutic agent such as cisplatin and etoposide in some types of cancer cells. The nuclear export of phosphorylated Gal-3 regulates its anti-apoptotic activity in response to chemotherapeutic drugs. It is suggested that targeting Gal-3 could improve the efficacy of anticancer drug chemotherapy in several types of cancer. Gal-3, which plays an important role in the biology of angiosarcoma (ASA) in humans and hemangiosarcoma (HSA) in dogs is identified as a potential therapeutic target in tumors arising from malignant endothelial cells (Fukumori et al. 2007; Johnson et al. 2007) (See Chap, 13).

12.7 Xenopus-Cortical Granule Lectin: A Human Homolog of Gal-3

Cortical granule lectin (xCGL) is a candidate target glycoprotein of Xenopus galectin-VIIa (xgalectin-VIIa) in Xenopus embryos. In addition, another member of the xCGL family is xCGL2. Expression of the mRNAs of xCGL and xCGL2, as well as that of xgalectin-VIIa occurs throughout early embryogenesis. Two and three potential N-glycosylation sites were deduced from the amino acid sequences of xCGL and xCGL2, respectively. The xgalectin-VIIa recognizes N-glycans linked to a common site in xCGL and xCGL2 in addition to N-glycans linked to a site specific to xCGL2. However, interaction between xgalectin-Ia and xCGLs was not detectable. The oligosaccharide specificity pattern of xgalectin-VIIa was similar to that of human homolog Gal-3. The N-acetyllactosamine type, biantennary N-glycans exhibit high affinity for xgalectin-VIIa (K_D = 11 μM) families (Shoji et al. 2005).

References

Ahmad N, Gabius HJ, Andre S et al (2004a) Galectin-3 precipitates as a pentamer with synthetic multivalent carbohydrates and forms heterogeneous cross-linked complexes. J Biol Chem 279:10841–10847

Ahmad N, Gabius HJ, Sabesan S et al (2004b) Thermodynamic binding studies of bivalent oligosaccharides to Gal-1, galectin-3, and the carbohydrate recognition domain of galectin-3. Glycobiology 14:817–825

Ahmed H, Du S-J, O'Leary N, Vasta GR (2004) Biochemical and molecular characterization of galectins from zebrafish (Danio rerio): notochord-specific expression of a prototype galectin during early embryogenesis. Glycobiology 14:219–232

Akahani S, Nangia-Makker P, Inohara H et al (1997) Galectin-3: a novel antiapoptotic molecule with a functional BH1 (NWGR) domain of Bcl-2 family. Cancer Res 57:5272–5276

Almkvist J, Dahlgren C, Leffler H, Karlsson A (2004) Newcastle disease virus neuraminidase primes neutrophils for stimulation by galectin-3 and formyl-Met-Leu-Phe. Exp Cell Res 298:74–82

Arata Y, Tamura M, Nonaka T, Kasai K (2006) Crosslinking of low-affinity glycoprotein ligands to galectin LEC-1 using a photoactivatable sulfhydryl reagent. Biochem Biophys Res Commun 350:185–190

Barboni E, Coade S, Fiori A (2005) The binding of mycolic acids to galectin-3: a novel interaction between a host soluble lectin and trafficking mycobacterial lipids? FEBS Lett 579:6749–6755

Bawumia S, Barboni EAM, Menon RP, Hughes RC (2003) Specificity of interactions of galectin-3 with Chrp, a cysteine- and histidine-rich cytoplasmic protein. Biochimie 85:189–194

Beatty WL, Rhoades ER, Hsu DK et al (2002) Association of a macrophage galactoside-binding protein with Mycobacterium-containing phagosomes. Cell Microbiol 4:167–176

Bernardes ES, Silva NM, Ruas LP et al (2006) *Toxoplasma gondii* infection reveals a novel regulatory role for galectin-3 in the interface of innate and adaptive immunity. Am J Pathol 168:1910–1920

Bichara M, Attmane-Elakeb A, Brown D et al (2006) Exploring the role of galectin 3 in kidney function: a genetic approach. Glycobiology 16:36–45

Bickle Q, Helmby H (2007) Lack of galectin-3 involvement in murine intestinal nematode and schistosome infection. Parasite Immunol 29:93–100

Birdsall B, Feeney J, Burdett IDJ et al (2001) NMR solution studies of hamster galectin-3 and electron microscopic visualization of surface-adsorbed complexes: evidence for interactions between the N- and C-terminal domains. Biochemistry 40:4859–4866

Bresalier RS, Mazurek N, Sternberg LR et al (1998) Metastasis of human colon cancer is altered by modifying expression of the β-galactoside-binding protein galectin 3. Gastroenterology 115: 287–296

Bresalier RS, Byrd JC, Tessler D, Lebel J, Koomen J, Hawke D, Half E, Liu KF (2004) A circulating ligand for galectin-3 is a haptoglobin-related glycoprotein elevated in individuals with colon cancer. Gastroenterology 127:741–748

Breuilh L, Vanhoutte F, Fontaine J et al (2007) Galectin-3 modulates immune and inflammatory responses during helminthic infection: impact of galectin-3 deficiency on the functions of dendritic cells. Infect Immun 75:5148–5157

Brewer CF, Miceli MC, Baum LG (2002) Clusters, bundles, arrays and lattices: novel mechanisms for lectin-saccharide-mediated cellular interactions. Curr Opin Struct Biol 12:616–623

Cao Z, Said N, Amin S et al (2002a) Galectins-3 and -7, but not Gal-1, play a role in re-epithelialization of wounds. J Biol Chem 277:42299–42305

Cao Z, Wu HK, Bruce A et al (2002b) Detection of differentially expressed genes in healing mouse corneas, using cDNA microarrays. Invest Ophthalmol Vis Sci 43:2897–2904

Cecchinelli B, Lavra L, Rinaldo C et al (2006) Repression of the antiapoptotic molecule galectin-3 by homeodomain-interacting protein kinase 2-activated p53 is required for p53-induced apoptosis. Mol Cell Biol 26:4746–4757

Chen H-Y, Fermin A, Vardhana S et al (2009) Galectin-3 negatively regulates TCR-mediated CD4+ T-cell activation at the immunological synapse. Proc Natl Acad Sci USA 106:14496–14501

Chen HY, Liu FT, Yang RY (2005) Roles of galectin-3 in immune responses. Arch Immunol Ther Exp (Warsz) 53:497–504

Chen HY, Sharma BB, Yu L et al (2006) Role of galectin-3 in mast cell functions: galectin-3-deficient mast cells exhibit impaired mediator release and defective JNK expression. J Immunol 177:4991–4997

Chiu MG, Johnson TM, Woolf AS (2006) Galectin-3 associates with the primary cilium and modulates cyst growth in congenital polycystic kidney disease. Am J Pathol 169:1925–1938

Collins PM, Hidari KI, Blanchard H (2007) Slow diffusion of lactose out of galectin-3 crystals monitored by x-ray crystallography: possible implications for ligand-exchange protocols. Acta Crystallogr D Biol Crystallogr 63:415–419

Dabelic S, Supraha S, Dumic J (2006) Galectin-3 in macrophage-like cells exposed to immunomodulatory drugs. Biochim Biophys Acta 1760:701–709

David A, Kopeckova P, Minko T et al (2004) Design of a multivalent galactoside ligand for selective targeting of HPMA copolymer-doxorubicin conjugates to human colon cancer cells. Eur J Cancer 40:148–157

Davidson PJ, Li SY, Lohse AG et al (2006) Transport of galectin-3 between the nucleus and cytoplasm. I. Conditions and signals for nuclear import. Glycobiology 16:602–611

Delacour D, Cramm-Behrens CI, Drobecq H et al (2006) Requirement for galectin-3 in apical protein sorting. Curr Biol 16:408–414

Demetriou M, Granovsky M, Quaggin S, Dennis JW (2001) Negative regulation of T-cell activation and autoimmunity by Mgat5 N-glycosylation. Nature 409:733–739

Deschildre C, Ji JW, Chater S et al (2007) Expression of galectin-3 and its regulation in the testes. Int J Androl 30:28–40

Devouassoux-Shisheboran M, Deschildre C et al (2006) Expression of galectin-3 in gonads and gonadal sex cord stromal and germ cell tumors. Oncol Rep 16:335–340

Dong S, Hughes RC (1997) Macrophage surface glycoproteins binding to galectin-3 (Mac-2-antigen). Glycoconj J 14:267–274

Dumic J, Dabelic S, Flogel M (2006) Galectin-3: an open-ended story. Biochim Biophys Acta 1760:616–635

Duneau M, Boyer-Guittaut M, Gonzalez P et al (2005) Galig, a novel cell death gene that encodes a mitochondrial protein promoting cytochrome c release. Exp Cell Res 302:194–205

Farnworth SL, Henderson NC, MacKinnon AC et al (2008) Galectin-3 reduces the severity of pneumococcal pneumonia by augmenting neutrophil function. Am J Pathol 172:395–405

Fernandez GC, Ilarregui JM, Rubel CJ et al (2005) Galectin-3 and soluble fibrinogen act in concert to modulate neutrophil activation and survival: involvement of alternative MAPK pathways. Glycobiology 15:519–527

Feuk-Lagerstedt E, Jordan ET, Leffler H, Dahlgren C, Karlsson A (1999) Identification of CD66a and CD66b as the major galectin-3 receptor candidates in human neutrophils. J Immunol 163:5592–5598

Forsman H, Islander U, Andréasson E et al (2011) Galectin 3 aggravates joint inflammation and destruction in antigen-induced arthritis. Arthritis Rheum 63:445–454

Fowler M, Thomas RJ, Atherton J, Roberts IS, High NJ (2006) Galectin-3 binds to *Helicobacter pylori* O-antigen: it is upregulated and rapidly secreted by gastric epithelial cells in response to *H. pylori* adhesion. Cell Microbiol 8:44–54

Fradin C, Poulain D, Jouault T (2000) β-1,2-linked oligomannosides from *Candida albicans* bind to a 32 kilodalton macrophage membrane protein homologous to the mammalian lectin galectin-3. Infect Immun 68:4391–4398

Friedrichs J, Manninen A, Muller DJ, Helenius J (2008) Galectin-3 regulates integrin _α2β1-mediated adhesion to collagen-I and -IV. J Biol Chem 283:32264–32272

Frigeri LG, Zuberi RI, Liu FT (1993) Epsilon BP, a beta-galactoside-binding animal lectin, recognizes IgE receptor (Fc epsilon RI) and activates mast cells. Biochemistry 32:7644–7649

Fukumori T, Kanayama HO, Raz A (2007) The role of galectin-3 in cancer drug resistance. Drug Resist Update 10:101–108

Fukumori T, Takenaka Y, Oka N et al (2004) Endogenous galectin-3 determines the routing of CD95 apoptotic signaling pathways. Cancer Res 64:3376–3379

Fukumori T, Takenaka Y, Yoshii T et al (2003) CD29 and CD7 mediate galectin-3-induced type II T-cell apoptosis. Cancer Res 63:8302–8311

Fukushi J, Makagiansar IT, Stallcup WB (2004) NG2 proteoglycan promotes endothelial cell motility and angiogenesis via engagement of galectin-3 and α3β1 integrin. Mol Biol Cell 15:3580–3590

Gebhardt A, Ackermann W, Unver N, Elsässer H-P. Expression of galectin-3 in the rat pancreas during regeneration following hormone-induced pancreatitis. Cell Tissue Res 2004; 315:321–9

Ghanooni R, Decaestecker C, Simon P et al (2006) Characterization of patterns of expression of protein kinase C- α, -δ, -η, -γ and -ζ and their correlations to p53, galectin-3, the retinoic acid receptor-β and the macrophage migration inhibitory factor (MIF) in human cholesteatomas. Hear Res 214:7–16

Gil CD, La M, Perretti M, Oliani SM (2006) Interaction of human neutrophils with endothelial cells regulates the expression of endogenous proteins annexin 1, Gal-1 and galectin-3. Cell Biol Int 30:338–344

Goletz S, Hanisch F-G, Karsten U (1997) Novel αGalNAc containing glycans on cytokeratins are recognized in vitro by galectins with type II carbohydrate recognition domains. J Cell Sci 110: 1585–1596

Gonen T, Grey AC, Jacobs MD, Donaldson PJ, Kistler J (2001) MP20, the second most abundant lens membrane protein and member of the tetraspanin superfamily, joins the list of ligands of galectin-3. BMC Cell Biol 2:17

Guevremont M, Martel-Pelletier J, Boileau C et al (2004) Galectin-3 surface expression on human adult chondrocytes: a potential substratem for collagenase-3. Ann Rheum Dis 63:636–643

Gupta SK, Masinick S, Garrett M, Hazlett LD (1997) Pseudomonas aeruginosa lipopolysaccharide binds galectin-3 and other human corneal epithelial proteins. Infect Immun 65:2747–2753

Henderson NC, Mackinnon AC, Farnworth SL et al (2006) Galectin-3 regulates myofibroblast activation and hepatic fibrosis. Proc Natl Acad Sci USA 103:5060–5065

Henrick K, Bawumia S, Barboni EAM et al (1998) Evidence for subsites in the galectins involved in sugar binding at the nonreducing end of the central galactose of oligosaccharide ligands: sequence analysis, homology modeling and mutagenesis studies of hamster galectin-3. Glycobiology 8:45–57

Hirabayashi J, Hashidate T, Arata Y et al (2002) Oligosaccharide specificity of galectins: a search by frontal affinity chromatography. Biochim Biophys Acta 1572:232–254

Hsu DK, Chen HY, Liu FT (2009) Galectin-3 regulates T-cell functions. Immunol Rev 230:114–127

Hsu DK, Yang RY, Pan Z et al (2000) Targeted disruption of the galectin-3 gene results in attenuated peritoneal inflammatory responses. Am J Pathol 156:1073–1083

Hudkins KL, Pichaiwong W, Wietecha T et al (2010) BTBR Ob/Ob mutant mice model progressive diabetic nephropathy. J Am Soc Nephrol 21:1533–1542

Huflejt ME, Turck CW, Lindstedt R et al (1993) L-29, a soluble lactose-binding lectin, is phosphorylated on serine 6 and serine 12 in vivo and by casein kinase I. J Biol Chem 268:26712–26718

Hughes RC (2001) Galectins as modulators of cell adhesion. Biochimie 83:667–676

Hughes RC (1994) Mac-2: a versatile galactose-binding protein of mammalian tissues. Glycobiology 4:5–12

Hughes RC (1999) Secretion of the galectin family of mammalian carbohydrate binding proteins. Biochim Biophys Acta 1473:172–185

Hughes RC (1997) The galectin family of mammalian carbohydrate-binding molecules. Biochem Soc Trans 25:1194–1198

Iacobini C, Amadio L, Oddi G et al (2003) Role of galectin-3 in diabetic nephropathy. J Am Soc Nephrol 14:S264–S270

Iacobini C, Menini S, Oddi G et al (2004) Galectin-3/AGE-receptor 3 knockout mice show accelerated AGE-induced glomerular injury: evidence for a protective role of galectin-3 as an AGE receptor. FASEB J 18:1773–1775

Iacobini C, Menini S, Ricci C et al (2009a) Accelerated lipid-induced atherogenesis in galectin-3-deficient mice: role of lipoxidation via receptor-mediated mechanisms. Arterioscler Thromb Vasc Biol 29:831–836

Iacobini C, Menini S, Ricci C et al (2009b) Advanced lipoxidation end-products mediate lipid-induced glomerular injury: role of receptor-mediated mechanisms. J Pathol 218:360–369

Iacobini C, Oddi G, Menini S et al (2005) Development of age-dependent glomerular lesions in galectin-3/AGE-receptor-3 knockout mice. Am J Physiol Renal Physiol 289:F611–F621

Iacovazzi PA, Notarnicola M, Caruso MG et al (2010) Serum levels of galectin-3 and its ligand 90 k/mac-2 bp in colorectal cancer patients. Immunopharmacol Immunotoxicol 32:160–164

Inohara H, Raz A (1994) Identification of human melanoma cellular and secreted ligands for galectin-3. Biochem Biophys Res Commun 201:1366–1375

Janeway CA Jr, Medzhitov R (2002) Innate immune recognition. Annu Rev Immunol 20:197–216

Jawhara S, Thuru X, Standaert-Vitse A et al (2008) Colonization of mice by *Candida albicans* is promoted by chemically induced colitis and augments inflammatory responses through galectin-3. J Infect Dis 197:972–980

Jeon SB, Yoon HJ, Chang CY et al (2010) Galectin-3 exerts cytokine-like regulatory actions through the JAK-STAT pathway. J Immunol 185:7037–7046

Jiang HR, Al Rasebi Z, Mensah-Brown E et al (2009) Galectin-3 deficiency reduces the severity of experimental autoimmune encephalomyelitis. J Immunol 182:1167–1173

Jin R, Greenwald A, Peterson MD, Waddell TK (2006) Human monocytes recognize porcine endothelium via the interaction of galectin 3 and α-GAL. J Immunol 177:1289–1295

Johnson KD, Glinskii OV, Mossine VV et al (2007) Galectin-3 as a potential therapeutic target in tumors arising from malignant endothelia. Neoplasia 9:662–670

Joo HG, Goedegebuure PS, Sadanaga N et al (2001) Expression and function of galectin-3, a beta-galactoside-binding protein in activated T lymphocytes. J Leukoc Biol 69:555–564

Jouault T, El Abed-El Behi M, Martinez-Esparza M et al (2006) Specific recognition of *Candida albicans* by macrophages requires galectin-3 to discriminate *Saccharomyces cerevisiae* and needs association with TLR2 for signaling. J Immunol 177:4679–4687

Kadrofske MM, Openo KP, Wang JL (1998) The human LGALS3 (galectin-3) gene: determination of the gene structure and functional characterization of the promoter. Arch Biochem Biophys 349:7–20

Kaji Y (2005) Expression and function of receptors for advanced glycation end products in bovine corneal endothelial cells. Nippon Ganka Gakkai Zasshi 109:691–699, Abstract

Karlsen AE, Storling ZM, Sparre T et al (2006) Immune-mediated β-cell destruction in vitro and in vivo-A pivotal role for galectin-3. Biochem Biophys Res Commun 344:406–415

Karlsson A, Christenson K, Matlak M, Björstad A, Brown KL, Telemo E, Salomonsson E, Leffler H, Bylund J (2009) Galectin-3 functions as an opsonin and enhances the macrophage clearance of apoptotic neutrophils. Glycobiology 19:16–20

Khoo KH, Dell A (2001) Glycoconjugates from parasitic helminths: structure diversity and immunobiological implications. Adv Exp Med Biol 491:185–205

Kim H, Kang TY, Joo HG, Shin T (2006) Immunohistochemical localization of galectin-3 in boar testis and epididymis. Acta Histochem 108:481–485

Kim HR, Lin HM, Biliran H, Raz A (1999) Cell cycle arrest and inhibition of anoikis by galectin-3 in human breast epithelial cells. Cancer Res 59:4148–4154

Kiwaki K, Novak CM, Hsu DK, Liu FT, Levine JA (2007) Galectin-3 stimulates preadipocyte proliferation and is up-regulated in growing adipose tissue. Obesity (Silver Spring) 15:32–39

Kleshchenko YY, Moody TN, Furtak VA et al (2004) Human galectin-3 promotes *Trypanosoma cruzi* adhesion to human coronary artery smooth muscle cells. Infect Immun 72:6717–6721

Kohatsu L, Hsu DK, Jegalian AG et al (2006) Galectin-3 induces death of *Candida* species expressing specific β-1,2-linked mannans. J Immunol 177:4718–4726

Koths K, Taylor E, Halenbeck R et al (1993) Cloning and characterization of a human Mac-2-binding protein, a new member of the superfamily defined by the macrophage scavenger receptor cysteine-rich domain. J Biol Chem 268:14245–14249

Krześlak A, Lipińska A (2004) Galectin-3 as a multifunctional protein. Cell Mol Biol Lett 9:305–328

Kuwabara I, Liu FT (1996) Galectin-3 promotes adhesion of human neutrophils to laminin. J Immunol 156:3939–3944

Lacina L, Plzakova Z, Smetana K Jr et al (2006) Glycophenotype of psoriatic skin. Folia Biol (Praha) 52:10–15

Li L, Liu CT, Wang K, Pang YM (2005) Inhibitory effects of galectin-3 on the inflammatory cytokines and chemokines in guinea pig asthma models. Sichuan Da Xue Xue Bao Yi Xue Ban 36:355–358

Li SY, Davidson PJ, Lin NY et al (2006) Transport of galectin-3 between the nucleus and cytoplasm II. Identification of the signal for nuclear export. Glycobiology 16:612–622

Li Y, Komai-Koma M, Gilchrist DS et al (2008) Galectin-3 is a negative regulator of lipopolysaccharide-mediated inflammation. J Immunol 181:2781–2789

Li YJ, Kukita A, Teramachi J et al (2009) A possible suppressive role of galectin-3 in upregulated osteoclastogenesis accompanying adjuvant-induced arthritis in rats. Lab Invest 89:26–37

Liao DI, Kapadia G, Ahmed H, Vasta GR, Herzberg O (1994) Structure of S-lectin, a developmentally regulated vertebrate β-galactoside-binding protein. Proc Natl Acad Sci USA 91:1428–1432

Liew FY, Xu D, Brint EK, O'Neill LA (2005) Negative regulation of toll-like receptor-mediated immune responses. Nat Rev Immunol 5:446–458

Lin H-M, Moon B-K, Yu F, Kim H-RC (2000) Galectin-3 mediates genistein–induced G2/M arrest and inhibits apoptosis. Carcinogenesis 21:1941–1945

Lin H-M, Pestell RG, Raz A, Kim H-RC (2002) Galectin-3 enhances cyclin D1 promoter activity through SP1 and a cAMP-responsive element in human breast epithelial cells. Oncogene 21:8001–8010

Lippert E, Falk W, Bataille F et al (2007) Soluble galectin-3 is a strong, colonic epithelial-cell-derived, lamina propria fibroblast-stimulating factor. Gut 56:43–51

Liu F-T, Patterson RJ, Wang JL (2002) Intracellular functions of galectins. Biochim Biophys Acta 1572:263–273

Liu F-T, Rabinovich GA (2005) Galectins as modulators of tumour progression. Nat Rev Cancer 5:29–41

Liu F-T (2005) Regulatory roles of galectins in the immune response. Int Arch Allergy Immunol 136:385–400

Liu L, Sakai T, Sano N, Fukui K (2004) Nucling mediates apoptosis by inhibiting expression of galectin-3 through interference with nuclear factor kappaB signalling. Biochem J 380:31–41

Lobsanov YD, Gitt MA, Leffler H et al (1993) Xray crystal structure of the human dimeric s-Lac lectin, L-14-II, in complex with lactose at 2.9 Å resolution. J Biol Chem 268:27034–27038

Lohr M, Kaltner H, Lensch M et al (2008) Cell-type-specific expression of murine multifunctional galectin-3 and its association with follicular atresia/luteolysis in contrast to pro-apoptotic galectins-1 and -7. Histochem Cell Biol 130:567–581

Lopez E, del Pozo V, Miguel T et al (2006) Inhibition of chronic airway inflammation and remodeling by galectin-3 gene therapy in a murine model. J Immunol 176:1943–1950

Lukyanov P, Furtak V, Ochieng J (2005) Galectin-3 interacts with membrane lipids and penetrates the lipid bilayer. Biochem Biophys Res Commun 338:1031–1036

MacKinnon AC, Farnworth SL, Hodkinson PS et al (2008) Regulation of alternative macrophage activation by galectin-3. J Immunol 180:2650–2658

Mandrell RE, Apicella MA, Lindstedt R, Leffler H (1994) Possible interaction between animal lectins and bacterial carbohydrates. Methods Enzymol 236:231–254

Matarrese P, Fusco O, Tinari N et al (2000) Galectin-3 overexpression protects from apoptosis by improving cell adhesion properties. Int J Cancer 85:545–554

Mazurek N, Conklin J, Byrd JC et al (2000) Phosphorylation of the β-galactoside-binding protein galectin-3 modulates binding to its ligands. J Biol Chem 275:36311–36315

McFarlane S, Glenn JV, Lichanska AM et al (2005) Characterisation of the advanced glycation endproduct receptor complex in the retinal pigment epithelium. Br J Ophthalmol 89:107–112

Menon RP, Hughes RC (1999) Determinants in the N-terminal domains of galectin-3 for secretion by a novel pathway circumventing the endoplasmic reticulum–golgi complex. Eur J Biochem 264:569–576

Menon RP, Strom M, Hughes RC (2000) Interaction of a novel cysteine and histidine-rich cytoplasmic protein with galectin-3 in a carbohydrateindependent manner. FEBS Lett 470:227–231

Mensah-Brown E, Shahin A, Parekh K et al (2006) Functional capacity of macrophages determines the induction of type 1 diabetes. Ann N Y Acad Sci 1084:49–57

Mensah-Brown EP, Al Rabesi Z, Shahin A et al (2009) Targeted disruption of the galectin-3 gene results in decreased susceptibility to multiple low dose streptozotocin-induced diabetes in mice. Clin Immunol 130:83–88

Mercer N, Ahmed H et al (2004) AGE-R3/galectin-3 expression in osteoblast-like cells: regulation by AGEs. Mol Cell Biochem 266:17–24

Mey A, Leffler H, Hmama Z et al (1996) The animal lectin galectin-3 interacts with bacterial lipopolysaccharides via two independent sites. J Immunol 156:1572–1577

Miller SI, Ernst RK, Bader MW (2005) LPS, TLR4 and infectious disease diversity. Nat Rev Microbiol 3:36–46

Mina-Osorio P, Soto-Cruz I, Ortega E (2007) A role for galectin-3 in CD13-mediated homotypic aggregation of monocytes. Biochem Biophys Res Commun 353:605–610

Mok SW, Riemer C, Madela K et al (2007) Role of galectin-3 in prion infections of the CNS. Biochem Biophys Res Commun 359:672–678

Moon B-K, Lee YJ, Battle P et al (2001) Galectin-3 protects human breast carcinoma cells against nitric oxideinduced apoptosis. Implication of galectin-3 function during metastasis. Am J Pathol 159:1055–1060

Morgan R, Gao G, Pawling J et al (2004) N-acetylglucosaminyl transferase V (Mgat5)-mediated N-glycosylation negatively regul ates Th1 cytokine production by T cells. J Immunol 173:7200–7208

Muller S, Schaffer T, Flogerzi B et al (2006) Galectin-3 modulates T cell activity and is reduced in the inflamed intestinal epithelium in IBD. Inflamm Bowel Dis 12:588–597

Nachtigal M, Ghaffar A, Mayer EP (2008) Galectin-3 gene inactivation reduces atherosclerotic lesions and adventitial inflammation in ApoE-deficient mice. Am J Pathol 172:247–255

Nagata S (2000) Apoptosis. Fibrinol Proteol 14:82–86

Nakahara S, Hogan V, Inohara H, Raz A (2006a) Importin-mediated nuclear translocation of galectin-3. J Biol Chem 281:39649–39659

Nakahara S, Oka N, Raz A (2005) On the role of galectin-3 in cancer apoptosis. Apoptosis 10:267–275

Nakahara S, Oka N, Wang Y et al (2006b) Characterization of the nuclear import pathways of galectin-3. Cancer Res 66:9995–10006

Narciso MS, Mietto Bde S, Marques SA et al (2009) Sciatic nerve regeneration is accelerated in galectin-3 knockout mice. Exp Neurol 217:7–15

Neidhart M, Zaucke F, von Knoch R et al (2005) Galectin-3 is induced in rheumatoid arthritis synovial fibroblasts after adhesion to cartilage oligomeric matrix protein. Ann Rheum Dis 64:419–424

Nickel W (2003) The mystery of nonclassical protein secretion. A current view on cargo proteins and potential export routes. Eur J Biochem 270:2109–2119

Nieminen J, Kuno A, Hirabayashi J, Sato S (2007) Visualization of galectin-3 oligomerization on the surface of neutrophils and endothelial cells using fluorescence resonance energy transfer. J Biol Chem 282:1374–1383

Nieminen J, St Pierre C, Sato S (2005) Galectin-3 interacts with naive and primed neutrophils, inducing innate immune responses. J Leukoc Biol 78:1127–1135

Nieminen J, St-Pierre C, Bhaumik P et al (2008) Role of galectin-3 in leukocyte recruitment in a murine model of lung infection by *Streptococcus pneumoniae*. J Immunol 180:2466–2473

Nio J, Iwanaga T (2007) Galectins in the mouse ovary: concomitant expression of galectin-3 and progesterone degradation enzyme (20αHSD) in the corpus luteum. J Histochem Cytochem 55:423–432

Nio J, Takahashi-Iwanaga H, Morimatsu M, Kon Y, Iwanaga T (2006) Immunohistochemical and in situ hybridization analysis of galectin-3, a β-galactoside binding lectin, in the urinary system of adult mice. Histochem Cell Biol 126:45–56

Nomoto K, Tsuneyama K, Abdel Aziz HO (2006) Disrupted galectin-3 causes non-alcoholic fatty liver disease in male mice. J Pathol 210:469–477

Ochieng J, Furtak V, Lukyanov P (2004) Extracellular functions of galectin-3. Glycoconj J 19:527–535

Ochieng J, Leite-Browning ML, Warfield P (1998) Regulation of cellular adhesion to extracellular matrix proteins by galectin-3. Biochem Biophys Res Commun 246:788–791

Ochieng J, Warfield P, Green-Jarvis B, Fentie I (1999) Galectin-3 regulates the adhesive interaction between breast carcinoma cells and elastin. J Cell Biochem 75:505–514

Ochieng J, Warfield P (1995) Galectin-3 binding potentials of mouse tumor RHS and human placental laminins. Biochem Biophys Res Commun 217:402–406

Ocklenburg F, Moharregh-Khiabani D, Geffers R et al (2006) UBD, a downstream element of FOXP3, allows the identification of LGALS3, a new marker of human regulatory T cells. Lab Invest 86:724–737

Oda Y, Herrmann J, Gitt MA et al (1993) Soluble lactose-binding lectin from rat intestine with two different carbohydrate-binding domains in the same peptide chain. J Biol Chem 268:5929–5939

Oguz E, Tsai TT, Di Martino A et al (2007) Galectin-3 expression in the intervertebral disc: a useful marker of the notochord phenotype? Spine 32:9–16

Ohannesian DW, Lotan D, Thoman P et al (1995) Carcinoembryonic antigen and other glycoconjugates act as ligands for galectin-3 in human colon carcinoma cells. Cancer Res 55:2191–2199

Okamura DM, Pasichnyk K, Lopez-Guisa JM et al (2010) Galectin-3 preserves renal tubules and modulates extracellular matrix remodeling in progressive fibrosis. Am J Physiol Renal Physiol 300:F245–F253

Oliveira FL, Chammas R, Ricon L et al (2009) Galectin-3 regulates peritoneal B1-cell differentiation into plasma cells. Glycobiology 19:1248–1258

Oliveira FL, Frazao P, Chammas R et al (2007) Kinetics of mobilization and differentiation of lymphohematopoietic cells during experimental murine schistosomiasis in galectin-3$^{-/-}$ mice. J Leukoc Biol 82:300–310

Ortega N, Behonick DJ, Colnot C et al (2005) Galectin-3 is a downstream regulator of matrix metalloproteinase-9 function during endochondral bone formation. Mol Biol Cell 16:3028–3039

Pace K, Lee C, Stewart PL, Baum LG (1999) Restricted receptor segregation into membrane microdomains occurs on human T cells during apoptosis induced by galectin-1. J Immunol 163:3801–3811

Park JW, Voss PG, Grabski S et al (2001) Association of Gal-1 and galectin-3 with Gemin4 in complexes containing the SMN protein. Nucl Acids Res 27:3595–3602

Patnaik SK, Potvin B, Carlsson S et al (2006) Complex N-glycans are the major ligands for Gal-1, -3, and -8 on Chinese hamster ovary cells. Glycobiology 16:305–317

Pelletier I, Sato S (2002) Specific recognition and cleavage of galectin-3 by *Leishmania major* through species-specific polygalactose epitope. J Biol Chem 277:17663–17670

Pilette C, Colinet B, Kiss R et al (2007) Increased galectin-3 expression and intraepithelial neutrophils in small airways in severe chronic obstructive pulmonary disease. Eur Respir J 29:914

Pricci F, Leto G, Amadio L et al (2000) Role of galectin 3 as a receptor for advanced glycosylation end products. Kidney Int 58:S31–S39

Probstmeier R, Montag D, Schachner M (1995) Galectin-3, a β- galactoside-binding animal lectin, binds to neural recognition molecules. J Neurochem 64:2465–2472

Pugliese G, Pricci F, Iacobini C et al (2001) Accelerated diabetic glomerulopathy in galectin-3/AGE receptor 3 knockout mice. FASEB J 15:2471–2479

Rabinovich GA, Baum LG et al (2002) Galectins and their ligands: amplifiers, silencers or tuners of the inflammatory response? Trends Immunol 23:313–320

Rosenberg I, Cherayil BJ et al (1991) Mac-2-binding glycoproteins. Putative ligands for a cytosolic β-galactoside lectin. J Biol Chem 266:18731–18736

Rosenberg IM, Iyer R, Cherayil B, Chiodino C, Pillai S (1993) Structure of the murine Mac-2 gene. Splice variants encode proteins lacking functional signal peptides. J Biol Chem 268:12393–12400

Rousseau C, Muriel M-P, Musset M, Botti J, Sève A-P (2000) Glycosylated nuclear lectin CBP70 also associated with endoplasmic reticulum and the golgi apparatus: does the "classic pathway" of glycosylation also apply to nuclear glycoproteins? J Cell Biochem 78:638–649

Saegusa J, Hsu DK, Chen H-Y et al (2009) Galectin-3 is critical for the development of the allergic inflammatory response in a mouse model of atopic dermatitis. Am J Pathol 174:922–931

Saegusa J, Hsu DK, Liu W et al (2008) Galectin-3 protects keratinocytes from UVB-induced apoptosis by enhancing AKT activation and suppressing ERK activation. J Invest Dermatol 128:2403–2411

Sano H, Hsu DK, Apgar JR et al (2003) Critical role of galectin-3 in phagocytosis by macrophages. J Clin Invest 112:389–397

Sano H, Hsu DK, Yu L et al (2000) Human galectin-3 is a novel chemoattractant for monocytes and macrophages. J Immunol 165:2156–2164

Sarafian V, Jadot M et al (1998) Expression of Lamp-1 and Lamp-2 and their interactions with galectin-3 in human tumor cells. Int J Cancer 75:105–111

Sato S, Burdett I, Hughes RC (1993) Secretion of the baby hamster kidney 30-kDa galactose-binding lectin from polarized and nonpolarized cells: a pathway independent of the endoplasmic reticulum-golgi complex. Exp Cell Res 207:8–18

Sato S, Hughes RC (1992) Binding specificity of a baby hamster kidney lectin for H type I and II chains, polylactosamine glycans, and appropriately glycosylated forms of laminin and fibronectin. J Biol Chem 267:6983–6990

Sato S, Nieminen J (2004) Seeing strangers or announcing "danger": galectin-3 in two models of innate immunity. Glycoconj J 19:583–591

Sato S, Ouellet N, Pelletier I et al (2002) Role of galectin-3 as an adhesion molecule for neutrophil extravasation during *Streptococcal pneumonia*. J Immunol 168:1813–1822

Schwaderer AL, Vijayakumar S, Al-Awqati Q, Schwartz GJ (2006) Galectin-3 expression is induced in renal β-intercalated cells during metabolic acidosis. Am J Physiol Renal Physiol 290:F148–F158

Seetharaman J, Kanigsberg A, Slaaby R et al (1998) X-ray crystal structure of the human galectin-3 carbohydrate recognition domain at 2.1-Å resolution. J Biol Chem 273:13047–13052

Sève A-P, Felin M et al (1993) Evidence for a lactose-mediated association between two nuclear carbohydrate-binding proteins. Glycobiology 3:23–30

Sharma UC, Pokharel S, van Brakel TJ et al (2004) Galectin-3 marks activated macrophages in failure-prone hypertrophied hearts and contributes to cardiac dysfunction. Circulation 110:3121–3128

Shimura T, Takenaka Y, Fukumori T et al (2005) Implication of galectin-3 in Wnt signaling. Cancer Res 65:3535–3537

Shimura T, Takenaka Y, Tsutsumi S et al (2004) Galectin-3, a novel binding partner of β-catenin. Cancer Res 64:6363–6367

Shoji H, Ikenaka K, Nakakita S et al (2005) Xenopus galectin-VIIa binds N-glycans of members of the cortical granule lectin family (xCGL and xCGL2). Glycobiology 15:709–720

Shou J, Bull CM, Li L et al (2006) Identification of blood biomarkers of rheumatoid arthritis by transcript profiling of peripheral blood mononuclear cells from the rat collagen-induced arthritis model. Arthritis Res Ther 8:R28

Silva-Monteiro E, Reis Lorenzato L, Kenji Nihei O et al (2007) Altered expression of galectin-3 induces cortical thymocyte depletion and premature exit of immature thymocytes during *Trypanosoma cruzi* infection. Am J Pathol 170:546–556

Srivatsan J, Smith DF, Cummings RD (1992) *Schistosoma mansoni* synthesizes novel biantennary Asn-linked oligosaccharides containing terminal beta-linked N-acetylgalactosamine. Glycobiology 2:445–452

Stillman BN, Hsu DK et al (2006) Galectin-3 and Gal-1 bind distinct cell surface glycoprotein receptors to induce T cell death. J Immunol 176:778–789

Stitt AW, McGoldrick C, Rice-McCaldin A et al (2005) Impaired retinal angiogenesis in diabetes: role of advanced glycation end products and galectin-3. Diabetes 54:785–794

Stowell SR, Arthur CM, Mehta P et al (2008) Gal-1, -2, and -3 exhibit differential recognition of sialylated glycans and blood group antigens. J Biol Chem 283:10109–10123

Takeda K, Kaisho T, Akira S (2003) Toll-like receptors. Annu Rev Immunol 21:335–376

van den Berg TK, Honing H, Franke N et al (2004) LacdiNAc-glycans constitute a parasite pattern for galectin-3-mediated immune recognition. J Immunol 173:1902–1907

van Kimmenade RR, Januzzi JL Jr, Ellinor PT et al (2006) Utility of amino-terminal pro-brain natriuretic peptide, galectin-3, and apelin for the evaluation of patients with acute heart failure. J Am Coll Cardiol 48:1217–1224

Vito P, Pellegrini L, Guiet C, D'Adamio L (1999) Cloning of AIP1, a novel protein that associates with the apoptosis-linked gene ALG-2 in a Ca2 + –dependent reaction. J Biol Chem 274:1533–1540

Vlassara H, Li YM, Imani F et al (1995) Identification of galectin 3 as a high affinity binding protein for advanced glycation end products (AGE): a new member of the AGEreceptor complex. Mol Med 1:634–646

Vray B, Camby I, Vercruysse V et al (2004) Up-regulation of galectin-3 and its ligands by *Trypanosoma cruzi* infection with modulation of adhesion and migration of murine dendritic cells. Glycobiology 14:647–657

Walzel H, Blach M, Hirabayashi J, Kasai KI, Brock J (2000) Involvement of CD2 and CD3 in galectin-1 induced signaling in human Jurkat T-cells. Glycobiology 10:131–140

Wang J, Barke RA, Charboneau R, Roy S (2005) Morphine impairs host innate immune response and increases susceptibility to *Streptococcus pneumoniae* lung infection. J Immunol 174:426–434

Wang JL, Gray RM, Haudek KC, Patterson RJ (2004) Nucleocytoplasmic lectins. Biochim Biophys Acta 1673:75–93

Wang W, Park JW et al (2006) Immunoprecipitation of spliceosomal RNAs by antisera to Gal-1 and galectin-3. Nucleic Acids Res 34:5166–5174

Weigert J, Neumeier M, Wanninger J et al (2010) Serum galectin-3 is elevated in obesity and negatively correlates with glycosylated hemoglobin in type 2 diabetes. J Clin Endocr Metab 9:1404–1411

Woldesenbet S, Garcia R, Igbo N et al (2004) Lectin receptors for endometrial H-type 1 antigen on goat conceptuses. Am J Reprod Immunol 52:74–80

Woo H-J, Shaw LM, Messier JM, Mercurio AM (1990) The major nonintegrin laminin binding protein of macrophages is identical to carbohydrate binding pro-tein 35 (Mac-2). J Biol Chem 265:7097–7099

Yang RY, Hsu DK, Liu FT (1996) Expression of galectin-3 modulates T-cell growth and apoptosis. Proc Natl Acad Sci USA 93:6737–6742

Yoshii T, Fukumori T et al (2002) Galectin phosphorylation is required for its anti-apoptotic function and cell cycle arrest. J Biol Chem 277:6852–6857

Yu F, Finley RL Jr, Raz A, Kim H-RC (2002) Galectin-3 translocates to the perinuclear membranes and inhibits cytochrome c release from the mitochondria. A role for synexin in galectin-3 translocation. J Biol Chem 277:15819–15827

Zhu W, Sano H, Nagai R et al (2001) The role of galectin-3 in endocytosis of advanced glycation end products and modified low density lipoproteins. Biochem Biophys Res Commun 280:1183–1188

Zlatkine P, Mehul B, Magee AI (1997) Retargeting of cytosolic proteins to the plasma membrane by the Lck protein tyrosine kinase dual acylation motif. J Cell Sci 110:673–679

Zörnig M, Hueber A-O, Baum W, Evan G (2001) Apoptosis regulators and their role in tumorigenesis. Biochim Biophys Acta 1551:F1–F37

Zuberi RI, Hsu DK, Kalayci O et al (2004) Critical role for galectin-3 in airway inflammation and bronchial hyperresponsiveness in a murine model of asthma. Am J Pathol 165:2045–2053

13 Galectin-3: A Cancer Marker with Therapeutic Applications

Anita Gupta

13.1 Galectin-3: A Prognostic Marker of Cancer

Galectin-3 (Gal-3) is a pleiotropic carbohydrate-binding protein involved in a variety of normal and pathological biological processes. Its carbohydrate-binding properties constitute the basis for cell-cell and cell-matrix interactions (Chap. 12) and cancer progression. Gal-3 is known to be expressed in various neoplasms including thyroid tumors. Gal-3 is widely spread among different types of cells and tissues, found intracellularly in nucleus and cytoplasm or secreted via non-classical pathway outside the cell, thus being found on the cell surface or in the extracellular space. Gal-3 is involved in RNA processing and cell cycle regulation through activation of transcription factors when translocated to the nucleus. It affects numerous biological processes and seems to be involved in different physiological and pathophysiological conditions. Most members of the galectin family including galectin-1 possess apoptotic activities, whereas Gal-3 possesses anti-apoptotic activity (Dumic et al. 2006). Studies lead to the recognition of Gal-3 as a diagnostic/prognostic marker for specific cancer types, such as thyroid and prostate (Califice et al. 2004b). Though the immune system recognizes diverse cancer antigens, tumors can evade the immune response, therefore growing and progressing. However, there is strong evidence indicating that the regulation of galectins function in the human tumor microenvironment is a complex process that is influenced by diverse biological circumstances.

13.2 Discriminating Malignant Tumors from Benign Nodules of Thyroid

Galectin-3 has been regarded as a useful tool for discriminating malignant tumors from benign nodules of the thyroid, including the distinction between follicular carcinoma and adenoma. Gal-3 is over-expressed in neoplastic human thyroid, while its expression in normal tissue and adenomas was absent or weak (Cvejic et al. 2005a, c; Kawachi et al. 2000; Yoshii et al. 2001; Coli et al. 2002; Faggiano et al. 2002; Herrmann et al. 2002; Nucera et al. 2005; Oestreicher-Kedem et al. 2004; Pisani et al. 2004; Takenaka et al. 2003). Gal-3 may potentially serve as a marker in difficult differential diagnosis cases involving Hürthle cell adenomas and Hürthle cell carcinomas (Nascimento et al. 2001). In addition, Gal-3 mRNA was observed in all malignant thyroid lesions, while in normal and non-malignant tissues, it was not detectable (Inohara et al. 1999). Gal-3 was predominantly found in the cytoplasm of follicular and parafollicular cells; a nuclear location was also sometimes observed. Gal-3 was significantly higher in papillary carcinomas with metastases than those without metastases (Kawachi et al. 2000). However, expression of Gal-3 in lymphatic metastases of papillary carcinoma and in lymph node metastases of medullary thyroid carcinoma appeared to be lower than in primary lesions (Faggiano et al. 2002; Kawachi et al. 2000). The antisense inhibition of Gal-3 expression in thyroid papillary carcinoma resulted in a marked reduction of malignant phenotype (Yoshii et al. 2001). Takenaka et al. (2003) gave evidence, which suggested that over-expression of Gal-3 in transfected Gal-3 cDNA normal thyroid follicular cells leads to the acquisition of malignant phenotype. The genes with increased expression include: retinoblastoma (RB), proliferating cell nuclear antigen (PCNA) and replication factor C (RCF), all of which are involved in the G1-S transition of cell cycle. Therefore, the possibility of Gal-3 involvement in the cell cycle regulation exists. From a biochemical point of view, the evaluation of cytoplasmic Gal-3 expression in epithelial cells isolated from FNAB should serve to make a differential presurgical diagnosis between benign follicular adenomas and differentiated carcinomas (i.e., papillary and widely and minimally invasive follicular carcinomas). Serum levels of galectins-1 and -3 are relatively high in patients with thyroid malignancy but there is

G.S. Gupta et al., *Animal Lectins: Form, Function and Clinical Applications*,
DOI 10.1007/978-3-7091-1065-2_13, © Springer-Verlag Wien 2012

considerable overlap in serum Gal-3 concentrations between those with benign and malignant nodular thyroid disease and, to a lesser extent, between those with and without nodular thyroid disease (Saussez et al. 2008c).

However, there are follicular tumors with unclear vascular or capsular invasion, which makes diagnosis more difficult. To find a relationship between Gal-3 expression and the degree of vascular or capsular invasion of follicular tumors, Ito et al. (2005) suggested that Gal-3 plays a role in the transformation of follicular tumors from benign to malignant. Moreover, in follicular tumors, the presence of this protein should not be required for diagnosing malignant transformation in all cases. Hence, Gal-3 should only be considered as an adjuvant marker for follicular carcinoma (Ito et al. 2005). Analyses of 202 specimens of papillary thyroid carcinoma (PTC) in relation to histomorphologic subtypes and clinicopathologic data showed that Gal-3 itself is not an indicator of local metastatic spread or extrathyroid invasion of PTC (Cvejic et al. 2005c). Another study indicated that Gal-3 gene is expressed at the protein level in most papillary microcarcinomas, a variant of PTC, although with slightly lower frequency than that reported for clinically evident PTC. The presence of Gal-3 in clinically silent microcarcinomas may indicate that Gal-3 is not related to growth or aggressiveness of papillary thyroid microcarcinomas but rather plays some other role in thyroid tumor biology. Results on the relationship between Gal-3 and proliferating cell nuclear antigen (PCNA) suggested that overexpression of Gal-3 is not clearly related to proliferative activity of PTC cells as assessed by PCNA immunostaining (Cvejic et al. 2005a, b; Pisani et al. 2004). In addition to 12/15 papillary carcinomas, among the malignant thyroid lesions, Gal-3 was also detected in 3/4 Hurthle cell carcinomas, 4/6 follicular carcinomas and 0/3 anaplastic carcinomas. Conversely, Gal-3 expression was absent in NT and in all benign thyroid lesions, but 1/15 Hashimoto's thyroiditis (HT) and 10/22 follicular adenomas (FA). Gal-3 cytoplasmic-perinuclear immunolocalization was observed in the majority of thyroid carcinomas and in more than half of the FA, theoretically suggesting an involvement of this protein in thyroid tumorigenesis throughout an antiapoptotic activity. Moreover, Gal-3 expression in FA might anticipate the likelihood of evolution of these benign lesions towards malignancy (Nucera et al. 2005; Oestreicher-Kedem et al. 2004). Diagnostic problems may arise in the presence of Hurthle cell proliferation or minimally invasive follicular carcinoma.

13.2.1 Large-Needle Aspiration Biopsy

Gasbarri et al. (2004a) illustrated the clinical impact of a new diagnostic test-method, named "Gal-3 thyrotest", which is based on the immunodetection of Gal-3 in specimens derived from thyroid nodular lesions. This diagnostic test method, which consistently improves the accuracy of conventional cytology, has been validated in a large multicenter study and is going to impact hardly the clinical management of patients bearing thyroid nodular diseases. The rationale of this diagnostic approach, the possibility to improve its performance in selected cases by using large needle aspiration biopsy (LNAB) together with technical details have been presented and discussed (Gasbarri et al. 2004a). Among 85 thyroid specimens, 42 Gal-3-positive cases were discovered preoperatively (11 thyroid cancers and 3 adenomas confirmed at the final histology), whereas Gal-3-negative cases were 71 (1 carcinoma and 70 benign proliferations at the final histology). Sensitivity, specificity and diagnostic accuracy of this integrated morphologic and phenotypic diagnostic approach were 91.6%, 97.2% and 95.3%, respectively. In conclusion, LNAB plus Gal-3 expression analysis when applied preoperatively to selected thyroid nodules candidate to surgery can potentially reduce unnecessary thyroid resections (Carpi et al. 2006). The routine correct use of Gal-3 can lead to a sensitive reduction of useless thyroid surgeries.

13.2.2 Fine-Needle Aspiration Biopsy

Among diagnostic modalities, fine-needle aspiration biopsy (FNAB) of clinically suspicious thyroid nodules is becoming increasingly popular. Assays using tumor-specific markers may improve the sensitivity and accuracy of FNA and so may be expected to reduce the frequency of open surgical procedures by identifying those patients with demonstrably benign lesions who do not require definitive surgical excision of their lesions for diagnosis. At the same time, thyroid-specific mRNA assays (especially thyroglobulin mRNA testing) have been used by investigators in the postoperative follow-up of patients with thyroid cancer as a potential means of detecting tumor recurrence in the peripheral blood (Rodrigo et al. 2006). Mills et al. (2005) suggested that Gal-3 does not appear to be a useful adjunct to diagnosis in thyroid FNA as it does not reliably distinguish malignant and benign lesions. Many thyroid aspirates are of low cellularity and are not suitable for cell block immunohistochemistry (Mills et al. 2005). Moreover, Gal-3 in the tissue of follicular adenomas with grave dysplasia and follicular carcinoma is an unfavourable prognostic sign and not a reliable immunohistochemical marker to distinguish benign from malignant thyroid follicular lesions (Mehrotra et al. 2004).

Although FNAB is the most reliable preoperative diagnostic procedure, it shows inherent limitations in differentiating adenoma from follicular carcinoma and, sometimes, follicular variants of papillary carcinoma. Gal-3 cytoplasmic neoexpression has been proposed as a peculiar feature of thyroid malignant cells, easily detectable in cytological and histological samples. Among 39 follicular carcinomas, 26 papillary carcinomas (PTC), and 105 adenomas in both cell-block samples and their histological counterparts, all papillary carcinoma samples showed high levels of Gal-3 immunoreactivity. Thirty-four follicular carcinomas were positive, whereas five were negative in cell-blocks but positive in their histological counterparts (Saggiorato et al. 2004). In Indian population, Gal-3 positivity was seen in 80% of papillary carcinomas, 13.5% of follicular neoplasms and in 60% of benign nodules. Study showed that Gal-3 was strongly expressed in smears of papillary carcinoma. However, since it is also expressed in a variety of benign lesions, its role as a pre-surgical markerfor differentiating benign from malignant thyroid nodules is limited (Aron et al. 2006).

From a total of 426 follicular lesions from patients who had undergone thyroidectomy for either benign or malignant nodules, Gal-3 could be used as a useful supplementary marker for cytologic diagnosis, although it was not an absolute marker in determining whether a lesion was benign or malignant (Bartolazzi and Bussolati 2006; Kim et al. 2006). Collet et al. (2005) suggested that Gal-3 constitutes a useful marker in the diagnosis of thyroid lesions classified as undeterminate by conventional cytology. Torres-Cabala et al. (2006) concluded that gal-3, gal-1 and S100C can be used to help in discriminating benign and malignant thyroid lesions.

Papillary carcinomas showed a strong, but cytoplasmic pattern of staining. Gal-3 was strongly positive in papillary carcinomas, and negative in benign lesions, confirming its value in differential diagnosis. The S100C, highly expressed in papillary carcinomas, is expressed in the nuclei of normal tissue, hyperplastic nodules, and follicular adenomas and carcinomas. It is helpful in pathological study of thyroid lesions, especially in cases in which follicular variants of papillary carcinoma and follicular carcinoma are considered in differential diagnosis (Torres-Cabala et al. 2004). Quantification of the TFF3/Gal-3 mRNA ratio (T/G ratio) may be a useful tool for the distinction between follicular adenomas and carcinomas, which is most difficult in thyroid pathology (Takano et al. 2005).

13.2.3 Combination of Markers

Gal-3 and HBME-1 Markers: Oncocytic cell tumors (OCTs) of the thyroid include oncocytic cell adenomas (OCAs) and oncocytic cell carcinomas (OCCs). Oncocytic variant of papillary carcinoma (OVPC) has also been described. These tumors may present similar diagnostic problems as their non-oncocytic counterparts, in both conventional histology and FNAB. Several markers can distinguish benign from malignant thyroid follicular tumors, Gal-3 and HBME-1 being the most promising ones. Though controversial data have been reported on their discriminatory potential in the small series of OCTs, Volante et al. (2004) assessed the role of Gal-3 and HBME-1 in a large series of 152 OCTs (including 50 OCAs, 70 OCCs and 32 OVPCs). Using a biotin-free detection system, the sensitivity of Gal-3 was 95.1%, while that for HBME-1 was nearly 53%. The combination of Gal-3 and HBME-1 increased the sensitivity up to 99%. However, for both markers, the specificity was 88%, lower than that reported for non-oncocytic follicular tumors. The relationship between the markers investigated and the nuclear changes suggest that the tumors containing them are pathogenetically linked to papillary carcinomas (Papotti et al. 2005). Immunochemical stains of cytokeratin (CK19), Gal-3 and HBME-1, especially when used in combination, can be an important adjunct to the histopathological diagnoses of thyroid lesions (Teng et al. 2004; Rossi et al. 2006). According to de Matos et al. (2005), from a panel of thyroid malignancy markers including HBME-1, CK19 and Gal-3, HBME-1 is the most sensitive marker though the three markers may be useful in specific cases (de Matos et al. 2005; Prasad et al. 2005).

The combined panel of antibodies against RET, HBME-1, and Gal-3 and the nuclear pleomorphism of follicular cells were effective in distinguishing between thyroid nodules requiring surgery from thyroid nodules requiring just follow-up (Rossi et al. 2005). In the group of nonoxyphilic tumors positive reaction with HBME-1 was more common in adenomas with intracapsular invasion and carcinomas, but positive reaction with anti-CD15 – only in carcinomas. Reactivity with these antibodies could mark malignancy. Barroeta et al. (2006) analyzed the diagnostic efficacy of a panel of antibodies to CK19, Gal-3, HBME-1, anti-MAP kinase (ERK), RET, and p16 using a tissue microarray consisting of both benign and malignant FDLT. HBME-1, ERK, and p16 were more specific for malignancy, whereas CK19 and Gal-3 stained benign lesions with a higher frequency and were not specific for malignant FDLT. RET-oncoprotein showed poor sensitivity and specificity.

Gal-3 and CD44v6: In FNAC of thyroid follicular tumors, the positive rates of Gal-3 and CD44v6 were 89% and 74% in follicular carcinoma, respectively, 25% and 30% in follicular adenoma, respectively. Positive staining of either Gal-3 or CD44v6 resulted in a diagnostic sensitivity of 97% and a specificity of 52% for follicular carcinoma among follicular tumors. Immunostaining of Gal-3 or CD44v6 on cytological samples can provide independent information to distinguish follicular carcinoma from adenoma (Maruta et al. 2004; Weber et al. 2004).

13.2.4 Hashimoto's Thyroiditis

Hashimoto's thyroiditis (HT) represents the most common cause of hypothyroidism and nonendemic goiter. In a study of 133 cases of HT, an unexpected expression of Gal-3 was observed in a subset of HT together with the presence of HBME-1, c-met and cyclin-D1 that are also involved in malignant transformation and deregulated cell growth (Gasbarri et al. 2004b). Furthermore, a loss of allelic heterozygosity in a specific cancer-related chromosomal region was demonstrated in some HT harbouring Gal-3-positive follicular cells. This study provided a well-substantiated demonstration that HT may include a spectrum of different thyroid conditions ranging from chronic autoimmune thyroiditis to thyroiditis triggered by specific immune-response to cancer-related antigens (Gasbarri et al. 2004b). Focal PTC like nuclear alterations have been documented in HT. The expression of four genes known to be up-regulated in PTC [*LGALS3* (galectin3), *CITED1, CK19* (cytokeratin 19) and FN1] and HBME1 were re-evaluated. Focal expression of galectin3 (GAL3), CITED1, CK19, HBME1 and FN1 was seen in 87%, 65%, 43%, 26% and 17% of HT, respectively, only in thyrocytes showing PTC-like nuclear alterations. Focal PTC-like immunophenotypic changes in HT suggested the possibility of early, focal premalignant transformation in some cases of HT (Prasad et al. 2004).

13.3 Breast Cancer

Breast tissue expresses a high level of Gal-3 whose expression is down-regulated in breast cancer. The reduced expression of Gal-3 was associated with increasing histologic grade, and thus with the acquisition of invasive and metastatic potential which possibly resulted from reduced extracellular matrix binding and increased cell motility (Idikio 1998).

Galectins in Mammary Carcinoma Cell Lines: Gal-3 is an anti-apoptotic protein that protects T cells, macrophages, and breast carcinoma cells from death triggered by a variety of agents. High levels of Gal-3 are present in a subset of B-cell neoplasms including diffuse large B-cell lymphoma (DLBCL), primary effusion lymphoma (PEL), and multiple myeloma (MM), in both cell lines and patient samples. Galectins are present on mouse mammary carcinoma cells in vitro and in vivo unlike non-malignant cells from the several tissues; and asialo-GM1 ganglioside carbohydrate part – containing probe was the most specific one. Different results concerning the role of Gal-3 in breast cancer malignancy were obtained in studies using cell lines. Galectin expression was reduced during tumor progression in more aggressive forms of spontaneous BLRB mammary carcinomas similarly shown in human breast carcinoma samples. Honjo et al. (2001) showed that the blocking of Gal-3 expression in highly malignant human breast carcinoma MDA-MB-435 cells led to the reversion of transformed cellular phenotype and to significant suppression of tumor growth in nude mice. It was suggested that the expression of Gal-3 is necessary for the maintenance of transformed and tumorigenic phenotype of MDA-MB-435 breast carcinoma cells. Furthermore, Song et al. (2002) showed that Gal-3 enhances the metastatic potential of human breast carcinoma BT 549 cells. It was established that Gal-3 in tumor cells can play the role of survival factor against. Histopathological data, however, led to suggest that galectin expression is hardly a suitable marker of aggressiveness of heterogeneous mammary. It was proposed that galectins that are selectively expressed during mouse mammary carcinoma progression, similarly to human breast carcinomas, seem to be proper targets for asialo-GM1-vectored cytotoxics and the present mouse model might be a relevant model to test novel modes of anti-breast cancer therapy (Moiseeva et al. 2005).

Uptake of Gal-3 in Breast Carcinoma Cells: Mazurek et al. (2005) found that the introduction of wild-type Gal-3 into nontumorigenic, Gal-3-null BT549 human breast epithelial cells conferred tumorigenicity and metastatic potential in nude mice, and that Gal-3 expressed by the cells was phosphorylated. In contrast, BT549 cells expressing Gal-3 incapable of being phosphorylated (Ser6—>Glu Ser6—> Ala) were nontumorigenic. Differentially regulated genes in breast cancers that are also predicted to be associated with phospho-Gal-3 in transformed BT549 cells include C-type lectin 2, insulin-like growth factor-binding protein 5, cathepsins L2, and cyclin D1. These results suggest that phosphorylation of the protein is necessary for regulation of unique sets of genes that play a role in malignant transformation (Mazurek et al. 2005). Gal-3 uptake modulates the adhesion plaques in those cells which express high levels of Gal-3 have thin-dot like plaques that may be suited for rapid adhesion and spreading while cells in which Gal-3 expression is reduced or knocked-down, have thick and elongated plaques which may be suited for a firmer adhesion to the substratum. Recombinant Gal-3 added exogenously reduced the thickness of the adhesion plaques of tumor cells with reduced Gal-3 expression. The report suggested that Gal-3 once externalized, is a powerful modulator of cellular adhesion and spreading in breast carcinoma cells (Baptiste et al. 2007).

Gal-3 in Breast Epithelial-Endothelial Interactions: Gal-3 is required for the stabilization of epithelial-endothelial interaction networks. Co-culture of epithelial cells with endothelial cells results in increase in levels of secreted Gal-3 and presence of proteolytically processed form of Gal-3 in the conditioned media. Data suggested that Gal-3 expression is associated with specific morphological precursor subtypes of breast cancer and undergoes a transitional shift in expression from luminal to peripheral cells as tumors progressed to comedo-DCIS or invasive carcinomas. Such a localized expression of Gal-3 in cancer cells proximal to the stroma could lead to increased invasive potential by inducing novel or better interactions with the stromal counterparts (Shekhar et al. 2004). In the majority of adenomas, the neoplastic cells show moderate to strong Gal-3 immunoreactivity, but in the majority of adenocarcinomas such immunoreactivity was weak. It was revealed that the progression of canine mammary tumors is associated with low Gal-3 expression (Choi et al. 2004; Johnson et al. 2007). Patients with metastatic cancer commonly show increased serum Gal-3 concentrations, which plays a critical role in cancer metastasis. It highlights the functional importance of altered cell surface glycosylation in cancer progression (Yu et al. 2007).

13.4 Tumors of Nervous System

13.4.1 Galectins and Gliomas

A relationship is assumed to exist between the levels of expression of Gal-3 and the level of malignancy in human gliomas (Bresalier et al. 1997; Gordower et al. 1999; Camby et al. 2001). Bresalier et al. (1997) showed that normal brain tissue and benign tumors did not express Gal-3 but anaplastic astrocytomas (grade 3) and glioblastomas (grade 4 astrocytomas) respectively exhibit intermediate and high level of expression of Gal-3. Moreover, a more significant expression of Gal-3 was associated in metastases than in the primary tumors from which they derived (Bresalier et al. 1997). Though, astrocytic tumor progression is known to be associated with an increased expression of Gal-3, Gordower et al. (1999) showed that the level of Gal-3 expression significantly is decreased in majority of astrocytic tumors. However, Gordower et al. (1999) suggested that human astrocytic tumors are very heterogenous, and in spite of the general decrease in the level of Gal-3 expresssion, some tumor cell clones express a higher level of Gal-3 with increasing level of malignancy (Gordower et al. 1999). Camby et al. (2001) assayed the levels of expression of galectin-1, -3 and -8 in human astrocytic tumors of grades 1–4. The levels of galectin-1 and Gal-3 expression significantly changed during the progression of malignancy in human astrocytic tumors, while that of galectin-8 remains unchanged (Camby et al. 2001). Colonic surgical resections were studied for gal-1 and gal-3. Gal-1 and gal-3 expressed in variable amounts in epithelial cells and the connective tissue of normal colon and the expression of both is significantly increased with degree of dysplasia, suggesting that gal-1 and gal-3 and their binding sites are related to malignant progression, while gal-8 was associated with suppressor activity (Hittelet et al. 2003). It seems that the expression of Gal-3 is highly dependent on non-tumor cells such as endothelial cells or macrophages/microglial cells. This feature can partly explain the conflicting results that have been published on Gal-3 expression in human gliomas (Deininger et al. 2002; Strik et al. 2001; Le Mercier et al. 2010). The regulation of Gal-3 expression by Runx-2, a transcription factor, has been recently suggested to contribute to the malignant progression of glial tumor. Knockdown of Runx2 was shown to be accompanied by a reduction in both Gal-3 mRNA and protein levels, dependent on the glial tumor cell line tested (Vladimirova et al. 2008).

Gal-3 in Distinguishing Gliomas: Cultured Gal-3 deficient glioblastoma cells showed increased motility potential on laminin and modifications in the cytoskeleton reorganization. In addition, c-DNA microarrays and quantitative immunofluorescence analysis showed that Gal-3 deficient U373 cells had an increased expression of integrins-$\alpha 6$ and -$\beta 1$ proteins known to be implicated in the regulation of cell adhesion (Debray et al. 2004). The distinction of astrocytomas and oligodendrogliomas, mainly pilocytic astrocytomas (PILOs) from infiltrating astrocytomas and oligodendrogliomas (ODs), and high-grade oligodendrogliomas from glioblastomas (GBMs), poses a serious clinical problem. Neder et al. (2004) identified Gal-3 as a possible tool to differentiate them based on gene expression profiles of GBMs. Higher expression of Gal-3 was observed in GBMs and PILOs than in OD, AODs and ASTs. Gal-3 appeared to be differentially expressed in central nervous system tumors, making IHC detection of Gal-3 a useful tool in distinguishing between these gliomas (Neder et al. 2004).

The Interactions Between Galectins and Integrins: Galectins are components of the ECM. The ECM comprises all secreted soluble and insoluble molecules found within the extracellular fluid of the extracellular space. The ECM is not only a static scaffolding for tissue organization but it is involved as well in many regulatory functions like modulation of migration, guidance of axonal growth, synapse formation and cell proliferation. Several reviews have already addressed an in-depth analysis of glioma ECM (Rutka et al. 1999; Wang et al. 2005a; Le Mercier et al. 2010). As emphasized by Uhm et al. (1999), integrins are cell-surface receptors that mediate the physical and functional interactions between a cell and its surrounding ECM. Although classically the role ascribed to integrins has been that of anchoring cells to ECM, the functions of integrins greatly exceed that of mere cell adhesion (Uhm et al. 1999). Within this multifaceted role, integrins have been shown to be molecular determinants of glioma invasion (Le Mercier et al. 2010; Natarajan et al. 2003; Rutka et al. 1999).

Galectins and integrins closely interact when modulating cell adhesion and/or cell migration. For example, Moiseeva et al. have shown that galectin-1 interacts with the integrin $\beta 1$ subunit in vascular smooth muscle cells (Fig. 13.1). Gal-3 was also shown to bind to $\alpha 1\beta 1$ integrin and it was suggested that this interaction regulates cell adhesion of various tumor cell lines by preventing $\alpha 1\beta 1$ integrin interaction with the ECM proteins (Ochieng et al. 1998) (Fig. 13.1). Gal-3 also forms a complex with $\alpha 3\beta 1$ integrins and NG2 on the surface of endothelial cells. The subsequent transmembrane signaling via $\alpha 3\beta 1$ has been shown to be responsible for endothelial cell motility and angiogenesis (Fukushi et al. 2004; Deininger et al. 2002). Among other galectins, Gal-8 was also reported to interact with a subgroup of integrins that include $\alpha 3\beta 1$, $\alpha 6\beta 1$, and to a lesser extent with the $\alpha 4$ and the $\beta 3$ subunits in human carcinoma (1,299) cells. Gal-8 binds $\alpha 1\beta 1$, $\alpha 3\beta 1$ and $\alpha 5\beta 1$ integrins in Jurkat T cells (Carcamo et al. 2006) (Fig. 13.1). These interactions were shown to inhibit cell adhesion and to induce apoptosis (Hadari et al. 2000).

Integrins play a significant role in the malignant progression of cancer through their involvement in cell adhesion, motility and intracellular signaling (Hynes 2002), with an emphasis on the role of $\beta 1$ integrin subunit in gliomas (Bartik et al. 2008; D'Abaco and Kaye 2007; Le Mercier et al. 2010). As galectins bind integrins, with galectin-1, Gal-3 and galectin-8 all known specifically to modulate $\beta 1$ integrin function, the understanding of molecular mediators such as galectins and the pathways through which they drive the cell invasion so descriptive of glioblastoma multiforme (GBM) is anticipated to reveal potential therapeutic targets (Fortin et al. 2004).

Galectins and Glioma Cell Migration: Cell migration involves at least three independent but highly coordinated biological processes: (1) cell adhesion to numerous components of the ECM; (2) cell motility, which involves the reorganization of the actin cytoskeleton mainly through modification of the components of the adhesion complex; and (3) invasion that involves the degradation of matrix proteins by tumor-secreted proteolytic enzymes, mainly serine proteases, cathepsins and metalloproteinases (MMPs) (Decaestecker et al. 2007; Lefranc et al. 2005; Rao 2003). Galectins are involved in cell migration at several steps (Stillman et al. 2005). Galectin-1, Gal-3 and to a lesser extent galectin-8, markedly stimulate the migration of glioma cells in vitro. Moreover, biological functions of Gal-3 were modulated by MMPs (Ochieng et al. 1998), which play crucial roles in glioma cell motility and invasion (Rao 2003). However, in contrary to the aforementioned, cultured Gal-3 deficient U373 glioblastoma cells had been shown to have both increased motility potential on laminin and displayed modifications in cytoskeleton reorganization. Indeed, Debray et al. have shown an increased motility of Gal-3 deficient cells cultivated on laminin (Debray et al. 2004; Le Mercier et al. 2010) whereas Camby et al. (2001) had observed an increase of motility when glioma cells were cultivated on plastic pre-coated with Gal-3. Thus, targeting both integrins and galectins represents is a feasible proposition in the future treatment of gliomas. Moreover, impairing galectin-1 expression in vivo in experimental gliomas through the delivery of anti-galectin-1 siRNA augments the therapeutic benefits contributed by temozolomide (Le Mercier et al. 2008).

Gal-3 Ligands in Meningiomas: Hancq et al. (2004) evaluated the discriminatory value of S100 proteins and Gal-3 and its ligand profile with respect to benign and atypical meningiomas. The combination of these three markers enabled an improved discriminatory criterion to be established between the benign and the atypical meningiomas. Study suggested that the Gal-3-binding sites and S100B (and S100A6 to a lesser extent) could play a role in the aggressiveness characterizing atypical meningiomas. During development and progression of pituitary tumors Riss et al. (2003) showed that only lactotroph (PRL) and corticotroph (ACTH) hormone-producing cells and tumors expressed Gal-3. Gal-3 was present in 63.2% PRL adenomas, 83.3% PRL carcinomas, 46.3% ACTH adenomas, and 87.5% ACTH carcinomas, but not in other pituitary adenomas and carcinomas. Riss et al. (2003) suggest that Gal-3 has an important role in pituitary cell proliferation and tumor progression.

13.5 Diffuse Large B-Cell Lymphoma

The WHO classification of lymphomas is based on clinical, morphological, immunohistochemical and genetic criteria. However, each entity displays its own spectrum of clinical aggressiveness. Treatment success varies widely and is not predictable. Gal-3 is an anti-apoptotic protein that protects T cells, macrophages, and breast carcinoma cells from death triggered by a variety of agents. High levels of Gal-3 are present in a subset of B-cell neoplasms including diffuse large B-cell lymphoma (DLBCL), primary effusion lymphoma (PEL), and multiple myeloma (MM), in both cell lines and patient samples. However, Gal-3 could not be detected in Burkitt lymphoma (BL), follicular lymphoma

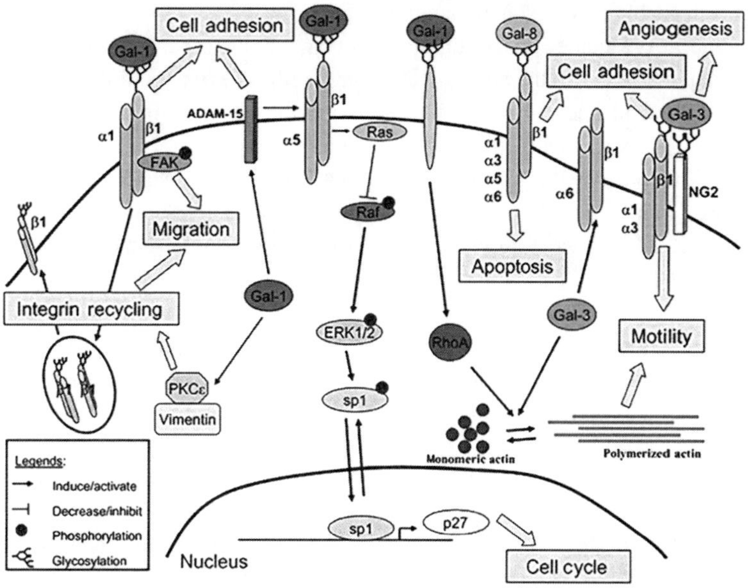

Fig. 13.1 Galectins, integrins and cell migration. The interaction of galectins with integrins modulates cell migration as well as other processes. Galectin-1 (Gal-1) interacts with the β1 integrin subunit inducing the phosphorylation of FAK, which modulates cell migration (Moiseeva et al. 2003). Binding of Gal-1 to integrin is involved in cell adhesion (Moiseeva et al. 1999) and also induces growth inhibition via its interaction with α5β1 (Fischer et al. 2005). Involvement of Ras–MEK–ERK pathway and the consecutive transactivation of Sp1, which induces p27 transcription is shown (Le Mercier et al. 2010). Gal-3 also forms a complex with α3β1 and the proteoglycan NG2 (Fukushi et al. 2004). This interaction regulates endothelial cell motility and angiogenesis. Finally, Gal-3 has been shown to regulate the expression of integrin α6β1 and actin cytoskeleton organization (Debray et al. 2004). However, it is not known with which molecule (s) Gal-3 is interacting to initiate this signaling. Galectin-8 (Gal-8) interacts with several integrins including α1β1, α3β1, α5β1 and α6β1. These interactions are involved in cell adhesion and apoptosis (Hadari et al. 2000). *ERK* extracellular signal-regulated kinase, *FAK* focal adhesion kinase, *MEK* MAP kinase/extracellular signal-regulated kinase kinase (MAPK/ERK Kinase), *PKCε* protein kinase Cε (Adapted with permission from Le Mercier et al. 2010 © John Wiley and Sons)

(FL), marginal zone lymphoma (MZL), MALT lymphoma or B-small lymphocytic lymphoma (B-SLL) cell lines or patient samples. The pattern of expression suggests that aberrantly increased Gal-3 levels in specific B-cell populations may yield a protective advantage during transformation and/or progression of certain B-cell neoplasms (Hoyer et al. 2004).

Non-Hodgkin's and Hodgkin's Lymphomas: Since galectins are involved in oncogenesis and the physiology of immune cells, D'Haene et al. (2005) investigated the expression of galectin-1 and Gal-3 in 25 normal lymphoid tissues, 42 non-Hodgkins and 42 Hodgkins lymphomas. Results showed that in normal lymphoid tissue, lymphocytes do not express galectin-1 and rarely express Gal-3. In contrast, Gal-3 was expressed in 8 of the 16 DLBCL cases and in 1 of the 8 FL cases. Furthermore, Gal-3 mRNA was expressed three times more in the DLBCLs than in the FLs. While the blood vessel walls of the lymphomas expressed galectin-1, the vessel walls of normal lymphoid tissues did not. This expression of galectin-1 in blood vessel walls was correlated with vascular density. This study shows that DLBCL can be distinguished from normal lymphoid tissue and other lymphomas on the basis of Gal-3 expression.

Karpas 299 T-Lymphoma Cells: PUVA treatment first induced G2/M cell cycle arrest resulting in a decrease in the cell proliferation rate in Karpas 299 T-lymphoma cells. Immediately following PUVA treatment, PUVA triggered mitochondrial apoptosis and enhanced the expression of peroxiredoxin, stress protein endoplasmin and Gal-3. Gal-3 was shown to protect mitochondrial membrane integrity and prevent cytochrome C release thereby blocking the effector stage of apoptosis. It was suggested that the elevated level of Gal-3 following PUVA treatment acts in synergy with the constitutively expressed chimeric kinase NPM/ALK to block the apoptosis (Bartosova et al. 2006).

13.6 Gal-3 in Melanomas

Pathogenesis of melanoma is a multi-step process that may include the phases of benign nevi and dysplastic nevi, melanoma, and metastatic melanoma. Dysregulation of cellular proliferation and apoptosis is probably involved in melanoma progression and response to therapy. Melanocytes accumulate Gal-3 with tumor progression, particularly in the nucleus. The strong association of cytoplasmic and nuclear expression in lesions of sun-exposed areas suggests an involvement of UV light in activation of Gal-3 (Prieto et al. 2006). An increase in serum Gal-3 production has been found in patients with advanced metastatic melanoma. Gal-3 concentration was significantly correlated with both LDH and CRP in the melanoma group. It seemed that at least part of serum Gal-3 might be produced by metastatic melanoma tissue. Gal-3 might play a role in melanoma progression and/or inflammation, and warrants further study (Vereecken et al. 2005, 2006; Vereecken and Heenen 2006). Zubieta et al. (2006) suggested that the range of Gal-3-positive tumor cells in melanoma biopsies varied between 0% and 93% and that of galectin-1-positive tumor cells varied between 5% and 97%. In addition, 23 ± 27% of tumor-associated lymphocytes were apoptotic. Although these results showed a correlation between Gal-3 expression and apoptosis of tumor-associated lymphocytes, such could not be found with galectin-1. Expression profiling of Gal-3-depleted melanoma cells reveals its major role in melanoma cell plasticity and vasculogenic mimicry (Mourad-Zeidan et al. 2008).

13.7 Head and Neck Carcinoma

Choufani et al. (1999) reported a significantly lower level of Gal-3 and its ligands in head and neck carcinoma (HNSCC) patients than their normal counterparts and the extent of decrease of Gal-3 expression correlated with the increasing level of clinically detectable HNSCCs aggressiveness. This was confirmed with a low differentiation status which is known as an indicator of recurrence rate in HNSCCs (Delorge et al. 2000; Lefranc et al. 2003). With 61 cases of HNSCC as their basis, including 31 oral, 20 laryngeal, and 10 hypopharyngeal lesions, Delorge et al. (2000) showed that the main modifications observed in connection with a loss of differentiation were related to a modification in the levels of both Gal-3/Gal-3-binding site and T-antigen/T-antigen-binding site expressions. Results also suggested that Gal-3 could act as an acceptor site for the T antigen (Delorge et al. 2000).

On the other hand, expression of Gal-3 correlated highly and positively with the level of apoptosis in human cholesteatomas, which is a benign disease, characterized by unstrained growth and accumulation of keratin debris in middle ear cavity. An up-regulation of Gal-3 expression, which is associated with pronounced apoptotic activity, could have a physiologically protective effect against the substantial apoptotic features occurring in recurrent cholestestomas (Sheikholeslam-Zadeh et al. 2001). Honjo et al. (2000) reported that the nuclear expression of Gal-3 markedly decreases during progression of squamous cell carcinomas of tongue from normal to cancerous states, while cytoplasmic expression increased. It was suggested that Gal-3 translocates from nucleus to cytoplasm during neoplastic progression and may serve as a prognostic factor for tongue cancer (He et al. 2004).

The expression of Gal-3 was examined in relation to neoplastic progression of HSCCs and LSCCs. Gal-3 positivity expressed as percentage of cells was significantly higher in LSCCs and HSCCs than in Low_D or High_D, respectively. Increased expression of Gal-3 in HSCCs was accompanied by a shift from the cytoplasmic compartment to the nucleus. In intertumor-type comparison, laryngeal carcinomas presented nuclear presence of Gal-3 only rarely (1 of 58 cases in laryngeal cancer vs. 27 of 79 cases in hypopharyngeal cancer), and a comparatively low labeling index. Results reveal an association between level of presence of Gal-3 and neoplastic progression of HSCCs and LSCCs (Saussez et al. 2008a, b).

Detection of accessible Gal-3-specific ligands is an independent prognostic marker in advanced head and neck squamous cell cancer with therapeutic potential (Plzak et al. 2004). Adenoid cystic carcinoma (ACC) is one of the most common malignant tumors of the salivary glands characterized by multiple recurrences and distant metastasis resulting in significantly worsening prognosis. Seventeen (48.6%) ACC tumor specimens were found Gal-3-positive. Gal-3 reactivity was significantly associated with regional and distant metastasis. There was no statistical significance in the correlation of Gal-3 expression and disease-free survival and overall survival rate. Gal-3 may be used as an indicator in the prediction of metastatic spread in ACC (Teymoortash et al. 2006).

Esophageal Squamous Carcinoma: In patients with esophageal squamous cell carcinoma (ESCC), neither nuclear nor cytoplasmic expression of Gal-3 was a prognostic indicator in ESCC. But elevated expression of Gal-3 in the nuclei but not the cytoplasm may be an important biological parameter related to histological differentiation and vascular invasion in patients with ESCC (Shibata et al. 2005; Saussez et al. 2007a, b).

13.8 Lung Cancer

Lung cancer is the leading cause of cancer deaths in the world. The tumor stage is the most powerful prognostic tool for predicting the survival rates of lung carcinoma patients.

Prognosis of individual patients is difficult in part because of the marked clinical heterogeneity among such patients. The histological expression of Gal-3 was assessed in a panel of lung tumor specimens including small-cell lung cancer (SCLC) and non-small-cell lung cancer (NSCLC). Among new prognostic markers, a striking difference in Gal-3 expression was observed between tumors, with high expression in NSCLC (42/47 samples) and low expression in SCLC (negative in 13/18, weak in 5/18). This differential expression of Gal-3 between histological types of lung carcinoma suggests that Gal-3 may have an important influence on tumor cell adhesion and apoptosis (Buttery et al. 2004). Prognostic value of Gal-3 expression in lung adenocarcinomas and squamous cell carcinomas has been studied by Mathieu et al. (2005). In 165 squamous cell carcinomas and 121 adenocarcinomas, immunostained for Gal-3, a large majority of cases displayed Gal-3 expression. Though, the Gal-3 immunohistochemical expression differs between squamous cell carcinomas and adenocarcinomas, the nuclear expression of Gal-3 behaved as a significant prognostic predictor for all the cases as a group (Mathieu et al. 2005).

However, binding of galectin-1 and its expression tended to increase, whereas the parameters for Gal-3 decreased in advanced pT and pN stages. The number of positive cases was considerably smaller among the cases with small cell lung cancer than in the group with non-small-cell lung cancer, among which adenocarcinomas figured prominently with the exception of galectin-1 expression. The survival rate of patients with Gal-3-binding or galectin-1-expressing tumors was significantly poorer than that of the negative cases. The expression and the capacity to bind the adhesion/growth regulatory Gal-3 were defined as an unfavorable prognostic factor not correlated with the pTN stage (Szoke et al. 2005).

13.9 Colon Neoplastic Lesions

Conflicting reports are available regarding expression of Gal-3 in human colonic mucosa and colonic tumors. Some studies suggest decreasing Gal-3 level in colon carcinoma progression (Castronovo et al. 1992; Lotz et al. 1993), whereas other reports presented opposite results (Irimura et al. 1991; Sanjuán et al. 1997; Schoeppner et al. 1995). Castronovo et al. (1992) and Lotz et al. (1993) found low levels of Gal-3 mRNA in human colon cancer tissues relative to levels in normal colonic mucosa. Castronovo et al. (1992) also stated that patients with Dukes' C and D tumors had a lower ratio of primary tumor Gal-3 mRNA to normal tissue mRNA than the patients with Dukes' B tumors. On the other hand, a higher content of Gal-3 was reported by Irimura et al. (1991) in advanced stage of Dukes' D colorectal cancer than in samples of early-stage disease. Though the normal mucosa distant from areas of neoplasia was characterized by weak or negative expression of Gal-3, Schoeppner et al. (1995) demonstrated that cytoplasmic levels of Gal-3 correlated with progression from adenoma to carcinoma. Thus, Gal-3 expression in invasive cancers varied according to the stage in Dukes' scale (Irimura et al. 1991; Schoeppner et al. 1995). Sanjuan et al. (1997) and Schoeppner et al. (1995) observed that normal colonic mucosa usually strongly expressed Gal-3 both in the nucleus (100% of cases) and in cytoplasm (77%). Nuclear and cytoplasmic expression was significantly down-regulated in adenomas (60%, 16%, respectively), whereas cytoplasmic expression of Gal-3 increased in carcinomas again (64%) although it usually did not reach the level of normal mucosa expression. A correlation between increased levels of Gal-3 expression and shorter survival periods, particularly in the case of patients with Dukes' A and B colon tumors, was observed (Nagy et al. 2003; Sanjuán et al. 1997; Schoeppner et al. 1995).

Gal-3 expression has been correlated with progression and metastasis in colon cancer. Results provide strong evidence that Gal-3 plays a role in the ability of colon cancer cells to metastasize to distant sites. Greco et al. (2004) found that (1) the expression of Gal-3 was significantly increased on the surface of cells from adenomas with respect to normal mucosa from the same patient; and (2) Gal-3 overexpression was not related with the presence of K-ras mutation. These results indicated that the evaluation of Gal-3 expression (and of its ligand, 90 kDa) can be of interest in the characterization of nonmalignant and malignant colon cancers (Greco et al. 2004; Endo et al. 2005). The incidence of lymph node and distant metastasis in galectin 3-positive cancer was significantly higher than that in Gal-3-negative cases. It was proposed that Gal-3 expression is an independent factor for prognosis in colorectal cancer (Endo et al. 2005). In colon cancer sera, the major Gal-3 ligand was a 40-kDa band distinct from mucin, carcinoembryonic antigen, and Mac-2 binding protein. The major circulating ligand for Gal-3, which is elevated in the sera of patients with colon cancer, is a cancer-associated glycoform of haptoglobin of 40-kDa which was 10 to 30-fold higher in patients than in healthy subjects (Bresalier et al. 2004). Greco et al. (2004) suggested that a 90 kDa ligand molecule of Gal-3 was increased in the blood from patients with both adenomatous and adenocarcinomatous lesions;

Gal-3 Up-Regulates MUC2 Transcription: Gal-3 and MUC2 in intestine have been correlated with the malignant behavior of colon cancer cells. Down-regulation of Gal-3 expression by antisense transfection resulted in a significant decrease in liver colonizing ability, whereas up-regulation of Gal-3 increased metastatic potential. Moreover, the alterations in Gal-3 levels resulted in parallel changes in the level of Muc 2 mucin – a major ligand for Gal-3 (Dudas et al. 2002). Gal-3 up-regulates MUC2 protein at the level of transcription through AP-1 activation. Gal-3 responsiveness was found between 1,500 and 2,186 bp upstream of the translation start site, a region that contains one consensus AP-1 binding site in MUC2 promoter constructs. Mutation in the AP-1 site markedly decreased MUC2 promoter activity. Analyses suggested an association between Gal-3, c-Jun, and Fra-1 in forming a complex at the AP-1 site on the MUC2 promoter (Song et al. 2005).

13.10 Expression of Gal-3 in Other Tumors

Vulvar Squamous Lesions: The expression patterns of Gal-3 and the frequency of infiltrating CD1a positive DCs were determined in 82 cases of vulvar tissues, consisting of normal squamous epithelia (NE), vulvar condylomas (VC), high grade vulvar intraepithelial neoplasias (HG-VIN) of common type, and invasive keratinizing squamous cell carcinomas (SCC). Gal-3 expression was cytoplasmic, nuclear or membranous in NE, VCs, and HG-VINs, with negative or weak and occasionally moderate reactivities. In keratinizing SCC, exclusively cytoplasmic staining patterns with moderate reactivity were observed in 59% of cases. Data indicated that qualitative and quantitative changes of Gal-3 expression and infiltration by CD1a positive DCs in vulvar NE, VCs, and HG-VIN lesions, respectively, compared with SCCs play a role in the development of an infiltrative phenotype, and may provide adjunctive criteria in the diagnosis of invasion of vulvar squamous epithelia (Brustmann 2006).

KSHV Downregulation of Gal-3 in Kaposi's Sarcoma: In prostate, ovarian and breast cancer, down regulation of Gal-3 is associated with malignancy and tumor progression. Kaposi's sarcoma (KS) is an angioproliferative tumor of vascular endothelial cells and produces rare B cell lymphoproliferative diseases in the form of primary effusion lymphomas and some forms of multicentric Castleman's disease. Alcendor et al. (2009) found reduced levels of gal-3 expression in a significant fraction of latency associated nuclear antigen (LANA) positive spindle cell regions in human archival KS tissue. Gal-3 protein expression is downregulated 10-fold in 10-day Kaposi's sarcoma-associated herpesvirus (KSHV) infected dermal microvascular endothelial cells (DMVEC) accompanied by down-regulation of mRNA with a consistent downregulation of Gal-3. Of the galectins assayed, only galectin-1 was also downregulated in KSHV infected DMVEC. Data suggest that KSHV vFLIP and LANA are the viral genes targeting Gal-3 down regulation.

Parathyroid Carcinoma: The diagnosis of parathyroid carcinoma (PC) is difficult and based on morphological features, which are not totally reliable. Hyperplastic and neoplastic parathyroid lesions may present overlapping morphologic features, and several markers have been proposed to distinguish benign from malignant growths. The expressin of Gal-3 in several malignant PC tumors, including follicular carcinomas of the thyroid indicated that Gal-3 immunostaining is a valuable tool to support a diagnosis of PC in highly proliferating tumors affecting a single parathyroid gland (Bergero et al. 2005). Hyperplastic lesions responsible for primary nonfamilial or tertiary hyperparathyroidism, as well as parathyroid adenomas, were negative for Gal-3, as opposed to carcinomas. In addition, secondary and familial primary hyperplasia cases were positive for Gal-3 in approximately two thirds of cases. All hyperplastic lesions (positive or negative for Gal-3) had a low Ki-67 index. Based on these findings, secondary hyperplasia has a low proliferative potential but an unexplained Gal-3 reactivity, which reduces its diagnostic role in differentiating benign from malignant nodules in the context of multiglandular parathyroid diseases (Saggiorato et al. 2006).

Prostate Cancer: Expression of Gal-3 is generally reduced in prostate cancer relative to the level in normal human prostate tissue (Ellerhorst et al. 1999; Pacis et al. 2000). Van den Brûle et al. (2000) demonstrated a clear change in location of Gal-3 in prostate carcinoma cells as compared with non-tumor cells. In general, normal glandular cells expressed Gal-3 in both the nucleus and cytoplasm. Studies suggest that Gal-3 might have anti-tumor activities when present in the nucleus, whereas it could favor tumor progression when expressed in the cytoplasm (Van den Brûle et al. 2000). Study on LNCaP, a human prostate cancer cell line, used to generate transfectants expressing Gal-3 either in the nucleus or in the cytosol, demonstrated that Gal-3 exerts opposite biological activities according to its cellular localization: nuclear Gal-3 plays antitumor functions and cytoplasmic Gal-3 promotes tumor progression (Califice et al. 2004a). He and Baum (2006) observed reduced T-cell migration across endothelial cells induced to increase galectin-1 expression by exposure to prostate cancer cell conditioned medium, compared to T-cell migration across control-treated endothelial cells; the inhibitory effect of galectin-1 on T-cell migration was reversed by specific antiserum. It was indicated that galectin-1-mediated clustering of CD43 contributes to the inhibitory effect on T-cell migration. Inhibition of T-cell migration is an anti-inflammatory activity of galectin-1 (He and Baum 2006).

Gal-3 Inhibits Anticancer Drug-Induced Apoptosis Through Regulation of Bad Protein: Prostate cancer exhibits resistance to anticancer drugs, at least in part due to enhanced antiapoptotic mechanisms. The expression of exogenous Gal-3 in LNCaP cells, which do not express Gal-3 constitutively, inhibits anticancer drug-induced apoptosis by stabilizing mitochondria. The expression of Gal-3 stimulated the phosphorylation of Ser(112) of Bcl-2-associated death (Bad) protein and down-regulated Bad expression after treatment with cis-diammine-dichloroplatinum. Findings indicated that Gal-3 inhibits anticancer drug-induced apoptosis through regulation of Bad protein and suppression of mitochondrial apoptosis pathway. Therefore, targeting Gal-3 could improve the efficacy of anticancer drug chemotherapy in prostate cancer (Fukumori et al. 2006).

DES-Induced Renal Tumors in Syrian Hamster: The diethylstilbestrol (DES)-induced renal tumors in male Syrian hamster kidney (SHKT) represent a unique animal model for the study of estrogen-dependent renal malignancies. Except galectin-4, all galectins (1, -3, -7, and -8) were expressed in kidney tumors. Small clusters of galectin-1-positive, most likely preneoplastic cells at the corticomedullary junction were already evident 1 week after DES administration. Expression of Galectin-1 and -3 was apparently associated with the first steps of the neoplastic transformation, because small tumorous buds were found to be positive after 1 month of treatment. In contrast, galectins-7 and -8 were detected in large tumors and medium-sized tumors, respectively, thereby indicating an involvement in later stages of DES-induced SHKT (Saussez et al. 2005, 2006).

Gal-3 Inhibits Apoptosis in Bladder Carcinoma Cells: Gal-3 has been shown to regulate CD95, a member of TNF family of proteins in the apoptotic signaling pathway. The generality of this phenomenon has been questioned by studying a different protein [e.g., TNF-related apoptosis-inducing ligand (TRAIL), which induces apoptosis in a wide variety of cancer cells]. Overexpression of Gal-3 in J82 human bladder carcinoma cells rendered them resistant to TRAIL-induced apoptosis, whereas phosphatidylinositol 3-kinase (PI3K) inhibitors (wortmannin and LY-294002) blocked the Gal-3 protecting effect. Because Akt is a major downstream PI3K target reported to play a role in TRAIL-induced apoptosis, Oka et al. (2005) questioned the possible relationship between Gal-3 and Akt. Overall results suggested that Gal-3 involves Akt as a modulator molecule in protecting bladder carcinoma cells from TRAIL-induced apoptosis (Oka et al. 2005).

Solid and Pseudopapillary Tumor of Pancreas: Berberat et al. (2001) showed that Gal-3 was strongly overexpressed at mRNA and protein level in human pancreatic cancer compared to its expression in normal human pancreas cells. On the other hand, metastatic pancreatic cancer cells in lymph nodes and in liver showed strong Gal-3 immunoreactivity, indicating that Gal-3 might have an impact on metastasis formation. Gal-3, containing NWGR antideath motif of the Bcl-2 protein family, is involved in various aspects of cancer progression. Solid pseudopapillary tumors (SPT) of the pancreas are rare neoplasms that occur mostly in young women. Gal-3 is a major factor in the carcinogenesis of pancreatic ductal adenocarcinoma in SPT. Gal-3 is strongly expressed in all SPTs, whereas its level was lowered in metastatic nodules. In contrast, Gal-3 expression was not found in normal pancreatic endocrine cells or in neuroendocrine tumors. Thus, Gal-3 is a useful marker to distinguish SPT from neuroendocrine tumor, and also indicator of behavior because its low expression is associated with metastatic spreading (Geers et al. 2006).

Hepatocellular Carcinoma: Studies demonstrated that normal hepatocytes do not express Gal-3, but it expresses in hepatocellular carcinoma (HCC), independent of prior hepatitis B virus (HBV) infection. However, Gal-3 expression in HCC can be positively influenced by HBV infection through a mechanism that may include the transactivation of the Gal-3 gene promoter. The focal regenerating nodules of cirrhotic tissue also express Gal-3. It is possible that these cells indicate early neoplastic events (Hsu et al. 1999).

Gastric Cancer: In patients with gastric cancer, Gal-3 expression was correlated with nodal status, lymphatic inivasion, pathological stage and histological parameters. On the other hand, Gal-3 expression did not correlate with the expression of Ki-67. By PCR-SSCP-sequence analysis, two single nucleotide polymorphisms (SNPs) were detected in the Gal-3 gene, but none showed mutations. The reduced Gal-3 expression was associated with lymph node metastasis, advanced stage and tumor differentiation in gastric cancer. Gal-3 expression could be a useful prognostic factor in gastric cancer (Okada et al. 2006).

ACTH-Producing Adenomas and Prolactinomas: Gal-3 is expressed in a subset of normal pituitary cells and tumors including PRL, ACTH, and in folliculo-stellate (FS) cells and tumors. Gal-3 has an important regulatory role in pituitary cell proliferation. Gal-3 is associated with functioning ACTH and PRL tumors and is expressed infrequently in silent ACTH adenomas, suggesting that Gal-3 protein and/or gene is altered in non-functioning ACTH tumors. The use of ACTH and Gal-3 immunostaining should help in the diagnosis of silent ACTH adenomas (Jin et al. 2005). Gal-3 is expressed both in invasive prolactinomas and noninvasive prolactinomas. Since significantly higher expression seen in the invasive prolactinomas, Gal-3 expression may be used as a useful indicator to determine the invasiveness and prognosis of prolactinomas (Wang et al. 2005b).

Pheochromocytomas: Malignant and benign pheochromocytomas were analyzed for the expression of Gal-3. One malignant pheochromocytoma with distant metastases showed strong and one malignant undifferentiated pheochromocytoma with local invasion showed partly strong cytoplasmic staining. Nine of ten sporadic and all hereditary benign pheochromocytomas had absent/weak staining. One benign sporadic pheochromocytoma had moderate cytoplasmic staining. The distinct expression in various types of pheochromocytomas is intriguing and requires further investigation (Gimm et al. 2006).

Testicular Tumors: In testicular tumorigenesis, Gal-3 has a dual function according to the histological type of tumors and their hormone dependency. In malignant testicular Sertoli cell tumors, the expression of Gal-3 is down-regulated while, in benign Leydig cell tumors, this expression is maintained, indicating the possible implication of this gene in the development of more aggressive testicular sex cord stromal tumors. In contrast to sex cord stromal tumors, Gal-3 expression is up-regulated in testicular germ cell tumors. A significant elevation of the Gal-3 mRNA level occurs in non-seminomatous testicular germ cell tumors and cell line as compared to normal testes and seminomas, indicating the possible role of this gene in the non-seminomatous differentiation of germ cell tumors (Devouassoux-Shisheboran et al. 2006).

13.11 Gal-3 in Metastasis

Adhesive interactions between the molecules on cancer cells and the target organ are one of the key determinants of the organ specific metastasis. It was shown that β1,6 branched N oligosaccharides which are expressed in a metastasis-dependent manner on B16-melanoma metastatic cell lines, participate in the adhesion process. High metastatic cells showed increased translocation of lysosome associated membrane protein (LAMP1), to the cell surface. LAMP1 on high metastatic cells, carries very high levels of these oligosaccharides, which are further substituted with poly N-acetyl lactosamine (polylacNAc), resulting in the expression of high density of high affinity ligands for Gal-3 on the cell surface. Krishnan et al. (2005) showed that Gal-3 is expressed in highest amount in the lungs as compared to other representative organs and that the lung vascular endothelial cells expressed Gal-3 constitutively on their surface. Gal-3 on the organ endothelium could thus serve as one of the anchors for the circulating cancer cells, expressing high density of very high affinity ligands on their surface, and facilitate organ specific metastasis. However, in liver and lung metastatic cells galectins seem to be expressed within cytoplasm and/or nuclei. Galectin expression correlated directly with aggressive tumor potential in the A/Sn transplantable animal model similar to findings in several human breast carcinoma cell lines.

C4.4A with Gal-3 Influences Laminin Adhesion: C4.4A is a member of the Ly6 family, with low homology to uPAR. It has been detected mainly on metastasizing carcinoma cells and proposed to be involved in wound healing. C4.4A has been regarded as an orphan receptor, whose functional activity has not been fully explored. C4.4A ligands are strongly expressed in tissues adjacent to squamous epithelia. For example, in tongue and esophagus, the expression pattern partly overlaps with laminin (LN) and complements the C4.4A expression that is found predominantly on the basal layers of squamous epithelium. Evidence suggests that association of C4.4A with Gal-3 influences LN (LN1 and LN5) adhesion. C4.4A was described originally as a metastasis-associated molecule (Paret et al. 2005).

Serum Gal-3 in Cancer Cell Endothelial Adhesion: Patients with metastatic cancer commonly show increased serum Gal-3 concentrations, which plays a critical role in cancer metastasis. It highlights the functional importance of altered cell surface glycosylation in cancer progression (Yu et al. 2007). MUC1, a large transmembrane mucin protein that is overexpressed and aberrantly glycosylated in epithelial cancer, is a natural ligand for Gal-3. Recombinant Gal-3 at concentrations similar to those found in the sera of patients with metastatic cancer increased adhesion of MUC1-expressing human breast (ZR-75-1) and colon (HT29-5F7) cancer cells to human umbilical vein endothelial cells (HUVEC). It was indicated that Gal-3, by interacting with cancer-associated MUC1 via oncofetal Thomsen-Friedenreich antigen (Galβ1,3 GalNAc-α(TF)), promotes cancer cell adhesion to endothelium by revealing epithelial adhesion molecules that are otherwise concealed by MUC1 (Yu et al. 2007).

13.12 β1,6 N-acetylglucosaminyltransferase V in Carcinomas

The Golgi enzyme β1,6 N-acetylglucosaminyltransferase V (Mgat5) is up-regulated in carcinomas and promotes the substitution of N-glycan with poly N-acetyllactosamine, the preferred ligand for Gal-3. Transformation is associated with increased expression of β1,6GlcNAc-branched N-glycans, products of Mgat5. The expression of Mgat5 sensitizes mouse cells to multiple cytokines. Gal-3 cross-linked Mgat5-modifies N-glycans on epidermal growth factor and transforming growth factor-β receptors at the cell surface and delayed their removal by constitutive endocytosis. Mgat5 also promoted cytokine-mediated leukocyte signaling, phagocytosis, and extravasation in vivo. Thus, conditional regulation of N-glycan processing drives synchronous modification of cytokine receptors, which balances their surface retention against loss via endocytosis (Partridge et al. 2004). Lagana et al. (2006) reported that fibronectin fibrillogenesis and fibronectin-dependent cell spreading are deficient in Mgat5$^{-/-}$ mammary epithelial tumor cells and inhibited in Mgat5$^{+/+}$ cells by blocking Golgi N-glycan processing with swainsonine or by competitive inhibition of galectin binding. It appeared that fibronectin polymerization and tumor cell motility are regulated by Gal-3 binding to branched N-glycan ligands that stimulate focal adhesion remodeling, FAK and PI3K activation, local F-actin instability, and α5β1 translocation to fibrillar adhesions.

Both tyrosine-phosphorylated caveolin-1 (pY42Cav1) and Mgat5 are linked with focal adhesions (FAs); their function in this context is unknown. Gal-3 binding to Mgat5-modified N-glycans functions together with pY42Cav1 to stabilize focal adhesion kinase (FAK) within FAs, and thereby promotes FA disassembly and turnover. Expression of the Mgat5/galectin lattice alone induces FAs and cell spreading. Results suggest that transmembrane crosstalk between the galectin lattice and pY42Cav1 promotes FA turnover by stabilizing FAK within FAs previously unknown, interdependent roles for Gal-3 and pY42Cav1 in tumor cell migration (Goetz et al. 2008).

13.13 Macrophage Binding Protein

The 90 K/Mac-2 binding protein (M2BP) is a member of the macrophage scavenger receptor cysteine-rich domain superfamily. Systemic levels of 90 K protein have been correlated with inflammation in many diseases, including asthma. 90 K protein is increased in asthma in blood. Its inhibitory effect on TH2 cytokine transcription suggests that increased 90 K protein expression is an attempt to limit the ongoing inflammation in asthma (Kalayci et al. 2004). Biliary M2BP levels, especially when used in conjunction with biliary CA19-9 levels, showed promise as a novel diagnostic marker for biliary tract carcinoma (Koopmann et al. 2004). The tumor-associated M2BP is highly expressed in lung cancer and the M2BP-specific immunity was observed in many patients with lung cancer. M2BP can be used as a target antigen in cancer immunotherapy. Hence peptides derived from M2BP with an HLA-A24 binding motif were analyzed for their ability to induce M2BP-specific cytotoxic T lymphocytes (CTL). Two CTLs, one induced with M2BP(241–250) (GYCASLFAIL) and the other with M2BP(568–576) (GFRTVIRPF), produced interferon-γ in response to HLA-A24-positive TISI cells pulsed with the same peptide in vitro. Although the CTLs induced with M2BP(241–250) reacted with both peptide-pulsed TISI cells and BT20 cells expressing both M2BP and HLA-A24, the CTLs induced with M2BP(568–576) did not react with BT20 cells. Findings suggested that M2BP (241–250) is naturally processed from the native M2BP molecule in cancer cells and recognized by M2BP-specific CTLs in an HLA-A24 restriction. The M2BP-derived CTL epitope with an HLA-A24 binding motif is expected to be useful as a target antigenic epitope in clinical immunotherapy for lung cancer (Kontani et al. 2004).

M2BP implicated in cancer progression and metastasis is modified by β1-6 branched N-linked oligosaccharides in colon cancer cells; glycans shown to contribute to cancer metastasis. M2BP-His bound to fibronectin, collagen IV, laminins-1, -5, and -10 and Gal-3 (Mac-2) but poorly to collagen I and galectin-1. As expected, binding of M2BP to Gal-3 was dependent on carbohydrate since it was inhibitable by lactose and asiolofetuin. Thus, a possible mechanism by which M2BP may contribute to colon cancer progression is by modulating tumor cell adhesion to extracellular proteins, including Gal-3 (Ulmer et al. 2006).

13.14 Galectinomics

Importantance of Galectins in Malignancy-Associated Processes: Gene expression pattern of human galectins have been demonstrated in tumor cell lines of various histogenetic origin (galectinomics). The presence of mRNAs for human galectins-1, -2, -3, -4, -7, -8, and -9 was monitored in a panel of 61 human tumor cell lines of different origin (breast, colon, lung, brain, skin, kidney, urogenital system, hematopoietic system). The results clearly demonstrate that human tumor cells express more mRNA species for galectins than those for galectins-1 and -3. To derive unequivocal diagnostic and prognostic information, additional monitoring of these so far insufficiently studied family members is essential (Lahm et al. 2001).

The presence and evidence of tumor-associated up-regulation were shown for galectin-1 and -3. This was less clear-cut for galectin-4 and -8. Galectin-7 was expressed in all cell lines; galectin-2 and -9 were detected at comparatively low levels. Galectin-2, -3 and -8 up-regulation was observed in superficial tumors, but not in muscle-invasive tumors. Immunoreactivity correlated with tumor grading for galectin-1, -2 and -8, and disease-dependent mortality correlated with galectin-2 and -8 expression. Binding sites were visualized using labelled galectins. Langbein et al. (2007) demonstrated a complex expression pattern of the galectin network in urothelial carcinomas. Galectin-1, -2, -3 and -8 are both potential disease markers and also possible targets for bladder cancer therapy.

Gastric Tumors: The expression of lactoside-binding lectin L-31 was higher in malignant gastric tumor than in normal tissue in 9/26 cases, similar in 42/26 cases, and lower in 3/26 cases. The higher expression of L-31 in primary cancers and metastases of certain types implicates this lectin in metastatic phenotype, but the presence of L-31 in primary cancer is not sufficient to allow the metastatic propensity of the tumor to be predicted (Lotan et al. 1994). In mucinous carcinomas of the ovary and gastrointestinal tract, cytoplasmic galectin 4 expression was relatively consistent. Meprin-α is an additional useful marker in differentiating primary from secondary mucinous adenocarcinomas of the ovary (Heinzelmann-Schwarz et al. 2007).

Reports suggest that a lack of RUNX3 function contributes to human gastric carcinogenesis. Sakakura et al. (2005) examined RUNX3 expression in clinical samples of peritoneal metastases in gastric cancers. Significant down-regulation of RUNX3 through methylation on the promoter region was observed in primary tumors (75%) as well as in clinical peritoneal metastases of gastric cancers (100%) as compared with normal gastric mucosa. Stable transfection of RUNX3 inhibited cell proliferation slightly, and modest TGF-beta-induced antiproliferative and apoptotic effects. It strongly inhibited peritoneal metastases of gastric cancers in animal model. Microarray analysis identified 28 candidate genes under the possible downstream control of RUNX3 and indicated that silencing of RUNX3 affects the expression of important genes involved in aspects of metastasis including cell adhesion (sialyltransferase 1 and galectin 4), proliferation, apoptosis, and promoting the peritoneal metastasis of gastric cancer. Identification of such genes could suggest new therapeutic modalities and therapeutic targets (Sakakura et al. 2005).

Colorectal cancer: The levels of Gal-3 (L31) in colorectal cancer specimens from primary tumors of patients with distant metastases (Dukes' stage D) were significantly higher than those from patients without detectable metastases (Dukes' stages B1 and B2). Results indicated that the relative amount of the L-31 lectin increases as the colorectal cancer progresses to a more malignant stage (Lotan et al. 1991). Wollina et al. (2002) suggest that galectins-1, -3, and -4 may be involved in early stages of human colon carcinoma development and that galectin-8 is involved in the later stages (Nagy et al. 2003; Wollina et al. 2002). In view of relevant ligands such as Bcl-2 or integrins, the presence of galectins-3 and -8 seems to be related to the loss of proliferation control and change in cell adhesion properties that are involved in clonal expansion and epidermal spread of malignant T cell clones. Successful chemotherapy of CTCL alters galectin expression selectively as shown for liposomal doxorubicin (Wollina et al. 2002). It has been noted that the majority of colon cancers develop from pre-existing adenomas. Differentially expressed genes were detected between normal-adenoma and adenoma-carcinoma, and were grouped according to the patterns of expression changes. Down-regulated genes in the sequence included galectin 4. Up-regulated genes included matrix metalloproteinase 23B in carcinoma but not in adenoma, supporting the pathobiological roles in malignant transformation (Lee et al. 2006).

Tumor vasculature: Little is known about galectin expression and regulation in tumor vasculature. Galectin-1/-3/-8/-9 are overexpressed in endothelium. Galectin-2/-4/-12 were detectable at mRNA level, albeit very low. Galectin-8 and -9 displayed alternative splicing. Endothelial cell activation in vitro significantly increased the expression of galectin-1 and decreased the expression of both galectin-8 and galectin-9. Gal-3 expression was unaltered. Although a portion of these proteins is expressed intracellularly, the membrane protein level of galectin-1/-8/-9 was significantly increased on cell activation in vitro, 6-fold, 3-fold, and 1.4-fold, respectively. Data showed that endothelial cells express several members of the galectin family and that their expression and distribution changes on cell activation, resulting in a different profile in the tumor vasculature (Thijssen et al. 2008).

13.15 Mechanism of Malignant Progression by Galectin-3

Galectin-3 Phosphorylation at Ser-6 by Casein Kinase 1: The mechanisms by which Gal-3 contributes to malignant progression are not fully understood. It has been shown that the antiapoptotic activity of Gal-3 is regulated by the phosphorylation at Ser-6 by casein kinase 1 (CK1). How phosphorylation at Ser-6 regulates Gal-3 function, was explored by Takenaka et al. (2004) who generated serine-to-alanine (S6A) and serine-to-glutamic acid (S6E) Gal-3 mutants and transfected them into the BT-549 human breast carcinoma cell line, which does not express Gal-3. BT-549 cell clones expressing wild-type (wt) and mutant Gal-3 were exposed to chemotherapeutic anticancer drugs. Perhaps Ser-6 phosphorylation acts as a molecular switch for its cellular translocation from the nucleus to the cytoplasm and, as a result, regulates the antiapoptotic activity of Gal-3 (Takenaka et al. 2004).

DNA analysis of a human pituitary tumor, breast carcinoma cell lines, and thyroid carcinoma cell lines showed that in cells expressing Gal-3 protein, the *LGALS3* gene was unmethylated, whereas in Gal-3 null cells, the promoter of the *LGALS3* gene was methylated. It indicated that Gal-3 expression is regulated in part by methylation of promoter in pituitary as well as in other tumors (Ruebel et al. 2005). The combination of 5-mc with Gal-3 led to an excellent accuracy level of 96%. Among follicular neoplasia 5-mc, accuracy to differentiate malignant tumors tends to be higher than Gal-3. These data stress the necessity of epigenetic events evaluation among thyroid nodules and propose global DNA methylation assessment as a potential diagnostic tool to combine with other valuable markers (Galusca et al. 2005). Since Gal-3 is functionally involved in cancer progression and metastasis, it may serve as a possible therapeutic target in the treatment of pituitary tumors.

Gal-3 Regulates a Molecular Switch from N-Ras to K-Ras: Depending on the cellular context, Ras can activate characteristic effectors by mechanisms still poorly understood. Promotion by galectin-1 of Ras activation of Raf-1 but not of phosphoinositide 3-kinase (PI3-K) is one such mechanism. Elad-Sfadia et al. (2004) described a mechanism controlling selectivity of K-Ras4B (K-Ras), an important Ras oncoprotein. It was demonstrated that Gal-3 acts as a selective binding partner of activated K-Ras. Unlike galectin-1, which prolongs Ras activation of ERK and inhibits PI3-K, K-Ras-GTP/Gal-3 interactions promote, in addition to PI3-K and Raf-1 activation, a third inhibitory signal that attenuates active ERK (Elad-Sfadia et al. 2004). Though, the Gal-3 is a selective binding partner of activated K-Ras-GTP and since both proteins are antiapoptotic and associated with cancer progression, Shalom-Feuerstein et al. (2005) questioned the possible functional role of Gal-3 in K-Ras activation. Shalom-Feuerstein et al. (2005) found that overexpression of Gal-3 in human breast cancer cells (BT-549/Gal-3) coincided with a significant increase in wild-type (wt) K-Ras-GTP coupled with loss in wt N-Ras-GTP, whereas the nononcogenic Gal-3 mutant proteins [Gal-3(S6E) and Gal-3(G182A)] failed to induce the Ras isoform switch. Only wt Gal-3 protein co-precipitated and colocalized with oncogenic K-Ras, resulting in its activation with radical alterations in Ras signaling pathway, whereby the activation of AKT and Ral was suppressed and shifted to the activation of extracellular signal-regulated kinase (ERK). These workers suggested that Gal-3 confers on BT-549 human breast carcinoma cells several oncogenic functions by binding to and activation of wt K-Ras, suggesting that some of the molecular functions of Gal-3 are, at least in part, a result of K-Ras activation (Shalom-Feuerstein et al. 2005). Eude-Le Parco et al. (2009) made genetic assessment of importance of Gal-3 in cancer initiation, progression, and dissemination in mice.

13.16 Anti-Galectin Compounds as Anti-Cancer Drugs

To interfere with galectin-carbohydrate interactions during tumor progression, a current challenge is the design of specific galectin inhibitors for therapeutic purposes. Certain galectins directly involved in cancer progression seem to be promising targets for the development of novel therapeutic strategies to combat cancer. Indeed, migrating cancer cells resistant to apoptosis still constitute the principal target for the cytotoxic drugs used to treat cancer patients. Reducing the levels of migration in apoptosis-resistant cancer cells can restore certain levels of sensitivity to apoptosis in restricted-migration cancer cells. Anti-galectin agents can restrict the levels of migration of several types of cancer cell and should therefore be used in association with cytotoxic drugs to combat metastatic cancer. Experimental proof in support of this concept with particular attention to glioblastomas has been provided (Ingrassia et al. 2006). Glioblastomas form the most common type of malignant brain tumor in children and adults, and no glioblastoma patient has been cured to date (Ingrassia et al. 2006).

Galactosides and Lactosides as Inhibitors of Gal-1 and -3: Among aryl 1-thio-β-d-galacto- and lacto-pyranosides carrying a panel of substituents on the phenyl groups, best galectin-1 inhibitors were p-nitrophenyl thiogalactoside 5a for the monosaccharide and o-nitrophenyl thiolactoside 6f or napthylsulfonyl lactoside 8c, both being 20 times better relative to natural ligands. Relative inhibitory properties of these were as low as 2,500 and 40 μM, respectively (Giguere et al. 2006a).

Evaluation of the N acetyllactosamine thioureas as inhibitors against galectins-1, 3, 7, 8N (N-terminal domain), and 9N (N-terminal domain) revealed thiourea-mediated affinity enhancements for galectins-1, 3, 7, and 9N and in particular, good inhibitors against galectin-7 and 9N (K_D 23 and 47 μM, respectively, for a 3-pyridylmethylthiourea derivative), representing more than an order of magnitude affinity enhancement over the parent natural N-acetyllactosamine (Salameh et al. 2006).

Copper(I)-catalyzed addition of alkynes to methyl 3-azido-3-deoxy-1-thio-β-D-galactopyranoside afforded stable and structurally simple 3-deoxy-3-(1H-1,2,3-triazol-1-yl)-1-thio-galactosides carrying a panel of substituents at the triazole C4 in high yields. The 3-(1H-[1,2,3]-triazol-1-yl)-1-thio-galactoside collection synthesized contained inhibitors of the tumor- and inflammation-related Gal-3 with K_D values as low as 107 μM, which is as potent as the natural disaccharide inhibitors lactose and N-acetyllactosamine (Salameh et al. 2005). Among anomeric oxime ether derivatives of β-galactose, the best inhibitor, [E]-O-(β-D galactopyranosyl)-indole-3-carbaldoxime (E-52), had a K_D value of 180 μM, which is 24 times better than methyl β-D-galactopyranoside (K_D = 4,400 μM) and in the same range as methyl lactoside (K_D = 220 μM) (Tejler et al. 2005). Galactosides and lactosides bearing triazoles or isoxazoles, provided specific galectin-1 and -3 inhibitors with potencies as low as 20 μM (Giguere et al. 2006b).

Cumpstey et al. (2005) synthesized compounds that can bind galectins-1, -3, -7, -8N and -9N. An aromatic nucleophilic substitution reaction between 1,5-difluoro-2,4-dinitrobenzene and a galacto thiol gave 5-fluoro-2,4-dinitrophenyl 2,3,4,6-tetra-O-acetyl-1-thio-β-D-galactopyranoside. The modified forms of these compounds were efficient inhibitors against galectin-7. The best inhibitors against galectin-7 were poor against the other galectins and thus have potential as simple and selective tools for dissecting biological functions of galectin-7.

Synthetic Lactulose Amines: Anticancer Agents: Rabinovich et al. (2006) reported the synthesis of three low molecular weight synthetic lactulose amines (SLA): (1) N-lactulose octamethylenediamine (LDO), (2) N, N′-dilactulose-octamethylenediamine (D-LDO), and (3) N, N′-dilactulose-dodecamethylenediamine (D-LDD). These galectin inhibitors with subtle differences in their carbohydrate structures may be potentially used to specifically block different steps of tumor growth and metastasis (Rabinovich et al. 2006).

In an attempt to block the Gal-3 carbohydrate recognition domain with synthetic peptides and reduce metastasis-associated carcinoma cell adhesion, Zou et al. (2005) demonstrated that carbohydrate-mediated, metastasis-associated tumor cell adhesion could be inhibited efficiently with short synthetic peptides which do not mimic naturally occurring glycoepitopes yet bind to the Gal-3 CRD with high affinity and specificity (Zou et al. 2005). Protein-ligand interactions can be significantly enhanced by the fine-tuning of arginine-arene interactions (Sorme et al. 2005).

References

Alcendor DJ, Knobel SM, Desai P et al (2009) KSHV downregulation of galectin-3 in Kaposi's sarcoma. Glycobiology 20:521–532.

Aron M, Kapila K, Verma K (2006) Utility of galectin 3 expression in thyroid aspirates as a diagnostic marker in differentiating benign from malignant thyroid neoplasms. Indian J Pathol Microbiol 49: 376–380

Baptiste TA, James A, Saria M, Ochieng J (2007) Mechano-transduction mediated secretion and uptake of galectin-3 in breast carcinoma cells: implications in the extracellular functions of the lectin. Exp Cell Res 313:652–664

Barroeta JE, Baloch ZW, Lal P et al (2006) Diagnostic value of differential expression of CK19, galectin-3, HBME-1, ERK, RET, and p16 in benign and malignant follicular-derived lesions of the thyroid: an immunohistochemical tissue microarray analysis. Endocr Pathol 17:225–234, Fall

Bartik P, Maglott A, Entlicher G et al (2008) Detection of a hypersialylated beta1 integrin endogenously expressed in the human astrocytoma cell line A172. Int J Oncol 32:1021–1031

Bartolazzi A, Bussolati G (2006) Galectin-3 does not reliably distinguish benign from malignant thyroid neoplasms. Histopathology 48:212–213

Bartosova J, Kuzelova K, Pluskalova M et al (2006) UVA-activated 8-methoxypsoralen (PUVA) causes G2/M cell cycle arrest in Karpas 299 T-lymphoma cells. J Photochem Photobiol B 85:39–48

Berberat PO, Friess H, Wang,L, et al. (2001) Comparative analysis of galectins in primary tumors and tumor metastasis in human pancreatic cancer. J Histochem Cytochem. 49:539–49

Bergero N, De Pompa R, Sacerdote C et al (2005) Galectin-3 expression in parathyroid carcinoma: immunohistochemical study of 26 cases. Hum Pathol 36:908–914

Bresalier RS, Yan P-S, Byrd JC et al (1997) Expression of the endogenous galactose-binding protein galectin-3 correlates with the malignant potential of tumors in the central nervous system. Cancer 80:776–787

Bresalier RS, Byrd JC, Tessler D et al (2004) A circulating ligand for galectin-3 is a haptoglobin-related glycoprotein elevated in individuals with colon cancer. Gastroenterology 127:741–748

Brustmann H (2006) Galectin-3 and CD1a-positive dendritic cells are involved in the development of an invasive phenotype in vulvar squamous lesions. Int J Gynecol Pathol 25:30–37

Buttery R, Monaghan H, Salter DM, Sethi T (2004) Galectin-3: differential expression between small-cell and non-small-cell lung cancer. Histopathology 44:339–344

Califice S, Castronovo V, Bracke M et al (2004a) Dual activities of galectin-3 in human prostate cancer: tumor suppression of nuclear galectin-3 vs tumor promotion of cytoplasmic galectin-3. Oncogene 23:7527–7536

Califice S, Castronovo V, Van Den Brule F (2004b) Galectin-3 and cancer (review). Int J Oncol 25:983–992

Camby I, Belot N, Rorive S et al (2001) Galectins are differentially expressed in supratentorial pilocytic astrocytomas, astrocytomas, anaplastic astrocytomas and glioblastomas, and significantly modulate tumor astrocytes migration. Brain Pathol 11:12–26

Carcamo C, Pardo E, Oyanadel C et al (2006) Galectin-8 binds specific beta1 integrins and induces polarized spreading highlighted by asymmetric lamellipodia in Jurkat T cells. Exp Cell Res 312:374–386

Carpi A, Naccarato AG, Iervasi G et al (2006) A. Large needle aspiration biopsy and galectin-3 determination in selected thyroid nodules with indeterminate FNA-cytology. Br J Cancer 95:204–209

Castronovo V, Campo E, van den Brûle FA et al (1992) Inverse modulation of steady-state messenger RNA levels of two non-integrin laminin-binding proteins in human colon carcinoma. J Natl Cancer Inst 84:1161–1169

Choi YK, Hong SH, Kim BH et al (2004) Immunohistochemical expression of galectin-3 in canine mammary tumours. J Comp Pathol 131:242–245

Choufani G, Nagy N, Saussez S et al (1999) The levels of expression of galectin-1, galectin-3, and the Thomsen-Friedenreich antigen and their binding sites decrease as clinical aggressiveness increases in head and neck cancers. Cancer 86:2353–2363

Coli A, Bigotti G, Zuchetti F et al (2002) Galectin-3, a marker of well-differentiated thyroid carcinoma, is expressed in thyroid nodules with cytological atypia. Histopathology 40:80–87

Collet JF, Hurbain I, Prengel C et al (2005) Galectin-3 immunodetection in follicular thyroid neoplasms: a prospective study on fine-needle aspiration samples. Br J Cancer 93:1175–1181

Cumpstey I, Carlsson S, Leffler H, Nilsson UJ (2005) Synthesis of a phenyl thio-β-D galactopyranoside library from 1,5-difluoro-2,4-dinitrobenzene: discovery of efficient and selective monosaccharide inhibitors of galectin-7. Org Biomol Chem 3:1922–1932

Cvejic D, Savin S, Petrovic I et al (2005a) Galectin-3 and proliferating cell nuclear antigen (PCNA) expression in papillary thyroid carcinoma. Exp Oncol 27:210–214

Cvejic D, Savin S, Petrovic I et al (2005b) Galectin-3 expression in papillary microcarcinoma of the thyroid. Histopathology 47:209–214

Cvejic DS, Savin SB, Petrovic IM et al (2005c) Galectin-3 expression in papillary thyroid carcinoma: relation to histomorphologic growth pattern, lymph node metastasis, extrathyroid invasion, and tumor size. Head Neck 27:1049–1055

D'Abaco GM, Kaye AH (2007) Integrins: molecular determinants of glioma invasion. J Clin Neurosci 14:1041–1048

D'Haene N, Maris C, Sandras F, et al. (2005) The differential expression of galectin-1 and galectin-3 in normal lymphoid tissue and non-Hodgkins and Hodgkins lymphomas. Int J Immunopathol Pharmacol. 18:431–43

de Matos PS, Ferreira AP, de Oliveira Facuri F et al (2005) Usefulness of HBME-1, cytokeratin 19 and galectin-3 immunostaining in the diagnosis of thyroid malignancy. Histopathology 47:391–401

Debray C, Vereecken P, Belot N et al (2004) Multifaceted role of galectin-3 on human glioblastoma cell motility. Biochem Biophys Res Commun 325:1393–1398

Decaestecker C, Debeir O, Van Ham P, Kiss R (2007) Can anti-migratory drugs be screened in vitro? A review of 2D and 3D assays for the quantitative analysis of cell migration. Med Res Rev 27:149–176

Deininger MH, Trautmann K, Meyermann R et al (2002) Galectin-3 labeling correlates positively in tumor cells and negatively in endothelial cells with malignancy and poor prognosis in oligodendroglioma patients. Anticancer Res 22:1585–1592

Delorge S, Saussez S, Pelc P et al (2000) Correlation of galectin-3/galectin-3-binding sites with low differentiation status in head and neck squamous cell carcinomas. Otolaryngol Head Neck Surg 122:834–841

Devouassoux-Shisheboran M, Deschildre C, Mauduit C et al (2006) Expression of galectin-3 in gonads and gonadal sex cord stromal and germ cell tumors. Oncol Rep 16:335–340

Dudas SP, Yunker CK, Sternberg LR et al (2002) Expression of human intestinal mucin is modulated by the β-galactoside binding protein galectin-3 in colon cancer. Gastroenterology 123:817–826

Dumic J, Dabelic S, Flogel M (2006) Galectin-3: an open-ended story. Biochim Biophys Acta 1760:616–635

Elad-Sfadia G, Haklai R, Balan E, Kloog Y (2004) Galectin-3 augments K-Ras activation and triggers a Ras signal that attenuates ERK but not phosphoinositide 3-kinase activity. J Biol Chem 279:34922–34930

Ellerhorst J, Troncoso P, Xu XC et al (1999) Galectin-1 and galectin-3 expression in human prostate tissue and prostate cancer. Urol Res 27:362–367

Endo K, Kohnoe S, Tsujita E, Watanabe A et al (2005) Galectin-3 expression is a potent prognostic marker in colorectal cancer. Anticancer Res 25:3117–3121

Eude-Le Parco I, Gendronneau G, Dang T et al (2009) Genetic assessment of the importance of galectin-3 in cancer initiation, progression, and dissemination in mice. Glycobiology 19:68–75

Faggiano A, Talbot M, Lacroix L et al (2002) Differential expression of galectin-3 in medullary thyroid carcinoma and C-cell hyperplasia. Clin Endocrinol 57:813–819

Fischer C, Sanchez-Ruderisch H, Welzel M et al (2005) Galectin-1 interacts with the $\alpha_5\beta_1$ fibronectin receptor to restrict carcinoma cell growth via induction of p21 and p27. J Biol Chem 280:37266–37277

Fortin S, Le Mercier M, Camby I, et al (2010) Galectin-1 is implicated in the protein kinase C epsilon/vimentin-controlled trafficking of integrin-beta1 in glioblastoma cells. Brain Pathol 20:39–49

Fukumori T, Oka N, Takenaka Y et al (2006) Galectin-3 regulates mitochondrial stability and antiapoptotic function in response to anticancer drug in prostate cancer. Cancer Res 663:114–119

Fukushi J, Makagiansar IT, Stallcup WB (2004) NG2 proteoglycan promotes endothelial cell motility and angiogenesis via engagement of galectin-3 and alpha3beta1 integrin. Mol Biol Cell 15: 3580–3590

Galusca B, Dumollard JM, Lassandre S et al (2005) Global DNA methylation evaluation: potential complementary marker in differential diagnosis of thyroid neoplasia. Virchows Arch 447: 18–23

Gasbarri A, Marchetti C, Iervasi G et al (2004a) From the bench to the bedside. Galectin-3 immunodetection for improving the preoperative diagnosis of the follicular thyroid nodules. Biomed Pharmacother 58:356–359

Gasbarri A, Sciacchitano S, Marasco A et al (2004b) Detection and molecular characterisation of thyroid cancer precursor lesions in a specific subset of Hashimoto's thyroiditis. Br J Cancer 91:1096–1104

Geers C, Moulin P, Gigot JF et al (2006) Solid and pseudopapillary tumor of the pancreas–review and new insights into pathogenesis. Am J Surg Pathol 30:1243–1249

Giguere D, Patnam R, Bellefleur MA et al (2006a) Carbohydrate triazoles and isoxazoles as inhibitors of galectins-1 and -3. Chem Commun (Camb) 22:2379–2381

Giguere D, Sato S, St-Pierre C et al (2006b) Aryl O- and S-galactosides and lactosides as specific inhibitors of human galectins-1 and -3: role of electrostatic potential at O-3. Bioorg Med Chem Lett 16:1668–1672

Gimm O, Krause U, Brauckhoff M et al (2006) Distinct expression of galectin-3 in pheochromocytomas. Ann N Y Acad Sci 1073:571–577

Goetz JG, Joshi B, Lajoie P et al (2008) Concerted regulation of focal adhesion dynamics by galectin-3 and tyrosine-phosphorylated caveolin-1. J Cell Biol 180:1261–1275

Gordower L, Decaestecker C, Kacem Y et al (1999) Galectin-3 and galectin-3-binding site expression in human adult astrocytic tumours and related angiogenesis. Neuropathol Appl Neurobiol 25:319–330

Greco C, Vona R, Cosimelli M, Matarrese P et al (2004) Cell surface overexpression of galectin-3 and the presence of its ligand 90 k in the blood plasma as determinants in colon neoplastic lesions. Glycobiology 14:783–792

Hadari YR, Arbel-Goren R, Levy Y, Amsterdam A, Alon R, Zakut R, Zick Y et al (2000) Galectin-8 binding to integrins inhibits cell adhesion and induces apoptosis. J Cell Sci 113:2385–2397

Hancq S, Salmon I, Brotchi J et al (2004) Detection of S100B, S100A6 and galectin-3 ligands in meningiomas as markers of aggressiveness. Int J Oncol 25:1233–1240

He J, Baum LG (2006) Endothelial cell expression of galectin-1 induced by prostate cancer cell inhibits T-cell transendothelial migration. Lab Invest 86:578–590

He QY, Chen J, Kung HF et al (2004) Identification of tumor-associated proteins in oral tongue squamous cell carcinoma by proteomics. Proteomics 4:271–278

Heinzelmann-Schwarz VA, Scolyer RA, Scurry JP et al (2007) Low meprin alpha expression differentiates primary ovarian mucinous carcinoma from gastrointestinal cancers that commonly metastasise to the ovaries. J Clin Pathol 60:622–626

Herrmann ME, LiVolsi VA, Pasha TL et al (2002) Immunohistochemical expression of galectin-3 in benign and malignant thyroid lesions. Arch Pathol Lab Med 126:710–713

Hittelet A, Legendre H, Nagy N et al (2003) Upregulation of galectins-1 and -3 in human colon cancer and their role in regulating cell migration. Int J Cancer 103:370–379

Honjo Y, Inohara H, Akahani S et al (2000) Expression of cytoplasmic galectin-3 as a prognostic marker in tongue carcinoma. Clin Cancer Res 6:4635–4640

Honjo Y, Nangia-Makker P, Inohara H, Raz A (2001) Down-regulation of galectin-3 suppresses tumorigenecity of human breast carcinoma cells. Clin Cancer Res 7:661–668

Hoyer KK, Pang M, Gui D et al (2004) An anti-apoptotic role for galectin-3 in diffuse large B-cell lymphomas. Am J Pathol 164:893–902

Hsu DK, Dowling CA, Jeng K-CG et al (1999) Galectin-3 expression is induced in cirrhotic liver and hepatocellular carcinoma. Int J Cancer 81:519–526

Hynes RO (2002) Integrins: bidirectional, allosteric signaling machines. Cell 110:673–687

Idikio H (1998) Galectin-3 expression in human breast carcinoma: correlation with cancer histologic grade. Int J Oncol 12:1287–1290

Ingrassia L, Camby I, Lefranc F et al (2006) Anti-galectin compounds as potential anti-cancer drugs. Curr Med Chem 13:3513–3527

Inohara H, Honjo Y, Yoshii T, Akahani S, et al. (1999) Expression of galectin-3 in fine-needle aspirates as a diagnostic marker differentiating benign from malignant thyroid neoplasms. Cancer. 85:2475–84

Irimura T, Matsushita Y, Sutton RC et al (1991) Increased content of a endogenous lactose-binding lectin in human colorectal carcinoma progressed to metastatic stages. Cancer Res 51:387–393

Ito Y, Yoshida H, Tomoda C et al (2005) Galectin-3 expression in follicular tumours: an immunohistochemical study of its use as a marker of follicular carcinoma. Pathology 37:296–298

Jin L, Riss D, Ruebel K, Kajita S, Scheithauer BW, Horvath E, Kovacs K, Lloyd RV (2005) Galectin-3 expression in functioning and silent ACTH-producing adenomas. Endocr Pathol 16:107–114, Summer

Johnson KD, Glinskii OV, Mossine VV et al (2007) Galectin-3 as a potential therapeutic target in tumors arising from malignant endothelia. Neoplasia 9:662–670

Kalayci O, Birben E, Tinari N et al (2004) Role of 90 K protein in asthma and TH2-type cytokine expression. Ann Allergy Asthma Immunol 93:485–492

Kawachi K, Matsushita Y, Yonezawa S et al (2000) Galectin-3 expression in various thyroid neoplasms and its possible role in metastasis formation. Hum Pathol 31:428–433

Kim MJ, Kim HJ, Hong SJ, Shong YK, Gong G (2006) Diagnostic utility of galectin-3 in aspirates of thyroid follicular lesions. Acta Cytol 50:28–34

Kontani K, Teramoto K, Ozaki Y et al (2004) Identification of antigenic epitopes recognized by Mac-2 binding protein-specific cytotoxic T lymphocytes in an HLA-A24 restricted manner. Int J Oncol 25:1537–1542

Koopmann J, Thuluvath PJ, Zahurak ML et al (2004) Mac-2-binding protein is a diagnostic marker for biliary tract carcinoma. Cancer 101:1609–1615

Krishnan V, Bane SM, Kawle PD et al (2005) Altered melanoma cell surface glycosylation mediates organ specific adhesion and metastasis via lectin receptors on the lung vascular endothelium. Clin Exp Metastasis 22:11–24

Lagana A, Goetz JG, Cheung P et al (2006) Galectin binding to Mgat5-modified N-glycans regulates fibronectin matrix remodeling in tumor cells. Mol Cell Biol 26:3181–3193

Lahm H, Andre S, Hoeflich A et al (2001) Comprehensive galectin fingerprinting in a panel of 61 human tumor cell lines by RT-PCR and its implications for diagnostic and therapeutic procedures. J Cancer Res Clin Oncol 127:375–386

Langbein S, Brade J, Badawi JK et al (2007) Gene-expression signature of adhesion/growth-regulatory tissue lectins (galectins) in transitional cell cancer and its prognostic relevance. Histopathology 51:681–690

Le Mercier M, Mathieu V, Haibe-Kains B, Bontempi G, Mijatovic T, Decaestecker C et al (2008) Knocking down galectin 1 in human hs683 glioblastoma cells impairs both angiogenesis and endoplasmic reticulum stress responses. J Neuropathol Exp Neurol 67:456–469

Le Mercier M, Fortin S, Mathieu V et al (2010) Galectins and gliomas. Brain Pathol 20:17–27

Lee S, Bang S, Song K, Lee I (2006) Differential expression in normal-adenoma-carcinoma sequence suggests complex molecular carcinogenesis in colon. Oncol Rep 16:747–754

Lefranc F, Chevalier C, Vinchon M et al (2003) Characterization of the levels of expression of retinoic acid receptors, galectin-3, macrophage migration inhibiting factor, and p53 in 51 adamantinomatous craniopharyngiomas. J Neurosurg 98:145–153

Lefranc F, Brotchi J, Kiss R (2005) Possible future issues in the treatment of glioblastomas: special emphasis on cell migration and the resistance of migrating glioblastoma cells to apoptosis. J Clin Oncol 23:2411–2422

Lotan R, Matsushita Y, Ohannesian D et al (1991) Lactose-binding lectin expression in human colorectal carcinomas. Relation to tumor progression. Carbohydr Res 213:47–57

Lotan R, Ito H, Yasui W et al (1994) Expression of a 31-kDa lactoside-binding lectin in normal human gastric mucosa and in primary and metastatic gastric carcinomas. Int J Cancer 56:474–480

Lotz MM, Andrews CW Jr, Korzelius CA et al (1993) Decreased expression of Mac-2 (carbohydrate binding protein 35) and loss of its nuclear localization are associated with the neoplastic progression of colon carcinoma. Proc Natl Acad Sci USA 90:3466–3470

Maruta J, Hashimoto H, Yamashita H et al (2004) Immunostaining of galectin-3 and CD44v6 using fine-needle aspiration for distinguishing follicular carcinoma from adenoma. Diagn Cytopathol 31:392–396

Mathieu A, Saal I, Vuckovic A, Ransy V et al (2005) Nuclear galectin-3 expression is an independent predictive factor of recurrence for adenocarcinoma and squamous cell carcinoma of the lung. Mod Pathol 18:1264–1271

Mazurek N, Sun YJ, Price JE et al (2005) Phosphorylation of galectin-3 contributes to malignant transformation of human epithelial cells via modulation of unique sets of genes. Cancer Res 65:10767–10775

Mehrotra P, Okpokam A, Bouhaidar R et al (2004) Galectin-3 does not reliably distinguish benign from malignant thyroid neoplasms. Histopathology 45:493–500

Mills LJ, Poller DN, Yiangou C (2005) Galectin-3 is not useful in thyroid FNA. Cytopathology 16:132–138

Moiseeva EP, Spring EL, Baron JH, de Bono DP (1999) Galectin 1 modulates attachment, spreading and migration of cultured vascular smooth muscle cells via interactions with cellular receptors and components of extracellular matrix. J Vasc Res 36:47–58

Moiseeva EP, Williams B, Goodall AH, Samani NJ (2003) Galectin-1 interacts with beta-1 subunit of integrin. Biochem Biophys Res Commun 310:1010–1016

Moiseeva EV, Rapoport EM, Bovin NV et al (2005) Galectins as markers of aggressiveness of mouse mammary carcinoma: towards a lectin target therapy of human breast cancer. Breast Cancer Res Treat 91:227–241

Mourad-Zeidan AA, Melnikova VO, Wang H, Raz A et al (2008) Expression profiling of galectin-3-depleted melanoma cells reveals its major role in melanoma cell plasticity and vasculogenic mimicry. Am J Pathol 173:1839–1852

Nagy N, Legendre H, Engels O et al (2003) Refined prognostic evaluation in colon carcinoma using immunohistochemical galectin fingerprinting. Cancer 97:1849–1858

Nascimento MC, Bisi H, Alves VA, Medeiros-Neto G et al (2001) Differential reactivity for galectin-3 in Hürthle cell adenomas and carcinomas. Endocr Pathol 12:275–279, Fall

Natarajan M, Hecker TP, Gladson CL (2003) FAK signaling in anaplastic astrocytoma and glioblastoma tumors. Cancer J 9:126–133

Neder L, Marie SK, Carlotti CG Jr et al (2004) Galectin-3 as an immunohistochemical tool to distinguish pilocytic astrocytomas from diffuse astrocytomas, and glioblastomas from anaplastic oligodendrogliomas. Brain Pathol 14:399–405

Nucera C, Mazzon E, Caillou B et al (2005) Human galectin-3 immunoexpression in thyroid follicular adenomas with cell atypia. J Endocrinol Invest 28:106–112

Ochieng J, Leite-Browning ML, Warfield P (1998) Regulation of cellular adhesion to extracellular matrix proteins by galectin-3. Biochem Biophys Res Commun 246:788–791

Oestreicher-Kedem Y, Halpern M, Roizman P et al (2004) Diagnostic value of galectin-3 as a marker for malignancy in follicular patterned thyroid lesions. Head Neck 26:960–966

Oka N, Nakahara S, Takenaka Y et al (2005) Galectin-3 inhibits tumor necrosis factor-related apoptosis-inducing ligand-induced apoptosis by activating Akt in human bladder carcinoma cells. Cancer Res 65:7546–7553

Okada K, Shimura T, Suehiro T et al (2006) Reduced galectin-3 expression is an indicator of unfavorable prognosis in gastric cancer. Anticancer Res 26:1369–1376

Pacis RA, Pilat MJ, Pienta KJ et al (2000) Decreased galectin-3 expression in prostate cancer. Prostate 44:118–123

Papotti M, Rodriguez J, De Pompa R et al (2005) Galectin-3 and HBME-1 expression in well-differentiated thyroid tumors with follicular architecture of uncertain malignant potential. Mod Pathol 18:541–546

Paret C, Bourouba M, Beer A et al (2005) Ly6 family member C4.4A binds laminins 1 and 5, associates with galectin-3 and supports cell migration. Int J Cancer 115:724–733

Partridge EA, Le Roy C et al (2004) Regulation of cytokine receptors by golgi N-glycan processing and endocytosis. Science 306 (5693):120–124

Pisani T, Vecchione A, Giovagnolii MR (2004) Galectin-3 immunodetection may improve cytological diagnosis of occult papillary thyroid carcinoma. Anticancer Res 24:1111–1112

Plzak J, Betka J, Smetana K Jr et al (2004) Galectin-3: an emerging prognostic indicator in advanced head and neck carcinoma. Eur J Cancer 40:2324–2330

Prasad ML, Huang Y, Pellegata NS et al (2004) Hashimoto's thyroiditis with papillary thyroid carcinoma (PTC)-like nuclear alterations express molecular markers of PTC. Histopathology 45:39–46

Prasad ML, Pellegata NS, Huang Y et al (2005) Galectin-3, fibronectin-1, CITED-1, HBME1 and cytokeratin-19 immunohistochemistry is useful for the differential diagnosis of thyroid tumors. Mod Pathol 18:48–57

Prieto VG, Mourad-Zeidan AA, Melnikova V et al (2006) Galectin-3 expression is associated with tumor progression and pattern of sun exposure in melanoma. Clin Cancer Res 12:6709–6715

Rabinovich GA, Cumashi A, Bianco GA et al (2006) Synthetic lactulose amines: novel class of anticancer agents that induce tumor-cell apoptosis and inhibit galectin-mediated homotypic cell aggregation and endothelial cell morphogenesis. Glycobiology 16:210–220

Rao JS (2003) Molecular mechanisms of glioma invasiveness: the role of proteases. Nat Rev Cancer 3:489–501

Riss D, Jin L, Qian X et al (2003) Differential expression of galectin-3 in pituitary tumors. Cancer Res 63:2251–2255

Rodrigo JP, Rinaldo A, Devaney KO et al (2006) Molecular diagnostic methods in the diagnosis and follow-up of well-differentiated thyroid carcinoma. Head Neck 28:1032–1039

Rossi ED, Raffaelli M, Minimo C et al (2005) Immunocytochemical evaluation of thyroid neoplasms on thin-layer smears from fine-needle aspiration biopsies. Cancer 105:87–95

Rossi ED, Raffaelli M, Mule' A et al (2006) Simultaneous immunohistochemical expression of HBME-1 and galectin-3 differentiates papillary carcinomas from hyperfunctioning lesions of the thyroid. Histopathology 48:795–800

Ruebel KH, Jin L, Qian X et al (2005) Effects of DNA methylation on galectin-3 expression in pituitary tumors. Cancer Res 65:1136–1140

Rutka JT, Muller M, Hubbard SL et al (1999) Astrocytoma adhesion to extracellular matrix: functional significance of integrin and focal adhesion kinase expression. J Neuropathol Exp Neurol 58:198–209

Saggiorato E, Aversa S, Deandreis D et al (2004) Galectin-3: presurgical marker of thyroid follicular epithelial cell-derived carcinomas. J Endocrinol Invest 27:311–317

Saggiorato E, Bergero N, Volante M et al (2006) Galectin-3 and Ki-67 expression in multiglandular parathyroid lesions. Am J Clin Pathol 126:59–66

Sakakura C, Hasegawa K, Miyagawa K et al (2005) Possible involvement of RUNX3 silencing in the peritoneal metastases of gastric cancers. Clin Cancer Res 11:6479–6488

Salameh BA, Leffler H, Nilsson UJ (2005) 3-(1,2,3-Triazol-1-yl)-1-thio-galactosides as small, efficient, and hydrolytically stable inhibitors of galectin-3. Bioorg Med Chem Lett 15:3344–3346

Salameh BA, Sundin A, Leffler H, Nilsson UJ (2006) Thioureido N-acetyllactosamine derivatives as potent galectin-7 and 9N inhibitors. Bioorg Med Chem 14:1215–1220

Sanjuán X, Fernández PL, Castells A et al (1997) Differential expression of galectin 3 and galectin 1 in colorectal cancer progression. Gastroenterology 113:1906–1915

Saussez S, Nonclercq D, Laurent G et al (2005) Toward functional glycomics by localization of tissue lectins: immunohistochemical galectin fingerprinting during diethylstilbestrol-induced kidney tumorigenesis in male Syrian hamster. Histochem Cell Biol 123:29–41

Saussez S, Lorfevre F, Nonclercq D et al (2006) Towards functional glycomics by localization of binding sites for tissue lectins: lectin histochemical reactivity for galectins during diethylstilbestrol-induced kidney tumorigenesis in male Syrian hamster. Histochem Cell Biol 126:57–69

Saussez S, Camby I, Toubeau G, Kiss R (2007a) Galectins as modulators of tumor progression in head and neck squamous cell carcinomas. Head Neck 29:874–884

Saussez S, Decaestecker C, Lorfevre F et al (2007b) High level of galectin-1 expression is a negative prognostic predictor of recurrence in laryngeal squamous cell carcinomas. Int J Oncol 30:1109–1117

Saussez S, Lorfevre F, Lequeux T et al (2008a) The determination of the levels of circulating galectin-1 and -3 in HNSCC patients could be used to monitor tumor progression and/or responses to therapy. Oral Oncol 44:86–93

Saussez S, Decaestecker C, Mahillon V et al (2008b) Galectin-3 upregulation during tumor progression in head and neck cancer. Laryngoscope 118:1583–1590

Saussez S, Glinoer D, Chantrain G et al (2008c) Serum galectin-1 and galectin-3 levels in benign and malignant nodular thyroid disease. Thyroid 18:705–712

Schoeppner HL, Raz A, Ho SB, Bresalier RS (1995) Expression of an endogenous galactose-binding lectin correlates with neoplastic progression in the colon. Cancer 75:2818–2826

Shalom-Feuerstein R, Cooks T, Raz A, Kloog Y (2005) Galectin-3 regulates a molecular switch from N-Ras to K-Ras usage in human breast carcinoma cells. Cancer Res 65:7292–7300

Sheikholeslam-Zadeh R, Decaestecker C, Delbrouck C et al (2001) The levels of expression of galectin-3, but not of galectin-1 and galectin-8, correlate with apoptosis in human cholesteatomas. Laryngoscope 111:1042–1047

Shekhar MP, Nangia-Makker P, Tait L et al (2004) Alterations in galectin-3 expression and distribution correlate with breast cancer progression: functional analysis of galectin-3 in breast epithelial-endothelial interactions. Am J Pathol 165:1931–1941

Shibata T, Noguchi T, Takeno S et al (2005) Impact of nuclear galectin-3 expression on histological differentiation and vascular invasion in patients with esophageal squamous cell carcinoma. Oncol Rep 13:235–239

Song YK, Billiar TR, Lee YJ (2002) Role of galectin-3 in breast cancer metastasis. Involvement of nitric oxide. Am J Pathol 160:1069–1075

Song S, Byrd JC, Mazurek N et al (2005) Galectin-3 modulates MUC2 mucin expression in human colon cancer cells at the level of transcription via AP-1 activation. Gastroenterology 129:1581–1591

Sorme P, Arnoux P, Kahl-Knutsson B et al (2005) Structural and thermodynamic studies on cation-Pi interactions in lectin-ligand complexes: high-affinity galectin-3 inhibitors through fine-tuning of an arginine-arene interaction. J Am Chem Soc 127:1737–1743

Stillman BN, Mischel PS, Baum LG (2005) New roles for galectins in brain tumors–from prognostic markers to therapeutic targets. Brain Pathol 15:124–132

Strik HM, Deininger MH, Frank B et al (2001) Galectin-3: cellular distribution and correlation with WHO-grade in human gliomas. J Neurooncol 53:13–20

Szoke T, Kayser K, Baumhakel JD et al (2005) Prognostic significance of endogenous adhesion/growth-regulatory lectins in lung cancer. Oncology 69:167–174

Takano T, Miyauchi A, Yoshida H et al (2005) Decreased relative expression level of trefoil factor 3 mRNA to galectin-3 mRNA distinguishes thyroid follicular carcinoma from adenoma. Cancer Lett 219:91–96

Takenaka Y, Inohara H, Yoshii T et al (2003) Malignant transformation of thyroid follicular cells by galectin-3. Cancer Lett 195: 111–119

Takenaka Y, Fukumori T, Yoshii T et al (2004) Nuclear export of phosphorylated galectin-3 regulates its antiapoptotic activity in response to chemotherapeutic drugs. Mol Cell Biol 24:4395–4406

Tejler J, Leffler H, Nilsson UJ (2005) Synthesis of O-galactosyl aldoximes as potent LacNAc-mimetic galectin-3 inhibitors. Bioorg Med Chem Lett 15:2343–2345

Teng XD, Wang LJ, Yao HT et al (2004) Expression of cytokeratin19, galectin-3 and HBME-1 in thyroid lesions and their differential diagnoses. Zhonghua Bing Li Xue Za Zhi 33:212–216, Abstract

Teymoortash A, Pientka A, Schrader C et al (2006) Expression of galectin-3 in adenoid cystic carcinoma of the head and neck and its relationship with distant metastasis. J Cancer Res Clin Oncol 132: 51–56

Thijssen VL, Hulsmans S, Griffioen AW (2008) The galectin profile of the endothelium: altered expression and localization in activated and tumor endothelial cells. Am J Pathol 172:545–553

Torres-Cabala C, Panizo-Santos A, Krutzsch HC et al (2004) Differential expression of S100C in thyroid lesions. Int J Surg Pathol 12:107–115

Torres-Cabala C, Bibbo M, Panizo-Santos A et al (2006) Proteomic identification of new biomarkers and application in thyroid cytology. Acta Cytol 50:518–528

Uhm JH, Gladson CL, Rao JS (1999) The role of integrins in the malignant phenotype of gliomas. Front Biosci 4:D188–D199

Ulmer TA, Keeler V, Loh L et al (2006) Tumor-associated antigen 90 K/Mac-2-binding protein: possible role in colon cancer. J Cell Biochem 98:1351–1366

Van den Brûle FA, Waltregny D, Liu FT, Castronovo V (2000) Alteration of the cytoplasmic/nuclear expression pattern of galectin-3 correlates with prostate carcinoma progression. Int J Cancer 89:361–367

Vereecken P, Heenen M (2006) Serum galectin-3 in advanced melanoma patients: a hypothesis on a possible role in melanoma progression and inflammation. J Int Med Res 34:119–120

Vereecken P, Debray C, Petein M et al (2005) Expression of galectin-3 in primary and metastatic melanoma: immunohistochemical studies on human lesions and nude mice xenograft tumors. Arch Dermatol Res 296:353–358

Vereecken P, Zouaoui Boudjeltia K et al (2006) High serum galectin-3 in advanced melanoma: preliminary results. Clin Exp Dermatol 31:105–109

Vladimirova V, Waha A, Luckerath K et al (2008) Runx2 is expressed in human glioma cells and mediates the expression of galectin-3. J Neurosci Res 86:2450–2461

Volante M, Bozzalla-Cassione F, DePompa R et al (2004) Galectin-3 and HBME-1 expression in oncocytic cell tumors of the thyroid. Virchows Arch 445:183–188

Wang D, Anderson JC, Gladson CL (2005a) The role of the extracellular matrix in angiogenesis in malignant glioma tumors. Brain Pathol 15:318–326

Wang H, Wang MD, Ma WB et al (2005b) Expression of galectin-3 in invasive prolactinomas. Zhongguo Yi Xue Ke Xue Yuan Xue Bao 27:380–381, Abstract

Weber KB, Shroyer KR, Heinz DE et al (2004) The use of a combination of galectin-3 and thyroid peroxidase for the diagnosis and prognosis of thyroid cancer. Am J Clin Pathol 122:524–531

Wollina U, Graefe T, Feldrappe S et al (2002) Galectin fingerprinting by immuno- and lectin histochemistry in cutaneous lymphoma. J Cancer Res Clin Oncol 128:103–110

Yoshii T, Inohara H, Takenaka Y et al (2001) Galectin-3 maintains the transformed phenotype of thyroid papillary carcinoma cells. Int J Oncol 18:787–792

Yu LG, Andrews N, Zhao Q et al (2007) Galectin-3 interaction with Thomsen-Friedenreich disaccharide on cancer-associated MUC1 causes increased cancer cell endothelial adhesion. J Biol Chem 282:773–781

Zou W (2005) Immunosuppressive networks in the tumour environment and their therapeutic relevance. Nat Rev Cancer 5: 263–274

Zubieta MR, Furman D, Barrio M et al (2006) Galectin-3 expression correlates with apoptosis of tumor-associated lymphocytes in human melanoma biopsies. Am J Pathol 168:1666–1675

Part V

R-Type Animal Lectins

R-Type Lectin Families

14

Rajesh K. Gupta and G.S. Gupta

14.1 Ricinus Communis Lectins

In 1888, Peter Hermann Stillmark reported that seed extracts of the poisonous plant Ricinus communis (castor bean) contain a toxin that can agglutinate erythrocytes. This agglutinin was named as "ricin." Other agglutinins were soon discovered in the seed extracts of other species of poisonous plants. In twentieth century, these proteins were recognized as important members of the glycan-binding proteins known as "lectins." The CRD of this lectin is closely related in sequence and three-dimensional structure to the CRDs of a number of other plant lectins. Some of these structural similarities were also noted in a variety of animal and bacterial glycan-binding proteins discussed in this chapter, causing all proteins containing this ricin-type CRD to be classified as R-type lectins Section 14.2.

The R-type domain is an ancient type of protein fold that is found in many glycosyltransferases as well as in bacterial and fungal hydrolases. Interestingly, the R-type CRD is the only one conserved between animal and bacterial lectins (Sharon and Lis 2004). Ricin was the first lectin discovered and it is the prototypical lectin in this category. Two different lectins have been purified from *R. communis* seeds, and in the original nomenclature they were termed RCA-I and RCA-II. RCA-I is an agglutinin but a very weak toxin. RCA-II is commonly called ricin, and it is both an agglutinin and a very potent toxin. The designation RCA-II has now been dropped, but the original name for the agglutinin RCA-I has been retained. The molecular mass of RCA-I is approximately 120 kDa and that of ricin is approximately 60-kDa. Ricin is a type II ribosome-inactivating protein (RIP-II). Although one might predict that RCA-I also would be highly toxic, it has weak activity compared to ricin because it lacks a separate A chain (Fig. 14.1).

14.1.1 Properties of Ricin

14.1.1.1 Ricin Is a Ribosome Inactivating Protein

Ricin is classified as a Type 2 ribosome inactivating protein (RIP). Whereas Type 1 RIPs consist of a single enzymatic protein chain, Type 2 RIPs, also known as holotoxins, are heterodimeric glycoproteins. Type 2 RIPs consist of an A chain that is functionally equivalent to a Type 1 RIP, covalently connected by a single disulfide bond to a B chain that is catalytically inactive, but serves to mediate entry of the A-B protein complex into the cytosol. Both Type 1 and Type 2 RIPs are functionally active against ribosomes in vitro, however only Type 2 RIPs display cytoxicity due to the lectin properties of the B chain. In order to display its ribosome inactivating function, the ricin disulfide bond must be reductively cleaved.

14.1.1.2 Ricin (RCA-II) Is Toxic and Differs from RCA-I

Ricinus communis agglutinin-I (RCA-I) is a tetramer with two ricin heterodimer-like proteins that are noncovalently associated. Each heterodimer in RCA-I contains an A-chain disulfide linked to the galactose-binding B chain. The sequences of the A chain of ricin and the A chain of RCA-I differ in 18 of 267 residues and are 93% identical, whereas the B chains differ in 41 of 262 residues and are 84% identical. All the subunits are N-glycosylated and usually express oligomannose-type N-glycans. The genome of R. communis also encodes several other proteins that have high homology with ricin, and some of these lectins have been designated ricin-A, -B, -C, -D, and -E.

Ricin is synthesized as a single prepropolypeptide with a 35-amino-acid amino-terminal signal sequence, followed by the A-chain domain, a 12-amino-acid linker region, and the B chain. Proteolysis results in cleavage between the A and B chains. Mature ricin contains four intrachain disulfide bonds; a single interchain disulfide bond links A chain to B chain.

G.S. Gupta et al., *Animal Lectins: Form, Function and Clinical Applications*,
DOI 10.1007/978-3-7091-1065-2_14, © Springer-Verlag Wien 2012

The A chain contains the catalytic activity responsible for the toxicity and the B chain has the glycan-binding activity. The mature A chain has 267 amino acids and the B chain has 262 amino acids. Each B subunit is a product of gene duplications and has two CRDs, each of which is composed of an ancient 40-amino-acid galactose-binding polypeptide region.

The B chain of ricin has two binding sites for sugars, which are about 35 Å apart. Ricin binds to β-linked galactose and N-acetylgalactosamine, whereas RCA-I prefers β-linked galactose. The affinities of these lectins for monosaccharides are quite low (K_D in the range 10^{-3}–10^{-4} M). In contrast, the binding to cells is of much higher affinity (K_D in the range 10^{-7}–10^{-8} M), owing to both increased avidity from the multivalency and enhanced binding to glycans terminating in the sequence Galβ1–4GlcNAc-R. Such sequences represent higher-affinity determinants for lectin binding. In general, ricin and RCA-I both preferentially bind to glycans containing nonreducing terminal Galβ1–4GlcNAc-R or GalNAcβ1–4GlcNAc-R, although they bind weakly to Galβ1–3GlcNAc-R. Neither ricin nor RCA-I binds well to glycoconjugates containing nonreducing terminal α-linked Gal residues. RCA-I is commonly used for glycan isolation and characterization because it is safer than ricin and has higher avidity because of its tetrameric nature. Ricin is often used as a toxin in cell selection for glycosylation mutants, and the A chain of ricin is often used in chimeric proteins as a toxin for specific-cell killing (Stirpe et al. 1992; Lord et al. 1994; Sharon and Lis 2004).

- The tertiary structure of ricin is a globular, glycosylated heterodimer of approximately 60–65 kDa. Ricin toxin A chain and ricin toxin B chain are of similar molecular weight, approximately 32 and 34 kDA respectively.
- **Ricin A Chain** (RTA) is an N-glycoside hydrolase composed of 267 amino acids. It has three structural domains with approximately 50% of the polypeptide arranged into α-helices and β-sheets. The three domains form a pronounced cleft that is the active site of RTA.
- **Ricin B Chain** (RTB) is a lectin composed of 262 amino acids that is able to bind terminal galactose residues on cell surfaces. RTB form a bilobal, barbell-like structure lacking α-helices or β-sheets where individual lobes contain three subdomains. At least one of these three subdomains in each homologous lobe possesses a sugar-binding pocket that gives RTB its functional character.

The gene for ricin A chain has been cloned (Halling et al. 1985). Because ricin is a multimeric glycoprotein, recombinant forms produced in Escherichia coli (such as ricin B chain) are usually poorly active. A chain expressed from recombinant sources will be referred to here as rRTA. The rRTA derived from an *E. coli* expression system was crystallized and refined by X-ray structure (Kim et al. 1992; Mlsna et al. 1993). The structure of the heterodimeric ricin has been solved and refined to 2.5 A resolution (Rutenber et al. 1991). The structures of both the -A (Katzin et al. 1991) and the lectin B chain (Rutenber and Robertus 1991) have been described. The RTA is linked by a disulfide bond to the B chain (RTB). The RTB binds to target cell surfaces via its lectin action, and the disulfide bond keeps RTA, the toxic moiety, tethered until it is taken up by endocytosis. The bond is reduced in the cell permitting enzyme action; it is known that the heterodimer is inactive against ribosomes.

14.1.1.3 Carbohydrate-Binding Module

The R-type domain contained in these proteins is the CRD, which is also termed a carbohydrate-binding module (CBM) and has been placed in the CBM13 family in the CAZy database (carbohydrate-active enzymes database). Figure 14.1 shows that the A chain has eight α-helices and eight β-strands and it is the catalytic subunit, as discussed above. The B chain, which contains R-type lectin domains, has two tandem CRDs that are about 35 Å apart and have a shape resembling a barbell, with one binding domain at each end. Each R-type domain has a three-lobed organization that is a β-trefoil structure (from the Latin trifolium meaning "three-leaved plant") (Fig. 14.2). The β-trefoil structure probably arose evolutionarily through gene fusion events linking a 42-amino-acid peptide subdomain that has galactose-binding activity. The three lobes are termed α, β, and γ and are arranged around a threefold axis. Conceivably, each lobe could be an independent binding site, but in most R-type lectins only one or two of these lobes retain the conserved amino acids required for sugar binding. Sugar binding is relatively shallow in these loops and arises from aromatic amino acid stacking against the Gal/GalNAc residues and from hydrogen bonding between amino acids and hydroxyl groups of the sugar ligands. A characteristic feature of these loops in the R-type domain in ricin is the presence of (QxW)3 repeats (where x is any amino acid), which are found in many, but not all, R-type family members (Hazes 1996).

14.1.2 Other R-Type Plant Lectins

In addition to RCA-I and ricin, other plant lectins with R-type domains include the ricin homolog from *Abrus precatorius* and the bark lectins from the elderberry plant, *Sambucus sieboldiana* lectin (SSA) and *Sambucus nigra* agglutinin (SNA). The SSA and SNA are unusual in that they are the only R-type lectins that bind well to α2–6-linked sialic acid–containing ligands and they do not bind to α2–3-linked sialylated ligands. SSA and SNA are heterotetramers (~140 kD) composed of two heterodimers each containing an A chain (which resembles ricin A chain) disulfide-bonded

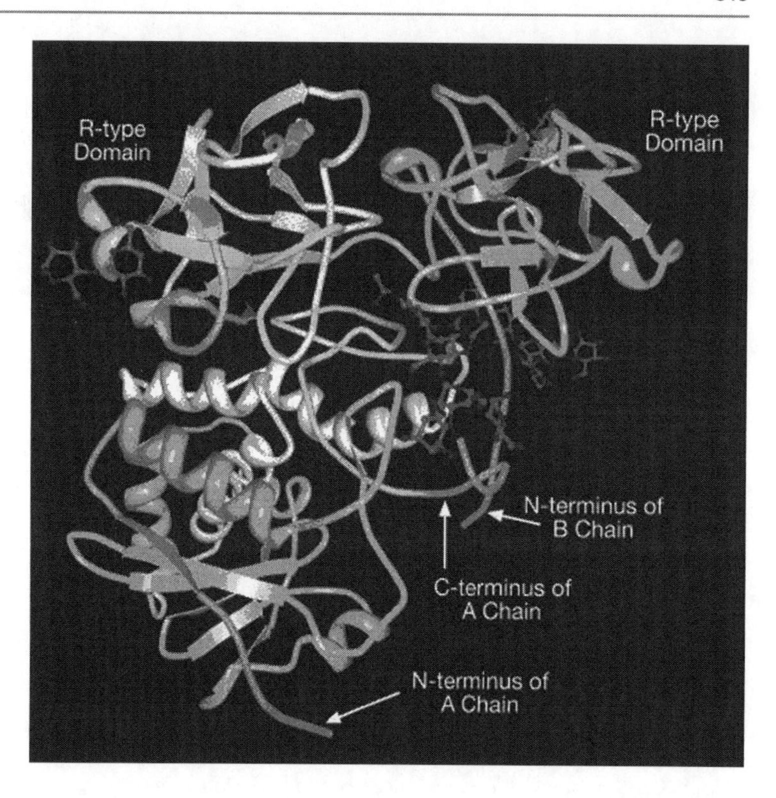

Fig. 14.1 The crystal structure of ricin refined to 2.5 Å. (Adapted by permission from Rutenber et al. 1991© John Wiley and Sons. PDB ID: 2AAI.)

to a B chain (which binds glycans and is an R-type lectin). The A chain in these proteins has very weak RIP-II activity in vitro. SSA and SNA may have the same overall organization as RCA-I (see Fig. 14.1). The toxins abrin, modeccin from Adenia digitata, Viscum album agglutinin (VAA or mistletoe lectin), and volkensin also have R-type domains, belong to the RIP-II class, and kill cells in a manner similar to ricin. There are other R-type plant lectins in the RIP-II class that are not toxic, and these include several proteins from the genus Sambucus (elderberry), such as nigrin-b, sieboldin-b, ebulin-f, and ebulin-r. All of the B subunits of these proteins appear to bind Gal/GalNAc, but they may have some differences in affinity and may recognize different Gal/GalNAc-containing glycoconjugates.

14.2 R-Type Lectins in Animals

The R-type lectin domain is found in several animal lectins, including the mannose receptor (MR) family, discussed in subsequent chapter, and in some invertebrate lectins. EW29 is a galactose-binding lectin from the annelid (earthworm) Lumbricus terrestris. The R-type domain is also found in pierisin-1, which is a cytotoxic protein from the cabbage butterfly Pieris rapae, and in the homologous protein pierisin-2, from Pieris brassicae. Tandem R-type motifs are found in some other R-type family members, including ricin, but the presence of four such motifs is unique, thus far, to pierisin-1. Some proteins with the R-type lectin domain are also enzymes and these are found in both animals and microbes. For example, Limulus horseshoe crab coagulation factor G has a central R-type lectin domain, which is flanked at the amino terminus by a xylanase Z-like domain and at the carboxyl terminus by a glucanase-like domain. This protein also has a subunit that is a serine protease.

14.3 Mannose Receptor Family

There are four known members of MR family in humans, all of which contain an R-type lectin domain; other family members are not predicted. The MR family includes the MR, the phospholipase A2 (PLA2) receptor, DEC-205/MR6-gp200, and Endo180/urokinase plasminogen activator receptor-associated protein (Fig. 14.3). All of these proteins are large type I transmembrane glycoproteins and they contain a single fibronectin type II domain similar to R-type CRDs of ricin, 8–10 C-type lectin domains (CTLDs), and an amino-terminal cysteine-rich domain (East et al. 2002;

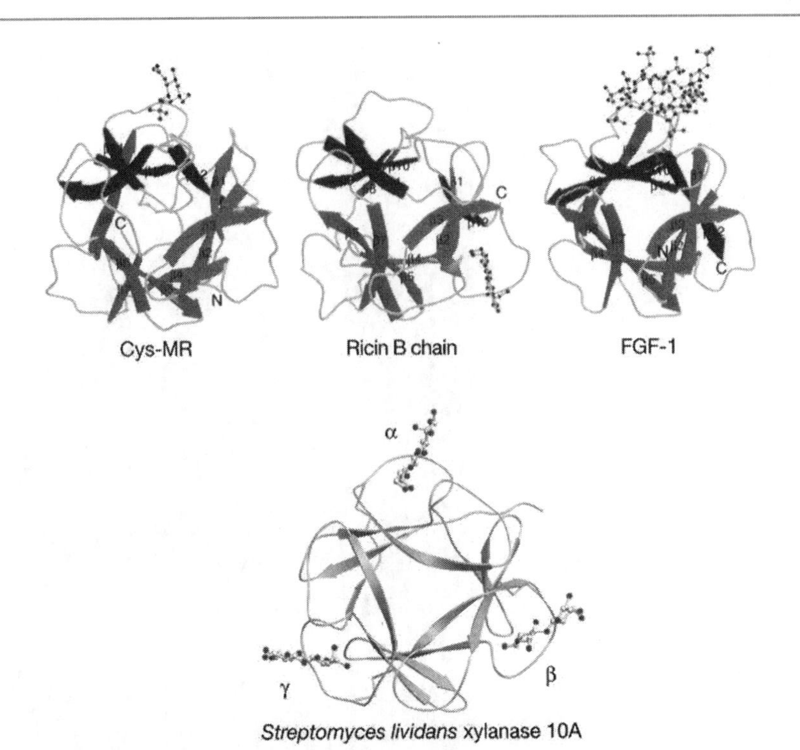

Fig. 14.2 Structures of the β-trefoil R-type domains in different proteins. (*Top*) Cysteine-rich R-type domain of the mannose receptor (MR) in complex with 4-O-sulfated GalNAc; ricin B chain in complex with galactose; and acidic fibroblast growth factor (FGF-1) in complex with sulfated heparan decasaccharide. N and C-termini are labeled, and lobes I, II, and III are indicated in different shades. Each structure is depicted with a bound ligand (4-SO_4-GalNAc for Cys-MR, galactose for ricin B chain, and sulfated heparin decasaccharide for aFGF) (Adapted with permission from Liu et al. 2000 © The Rockefeller University Press). (*Bottom*) *Streptomyces lividans* endo-β1-4xylanase in complex with lactose (Adapted by permission from Notenboom et al. 2002 © American Chemical Society)

Llorca 2008). However, despite the common presence of multiple lectin-like domains, these four endocytic receptors have divergent ligand binding activities, and it is clear that the majority of these domains do not bind sugars. Endo180 binds in a Ca^{2+}-dependent manner to mannose, fucose, and N-acetylglucosamine but not to galactose. This activity is mediated by one of the eight CTLDs, CTLD2. Monosaccharide binding specificity of Endo180 CTLD2 is similar to that of MR CTLD4. However, additional experiments indicate that, unlike the cysteine-rich domain of the MR, the cysteine-rich domain of Endo180 does not bind sulfated sugars. Thus, although Endo180 and the MR are now both known to be mannose binding lectins, each receptor is likely to have a distinct set of glycoprotein ligands in vivo (East et al. 2002; Yan et al., 1997). The mannose receptor acts as a molecular scavenger, clearing harmful glycoconjugates or micro-organisms through recognition of their defining carbohydrate structures. The MR can also bind collagen and that the fibronectin type II domain mediates this activity. Neither of the two types of sugar-binding domain in MR is involved in collagen binding. The fibronectin type II domain shows the same specificity for collagen as the whole receptor, binding to type I, type III and type IV collagens. These are additional roles for this multifunctional receptor which mediates collagen clearance or cell-matrix adhesion (Napper et al. 2006). Details of this family have been discussed in next Chap. 15.

14.4 UDP-Galnac: Polypeptide α-N-Acetyl-galactosaminyltransferases

14.4.1 Characteristics of UDP-GalNAc: α-N-Acetylgalactosaminyltransferases

A large homologous family of uridine diphosphate (UDP)-*N*-acetyl-α-D galactosamine(GalNAc): poly-peptide *N*-acetylgalactosaminyltransferases (ppGalNAc-Ts, EC 2.4.1.41) initiate mucin-type *O*-glycosylation by transferring GalNAc to the hydroxyl group of serine and threonine residues (GalNAcα1-*O*-Ser/Thr) (Fig. 14.3). In *Caenorhabditis elegans*, a total of 11

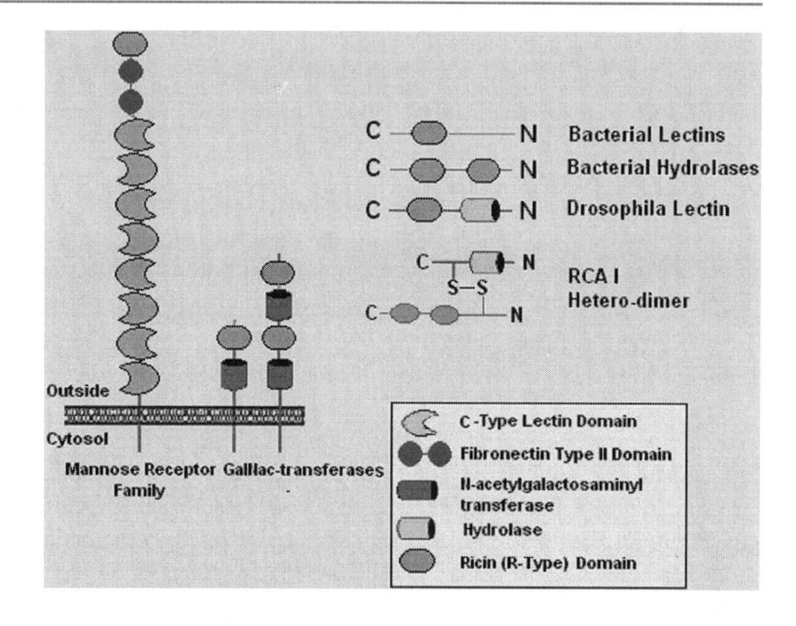

Fig. 14.3 The R-type lectin superfamily. Different groups within the family of animal, bacterial and ricin (plant) are indicated with the domain structures shown In mannose recepror family, DEC-205 contains ten lectin domains where as others contain eight CTLDs

distinct sequence homologs of ppGalNAc-T gene family were cloned, sequenced, and expressed. All clones encoded type II membrane proteins that shared 60–80% amino acid sequence similarity with the catalytic domain of mammalian ppGalNAc-Ts enzymes. Two sets of cDNA clones contained variants that appeared to be produced by alternative message processing. In addition to the existence of ppGalNAc-T enzymes in a nematode organism, the substantial diversity of these isoforms in *C. elegans* suggests that mucin O-glycosylation is catalyzed by a complex gene family, which is conserved among evolutionary-distinct organisms (Hagen and Nehrke 1998). Human and rodent ppGalNAc-T families are said to include 20 distinct isoforms, of which 16 have been characterized (Ten Hagen et al. 2003). Despite the seeming simplicity of ppGalNAc-T catalytic function, it is estimated that there are 24 unique ppGalNAc-T human genes. These ppGalNAcTs may be separated into two general classes: those that can transfer N-acetyl-galactosamine from UDP-GalNAc to unmodified polypeptide acceptors, and those that prefer acceptor glycopeptides containing GalNAc-Ser/Thr (i.e., a set predicted to have a CRD). Cloning of the genes encoding these enzymes revealed that they are large proteins and have a unique multidomain structure. All the enzymes are type II transmembrane proteins and have a carboxy-terminal R-type domain of about 130 amino acid residues and an amino-terminal catalytic domain (Ten Hagen et al. 2001, 2003). The R-type domain has the QxW repeat, and recent evidence shows that these domains in the ppGalNAcTs function as lectins during the catalytic process.

The ppGalNAc-T isoforms display tissue-specific expression in adult mammals as well as unique spatial and temporal patterns of expression during murine development. In vitro assays suggest that a subset of the ppGalNAc-Ts have overlapping substrate specificities, but at least two ppGalNAc-Ts (ppGalNAc-T-T7 and -T9 [now designated -T10]) appear to require the prior addition of GalNAc to a synthetic peptide before they can catalyze sugar transfer to this substrate. Site-specific O-glycosylation by several ppGalNAc-Ts is influenced by the position and structure of previously added O-glycans. Collectively, these observations argue in favor of a hierarchical addition of core GalNAc residues to the apomucin. Various forms of O-glycan pathobiology may be reexamined in light of the existence of an extensive ppGalNAc-T family of enzymes. Studies have demonstrated that at least one ppGalNAc-T isoform is required for normal development in *Drosophila melanogaster* (Ten Hagen et al. 2003). The ppGalNAc-T family is conserved in evolution, and distinct subfamilies of orthologous isoforms with conserved kinetic properties have been identified in vertebrates and invertebrates (Schwientek et al. 2002; Stwora-Wojczyk et al. 2004). Differences in kinetic properties, substrate specificities, and expression patterns of these isoenzymes provide for differential regulation of O-glycan attachment sites and density (Hassan et al. 2000; Ten Hagen et al. 2003). Although ablation of several ppGalNAc-T isoforms in mice have not demonstrated a phenotype (Ten Hagen et al. 2003), the finding that impairment of a single isoform in *Drosophila melanogaster* disrupts development (Ten Hagen and Tran 2002; Schwientek et al. 2002), and that the human ppGalNAc-T isoform, GalNAc-T3, is implicated in the disease familial tumoral calcinosis (Topaz et al. 2004) demonstrate the nonredundant function of some ppGalNAc-T isoforms.

The GalNAc-glycopeptide substrate specificity exhibited by some isoforms has been associated with a ricin-like lectin domain found in the C-terminal region of most ppGalNAc-T isoforms (Hassan et al. 2000). This distinct lectin-like domain was originally identified by Hazes (1996) and Imberty et al. (1997). Studies have demonstrated that the catalytic and lectin domains of ppGalNAc-Ts fold into separate domains (Fritz et al. 2004, 2006; Kubota et al. 2006). In agreement with predictions from studies of the role of the lectin domain in directing GalNAc-glycopeptide substrate specificities, Fritz and associates were able to model a glycopeptide into the catalytic pocket and show potential interactions of the GalNAc residues with the carbohydrate-binding sites of the lectin domain.

The ppGalNAc-Ts can be subdivided into three putative domains, each containing a characteristic sequence motif. The 112-amino acid glycosyltransferase 1 (GT1) motif represents the first half of the catalytic unit and contains a short aspartate-any residue-histidine (DXH) or aspartate-any residue-aspartate (DXD)-like sequence. Secondary structure predictions suggest that the GT1 motif forms a 5-stranded parallel β-sheet flanked by 4 α-helices, which resembles the first domain of the lactose repressor. Four invariant carboxylates and a histidine residue are predicted to lie at the C-terminal end of three β-strands and line the active site cleft. Site-directed mutagenesis of murine ppGalNAc-T1 reveals that conservative mutations at these five positions result in products with no detectable enzyme activity. The second half of the catalytic unit contains a motif (positions 310–322) which is also found in β1,4-galactosyltransferases (termed the Gal/GalNAc-T motif). Mutants of carboxylates within this motif express either no detectable activity. Mutagenesis of highly conserved (but not invariant) carboxylates produces only modest alterations in enzyme activity. Mutations in the C-terminal 128-amino acid ricin-like lectin motif do not alter the enzyme's catalytic properties (Amadoetal 1999; Hagen et al. 1999; Ten Hagen et al. 2001).

Hagen et al. (1999) initially investigated the function of the lectin domain of GalNAc-T1, and found that selective mutational disruption of the lectin domain of GalNAc-T1 did not severely affect the catalytic function of the enzyme with peptide substrates. But Bennett et al. (1998) found that the GalNAc-T4 isoform exhibited a unique GalNAc-glycopeptide substrate specificity. Mutational analysis of GalNAc-T4 demonstrated that a mutation in the α-repeat of the C-terminal lectin domain of GalNAc-T4 selectively inactivated the GalNAc-glycopeptide catalytic function of the enzyme, whereas activity with unglycosylated peptides was unaffected, in accordance with the original study of GalNAc-T1 (Hassan et al. 2000). Now, it has emerged that some ppGalNAc-T isoforms in vitro selectively function with partially GalNAc O-glycosylated acceptor peptides rather than with the corresponding unglycosylated peptides. O-Glycan attachment to selected sites, most notably two sites in the MUC1 tandem repeat, is entirely dependent on the glycosylation-dependent function of GalNAc-T4. Furthermore, results suggest that the GalNAc-T4 lectin domain modulates the function of the enzyme through interaction with the GalNAc residues of the glycopeptide substrate. Similar studies with GalNAc-T1 and -T2 suggested that the lectin domains of these may be important for the glycosylation of partially GalNAc-glycosylated peptide substrates (Tenno et al. 2002; Fritz et al. 2006). Direct carbohydrate binding of two ppGalNAc-T lectin domains, GalNAc-T4 and GalNAc-T2, representing isoforms with distinct glycopeptide activity (GalNAc-T4) and isoforms without apparent distinct GalNAc-glycopeptide specificity (GalNAc-T2) suggests that ppGalNAc-T lectins serve to modulate the kinetic properties of the enzymes in the late stages of the initiation process of O-glycosylation to accomplish dense or complete O-glycan occupancy (Wandall et al. 2007).

Using a set of synthetic peptides and glycopeptides it was demonstrated that the lectin domain of ppGalNAcT-2 (hT2) directs glycosylation site selection for glycopeptide substrates. It was found that glycosylation of peptide substrates by glycopeptide transferase ppGalNAcT-10 (hT10) requires binding of existing GalNAcs on the substrate to either its catalytic or lectin domain, thereby resulting in its apparent strict glycopeptide specificity. These results highlight the existence of two modes of site selection used by these ppGalNAcTs: local sequence recognition by the catalytic domain and the concerted recognition of distal sites of prior glycosylation together with local sequence binding mediated, respectively, by the lectin and catalytic domains (Raman et al. 2008).

14.4.2 The Crystal Structure of Murine ppGalNAc-T-T1

The family of ppGalNAcTs is unique among glycosyltransferases, containing a catalytic and a C-terminal lectin domain that were shown to be closely associated. The x-ray crystal structure of a ppGalNAc-T, murine ppGalNAc-T-T1, showed that the enzyme folds to form distinct catalytic and lectin domains. The association of two domains forms a large cleft in the surface of the enzyme that contains a Mn^{2+} ion complexed by invariant D209 and H211 of the "DXH" motif and by invariant H344. Each of the three potential lectin domain carbohydrate-binding sites (α, β, and γ) is located on the active-site face of the enzyme, suggesting a mechanism by which the transferase may accommodate multiple conformations of glycosylated acceptor substrates. A model of a mucin 1 glycopeptide substrate bound to the enzyme shows that the spatial separation between the lectin α site and a modeled active site UDP-GalNAc is consistent with the in vitro pattern of glycosylation observed for this peptide catalyzed by ppGalNAc-T-T1. The structure also provides a template for the larger ppGalNAc-T family, and homology models of several ppGalNAc-T isoforms predict dramatically different surface chemistries consistent with isoform-selective acceptor substrate recognition (Fritz et al. 2004).

Fritz et al. (2006) described the x-ray crystal structures of human ppGalNAcT-2 (hT2) bound to the product UDP and to UDP and an acceptor peptide substrate EA2 (PTTDSTTPAPTTK). The conformations of both UDP and residues Arg362-Ser372 vary greatly between the two structures. In the hT2-UDP-EA2 complex, residues Arg362-Ser373 comprise a loop that forms a lid over UDP, sealing it in the active site, whereas in the hT2-UDP complex this loop is folded back, exposing UDP to bulk solvent. EA2 binds in a shallow groove with threonine 7 positioned consistent with in vitro data showing it to be the preferred site of glycosylation. The relative orientations of the hT2 catalytic and lectin domains differ dramatically from that of murine ppGalNAcT-1 and also vary considerably between the two hT2 complexes. Indeed, in the hT2-UDP-EA2 complex essentially no contact is made between the catalytic and lectin domains except for the peptide bridge between them. Thus, the hT2 structures reveal an unexpected flexibility between the catalytic and lectin domains and suggest a new mechanism used by hT2 to capture glycosylated substrates. Kinetic analysis of hT2 lacking the lectin domain confirmed the importance of this domain in acting on glycopeptide but not peptide substrates. The structure of the hT2-UDP-EA2 complex also resolves long standing questions regarding ppGalNAcT acceptor substrate specificity (Fritz et al. 2006).

The murine enzyme MT1 does not require the R-type domain for catalysis with either glycopeptide (GalNAc-peptide) or peptide acceptors. In contrast, in human hT2, the lectin domain is important in catalysis with GalNAc-peptide acceptors, but not peptide acceptors, and the lectin domains of both hT2 and hT4 can directly bind N-acetylgalactosamine. In mT1, the R-type domain has the β-trefoil structure and the three-lobe repeating α, β, and γ loops. Mutations in the repeats can abolish binding to GalNAc-peptide acceptors. The ppGalNAcTs are the only known glycosyltransferases that have a lectin and catalytic domain conjoined. The available models suggest that each of the ppGalNAcTs has a somewhat different peptide or glycopeptide acceptor specificity, allowing the assortment of enzymes to be highly efficient at adding N-acetylgalactosamine to a tremendous variety of polypeptide substrates, including long mucin polypeptides of thousands of amino acids, and to specific single serine or threonine residues on membrane and secreted glycoproteins. The presence of the two domains in these enzymes and the multiple family members may promote an efficient processive activity and association of these enzymes with acceptor proteins (Fig. 14.4).

14.4.3 Parasite ppGalNAc-Ts

The ppGalNAc-T from human disease-causing parasite, *Toxoplasma gondii* catalyzes the initial step of mucin-type O-glycosylation, the transfer of GalNAc in O-glycosidic linkage to serine and threonine residues in polypeptides. The 84-kDa type II membrane protein contains a 49-amino acid N-terminal cytoplasmic domain, a 22-amino acid hydrophobic transmembrane domain, and a 680-amino acid C-terminal lumenal domain. Sequence motifs include a glycosyltransferase 1 (GT1) motif containing a DXH sequence, a Gal/GalNAc-T motif, and a region homologous to ricin lectin in a single 5.5-kb ppGalNAc-T transcript. Genomic DNA sequences revealed that this transferase is encoded by 10 exons in a 10 kb region. *T. gondii* demonstrates that this human parasite has its own enzymatic machinery for the O-glycosylation of toxoplasmal proteins (Wojczyk et al. 2003).

A full-length cDNA for ppGalNAc-T from the cestode *Echinococcus granulosus* (Eg-ppGalNAc-T1) was found to code for a 654-amino-acid protein containing all the structural features of ppGalNAc-Ts. Interestingly the C-terminal region of Eg-ppGalNAc-T1 bears a highly unusual lectin domain, considerably longer than the one from other members of the family, and including only one of the three ricin B repeats generally present in ppGalNAc-Ts. The role of the lectin domain in the determination of the substrate specificity of these enzymes suggests that Eg ppGalNAc-T1 would be involved in the glycosylation of a special type of substrate. This transferase is expressed in the hydatid cyst wall and the subtegumental region of larval worms. Therefore it seems to participate in the biosynthesis of O-glycosylated parasite proteins exposed at the interface between E. granulosus and its hosts (Freire et al. 2004).

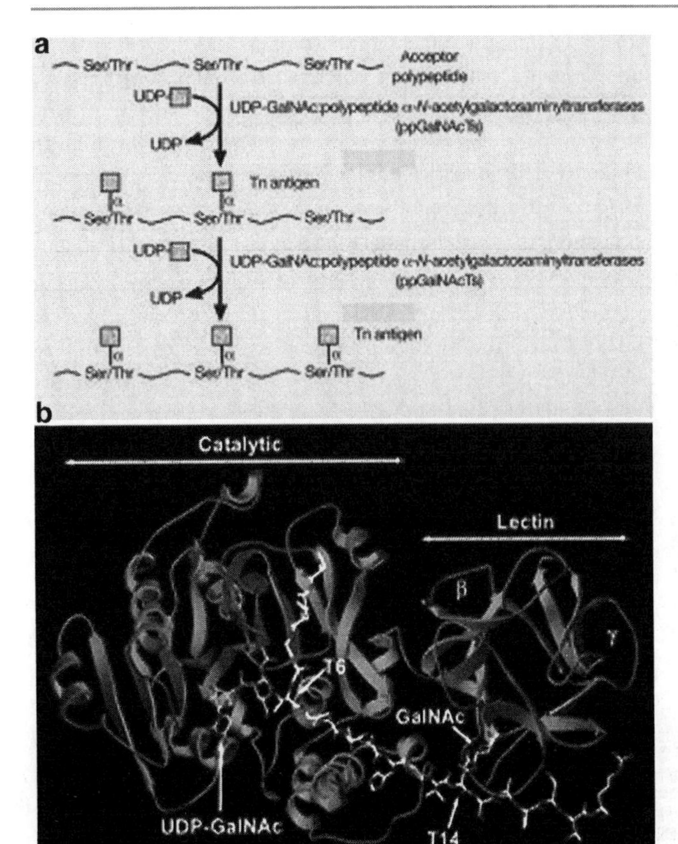

Fig. 14.4 Structure and function of UDP-GalNAc:polypeptide α-*N*-acetylgalactosaminyltransferases (ppGalNAcTs). (a) The N-acetylgalactosamine (GalNAc, *yellow* square) transfer reaction of the ppGalNAcT showing an acceptor peptide with Ser and Thr residues and UDP-GalNAc as the donor. Some of the ppGalNAcTs may also prefer to act on the product of this reaction and use peptides with attached *N*-acetylgalactosamine as the acceptor (Cummings and Etzler 2009). **(b) The crystal structure of murine ppGalNAc-T-T1.** Modeled binding of a MUC1 glycopeptide to ppGaNTase-T1 is consistent with its in vitro pattern of glycosylation. Structural alignment of the ppGaNTase-T1 and *S. olivaceoviridis* β-xylanase (PDB ID code 1XYF) lectin domains was used to model GalNAc covalently attached to Thr-14 of a MUC1 peptide (PAPGSTAPPAHGVTSAPDTR, white carbons) into the α site of the ppGaNTase-T1 lectin domain. This docking allowed Thr-6 of the peptide to be positioned within 2.5 Å of the anomeric carbon of the modeled UDP-GalNAc (*gray* carbons). The remainder of the peptide was positioned to avoid overlap with the enzyme. All peptide ϕ and ψ angles are in allowed regions of the Ramachandran plot (Adapted by permission from Fritz et al. 2004 © Nationl Academy of Sciences, USA)

N-Acetylgalactosamine (GalNAc)

14.4.4 Crystal Structure of CEL-III from *Cucumaria echinata* Complexed with GalNAc

CEL-III is a Ca^{2+}-dependent and galactose-specific lectin with two β-trefoil folds from sea cucumber, *Cucumaria echinata*, which exhibits hemagglutinating activity. Six molecules of CEL-III are assumed to oligomerize to form an ion-permeable pore in cell membrane. CEL-III consists of three distinct domains: two carbohydrate-binding domains (1 and 2) at N-terminus that adopt β-trefoil folds as in B-chain of ricin and are members of $(QXW)_3$ motif family; and domain 3, which is a novel fold composed of two α-helices and one β-sandwich. Despite sharing the structure of the B-chain of ricin, CEL-III binds five Ca^{2+} ions at five of the six sub-domains in both domains 1 and 2. Considering the relatively high similarity among the five sub-domains, they are putative binding sites for galactose-related carbohydrates. The paucity of hydrophobic interactions in the interfaces between the domains and biochemical data suggest that these domains rearrange upon carbohydrate binding in erythrocyte membrane. This conformational change may be responsible for oligomerization of CEL-III molecules and hemolysis in the erythrocyte membranes (Uchida et al. 2004; Hatakeyama et al. 2007).

The three-dimensional structure of CEL-III/GalNAc and CEL-III/methyl α-galactoside complexes was solved by X-ray crystallography. In these complexes, five carbohydrate molecules were found to be bound to two carbohydrate-binding domains (domains 1 and 2) located in the N-terminal 2/3 portion of the polypeptide and that contained β-trefoil folds similar to ricin B-chain. The 3-OH and 4-OH of bound carbohydrate molecules were coordinated with Ca^{2+} located at the subdomains 1α, 1γ, 2α, 2β, and 2γ, simultaneously forming hydrogen bond networks with nearby amino acid side chains, which is similar to carbohydrate binding in C-type lectins. The binding of carbohydrates was further stabilized by aromatic amino acid residues, such as tyrosine and tryptophan, through a stacking interaction with the hydrophobic face of carbohydrates. The orientation of bound GalNAc and methyl α-galactoside was similar to the galactose moiety of lactose bound to the carbohydrate-binding site of the ricin B-chain, although the ricin B-chain does not require Ca^{2+} ions for carbohydrate binding. The binding of the carbohydrates induced local structural changes in carbohydrate-binding sites in subdomains 2α and 2β (Hatakeyama et al. 2007).

14.5 Microbial R-Type Lectins

A feature of many microbial glycosidases is the presence of both a catalytic domain and a carbohydrate-binding module (CBM). *Streptomyces lividans* endo-β1–4xylanase 10A (Xyn10A) is a good example of such an enzyme. Xyn10A catalyzes the cleavage of β1–4xylans and can bind to xylan and a variety of small soluble sugars, including galactose, lactose, and xylo- and arabino-oligosaccharides. The catalytic domain is at the amino terminus and the carboxyl terminus has an R-type β-trefoil motif. As mentioned above, the R-type domain CBM represents the CBM13 family in the CAZy database. In Xyn10A, all of the original β-trefoil sugar-binding motifs are retained, along with the conserved disulfide bridges, and evidence suggests that each of the three potential sugar-binding sites in β-trefoil structure interact with sugars and each site may span up to four xylose residues. The binding to monosaccharides is very weak (K_D in the range of 10^{-2}–10^{-3} M), but multivalent binding to polysaccharides can be of very high affinity.

14.5.1 *S. olivaceoviridis* E-86 Xylanase: Sugar Binding Structure

Endo-β-1,4-xylanase 10A (Xyn10A) from *Streptomyces lividans* includes an N-terminal catalytic module and a 130-residue C-terminal family 13 carbohydrate-binding module(CBM13). This latter domain adopts a β-trefoil structure with three potential binding sites (α, β, and γ) for a variety of small sugars, xylooligosaccharides, and xylan polymers. CBM13 binds mono- and oligo-saccharides with association constants of $1-10 \times 10^2$ M^{-1}. The primary function of CBM13 is to bind the polysaccharide xylan, but it retains the ability of R-type lectins to bind small sugars such as lactose and galactose. The association of CBM13 with xylan appears to involve cooperative and additive participation of three binding pockets in each of the three trefoil domains of CBM13, suggesting a novel mechanism of CBM-xylan interaction. It appears to be specific only for pyranose sugars. CBM13 binds insoluble and soluble xylan, holocellulose, pachyman, lichenan, arabinogalactan and laminarin. Site-directed mutation indicates the involvement of three functional sites on CBM13 in binding to soluble xylan. The sites are similar in sequence, and are predicted to have similar structures, to α, β and γ sites of ricin toxin B-chain, which is also in family 14. The binding of CBM13 to soluble xylan involves additive and co-operative interactions between the three binding sites. Analysis of ^{15}N NMR relaxation data revealed that CBM13 tumbles as an oblate ellipsoid and that its backbone is relatively rigid on the sub-nanosecond time scale. In particular, the three binding sites show no distinct patterns of increased internal mobility (Boraston et al. 2000). Chemical shift changes in spectra of CBM13 demonstrated that sugars (L-arabinose, lactose, D-xylose, xylobiose, xylotetraose, and xylohexaose) associate independently with the three binding sites of CBM14. The site-specific association constants showed that L-arabinose, lactose, and D-xylose preferentially bind to α site of CBM13, xylobiose binds equally well to all three sites, and xylotetraose and xylohexaose prefer binding to the β site. Inspection of the crystallographic structure of CBM13 provides a rationalization for these results (Schärpf et al. 2002). Crystal structures of CBM13 in complex with lactose and xylopentaose revealed two distinct mechanisms of ligand binding. CBM13 has retained its specificity for lactose via Ricin-like binding in all of the three classic trefoil binding pockets. However, CBM13 has the ability to bind either the nonreducing galactosyl moiety or the reducing glucosyl moiety of lactose. The mode of xylopentaose binding suggests adaptive mutations in the trefoil sugar binding scaffold to accommodate internal binding on helical polymers of xylose (Notenboom et al. 2002).

In addition to CBM13 as a xylan binding domain (XBD), *S. olivaceoviridis* E-86 contains a $(\beta/\alpha)_8$-barrel as a catalytic domain and a Gly/Pro-rich linker between them. The crystal structure of this enzyme showed that XBD has three similar subdomains, as indicated by the presence of a triple-repeated sequence, forming a galactose binding lectin fold similar to that found in Ricin toxin B-chain. Comparison with the structure of ricin/lactose complex suggests three potential sugar binding sites in XBD. In the catalytic cleft, bound sugars were observed in the xylobiose and xylotriose complex structures. In the XBD, bound sugars were identified in subdomains α and γ in all complexes with xylose, xylobiose, xylotriose, glucose, galactose and lactose. XBD binds xylose or xylooligosaccharides at same sugar binding sites as in Ricin/lactose complex but its binding manner for xylose and xylooligosaccharides is different from the galactose binding mode in ricin, even though XBD binds galactose in the same manner as in the ricin/galactose complex. These different binding modes are utilized efficiently and differently to bind the long substrate to xylanase and ricin-type lectin. Family 13 CBM has rather loose and broad sugar specificities and is used by some kinds of proteins to bind their target sugars. In such enzyme, XBD binds xylan, and the catalytic domain may assume a flexible position with respect to XBD/xylan complex, in as much as the linker region is unstructured (Fujimoto et al. 2002).

14.5.2 The Mosquitocidal Toxin (MTX) from *Bacillus sphaericus*

The mosquitocidal toxin (MTX) from *Bacillus sphaericus* and the apoptosis-inducing pierisin-1 from the cabbage butterfly Pieris rapae are two of the most intriguing members of the family of ADP-ribosyltransferases. Both are approximately 100 kDa proteins, composed of an N-terminal ADP-ribosyltransferase (~27 kDa) and a C-terminal putative binding and translocation domain (~70 kDa) consisting of four ricin-B-like domains. They both share structural homologies, with an overall amino acid sequence identity of approximately 30% and seem to largely differ with regard to their targets or cell internalization mechanisms. MTX ADP-ribosylates numerous proteins in lysates of target insect cells at arginine residues, whereas pierisin-1 modifies DNA of insect and mammalian cells by ADP-ribosylation at 2′-deoxyguanosine residues resulting in DNA adducts, mutations and eventually apoptosis (Carpusca et al. 2006).

The crystal structure of mosquitocidal toxin from *Bacillus sphaericus* (MTX), determined at 2.5 A resolution, revealed essentially a chain consisting of four ricin B-type domains curling around the catalytic domain in a hedgehog-like assembly. The structure is probably not affected by packing contacts and explains autoinhibition data reported earlier. An analysis of ricin B-type lectin complexes and sugar molecules shows that the general construction principle applies to all four lectin domains of MTX, indicating 12 putative sugar-binding sites. These sites are sequence-related to pierisin, which is known to bind glycolipids. It seems therefore likely that MTX also binds glycolipids. The seven contact interfaces between the five domains are predominantly polar and not stronger than common crystal contacts so that in an appropriate environment, the multi-domain structure would likely uncurl into a string of single domains. The structure of the isolated catalytic domain plus an extended linker was established earlier in three crystal packings, two of which showed a peculiar association around a 7-fold axis. The catalytic domain of the reported MTX closely resembles all three published structures, except one with an appreciable deviation of the 40 N-terminal residues. A comparison of all structures suggests a possible scenario for the translocation of the toxin into the cytosol (Treiber et al. 2008). The crystal structure of MTX catalytic domain is helpful to reveal new insights into structural organization, catalytic mechanisms, and autoinhibition of both enzymes.

14.6 R-Type Lectins in Butterflies

14.6.1 Pierisin-1

Pierisin-like proteins are found in subtribes Pierina, Aporiina and Appiadina. Pierisin from P. rapae is called pierisin-1, and that from P. brassicae is called pierisin-2. Pierisin-1 is a 98-kDa protein comprising 850 aa consisting of N-terminal region (27 kDa) and C-terminal region (71 kDa). The N-terminal region of pierisin-1 has a partial regional sequence similarity with ADP-ribosylating toxins such as the A-subunit of cholera toxin, and disruption of this possible NAD-binding site by site-directed mutagenesis abolishes its apoptosis-inducing activity (Watanabe et al. 1999). Unlike other ADP-ribosyltransferases, the N-terminal region of pierisin-1 targets the N2 amino groups of guanine residues in DNA to yield N2-(ADP-ribos-1-yl)-2′-deoxyguanosine (Takamura-Enya et al. 2001). The C-terminal region of pierisin-1 shares sequence similarity with HA-33, a subcomponent of hemagglutinin of botulinum toxin that binds to sialic acid or galactose moieties on surfaces of neuronal cells (Inoue et al. 1999; Lord et al. 2003). Receptors for pierisin-1 on mammalian cells have been found to be the neutral glycosphingolipids, including globotriaosylceramide and globotetraosylceramide, and their expression levels largely determine sensitivity to the toxic protein (Matsushima-Hibiya et al. 2003). Pierisin-1 is distributed in fat bodies during the final larval instar and is abundantly expressed in fifth instar larvae and early pupae in the cabbage butterfly, Pieris rapae (Watanabe et al. 1998, 2004a). It appears that pierisin-1 may play important roles in induction of apoptosis to remove larval cells in the pupation of Pieris rapae. This protein has potent cytotoxic activity against TMK-1 human gastric cancer cells and various human cancer cell lines, inducing typical apoptotic cell death with characteristic morphological features, DNA fragmentation, and cleavage of poly(ADP-ribose) polymerase (Watanabe et al. 1998; Kono et al. 1999). Among mammalian cell lines tested, human HeLa cells were the most sensitive to the cytotoxic effects of pierisin-1 (Kono et al. 1999). The other cabbage white butterfly, Pieris brassicae, also contains the cytotoxic protein and named pierisin-2. Its amino acid sequence is 91% identical to that of pierisin-1. Pierisin-2 targets DNA, and the structure of the DNA adduct produced by pierisin-2 is the same as that produced by pierisin-1 (Takamura-Enya et al. 2004).

Globotriaosylceramide (Gb3) and globotetraosylceramide (Gb4), two neutral glycosphingolipids showed receptor activities for pierisin-1. Alteration of QXW by site-directed mutagenesis caused marked reduction of pierisin-1 cytotoxicity. Study suggests that pierisin-1 binds to Gb3 and Gb4 receptors at C-terminal region, in a manner similar to ricin, and then exhibits cytotoxicity after incorporation into the cell (Matsushima-Hibiya et al. 2003).

14.6.1.1 Cytotoxic and Apoptotic Activity in Pierisin

ADP-ribosylation is generally known to be a posttranslational modification where the ADP-ribose moiety of b-NAD is transferred to specific proteins. Several types of bacteria have been shown to produce mono(ADP-ribosyl) transferase the acceptors of which are usually specific amino acid residues in proteins in eukaryotic cells. Cholera toxin ADP-ribosylates arginine residues in G proteins, whereas pertussis toxin ADP-ribosylates a cysteine residue. Diphtheria toxin and Pseudomonas aeruginosa exotoxin A use diphthamide, a modified histidine, as the acceptor amino acid. Clostridium botulinum C3 exoenzyme is an asparagine-specific ADP-ribosyltransferase. Thus, mono(ADPribosyl) ation reactions occur at nitrogen or sulfur atoms in different amino acids to produce N- or S-glycosides.

Pierisin-1 is a potent inducer of apoptosis of mammalian cells; apoptosis is accompanied by cleavage of DNA to nucleosome units and of poly(ADP-ribose) polymerase (Watanabe et al. 1998; Kono et al. 1999). Like diphtheria toxin, Pierisin-1 is also considered to be an ADP-ribosylating toxin. Several spectral analyses and independent syntheses indicated that the acceptor site for ADP-ribosylation is N-2 of guanine base. Cytotoxic activity in extracts of pupae and adults of various kinds of butterflies and moths was tested in vitro against the human gastric carcinoma cell line, TMK-1. Among species examined, cytotoxicity was limited to Pieris rapae, Pieris napi and Pieris brassicae, while with the other butterflies and moths no activity was observed, even at high concentration. The pupae showed the strongest activity. The active principle in the pupae of Pieris rapae was heat-labile and not extractable with organic solvents. This cytotoxic factor was named pierisin (Watanabe et al. 1998) and later Pierisin-1. In addition to human gastric cancer TMK-1 cell line, Pierisin-1 showed cytotoxic effects in nine other human cancer cell lines and human umbilical vein endothelial cells (HUVECs) of the human cells (Kono et al. 1999). After incorporation of pierisin-1 into the cell by interaction of its C-terminal region with the receptor in the cell membrane, the entire protein is cleaved into the N- and C-terminal fragments with intracellular protease, and the N-terminal fragment then exhibits cytotoxicity (Kanazawa et al. 2001). Pierisin-1 is a toxic protein in mice and rats and results in a gradual decrease in body weight due to decreased food intake, relative polycythemia with low serum albumin concentration and atrophy of the thymus, spleen, seminal vesicles and adipose tissue after i.p. administration. It induced diarrhea, fusion and atrophy of the villi and dilatation of the crypts in the small intestine of BALB/c mice. However, oral administration of pierisin-1 at a dose of 10,000 μ/kg body weight did not exert any obvious effects (Shiga et al. 2006). Pierisin-1 has an A small middle dotB structure-function organization like cholera or diphtheria toxin, where the "A" domain (N-terminal) exhibits ADP-ribosyltransferase activity. The target molecule for ADP-ribosylation by pierisin-1 in the presence of β-[adenylate-32P]NAD was found DNA as the acceptor, but not protein as is the case with other bacteria-derived ADP-ribosylating toxins such as such as cholera toxin and pertussis toxin. Thus, the targets for ADP-ribosylation by pierisin-1 were concluded to be 2′-deoxyguanosine residues in DNA. Pierisin-1 efficiently catalyzes the ADP-ribosylation of double-stranded DNA. The ADP-ribose moiety of NAD is transferred by pierisin-1 to the amino group at N2 of the deoxyguanosine base (Takamura-Enya et al. 2001; Kanazawa et al. 2001). These findings opened a new field regarding the biological significance of ADP-ribosylation (Takamura-Enya et al. 2001).

Pierisin-1 induced apoptosis in mammalian cells is accompanied by a release of cytochrome C and activation of a variety of caspases, and this apoptosis was inhibited by over-expression of Bcl-2 (Watanabe et al. 2002; Kanazawa et al. 2002). Pierisin-1 treatment primarily activates ATR pathway and eventually activates ATM pathway as a result of the induction of apoptosis. It was suggested that mono-ADP-ribosylation of DNA causes a specific type of fork blockage that induces checkpoint activation and signaling (Shiotani et al. 2006).

14.6.1.2 Cytotoxicity of Butterflies Extracts Against Cancer Cells

Cells cytotoxicity has been studied in 18 kinds of butterflies against TMK-1 cells. Positive results have been obtained with extracts from Pieris rapae, Pieris brassicae, and Pieris napi among the genus Pieris. However, no cytotoxicity was observed in the other extracts from examined butterflies: Eurema hecabe, Colias erate, and Hebomoia glaucippe of the family Pieridae; Papilio bianor, Papilio helenus, Papilio maackii, Papilio machaon, Papilio protenor, and Papilio xuthus of the family Papilionidae; Dichorragia nesimachus, Vanessa indica, Sasakia charonda, and Hestina japonica of the family Nymphalidae; Celastrina argiolus and Lycaena phlaeas of the family Lycaenidae (Matsumoto et al. 2008). In further study (Matsumoto et al. 2008), crude extracts from 20 other species of Pieridae family were examined for cytotoxicity in HeLa cells and DNA ADP-ribosylating activity. Both activities were detected in extracts from 13 species: subtribes Pierina. All of these extracts contained substances recognized by anti-pierisin-1 antibodies, with a molecular mass of ≈100 kDa close to pierisin-1. Extracts from seven species, Appias lyncida, Leptosia nina, Anthocharis scolymus, Eurema hecabe, Catopsilia pomona, Catopsilia scylla, and Colias erate, showed neither cytotoxicity nor DNA ADP-ribosylating activity, anti-pierisin-1 immune activity. Thus, pierisin-like proteins, showing cytotoxicity and DNA ADP-ribosylating activity, are suggested to be present in the extracts from butterflies not only among the subtribe Pierina, but also among the subtribes Aporiina and Appiadina. These findings offer insight to understanding the nature of DNA ADP-ribosylating activity in the butterfly.

14.6.1.3 Molecular Cloning of Pierisin-1

The pierisin gene encodes an 850-amino acid protein with a molecular weight of 98,081. The expressed protein induced apoptosis in human gastric carcinoma TMK-1 and cervical carcinoma HeLa cells, like the native protein, indicating functional activity. The deduced amino acid sequence of pierisin showed 32% homology with a 100-kDa mosquitocidal toxin from *Bacillus sphaericus* SSII-1. In addition, pierisin showed regional sequence similarities with ADP-ribosylating toxins, such as the A subunit of cholera toxin. A glutamic acid residue at the putative NAD-binding site, conserved in all ADP-ribosylating toxins, was also found in pierisin. Substitution of another amino acid for glutamic acid 165 resulted in a great decrease in cytotoxicity and induction of apoptosis. Moreover, inhibitors of ADP-ribosylating enzymes reduced pierisin-induced apoptosis. Hence, pierisin might possess ADP-ribosylation activity that leads to apoptosis of the cells (Watanabe et al. 1999).

14.6.1.4 Enzymatic Properties of Pierisin-1

ADP-ribosyltransferase catalyzes the transfer of an ADP-ribosyl moiety of NAD to specific proteins. ADP-ribosyltransferase showed a K_m for NAD of 0.17 mM and kcat of 55 s^{-1} (Watanabe et al. 2004a). Binding of C-terminal region of pierisin-1 to glycosphingolipid Gb3 and Gb4 receptors on cell membrane is necessary for incorporation into cells, while the N-terminal polypeptide catalyzes transfer of the ADP-ribose moiety of NAD at N2 of dG in DNA. Resulting DNA adducts cause mutation if they are present at low levels. If the DNA damage is more severe, the cells undergo apoptosis. Shiotani et al. (2005) examined the repair system for ADP-ribosylated dG adducts using nucleotide excision repair (NER) mutants of CHO cells. It is suggested that the NER system is involved in the repair of ADP-ribosylated dG adducts in DNA. Kawanishi et al. (2007) examined the involvement of NER system in the removal of N2-ADPR-dG in E. coli and human cells and suggested the involvement of the NER system in the repair of N2-ADPR-dG in both E. coli and human cells.

Similar to pierisin-1 and -2, crude extracts from the clams Meretrix lamarckii (shellfish), Ruditapes philippinarum, and Corbicula japonica incubated with calf thymus DNA and β-NAD results in production of N2-(ADP-ribos-1-yl)-2′-deoxyguanosine. The CARP-1 showed no homology with pierisin-1 or -2. However, a glutamic acid residue (E128) at NAD-binding site was conserved in in CARP-1. Although the CARP-1 in the culture medium showed no cytotoxicity against HeLa and TMK-1 cells, introduction of this protein by electroporation induced apoptosis in these cells (Nakano et al. 2006).

14.6.2 Pierisin-2, Pierisin-3 and -4

Pierisin from P. rapae is called pierisin-1, and that from P. brassicae is called pierisin-2. Pierisin-2 is a cytotoxic and apoptosis-inducing protein present in Pieris brassicae and purified from pupae. The cDNA encodes an 850-amino-acid protein with a molecular mass of 97,986. The deduced amino-acid sequence of pierisin-2 was 91% identical with that of pierisin-1. The results from site-directed mutagenesis at Glu165, a conserved residue among ADP-ribosylating enzymes necessary for NAD binding, and from experiments with ADP-ribosylating enzyme inhibitors suggested that pierisin-2 could be considered as an ADP-ribosylating toxin like pierisin-1 (Matsushima-Hibiya et al. 2000). Like pierisin-1, pierisin-2 also catalyzed ADP-ribosylation of dG in DNA to give the same reaction product as demonstrated for pierisin-1. With oligonucleotides as substrates, ADP-ribosylation by pierisin-2 was suggested to occur by one-side attack of the carbon atom at 1 position of the ribose moiety in NAD toward N2 of dG (Takamura-Enya et al. 2004).

Pierisin-like proteins are found in subtribes Pierina, Aporiina and Appiadina. The nucleotide sequences of Pierisin-3 and -4 encode an 850 and an 858 amino acid protein, respectively. The partial peptide sequences of Pierisin-3 and -4 purified from pupae were identical to the deduced amino acid sequence of ORF. Pierisin-3 showed 93% similarity to Pierisin-1 and 64% similarity to Pierisin-4 in amino acid sequences. Pierisin-3 and -4 synthesized in vitro exhibited apoptosis-inducing activity against human cervical carcinoma HeLa and human gastric carcinoma TMK-1 cells. Site-directed mutagenesis at a glutamic acid residue comprising the NAD-binding site resulted in a decrease in cytotoxicity of both proteins. Moreover, proteins with calf thymus DNA and β-NAD resulted in the formation of N2-(ADP-ribos-1-yl)-2′-deoxyguanosine, as in Pierisin-1 and -2. Results suggest apoptosis-inducing ability and molecular evolution of Pierisin-like proteins in family Pieridae (Yamamoto et al. 2009).

14.7 Discoidin Domain and Carbohydrate-Binding Module

14.7.1 The Discoidin Domain

Discoidin domain (DS) (also known as F5/8 type C domain, or C2-like domain) was first identified in discoidin proteins of *Dictyostelium discoideum* and subsequently found in a variety of extracellular and membrane proteins including the blood coagulation Factor V and Factor VIII, milk fat globule protein, neuropilins, neurexin IV, and discoidin domain receptor proteins (Kiedzierska et al. 2007; Baumgartner et al. 1998; Pratt et al. 1999). The DS domain is a structural and functional motif that is appended, singly or in tandem, to various eukaryotic and prokaryotic proteins. The first DS domain in the amoeba *Dictyostelium discoideum* was described as a lectin

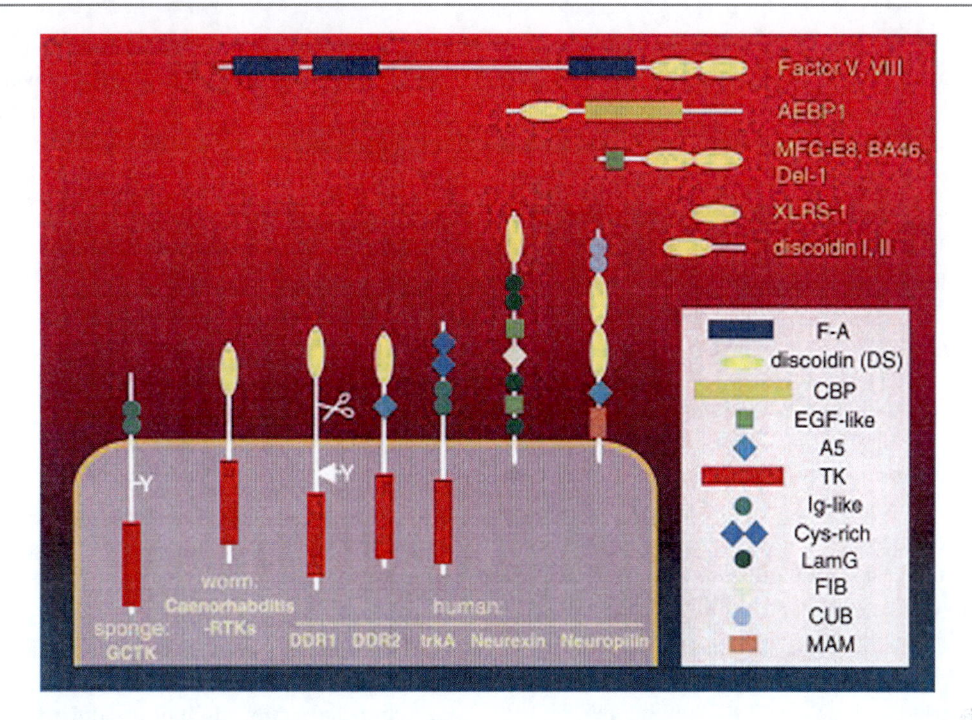

Fig. 14.5 The family of proteins with DS domains. Schematic representation of transmembrane and secreted proteins. Abbreviations for the domains shown are as follows: *F-A* A-domain in Factor V and VIII, *CBP* carboxypeptidase, *EGF* epidermal growth factor, *TK* tyrosine kinase, *A5* homology to A5 antigen, *Ig* immunoglobulin, *LamG* laminin-G, *FIB* fibronectin-like, *CUB* complement binding, *MAM* meprin/A5/PTPmu. The tyrosines in the N-P-X-Y motives of GCTK and DDR1b are highlighted. Proteolytic processing of DDR1 is indicated (Adapted by permission from Vogel 1999 © The Federation of American Societies for Experimental Biology)

with high affinity for galactose and galactose derivatives (Sauer et al. 1997). Since then, high resolution structures of a number of discoidin domains have been determined (Pratt et al. 1999; Macedo-Ribeiro et al. 1999; Lee et al. 2003). In all cases, the domain consists of a 5-strand antiparallel β-sheet that packs against a 3-strand antiparallel β-sheet to form a core barrel-like structure. An intramolecular disulfide bond between conserved cysteine residues at the beginning and end of the discoidin domain stabilizes the three dimensional structure. At the opposite end spikes or loops protrude from the core β-barrel structure to define a groove or cleft that serves as the ligand binding site (Fig. 14.5).

Kiedzierska et al. (2007) outlined the biological role of this module in various eukaryotic proteins. The DS domain binds a wide variety of ligand molecules, including phospholipids, carbohydrates, and partner proteins, thus enabling its cognate protein to participate in various physiological functions such as cellular adhesion, migration, neural development, and nutrition assimilation. Discoidin domain receptors interact with collagen to regulate cell proliferation and extracellular matrix modeling via activation of their tyrosine kinase activities (Vogel et al. 2006). A subgroup of the domain possessing carbohydrate-binding ability is also classified as the carbohydrate-binding module family 32 (CBM32). Co-crystallizations of CBM32 members and their ligands, such as the module of Clostridium perfringens N-acetylglucosaminidase with β-galactosyl-1,4-β-N-acetylglucosamine or the module of Micromonospora viridifaciens sialidase with lactose, demonstrate that the protruding loops form the ligand binding site (Macedo-Ribeiro et al. 1999; Lee et al. 2003). Discoidin proteins from *Dictyostelium discoideum* function as lectins with a high affinity for galactose residues to promote cell aggregation (Poole et al. 1981). Factors, V and -VIII bind phosphatidylserine on the surface of platelets and endothelial cells through their discoidin domains as a crucial step in initiating the blood coagulation cascade (Fuentes-Prior et al. 2002). The discoidin domains of SED1 mediate the interaction of sperm and egg as part of the fertilization process (Ensslin and Shur 2003).

14.7.2 Discoidins from *Dictyostelium discoideum* (DD)

The cell adhesion protein discoidin from cellular slime mould *Dictyostelium discoideum* has a binding site for carbohydrate residues related to galactose. The lectin, that consists of two distinct species (discoidins I and II), is synthesised as the cells differentiate from vegetative to aggregation phase and was originally thought to be involved in intercellular adhesion, but discoidin I is now thought to be involved in adhesion to the substratum by a mechanism resembling that of fibronectin in animals. The social amoeba *D. discoideum* (Slime mold) adopts a cohesive stage upon starvation and then produces Discoidin I and II, two proteins able to bind galactose and N-acetyl-galactosamine. The carbohydrate-binding protein (discoidin) agglutinated formalinized sheep erythrocytes and synthesized by *D. discoideum* cells when cells were deprived of food. Agglutination of erythrocytes by this protein was inhibited by N-acetyl-D-galactosamine, D-galactose, and Lfucose, but other monosaccharides had little or no effect. It appears to be present on the surface of cohesive but not vegetative slime-mold cells. The possibility that this protein may mediate intercellular adhesion in Dictyostelium has been suggested (Rosen et al. 1973; Kamohara et al. 2001). The discoidin domain is a approximately 150 amino acid motif common in both eukaryotic and prokaryotic proteins and arranged into a β-sandwich fold with several flexible loops. Presumably, the β-sandwich fold is stabilized predominantly by hydrophobic interactions. The variability within the loops has been suggested to account for the diverse binding spectrum of the DS domain. The cell adhesion protein discoidin, a related domain, named discoidin I-like domain (DLD, or D) has been found to share a common C-terminal region of about 110 amino acids with the FA5/8C domain, but whose N-terminal 40 amino acids are much less conserved. Discoidin domain is major domain of many blood coagulation factors, as blood coagulation factors V and VIII. In coagulation factors V and VIII the repeated domains compose part of a larger functional domain which promotes binding to anionic phospholipids on the surface of platelets and endothelial cells. The C-terminal domain of the second FA5/8C repeat (C2) of coagulation factor VIII has been shown to be responsible for phosphatidylserine-binding and essential for activity. It forms an amphipathic α-helix, which binds to the membrane. FA5/8C contains two conserved cysteines in most proteins, which link the extremities of the domain by a disulfide bond. A further disulfide bond is located near C-terminal of second FA58C domain in MFGM Q08431. Ricin B chain and discoidin I share a common primitive protein fold (Robertus and Ready 1984).

Similar domains have been detected in other extracellular and membrane proteins. The X-ray structures of the wild-type and recombinant Discoidin II in unliganded state and in complex with monosaccharides revealed that the protein forms a homotrimer which presents two binding surfaces situated on the opposite boundaries of the structure. The binding sites of N-terminal domain contain PEG molecules that could mimic binding of natural ligand. The C-terminal lectin domain interactions with N-acetyl-D-galactosamine and methyl-β-galactoside have been reported. The carbohydrate binding sites are located at the interface between monomers. Specificity for galacto configuration can be rationalized since the axial O_4 hydroxyl group is involved in several hydrogen bonds with protein side chains. Results highlight the structural differentiation of the DS domain involved in many cell-adhesion processes from the lectin activity of *Dictyostelium discoidins (*Aragão et al. 2008).

14.7.3 Discoidin Domain Receptors (DDR1 and DDR2)

Two mammalian receptor tyrosine kinases (DDR1 and DDR2) have extracellular domains closely related to lectin, discoidin, required for cell aggregation. The subgroup of DDRs is distinguished from other members of the receptor tyrosine kinase (RTK) family by a discoidin homology repeat in their extracellular domains that is also found in a variety of other transmembrane and secreted proteins. Two mammalian receptor tyrosine kinases (DDR1 and DDR2) have extracellular domains closely related to a *D. discoideum* lectin, discoidin, required for cell aggregation. DDR1 and DDR2 are detected as 125- and 130-kDa glycosylated proteins in a Western blot of lysates from overexpressing cells (Alves et al. 1995). The mammalian DDR receptors bind and get activated by specific types of collagen. Stimulation of DDR receptor tyrosine kinase requires the native triple-helical structure of collagen and occurs over an extended period of time. Collagen activation of DDR1 induces phosphorylation of a docking site for Shc phosphotyrosine binding domain, whose presence is controlled by alternative splicing. Activation of DDR2 by collagen results in the up-regulation of matrix metalloproteinase-1 expression. Thus the discoidin-related DDR tyrosine kinases are novel collagen receptors with the potential to control cellular responses to the extracellular matrix (Vogel et al. 1997).

The cDNAs of RTK subfamily have been cloned from different species by many laboratories. They represent two distinct genes, which have now been renamed DDR1 (Vogel et al. 1997) and DDR2 (Alves et al. 1995; Lai and Lemke 1994; Karn et al. 1993; Playford et al. 1996). It is surprising to note that the closest relative to the tyrosine kinase domain of DDR1 is found in the genome of the marine sponge *Geodia cydonium*. The multiple sequence alignment shows that the catalytic core region of DDR1 is 59% identical to the Geodia tyrosine kinase called GCTK (61% for DDR2), whereas the closest mammalian RTK subfamily, the neurotrophin receptors, are 55–58% related (Gamulin et al. 1997). A similar Ig-domain repeat is found in the mammalian nerve growth factor receptors, which are the second-most-related RTK subfamily to the Geodia sequence. Thus far, five isoforms of the DDR1 protein have been characterized, that arise by alternative splicing (Alves et al. 2001). The longest DDR1 transcript codes for the full-length 919 amino acid long receptor (c-isoform), whereas the a- or b-receptor isoforms lack 37 or 6 amino acids in the juxtamembrane or kinase domain, respectively (Alves et al. 1995). The DDR1b protein is the predominant isoform expressed during embryogenesis, whereas the a-isoform is upregulated in certain human mammary carcinoma cell lines (Johnson et al. 1993; Perez et al. 1996). Results suggest that the discoidin-related DDR tyrosine kinases are novel collagen receptors with the potential to control cellular responses to the extracellular matrix (Vogel 1999, 2001). The revelation that collagens function as ligands for DDR1 and DDR2 suggests that some of the other mammalian DS domains may interact with matrix proteins as well.

14.7.4 Earth Worm (EW)29 Lectin

The galactose-binding lectin EW29 from the earthworm Lumbricus terrestris is composed of two homologous domains, both of which are members of the R-type lectin family. The EW29 shows haemagglutination activity and is composed of a single peptide chain that includes two homologous domains: N-terminal and C-terminal domains (14,500 Da) and show 27% identity with each other. A truncated mutant of EW29 comprising the C-terminal domain (rC-half) has haemagglutination activity by itself. In order to clarify how rC-half recognizes ligands and shows haemagglutination activity, X-ray crystal structures of rC-half in complex with D-lactose and N-acetyl-D-galactosamine have been determined. The structure of rC-half is similar to that of the ricin B chain and consists of a β-trefoil fold; the fold is further divided into three similar subdomains referred to as subdomains α, β and γ, which are gathered around the pseudo-threefold axis. The structures of sugar complexes demonstrated that subdomains α and γ of rC-half bind terminal galactosyl and N-acetylgalactosaminyl glycans. The sugar-binding properties are common to both ligands in both subdomains and are quite similar to those of ricin B chain-lactose complexes. These results indicate that the C-terminal domain of EW29 uses these two galactose-binding sites for its function as a single-domain-type haemagglutinin (Suzuki et al. 2009; Hemmi et al. 2009). The truncated mutant rC-half comprising C-terminal domain was crystallized by the hanging-drop vapour-diffusion method. The crystal belonged to space group P4(3)2(1)2, with unit-cell parameters a = b = 61.2, c = 175.6 A, and diffracted to beyond 1.9 A resolution. Matthews coefficient calculations suggested that this crystal contained two molecules per asymmetric unit (Suzuki et al. 2004). Since ricin B chain and discoidin I share a common primitive protein fold (Robertus and Ready 1984), presence of β-trefoil fold in discoidin domain containing lectins is subject of further investigations.

References

Alves F, Vogel W, Mossie K et al (1995) Distinct structural characteristics of discoidin I subfamily receptor tyrosine kinases and complementary expression in human cancer. Oncogene 10:609–618

Alves F, Saupe S, Ledwon M et al (2001) Identification of two novel, kinase-deficient variants of discoidin domain receptor 1: differential expression in human colon cancer cell lines. FASEB J 15: 1321–1323

Amado M, Almeida R, Schwientek T, Clausen H (1999) Identification and characterization of large galactosyltransferase gene families: Galactosyltransferases for all functions. Biochim Biophys Acta 1473:35–53

Aragão KS, Satre M, Imberty A, Varrot A (2008) Structure determination of Discoidin II from *Dictyostelium discoideum* and carbohydrate binding properties of the lectin domain. Proteins 73:43–52

Baumgartner S, Hofmann K, Chiquet-Ehrismann R, Bucher P (1998) The discoidin domain family revisited: new members from prokaryotes and a homology-based fold prediction. Protein Sci 7:1626–1631

Bennett EP, Hassan H, Mandel U et al (1998) Cloning of a human UDP-N-acetyl-α-D-galactosamine: polypeptide N-acetylgalactosaminyl transferase that complements other PpGalNAc-Ts in complete O-glycosylation of the MUC1 tandem repeat. J Biol Chem 273: 30472–30481

Boraston AB, Tomme P, Amandoron EA, Kilburn DG (2000) A novel mechanism of xylan binding by a lectin-like module from *Streptomyces lividans* xylanase 10A. Biochem J 350(Pt 3):933–941

Carpusca I, Jank T, Aktories K (2006) *Bacillus sphaericus* mosquitocidal toxin (MTX) and pierisin: the enigmatic offspring from the family of ADP-ribosyltransferases. Mol Microbiol 62:621–630

Cummings RD (2009) Etzler ME. In: Varki A, Cummings RD, Esko JD, Freeze HH, Stanley P, Bertozzi CR, Hart GW, Etzler ME (eds) Essentials of glycobiology, 2nd edn. Cold Spring Harbor Laboratory Press, Cold Spring Harbor

East L, Rushton S, Taylor ME, Isacke CM (2002) Characterization of sugar binding by the mannose receptor family member, Endo180. J Biol Chem 277:50469–50475

Ensslin MA, Shur BD (2003) Identification of mouse sperm SED1, a bimotif EGF repeat and discoidin-domain protein involved in sperm-egg binding. Cell 114:405–417

Freire T, Fernández C, Chalar C et al (2004) Characterization of a UDP-N-acetyl-D-galactosamine:polypeptide N-acetylgalactosaminyltransferase with an unusual lectin domain from the platyhelminth parasite *Echinococcus granulosus*. Biochem J 382:501–510

Fritz TA, Hurley JH, Trinh LB, Shiloach J, Tabak LA (2004) The beginnings of mucin biosynthesis: the crystal structure of UDP-GalNAc:polypeptide α-N-acetylgalactosaminyltransferase-T1. Proc Natl Acad Sci USA 101:15307–15312

Fritz TA, Raman J, Tabak LA (2006) Dynamic association between the catalytic and lectin domains of human UDP-GalNAc:polypeptide α-N-acetylgalactosaminyltransferase-2. J Biol Chem 281:8613–8619

Fuentes-Prior P, Fujikawa K, Pratt KP (2002) New insights into binding interfaces of coagulation factors V and VIII and their homologues lessons from high resolution crystal structures. Curr Protein Pept Sci 3:313–339

Fujimoto Z, Kuno A, Kaneko S et al (2002) Crystal structures of the sugar complexes of Streptomyces olivaceoviridis E-86 xylanase: sugar binding structure of the family 13 carbohydrate binding module. J Mol Biol 316:65–78

Gamulin V, Skorokhod A, Kavsan V et al (1997) Experimental indication in favor of the introns-late theory: the receptor tyrosine kinase gene from the sponge *Geodia cydonium*. J Mol Evol 44: 242–252

Hagen FK, Nehrke K (1998) cDNA cloning and expression of a family of UDP-N-acetyl-D-galactosamine:polypeptide N-acetylgalactosaminyltransferase sequence homologs from *Caenorhabditis elegans*. J Biol Chem 273:8268–8277

Hagen FK, Hazes B, Raffo R et al (1999) Structure-function analysis of the UDP-N-acetyl-D-galactosamine:polypeptide N-acetylgalactosaminyltransferase. Essential residues lie in a predicted active site cleft resembling a lactose repressor fold. J Biol Chem 274: 6797–6803

Halling KC, Halling AC, Murray EE et al (1985) Genomic cloning and characterization of a ricin gene from *Ricinus communis*. Nucleic Acids Res 1:8019–8033

Hassan H, Reis CA, Bennett EP et al (2000) The lectin domain of UDP-N-acetyl-D-galactosamine: polypeptide N-acetylgalactosaminyl - transferase-T4 directs its glycopeptide specificities. J Biol Chem 275:38197–38205

Hatakeyama T, Unno H, Kouzuma Y, Uchida T, Eto S, Hidemura H, Kato N, Yonekura M, Kusunoki M (2007) C-type lectin-like carbohydrate recognition of the hemolytic lectin CEL-III containing ricin-type -trefoil folds. J Biol Chem 282:37826–37835

Hazes B (1996) The (QxW)3 domain: a flexible lectin scaffold. Protein Sci 5:1490–1501

Hemmi H, Kuno A, Ito S et al (2009) NMR studies on the interaction of sugars with the C-terminal domain of an R-type lectin from the earthworm Lumbricus terrestris. FEBS J 276:2095–2105

Imberty A, Piller V, Piller F, Breton C (1997) Fold recognition and molecular modeling of a lectinlike domain in UDP-GalNac:polypeptide N-acetylgalactosaminyltransferases. Protein Eng 10:1353–1356

Inoue K, Fujinaga Y, Honke K et al (1999) Characterization of haemagglutinin activity of Clostridium botulinum type C, D 16S toxins, and one subcomponent of haemagglutinin (HA1). Microbiology 145:2533–2542

Johnson JD, Edman JC, Rutter WJ (1993) A receptor tyrosine kinase found in breast carcinoma cells has an extracellular discoidin I-like domain. Proc Natl Acad Sci USA 90:5677–5681

Kamohara H, Yamashiro S, Galligan C, Yoshimura T (2001) Discoidin domain receptor 1 isoform-a (DDR1α) promotes migration of leukocytes in three-dimensional collagen lattices. FASEB J 15:2724–2726

Kanazawa T, Watanabe M, Matsushima-Hibiya Y et al (2001) Distinct roles for the N- and C-terminal regions in the cytotoxicity of pierisin-1, a putative ADP-ribosylating toxin from cabbage butterfly, against mammalian cells. Proc Natl Acad Sci USA 98:2226–2231

Kanazawa T, Kono T, Watanabe M et al (2002) Bcl-2 blocks apoptosis caused by pierisin-1, a guanine-specific ADP-ribosylating toxin from the cabbage butterfly. Biochem Biophys Res Commun 296:20–25

Karn T, Holtrich U, Bräuninger A et al (1993) Structure, expression and chromosomal mapping of TKT from man and mouse: a new subclass of receptor tyrosine kinases with a factor VIII-like domain. Oncogene 8:3433–3440

Katzin BJ, Collins EJ, Robertus JD (1991) Structure of ricin A-chain at 2.5 A. Proteins 10:251–259

Kawanishi M, Matsukawa K, Kuraoka I et al (2007) Molecular evidence of the involvement of the nucleotide excision repair (NER) system in the repair of the mono(ADP-ribosyl)ated DNA adduct produced by pierisin-1, an apoptosis-inducing protein from the cabbage butterfly. Chem Res Toxicol 20:694–700

Kiedzierska A, Smietana K, Czepczynska H, Otlewski J (2007) Structural similarities and functional diversity of eukaryotic discoidin-like domains. Biochim Biophys Acta 1774:1069–1078

Kim Y, Mlsna D, Monzingo AF et al (1992) The structure of a ricin mutant showing rescue of activity by a noncatalytic residue. Biochembtry 31:3294–3296

Kono T, Watanabe M, Koyama K et al (1999) Cytotoxic activity of pierisin, from the cabbage butterfly, Pieris rapae, in various human cancer cell lines. Cancer Lett 137:75–81

Kubota T, Shiba T, Sugioka S et al (2006) Structural basis of carbohydrate transfer activity by human UDP-GalNAc: polypeptide α-N-Acetylgalactosa-minyltransferase (pp-GalNAc-T10). J Mol Biol 359:708–727

Lai C, Lemke G (1994) Structure and expression of the tyro 10 receptor tyrosine kinase. Oncogene 9:877–883

Lee CC, Kreusch A, McMullan D et al (2003) Crystal structure of the human neuropilin-1 b1 domain. Structure 11:99–108

Liu Y, Chirino AJ, Misulovin Z, Leteux C, Feizi T, Nussenzweig MC, Bjorkman PJ (2000) Crystal structure of the cysteine-rich domain of mannose receptor complexed with a sulfated carbohydrate ligand. J Exp Med 191:1105–1116

Llorca O (2008) Extended and bent conformations of the mannose receptor family. Cell Mol Life Sci 65:1302–1310

Lord JM, Roberts LM, Robertus JD (1994) Ricin: structure, mode of action, and some current applications. FASEB J 8:201–208

Lord MJ, Jolliffe NA, Marsden CJ et al (2003) Ricin. Mechanisms of cytotoxicity. Toxicol Rev 22:53–64

Macedo-Ribeiro S, Bode W, Huber R et al (1999) Crystal structures of the membrane-binding C2 domain of human coagulation factor V. Nature 402:434–439

Matsumoto Y, Nakano T, Yamamoto M et al (2008) Distribution of cytotoxic and DNA ADP-ribosylating activity in crude extracts from butterflies among the family Pieridae. Proc Natl Acad Sci USA 105:2516–2520

Matsushima-Hibiya Y, Watanabe M, Kono T et al (2000) Purification and cloning of pierisin-2, an apoptosis-inducing protein from the cabbage butterfly, Pieris brassicae. Eur J Biochem 267:5742–5750

Matsushima-Hibiya Y, Watanabe M, Hidari KI et al (2003) Identification of glycosphingolipid receptors for pierisin-1, a guanine-specific ADP-ribosylating toxin from the cabbage butterfly. J Biol Chem 278:9972–9978

Mlsna D, Monzingo AF, Katzin BJ et al (1993) Structure of recombinant ricin A chain at 2.3 A. Protein Sci 2:429–435

Nakano T, Matsushima-Hibiya Y, Yamamoto M et al (2006) Purification and molecular cloning of a DNA ADP-ribosylating protein, CARP-1, from the edible clam Meretrix lamarckii. Proc Natl Acad Sci USA 103:13652–13657

Napper CE, Drickamer K, Taylor ME (2006) Collagen binding by the mannose receptor mediated through the fibronectin type II domain. Biochem J 395:579–586

Notenboom V, Boraston AB, Williams SJ, Kilburn DG, Rose DR (2002) High-resolution crystal structures of the Lectin-like Xylan binding domain from *Streptomyces lividans* Xylanase 10A with bound substrates reveal a novel mode of Xylan binding. Biochemistry 4:4246–4254

Perez JL, Jing SQ, Wong TW (1996) Identification of two isoforms of the Cak receptor kinase that are coexpressed in breast tumor cell lines. Oncogene 12:1469–1477

Playford MP, Butler RJ, Wang XC et al (1996) The genomic structure of discoidin receptor tyrosine kinase. Genome Res 6:620–627

Poole S, Firtel RA, Lamar E, Rowekamp W (1981) Sequence and expression of the discoidin I gene family in *Dictyostelium discoideum.* J Mol Biol 153:273–289

Pratt KP, Shen BW, Takeshima K et al (1999) Structure of the C2 domain of human factor VIII at 1.5 A resolution. Nature 402:439–442

Raman J, Fritz TA, Gerken TA et al (2008) The catalytic and lectin domains of UDP-GalNAc:polypeptide α-N Acetylgalactosaminyltransferase function in concert to direct glycosylation site selection. J Biol Chem 283:22942–22951

Robertus JD, Ready MP (1984) Ricin B chain and discoidin I share a common primitive protein fold. J Biol Chem 259:13953–13956

Rosen SD, Kafka JA, Simpson DL et al (1973) Developmentally regulated, carbohydrate-binding protein in *dictyostelium discoideum.* Proc Nat Acad Sci USA 70:2554–2557

Rutenber E, Robertus JD (1991) The structure of ricin B chain at 2.5 A resolution. Proteins Struct Funct Genet 10:260–269

Rutenber E, Katzin BJ, Collins EJ et al (1991) The crystallographic refinement of ricin at 2.5 A resolution. Proteins Struct Funct Genet 10:240–250

Sauer CG, Gehrig A, Warneke-Wittstock R et al (1997) Positional cloning of the gene associated with X-linked juvenile retinoschisis. Nat Genet 17:164–170

Schärpf M, Connelly GP, Lee GM et al (2002) Site-specific characterization of the association of xylooligosaccharides with the CBM13 lectin-like xylan binding domain from *Streptomyces lividans* xylanase 10A by NMR spectroscopy. Biochemistry 41:4255–4263

Schwientek T, Bennett EP, Flores C et al (2002) Functional conservation of subfamilies of putative UDP-N-acetylgalactosamine: polypeptide N-acetylgalacto-saminyltransferases in *Drosophila, Caenorhabditis elegans*, and mammals—One subfamily composed of l(2)35Aa is essential in *Drosophila.* J Biol Chem 277:22623–22638

Sharon N, Lis H (2004) History of lectins: from hemagglutinins to biological recognition molecules. Glycobiology 14:53R–62R

Shiga A, Kakamu S, Sugiyama Y et al (2006) Acute toxicity of pierisin-1, a cytotoxic protein from Pieris rapae, in mouse and rat. J Toxicol Sci 31:123–137

Shiotani B, Watanabe M, Totsuka Y et al (2005) Involvement of nucleotide excision repair (NER) system in repair of mono ADP-ribosylated dG adducts produced by pierisin-1, a cytotoxic protein from cabbage butterfly. Mutat Res 572:150–155

Shiotani B, Kobayashi M, Watanabe M et al (2006) Involvement of the ATR- and ATM-dependent checkpoint responses in cell cycle arrest evoked by pierisin-1. Mol Cancer Res 4:125–133

Stirpe F, Barbieri L, Battelli MG, Soria M, Lappi DA (1992) Ribosome-inactivating proteins from plants: present status and future prospects. Biotechnology 10:405–412

Stwora-Wojczyk MM, Dzierszinski F, Roos DS et al (2004) Functional characterization of a novel *Toxoplasma gondii* glycosyltransferase: UDP-N-acetyl-D-galactosamine: polypeptide N-acetylgalactosaminyltransferase-T3. Arch Biochem Biophys 426:231–240

Suzuki R, Fujimoto Z, Kuno A et al (2004) Crystallization and preliminary X-ray crystallographic studies of the C-terminal domain of galactose-binding lectin EW29 from the earthworm Lumbricus terrestris. Acta Crystallogr D Biol Crystallogr 60(Pt 10):1895–1896

Suzuki R, Kuno A, Hasegawa T et al (2009) Sugar-complex structures of the C-half domain of the galactose-binding lectin EW29 from the earthworm Lumbricus terrestris. Acta Crystallogr D Biol Crystallogr 65(Pt 1):49–57

Takamura-Enya T, Watanabe M, Totsuka Y et al (2001) Mono(ADP-ribosyl)ation of 2′-deoxyguanosine residue in DNA by an apoptosis-inducing protein, pierisin-1, from cabbage butterfly. Proc Natl Acad Sci USA 98:12414–12419

Takamura-Enya T, Watanabe M, Koyama K et al (2004) Mono(ADP-ribosyl)ation of the N2 amino groups of guanine residues in DNA by pierisin-2, from the cabbage butterfly, Pieris brassicae. Biochem Biophys Res Commun 323:579–582

Ten Hagen KG, Tran DT (2002) A UDP-GalNAc:polypeptide N-acetylgalactosaminyltransferase is essential for viability in *Drosophila melanogaster.* J Bio Chem 277:22616–22622

Ten Hagen KG, Bedi GS, Tetaert D et al (2001) Cloning and characterization of a ninth member of the UDPGalNAc: polypeptide N-acetylgalactosaminyltransferase family, ppGalNAc-T-T9. J Biol Chem 276:17395–17404

Ten Hagen KG, Fritz TA, Tabak LA (2003) All in the family: the UDP-GalNAc:polypeptide N-acetylgalactosaminyl- transferases. Glycobiology 13:1R–16R

Tenno M, Saeki A, Kezdy FJ et al (2002) The lectin domain of UDPGalNAc: polypeptide N-acetylgalactosaminyltransferase 1 is involved in O-glycosylation of a polypeptide with multiple acceptor sites. J Biol Chem 277:47088–47096

Topaz O, Shurman DI, Bergman R et al (2004) Mutations in GALNT3, encoding a protein involved in O-linked glycosylation, cause familial tumoral calcinosis. Nat Genet 36:579–581

Treiber N, Reinert DJ, Carpusca I, Aktories K, Schulz GE (2008) Structure and mode of action of a mosquitocidal holotoxin. J Mol Biol 381:150–159

Uchida T, Yamasaki T, Eto S et al (2004) Crystal structure of the hemolytic lectin CEL-III isolated from the marine invertebrate Cucumaria echinata: implications of domain structure for its membrane pore-formation mechanism. J Biol Chem 279:37133–37141

Vogel W (1999) Discoidin domain receptors: structural relations and functional implications. FASEB J 13:S77–S82

Vogel WF (2001) Collagen-receptor signaling in health and disease. Eur J Dermatol 11:506–514

Vogel W, Gish GD, Alves F et al (1997) The discoidin domain receptor tyrosine kinases are activated by collagen. Mol Cell 1:13–23

Vogel WF, Abdulhussein R, Ford CE (2006) Sensing extracellular matrix: an update on discoidin domain receptor function. Cell Signal 18:1108–1116

Wandall HH, Irazoqui F, Tarp MA et al (2007) The lectin domains of polypeptide PpGalNAc-Ts exhibit carbohydrate-binding specificity for GalNAc: lectin binding to GalNAc-glycopeptide substrates is required for high density GalNAc-O-glycosylation. Glycobiology 17:374–387

Watanabe M, Kono T, Koyama K et al (1998) Purification of pierisin, an inducer of apoptosis in human gastric carcinoma cells, from cabbage butterfly, Pieris rapae. Jpn J Cancer Res 89:556–561

Watanabe M, Kono T, Matsushima-Hibiya Y et al (1999) Molecular cloning of an apoptosis-inducing protein, pierisin, from cabbage butterfly: possible involvement of ADP-ribosylation in its activity. Proc Natl Acad Sci USA 96:10608–10613

Watanabe M, Takamura-Enya T, Kanazawa T et al (2002) Mono(ADP-ribosyl)ation of DNA by apoptosis-inducing protein, pierisin. Nucleic Acids Res Supp 2:243–244

Watanabe M, Enomoto S, Takamura-Enya T et al (2004a) Enzymatic properties of pierisin-1 and its N-terminal domain, a guanine-specific

ADP ribosyltransferase from the cabbage butterfly. J Biochem 135:471–477
Watanabe M, Nakano T, Shiotani B et al (2004b) Developmental stage-specific expression and tissue distribution of pierisin-1, a guanine-specific ADP-ribosylating toxin, in Pieris rapae. Comp Biochem Physiol A Mol Integr Physiol 139:125–131
Wojczyk BS, Stwora-Wojczyk MM et al (2003) cDNA cloning and expression of UDP-N-acetyl-D-galactosamine:polypeptide N-acetylgalactosaminyl -transferase T1 from *Toxoplasma gondii*. Mol Biochem Parasitol 131:93–107
Yamamoto M, Nakano T, Matsushima-Hibiya Y et al (2009) Molecular cloning of apoptosis-inducing Pierisin-like proteins, from two species of white butterfly, Pieris melete and Aporia crataegi. Comp Biochem Physiol B Biochem Mol Biol 154:326–333
Yan X, Hollis T, Svinth M et al (1997) Structure-based identification of a ricin inhibitor. J Mol Biol 266:1043–1049

Mannose Receptor Family: R-Type Lectins 15

Rajesh K. Gupta and G.S. Gupta

15.1 R-Type Lectins in Animals

R-type lectins exist ubiquitously in nature and mainly bind to galactose unit of sugar chains. Originally found in plant lectin, Ricin, the R-type lectin domain is found in several animal lectins, including the members of mannose receptor (MR) family, and in some invertebrate lectins (discussed in Chap. 14). The R-type domain contained in these proteins is the CRD, which is also termed a carbohydrate-binding module (CBM) and has been placed in the CBM13 family in the CAZy database (carbohydrate-active enzymes database). While the A chain in ricin has eight α-helices and eight β-strands, and is the catalytic subunit, the B chain contains R-type lectin domains, has two tandem CRDs that are about 35 Å apart and have a shape resembling a barbell, with one binding domain at each end. Each R-type domain has a three-lobed organization that is a β-trefoil structure (from the Latin trifolium meaning "three-leaved plant"). The β-trefoil structure probably arose evolutionarily through gene fusion events linking a 42-amino-acid peptide subdomain that has galactose-binding activity. The three lobes are termed α, β, and γ and are arranged around a threefold axis (See Fig. 14.2; Chap. 14). Conceivably, each lobe could be an independent binding site, but in most R-type lectins only one or two of these lobes retain the conserved amino acids required for sugar binding. The R-type domain is also found in pierisin-1, which is a cytotoxic protein from the cabbage butterfly Pieris rapae, and in the homologous protein pierisin-2, from Pieris brassicae. Tandem R-type motifs are found in some other R-type family members. For example, Limulus horseshoe crab coagulation factor G has a central R-type lectin domain, which is flanked at the amino terminus by a xylanase Z-like domain and at the carboxyl terminus by a glucanase-like domain. In this chapter we will restrict our discussion to R-type lectins of mannose receptor family, which comprises also of endocytic receptors.

15.2 Mannose Receptor Lectin Family

There are four known members of the MR family in humans, all of which contain an R-type lectin domain. In addition, the members of mannose receptor family have a unique structural composition due to the presence of multiple C-type lectin-like domains within a single polypeptide backbone. The four members of the mannose receptor family [the mannose receptor, the M-type phospholipase A_2 receptor, DEC-205 and Endo180 (or urokinase plasminogen activator receptor-associated protein)] share a common extracellular arrangement of an amino-terminal cysteine-rich domain related to R-type domain, followed by a fibronectin type II domain and 8–10 C-type lectin-like CRD domains within a single polypeptide. In addition, all have a short cytoplasmic domain, which mediates their constitutive recycling between the plasma membrane and the endosomal apparatus, suggesting that these receptors function to internalize ligands for intracellular delivery (See Fig. 14.3; Chap 14). However, despite the common presence of multiple lectin-like domains, these four endocytic receptors have divergent ligand binding activities, and it is clear that the majority of these domains do not bind sugars. All of the MR family members except DEC-205 recycle back to the cell surface from early endosomes, but DEC-205 recycles from late endosomes. However, each receptor has evolved to have distinct functions and distributions. These receptors are unusual among animal lectins in that they can bind ligands in either a "cis" or "trans" fashion, which means they can bind to cell-surface glycoconjugates on the same cell or to those on other cells and to soluble ligands.

G.S. Gupta et al., *Animal Lectins: Form, Function and Clinical Applications*,
DOI 10.1007/978-3-7091-1065-2_15, © Springer-Verlag Wien 2012

15.3 The Mannose Receptor (CD206)

15.3.1 Human Macrophage Mannose Receptor (MMR)

The CD206, also known as macrophage MR (MMR) is the best characterised member of the family of four endocytic molecules that share a common domain structure; a cysteine-rich (CR) domain related to the R-type CRD of ricin, a fibronectin-type II (FNII) domain and tandemly arranged C-type lectin-like domains (CTLD, eight in the case of MR). The MR and other members of this family are among the few mammalian glycan-binding proteins that have two separate lectin motifs (C-type and R-type) in the same molecule. This group is also unusual in that it is the only known lectin group in mammals with more than two C-type lectin domains in the same molecule. Only CTLDs 4 and 5 of the MR have been shown to bind glycans in a Ca^{2+}-dependent manner and to bind mannose, N-acetylglucosamine, and fucose. Each protein is a recycling plasma membrane receptor with a cytoplasmic domain that mediates clathrin-dependent endocytosis and uptake of extracellular glycan-containing ligands. The MR is the heavily glycosylated endocytic receptor. Glycosylation differentially affects both MR lectin activities. The glycosylation of MR, terminal sialylation in particular, could influence its binding properties at two levels: (1) it is required for mannose recognition; and (2) it modulates the tendency of MR to self-associate, effectively regulating the avidity of the CR domain for sulfated sugar ligands (Su et al. 2005). The MR is unusual protein in sense that it is the only member of the MR family that can function both in clathrin-dependent endocytosis and in the phagocytosis of nonopsinized microbes and large ligands.

The MR of macrophages (MMR), epithelial, and endothial cells acts as a molecular scavenger, binding to and internalizing a variety of pathogenic microorganisms and harmful glycoproteins. The macrophage MR is a 175–180 kDa type I transmembrane glycoprotein. The MR is expressed at high levels on hepatic endothelial cells and Kupffer cells as well as on many other endothelial and epithelial cells, macrophages, and immature dendritic cells (DCs). The MR was originally discovered in the 1980s in rabbit alveolar macrophages as a membrane protein that bound mannose-containing ligands as well as pituitary hormones such as lutropin and thyrotropin, which have 4-O-sulfated N-acetylgalactosamine residues on N-glycans (Leteux et al. 2000). The MR is part of the innate immune system and facilitates the phagocytosis of mannose-rich pathogens. It also assists leukocytes in responding appropriately to antigens by promoting trafficking to the germinal center and is also involved in antigen presentation.

The primary structure of the mannose receptor reflects its diverse carbohydrate specificity. Size and shape parameters indicate that the receptor is a monomeric, elongated and asymmetric molecule. Domain organization of the CRD-4 monomer in MR represents two possible conformations: extended and U-shaped. Hydrodynamic coefficients predicted for modeled receptor conformations are consistent with an extended conformation with close contacts between three pairs of CRDs. The N-terminal cysteine-rich domain and the fibronectin type II repeat appear to increase the rigidity of the molecule. The rigid, extended conformation of the receptor places domains with different functions at distinct positions with respect to the membrane. An N-terminal cysteine-rich domain mediates recognition of sulfated *N*-acetylgalactosamine, which is the terminal sugar of the unusual oligosaccharides present on pituitary hormones (Fiete et al. 1998; Napper et al. 2001). The extracellular domains are linked to a transmembrane region and a small cytoplasmic domain. The CRDs of the extracellular region mediate calcium-dependent binding to sugars that are commonly found on microorganisms, but rarely seen in sufficient density in terminal positions of mammalian oligosaccharides (Weis et al. 1998; Drickamer and Taylor 1993).

The gene for the human MMR is divided into 30 exons. The first three exons encode the signal sequence, the NH2-terminal cysteine-rich domain, and the fibronectin type II repeat, while the final exon encodes the transmembrane anchor and the cytoplasmic tail. The intervening 26 exons encode the eight CRDs and intervening spacer elements. The pattern of intron positions and comparison of sequences of CRDs suggests that these domains evolved by duplication (Kim et al. 1992).

15.3.2 Structure-Function Relations

Understanding the molecular basis of cell surface ligand recognition and endosomal release by the MR requires information about how individual domains interact with sugars as well as the structural arrangement of the multiple domains. The NH2-terminal cysteine-rich domain and the fibronectin type II repeat are not necessary for endocytosis of mannose-terminated glycoproteins. The CRDs 1–3 have at most very weak affinity for carbohydrate, where as, of the eight C-type CRDs, CRDs 4–8 are required for binding and endocytosis of mannose/GlcNAc/fucose-terminated ligands, but only CRD-4 has demonstrable sugar binding activity in isolation. CRD 4 shows the highest affinity binding and has multispecificity for a variety of monosaccharides. As the main mannose-recognition domain of MR (CRD4) is the central ligand binding domain of the receptor, analysis of this domain suggests ways in which multiple CRDs in whole receptor might interact with each other (Feinberg et al. 2000;

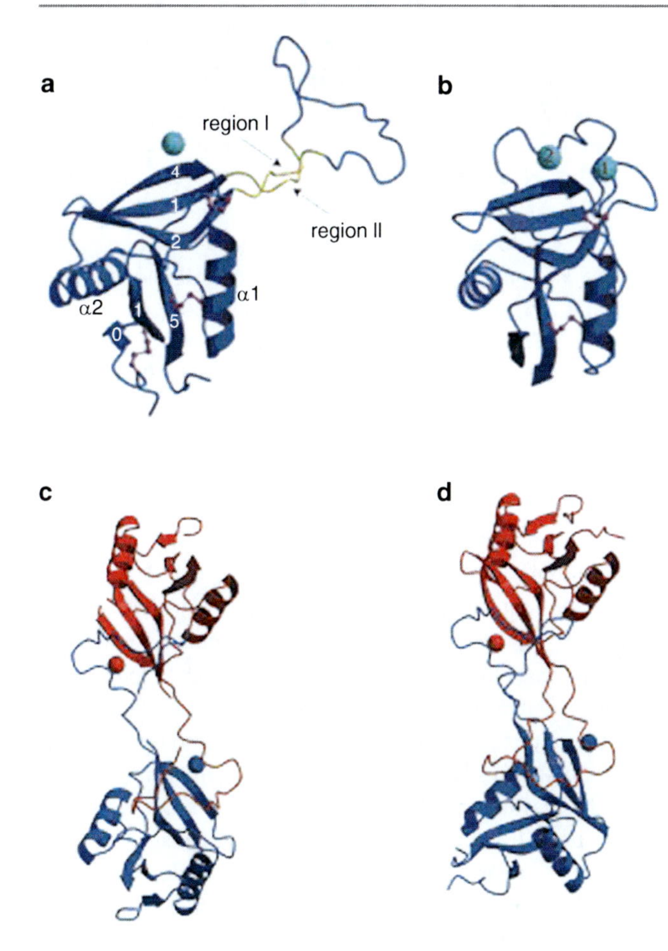

Fig. 15.1 Structure of CRD-4 monomer and comparison to rat Mannose-binding protein A. (a), Ribbon diagram of CRD-4. Disulfide bonds are shown in *pink ball-and-stick* representation, and the Ca^{2+} is shown as a *bluegreen sphere*. The two segments that connect the extended loop to the core of the CRD, region I (residues 701–708) and region II (residues 729–734), are shown in *yellow*. (**b**). Ribbon diagram of the MBP-A CRD. Ca^{2+} site 1 is the auxiliary site, and Ca^{2+} site 2 is the principal site. **Figs 14.1c and 14.1d: Domain-swapped dimer structure.** (**c**). Two molecules of copy A of CRD-4, related by a crystallographic twofold rotational symmetry axis in the lattice. (**d**). crystallographic dimer of copy **C**. The two protomers are shown in *blue* and *red*. The distal part of the extended loop of the partner protomer in the dimer, which forms part of the Ca^{2+}-binding site, is shown in *red*, next to the *blue* core of the other protomer (Adapted by permission from Feinberg et al. (2000) © The American Society for Biochemistry and Molecular Biology)

Mullin et al. 1997). However, CRD 4 alone cannot account for the binding of the receptor to glycoproteins. At least 3 CRDs (4, 5, and 7) are required for high affinity binding and endocytosis of multivalent glycoconjugates. In this respect, the MR is like other carbohydrate-binding proteins, in which several CRDs, each with weak affinity for single sugars, are clustered to achieve high affinity binding to oligosaccharides (Taylor et al. 1992). The overall structure of CRD-4 (Fig. 15.1a) is similar to other C-type CRDs, containing two α helices and two small antiparallel β sheet in MBP (Fig. 15.1b). The core region of the CRD-4 domain, consisting of β strands 1–5 and the two α helices, superimposes on the equivalent residues of the rat MBP-A CRD. The principal difference resides in the position of helix α2, which is the most variable element of secondary structure among the C-type lectin-like folds.

15.3.2.1 Domain-Swapped Dimer Structure

Extensive interactions occur between the extended loops formed by residues 701–734 of two CRD-4 molecules related by a crystallographic twofold rotational symmetry axis (Fig. 15.1c, d). Most strikingly, the most distal portion of the loop (residues 708–728) forms the upper portion of the Ca^{2+}- and sugar-binding site of the partner molecule. This domain swapping (3D domain swapping) is a mechanism for forming oligomeric proteins from their monomers), Bennett et al. (1995) produces a dimer in which each end, comprised of residues 625–700 and 735–768 from one polypeptide chain and residues 708–728 from a crystal symmetry-related molecule, has the compact fold typical of C-type CRDs (Fig. 15.1a). This "hybrid molecule" was referred to as CRD4 monomer-like (CRD4-M) (Feinberg et al. 2000). Comparison of CRD4-M with MBP (Fig. 15.1b) and E-selectin and other C-type CRDs reveals a remarkable similarity in the structure of these loops. The two crossover segments that connect residues 708–728 to the core of the CRD, residues 701–707 and residues 729–734, referred to as regions I and II, respectively, are examples of "hinge loops," which are segments of polypeptide that link the swapped domain to the remainder of the molecule and which have different conformations in the monomer and dimer (Feinberg et al. 2000).

Although the basic C-type lectin fold is preserved, a loop extends away from the core of the domain to form a domain-swapped dimer in the crystal (Fig. 15.1c, d). Combined studies on ligand binding, site-directed mutagenesis, and NMR indicated that CRD-4 of the mannose receptor has specificity for mannose, GlcNAc, and fucose like the C-type CRDs of rat serum and liver mannose-binding proteins (MBPs) (Drickamer and Taylor 1993; Weis et al. 1998; Weis et al. 1991; Wu et al. 1996). Some aspects of binding of sugar and Ca^{2+} by CRD-4 are similar to those of the mannose binding protein (MBP) CRDs, but others are different (Wu et al. 1996) (Fig. 15.1a). The structure likely represents an endosomal form of the domain formed when Ca^{2+} is lost from the auxiliary calcium site3 (Feinberg et al. 2000). Transfection of the mannose receptor cDNA into Cos-I cells is necessary for receptor-mediated endocytosis of mannose-rich glycoconjugate as well as phagocytosis of yeasts. Deletion of the cytoplasmic tail results in a mutant receptor that is able to bind but not ingest the ligated pathogens, suggesting that the signal for phagocytosis is contained in the cytoplasmic tail (Ezekowitz et al. 1990).

Size and shape parameters derived from sedimentation and diffusion coefficients suggest that the receptor is a monomeric, elongated and asymmetric molecule. Proteolysis experiments indicated the presence of close contacts between several pairs of domains and exposed linker regions separating CRDs 3 and 6 from their neighboring domains. Hydrodynamic coefficients predicted for modeled receptor conformations are consistent with an extended conformation with close contacts between three pairs of CRDs. The N-terminal cysteine-rich (CR) domain and the fibronectin type II repeat appear to increase the rigidity of the molecule. The rigid, extended conformation of the receptor places domains with different functions at distinct positions with respect to the membrane (Napper et al. 2001). The crystal structures of CR alone and complexed with 4-sulfated-*N*-acetylgalactosamine showed that CR folds into threefold symmetric ß-trefoil shape resembling fibroblast growth factor. The sulfate portion of 4-sulfated-*N*-acetylgalactosamine and an unidentified ligand found in the native crystals binds in another pocket in the third lobe (Liu et al. 2000).

15.3.2.2 Ca^{2+} and Monosaccharide Binding at CRD-4 in MR

CRD-4 requires two Ca^{2+} for sugar binding, like the CRD of rat MBP-A. The binding site for one Ca^{2+} which ligates to the bound sugar in MBP-A, is conserved in CRD-4 where as the auxiliary Ca^{2+} binding site is not. Mutation of the four residues at positions in CRD-4 equivalent to the auxiliary Ca^{2+} binding site in MBP-A indicated that only one, Asn^{728}, is involved in the ligation of Ca^{2+}. Sequence comparisons with other C-type CRDs suggest that the proposed binding site for the auxiliary Ca^{2+} in CRD-4 of the MR is unique. Proton NMR spectra of methyl glycosides of mannose, GlcNAc, and fucose in the presence of CRD-4 and site-directed mutagenesis indicated that a stacking interaction with Tyr^{729} is also involved in binding of sugars to CRD-4. C-5 and C-6 of mannose interact with Tyr^{729}, whereas C-2 of GlcNAc is closest to this residue, indicating that these two sugars bind to CRD-4 in opposite orientations. Sequence comparisons with other mannose/GlcNAc-specific C-type CRDs suggest that use of a stacking interaction in the binding of these sugars is probably unique to CRD-4 of the MR (Mullin et al. 1997).

15.3.3 Cell and Tissue Distribution

Immune Cells: MR expression is detectable on monocytes cultured for 3 days (macrophages). In the thymus and lymph node, MR-positive branched cells (macrophages and DCs) were detected in connective tissue, thymus cortex (not medulla), and in the T cell area (not the B cell area) of lymph nodes (Noorman et al. 1997). Mononuclear phagocytes comprise the majority of interstitial cells in the mouse dermis. These cells express the mouse macrophage galactose-/N-acetylgalactosamine-specific-lectin (mMGL)/CD301 as well as the MR/CD206 (Dupasquier et al. 2006; van Vliet et al. 2005). MR positive APCs are present in several peripheral organs: skin, liver, cardiac and skeletal muscle and tongue. MR positive cells in salivary gland, thyroid and pancreas co-express MHC class II and the myeloid markers, macrosialin and sialoadhesin, but not the DC markers CD11c or DEC-205. MR and MHC class II co-localized, implying that antigen capture may be the primary role of MR in these cells. The tissue and sub-cellular distribution of MR suggest that it is appropriately located to serve as a high efficiency antigen uptake receptor of APC (Linehan 2005). The macrophage MR is significantly upregulated in nasal polyposis (NP) compared to patients with chronic rhinosinusitis (CS) without NP or turbinate tissue of controls (Claeys et al. 2004). The MR is up-regulated on dexamethasone-treated (immunosuppressed) macrophages, and down-regulated on LPS-activated macrophages (Dupasquier et al. 2006). Embryonic and adult mouse tissues express MR from early embryogenesis till adulthood. The MR is first observed on embryonic day 9 on cells that line blood island vessel walls in the yolk sac. Thus, the MR, expressed on tissue macrophages, is also expressed on subsets of vascular and lymphatic endothelial cells. The MR may be a marker of the so-called reticuloendothelial system (Takahashi et al. 1998).

Non-Immune Organs: Studies demonstrate that Pneumocystis (Pc) -mediated IL-8 release by human alveolar macrophages (AM) requires the coexpression of MR and toll likr receptor 2 (TLR2) and supports the concept that combinatorial interactions of macrophage innate receptors provide specificity of host defense cell responses to infectious challenge (Tachado et al. 2007). A mannose-specific receptor has also been found on retinal pigment epithelial cells (RPE). Human RPE expresses a MR on its apical surface (as does the rat RPE) and that this receptor is similar to the human macrophage MR. It seems that a MR is involved in the phagocytosis of rod outer segments by rat and human retinal pigment epithelium. Glycoproteins having oligosaccharides with terminal sequence SO_4-4-GalNAcβ1,4GlcNAcβ1,4Man-(S4GGnM) are rapidly removed from circulation by a S4GGnM-specific receptor (S4GGnM-R) expressed at the surface of hepatic endothelial cells. The S4GGnM-R from rat liver is closely related to the macrophage MR from rat lung both antigenically and structurally (Fiete et al. 1997). Squamous cell cancers of the head and neck and breast cancer specimens containing peritumoral vessels express MR, and the intratumoral lymph vessels often expressed them in both tumor types (Irjala et al. 2003).

Astrocytes and Microglia: Expression of MR has been demonstrated in brain, both in its soluble and membrane forms. Astrocytes and microglia, two types of glial cells that can be turned into immune-competent cells, are the main sites of expression in vivo and in vitro. Rodent Schwann cells (SC) in primary cultures take up MR ligand mannosyl/man-BSA-FITC in a highly specific manner which suggest that SC express MR in a prospectively functional state and suggest an antigen-presenting function of SC, compatible with a role in infectious/inflammatory states of the peripheral nervous system (Baetas-da-Cruz et al. 2009). MR mediates in vitro pinocytosis by astrocytes and microglia and phagocytosis by microglia. Regional expression of MR in glial and neuronal cells strongly suggests that this receptor plays an important role in homeostasis during brain development and/or neuronal function (Régnier-Vigouroux 2003). Being a differentiation marker and a relevant glycoprotein for the phagocytic and endocytic function of macrophages, the presence of MR in microglia suggests that MR could participate in multiple physiologic and pathologic conditions in the CNS, including inflammation, ischaemia, and neurodegenerative diseases.

Studies demonstrate that the expression and function of MR are inversely regulated by anti-and pro-inflammatory compounds. As observed for macrophages, IFNγ decreases, where as IL-4 increases MR expression. Consequently, the rates of pinocytosis were strongly up-regulated by IL-4 and inhibited by IFNγ. Down-regulation of the MR by IFN-γ is concomitant with the induction of the invariant chain, which is also induced by GM-CSF + IL-4. MR-expressing astrocytes may act as scavenger not only in CNS development but also in defense, against soluble and particulate mannosylated pathogens, presenting fragments there of at strategic locations in the CNS (Burudi et al. 1999; Zimmer et al. 2003).

Role in HIV-1 Neuropathogenesis: Both microglia and astrocytes are susceptible to HIV-1 infection. The HIV-1 in the CNS causes a variety of neurobehavioral and neuropathological disorders. Microglial MR is down-regulated by LPS and up-regulated by dexamethasone, as described for peripheral macrophages (Marzolo et al. 1999). Unlike microglia that express and utilize CD4 and chemokine co-receptors CCR5 and CCR3 for HIV-1 infection, astrocytes fail to express CD4. Results demonstrate the direct involvement of human MR in HIV-1 infection of astrocytes and suggest that HIV-1 interaction with MR plays an important role in HIV-1 neuropathogenesis. It gives a notion that humanMR is capable of eliciting intracellular signaling upon ligand binding (Liu et al. 2004; Lopez-Herrera et al. 2005).

15.3.4 Ligands

Multiple exogenous ligands (e.g. viruses, bacteria and fungi) as well as endogenous ligands (e.g. lutropin, myeloperoxidase and thyroglobulin) have been identified for the mannose-binding region of MR. Two distinct lectin activities have been described: (1) the cysteine-rich (CR) domain recognises sulphated carbohydrates while (2) the CTLDs mediate binding to mannose, fucose or N-acetylglucosamine. FNII domains are known to be important for collagen binding and this has been studied in the context of two members of the MR family, Endo180 and the phospholipase A2 receptor. Whilst no exogenous ligands have been reported for CR domain, the cysteine-rich R-type domain of the MR binds sulfated glycans and also N-glycans on pituitary glyco-hormones (e.g. lutropin) containing 4-SO4-GalNAcβ1–4GlcNAcβ1–2Manα1-R. Other ligands for the MR include chondroitin-4-sulfate proteoglycans on leukocytes that also contain 4-SO4-GalNAc β1-R residues and perhaps sulfated glycans, such as those containing 3-O-sulfated galactose, blood group 3-O-sulfated Le^x, and 3-O-sulfated Le^a. The cysteine-rich domain in MR also mediates recognition of selected glycoforms of membrane receptors expressed by metallophili macrophages in the spleen. The cysteine-rich R-type domain in the MR binds with relatively low affinity to sulfated monosaccharides (K_D in the range of 10^{-3}–10^{-4} M), but oligomeric forms of the protein probably display much higher avidity for glycoproteins with multiple sulfated glycans. Although the cysteine-rich R-type domain binds sulfated glycans, the C-type lectin domains bind unsulfated glycans and may also play a role in glycoprotein homeostasis and clearance. The full-length soluble form of MR was able to bind simultaneously polysaccharide via the carbohydrate recognition domains and sulfated oligosaccharide via the cysteine-rich domain (Zamze et al. 2002).

The MR can also bind collagen and the fibronectin type II domain mediates this activity. Neither of the two types of sugar-binding domain in the receptor is involved in collagen binding. Fibroblasts expressing the mannose receptor adhere to type I, type III and type IV collagens, but not to type V collagen, and the adherence is inhibited by isolated MR fibronectin type II domain. The fibronectin type II domain shows the same specificity for collagen as the whole receptor, binding to type I, type III and type IV collagens. MR is able to bind and internalise collagen in a carbohydrate-independent manner and that MR deficient macrophages have a marked defect in collagen IV and gelatin internalisation. These results have major implications at the molecular level as there are now three distinct ligand-binding sites described for MR (Martinez-Pomares et al.

2006). These results suggest additional roles for this multi-functional receptor in mediating collagen clearance or cell-matrix adhesion.

The *Mycobacterium tuberculosis* (M.tb) envelope is highly mannosylated with phosphatidyl-myo-inositol mannosides (PIMs), lipomannan, and mannose-capped lipoarabinomannan (ManLAM). It was demonstrated that recognition of M.tb PIMs by host cell C-type lectins is dependent on both the nature of the terminal carbohydrates and degree of acylation. Subtle structural differences among the PIMs impact host cell recognition and response and are predicted to influence the intracellular fate of M.tb. For example, higher-order PIMs preferentially associated with MR. In contrast, the lower-order PIM(2)f associated poorly with MDMs and did not bind to COS-1-MR. In contrast with the MR, the PIM(2)f and lipomannan were recognized by DC-SIGN comparable to higher-order PIMs and ManLAM, and the association was independent of their degree of acylation (Torrelles et al. 2006). Glyco-peptidolipids (GPLs) can function to delay phagosome-lysosome fusion and GPLs, like ManLAM, work through the MR to mediate this activity (Sweet et al. 2010; Torrelles et al. 2006).

15.3.5 Functions of Mannose Receptor

15.3.5.1 Role in Immunity

Genetic deletion in mice has highlighted the role of MR in the clearance of endogenous glycoproteins. Therefore MR has been implicated in both homeostatic processes and pathogen recognition. However, the function of MR in host defence is not yet clearly understood as MR-deficient animals do not display enhanced susceptibility to pathogens bearing MR ligands. Moreover, the MR mediates uptake of soluble but not of cell-associated antigen for cross-presentation. In vivo, MR deficiency impaired endocytosis of soluble OVA by DC and concomitant OT-I cell activation. Findings demonstrate that DC use the MR for endocytosis of a particular Ag type intended for cross-presentation (Burgdorf et al. 2006). This scenario is even more complex when considering the role of MR in innate immune activation as, even though no intracellular signaling motif has been identified at its cytoplasmic tail, MR has been shown to be essential for cytokine production, both pro-inflammatory and anti-inflammatory. Furthermore, MR might interact with other canonical pattern recognition receptors in order to mediate intracellular signaling. Gazi and Martinez-Pomares (2009) have summarised recent observations relating to MR function in immune responses and focused on its participation in phagocytosis, antigen processing and presentation, cell migration and intracellular signaling. The MR also functions in adaptive immunity through its ability to deliver antigens to MHC class II compartments and through its cleavage and release as a soluble protein into blood. It is speculated that the membrane-bound MR may bind antigens through its C-type lectin domains and then, following proteolytic cleavage, the soluble MR bound to its cargo may move to germinal centers where it may bind via its R-type domain to macrophages and dendritic cells expressing ligands such as sialoadhesion or CD45 that may contain sulfated glycans.

15.3.5.2 Phagocyte-Bacteria Interactions

The macrophage MR recognizes carbohydrates on the cell walls of infectious organisms. After ligation of mannose-rich glycoconjugates or pathogens, the macrophage MR mediates endocytosis and phagocytosis of the bound ligands by macrophages. The receptor-mediated uptake may either arm the macrophage to contribute to oxidant-mediated tissue damage or may function to clear extracellular myeloperoxidase during the resolution phase of the inflammatory process. Where as human keratinocytes appear to kill *Candida* in presence of MR (Szolnoky et al. 2001), Lee et al. (2003) suggested that MR is not required for the normal host defense during disseminated candidiasis or for the phagocytosis of *C. albicans*.(Lee et al. 2003; Le Cabec et al. 2005). One proposed function for MR found on macrophages and hepatic endothelial cells is to enhance the uptake and process glycoprotein antigens for presentation by MHC class II molecules. Incubation of RNase A (non-glycosylated) or RNAse B (mannosylated) with the transfected cells resulted in identical stimulation of ribonuclease-specific T cells, indicating that endocytosis of the glycosylated protein by the MR does not enhance presentation of this antigen (Napper and Taylor 2004).

A nonredundant role of MR has been demonstrated in the development of $CD4^+$ T-cell responses to *Cryptococcal* mannoproteins (MP) and protection from *C. neoformans* (Dan et al. 2008). *Trichuris muris* is a natural mouse model of the human gastrointestinal nematode parasite *Trichuris trichiura*. Macrophages accumulate in the large intestine of mice during infection and known to express MR. Infection of MR knockout mice with *T. muris* reveals that this receptor is not necessary for the expulsion of the parasite because MR knockout mice expel parasites with the same kinetics as wild-type animals and have similar cytokine responses in the mesenteric lymph nodes. This work suggests that, despite binding components of *T. muris* secretory products, the MR is not critically involved in the generation of the immune response to this parasite (deSchoolmeester et al. 2009).

MR, specifically, recognizes mannose residues on the surface of *Leishmania* parasites. Infection of mouse peritoneal macrophages with *Leishmania donovani* results in decrease in MR activity (Basu et al. 1991). Akilov et al. (2007) demonstrated that host MR is not essential for blocking IFN-γ/LPS-induced IL-12 production and MAPK activation by *Leishmania*. Thus, the MR is not essential for host defense against *Leishmania* infection or regulation of IL-12 production (Akilov et al. 2007). It was proposed that MR is a binding receptor, which requires a partner to trigger phagocytosis in some specialized cells such as macrophages.

Although alveolar macrophage MR activity is down-regulated in individuals infected with HIV, and that functional MR is shed from the macrophage cell surface, *P. carinii* enhances the formation of soluble MR by macrophages in vitro. Soluble MR was detected in cell-free alveolar fluid from humans infected with HIV and/or *P. carinii* (Camner et al. 2002; Swain et al. 2003). Macrophages recognize and subsequently kill *Klebsiella* expressing Man-α 2/3-Man or Rha-α 2/3-Rha sequences in their capsular polysaccharides by (a) MMRs, and (b) opsonization by the lung surfactant protein A (SP-A), which binds to the capsular polysaccharides of *Klebsiella* and to SP-A receptors on the macrophages (Keisari et al. 1997). MR contributes equally to TLR2 in proinflammatory cytokine production by human monocytes in response to *P. aeruginosa* infection. MR follows the same kinetics and colocalizes with TLR2 in the endosome during in vivo infection of human macrophages with *P. aeruginosa* (Xaplanteri et al. 2009).

15.3.5.3 Macrophage MR in HIV Binding

A role for MMR in the binding and transmission of HIV-1 by macrophages has been suggested (Nguyen and Hildreth 2003). Nonetheless, alveolar macrophage MR activity is down-regulated in individuals infected with HIV, and that functional MR is shed from the macrophage cell surface (Camner et al. 2002). As expected, mannan bound to the CRDs of MR dimers mostly in a calcium-dependent fashion. Surprisingly, gp120-mediated binding of HIV to dimers on MR-transfected Rat-6 cells and macrophages was calcium-independent, and only partially blocked by mannan and partially inhibited by *N*-acetylgalactosamine 4-sulfate. Thus gp120-mediated HIV binding occurs via calcium-dependent, calcium-independent CRDs and CR domain at the C terminus of MR dimers, presenting a much broader target for potential inhibitors of gp120-MR binding (Lai et al. 2009).

15.3.5.4 MR as Endocytic Receptor During Entry of Other Viruses

Influenza viruses showed marked differences in their ability to infect murine macrophages, including resident alveolar and peritoneal macrophages. The hierarchy in infectivity of the viruses resembles their reactivity with mannose-binding lectins. The possible involvement of MR in infection of macrophages by influenza virus has been suggested (Reading et al. 2000). Macrophages (MΦ) and mononuclear phagocytes are major targets of infection by dengue virus (DV), The MMR binds to all four serotypes of DV and specifically to the envelope glycoprotein. Pre-treatment of human monocytes or MΦ with type 2 cytokines (IL-4 or IL-13) enhances their susceptibility to productive DV infection. These findings indicate a new functional role for the MR in DV infection (Miller et al. 2008).

When human MR cDNA is transfected into Cos cells, these usually non-phagocytic cells express cell surface MR and bind and ingest MR ligands such as zymosan, yeast, and *Pneumocystis carinii* (Kruskal et al. 1992). Interaction of MR with Fc receptors is critical for the development of crescentic glomerulonephritis (CGN) in mice. In mouse model of CGN, MR-deficient ($MR^{-/-}$) mice were protected from CGN despite generating humoral and T cell responses similar to those of WT mice, but they demonstrated diminished macrophage and MC Fc receptor-mediated functions, including phagocytosis and Fc-mediated oxygen burst activity. Results demonstrate that MR augments Fc-mediated function and promotes MC survival (Chavele et al. 2010)

15.3.5.5 MR Regulates Cell Migration

MR has been shown to bind and internalize carbohydrate and collagen ligands and to have a role in myoblast motility and muscle growth. Since the related Endo180 (CD280) receptor has also been shown to have a promigratory role, it was likely that MR may be involved in regulating macrophage migration and/or chemotaxis. Contrary to expectation, bone marrow-derived macrophages (BMM) from MR-deficient mice showed an increase in random cell migration and no impairment in chemotactic response to a gradient of CSF-1 (Sturge et al. 2007).

Marttila-Ichihara et al. (2008) showed that migration of lymphocytes from the skin into the draining lymph nodes through the afferent lymphatics is reduced in MR-deficient mice, while the structure of lymphatic vasculature remains normal in these animals. Moreover, in a tumor model the primary tumors grow significantly bigger in $MR^{-/-}$ mice than in the wild-type (WT) controls, whereas the regional lymph node metastases are markedly smaller. Adhesion of both normal lymphocytes and tumor cells to lymphatic vessels is significantly decreased in MR-deficient mice. Thus, MR on lymphatic endothelial cells is involved in leukocyte trafficking and contributes to the metastatic behavior of cancer cells. Blocking of MR may provide a new approach to controlling inflammation and cancer metastasis by targeting the lymphatic vasculature.

15.3.6 Mouse Mannose Receptor

In mouse macrophages, the newly synthesized receptor has a Mr of 157 kDa that rapidly matures to a protein with a Mr of

172 kDa. Both forms of receptors are tightly associated with cell membranes. The receptor is found in a number of mouse macrophage cell types but not found in mouse fibroblasts. The transcript encoding this lectin is present in a number of highly endothelialized sites as well as in chondrocytes in cartilaginous regions of the embryo. Receptor-ligand binding was Ca^{2+} and pH dependent. D-mannose and L-fucose partially inhibit receptor binding to the ligand D-mannose-BSA (Blum et al. 1991; Wu et al. 1996). Cysteine-rich domain of the murine MR binds to macrophages from splenic marginal zone and lymph node subcapsular sinus and to germinal centers: Ligands for cysteine-rich (CR) domain of the murine MR have been detected in murine tissues. In naive mice, the CR-Fc, a Fc chimeric protein bound to sialoadhesin^{+}, F_4/80low/$^{-}$, macrosialin^{+} macrophages (Mφ) in spleen marginal zone (metallophilic Mφ) and lymph node subcapsular sinus. These results confirmed the identification of the CR region of the mannose receptor as a lectin (Martinez-Pomares et al. 1996, 1999). MR is localized with CR ligands in Lyve-1^{+} cells lining venous sinuses. These cells form a physical barrier for blood cells as they need to migrate through sinuses in order to exit the splenic parenchyma and, in this way, contribute to the unique filtration function of this organ. Furthermore, unlike murine spleen, CD68^{+} red pulp MΦ lack MR expression. However, results of Martinez-Pomares et al. (2005) suggest an unexpected contribution of MR to splenic function through the recognition of sulphated ligands that could influence the filtering capability of this organ (Martinez-Pomares et al. 2005).

Murine MR is a member of type C lectins family with a cysteine-rich domain, a fibronectin type 2 domain, eight type C lectin domains, a transmembrane domain, and a short cytoplasmic carboxyl terminus. Genomic analysis suggests it to be a conserved protein with human sequence homology with the murine form. It expresses as a large transcript in a number of human and murine tissues and tumor cells, and an alternatively spliced smaller transcript with a divergent 5′ sequence, expressed specifically in the human fetal liver. The gene encoding this lectin is interrupted by a large number of introns; the intron structure was similar to macrophage MR (Wu et al. 1996). The 854 bp of rat MR promoter sequence revealed one Sp1 site, three PU.1 sites, and a potential TATA box (TTTAAA)-33 bp 5′ of the transcriptional start site. The promoter was most active in the mature macrophage cell line NR8383 although the promoter also showed activity in the monocytic cell line RAW. Mutation in TTTAAA sequence to TTGGAA, resulting in decrease in activity, suggested that the promoter contains a functional TATA box. Though, the transcription factors Sp1, PU.1, and USF bound to the MR promoter, but only PU.1 and USF contributed to activation. Comparison of the rat, mouse, and human promoter sequences demonstrated that some binding sites are not conserved. Transcription of the rat MR is regulated by binding of PU.1 and a ubiquitous factor at an adjacent site, similar to other myeloid promoters (Egan et al. 1999).

Paracrine Control of Macrophage MR: The macrophage MR is progressively up-regulated as bone marrow precursor cells mature into macrophages and thus serves as a marker of differentiation. Prostaglandins E (PGE) are known inhibitors of monocyte and macrophage precursor proliferation, an effect often associated with cellular maturation. Prostaglandins accelerate macrophage MR expression and hence the differentiation of macrophage precursor cells, suggesting that a paracrine mechanism may exist to regulate macrophage MR expression and function (Schreiber et al. 1990). IL-4 up-regulates total cell-associated MR (cMR), correlating with enhanced surface expression. The influence of IL-10 showed an effect similar to IL-4. In both cases, enhanced cMR levels translated into increased production of the soluble form of the receptor (sMR). These data support a role for MR in T helper cell type 2 cytokine-driven, immune responses and suggest a non-macrophage contribution to sMR production in vivo (Martinez-Pomares et al. 2003). The induction of cyclooxygenase-2 (COX-2) and the production of PGE_2 in response to pathogen-associated decorated with mannose moieties were studied in human monocytes and monocyte-derived macrophages (MDM). It showed that mannan is a strong inducer of COX-2 expression in human MDM, most likely by acting through the MR route. Because COX-2 products can be both, proinflammatory and immunomodulatory, these results disclose a signaling route triggered by mannose-decorated pathogen-associated molecular patterns, which can be involved in both the response to pathogens and the maintenance of homeostasis (Fernández et al 2005).

PPARγ Promotes MR Gene Expression in Murine Macrophages: The Macrophage MR expression is induced in mouse peritoneal macrophages following exposure to PPARγ ligands or to IL-13 via a PPARγ signaling pathway. This novel signaling pathway controlling the macrophage MR surface expression involves the endogenous PPARγ ligand produced by phospholipase A2 activation that may be an important regulator of macrophage MR expression by IL-13 (Coste et al. 2003). Ligand activation of the PPARγ in macrophages promotes uptake, killing of *C. albicans*, and reactive oxygen intermediates production triggered by the yeasts through macrophage MR over-expression.

15.3.7 Interactions of MR with Branched Carbohydrates

15.3.7.1 Inhibitory Potency of Monosaccharides of MR

The relative inhibitory potency of monosaccharides of human placental MR was found as: L-Fuc > D-Man > D-Glc > D-GlcNAc > Man-6-P >> D-Gal >> L-Rha >> GalNAc. The inhibitory potency of mannose, however, increased by two orders of magnitude when linear oligomer was used. Oligomers containing α1-3- and α1-6-linked mannose residues were more inhibitory than oligomers, which contained α1-2- and α1-4-linked mannoses. Linear or branched oligomannosides larger than three units did not have a significant influence on their inhibitory potency; rather, potency was found to decrease in comparison with oligomannosides with three units. Compared to linear oligomers, inhibition of binding was maximum using branched mannose oligosaccharides, α-D-Man-bovine serum albumin conjugates, or mannan. A model is discussed in which branched ligand is bound to spatially distinct sites on the human MR (Kery et al. 1992).

15.3.7.2 Di-Mannoside Clusters Target MR Whereas Lewis Clusters Target DC-SIGN

Dimannoside clusters, recognized by the MR with an affinity constant close to 10^6 liter mol^{-1} have very low affinity for DC-SIGN (less than 10^4 liter mol^{-1}). Conversely, Lewis clusters had higher affinity toward DC-SIGN than toward the MR. Dimannoside clusters are efficiently taken up by human DCs as well as by rat fibroblasts expressing the MR but not by HeLa cells or rat fibroblasts expressing DC-SIGN; DC-SIGN-expressing cells take up Lewis clusters. This suggested that ligands containing di-mannoside clusters can specifically target the MR, whereas ligands containing Lewis clusters will target DC-SIGN (Frison et al. 2003; Guo et al., 2004).

The MR is involved in prevention of LPS-induced acute lung injury after administration of mannose in mice. Mannose also prevented the inflammatory cell accumulation, and inhibited production of cytokines. Furthermore, mannose receptor was up-regulated after mannose administration. Studies reveal involvement of MR and impaired NF-kB activation in the mannose prevention of acute lung injury, and implicate MR as a potential therapeutic target during acute lung injury (Xu et al. 2010).

15.3.7.3 Tissue-Type Plasminogen Activator-Receptor Interaction

Tissue-type plasminogen activator (t-PA) in blood is cleared by the liver partially through a mannose-specific uptake system. The MR (175 kDa) from bovine AM is specifically bound to t-PA and showed saturable binding in presence of Ca^{2+}. Mannose-albumin was an effective inhibitor, whereas galactose-albumin did not have a significant effect. Among monosaccharides, D-mannose and L-fucose were the most potent inhibitors whereas D-galactose and N-acetyl-D-galactosamine were ineffective. t-PA, deglycosylated by endoglycosidase H, did not interact with the receptor. The MR binding to t-PA, probably occurs through its high mannose-type oligosaccharide (Otter et al. 1991). The interaction of ^{125}I-t-PA with isolated rat parenchymal and endothelial liver cells was studied by Otter et al. (1992). It was concluded that (1) the mannose receptor and LDL receptor-related protein (LRP) appear to be the sole major receptors responsible for tPA clearance and (2) therapeutic levels of tPA can be maintained for a prolonged time span by co-administration of the receptor antagonists (Biessen et al. 1997; Otter et al. 1992).

15.3.8 Mannose Receptor-Targeted Drugs and Vaccines

Targeting antigens to endocytic receptors on professional antigen-presenting cells (APCs) represents an attractive strategy to enhance the efficacy of vaccines. Such APC-targeted vaccines have an exceptional ability to guide exogenous protein antigens into vesicles that efficiently process the antigen for MHC class I and class II presentation. The MR and related C-type lectin receptors are particularly designed to sample antigens, much like pattern recognition receptors, to integrate the innate with adaptive immune responses. In fact, a variety of approaches involving delivery of antigens to the MR have demonstrated effective induction of potent cellular and humoral immune responses. Yet, although several lines of evidence in diverse experimental systems attest to the efficacy of targeted vaccine strategies, it is becoming increasingly clear that additional signals, such as those afforded by adjuvants, may be critical to elicit sustained immunity. Therefore, MR-targeted vaccines are likely to be most efficacious in vivo when combined with agents that elicit complementary activation signals. A better understanding of the mechanism associated with the induction of immune responses as a result of targeting antigens to the MR, will be important in exploiting MR-targeted vaccines not only for mounting immune defenses against cancer and infectious disease, but also for specific induction of tolerance in the treatment of autoimmune disease (Gupta et al. 2009; Irache et al. 2008; Keler et al. 2004).

15.4 Phospholipase A2-Receptors

The PLA2 receptor was discovered as a receptor for phospholipase A2 neurotoxins in snake venoms and was referred to as the M-type PLA2 receptor to distinguish it from the neuronal or N-type PLA2 receptor. It can occur as both a long form that is a type I trans-membrane glycoprotein with a domain structure like the MR and a shorter form that is secreted.

15.4.1 The Muscle (M)-Type sPLA2 Receptors

15.4.1.1 Molecular Cloning and Chromosomal Localization

Cloning of 180 kDa PLA2 receptor from rabbit skeletal muscle revealed that it is a membrane protein with a N-terminal cysteine-rich domain, a fibronectin type II domain, eight repeats of a carbohydrate recognition domain, a unique transmembrane domain, and a intracellular C-terminal domain. The 1458-residue PLA2 receptor, expressed in transfected cells, binds venom PLA2 (vPLA2) with very high affinity (K_D = 10–20 pM). It also tightly binds the two structural types of sPLA2s, i.e. pancreatic PLA2 and synovial PLA2 (K_D = 1–10 nM) (Lambeau and Lazdunski 1999). Cloning of sPLA2 receptors from human kidney revealed two transcripts. One encodes for a transmembrane form of the sPLA2 receptor and the other one is an alternatively processed transcript, caused by polyadenylation occurring at a site within an intron in the C terminus part of the transcriptional unit. This transcript encodes for a shortened secreted soluble sPLA2 receptor lacking the coding region for the transmembrane segment. Soluble and membrane-bound human sPLA2 receptors both bind sPLA2 with high affinities, although the binding properties of the human receptors are different from those obtained with the rabbit membrane-bound sPLA2 receptor. The 180-kDa human sPLA2 receptor gene has been mapped in the q23–q24 bands of chromosome 2. Cloned 180-kDa sPLA2 receptors have the same structural organization as the macrophage mannose receptor, a membrane protein involved in the endocytosis of glycoproteins, and the phagocytosis of microorganisms bearing mannose residues on their surface (Ancian et al. 1995b).

The M-type 180 kDa sPLA2 receptor has been cloned in rabbit, bovine, and humans (Lambeau and Lazdunski 1999; Ancian et al. 1995a). The cloned receptors are homologous to the macrophage MR (Ezekowitz et al. 1990), as well as to DEC-205, a protein involved in the presentation of antigens. Interestingly, all of these proteins are predicted to share the same structural organization, *i.e.* a large extracellular region composed of an N-terminal cysteine-rich domain, a fibronectin like type II domain, eight (Lambeau et al. 1994; Taylor et al. 1992) or ten repeats of a carbohydrate recognition domain (CRD), followed by a unique transmembrane domain and a short intracellular C-terminal domain (Jiang et al. 1995). This latter domain contains a consensus sequence for the internalization of ligand-receptor complexes and is thought to confer to these receptors their endocytic properties (Ancian et al. 1995a; Kruskal et al. 1992).

15.4.1.2 Structural Elements

Specific membrane receptors for sPLA2s were initially identified with snake venom sPLA2s called OS1 and OS2 (Lambeau et al. 1991). One of these sPLA2 receptors (muscle M-type, 180 kDa) has a very high affinity for OS1 and OS2 and a high affinity for pancreatic and inflammatory-type mammalian sPLA2s, which might be the natural endogenous ligands for PLA2 receptors. The binding of pancreatic sPLA2 mutants to the M-type receptor showed that residues within or close to the Ca^{2+}-binding loop of pancreatic sPLA2 are crucially involved in the binding step. Although the presence of Ca^{2+} is essential for the enzymatic activity, it is not required for binding to the receptor. These residues include Gly-30 and Asp-49, which are conserved in all sPLA2s. Leu-31 is also essential for binding of pancreatic sPLA2 to its receptor. Conversion of pancreatic prophospholipase to phospholipase is essential for the acquisition of binding properties to the M-type receptor (Lambeau et al. 1995).

Binding Domain for sPLA2: The rabbit M-type receptor for sPLA2s has a large extracellular domain of 1,394 amino acids, composed of an N-terminal cysteine-rich domain, a fibronectin-like type II domain, and eight carbohydrate recognition domains (CRDs). It is thought to mediate some of the physiological effects of mammalian sPLA2s, including vascular smooth muscle contraction and cell proliferation, and is able to internalize sPLA2s. Site-directed mutagenesis studies showed that a snake venom sPLA2 (OS1), binds to the receptor via its CRDs and that deletion of CRD 5 completely abolishes the binding of sPLA2s. Moreover, a receptor lacking all CRDs but CRD 5 was still able to bind OS1 although with a lower affinity. Deletion of CRDs 4 and 6, surrounding the CRD 5, slightly reduced the affinity for OS1, thus suggesting that these CRDs are also involved in the binding of OS1. The M-type sPLA2 receptor and the macrophage MR are predicted to share the same tertiary structure. The p-Aminophenyl-α-D-mannopyranoside bovine serum albumin, a known ligand of the macrophage MR, binds to the M-type sPLA2 receptor essentially via CRDs 3–6 (Nicolas et al. 1995).

Extended and Bent Conformations of MR Family: M-type receptor for sPLA2s is a membrane protein with a N-terminal cystein-rich domain, a fibronectin-like type II domain, eight repeats of a carbohydrate recognition domain, a unique transmembrane and an intracellular C-terminal. When expressed in transfected cells, the rabbit M-type receptor binds both the inflammatory-type and the pancreatic-type msPLA2′s with fairly high affinities (K_D ~ 1–10 nM) suggesting that the sPLA2 receptors for vPLA2′s are the normal targets of endogenous msPLA2′s involved in a variety of diseases. Residues within or close to the Ca^{2+} binding loop of pancreatic-type PLA2 are crucially involved in the binding step although the presence of Ca^{2+} which is essential for the enzymatic activity is not required for binding to the receptor. The domain in charge of sPLA2 binding in the M-type receptor has been identified (Llorca 2008). Llorca (2008) reviewed three dimensional structures of the receptors of MR family and their functional implications. Recent research has revealed that several members of this family can exist in at least two configurations: an extended conformation with the N-terminal cysteine rich domain pointing outwards from the cell membrane and a bent conformation where the N-terminal domains fold back to interact with C-type lectin-like domains at the middle of the structure. Conformational transitions between these two states seem to regulate the interaction of these receptors with ligands and their oligomerization.

15.4.1.3 Phospholipases A2 as Ligand of sPLA2 Receptors

Phospholipases A2 (PLA2s) form a large family of structurally related enzymes that catalyse the hydrolysis of the sn-2 acyl bond of glycerophospholipids to produce free fatty acids and lysophospholipids. Venom vPLA2s show a variety of toxic effects in animals including neurotoxicity, myotoxicity, hypotensive, anticoagulant, and proinflammatory effects. Snake venoms are known for decades to contain a tremendous molecular diversity of PLA2s, which can exert a myriad of toxic and pharmacological effects. Because PLA2 products are important for cell signaling and the biosynthesis of biologically active lipids, including eicosanoids and platelet-activating factor, PLA2s are generally considered as key enzymes that control the release of lipid mediator precursors. Secretory sPLA2s form a large family of structurally related enzymes which are widespread in nature. Mammalian cells also express a variety of sPLA2s with 12 distinct members identified so far, in addition to the various other intracellular PLA2s. On the other hand, mammalian secretory PLA2′s (sPLA2′s) are now implicated in many biological functions besides digestion, such as airway and vascular smooth muscle contraction, cell proliferation, and in a variety of diseases associated with inflammation such as rheumatoid arthritis, endotoxic shock, and respiratory distress syndrome. More speculative results suggest the involvement of one or more sPLA2s in promoting atherosclerosis and cancer.

High level PLA2 activity is found in serum and biological fluids during the acute-phase response (APR). Extracellular PLA2 in fluids of patients with inflammatory diseases such as sepsis, acute pancreatitis or rheumatoid arthritis is also associated with propagation of inflammation. PLA2 activity is involved in the release of both pro- and anti-inflammatory lipid mediators from phospholipids of cellular membranes or circulating lipoproteins. PLA2 may thus generate signals that influence immune responses. PLA2 treatment of differentiating monocytes in the presence of granulocyte/macrophage colony-stimulating factor and IL-4 yielded cells with phenotypical and functional characteristics of mature DC. The effects of PLA2 on DC maturation were mainly dependent on enzyme activity and correlated with the activation of NF-kB, AP-1 and NFAT. The transient increase in PLA2 activity seems to generate signals that promote transition of innate to adaptive immunity during the APR (Perrin-Cocon et al. 2004). Scanning of nucleic acid databases indicated that several sPLA2s are also present in invertebrate animals like *Drosophila melanogaster* as well as in plants.

The variety of effects, so mentioned, are apparently linked to the existence of a diversity of very high affinity receptors (K_D values as low as 1.5 pM) for these toxic PLA2s (Lambeau and Gelb 2008). Several different types of receptors (N and M) have been identified for vPLA2′s. The identification of a variety of membrane and soluble proteins that bind to sPLA2s suggests that the sPLA2 enzymes also function as high affinity ligands. Venom sPLA2s and group IB and IIA mammalian sPLA2s have been shown to bind to membrane and soluble mammalian proteins of the C-type lectin superfamily (M-type sPLA2 receptor and lung surfactant proteins), to pentraxin and reticulocalbin proteins, to factor Xa, and proteoglycans including glypican and decorin, a mammalian protein containing a leucine-rich repeat and to N-type receptors. Venom sPLA2s also associate with three distinct types of sPLA2 inhibitors purified from snake serum that belong to the C-type lectin superfamily, to the three-finger protein superfamily and to proteins containing leucine-rich repeats (Cupillard et al. 1999; Valentin and Lambeau 2000; Wu et al. 2006). On the basis of their different molecular properties and tissue distributions, each sPLA2 is likely to exert distinct functions by acting as an enzyme or ligand for specific soluble proteins or receptors, among which the M-type receptor is the best-characterized target. Results reaffirmed that the mouse M-type receptor is selective for only a subset of mouse sPLA2s from the group I/II/V/X structural collection. Binding of mouse sPLA2s to a recombinant soluble mouse M-type receptor leads in all cases to inhibition of enzymatic activity, and the extent of deglycosylation of the receptor decreases yet does not abolish sPLA2 binding. The two mouse sPLA2s (group IB and group IIA) have relatively high affinities for the mouse M-type receptor, but they can have much lower affinities for receptors from other animal species, indicating species specificity for sPLA2 binding to M-type receptors (Cupillard et al. 1999; Rouault et al. 2007).

A number of competitive inhibitors have been developed against the inflammatory-type human group IIA (hGIIA) sPLA2 with the aim of specifically blocking its catalytic activity and pathophysiological functions. The sPLA2:M-receptor interactions suggest that the therapeutic effects of sPLA2 inhibitors may be due not only to inhibition of enzymatic activity but also to modulation of binding of sPLA2 to the M-type receptor or other protein targets (Boilard et al. 2006).

15.4.1.4 Functions of M-type sPLA2 Receptor

The rabbit M-type sPLA2 receptor is a multifunctional protein, which seems to mediate the physiological effects of group I sPLA2, including smooth muscle contraction and cell proliferation. This receptor binds with high affinity to pancreatic group I and inflammatory group II sPLA2s as well as various sPLA2s from snake venoms. The rabbit M-type sPLA2 receptor is able to promote cell adhesion on type I and IV collagens most probably via its N-terminal fibronectin-like type II domain. It also shows that binding of sPLA2s to a recombinant soluble form of this receptor is associated with a noncompetitive inhibition of phospholipase A2 activity (Ancian et al. 1995a).

Endocytic Properties: Rabbit myocytes express M-type sPLAs receptor at high levels. Internalization of the receptor was shown to be clathrin-coated pit-mediated, rapid ($ke = 0.1\ min^{-1}$), and ligand-independent. Analysis of the internalization efficiency of the mutants suggested that the NSYY motif encodes the major endocytic signal, with the distal tyrosine residue playing the key role (Zvaritch et al. 1996). The membrane-bound form can function as an endocytic receptor to internalize phospholipase A2 ligands.

Regulation of Proinflammatory Cytokines: In animal cells, secreted phospholipase A2s are important in the degradation of phospholipids and the release of arachidonic acid, which is the precursor for prostaglandins, leukotrienes, thromboxanes, and prostacyclins. The murine PLA2 receptor binds to sPLA2-X enzyme, where as the human PLA2 receptor may bind to several different phospholipase A2 isozymes. Although from early studies it was thought that the PLA2 receptor might be involved in clearance of phospholipase A2s, murine knockouts for PLA2 receptor showed unusual phenotype of resistance to endotoxic shock. This suggested that the PLA2 receptor might be important in regulating production of proinflammatory cytokines by soluble phospholipase A2s. Thus, the PLA2 receptor might function in signal transduction mediated by phospholipase A2 binding. Some of the C-type lectin domains in the PLA2 receptor function by binding phospholipase A2 ligands, rather than binding directly to glycan ligands. The fibronectin type II domain binds to collagen, which is a feature shared by this domain in other MR family members, except DEC-205.

Upregulation of PLA2 and M-Type Receptor in Rat ANTI-THY-1 Glomerulonephritis: Treatment of rat glomerular mesangial cell (GMC) cultures with pancreatic sPLA2-IB results in an enhanced expression of sPLA2-IIA and COX-2, possibly via binding to its specific M-type sPLA2 receptor. During glomerulonephritis (GN), shortly after induction of anti-Thy 1.1-GN, expression of sPLA2-IB was up-regulated in the kidney with strongest sPLA2-IB protein expression on infiltrated granulocytes and monocytes, and markedly upregulated expression of M-type receptor on resident glomerular cells. It suggested that both sPLA2-IB and the M-type sPLA2 receptors are involved in the autocrine and paracrine amplification of the inflammatory process in different resident and infiltrating cells (Beck et al. 2006).

15.4.2 Neuronal or N-Type PLA2 Receptor

The sPLA2 found in PLA2 honey bee venom is neurotoxic and binds to N-type PLA2 receptor with high affinity. Mutations in the interfacial binding surface, in the Ca^{2+}-binding loop and in the hydrophobic channel lead to a dramatic decrease in binding to N-type receptors, whereas mutations of surface residues localized in other parts of the sPLA2 structure do not significantly modify the binding properties. Neurotoxicity experiments show that mutants with low affinity for N-type receptors are devoid of neurotoxic properties, even though some of them retain high enzymatic activity. These results provide evidence for the surface region surrounding the hydrophobic channel of bee venom sPLA2 as the N-type receptor recognition domain (Nicolas et al. 1997). The bee venom PLA2 specifically binds to a single polypeptide with a mass of approximately 180 kDa. Moreover, mannose-BSA and the bee venom PLA2 bound to the same site on macrophages. Results suggested that bee venom PLA2 binding to macrophages is mediated through MR (Mukhopadhyay and Stahl 1995).

15.5 DEC-205 (CD205)

15.5.1 Characterization

DEC-205 (also known as Ly75 and designated as CD205) is an endocytic receptor with ten membrane-external, contiguous C-type lectin domains. DEC-205 is a 205-kDa protein of MR family that is expressed by dermal dendritic cells and, at a lower level, by epidermal Langerhans cells (Hawiger et al.

2001; Witmer-Pack et al. 1995). It is also expressed on some epithelial cells, on bone marrow stroma, and by endothelial cells. Cloning of DEC-205 gene revealed its relationship to MR and its unique characteristic of having ten tandem C-type lectin domains, rather than the eight found in other MR family members. Anti-DEC-205/CD205 antibodies are useful for identifying DCs in human splenic white pulp and its border region with the red pulp (Pack et al. 2008). Unlike mouse DEC-205, which is reported to have predominant expression on DC, human DEC-205 was detected at relatively high levels on myeloid blood DC and monocytes, at moderate levels on B lymphocytes and at low levels on NK cells, plasmacytoid blood DC and T lymphocytes (Kato et al. 2006). None of the ten CTLDs in DEC-205 have the conserved amino acids known to be important in carbohydrate binding, and thus far there is no evidence that these domains bind glycans. There is also no evidence that the cysteine-rich R-type domain at the amino terminus binds to glycan ligands.

The mouse DEC-205 cDNA predicts a molecular structure which has a marked similarity to the MMR. The full coding region of human DEC-205 cDNA from the Hodgkin's disease-derived L428 cell line predicted protein structure of 1722 amino acids consisting of a signal peptide, cysteine-rich domain, fibronectin type II domain, ten carbohydrate recognition-like domains, transmembrane domain, and a cytoplasmic tail. Human DEC-205 is 77% identical to the mouse protein with completely conserved cysteines. The DEC-205 gene was mapped to chromosome band 2q24. The 7.8 and 9.5 kb DEC-205 transcripts are present in myeloid, B lymphoid and Hodgkin's disease-derived cell lines. Immature blood DC contained a barely detectable amount of DEC-205 transcripts but these were markedly increased upon differentiation/activation (Kato et al. 1998). The multi-domain structure of mouse and human DEC-205 was completely conserved in hamster with the overall identity of approximately 80%. DEC-205 transcripts were detected in the thymus and bone marrow cells cultured in the presence of mouse GMC-SF and interleukin-4 in which the DEC-205 expression was up-regulated in the course of cultures. Hamster DEC-205 was mainly detected on cell membrane and shown to mediate the uptake of FITC-conjugated ovalbumin (Maruyama et al. 2002).

15.5.2 Functions of DEC-205

Studies suggest that CD205 has two distinct functions – one as an endocytic receptor on immature dendritic cells and a second as a non-endocytic molecule on mature dendritic (Butler et al. 2007). DEC-205 is important in the recognition and internalization of antigens for presentation to T cells. Upon endocytosis, DEC-205 internalizes to late endosomes/lysosomes and recycles to the surface. Expression of DEC-205 is enhanced in both types of cells upon cell maturation induced by inflammatory stimuli (Hawiger et al. 2001; Jiang et al. 1995; Mahnke et al. 2000). The DEC-205 is rapidly taken up by means of coated pits and vesicles, and is delivered to a multi-vesicular endosomal compartment that resembles the MHC class II-containing vesicles implicated in antigen presentation (Mahnke et al. 2000).

In the steady state, DEC-205 represents a specific receptor for DCs to induce peripheral tolerance to soluble antigens for both $CD4^+$ (Hawiger et al. 2001) and $CD8^+$ T cells (Bonifaz et al. 2002). The key observations are that DCs have efficient receptor based mechanisms to enhance presentation on MHC class I products in vivo, and that these operate in the steady state, and finally, the consequence of presentation is peripheral tolerance in the $CD8^+$ compartment by a deletional mechanism (Bonifaz et al. 2002, 2004; Steinman et al. 2003). Although its ligands await identification, the endocytic properties of CD205 make it an ideal target for those wishing to design vaccines and targeted immunotherapies. Unlike other members of the MMR family, CD205 was up-regulated upon dendritic cell maturation. Furthermore, a small amount of the CD205-DCL-1 fusion protein was detected in mature DC.

15.6 ENDO 180 (CD280)/uPARAP

15.6.1 Urokinase Receptor (uPAR)-Associated Protein

Urokinase receptor (uPAR)-associated protein is a member of macrophage MR protein family and known as Endo180 or urokinase receptor-associated protein (uPARAP) and was discovered independently by several groups. It was found to be part of a trimolecular cell-surface complex with urokinase plasminogen activator (uPA) and its receptor (uPAR). It was also discovered as a novel antigen on macrophages and human fibroblasts. The plasminogen activation cascade system, directed by urokinase and the urokinase receptor, plays a key role in extracellular proteolysis during tissue remodeling. To identify molecular interaction partners of these trigger proteins on the cell, Behrendt et al. (2000) observed a specific tri-molecular complex on addition of pro-urokinase to human U937 cells. This complex included the urokinase receptor, pro-urokinase, and an unknown, high molecular weight urokinase receptor-associated protein. Further analysis identified the novel protein as the human homologue of a murine membrane-bound lectin with hitherto unknown function. The protein, designated uPARAP, is the fourth member of the macrophage MR protein family. The large extracellular domain of Endo 180 contains an N-terminal cysteine-rich domain, a single fibronectin type

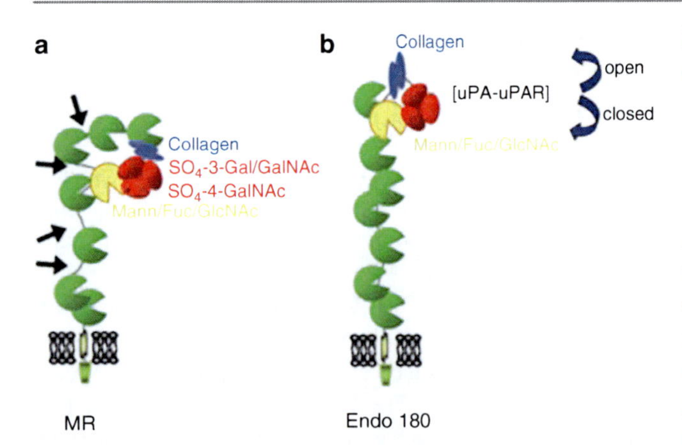

Fig. 15.2 The structural organization of mannose receptor (a) and Endo 180 (b). Globular conformation of the mannose receptor family members where the CysR domain (*red*) interacts with the physiologically active CTLD (*yellow*) and corresponding to CTLD4 in theMR (**a**) and CTLD2 in Endo180 (**b**). The FNII domain has been colored *blue*, while the non-functional CTLDs are shown in *green*. This conformation seems to be receptive to ligand binding and it can interact with several substrates, which are indicated with the corresponding color beside the domain responsible for the interaction. Regions accessible for proteolytic cleavage for the MR have been identified by Napper et al, (2001) and they are highlighted with *black arrows*. In an acidic environment it is predicted that Endo180 switches to a more conformation. For the MR, conformational changes could modify its overall compactness by either an opening or closing of the observed conformation (Adapted by permission from Boskovic et al. (2006) © The American Society for Biochemistry and Molecular Biology)

II domain (a putative collagen-binding site) and eight C-type lectin-like domains. The second of these lectin-like domains has been shown to mediate Ca^{2+} –dependent mannose binding. Endo180 like MR is expressed on macrophages. But it is also expressed on fibroblasts and chondrocytes, some endothelial cells, and tissues undergoing ossification. The cysteine-rich R-type domain at the amino terminus of Endo180 has been shown to be unable to bind sulfated glycans. The function of cysteine-rich domain is unknown. Targeted deletion of Endo180 exons 2–6 in mice showed that this mutation results in the efficient expression of a truncated Endo180 protein that lacks the cysteine-rich domain, the FNII domain and CTLD1. This mutation does not disrupt the C-type lectin activity that is mediated by CTLD2, but results in cells that have a defect in collagen binding and internalization and an impaired migratory phenotype (East et al. 2003).

Despite similarities between Endo180 and the MR, conflicting reports have been published on the three-dimensional arrangement of these receptors. Single particle electron microscopy of the MR and Endo180 display distinct three-dimensional structures, which are, however, conceptually very similar: a bent and compact conformation built upon interactions of the CysR domain and the lone functional CTLD. Studies indicated that, under a low pH mimicking the endosomal environment, both MR and Endo180 experience large conformational changes (Fig. 15.2) (Boskovic et al. 2006).

15.6.2 Interactions of Endo180

Although Endo180 and the MR are now both known to be mannose binding lectins, each receptor is likely to have a distinct set of glycoprotein ligands in vivo (East and Isacke 2002). The distribution and post-translational processing is consistent with Endo180 functioning to internalize glycosylated ligands from the extracellular milieu for release in an endosomal compartment (Sheikh et al. 2000).

The Endo180 interacts with ligand-bound uPAR, uPA, matrix metalloprotease-13 (MMP-13), and collagen V on the cell surface. Endo180 binds to the carboxy-terminal region of type I collagen, and collagens type II, IV, and V. In vitro assays showed that Endo180 binds both to native and denatured collagens and the binding is mediated by the fibronectin type II domain. The restricted expression of Endo180 in both embryonic and adult tissue indicates that Endo180 plays a physiological role in mediating collagen matrix remodeling during tissue development and homeostasis; and the observed receptor upregulation in pathological conditions may contribute to disease progression (Thomas et al. 2005; Wienke et al. 2003). Cross-linking studies have identified Endo180 as an uPAR associated protein and this interaction could be blocked by collagen V. This collagen binding reaction at the exact site of plasminogen activation on the cell may lead to adhesive functions as well as a contribution to cellular degradation of collagen matrices (Behrendt et al. 2000).

The uptake and lysosomal degradation of collagen by fibroblasts constitute a major pathway in the turnover of connective tissue. Studies suggest a central function of Endo180 in cellular collagen interactions (Engelholm et al. 2003). During post-implantation development in mice, Endo180 was expressed in all tissues undergoing primary ossification, including the developing bones that ossify intramembranously, and developing long bones undergoing endochondral ossification. Osteoblasts also expressed uPAR. Besides bone-forming tissues, uPARAP/Endo180 was detected in a mesenchymal condensation of the midbrain and in the developing lungs. Endo180 receptor is strongly expressed in chondrocytes in the articular cartilage of young mice. Expression of Endo180 in articular cartilage chondrocytes of young, but not old, mice and the reciprocal expression of Endo180 and its ligands in the growth plate suggest that this receptor is involved in cartilage development but not in cartilage homeostasis. Endo180 does not appear to play a role in the development or progression of murine osteoarthritis (Howard et al. 2004; Howard and Isocke 2002), though it is involved in the clearance of uPA:uPAR complex as well as other possible ligands during benign and malignant tissue remodeling (Schnack Nielsen et al. 2002).

References

Akilov OE, Kasuboski RE, Carter CR, McDowell MA (2007) The role of mannose receptor during experimental leishmaniasis. J Leuk Biol 81:1188–1196

Ancian P, Lambeau G, Lazdunski M (1995a) Multifunctional activity of the extracellular domain of the M-type (180 kDa) membrane receptor for secretory phospholipases A2. Biochemistry 34: 13146–13151

Ancian P, Lambeau G, Mattéi MG, Lazdunski M (1995b) The human 180-kDa receptor for secretory phospholipases A2. Molecular cloning, identification of a secreted soluble form, expression, and chromosomal localization. J Biol Chem 270:8963–8970

Baetas-da-Cruz W, Alves L, Pessolani MC et al (2009) Schwann cells express the macrophage mannose receptor and MHC class II. Do they have a role in antigen presentation? J Peripher Nerv Syst 14: 84–92

Basu N, Sett R, Das PK (1991) Down-regulation of mannose receptors on macrophages after infection with *Leishmania donovani*. Biochem J 277:451–456

Beck S, Beck G, Ostendorf T et al (2006) Upregulation of group IB secreted phospholipase A(2) and its M-type receptor in rat ANTI-THY-1 glomerulonephritis. Kidney Int 70:1251–1260

Behrendt N, Jensen ON, Engelholm LH et al (2000) A urokinase receptor-associated protein with specific collagen binding properties. J Biol Chem 275:1993–2002

Bennett MJ, Schlunegger MP, Eisenberg D (1995) 3D domain swapping: a mechanism for oligomer assembly. Protein Sci 4:2455–2468

Biessen EA, van Teijlingen M, Vietsch H et al (1997) Antagonists of the mannose receptor and the LDL receptor-related protein dramatically delay the clearance of tissue plasminogen activator. Circulation 95:46–52

Blum JS, Stahl PD, Diaz R, Fiani ML (1991) Purification and characterization of the D-mannose receptor from J774 mouse macrophage cells. Carbohydr Res 213:145–153

Boilard E, Rouault M, Surrel F et al (2006) Secreted phospholipase A2 inhibitors are also potent blockers of binding to the M-type receptor. Biochemistry 45:13203–13218

Bonifaz L, Bonnyay D, Mahnke K et al (2002) Efficient targeting of protein antigen to the dendritic cell receptor DEC-205 in the steady state leads to antigen presentation on major histocompatibility complex class I products and peripheral CD8+ T cell tolerance. J Exp Med 196:1627–1638

Bonifaz LC, Bonnyay DP, Charalambous A et al (2004) In vivo targeting of antigens to maturing dendritic cells via the DEC-205 receptor improves T cell vaccination. J Exp Med 199:815–824

Boskovic J, Arnold JN, Stilion R et al (2006) Structural model for the mannose receptor family uncovered by electron microscopy of Endo180 and the mannose receptor. J Biol Chem 28:8780–8787

Burgdorf S, Lukacs-Kornek V, Kurts C (2006) The mannose receptor mediates uptake of soluble but not of cell-associated antigen for cross-presentation. J Immunol 176:6770–6776

Burudi EM, Riese S, Stahl PD et al (1999) Identification and functional characterization of the mannose receptor in astrocytes. Glia 25: 44–55

Butler M, Morel AS, Jordan WJ et al (2007) Altered expression and endocytic function of CD205 in human dendritic cells, and detection of a CD205 DCL 1 fusion protein upon dendritic cell maturation. Immunology 120:362–371

Camner P, Lundborg M, Lastbom L et al (2002) Experimental and calculated parameters on particle phagocytosis by alveolar macrophages. J Appl Physiol 92:2608–2616

Chavele KM, Martinez-Pomares L, Domin J et al (2010) Mannose receptor interacts with Fc receptors and is critical for the development of crescentic glomerulonephritis in mice. J Clin Invest 120: 1469–1478

Claeys S, De Belder T, Holtappels G et al (2004) Macrophage mannose receptor in chronic sinus disease. Allergy 59:606–612

Coste A, Dubourdeau M, Linas MD et al (2003) PPARγ promotes mannose receptor gene expression in murine macrophages and contributes to the induction of this receptor by IL-13. Immunity 19:329–339

Cupillard L, Mulherkar R, Gomez N et al (1999) Both group IB and group IIA secreted phospholipases A2 are natural ligands of the mouse 180-kDa M-type receptor. J Biol Chem 274:7043–7051

Dan JM, Kelly RM, Lee CK et al (2008) Role of the mannose receptor in a murine model of *Cryptococcus* neoformans infection. Infect Immun 76:2362–2367

deSchoolmeester ML, Martinez-Pomares L, Gordon S, Else KJ (2009) The mannose receptor binds *Trichuris muris* excretory/secretory proteins but is not essential for protective immunity. Immunology 126:246–255

Drickamer K, Taylor ME (1993) Biology of animal lectins. Annu Rev Cell Biol 9:237–264

Dupasquier M, Stoitzner P, Wan H et al (2006) The dermal microenvironment induces the expression of the alternative activation marker CD301/mMGL in mononuclear phagocytes, independent of IL-4/IL-13 signaling. J Leukoc Biol 80:838–849

East L, Isacke CM (2002) The mannose receptor family. Biochim Biophys Acta 1572:364–386

East L, McCarthy A, Wienke D et al (2003) A targeted deletion in the endocytic receptor gene Endo180 results in a defect in collagen uptake. EMBO Rep 4:710–716

Egan BS, Lane KB, Shepherd VL (1999) PU.1 and USF are required for macrophage-specific mannose receptor promoter activity. J Biol Chem 274:9098–9107

Engelholm LH, List K, Netzel-Arnett S et al (2003) uPARAP/Endo180 is essential for cellular uptake of collagen and promotes fibroblast collagen adhesion. J Cell Biol 160:1009–1015

Ezekowitz RA, Sastry K, Bailly P, Warner A (1990) Molecular characterization of the human macrophage mannose receptor: demonstration of multiple carbohydrate recognition-like domains and phagocytosis of yeasts in Cos-1 cells. J Exp Med 172:1785–1794

Feinberg H, Park-Snyder S, Kolatkar AR et al (2000) Structure of a C-type carbohydrate recognition domain from the macrophage mannose receptor. J Biol Chem 275:21539–21548

Fernández N, Alonso S, Valera I et al (2005) Mannose-containing molecular patterns are strong inducers of cyclooxygenase-2-expression and prostaglandin E_2 production in human macrophages. J Immunol 174:8154–8162

Fiete D, Beranek MC, Baenziger JU (1997) The macrophage/endothelial cell mannose receptor cDNA encodes a protein that binds oligosaccharides terminating with SO4-4 GalNAcβ1,4GlcNAcβ or Man at independent sites. Proc Natl Acad Sci USA 94:11256–11261

Fiete DJ, Beranek MC, Baenziger JU (1998) A cysteine-rich domain of the "mannose" receptor mediates GalNAc-4-SO_4 binding. Proc Natl Acad Sci USA 95:2089–2093

Frison N, Taylor ME, Soilleux E et al (2003) Oligolysine-based oligosaccharide clusters: selective recognition and endocytosis by the mannose receptor and dendritic cell-specific intercellular adhesion molecule 3 (ICAM-3)-grabbing nonintegrin. J Biol Chem 278: 23922–23929

Gazi U, Martinez-Pomares L (2009) Influence of the mannose receptor in host immune responses. Immunobiology 214:554–561

Guo Y, Feinberg H, Conroy E et al (2004) Structural basis for distinct ligand-binding and targeting properties of the receptors DC-SIGN and DC-SIGNR. Nat Struct Mol Biol 11:591–598

Gupta A, Gupta RK, Gupta GS (2009) Targeting cells for drug and gene delivery: emerging applications of mannans and mannan binding lectins. J Sci Ind Res 68:465–483

Hawiger D, Inaba K, Dorsett Y et al (2001) Dendritic cells induce peripheral T cell unresponsiveness under steady state conditions in vivo. J Exp Med 194:769–779

Howard MJ, Isacke CM (2002) The C-type lectin receptor Endo180 displays internalization and recycling properties distinct from other members of the mannose receptor family. J Biol Chem 277: 32320–32331

Howard MJ, Chambers MG, Mason RM et al (2004) Distribution of Endo180 receptor and ligand in developing articular cartilage. Osteoarthr Cartil 12:74–82

Irache JM, Salman HH, Gamazo C, Espuelas S (2008) Mannose-targeted systems for the delivery of therapeutics. Expert Opin Drug Deliv 5:703–724

Irjala H, Alanen K, Grénman R, Heikkilä P, Joensuu H, Jalkanen S (2003) Mannose receptor (MR) and common lymphatic endothelial and vascular endothelial receptor (CLEVER)-1 direct the binding of cancer cells to the lymph vessel endothelium. Cancer Res 63: 4671–4676

Jiang WP, Swiggard WJ, Heufler C et al (1995) The receptor DEC-205 expressed by dendritic cells and thymic epithelial cells is involved in antigen processing. Nature 375:151–155

Kato M, McDonald KJ, Khan S et al (2006) Expression of human DEC-205 (CD205) multilectin receptor on leukocytes. Int Immunol 18:857–869

Kato M, Neil TK, Clark GJ, Morris CM, Sorg RV, Hart DN (1998) cDNA cloning of human DEC-205, a putative antigen-uptake receptor on dendritic cells. Immunogenetics 47:442–450

Keisari Y, Kabha K, Nissimov L, Schlepper-Schafer J, Ofek I (1997) Phagocyte-bacteria interactions. Adv Dent Res 11:43–49

Keler T, Ramakrishna V, Fanger MW (2004) Mannose receptor-targeted vaccines. Expert Opin Biol Ther 4:1953–1962

Kery V, Krepinsky JJ, Warren CD et al (1992) Ligand recognition by purified human mannose receptor. Arch Biochem Biophys 298: 49–55

Kim SJ, Ruiz N, Bezouska K et al (1992) Organization of the gene encoding the human macrophage mannose receptor (MRC1). Genomics 14:721–727

Kruskal BA, Sastry K, Warner AB et al (1992) Mannose receptor phagocytic chimeric receptors require both transmembrane and cytoplasmic domains from the mannose receptor. J Exp Med 176:1673–1680

Lai J, Bernhard OK, Turville SG et al (2009) Oligomerization of the macrophage mannose receptor enhances gp120-mediated binding of HIV-1. J Biol Chem 284:11027–11038

Lambeau G, Gelb MH (2008) Biochemistry and physiology of mammalian secreted phospholipases A_2. Annu Rev Biochem 77:495–520

Lambeau G, Lazdunski M (1999) Receptors for a growing family of secreted phospholipases A2. Trends Pharmacol Sci 20:162–170

Lambeau G, Barhanin J, Lazdunski M (1991) Identification of different receptor types for toxic phospholipases A2 in rabbit skeletal muscle. FEBS Lett 293:29–33

Lambeau G, Ancian P, Barhanin J et al (1994) Cloning and expression of a membrane receptor for secretory phospholipases A. J Biol Chem 269:1575–1578

Lambeau G, Ancian P, Nicolas JP et al (1995) Structural elements of secretory phospholipases A2 involved in the binding to M-type receptors. J Biol Chem 270:5534–5540

Le Cabec V, Emorine LJ, Toesca I et al (2005) The human macrophage mannose receptor is not a professional phagocytic receptor. J Leukoc Biol 77:934–943

Lee SJ, Zheng NY, Clavijo M et al (2003) Normal host defense during systemic candidiasis in mannose receptor-deficient mice. Infect Immun 71:437–445

Leteux C, Chai W, Loveless RW et al (2000) The cysteine-rich domain of the macrophage mannose receptor is a multispecific lectin that recognizes chondroitin sulfates A and B and sulfated oligosaccharides of blood group Lewis[a] and Lewis[x] types in addition to the sulfated N-glycans of lutropin. J Exp Med 191:1117–1126

Linehan SA (2005) The mannose receptor is expressed by subsets of APC in non-lymphoid organs. BMC Immunol 6:4

Liu Y, Arthur J, Chirino AJ et al (2000) Crystal structure of the cysteine-rich domain of mannose receptor complexed with a sulfated carbohydrate ligand. J Expt Med 191:1105–1116

Liu Y, Liu H, Kim BO et al (2004) CD4-independent infection of astrocytes by human immuno deficiency virus type 1: requirement for the human mannose receptor. J Virol 78:4120–4133

Llorca O (2008) Extended and bent conformations of the mannose receptor family. Cell Mol Life Sci 65:1302–1310

Lopez-Herrera A, Liu Y, Rugeles MT, He JJ (2005) HIV-1 interaction with human mannose receptor (hMR) induces production of matrix metalloproteinase 2 (MMP-2) through hMR-mediated intracellular signaling in astrocytes. Biochim Biophys Acta 1741:55–64

Mahnke K, Guo M, Lee S et al (2000) The dendritic cell receptor for endocytosis, DEC-205, can recycle and enhance antigen presentation via major histocompatibility complex class II-positive lysosomal compartments. J Cell Biol 151:673–684

Martinez-Pomares L, Kosco-Vilbois M, Darley E et al (1996) Fc chimeric protein containing the cysteine-rich domain of the murine mannose receptor binds to macrophages from splenic marginal zone and lymph node subcapsular sinus and to germinal centers. J Exp Med 184:1927–1937

Martinez-Pomares L, Crocker PR, Da Silva R et al (1999) Cell-specific glycoforms of sialoadhesin and CD45 are counter-receptors for the cysteine-rich domain of the mannose receptor. J Biol Chem 274: 35211–35218

Martinez-Pomares L, Reid DM, Brown GD et al (2003) Analysis of mannose receptor regulation by IL-4, IL-10, and proteolytic processing using novel monoclonal antibodies. J Leukoc Biol 73:604–613

Martinez-Pomares L, Hanitsch LG, Stillion R, Keshav S, Gordon S (2005) Expression of mannose receptor and ligands for its cysteine-rich domain in venous sinuses of human spleen. Lab Invest 85:1238–1249

Martinez-Pomares L, Wienke D, Stillion R et al (2006) Carbohydrate-independent recognition of collagens by the macrophage mannose receptor. Eur J Immunol 36:1074–1082

Marttila-Ichihara F, Turja R, Miiluniemi M et al (2008) Macrophage mannose receptor on lymphatics controls cell trafficking. Blood 112:64–72

Maruyama K, Akiyama Y, Cheng J et al (2002) Hamster DEC-205, its primary structure, tissue and cellular distribution. Cancer Lett 181:223–232

Marzolo MP, von Bernhardi R, Inestrosa NC (1999) Mannose receptor is present in a functional state in rat microglial cells. J Neurosci Res 58:387–395

Miller JL, de Wet BJ, Martinez-Pomares L et al (2008) The mannose receptor mediates dengue virus infection of macrophages. PLoS Pathog 4:e17

Mullin NP, Hitchen PG, Taylor ME (1997) Mechanism of Ca2+ and monosaccharide binding to a C-type carbohydrate-recognition domain of the macrophage mannose receptor. J Biol Chem 272: 5668–5681

Mukhopadhyay A, Stahl P (1995) Bee venom phospholipase A2 is recognized by the macrophage mannose receptor. Arch Biochem Biophys 324:78–84

Napper CE, Taylor ME (2004) The mannose receptor fails to enhance processing and presentation of a glycoprotein antigen in transfected fibroblasts. Glycobiology 14:7C–12C

Napper CE, Dyson MH, Taylor ME (2001) An extended conformation of the macrophage mannose receptor. J Biol Chem 276:14759–14766
Napper CE, Drickamer K, Taylor ME (2006) Collagen binding by the mannose receptor mediated through the fibronectin type II domain. Biochem J 395:579–586
Nguyen DG, Hildreth JE (2003) Involvement of macrophage mannose receptor in the binding and transmission of HIV by macrophages. Eur J Immunol 33:483–493
Nicolas JP, Lambeau G, Lazdunski M (1995) Identification of the binding domain for secretory phospholipases A2 on their M-type 180-kDa membrane receptor. J Biol Chem 270:28869–28873
Nicolas JP, Lin Y, Lambeau G et al (1997) Localization of structural elements of bee venom phospholipase A2 involved in N-type receptor binding and neurotoxicity. J Biol Chem 272:7173–7181
Noorman F, Braat EA, Barrett-Bergshoeff M et al (1997) Monoclonal antibodies against the human mannose receptor as a specific marker in flow cytometry and immunohistochemistry for macrophages. J Leukoc Biol 61:63–72
Otter M, Barrett-Bergshoeff MM, Rijken DC (1991) Binding of tissue-type plasminogen activator by the mannose receptor. J Biol Chem 266:13931–13935
Otter M, Kuiper J, Bos R, Rijken DC, van Berkel TJ (1992) Characterization of the interaction both in vitro and in vivo of tissue-type plasminogen activator (t-PA) with rat liver cells. Effects of monoclonal antibodies to t-PA. Biochem J 284:545–550
Pack M, Trumpfheller C, Thomas D et al (2008) DEC-205/CD205+ dendritic cells are abundant in the white pulp of the human spleen, including the border region between the red and white pulp. Immunology 123:438–446
Perrin-Cocon L, Agaugué S, Coutant F et al (2004) Secretory phospholipase A2 induces dendritic cell maturation. Eur J Immunol 34: 2293–2302
Reading PC, Miller JL, Anders EM (2000) Involvement of the mannose receptor in infection of macrophages by influenza virus. J Virol 74:5190–5197
Régnier-Vigouroux A (2003) The mannose receptor in the brain. Int Rev Cytol 226:321–342
Rouault M, Le Calvez C, Boilard E et al (2007) Recombinant production and properties of binding of the full set of mouse secreted phospholipases A2 to the mouse M-type receptor. Biochemistry 46:1647–1662
Schnack Nielsen B, Rank F, Engelholm LH et al (2002) Urokinase receptor-associated protein (uPARAP) is expressed in connection with malignant as well as benign lesions of the human breast and occurs in specific populations of stromal cells. Int J Cancer 98: 656–664
Schreiber S, Blum JS, Chappel JC et al (1990) Prostaglandin E specifically upregulates the expression of the mannose-receptor on mouse bone marrow-derived macrophages. Cell Regul 1:403–413
Sheikh H, Yarwood H, Ashworth A, Isacke CM (2000) Endo180, an endocytic recycling glycoprotein related to the macrophage mannose receptor is expressed on fibroblasts, endothelial cells and macrophages and functions as a lectin receptor. J Cell Sci 113(Pt 6):1021–1032
Steinman RM, Hawiger D, Nussenzweig MC (2003) Tolerogenic dendritic cells. Annu Rev Immunol 21:685–711
Sturge J, Todd SK, Kogianni G, McCarthy A, Isacke CM (2007) Mannose receptor regulation of macrophage cell migration. J Leukoc Biol 82:585–593
Su Y, Bakker T, Harris J et al (2005) Glycosylation influences the lectin activities of the macrophage mannose receptor. J Biol Chem 280: 32811–32820
Swain SD, Lee SJ, Nussenzweig MC, Harmsen AG (2003) Absence of the macrophage mannose receptor in mice does not increase susceptibility to *Pneumocystis carinii* infection in vivo. Infect Immun 71:6213–6221
Sweet L, Singh PP, Azad AK et al (2010) Mannose receptor-dependent delay in phagosome maturation by Mycobacterium avium glycopeptidolipids. Infect Immun 78:518–526
Szolnoky G, Bata-Csörgö Z, Kenderessy AS et al (2001) A mannose-binding receptor is expressed on human keratinocytes and mediates killing of Candida albicans. J Invest Dermatol 117:205–213
Tachado SD, Zhang J, Zhu J, Patel N, Cushion M, Koziel H (2007) Pneumocystis-mediated IL-8 release by macrophages requires coexpression of mannose receptors and TLR2. J Leukoc Biol 81: 205–211
Takahashi K, Donovan MJ, Rogers RA, Ezekowitz RAB (1998) Distribution of murine mannose receptor expression from early embryogenesis through to adulthood. Cell Tissue Res 292:311–323
Taylor ME, Bezouska K, Drickamer K (1992) Contribution to ligand binding by multiple carbohydrate-recognition domains in the macrophage mannose receptor. J Biol Chem 267:1719–1726
Thomas EK, Nakamura M, Wienke D et al (2005) Endo180 binds to the C-terminal region of type I collagen. J Biol Chem 280: 22596–22605
Torrelles JB, Azad AK, Schlesinger LS (2006) Fine discrimination in the recognition of individual species of phosphatidyl-myo-inositol mannosides from Mycobacterium tuberculosis by C-type lectin pattern recognition receptors. J Immunol 177:1805–1816
Valentin E, Lambeau G (2000) Increasing molecular diversity of secreted phospholipases A(2) and their receptors and binding proteins. Biochim Biophys Acta 1488:59–70
van Vliet SJ, van Liempt E, Saeland E et al (2005) Carbohydrate profiling reveals a distinctive role for the C-type lectin MGL in the recognition of helminth parasites and tumor antigens by dendritic cells. Intern Immunol 17:661–669
Weis WI, Kahn R, Fourme R, Drickamer K, Hendrickson WA (1991) Structure of the calcium-dependent lectin domain from a rat mannose-binding protein determined by MAD phasing. Science 254:1608–1615
Weis WI, Taylor ME, Drickamer K (1998) The C-type lectin superfamily in the immune system. Immunol Rev 163:19–34
Wienke D, MacFadyen JR, Isacke CM (2003) Identification and characterization of the endocytic transmembrane glycoprotein Endo180 as a novel collagen receptor. Mol Biol Cell 14:3592–3604
Witmer-Pack MD, Swiggard WJ, Mirza A et al (1995) Tissue distribution of the DEC-205 protein that is detected by the monoclonal antibody NLDC-145. II. Expression in situ in lymphoid and nonlymphoid tissues. Cell Immunol 163:157–162
Wu K, Yuan J, Lasky LA (1996) Characterization of a novel member of the macrophage mannose receptor type C lectin family. J Biol Chem 271:21323–21330
Wu YZ, Manevich Y, Baldwin JL et al (2006) Interaction of surfactant protein A with peroxiredoxin 6 regulates phospholipase A2 activity. J Biol Chem 281:7515–7525
Xaplanteri P, Lagoumintzis G, Dimitracopoulos G, Paliogianni F (2009) Synergistic regulation of Pseudomonas aeruginosa-induced cytokine production in human monocytes by mannose receptor and TLR2. Eur J Immunol 39:730–740
Xu X, Xie Q, Shen Y et al (2010) Involvement of mannose receptor in the preventive effects of mannose in lipopolysaccharide-induced acute lung injury. Eur J Pharmacol 641:229–237
Zamze S, Martinez-Pomares L, Jones H et al (2002) Recognition of bacterial capsular polysaccharides and lipopolysaccharides by the macrophage mannose receptor. J Biol Chem 277:41613–41623
Zimmer H, Riese S, Regnier-Vigouroux A (2003) Functional characterization of mannose receptor expressed by immunocompetent mouse microglia. Glia 42:89–100
Zvaritch E, Lambeau G, Lazdunski M (1996) Endocytic properties of the M-type 180-kDa receptor for secretory phospholipases A2. J Biol Chem 271:250–257

Part VI

I-Type Lectins

I-Type Lectins: Sialoadhesin Family

16

G.S. Gupta

16.1 Sialic Acids

Sialic acids are a family of α-keto acids with nine-carbon backbones. Sialic acids are acidic monosaccharides typically found at the outermost ends of the sugar chains of animal glycoconjugates. Sialic acids belong to most important molecules of higher animals and also occur in some microorganisms. Their structural diversity is high and, correspondingly, the mechanisms for their biosynthesis complex. Sialic acids are involved in a great number of cell functions. They are bound to complex carbohydrates and occupy prominent positions, especially in cell membranes. However, receptors or adhesion molecules mediating such functions between eukaryotic cells were unknown until 1985, when it was found that the members of the selectin family mediate adhesion of leukocytes to specific endothelia through binding to sialylated glycans like sialyl Lewis. Sialic acids are expressed abundantly in animals of deuterostome lineage (primarily in echinoderms and vertebrates), but their expression in another major group of animals, the protostomes (including nematodes, arthropods, and mollusks), is inconspicuous. They are found mostly at distal positions of oligosaccharide chains of glycoproteins and glycolipids and are thus exposed to the extracellular environment, allowing them to be recognized during initial contact of cells with various pathogenic agents such as viruses, bacteria, protozoa, and toxins. The marked structural complexity of sialic acids can thus be interpreted as a result of evolutionary arms race between the hosts and the pathogens.

Various proteins can recognize and bind to Sialic acids family of monosaccharides. Particular attention is focused on the evolving information about sialic acid recognition by certain C-type lectins (the selectins), I-type lectins, and a complement regulatory protein (the H protein). The last two instances are examples of the importance of the side chain of sialic acids and the effects of natural substitutions (e.g., 9-O-acetylation) of this part of the molecule. Accumulated evidence show that sialic acids function in cellular interactions either by masking or as a recognition site. Due to their cell surface location these acidic molecules shield macromolecules and cells from enzymatic and immunological attacks and thus contribute to innate immunity. In comparison to the masking role, sialic acids also represent recognition sites for various physiological receptors, such as selectins and siglecs, as well as for toxins and microorganisms. The recognition function of sialic acids can be masked by O-acetylation, which modifies the interaction with receptors. Many viruses use sialic acids for the infection of cells. Since sialic acids play also a decisive role in tumor biology, they modulate biological and pathological cellular events in a sensitive way. Thus, they are most prominent representatives of mediators of molecular and cellular recognition (Schauer 2004).

Most mammalian cell surfaces display two major sialic acids (Sias), N-acetylneuraminic acid (Neu5Ac) and N-glycolylneuraminic acid (Neu5Gc). Humans lack Neu5Gc due to a mutation in cytidine-5′-monophosphate (CMP)-Neu5Ac hydroxylase, which occurred after evolutionary divergence from great apes. Sonnenburg et al. (2004) described an apparent consequence of human Neu5Gc loss: domain-specific functional adaptation of Siglec-9 (CD33-related Siglecs = CD33rSiglecs). While recombinant human Siglec-9 showed recognition of both Neu5Ac and Neu5Gc, in striking contrast, chimpanzee and gorilla Siglec-9 strongly preferred binding Neu5Gc. Reports also indicated that endogenous Sias (rather than surface Sias of bacterial pathogens) are the functional ligands of CD33r-Siglecs and suggested that the endogenous Sia landscape is the major factor directing evolution of CD33rSiglec binding specificity. Such domain-specific divergences should be taken into consideration in upcoming comparisons of human and chimpanzee genomes (Sonnenburg et al. 2004).

G.S. Gupta et al., *Animal Lectins: Form, Function and Clinical Applications*,
DOI 10.1007/978-3-7091-1065-2_16, © Springer-Verlag Wien 2012

16.2 Sialic Acid-Binding Ig-Like Lectins (I-Type Lectins)

The immunoglobulin superfamily (IgSF), integrins, cadherins, and selectins comprise distinct categories of cell adhesion molecules. Of these IgSFs, the selectins function through glycoconjugate-mediated interactions, and members of the other families function either through homophilic or heterophilic protein–protein recognition. Proteins other than antibodies and T-cell receptors that mediate glycan recognition via Ig-like domains are called "I-type lectins or Siglecs (Sia-recognizing Ig-superfamily lectins)." The Ig superfamily is defined by their structural similarity to immunoglobulins. The siglecs mediate cell–cell interactions through recognition of specific sialylated glycoconjugates as their counter receptors (Gabius et al. 2002; May et al. 1998). Sialoadhesins recognizing sialylated glycan structures represent the best characterized subgroup. The majority of these proteins are involved in protein-protein binding as receptors, antibodies or cell adhesion molecules and capable of carbohydrate-protein interactions. In contrast to the selectins, these proteins are associated with diverse biological processes, i.e. hemopoiesis, neuronal development and immunity. Their properties, carbohydrate specificities and potential biological functions have been vividly discussed.

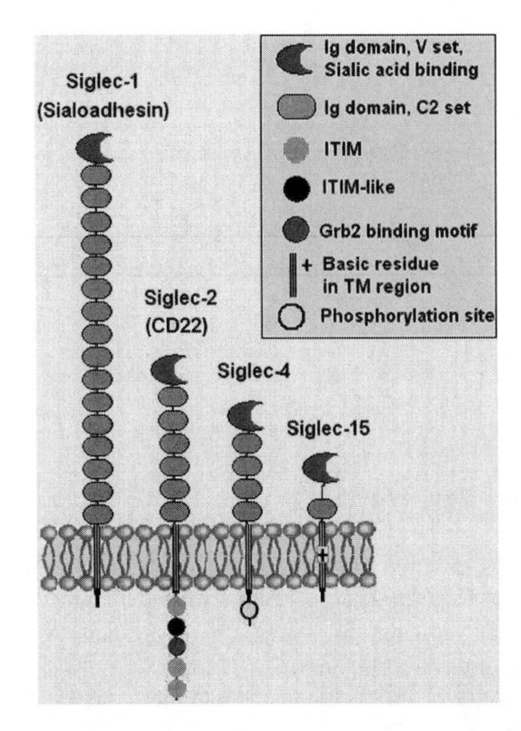

Fig. 16.1 Domain organization of known Siglecs of Sialoadhesin family in humans: Sialoadhesin (Siglec-1), Siglec-2 (CD22), Siglec-4 (Mylin-associated glycoprotein: MAG), and Siglec-15. Most Siglecs have one or more tyrosine-based signaling motifs. Exceptions are Sn, Siglec-14, Siglec-15. Some siglecs having no defined cytoplasmic domain have positively charged amino acid within TM domain, that can associate with DAP12 (DNAX activation protein-12) ITAM containing adaptor

16.2.1 Two Subsets of Siglecs

Siglecs are type 1 transmembrane proteins which comprise a sialic acid-binding N-terminal V-set domain, variable numbers of C2-set Ig domains, a transmembrane region and a cytosolic tail. Over the past few years, several novel siglecs have been discovered through genomics and functional screens. On the basis of their sequence similarities and evolutionary conservation, two primary subsets of siglecs have been identified:

1. **First subset** of the sialoadhesin (Sn) family, classified by sequence homology to members of IgSF, exhibits protein–carbohydrate recognition. In addition to Sn (Siglec-1), the subset includes CD22 (Siglec-2), MAG (myelin-associated glycoprotein) (Siglec-4a), Schwann cell myelin protein (Siglec 4b), Siglec-15, all of which are well-conserved in mammals (Fig. 16.1).
2. **Second**, rapidly evolving, subset is designated the CD33-related siglecs. In humans, these include CD33 and siglecs-5, -6, -7 (7/p75/AIRM1), -8, -9, 10, -11, -14 and -16, whereas, in mice, they comprise murine CD33 and siglecs-E, -F, -G and –H (Angata 2006; Floyd et al. 2000; Crocker and Redelinghuys 2008; Varki and Angata 2006; Yu et al. 2001). Thirteen functional Siglecs [sialoadhesin (siglec-1), CD22 (siglec-2), MAG, CD33 (Siglecs-3) and Siglecs-5-13 (CD33-related-Siglecs)] have been identified in great apes, of which humans lack Siglec-13 (Angata et al. 2004). Along with sialoadhesin, Siglec-2–4 and Siglec 5–16 molecules form the Siglec family of sialic acid-binding lectins (van den Berg et al. 2001; Crocker and Redelinghuys 2008) (see Chap. 17; Fig. 17.1).

The "CD33-related" siglecs have molecular features of inhibitory receptors and may be important in regulating leucocyte activation during immune and inflammatory responses (Crocker 2002; Varki and Angata 2006). Siglecs from both types show specificity for sialic acid-containing glycans that are found on the nonreducing ends of oligosaccharide chains of *N*- and *O*-linked carbohydrates or on glycosphingolipids. Siglecs are further distinguished by their specificity for sialic acids that possess α(2,3), α(2,6), or α(2,8) linkages to an extended oligosaccharide structure, as well as distinguish the structures of the extended glycans themselves (Crocker et al. 2007). With the notable exception of MAG, which is expressed in the nervous system, siglecs are differentially expressed on various subsets of leucocytes where they play a role in the positive and negative regulation of immune and inflammatory responses (McMillan and Crocker 2008; Crocker et al. 2007).

Most Siglecs with two exceptions (Siglec-4/MAG and Siglec-6) are expressed on various white cells of immune system; Siglec-4/MAG is expressed exclusively in the nervous system (Crocker et al. 2007; Lock et al. 2004; Varki and Angata 2006; Zhang et al 2004) . In particular, CD33-related Siglecs show marked inter-species differences even within the same order of mammals (Angata and Binkman-vander Linden 2004). Although the biological functions of Siglecs expressed in the immune system are not yet fully understood, in vitro and in vivo analyses suggest that these lectins are involved in coupling glycan recognition to immunological regulation. On the other hand, the precise functional importance of sialic acid-binding property of Siglecs is not well understood, although the phenotypic similarity between Siglec-2 null mice and ST6Gal-I (Galβ1–4GlcNAc α2–6 sialyltransferase) null mice suggests that sialic acid binding does affect the signaling activity of this Siglec

I-type lectins are abundant in nervous system and have been implicated in a number of morphogenetic processes as fundamental as axon growth, myelin formation and growth factor signaling. The structural and functional properties of I-type lectins expressed in neural tissues have been reviewed with a main focus on MAG/Siglec-4 (Angata and Binkman-Vander Linden 2002). As endocytic receptors, siglecs provide portals of entry for certain viral and bacterial pathogens, as well as therapeutic opportunities for targeting innate immune cells in disease. Three Siglec family members appear to be tightly restricted in expression to specific cell populations within hematopoietic lineage: sialoadhesin/Siglec-1 to macrophages, CD33/Siglec-3 to cells of the myelomonocytic lineage, and CD22/Siglec-2 to B cells. Likewise MAG/Siglec-4 is only expressed on oligodendrocytes in the central nervous system and on Schwann cells in the peripheral nervous system. For Sn and MAG, specific cell populations which bear the cognate "ligand" have been identified: Sn preferentially interacts with cells of the granulocytic lineage. Additionally, for each Siglec family member, certain sialic acid ligand preferences have been determined (Patel et al. 1999) (Fig. 16.2).

Human mast cells (MC) derived from $CD34^+$ peripheral blood precursors express mRNA for CD22 (Siglec-2), CD33 (Siglec-3), Siglec-5, Siglec-6, Siglec-8 and Siglec-10 and surface expression of these proteins except CD22 and Siglec-10, whose levels were low or undetectable. However, expression of CD22 and Siglec-10 was mostly cytoplasmic. $CD34^+$ precursor cells from peripheral blood constitutively expressed surface CD33, Siglec-5 and Siglec-10. Phenotypic analysis of LAD-2 MC yielded a similar pattern of Siglec expression except that CD22 expression was particularly prominent (Yokoi et al. 2006).

16.2.2 Siglecs as Inhibitory Receptors

Each Siglec has a distinct expression pattern suggesting that these molecules play unique roles in the cells expressing them. Siglecs differ from traditional Ig superfamily members in several ways. Although their extracellular domains contain a variable number of C2-set Ig domains, unlike other Ig superfamily members, Siglecs possess an NH_2-terminal V-set Ig domain that binds sialylated structures (May et al. 1998; Yamaji et al. 2002, Zaccai et al. 2003; Alphey et al. 2003; Dimasi et al. 2004). In addition, many Siglecs have potential tyrosine phosphorylation sites in the context of an immunoreceptor tyrosine-based inhibitory motif (ITIM) in their cytoplasmic tails, suggesting their involvement in intracellular signaling pathways and in endocytosis. In fact, Siglecs-3 and -7 have been shown to be capable of transmitting negative regulatory signals upon cross-linking by specific antibodies. Although the function of other family members remains unknown, the presence of ITIM sequences in the cytoplasmic tails of some of human siglecs renders them potential inhibitory receptors. Human CD33 has been characterized as an inhibitory receptor by virtue of its ability to bind tyrosine phosphatases SHP-1 and SHP-2 (Ulyanova et al. 1999, 2001). In case of CD22/Siglec-2, a negative regulatory role was further proven by the studies using genetically engineered mice, which suggests that Siglecs possess signal transduction activity. Evidence of signaling has been shown for several human Siglecs (Crocker and Varki 2001; Ikehara et al. 2004; Mingari et al. 2001).

16.2.3 Binding Characteristics of Siglecs

The common feature of Siglec proteins is their recognition of sialic acid residues on cell surface glycoproteins and glycolipids, which is mediated by the amino-terminal V-set domain of a siglec (May et al. 1998). Although recognition of sialic acid residues is a hallmark feature of all siglecs, the specificity of binding has been shown for certain family members in terms of their preferences for the position of a sialic acid residue on N-linked oligosaccharides as well as in the binding of ligands expressed on different cell types (Crocker et al. 1997). Each siglec is expressed in a highly restricted fashion (Angata et al. 2001; Yu et al. 2001; Floyd et al. 2000; Patel et al. 1999; Crocker and Redelinghuys 2008), implying a specific function for each family member. Indeed, it is believed that sialoadhesin is involved in regulation of macrophage function (Crocker et al. 1997); MAG is implicated in myelinogenesis (Umemori et al. 1994), while CD22 serves to inhibit signaling through B-cell antigen receptor via binding of SHP-1 tyrosine phosphatase to its ITIMs (Cyster and Goodnow 1997; Doody et al. 1995), and

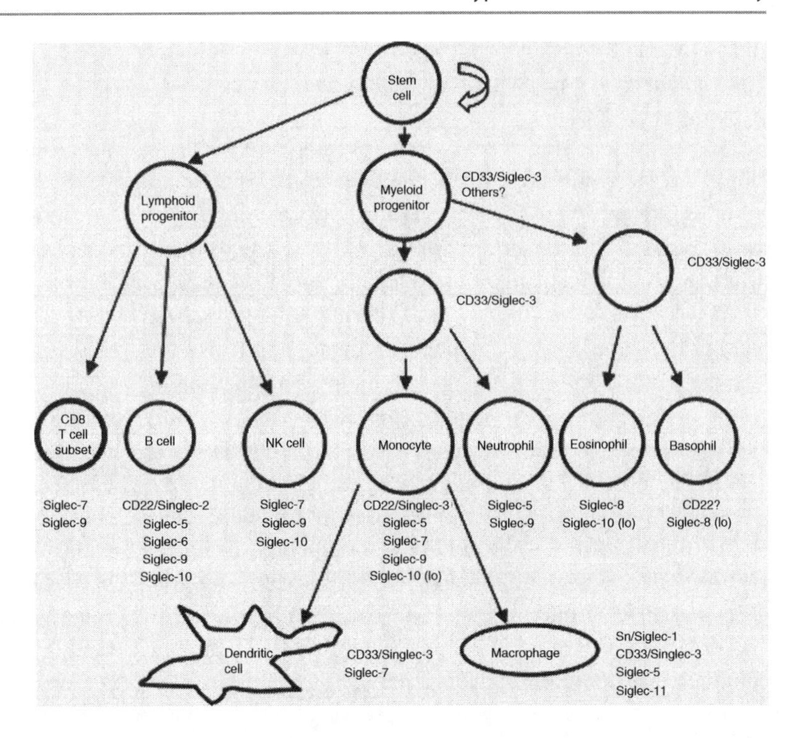

Fig. 16.2 Cell-type specific expression of Siglecs in the hematopoietic and immune cells of humans. The figure shows the distribution of human Siglecs on various cell types (Modified from Crocker and Varki (2001)). Note that expression patterns of Siglecs on bone marrow precursor cell types has not been well studied, except in the case of CD33/Siglec-3. In a few instances, expression is not found in all humans studied (e.g., Siglec-7 and -9 on a small subset of T cells) (Adapted by permission from Varki and Angata (2006) © Oxford University Press)

the lectin function of CD22 is thought to be important in recruiting CD22 to the B-cell antigen receptor (Tedder et al. 1997). Mouse Siglecs (mSiglecs) orthologous to hSiglec-1 (Sn), hSiglec-2 (CD22), and hSiglec-4 (MAG) have been characterized, with latter two including phenotypes with gene disruptions (Otipoby et al. 1996; O'Keefe et al. 1996; Nitschke et al. 2001) .

Siglecs have a different binding pattern and each member shows a distinct specificity for the type of sialic acid with which it interacts. Sialic acids occur naturally in about 30–40 forms, and it has been shown that Sn and MAG are specific for N-acetyl neuraminic acid (Collins et al. 1997), whereas CD22 can bind either N-glycolyl or N-acetyl neuraminic acid (Sjoberg et al. 1994). The glycerol and carboxylate side chains are also essential for binding by Sn and MAG (Collins et al. 1997). The linkage of sialic acid to underlying oligosaccharide chain defines another level of specifity for siglec family members. CD22 is highly specific for α2,6-linked sialic acids (Powell and Varki 1994), whereas sialoadhesin, MAG, and CD33 prefer glycans terminated in α2,3-linked sialic acids (Kelm et al. 1994). The affinities of members of the siglec family for free oligosaccharides are weak (32 μM for CD22; Powell et al. 1995); ~1 mM [Sn and MAG) and for many other lectins (May et al. 1998). Unlike L-selectin and Langerin, which also bind to sulphated analogues of sialyl-Le^x, the siglecs do not give detectable binding signals with sulphated analogues that are lacking sialic acid. The sulphate groups, however, modulate the Siglec binding to the sialyl-Le^x sequence in positive or negative fashion (Campanero-Rhodes et al. 2006).

16.2.4 Siglecs of Sialoadhesin Family

In addition to sialoadhesin or Siglec-1, the subset includes CD22 (Siglec-2), MAG (myelin-associated glycoprotein) (Siglec-4a), Schwann cell myelin protein (Siglec 4b), Siglec-15, all of which are well-conserved in mammals. Sialoadhesin is a macrophage receptor where as CD22 is expressed by B lymphocytes. Both proteins share sequence similarity with MAG, an adhesion molecule of oligodendrocytes and Schwann cells that has been implicated in the process of myelination. Sialoadhesin, CD22, and MAG mediate cell adhesion by binding to cell-surface glycans that contain sialic acid. Whereas sialoadhesin binds equally to the sugar moieties NeuAc α 2—> 3Gal β1—> 3(4) GlcNAc or NeuAc α2—> 3Gal β1—> 3GalNAc, MAG recognizes only NeuAc α2—> 3Gal β1—> 3GalNAc, where as CD22 binds specifically to NeuAc α2—> 6Gal β1—> 4GlcNAc. Moreover, the recognition of sialylated glycans on the surface of particular cell types leads to the selective binding: such as Sn to neutrophils, MAG to neurons and CD22 to lymphocytes (Kelm et al. 1994; Angata 2006; Crocker 2005; Crocker and Zhang 2002). CD22 is a well-characterised B cell restricted siglec that has been shown to mediate both sialic acid-dependent and -independent signaling functions in B cell regulation.

16.3 Sialoadhesin (Sn)/Siglec-1 (CD169)

16.3.1 Characterization of Sialoadhesin/ Siglec-1

Sialoadhesin is a member of Ig superfamily with 17 extracellular domains. The rodent sialoadhesin gene, *Sn*, is located to chromosome 2 F-H1. The human sialoadhesin gene, *Sn*, is localized to the conserved syntenic region on human chromosome 20p13. Hence, the sialoadhesin gene is not linked to the other members of the Sialoadhesin family, CD22, MAG, and CD33, which have been independently mapped to the distal region of mouse chromosome 7 and to human chromosome 19q13.1–3 (Mucklow et al. 1995). The predicted protein sequences of human and mouse Sn are about 72% identical, with the greatest similarity in the extracellular region, which comprises 17 Ig domains (1 variable domain and 16 constant domains) in both species. The expression pattern of human Sn was found to be similar to that of the mouse receptor, being absent from monocytes and other peripheral blood leukocytes, but expressed strongly by tissue macrophages in the spleen, lymph node, bone marrow, liver, colon, and lungs. High expression was also found on inflammatory macrophages present in affected tissues from patients with rheumatoid arthritis (Hartnell et al. 2001).

N-Terminal Sialoadhesin Glycopeptide Domain Mediates Ligand Binding: In Sn, sialic acid binding region has been characterized in detail by X-ray crystallography, nuclear magnetic resonance, and site-directed mutagenesis. Studies indicate that this receptor is likely to function as a macrophage accessory molecule in a variety of cell-cell and cell-extracellular matrix interactions and mediates cell surface interactions through binding of sialylated glycoconjugates. Studies have shown that the membrane distal V-set Ig domain of sialoadhesin contains sialic acid binding site (Vinson et al. 1996). Site-directed mutagenesis of a subset of non-conservative mutations disrupted sialic acid-dependent binding without affecting binding of mAbs directed to two distinct epitopes of Sn. A CD8α-based molecular model predicts that these residues form a contiguous binding site on GFCC'C'' β-sheet of the V-set domain centered around an arginine in F strand. A conservative mutation of this arginine to lysine also abolished binding. This amino acid is conserved among all members of Sn family and is therefore likely to be a key residue in mediating sialic acid-dependent binding of sialoadhesins to cells (Vinson et al. 1996). However, the N-terminal Sn domain can mediate sialic acid-binding on its own. The structure of N-terminal Sn domain, in complex with a sialic acid-contains heptapeptide, (Ala-Gly-His-Thr (Neu5Ac)-Trp-Gly-His). The affinity of Sn for this ligand is four times higher than the affinity for the natural linkage 2,3′-sialyllactose (Bukrinsky et al. 2004).

There is a human Siglec-like molecule (Siglec-L1) that lacks a conserved arginine residue known to be essential for optimal sialic acid recognition by known Siglecs. Loss of arginine from an ancestral molecule was caused by a single nucleotide substitution that occurred after common ancestor of humans with great apes but before the origin of modern humans. The chimpanzee Siglec-L1 ortholog preferentially recognizes N-glycolylneuraminic acid, which is a common sialic acid in great apes and other mammals. Reintroducing the ancestral arginine into human molecule regenerates the same properties. Thus, a single base pair mutation that replaced arginine on human Siglec-L1 is likely to be evolutionarily relative to the previously reported loss of N glycolylneuraminic acid expression in the human lineage. It seems that additional changes in the biology of sialic acids have taken place during human evolution (Angata et al. 2001).

16.3.2 Cellular Expression of Sialoadhesin

Stromal macrophages in lymphohemopoietic tissues express macrophage-restricted plasma membrane receptors involved in nonphagocytic interactions with other hemopoietic cells (originally named sheep erythrocyte receptor). Sn is first detected on fetal liver macrophages on day 18 of development. In spleen and bone marrow, Sn appears between day 18 and birth, in parallel with myeloid development. Sialoadhesin is differentially regulated compared with the erythroblast receptor and F4/80 antigen, that it is not required for fetal erythropoiesis, and that its induction on stromal macrophages is delayed until the onset of myeloid and lymphoid development. Resident peritoneal macrophages do not express high levels of Sn in vitro unless an inducing element found in normal mouse serum is present, but such macrophages did not acquire Sn-independent EbR activity (Van den Berg et al. 1996; Hartnell et al. 2001; van den Berg et al. 2001a). Sn can also be induced on macrophages present at sites of inflammation in both humans and rodents (Jiang et al. 2006; Lai et al. 2001). During chronic inflammation, as occurs during autoimmune disease, high levels of Sn are found on macrophages in inflammatory infiltrates, where it is suggested to mediate local cell-cell interactions (van den Berg et al. 2001a). Among haemopoetic cells, Sn binds preferentially to mature granulocytes (Crocker et al. 1997). It has been proposed that in lymphoid tissues Sn may act as a lymphocyte adhesion molecule (Van den Berg et al. 1992), and its selective expression on macrophages in the marginal zone of the spleen suggests a role in antigen presentation to B cells (Steiniger et al. 1997). However, in chronic inflammatory conditions, such as atherosclerosis and rheumatoid arthritis, Sn is expressed at high levels on inflammatory macrophages. Sn is also detected in intracellular vesicles that were apparently taken up by macrophages. Surprisingly, Sn is also found at contact points of macrophages with other macrophages, sinus-lining cells and reticulum cells, suggesting that it also mediates interactions with these cell types (Schadee-Eestermans et al. 2000). Transcriptional and protein levels of Sn on monocytes in coronary artery disease patients are significantly increased compared with healthy controls, but increased Sn had no correlation with level of serum lipids. Sialoadhesin may be considered as a potential non-invasive indicator for monitoring disease severity and a biomarker for predicting the relative risk of cardiovascular events (Xiong et al. 2009).

Microglia, the resident macrophages of the CNS, reside behind blood–brain barrier and do not express Sn. Microglia and macrophage population in cicumventricular organs, choroid plexus and leptomeninges are exposed to plasma proteins and some macrophages express Sn at these sites. Injury to the CNS, which damages the blood–brain barrier, induces Sn expression on a proportion of macrophages and microglia within the parenchyma (Perry et al. 1992). Macrophages from mesenteric and axillar lymph nodes exhibited higher activity than those from spleen. Quantity of Sn present in lymph node macrophages was 25-fold higher than in splenic macrophages. It suggested that macrophages express high levels of unmasked Sn in lymph nodes. The unmasked forms on these macrophages are available for Sn-dependent adhesive functions, unlike the masked forms on the majority of splenic macrophages (Nakamura et al. 2002).

16.3.3 Ligands for Sialoadhesin

16.3.3.1 Specificity for Sialylated Glycans

Surface proteins on tolerogenic, immature dendritic cells and regulatory T cells are highly α2,6-sialylated, suggesting a glycan motif of tolerogenic cells which might serve as ligand for inhibitory siglecs on the surface of effector cells (Jenner et al. 2006). Sn contains 17 extracellular Ig-like domains, which recognizes oligosaccharides/glycoconjugates containing terminal oligosaccharides NeuNAc2,3-Gal 1,3-GalNAc (disaccharide α-D-Neu5Ac-(2—> 3)-β-D-Gal) or NeuNAc2,3-Gal 1,3-GlcNAc in N- and O-linked glycans, and as such mediates adhesive interactions with lymphoid and myeloid cells (Van den Berg et al. 1992; Crocker et al. 1995). Sn specifically binds to α-2—> 3-sialylated N-acetyl lactosamine residues of glycan chains. The experimental and theoretical STD values indicate that a combined modeling/STD NMR approach yields a reliable structural model for the complex of Sn with α-D-Neu5Ac-(2—> 3)-β-D-Gal-(1—> 4)-D-Glc 1 in aqueous solution (Bhunia et al. 2004). Since, each Siglec exhibits a unique specificity for sialylated glycans, Sn prefers 2,3-linked sialic acids of the Neu5Ac rather than the Neu5,9(Ac)2 or Neu5Gc types (Angata and Brinkman-Vanden Linden 2002; Bhunia et al. 2004). On the other hand, CD22, a B cell-restricted receptor with seven Ig-like domains, selectively recognizes oligosaccharides terminating in NeuAc α2-6Gal in N-glycans. In humans, the amino-terminal V-set Ig-like domain in both proteins is both necessary and sufficient to mediate sialic acid-dependent adhesion of correct specificity. In contrast in murine CD22, only constructs containing both the V-set domain and the adjacent C2-set domain were able to mediate sialic acid-dependent binding (Nath et al. 1995).

The structural diversity of sialic acids influences cell adhesion mediated by molecules like Sn and CD22 in murine macrophages and B-cells. It was shown that the 9-O-acetyl group of Neu5,9Ac2 and the N-glycoloyl residue of Neu5Gc interfere with Sn binding. In contrast, CD22 binds more strongly to Neu5Gc compared to Neu5Ac. Of two synthetic sialic acids tested, only CD22 bound the N-formyl derivative, whereas a N-trifluoroacetyl residue was accepted by Sn (Kelm et al. 1994). Sn-deficient ($Sn^{-/-}$) mice suggest a role for Sn in regulating cells of the immune system rather than in influencing steady-state hematopoiesis (Oetke et al. 2006).

Using a series of synthetic sialic-acid analogues either on resialylated human erythrocytes or as free α-glycosides in hapten inhibition, siglecs required hydroxy group at C-9 for binding, suggesting hydrogen bonding of this substituent with the binding site. Besides, remarkable differences were found among the proteins in their specificity for modifications of N-acetyl group. Whereas Sn, MAG and SMP do not tolerate a hydroxy group as in N-glycolylneuraminic acid, they bind to halogenated acetyl residues. Study indicates that interactions of hydroxy group at position 9 and the N-acyl substituent contribute significantly to the binding strength (Kelm et al. 1998).

16.3.3.2 CD43 as T Cell Counter-Receptor for Sn

Evidences indicate that cell adhesion molecules of Ig family use GFCC'C β-sheet of membrane-distal V-set domains that bind structurally different ligands. Such surface is favored for cell-cell recognition (van der Merwe et al. 1996). Sn has been shown to bind several membrane receptors via both sialic acid-dependent and -independent mechanisms; for instance, the sialomucins leukosialin (CD43) on T cells and MUC-1 on breast cancer cells are putative sialic acid-dependent counter receptors, whereas the macrophage mannose receptor, which is present on several types of myeloid cells (Martinez-Pomares et al. 1996), and the mouse macrophage galactose-type C-type lectin 1 (Kumamoto et al. 2004) binds Sn in a sialic acid-independent manner. Among major glycoproteins (85, 130, 240 kDa) from a murine T cell line (TK-1), CD43 from COS cells supported increased binding to immobilized Sn. Sn-binding glycoproteins were identified as the sialomucins CD43 and P-selectin glycoprotein ligand 1 (PSGL-1 or CD162), corresponding to 130- and 240-kDa respectively. Further more, Sn bound to different glycoforms of CD43 expressed in CHO cells, including unbranched (core 1) and branched (core 2) *O*-linked glycans, that are normally found on CD43 in resting and activated T cells. Results suggest CD43 as a T cell counter receptor for Sn (van den Berg et al. 2001b). The nature of the sialoglycoprotein recognized by Sn on breast cancer cells was a major band of ~240 kDa, and was shown to be the epithelial mucin, MUC1 (Nath et al. 1999).

16.3.4 Sialoadhesin Structure

Domain deletion and site-directed mutagenesis studies (Vinson et al. 1996) have indicated that functionally important sialic acid–binding portion of the molecule is localized on the N-terminal V-set Ig domain. Molecular cloning of murine Sn showed that it has 17 Ig-like domains. The most similar proteins in the database were CD22, MAG, Schwann cell myelin protein and CD33. Low angle shadowing and electron microscopy showed that Sn consisted of a globular head region of approximately 9 nm and an extended tail of approximately 35 nm. Sialoadhesin specifically recognizes oligosaccharide sequence Neu5Ac α2,3Gal β1,3GalNAc in either sialoglycoproteins or gangliosides. Findings imply that specific sialoglycoconjugates carrying this structure may be involved in cellular interactions between stromal macrophages and subpopulations of haemato-poietic cells and lymphocytes. (Crocker et al. 1991, 1994).

1. Proton NMR analysis of Sn: The molecular interactions between Sn and sialylated ligands have been investigated by using proton NMR. Addition of ligands to 12 kDa N-terminal Ig-like domain of Sn results in resonance shifts in the protein. The results indicated that α2, 3-sialyllactose and α2,6-sialyl-lactose bind respectively 2- and 1.5-fold more strongly than does α-methyl-N-acetylneuraminic acid (α-Me-NeuAc). The resonances corresponding to the methyl protons within the N-acetyl moiety of sialic acid undergo up-field shifting and broadening, reflecting an interaction of this group with Trp-2 in Sn as observed in co-crystals of the terminal domain with bound ligand. Affinities of mutant and wild-type forms of Sn in which the first three domains were fused to the Fc region of human IgG, revealed that substitution of Arg^{97} by alanine completely abrogates interaction with α-Me-NeuAc, whereas a conservative substitution with lysine resulted in a 10-fold decrease in affinity. These results confirm the critical importance of conserved arginine in interactions between sialosides and members of Ig-like lectins (Crocker et al. 1999).

2. Crystallographic and in silico analysis of sialoadhesin: The X-ray crystal structures of N-terminal domain of Sn provides important insights into how this transmembrane-spanning receptors functions. A functional fragment of Sn, comprising the N-terminal Ig domain, was expressed in CHO cells as both native (SnD1) and selenomethionyl (Se-SnD1) stop protein. SnD1 in absence and presence of its ligand, 2,3 sialyllactose and Se-SnD1 in absence of ligand have been crystallized. The ligand-free crystals of SnD1 and Se-SnD1 were isomorphous, of space group P3(1)21 or P3(2)21, with unit cell dimensions a = b 38.9 Å, c = 152.6 Å, α = β = 90º, γ = 120º, and diffracted to a maximum resolution of 2.6 A. Cocrystals containing 2,3 sialyllactose diffracted to 1.85 A at a synchrotron source and belong to space group P2(1)2(1)2(1), with unit cell dimensions a = 40.9 Å, b = 97.6 Å, c = 101.6 Å, α = β = γ = 90º (May et al. 1997).

The predicted V-set N-terminal domain of sialoadhesin (SnD1), defined as residues 1–119, contains no O- or N-linked glycosylation sites, and in isolation as a soluble fragment remains competent to bind sialic acids. The structure of the functional N-terminal domain from the extracellular region of sialoadhesin SnD1 was solved in presence of its ligand 3′ sialyllactose to a resolution of 1.85 Å. The structure conforms to the V-set Ig-like fold but contains several distinctive features, including an intra–β sheet disulphide and a splitting of the standard β strand G into two shorter strands. These features appear important in adapting the V-set fold for sialic acid–mediated recognition. Analysis of the complex with 3′ sialyllactose highlights three residues, conserved throughout the siglec family, as key features of sialic acid–binding template. The complex provides information for a heterotypic cell adhesion interaction (May et al. 1998) (Fig. 16.3).

The crystal structure of SnD1 in complex with 2,3-sialyllactose has informed the design of sialic acid analogs (sialosides) that bind Siglecs with significantly enhanced affinities and specificities. Binding assays against sialoadhesin (Siglec-1), CD22 (Siglec-2), and MAG (Siglec-4) showed a 10–300-fold reduction in IC_{50} values (relative to methyl-α-Neu5Ac) for three sialosides bearing aromatic group modifications of the glycerol side chain: Me-α-9-N-benzoyl-amino-9-deoxy-Neu5Ac (BENZ), Me-α-9-N-(naphthyl-2-carbonyl)-amino-9-deoxy-Neu5Ac (NAP), and Me-α-9-N-(biphenyl-4-carbonyl)-amino-9-deoxy-Neu5Ac (BIP). Zaccai et al. (2007) determined the crystal structures of SnD1 in complex with 2-benzyl-Neu5NPro and 2-benzyl-Neu5NAc. These structures reveal that SnD1 undergoes very few structural changes on ligand binding and detail how two novel classes of sialic acid

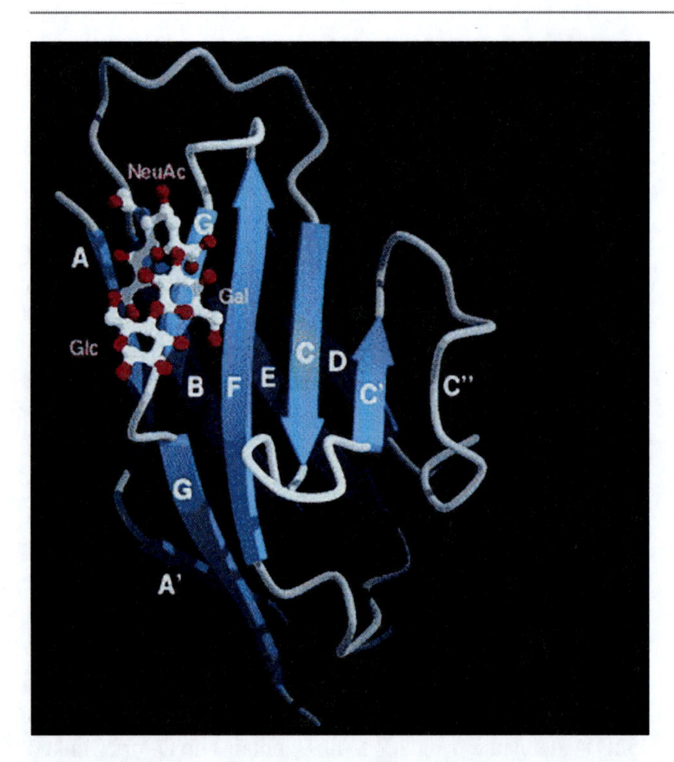

Fig. 16.3 Structure of N-Terminal domain of Sialoadhesin in complex with 3′ Sialyllactose. Each strand is labeled. The 3′ sialyllactose lies along strand G and makes interactions with residues from the A, G, and F strands (Adapted by permission from May et al. (1998) © Elsevier)

analogs bind, one of which unexpectedly can induce Siglec dimerization. In conjunction with in silico analysis, crystal structures of sialosides in complex with SnD1 suggest explanations for the differences in specificity and affinity, providing further ideas for compound design of physiological and potentially therapeutic relevance (Zaccai et al. 2003; 2007).

16.3.5 Regulation of Sialoadhesin

In bone marrow, IL-4 is a potent regulator of sialic acid-specific receptor implicated in macrophage-hemopoietic cell interactions (McWilliam et al. 1992). In bone marrow stromal macrophages, Sn promotes the interactions with developing myeloid cells, and by a subset of tissue macrophages helps in antigen presentation and activation of tumor-reactive T cells. Growth of Ehrlich tumor (ET), a murine mammary carcinoma - may modify the Sn expression by bone marrow macrophages (Kusmartsev et al. 1999). Treatment of bone marrow cells with IFN-γ improved adhesive properties of macrophages, but did not modulate expression of Sn. IL-1 and TNF-α had no effect while combined treatment with these cytokines enhanced binding of sheep erythrocytes to macrophages. Administration of LPS or combined with IFN-γ and TNF-α increased the number of macrophages adhering to plastic and stimulated expression of Sn. Combined treatment with IFN-γ and TNF-α stimulated production of NO by bone marrow macrophages. Blockade of NO synthesis had no effect on adhesive properties of macrophages and expression of Sn (Kusmartsev et al. 2003). IL-13, expressed by activated lymphocytes, markedly suppressed NO release and to a lesser extent secretion of TNF-α by macrophages (Doyle et al. 1994). Glucocorticoids (GC) induce Sn expression on freshly isolated rat macrophages and macrophage cell line R2 in vitro. The cytokines IFN-β, IFN-γ, IL-4, and LPS, although unable to induce Sn expression by themselves, were able to enhance GC-mediated induction of Sn. Effect of GC on Sn expression on rat macrophages can be enhanced by IFN-β, T cell-derived cytokines, or LPS (van den Berg et al. 1996).

16.3.6 Functions of Sialoadhesin

Siglecs are expressed in a highly cell type–specific fashion and appear to be involved in discrete functions ranging from control of myeloid cell interactions (sialoadhesin/CD33; Freeman et al. 1995) to activation of B cells (CD22, Cyster and Goodnow 1997) and regulation of neuronal cell growth and maintenance of myelination in nervous system (MAG) (Li et al. 1994).

Cell-Cell Interactions: The affinity of Sn for sialosides is low (10^{-3} M range), and high-avidity binding requires receptor and ligand clustering. This low affinity, together with a highly extended extracellular domain are key features in permitting Sn to mediate cell–cell interactions, particularly in the plasma microenvironment where the poorly clustered glycans of plasma glycoproteins are unable to compete efficiently with the highly clustered cell-associated glycans involved in avid Sn-binding. Furthermore, cell–cell and cell–matrix interactions are accentuated further by the extension of the N-terminal V-set domain beyond the reach of shorter *cis*-interacting inhibitory siglecs closer to the plasma membrane. Although the expression and features of Sn could potentially influence many macrophage cellular interactions relating to homoeostasis and immunity (Munday et al. 1999), studies on Sn-deficient mice suggest that these may be more important in the regulation of adaptive immune responses. Crocker et al. (2007) reviewed and indicated that Sn-deficient mice exhibit reduced $CD4^+$ T-cell and inflammatory responses in a model of autoimmune uveoretinitis and reduced $CD8^+$ T-cell and macrophage recruitment in models of inherited demyelinating neuropathy in both the central and peripheral nervous systems (Jiang, et al. 2006). Subtle effects on T-cells and IgM antibody levels were also seen in Sn-deficient mice maintained under specific pathogen-free conditions (Oetke et al. 2006). The sialomucin CD43 on T-cells (van den Berg et al. 2001b) and mucin-1 on breast cancer cells have both been identified as putative Sn counter-receptors.

The rat murine and rat alveolar macrophages express Sn and that sialic acid-dependent receptor (SAR) and sheep erythrocyte receptor (SER)-like activities are mediated by Sn. The mouse and rat Sn on macrophages can function as a lymphocyte adhesion molecule. Sn has been implicated in cellular interactions of stromal macrophages with developing myeloid cells. In all assays, Sn exhibited specific, differential binding to various murine cell populations of hemopoietic origin. In rank order, sialoadhesin bound neutrophils > bone marrow cells = blood leukocytes > lymphocytes > thymocytes. Single-cell analyses confirmed that Sn selectively bound myeloid cells in complex cell mixtures obtained from the bone marrow and blood. This gives the notion that Sn is involved in interactions with granulocytes at different stages of their life histories (Crocker et al. 1995). Thus, Sn provides the example of a macrophage-restricted lymphocyte adhesion molecule (van den Berg et al. 1992). Umansky et al. (1996) described the functional role of SER^+ spleen macrophages in antigen processing and presentation to T lymphocytes. In two syngeneic murine tumor systems (ESb-MP and lacZ transduced ESbL T-lymphoma cells), it was suggested that in situ-activated SER^+ macrophages contribute to host resistance against metastasis (Umansky et al. 1996). The in vivo interactions of T lymphocytes in the Tn syndrome with CD22 are not likely to be affected, whereas adhesion mediated by Sn or MAG could be strongly reduced (Mrkoci et al. 1996).

Importantly, Sn-mediated interactions appear to be important for effective killing of tumor cells by CTLs in a murine tumor model in vivo (van den Berg et al. 1996). Although Sn function is not known, it is involved in the attachment and internalization of certain viruses (Hartnell et al. 2001; Vanderheijden et al. 2003), and has the potential to endocytose sialylated bacteria such as *Neisseria meningitidis* (Jones et al. 2003). In addition, Sn has been known to promote adhesion of macrophages to T cells and to other cell types such as neutrophils and macrophages (Muerkoster et al. 1999). Because Sn is expressed on subsets of inflammatory macrophages, it can serve as a marker of activation. Anti-inflammatory treatments, for instance with IL-11, are associated with reduction in overall tissue damage with a selective decrease in the number of Sn^+ macrophages (Lai et al. 2001).

Generation of Activated Sn-Positive Microglia During Retinal Degeneration: The retina contains a rich network of myeloid-derived cells (microglia) within the retinal parenchyma and surrounding vessels. Their response and behavior during inflammation and neurodegeneration has been examined during the onset of photoreceptor degeneration in the rods of mouse and to assess their role in photoreceptor apoptosis. During retinal degeneration, activated microglia expressed Sn. The temporal relationship between photoreceptor apoptosis and microglial response suggests that microglia are not responsible for the initial wave of photoreceptor death, and this was corroborated by the absence of iNOS and nitrotyrosine. Expression of Sn may indicate blood-retinal barrier breakdown, which has immune implications for subretinal gene therapeutic strategies (Hughes et al. 2003).

16.3.7 Lessons from Animal Experiments

Sn is expressed at high levels on discrete subsets of tissue macrophages, particularly those found in secondary lymphoid tissues (Munday et al. 1999). High expression is also seen in chronic inflammatory diseases such as rheumatoid arthritis (Hartnell et al. 2001), atherosclerosis (Gijbels et al. 1999) and models of inherited demyelinating diseases of the nervous system (Ip et al. 2007). Sn-deficient mice exhibit changes in B- and T-cell populations suggesting that sialoadhesin regulates cells of immune system rather than influencing steady-state hematopoiesis (Oetke et al. 2006). In experimental autoimmune uveoretinitis (EAU), different macrophage surface markers are expressed during different stages of EAU. Sn expression occurs at peak and later stages of the disease, not during initiation of EAU Jiang et al. (1999). In Sn-deficient (Sn-KO) mice model, EAU was reduced in severity in the initial stages. Furthermore, activated T cells from the draining lymph nodes of Sn-KO mice secreted lower levels of IFN-γ. It suggested that Sn plays a role in “fine tuning” of the immune response to autoantigens by modulating T cell priming (Avichezer et al. 2000; Jiang et al. 2006).

Role in Renal Disease: Accumulation of Sn^+ macrophages is a marker of disease progression versus remission in rat mesangial proliferative nephritis. Sn^+ macrophages were localized in areas of focal glomerular and interstitial damage. Accumulation of Sn^+ macrophage subset in the kidney correlated with proteinuria and histologic damage. It suggests that Sn^+ macrophages may play an important role in progressive renal disease (Ikezumi et al. 2005). IL-11, a cytokine with anti-inflammatory activity reduced the number of Sn^+ macrophages in nephrotoxic nephritis rats and markedly reduced glomerular injury and macrophage Sn expression, but without an alteration of macrophage numbers, suggesting that IL-11 may be acting in part to reduce macrophage activation (Lai et al. 2001).

Sialoadhesin Deficiency and Myelin Degeneration: Mouse mutants heterozygously deficient for the myelin component P0 mimic some forms of inherited neuropathies and offer future treatment strategies for inherited demyelinating neuropathies in humans. During search of possible role of macrophage-restricted Sn in the pathogenesis of inherited demyelination in $P0^{+/-}$ mice, it was found that most peripheral nerve macrophages express Sn in the mutants. Myelin mutants devoid of Sn showd reduced numbers of $CD8^+$ T lymphocytes and macrophages in peripheral nerves and less severe demyelination, resulting in improved nerve conduction properties (Kobsar et al. 2006). By cross-breeding these mutants with RAG-1-deficient mice lacking mature lymphocytes, pathogenetic impact of the $CD8^+$ cells was demonstrated. Ip et al. (2007) investigated the pathogenetic impact of $CD11b^+$ macrophages by cross-breeding the myelin mutants mice deficient for Sn. In the wild-type mice, Sn is barely detectable on $CD11b^+$ cells, whereas in the myelin mutants, almost all $CD11b^+$ cells expressed Sn. In the double mutants, upregulation of $CD8^+$ T-cells and $CD11b^+$ macrophages was reduced and pathological alterations were ameliorated. These results suggest that in a genetically caused myelin disorder of CNS, macrophages expressing Sn partially mediate pathogenesis. These results have substantial impact on treatment strategies for leukodystrophic disorders and some forms of multiple sclerosis.

16.3.8 Interactions with Pathogens

Candida albicans yeast cells specifically adhere to mouse macrophages in the splenic marginal zone and in lymph node subcapsular and medullary sinuses. These macrophages express Sn that binds erythrocytes, but binding of yeast cells is not mediated by Sn (Han et al. 1994). *Trypanosoma cruzi* is a parasite with large amounts of sialic acid residues exposed at its surface that seems to be involved in macrophages infection. Some macrophages, present in *T. cruzi* infected tissues, expresses Sn. Sn was induced in mice peritoneal macrophages by homologous serum (HS) cultivation. Desialylation reduced the association of parasites to HS cultured macrophages indicating importance of Sn. Sn positive macrophages seem to be important in the initial trypomastigote infection and in the establishment of Chagas disease (Monteiro et al. 2005).

Internalization of Porcine Arterivirus: Porcine Sn mediates internalization of the arterivirus porcine reproductive and respiratory syndrome virus (PRRSV) in alveolar macrophages (Vanderheijden et al. 2003). α2-3- and α2-6-linked sialic acids on the virion are important for PRRSV infection of porcine alveolar macrophages (PAM). It suggested that pSn is a sialic acid binding lectin which interacts with sialic acid on PRRS virion and essential for PRRSV infection of PAM (Delputte and Nauwynck 2004, 2006; Genini et al. 2008). The p210 protein involved in infection of PAM showed sequence identities ranging from 56% to 91% with mouse Sn. The full p210 cDNA sequence (5,193 bp) shared 69 and 78% amino acid identity, respectively, with mouse and human sialoadhesins. Results show that sialoadhesin is involved in the entry process of PRRSV in PAM (Vanderheijden et al. 2003). Study of attachment kinetics of PRRSV to macrophages revealed that early attachment is mediated mainly via an interaction with heparan sulphate, followed by a gradual interaction with Sn. By using wild-type CHO and CHO deficient in heparan sulphate expression, it was shown that heparan sulphate alone is sufficient to mediate PRRSV attachment, but not entry, and that heparin sulphate is not necessary for Sn to function as a PRRSV internalization receptor, but enhances the interaction of the virus with Sn (Delputte et al. 2005).

In addition to its role in cell adhesion, Sn has also been shown to facilitate pathogen interactions. For example, Sn can promote macrophage uptake of sialylated strains of *Neisseria meningitidis* (Jones et al. 2003) and functions in endocytosis of the macrophage/monocyte-tropic PRRSV (porcine reproductive and respiratory syndrome virus) (Delputte et al. 2005) (Fig. 16.4) and internalization is thought to be triggered via interactions between Sn and sialylated N-linked glycans present on the four structural viral glycoproteins GP_2–GP_5 of PRRSV (Vanderheijden et al. 2003; Delputte and Nauwynck 2004). HIV-1 has also been shown to interact with Sn (Rempel et al. 2008). During acute period of HIV-1 infection, IFNγ is produced by NK cells and T-cells, and IFNα released by pDCs (plasmacytoid dendritic cells) as part of antiviral response may lead to induction of Sn on monocytes. This in turn binds to the virus in a sialic acid-dependent manner and may permit *trans*-infection of permissive cells and the delivery and distribution of HIV-1 to target cells. This suggests that Sn may be involved in both capture of free virus through sialylated glyconjugates on target cells such as T-cells, and enhancing the binding of HIV-1 gp120 to its cognate receptors (Rempel et al. 2008).

16.4 CD22 (Siglec-2)

16.4.1 Characterization and Gene Organization

The human CD22 gene is expressed specifically in B lymphocytes and likely has an important function in cell-cell interactions. Sequence analysis of a full-length B cell cDNA clone revealed an open reading frame of 2,541 bases coding for a predicted protein of 847 amino acids with a molecular mass of 95 kDa. The B lymphocytes-CAM cDNA is nearly identical to a cDNA clone for CD22, with the exception of an additional 531 bases in the coding region of BL-CAM. BL-CAM has a predicted transmembrane spanning region and a 140-amino acid intracytoplasmic domain. This protein had significant homology with three homotypic cell adhesion proteins: carcinoembryonic antigen (29% identity over 460 amino acids), MAG, and neural cell adhesion molecule (NCAM) (21.5% over 274 amino acids). BL-CAM mRNA expression was increased after B cell activation with *S. aureus* Cowan strain 1 and phorbol myristate acetate, but not by various cytokines. An antisense BL-CAM RNA probe revealed expression in B cell-rich areas in tonsil and lymph node, although the most striking hybridization was in the germinal centers (Wilson et al. 1992).

Genomic Structure: A full length human CD22 cDNA clone, spread over 22 kb of DNA, is composed of 15 exons. The first exon contains the major transcriptional start sites. The translation initiation codon is located in exon 3, which also encodes a portion of signal peptide. Exons 4–10 encode the seven Ig domains of CD22, exon 11 encodes the transmembrane domain, exons 12–15 encode the intracytoplasmic domain of CD22, and exon 15 also contains the 3' untranslated region. A minor form of CD22 mRNA results from splicing of exon 5 to exon 8, skipping exons 6 and 7. TheCD22 gene is located within the band region q13.1 of chromosome 19. Two closely clustered major transcription start sites and several minor start sites were mapped by primer extension. Similarly to many other lymphoid-specific genes, the CD22 promoter lacks an obvious TATA box. Approximately 4 kb of DNA 5' of the transcription start sites were sequenced and found to contain multiple Alu elements. Potential binding sites for the transcriptional factors NF-kB, AP-1, and Oct-2 are located within 300 bp 5′ of the major transcription start sites. A 400-bp fragment (bp −339 through +71) of CD22 promoter region was found to be active in both B and T cells (Wilson et al. 1991, 1993).

Gene Variations in Autoimmune Diseases: Investigations of the pathogenic role of autoantibodies in rheumatic diseases, studies suggest a more central role of B cells in the maintenance of the disease beyond just being precursors of (auto) antibody-producing plasma cells. Detailed analyses have implicated a number of surface molecules and subsequent downstream signaling pathways in regulation of events induced by BCR engagement. After screening of CD22 coding region, seven non-synonymous and four synonymous substitutions were found associated with rheumatic diseases in Japanese patients with systemic lupus erythematosus (SLE), and patients with rheumatoid arthritis (RA). In addition, single base substitutions were found in two introns flanking exon-intron junctions. Among these variations, Q152E substitution within the second extracellular domain was observed with a marginally higher frequency in patients with SLE (3/68, 4.4%) than in healthy individuals (1/207, 0.5%), although this difference was no longer significant after correction for number of comparisons. No significant association was observed between any of the variations and RA (Hatta et al. 1999).

Activating and inhibitory receptors have been implicated both in human systemic sclerosis (SSc) and tight-skin mouse, a model for SSc. A SNP in CD22 were genotyped in Japanese patients with SSc. At c.2304 C > A SNP coding for a synonymous substitution in exon 13, A/A genotype was observed in six patients with SSc (4.8%) but none in the controls. All six patients with A/A genotype belonged to lcSSc subgroup (7.6%). Surface expression level of CD22 tended to be lower in B cells from the patients with A/A genotype as compared with C/A or C/C genotype (17% decrease). Taken together with observation on CD19 polymorphism, the expression level of CD22 was suggested to play a causative role in a proportion of patients with lcSSc (Hitomi et al. 2007).

16.4.2 Functional Characteristics

CD22/Siglec-2 is a B cell–specific glycoprotein expressed in the cytosol of pre- and pro-B cells, and on the plasma membrane of mature B cells (Cyster and Goodnow 1997; Law et al. 1994, 1996; Tedder et al. 1997). The predominant form of cell surface CD22 (CD22ß) is a 140-kDa type I transmembrane protein. The CD22 associates with B cell receptor (BCR) both physically and functionally (Leprince et al. 1993; Peaker and Neuberger 1993) and negatively regulates BCR signaling (Doody et al. 1995). Six tyrosine residues are located in the cytosolic portion of CD22, and BCR stimulation can induce phosphorylation on some of them. CD22-deficient mice generally show a hyperreactive B cell phenotype (Nitschke et al. 1997; O'Keefe et al. 1996; Otipoby et al. 1996), affirming that the primary function of CD22 is to dampen BCR signaling.

Studies suggest that cell surface CD22 undergoes constitutive endocytosis and degradation (Chan et al. 1998; Shan and Press 1995). Because CD22 expression level on B cells is significantly reduced in ST6Gal-I knockout mouse, where

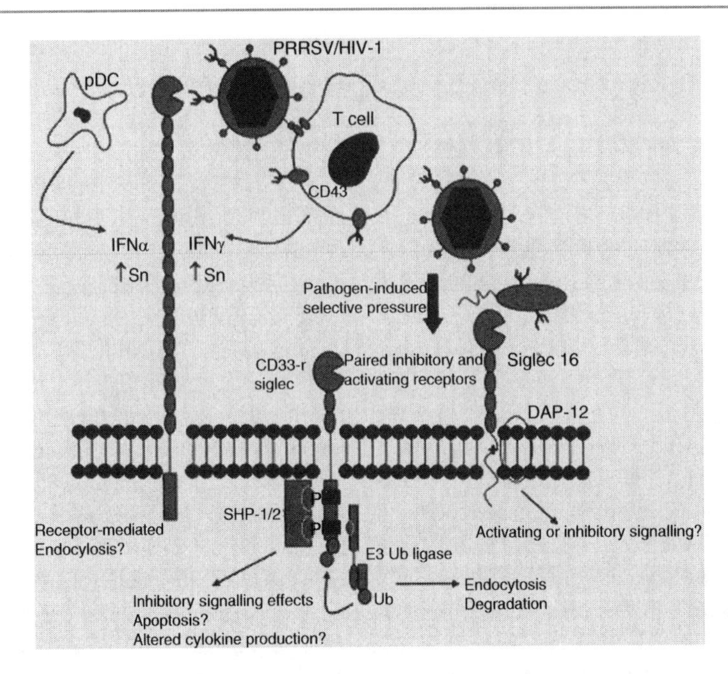

Fig. 16.4 Sn is a highly extended siglec which mediates cell–cell and cell-matrix interactions. Sn expression is induced by IFNs produced by pDCs and T-cells, and it acts both as an endocytic receptor for PRRSV as well as a potential mediator of the *trans*-infection of T-cells by HIV-1. Sialylated pathogens may have driven the evolution of siglecs such as siglec-14 and -16 into DAP-12-associated activating receptors which are paired with inhibitory counterparts such as siglec-5 and -11. Tyrosyl phosphorylation, ubiquitination and internalization are mechanisms by which siglecs transmit their inhibitory and activating signals in response to ligand binding (*Ub* ubiquitin) (Adapted by permission from Crocker and Redelinghuys (2008) © Biochemical Society)

CD22 ligand formation is abolished (Hennet et al. 1998), it is possible that cis binding sialic acid ligands might help maintain optimal CD22 levels on the cell surface by restricting its rate of endocytosis. Zhang and Varki (2004), while examining antibody-mediated endocytosis of CD22, an issue that is important in anti-CD22-based immunotherapy of B cell leukemias and lymphomas (Tuscano et al. 2003), suggested a mechanism how sialic acid-binding and sIgM ligation affect these processes. An approach for simultaneous biotinylation and cross-linking showed that CD22 associates with CD45 and sIgM, possibly involving cell surface multimers of CD22. Sialic acid removal or mutation of a CD22 arginine residue required for sialic acid recognition did not affect these associations even in human:mouse heterologous systems, indicating that they are primarily determined by evolutionarily conserved protein–protein interactions. Thus masking of the sialic acid-binding site of CD22 involves many cell surface sialoglycoproteins, without requiring specific ligand(s) and/or is mediated by secondary interactions with sialic acids on CD45 and sIgM (Zhang and Varki 2004).

N-Linked Glycosylation Site in Ligand Recognition: Site-specific mutagenesis of potential N-linked glycosylation sites on CD22 showed that mutation of a single potential N-linked glycosylation site in first Ig-domain of CD22 completely abrogates ligand recognition. Interestingly, this site is characterized by the sequence NCT, where the cysteine is thought to be involved in an intrachain disulfide bond. Site-directed mutagenesis of similar NC(T/S) motifs in the first or second Ig domains of MAG, and Sn did not disrupt their ability to mediate sialic acid binding. In contrast, mutation of a NCS motif in first Ig domain of CD33 (Siglec-3) unmasked its sialic acid binding activity. Thus, a single N-linked glycosylation site located at a similar position in the CD22 and CD33 glycoproteins is critical for regulating ligand recognition by both receptors (Nath et al. 1995; Sgroi et al. 1996).

16.4.3 Ligands of CD22

As in other members of the family, the extracellular N-terminal Ig domain of CD22 binds to glycan ligands containing sialic acid, which are highly expressed on B-cell glycoproteins, in a highly linkage specific manner. Human and mouse CD22 bind selectively to 2-6-linked sialic acids (Engel et al. 1993; Kelm et al. 1994; Powell et al. 1995; Powell and Varki 1994; Sgroi et al. 1996). The amino-terminal Ig-like V-set domain is critical in this binding (Engel et al. 1995), requiring a conserved arginine residue that likely forms a salt bridge with the carboxylate groups of sialic acid ligand (Van der Merwe et al. 1996). Like most of the other Siglecs, CD22 is natively bound to sialylated cis ligands on the same cell surface (Razi and Varki 1998, 1999). This "masking" effect is abolished by sialidase treatment, and a small amount of "unmasking" has been found on activated human B cells (Razi and Varki 1998). Desialylation of in vivo macrophage sialylconjugates enhances Sn-mediated lectin activity. Receptor sialylation of soluble Sn inhibits its binding to Jurkat cell ligands, and that charge-dependent repulsion alone cannot explain this inhibition. Moreover, the inhibitory effect of sialic acid is partially dependent on the presence of an intact exocyclic side chain. Sialylation of siglecs by specific glycosyltransferases may be a common mechanism by which siglec-mediated adhesion is regulated (Barnes et al. 1999). Thus, different studies showed that cis binding of sialic acid by CD22 is required for its optimal function as an inhibitory regulator of BCR (Jin et al. 2002; Kelm et al. 2002). Notably, CD45, surface IgM and other glycoproteins that bind to CD22 in vitro do not appear to be important cis ligands of CD22 in situ. Instead, CD22 seems to recognize glycans of neighboring CD22 molecules as cis ligands, forming homomultimeric complexes (Han et al. 2005).

The physiological cis ligands for CD22 are not well defined (Tedder et al. 1997). Several molecules that carry 2-6-linked sialic acids, for example, cell surface IgM (sIgM) or CD45, as well as circulating glycoproteins such as IgM and haptoglobin have been suggested as candidate ligands (Hanasaki et al. 1995a; Sgroi et al. 1996). Studies hypothesize that associations of CD22 with sIgM and CD45 are mediated by CD22 recognition of their sialic acid residues (Collins et al. 2002; Cyster and Goodnow 1997). Support for this hypothesis comes from the finding that B cells from mice deficient in sialylated CD22 ligands show constitutive unmasking of CD22 and altered sIgM signaling responses (Hennet et al. 1998). However, there is as yet no direct proof for a role of sialic acids in forming or maintaining specific interactions of other proteins with CD22. It was shown that although a low-density CD22-Fc column selectively interacted with sIgM and haptoglobin from blood plasma, a high-density column could bind most of the 2–6 sialylated glycoproteins in the sample (Hanasaki et al. 1995a, b). Mild detergent extracts detected very limited (1–2%) interactions of CD22 with sIgM (Law et al. 1994, 1996; Leprince et al. 1993; Peaker and Neuberger 1993).

Binding to CD45 and Synthetic Oligosaccharide: One prominent ligand for CD22 is the highly glycosylated leukocyte surface protein CD45, which carries α2-6-linked sialic acid on N-glycans. In situ desialylation and resialylation of immobilized CD45-thy, mouse CD22 binds to the sialoglycoconjugate NeuGc α2-6Gal β1-4GlcNAc on CD45-thy N-glycans. Evidences indicate that cell adhesion molecules (CAMs) of Ig family use GFCC'C β-sheet of membrane-distal V-set domains that bind structurally different ligands. Such surface is favored for cell-cell recognition (van der Merwe et al. 1996). Using surface plasmon resonance, thermodynamic analysis of CD22 binding to native CD45 showed a low affinity ($K_D = 130$ μM at 25 º C) with very fast kinetics. Binding reaction was enthalpically driven at physiological temperatures, as found in most lectin-carbohydrate interactions. Since CD22 binds preferably to CD45, even though many cell surface proteins carry α2-6-linked sialic acid, comparison of affinities of CD22 to CD45, with other sialoglyco-conjugates carrying α2-6-linked sialic acid as CD4, and to a synthetic sialoglyco-conjugate, did not differ significantly. This suggested that CD22 binds preferentially to CD45 not because the latter presents higher affinity ligands but because it carries multiple copies of CD45 (Bakker et al. 2002).

Surface 9-O-Acetylation and Recognition Processes: O-Acetylation of 9-hydroxyl group of sialic acids has been suggested to modify various recognition phenomena involving these molecules. The extent of 9-O-acetylation of surface sialic acids on murine erythroleukemia (MEL) cells can be modified by various manipulations, including differentiation, nocodazole treatment, and 9-O-acetyl esterase treatment (Shi et al. 1996a). Induced differentiation of MEL cells causes resistance to lysis, and this correlates directly with extent of decrease in 9-O-acetylation. A similar resistance to alternative pathway lysis can be obtained by selective enzymatic removal of 9-O-acetyl groups from sialic acids. Thus, a 9-O-acetyl group added to the side chain of cell surface sialic acids may abrogate its normal function in restricting alternative pathway activation. MEL cells are also known to have cell surface ligands for Sn and CD22. Sialoadhesin (but not CD22) binding is selectively enhanced by differentiation-induced loss of cell surface 9-O-acetylation and by direct enzymatic removal of ester groups. Since Sn is expressed on some macrophages in vivo, it was reasoned and proved that tissue homing of MEL cells might be affected by O-acetylation. In particular, de-O-acetylation caused significant increase in homing to the liver and spleen. These results indicate that cell surface 9-O-acetylation can affect a variety of biological recognition phenomena and provide a system for further exploration of molecular mechanisms involved (Shi et al. 1996b).

Sialylated Multivalent Antigens Inhibit B Cell Activation: Although antigens can display CD22 ligands, the receptor is known to bind to α2-6-sialylated glycan as a specific ligand on the cell surface. Kimura et al. (2007) proposed that α2-6-sialylated and 6-GlcNAc-sulfated determinant serve as a preferred ligand for CD22. The α2-6-sialylated 6-sulfo-LacNAc determinant serves as an endogenous ligand for human CD22 that suggests the possibility that 6-GlcNAc sulfation as well as α2-6-sialylation may regulate CD22/Siglec-2 functions in humans.

A ligand for CD22 was identified on human T cells as a low molecular mass isoform of leukocyte common Ag, CD45RO. Murine and human sequences overall have 62% identity, which includes 18 of 20 extracellular cysteines and six of six cytoplasmic tyrosines. BHK cells transfected with mCD22 cDNA specifically adhere to resting and activated T lymphocytes and in addition bound activated, but not resting, B cells. The propinquity of CD22 and cell-surface glycoprotein ligands has led to the conclusion that the inhibitory properties of the receptor are due to cis interactions. Courtney et al. (2009) examined the functional consequences of trans interactions by employing sialylated multivalent antigens that can engage both CD22 and the BCR. Exposure of B cells to sialylated antigens results in the inhibition of key steps in BCR signaling. These results reveal that antigens bearing CD22 ligands are powerful suppressors of B cell activation. The ability of sialylated antigens to inhibit BCR signaling through trans CD22 interactions reveals a previously unrecognized role for the Siglec-family of receptors as modulators of immune signaling.

Loss of N-Glycolylneuraminic Acid in Human Evolution: The common sialic acids of mammalian cells are N-acetylneuraminic acid (Neu5Ac) and N-glycolylneuraminic acid (Neu5Gc). Humans are exception, because of a mutation in cytidine-50-monophosphate (CMP)-sialic acidhydroxylase, which occurred after our ancestor, great apes. The resulting loss of Neu5Gc and increase in Neu5Ac in humans altered the biology of the siglecs, which recognize sialic acids. Human siglec-1 (Sn) strongly prefers Neu5Ac over Neu5Gc. Thus, humans have a higher density of siglec-1 ligands than great apes. Siglec-1-positive macrophages in humans are found primarily in perifollicular zone, whereas in chimpanzees they also occur in the marginal zone and surrounding the periarteriolar lymphocyte sheaths. Although only a subset of chimpanzee macrophages expresses siglec- 1, most human macrophages are positive. A known evolutionary difference is the strong preference of mouse siglec-2 (CD22) for Neu5Gc, contrasting with human siglec-2, which binds Neu5Ac equally well. In fact, siglec-2 had evolved a higher degree of recognition flexibility before Neu5Gc was lost in humans. Human siglec-3 (CD33) and siglec-6 (obesity-binding protein 1) also recognize both Neu5Ac and Neu5Gc, and siglec-5 may have some preference for Neu5Gc. Others showed that siglec-4a (MAG) prefers Neu5Ac over Neu5Gc. In fact, siglec-2 had evolved a higher degree of recognition flexibility before Neu5Gc was lost in humans. Thus, the human loss of Neu5Gc may alter biological processes involving siglec-1, and possibly, siglec-4a or -5 (Brinkman-Van der Linden et al. 2000).

16.4.4 Regulation of CD22

IFNα, a potent antiviral cytokine and immune modulator, induces Sn expression in monocytes which normally do not express the receptor and also to increase Sn expression in macrophages (York et al. 2007). IFNγ produced by activated T-cells and NK cells has also been shown to induce Sn expression on monocytes (Hartnell et al. ; Rempel et al. 2008). The induction of Sn expression in cells of the monocyte–macrophage lineage by IFNs may play a role in the potentiation of inflammatory diseases including rheumatoid arthritis (where Sn serves as a restricted inflammatory marker for tissue macrophages), systemic sclerosis (York et al. 2007), SLE (systemic lupus erythematosus) (Biesen et al. 2008) and proliferative glomerulonephritis (Ikezumi et al. 2005).

Inhibitory receptors are involved in negatively regulating B cell immune response and in preventing autoimmunity. The IL-4 reduces expression of CD22, FcγRII (CD32), CD72, and paired Ig-like receptor (PIR)-B on activated B cells at m RNA and protein level. This reduced expression is dependent on continuous exposure to IL-4 and is mediated through Stat6 (Rudge et al. 2002). Furthermore, treatment of SLE patients with glucocorticoids resulted in a strong decrease in Sn^+ monocytes, suggesting that Sn may be a useful biomarker for disease monitoring in response to therapeutic treatments. High expression is seen in chronic inflammatory diseases such as rheumatoid arthritis (Hartnell et al. 2001), atherosclerosis (Gijbels et al. 1999) and models of inherited demyelinating diseases of the nervous system (Ip et al. 2007). Moreover, a correlation between the frequency of circulating Sn^+ monocytes and the titres of anti-dsDNA (double-stranded DNA) auto-antibodies indicates that the expression of Sn closely parallels extent of SLE disease (Biesen et al. 2008).

Mucins isolated from colon cancer cells and bovine submaxillary mucins bound to CD22 cDNA transfectants and a human B cell line, Daudi cell. Results suggest that in the tumor-bearing state a portion of the mucins in the bloodstream was taken up by the spleen and ligated to CD22 expressed on splenic B cells, which may have led to down-regulation of signal transduction (Toda et al. 2008).

16.4.5 Functions of CD22

B cells express two members of the Siglec family, CD22 (Siglec-2) and Siglec-G, both of which have been shown to inhibit B-cell signaling. Interaction with antigen-presenting accessory cells is thought to be an important step in B-cell activation, and the B-cell receptor CD22, which is coordinately expressed with surface immunoglobulin, has been proposed to participate in the antigen response. Stamenkovic and Seed (1990) showed that CD22, a structure related to MAG (a NCAM), mediates monocyte and erythrocyte adhesion. Like CD2, the CD22 may facilitate antigen recognition by promoting antigen-nonspecific contacts with accessory cells.

Regulation of B Cell Development and B Cell Signaling by CD22: Engagement of CD22 with a mAb (HB22.23) results in rapid CD22 tyrosine phosphorylation and in increased association of CD22 with p53/56lyn kinase, p85 phosphatidyl inositol-3 kinase, and p72syk kinase. Synthetic peptides that span various regions of CD22 showed that these kinases associated with a tyrosine-phosphorylated peptide, which spans tyrosine amino acid residues 822 and 842, and implicate this as an important region in mediating CD22 signal transduction. Engagement of CD22 with HB22.23 was sufficient to stimulate normal B cell proliferation and indicates CD22 as a B lymphocyte signaling molecule (Tuscano et al. 1996). CD22 is an inhibitory co-receptor of BCR-mediated signaling and binds specifically to glycan ligands containing α2,6-linked sialic acids. B cells deficient in enzyme (ST6Gal I) that forms the CD22 ligand (α2,6-linked sialic acids) show suppressed BCR signaling. Mice deficient in receptor/ligand pair (double mutant) in both ST6GalI-deficient and ST6GalI x CD22 showed normal B cell development, but an impaired marginal zone B cell population in the spleen. Both types of mutant mice also showed a reduced population of bone marrow recirculating B cells, a defect previously detected in $CD22^{-/-}$ mice. This suggests a direct involvement of CD22 and its ligands α2,6-linked sialic acid in a homing process of recirculating B cells to the bone marrow. Interestingly, defective B cell Ca^{2+} signaling and proliferation of $ST6Gal^{-/-}$ mice was rescued in ST6GalI x CD22-deficient mice. These studies suggest an important function for CD22-ligand interaction in regulating BCR signal and microdomain localization (Collins et al. 2006b; Ghosh et al. 2006). It has been observed that CD22 recruits the tyrosine phosphatase SHP-1 ITIMs and inhibits BCR-induced Ca^{2+} signaling on normal B cells. CD22 interacts specifically with ligands carrying α2-6-linked sialic acids. Interaction with these ligands in cis regulates the association of CD22 with BCR and thereby modulates the inhibitory function of CD22. Interaction of CD22 to ligands in trans can regulate both B-cell migration as well as BCR signaling threshold (Nitschke 2009).

Siglec-G is a recently identified protein with an inhibitory function restricted to a B-cell subset, the B1 cells. Siglec-G inhibits Ca^{2+} signaling specifically in these cells. In addition, it controls the cellular expansion and antibody secretion of B1 cells. Nitschke (2009) indicated that both Siglecs CD22 and Siglec-G modulate BCR signaling on different B-cell populations in a mutually exclusive fashion.

Recognition of Pathogens and Endocytosis: Antibody ligation of siglec proteins initiates their endocytosis, suggesting that endocytic activity is also a general property of this subfamily (Biedermann et al. 2007; Jones et al. 2003; Nguyen et al. 2006; Walter et al. 2005; Zhang et al. 2006). Over 20 pathogenic microorganisms express sialic acid-containing glycans on their surface (Crocker 2005). Demonstration of the binding or uptake of several sialylated pathogens, including *Neisseria meningitidis*, *Trypanosoma cruzi*, *Campylobacter jejuni*, and group B *Streptococcus*, by Sn and Siglec-5, -7, and -9 has suggested various roles of these siglec proteins in the immune responses to these organisms (Avril et al. 2006; Crocker 2005; Delputte et al. 2005; Jones et al. 2003; Monteiro et al. 2005). Since mechanisms of pathogen entry and engulfment are increasingly recognized to involve the host cells' endocytic machinery, the endocytic functions of the siglec proteins are also relevant to their roles in pathogen recognition and uptake.

16.4.6 Signaling Pathway of Human CD22 and Siglec-F in Murine

16.4.6.1 Many siglec proteins contain one or more ITIMs

Many siglec proteins contain one or more immunoreceptor tyrosine-based inhibitory motifs (ITIMs), (I/L/V)XYXX (L/V), suggesting that they play important roles as inhibitory receptors of cell signaling (Crocker 2005; Varki and Angata 2006), as exemplified by CD22, which is well documented as a regulator of BCR signaling. Upon antigen binding to the BCR, the ITIMs of CD22 are quickly tyrosine phosphorylated and recruit protein tyrosine phosphatase SHP-1, which dephosphorylates the BCR and dampens the B-cell response, setting a threshold for B-cell activation (Crocker et al. 2007; Tedder et al. 2005).

CD22 is also known to undergo endocytosis following ligation by anti-CD22 antibody (Jones et al. 2003) or high-affinity multivalent-sialoside ligands (Collins et al. 2006a). Sn and most CD33-related siglec proteins are expressed on cells of the innate immune system, including monocytes, macrophages, neutrophils, eosinophils, and dendritic cells (Crocker 2005; Lock et al. 2004; Nguyen et al. 2006; Zhang et al. 2004). Like CD22, ligation of CD33-related siglec proteins (CD33 and Siglec-5, -7, and -9) also induces recruitment of SHP-1 via phosphorylated ITIMs (c/r Avril et al. 2005).

Murine Siglec-F is predominately expressed on eosinophils (Zhang et al. 2004). Eosinophils are best known for their role in allergic diseases. Siglec-F null mice revealed that Siglec-F is a negative regulator of eosinophil response to allergens (Zhang et al. 2007). Eosinophils also contribute to an immune response against foreign pathogens, including binding, engulfment, and killing of microbes (Galioto et al. 2006; Inoue et al. 2005). Tateno et al. (2007) investigated the endocytic pathways of CD22 (Siglec-2) and mouse eosinophil Siglec-F that are expressed on cells of the innate immune system. Though both the Siglecs required intact cytoplasmic ITIM motifs, they differed in molecular mechanisms in endocytic pathways, suggesting that these two siglec proteins evolved distinctly (Tateno et al. 2007).

During endocytosis following ligation by anti-CD22 antibody (Jones et al. 2003) or high-affinity multivalent-sialoside ligands (Collins et al. 2006a), the tyrosine-based ITIMs of CD22 also fit the sorting signal YXXØ (where Ø is a hydrophobic residue) for association with the adaptor complex 2 (AP2), which directs recruitment of receptors into clathrin-coated pits (Bonifacino and Traub 2003). John et al. reported that CD22 associates with the AP50 subunit of AP2 through these tyrosine-based motifs and that they are required for endocytosis (John et al. 2003). Consistent with this observation, CD22 is predominantly localized in clathrin-rich domains (Collins et al. 2006b; Grewal et al. 2006). Antigen ligation of the BCR results in mobilization of the BCR to "activation rafts," which subsequently fuse with clathrin domains prior to endocytosis (Stoddart et al. 2002; Stoddart et al. 2005). Since CD22 is specifically excluded from activation rafts (Pierce 2002), the negative regulatory effect of CD22 on BCR signaling has been proposed to occur following its movement to clathrin domains (Collins et al. 2006b), linking the endocytic activity of CD22 to its role in regulation of BCR signaling.

16.4.6.2 PTKs, PTPs and PLCγ1 in signal transduction in B cell activation

CD22 interacts specifically with ligands carrying α2-6-linked sialic acids. Interaction with these ligands in cis regulates the association of CD22 with the BCR and thereby modulates the inhibitory function of CD22. Interaction of CD22 to ligands in trans can regulate both B-cell migration as well as the BCR signaling threshold (Nitschke 2009).

Cross-linking BCR elicits early signal transduction events, including activation of protein tyrosine kinases, phosphorylation of receptor components, activation of phospholipase C-γ (PLC-γ), and increase in intracellular Ca^{2+}. Cross-linking of BCR leads to rapid translocation of cytosolic protein tyrosine phosphatase (PTP)1 C to the particulate fraction, where it became associated with CD22. The association of PTP-1 C with CD22 was mediated by the NH2-terminal SH2 domain of PTP-1 C. Complexes of either CD22/PTP-1 C/Syk/PLC-γ could be isolated from B cells stimulated by BCR engagement or a mixture of hydrogen peroxidase and sodium orthovanadate, respectively. The binding of PLC-γ1 and Syk to tyrosyl-phosphorylated CD22 was mediated by NH2-terminal SH2 domain of PLC-γ1 and the COOH-terminal SH2 domain of Syk, respectively. Results suggest that tyrosyl-phosphorylated CD22 may down-regulate the activity of this complex by dephosphorylation of CD22, Syk, and/or PLC-γ1. Transient expression of CD22 and a null mutant of PTP-1 C (PTP-1CM) in COS cells resulted in an increase in tyrosyl phosphorylation of CD22 and its interaction with PTP-1CM. By contrast, CD22 was not tyrosyl phosphorylated or associated with PTP-1CM in presence of wild-type PTP-1 C. These results suggest that tyrosyl-phosphorylated CD22 may be a substrate for PTP-1 C regulates tyrosyl phosphorylation of CD22.

Binding Site for SH-2-Containing Protein-Tyrosine Phosphatase-1 in CD22: CD22 contains ITIMs in the cytoplasmic region and recruits SHP-1 to the phosphorylated ITIMs upon ligation of BCR, thereby negatively regulating BCR signaling. Among three identified ITIMs, two ITIMs containing tyrosine residues at position 843 (Tyr^{843}) and 863 (Tyr^{863}), respectively, are required for CD22 to recruit SHP-1 and regulate BCR signaling upon BCR ligation by anti-Ig Ab, indicating that CD22 has the SHP-1-binding domain at the region containing (Tyr^{843}) and (Tyr^{863}). Further, the CD22 mutant in which both Tyr^{843} and Tyr^{863} are replaced by phenylalanine (CD22F5/6) recruits SHP-1 and regulates BCR signaling upon stimulation with antigen but not anti-Ig Ab. This suggests that CD22 contains another SHP-1 binding domain that is specifically activated upon stimulation with antigen. Both of the flanking sequences of Tyr^{783} and Tyr^{817} fit the consensus sequence of ITIM, and the CD22F5/6 mutant requires these tyrosine residues for SHP-1 binding and BCR regulation. Thus, these ITIMs constitute a novel conditional SHP-1-binding site of CD22 that is activated upon BCR ligation by antigen but not by anti-Ig Ab (Zhu et al. 2008).

16.4.6.3 Endocytic mechanisms of CD22 and Siglec-F in cell signaling

Tateno et al. (2007) investigated the endocytic pathways of CD22 and mouse eosinophil Siglec-F that are expressed on cells of the innate immune system. Siglec-F showed efficient endocytosis of anti-Siglec antibody, and *Neisseria meningitidis* bearing sialylated glycans. Like CD22, endocytosis was dependent on its cytoplasmic ITIM and ITIM-like motifs. While endocytosis of CD22 was mediated by a clathrin-dependent mechanism and was sorted to early endosome and recycling compartments, Siglec-F endocytosis was directed to lysosomal compartments and was mediated by a mechanism that was independent of clathrin and dynamin. Like CD22, Siglec-F mediated endocytosis of anti-Siglec-F and sialoside ligands, a function requiring intact tyrosine-based motifs. In contrast, however, Siglec-F endocytosis was clathrin and dynamin independent, required ADP ribosylation factor 6, and trafficed to lysosomes. Comparative results suggest that these two siglec proteins have evolved distinct endocytic mechanisms consistent with roles in cell signaling and innate immunity (Tateno et al. 2007).

16.4.7 CD22 as Target for Therapy

The restricted expression of several siglecs to one or a few cell types makes them attractive targets for cell-directed therapies. The anti-CD33 (also known as Siglec-3) antibody gemtuzumab (Mylotarg) is approved for the treatment of acute myeloid leukemia, and antibodies targeting CD22 (Siglec-2) are currently in clinical trials for treatment of B cell non-Hodgkins lymphomas and autoimmune diseases. Because siglecs are endocytic receptors, they are well suited for a 'Trojan horse' strategy, whereby therapeutic agents conjugated to an antibody, or multimeric glycan ligand, bind to the siglec and are efficiently carried into the cell. Although the rapid internalization of unmodified siglec antibodies reduces their utility for induction of antibody-dependent cellular cytotoxicity or complement-mediated cytotoxicity, antibody binding of Siglec-8, Siglec-9 and CD22 has been demonstrated to induce apoptosis of eosinophils, neutrophils and depletion of B cells, respectively. Properties of siglecs that make them attractive for cell-targeted therapies have been reviewed (Kreitman 2006; O'Reilly and Paulson 2009). Anti-CD22 antibodies are theoretically good candidates alone and in combination with other drugs in the treatment of B cell malignancies. Strategies include targeting B-cell surface markers such as CD22, as well as blocking B-cell-activating factors or their receptors. Refinement of existing immunotoxins and development of new immunotoxins are underway to improve the treatment of cancer (Kreitman 2006). CMC-544 (inotuzumab ozogamicin), a CD22-specific cytotoxic immunoconjugate of calicheamicin is intended for treatment of B-lymphoid malignancies. CMC-544 targets CD22 expressed by B-lymphoid malignancies. CMC-544 comprises a humanized IgG4 anti-CD22 mAb, G5/44, covalently linked to CalichDMH via an acid-labile linker. CMC-544 inhibited in vitro growth of acute lymphoblastic leukemia (ALL) cell lines more potently than that of Ramos B-lymphoma cells. In nude mice with established sc xenografts of REH ALL, CMC-544 caused dose-dependent inhibition of xenograft growth producing complete tumor regression and cures in tumor-bearing mice after treatment of conjugated calicheamicin. The anti-leukemia activity of CMC-544 supports clinical evaluation of CMC-544 for the treatment of CD22+ leukemia (Dijoseph et al. 2004, 2007).

B cells play an important role in the pathogenesis of many autoimmune diseases. Different approaches targeting B cell compartment are under investigation. Selective modulation of B cells has been achieved using a humanized mAb against CD22. The antibody (epratuzumab), originally developed for treatment of non-Hodgkin's lymphoma, was found to be effective, with a very good safety profile. Studies have demonstrated the efficacy and safety of epratuzumab in several autoimmune diseases, including systemic lupus erythematosus and primary SjÃ gren's syndrome (Steinfeld and Youinou 2006). Anti-CD22/cal mAb therapy resulted in early and prolonged B-cell depletion and delayed disease in pre-diabetic mice. Importantly, when new-onset hyperglycemic mice were treated with anti-CD22/cal mAb, 100% of B-cell-depleted mice became normoglycemic by 2 days, and 70% of them maintained a state of long-term normoglycemia. Targeting CD22 depletes and reprograms B-cells and reverses autoimmune diabetes, thereby providing a blueprint for development of novel therapies to cure autoimmune diabetes (Fiorina et al. 2008).

Newer methods are in progress to generate multivalent antibodies. The dock and lock (DNL) method is a new technology for generating multivalent antibodies. Rossi et al. (2009) reported characterizations of 20–22 and 22–20, a pair of humanized hexavalent anti-CD20/22 bispecific antibodies (bsAbs) derived from veltuzumab (v-mab) and epratuzumab (e-mab). Each bsAb translocates both CD22 and CD20 into lipid rafts, induces apoptosis and growth inhibition without second-antibody cross-linking, and is significantly more potent in killing lymphoma cells in vitro than their parental antibodies. Results suggest multiple advantages of hexavalent anti-CD20/22 bsAbs over the individual parental antibodies and suggest that these may represent a new class of cancer therapeutics (Rossi et al. 2009).

16.5 Siglec-4 [Myelin-Associated Glycoprotein, (MAG)]

16.5.1 MAG and Myelin Formation

Several glia-associated cell surface molecules have been implicated in myelin formation in CNS and peripheral nervous system (PNS). In PNS, the major peripheral myelin protein PO and the peripheral myelin protein (PMP) 22 are involved in spiral formation as reflected by retarded myelin formation in mice deficient for the respective molecules. The involvement of myelin-associated glycoprotein (MAG) in this process was detected in mice deficient in both PO and MAG, suggesting that PO can replace MAG during the formation of spiraling loops. For the maintenance of the association of Schwann cell and myelin with its ensheathed axon, the myelin components PO, MAG, and Connexin 32 are crucial. In CNS, recognition of oligodendrocytes and axons and the formation of spiraling loops is mediated by MAG (Martini and Schachner 1997).

16.5.2 Characteristics of MAG

MAG, membrane glycoprotein of 100 kDa, is expressed abundantly early in the myelination process, indicating an important role for MAG in initial stages of myelination. Reports suggest that CNS nerves may be coaxed into full functional regeneration in a clinical setting. However, only a small number of axons regenerate in these studies, in part because of other inhibitory proteins associated with myelin. One candidate for such a protein is myelin-associated glycoprotein (MAG). The MAG is a 100-kD type I transmembrane integral membrane glycoprotein, which is a member of Siglec family. It makes up ~1% of CNS and ~0.1% of PNS myelin proteins (Trapp 1990). The MAG is localized in periaxonal Schwann cell and oligodendroglial membranes of myelin sheaths where it functions in interactions between myelin-forming cells (both oligodendrocytes and Schwann cells) and the axolemma in PNS and CNS.

MAG contains five Ig-like domains and belongs to the Siglec subgroup of Ig superfamily and shares significant homology with neural cell adhesion molecule (N-CAM) (Salzer et al. 1987). MAG is expressed as two developmentally regulated isoforms with different cytoplasmic domains that may activate different signal transduction pathways in myelin-forming cells. MAG contains a carbohydrate epitope shared with other glycoconjugates that is a target antigen in autoimmune peripheral neuropathy associated with IgM γ-pathy and has been implicated in a dying back oligodendrogliopathy in multiple sclerosis (Quarles 2007).

16.5.3 MAG Isoforms

Myelin-associated Glycoprotein can be obtained from adult mouse brain from detergent-lysates of a crude membrane fraction as a 96–100 kDa form (detergent solubilized MAG or L-MAG), and from 100,000 g supernatants of homogenates as a 90–96 kDa form (soluble MAG or S-MAG). Both molecular forms bind to heparin in hypo- and isotonic buffers. Soluble MAG binds to several collagens (type G, I, II, III, IV, V, VI, IX) with a K_D of 5.7×10^{-8} M for collagen type IX and 2.0×10^{-7} for collagen type IV. The MAG is localized in basal lamina and interstitial collagens of the sciatic nerve in situ (Fahrig et al. 1987). It is heavily glycosylated containing 30% carbohydrate by weight. Of the 9 MAG sequons 7 were glycosylated and 1 was partially glycosylated at Asn106. Asn332, which was not recovered in the glycopeptide fractions and one was probably not glycosylated. All MAG glycosylated sequons might bear the L2/HNK-1 epitope (Burger et al. 1993). Two isoforms of the MAG are the result of alternative splicing of the primary MAG transcript. The small (S-MAG) and large (L-MAG) isoforms are identical in their extracellular and transmembrane domains but are distinct at their C-terminal ends. Early in the myelination process expression of L-MAG predominates, whereas S-MAG accumulates in later stages (Inuzuka et al. 1991; Pedraza et al. 1991).

The functions of two MAG isoforms, which differ only in their cytoplasmic domains, are not well known. In rat and mouse, expression of the two forms of mRNA is developmentally regulated; the mRNA without exon 12 portion is expressed mainly in the actively myelinating stage of development. In quaking mouse, the mRNA without a 45-nucleotide exon portion was scarcely expressed throughout development (Fujita et al. 1989). Fujita et al. (1989) determined the structures of three forms of mouse MAG mRNAs. Two forms of mRNAs were reported to be different by alternate inclusion of exon 2 and 12 in rat brain. One of the three forms of clones appeared to be mRNA, which lacked both the exon 2 and 12 portions, although others were identical splicing patterns to those of rat. Northern blot analysis using specific probes to mRNAs with or without the exon 2 portion in normal and quaking mouse confirmed that the splicing of exon 2 and 12 occurred independently (Fujita et al. 1989; Sato et al. 1989).

On human locus MAG is assigned to chromosome 19 and the mouse locus to chromosome 7. Since the region of mouse chromosome 7-known to contain several other genes that are homologous to genes on human chromosome 19-also carries the quivering (qv) locus; the possibility that a mutation in the MAG gene could be responsible for this neurological disorder. While MAG-specific DNA restriction fragments, mRNA, and protein from qv/qv mice were apparently normal in size and abundance, the possibility was not ruled out that qv could be caused by a point mutation in the MAG gene (Barton et al. 1987). The human MAG sequence provides an open reading frame of 1,878 nt encoding a peptide of 626 amino acids with a molecular mass of 69.1 kD. It is 89% homologous to nt sequence to the large isoform of rat MAG, with 95% homology in the amino acid sequence. It contains 9 potential glycosylation sites, one more than in rat, and shares other key features with rat MAG, including 5 Ig-like regions of internal homology, an RGD sequence, and potential phosphorylation sites. Its structure appears to be highly conserved in evolution, possibly suggesting a close interdependence between its structure and function. The human gene is located on the proximal long arm of chromosome 19 (19q12–q13.2) (Spagnol et al. 1989). In transgenic mouse line that specifically expresses GFP-tagged S-MAG, Erb et al. (2006) reported differential expression pattern and spatial distribution of L- and S-MAG during development as well as in the adult central and peripheral nervous system. In peripheral nerves, where S-MAG is the sole isoform, S-MAG concentrated in different ring-like structures such as periaxonal and abaxonal rings, and discs spanning through the compact myelin sheath perpendicular to the axon. This provides a new insight in the subcellular distribution of MAG isoforms for the understanding of their specific functions in myelin formation.

Interestingly, cytoplasmic region unique to L-MAG contains a tyrosine phosphorylation site, suggesting a role in the regulation of MAG function (Edwards et al. 1988; Jaramillo et al. 1994). MAG is tyrosine-phosphorylated in the developing brain. The major tyrosine phosphorylation at residue 620 interacts specifically with the SH2 domains of phospholipase C (PLCγ). This domain may represent a protein binding motif that can be regulated by tyrosine phosphorylation. MAG also specifically bound the Fyn tyrosine kinase, suggesting that MAG serves as a docking protein that allows the interaction between different signaling molecules (Jaramillo et al. 1994). Gene-targeted mutant mice express a truncated form of L-MAG isoform, eliminating the unique portion of its cytoplasmic domain, but they continue to express S-MAG. Similar to the total MAG knockouts, these animals do not express an overt clinical phenotype. CNS myelin of L-MAG mutant mice displays most of the pathological abnormalities reported for the total MAG knockouts. In contrast to the null MAG mutants, however, PNS axons and myelin of older L-MAG mutant animals do not degenerate, indicating that S-MAG is sufficient to maintain PNS integrity. These observations demonstrate a differential role of L-MAG isoform in CNS and PNS myelin (Fujita et al. 1998).

16.5.4 Ligands of MAG: Glycan Specificity of MAG

Using oligosaccharides with modifications in the sialic acid, galactose or N-acetylglucosamine moieties, it was demonstrated that both MAG and Sn bind with high preference to α2,3-linked sialic acid and interact at least with the three terminal monosaccharide units. An additional sialic acid at position six of the third-terminal monosaccharide unit enhances binding to MAG, whereas it does not influence binding to Sn significantly. The hydroxy groups at positions 8 and 9 are required for binding to both proteins. Surprisingly, MAG binds 2-keto-3-deoxy-D-glycero-D-galacto-nononic acid significantly better than N-acetylneuraminic acid, whereas Sn prefers the latter structure. Thus the interactions of MAG and Sn are mainly with sialic acid and that additional contacts with the subterminal galactose and N-acetylglucosamine residues also contribute to the binding strength, although to a lesser degree (Strenge et al. 1998).

Neuronal Ligands: The carbohydrate binding specificities of three sialoadhesins were compared by measuring lectin-transfected COS cell adhesion to natural and synthetic gangliosides. The neural sialoadhesins, MAG and Schwann cell myelin protein (SMP) had similar and stringent binding specificities. Each required an α2,3-linked sialic acid on the terminal galactose of a neutral saccharide core, and they shared the following rank-order potency of binding: GQ1bα >> GD1a = GT1b >> GM3 = GM4 >> GM1, GD1b, GD3, GQ1b (nonbinders). In contrast, sialoadhesin had less specificity, binding to gangliosides that bear either terminal α2,3- or α2,8-linked sialic acids with the following rank-order potency of binding: GQ1bα > GD1a = GD1b = GT1b = GM3 = GM4 > GD3 = GQ1b >> GM1 (nonbinder). Binding of MAG, SMP, and sialoadhesin was abrogated by chemical modification of either the sialic acid carboxylic acid group or glycerol side chain on a target ganglioside. These results are consistent with sialoadhesin binding to one face of the sialic acid moiety, whereas MAG (and SMP) may have more complex binding sites or may bind sialic acids only in the context of more restricted oligosaccharide conformations (Collins et al. 1997).

Gangliosides, the most abundant sialylated glycoconjugates in brain, may be the functional neuronal ligands for MAG. Cells engineered to express MAG on their surface adhered specifically to gangliosides bearing an α2,3-linked N-acetylneuraminic acid on a terminal galactose, with the following relative potency: GQ1b α >> GD1a, GT1b >> GM3, GM4 (GM1, GD1b, GD3, and GQ1b did not support adhesion). MAG binding was abrogated by modification of the carboxylic acid, any hydroxyl, or the N-acetyl group of the ganglioside's N-acetylneuraminic acid moiety. Related Ig superfamily members either failed to bind gangliosides (CD22) or bound with less stringent specificity (sialoadhesin), whereas a modified form of MAG (bearing three of its five extra-cellular Ig-like domains) bound only GQ1b α. Enzymatic removal of sialic acids from the surface of intact nerve cells altered their functional interaction with myelin. These data are consistent with a role for gangliosides in MAG-neuron interactions (Schnaar et al. 1998).

MAG binds with high affinity and specificity to two major brain gangliosides, GD1a and GT1b, that are expressed prominently on axons and that bear the MAG-binding terminal sequence NeuAc1-3Galß1-3GalNAc (Collins et al. 1997; Yang et al. 1996). Mice lacking a key enzyme involved in ganglioside biosynthesis, UDP-N-acetyl-D-galactosamine:GM3/GD3 N-acetyl-D-galactosaminyltransferase (EC 2.4.1.92), do not express the NeuAc1-3Galß1-3GalNAc terminus, and display axon degeneration and dysmyelination similar to Mag-null mice (Sheikh et al. 1999), as well as progressive motor behavioral deficits (Chiavegatto et al. 2000). Furthermore, nerve cells from these mice are less responsive to MAG as an inhibitor of neurite outgrowth (Vyas et al. 2002). These and other studies (Vinson et al. 2001; Yamashita et al. 2002) implicate complex brain gangliosides, particularly GD1a and GT1b, as functional MAG ligands.

MAG expression was also decreased in mice lacking complex brain gangliosides (Sheikh et al. 1999). Mice with a disrupted Galgt1 gene lack UDP-GalNAc:GM3/GD3 N-acetylgalactosaminyltransferase (GM2/GD2 synthase) and fail to express complex brain gangliosides, including GD1a and GT1b, instead expressing a comparable amount of the simpler gangliosides GM3, GD3, and O-acetyl-GD3. Sun et al. (2004) indicated that the maintenance of MAG protein levels depends on the presence of complex gangliosides, perhaps due to enhanced stability when MAG on myelin binds to its complementary ligands, GD1a and GT1b, on the apposing axon surface. Data support the conclusion that MAG interaction with complex brain gangliosides markedly affects its steady-state expression.

16.5.5 Functions of MAG

MAG is a bifunctional molecule, which has been implicated in the formation and maintenance of myelin sheaths. CNS myelin formation is delayed in MAG gene mutant mice (Montag et al. 1994; Bartsch et al. 1997). Moreover, oligodendrocytic cytoplasmic collars of mature CNS myelin are frequently missing or reduced, whereas controversy persists with regard to the effect of MAG deficiencies on periaxonal spacing (see Bartsch 1996). Compact myelin of MAG mutants also contains an increased presence of cytoplasmic loops of oligodendrocytes. In contrast, peripheral nervous system (PNS) myelin formation proceeds normally in MAG-deficient animals. Older mutants, however, display PNS axonal and myelin degeneration with the presence of superfluous Schwann cell processes, indicating that MAG plays a critical role in maintaining PNS integrity (Fruttiger et al. 1995).

The axon is dependent on signals from myelin, specifically MAG, for its cytoarchitecture, structure, and long-term stability (Bjartmar et al. 1999). MAG appears to function both as a ligand for an axonal receptor that is needed for the maintenance of myelinated axons and as a receptor for an axonal signal that promotes the differentiation, maintenance and survival of oligodendrocytes. Its function in the maintenance of myelinated axons may be related to its role as one of the white matter inhibitors of neurite outgrowth acting through a receptor complex involving Nogo receptor and/or gangliosides containing 2,3-linked sialic acid. Genetic ablation of MAG results in reduced axon caliber, reduced axon neurofilament spacing and phosphorylation, and progressive axon degeneration (Fruttiger et al. 1995; Li et al. 1994; Montag et al. 1994; Yin et al. 1998). These observations led to conclusion that MAG is an important signaling molecule in myelin–axon interactions and is required for optimal long-term axon stability (Schachner and Bartsch 2000).

MAG Inhibits Neurite Outgrowth in Cell Culture: MAG and Nogo are potent inhibitors of neurite outgrowth from a variety of neurons, and they have been identified as possible components of CNS myelin that prevents axonal regeneration in the adult vertebrate CNS. Nogo, MAG, and -oligodendrocyte-myelin glycoprotein act on neurons through 75 receptor (p75) (Wang et al. 2002a*)* in complex with Nogo receptor (Mimura et al. 2006). Although also found in PNS myelin and abundant in the CNS, MAG inhibits neurite outgrowth in cell culture, and its role in regeneration is controversial (David et al. 1995; Bartsch 1996). It is present in a neurite outgrowth-inhibitory fraction of CNS myelin (David et al. 1995; Bartsch 1996; McKerracher et al. 1994). MAG causes growth cone collapse and inhibits neurite outgrowth from various vertebrate neuronal cell types, including retinal ganglion cells (RGCs) (Li et al. 1996; McKerracher et al. 1994; Mukhopadhyay et al. 1994; Song et al. 1998). However, the inhibitory activity of CNS myelin from MAG-deficient transgenic mice was not significantly diminished, and axon regeneration in CNS of these MAG knock-out mice was improved only slightly (David et al. 1995) or not at all (Bartsch et al. 1995). In addition, MAG inhibits axon regeneration after injury (Li et al. 1996; McKerracher et al. 1994; Mukhopadhyay et al. 1994). Along with Nogo, oligodendrocyte myelin glycoprotein, and chondroitin sulfate proteoglycans, MAG also contributes to CNS an environment that is highly inhibitory for nerve regeneration (Sandvig et al. 2004; Wang et al. 2002b). It has been suggested that MAG prevents axonal regeneration in lesioned nervous tissue. MAG is now well known as one of several white matter inhibitors of neurite outgrowth in vitro and axonal regeneration in vivo. MAG knockout mice revealed that MAG is not essential for the initiation of myelination; but, it plays an important role in maintaining a stable interaction between axons and myelin (Filbin 1995). By acute inactivation of MAG in situ in chick retina-optic nerve cultures, Wong et al. (2003) suggested that the acute loss of MAG function can promote significant axon growth across a site of CNS nerve damage.

In contrast to these studies, $MAG^{-/-}$ mice cross-bred with C57BL/Wld^{S} showed improved PNS nerve regeneration in vivo (Schafer et al. 1996). The loss of inhibitory MAG activity may be compensated by expression of other myelin inhibitory proteins (depending on the genetic background). Analysis of optic nerves from mutant mice revealed that MAG is functionally involved in the recognition of axons by oligodendrocytes and in the morphological maturation of myelin sheaths. However, results did not support a role of MAG as a potent inhibitor of axonal regeneration in the adult mammalian CNS (Bartsch 1996). The MAG has been implicated in the formation and maintenance of myelin. Although the analysis of MAG null mutants confirmed this view, the phenotype of this mutant is surprisingly subtle. In the CNS of MAG-deficient mice, initiation of myelination, formation of morphologically intact myelin sheaths and to a minor extent, integrity of myelin is affected. In PNS, in comparison, only maintenance of myelin was impaired. Observations also suggest that other molecules performing similar functions as MAG might compensate, at least partially, for the absence of MAG in the null mutant (Schachner and Bartsch 2000; Tang et al. 2001). Thus, there is still no clear role for MAG in inhibiting nerve regeneration in the CNS in situ. Additional research is needed to determine if receptors and signaling systems are similar to those responsible for MAG inhibition of neurite outgrowth (Quarles 2009).

MAG Induces Intramembrane Proteolysis of p75 Neurotrophin Receptor to Inhibit Neurite Outgrowth: The three known inhibitors of axonal regeneration present in myelin—MAG, Nogo, and OMgp—all interact with the same receptor complex to effect inhibition via PKC-dependent activation of the small GTPase Rho. The activation of RhoA and Rho-kinase is reported to be an essential part of signaling mechanism of MAG. Data suggest the important roles of collapsing response mediator protein-2 (CRMP-2) and microtubules in the inhibition of the axon regeneration by the myelin-derived inhibitors. The transducing component of this receptor complex is the p75 neurotrophin receptor. It was shown that MAG binding to cerebellar neurons induces α- and then γ-secretase proteolytic cleavage of p75, in a protein kinase C-dependent manner, and that this cleavage is necessary for both activation of Rho and inhibition of neurite outgrowth (Domeniconi et al. 2005).

16.5.6 MAG in Demyelinating Disorders

The early loss of MAG in the development of multiple sclerosis plaques in comparison to other myelin proteins suggests that it plays a key role in the pathogenesis of this disease. The selective loss of MAG may relate to the high susceptibility of human MAG to cleavage by a Ca^{2+}-activated neutral protease. Human MAG also contains a highly immunogenic carbohydrate determinant that is also expressed on other neural glycoconjugates and is the antigen recognized by many human IgM paraproteins that occur in patients with peripheral neuropathy.

Association of the MAG Locus with Schizophrenia in Chinese Population: Neurotransmitter-based hypotheses have so far led to only moderate success in predicting new pathogenetic findings in etiology of schizophrenia. On the other hand, the more recent oligodendroglia hypotheses of this disorder have been supported by an increasing body of evidence. The expression MAG gene has been shown to be significantly lower in schizophrenia patient groups compared to control groups. Such an effect might be a result of genetic variations of the MAG gene. Genotyping of four markers within the MAG locus in 413 trios sample of the Han Chinese using allele-specific PCR demonstrated that MAG might play a role in genetic susceptibility to schizophrenia (Yang et al. 2005). In order to further assess the role of MAG in schizophrenia, Wan et al. (2005) examined four single nucleotide polymorphisms (SNPs), namely rs2301600, rs3746248, rs720309 and rs720308, of this gene in Chinese schizophrenic patients and healthy controls. The distribution of rs720309 T/A genotypes showed a strong association with schizophrenia. A haplotype constructed of rs720309-rs720308 also revealed a significant association with schizophrenia. The finding of a possible association between the MAG locus and schizophrenia is in agreement with the hypotheses of oligodendrltic and myelination dysfunction in the Chinese Han population (Wan et al. 2005).

Loss of MAG Reflects Hypoxia-Like White Matter Damage in Brain Diseases: Destruction of myelin and oligodendrocytes leading to the formation of large demyelinated plaques is the hallmark of multiple sclerosis (MS) pathology. In a subset of MS patients termed pattern III, actively demyelinating lesions show preferential loss of MAG and apoptotic-like oligodendrocyte destruction, whereas other myelin proteins remain well preserved. MAG is located in the most distal periaxonal oligodendrocyte processes and primary "dying back" oligodendrogliopathy may be the initial step of myelin degeneration in pattern III lesions. In addition to a subset of MS cases, a similar pattern of demyelination was found in some cases of virus encephalitis as well as in all lesions of acute white matter stroke. Brain white matter lesions presenting with MAG loss and apoptotic-like oligodendrocyte destruction, irrespective of their primary disease cause, revealed a prominent nuclear expression of hypoxia inducible factor-1α in various cell types, including oligodendrocytes. Data suggest that a hypoxia-like tissue injury may play a pathogenetic role in a subset of inflammatory demyelinating brain lesions (Aboul-Enein et al. 2003).

Shear Stress Alters the Expression of MAG: Reports revealed that Schwann cells undergo concurrent proliferation and apoptosis after a chronic nerve injury that is independent of axonal pathology. Gupta et al (2005) postulated that this response may be triggered directly by mechanical stimuli. Immunochemical analysis showed that the Schwann cells are positive for S-100, MAG, and MBP in greater than 99% of the experimental cells. Stimulated cells also revealed an increased rate of proliferation by as much as 100%. The mRNA expression of MAG and MBP was down-regulated by 21% and 18%, respectively, in experimental cells, while protein was down-regulated by 29% and 35%, respectively. This study provides information regarding Schwann cell direct response to physical stimulus, which is not secondary to an axonal injury (Gupta et al. 2005).

16.5.7 Inhibitors of Regeneration of Myelin

The lack of axonal growth after injury in the adult CNS is due to several factors including the formation of a glial scar, the absence of neurotrophic factors, the presence of growth-inhibitory molecules associated with myelin and the intrinsic growth-state of the neurons. The proteolytic fragment of MAG (dMAG), consisting of entire extracellular domain, is readily released from myelin and is found in vivo. Three inhibitors of axonal growth have been identified in myelin: MAG, Nogo-A, and Oligodendrocyte-Myelin glycoprotein (OMgp). MAG inhibits axonal regeneration by high affinity interaction ($K_d = 8$ nM) with the Nogo66 receptor (NgR) and activation of a p75 neurotrophin receptor (p75NTR)-mediated signaling pathway. Two myelin inhibitors, MAG and Nogo, both transmembrane proteins, have been identified. MAG, a sialic acid binding protein and a component of myelin, is a potent inhibitor of neurite outgrowth from a variety of neurons both in vitro and in vivo. The MAG's sialic acid binding site is distinct from its neurite inhibitory activity. Alone, sialic acid-dependent binding of MAG to neurons is insufficient to affect inhibition of axonal growth. Thus, while soluble MAG-Fc (MAG extracellular domain fused to Fc), a truncated form of MAG-Fc missing Ig-domains 4 and 5, MAG(d1-3)-Fc, and another sialic acid binding protein, sialoadhesin, each bind to neurons in a sialic acid- dependent manner, only full-length MAG-Fc inhibits neurite outgrowth. Results indicated that a second site must exist on MAG which elicits this response. Consistent with this model, mutation of Arg^{118} in MAG to either alanine or aspartate abolishes its sialic acid-dependent binding. However, when expressed at the surface of either CHO or Schwann cells, Arg^{118}-mutated MAG retains the ability to inhibit axonal outgrowth. Hence, MAG has two recognition sites for neurons, the sialic acid binding site at Arg^{118}8 and a distinct inhibition site which is absent from the first three Ig domains (Tang et al. 1997).

Tang et al. (2001) showed that a soluble MAG-Fc, when secreted from CHO cells in a collagen gel inhibits/deflects neurite outgrowth from P6 dorsal root ganglion (DRG) neurons. Using the same assay system, results showed that factors secreted from damaged white matter inhibit axonal regeneration and that the majority of inhibitory activity could be accounted for by the proteolytic fragment of MAG (dMAG). Thus, released dMAG is likely to play an important role in preventing regeneration, immediately after injury before the glial scar forms (Tang et al. 2001).

16.5.8 Axonal Regeneration by Overcoming Inhibitory Activity of MAG

The interaction of MAG and its neuronal receptors mediates bidirectional signaling between neurons and oligodendrocytes. When cultured on MAG-expressing cells, dorsal root ganglia neurons (DRG) older than post-natal day 4 (PND4) extend neuritis 50% shorter on average than when cultured on control cells. In contrast, MAG promotes neurite outgrowth from DRG neurons from animals younger than PND4. The response switch, which is also seen in retinal ganglia (RGC) and Raphe nucleus neurons, is concomitant with a developmental decrease in the endogenous neuronal cAMP levels. The artificially increasing cAMP levels in older neurons can alter their growth-state and induce axonal growth in the presence of myelin-associated inhibitors (Domeniconi and Filbin 2005).

Glycan Inhibitors of MAG Enhance Axon Outgrowth In Vitro: MAG is one of several endogenous axon regeneration inhibitors that limit recovery from central nervous system injury and disease. Molecules that block such inhibitors may enhance axon regeneration and functional recovery. Blixt et al. (2003) evaluated ten known human siglecs and their murine orthologs for their specificity for more than 25 synthetic sialosides. Among these siglecs, Siglec-4 binds with 500–10,000-fold higher affinity to a series of mono- and di-sialylated derivatives of the O-linked T-antigen (Galβ (1–3)-GalNAc(α)OThr) as compared with α-methyl-NeuAc.

Potent monovalent sialoside inhibitors of MAG have been identified (Blixt et al. 2003). The potent of these were tested for their ability to reverse MAG-mediated inhibition of axon outgrowth from rat cerebellar granule neurons in vitro. It was found that monovalent sialoglycans enhance axon regeneration in proportion to their MAG binding affinities. The most potent glycoside was disialyl T antigen (NeuAcα2-3Galβ1-3[NeuAcα2-6]GalNAc-R), followed by 3-sialyl T antigen (NeuAcα2-3Galβ1-3GalNAc-R), structures expressed on O-linked glycoproteins as well as on gangliosides. Blocking gangliosides reversed MAG inhibition. But blocking O-linked glycoprotein sialylation with benzyl-α-GalNAc had no effect. The ability to reverse MAG inhibition with monovalent glycosides suggests further exploration of glycans as blockers of MAG-mediated axon outgrowth inhibition (Vyas et al. 2005).

Immunization with r-Nogo-66/MAG Promotes Axon Regeneration: Immunization with myelin is able to promote robust regeneration of corticospinal tract fibers in adult mice. The effectiveness of such immunization with myelin was compared to that of a combination of two axon growth inhibitors in myelin, Nogo-66 (the 66-amino-acid inhibitory region of Nogo-A) and MAG in SJL/J mice, a strain that is susceptible to autoimmune experimental allergic encephalomyelitis (EAE). None of the immunized mice showed EAE. Long-distance axon regeneration and sprouting of the corticospinal tract was observed in myelin and Nogo-66/MAG immunized mice. Thus, this work shows that axon growth inhibitors in myelin can be selectively blocked using this immunization approach to promote long-distance axon regeneration in the spinal cord (Sicotte et al. 2003). The finding that an anti-MAG monoclonal antibody not only possesses the ability to neutralise the inhibitory effect of MAG on neurons but also directly protects oligodendrocytes from glutamate-mediated oxidative stress-induced cell death. Administration of anti-MAG antibody (centrally and systemically) starting 1 h after middle cerebral artery occlusion in the rat significantly reduced lesion volume at 7 days. This neuroprotection was associated with a robust improvement in motor function compared with animals receiving control IgG1, highlighting the potential for the use of anti-MAG antibodies as therapeutic agents for the treatment of stroke (Irving et al. 2005).

16.5.9 Fish Siglec-4

The combination of microarray technology with the zebrafish model system can provide useful information on how genes are coordinated in a genetic network to control zebrafish embryogenesis and can help to identify novel genes that are important for organogenesis (Lo et al. 2003). To understand the evolution of siglecs, in particular the origin of this family, Lehmann et al. (2004) investigated the occurrence of corresponding genes in bony fish. Interestingly, only unambiguous orthologs of mammalian siglec-4, a cell adhesion molecule expressed exclusively in the nervous system, could be identified in the genomes of fugu and zebrafish, whereas no obvious orthologs of the other mammalian siglecs were found. As in mammals, fish siglec-4 expression is restricted to nervous tissues. Expressed as recombinant protein, fish siglec-4 binds to sialic acids with a specificity similar to the mammalian orthologs. Relatively low sequence similarities in the cytoplasmic tail as well as an additional splice variant found in fish siglec-4 suggest alternative signaling pathways compared to mammalian species. Observations suggest that this siglec occurs at least in the nervous system of all vertebrates (Lehmann et al. 2004).

16.6 Siglec-15

Siglec-15 is a type-I transmembrane protein consisting of: two Ig-like domains, a transmembrane domain containing a lysine residue, and a short cytoplasmic tail. Siglec-15 is expressed on the cells of immune system, can recognize sialylated ligands. The extracellular domain of Siglec-15 preferentially recognizes the Neu5Acα2–6GalNAcα– structure. Siglec-15 associates with the activating adaptor proteins DNAX activation protein (DAP)12 and DAP10 via its lysine residue in the transmembrane domain, implying that it functions as an activating signaling molecule. Siglec-15 is another human Siglec identified to have an activating signaling potential. However, unlike Siglec-14, it does not have an inhibitory counterpart. Orthologs of Siglec-15 are present not only in mammals but also in other branches of vertebrates (fish); in contrast, no other known Siglec expressed in the immune system has been conserved throughout vertebrate evolution. Probably Siglec-15 plays a conserved, regulatory role in the immune system of vertebrates (Angata et al. 2007).

References

Aboul-Enein F, Rauschka H, Kornek B et al (2003) Preferential loss of myelin-associated glycoprotein reflects hypoxia-like white matter damage in stroke and inflammatory brain diseases. J Neuropathol Exp Neurol 62:25–33

Alphey MS, Attrill H, Crocker PR, Van Aalten DM (2003) High resolution crystal structures of Siglec-7. Insights into ligand specificity in the Siglec family. J Biol Chem 278:3372–3377

Angata T (2006) Molecular diversity and evolution of the Siglec family of cell-surface lectins. Mol Divers 10:555–566

Angata T, Brinkman-Van der Linden E (2002) I-type lectins. Biochim Biophys Acta 1572:294–316

Angata T, Varki NM, Varki A (2001) A second uniquely human mutation affecting sialic acid biology. J Biol Chem 276: 40282–40287

Angata T, Margulies EH, Green ED, Varki A (2004) Large-scale sequencing of the CD33-related Siglec gene cluster in five mammalian species reveals rapid evolution by multiple mechanisms. Proc Natl Acad Sci USA 101:13251–13256

Angata T, Tabuchi Y, Nakamura K, Nakamura M (2007) Siglec-15: an immune system Siglec conserved throughout vertebrate evolution. Glycobiology 17:838–846

Avichezer D, Silver PB, Chan CC et al (2000) Identification of a new epitope of human IRBP that induces autoimmune uveoretinitis in mice of the H-2b haplotype. Invest Ophthalmol Vis Sci 41:127–131

Avril T, Freeman SD, Attrill H et al (2005) Siglec-5 (CD170) can mediate inhibitory signaling in the absence of immunoreceptor tyrosine-based inhibitory motif phosphorylation. J Biol Chem 280: 19843–19851

Avril T, Wagner ER, Willison HJ, Crocker PR (2006) Sialic acid-binding immunoglobulin-like lectin 7 mediates selective recognition of sialylated glycans expressed on Campylobacter jejuni lipooligosaccharides. Infect Immun 74:4133–4141

Bakker TR, Piperi C, Davies EA, Merwe PA (2002) Comparison of CD22 binding to native CD45 and synthetic oligosaccharide. Eur J Immunol 32:1924–1932

Barnes YC, Skelton TP, Stamenkovic I, Sgroi DC (1999) Sialylation of the sialic acid binding lectin sialoadhesin regulates its ability to mediate cell adhesion. Blood 93:1245–1252

Barton DE, Arquint M, Roder J et al (1987) The myelin-associated glycoprotein gene: mapping to human chromosome 19 and mouse chromosome 7 and expression in quivering mice. Genomics 1:107–112

Bartsch U (1996) Myelination and axonal regeneration in the central nervous system of mice deficient in the myelin-associated glycoprotein. J Neurocytol 25:303–313

Bartsch U, Montag D, Bartsch S, Schachner M (1995) Multiply myelinated axons in the optic nerve of mice deficient for the myelin-associated glycoprotein. Glia 14:115–122

Bartsch S, Montag D, Schachner M, Bartsch U (1997) Increased number of unmyelinated axons in optic nerves of adult mice deficient in the myelin-associated glycoprotein (MAG). Brain Res 762:231–234

Bhunia A, Jayalakshmi V, Benie AJ et al (2004) Saturation transfer difference NMR and computational modeling of a sialoadhesin-sialyl lactose complex. Carbohydr Res 339:259–267

Biedermann B, Gil D, Bowen DT, Crocker PR (2007) Analysis of the CD33-related siglec family reveals that Siglec-9 is an endocytic receptor expressed on subsets of acute myeloid leukemia cells and absent from normal hematopoietic progenitors. Leuk Res 31: 211–220

Biesen R, Demir C, Barkhudarova F et al (2008) Sialic acid-binding Ig-like lectin 1 expression in inflammatory and resident monocytes is a potential biomarker for monitoring disease activity and success of therapy in systemic lupus erythematosus. Arthritis Rheum 58:1136–1145

Bjartmar C, Yin X, Trapp BD (1999) Axonal pathology in myelin disorders. J Neurocytol 28:383–395

Blixt O, Collins BE, van den Nieuwenhof IM et al (2003) Sialoside specificity of the siglec family assessed using novel multivalent probes: identification of potent inhibitors of myelin-associated glycoprotein. J Biol Chem 278:31007–31019

Bonifacino JS, Traub LM (2003) Signals for sorting of transmembrane proteins to endosomes and lysosomes. Annu Rev Biochem 72: 395–447

Brinkman-Van der Linden EC, Sjoberg ER et al (2000) Loss of N-glycolylneuraminic acid in human evolution. Implications for sialic acid recognition by siglecs. J Biol Chem 275:8633–8640

Bukrinsky JT, St Hilaire PM, Meldal M et al (2004) Complex of sialoadhesin with a glycopeptide ligand. Biochim Biophys Acta 1702:173–179

Burger D, Pidoux L, Steck AJ (1993) Identification of the glycosylated sequence of human myelin-associated glycoprotein. Biochem Biophys Res Commun 197:457–464

Campanero-Rhodes MA, Childs RA, Kiso M et al (2006) Carbohydrate microarrays reveal sulphation as a modulator of siglec binding. Biochem Biophys Res Commun 344:1141–1146

Chan CH, Wang J, French RR, Glennie MJ (1998) Internalization of the lymphocytic surface protein CD22 is controlled by a novel membrane proximal cytoplasmic motif. J Biol Chem 273:27809–27815

Chiavegatto S, Sun J, Nelson RJ, Schnaar RL (2000) A functional role for complex gangliosides: motor deficits in GM2/GD2 synthase knockout mice. Exp Neurol 166:227–234

Collins BE, Kiso M, Hasegawa A et al (1997) Binding specificities of the sialoadhesin family of I-type lectins. Sialic acid linkage and substructure requirements for binding of myelin-associated glycoprotein, Schwann cell myelin protein, and sialoadhesin. J Biol Chem 272:16889–16895

Collins BE, Blixt O, Bovin NV et al (2002) Constitutively unmasked CD22 on B cells of ST6Gal I knockout mice: novel sialoside probe for murine CD22. Glycobiology 12:563–571

Collins BE, Smith BA, Bengtson P, Paulson JC (2006a) Ablation of CD22 in ligand-deficient mice restores B cell receptor signaling. Nat Immunol 7:199–206

Collins BE, Blixt O, Han S et al (2006b) High-affinity ligand probes of CD22 overcome the threshold set by cis ligands to allow for binding, endocytosis, and killing of B cells. J Immunol 177:2994–3003

Courtney AH, Puffer EB, Pontrello JK et al (2009) Sialylated multivalent antigens engage CD22 in trans and inhibit B cell activation. Proc Natl Acad Sci USA 106:2500–2505

Crocker PR (2002) Siglecs: sialic-acid-binding immunoglobulin-like lectins in cell-cell interactions and signaling. Curr Opin Struct Biol 12:609–615

Crocker PR (2005) Siglecs in innate immunity. Curr Opin Pharmacol 5:431–437

Crocker PR, Redelinghuys P (2008) Siglecs as positive and negative regulators of the immune system. Biochem Soc Trans 36: 1467–1471

Crocker PR, Varki A (2001) Siglecs, sialic acids and innate immunity. Trends Immunol 22:337–342

Crocker PR, Zhang J (2002) New I-type lectins of the CD 33-related siglec subgroup identified through genomics. Biochem Soc Symp 69:83–94

Crocker PR, Kelm S, Dubois C et al (1991) Purification and properties of sialoadhesin, a sialic acid-binding receptor of murine tissue macrophages. EMBO J 10:1661–1669

Crocker PR, Mucklow S, Bouckson V et al (1994) Sialoadhesin, a macrophage sialic acid binding receptor for haemopoietic cells with 17 immunoglobulin-like domains. EMBO J 13:4490–4503

Crocker PR, Freeman S, Gordon S, Kelm S (1995) Sialoadhesin binds preferentially to cells of the granulocytic lineage. J Clin Invest 95:635–643

Crocker PR, Hartnell A, Munday J, Nath D (1997) The potential role of sialoadhesin as a macrophage recognition molecule in health and disease. Glycocon J 14:601–609

Crocker PR, Vinson M, Kelm S, Drickamer K (1999) Molecular analysis of sialoside binding to sialoadhesin by NMR and site-directed mutagenesis. Biochem J 341:355–361

Crocker PR, Paulson JC, Varki A (2007) Siglecs and their roles in the immune system. Nat Rev Immunol 7:255–266

Cyster JG, Goodnow CC (1997) Tuning antigen receptor signaling by CD22: integrating cues from antigens and the microenvironment. Immunity 6:509–517

David S, Braun PE, Jackson DL et al (1995) Laminin overrides the inhibitory effects of peripheral nervous system and central nervous system myelin-derived inhibitors of neurite growth. J Neurosci Res 42:594–602

Delputte PL, Nauwynck HJ (2004) Porcine arterivirus infection of alveolar macrophages is mediated by sialic acid on the virus. J Virol 78:8094–8101

Delputte PL, Nauwynck HJ (2006) Porcine arterivirus entry in macrophages: heparan sulfate-mediated attachment, sialoadhesin-mediated internalization, and a cell-specific factor mediating virus disassembly and genome release. Adv Exp Med Biol 581:247–252

Delputte PL, Costers S, Nauwynck HJ (2005) Analysis of porcine reproductive and respiratory syndrome virus attachment and internalization: distinctive roles for heparan sulphate and sialoadhesin. J Gen Virol 86:1441–1445

DiJoseph JF, Armellino DC, Boghaert ER et al (2004) Antibody-targeted chemotherapy with CMC-544: a CD22-targeted immunoconjugate of calicheamicin for the treatment of B-lymphoid malignancies. Blood 103:1807–1814

Dijoseph JF, Dougher MM, Armellino DC et al (2007) Therapeutic potential of CD22-specific antibody-targeted chemotherapy using inotuzumab ozogamicin (CMC-544) for the treatment of acute lymphoblastic leukemia. Leukemia 21:2240–2245

Dimasi N, Moretta A, Moretta L et al (2004) Structure of the saccharide-binding domain of the human natural killer cell inhibitory receptor p75/AIRM1. Acta Crystallogr Sect D Biol Crystallogr 60:401–403

Domeniconi M, Filbin MT (2005) Overcoming inhibitors in myelin to promote axonal regeneration. J Neurol Sci 233:43–47

Domeniconi M, Zampieri N, Spencer T et al (2005) MAG induces regulated intramembrane proteolysis of the p75 neurotrophin receptor to inhibit neurite outgrowth. Neuron 46:849–855

Doody GM, Justement LB, Delibrias CC et al (1995) A role in B cell activation for CD22 and the protein tyrosine phosphatase SHP. Science 269:242–244

Doyle AG, Herbein G, Montaner LJ et al (1994) Interleukin-13 alters the activation state of murine macrophages in vitro: comparison with interleukin-4 and interferon-gamma. Eur J Immunol 24: 1441–5

Edwards AM, Arquint M, Braun PE et al (1988) Myelin-associated glycoprotein, a cell adhesion molecule of oligodendrocytes, is phosphorylated in brain. Mol Cell Biol 8:2655–2658

Engel P, Nojima Y, Rothstein D et al (1993) The same epitope on CD22 of B lymphocytes mediates the adhesion of erythrocytes, T and B lymphocytes, neutrophils, and monocytes. J Immunol 150: 4719–4732

Engel P, Wagner N, Miller AS, Tedder TF (1995) Identification of the ligand-binding domains of CD22, a member of the immunoglobulin superfamily that uniquely binds a sialic acid-dependent ligand. J Exp Med 181:1581–1586

Erb M, Flueck B, Kern F et al (2006) Unraveling the differential expression of the two isoforms of myelin-associated glycoprotein in a mouse expressing GFP-tagged S-MAG specifically regulated and targeted into the different myelin compartments. Mol Cell Neurosci 31:613–627

Fahrig T, Landa C, Pesheva P et al (1987) Characterization of binding properties of the myelin-associated glycoprotein to extracellular matrix constituents. EMBO J 6:2875–2883

Filbin MT (1995) Myelin-associated glycoprotein: a role in myelination and in the inhibition of axonal regeneration? Curr Opin Neurobiol 5:588–595

Fiorina P, Vergani A, Dada S et al (2008) Targeting CD22 reprograms B-cells and reverses autoimmune diabetes. Diabetes 57:3013–3024

Floyd H, Ni J, Cornish AL et al (2000) Siglec-8: a novel eosinophil-specific member of the immunoglobulin superfamily. J Biol Chem 275:861–866

Freeman SD, Kelm S, Barber EK et al (1995) Characterisation of CD33 as a new member of the sialoadhesin family of cellular interaction molecules. Blood 84:2005–2012

Fruttiger M, Montag D, Schachner M, Martini R (1995) Crucial role for the myelin-associated glycoprotein in the maintenance of axon-myelin integrity. Eur J Neurosci 7:511–515

Fujita N, Sato S, Kurihara T et al (1989) cDNA cloning of mouse myelin-associated glycoprotein: a novel alternative splicing pattern. Biochem Biophys Res Commun 165:1162–1169

Fujita N, Kemper A, Dupree J et al (1998) The cytoplasmic domain of the large myelin-associated glycoprotein isoform is needed for proper cns but not peripheral nervous system myelination. J Neurosci 18:1970–1978

Gabius HJ, Andre S, Kaltner H, Siebert HC (2002) The sugar code: functional lectinomics. Biochim Biophys Acta 1572:165–177

Galioto AM, Hess JA, Nolan TJ et al (2006) Role of eosinophils and neutrophils in innate and adaptive protective immunity to larval *Strongyloides stercoralis* in mice. Infect Immun 74:5730–5738

Genini S, Malinverni R, Delputte PL et al (2008) Gene expression profiling of porcine alveolar macrophages after antibody-mediated cross-linking of Sialoadhesin (Sn, Siglec-1). J Recept Signal Transduct Res 28:185–243

Ghosh S, Bandulet C, Nitschke L (2006) Regulation of B cell development and B cell signaling by CD22 and its ligands α2,6-linked sialic acids. Int Immunol 18:603–611

Gijbels MJ, van der Cammen M, van der Laan LJ et al (1999) Progression and regression of atherosclerosis in APOE3-Leiden transgenic mice: an immunohistochemical study. Atherosclerosis 143:15–25

Grewal PK, Boton M, Ramirez K, Collins BE et al (2006) ST6Gal-I restrains CD22-dependent antigen receptor endocytosis and Shp-1 recruitment in normal and pathogenic immune signaling. Mol Cell Biol 26:4970–4981

Gupta R, Truong L, Bear D et al (2005) Shear stress alters the expression of myelin-associated glycoprotein (MAG) and myelin basic protein (MBP) in Schwann cells. J Orthop Res 23:1232–1239

Han Y, Kelm S, Riesselman MH (1994) Mouse sialoadhesin is not responsible for Candida albicans yeast cell binding to splenic marginal zone macrophages. Infect Immun 62:2115–8

Han S, Collins BE, Bengtson P, Paulson JC (2005) Homomultimeric complexes of CD22 in B cells revealed by protein-glycan cross-linking. Nat Chem Biol 1:93–97

Hanasaki K, Powell LD, Varki A (1995a) Binding of human plasma sialoglycoproteins by the B cell-specific lectin CD22. Selective recognition of immunoglobulin M and haptoglobin. J Biol Chem 270:7543–7550

Hanasaki K, Varki A, Powell LD (1995b) CD22-mediated cell adhesion to cytokine-activated human endothelial cells. Positive and negative regulation by α2–6-sialylation of cellular glycoproteins. J Biol Chem 270:7533–7542

Hartnell A, Steel J, Turley H et al (2001) Characterization of human sialoadhesin, a sialic acid binding receptor expressed by resident and inflammatory macrophage populations. Blood 97:288–296

Hatta Y, Tsuchiya N, Matsushita M et al (1999) Identification of the gene variations in human CD22. Immunogenetics 49:280–286

Hennet T, Chui D, Paulson JC, Marth JD (1998) Immune regulation by the ST6Gal sialyltransferase. Proc Nat Acad Sci USA 95: 4504–4509

Hitomi Y, Tsuchiya N, Hasegawa M et al (2007) Association of CD22 gene polymorphism with susceptibility to limited cutaneous systemic sclerosis. Tissue Antigens 69:242–249

Hughes EH, Schlichtenbrede FC, Murphy CC et al (2003) Generation of activated sialoadhesin-positive microglia during retinal degenera- tion. Invest Ophthalmol Vis Sci 44:2229–2234

Ikehara Y, Ikehara SK, Paulson JC (2004) Negative regulation of T cell receptor signaling by Siglec-7 (p70/AIRM) and Siglec-9. J Biol Chem 279:43117–43125

Ikezumi Y, Suzuki T, Hayafuji S et al (2005) The sialoadhesin (CD169) expressing a macrophage subset in human proliferative glomerulonephritis. Nephrol Dial Transplant 20:2704–2713

Inoue Y, Matsuwaki Y, Shin SH et al (2005) Nonpathogenic, environmental fungi induce activation and degranulation of human eosinophils. J Immunol 175:5439–5447

Inuzuka T, Fujita N, Sato S et al (1991) Expression of the large myelin-associated glycoprotein isoform during the development in the mouse peripheral nervous system. Brain Res 562:173–175

Ip CW, Kroner A, Crocker PR et al (2007) Sialoadhesin deficiency ameliorates myelin degeneration and axonopathic changes in the CNS of PLP overexpressing mice. Neurobiol Dis 25:105–111

Irving EA, Vinson M, Rosin C et al (2005) Identification of neuroprotective properties of anti-MAG antibody: a novel approach for the treatment of stroke? J Cereb Blood Flow Metab 25:98–107

Jaramillo ML, Afar DE, Almazan G, Bell JC (1994) Identification of tyrosine 620 as the major phosphorylation site of myelin-associated glycoprotein and its implication in interacting with signaling molecules. J Biol Chem 269:27240–27245

Jenner J, Kerst G, Handgretinger R, Müller I (2006) Increased alpha2,6-sialylation of surface proteins on tolerogenic, immature dendritic cells and regulatory T cells. Exp Hematol 34:1212–1218

Jiang HR, Lumsden L, Forrester JV (1999) Macrophages and dendritic cells in IRBP-induced experimental autoimmune uveoretinitis in B10RIII mice. Invest Ophthalmol Vis Sci 40:3177–3185

Jiang HR, Hwenda L, Makinen K et al (2006) Sialoadhesin promotes the inflammatory response in experimental autoimmune uveoretinitis. J Immunol 177:2258–2264

Jin L, McLean PA, Neel BG, Wortis HH (2002) Sialic acid binding domains of CD22 are required for negative regulation of B cell receptor signaling. J Exp Med 195:1199–1205

John B, Herrin BR, Raman C et al (2003) The B cell coreceptor CD22 associates with AP50, a clathrin-coated pit adapter protein, via tyrosine-dependent interaction. J Immunol 170:3534–3543

Jones C, Virji M, Crocker PR (2003) Recognition of sialylated meningococcal lipopolysacch aride by siglecs expressed on myeloid cells leads to enhanced bacterial uptake. Mol Microbiol 49:1213–1225

Kelm S, Schauer R, Manuguerra JC et al (1994) Modifications of cell surface sialic acids modulate cell adhesion mediated by sialoadhesin and CD22. Glycocon J 11:576–585

Kelm S, Brossmer R, Isecke R et al (1998) Functional groups of sialic acids involved in binding to siglecs (sialoadhesins) deduced from interactions with synthetic analogues. Eur J Biochem 255:663–672

Kelm S, Gerlach J, Brossmer R et al (2002) The ligand-binding domain of CD22 is needed for inhibition of the B cell receptor signal, as demonstrated by a novel human CD22-specific inhibitor compound. J Exp Med 195:1207–1213

Kimura N, Ohmori K, Miyazaki K et al (2007) Human B-lymphocytes express α2-6-sialylated 6-sulfo-N-acetyllactosamine serving as a preferred ligand for CD22/Siglec-2. J Biol Chem 282: 32200–32207

Kobsar I, Oetke C, Kroner A et al (2006) Attenuated demyelination in the absence of the macrophage-restricted adhesion molecule sialoadhesin (Siglec-1) in mice heterozygously deficient in P0. Mol Cell Neurosci 31:685–691

Kreitman RJ (2006) Immunotoxins for targeted cancer therapy. AAPS J 8:E532–E551

Kumamoto Y, Higashi N, Denda-Nagai K et al (2004) Identification of sialoadhesin as a dominant lymph node counter-receptor for mouse macrophage galactose-type C-type lectin 1. J Biol Chem 279: 49274–49280

Kusmartsev S, Ruiz de Morales JM (1999) Sialoadhesin expression by bone marrow macrophages derived from Ehrlich-tumor-bearing mice. Cancer Immunol Immunother 48:493–498

Kusmartsev SA, Danilets MG, Bel'skaya NV et al (2003) Effect of individual and combination treatment with cytokines on expression of sialoadhesin by bone marrow macrophages. Bull Exp Biol Med 136:139–141

Lai PC, Cook HT, Smith J et al (2001) Interleukin-11 attenuates nephrotoxic nephritis in Wistar Kyoto rats. J Am Soc Nephrol 12: 2310–2320

Law C-L, Sidorenko SP, Clark EA (1994) Regulation of lymphocyte activation by the cell-surface molecule CD22. Immunol Today 15:442–449

Law CL, Sidorenko SP, Chandran KA et al (1996) CD22 associates with protein tyrosine phosphatase 1C, Syk, and phospholipase C-γ1 upon B cell activation. J Exp Med 183:547–560

Lehmann F, Gathje H, Kelm S, Dietz F (2004) Evolution of sialic acid-binding proteins: molecular cloning and expression of fish siglec-4. Glycobiology 14:959–968

Leprince C, Draves KE, Geahlen RL et al (1993) CD22 associates with the human surface IgM-B-cell antigen receptor complex. Proc Natl Acad Sci USA 90:3236–3240

Li C, Tropak MB, Gerlai R et al (1994) Myelination in the absence of myelin-associated glycoprotein. Nature 369:747–750

Li M, Shibata A, Li C et al (1996) Myelin-associated glycoprotein inhibits neurite/axon growth and causes growth cone collapse. J Neurosci Res 46:404–414

Lo J, Lee S, Xu M et al (2003) 15000 unique zebrafish EST clusters and their future use in microarray for profiling gene expression patterns during embryogenesis. Genome Res 13:455–466

Lock K, Zhang J, Lu J et al (2004) Expression of CD33-related siglecs on human mononuclear phagocytes, monocyte-derived dendritic cells and plasmacytoid dendritic cells. Immunobiology 209: 199–207

Martinez-Pomares L, Kosco-Vilbois M, Darley E et al (1996) Fc chimeric protein containing the cysteine-rich domain of the murine mannose receptor binds to macrophages from splenic marginal zone and lymph node subcapsular sinus and to germinal centers. J Exp Med 184:1927–1937

Martini R, Schachner M (1997) Molecular bases of myelin formation as revealed by investigations on mice deficient in glial cell surface molecules. Glia 19:298–310

May AP, Robinson RC, Aplin RT et al (1997) Expression, crystallization, and preliminary X-ray analysis of a sialic acid-binding fragment of sialoadhesin in the presence and absence of ligand. Protein Sci 6:717–721

May AP, Robinson RC, Vinson M et al (1998) Crystal structure of the N-terminal domain of sialoadhesin in complex with 3' sialyllactose at 1.85 Å resolution. Mol Cell 1:719–728

McKerracher L, David S, Jackson DL et al (1994) Identification of myelin-associated glycoprotein as a major myelin-derived inhibitor of neurite growth. Neuron 13:805–811

McMillan SJ, Crocker PR (2008) CD33-related sialic-acid-binding immunoglobulin-like lectins in health and disease. Carbohydr Res 343:2050–2056

McWilliam AS, Tree P, Gordon S (1992) Interleukin 4 regulates induction of sialoadhesin, the macrophage sialic acid-specific receptor. Proc Natl Acad Sci USA 89:10522–6

Mimura F, Yamagishi S, Arimura N et al (2006) Myelin-associated glycoprotein inhibits microtubule assembly by a Rho-kinase-dependent mechanism. J Biol Chem 281:15970–15979

Mingari MC, Vitale C, Romagnani C et al (2001) p75/AIRM1 and CD33, two sialoadhesin receptors that regulate the proliferation or the survival of normal and leukemic myeloid cells. Immunol Rev 181:260–268

Montag D, Giese KP, Bartsch U et al (1994) Mice deficient for the myelin-associated glycoprotein show subtle abnormalities in myelin. Neuron 13:229–246

Monteiro VG, Lobato CS, Silva AR et al (2005) Increased association of Trypanosoma cruzi with sialoadhesin positive mice macrophages. Parasitol Res 97:380–385

Mrkoci K, Kelm S, Crocker PR et al (1996) Constitutively hyposialylated human T-lymphocyte clones in the Tn-syndrome: binding characteristics of plant and animal lectins. Glycoconj J 13:567–573

Mucklow S, Hartnell A, Mattei MG et al (1995) Sialoadhesin (Sn) maps to mouse chromosome 2 and human chromosome 20 and is not linked to the other members of the sialoadhesin family, CD22, MAG, and CD33. Genomics 28:344–346

Muerkoster S, Rocha M, Crocker PR et al (1999) Sialoadhesin-positive host macrophages play an essential role in graft-versus-leukemia reactivity in mice. Blood 93:4375–4386

Mukhopadhyay G, Doherty P, Walsh FS et al (1994) A novel role for myelin-associated glycoprotein as an inhibitor of axonal regeneration. Neuron 13:757–767

Munday J, Floyd H, Crocker PR (1999) Sialic acid binding receptors (siglecs) expressed by macrophages. J Leukoc Biol 66: 705–711

Nakamura K, Yamaji T, Crocker PR et al (2002) Lymph node macrophages, but not spleen macrophages, express high levels of unmasked sialoadhesin: implication for the adhesive properties of macrophages in vivo. Glycobiology 12:209–216

Nath D, van der Merwe PA, Kelm S et al (1995) The amino-terminal immunoglobulin-like domain of sialoadhesin contains the sialic acid binding site. Comparison with CD22. J Biol Chem 270: 26184–26191

Nath D, Hartnell A, Happerfield L, Crocker PR et al (1999) Macrophage-tumour cell interactions: identification of MUC1 on breast cancer cells as a potential counter-receptor for the macrophage-restricted receptor, sialoadhesin. Immunology 98: 213–219

Nguyen DH, Ball ED, Varki A (2006) Myeloid precursors and acute myeloid leukemia cells express multiple CD33-related Siglecs. Exp Hematol 34:728–735

Nitschke L (2009) CD22 and Siglec-G: B-cell inhibitory receptors with distinct functions. Immunol Rev 230:128–143

Nitschke L, Carsetti R, Ocker B et al (1997) CD22 is a negative regulator of B-cell receptor signaling. Curr Biol 7:133–143

Nitschke L, Floyd H, Crocker PR (2001) New functions for the sialic acid-binding adhesion molecule CD22, a member of the growing family of Siglecs. Scand J Immunol 53:227–234

Oetke C, Vinson MC, Jones C, Crocker PR (2006) Sialoadhesin-deficient mice exhibit subtle changes in B- and T-cell populations and reduced immunoglobulin M levels. Mol Cell Biol 26: 1549–1557

O'Keefe TL, Williams GT, Davies SL et al (1996) Hyperresponsive B cells in CD22-Deficient mice. Science 274:798–801

O'Reilly MK, Paulson JC (2009) Siglecs as targets for therapy in immune-cell-mediated disease. Trends Pharmacol Sci 30: 240–248

Otipoby KL, Andersson KB, Draves KE et al (1996) CD22 regulates thymus-independent responses and the lifespan of B cells. Nature 384:634–637

Patel N, Brinkman-Van der Linden ECM, Altmann SW et al (1999) OB-BP1/Siglec-6—a leptin- and sialic acid-binding protein of the immunoglobulin superfamily. J Biol Chem 274:22729–22738

Peaker CJ, Neuberger MS (1993) Association of CD22 with the B cell antigen receptor. Eur J Immunol 23:1358–1363

Pedraza L, Frey AB, Hempstead BL et al (1991) Differential expression of MAG isoforms during development. J Neurosci Res 29:141–148

Perry VH, Crocker PR, Gordon S (1992) The blood–brain barrier regulates the expression of a macrophage sialic acid-binding receptor on microglia. J Cell Sci 101:201–207

Pierce SK (2002) Lipid rafts and B-cell activation. Nat Rev Immunol 2:96–105

Powell LD, Varki A (1994) The oligosaccharide binding specificities of CD22b, a sialic acid-specific lectin of B cells. J Biol Chem 269: 10628–10636

Powell LD, Jain RK, Matta KI et al (1995) Characterization of sialyloligosaccharide binding by recombinant soluble and native cell-associated CD22. J Biol Chem 270:7523–7532

Quarles RH (2007) Myelin-associated glycoprotein (MAG): past, present and beyond. J Neurochem 100:1431–1448

Quarles RH (2009) A hypothesis about the relationship of myelin-associated glycoprotein's function in myelinated axons to its capacity to inhibit neurite outgrowth. Neurochem Res 34:79–86

Razi N, Varki A (1998) Masking and unmasking of the sialic acid-binding lectin activity of CD22 (Siglec-2) on B lymphocytes. Proc Natl Acad Sci USA 95:7469–7474

Razi N, Varki A (1999) Cryptic sialic acid binding lectins on human blood leukocytes can be unmasked by sialidase treatment or cellular activation. Glycobiology 9:1225–1234

Rempel H, Calosing C, Sun B, Pulliam L (2008) Sialoadhesin expressed on IFN-induced monocytes binds HIV-1 and enhances infectivity. PLoS One 3:e1967

Rossi EA, Goldenberg DM, Cardillo TM et al (2009) Hexavalent bispecific antibodies represent a new class of anticancer therapeutics: 1. Properties of anti-CD20/CD22 antibodies in lymphoma. Blood 113:6161–6171

Rudge EU, Cutler AJ, Pritchard NR, Smith KG (2002) Interleukin 4 reduces expression of inhibitory receptors on B cells and abolishes CD22 and Fcγ RII-mediated B cell suppression. J Exp Med 195:1079–1085

Salzer JL, Holmes WP, Colman DR (1987) The amino acid sequences of the myelin-associated glycoproteins: homology to the immunoglobulin gene superfamily. J Cell Biol 104:957–965

Sandvig A, Berry M, Barrett LB et al (2004) Myelin-, reactive glia-, and scar-derived CNS axon growth inhibitors: expression, receptor signaling, and correlation with axon regeneration. Glia 46:225–251

Sato S, Fujita N, Kurihara T et al (1989) cDNA cloning and amino acid sequence for human myelin-associated glycoprotein. Biochem Biophys Res Commun 163:1473–1480

Schachner M, Bartsch U (2000) Multiple functions of the myelin-associated glycoprotein MAG (siglec-4a) in formation and maintenance of myelin. Glia 29:154–165

Schadee-Eestermans IL, Hoefsmit EC, van de Ende M et al (2000) Ultrastructural localisation of sialoadhesin (siglec-1) on macrophages in rodent lymphoid tissues. Immunobiology 202: 309–325

Schafer M, Fruttiger M, Montag D et al (1996) Disruption of the gene for the myelin-associated glycoprotein improves axonal regrowth along myelin in C57BL/Wlds mice. Neuron 16:1107–1113

Schauer R (2004) Victor Ginsburg's influence on my research of the role of sialic acids in biological recognition. Arch Biochem Biophys 426:132–141

Schnaar RL, Collins BE, Wright LP et al (1998) Myelin-associated glycoprotein binding to gangliosides. Structural specificity and functional implications. Ann NY Acad Sci 845:92–105

Sgroi D, Nocks A, Stamenkovic I (1996) A single N-linked glycosylation site is implicated in the regulation of ligand recognition by the I-type lectins CD22 and CD33. J Biol Chem 271:18803–18809

Shan D, Press OW (1995) Constitutive endocytosis and degradation of CD22 by human B cells. J Immunol 154:4466–4475

Sheikh KA, Sun J, Liu Y et al (1999) Mice lacking complex gangliosides develop Wallerian degeneration and myelination defects. Proc Natl Acad Sci USA 96:7532–7537

Shi WX, Chammas R, Varki A (1996a) Regulation of sialic acid 9-*O*-acetylation during the growth and differentiation of murine erythroleukemia cells. J Biol Chem 271:31517–31525

Shi WX, Chammas R, Varki NM et al (1996b) Sialic acid 9-O-acetylation on murine erythroleukemia cells affects complement activation, binding to I-type lectins, and tissue homing. J Biol Chem 271:31526–31532

Sicotte M, Tsatas O, Jeong SY et al (2003) Immunization with myelin or recombinant Nogo-66/MAG in alum promotes axon regeneration and sprouting after corticospinal tract lesions in the spinal cord. Mol Cell Neurosci 23:251–263

Sjoberg ER, Powell LD, Klein A, Varki A (1994) Natural ligands of the B-cell adhesion molecule CD22β can be masked by 9-O-acetylation of sialic acids. J Cell Biol 126:549–562

Song H, Ming G, He Z et al (1998) Conversion of neuronal growth cone responses from repulsion to attraction by cyclic nucleotides. Science 281:1515–1518

Sonnenburg JL, Altheide TK, Varki A (2004) A uniquely human consequence of domain-specific functional adaptation in a sialic acid-binding receptor. Glycobiology 14:339–346

Spagnol G, Williams M, Srinivasan J et al (1989) Molecular cloning of human myelin-associated glycoprotein. J Neurosci Res 24:137–142

Stamenkovic I, Seed B (1990) The B-cell antigen CD22 mediates monocyte and erythrocyte adhesion. Nature 345(6270):74–77

Steinfeld SD, Youinou P (2006) Epratuzumab (humanised anti-CD22 antibody) in autoimmune diseases. Expert Opin Biol Ther 6: 943–949

Steiniger B, Barth P, Herbst B et al (1997) The species-specific structure of microanatomical compartments in the human spleen: strongly sialoadhesin-positive macrophages occur in the perifollicular zone, but not in the marginal zone. Immunology 92:307–316

Stoddart A, Dykstra ML, Brown BK et al (2002) Lipid rafts unite signaling cascades with clathrin to regulate BCR internalization. Immunity 17:451–462

Stoddart A, Jackson AP, Brodsky FM (2005) Plasticity of B cell receptor internalization upon conditional depletion of clathrin. Mol Biol Cell 16:2339–2348

Strenge K, Schauer R, Bovin N et al (1998) Glycan specificity of myelin-associated glycoprotein and sialoadhesin deduced from interactions with synthetic oligosaccharides. Eur J Biochem 258: 677–685

Sun J, Shaper NL, Itonori S et al (2004) Myelin-associated glycoprotein (Siglec-4) expression is progressively and selectively decreased in the brains of mice lacking complex gangliosides. Glycobiology 14:851–857

Tang S, Shen YJ, DeBellard ME et al (1997) Myelin-associated glycoprotein interacts with neurons via a sialic acid binding site at ARG118 and a distinct neurite inhibition site. J Cell Biol 138: 1355–1366

Tang S, Qiu J, Nikulina E, Filbin MT (2001) Soluble myelin-associated glycoprotein released from damaged white matter inhibits axonal regeneration. Mol Cell Neurosci 18:259–269

Tateno H, Li H, Schur MJ, Bovin N et al (2007) Distinct endocytic mechanisms of CD22 (Siglec-2) and Siglec-F reflect roles in cell signaling and innate immunity. Mole Cell Biol 27:5699–5710

Tedder TF, Tuscano J, Sato S, Kehrl JH (1997) CD22, A B lymphocyte-specific adhesion molecule that regulates antigen receptor signaling. Annu Rev Immunol 15:481–504

Tedder TF, Poe JC, Haas KM (2005) CD22: a multifunctional receptor that regulates B lymphocyte survival and signal transduction. Adv Immunol 88:1–50

Toda M, Akita K, Inoue M et al (2008) Down-modulation of B cell signal transduction by ligation of mucins to CD22. Biochem Biophys Res Commun 372:45–50

Trapp BD (1990) Myelin-associated glycoprotein. Location and potential functions. Ann NY Acad Sci 605:29–43

Tuscano JM, Engel P, Tedder TF et al (1996) Involvement of p72syk kinase, p53/56lyn kinase and phosphatidyl inositol-3 kinase in signal transduction via the human B lymphocyte antigen CD22. Eur J Immunol 26:1246–1252

Tuscano JM, O'Donnell RT, Miers LA et al (2003) Anti-CD22 ligand-blocking antibody HB22.7 has independent lymphomacidal properties and augments the efficacy of 90Y- DOTA-peptide-Lym-1 in lymphoma xenografts. Blood 101:3641–3647

Ulyanova T, Blasioli J, Woodford-Thomas TA et al (1999) The sialoadhesin CD33 is a myeloid-specific inhibitory receptor. Eur J Immunol 29:3440–3449

Ulyanova T, Shah DD, Thomas ML (2001) Molecular cloning of MIS, a myeloid inhibitory siglec, that binds protein-tyrosine phosphatases SHP-1 and SHP-2. J Biol Chem 276:14451–14458

Umansky V, Beckhove P, Rocha M et al (1996) A role for sialoadhesin-positive tissue macrophages in host resistance to lymphoma metastasis in vivo. Immunology 87:303–309

Umemori H, Sato S, Yagi T et al (1994) Initial events of myelination involve Fyn tyrosine kinase signaling. Nature 367:572–576

Valent P, Ghannadan M, Akin C et al (2004) On the way to targeted therapy of mast cell neoplasms: identification of molecular targets in neoplastic mast cells and evaluation of arising treatment concepts. Eur J Clin Invest 34(Suppl 2):41–52

van den Berg TK, Breve JJ, Damoiseaux JG et al (1992) Sialoadhesin on macrophages: its identification as a lymphocyte adhesion molecule. J Exp Med 176:647–655

van den Berg TK, van Die I, de Lavalette CR et al (1996) Regulation of sialoadhesin expression on rat macrophages. Induction by glucocorticoids and enhancement by IFN-β, IFN-γ, IL-4, and lipopolysaccharide. J Immunol 157:3130–3138

van den Berg TK, Dopp EA, Dijkstra CD (2001a) Rat macrophages: membrane glycoproteins in differentiation and function. Immunol Rev 184:45–57

van den Berg TK, Nath D, Ziltener HJ et al (2001b) Cutting edge: CD43 functions as a T cell counterreceptor for the macrophage adhesion receptor sialoadhesin (Siglec-1). J Immunol 166:3637–3640

van der Merwe PA, Crocker PR, Vinson M et al (1996) Localization of the putative sialic acid-binding site on the immunoglobulin superfamily cell-surface molecule CD22. J Biol Chem 271:9273–9280

Vanderheijden N, Delputte PL, Favoreel HW et al (2003) Involvement of Sn in entry of porcine reproductive and respiratory syndrome virus into porcine alveolar macrophages. J Virol 77:8207–8215

Varki A, Angata T (2006) Siglecs—the major subfamily of I-type lectins. Glycobiology 16:1R–27R

Vinson M, van der Merwe PA, Kelm S et al (1996) Characterization of the sialic acid-binding site in sialoadhesin by site-directed mutagenesis. J Biol Chem 271:9267–9272

Vinson M, Strijbos PJ, Rowles A et al (2001) Myelin-associated glycoprotein interacts with ganglioside GT1b: a mechanism for neurite outgrowth inhibition. J Biol Chem 276:20280–20285

Vyas AA, Patel HV, Fromholt SE et al (2002) Gangliosides are functional nerve cell ligands for myelin-associated glycoprotein (MAG), an inhibitor of nerve regeneration. Proc Natl Acad Sci USA 99:8412–8417

Vyas AA, Blixt O, Paulson JC et al (2005) Potent glycan inhibitors of myelin-associated glycoprotein enhance axon outgrowth in vitro. J Biol Chem 280:16305–16310

Walter RB, Raden BW, Kamikura DM et al (2005) Influence of CD33 expression levels and ITIM-dependent internalization on gemtuzumab ozogamicin-induced cytotoxicity. Blood 105:1295–1302

Wan C, Yang Y, Feng G et al (2005) Polymorphisms of myelin-associated glycoprotein gene are associated with schizophrenia in the Chinese Han population. Neurosci Lett 388:126–131

Wang KC, Kim JA, Sivasankaran R et al (2002a) P75 interacts with the Nogo receptor as a co-receptor for Nogo, MAG and OMgp. Nature 420:74–78

Wang KC, Koprivica V, Kim JA et al (2002b) Oligodendrocyte-myelin glycoprotein is a Nogo receptor ligand that inhibits neurite outgrowth. Nature 417:941–944

Wilson GL, Fox CH, Fauci AS, Kehrl JH (1991) cDNA cloning of the B cell membrane protein CD22: a mediator of B-B cell interactions. J Exp Med 173:137–146

Wilson GL, Najfeld V, Kozlow E et al (1993) Genomic structure and chromosomal mapping of the human CD22 gene. J Immunol 150:5013–5024

Wong EV, David S, Jacob MH, Jay DG (2003) Inactivation of myelin-associated glycoprotein enhances optic nerve regeneration. J Neurosci 23:3112

Xiong YS, Zhou YH, Rong GH et al (2009) Siglec-1 on monocytes is a potential risk marker for monitoring disease severity in coronary artery disease. Clin Biochem 42:1057–1063

Yamaji T, Teranishi T, Alphey MS et al (2002) A small region of the natural killer cell receptor, Siglec-7, is responsible for its preferred binding to α 2,8-disialyl and branched α 2,6-sialyl residues. A comparison with Siglec-9. J Biol Chem 277:6324–6332

Yamashita T, Higuchi H, Tohyama M (2002) The p75 receptor transduces the signal from myelin-associated glycoprotein to Rho. J Cell Biol 157:565–570

Yang LJS, Zeller CB, Shaper NL et al (1996) Gangliosides are neuronal ligands for myelin-associated glycoprotein. Proc Natl Acad Sci USA 93:814–818

Yang YF, Qin W, Shugart YY et al (2005) Possible association of the MAG locus with schizophrenia in a Chinese Han cohort of family trios. Schizophr Res 75:11–19

Yin X, Crawford TO, Griffin JW et al (1998) Myelin-associated glycoprotein is a myelin signal that modulates the caliber of myelinated axons. J Neurosci 18:1953–1962

Yokoi H, Myers A, Matsumoto K et al (2006) Alteration and acquisition of Siglecs during in vitro maturation of CD34+ progenitors into human mast cells. Allergy 61:769–776

York MR, Nagai T, Mangini AJ et al (2007) A macrophage marker, Siglec-1, is increased on circulating monocytes in patients with systemic sclerosis and induced by type I interferons and Toll-like receptor agonists. Arthritis Rheum 56:1010–1020

Yu Z, Lai CM, Maoui M et al (2001) Identification and characterization of S2V, a novel putative siglec that contains two V set Ig-like domains and recruits protein-tyrosine phosphatases SHPs. J Biol Chem 276:23816–23824

Zaccai NR, Maenaka K, Maenaka T et al (2003) Structure-guided-design of sialic acid-based Siglec inhibitors and crystallographic analysis in complex with sialoadhesin. Structure 11:557–567

Zaccai NR, May AP, Robinson RC et al (2007) Crystallographic and in Silico analysis of the sialoside-binding characteristics of the siglec sialoadhesin. J Mol Biol 365:1469–1479

Zhang M, Varki A (2004) Cell surface sialic acids do not affect primary CD22 interactions with CD45 and surface IgM nor the rate of constitutive CD22 endocytosis. Glycobiology 14:939–949

Zhang JQ, Biedermann B, Nitschke L, Crocker PR (2004) The murine inhibitory receptor mSiglec-E is expressed broadly on cells of the innate immune system whereas mSiglec-F is restricted to eosinophils. Eur J Immunol 34:1175–1184

Zhang J, Raper A, Sugita N et al (2006) Characterization of Siglec-H as a novel endocytic receptor expressed on murine plasmacytoid dendritic cell precursors. Blood 107:3600–3608

Zhang M, Angata T, Cho JY et al (2007) Defining the in vivo function of Siglec-F, a CD33-related Siglec expressed on mouse eosinophils. Blood 109:4280–4287

Zhu C, Sato M, Yanagisawa T et al (2008) Novel binding site for Src homology 2-containing protein-tyrosine phosphatase-1 in CD22 activated by B lymphocyte stimulation with antigen. J Biol Chem 283:1653–1659

CD33 (Siglec 3) and CD33-Related Siglecs

17

G.S. Gupta

Sialic-acid-binding immunoglobulin-like lectins (Siglecs) bind sialic acids in different linkages in a wide variety of glycoconjugates. These membrane receptors are expressed in a highly specific manner, predominantly within haematopoietic system. Activating and inhibitory receptors act in concert to regulate cellular activation. Second sub-set of rapidly evolving siglecs is designated the CD33-related siglecs. In humans, these include CD33 and siglecs-5, -6, -7 (7/p75/AIRM1), -8, -9, 10, -11, -14 and -16, whereas, in mice, they comprise murine CD33 and siglec-E, -F, -G and -H. CD33 is a myeloid-specific inhibitory receptor, which along with CD33-related Siglecs represent a distinct subgroup that is undergoing rapid evolution. The structural features of CD33-related Siglecs and the frequent presence of conserved cytoplasmic signaling motifs point to their roles in regulating leukocyte functions that are important during inflammatory and immune responses. McMillan and Crocker (2008) reviewed ligand binding preferences and described the functional roles of CD33-related Siglecs in the immune system and the potential for targeting novel therapeutics against these surface receptors. Siglec-4 belonging to subset 1 of sialoadhesin family is discussed in Chap. 16.

17.1 Human CD33 (Siglec-3)

17.1.1 Human CD33 (Siglec-3): A Myeloid-specific Inhibitotry Receptor

Human CD33 (Siglec-3) is a 67-kDa transmembrane lectin-like glycoprotein that contains one V-set and one C2-set Ig-like domain (Freeman et al. 1995). In addititio to two Ig domains, it contains a transmembrane region and a cytoplasmic tail that has two potential ITIM sequences (Fig. 17.1). CD33 is a marker of myeloid progenitor cells, mature myeloid cells, and most myeloid leukemias (see Fig. 16.2, Chap. 16). Although its biologic function remains unknown, it functions as a sialic acid-specific lectin and a cell adhesion molecule. The CD33 binds sialic acid residues in N- and O-glycans on cell surfaces. CD33 is also a serine/threonine phosphoprotein, containing at least 2 sites of serine phosphorylation in its cytoplasmic domain, catalyzed by protein kinase C (PKC) Many of the Siglecs have been reported to be tyrosine phosphorylated in the cytosolic tails under specific stimulation conditions.

CD33 is an and inhibitory receptor that acts in concert to regulate cellular activation. Inhibitory receptors are characterized by the presence of a characteristic sequence known as an immunoreceptor tyrosine-based inhibitory motif (ITIM) in their cytoplasmic tail. Phosphorylated ITIM serves as docking sites for the SH2-containing phosphatases which then inhibit signal transduction. The proximal ITIM is necessary and sufficient for SHP-1 binding which is mediated by the amino-terminal SH2 domain. Treatment of SHP-1 with a phosphopeptide representing the proximal CD33 ITIM results in increased SHP-1 enzymatic activity. CD33 exerts an inhibitory effect on tyrosine phosphorylation and Ca^{2+} mobilization (Taylor et al. 1999) when co-engaged with activating FcγRI receptor. Thus, CD33 is an inhibitory receptor that may regulate FcγRI signal transduction. Engagement of CD33 on chronic and acute myeloid leukemias (AML) inhibits the proliferation of these cells and activates a process leading to apoptotic cell death on AML cells (Vitale et al. 2001). Phosphorylation could be augmented in presence of IL-3, erythropoietin, or GM-CS factor, in a cytokine-dependent cell line, TF-1. The phosphorylation of CD33 is cross-regulated with its lectin activity. Although this is one example of serine/threonine phosphorylation in the subfamily of CD33-related Siglecs, some of the other members also have putative sites in their cytoplasmic tails (Grobe and Powell 2002).

G.S. Gupta et al., *Animal Lectins: Form, Function and Clinical Applications*,
DOI 10.1007/978-3-7091-1065-2_17, © Springer-Verlag Wien 2012

17.2 CD33-Related Siglecs (CD33-rSiglecs)

17.2.1 CD33-Related Siglecs Family

Within Siglec family there exists a subgroup of molecules which bear a very high degree of homology with CD33/Siglec-3, and has thus been designated Siglec-3-like subgroup of Siglecs. The CD33-related siglecs show complex recognition patterns for sialylated glycans. The CD33 and CD33-related type 1 transmembrane proteins are thought to exert their functions through glycan recognition (Crocker et al. 2007; Varki and Angata 2006). This subset of Siglec proteins shows relatively a high sequence similarity to CD33 and many members have been shown to associate with protein tyrosine phosphatase SHP-1 via ITIM. These include CD33/Siglec-3, Siglec-5, Siglec-6/OB-BP1, Siglec-7/AIRM1, Siglec-8, Siglec-9, Siglec-10, Siglec-11, Siglec-14, Siglec-16, and a Siglec-like molecule (Siglec-L1) in humans, as well as five confirmed or putative Siglec proteins in mice (CD33, Siglec-E, Siglec-F, Siglec-G, and Siglec-H) and shares 50–99% sequence identity. This group is rapidly evolving, and has poorly conserved binding specificities and domain structures (Crocker et al. 2007; Crocker and Redelinghuys 2008; Varki and Angata 2006). Many CD33-rSiglecs are expressed on cells involved in innate immunity. For example, hSiglec-7/AIRM-1 is expressed on NK cells and monocytes (Nicoll et al. 1999, 2003); hSiglec-8 is expressed on eosinophils (Floyd et al. 2000), and mSiglec-F is expressed on immature cells of myeloid lineage (Angata et al. 2001a).

17.2.1.1 Differential Expression of Siglecs on Mononuclear Phagocytes and DCs

CD33-rSiglecs are expressed mostly in hematopoietic and immune systems with a highly cell type-specific expression pattern; for example, CD33 is found mainly on immature and mature myeloid cells (Crocker et al. 2007; Varki and Angata 2006; Crocker and Varki 2001a, b). Monocytes cultured to differentiate into macrophages using either GM-CSF or M-CSF retain the expression of siglec-3, -5, -7, -9 and -10 and their levels remain unaffected following stimulation with LPS. In comparison, monocyte-derived dendritic cells down-modulated siglec-7 and -9 following maturation with LPS. Plasmacytoid DCs in human blood expressed siglec-5 only. On monocytes, siglec-5 was shown to mediate rapid uptake of anti-siglec-5 (Fab)2 fragments into early endosomes. This suggests, in addition to inhibitory signaling, a potential role in endocytosis for siglec-5 and the other CD33-related siglecs. Studies show that siglecs are differentially expressed on mononuclear phagocytes and DCs and that some can be modulated by stimuli that promote maturation and differentiation (Lock et al. 2004). In order to characterize the spectrum of expression of other CD33rSiglecs on bone marrow precursors and AML cells, Nguyen et al. (2006a) demonstrated that Siglecs-3, -5, -6, -7, and -9 are expressed on subsets of normal bone marrow precursors, including promonocytes and myelocytes. Furthermore, most AML (but not ALL) cells express these Siglecs.

17.2.2 CD33-rSiglec Structures

The structurally CD33-rSiglecs are characterized by one N-terminal V-set Ig domain mediating sialic acid binding, followed by a variable number of C2-set Ig domains (1–16), ranging from 4 (in Siglec-10 and -11) to 1 (in CD33) (Crocker et al. 2007; Crocker 2002; Crocker 2005; Varki and Angata 2006) and a transmembrane domain, followed by a short cytoplasmic tail. Each siglec exhibits distinct and varied specificity for sialoside sequences on glycoprotein and glycolipid glycans that are expressed on the same cell (in *cis*) or on adjacent cells (in *trans*) (Crocker et al. 2007). Siglec-3-rSiglecs are characterized by their sequence similarity in first two Ig-like domains. The cytoplasmic domains of CD22 and most CD33-related siglecs contain ITIM and ITIM-like motifs involved in regulation of cell signaling. Several other siglecs (Siglecs-14-16 and murine Siglec-H) have no tyrosine motifs, but contain a positively charged trans-membrane spanning region. A charged residue permits association with the adapter protein DAP12 (12 kDa DNAX-activating protein), which bears a cytoplasmic ITAM that imparts both positive and negative signals (Crocker et al. 2007; Crocker and Redelinghuys 2008) and another putative signaling motif situated nearby (Crocker and Varki 2001a, b; Crocker 2002; Ravetch and Lanier 2000) (Fig. 17.1).

Paired Receptors: Immune cell surface receptors sharing similar sequences and having counteracting signaling properties are called "paired receptors." Known paired receptors belong to two classes of molecular families, namely, the Ig superfamily and the C-type lectin family (Long 1999). These receptor pairs (or receptor families) include the killer cell Ig-like receptors (KIR) of primates (Vilches and Parham 2002), the leukocyte Ig-like receptors (LILR) of primates (Martin et al. 2002), the paired Ig-like receptors (PIR) of rodents (Kubagawa et al. 1997), the myeloid-associated Ig-like receptors (MAIR) of rodents (Yotsumoto et al. 2003), the Ly49 family of rodents (Yokoyama and Plougastel 2003), and the CD94/NKG2 family of both primates and rodents (Gunturi et al. 2004). Although Ly49 and CD94/NKG2 families belong to C-type lectin family, none of them has been unequivocally shown to recognize glycan ligands (Angata et al. 2006). Rather, their ligands have been shown to be MHC class I proteins. The functional significance of the presence of paired inhibitory

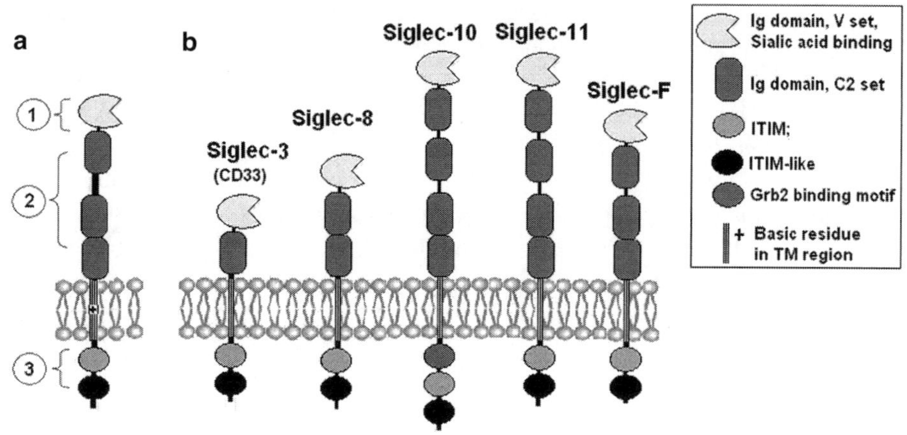

Fig. 17.1 (**a**) *Common structural features of siglecs:* The N-terminal 'V-set' Ig domain (1) contains a conserved arginine residue that confers sialic acid-binding ability. This domain is followed by a variable number (1–16) of 'C2-set' Ig domains (2). In the cytosolic domain, most siglecs contain some combination of tyrosine motifs, including ITIM, ITIM-like, Grb2-binding, and Fyn kinase sites (3). Siglecs-14, -15, and -16 contain a positively charged residue in the transmembrane spanning region (4) that enables association with the ITAM-bearing adaptor protein, DAP-12. It is speculated that these may have evolved to counteract ITIM-bearing siglecs (Crocker et al. 2007). With 99% sequence identity in the two first N-terminal Ig domains, Siglecs-5 and -14 are believed to be such paired receptors (O'Reilly and Paulson 2009). (**b**). Structural characteristics of human Siglec-3 (CD33), human Siglec-8, 10, 11 and mouse Siglec-F. For the cytoplasmic portions of each Siglec, the light gray and black circles represent ITIM and ITIM-like motifs, respectively. Siglec-6, -7, -8 and -9 have same number of Ig domain C2 set (Varki and Crocker 2009). With 99% sequence identity in the two first N-terminal Ig domains, Siglecs-5 and -14 are believed to be such paired receptors (O'Reilly and Paulson 2009)

and activating receptors remains elusive, but it is proposed that these paired receptors are involved in fine-tuning of immune responses (Lanier 2001) . Human Siglec-5, encoded by *SIGLEC5* gene, has four extracellular Ig-like domains and a cytosolic inhibitory motif. Human Siglec-14 has three Ig-like domains, encoded by the *SIGLEC14* gene, adjacent to *SIGLEC5*. Human Siglec-14 has almost complete sequence identity with human Siglec-5 at the first two Ig-like domains, shows a glycan binding preference similar to that of human Siglec-5, and associates with the activating adapter protein DAP12. Thus, Siglec-14 and Siglec-5 appear to be the first glycan binding paired receptors. Near-complete sequence identity of the amino-terminal part of human Siglec-14 and Siglec-5 indicates partial gene conversion between *SIGLEC14* and *SIGLEC5*. *SIGLEC14* and *SIGLEC5* in other primates also show evidence of gene conversions within each lineage. Evidently, balancing the interactions between Siglec-14, Siglec-5 and their common ligand(s) had selective advantage during the course of evolution. The "essential arginine" critical for sialic acid recognition in both Siglec-14 and Siglec-5 is present in humans but mutated in almost all great ape alleles (Angata et al. 2006b).

Activatory and Inhibitory Signals: Regulation of responses by cells in hemopoietic and immune systems depends on a balance between activatory and inhibitory signals. Their relative strengths set an appropriate activation threshold that helps fine-tune the response. CD33-rSiglecs have been shown to regulate negatively the cells, which express them (Mingari et al. 2001; Ulyanova et al. 2001; Nutku et al. 2005; von Gunten et al. 2005). Ab-mediated cross-linking of CD33rSiglecs leads to negative regulation of cellular activities, such as cell death or reduced cell proliferation (Von Gunten et al. 2005). Inhibitory signals are typically initiated by receptors containing one or more cytoplasmic ITIMs (Ravetch and Lanier 2000). ITIM in the intracytoplasmic domain of inhibitory receptors acts as a regulatory molecule to inhibit activation. Among 11 human CD33-related siglecs, most of them contain a membrane proximal ITIM and a membrane distal ITIM-like motif in their cytoplasmic tails. Where studied, CD33-rSiglecs can become tyrosine phosphorylated and recruit SHP-1 and SHP-2 after treatment of cells with pervanadate, a protein-tyrosine phosphatase inhibitor (Avril et al. 2004; Ikehara et al. 2004; Taylor et al. 1999; Ulyanova et al. 2001). In addition, CD33 and Siglecs-7 and -9 have been shown to exhibit enhanced siglec-dependent adhesion after tyrosine mutation of proximal ITIM (Taylor et al. 1999); antibody-mediated cross-linking of some CD33-rSiglecs results in inhibition of cell proliferation and function, and/or induction of apoptosis (Nutku et al. 2003; von Gunten et al. 2005). While these in vitro results suggest that CD33-rSiglecs are inhibitory signaling molecules that dampen immune-cell functions, in vivo proof is lacking. Anti-Siglec antibodies also tend to induce rapid endocytic clearing of cognate Siglec from cell surfaces (Nguyen et al. 2005, 2006a), complicating interpretation of the observed effects. The CD33-rSiglecs contain not only a cytoplasmic ITIM but also an immunoreceptor tyrosine-based switch–like motif (ITSM). While ligand-induced clustering of inhibitory receptors results in tyrosine phosphorylation of these ITIMs and recruitment of Src homology 2 (SH2) containing phosphatases (SHP-1/2) and inositol phosphatase (SHIP) (Malbec et al. 1998), the ITSM has been shown to switch between binding to signaling lymphocyte activated molecule (SLAM) associated protein (SAP) and EAT-2 or between SAP and SHP-2 in other receptors (Howie et al. 2002).

The consensus sequence for ITIMs is (V/I/L)XYXX (L/V), where X is any amino acid. A bioinformatics study by Staub et al. (2004) showed that the human proteome contains ~ 109 membrane proteins with cytoplasmic ITIMs, and many of these have been shown already to function as inhibitory receptors. Generally, when ITIM-containing receptors are engaged, they can become tyrosine-phosphorylated and then transmit inhibitory signals by binding and activating Src homology-2 domain (SH2)-containing tyrosine phosphatases (SHP1 and SHP2) and/or the SH2-containing inositol polyphosphate 5'-phosphatase (SHIP). In the case of CD33, several studies have demonstrated that SHP1 and SHP2, but not SHIP, are recruited and activated once the ITIMs are phosphorylated (Taylor et al. 1999).

ITAM Motif: Recent discoveries of Siglecs that associate with the activating adaptor molecule DAP12, which has an ITAM, have shed light on a subgroup of Siglecs that have an activating signaling potential (Angata et al. 2006; Blasius et al. 2006). The presence of both activating and inhibitory members within a family is a feature that is also found in other immunity-related cell-surface receptor families such as killer cell Ig-like receptors of primates (Vilches and Parham 2002), leukocyte Ig-like receptors of primates (Martin et al. 2002), and the Ly49 family of rodents (Yokoyama and Plougastel 2003). Although the functional significance of the presence of counteracting members within a family is not fully understood, it has been proposed that such members act cooperatively to fine-tune cellular responses (Lanier 2001), or that the activating members function to counteract pathogens that exploit their inhibitory counterparts (Arase and Lanier 2004).

17.2.3 Organization of CD33-rSiglec Genes on Chromosome 19q13.4

Genes of Siglec-3-like subgroup of Siglecs were mapped to a region of ~ 500 kb of chromosome 19q (cytological band 19q13.3–13.4) near the gene clusters containing KIR and LILR genes, suggesting their origin through repeated gene duplications (Crocker and Varki 2001a,b). The detailed map of the locus contains 8 Siglec genes and one Siglec-like gene (Siglec-L) and at least 13 Siglec-like pseudogenes (Angata et al. 2001b; Angata et al. 2004). Members of this subfamily exhibit two patterns of organization of the signal peptide, which is followed by one V-set domain (except for the long form of the siglecL1 gene). Exons containing the C2-set domains are all comparable in size and are separated by linker exons. The transmembrane domain is encoded for by a separate exon of almost the same size in all genes. The total number of exons differs according to the number of C2-set Ig domains, but intron phases are identical. The cytoplasmic domain is always encoded by two exons. Yousef et al. (2002) further identified two new Siglec pseudogenes in this locus, and analyzed their tissue expression pattern and their structural features. Although a few Siglecs are well-conserved throughout vertebrate evolution and show similar binding preference regardless of species of origin, CD33-related subfamily of Siglecs show marked inter-species differences in repertoire, sequence, and binding preference. Hayakawa et al. (2005) demonstrated human-specific gene conversion between *SIGLEC11* and the adjacent pseudogene *SIGLECP16*, resulting in human-specific expression of Siglec-11 in brain microglia (Hayakawa et al. 2005).

17.2.3.1 Loss of Siglec Expression on T Lymphocytes During Human Evolution

Human T cells give much stronger proliferative responses to specific activation via TCR than those from chimpanzees, our closest relatives. Among human immune cells, T lymphocytes are a striking exception, expressing little to none of these CD33-related Siglecs. In sharp contrast, T lymphocytes from chimpanzees as well as from "great apes" (bonobos, gorillas, and orangutans) express several CD33-related Siglecs on their surfaces. This suggests that human-specific loss of T cell Siglec expression occurred after our last common ancestor with great apes, potentially resulting in an evolutionary difference with regard to inhibitory signaling. This was confirmed by studying Siglec-5, which is prominently expressed on chimpanzee lymphocytes, including CD4 T cells. The human-specific loss of T cell Siglec expression associated with T cell hyperactivity may help explain the strikingly disparate prevalence and severity of T cell-mediated diseases such as AIDS and chronic active hepatitis between humans and chimpanzees (Nguyen et al. 2006b).

17.3 Siglec-5 (CD170)

Human Siglec-5, also known as CD170, encoded by *SIGLEC5* gene, has four extracellular Ig-like domains and a cytosolic inhibitory motif. Siglec-5 also known as obesity binding protein-2 (OB-BP2) shows no similarity to leptin receptor (Ob-R). Angata et al. (2006) discovered human Siglec-14 with three Ig-like domains, encoded by the

SIGLEC14 gene, adjacent to *SIGLEC5*. Human Siglec-14 has almost complete sequence identity with human Siglec-5 at the first two Ig-like domains, shows a glycan binding preference similar to that of human Siglec-5, and associates with the activating adapter protein DAP12. Thus, Siglec-14 and Siglec-5 appear to be first among glycan binding paired receptors.

17.3.1 Characterization

Siglec-5, a human CD33-related siglec, is expressed on granulocytes and monocytes (Cornish et al. 1998), as well as on plasmacytoid dendritic cells, monocyte-derived dendritic cells, and macrophages (Lock et al. 2004). A full-length cDNA encoding siglec-5 predicted that siglec-5 contains four extracellular Ig-like domains, the N-terminal two of which are 57% identical to the corresponding region of CD33. The cytoplasmic tail is also related to that of CD33, containing two tyrosine residues embodied in ITIM-like motifs. The siglec-5 gene was shown to map to chromosome 19q13.41-43, closely linked to the CD33 gene. Siglec-5 may be involved in cell-cell interactions. However, siglec-5 was found to have an expression pattern distinct from that of CD33, being present at relatively high levels on neutrophils but absent from leukemic cell lines representing early stages of myelomonocytic differentiation. Siglec-5 exists as a disulfide-linked dimer of approximately 140 kD (Cornish et al. 1998).

Angata et al. (2001a, 2004) identified *SIGLEC5**, a genomic segment adjacent to *SIGLEC5* gene in the Siglec gene cluster that showed extreme sequence identity with a part of *SIGLEC5* (Angata et al. 2004; Angata et al. 2001a). However, it was not clear if this genomic segment was a part of an active genetic element. It was evident that the genomic region downstream of *SIGLEC5** lacked sequences similar to the exon containing ITIM typically found in CD33rSiglecs, suggesting a possibility that the Siglec protein encoded by *SIGLEC5**, if any, may have a different signaling potential from that of Siglec-5, or any other human CD33rSiglecs. Patel et al. (1999) reported cloning of a leptin-binding protein of IgSF (OB-BP1/Siglec-6) and a cross-hybridizing clone (OB-BP2) that is identical to Siglec-5. Human tissues showed OB-BP2/Siglec-5 mRNA in peripheral blood leukocytes, lung, spleen, and placenta. Angata et al. (2006) proposed that Siglec-5 and Siglec-14 are paired inhibitory and activating receptors. It was also demonstrated that *SIGLEC14* and *SIGLEC5* have undergone concerted evolution via gene conversion in multiple primate species. Based on these results, the evolutionary dynamics behind the birth of Siglec-14 and its concerted evolution with Siglec-5 was suggested.

17.3.2 Siglec-5: An Inhibitory Receptor

Siglec-5 can recruit SHP-1 and SHP-2 after tyrosine phosphorylation and mediate inhibitory signaling, as measured by calcium flux and serotonin release after co-ligation with the activatory FcγRI. Surprisingly, however, mutagenesis studies showed that inhibition of serotonin release could still occur efficiently after a double tyrosine to alanine substitution, whereas suppression of Siglec-5-dependent adhesion required an intact tyrosine residue at the membrane proximal ITIM. A potential mechanism for tyrosine phosphorylation-independent inhibitory signaling was provided by results of in vitro phosphatase assays, which demonstrated low level activation of SHP-1 by the Siglec-5 cytoplasmic tail in the absence of tyrosine phosphorylation (Avril et al. 2005). Based on these findings, Siglec-5 can be classified as an inhibitory receptor with the potential to mediate SHP-1 and/or SHP-2-dependent signaling in the absence of tyrosine phosphorylation (Avril et al. 2005; Angata et al. 2006).

Siglec-5 was not expressed at significant levels by CD34+ progenitors either from bone marrow or mobilized peripheral blood. Siglec-5 expression remained absent or very low on cultured CD34+ cells, unlike CD33, which was present on almost all CD34+ cells by day 4. However, analysis of blasts from patients with AML revealed aberrant expression of Siglec-5 with CD34 in 50% of patients with CD34+ AML; 61% of AML cases were positive for Siglec-5 with an increased frequency in the French-American-British subtypes M3-5. All 13 acute lymphoblastic leukaemic (ALL) samples tested, were Siglec-5 negative (Virgo et al. 2003).

17.3.3 Siglec-5-Mediated Sialoglycan Recognition

The crystal structure of two N-terminal extracellular domains of human Siglec-5 and its complexes with two sialylated carbohydrates was determined. The native structure revealed an unusual conformation of the CC' ligand specificity loop and a unique interdomain disulfide bond (Fig. 17.2). The α(2,3)- and α(2,6)-sialyllactose complexed structures showed a conserved Sia recognition motif that involves both Arg124 and a portion of the G-strand in the V-set domain forming β-sheet-like hydrogen bonds with the glycerol side chain of the Sia. Only few protein contacts to the subterminal sugars are observed and mediated by the highly variable GG' linker and CC' loop. Structural observations provided mechanistic insights into linkage-dependent Siglec carbohydrate recognition and suggested that Siglec-5 and other CD33-related Siglec receptors are more promiscuous in sialoglycan recognition than previously understood (Zhuravleva et al. 2008)

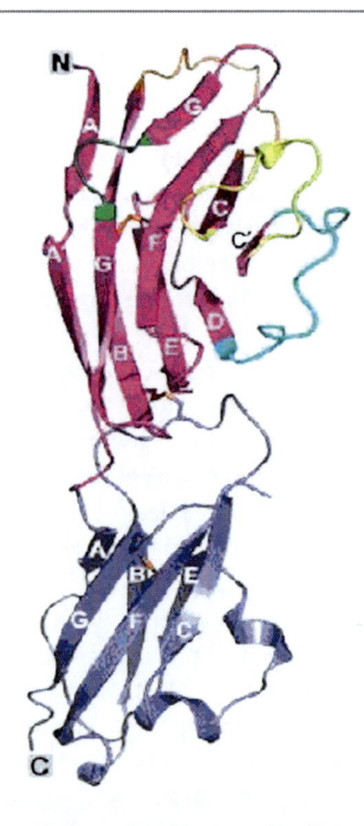

Fig. 17.2 Ribbon diagram showing the native Siglec-5 structure with V-domain in magenta and C2-domain in blue. The disulfides are depicted with sticks. BC and C′D loops and the specificity-determining GG′ and CC′ regions are depicted in orange, cyan, green and yellow, respectively (Adapted by permission from Zhuravleva et al. 2008 © Elsevier)

17.4 Siglec-6

17.4.1 Cloning and Gene Organization of Siglec-6 (OB-BP1)

Siglec-6 was cloned independently by two groups: in one instance from a human placental library (Takei et al. 1997) and in another as a leptin-binding protein (Patel et al. 1999). While Takei et al. (1997) noted an alternately spliced form of obesity binding protein-1 (OB-BP1) (CD33L2) that is predicted to encode a soluble form of protein, Siglec-6 shows strong expression in human placental trophoblast and also at variable levels on human B cells (Patel et al. 1999). Placental trophoblast expression of Siglec-6 is a human-specific phenomenon, which is associated with presence of Siglec-6 ligands on cells of placental and endometrial (uterine epithelial) origin. There is a relationship between the tempo of Siglec-6 expression and the onset and progression of labor. Given the presence of inhibitory signaling motifs on the cytosolic tail of Siglec-6 and its leptin-binding ability, one can speculate about mechanistic connections to the unusually prolonged nature of the human birth process. Human siglec-6 (OB-BP1), like siglec-3 (CD33) recognize both Neu5Ac and Neu5Gc. Siglec-6 shows significant binding to sialyl-Tn (Neu5Acα2-6-GalNAc), a tumor marker associated with poor prognosis. Siglec-6 is an exception among siglecs in not requiring the glycerol side chain of sialic acid for recognition.

The complete cDNA contains an open reading frame of 1,326 nt encoding 442 amino acids and belongs to the IgSF. It is composed of three Ig-like domains; its high degree of similarity to CD33 (70% identity) in the first and second of these domains implies that the placenta-specific gene product is likely to be associated with cell-cell interaction. Northern-blot analysis revealed transcripts of four distinct sizes, 7.5 kb, 5.0 kb, 4.1 kb, and 2.0 kb, specifically in placenta. Consequently, this gene was termed CD33-antigen-like. An alternatively-spliced transcript encoding a 342-amino-acid peptide which lacked the transmembrane region and the cytoplasmic tail was also isolated. It was localized to human chromosome 19q13.3, where the CD33 gene is also located (Takei et al. 1997).

The leptin-binding protein of IgSF (OB-BP1/Siglec-6) along with OB-BP2/Siglec-5, and CD33/Siglec-3 constitute a unique related subgroup with a high level of overall amino acid identity. The cytoplasmic domains are not as highly conserved, but display motifs which are putative sites of tyrosine phosphorylation, including an ITIM and a motif found in SLAM and SLAM-like proteins. Human tissues showed high levels of OB-BP1 mRNA in placenta and moderate expression in spleen, peripheral blood leukocytes, and small intestine. OB-BP2/Siglec-5 mRNA was detected in peripheral blood leukocytes, lung, spleen, and placenta. OB-BP1 showed high expression in placental cyto- and syncytiotrophoblasts. While OB-BP1 exhibited tight binding (K_D = 91 nM), the other two showed weak binding with K_D values in 1–2 μM range. OB-BP1 specifically bound Neu5Acα2–6GalNAcα (sialyl-Tn) allowing its formal designation as Siglec-6. The OB-BP1/Siglec-6 as a Siglec may mediate cell-cell recognition events by interacting with sialylated glycoprotein ligands expressed on specific cell populations (Patel et al. 1999). OB-BP1 and 2 display no similarity to leptin receptor (Ob-R). In surface plasmon resonance studies, OB-BP1 bound leptin with a moderate affinity, while OB-BP2 and CD33 bound with low affinities. The three exhibited binding kinetics with relatively slow on and off rates, which differed significantly from typical receptor-cytokine kinetics in which both on and off rates are fast (Patel et al. 1999). If leptin is an endogenous ligand of OB-BP1, role in leptin physiology could be speculated. Patel et al. (1999) hypothesized that OB-BP1 regulates circulating levels of leptin or acts as a leptin carrier in blood via B cells.

17.4.2 Siglec-6 (OB-BP1) and Reproductive Functions

Siglec-6/OB-BP1 is expressed on immune cells of both humans and the closely related great apes. Placental trophoblast expression is human-specific, with little or no expression in ape placentae. Human placenta also expresses natural ligands for Siglec-6 (a mixture of glycoproteins carrying cognate sialylated targets), in areas adjacent to Siglec-6 expression. Ligands were also found in uterine endometrium and on cell lines of trophoblastic or endometrial origin. Thus, Siglec-6 was recruited to placental expression during human evolution, presumably to interact with sialylated ligands for specific negative signaling functions and/or to regulate leptin availability. The control of human labor is poorly understood, but involves multiple cues, including placental signaling. Human birthing is also prolonged in comparison to that in our closest evolutionary relatives, the great apes. Siglec-6 levels are generally low in placentae from elective surgical deliveries without known labor and the highest following completion of labor. It is speculated that the negative signaling potential of Siglec-6 was recruited to human-specific placental expression, to slow the tempo of human birth process. The leptin-binding ability of Siglec-6 is also consistent with this hypothesis, as leptin-deficient mice have increased parturition times (Brinkman-Van der Linden et al. 2007).

Preeclampsia (PE), which affects 4–8% of human pregnancies, causes significant maternal and neonatal morbidity and mortality. Within the basal plate, placental cytotrophoblasts (CTBs) encode proteins associated with PE and molecules like Siglec-6 and pappalysin-2, localized to invasive CTBs and syncytiotrophoblasts. Alterations in Siglec-6 have been reported in PE. Siglec-6 placental expression is unique to human, as is spontaneous PE (Winn et al. 2009).

17.5 Siglec-7 (p75/AIRM1)

17.5.1 Characterization

Siglec-7, the adhesion inhibitory receptor molecule 1 (p75/AIRM1), is a 75-kD surface glycoprotein that displays homology with the myeloid cell antigen CD33. In lymphoid cells, p75/AIRM1 is confined to NK cells and mediates inhibition of their cytolytic activity. p75/AIRM1 is also expressed by cells of the myelomonocytic cell lineage, in which it appears at a later stage as compared with CD33 (Vitale et al. 2001). The p75/AIRM1 gene is located on human chromosome 19 and encodes a member of the sialoadhesin family characterized by three Ig-like extracellular domains (one NH_2-terminal V-type and two C2-type) and a classical ITIM in the cytoplasmic portion. The highest amino acid sequence similarity has been found with the myeloid-specific CD33 molecule and the placental CD33L1 protein. Similar to other sialoadhesin molecules, p75/AIRM1 appears to mediate sialic acid-dependent ligand recognition (Nicoll et al. 1999; Angata and Varki 2000b).

Sugar-binding specificity of Siglec-7 expressed on CHO cells was characterized. Glyco-probes carrying unique oligosaccharide structures such as GD3 (NeuAc α2,8NeuAc α2,3Gal β 1,4Glc) and LSTb (Gal β1,3[NeuAc α 2,6]GlcNAc β1,3Gal β1,4Glc) oligosaccharides bound to Siglec-7 better than those carrying LSTc (NeuAc α2,6Gal β1,4GlcNAc β1,3Gal β1,4Glc) or GD1a (NeuAc α 2,3Gal β1,3GalNAc β1,4[NeuAc α2,3]Gal β1,4Glc) oligosaccharides. In contrast, Siglec-9, which is 84% identical to Siglec-7, did not bind to the GD3 and LSTb probes but did bind to the LSTc and GD1a probes. Substitution of a small region, Asn70-Lys75, of Siglec-7 with the equivalent region of Siglec-9 resulted in loss of Siglec-7-like binding specificity and acquisition of Siglec-9-like binding properties. In comparison, a Siglec-9-based chimera, which contains Asn70-Lys75 with additional amino acids derived from Siglec-7, exhibited Siglec-7-like specificity. These results with molecular modeling suggest that the C-C' loop in the sugar-binding domain plays a major role in determining the binding specificities of Siglecs-7 and -9 (Yamaji et al. 2002).

17.5.2 Cytoplasmic Domain of Siglec-7 (p75/AIRM1)

The cytoplasmic domain of Siglec-7 contains two signaling motifs: a membrane-proximal ITIM (Ile435-Gln-Tyr-Ala-Pro-Leu440) and a membrane-distal motif (Asn458-Glu-Tyr-Ser-Glu-Ile463). Upon pervanadate (PV) treatment, Siglec-7 recruited the protein tyrosine phosphatases Src homology-2 (SH2) domain-containing protein-tyrosine phosphatase-1 (SHP-1) and SHP-2 less efficiently than did other inhibitory receptors such as Siglec-9. Alignment of the amino acid sequences of the two Siglecs revealed only three amino acids difference in these motifs. These amino acids appeared to affect not only phosphatase recruitment but also the subsequent attenuation of Syk phosphorylation (Yamaji et al. 2005).

Siglec-7 shows a preference for α(2,8)-disialylated ligands and provides a structural template for studying the key interactions that drive this selectivity. The crystal

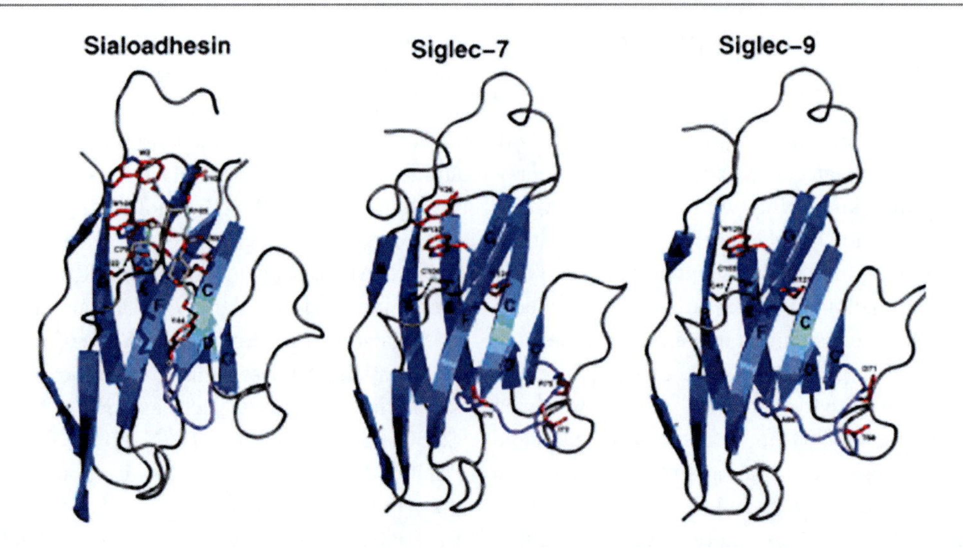

Fig. 17.3 *Comparison of sialoadhesin, Siglec-7, and Siglec-9 structures.* The crystal structures of sialoadhesin in complex with sialyllactose, the crystal structure of Siglec-7, and a model for Siglec-9 are shown in two representations. The ackbones are shown as *ribbons* with the side chains surrounding the ligand-binding site shown as *sticks* with *orange* carbons. The C-C' loop is *highlighted* in *magenta.* The two pyranose sugars from the sialyllactose ligand are shown as *sticks* with *green* carbons. Hydrogen bonds with the protein are shown as *black dotted lines.* For sialoadhesin, the two pyranose sugars of the ligand are shown in a *sticks* representation (Adapted by permission from Alphey et al. 2003 © The American Society for Biochemisty and Molecular Biology)

structure of the co-crystallized Siglec-7 with a synthetic oligosaccharide corresponding to the α(2,8)-disialylated ganglioside GT1b offers a first glimpse into how this important family of lectins binds the structurally diverse gangliosides. The complex structure revealed that the C-C' loop, a region implicated in previous studies as driving siglec specificity, undergoes a dramatic conformational shift, allowing it to interact with the underlying neutral glycan core of the ganglioside. Studies showed that binding of the ganglioside is driven by extensive hydrophobic contacts together with key polar interactions and that the binding site structure is complementary to preferred solution conformations of GT1b (Attrill et al. 2006a).

17.5.3 Crystallographic Analysis

The expression, crystallization and preliminary X-ray characterization of the Ig-V like domain of p75/AIRM1 have been reported. X-ray data were collected from a single crystal at 100 K at 1.45 A resolution showed that the crystal belongs to a primitive monoclinic space group, with unit-cell parameters a = 32.65, b = 49.72, c = 39.79 A, $\alpha = \gamma = 90$, $\beta = 113$. The systematic absences indicate that the space group is P2(1). Assuming one molecule per asymmetric unit, V(M) (Matthews coefficient) was calculated to be 1.879 $A^3\ Da^{-1}$ and the solvent content was estimated to be 32.01%. The structure belongs to a different space group than Siglec-7 structure and was obtained using a bacterial expression system. The structure unveils the fine structural requirements adopted by a natural killer cell inhibitory receptor of the Siglec family in target-cell recognition and binding (Dimasi et al. 2004).

Alphey et al. (2003) described the high resolution structures of the N-terminal V-set Ig-like domain of Siglec-7 in two crystal forms. The latter crystal form reveals the full structure of this domain and allows us to speculate on the differential ligand binding properties displayed by members of the Siglec family. A fully ordered N-linked glycan is observed, tethered by tight interactions with symmetry-related protein molecules in the crystal. Comparison of the structure with that of sialoadhesin and a model of Siglec-9 shows that the unique preference of Siglec-7 for α(2,8)-linked disialic acid is likely to reside in the C-C' loop, which is variable in the Siglec family (Fig. 17.3). In the Siglec-7 structure, the ligand-binding pocket is occupied by a loop of a symmetry-related molecule, mimicking the interactions with sialic acid (Alphey et al. 2003).

17.5.4 Interactions of Siglec-7

Avril et al. (2006) examined the interaction of 10 siglecs with lipooligosaccharides (LOS) purified from four different *C. jejuni* isolates expressing GM1-like, GD1a-like, GD3-like, and GT1a-like oligosaccharides. Of all siglecs examined, only Siglec-7 exhibited specific, sialic acid-dependent interactions with *C. jejuni* LOS. Binding of Siglec-7 was also observed with intact bacteria expressing these LOS structures. Specific binding of HS:19(GM1$^+$ GT1a$^+$) bacteria was demonstrated with Siglec-7 expressed on transfected CHO cells and with peripheral blood leukocytes, among which HS:19(GM1$^+$ GT1a$^+$) bacteria bound selectively to both NK cells and monocytes which naturally express Siglec-7. These results raise the possibility that, in addition to their role in generating autoimmune antibody responses, *C. jejuni* LOS could interact with Siglec-7 expressed by leukocytes, modulate the host-pathogen interaction, and contribute to the clinical outcome and the development of secondary complications such as Guillain-Barre syndrome (Avril et al. 2006).

In addition to their ability to recognize sialic acid residues, Siglec-7 and Siglec-9 display two conserved tyrosine-based motifs in their cytoplasmic region similar to those found in inhibitory receptors of the immune system. Siglecs-7 and -9 are able to inhibit the FcεRI-mediated serotonin release from RBL cells following co-crosslinking. In addition, under these conditions or after pervanadate treatment, Siglecs-7 and -9 associate with the Src homology region 2 domain-containing phosphatases (SHP), SHP-1 and SHP-2. Site-directed mutagenesis showed that the membrane-proximal tyrosine motif is essential for the inhibitory function of both Siglec-7 and -9, and is also required for tyrosine phosphory lation and recruitment of SHP-1 and SHP-2 phosphatases. The mutation of the membrane-proximal motif increased the sialic acid binding activity of Siglecs-7 and -9, suggesting the possibility that "inside-out" signaling may occur to regulate ligand binding (Angata et al. 2006). The crystal structure of siglec-7 in complex with a sialylated ligand, the ganglioside analogue DSLc4 [α(2,3)/α (2,6) disialyl lactotetraosyl 2-(trimethylsilyl)ethyl], allows for a detailed description of the binding site, required for structure-guided inhibitor design. Mutagenesis and binding assays demonstrated the key structural role for Lys131, a residue that changes conformation upon sialic acid binding. Differences between the binding sites of siglec family members were then exploited using α-methyl Neu5Ac (N-acetylneuraminic acid) as a basic scaffold. A co-crystal of siglec-7 in complex with the sialoside inhibitor, oxamido-Neu5Ac [methyl α-9-(amino-oxalyl-amino)-9-deoxy-Neu5Ac] and inhibition data for the sialosides gives clear leads for future inhibitor design (Attrill et al. 2006b).

17.5.5 Functions of Siglec-7

Negative Regulation of T Cell Receptor Signaling: Siglec-7 and Siglec-9 are capable of modulating T cell receptor (TCR) signaling in Jurkat T cells stably and transiently transfected with Siglec-7 or Siglec-9. Following either pervanadate stimulation or TCR engagement, both Siglecs exhibited increased tyrosine phosphorylation and recruitment of SHP-1. There was also a corresponding decreased transcriptional activity of nuclear factor of activated T cells (NFAT) as determined using a luciferase reporter gene. Like all siglecs, Siglec-7 and -9 recognize sialic acid-containing glycans of glycoproteins and glycolipids as ligands. Mutation of the conserved Arg in the ligand binding site of Siglec-7 (Arg^{124}) or Siglec-9 (Arg^{120}) resulted in reduced inhibitory function in the NFAT/luciferase transcription assay, suggesting that ligand binding is required for optimal inhibition of TCR signaling. The combined results demonstrate that both Siglec-7 and Siglec-9 are capable of negative regulation of TCR signaling and that ligand binding is required for optimal activity (Ikehara et al. 2004).

SOCS3 Targets Siglec 7 for Proteasomal Degradation: SOCS3 (Suppressor of Cytokine Signaling 3) is up-regulated during inflammation and competes with SHP-1/2 for binding to ITIM-like motifs on various cytokine receptors resulting in inhibition of signaling. Orr et al. (2007a) showed that SOCS3 binds the phosphorylated ITIM of Siglec 7 and targets it for proteasomal-mediated degradation suggesting that the Siglec 7 receptor is a SOCS target. In addition, SOCS3 expression blocks Siglec 7 mediated inhibition of cytokine-induced proliferation. It seems that SOCS target degrades simultaneously with the SOCS protein and that inhibitory receptors may be degraded in this way. This may be a mechanism by which the inflammatory response is potentiated during infection (Orr et al. 2007a).

17.6 Siglec-8

17.6.1 Characteristics and Cellular Specificity

Siglec-8 (SAF-2) is selectively expressed on human eosinophils, basophils, and mast cells, where it regulates their function and survival (Kikly et al. 2000; Floyd et al. 2000). SAF-2, identified by Kikly et al. (2000) was homologous with CD33 and siglec-5. SAF-2 is a 431-amino acid protein composed of 3 Ig domains with a 358-amino acid extracellular domain and a 47-amino acid tail. SAF-2 is useful in the detection and/or modulation of allergic cells (Bochner 2009; Aizawa et al. 2003). Siglec-8 was also found to be expressed on human mast cells and to a weak but consistent degree on human basophils (Kikly et al. 2000). Siglec-8 exists in 2 isoforms with identical extracellular and transmembrane sequences. A splice variant of Siglec-8, termed Siglec-8L, which contains an identical extracellular domain but a longer cytoplasmic tail possessing two ITIMs, was discovered from human genomic DNA (Bochner 2009; Foussias et al. 2001; Yousef et al. 2002). One isoform (Siglec-8) has a short cytoplasmic tail with no known signaling sequences, while the other, Siglec-8 long form (Siglec-8L) has a longer cytoplasmic tail containing 2 - tyrosine-based signaling motifs (Foussias et al. 2000; Munday et al. 2001). The cytoplasmic region of Siglec-8L contains one consensus ITIM and a signaling lymphocyte activation molecule (SLAM)–like motif, suggesting that Siglec-8L may possess signal transduction activity (Crocker and Varki 2001a, b; Foussias et al. 2000; Munday et al. 2001). The highest levels of homology are found between Siglec-8 and Siglec-3 (49%), Siglec-5 (42%), and Siglec-7 (68%), with virtually all homology due to similarities in the extracellular and transmembrane regions. Siglec-8 is expressed not only on the surface of eosinophils but also on basophils and mast cells (Kikly et al. 2000), and the existence of both the Siglec-8 and Siglec-8L isoforms was verified in human eosinophils, basophils, and mast cells (Aizawa et al. 2002; Bochner 2009). Antibody cross-linking of Siglec-8 on human eosinophils induces caspase-dependent apoptosis in vitro (Nutku et al. 2003). Like Siglec-3 and Siglec-7 (p75/AIRM-1), ligation of Siglec-8 inhibits eosinophil survival. Cross-linking Siglec-8 with antibodies rapidly generated caspase-3–like activity and reduced eosinophil viability through induction of apoptosis. Siglec-8 activation may provide a useful therapeutic approach to reduce numbers of eosinophils in disease states where these cells are important (Nutku et al. 2008).

The siglec-8 gene mapped on chromosome 19q13.33-41, ~330 kb down-stream of Siglec-9 gene, is closely linked to genes encoding CD33, siglec-5, siglec-6, and siglec-7. It mediates sialic acid-dependent binding to human erythrocytes and to soluble sialoglycoconjugates (Floyd et al. 2000). Both Siglec-8 and Siglec8-L comprise of seven exons, of which the first five are identical, followed by marked differences in exon usage and mRNA splicing. The 499 amino acid protein encoded by the Siglec8-L open reading frame has a molecular weight of 54 kDa (Foussias et al. 2000) (Fig. 17.1).

17.6.2 Ligands for Siglec-8

Red blood cell rosettes are formed in presence of Siglec-8, and neuraminidase treatment alters rosette formation (Kikly et al. 2000; Floyd et al. 2000). Specific structures shown to bind Siglec-8 include forms of sialic acid that are linked α2–3 or α2–6 to Galβ1–4GlcNAc (Floyd et al. 2000). The binding of Siglec-8 is sialic acid-dependent. A Siglec-8-Ig chimeric protein revealed that of 172 glycan structures, ~40 structures were sialylated. Among these, avid binding was detected to a single defined glycan, NeuAcα2–3(6-*O*-sulfo)Galβ1–4[Fucα1–3]GlcNAc, also referred to as 6' sulphated sialyl Lewis X (6′-sulfo-sLe^x). Notably, neither unsulfated sLe^x (NeuAcα2–3Galβ1–4[Fucα1–3]GlcNAc), a known ligand for E-, P- and L-selectin nor an isomer with the sulfate on the 6-position of the GlcNAc residue (6-sulfo-sLe^x, NeuAcα2–3Galβ1–4[Fucα1–3](6-*O*-sulfo)GlcNAc) supported detectable binding. Whereas surfaces derivatized with sLe^x and 6-sulfo-sLe^x failed to support detectable Siglec-8 binding, 6′-sulfo-sLe^x supported significant binding with a K_D of 2.3 μM. Siglec-8 binds preferentially to the sLe^x structure bearing an additional sulfate ester on the galactose 6-hydroxyl (Bochner et al. 2005; Bochner 2009). Bochner et al. (2005) indicated that Siglec-8 is a highly specific lectin, binding preferentially to the sLe^x structure bearing an additional sulfate ester on the galactose 6-hydroxyl, namely NeuAcα2–3(6-*O*-sulfo)Galβ1–4[Fucα1–3]GlcNAc, also referred to in the literature as 6′-sulfo-sLe^x, which is a structure closely related to 6-sulfo-sLe^x, a candidate ligand for L-selectin. However, Blixt et al. (2003) suggested that Siglec-Ig chimeras showed a wide range of binding patterns with no clear binding preference for Siglec-8. Thus, Siglec-8 requires the presence of the sulphate on the sixth position of the galactose residue (Bochner 2009), though, the exact biochemical identity of natural tissue ligands for Siglec-8 remains unknown. Based on the occurrence of sulphated sLex-like structures made by airway cells, bronchial (and perhaps mononuclear cell) mucins have been suggested as candidate ligands for Siglec-F and Siglec-8 (Bochner 2009; Zhang et al. 2007).

17.6.3 Functions in Apoptosis

Recent work suggests that Siglecs are also empowered to transmit death signals, at least in myeloid cells. Strikingly, death induction by Siglecs is enhanced when cells are exposed to proinflammatory survival cytokines. Based on these insights, von Gunten and Simon (2006) hypothesized that at least some members of the Siglec family regulate immune responses via the activation of caspase-dependent and caspase-independent cell death pathways. Siglec-8 cross-linking with antibodies rapidly generated caspase-3-like activity and reduced eosinophil viability through induction of apoptosis. Siglec-8 crosslinking on eosinophils increased dissipation of mitochondrial membrane potential upstream of caspase activation. Inhibitors of mitochondrial respiratory chain components completely inhibited apoptosis. Additional experiments demonstrated that ROS was also essential for Siglec-8-mediated apoptosis and preceded Siglec-8-mediated mitochondrial dissipation (Nutku et al. 2003; 2005).

Based on structural homology alone, there was no clear mouse ortholog of Siglec-8. Therefore, using different strategies, it is now clear that Siglec-G is expressed on B lymphocytes (Hoffmann et al. 2007) while Siglec-F is most prominently expressed by mouse eosinophils (Angata et al. 2001a; Zhang et al. 2004; Tateno et al. 2005). This was somewhat unexpected based on sequence homology alone, because Siglec-F more closely resembles human Siglec-5, which is not expressed by human eosinophils (Aizawa et al. 2003). Siglec-F is not expressed on mouse mast cells and instead is expressed on a wider range of cells including alveolar macrophages and at very low levels on T cells and neutrophils, none of which express Siglec-8 in human; and unlike Siglec-8, surface levels of Siglec-F on eosinophils and other cells increase during allergic inflammatory responses (Bochner 2009). Therefore, because of its preferential expression on eosinophils and preference for binding the same ligand, Siglec-F and Siglec-8 are best thought of as functionally convergent paralogs.

17.6.4 Siglec-8 in Alzheimer's Disease

Recent progress in pattern recognition receptors of monocytes and macrophages has revealed that Siglec family of receptors is an important recognition receptor for sialylated glycoproteins and glycolipids. Studies revealed that microglial cells contain only one type of Siglec receptors, Siglec-11, which mediates immunosuppressive signals and thus inhibits the function of other microglial pattern recognition receptors, such as TLRs, NLRs, and RAGE receptors. Salminen and Kaarniranta (2009) reviewed and indicated that aggregating amyloid plaques are masked in AD by sialylated glycoproteins and gangliosides. Sialylation and glycosylation of plaques, mimicking the cell surface glycocalyx, can activate the immunosuppressive Siglec-11 receptors, as well as hiding the neuritic plaques, allowing them to evade the immune surveillance of microglial cells. This kind of immune evasion can prevent the microglial cleansing process of aggregating amyloid plaques in AD.

17.7 Siglec-9

17.7.1 Characterization and Phylogenetic Analysis

The cDNA of human Siglec-9 encodes a type 1 transmembrane protein with three extracellular immunoglobulin-like domains and a cytosolic tail containing two tyrosines, one within a typical ITIM. The N-terminal V-set Ig domain has most amino acid residues typical of Siglecs. Expression of the full-length cDNA in COS cells induces sialic-acid dependent erythrocyte binding. A recombinant soluble form of the extracellular domain binds to α2-3 and α2-6-linked sialic acids. The carboxyl group and side chain of sialic acid are essential for recognition, and mutation of a critical arginine residue in domain 1 abrogates binding. The underlying glycan structure also affects binding, with Galβ1-4Glc[NAc] being preferred. Siglec-9 shows closest homology to Siglec-7 and both belong to a Siglec-3/CD33-related subset of Siglecs (with Siglecs-5, -6, and -8). The Siglec-9 gene is on chromosome 19q13.3-13.4, in a cluster with all Siglec-3/CD33-related Siglec genes, suggesting their origin by gene duplications. A homology search of the *Drosophila melanogaster* and *C.elegans* genomes suggests that Siglec expression may be limited to animals of deuterostome lineage, coincident with the appearance of the genes of the sialic acid biosynthetic pathway (Angata and Varki 2000a).

A full-length cDNA encoding Siglec-9, isolated from a dibutyryl cAMP-treated HL-60 cell cDNA library, is predicted to contain three extracellular Ig-like domains that comprise an N-terminal V-set domain and two C2-set domains, a transmembrane region and a cytoplasmic tail containing two putative tyrosine-based signaling motifs. Overall, Siglec-9 is approximately 80% identical in amino acid sequence to Siglec-7, suggesting that the genes encoding these two proteins arose relatively recently by gene duplication. Siglec-9 was found to be expressed by monocytes, neutrophils, and $CD16^+$, $CD56^-$ cells. Weaker expression was observed on app. 50% of B cells and NK cells and minor subsets of $CD8^+$ T cells and $CD4^+$ T cells (Zhang et al. 2000).

17.7.2 Functions of Siglec-9

17.7.2.1 Siglec-9 Transduces Apoptotic and Nonapoptotic Death Signals

In normal neutrophils, Siglec-9 ligation induces apoptosis. The increased Siglec-9-mediated death was mimicked in vitro by proinflammatory cytokines, such as GM-CSF, IFN-α, and IFN-γ, and was demonstrated to be caspase independent. Study suggested that apoptotic (ROS- and caspase-dependent) and nonapoptotic (ROS-dependent) death pathways are initiated in neutrophils via Siglec-9. The new insights have important implications for the pathogenesis, diagnosis, and treatment of inflammatory diseases such as sepsis and rheumatoid arthritis (von Gunten et al. 2005). Siglec-9, present on neutrophils is activated by intravenous (iv) administration of Ig during immuno-therapy, resulting in caspase-dependent and caspase-independent forms of cell death, resulting in neutropenia that is sometimes seen in association with iv-Ig therapy Neutrophil death was mediated by naturally occurring anti-Siglec-9 autoantibodies present in iv-Ig. Anti- Siglec-9 autoantibody-depleted iv-Ig failed to induce this caspase-independent neutrophil death. Results explain the cause of neutropenia that is seen in association with iv Ig therapy (von Gunten et al. 2006).

17.7.2.2 Bacterial Pathogen Lower Siglec-9 Innate Immune Response

Human neutrophil Siglec-9 recognizes host Sias as ""self,"" including *in cis* interactions with Sias on the neutrophil's own surface, thereby dampening unwanted neutrophil reactivity. Neutrophils presented with immobilized multimerized Sia α2-3Galβ1-4GlcNAc units engage them *in trans* via Siglec-9. The sialylated capsular polysaccharide of group B *Streptococcus* (GBS) also presents terminal Sia α2-3Galβ1-4GlcNAc units, and similarly engages neutrophil Siglec-9, dampening neutrophil responses in a Sia- and Siglec-9–dependent manner. GBS can impair neutrophil defense functions by coopting a host inhibitory receptor via sialoglycan molecular mimicry, a novel mechanism of bacterial immune evasion (Carlin et al. 2009). Experiments using a pan-caspase-inhibitor provided evidence for caspase-independent neutrophil death in Siglec-9 responders upon Siglec-9 ligation. Septic shock patients exhibit different ex vivo death responses of blood neutrophils after Siglec-9 ligation early in shock (von Gunten et al. 2009).

17.7.2.3 Siglec-9 Enhances IL-10 Production in Macrophages

Siglec-9 modulates cytokine production in macrophage cell line RAW264. Overexpression of siglec-9 in macrophage cell lines inhibited production of pro-inflammatory cytokines such as TNFα and enhanced the production of IL-10 in an ITIM-dependent manner in response to Toll-like receptor signaling (Ando et al. 2008). Cells expressing Siglec-9 produced low levels of TNF-α upon stimulation with LPS, peptidoglycan, unmethylated CpG DNA, and double-stranded RNA. On the other hand, IL-10 production was strongly enhanced in Siglec-9-expressing cells. Similar activity was also demonstrated by Siglec-5. However, the up-regulation of IL-10 as well as the down-regulation of TNF-α was abrogated when two tyrosine residues in the cytoplasmic tail of Siglec-9 were mutated to phenylalanine (Ando et al. 2008).

17.7.2.4 Siglec-9 is Expressed on Subsets of Acute Myeloid Leukemia Cells

Like Siglec-3, Siglec-9 is also expressed on subsets of AML and may provide additional therapeutic targets in the future (Biedermann et al. 2007). Siglec-9 was absent from normal bone marrow myeloid progenitors but present on monocytic precursors. Using primary AML cells or transfected rat basophilic leukemia cells, Siglec-9 mediated rapid endocytosis of anti-Siglec-9 mAb. Siglec-9 is not only a useful marker for certain subsets of AML, but also presents as a potential therapeutic target (Biedermann et al. 2007).

17.8 Siglec-10, -11, -12, and -16

17.8.1 Siglec-10

The full-length-Siglec-10 cDNA encodes a type 1 transmembrane protein containing four extracellular Ig-like domains, a transmembrane region, and a cytoplasmic tail with two ITIMs. The N-terminal V-set Ig domain has most of the amino acid residues typical of the Siglecs. Siglec-10 shows the closest homology to Siglec-5 and Siglec-3/CD33 and mapped to the same region, on chromosome 19q13.3. Various cells and cell lines including monocytes and dendritic cells express Siglec-10. Siglec-10 is an immune system-restricted membrane-bound protein that is highly expressed in peripheral blood leukocytes, spleen, and liver. The expressed protein was able to mediate sialic acid-dependent binding to human erythrocytes and soluble sialoglycoconjugates. Siglec-10 was detected on subsets of human leucocytes including eosinophils, monocytes and a minor population of NK-like cells. The molecular properties and expression pattern suggest that Siglec-10 may function as an inhibitory receptor within the innate immune system (Li et al. 2001; Munday et al. 2001). Genomic sequence of siglec-10 is localized within the cluster of genes on chromosome 19q13.3-4 that encodes other siglec family members. The extracellular domain of siglec-10 was capable of binding to peripheral blood leukocytes. The cytoplasmic tail of siglec-10 contains four tyrosines, two of which are embedded in ITIM-signaling motifs (Y^{597} and Y^{667}) and are likely involved in intracellular signaling. The ability of tyrosine kinases to phosphorylate the cytoplasmic tyrosines was evaluated by kinase assay using wild-type siglec-10 cytoplasmic domain and Y—>F mutants. The majority of the phosphorylation could be attributed to Y^{597} and Y^{667}. Further experiments with cell extracts suggest that SHP-1 interacts with Y^{667} and SHP-2 interacts with Y^{667} in addition to another tyrosine. Therefore siglec-10, as CD33, may be characterized as an inhibitory receptor (Whitney et al. 2001). A splice variant of Siglec-10, called Siglec-10Sv3 expresses in T- and B-cells. Moreover, another splicing form of Siglec-10, named Siglec-10Sv4, was also identified. One common characteristic of all Siglec-10 splice forms (except for Siglec-10Sv2) is their cytoplasmic tail with two ITIMs and one CD150-like sequence (Kitzig et al. 2002).

17.8.2 Siglec-11

As with others in this subgroup, the cytosolic domain of Siglec-11 is phosphorylated at tyrosine residue(s) upon pervanadate treatment of cells and then recruits the protein-tyrosine phosphatases SHP-1 and SHP-2. However, Siglec-11 has several novel features relative to the other CD33/Siglec-3-related Siglecs. First, it binds specifically to α2-8-linked sialic acids. Second, unlike other CD33/Siglec-3-related Siglecs, Siglec-11 was not found on peripheral blood leukocytes. Instead, it was expressed on macrophages in various tissues, such as liver Kupffer cells. Third, it was also expressed on brain microglia, thus becoming the second Siglec to be found in the nervous system. Fourth, whereas the Siglec-11 gene is on human chromosome 19, it lies outside the previously described CD33/Siglec-3-related Siglec cluster on this chromosome. Fifth, analyses of genome data bases indicate that Siglec-11 has no mouse ortholog and that it is likely to be the last canonical human Siglec to be reported. Finally, although Siglec-11 showed marked sequence similarity to human Siglec-10 in its extracellular domain, the cytosolic tail appeard only distantly related. Analysis of genomic regions surrounding the Siglec-11 gene suggests that it is actually a chimeric molecule that arose from relatively recent gene duplication and recombination events, involving the extracellular domain of a closely related ancestral Siglec gene (which subsequently became a pseudogene) and a transmembrane and cytosolic tail derived from another ancestral Siglec (Angata et al. 2002b). Human Siglec-11 was expressed by tissue macrophages, including brain microglia (Angata et al. 2004). Comparisons with the chimpanzee genome indicate that human Siglec-11 emerged through human-specific gene conversion by an adjacent pseudogene. Conversion involved 5 cent untranslated sequences and the Sia-recognition domain. This human protein shows reduced binding relative to the ancestral form but recognizes oligosialic acids, which are enriched in the brain. Siglec-11 is expressed in human but not in chimpanzee brain microglia. Further studies will determine if this event was related to the evolution of Homo (Hayakawa et al. 2005).

17.8.3 Siglec-12

Analyses of genomic *SIGLEC* sequences across humans, chimpanzees, baboons, rats, and mice showed that CD33rSiglecs are evolving rapidly. This is particularly pronounced in the Sia-recognizing V-set domain, suggesting that this domain is under the greatest selection pressure. The human ortholog of Siglec-12 (formerly Siglec-L1) has an Arg Cys (R122C) substitution resulting in a protein unable to bind Sias (Angata et al. 2001). This protein is referred to as Siglec-XII in humans and differentially named as Siglec-12 in primates, where the Sia-binding arginine is present. The gene in both cases is referred to as *SIGLEC12*. Reversing this mutation *in vitro* restored Sia binding. The R122C mutation of the Siglec-XII protein is fixed in the human population, i.e. it occurred prior to the origin of modern humans. Additional mutations have since completely inactivated the *SIGLEC12* gene in some but not all humans. The chimpanzee Siglec-12 is fully functional and preferentially recognizes N-glycolylneuraminic acid, which is a common sialic acid in great apes and other mammals. Reintroducing the ancestral arginine into the human molecule regenerates the same properties. Thus, the single base pair mutation that replaced the arginine on human Siglec-L1 is likely to be evolutionarily related to the previously reported loss of N-glycolylneuraminic acid expression in the human lineage. The human genome contains several Siglec-like pseudogenes that have independent mutations that would have replaced the arginine residue required for optimal sialic acid recognition during human evolution (Angata et al, 2001). Unlike other CD33-rSiglecs that are primarily found on immune cells, Siglec-XII protein is expressed on some macrophages and also on various epithelial cell surfaces in humans and chimpanzees. *SIGLEC12* gene also expresses on certain human prostate epithelial carcinomas and carcinoma cell lines. This expression correlates with the presence of the non-frame-shifted, intact *SIGLEC12* allele. Polymorphic expression of Siglec-XII in humans thus has implications for prostate cancer biology and therapeutics (Mitra et al. 2011)

17.8.3.1 Siglec-like Gene (SLG)/S2V

Foussias et al. (2001) identified the complete genomic structure of Siglec-like gene (SLG), a putative member of Siglec-3-like subgroup of Siglecs, as well as two alternative splice variants. The SLG gene is localized 32.9 kb downstream of Siglec-8 on chromosome 19q13.4. The putative SLG-S and SLG-L proteins, of 477 and 595 amino acid residues, respectively, show extensive homology to many members of the Siglec-3-like subgroup. This homology is conserved in the extracellular Ig-like domains, as well as in the cytoplasmic tyrosine-based motifs. However, SLG-L protein contains two N-terminal V-set Ig-like domains, as opposed to SLG-S and other Siglec-3-like subgroup members which contain only one such domain. The SLG-S is highly expressed in spleen, small intestine and adrenal gland, while SLG-L exhibits high levels of expression in spleen, small intestine, and bone marrow (Foussias et al. 2001).

The Siglec-like gene which comprises of 11 exons, with 10 intervening introns, is localized 278 kb telomeric to Siglec-9 and 35 kb centromeric to Siglec-8 and on chromosome 19q13.4. The coding region consists of 2,094 bp, and encodes for a putative 76.6 kDa protein. All Siglec-conserved structural features, including V-set domain, three C-set domains, transmembrane domain, ITIM and SLAM motifs, were found in this Siglec-like gene. Also, it has the conserved amino acids essential for sialic acid binding (Yousef et al. 2002). The cDNA of S2V Siglec encodes a type 1 transmembrane protein with four extracellular Ig-like domains and a cytoplasmic tail bearing a typical ITIM and an ITIM-like motif. A unique feature of S2V is the presence of two V-set Ig-like domains responsible for the binding to sialic acid, whereas all other known siglecs possess only one. S2V is predominantly expressed in macrophage. S2V is involved in the negative regulation of signaling in macrophage by functioning as an inhibitory receptor (Yu et al. 2001a).

17.8.4 Siglec-16

The lineages of *SIGLEC11* genes in human, rodent, dog, cow and non-human primates have undergone dynamic gene duplication and conversion, forming a potential inhibitory (Siglec-11)/activating (Siglec-16) receptor pair in chimpanzee and humans. A cDNA encoding human Siglec-16, classed as a pseudogene in databases (*SIGLECP16*), is expressed in various cell lines and tissues. A polymorphism screen for the two alleles (wild type and four-base pair deletion) of *SIGLEC16* found their frequencies to be 50% amongst the UK population. A search for donor sequences for Siglec-16 revealed a subfamily of activating Siglec with charged transmembrane domains predicted to associate with ITAM-encoding adaptor proteins (Cao et al. 2008). Using antisera specific to the cytoplasmic tail of Siglec-16, Cao et al. (2008) identified Siglec-16 expression in $CD14^{+}$ tissue macrophages and in normal human brain, cancerous oesophagus and lung. Although, most CD33rSiglecs have immune receptor tyrosine-based inhibitory motifs and signal negatively, novel DAP-12-coupled 'activating' CD33rSiglecs have been identified, such as siglec-14 and siglec-16, which are paired with the inhibitory receptors, siglec-5 and siglec-11, respectively (Cao and Crocker 2011). Siglec-16 was expressed at cell surface in the presence of DAP12, but not the FcRγ chain.

17.9 Mouse Siglecs

17.9.1 Evolution of Mouse and Human CD33-rSiglec Gene Clusters

Unlike most human Siglec-3 (hSiglec-3)-related Siglecs with promiscuous linkage specificity, mouse Siglec (mSiglec-F) shows a strong preference for α2–3-linked sialic acids. It is predominantly expressed in immature cells of the myelomonocytic lineage and in a subset of CD11b (Mac-1)-positive cells in some tissues. A comprehensive comparison of Siglecs between human and mouse genomes suggests that mouse genome contains eight Siglec genes, whereas the human genome contains 14 Siglec genes and a Siglec-like gene. Although a one-to-one orthologous correspondence between human and mouse Siglecs 1, 2, and 4 is confirmed, the Siglec-3-related Siglecs showed marked differences between human and mouse. Angata et al. (2001b) identified only four Siglec genes and two pseudogenes in the mouse chromosome 7 region syntenic to the Siglec-3-related gene cluster on human chromosome 19, which, in contrast, contains seven Siglec genes, a Siglec-like gene, and thirteen pseudogenes. Although analysis of gene maps and exon structures allows tentative assignments of mouse-human Siglec ortholog pairs, the possibility of unequal genetic recombination makes the assignments inconclusive. Current information suggests that mSiglec-F is likely a hSiglec-5 ortholog. The previously reported mSiglec-3/CD33 and mSiglec-E/MIS are likely orthologs of hSiglec-3 and hSiglec-9, respectively. The other Siglec-3-like gene in the cluster (mSiglec-G) is probably a hSiglec-10 ortholog. Another mouse gene (mSiglec-H), without an apparent human ortholog, lies outside of the cluster. Thus, although some duplications of Siglec-3-related genes predated separation of the primate and rodent lineages (about 80–100 million years ago), this gene cluster underwent extensive duplications in the primate lineage thereafter (Angata et al. 2001b).

17.9.2 Mouse CD33/Siglec-3

A cDNA for the putative mouse ortholog of hCD33 was cloned in 1994 (Tchilian et al. 1994). Similarities within the extracellular domain between hCD33 and mouse CD33 (mCD33; 62% identity in amino acid sequence) (Tchilian et al. 1994) and similar gene structure and chromosomal position relative to adjacent genes (Angata et al. 2001b) warrant its designation as the mouse ortholog of hCD33. However, the lack of sequence similarity in the cytosolic domains and difficulties in resolving phylogenetic relationships among the related Siglec molecules in humans and mice have raised questions about its functional equivalence to hCD33 (Angata et al. 2000, 2001a).

Differences with Human CD33: Although 16 human siglec proteins have been discovered, the mouse siglec family has fewer members (Yu et al. 2001b; Ulyanova et al. 2001): sialoadhesin, MAG, SMP, CD22, and CD33. Structurally, sialoadhesin, MAG, SMP, and CD22 are very similar in mice and humans. In contrast, human and mouse CD33, although somewhat similar in their extracellular domains, are strikingly different in their transmembrane and cytoplasmic regions. Human CD33 contains two ITIM sequences in the cytoplasmic tail, whereas mouse CD33 does not. Characterization of several ITIM-containing human siglecs implies that there must be murine ITIM-containing siglecs. Ulyanova et al. (2001) described the molecular cloning and characterization of a murine ITIM-containing myeloid cell-restricted inhibitory siglec, MIS. None of the two Siglec-3-related Siglecs in mouse, i.e. mSiglec-3/CD33 (Tchilian et al. 1994) and mSiglec-E/MIS (Yu et al. 2001b) shows a clear one-to-one orthologous correspondence to hSiglecs (Angata and Varki 2000; Ulyanova et al. 2001). Although it is likely that these mSiglec-3-related Siglecs, as well as one reported by Dehal et al. (2001) are present in a similar gene cluster in mouse chromosome 7 (Dehal et al. 2001), there have been few reports of a comprehensive analysis of Siglec-3-related mouse Siglecs. Like hCD33, mCD33 is expressed on myeloid precursors in the bone marrow, albeit mostly in the more mature stages of the granulocytic lineage. Moreover, unlike hCD33, mCD33 in peripheral blood is primarily expressed on granulocytes. Also, unlike hCD33, mCD33 did not bind to α2-3- or α2-6-linked sialic acids on lactosamine units. Instead, it showed distinctive sialic acid-dependent binding only to the short O-linked glycans of certain mucins and weak binding to the sialyl-Tn epitope. Results indicate substantial species differences in CD33 expression patterns and ligand recognition and suggest functional degeneracy between mCD33 and the other CD33-related Siglec proteins expressed on cells of the myeloid lineage (Brinkman-Van der Linden et al. 2003).

17.9.3 Siglec-3-Related Siglecs in Mice

17.9.3.1 mSiglec-E

The mouse siglec E (mSiglec-E)- cDNA encodes a protein of 467 amino acids that contains three extracellular Ig-like domains, a transmembrane region and a cytoplasmic tail bearing two ITIMs. mSiglec-E is highly expressed in mouse spleen, which is rich in leucocytes. The ITIMs of mSiglec-E can recruit SHP-1 and SHP-2, two inhibitory regulators of immunoreceptor signal transduction. mSiglec-E, through cell-cell interactions, is probably involveed in haematopoietic cells and the immune system as an inhibitory receptor. In comparison with the known members, mSiglec-E exhibits a high degree of sequence similarity to both human siglec-7 and siglec-9. The gene encoding mSiglec-E is localized in the same chromosome as that encoding mouse CD33. Phylogenetically, though Yu et al. (2001a,b) did not show clear relation between mouse mSiglec-E with any human siglecs, Angata et al. (2001b) suggested that mSiglec-E is likely an ortholog of hSiglec-9.

17.9.3.2 mSiglec-F

Siglec-F is a CD33-rSiglec prominently expressed on mature circulating mouse eosinophils, and on some myeloid precursors in bone marrow (Zhang et al. 2004; Angata et al. 2001a). It has a binding preference for α2-3–linked Sias (Angata et al. 2001), with the best-known ligand being 6'sulfo-sialyl-Le^X (Tateno et al. 2005). Of interest, this structure is also the preferred ligand for human Siglec-8- (Bochner et al. 2005), a molecule also specifically expressed on human eosinophils (Floyd et al. 2000; Kikly et al. 2000). Although mouse Siglec-F is not the true ortholog of human Siglec-8 (Angata et al. 2001a), their marked similarities in expression patterns and ligand preferences suggest that they play equivalent roles.

The elevated eosinophil count in allergic conditions is well known as is a critical role for CD4$^+$ Th2 cells in regulating allergic inflammatory responses involving eosinophils. Zhang et al. (2007) investigated the biologic roles of Siglec-F in Siglec-F–null mice in an induced lung allergic response model. This model also mimics some other features of bronchial asthma in humans, such as IgE-mediated mast-cell activation and degranulation, airway inflammation and hyperreactivity, CD4$^+$ T-cell infiltration and cytokine production, goblet-cell hyperplasia, and mucus overproduction. Data with WT mice using this model suggested a negative feedback loop involving Siglec-F in controlling eosinophilic responses, a hypothesis confirmed by studies of Siglec-F–null mice. Zhang et al. (2007) studied in vivo functions of Siglec-F, expressed on mouse eosinophils, which are prominent in allergic processes. Induction of allergic lung inflammation in mice caused up-regulation of Siglec-F on blood and bone marrow eosinophils, accompanied by newly induced expression on some CD4$^+$ cells, as well as quantitative up-regulation of endogenous Siglec-F ligands in the lung tissue and airways. With the ITIM in the cytosolic tail of Siglec-F, Zhang et al. (2007) suggested a negative feedback loop, controlling allergic responses of eosinophils and helper T cells, via Siglec-F and Siglec-F ligands. Siglec-F–null mice, allergen-challenged null mice showed increased lung eosinophil infiltration, enhanced bone marrow and blood eosinophilia, delayed resolution of lung eosinophilia, and reduced peribronchial-cell apoptosis. Data supporting the proposed negative feedback role for Siglec-F, represent the in vivo demonstration of biologic functions for any CD33rSiglec, and predict a role for human Siglec-8 (the isofunctional paralog of mouse Siglec-F) in regulating the pathogenesis of human eosinophil-mediated disorders. This demonstration suggests an in vivo biologic role for a CD33rSiglec, and also potential role for CD33rSiglecs in regulating T-cell induction of eosinophilic responses. To search for a mouse Siglec (mSiglec) ortholog of Siglec-8 and other mouse Siglec paralogs, Aizawa et al. (2003) searched public database with cDNA sequences of human

Siglec-5 to -10 and identified two novel mSiglecs. One has significant sequence identity to human Siglec-5 and is a splice variant of mSiglec-F. The other has greatest sequence identity to human Siglec-10 (mSiglec-G). Both mSiglecs have extracellular Ig-like domains and intracellular tyrosine-based motifs. A expression of mSiglec-5 (or -F), -10, and -E mRNA was detected in purified mouse eosinophils, but analyses suggested that mSiglec-10 is probably most relevant to mouse eosinophils.

Mouse Siglec-F and Human Siglec-8 are Functionally Convergent Paralogs: Unlike most human hSiglec-3rSiglecs with promiscuous linkage specificity, mSiglec-F shows a strong preference for α2-3-linked sialic acids. It is predominantly expressed in immature cells of the myelomonocytic lineage and in a subset of CD11b (Mac-1)-positive cells in some tissues. The sialoside sequence 6'-sulfo-sLe^x (Neu5Ac α2-3 [6-SO4]Galβ1-4[Fuc α1-3]GlcNAc) is a preferred ligand for Siglec-F. The lectin activity of Siglec-F on mouse eosinophils was "masked" by endogenous cis ligands and could be unmasked by treatment with sialidase. Unmasked Siglec-F mediated mouse eosinophil binding and adhesion to multivalent 6'-sulfo-sLe^x structure. Although there is no clear-cut human ortholog of Siglec-F, Siglec-8 is encoded by a paralogous gene that is expressed selectively by human eosinophils and has been found to recognize 6'-sulfo-sLe^x. It seems that mouse Siglec-F and human Siglec-8 have undergone functional convergence during evolution and implicate a role for the interaction of these siglecs with their preferred 6'-sulfo-sLe^x ligand in eosinophil biology (Tateno et al. 2005).

Siglec-F showed efficient endocytosis of anti-Siglec antibody, and *Neisseria meningitidis* bearing sialylated glycans. Like CD22, endocytosis was dependent on ITIM and ITIM-like motifs. While endocytosis of CD22 was mediated by a clathrin-dependent mechanism and was sorted to early endosome and recycling compartments, Siglec-F endocytosis was directed to lysosomal compartments and was mediated by a mechanism that was independent of clathrin and dynamin. Like CD22, Siglec-F mediated endocytosis of anti-Siglec-F and sialoside ligands, a function requiring intact tyrosine-based motifs. In contrast, however, Siglec-F endocytosis was clathrin and dynamin independent, required ADP ribosylation factor 6, and trafficed to lysosomes (Tateno et al. 2007).

17.9.3.3 mSiglec-G

It is now well accepted that the innate immune system recognizes both damage (or danger)- and pathogen-associated molecular patterns (DAMP and PAMP, respectively) through pattern recognition receptors, such as Toll-like receptors (TLR) and/or Nod-like receptors (NLR). Less clear are whether and how the response to PAMP and DAMP are regulated differentially. The answers may reveal whether the primary goal of the immune system is to defend against infections or to alert the host of tissue injuries. As demonstrated the host response to DAMP is controlled by a DAMP-CD24-Siglec axis. CD24-Siglec G/10 pathway plays a key role in discriminating between DAMPs and PAMPs (Liu et al. 2009). Siglec-G is a recently identified protein with an inhibitory function restricted to a B-cell subset, the B1 cells and is an ortholog of Siglec-10. Siglec-G inhibits Ca2+ signaling specifically in these cells. In addition, it controls the cellular expansion and antibody secretion of B1 cell population (Hoffmann et al. 2007). Thus, both Siglecs, Siglec-G/Siglec-10 and CD22 modulate BCR signaling on different B-cell populations in a mutually exclusive fashion (Nitschke 2009).

17.9.3.4 mSiglec-H as an Endocytic Receptor

Siglec-H is a murine CD33-related endocytic receptor that lacks intrinsic tyrosine-based signaling motifs. Although Siglec-H has the typical structural features required for sialic acid binding, no evidence for carbohydrate recognition has been obtained. Siglec-H is expressed specifically on plasmacytoid dendritic cell (pDC) precursors in bone marrow, spleen, blood, and lymph nodes. It is also expressed in a subset of marginal zone macrophages in spleen and in medullary macrophages in lymph nodes. Siglec-H functions as an endocytic receptor and mediates efficient internalization of anti-Siglec-H Abs. Immunizing mice with ovalbumin-conjugated anti-Siglec-H Ab in presence of CpG generated antigen-specific CD8 T cells in vivo. Targeting Siglec-H may therefore be a useful way of delivering antigens to pDC precursors for cross-presentation (Zhang et al. 2006). Siglec-H depends on DAP12 for surface expression; and cross-linking with anti-siglec-H antibodies can selectively inhibit interferon-α production by pDCs following TLR9 (Toll-like receptor 9) ligation.

17.9.3.5 Myeloid Inhibitory Siglec in Mice

Myeloid inhibitory siglec (MIS) belongs to the family of sialic acid-binding Ig-like lectins. The full-length MIS cDNA from murine bone marrow cells is predicted to contain an extracellular region comprising three Ig-like domains (V-set amino-terminal domain followed by two C-set domains), a transmembrane domain and a cytoplasmic tail with two ITIM-like sequences. The closest relative of MIS in the siglec family is human siglec 8. Extracellular regions of these two siglecs share 47% identity at the amino acid level. Southern blot analysis suggests the presence of one MIS gene. MIS is expressed in the spleen, liver, heart, kidney, lung and testis tissues. Several isoforms of MIS protein exist due to the alternative splicing. In a human promonocyte cell line, MIS was able to bind Src homology 2-containing protein-tyrosine phosphatases, SHP-1 and SHP-2. This binding was mediated by the membrane-proximal ITIM of MIS. Moreover, MIS

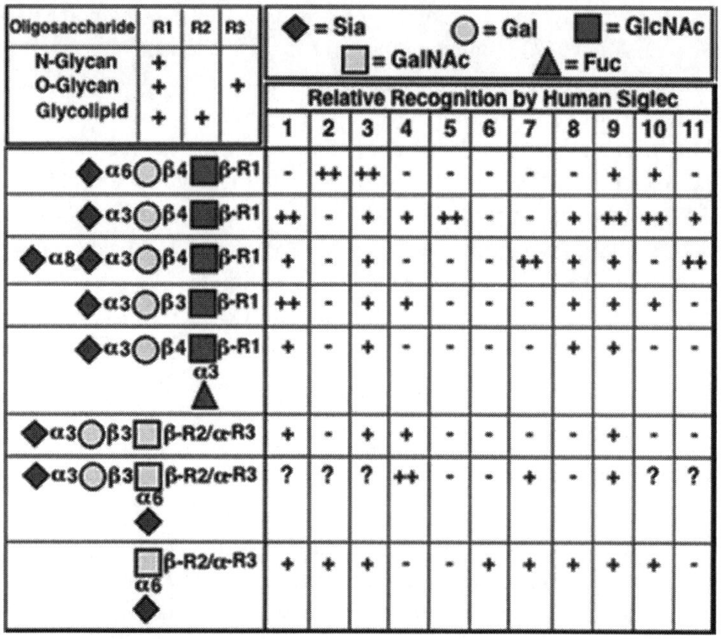

Oligosaccharide	R1	R2	R3
N-Glycan	+		
O-Glycan	+		+
Glycolipid	+	+	

◆ = Sia ○ = Gal ■ = GlcNAc □ = GalNAc ▲ = Fuc

	Relative Recognition by Human Siglec										
	1	2	3	4	5	6	7	8	9	10	11
◆α6○β4■β-R1	-	++	++	-	-	-	-	-	+	+	-
◆α3○β4■β-R1	++	-	+	+	++	-	-	+	++	++	+
◆α8◆α3○β4■β-R1	+	-	+	-	-	-	++	+	+	-	++
◆α3○β3■β-R1	++	-	+	+	-	-	-	+	+	+	-
◆α3○β4■β-R1 (■ α3 ▲)	+	-	+	-	-	-	-	+	+	-	-
◆α3○β3□β-R2/α-R3	+	-	+	+	-	-	-	-	+	-	-
◆α3○β3□β-R2/α-R3 (□ α6 ◆)	?	?	?	++	-	-	+	-	+	?	?
□β-R2/α-R3 (□ α6 ◆)	+	+	+	-	-	+	+	+	+	+	-

Fig. 17.4 *Glycan-binding specificities of human Siglecs.* With a few exceptions (CD22 and MAG), binding specificities of human Siglecs have varied significantly. In addition to assay formats and glycan linker issues, the density and arrangement of the ligands studied could be responsible for this variation. The most commonly reported specificities for the most commonly studied sialylated glycans are shown. Relative binding within studies of each Siglec is indicated as ++, strong binding; +, detectable binding; and –, very weak or undetectable binding. Not shown is the recently reported strong-binding preference of hSiglec-8 and mSiglec-F for 6′-sulfated-sialyl-Lewis x (sLex) and of hSiglec-9 for 6-sulfated-sLex. See text for the discussion (Reprinted with permission from Varki and Angata 2006 © Oxford University Press)

exerted an inhibitory effect on FcγRI receptor-induced calcium mobilization (Ulyanova et al. 2001).

17.10 Glycoconjugate Binding Specificities of Siglecs

Siglecs exhibit specificities for both the linkage and the nature of sialic acids in N-glycans, O-glycans and glycolipids. Many Siglecs recognize α2–3- and α2–6-linked sialic acids (Freeman et al. 1995; Brinkman-Van der Linden and Varki 2000), whereas others bind to other sialylated structures (Fig. 17.4). For example, Siglec-1 has been shown to bind the highly glycosylated surface protein CD43 (van den Berg et al. 2001), the epithelial mucin MUC-1 (Nath et al. 1999), and sialylated LPS of *Neisseria meningitidis* (Nm) (Jones et al. 2003). The Nm LPS sialylation can lead to increased bacterial susceptibility to phagocytic uptake, a phenomenon in direct contrast to earlier reported protective effects of LPS sialylation (Jones et al. 2003). Earlier studies of siglec specificities focused on α2-3- and α2-6-sialyllactos(amin)es and on one or two of the siglecs at a time. The binding of siglec-1, siglec-3 (CD33), siglec-4a (MAG), and siglec-5 to α2-3 sialyllactosamine is affected markedly by the presence of an α1-3-linked fucose. Studies revealed that: (1) while siglecs may not interfere with selectin-mediated recognition, fucosylation could negatively regulate siglec binding; (2) In contrast to earlier studies, siglec-3 prefers α2-6-sialyllactose; (3) siglec-5 binds α2-8-linked sialic acid, making it least specific for linkage recognition; (4) siglecs-2 (CD22), -3, -5, and -6 (obesity-binding protein-1) showed significant binding to sialyl-Tn (Neu5Acα2-6-GalNAc), a tumor marker associated with poor prognosis; (5) siglec-6 is an exception among siglecs in not requiring the glycerol side chain of sialic acid for recognition; (6) all siglecs require the carboxyl group of sialic acid for binding; (7) the presentation of sialyl-Tn epitope and/or more extended structures that include this motif may be important for optimal recognition by siglecs (Brinkman-Van der Linden and Varki 2001). Among a panel of glycans tested, Siglec-3 showed enhanced binding to a multivalent form of sialyl-Tn (NeuAcα2–6GalNAc) disaccharides (Brinkman-Van der Linden et al. 2003).

To determine if differences in glycan composition of siglec-5, siglec-7 and siglec-8 that may modify their function, Freeman et al. (2001) characterized N-linked oligosaccharide distribution in these three glycoproteins. The glycan pools from siglec-5 and siglec-7 contained a larger proportion of sialylated and core-fucosylated biantennary, triantennary and tetra-antennary oligosaccharides, whereas carbohydrate mixture released from siglec-8 is noticeably less sialylated and is more abundant in 'high-mannose'-type glycans. Moreover, it was found that, in contrast to CD22 and CD33, mutating the conserved potentially N-linked glycosylation site in first domain has no effect on binding mediated by siglec-5 or siglec-7 (Freeman et al. 2001). Using recombinant chimeric soluble receptors, siglec-transfected cell lines and macrophages from wild-type and Sn-deficient mice, it was observed that sialylated but not non-sialylated variants of either genetic background were specifically recognized by Sn and siglec-5, whereas other siglecs examined were ineffective. In addition, macrophages expressing Sn, as well as transfectants expressing Sn or siglec-5, bound and phagocytosed sialylated bacteria in a siglec- and sialic acid-dependent manner.

Most mammalian cell surfaces display two major sialic acids (Sias), N-acetylneuraminic acid (Neu5Ac) and N-glycolylneuraminic acid (Neu5Gc). Humans lack Neu5Gc due to a mutation in CMP-Neu5Ac hydroxylase, which occurred after evolutionary divergence from great apes. Sonnenburg et al. (2004) described an apparent consequence of human Neu5Gc loss: domain-specific functional adaptation of Siglec-9. While recombinant human Siglec-9 showed recognition of both Neu5Ac and Neu5Gc, in striking contrast, chimpanzee and gorilla Siglec-9 strongly preferred binding Neu5Gc. Results also indicated that endogenous Sias (rather than surface Sias of bacterial pathogens) are the functional ligands of CD33rSiglecs and suggested that the endogenous Sia landscape is the major factor directing evolution of CD33rSiglec binding specificity. Findings suggested ongoing adaptive evolution specific to the Sia-binding domain, possibly of an episodic nature. Such domain-specific divergences should also be considered in upcoming comparisons of human and chimpanzee genomes (Sonnenburg et al. 2004; Varki and Crocker 2009). Recent work has also shown that engagement of neutrophil-expressed siglec-9 by certain strains of sialylated Group B streptococci can suppress killing responses, thereby providing experimental support for pathogen exploitation of host CD33rSiglecs (Cao and Crocker 2011).

Siglec-Mediated Cell Adhesion to Gangliosides: Binding specificities of MAG, SMP, and sialoadhesin were compared by measuring siglec-mediated cell adhesion to immobilized gangliosides. The α-series gangliosides displayed enhanced potency for MAG- and SMP-mediated cell adhesion, whereas sialoadhesin-mediated adhesion was comparable with α-series and non-α-series gangliosides. GD1α derivatives with modified sialic acids (7-, 8-, or 9-deoxy) or sulfate (instead of sialic acid) at the III(6)-position supported adhesion comparable with that of GD1α. A novel GT1aα analog with sulfates at two internal sites of sialylation (NeuAcα 2,3Galβ1,4GalNAc-6-sulfateβ1, 4Gal3-sulfateβ1, 4Glcβ1,1'ceramide) was the most potent siglec-binding structure. Compiled reports indicated that MAG and SMP display an extended structural specificity with a requirement for a terminal α2, 3-linked NeuAc and great enhancement by nearby precisely spaced anionic charges (Collins et al. 1999). Studies on the interaction of CD33 related-siglecs-5,-7,-8,-9,-10 with gangliosides GT1b, GQ1b, GD3, GM2, GM3 and GD1a revealed that Siglec-5 bound preferentially to GQ1b, but weakly to GT1b, whereas siglec-10 interacted only with GT1b ganglioside. Siglec-7 and siglec-9 displayed binding to gangliosides GD3, GQ1b and GT1b bearing a disialoside motif, though siglec-7 was more potent; besides, siglec-9 interacted also with GM3. Siglec-8 demonstrated low affinity to the gangliosides tested compared with other siglecs. Despite high structural similarity of CD33 related siglecs, they demonstrated different ganglioside selectivity, in particular to the Neu5Acα2-8Neu5Ac motif (Rapoport et al. 2003a). Siglec-7 binds to GD3 ganglioside, LSTb oligosaccharide, sialyl Le^a, and NeuAcα2–8NeuAc, whereas Siglec-9 preferentially binds GD1a ganglioside and LSTc oligosaccharide (Yamaji et al. 2002; Rapoport et al. 2003a; Miyazaki et al. 2004).

Sulfated Gangliosides, High-Affinity Ligands for Neural Siglecs, Inhibit NADase Activity of CD38: Three kinds of novel sulfated gangliosides structurally related to the Chol-1 (α-series) ganglioside GQ1bα were synthesized. These sulfated gangliosides were potent inhibitors of NADase activity of leukocyte cell surface antigen CD38. Among the synthetic gangliosides, GSC-338 (II(3)III(6)-disulfate of iso-GM1b) was surprisingly found to be the most potent structure in both the NADase inhibition and MAG-binding activity. The study indicated that the sulfated gangliosides are useful to study the recognition of the internal tandem sialic acid residues α2-3-linked to Gal(II(3)) as well as the siglec-dependent recognition including a terminal sialic acid residue (Hara-Yokoyama et al. 2003; Ito et al. 2003).

Binding of Soluble Siglecs with Sulfated Oligosaccharides: Soluble siglecs were studied with polyacrylamide glycoconjugates in which: (1) the Neu5Ac residue was substituted by a sulfate group (Su); (2) glycoconjugates contained both Su and Neu5Ac; (3) sialoglycoconjugates contained a tyrosine-O-sulfate residue. Sulfate derivatives of LacNAc did not bind siglecs-1, -4, -5, -6, -7, -8, -9, and -10; binding of 6'-O-Su-LacNAc to siglec-8 was stronger than binding of 3'SiaLacNAc. The relative affinity of 3'-O-Su-TF binding to siglecs-1, -4, and -8 was similar to that of 3'SiaTF. 3'-O-Su-Le(c) displayed two-fold weaker binding to siglec-1 and siglec-4 than 3'SiaLe(c). The interaction of soluble siglecs with sulfated oligosaccharides containing sialic acid showed that siglecs-1, -4, -5, -6, -7, -9, and -10 did not interact with these compounds; binding of 6-O-Su-3'SiaLacNAc and 6-O-Su-3'SiaTF to siglec-8 was weaker than that of the corresponding sulfate-free sialoside probes. Siglec-8 displayed affinity to 6'-O-Su-LacNAc and 6'-O-Su-SiaLe^x, and defucosylation of the latter compound led to an increase in the binding. Sialoside probes containing tyrosine-O-sulfate residue did not display increased affinity to siglecs-1 and -5 compared with glycoconjugates containing only sialoside. Cell-bound siglecs-1, -5, -7, and -9 did not interact with 6-O-Su-3'SiaLacNAc, whereas the sulfate-free probe 3'SiaLacNAc demonstrated binding. In contrast, the presence of sulfate in 6-O-Su-6'SiaLacNAc did not affect binding of the sialoside probe to siglecs. 6'-O-Su-SiaLe^x displayed affinity to cell-bound siglecs-1 and -5; its isomer 6-O-Su-SiaLe^x bound more strongly to siglecs-1, -5, and -9 than SiaLe^x (Rapoport et al. 2006).

Potent Inhibitors of MAG: Ten of the known human siglecs or their murine orthologs were evaluated for their specificity for over 25 synthetic sialosides representing most of the major sequences terminating carbohydrate groups of glycoproteins and glycolipids. Each siglec was found to have a unique specificity for binding 16 different sialoside-streptavidin-alkaline phosphatase probes. Competitive inhibition studies revealed that Siglec-4 binds with 500–10,000-fold higher affinity to a series of mono- and di-sialylated derivatives of the O-linked T-antigen (Galβ1-3 GalNAc α OThr) as compared with α-methyl-NeuAc (Blixt et al. 2003).

Group B Streptococcal Capsular Sialic Acids Interact with Siglecs: Group B Streptococcus (GBS) is classified into nine serotypes that vary in capsular polysaccharide (CPS) architecture, but share the presence of terminal sialic acid (Sia) residue. The position and linkage of GBS Sia closely resembles that of cell surface glycans found abundantly on human cells. CD33-rSiglecs are expressed on host leukocytes that engage host Sia-capped glycans and send signals that dampen inflammatory gene activation. GBS evolved to display CPS Sia as a form of molecular mimicry limiting the activation of an effective innate immune response. Carlin et al. (2007) demonstrated that GBS of several serotypes interact in a Sia- and serotype-specific manner with certain human CD33-rSiglecs, including hSiglec-9 and hSiglec-5 expressed on neutrophils and monocytes. Modification of GBS CPS Sia by O-acetylation (OAc) has been recognized, and the degree of OAc can markedly impact the interaction between GBS and hSiglecs-5, -7, and -9. Thus, production of Sia-capped bacterial polysaccharide capsules that mimic human cell surface glycans in order to engage CD33-rSiglecs may be an example of bacterial mechanism of leukocyte manipulation (Carlin et al. 2007).

17.11 Functions of CD33-Related Siglecs

17.11.1 Endocytosis

Although it is assumed that CD33-related Siglecs have important roles in modulating leukocyte behavior, including inhibition of proliferation or cellular activation, modulation of cytokine secretion, and induction of apoptosis, their precise signaling functions remain unknown (Crocker et al. 2007; Crocker 2005; Varki and Angata 2006). Nevertheless, studies have demonstrated their interactions with various sialylated pathogens and suggested their potential importance in host defense and pathogenicity. For example, several pathogens, including *Neisseria meningitides*, Group B Streptococci, and *Campylobacter jejuni*, have been shown to bind to CD33-related Siglecs (Jones et al.; 11 2003; Avril et al. 2006; Carlin et al. 2007). Furthermore, there is increasing evidences that indicate endocytic capacities of human CD33-related Siglecs (Lock et al. 2004; Walter et al. 2005; Biedermann et al. 2007). This might be of physiological relevance for clearance of sialylated antigens and modulation of antigen presentation. Endocytosis of CD33 based immunotherapy results in cellular uptake of the drug (GO; Mylotarg™), which is cleaved intracellularly to release the toxic moiety (Linenberger 2005; Damle and Frost 2003). CD33 is present on tumor cells of 85–90% of adult and pediatric patients with acute myeloid leukemia (AML), and the therapeutic success of GO depends largely on its CD33-dependent uptake. Other CD33-related Siglecs, such as Siglec-8 for targeting eosinophils in allergic inflammation or other disease states, or Siglec-9 for treatment of AML or inflammatory diseases suggest are potential targets in clinical exploitation (Biedermann et al. 2007; Nutku et al. 2003; Von Gunten et al. 2005). The mechanisms underlying the endocytosis of CD33-related Siglecs remain elusive. Studies of CD33, Siglec-5, Siglec-F, and Siglec-9 have shown that they all endocytose slowly when bound to antibody. Also, CD33, Siglec-F, and Siglec-9 use their ITIMs for endocytosis (Walter et al. 2005; Biedermann et al. 2007; Tateno et al. 2007). However, it is not clear whether the ITIMs stimulate endocytosis when they are phosphorylated or when they are not phosphorylated. Walter et al. (2008b) identified proteins that bind to CD33 in an ITIM-dependent manner and assessed their importance for CD33 internalization in human myeloid cells by specific silencing of target genes through expression of siRNA.

It was shown that endocytosis is largely limited and determined by the intracellular domain while the extracellular and transmembrane domains play a minor role. Tyrosine phosphorylation, most likely through Src family kinases, increases uptake of CD33 depending on the integrity of two cytoplasmic ITIMs. Simultaneous depletion of protein tyrosine phosphatases (SHP1 and SHP2), which bound to phosphorylated CD33, increased internalization of CD33 slightly in some cell lines, whereas depletion of spleen tyrosine kinase (Syk) had no effect, implied that SHP1 and SHP2 can dephosphorylate the ITIMs or mask binding of phosphorylated ITIMs to an endocytic adaptor. Studies indicated that restraint of CD33 internalization through intracellular domain is relieved partly when the ITIMs are phosphorylated and thus showed that Shp1 and Shp2 can modulate this process (Walter et al. 2008c). Walter et al. (2008b) indicated that phosphorylation-dependent ubiquitylation regulates cell surface expression and internalization, and thus possibly function, of CD33/Siglec-3, suggesting an important role of ubiquitin in endocytosis of ITIM-bearing inhibitory immunoreceptors.

17.11.2 Phagocytosis of Apoptotic Bodies

Elimination of apoptotic bodies is one of the important functions of macrophages. A specific apoptotic glycosylation pattern may play an assistant or even a causative role in phagocytosis of apoptotic bodies. Taking into account that siglecs, a mannose receptor and galectins expressed on macrophages could be involved in engulfment of apoptotic bodies, their potential expression on THP-1 cells was assessed by means of polyacrylamide glycoconjugates. A strong binding of cells occurred to siglec ligands (3'SiaLac, 6'SiaLac, [Neu5Acα2-8]2) and galectin ligands (LacNAc, GalNAcβ1 - 4GlcNAc, Galβ1 - 3GalNAcβ and asialoGM1), where as Galβ1 - 3GalNAcβ-terminated chains represented the apoptotic bodies; the other tested galectin ligands did not appear to be the target for THP-1 cells (Rapoport et al. 2003b).

Macrophage lectins were probed with neoglycoconjugates. Glyc-PAA-fluo where carbohydrate is linked to fluorescein-labeled polyacrylamide. The neoglycoconjugates containing a Neu5Acα2-3Gal fragment bound to macrophages isolated from blood of healthy donors. Besides, carbohydrate chains containing the same fragment were revealed on apoptotic bodies. Phagocytosis of apoptotic bodies by macrophages was inhibited with sialooligosaccharide ligands of siglec-5 and mAbs to siglec-5. Thus, siglec-5 expressed on macrophages could participate in phagocytosis of apoptotic bodies. In addition, siglecs of tumor-associated macrophages modify engulfment of apoptotic bodies. The phagocytic potency of macrophages isolated from blood of breast cancer patients was lower than engulfment ability of macrophages obtained from healthy donors. Staining of macrophages from blood of tumor patients was more intense than that of macrophages from healthy donors; phagocytosis of apoptotic bodies by tumor-associated macrophages was inhibited by carbohydrates that are known to be ligands for siglecs (Rapoport et al. 2005).

17.12 Siglecs as Targets for Immunotherapy

The restricted expression of several siglecs to one or a few cell types makes them attractive targets for cell-directed therapies. Anti-CD33 antibodies have been used alone-and more effectively, attached to chemotherapy agents or radioisotopes-to treat those with AML (Nemecek and Matthews 2002). The anti-CD33 (Siglec-3) antibody gemtuzumab (Mylotarg) is approved for treatment of acute myeloid leukemia (AML) and antibodies targeting CD22 (Siglec-2) are currently in clinical trials for treatment of B cell non-Hodgkins lymphomas and autoimmune diseases. O'Reilly and Paulson (2009) reviewed the properties of siglecs that make them attractive for cell-targeted therapies. In normal myelopoiesis, expression of CD33 is restricted to advanced stages of differentiation, whereas primitive stem cells do not express CD33. It was shown that leukaemic stem cells in patients with $CD33^+$ AML express CD33 (Hauswirth et al. 2007). Antibody-targeted chemotherapy is a therapeutic strategy in cancer therapy that involves a monoclonal antibody specific for a tumor-associated antigen, covalently linked via a suitable linker to a potent cytotoxic agent. Tumor-targeted delivery of a cytotoxic agent in the form of an immunoconjugate is expected to improve its antitumor activity and safety.

Antibody-targeted chemotherapy with gemtuzumab ozogamicin (GO) (CMA-676, a CD33-targeted immunoconjugate of N-acetyl-γ-calicheamicin dimethyl hydrazide [CalichDMH], a potent DNA-binding cytotoxic antitumor antibiotic) is a clinically validated therapeutic option for patients with AML. Calicheamicin is a cytotoxic natural product isolated from Micromonospora echinospora that is at least 1000-fold more potent than conventional cytotoxic chemotherapeutics. Gemtuzumab ozogamicin is indicated for the treatment of elderly patients with relapsed AML (Damle 2004).

Immunoconjugates of calicheamicin targeted against tumor-associated antigens exhibit tumour-specific cytotoxic effects and cause regression of established human tumor xenografts in nude mice. CD33-specific binding triggers internalization of GO and subsequent hydrolytic release of calicheamicin. Calicheamicin then translocates to the nucleus, where calicheamicin binds DNA in the minor groove and causes double-strand DNA breaks, leading to cell death. GO is part of clinical practice for AML, but is frequently associated with severe side effects. The histone deacetylase inhibitor valproic acid potently augments gemtuzumab ozogamicin-induced apoptosis in acute myeloid leukemic cells. The synergistic proapoptotic activity of cotreatment of AML cells with VPA and GO indicates the potential value of this strategy for AML (Ten Cate et al. 2007). Finally, compared with either agent alone, antibody BC8 combined with GO resulted in marked tumor growth inhibition and superior survival rates of mice bearing human AML xenografts. Further study of this antibody combination for clinical use in AML is warranted (Walter et al. 2008a). Lintuzumab (HuM195) is an unconjugated humanized murine mAb directed against cell surface myelomonocytic CD33 (Feldman et al. 2005).

17.13 Molecular Diversity and Evolution of Siglec Family

The Siglecs can be divided into two groups: an evolutionarily conserved subgroup (Siglecs-1, -2, and -4) and a CD33/Siglec-3-related subgroup (Siglecs-3 and -5-13 in primates), which appear to be rapidly evolving (Varki et al. 2006). A human Siglec-like molecule (Siglec-L1) lacks a conserved arginine residue known to be essential for optimal sialic acid recognition by previously known Siglecs. Loss of the arginine from an ancestral molecule was caused by a single nucleotide substitution that occurred after the common ancestor of humans with the great apes but before the origin of modern humans. The chimpanzee Siglec-L1 ortholog remains fully functional and preferentially recognizes N-glycolylneuraminic acid, which is a common sialic acid in great apes and other mammals. Reintroducing the ancestral arginine into the human molecule regenerates the same properties. Thus, the single base pair mutation that replaced the arginine on human Siglec-L1 is likely to be evolutionarily related to the previously reported loss of N-glycolylneuraminic acid expression in the human lineage suggesting additional changes in the biology of sialic acids that may have taken place during human evolution (Angata et al. 2001b). Although a few Siglecs are well-conserved throughout vertebrate evolution and show similar binding preference regardless of the species of origin, most others, particularly the CD33-related subfamily of Siglecs, show marked inter-species differences in repertoire, sequence, and binding preference. The diversification of CD33-related Siglecs may be driven by direct competition against pathogens, and/or by necessity to catch up with the changing landscape of endogenous glycans, which may in turn be changing to escape exploitation by other pathogens (Angata 2006).

Comparison of different mammalian species has revealed differential and complex evolutionary paths for Siglec protein family, even within the primate lineage. The combination of microarray technology with the zebrafish model system can provide useful information on how genes are coordinated in a genetic network to control zebrafish embryogenesis and can help to identify novel genes that are important for organogenesis (Lo et al. 2003). Lehmann et al. (2004) investigated the occurrence of corresponding genes in bony fish. Interestingly, only unambiguous orthologs of mammalian siglec-4, a cell adhesion molecule expressed exclusively in the nervous system, could be identified in the genomes of fugu and zebrafish, whereas no obvious orthologs of the other mammalian siglecs were found. As in mammals, fish siglec-4 expression is restricted to nervous tissues. Expressed as recombinant protein, fish siglec-4 binds to sialic acids with a specificity similar to the mammalian orthologs. Relatively low sequence similarities in the cytoplasmic tail as well as an additional splice variant found in fish siglec-4 suggest alternative signaling pathways compared to mammalian species. Observations suggest that this siglec occurs at least in the nervous system of all vertebrates (Lehmann et al. 2004).

A comparison of the CD33rSiglec gene cluster in different mammalian species showed that it can be divided into subclusters, A and B. The two subclusters, inverted in relation to each other, each encodes a set of CD33rSiglec genes arranged head-to-tail. Two regions of strong correspondence provided evidence for a large-scale inverse duplication, encompassing the framework CEACAM-18 (CE18) and ATPBD3 (ATB3) genes that seeded the mammalian CD33-rSiglec cluster. Phylogenetic analysis was consistent with the predicted inversion. Rodents appear to have undergone wholesale loss of CD33rSiglec genes after the inverse duplication. In contrast, CD33rSiglecs expanded in primates and many are now pseudogenes with features consistent with activating receptors. In contrast to mammals, the fish CD33rSiglecs clusters show no evidence of an inverse duplication. They display greater variation in cluster size and structure than mammals. The close arrangement of other Siglecs and CD33rSiglecs in fish is consistent with a common ancestral region for Siglecs. Expansion of mammalian CD33rSiglecs appears to have followed a large inverse duplication of a smaller primordial cluster over 180 million years ago, prior to eutherian/marsupial divergence. Inverse duplications in general could potentially have a stabilizing effect in maintaining the size and structure of large gene clusters, facilitating the rapid evolution of immune gene families (Cao et al. 2009).

A comprehensive comparative study of Siglecs between the human and mouse genomes suggests that the mouse genome contains eight Siglec genes, whereas the human genome contains 11 Siglec genes and a Siglec-like gene. Although a one-to-one orthologous correspondence between human and mouse Siglecs 1, 2, and 4 is confirmed, the Siglec-3-related Siglecs showed marked differences between human and mouse. Angata et al. (2001a) found only four Siglec genes and two pseudogenes in the mouse chromosome 7 region syntenic to the Siglec-3-related gene cluster on human chromosome 19, which, in contrast, contains seven Siglec genes, a Siglec-like gene, and thirteen pseudogenes. Although some duplications of Siglec-3-related genes predated separation of the primate and rodent lineages (about 80–100 million years ago), this gene cluster underwent extensive duplications in the primate lineage thereafter (Angata et al. 2001a). A temporary lettered nomenclature for additional mouse Siglecs suggests that mSiglec-F is likely a hSiglec-5 ortholog and the previously reported mSiglec-3/CD33 and mSiglec-E/MIS are likely orthologs of hSiglec-3 and hSiglec-9, respectively. The other Siglec-3-like gene in the cluster (mSiglec-G) is probably a hSiglec-10 ortholog. Another mouse gene (mSiglec-H), without an apparent human ortholog, lies outside of the cluster. Thus, although some duplications of Siglec-3-related genes predated separation of the primate and rodent lineages (about 80–100 million years ago), this gene cluster underwent extensive duplications in the primate lineage thereafter (Angata et al. 2001a).

Humans are genetically very similar to "great apes", (chimpanzees, bonobos, gorillas and orangutans), our closest evolutionary relatives. The human T cells give much stronger proliferative responses to specific activation via the T cell receptor (TCR) than those from chimpanzees. Nonspecific activation using phytohemagglutinin was robust in chimpanzee T cells, indicating that the much lower response to TCR simulation is not due to any intrinsic inability to respond to an activating stimulus. CD33-related Siglecs are inhibitory signaling molecules expressed on most immune cells and are thought to down-regulate cellular activation pathways via cytosolic ITIMs. Among human immune cells, T lymphocytes are a striking exception, expressing little to none of these molecules. In stark contrast, T lymphocytes from chimpanzees as well as the other closely related "great apes" (bonobos, gorillas, and orangutans) express several CD33-related Siglecs on their surfaces. Thus, human-specific loss of T cell Siglec expression occurred after our last common ancestor with great apes, potentially resulting in an evolutionary difference with regard to inhibitory signaling. This was confirmed by studying Siglec-5, which is prominently expressed on chimpanzee lymphocytes, including CD4 T cells (Nguyen et al. 2006b).

Altheide et al. (2006) presented further evidence for accelerated evolution in Sia-binding domains of CD33-related Sia-recognizing Ig-like lectins. Other gene classes are more conserved, including those encoding the sialyltransferases that attach Sia residues to glycans. Despite this conservation, tissue sialylation patterns are shown to differ widely among these species, presumably because of rapid evolution of sialyltransferase expression patterns. Sia modifications on these glycopeptides also appear to be undergoing rapid evolution. This rapid evolution of the sialome presumably results from the ongoing need of organisms to evade microbial pathogens that use Sia residues as receptors. The rapid evolution of Sia-binding domains of the inhibitory CD33-related Sia-recognizing Ig-like lectins is likely to be a secondary consequence, as these inhibitory receptors presumably need to keep up with recognition of the rapidly evolving "self"-sialome.

References

Aizawa H, Plitt J, Bochner BS (2002) Human eosinophils express two siglec-8 splice variants. J Allergy Clin Immunol 109:176

Aizawa H, Zimmermann N, Carrigan PE et al (2003) Molecular analysis of human Siglec-8 orthologs relevant to mouse eosinophils: identification of mouse orthologs of Siglec-5 (mSiglec-F) and Siglec-10 (mSiglec-G). Genomics 82:521–530

Alphey MS, Attrill H, Crocker PR, van Aalten DM (2003) High resolution crystal structures of Siglec-7. Insights into ligand specificity in the Siglec family. J Biol Chem 278:3372–3377

Altheide TK, Hayakawa T, Mikkelsen TS et al (2006) System-wide genomic and biochemical comparisons of sialic acid biology among primates and rodents: evidence for two modes of rapid evolution. J Biol Chem 281:25689–25702

Ando M, Tu W, Nishijima K, Iijima S (2008) Siglec-9 enhances IL-10 production in macrophages via tyrosine-based motifs. Biochem Biophys Res Commun 369:878–883

Angata T (2006) Molecular diversity and evolution of the Siglec family of cell-surface lectins. Mol Divers 10:555–566

Angata T, Varki A (2000a) Cloning, characterization, and phylogenetic analysis of siglec-9, a new member of the CD33-related group of siglecs. Evidence for co-evolution with sialic acid synthesis pathways. J Biol Chem 275:22127–22135

Angata T, Varki A (2000b) Siglec-7: a sialic acid-binding lectin of the immunoglobulin superfamily. Glycobiology 10:431–438

Angata T, Hingorani R et al (2000) Cloning and characterization of siglec binding specificities, including the significance of fucosylation and of the sialyl-Tn epitope. Sialic acid-binding immunoglobulin superfamily lectins. J Biol Chem 275:8625–8632

Angata T, Hingorani R, Varki NM, Varki A (2001a) Cloning and characterization of a novel mouse siglec, mSiglec-F. Differential evolution of the mouse and human (CD33) siglec-3-related gene clusters. J BiolChem 276:45128–45136

Angata T, Varki NM, Varki A (2001b) A second uniquely human mutation affecting sialic acid biology. J Biol Chem 276:40282–40287

Angata T, Kerr SC et al (2002) Cloning and characterization of human Siglec-11. A recently evolved signaling that can interact with SHP-1 and SHP-2 and is expressed by tissue macrophages, including brain microglia. J Biol Chem 277:24466–24474

Angata T, Margulies EH, Green ED, Varki A (2004) Large-scale sequencing of the CD33-related Siglec gene cluster in five mammalian species reveals rapid evolution by multiple mechanisms. Proc Natl Acad Sci USA 101:13251–13256

Angata T, Hayakawa T, Yamanaka M et al (2006) Discovery of Siglec-14, a novel sialic acid receptor undergoing concerted evolution with Siglec-5 in primates. FASEB J 20:1964–1973

Arase H, Lanier LL (2004) Specific recognition of virus-infected cells by paired NK receptors. Rev Med Virol 14:83–93

Attrill H, Imamura A, Sharma RS et al (2006a) Siglec-7 undergoes a major conformational change when complexed with the α(2,8)-disialylganglioside GT1b. J Biol Chem 281:32774–32783

Attrill H, Takazawa H, Witt S et al (2006b) The structure of siglec-7 in complex with sialosides: leads for rational structure-based inhibitor design. Biochem J 397:271–278

Avril T, Floyd H, Lopez F et al (2004) The membrane-proximal immunoreceptor tyrosine-based inhibitory motif is critical for the inhibitory signaling mediated by Siglecs-7 and -9, CD33-related Siglecs expressed on human monocytes and NK cells. J Immunol 173:6841–6849

Avril T, Freeman SD, Attrill H et al (2005) Siglec-5 (CD170) can mediate inhibitory signaling in the absence of immunoreceptor tyrosine-based inhibitory motif phosphorylation. J Biol Chem 280: 19843–19851

Avril T, Wagner ER, Willison HJ, Crocker PR (2006) Sialic acid-binding immunoglobulin-like lectin 7 mediates selective recognition of sialylated glycans expressed on Campylobacter jejuni lipooligosaccharides. Infect Immun 74:4133–4141

Biedermann B, Gil D et al (2007) Analysis of the CD33-related siglec family reveals that Siglec-9 is an endocytic receptor expressed on subsets of acute myeloid leukemia cells and absent from normal hematopoietic progenitors. Leuk Res 31:211–220

Blasius AL, Cella M, Maldonado J et al (2006) Siglec-H is an IPC-specific receptor that modulates type I IFN secretion through DAP12. Blood 107:2474–2476

Blixt O, Collins BE, van den Nieuwenhof IM et al (2003) Sialoside specificity of the siglec family assessed using novel multivalent probes: identification of potent inhibitors of myelin-associated glycoprotein. J Biol Chem 278:31007–31019

Bochner BS (2009) Siglec-8 on human eosinophils and mast cells, and Siglec-F on murine eosinophils, are functionally related inhibitory receptors. Clin Exp Allergy 39:317–324

Bochner BS, Alvarez RA, Mehta P et al (2005) Glycan array screening reveals a candidate ligand for Siglec-8. J Biol Chem 280:4307–4312

Brinkman-Van der Linden EC, Varki A (2000) New aspects of siglec binding specificities, including the significance of fucosylation and of the sialyl-Tn epitope. Sialic acid-binding immunoglobulin superfamily lectins. J Biol Chem 275:8625–8632

Brinkman-Van der Linden EC, Varki A (2001) New aspects of a novel mouse Siglec, mSiglec-F: differential evolution of the mouse and human (CD33) Siglec-3-related gene clusters. J Biol Chem 276: 45128–45136

Brinkman-Van der Linden EC, Angata T, Reynolds SA et al (2003) CD33/Siglec-3 binding specificity, expression pattern, and consequences of gene deletion in mice. Mol Cell Biol 23:4199–4206

Brinkman-Van der Linden ECM, Hurtado-Ziola N et al (2007) Human-specific expression of Siglec-6 in the placenta. Glycobiology 17: 922–931

Cao H, Crocker PR (2011) Evolution of CD33-related siglecs: regulating host immune functions and escaping pathogen exploitation? Immunology 132:18–26

Cao H, Lakner U, de Bono B et al (2008) SIGLEC16 encodes a DAP12-associated receptor expressed in macrophages that evolved from its inhibitory counterpart SIGLEC11 and has functional and non-functional alleles in humans. Eur J Immunol 38:2303–2315

Cao H, de Bono B, Belov K et al (2009) Comparative genomics indicates the mammalian CD33rSiglec locus evolved by an ancient large-scale inverse duplication and suggests all Siglecs share a common ancestral region. Immunogenetics 61:401–417

Carlin AF, Lewis AL, Varki A, Nizet V (2007) Group B streptococcal capsular sialic acids interact with Siglecs (immunoglobulin-like lectins) on human leukocytes. J Bacteriol 189:1231–1237

Carlin AF, Uchiyama S, Chang Y-C et al (2009) Molecular mimicry of host sialylated glycans allows a bacterial pathogen to engage neutrophil Siglec-9 and dampen the innate immune response. Blood 113:3333–3336

Collins BE, Ito H, Sawada N et al (1999) Enhanced binding of the neural siglecs, myelin-associated glycoprotein and Schwann cell myelin protein, to Chol-1 (α-series) gangliosides and novel sulfated Chol-1 analogs. J Biol Chem 274:37637–37643

Cornish AL, Freeman S, Forbes G et al (1998) Characterization of siglec-5, a novel glycoprotein expressed on myeloid cells related to CD33. Blood 92:2123–2132

Crocker PR (2002) Siglecs: sialic-acid-binding immunoglobulin-like lectins in cell-cell interactions and signaling. Curr Opin Struct Biol 12:609–615

Crocker PR (2005) Siglecs in innate immunity. Curr Opin Pharmacol 5:431–437

Crocker PR, Varki A (2001a) Siglecs in the immune system. Immunology 103:137–145

Crocker PR, Varki A (2001b) Siglecs, sialic acids and innate immunity. Trends Immunol 22:337–342

Crocker PR, Paulson JC, Varki A (2007) Siglecs and their roles in the immune system. Nat Rev Immunol 7:255–266

Crocker PR, Redelinghuys P (2008) Siglecs as positive and negative regulators of the immune system. Biochem Soc Trans 36(Pt 6): 1467–1471

Damle NK (2004) Tumour-targeted chemotherapy with immunoconjugates of calicheamicin. Expert Opin Biol Ther 4:1445–1452

Damle NK, Frost P (2003) Antibody-targeted chemotherapy with immunoconjugates of calicheamicin. Curr Opin Pharmacol 3: 386–390, 20

Dehal P, Predki P, Olsen AS, Kobayashi A et al (2001) Human chromosome 19 and related regions in mouse: conservative and lineage-specific evolution. Science 293:104–111

Dimasi N, Moretta A, Moretta L et al (2004) Structure of the saccharide-binding domain of the human natural killer cell inhibitory receptor p75/AIRM1. Acta Crystallogr D Biol Crystallogr 60 (Pt 2):401–403

Feldman EJ, Brandwein J, Stone R et al (2005) Phase III randomized multicenter study of a humanized anti-CD33 monoclonal antibody, lintuzumab, in combination with chemotherapy, versus chemotherapy alone in patients with refractory or first-relapsed acute myeloid leukemia. J Clin Oncol 23:4110–4116

Floyd H, Ni J, Cornish AL et al (2000) Siglec-8: a novel eosinophil-specific member of the immunoglobulin superfamily. J Biol Chem 275:861–866

Foussias G, Yousef GM, Diamandis EP (2000) Molecular characterization of a Siglec8 variant containing cytoplasmic tyrosine-based motifs, and mapping of the Siglec8 gene. Biochem Biophys Res Commun 278:775–781

Foussias G, Taylor SM, Yousef GM et al (2001) Cloning and molecular characterization of two splice variants of a new putative member of the Siglec-3-like subgroup of Siglecs. Biochem Biophys Res Commun 284:887–899

Freeman SD, Kelm S, Barber EK, Crocker PR (1995) Characterization of CD33 as a new member of the sialoadhesin family of cellular interaction molecules. Blood 85:2005–2012

Freeman S, Birrell HC, D'Alessio K et al (2001) A comparative study of the asparagine-linked oligosaccharides on siglec-5, siglec-7 and siglec-8, expressed in a CHO cell line, and their contribution to ligand recognition. Eur J Biochem 268:1228–1237

Grobe K, Powell LD (2002) Role of protein kinase C in the phosphorylation of CD33 (Siglec-3) and its effect on lectin activity. Blood 99:3188–3196

Gunturi A, Berg RE, Forman J (2004) The role of CD94/NKG2 in innate and adaptive immunity. Immunol Res 30:29–34

Hara-Yokoyama M, Ito H, Ueno-Noto K et al (2003) Novel sulfated gangliosides, high-affinity ligands for neural siglecs, inhibit NADase activity of leukocyte cell surface antigen CD38. Bioorg Med Chem Lett 13:3441–3445

Hauswirth AW, Florian S, Printz D et al (2007) Expression of the target receptor CD33 in CD34/CD38/CD123 AML stem cells. Eur J Clin Invest 37:73–82

Hayakawa T, Angata T, Lewis AL et al (2005) A human specific gene in microglia. Science 309(5741):1693, Comment in: Science. 2005; 309(5741):1662–1663

Hoffmann A, Kerr S, Jellusova J et al (2007) Siglec-G is a B1 cell-inhibitory receptor that controls expansion and calcium signaling of the B1 cell population. Nat Immunol 8:695–704

Howie D, Simarro M, Sayos J et al (2002) Molecular dissection of the signaling and costimulatory functions of CD150 (SLAM): CD150/SAP binding and CD150-mediated costimulation. Blood 99:957–965

Ikehara Y, Ikehara SK, Paulson JC (2004) Negative regulation of T cell receptor signaling by Siglec-7 (p70/AIRM) and Siglec-9. J Biol Chem 279:43117–43125

Ito H, Ishida H, Collins BE et al (2003) Systematic synthesis and MAG-binding activity of novel sulfated GM1b analogues as mimics of Chol-1 (α-series) gangliosides: highly active ligands for neural siglecs. Carbohydr Res 338:1621–1639

Jones C, Virji M, Crocker PR (2003) Recognition of sialylated meningococcal lipopolysaccharide by siglecs expressed on myeloid cells leads to enhanced bacterial uptake. Mol Microbiol 49:1213–1225

Kikly KK, Bochner BS, Freeman SD et al (2000) Identification of SAF-2, a novel siglec expressed on eosinophils, mast cells, and basophils. J Allergy Clin Immunol 105:1093–1100

Kitzig F, Martinez-Barriocanal A, López-Botet M, Sayós J (2002) Cloning of two new splice variants of Siglec-10 and mapping of the interaction between Siglec-10 and SHP-1. Biochem Biophys Res Commun 296:355–362

Kubagawa H, Burrows PD, Cooper MD (1997) A novel pair of immunoglobulin-like receptors expressed by B cells and myeloid cells. Proc Natl Acad Sci USA 94:5261–5266

Lanier LL (2001) Face off—the interplay between activating and inhibitory immune receptors. Curr Opin Immunol 13:326–331

Lehmann F, Gäthje H, Kelm S, Dietz F (2004) Evolution of sialic acid-binding proteins: molecular cloning and expression of fish siglec-4. Glycobiology 14:959–968

Li N, Zhang W, Wan T et al (2001) Cloning and characterization of Siglec-10, a novel sialic acid binding member of the Ig superfamily, from human dendritic cells. J Biol Chem 276:28106–28112

Linenberger ML (2005) CD33-directed therapy with gemtuzumab ozogamicin in acute myeloid leukemia: progress in understanding cytotoxicity and potential mechanisms of drug resistance. Leukemia 19:176–182

Liu Y, Chen GY, Zheng P (2009) CD24-Siglec G/10 discriminates danger- from pathogen-associated molecular patterns. Trends Immunol 30:557–561

Lock K, Zhang J, Lu J et al (2004) Expression of CD33-related siglecs on human mononuclear phagocytes, monocyte-derived dendritic cells and plasmacytoid dendritic cells. Immunobiology 209: 199–207

Long EO (1999) Regulation of immune responses through inhibitory receptors. Annu Rev Immunol 17:875–904

Lo J, Lee S, Xu M et al (2003) Unique zebrafish EST clusters and their future use in microarray for profiling gene expression patterns during embryogenesis. Genome Res 13:455–466

Malbec O, Fong DC, Turner M et al (1998) Fcε receptor I-associated lyn-dependent phosphorylation of Fc gamma receptor IIB during negative regulation of mast cell activation. J Immunol 160:1647–1658

Martin AM, Kulski JK, Witt C et al (2002) Leukocyte Ig-like receptor complex (LRC) in mice and men. Trends Immunol 23:81–88

McMillan SJ, Crocker PR (2008) CD33-related sialic-acid-binding immunoglobulin-like lectins in health and disease. Carbohydr Res 343:2050–2056

Mingari MC, Vitale C, Romagnani C et al (2001) p75/AIRM1 and CD33, two sialoadhesin receptors that regulate the proliferation or the survival of normal and leukemic myeloid cells. Immunol Rev 181:260–268

Miyazaki K, Ohmori K, Izawa M et al (2004) Loss of disialyl Lewis(a), the ligand for lymphocyte inhibitory receptor sialic acid-binding immunoglobulin-like lectin-7 (Siglec-7) associated with increased sialyl Lewis(a) expression on human colon cancers. Cancer Res 64:4498–4505

Mitra N, Banda K, Altheide TK et al (2011) SIGLEC12, a human-specific segregating (pseudo)gene, encodes a signaling molecule expressed in prostate carcinomas. J Biol Chem 286:23003–23011

Munday J, Kerr S, Ni J et al (2001) Identification, characterization and leucocyte expression of Siglec-10, a novel human sialic acid-binding receptor. Biochem J 355:489–497

Nath D, Hartnell A, Happerfield L et al (1999) Macrophage-tumour cell interactions: identification of MUC1 on breast cancer cells as a potential counter-receptor for the macrophage-restricted receptor, sialoadhesin. Immunology 98:213–219

Nemecek ER, Matthews DC (2002) Antibody-based therapy of human leukemia. Curr Opin Hematol 9(4):316–321

Nguyen DH, Tangvoranuntakul P, Varki A (2005) Effects of natural human antibodies against a nonhuman sialic acid that metabolically incorporates into activated and malignant immune cells. J Immunol 175:228–236

Nguyen DH, Ball ED, Varki A (2006a) Myeloid precursors and acute myeloid leukemia cells express multiple CD33-related Siglecs. Exp Hematol 34:728–735

Nguyen DH, Hurtado-Ziola N, Gagneux P, Varki A (2006b) Loss of Siglec expression on T lymphocytes during human evolution. Proc Natl Acad Sci USA 103:7765–7770

Nicoll G, Ni J, Liu D et al (1999) Identification and characterization of a novel siglec, siglec-7, expressed by human natural killer cells and monocytes. J Biol Chem 274:34089–34095

Nicoll G, Avril T, Lock K et al (2003) Ganglioside GD3 expression on target cells can modulate NK cell cytotoxicity via siglec-7-dependent and – independent mechanisms. Eur J Immunol 33: 1642–1648

Nitschke L (2009) CD22 and Siglec-G: B-cell inhibitory receptors with distinct functions. Immunol Rev 230:128–143

Nutku E, Aizawa H, Hudson SA, Bochner BS (2003) Ligation of Siglec-8: a selective mechanism for induction of human eosinophil apoptosis. Blood 101:5014–5020

Nutku E, Hudson SA, Bochner BS (2005) Mechanism of Siglec-8-induced human eosinophil apoptosis: role of caspases and mitochondrial injury. Biochem Biophys Res Commun 336:918–924

Nutku E, Hudson SA, Bochner 1SB (2008) Interleukin-5 priming of human eosinophils alters Siglec-8-mediated apoptosis pathways. Am J Respir Cell Mol Biol 38:121–124

Oetke C, Vinson MC, Jones C, Crocker PR (2006b) Sialoadhesin-deficient mice exhibit subtle changes in B- and T-cell populations and reduced immunoglobulin M levels. Mol Cell Biol 26:1549–1557

O'Reilly MK, Paulson JC (2009) Siglecs as targets for therapy in immune-cell-mediated disease. Trends Pharmacol Sci 30:240–248

Orr SJ, Morgan NM, Buick RJ et al (2007) SOCS3 targets Siglec 7 for proteasomal degradation and blocks Siglec 7-mediated responses. J Biol Chem 282:3418–3422

Patel N, Brinkman-Van der Linden EC et al (1999) OB-BP1/Siglec-6. a leptin- and sialic acid-binding protein of the immunoglobulin superfamily. J Biol Chem 274:22729–22738

Rapoport E, Khaidukov S, Baidina O et al (2003a) Involvement of the Galβ1–3GalNAcβ structure in the recognition of apoptotic bodies by THP-1 cells. Eur J Cell Biol 82:295–302

Rapoport E, Mikhalyov I, Zhang J et al (2003b) Ganglioside binding pattern of CD33-related siglecs. Bioorg Med Chem Lett 13:675–678

Rapoport EM, Sapot'ko YB, Pazynina GV et al (2005) Sialoside-binding macrophage lectins in phagocytosis of apoptotic bodies. Biochemistry (Mosc) 70:330–338

Rapoport EM, Pazynina GV, Sablina MA et al (2006) Probing sialic acid binding Ig-like lectins (siglecs) with sulfated oligosaccharides. Biochemistry (Mosc) 71:496–504

Ravetch JV, Lanier LL (2000) Immune inhibitory receptors. Science 290:84–89

Salminen A, Kaarniranta K (2009) Siglec receptors and hiding plaques in Alzheimer's disease. Mol Med 87:697–701

Sonnenburg JL, Altheide TK, Varki A (2004) A uniquely human consequence of domain-specific functional adaptation in a sialic acid-binding receptor. Glycobiology 14:339–346

Staub E, Rosenthal A, Hinzmann B (2004) Systematic identification of immunoreceptor tyrosine-based inhibitory motifs in the human proteome. Cell Signal 16:435–456

Takei Y, Sasaki S, Fujiwara T et al (1997) Molecular cloning of a novel gene similar to myeloid antigen CD33 and its specific expression in placenta. Cytogenet Cell Genet 78:295–300

Tateno H, Crocker PR, Paulson JC (2005) Mouse Siglec-F and human Siglec-8 are functionally convergent paralogs that are selectively expressed on eosinophils and recognize 6'-sulfo-sialyl Lewis X as a preferred glycan ligand. Glycobiology 15:1125–1135

Tateno H, Li H, Schur MJ, Wakarchuk WW, Paulson JC et al (2007) Distinct endocytic mechanisms of CD22 (Siglec-2) and Siglec-F reflect roles in cell signaling and innate immunity. Mole Cell Biol 27:5699–5710

Taylor VC, Buckley CD, Douglas M et al (1999) The myeloid-specific sialic acid-binding receptor, CD33, associates with the protein-tyrosine phosphatases, SHP-1 and SHP-2. J Biol Chem 274:11505–11512

Tchilian EZ, Beverley PC, Young BD, Watt SM (1994) Molecular cloning of two isoforms of the murine homolog of the myeloid CD33 antigen. Blood 83:3188–3198

Ten Cate B, Samplonius DF, Bijma T et al (2007) The histone deacetylase inhibitor valproic acid potently augments gemtuzumab ozogamicin-induced apoptosis in acute myeloid leukemic cells. Leukemia 21:248–252

Ulyanova T, Shah DD, Thomas ML (2001) Molecular cloning of MIS, a myeloid inhibitory siglec, that binds protein-tyrosine phosphatases SHP-1 and SHP-2. J Biol Chem 276:14451–14458

van den Berg TK, Nath D, Ziltener HJ et al (2001) Cutting edge: CD43 functions as a T cell counterreceptor for the macrophage adhesion receptor sialoadhesin (Siglec-1). J Immunol 166:3637–3644

Varki A (2009) Multiple changes in sialic acid biology during human evolution. Glycoconj J 26:231–245

Varki A, Angata T (2006) Siglecs—the major subfamily of I-type lectins. Glycobiology 16:1R–27R

Varki A, Crocker PR. I-type lectins, in Essentials of Glycobiology, 2nd edition (Varki A, Cummings RD, Esko JD, et al., editors), Cold Spring Harbor (NY), 2009.

Vilches C, Parham P (2002) KIR: diverse, rapidly evolving receptors of innate and adaptive immunity. Annu Rev Immunol 20:217–251

Virgo P, Denning-Kendall PA, Erickson-Miller CL et al (2003) Identification of the CD33-related Siglec receptor, Siglec-5 (CD170), as a useful marker in both normal myelopoiesis and acute myeloid leukaemias. Br J Haematol 123:420–430

Vitale C, Romagnani C, Puccetti A et al (2001) Surface expression and function of p75/AIRM-1 or CD33 in acute myeloid leukemias: engagement of CD33 induces apoptosis of leukemic cells. Proc Natl Acad Sci USA 98:5764–5769

von Gunten S, Simon HU (2006) Sialic acid binding immunoglobulin-like lectins may regulate innate immune responses by modulating the life span of granulocytes. FASEB J 20:601–605

von Gunten S, Yousefi S, Seitz M et al (2005) Siglec-9 transduces apoptotic and nonapoptotic death signals into neutrophils depending on the proinflammatory cytokine environment. Blood 106:1423–1431

von Gunten S, Schaub A, Vogel M et al (2006) Immunologic and functional evidence for anti-Siglec-9 autoantibodies in intravenous immunoglobulin preparations. Blood 108:4255–4259

von Gunten S, Jakob SM, Geering B et al (2009) Different patterns of Siglec-9-mediated neutrophil death responses in septic shock. Shock 32:386–392

Walter RB, Raden BW, Kamikura DM et al (2005) Influence of CD33 expression levels and ITIM-dependent internalization on gemtuzumab ozogamicin-induced cytotoxicity. Blood 105:1295–1302

Walter RB, Boyle KM, Appelbaum FR et al (2008a) Simultaneously targeting CD45 significantly increases cytotoxicity of the anti-CD33 immunoconjugate, gemtuzumab ozogamicin, against acute myeloid leukemia (AML) cells and improves survival of mice bearing human AML xenografts. Blood 111:4813–4816

Walter RB, Häusermann P, Raden BW et al (2008b) Phosphorylated ITIMs enable ubiquitylation of an inhibitory cell surface receptor. Traffic 9:267–279

Walter RB, Raden BW, Zeng R et al (2008c) ITIM-dependent endocytosis of CD33-related Siglecs: role of intracellular domain, tyrosine phosphorylation, and the tyrosine phosphatases, Shp1 and Shp2. J Leukoc Biol 83:200–211

Whitney G, Wang S, Chang H et al (2001) A new siglec family member, siglec-10, is expressed in cells of the immune system and has signaling properties similar to CD33. Eur J Biochem 268: 6083–6096

Winn VD, Gormley M, Paquet AC et al (2009) Severe preeclampsia-related changes in gene expression at the maternal-fetal interface include sialic acid-binding immunoglobulin-like lectin-6 and pappalysin-2. Endocrinology 150:452–462

Yamaji T, Teranishi T, Alphey MS et al (2002) A small region of the natural killer cell receptor, Siglec-7, is responsible for its preferred binding to α 2,8-disialyl and branched α 2,6-sialyl residues. A comparison with Siglec-9. J Biol Chem 277:6324–6332

Yamaji T, Mitsuki M, Teranishi T et al (2005) Characterization of inhibitory signaling motifs of the natural killer cell receptor Siglec-7: attenuated recruitment of phosphatases by the receptor is attributed to two amino acids in the motifs. Glycobiology 15: 667–676

Yokoyama WM, Plougastel BF (2003) Immune functions encoded by the natural killer gene complex. Nat Rev Immunol 3:304–316

Yotsumoto K, Okoshi Y, Shibuya K et al (2003) Paired activating and inhibitory immunoglobulin-like receptors, MAIR-I and MAIR-II, regulate mast cell and macrophage activation. J Exp Med 198: 223–233

Yousef GM, Ordon MH, Foussias G, Diamandis EP (2002) Genomic organization of the siglec gene locus on chromosome 19q13.4 and cloning of two new siglec pseudogenes. Gene 286:259–270

Yu Z, Lai CM, Maoui M et al (2001a) Identification and characterization of S2V, a novel putative siglec that contains two V set Ig-like domains and recruits protein-tyrosine phosphatases SHPs. J Biol Chem 276:23816–23824

Yu Z, Maoui M et al (2001b) mSiglec-E, a novel mouse CD33-related siglec (sialic acid-binding immunoglobulin-like lectin) that recruits Src homology 2 (SH2)-domain-containing protein tyrosine phosphatases SHP-1 and SHP-2. Biochem J 353:483–492

Zhang JQ, Nicoll G, Jones C, Crocker PR (2000) Siglec-9, a novel sialic acid binding member of the immunoglobulin superfamily expressed broadly on human blood leukocytes. J Biol Chem 275:22121–22126

Zhang JQ, Biedermann B, Nitschke L, Crocker PR (2004) The murine inhibitory receptor mSiglec-E is expressed broadly on cells of the innate immune system whereas mSiglec-F is restricted to eosinophils. Eur J Immunol 34:1175–1184

Zhang J, Raper A, Sugita N et al (2006) Characterization of Siglec-H as a novel endocytic receptor expressed on murine plasmacytoid dendritic cell precursors. Blood 107:3600–3608

Zhang M, Angata T, Cho JY et al (2007) Defining the in vivo function of Siglec-F, a CD33-related Siglec expressed on mouse eosinophils. Blood 109:4280–4287

Zhuravleva MA, Trandem K, Sun PD (2008) Structural implications of Siglec-5-mediated sialoglycan recognition. J Mol Biol 375:437–447

Part VII

Novel Super-Families of Lectins

Fibrinogen Type Lectins

18

Anita Gupta

18.1 Ficolins

Ficolins are one of the most important groups of proteins capable of recognizing pathogens, and they function in the innate immune defence as pathogen-associated molecular pattern recognition molecules (Bohlson et al. 2007; Thiel et al. 1997; Runza et al. 2008). Ficolins comprises of a collagen like domain at the N-terminus and a FBG (fibrinogen-like domain), which is the ligand-binding site, at the C-terminus, and they form trimer-based multimers that are N-terminally linked by disulfide bonds (Ohashi and Erickson 2004; Hummelshoj et al. 2007). Three human ficolins (L-, M and H -ficolins) have been characterized. The amino acid sequence homologies between L-ficolin and M-ficolin, and between H-ficolin and either L-ficolin or M-ficolin, are 80% and 48% respectively. These ficolins are associated with the mannose-binding lectin-associated serine protease, and the complexes activate the lectin complement pathway. Interestingly, ficolins collaborate with CRP (C-reactive protein), which is highly up-regulated during the acute-phase response, and the interaction stabilizes CRP binding to bacteria, resulting in the activation of the lectin complement pathway (Ng et al. 2007). In addition to humans, ficolins have been identified in different mammalian species including rodents and pigs, which have two related ficolin genes called A and B and α and β, respectively, orthologous to the human L- and M-ficolin genes, respectively (Endo et al. 2004). To date, H-ficolin has only been identified in humans where as the mouse and rat homologues of H-ficolin gene are pseudogenes, which accounts for the absence of the corresponding protein in rodents (Endo et al. 2004).

18.1.1 Ficolins versus Collectins

Collectins and ficolins bind to oligosaccharide structures on the surface of microorganisms, leading to the killing of bound microbes through complement activation and phagocytosis. The human collectins, which are oligomeric proteins composed of CRDs attached to collagenous regions, are structurally similar to ficolins (L-ficolin, M-ficolin, and H-ficolin). However, they make use of different CRD structures: C-type lectin domains for the collectins and fibrinogen-like domains for the ficolins. Collectins and ficolins bear no significant sequence homology except for the presence of collagen-like sequences over the N-terminal halves of the polypeptides that enable the assembly of these molecules into oligomeric structures. Collectins and ficolins both contain lectin activities within the C-terminal halves of their polypeptides, the C-type CRDs and fibrinogen β/γ (homology) (FBG) domain, respectively. These domains form trimeric clusters at the ends of the collagen triple helices emanating from a central hub, where the N-terminal ends of the polypeptides merge. The collectins and ficolins seem to have evolved to recognize the surface sugar codes of microbes and their binding, to the arrays of cell surface carbohydrate molecules, targets the microbe for subsequent clearance by phagocytic cells (Holmskov et al. 2003).

18.1.2 Characterization of Ficolins

Ficolins resemble collectins in structure, but their collagen-like stalks are followed by a domain homologous to the fibrinogen β and γ chains (Matsushita and Fujita 2001). While a collagen-like domain is found at the N-terminus, the fibrinogen-like domain (FD1), which is the sugar-binding site, is present at the C-terminus (Le et al. 1998; Teh et al. 2000). Ficolins are assembled from homotrimeric subunits comprising a collagen-like triple helix and a lectin-like domain—composed of three fibrinogen-like (FBG)-domains. Two cysteines at the N-terminal end of the polypeptide chains form interchain disulfide bonds that mediate assembly into higher oligomerization structures (Ohashi and Erickson 2004; Hummelshoj et al. 2007). Ficolins are structurally and functionally similar to MBL. Ficolins present in serum have

G.S. Gupta et al., *Animal Lectins: Form, Function and Clinical Applications*,
DOI 10.1007/978-3-7091-1065-2_18, © Springer-Verlag Wien 2012

common binding specificity for N-acetylglucosamine (GlcNAc). MBL is also a collagenous lectin found in serum and specific for GlcNAc and mannose binding. Its domain organization is similar to that of ficolins, except that MBL has a CRD instead of a fibrinogen-like domain. Ficolin from monocytes is a polypeptide of ~ 42 kDa, which is similar in size to that of ficolin predicted from its cDNA-derived sequence and bound to *E. coli*. Protein(s) reactive with GlcNAc from porcine serum has M_r mainly 40 kDa and plays an important role(s) in innate immunity against microbial infection with Gram-positive and -negative bacteria (Nahid and Sugii 2006).

In humans, L-ficolin/P35 (or ficolin 2) and H-ficolin (or ficolin 3) are serum proteins, whereas M-ficolin (or ficolin 1) is a secretory protein produced by lung and blood cells (Matsushita et al. 1996; Sugimoto et al. 1998; Liu et al. 2005) and mainly expressed by the monocytic cell lineage (Matsushita et al. 1996; Sugimoto et al. 1998). In addition to humans, ficolins have been identified in different mammalian species including rodents and pigs (Ichijo et al. 1991; Fujimori et al. 1998), which have two related ficolin genes called A and B and α and β, respectively, orthologous to human L- and M-ficolin genes, respectively (Endo et al. 2004). Three human ficolins, ficolin-1 (M-ficolin), ficolin-2 (L-ficolin) and ficolin-3 (H-ficolin or Hakata antigen) are encoded by the *FCN1*, *FCN2* and *FCN3* genes, respectively. If MBL plays a role in innate immunity by acting as an opsonin and activating complement in association with MBL-associated serine protease (MASP) via lectin pathway, human serum ficolins, L-ficolin/P35 and H-ficolin (Hakata antigen), are also associated with MASPs and sMAP, a truncated protein of MASP-2, and activate complement. Thus, serum ficolins are structurally and functionally similar to MBL and have the capacity to activate the lectin pathway and have a role in innate Immunity (Matsushita and Fujita 2001; Krarup et al. 2004).

18.1.2.1 L-Ficolin (Ficolin-2)

L-ficolin (synonymous with ficolin-2 or Ficolin/P35) is synthesized in the liver and found in blood circulation (Matsushita et al. 1996). Adult plasma contains, on average, a level of L-ficolin that is threefold higher than found in cord blood, implying a protective role of this lectin. L-ficolin indeed binds to sugar structures via its FBG domains (Le et al. 1998) and, on binding to carbohydrates on bacteria, promotes clearance by phagocytosis (Matsushita et al. 1996). Like MBL, L-ficolin forms a complex with the MBL-associated proteases (MASPs) and binding of this complex to *Salmonella typhimurium* activated the complement system (Matsushita and Fujita 2001). L-ficolin binds to acetyl groups such as GlcNAc. This protein binds to clinically important bacteria, including *Salmonella typhimurium*, *Staphylococcus aureus*, *Streptococcus pneumoniae*, *Streptococcus pyogenes* and *Streptococcus agalactiae* (Krarup et al. 2005; Lynch et al. 2004; Aoyagi et al. 2005). L-ficolin/P35 binds to an Ra chemotype strain of *S. typhimurium* (TV119) which has an exposed GlcNAc at the non-reducing termini of the polysaccharide, where as it did not bind to LT2, a smooth type strain of *S. typhimurium* with additional O-polysaccharides covering GlcNAc (Taira et al. 2000).

18.1.2.2 H-Ficolin (Ficolin-3)

H-ficolin (synonymous with ficolin-3 or Hakata antigen) was initially identified as a serum auto-antigen, recognized by antibodies in patients suffering from systemic lupus erythematosus and other autoimmune diseases (Sugimoto et al. 1998). It is synthesized in the liver by hepatocytes and bile duct epithelial cells and is secreted into both blood circulation and bile (Akaiwa et al. 1999). It is also synthesized by ciliated bronchial and Type II alveolar epithelial cells and is secreted into bronchus and the alveolar space (Akaiwa et al. 1999). H-ficolin is a lectin that binds to carbohydrate structures found on bacteria and may therefore play an important role in both systemic and mucosal immune defence systems (Endo et al. 2007; Dahl et al. 2001, 1997) Both H- and L-ficolins have also been shown to bind to carbohydrate structures found on Gram-negative bacteria. (Akaiwa et al. 1999; Le et al. 1998; Matsushita et al. 1996) Binding of L-ficolin to *S. typhimurium* directly opsonizes the bacteria for enhanced phagocytosis by polymorphonuclear leucocytes (PMNs) (Matsushita et al. 1996) and this is probably mediated by an L-ficolin receptor(s) on PMNs. H-ficolin shows affinity for GlcNAc, GalNAc and D-fucose, and binds to *S. typhimurium*, *Salmonella minnesota* and *Aerococcus viridans* (Sugimoto et al. 1998; Krarup et al. 2005). The finding that L-ficolin formed a functional complex with the MASPs that bound to *S. typhimurium* and subsequently activated the complement system, implied that L-ficolin could mediate killing and clearance of pathogens more effectively through complement activation and complement receptor-mediated phagocytosis (Matsushita et al. 2000). L ficolin and H-ficolin recognize surface structures on apoptotic cells and initiate the activation of lectin complement pathway (Kuraya et al. 2005). H-ficolin has only been identified in humans where as the mouse and rat homologues of H-ficolin gene are pseudogenes, which account for the absence of corresponding protein in rodents (Endo et al. 2004).

18.1.2.3 M-Ficolin (Ficolin-1)

M-ficolin was initially identified from the membrane fraction of pig uterus (Ichijo et al. 1993). Significant M-ficolin has been detected on U937 cells and on monocytes, but not on lymphocytes and granulocytes. Association with the surface of monocytes or U937 cells is apparently an intrinsic property of M-ficolin instead of that of monocytes or U937 cells. M-ficolin was shown to mediate U937 cell adhesion to the immobilized F(ab′)2 fragment of the anti-FBG antibody and therefore M-ficolin was linked to the cell cytoskeleton either directly or indirectly through its cytoskeleton-linked receptor(s). M-ficolin-mediated U937 cell adhesion may therefore be a novel pathway of monocyte emigration into extravascular tissues.

M-ficolin is synthesized in peripheral blood monocytes. Ficolin mRNA is synthesized by a human monocyte cells, and thus blood monocytes also normally synthesize human ficolin. Peripheral blood monocytes from adult human donors showed that ficolin mRNA is highly expressed in monocytes throughout the first 20 h of adhesion. The origin of ficolin from monocytes, together with its structural similarity to C1q and the collectins, raised the possibility that ficolin is another plasma protein capable of binding to surface structures of micro-organisms (Lu et al. 1996b). However, its expression is down-regulated during monocyte differentiation and its mRNA is not detectable in mature macrophages (Lu et al. 1996b). M-ficolin mRNA has been found to be abundant in peripheral blood monocytes, accounting for 0·44% of the total mRNA in the cells (Hashimoto et al. 1999b). However, M-ficolin mRNA is not detectable in monocyte-derived DCs and was detected only at a very low level in monocyte-derived macrophages in vitro (Hashimoto et al. 1999a, 1999b). M-ficolin expression is a marker of circulating monocytes and has been reported to be located in secretory granules in the cytoplasm of neutrophils, monocytes and type II alveolar epithelial cells in the lung (Liu et al. 2005). M-ficolin shows affinity for GlcNAc, GalNAc and sialic acid, and it binds to *Staph. aureus* through GlcNAc (Liu et al. 2005). M-ficolin binds weakly to *S. typhimurium*, but this binding is not inhibited by GlcNAc (Liu et al. 2005). In addition, the peptide Gly-Pro-Arg-Pro, which mimics the N-terminal sequence of the fibrin α-chain and inhibits fibrin polymerization, prevents binding of the M-ficolin FBG domain to GlcNAc. M-ficolin is expressed by blood monocytes and type II alveolar epithelial cells (Lu et al. 1996a; Theil et al. 2000; Liu et al. 2005; Frederiksen et al. 2005). Interestingly, M-ficolin expression is downregulated when monocytes mature into macrophages, but can be re-induced in mature macrophages when they are treated with bacterial products such as lipopolysaccharide (Frankenberger et al. 2008). M-ficolin, unlike L-ficolin and H-ficolin, was not considered to be a serum protein, but it was recently demonstrated that M-ficolin exists in human plasma and serum under normal conditions (Honore et al. 2008).

M-ficolin is highly homologous to L-ficolin at the amino-acid sequence level and also shows a marked preference for acetylated compounds (Frederiksen et al. 2005; Liu et al. 2005). H-ficolin, in contrast, is less closely related to L-ficolin and shows no binding affinity for acetylated derivatives. So far, it has been only reported to bind to Aerococcus viridans (Tsujimura et al. 2001; Krarup et al. 2005). M-ficolin also has lectin activity and, like L-ficolin (Le et al. 1998) it binds to GlcNAc via the FBG domain. As GlcNAc is a common carbohydrate moiety on Gram-negative bacteria, M-ficolin may also recognize these pathogens.

18.1.2.4 Veficolins

Cerberus rynchops (dog-faced water snake) belongs to Homalopsidae of Colubroidea (rear-fanged snakes). In addition to C-type lectins, the venom gland of *C. rynchops* contains two proteins that showed sequence homology to ficolin. These proteins were named as ryncolin 1 and ryncolin 2 (rynchops ficolin) and this new family of snake venom proteins as veficolins (venom ficolins). On the basis of its structural similarity to ficolin, it is speculated that ryncolins may induce platelet aggregation and/or initiate complement activation (OmPraba et al. 2010).

18.1.3 Ligands of Ficolins

Both H- and L-ficolins have also been shown to bind to carbohydrate structures found on Gram-negative bacteria (Akaiwa et al. 1999). L-ficolin binds to GlcNAc (Matsushita et al. 1996; Le et al. 1997, 1998) and N-acetyl-d-galactosamine (GalNAc) (Le et al. 1998). The binding ability is inhibited by acetylated compounds, indicating that this protein specifically recognizes acetyl groups (Krarup et al. 2004). L-ficolin activates the lectin pathway after binding to various capsulated bacteria (Matsushita et al. 1996; Lynch et al. 2004; Aoyagi et al. 2005). A binding specificity for GlcNAc was initially characterized by Matsushita et al. (1996, 2000), in agreement with the finding that L-ficolin recognizes lipoteichoic acid, a GlcNAc-containing cell wall component characteristic of Gram-positive bacteria (Lynch et al. 2004). Binding to a fungal 1,3-β-D-glucan preparation was later reported (Ma et al. 2004), whereas further investigations revealed specificity for N-acetylated carbohydrates and other non-carbohydrate acetylated compounds such as acetylcholine (Krarup et al. 2004). Where as M-ficolin was initially identified as a membrane-associated protein in pig uterus, based on its affinity for TGF-β1 (Ichijo et al. 1993), L-ficolin was shown to bind to corticosteroids (Edgar 1995), and both L- and M-ficolin were found to be elastin-binding proteins (Harumiya et al. 1995, 1996).

L-ficolin activates the lectin-complement pathway upon binding to lipoteichoic acid, a cell component found in all Gram-positive bacteria (Lynch et al. 2004). H-ficolin (ficolin-3), the primary structure of which is 48% identical to that of L-ficolin, binds to GlcNAc, GalNAc and d-fucose (Sugimoto et al. 1998). Unlike L-ficolin, the GlcNAc-binding activity of H-ficolin is not inhibited by acetyl compounds (Krarup et al. 2004). M-ficolin, the primary structure of which is 80% and 48% identical to those of L-ficolin and H-ficolin, respectively (Sugimoto et al. 1998), binds to GlcNAc, GalNAc and sialic acid (Teh et al. 2000; Liu et al. 2005). This protein also recognizes acetyl groups, like L-ficolin (Frederiksen et al. 2005). Unlike the serum ficolins, M-ficolin has been detected on the surfaces of peripheral blood monocytes and promonocytic U937 cells (Lu et al. 1996a; Teh et al. 2000; Frederiksen et al. 2005). Gout et al. (2010) reported that L-ficolin preferentially recognized disulfated N-acetyllactosamine and tri- and tetrasaccharides containing terminal galactose or N-acetylglucosamine. Binding was sensitive to the position and orientation of the bond between N-acetyllactosamine and the adjacent carbohydrate. The crystal structure of the Y271F mutant fibrinogen domain showed that the mutation does not alter the structure of the ligand binding pocket. These analyses reveal ficolin ligands such as sulfated N-acetyllactosamine (L-ficolin) and gangliosides (M-ficolin) and provide precise insights into the sialic acid binding specificity of M-ficolin, emphasizing the essential role of Tyr^{271} in this respect.

18.1.4 X-ray Structures of M, L- and H-Ficolins

18.1.4.1 Homology to Tachylectin 5A

Several observations indicate that ficolins are not classical lectins and raise questions about their mechanism of action. To answer these questions at molecular level, the X-ray structures of their trimeric recognition domains, alone and in complex with various ligands, were solved to resolutions up to 1.95 and 1.7 Å, respectively. Both domains have three-lobed structures with clefts separating the distal parts of the protomers. Ca^{2+} ions are found at sites homologous to those described for *Tachypleus tridentatus* tachylectin-5A (TL5A), an invertebrate lectin (Kairies et al. 2001). Outer binding sites (S1) homologous to the GlcNAc-binding pocket of TL5A are present in ficolins but show different structures and specificities. In L-ficolin, three additional binding sites (S2–S4) surround the cleft (Garlatti et al. 2007a). The structures revealed that L-ficolin has evolved to be a versatile recognition protein able to recognize a variety of acetylated and carbohydrate targets through three different sites, suggesting that ficolins represent novel types of pattern recognition proteins. Results defined an unpredicted continuous recognition surface able to sense various acetylated and neutral carbohydrate markers in the context of extended polysaccharides such as 1,3-β-D-glucan, as found on microbial or apoptotic surfaces.

18.1.4.2 Domain Structures of M-Ficolin

The human M-ficolin FD1, which contains the ligand-binding site, was over-expressed in Pichia pastoris, purified and crystallized using the vapour-diffusion method at 293 K. The crystals belong to the monoclinic space group P21, with unit-cell parameters a = 55.16, b = 117.45, c = 55.19 Å, β = 99.88°, and contain three molecules per asymmetric unit (Tanio et al. 2006). To investigate the discrimination mechanism between self and non-self by ficolins, Tanio et al. (2007) determined the crystal structure of the human M-ficolin FD1 at 1.9A resolution. Although the FD1 monomer shares a common fold with the fibrinogen γ-fragment and tachylectin-5A, the Asp^{282}-Cys^{283} peptide bond, which is the predicted ligand-binding site on the C-terminal P domain, is a normal trans bond, unlike the cases of the other two proteins. The trimeric formation of FD1 results in the separation of the three P domains, and the spatial arrangement of the three predicted ligand-binding sites on the trimer is very similar to that of the trimeric collectin, indicating that such an arrangement is generally required for pathogen-recognition. The ligand binding study of FD1 in solution indicated that the recombinant protein binds to N-acetyl-d-glucosamine and the peptide Gly-Pro-Arg-Pro and suggested that the ligand-binding region exhibits a conformational equilibrium involving cis-trans isomerization of the Asp^{282}-Cys^{283} peptide bond. The crystal structure and the ligand binding study of FD1 provide an insight of the self- and non-self discrimination mechanism by ficolins (Tanio et al. 2007).

Since GlcNAc, the common ligand of ficolins from several species, is universally expressed on both pathogens and hosts, the discrimination mechanism between self (host) and nonself (pathogen) by ficolins is of particular interest. With regard to this, the trimer formation by the FD1 (M-ficolin FBG domain) is required to recognize pathogen surfaces with high ligand density. In addition, from structural and functional studies of M-ficolin FD1, Tanio et al. (2007) proposed that the ligand-binding region of FD1 exists in a conformational equilibrium between active and non-active states depending on three groups with a pK_a of 6.2, which are probably histidine residues, and suggested that the 2-state conformational equilibrium as well as the trimer formation contributes to the discrimination mechanism between self and non-self of FD1. The GlcNAc binding study of a series of single histidine mutants of FD1 demonstrated that His^{251}, His^{284} and His^{297} are required for the activity, and thus three histidines are the origins of pH dependency of FD1. The analyses of GlcNAc association and dissociation of FD1 provided evidence that FD1 always exchanges between the active and non-active states with the pH-dependent populations in solution. The biological roles of the histidine-regulated conformational equilibrium of M-ficolin are important in terms of self and non-self discrimination mechanism (Tanio and Kohno 2009).

The ligand-bound crystal structures of the CRD of M-ficolin, at high resolution provide the structural insights into its binding properties. Interaction with acetylated carbohydrates differs from the one described for L-ficolin (Fig. 18.1). Garlatti et al. (2007b) also revealed the structural determinants for binding to sialylated compounds, a property restricted to human M-ficolin and its mouse counterpart, ficolin B. Comparison between the ligand-bound structures obtained at neutral pH and nonbinding conformations observed at pH 5.6 showed how the ligand binding site is dislocated at acidic pH. This means that the binding function of M-ficolin is subject to a pH-sensitive conformational switch. Considering that the homologous ficolin B is found in the lysosomes of activated macrophages (Runza et al. 2006), it is proposed that this switch can play a physiological role in such acidic compartments (Garlatti et al. 2007b).

The structure of human M-ficolin FD1 has been compared with the human fibrinogen γ fragment, tachylectin-5A, L-ficolin and H-ficolin. The overall structure of FD1 is similar to that of other proteins, although the peptide bond between Asp^{282} and Cys^{283}, which is in a predicted ligand-binding site, is a normal *trans* bond, unlike the cases of the other proteins. Analysis of the pH-dependent ligand-binding activity of FD1 in solution suggested that a conformational equilibrium between active and non-active forms in the ligand-binding region, involving cis-trans isomerization of the Asp^{282}-Cys^{283} peptide bond, contributes to the discrimination between self and non-self, and that the pK_a values of His^{284} are 6.1 and 6.3 in the active and non-active forms, respectively (Tanio et al. 2008).

18.1.4.3 Ligand-binding site of Ficolins

The crystal structures of human fibrinogen γ fragment (Pratt et al. 1997) and of the FD1 of *Tachypleus tridentatus* tachylectin-5A (TL5A), an invertebrate lectin (Kairies et al. 2001), have shown that the P domain contributes to ligand binding. In angiopoietin-2, the P domain of the FD1 relates to receptor binding (Barton et al. 2005). These findings suggest that the P domain of ficolins includes the sugar-binding site. Interestingly, although the sugars recognized by ficolins exist on the surface of the host cell, these proteins can discriminate between pathogens and the host cell. The crystal structures of ficolins, complexed with ligands, should clarify the molecular mechanism of the discrimination and advance drug design for general pathogens. The FReD (fibrinogen-related domain) of FIBCD1 (Fibrinogen C domain containing 1) forms noncovalent tetramers and that the acetyl-binding site of FReDs of FIBCD1 is homologous to that of tachylectin 5A and M-ficolin but not to the FReD of L-ficolin (Thomsen et al. 2010).

18.1.5 Functions of Ficolins

The lectin pathway of complement activation in humans is triggered through the action of MBL-associated serine protease-2 in response to recognition of neutral carbohydrates and other motifs present on microbial surfaces. This pathway is initiated by MBL, L-ficolin, H-ficolin, and M-ficolin (Matsushita and Fujita 2001; Liu et al. 2005; Frederiksen et al. 2005). MBL, through its C-type lectin domain, recognizes terminal carbohydrates, provided that their hydroxyl groups at positions C3 and C4 are in equatorial orientation (Weis et al. 1992).

Upon recognition of the infectious agent, the ficolins act through two distinct routes: initiate the lectin pathway of complement activation through attached mannose-binding lectin-associated serine proteases (MASPs), and a primitive opsonophagocytosis thus limiting the infection and concurrently orchestrating the subsequent adaptive clonal immune response. Ficolins are associated with the MASPs and the complexes activate the lectin-complement pathway (Fujita et al. 2004; Liu et al. 2005; Frederiksen et al. 2005) in response to recognition of neutral carbohydrates and N-acetyl groups on pathogens and damaged cells. This results from the ability of MBL and ficolins to associate with and trigger activation of MBL-associated serine protease (MASP)-2. Activated MASP-2 cleaves the complement proteins C2 and C4, thereby triggering the complement cascade (Thiel et al. 1997; Matsushita, et al. 2000). Three other MBL/ficolins-associated proteins have been described, the MASP-1 and MASP-3 proteases (Matsushita et al. 2002; Dahl, et al. 2001) and a truncated form of MASP-2 called MAp19 (19-kDa MBL-associated protein) or sMAp (Stover et al. 1999; Takahashi et al. 1999). MASP-3 has no known physiological substrates whereas MASP-1 cleaves with a low efficiency a few protein substrates, among which are fibrinogen and coagulation factor XIII (Krarup et al. 2008). It has been proposed that MASP-1 might contribute to the activation of the lectin pathway, though this issue is still controversial (Takahashi et al. 2008; Rossi et al. 2001).

Complement activation results in opsonization of microbes and apoptotic cells with C3-derived fragments, thus promoting their clearance through interaction with C3 receptors on phagocytes (Aoyagi et al. 2005; Kuraya et al. 2005). In addition, ficolins may themselves function as opsonins, as suggested by the ability of L-ficolin to enhance phagocytosis of *Salmonella typhimurium* by neutrophils (Matsushita et al. 1996) and the ability of L- and H-ficolins to increase adhesion/uptake of late apoptotic cells by macrophages (Jensen 2007; Honoré et al. 2007; Aoyagi et al. 2005; Kuraya et al. 2005).

The opsonizing effects of ficolins are exerted through receptors present on phagocytic cells that are likely common to other defense collagens such as C1q and the collectins. A likely candidate is the receptor for the collagenous domain of C1q (cC1qR) or calreticulin (Crt) (Stuart et al. 1997), which is thought to function in complex with the endocytic receptor CD91 (Ogden et al. 2001; Vandivier et al. 2002). This hypothesis is supported by studies showing that L- and

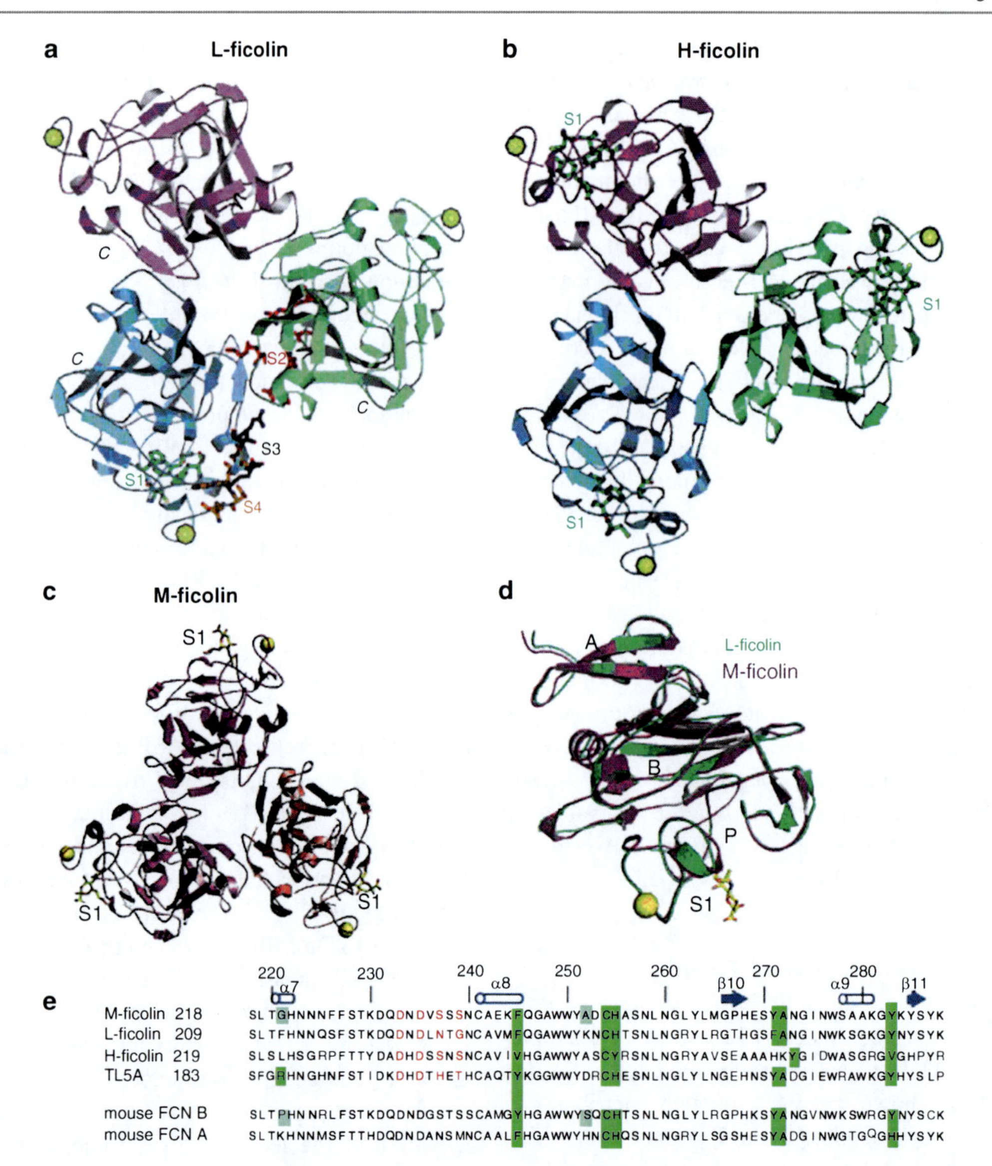

Fig. 18.1 *Homotrimeric structure of the recognition domains of human L-, H- and M-ficolins and location of their binding site(s)*: L-ficolin (**a**) and H-ficolin (**b**) structures seen from the target binding surface (*bottom* view). The side chains of the binding site residues are displayed as ball and sticks and colored *green* (S1), *red* (S2), *black* (S3), and *orange* (S4). To enhance clarity of the side view, only one of each representative binding sites is shown on the L-ficolin trimer. N and C indicate the N- and C-terminal ends of each protomer. Ca^{2+} ions are represented as golden spheres (Reprinted by permission from Macmillan Publishers Ltd: EMBO Journal, Garlatti. et al., © 2007a). (**c**) *Bottom* view of the homotrimeric structure of M-ficolin solved at neutral pH. Ca^{2+} ions are represented as golden spheres. The sialic acid ligand bound to site S1 is shown in a *yellow* ball and stick representation. (**d**) Superposition of the similar fibrinogen-like protomers of M-ficolin (*magenta*) and L-ficolin (*green*). Domains A, B, and P are labeled. (**e**) Sequence alignment of the P domains of human ficolins M, L, and H; mouse ficolins FCN B and FCN A; and TL5A. The residue numbering and the secondary structure elements apply to M-ficolin. Residues involved in the S1 binding site are colored green, and those involved in Ca^{2+} binding are colored red. Small residues allowing accommodation of sialic acid in site S1 are colored blue (Reprinted with permission from Garlatti et al. 2007b © American Society for Biochemistry and Molecular Biology)

H-ficolins bind to cC1qR/CRT (Kuraya et al. 2005) and that H-ficolin binding to Crt can be inhibited by MBL, suggesting that both proteins share a common binding site on CRT (Honoré et al. 2007). In addition, ficolins may themselves function as opsonins, as suggested by the ability of L-ficolin to enhance phagocytosis of *Salmonella typhimurium* by neutrophils and the ability of L- and H-ficolins to increase adhesion/uptake of late apoptotic cells by macrophages (Jensen et al. 2007; Honoré et al. 2007). The other putative function of monocyte surface M-ficolin was implicated by its affinity for GlcNAc, a common sugar structure on Gram-negative bacteria and other pathogens. M-ficolin might act as a phagocytic receptor or adaptor on circulating monocytes for micro-organism recognition and may potentially mediate monocyte adhesion (Teh et al. 2000).

Studies have shown that MASP-2 binds to a short segment of the collagen-like domain of MBL. Recent studies revealed that Lys^{55} in the collagenous region of MBL is important for interaction with MASPs and CRT. On the other hand residues Lys^{57} of L-ficolin and Lys^{47} of H-ficolin are key components of the interaction with MASPs and CRT, providing strong indication that MBL and the ficolins share homologous binding sites for both types of proteins (Lacroix et al. 2009). Girija et al. (2007) showed that the MASP-2 binding site on rat ficolin-A is also located within the collagen-like domain and encompasses a conserved motif that is present in both MBLs and ficolins. Site-directed mutagenesis revealed that a lysine residue in the X position of the Gly-X-Y collagen repeat, Lys^{56} in ficolin-A, which is present in all ficolins and MBLs known to activate complement, is essential for MASP-2 binding. Similar binding sites and activation kinetics of MASP-2 suggest that complement activation by ficolins and MBLs follows analogous mechanisms.

18.1.5.1 The C3b Deposited by L-Ficolin/MBL forms Alternative Pathway C3 Convertase C3bBb

Serotype III group *B streptococci* (GBS) are the common cause of neonatal sepsis and meningitis. Although deficiency in maternal capsular polysaccharide (CPS)-specific IgG correlates with susceptibility of neonates to the GBS infection, serum deficient in CPS-specific IgG mediates significant opsonophagocytosis. The IgG-independent opsonophagocytosis requires activation of the complement pathway, a process requiring the presence of both Ca^{2+} and Mg^{2+}, and is significantly reduced by chelating Ca^{2+} with EGTA. The role of L-ficolin/mannose-binding lectin-associated serine protease (MASP) complexes in Ca^{2+}-dependent, Ab-independent opsonophagocytosis of serotype III GBS has been defined. The binding of L-ficolin/MASP complexes to the CPS generates C3 convertase C4b2a, which deposits C3b on GBS. The C3b deposited by this lectin pathway forms alternative pathway C3 convertase C3bBb whose activity is enhanced by CPS-specific IgG2, leading to an increased opsonophagocytic killing by further deposition of C3b on the GBS (Aoyagi et al. 2005).

18.1.6 Pathophysiology of Ficolins

Like invertebrate innate immune reactions, the human PGN and 1, 3-β-D-glucan recognition proteins function as complement-activating lectins. The counterparts of these proteins from human serum are MBL and L-ficolin. The specific microbial cell component-coupled columns demonstrated that MBL and L-ficolin bind to PGN and 1, 3-β-D-D-glucan, respectively. MBL and L-ficolin were associated with MBL-associated serine proteases-1 and -2 (MASPs) and small MBL-associated protein. The binding of purified MBL/MASP and L-ficolin/MASP complexes to PGN and 1, 3-β-D-glucan, respectively, resulted in the activation of the lectin-complement pathway (Ma et al. 2004; Matsushita and Fujita 2001). A lot of reports showed that dysfunction or abnormal expressions of ficolins may play crucial roles in the pathogenesis of human diseases including: (1) infectious and inflammatory diseases, e.g., recurrent respiratory infections; (2) apoptosis, and autoimmune disease; (3) systemic lupus erythematosus; (4) IgA nephropathy; (5) clinical syndrome of preeclampsia; (6) other diseases associated factor e.g. C-reactive protein. Zhang and Ali (2008) reviewed the structures, functions, and clinical implications of ficolins and summarized the reports on the roles of ficolins in human diseases. Precise identification of ficolins functions will provide novel insight in the pathogenesis of these diseases and may provide novel innate immune therapeutic options to treat disease progression.

In clinical studies, like MBL, no absolute levels of deficiency have yet been defined. In more than 300 children with recurrent respiratory tract infections no correlation has been observed between L-ficolin levels and severity of disease (Atkinson et al. 2004). An association with MBL deficiency in the same patient cohort had already been reported (Cedzynski et al. 2004). In this study, low levels of L-ficolin were more common in patients than in controls and most common in patients with co-existing atopic disorders, suggesting a role for L-ficolin in protection from microorganisms complicating allergic disease. Polymorphisms in the ficolins have been identified, although their clinical significance is as yet unknown (Dommett et al. 2006).

18.2 Tachylectins

18.2.1 Horseshoe Crab Tachylectins

Tachylectin-related proteins belong to fibrinogen type lectins and function in innate immunity of various animals, from ancient sponges to vertebrates. The specific recognition by horseshoe crab tachylectins with a propeller-like fold or a propeller-like oligomeric arrangement is reinforced by the short distance between the individual binding sites that interact with pathogen-associated molecular patterns (PAMPs). There is virtually no conformational change in the main or side chains of tachylectins upon binding with the ligands. This low structural flexibility of the propeller structures must be very important for specific interaction with PAMPs. While MBL and ficolins trigger complement activation through the lectin pathway in the form of opsonins, tachylectins have no effector collagenous domains and no lectin-associated serine proteases found in the mammalian lectins. Furthermore, no complement-like proteins have been found in horseshoe crabs, except for α_2-macroglobulin. The mystery of the molecular mechanism of the scavenging pathway of pathogens in horseshoe crabs is to be solved (Kawabata and Tsuda 2002).

Kawabata and Iwanaga (1999) purified five types of tachylectins from circulating hemocytes and hemolymph plasma of the Japanese horseshoe crab, *Tachypleus tridentatus*. Tachylectin-1 interacts with Gram-negative bacteria probably through 2-keto-3-deoxyoctonate, one of the constituents of LPS. Tachylectin-1 also binds to polysaccharides such as agarose and dextran with broad specificity. Tachylectin-2 binds to D-GlcNAc or D-GalNAc and recognizes staphylococcal lipoteichoic acids and LPS from several Gram-negative bacteria. In contrast, tachylectins-3 and -4 specifically bind to S-type LPS from several Gram-negative bacteria through a certain sugar moiety on the O-specific polysaccharides (O-antigens). Tachylectin-5 identified in hemolymph plasma has the strongest bacterial agglutinating activity in the five types of tachylectins, and exhibits broad specificity against acetyl group-containing substances. Thus, the innate immune system of horseshoe crab may recognize invading pathogens through a combinatorial method using lectins with different specificities against carbohydrates exposed on pathogens. An encounter of these lectins derived from hemocytes and hemolymph plasma at injured sites, in response to the stimulation of LPS, suggests that they serve synergistically to accomplish an effective host defense against invading microbes and foreign substances (Kawabata and Iwanaga 1999). The structure of tachylectin-2, the first example of a fivefold symmetric β-propeller protein, sheds light on the role played by this lectin in horseshoe crab host defense (Rini and Lobsanov 1999).

18.2.1.1 Tachylectin 5A and 5B

Gokudan et al. (1999) characterized tachylectins 5A and 5B (TLs-5). TLs-5 agglutinated all types of human erythrocytes and Gram-positive and Gram-negative bacteria. TLs-5 specifically recognizes acetyl group-containing substances including noncarbohydrates; the acetyl group is required and is sufficient for recognition. TLs-5 enhanced the antimicrobial activity of a horseshoe crab-derived big defensin. cDNA sequences of TLs-5 indicated that they consist of a short N-terminal Cys-containing segment and a C-terminal fibrinogen-like domain with the highest sequence identity (51%) to that of mammalian ficolins. TLs-5, however, lack the collagenous domain found in a kind of "bouquet arrangement" of ficolins and collectins. Electron microscopy revealed that TLs-5 form two- to four-bladed propeller structures. Kairies et al. (2001) reported the crystal structure of TL5A, a lectin from hemolymph plasma of T. tridentatus. TL5A shares not only a common fold but also related functional sites with γ fragment of mammalian fibrinogen (Fig. 18.2). A fibrinogen-related lectin, named Dorin M, from the hemolymph plasma of soft tick, Ornithodoros moubata, is produced in the tick hemocytes. Dorin M is also expressed in salivary glands and has a fibrinogen-like domain. It exhibits similarity with tachylectins 5A and 5B from a horseshoe crab, Tachypleus tridentatus. In addition, Dorin M is closely related to tachylectins-5 (Rego et al. 2006).

18.2.2 X-ray Structure

TL5A is an ellipsoidal molecule (overall dimensions ~34 × 36 × 53 Å^3), subdivided into three distinct but interacting domains (Fig. 18.2 a and b). The N-terminal domain A (residues Asp^{45}–Trp^{89}) comprises two short helices and a small two-stranded antiparallel β-sheet. The N-terminal helix (α1) is diagonally twisted and anchored through a disulfide bond (Cys^{49}–Cys^{80}) to the second strand of β-sheet (β2). The second, short helix β2 connects strand β1 with strand β2. Structural elements N-terminal of Asp^{45} could not be resolved because of poorly defined electron density. Cysteine residues Cys^6 and Cys^{170} were expected to be involved in lectin homodimerization. However, electron density revealed Cys^{170} to be present in the free -SH form. In addition, chemical-labeling experiments also showed that Cys^6 and Cys^{170} are present in the free SH forms in native TL5A. The two disulfide bridges, Cys^{49}–Cys^{80} and Cys^{206} Cys^{219}, were confirmed by sequence analysis of peptides derived from TL5A (Kairies et al. 2001). The central and larger domain B (residues Thr^{90}–Ala^{180} and Pro^{253}–Phe^{264}) is clamped through domains A and P and predominantly made up by a twisted seven-stranded antiparallel β-sheet (strands β3–β7, β9 and β12). Helices β4 and β5 within this domain can be interpreted as a single helix divided by a loop, which participates with a small antiparallel two-stranded β-sheet (β4, β5) to the main seven-stranded β-sheet. The central strand β12 (residues Gln^{254}–Pro^{261}) extends the C terminus of domain P back to domain B, bringing both polypeptide termini in close proximity. The C-terminal domain P (residues Gly^{181}–Leu^{252}) possesses only a few short elements of regular secondary structure, and comprises two major functional sites within TL5A: the Ca^{2+}-binding site and the nearby acetyl group-binding pocket (Fig. 18.2b). The Ca^{2+} ion is bound in a loop region located ~11 Å away from the adjacent ligand binding site, coordinated by seven oxygen ligands in a pentagonal bipyramidal manner. Both carboxylate oxygen atoms of Asp^{198}, the main chain carbonyl oxygens of His^{202} and Thr^{204}, and

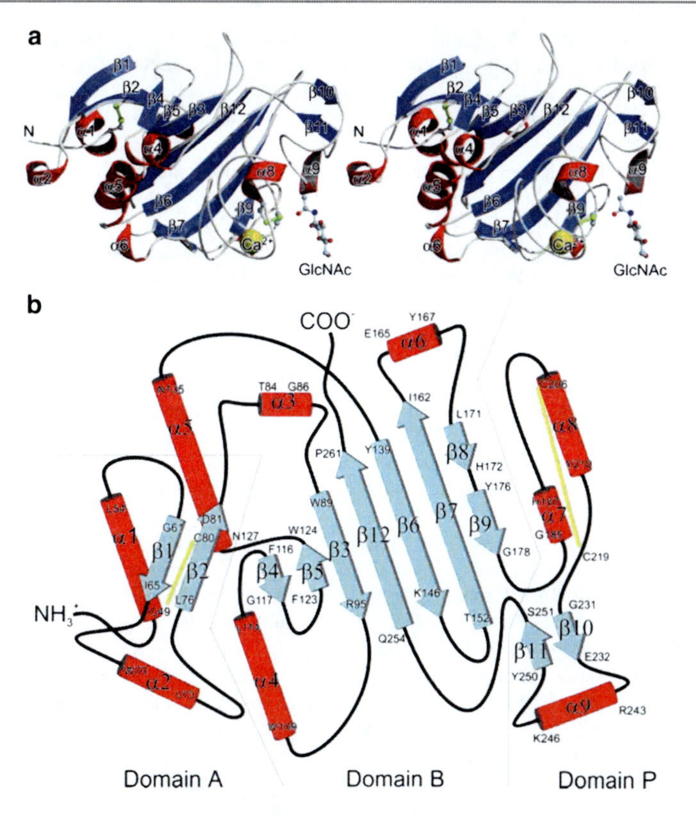

Fig. 18.2 *Crystal structure of TL5A from the Japanese horseshoe crab T. tridentatus*. (**a**) Stereo view of the TL5A ribbon plot. GlcNAc is represented by a ball-and-stick model and the Ca^{2+} ion is represented by a golden sphere. Disulfide bridges are colored yellow. (**b**) Topology diagram showing the arrangement of secondary-structure elements in TL5A. Disulfide bridges are indicated by yellow lines. Domains named in analogy to γ-fibrinogen fragment (Yee et al. 1997) (Reprinted with permission from Kairies et al. 2001© National Academy of Sciences USA)

one water molecule (WAT 40) form the pentagonal base, whereas WAT61 and the side chain oxygen atom OD1 (Asp^{200}) occupy the vertices of the bipyramide (Kairies et al. 2001).

References

Akaiwa M, Yae Y, Sugimoto R et al (1999) Hakata antigen, a new member of the ficolin/opsonin p35 family, is a novel human lectin secreted into bronchus/alveolus and bile. J Histochem Cytochem 47:777–786

Aoyagi Y, Adderson EE, Min JG et al (2005) Role of L-ficolin/mannose-binding lectin-associated serine protease complexes in the opsonophagocytosis of type III group B streptococci. J Immunol 174:418–425

Atkinson AP, Cedzynski M, Szemraj J et al (2004) L-ficolin in children with recurrent respiratory infections. Clin Exp Immunol 138:517–520

Barton WA, Tzvetkova D, Nikolov DB (2005) Structure of the angiopoietin-2 receptor binding domain and identification of surfaces involved in Tie2 recognition. Structure 13:825–832

Bohlson SS, Fraser DA, Tenner AJ (2007) Complement proteins C1q and MBL are pattern recognition molecules that signal immediate and long-term protective immune functions. Mol Immunol 44:33–43

Cedzynski M, Szemraj J, Swierzko AS et al (2004) Mannan-binding lectin insufficiency in children with recurrent infections of the respiratory system. Clin Exp Immunol 136:304–311

Dahl MR, Thiel S, Matsushita M, Fujita T et al (2001) MASP-3 and its association with distinct complexes of the mannan-binding lectin complement activation pathway. Immunity 15:127–135

Dommett RM, Klein N, Turner MW (2006) Mannose-binding lectin in innate immunity: past, present and future. Tissue Antigens 68: 193–209

Edgar PF (1995) Hucolin, a new corticosteroid-binding protein from human plasma with structural similarities to ficolins, transforming growth factor-β1-binding proteins. FEBS Lett 375:159–161

Endo Y, Sato Y, Matsushita M, Fujita T (1996) Cloning and characterization of the human lectin P35 gene and its related gene. Genomics 36:515–521

Endo Y, Liu Y, Kanno K et al (2004) Identification of the mouse H-ficolin gene as a pseudogene and orthology between mouse ficolins A/B and human L-/M-ficolins. Genomics 84:737–744

Endo Y, Matsushita M, Fujita T (2007) Role of ficolin in innate immunity and its molecular basis. Immunobiology 212:371–379

Endo Y, Takahashi M, Fujita T (2006) Lectin complement system and pattern recognition. Immunobiology 221:283–293

Frankenberger M, Schwaeble W, Ziegler-Heitbrock L (2008) Expression of M-ficolin in human monocytes and macrophages. Mol Immunol 45:1424–1430

Frederiksen PD, Thiel S, Larsen CB, Jensenius JC (2005) M-ficolin, an innate immune defence molecule, binds patterns of acetyl groups and activates complement. Scand J Immunol 62:462–473

Fujimori Y, Harumiya S, Fukumoto Y et al (1998) Molecular cloning and characterization of mouse ficolin-A. Biochem Biophys Res Commun 244:796–800

Fujita T (2002) Evolution of the lectin-complement pathway and its role in innate immunity. Nat Rev Immunol 2:346–353

Fujita T, Matsushita M, Endo Y (2004) The lectin-complement pathway–Its role in innate immunity and evolution. Immunol Rev 198:185–202

Garlatti V, Belloy N, Martin L et al (2007a) Structural insights into the innate immune recognition specificities of L- and H-ficolins. EMBO J 26:623–633

Garlatti V, Martin L, Gout E et al (2007b) Structural basis for innate immune sensing by M-ficolin and its control by a pH-dependent conformational switch. J Biol Chem 282:35814–35820

Girija UV, Dodds AW, Roscher S et al (2007) Localization and characterization of the mannose-binding lectin (MBL)-associated-serine protease-2 binding site in rat ficolin-A: equivalent binding sites within the collagenous domains of MBLs and ficolins. J Immunol 179:455–462

Gokudan S, Muta T, Tsuda R et al (1999) Horseshoe crab acetyl group-recognizing lectins involved in innate immunity are structurally related to fibrinogen. Proc Natl Acad Sci USA 6:10086–10091

Gout E, Garlatti V, Smith DF, Thielens NM et al (2010) Carbohydrate recognition properties of human ficolins: glycan array screening reveals the sialic acid binding specificity of M-ficolin. J Biol Chem 285:6612–6622

Harumiya S, Omori A, Sugiura T et al (1995) EBP-37, a new elastin-binding protein in human plasma: structural similarity to ficolins, transforming growth factor-β1-binding proteins. J Biochem 117:1029–1035

Harumiya S, Takeda K, Sugiura T et al (1996) Characterization of ficolins as novel elastin-binding proteins and molecular cloning of human ficolin-1. J Biochem 120:745–751

Hashimoto S, Suzuki T, Dong HY et al (1999a) Serial analysis of gene expression in human monocyte-derived dendritic cells. Blood 94:845–852

Hashimoto S, Suzuki T, Dong HY et al (1999b) Serial analysis of gene expression in human monocytes and macrophages. Blood 94:837–844

Holmskov U, Thiel S, Jensenius JC (2003) Collectins and ficolins: humoral lectins of the innate immune defense. Annu Rev Immunol 21:547–578

Honoré C, Hummelshoj T, Hansen BE et al (2007) The innate immune component ficolin 3 (Hakata antigen) mediates the clearance of late apoptotic cells. Arthritis Rheum 56:1598–1607

Honore C, Rorvig S, Munthe-Fog L et al (2008) The innate pattern recognition molecule Ficolin-1 is secreted by monocytes/macrophages and is circulating in human plasma. Mol Immunol 45:2782–2789

Hummelshoj T, Thielens NM, Madsen HO et al (2007) Molecular organization of human ficolin-2. Mol Immunol 44:401–411

Hummelshoj T, Fog LM, Madsen HO et al (2008) Comparative study of the human ficolins reveals unique features of Ficolin-3 (Hakata antigen). Mol Immunol 45:1623–1632

Ichijo H, Hellman U, Wernstedt C et al (1993) Molecular cloning and characterization of ficolin, a multimeric protein with fibrinogen- and collagen-like domains. J Biol Chem 268:14505–14513

Ichijo H, Rönnstrand L, Miyagawa K et al (1991) Purification of transforming growth factor-β1 binding proteins from porcine uterus membranes. J Biol Chem 266:22459–22464

Janeway CA, Medzhitov R (2002) Innate immune recognition. Annu Rev Immunol 20:197–216

Jensen ML, Honoré C, Hummelshøj T, Hansen BE, Madsen HO, Garred P (2007) Ficolin-2 recognizes DNA and participates in the clearance of dying host cells. Mol Immunol 44:856–865

Kairies N, Beisel HG, Fuentes-Prior P et al (2001) The 2.0-Å crystal structure of tachylectin 5A provides evidence for the common origin of the innate immunity and the blood coagulation systems. Proc Natl Acad Sci USA 98:13519–13524

Kawabata S, Iwanaga S (1999) Role of lectins in the innate immunity of horseshoe crab. Dev Comp Immunol 23:391–400

Kawabata S, Nagayama R, Hirata M et al (1996) Tachycitin, a small granular component in horseshoe crab hemocytes, is an antimicrobial protein with chitin-binding activity. J Biochem 120: 1253–1260

Kawabata S, Tsuda R (2002) Molecular basis of non-self recognition by the horseshoe crab tachylectins. Biochim Biophys Acta 1572: 414–421

Kim J-I, Lee CJ, Jin MS et al (2005) Crystal structure of CD14 and its implications for lipopolysaccharide signalling. J Biol Chem 280:11347–11351

Krarup A, Gulla KC, Gál P et al (2008) The action of MBL-associated serine protease 1 (MASP1) on factor XIII and fibrinogen. Biochim Biophys Acta 1784:1294–1300

Krarup A, Sorensen UB, Matsushita M et al (2005) Effect of capsulation of opportunistic pathogenic bacteria on binding of the pattern recognition molecules mannan-binding lectin, L-Ficolin and H-Ficolin. Infect Immun 73:1052–1060

Krarup A, Thiel S, Hansen A et al (2004) L-ficolin is a pattern recognition molecule specific for acetyl groups. J Biol Chem 279: 47513–47519

Kuraya M, Ming Z, Liu X et al (2005) Specific binding of L-ficolin and H-ficolin to apoptotic cells leads to complement activation. Immunobiology 209:689–697

Lacroix M, Dumestre-Pérard C, Schoehn G et al (2009) Residue Lys57 in the collagen-like region of human L-ficolin and its counterpart Lys47 in H-ficolin play a key role in the interaction with the mannan-binding lectin-associated serine proteases and the collectin receptor calreticulin. J Immunol 182:456–465

Le Y, Lee SH, Kon OL, Lu J (1998) Human L-ficolin: plasma levels, sugar specificity, and assignment of its lectin activity to the fibrinogen-like (FBG) domain. FEBS Lett 425:367–370

Le Y, Tan SM, Lee SH, Kon OL, Lu J (1997) Purification and binding properties of a human ficolin-like protein. J Immunol Methods 204:43–49

Liu Y, Endo Y, Iwaki D et al (2005) Human M-ficolin is a secretory protein that activates the lectin complement pathway. J Immunol 175:3150–3156

Lu J, Le Y, Kon OL et al (1996a) Biosynthesis of human ficolin, an Escherichia coli-binding protein, by monocytes: comparison with the synthesis of two macrophage-specific proteins, C1q and the mannose receptor. Immunology 89:289–294

Lu J, Teh C, Kishore U, Reid KB (2002) Collectins and ficolins: sugar pattern recognition molecules of the mammalian innate immune system. Biochim Biophys Acta 1572:387–400

Lu J, Tay PN, Kon OL, Reid KB (1996b) Human ficolin: cDNA cloning, demonstration of peripheral blood leucocytes as the major site of synthesis and assignment of the gene to chromosome 9. Biochem J 313:473–478

Lynch NJ, Roscher S, Hartung T et al (2004) L-ficolin specifically binds to lipoteichoic acid, a cell wall constituent of gram-positive bacteria, and activates the lectin pathway of complement. J Immunol 172:1198–1202

Ma YG, Cho MY, Zhao M et al (2004) Human mannose-binding lectin and L-ficolin function as specific pattern recognition proteins in the lectin activation pathway of complement. J Biol Chem 279:25307–25312

Matsushita M, Endo Y, Fujita T (2000) Complement-activating complex of ficolin and mannose binding lectin-associated serine protease. J Immunol 164:2281–2284

Matsushita M, Endo Y, Taira S et al (1996) A novel human serum lectin with collagen- and fibrinogen-like domains that functions as an opsonin. J Biol Chem 271:2448–2454

Matsushita M, Fujita T (2001) Ficolins and the lectin complement pathway. Immunol Rev 180:78–85

Matsushita M, Kuraya M, Hamasaki N et al (2002) Activation of the lectin complement pathway by H-ficolin (Hakata antigen). J Immunol 168:3502–3506

Matsushita M, Takahashi A, Hatsuse H et al (1992) Human mannose-binding protein is identical to a component of Ra-reactive factor. Biochem Biophys Res Commun 183:645–651

Nahid AM, Sugii S (2006) Binding of porcine ficolin-alpha to lipopolysaccharides from Gram-negative bacteria and lipoteichoic acids from Gram-positive bacteria. Dev Comp Immunol 30:335–343

Ng PM, Le Saux A et al (2007) C-reactive protein collaborates with plasma lectins to boost immune response against bacteria. EMBO J 26:3431–3440

Ogden CA, deCathelineau A, Hoffmann PR et al (2001) C1q and mannose binding lectin engagement of cell surface calreticulin and CD91 initiates macropinocytosis and uptake of apoptotic cells. J ExpMed 194:781–796

Ohashi T, Erickson HP (2004) The disulfide bonding pattern in ficolin multimers. J Biol Chem 279:6534–6539

OmPraba G, Chapeaurouge A, Doley R et al (2010) Identification of a novel family of snake venom proteins Veficolins from Cerberus rynchops using a venom gland transcriptomics and proteomics approach. J Proteome Res 9:1882–1893

Pratt KP, Cote HC, Chung DW et al (1997) The primary fibrin polymerization pocket: three-dimensional structure of a 30-kDa C-terminal gamma chain fragment complexed with the peptide Gly-Pro-Arg-Pro. Proc Natl Acad Sci USA 94:7176–7181

Rego RO, Kovár V, Kopácek P et al (2006) The tick plasma lectin, Dorin M, is a fibrinogen-related molecule. Insect Biochem Mol Biol 36:291–299

Rini JM, Lobsanov YD (1999) New animal lectin structures. Curr Opin Struct Biol 9:578–584

Rossi V, Cseh S, Bally I, Thielens NM, Jensenius JC, Arlaud GJ (2001) Substrate specificities of recombinant mannan-binding lectin-associated serine proteases-1 and –2. J Biol Chem 276:40880–40887

Runza VL, Hehlgans T, Echtenacher B, Zahringer U, Schwaeble WJ, Mannel DN (2006) Localization of the mouse defense lectin ficolin B in lysosomes of activated macrophages. J Endotoxin Res 12:120–126

Runza VL, Schwaeble W, Mannel DN (2008) Ficolins: novel pattern recognition molecules of the innate immune response. Immunobiology 213:297–306

Stover CM, Thiel S, Thelen M et al (1999) Two constituents of the initiation complex of the mannan-binding lectin activation pathway of complement are encoded by a single structural gene. J Immunol 162:3481–3490

Stuart GR, Lynch NJ, Day AJ, Schwaeble WJ, Sim RB (1997) The C1q and collectin binding site within C1q receptor (cell surface calreticulin). Immunopharmacology 38:73–80

Sugimoto R, Yae Y, Akaiwa M et al (1998) Cloning and characterization of the Hakata antigen, a member of the ficolin/opsonin p35 lectin family. J Biol Chem 273:20721–20727

Taira S, Kodama N, Matsushita M, Fujita T (2000) Opsonic function and concentration of human serum ficolin/P35. Fukushima J Med Sci 46:13–23

Takahashi M, Endo Y, Fujita T, Matsushita M (1999) A truncated form of mannose-binding lectin-associated serine protease (MASP)-2-expressed by alternative polyadenylation is a component of the lectin complement pathway. Int Immunol 11:859–863

Takahashi M, Iwaki D, Kanno K, Ishida Y, Xiong J, Matsushita M, Endo Y, Miura S, Ishii N, Sugamura K, Fujita T (2008) Mannose-binding lectin (MBL)-associated serine protease (MASP)-1 contributes to activation of the lectin complement pathway. J Immunol 180:6132–6138

Tanio M, Kohno T (2009) Histidine-regulated activity of M-ficolin. Biochem J 417:485–491

Tanio M, Kondo S, Sugio S, Kohno T (2006) Overexpression, purification and preliminary crystallographic analysis of human M-ficolin fibrinogen-like domain. Acta Crystallogr Sect F Struct Biol Cryst Commun 62:652–655

Tanio M, Kondo S, Sugio S, Kohno T (2008) Trimeric structure and conformational equilibrium of M-ficolin fibrinogen-like domain. J Synchrotron Radiat 15:243–245

Tanio M, Kondo S, Sugio S et al (2007) Trivalent recognition unit of innate immunity system: crystal structure of trimeric human M-ficolin fibrinogen-like domain. J Biol Chem 282:3889–3895

Teh C, Le Y, Lee SH, Lu J (2000) M-ficolin is expressed on monocytes and is a lectin binding to N-acetyl-d-glucosamine and mediates monocyte adhesion and phagocytosis of Escherichia coli. Immunology 101:225–232

Thiel S, Vorup-Jensen T, Stover CM et al (1997) A second serine protease associated with mannan-binding lectin that activates complement. Nature 386(6624):506–510

Thiel S (2007) Complement activating soluble pattern recognition molecules with collagen-like regions, mannan-binding lectin, ficolins and associated proteins. Mol Immunol 44:3875–3888

Thomsen T, Moeller JB, Schlosser A et al (2010) The recognition unit of FIBCD1 organizes into a noncovalently linked tetrameric structure and uses a hydrophobic funnel (S1) for acetyl group recognition. J Biol Chem 285:1229–1238

Tsujimura M, Ishida C, Sagara Y et al (2001) Detection of a serum thermolabile β-2 macroglycoprotein (hakata antigen) by enzyme-linked immunosorbent assay using polysaccharide produced by Aerococcus viridans. Clin Diagn Lab Immunol 8:454–459

Vandivier RW, Ogden CA, Fadok VA et al (2002) Role of surfactant proteins A, D, and C1q in the clearance of apoptotic cells in vivo and in vitro: calreticulin and CD91 as a common collectin receptor complex. JImmunol 169:3978–3986

Weis WI, Drickamer K, Hendrickson WA (1992) Structure of a C-type mannose-binding protein complexed with an oligosaccharide. Nature 360:127–134

Yee VC, Pratt KP, Côté HC, Trong IL, Chung DW, Davie EW, Stenkamp RE, Teller DC (1997) Crystal structure of a 30 kDa C-terminal fragment from the gamma chain of human fibrinogen. Structure 5:125–138

Zhang XL, Ali MA (2008) Ficolins: structure, function and associated diseases. Adv Exp Med Biol 632:105–115

Chi-Lectins: Forms, Functions and Clinical Applications

19

Rajesh K. Gupta and G.S. Gupta

19.1 Glycoside Hydrolase Family 18 Proteins in Mammals

19.1.1 Chitinases

Chitinases (EC.3.2.1.14) hydrolyze the β-1,4-linkages in chitin, an abundant N-acetylglucosamine (GlcNAc) polysaccharide that is a structural component of protective biological matrices of invertebrate such as insect exoskeletons and fungal cell walls. Chitinases cleave chitin and contain the conserved sequence motif DXXDXDXE, in which the glutamate is the catalytic residue. Chitinases are found in species including archaea, bacteria, fungi, plants, and animals. On the basis of sequence homologies, chitinases fall into two groups: families 18 and 19 of glycosyl hydrolases. Members of family 18 employ a substrate-assisted reaction mechanism (van Aalten et al. 2001), whereas those of family 19 adopt a fold-and-reaction mechanism similar to that of lysozyme (Monzingo et al. 1996), suggesting that these families evolved independently to deal with chitin. The glycoside hydrolase 18 (GH18) family of chitinases is an ancient gene family widely expressed in archea, prokaryotes and eukaryotes. Since chitin is an important structural component of pathogens like fungi as well as a constituent of the mammalian diet, a dual function for mammalian chitinases in innate immunity and food digestion has been envisioned (Suzuki et al. 2002; Boot et al. 2005a). Indeed, for human chitotriosidase, an enzyme predominantly expressed by phagocytes, a fungistatic effect has been demonstrated (van Eijk et al. 2005; Brunner et al. 2008). Several studies have tried to link a common chitotriosidase deficiency to susceptibility for infection by chitin-containing parasites (Bussink et al. 2006). The physiological function of the second mammalian chitinase, acidic mammalian chitinase (AMCase), has attracted considerable attention due to a report linking the protein to the pathophysiology of asthma (Zhu et al. 2004a).

19.2 Chitinase-Like Lectins: Chi-Lectins

In addition to active chitinases, highly homologous mammalian proteins lacking enzymatic activity due to substitution of active-site catalytic residues have been identified. Despite their lack of enzymatic activity, these proteins have retained active-site carbohydrate binding and hence have been named chi-lectins (Renkema et al. 1998; Houston et al. 2003; Bussink et al. 2006). Like the active chitinases, chi-lectins belong to family 18 of glycosyl hydrolases, consisting of a 39-kDa catalytic domain having a TIM-barrel structure, one of the most versatile folds in nature (Sun et al. 2001; Fusetti et al. 2003; Houston et al. 2003). In contrast to both chitinases, chi-lectins lack the conserved additional chitin-binding domain (Boot et al. 2001). Despite the detailed knowledge regarding structure, insight into the exact physiological function of the various chi-lectins is limited (Bussink et al. 2006). Similar to chitotriosidase and AMCase, chi-lectins are secreted locally or into the circulation and hence Chi-lectins play a significant role in inflammatory conditions. For example, human cartilage GP39 (HC-gp39/YKL-40/CHI3L1), a protein expressed by chondrocytes and phagocytes, has been implicated in arthritis, tissue remodeling, fibrosis, and cancer (Johansen 2006). Similarly, the human chi-lectin YKL-39 (CHI3L2) and the murine Ym1 (CHI3L3/ECF-L) have been associated with the pathogenesis of arthritis (Tsuruha et al. 2002) and allergic airway inflammation, respectively (Ward et al. 2001; Homer et al. 2006). Investigation of family 18 glycosyl hydrolases has revealed that active chitinases and chi-lectins are widespread and conserved in the mammalian kingdom. An ancient gene duplication first allowed the specialization of two active chitinases, chitotriosidase and AMCase, and subsequent gene duplications followed by loss-of-enzymatic-function mutations, have led to the evolution of a broad spectrum of chi-lectins in mammals.

G.S. Gupta et al., *Animal Lectins: Form, Function and Clinical Applications*,
DOI 10.1007/978-3-7091-1065-2_19, © Springer-Verlag Wien 2012

Chitin is not found in vertebrates, but vertebrates do possess a small number of closely-related active chitinases. In mammals these are represented primarily by AMCase and chitotriosidase, both of which appear to have roles in the immune system. Eukaryotes also possess di-N-acetylchitobiase, a lysosomal enzyme involved in the degradation of N-linked glycoproteins, which is distantly related to AMCase and chitotriosidase. In mammals, but not other vertebrates, there are non-enzymatic chitinase-like proteins very similar to AMCase and chitotriosidase, such as YKL-40. Preservation of the hydrophobic substrate binding cleft, and consequently of high affinity binding to chito-oligosaccharides, has been demonstrated in some of these proteins, which have been termed chitinase-like lectins or Chi-lectins. Stabilin-1 interacting chitinase-like protein (SI-CLP, CHID1) is a non-enzymatic protein of unknown function which is distantly related to other mammalian glycoside hydrolase family 18 proteins and is conserved in eukaryotes.

19.3 YKL-40 [Chitinase 3-Like Protein 1 (CHI3L1)]

19.3.1 The Protein

Johansen and associates identified a protein secreted in vitro by the human osteosarcoma cell line MG63 and named it 'YKL-40' based on its three NH2-terminal amino acids tyrosine (Y), lysine (K), and leucine (L) and its molecular weight of 40 kDa (Johansen et al. 1992). Chitinase 3-like protein 1 or YKL-40 contains a single polypeptide chain of 383 amino acids. The complete amino acid and cDNA sequence of human YKL-40 was deduced by Hakala et al. (1993). The sequence of YKL-40 from several other mammals is known. Amino acid sequence of YKL-40 suggested that YKL-40 belongs to the glycosyl hydrolase family 18. YKL-40 shares significant amino acid sequence homology to bacterial chitinases and to six "mammalian chitinase–like proteins": oviduct-specific glycoprotein, chitotriosidase, YKL-39, TSA 1902, inducible silicotic bronchoalveolar lavage protein-p58 also named acidic mammalian chitinase, and mouse Ym1 also named eosinophil chemotactic cytokine (Boot et al. 2005b; Johansen et al. 2006). Interestingly, *Drosophila melanogaster* secretes several proteins, DS47 and imaginal disc growth factors, with sequence homology to YKL-40 (Kawamura et al. 1999).

The gene for human CHI3L1 is located on chromosome 1q31–q32, and consists of 10 exons and spans about 8 kb of genomic DNA (Rehli et al. 1997). The genes of all other human mammalian chitinase–like proteins identified are also located on chromosome 1. The transcriptional regulation of YKL-40 during human macrophage differentiation has been described. There are probably two independent transcription start sites and the promoter sequence that contains binding sites for several known factors and specific binding of nuclear PU.1, Sp1, Sp3, USF, AML-1, and C/EBP proteins (Rehli et al. 2003). The Sp1-family transcription factors seem to have a predominating role in controlling YKL-40 promoter activity.

19.3.2 Cell Distribution and Regulation

CHI3L1 is expressed by a variety of cell types including synovial cells, chondrocytes and smooth muscle cells, neutrophils, macrophages, fibroblast-like synovial cells, chondrocytes, and hepatic stellate cells. The expression of CHI3L1 is totally absent in monocytes (Rehli et al. 1997) and marginally expressed in monocyte-derived dendritic cells (Krause et al. 1996), but is strongly induced during the late stages of human macrophage differentiation. Rehli et al. (1997) demonstrated that promoter elements (in particular, the proximal -377 base pairs of the CHI3L1 promoter region) control the expression of CHI3L1 in the macrophage. CHI3L1 is also expressed in neutrophils (Volck et al. 1998), chondrocytes (Hakala et al. 1993), fibroblast-like synoviocytes (De Ceuninck et al. 2001), vascular smooth muscle cells, vascular endothelial cells (Malinda et al. 1999), ductal epithelial cells (Qin et al. 2007), hepatic stellate cells (Johansen 2006), and colonic epithelial cells (Mizoguchi 2006). In physiological concentrations CHI3L1 tends to promote proliferation of these cell types.

In U87, hypoxia and ionizing radiation induced a significant increase in YKL-40 after 24–48 h. The hypoxic induction of YKL-40 was independent of HIF1. Etoposide, ceramide, serum depletion and confluence led to elevated YKL-40. Inhibition of p53 augmented the YKL-40 expression indicating that YKL-40 is attenuated by p53. In contrast, both basic fibroblast growth factor and tumor necrosing factor-α repressed YKL-40. Diverse types of stress resulted in YKL-40 elevation, which strongly supports an involvement of YKL-40 in the malignant phenotype as a cellular survival factor in an adverse microenvironment (Junker et al. 2005a, b).

CHI3L1 is related to several other mammalian chitinase-like proteins (CLPs): chitotriosidase, YKL-39, Ym1, AMCase, oviduct-specific glycoprotein,, human cartilage glycoprotein 39 (HC-gp39) and stabilin-1-interacting (SI)-CLP. These mammalian chitinase-like proteins possess a conserved sequence motif (DXXDXDXE) on strand 4, and catalytic activity in these chitinases is mediated by the glutamic acid (E), which protonates the glycosidic bond with chitin (van Aalten et al. 2001). AMCase and chitotriosidase have catalytic activity but CHI3L1, YKL-39, Ym1, oviductin, SI-CLP do not have this activity (Kzhyshkowska et al. 2006). Due to substitution of an essential glutamic acid residue to leucine, CHI3L1 loses chitinase activity, but still can bind to chitin and chito-oligosaccharides with high affinity through a preserved hydrophobic substrate binding cleft (Kawada et al. 2007). SI-CLP is upregulated by IL-4 as well as by glucocorticoids. This feature of SI-CLP makes it an attractive candidate for the examination of individual sensitivity of patients to glucocorticoid treatment and prediction of side effects of glucocorticoid therapy

Exposure to antigens containing chitin- or chitin-like structures sometimes induces strong T helper type-I. responses in mammals which may be associated with the induction of mammalian chitinases

19.3.3 Ligands of YKL-40

YKL-40 binds chitin of different lengths in a similar fashion as with other family 18 chitinases, but has no chitinase activity (Hakala et al. 1993; Renkema et al. 1998). The amino acids essential for the catalytic activity in chitinases are three acidic residues Asp, Glu, and Asp. The corresponding residues in human YKL-40 are Asp^{115}, Leu^{119}, and Asp^{186}. The recombinant chitin-binding domain binds to chitin, but not to glucan, xylan, or mannan and the binding of the recombinant chitin-binding domain to chitin was inhibited by N-acetylglucosamine, di-N-acetylchitobiose, and hyaluronan, but not by N-acetylgalactosamine or chondroitin. Furthermore, the recombinant domain interacts specifically with hyaluronan and hybrid-type N-linked oligosaccharide chains on glycoproteins, and that the oligosaccharide-binding characteristics are similar to those of wheat germ agglutinin, a lectin that binds to chitin (Ujita et al. 2003). YKL-40 binds specifically to collagen triple helices and regulates cleavage and fibril formation. Furthermore, Bigg et al., have reported an ability of YKL-40 to bind to collagen type 1, 2 and 3 (Bigg et al. 2006). YKL-40 binds chitin and chito oligosaccharides using 9 GlcNAc-binding subsites. The short chito-oligosaccharides which prime hyaluronic acid synthesis and are retained at reducing ends of hyaluronic acid molecules in mammals are suggested carbohydrate ligands for YKL-40; hyaluronic acid is associated with similar processes to YKL-40. Heparan sulphate, found on proteoglycans, is also a potential ligand and cell surface receptor for YKL-40 (Fusetti et al. 2003; Houston et al. 2003). The putative heparin-binding site is GRRDKQH (residues 143–149) (Fusetti et al. 2003). YKL-40 has therefore been defined as a CLP or chitinase-like lectin (Chi-lectin). YKL-39 is a protein of unknown function closely related to YKL-40 and present in humans and other primates.

YKL-40 is a glycoprotein in which most of the glucosamine is incorporated into N-linked complex oligosaccharides (De Ceuninck et al. 1998). Two possible sites of glycosylation are found in YKL-40, but only the NH2-terminal is glycosylated with two units of N-acetylglucosamine through β(1–4) linkage. Binding of either short or long oligosaccharides to human YKL-40 is possible, and the presence of two distinct binding sites with selective affinity for long and short oligosaccharides suggests that YKL-40 could function by cross-linking two targets. Glycosylation is a unique feature of the YKL-40 structure as the residue corresponding to Arg^{84} does not exist in chitinases and is mutated to Pro in other mammalian chitinase–like proteins. YKL-40 also binds heparin and amino acid sequence analysis reveals that YKL-40 contains one heparin binding motif (GRRDKQH at position 143–149) and two potential hyaluronan binding sites on the external face of the folded protein (Fusetti et al. 2003; Malinda et al. 1999). The putative heparin-binding site is located in a surface loop. It has been suggested that heparan sulfate is a more likely ligand of YKL-40, and unsulfated fragments of heparan sulfate can be accommodated in the binding groove of YKL-40 (Fusetti et al. 2003).

19.3.4 The Crystal Structure of YKL-40

Family 18 chitinases have catalytic domains of triosephosphate isomerase (TIM barrel) fold with a conserved DxDxE motif (Vaaje-Kolstad et al. 2004) and catalyze the hydrolytic reaction by substrate-assisted mechanism (Fig. 19.1a) (Terwisscha van Scheltinga et al. 1996; van Aalten et al. 2001), whereas family 19 chitinases have high percentage of α-helices and adopt the single displacement catalytic mechanism (Hoell et al. 2006). The crystallographic structures of human YKL-40 (Fusetti et al. 2003; Houston et al. 2003) and goat YKL-40 (Mohanty et al. 2003) display the typical fold of family 18 glycosyl hydrolases (Henrissat and Davies 1997). The structure is divided into two globular domains: a big core domain which consists of a $(\beta/\alpha)_8$ domain structure with a TIM barrel fold, and a small α/β domain, composed of five antiparallel β-strands and one α-helix, inserted in the loop between strand β7 and helix α7. This gives the active site of YKL-40 a groove-like character (Fig. 19.1b and Fig. 19.2). The TIM barrel domain consisting an $(\alpha/\beta)_8$-barrel fold has been found in many different proteins, most of which are enzymes. The TIM barrel domains share low sequence identity and have a diverse range of functions. The specific enzyme activity is determined by the eight loops at the carboxyl end of β-strands. In some TIM barrels, an additional loop

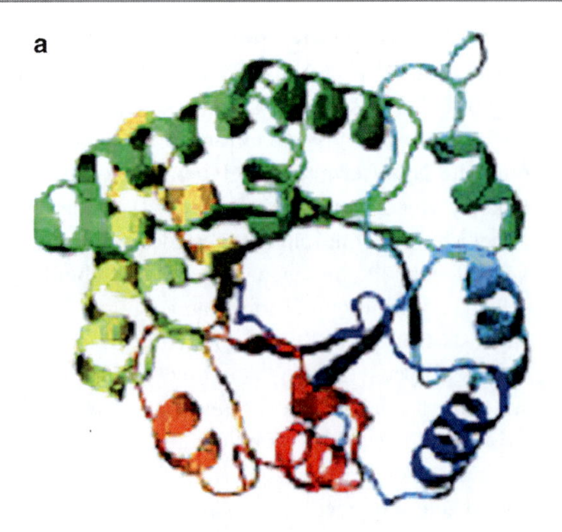
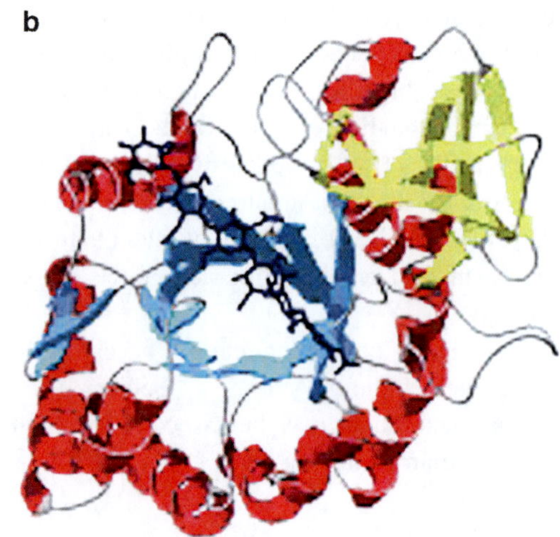

Fig. 19.1 (**a**) Top view of a triosephosphateisomerase (TIM) barrel (PDB code: 8TIM), colored from *blue* (N-terminus) to *red* (C-terminus). (**b**) Structure of human YKL-40 with bound GlcNAc oligomer. β-strands are colored as described in Fig. 19.1a. The GlcNAc oligomer is shown in *dark blue* (PDB code: 1HJW)

from a second domain approaches the active site of the TIM domain and participates in binding and catalysis (Li and Greene 2010).

Mammary Gland Protein-40 (MGP40): X-ray structure of chitolectin, mammary gland protein (MGP-40), is consistent with the $(\beta/\alpha)_8$ barrel topology of the Family 18 glycosidase proteins (Houston et al. 2003; Sun et al. 2001; van Aalten et al. 2001; Varela et al. 2002). MGP40 is an Asn-linked glycoprotein itself. The single disaccharide is covalently linked to the protein (at Asn^{39}) and forms hydrogen bonds with Arg^{84} and Ile^{40}, influencing the backbone conformation of loop Val^{75}-Phe^{85}, which in turn alters the disposition of Trp^{78}. Since Trp^{78} is an essential member of the binding site, the altered positioning of the site leads to constriction of the binding site, thereby leading to the inability of the MGP40 to bind sugar. Therefore, the oligosaccharides cannot bind to the putative binding site. Finally, Arg^{84}, which is hydrogen-bonded to the covalently bond sugar, is conserved in this particular class of chitolectins. The active chitinases in the Family 18 group which bind and cleave sugars possess a Pro in place of Arg^{84}, which in turn leads to inability of hydrogen bonding to the covalent linked sugar. Subsequently, there is a "relaxed" backbone conformation of the Val^{75}-Phe^{85} loop in the Family 18 chitinases and thereby no constriction of the binding site (Mohanty et al. 2003).

Human Cartilage Glycoprotein-39 (HUMGP39): The X-ray crystal structures of another chitolectin HumGP39 has been reported independently by two different groups (Houston et al. 2003; Fusetti et al. 2003). These structures are also consistent with (β/α)8 barrel topology. The HumGP39 protein structures show a disaccharide covalently linked to it at Asn^{19}. However, these structures depict in addition an oligosaccharide bound in the active site. Crystal structures of this protein show complexes with di, tri, tetra, penta and hexasaccharide bound at the active site. Based on these studies, it appears that the chitolectin group of proteins from Family 18 can bind chitin oligosaccharides. In comparison to MGP40 and HumGP39, the active site of glycosidases shows a difference of a glutamic acid, an essential acid/base residue for chitin cleavage and replaced with leucine or glutamine in chito-lectins.

The primary sequences of MGP40 and HumGP39 are highly identical (83 % identity). It was observed that since Arg^{84} is conserved between the two proteins, it is not the cause for the lack of oligosaccharide binding in MGP40 (Mohanty et al. 2003). Haq et al. (2007) compared the aromatic residues in both the proteins, focusing on Trp and Tyr residues present in the floor of the binding groove that are likely to interact strongly with the sugar molecules. Haq et al. (2007) suggested that MGP40 is capable of binding oligo-saccharides contrary to the conclusion of Mohanty

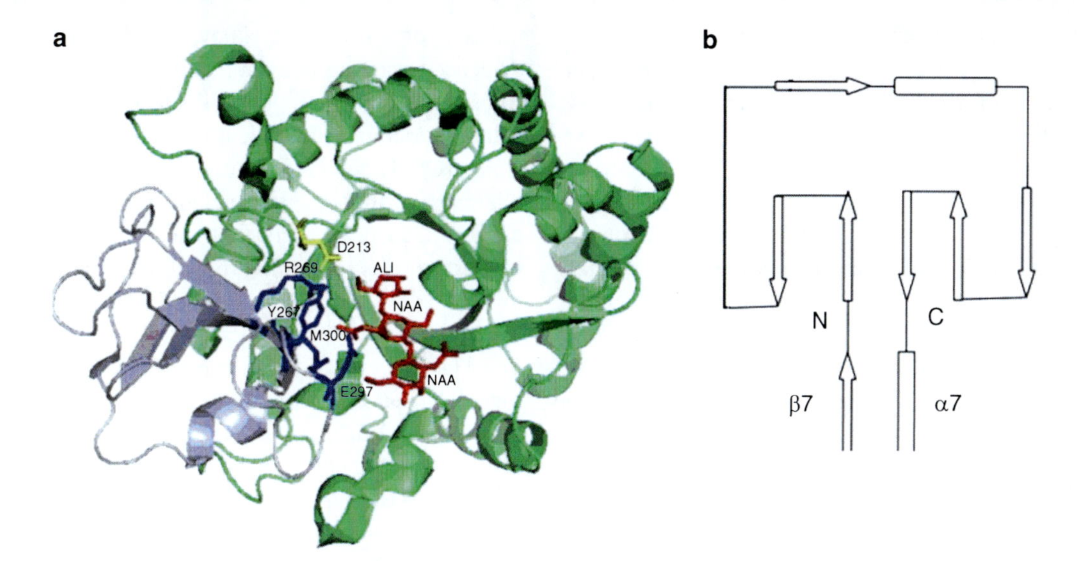

Fig. 19.2 *Structural analysis of Chitinase insertion domain (CID)*: (**a**) Ribbon model of human chitotriosidase (PDB: 1HKM) in complex with the substrate (NAA-NAA-ALI), showing the TIM barrel and CID. The helices and strands on the TIM barrel are coloured in *green* and those on the CID are coloured in *light blue*. Some residues (Tyr^{267}, Arg^{269}, Glu^{297}, and Met^{300}) in *blue* on the CID and Asp^{213} in *yellow* on the TIM barrel interact with the substrate in *red*. (**b**) Schematic representation of the CID between β7 and α7 on the TIM barrel, which is composed of two anti-parallel β-strands followed by one β-strand, one short α-helix, and lastly three anti-parallel β-strands. The arrows indicate β-strands and the rectangles are α-helices. The lines stand for the loops connecting α-helices or β-strands (Reprinted from Li and Greene 2010 © PLoS One)

et al. (2003) and that the conformation of residues 209–213 is not altered upon ligand binding as proposed by Mohanty et al. (2003). Based on the two HumGP39 structures, it was inferred that there are two distinct binding sites; a distal site for trisaccharides and the main site for tetra- and larger saccharides. It was proposed that the Trp in the +1 position functions as a "gate" to the main binding site that is in the "pinched" conformation when the oligosaccharide is not bound and is in the "stacked" conformation while interacting with the oligosaccharide.

Chitinase Insertion Domain (CID): While catalytic domain in chitinase activity has been studied extensively, the function of the chitinase insertion domain is least understood. For example, human chitotriosidase (PDB code: 1HKM), as a family 18 chitinase in the subfamily A, has a TIM domain and a chitinase insertion domain (CID), which is a module inserted into the TIM barrel (Fig. 19.2). Therefore the presence or absence of the insertion domain appears to be subfamily specific (Suzuki et al. 1999; Li and Greene 2010). Chitinase insertion domain sequences include four experimentally determined structures and span five kingdoms. The role of conserved residues was explored by conducting a structural analysis of a number of holo-enzymes. Hydrogen bonding and van der Waals calculations revealed a distinct subset of four conserved residues constituting two sequence motifs that interact with oligosaccharides. The other conserved residues may be key to the structure, folding, and stability of this domain. Sequence and structural studies of the chitinase insertion domains conducted within the framework of evolution identified four conserved residues which clearly interact with the substrates. Evolutionary studies propose a link between the appearance of chitinase insertion domain and the function of family 18 chitinases in the subfamily A (Li and Greene 2010).

19.4 Human Cartilage 39-KDA Glycoprotein (or YKL-39)/(CHI3L2)

The homologous human cartilage 39-kDa glycoprotein (HC gp-39) from articular cartilage chondrocytes primary culture has the N-terminal sequence YKL, which was termed YKL-19. The cDNA of HC gp-39 contained an open reading frame coding for a 383-amino acid long peptide which contains regions displaying significant homology with a group of bacterial and fungal chitinases and a similar enzyme found in the nematode, Brugia malayi. In addition significant homologies were observed with three mammalian secretory proteins, suggesting that a related protein family exists in mammals. The human protein does not possess any glycosidic activity against chitinase substrates. The mRNA of HC gp-39 is present in human articular chondrocytes as well is in liver, but undetectable in muscle tissues, lung, pancreas, mononuclear cells, or fibroblasts. Neither the protein nor mRNA for HC gp-39 was detectable in normal newborn or adult human articular cartilage obtained at surgery, while mRNA for HC gp-39 was detectable both in synovial specimens and in cartilage obtained from patients with rheumatoid arthritis (Hakala et al. 1993; Renkema et al. 1998). One major site of synthesis of HC gp-39is the involuting mammary gland upon cessation of lactation (Morrison and Leder 1994). The 1434-nt sequence of YKL-39 cDNA predicts a 385-residue initial translation product and a 364-residue mature YKL-19. The amino acid sequence of YKL-39 is most closely related to YKL-40, followed by macrophage chitotriosidase, oviductal glycoprotein, and macrophage YM-1. These proteins share significant sequence identity with bacterial chitinases and have the probable structure of an $(\alpha\beta)_8$ barrel. The YKL-39 lacks the active site glutamate, which is essential for the activity of chitinases, and as expected has no chitinase activity. The highest level of YKL-39 mRNA expression is seen in chondrocytes, followed by synoviocytes, lung, and heart. YKL-39 accounts for 4 % of the protein in chondrocyte-conditioned medium, prostromelysin accounts for 17 %, and YKL-40 accounts for 33 %. In contrast to YKL-40, YKL-39 is not a glycoprotein and does not bind to heparin.

HC gp-39 and chitotriosidase are expressed in lipid-laden macrophages accumulated in various organs during Gaucher disease. In addition, these proteins can be induced with distinct kinetics in cultured macrophages. Studies depict remarkable phenotypic variation among macrophages present in the atherosclerotic lesion. Furthermore, chitotriosidase enzyme activity was shown to be elevated up to 55-fold in extracts of atherosclerotic tissue. Although a function for chitotriosidase and HC gp-39 has not been identified, it was hypothesized that these proteins play a role in cell migration and tissue remodeling during atherogenesis (Boot et al. 1999). However, little is known about the distribution of HC-gp39 and its role in fetal development (Ling and Recklies 2004).

19.5 Ym1 and Ym2: Murine Proteins

19.5.1 The Protein

Ym1 (or CH13L3) and Ym2 (or CH13L4) are murine chi-lectins, which are produced by macrophages, dendritic cells and mast cells. Ym1 is a murine protein secreted in large quantities by peritoneal macrophages upon nematode infection and is a marker for alternatively activated macrophages, which exert anti-inflammatory effects and promote wound healing, as well as combating parasitic infections. It is a single chain polypeptide (45 kDa) with a strong tendency to crystallize at its isoelectric point (pI 5.7). Upon nematode infection, murine peritoneal macrophages synthesize and secrete Ym1 protein, which is a functional marker for alternatively activated macrophages in T_{H2}-mediated inflammatory responses. Ym1 shares significant structural similarity to the family 18 chitinases. The function of Ym1 is not known, and there is conflicting evidence regarding both chitinase activity (weak or absent) and sugar-binding activity. SPR studies suggested that Ym1 exhibits calcium-independent binding at low pH to glucosamine (GlcN) and galactosamine (GalN), and especially to oligomers of these sugars, including heparin, but not to the unsubstituted or N-acetylated sugar. Treatment of macrophages and mast cells in vitro with IL-4 induced expression of Ym1 and Ym2 mRNA (HogenEsch et al. 2006). While Ym2 mRNA expression increased 976-fold, the mRNA of Ym1 increased 24-fold in the skin of cpdm/cpdm mice. Macrophages, dendritic cells and mast cells are cellular sources of Ym1 and Ym2 proteins. Eosinophils and neutrophils did not contain detectable concentrations of these proteins (Boot et al. 2005b; HogenEsch et al. 2006). Ym2 is another murine protein of unknown function closely related to Ym1. The Ym1 gene is mapped to a central region of mouse chromosome 3 (in the region syntenic to human chromosome 1p13) and that of another murine chitinase-like gene, Brp39, to a central region of mouse chromosome 1 (in the region syntenic to human chromosome 1q31).

Ym1 and Ym2 (Ym1/2) are induced by IL-4 in mouse bone marrow-derived mast cells and highly induced in bronchoalveolar lavage fluid (BALF) and the lung. Ym1/2 expression was completely inhibited by dexamethasone (Dex) in BALF and weakly inhibited in the lung. Although Dex pretreatment inhibited the Ym1/2 expression level in an animal model, it did not reduce IL-4 induction of Ym1/2 expression in primary cultured macrophages. It had no inhibitory effect on the phosphorylation level of STAT6 in macrophages. The inhibitory effect of Dex on Ym1/2 protein expression in the murine model of asthma does not involve the STAT6 signaling pathway (Lee et al. 2008).

Oral infections of mice with *Trichinella spiralis* induce activation of peritoneal exudate cells to transiently express and secrete Ym1. Co-expression of Ym1 with Mac-1 and scavenger receptor pinpoints macrophages as its main producer. A single open reading frame of 398 amino acids with a leader peptide (21 residues) typical of secretory protein has been deduced. Ym1 exhibits binding specificity to saccharides with a free amine group, such as GlcN, GalN, or GlcN polymers, but it does not bind to other saccharides. The interaction is pH-dependent but Ca^{2+} and Mg^{2+} ion-independent. The binding avidity of Ym1 to GlcN oligosaccharides was enhanced due to the clustering effect. Specific binding of Ym1 to heparin suggests that heparin/heparan sulfate may be its physiological ligand in vivo during inflammation and/or tissue remodeling. Although it shares 30 % homology with microbial chitinases, no chitinase activity was found associated with Ym1 (Chang et al. 2001).

Crystals discovered within the aged lung at the sites of chronic inflammation in a mouse model of chronic

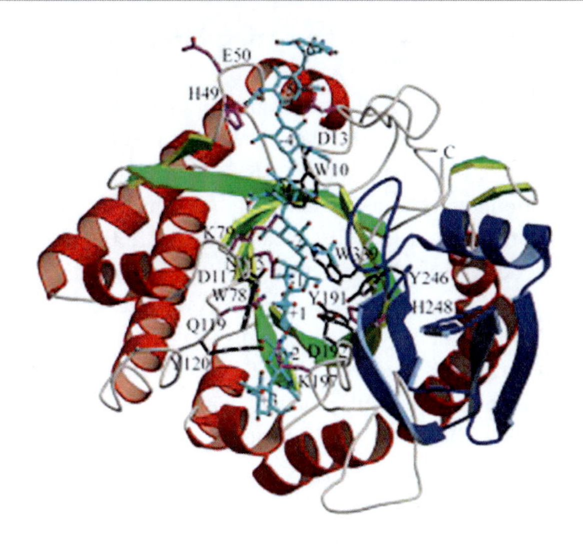

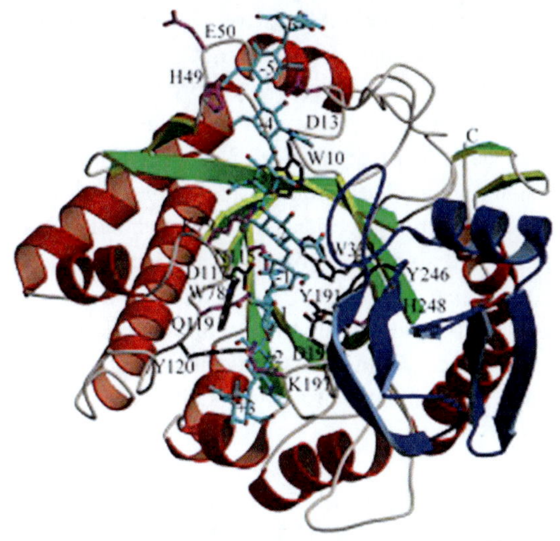

Fig. 19.3 Stereo view of the refined Ym1 structure with a model of N-acetylglucosamine$_9$ (NAG9). The NAG9 model was generated through superposition of the hCGP39–NAG8 complex, the SmChiA–NAG8 complex, and the SmChiB–NAG5 complex. The NAG9 model is labeled with the – 6 to +3 sugar-binding subsites and is shown as a stick drawing with cyan carbons. Helices in the TIM-barrel domain are colored in *red*, strands in *green*, and the additional α + β domain in *blue*. The conserved carbohydrate-interacting residues are shown in *magenta*, where non-conserved residues are shown in *black* (Reprinted with permission from Tsai et al. 2004 © Elsevier)

granulomatous disease, were identified as the chitinase-like protein Ym1. Ym1, found as a neutrophil granule protein, had weak β-N-acetylglucosaminidase activity, indicating that it might contribute to the digestion of glycosaminoglycans. Crystal formation is likely to be a function of excess neutrophil turnover at sites of inflammation in the chronic granulomatous disease mouse. Failure to remove subcutaneous Ym1 crystals injected into knockout mice indicates that a failure of digestion may also contribute to crystallization (Harbord et al. 2002).

19.5.2 Crystal Structure of Ym1

Co-crystallization and soaking experiments with various glucosamine or N-acetylglucosamine oligomers yield only the uncomplexed Ym1. The refined Ym1 structure at 1.31A resolution clearly displays a water cluster forming an extensive hydrogen bond network with the "active-site" residues (Fig. 19.3). This water cluster contributes notable electron density to lower resolution maps and this might have misled and given rise to a previous proposal for a monoglucosamine-binding site for Ym1. But crystals of Ym1 complexed with GlcN or GlcNAc polymers could not be obtained. A structural comparison of family 18 glycosidase like proteins reveals a lack of several conserved residues in Ym1, and illustrates the versatility of the divergent active sites. Therefore, Ym1 may lack N-acetylglucosamine-binding affinity, and this suggests that a new direction should be taken to unravel the function of Ym1 (Tsai et al. 2004).

Mouse Breast Regression Protein 39: Mouse breast regression protein 39 (BRP-39; CHI3l1) and its human homologue YKL-40 are chitinase-like proteins that lack chitinase activity. The biological properties of BRP-39/YKL-40 have only been rudimentarily defined. BRP-39$^{-/-}$ mice have markedly diminished antigen-induced T_{H2} responses and that epithelial YKL-40 rescued the T_{H2} responses in these animals. Mechanistic investigations demonstrated that BRP-39 and YKL-40 play an essential role in antigen sensitization and immunoglobulin E induction, stimulate dendritic cell accumulation and activation, and induce alternative macrophage activation. These proteins also inhibit inflammatory cell apoptosis/cell death while inhibiting Fas expression, activating protein kinase B/AKT, and inducing Faim 3. These proteins may prove therapeutic targets in T_{H2}- and macrophage-mediated disorders.

19.5.3 Oviductin

The high-molecular-weight oviductins, consisting of the amino-terminal 39-kDa catalytic domain followed by a heavily glycosylated serine/threonine-rich domain, are secreted by nonciliated oviductal epithelial cells and have been shown to play a role in fertilization and early embryo development (Buhi 2002). Among glycoside hydrolase family 18 proteins, it has a unique O-glycosylated mucin-like region following the chitinase-like domain. Carbohydrate binding has yet to be investigated in oviductin. The deduced amino acid sequence of an estrogen-dependent sheep oviductal glycoprotein (Mr 90,000–116,000) revealed the presence of several potential sites for glycan substitution on a protein backbone of Mr 66,500, and identity with chitinases. The oviductal glycoprotein contained N-acetylgalactosamine, N-acetylglucosamine, galactose, fucose, and sialic acid both in α2,3 and α2,6 linkages, typical of sialomucins. The oviductal glycoprotein was resistant to digestion with O-glycanase alone and chondroitinase ABC, with the latter indicating that it was not a proteoglycan. Results suggest that the secreted glycoprotein contains saccharide residues typical of sialomucins, and despite primary amino acid sequence identity, the oviductal glycoprotein does not share an enzymatic relationship with chitinases (DeSouza and Murray 1995).

19.6 Functions of CHI3L1 (YKL-40)

Although CHI3L1 was first identified in 1993 (Hakala et al. 1993), its biological function has been largely undetermined. CHI3L1 possesses a functional carbohydrate-binding motif which allows binding with a polymer or oligomer of GlcNAc, but CHI3L1 lacks enzymatic activity entirely. It is not yet known to have a specific receptor. Expression of YKL-40 appears to be induced by a change in the extracellular matrix environment, and is thought to have a role in development, tissue remodelling and inflammation. However, its pattern of expression is associated with pathogenic processes related to inflammation, extracellular tissue remodeling, fibrosis and solid carcinomas. It is assumed that YKL-40 plays a role in cancer cell proliferation, survival, invasiveness and in the regulation of cell-matrix interactions. YKL-40 is suggested as a marker associated with a poorer clinical outcome in genetically defined subgroups of different tumors (Kazakova and Sarafian 2009; Lee et al. 2011).

19.6.1 Role in Remodeling of Extracellular Matrix and Defense Mechanisms

Mammalian chitinase-like proteins have a postulated role in remodeling of extracellular matrix and defense mechanisms against chitin-containing pathogens. The expression of these proteins is increased in parasitic infections and allergic airway disease. YKL-40 has been indicated as a biomarker of sepsis (Hattori et al. 2009). YKL-40 is a growth factor for fibroblasts and has an anti-catabolic effect preserving extracellular matrix during tissue remodeling (Ling and Recklies 2004). YKL-40 stimulates migration and adhesion of endothelial cells and vascular smooth muscle cells (VSMCs), suggesting a role in angiogenesis (Malinda et al. 1999; Nishikawa and Millis 2003), and may play a role in regulating response of cancer cells to hypoxia (Saidi et al. 2008).

19.6.2 Growth Stimulating Effect

Since CHI3L regulates the growth of imaginal disc cells in *D.melanogaster*, it has been predicted that CHI3L1 can have a growth stimulating effect (Kawamura et al. 1999). Growth stimulating effects of CHI3L1 have been demonstrated on connective tissue cells (Recklies et al. 2002) and on endothelial cells (Malinda et al. 1999). CHI3L1 also stimulates angiogenesis and reorganization of vascular endothelial cells (Malinda et al. 1999). Insulin-like growth factor-1 works synergistically with CHI3L1 to enhance the response of human synovial cells isolated from patients with osteoarthritis (Recklies et al. 2002). In addition, CHI3L1 also promotes the activation of 2 major signaling pathways associated with mitogenesis and cell survival: MAPK (mitogen-activated protein kinase) pathways and PI-3 K (phosphoinositide 3-kinase)-mediated pathways in fibroblast cells. The purified human CHI3L1 efficiently phosphorylates MAPK p42/p44 in human synovial cells, fibroblasts, articular chondrocytes (Recklies et al. 2002) and human colonic epithelial cells (Eurich et al. 2009) in a dose-dependent manner. It has been suggested that G-protein-regulated MAPK networks are involved in the action of most non-nuclear oncogenes and subsequent carcinogenesis and tumor progression (Pearson et al. 2001). The networks are involved in the activation of MAPK p42/p44 which may enhance the carcinogenic change of epithelial cells during upregulated CHI3L1 expression under inflammatory conditions (Eurich et al. 2009).

19.7 Chi-Lectins As Markers of Pathogenesis

19.7.1 A Marker of Inflammation

The CHI3L1 plays a role in the pathological conditions leading to arthritis and tissue fibrosis (Recklies et al. 2002). Increased circulating levels of CHI3L1 have been reported in the serum of patients with several inflammatory conditions including IBD (Crohn's disease (CD) and ulcerative colitis (UC) (Vind et al. 2003), asthma (Chupp et al. 2007; Ober et al. 2008) and liver cirrhosis (Johansen et al. 1997). Serum CHI3L1 is rarely detectable in healthy individuals (Vind et al. 2003), and therefore CHI3L1 has been proposed as a useful marker for indicating inflammatory activity and poor clinical prognosis for IBD (Vind et al. 2003). CHI3L1 is strongly expressed by macrophages in the synovial membrane of patients with rheumatoid arthritis (RA) and a polarized IFNγ-mediated proinflammatory Th1-type immune response has been observed in half the patients with RA. CHI3L1 seems to be the cross-tolerance inducing protein in chronic arthritis which effectively downregulates the pathogenic immune responses. It is possible that nasal administration of

CHI3L1 represents an attractive approach for suppressing the clinical manifestation of chronic types of inflammation as well as autoimmune diseases.

Human cartilage GP39 (HC-gp39) has been implicated in arthritis, inflammatory bowel disease, tissue remodeling, fibrosis, and cancer (Hakala et al. 1993; Recklies et al. 2002; Johansen 2006). Prominent sites of HC-gp39 production are degenerate articular cartilage and inflamed or hyperplastic synovium, fibrotic liver tissue and gliomas, where a correlation of HC-gp39 production with malignancy has been reported. The presence of HC-gp39 at sites of inflammation has been described in several studies (Hormigo et al. 2006; Johansen et al. 1999a, 1999b), and the use of serum levels of this protein as a disease marker for the progression of joint erosion has been proposed. Cintin et al. (2002) have reported that serum levels of HC-gp39 correlate negatively with survival in patients with colorectal carcinomas; a similar correlation has been suggested for patients with breast carcinomas (Johansen et al. 1995). Similarly, the human chi-lectin YKL-39 (CHI3L2) and the murine Ym1 (CHI3L3/ECF-L) have been associated with the pathogenesis of arthritis (Tsuruha et al. 2002) and allergic airway inflammation, respectively (Chang et al. 2001; Ward et al. 2001; Homer et al. 2006).

Acidic mammalian chitinase (AMCase) and murine lectin Ym1 are upregulated in Th2-environment, and enzymatic activity of AMCase contributes to asthma pathogenesis. Chitotriosidase and YKL-40 reflect the macrophage activation in atherosclerotic plaques. Serum level of YKL-40 is a diagnostic and prognostic marker for numerous types of solid tumors. YKL-39 is a marker for the activation of chondrocytes and the progression of the osteoarthritis in human. The genetic contribution of YKL-40 gene to atopic susceptibility strongly suggests that the g.-247C/T polymorphism in the CHI3L1 promoter region is associated with the risk of atopy (Sohn and Wu 2009).

Expression in Osteoarthritis: Among numerous substances increasingly proposed for diagnostic purposes, very few may be considered as true disease markers in osteoarthritis (OA); COMP, antigenic keratin sulphate, hyaluronic acid, YKL-40, type III collagen N-propeptide and urinary glucosyl-galactosyl pyridinoline seem to be the most promising (Punzi et al. 2005). The presence of YKL-40 in cartilage and synovium in OA patients correlates with histopathological changes and may reflect local disease activity. In addition, the levels of YKL-40 in serum and synovial fluid also seem to correlate with disease severity. The functional role of YKL-40 is not yet clear, but its production as part of the inflammatory response in articular chondrocytes may modulate the cellular response to proinflammatory cytokines, acting to limit connective tissue degradation. Further elucidation of its roles and relationships may enable YKL-40 to act as a useful biomarker in the development of therapies for OA (Huang and Wu 2009).

Biomarkers for Asthma: Ober and Chupp (2009) provided an overview of genetic basis of asthma and immune-mediated diseases with polymorphisms in the genes encoding these proteins, CHIT1, CHI3L1, and CHIA, respectively. Polymorphisms in the CHIT1, CHIA, and CHI3L1 genes influence chitotriosidase enzyme activity, acid mammalian chitinase activity, and YKL-40 levels, respectively. Regulatory SNPs in CHI3L1 have been also associated with asthma, atopy, and immune-mediated diseases, and nonsynonymous SNPs in CHIA were associated with asthma. No CHIT1 polymorphisms, including a common nonfunctional 24-bp duplication allele, have been associated with asthma. These genes represent novel asthma susceptibility genes. Additional studies of this gene in populations of diverse ancestries are warranted (Ober et al. 2009). There is ample evidence to support an association of acidic mammalian chitinase (AMC) and YKL-40 with allergic bronchial asthma in patients. Recent studies in a mouse asthma model revealed that anti-inflammatory drugs like corticosteroid and cysteinyl leukotriene receptor antagonist were able to suppress elevated pulmonary levels of mammalian chitinases. Taken together, mammalian chitinases and chitinase-like proteins may be useful as biomarkers for asthma (Shuhui et al. 2009). Assessments of YKL-40 may allow physicians to more accurately diagnose and predict the course of asthma and thereby allow therapy to be appropriately tailored for a given patient (Hartl et al. 2009)

An overview of the chitinase and chitinase-like proteins, has been related to the genetic studies of asthma and immune-mediated diseases with polymorphisms in the genes encoding these proteins, CHIT1, CHI3L1, and CHIA, respectively. Polymorphisms in the genes of these proteins influence chitotriosidase enzyme activity, acid mammalian chitinase activity, and YKL-40 levels, respectively. Regulatory SNPs in CHI3L1 were also associated with asthma, atopy, and immune-mediated diseases, and nonsynonymous SNPs in CHIA were associated with asthma. No CHIT1 polymorphisms, including a common nonfunctional 24-bp duplication allele, have been associated with asthma. These genes represent novel asthma susceptibility genes (Ober and Chupp 2009).

Cardiovascular Diseases: Several inflammatory cytokines are involved in vascular inflammation resulting in endothelial dysfunction which is the earliest event in the atherosclerotic process leading to manifest cardiovascular disease. Human cartilage glycoprotein-39 (YKL-40) is secreted into circulation by macrophages, neutrophils, chondrocytes, vascular smooth muscle cells and cancer cells. The YKL-40 is a serum biomarker in diseases with fibrosis, inflammation and tissue remodelling. In contrast to C-reactive protein (CRP) produced in the liver in response to inflammation, YKL-40 is produced by lipid-laden macrophages inside the vessel wall. YKL-40 is an inflammatory glycoprotein involved in endothelial dysfunction by promoting chemotaxis, cell attachment and migration, reorganization and tissue remodeling as a response to endothelial damage. YKL-40 protein expression is seen in macrophages and smooth muscle cells in atherosclerotic plaques with the highest expression seen in macrophages in the early lesion of atherosclerosis. Several studies demonstrate that elevated serum YKL-levels are independently associated with the presence and extent of coronary artery disease and even higher YKL-40 levels are documented in patients with myocardial infarction. Moreover, elevated serum YKL-40 levels have also been found to be associated with all-cause as well as cardiovascular mortality. YKL-40 levels are elevated both in patients with type-1 and type-2 diabetes, known to be at high risk for the development of cardiovascular diseases, when compared to non-diabetic persons.

Atherosclerotic plaque macrophages express YKL-40, particularly macrophages that have infiltrated deeper into the lesion, and the highest YKL-40 mRNA expression is found in macrophages in early atherosclerotic lesion (Boot et al. 1999). Studies have suggested that high serum YKL-40 levels could be a prognostic biomarker of short survival. This is demonstrated in 80-year-old people and in patients with *Streptococcus pneumoniae* bacteraemia, and local or metastatic cancer (Johansen et al. 2009; Kastrup et al. 2009). Serum YKL-40 levels are elevated in patients with acute myocardial infarction (MI) ((Nøjgaard et al. 2008; Wang et al. 2008) and stable coronary artery disease (CAD),and associated with the number of diseased vessels assessed by coronary angiography. Elevated serum YKL-40 is a risk factor for acute coronary syndrome and death in patients with stable CAD (Kastrup et al. 2009).

The serum YKL-40 was increased in chronic heart failure (CHF), and YKL-40 detected high risk patients for adverse outcomes in CHF (Bilim et al. 2010; Johansen et al. 2010). The YKL-40 is elevated and associated with mortality in patients with stable coronary artery disease (CAD). The influence of statin treatment and lipid status has been studied on serum YKL-40 and Hs-CRP in patients with stable CAD. HsCRP, but not YKL-40, is associated with the cholesterol levels in statin treated patients. In general population, elevated plasma YKL-40 levels are associated with increased risk of ischemic stroke and ischemic cerebrovascular disease, independent of plasma CRP levels (Kjaergaard et al. 2010), indicating that YKL-40 could be a superior prognostic biomarker in patients with stable CAD, since it is independent of changes in cholesterol levels in both statin and non-statin treated patients (Mygind et al. 2011).

A positive association between elevated circulating YKL-40 levels and increasing levels of albumin-uria has been described in patients with type 1 diabetes indicating a role of YKL-40 in the progressing vascular damage resulting in microvascular disease. YKL-40 has been related to endothelial dysfunction, atherosclerosis, cardiovascular disease and diabetes and offers future perspectives in research (Catalán et al. 2011; Rathcke et al. 2009; Rathcke and Vestergaard 2009; Mathiasen et al. 2010). However, Plasma YKL-40 was identified as an obesity-independent marker of type 2 diabetes related to fasting plasma glucose and plasma IL-6 levels (Nielsen et al. 2008).

Neurological Diseases: Elevated levels of CHI3L1 (YKL-40) have been observed in the cerebrospinal fluid (CSF) of human and non-human primates with lentiviral encephalitis. Immunohistochemistry showed that CHI3L1 was associated with astrocytes. It was demonstrated that CHI3L1 is induced in astrocytes in a variety of neurological diseases but that it is most abundantly associated with astrocytes in regions of inflammatory cells (Bonneh-Barkay et al. 2010; Gaidamashvili et al. 2004; Nigro et al. 2005). Disease-modifying therapies for Alzheimer's disease (AD) would be most effective during the preclinical stage before significant neuronal loss occurs. Data demonstrate that YKL-40 is elevated in AD and, together with Aβ42, has potential prognostic utility as a biomarker for preclinical AD (Craig-Schapiro et al. 2010).

Respiratory Disorders: The exaggerated expression of YKL-40, the human homolog of BRP-39, has been reported in chronic obstructive pulmonary disease (COPD). We hypothesized and confirmed that BRP-39/YKL-40 plays an important role in the pathogenesis of cigarette smoke (CS)-induced emphysema. Studies demonstrate a regulatory role of BRP-39/YKL-40 in CS-induced inflammation and emphysematous destruction. Studies also highlight that maintaining the physiologic levels of YKL-40 in the lung will be therapeutically important to prevent excessive inflammatory responses or emphysematous alveolar destruction (Matsuura et al. 2011). Idiopathic pulmonary fibrosis (IPF) is a progressive interstitial lung disease that is hallmarked by fibrosis, inflammation and tissue remodelling. The -329 A/G polymorphism was associated with serum and BALF YKL-40 levels in IPF patients. High serum and BALF YKL-40 levels are associated with poor survival in IPF patients and could be useful prognostic markers for survival in IPF (Korthagen et al. 2011).

19.7.2 CHI3L1 as Biomarker in Solid Tumors

CHI3L1 or YKL-40 is expressed and secreted by several types of solid tumors including glioblastoma, colon cancer, breast cancer and malignant melanoma (Table 19.1). Although the exact function of CHI3L1 in inflammation and cancer is largely unknown, CHI3L1 plays a pivotal role in exacerbating the inflammatory processes and in promoting angiogenesis and remodeling of the extracellular matrix. CHI3L1 may be highly involved in the chronic engagement of inflammation which potentiates development of epithelial tumorigenesis presumably by activating the mitogen-activated protein kinase and the protein kinase B signaling pathways. Anti-CHI3L1 antibodies or pan-chitinase inhibitors may have the potential to suppress CHI3L1-mediated chronic inflammation and the subsequent carcinogenic change in epithelial cells (Eurich et al. 2009).

Clinical studies with different types of advanced tumors suggest that serum levels of CHI3L1 may be a new biomarker in cancer. In many cases, serum CHI3L1 provides independent information of survival. The highest serum CHI3L1 is detected in patients with advanced cancer and with the poorest prognosis. However, serum CHI3L1 cannot be used as a single screening test for cancer (Johansen et al. 2006). When evaluated with other prognostic factors of survival after recurrence of breast cancer, serum CHI3L1and serum LDH were the most significant independent factors. It was suggested that serum CHI3L1 may be of value in the follow-up of patients with breast cancer and in evaluating potential metastatic spread. CHI3L1 expression is strongly elevated in serum and biopsy material from glioblastomas patients. Though, the elevated plasma CHI3L1 is a biomarker of poor prognosis in cancer patients, but in general population, elevated plasma CHI3L1 predicts increased risk of gastrointestinal cancer and decreased survival after any cancer diagnosis (Johansen et al. 2009).

It has been strongly hypothesized that CHI3L1 plays a pivotal role as a growth stimulating factor for solid tumors or has a suppressive/protective effect in the apoptotic processes of cancer cells (Johansen et al. 2006) and inflammatory cells (Lee et al. 2009). In addition to CHI3L1 expression in a wide variety of human solid tumors, elevated levels of CHI3L1 in serum and/or plasma have been detected in patients with different types of solid tumors. Therefore, it is reasonable to predict that the serum level of CHI3L1 can be a reliable marker of progression of certain kinds of tumors and of a "bad prognosis" in patients with certain types of malignant tumors (Johansen et al. 2006).

The CHI3L1 can be used as a tumor marker for ovarian cancer (Høgdall et al. 2009), small cell lung cancer (Johansen et al. 2004), metastatic breast cancer (Johansen et al. 2009, 2010), and metastatic prostate cancer (Brasso et al. 2006). In addition, CHI3L1 is one of the most significant prognosis markers for cervical adenocarcinoma (Mitsuhashi et al. 2009), recurrent breast cancer (Johansen et al. 1995) and metastatic breast cancer (Jensen et al. 2003), as well as advanced stages of breast cancer (Coskun et al. 2007). Interestingly, the CHI3L1 serum level could be a useful and sensitive biomarker for recurrence in locally advanced breast carcinoma (Coskun et al. 2007), ovarian carcinoma (Gronlund et al. 2006), endometrial cancer (Diefenbach et al. 2007), squamous cell carcinoma of the head and neck (Roslind et al. 2007, 2008), metastatic prostate cancer or melanoma (Johansen et al. 2007; Schmidt et al. 2006), Hodgkin's lymphoma (Biggar et al. 2008) and colon cancer. Johansen and colleagues suggest that CHI3L1 may be used as a sensitive predictor of any cancer (Johansen

Table 19.1 *Expression of CHI3L1 (or YKL40) in solid tumors* (Eurich et al. 2009; Johansen et al. 2006)

Solid tumor	Location	Reference(s)
Glioma, Oligodendroglioma, glioblastoma	Brain	Tanwar et al. 2002; Nutt et al. 2005; Saidi et al. 2008; Bhat et al. 2008; Zhang et al. 2010
Squamous cell carcinoma of the head and neck	Head and neck	Johansen et al. 2006; Roslind et al. 2008.
Lung cancer (small cell carcinoma)	Lung	Junker et al. 2005
Breast cancer	Breast	Roslind et al. 2008; Qin et al. 2007; Johansen et al. 2006, 2008; Bhat et al. 2008, Junker et al. 2005; Cintin et al. 2002; Svane et al. 2007; Yamac et al. 2008
Hepatocellular carcinoma	Liver	Johansen et al. 2009
Colorectal cancer	Colon	Cintin et al. 2002; Bhat et al. 2008
Kidney tumor	Kidney	Bhat et al. 2008; Berntsen et al. 2008
Ovarian tumor, endometrial cancer	Ovary	Høgdall et al. 2009; Mitsuhashi et al. 2009; Coskun et al. 2007; Diefenbach et al. 2007; Peng et al. 2010; Zou et al. 2010
Primary prostate cancer, Metastatic prostate cancer	Prostate	Brasso et al. 2006; Johansen et al. 2007; Kucur et al. 2008
Papillary thyroid carcinoma, thyroid tumor	Thyroid	Huang et al. 2001
Extracellular myxoidchondrosarcoma	Bone	Sjögren et al. 2003
Multiple myeloma	Bone marrow	Mylin et al. 2006, 2009
Hodgkin's lymphoma	Lymph node	Biggar et al. 2008
Malignant melanoma	Melanocyte	Johansen et al. 2006; Schmidt et al. 2006
Myxoid liposarcoma	Fat cells	Sjögren et al. 2003

et al. 2009; Qin et al. 2007). They categorized patients into 5 distinct levels according to the amount of plasma CHI3L1 detected by ELISA, and found that participants with the highest level of plasma CHI3L1 had a median survival rate of only 1 year after the cancer diagnosis (Johansen et al. 2009). In contrast, the patients with the lowest level of plasma CHI3L1 had a survival rate of more than 4 years. Although the variation of CHI3L1 serum levels in healthy subjects in this study was relatively small, subsequent measurements would be required to determine cancer risk since the serum level of CHI3L1 could also be elevated in patients with other inflammatory diseases or autoimmune disorders (Johansen et al. 2008; van Aalten et al. 2001). From the results, it has been highly predicted that serum CHI3L1 levels seem to be a potential and promising biomarker for malignant tumors.

It is believed that IBD is a risk factor of cancer development based on the severity of the disease course. As previously reported, serum CHI3L1 concentration is elevated in patients with IBD (Vind et al. 2003) and primary colorectal cancer (Cintin et al. 2002). People with CD have a 5.6-fold increased risk of developing colon cancer; therefore screening for colon cancer by colonoscopy is strongly recommended for patients who have had CD for several years (Collins et al. 2006). Inflammation was recently recognized as an important factor in the pathogenesis of malignant tumors (Bromberg and Wang 2009).

Eurich et al. (2009) reviewed some examples of inflammation-mediated carcinogenesis and diseases which express CHI3L1 during the course of inflammation and the subsequent tumorigenesis. CHI3L1 protects cancer cells from undergoing apoptosis and also has an effect on extracellular tissue remodeling by binding specifically with collagen types I, II, and III (Bigg et al. 2006). Studies indicate that CHI3L1 is strongly associated with cell survival and cell migration during the drastic tissue remodeling processes by interacting with extracellular matrix components (Ling and Recklies 2004). The canonical (Wnt/β-catenin) pathway is known to play a crucial role in UC-related tumor progression (van Dekken et al. 2007). Reports strongly suggest that CHI3L1 may have a direct but not a secondary role for inflammation-associated tumorigenesis by continuously activating the Wnt/β-catenin canonical signaling pathway in CECs. As demonstrated, CHI3L1 expression is enhanced by proinflammatory cytokine interleukin-6 (De Ceuninck et al. 1998; Mizoguchi 2006), which also has a critical tumor-promoting effect during early colitis-associated cancer tumorigenesis (Grivennikov et al. 2009). The chitin-binding motif (CBM) of CHI3L1 is specifically associated with the CHI3L1-mediated activation of the Akt-signaling in CEC by transfecting the CBM-mutant CHI3L1 vectors in SW480 CECs. Downstream, CHI3L1 enhanced the secretion of IL-8 and TNFα in a dose-dependent manner. The 325 through 339 amino-acids in CBM are crucial for the biological function of CHI3L1 and is a critical region for the activation of Akt, IL-8 production, and for a specific cellular localization of CHI3L1. The activation may be associated with the development of chronic colitis (Chen et al. 2011).

19.7.3 Chitinase 3-Like-1 Protein (CHI3L1) or YKL-40 in Clinical Practice

Anti-Inflammatory Role of CHI3L1 (YKL-40): Chitinase 3-like-1 gene is significantly upregulated in inflamed colon of the dextran sulfate sodium-induced colitis model. CHI3L1 is mainly expressed in colonic epithelial cells and macrophages in the inflamed colon of dextran sulfate sodium-induced colitis. The CHI3L1 possesses an ability to enhance the adhesion and internalization of intracellular bacteria into colonic epithelial cells. In vivo neutralization of CHI3L1 significantly suppressed the development of dextran sulfate sodium-induced colitis by decreasing the bacterial adhesion and invasion into colonic epithelial cells. The CHI3L1 inhibited cellular responses to the inflammatory cytokines IL-1 and TNF-α. Stimulation of human skin fibroblasts or articular chondrocytes with IL-1 or TNF-α in presence of CHI3L1 resulted in a marked reduction of both p38 mitogen-activated protein kinase and stress-activated protein kinase/Jun N-terminal kinase phosphorylation, whereas nuclear translocation of NF-kB proceeded unimpeded. HC-gp39 suppressed the cytokine-induced secretion of MMP1, MMP3 and MMP13, as well as secretion of the chemokine IL-8 (Ling and Recklies 2004). Kawada et al. (2007) provided insight into the physiological role of mammalian chitinases in host/microbial interactions and suggested that inhibition of chitinase activity can be a novel therapeutic strategy of allergic and inflammatory disorder (Kawada et al. 2007). Unexpected roles have been identified for CHI3L1 in intestinal inflammation. While galectin-1 contributes to the suppression of intestinal inflammation by the induction of effector T cell apoptosis, in contrast, galectin-4 is involved in the exacerbation of this inflammation by specifically stimulating intestinal $CD4^+$ T cells to produce IL-6. CHI3L1 enhances the host/microbial interaction that leads to the exacerbation of intestinal inflammation (Mizoguchi and Mizoguchi 2007).

CHI3L1 plays a unique role during the development of intestinal inflammation, the protein is induced in both colonic lamina propria macrophages and CECs during the course of intestinal inflammation in experimental colitis models as well as in patients with IBD (Mizoguchi 2006). It has been suggested that a genetic defect against intracellular bacterial infection is strongly associated with the development of CD. CHI3L1 molecule particularly enhances the adhesion of chitin-binding protein-expressing bacteria to CECs through the conserved amino-acid residues (Kawada et al. 2008). Therefore, overexpression of CHI3L1 may be strongly associated with the intracellular bacterial adhesion and invasion on/into CECs in CD patients (Xavier and Podolsky 2007). In contrast, in an aseptic condition such as bronchial asthma, epithelium-expressing CHI3L1 seems to play a regulatory role by rescuing Th2-type immune responses (Lee et al. 2009). Further studies are required to understand the exact role of epithelium-expressing CHI3L1 in inflammatory conditions.

CHI3L1 was recently introduced into clinical practice, yet its application is still restricted (Kazakova and Sarafian 2009). CHI3L1 may participate in the innate immune response as it is regarded as an acute phase protein, since its plasma concentration is increased in inflammatory diseases (Chupp et al. 2007; Kronborg et al. 2002; Johansen 2006; Rathcke et al. 2006). The CHI3L1/YKL-40 is induced specifically during the course of inflammation in such disorders as inflammatory bowel disease, hepatitis and asthma and has been found to be either the cause or a biomarker for asthma. When the asthma subjects were stratified, serum CHI3L1 levels in the exacerbation group were higher than those in the stable and control groups (Chupp et al. 2007; Fontana et al. 2010). The CHI3L1 gene is abnormally expressed in the hippocampus of subjects with schizophrenia and may be involved in the cellular response to various environmental events that are reported to increase the risk of schizophrenia. The functional variants at the CHI3L1 locus influence the genetic risk of schizophrenia. CHI3L1 is a potential schizophrenia-susceptibility gene and suggest that the genes involved in the biological response to adverse environmental conditions are likely to play roles in the predisposition to schizophrenia (Zhao et al. 2007).

19.8 Evolution of Mammalian Chitinases (-Like) of GH18 Family

Homologous lectins to chitinases of the glycosyl hydrolase family 18 have been isolated from plants. Insects utilize multiple family 18 chitinolytic enzymes and also non-enzymatic chitinase-like proteins for degrading/remodeling/binding to chitin in different insect anatomical extracellular structures, such as the cuticle, peritrophic membrane, trachea and mouth parts during insect development, and possibly for other roles including chitin synthesis. Drosophila protein database revealed the presence of 18 chitinase-like proteins. Among these, seven are novel chitinase-like proteins that contain four signature amino acid sequences of chitinases, including both acidic and hydrophobic amino acid residues critical for enzyme activity (Zhu et al. 2004a). *D. melanogaster* imaginal disc growth factor proteins stimulate cell proliferation in common with mammalian chitinase-like protein YKL-40, but are only distantly related to mammalian glycoside hydrolase family 18 proteins, displaying around 25 % sequence identity. Chitinases may also affect gut physiology through their involvement in peritrophic membrane turnover. The chitinase from larvae of tomato moth (Lacanobia oleracea) shows 75–80 % identity with other Lepidopteran chitinases. L. oleracea chitinase caused chronic effects when fed, causing reductions in larval growth and food consumption by 60 % (Fitches et al. 2004). A chitinase-like protein from oyster, Clp1, which binds chitin and exhibits similar functional properties to YKL-40, has been described.

Gene duplication and loss according to a birth-and-death model of evolution is a feature of the evolutionary history of the GH18 family. Mammals are not known to synthesize chitin or metabolize it as a nutrient, yet the human genome encodes eight GH18 family members. Some GH18 proteins lack an essential catalytic glutamic acid and are likely to act as lectins rather than as enzymes. Both types of proteins widely occur in mammals although these organisms lack endogenous chitin. Molecular phylogenetic analyses suggest that both active chitinases (chitotriosidase and AMCase) result from an early gene duplication event. Further duplication events, followed by mutations leading to loss of chitinase activity, allowed evolution of chi-lectins. The homologous genes encoding chitinase (-like) proteins are clustered in two distinct loci that display a high degree of synteny among mammals. Despite the shared chromosomal location and high homology, individual genes have evolved independently. Orthologs are more closely related than paralogues, and calculated substitution rate ratios indicate that protein-coding sequences underwent purifying selection. Substantial gene specialization has occurred in time, allowing for tissue-specific expression of pH optimized chitinases and chi-lectins (Bussink et al. 2007). Bussink et al. (2007) reported that several family 18 chitinase-like proteins are present only in certain lineages of mammals, exemplifying evolutionary events in the chitinase protein family (Bussink et al. 2007).

The current human GH18 family likely originated from ancient genes present at the time of the bilaterian expansion (~550 M years). The family expanded in the chitinous protostomes *C. elegans* and *D. melanogaster*, declined in early deuterostomes as chitin synthesis disappeared, and expanded again in late deuterostomes with a significant increase in gene number after the avian/mammalian split. This comprehensive genomic study of animal GH18 proteins reveals three major phylogenetic groups in the family: chitobiases, chitinases/chitolectins, and stabilin-1 interacting chitolectins. Only the chitinase/chitolectin group is associated with expansion in late deuterostomes. Finding that the human GH18 gene family is closely linked to the human MHC paralogon on chromosome 1, together with the recent association of GH18 chitinase activity with Th2 cell inflammation, suggests that its late expansion could be related to an emerging interface of innate and adaptive immunity during early vertebrate history (Funkhouser and Aronson 2007).

References

Berntsen A, Trepiakas R, Wenandy L et al (2008) Therapeutic dendritic cell vaccination of patients with metastatic renal cell carcinoma: a clinical phase 1/2 trial. J Immunother 31:771–780

Bhat KP, Pelloski CE, Zhang Y, et al (2008) Selective repression of YKL-40 by NF-kB in glioma cell lines involves recruitment of histone deacetylase-1 and -2. FEBS Lett 582:3193–3200

Bigg HF, Wait R, Rowan AD, Cawston TE (2006) The mammalian chitinase-like lectin, YKL-40, binds specifically to type 1 collagen and modulates the rate of type 1 collagen fibril formation. J Biol Chem 281:21082–21095

Biggar RJ, Johansen JS, Smedby KE et al (2008) Serum YKL-40 and interleukin 6 levels in Hodgkin lymphoma. Clin Cancer Res 14:6974–6978

Bilim O, Takeishi Y et al (2010) Serum YKL-40 predicts adverse clinical outcomes in patients with chronic heart failure. J Card Fail 16:873–879

Bonneh-Barkay D, Wang G, Starkey A et al (2010) In vivo CHI3L1 (YKL-40) expression in astrocytes in acute and chronic neurological diseases. J Neuroinflammation 7:34

Boot RG, van Achterberg TA, van Aken BE et al (1999) Strong induction of members of the chitinase family of proteins in atherosclerosis: chitotriosidase and human cartilage gp-39 expressed in lesion macrophages. Arterioscler Thromb Vasc Biol 19:687–694

Boot RG, Blommaart EF, Swart E (2001) Identification of a novel acidic mammalian chitinase distinct from chitotriosidase. J Biol Chem 276:6770–6778

Boot RG, Bussink AP, Aerts JM (2005a) Human acidic mammalian chitinase erroneously known as eosinophil chemotactic cytokine is not the ortholog of mouse YM1. J Immunol 175:2041–2042

Boot RG, Bussink AP, Verhoek M et al (2005b) Marked differences in tissue-specific expression of chitinases in mouse and man. J Histochem Cytochem 53:1283–1292

Brasso K, Christensen IJ, Johansen JS et al (2006) Prognostic value of PINP, bone alkaline phosphatase, CTX-I, and YKL-40 in patients with metastatic prostate carcinoma. Prostate 66:503–513

Bromberg J, Wang TC (2009) Inflammation and cancer: IL-6 and STAT3 complete the link. Cancer Cell 15:79–80

Brunner JK, Scholl-Bürgi S, Hössinger D et al (2008) Chitotriosidase activity in juvenile idiopathic arthritis. Rheumatol Int 28:949–950

Buhi WC (2002) Characterization and biological roles of oviduct-specific, oestrogen-dependent glycoprotein. Reproduction 123:355–362

Bussink AP, van Eijk M, Renkema GH et al (2006) The biology of the Gaucher cell: the cradle of human chitinases. Int Rev Cytol 252:71–128

Bussink AP, Speijer D, Aerts JM et al (2007) Evolution of mammalian chitinase(-like) members of family 18 glycosyl hydrolases. Genetics 177:959–970

Catalán V, Gómez-Ambrosi J, Rodríguez A et al (2011) Icreased circulating and visceral adipose tissue expression levels of ykl-40 in obesity-associated type 2 diabetes are related to inflammation: impact of conventional weight loss and gastric bypass. J Clin Endocrinol Metab 96:200–209

Chang NC, Hung SI, Hwa KY et al (2001) A macrophage protein, Ym1, transiently expressed during inflammation is a novel mammalian lectin. J Biol Chem 276:17497–17506

Chen C-C, Llado V, Eurich K, et al (2011) Carbohydrate-binding motif in chitinase 3-like 1 (CHI3L1/YKL-40) specifically activates Akt signaling pathway in colonic epithelial cells. Clin Immunol 140:268–275

Chupp GL, Lee CG, Jarjour N et al (2007) A chitinase-like protein in the lung and circulation of patients with severe asthma. N Engl J Med 357:2016–2027

Cintin C, Johansen JS, Christensen IJ et al (2002) High serum YKL-40 level after surgery for colorectal carcinoma is related to short survival. Cancer 95:267–274

Collins PD, Mpofu C, Watson AJ, Rhodes JM (2006) Strategies for detecting colon cancer and/or dysplasia in patients with inflammatory bowel disease. Cochrane Database Syst Rev 336:CD000279

Coskun U, Yamac D, Gulbahar O et al (2007) Locally advanced breast carcinoma treated with neoadjuvant chemotherapy: are the changes in serum levels of YKL-40, MMP-2 and MMP-9 correlated with tumor response? Neoplasma 54:348–352

Craig-Schapiro R, Perrin RJ, Roe CM et al (2010) YKL-40: a novel prognostic fluid biomarker for preclinical Alzheimer's disease. Biol Psychiatry 68:903–912

De Ceuninck F, Pastoureau P, Bouet F et al (1998) Purification of guinea pig YKL40 and modulation of its secretion by cultured articular chondrocytes. J Cell Biochem 69:414–424

De Ceuninck F, Gaufillier S, Bonnaud A et al (2001) YKL-40 (cartilage gp-39) induces proliferative events in cultured chondrocytes and synoviocytes and increases glycosaminoglycan synthesis in chondrocytes. Biochem Biophys Res Commun 285:926–931

DeSouza MM, Murray MK (1995) An estrogen-dependent sheep oviductal glycoprotein has glycan linkages typical of sialomucins and does not contain chitinase activity. Biol Reprod 53: 1517–1526

Diefenbach CS, Shah Z, Iasonos A et al (2007) Preoperative serum YKL-40 is a marker for detection and prognosis of endometrial cancer. Gynecol Oncol 104:435–442

Eurich K, Segawa M, Toei-Shimizu S et al (2009) Potential role of chitinase 3-like-1 in inflammation-associated carcinogenic changes of epithelial cells. World J Gastroenterol 15:5249–5259

Fitches E, Wilkinson H, Bell H et al (2004) Cloning, expression and functional characterisation of chitinase from larvae of tomato moth (Lacanobia oleracea): a demonstration of the insecticidal activity of insect chitinase. Insect Biochem Mol Biol 34:1037–1050

Fontana RJ, Dienstag JL, Bonkovsky HL et al (2010) Serum fibrosis markers are associated with liver disease progression in non-responder patients with chronic hepatitis C. Gut 59:1401–1409

Funkhouser JD, Aronson NN Jr (2007) Chitinase family GH18: evolutionary insights from the genomic history of a diverse protein family. BMC Evol Biol 7:96

Fusetti F, Pijning T, Kalk KH et al (2003) Crystal structure and carbohydrate-binding properties of the human cartilage glycoprotein-39. J Biol Chem 278:37753–37760

Gaidamashvili M, Ohizumi Y, Iijima S et al (2004) Characterization of the yam tuber storage proteins from *Dioscorea batatas* exhibiting unique lectin activities. J Biol Chem 279:26028–26035

Grivennikov S, Karin E, Terzic J et al (2009) IL-6 and Stat3 are required for survival of intestinal epithelial cells and development of colitis-associated cancer. Cancer Cell 15:103–113

Gronlund B, Høgdall EV, Christensen IJ et al (2006) Pre-treatment prediction of chemoresistance in second-line chemotherapy of ovarian carcinoma: value of serological tumor marker determination (tetranectin, YKL-40, CASA, CA 125). Int J Biol Markers 21:141–148

Hakala BE, White C, Recklies AD (1993) Human cartilage gp-39, a major secretory product of articular chondrocytes and synovial cells, is a mammalian member of a chitinase protein family. J Biol Chem 268:25803–25810

Harbord M, Novelli M, Canas B et al (2002) Ym1 is a neutrophil granule protein that crystallizes in p47phox-deficient mice. J Biol Chem 277:5468–5475

Hartl D, Lee CG, Da Silva CA et al (2009) Novel biomarkers in asthma: chemokines and chitinase-like proteins. Curr Opin Allergy Clin Immunol 9:60–66

Hattori N, Oda S, Sadahiro T et al (2009) YKL-40 identified by proteomic analysis as a biomarker of sepsis. Shock 32:393–400

Henrissat B, Davies G (1997) Structural and sequence-based classification of glycoside hydrolases. Curr Opin Struct Biol 7:637–644

Hoell IA, Dalhus B, Heggset EB et al (2006) Crystal structure and enzymatic properties of a bacterial family 19 chitinase reveal differences from plant enzymes. FEBS J 273:4889–4900

Høgdall EV, Ringsholt M, Høgdall CK et al (2009) YKL-40 tissue expression and plasma levels in patients with ovarian cancer. BMC Cancer 9:8

HogenEsch H, Dunham A, Seymour R et al (2006) Expression of chitinase-like proteins in the skin of chronic proliferative dermatitis (cpdm/cpdm) mice. Exp Dermatol 15:808–814

Homer RJ, Zhu Z, Cohn L et al (2006) Differential expression of chitinases identify subsets of murine airway epithelial cells in allergic inflammation. Am J Physiol Lung Cell Mol Physiol 291:502–511

Hormigo A, Gu B, Karimi S et al (2006) YKL-40 and matrix metalloproteinase-9 as potential serum biomarkers for patients with high-grade gliomas. Clin Cancer Res 12:5698–5704

Houston DR, Recklies AD, Krupa JC et al (2003) Structure and ligandinduced conformational change of the 39-kDa glycoprotein from human articular chondrocytes. J Biol Chem 278:30206–30212

Huang K, Wu LD (2009) YKL-40: a potential biomarker for osteoarthritis. J Int Med Res 37:18–24

Huang Y, Prasad M, Lemon WJ et al (2001) Gene expression in papillary thyroid carcinoma reveals highly consistent profiles. Proc Natl Acad Sci USA 98:15044–15049

Jensen BV, Johansen JS, Price PA (2003) High levels of serum HER-2/neu and YKL-40 independently reflect aggressiveness of metastatic breast cancer. Clin Cancer Res 9:4423–4434

Johansen JS (2006) Studies on serum YKL-40 as a biomarker in diseases with inflammation, tissue remodelling, fibroses and cancer. Dan Med Bull 53:172–209

Johansen JS, Williamson MK, Rice JS et al (1992) Identification of proteins secreted by human osteoblastic cells in culture. J Bone Miner Res 7:501–512

Johansen JS, Cintin C, Jørgensen M et al (1995) Serum YKL-40: a new potential marker of prognosis and location of metastases of patients with recurrent breast cancer. Eur J Cancer 31A:1437–1442

Johansen JS, Møller S, Price PA et al (1997) Plasma YKL-40: a new potential marker of fibrosis in patients with alcoholic cirrhosis? Scand J Gastroenterol 32:582–590

Johansen JS, Baslund B, Garbarsch C et al (1999a) YKL-40 in giant cells and macrophages from patients with giant cell arthritis. Arthritis Rheum 42:2624–2630

Johansen JS, Stoltenberg M, Hansen M et al (1999b) Serum YKL-40 concentrations in patients with rheumatoid arthritis: relation to disease activity. Rheumatology (Oxford) 38:618–626

Johansen JS, Drivsholm L, Price PA, Christensen IJ (2004) High serum YKL-40 level in patients with small cell lung cancer is related to early death. Lung Cancer 46:333–340

Johansen JS, Jensen BV, Roslind A et al (2006) Serum YKL-40, a new prognostic biomarker in cancer patients? Cancer Epidemiol Biomarkers Prev 15:194–202

Johansen JS, Brasso K, Iversen P et al (2007) Changes of biochemical markers of bone turnover and YKL-40 following hormonal treatment for metastatic prostate cancer are related to survival. Clin Cancer Res 13:3244–3249

Johansen JS, Pedersen AN, Schroll M et al (2008) High serum YKL-40 level in a cohort of octogenarians is associated with increased risk of all-cause mortality. Clin Exp Immunol 151:260–266

Johansen JS, Bojesen SE, Mylin AK et al (2009) Elevated plasma YKL-40 predicts increased risk of gastrointestinal cancer and decreased survival after any cancer diagnosis in the general population. J Clin Oncol 27:572–578

Johansen JS, Bojesen SE, Tybjaerg-Hansen A, Nordestgaard BG et al (2010) Plasma YKL-40 and total and disease-specific mortality in the general population. Clin Chem 56:1580–1591

Junker N, Johansen JS, Andersen CB, Kristjansen PE (2005a) Expression of YKL-40 by peritumoral macrophages in human small cell lung cancer. Lung Cancer 48:223–231

Junker N, Johansen JS, Hansen LT et al (2005b) Regulation of YKL-40 expression during genotoxic or microenvironmental stress in human glioblastoma cells. Cancer Sci 96:183–190

Kastrup J, Johansen JS, Winkel P et al (2009) High serum YKL-40 concentration is associated with cardiovascular and all-cause mortality in patients with stable coronary artery disease. Eur Heart J 30:1066–1072

Kawada M, Hachiya Y, Arihiro A, Mizoguchi E (2007) Role of mammalian chitinases in inflammatory conditions. Keio J Med 56:21–27

Kawada M, Chen CC, Arihiro A et al (2008) Chitinase 3-like-1 enhances bacterial adhesion to colonic epithelial cells through the interaction with bacterial chitin-binding protein. Lab Invest 88:883–895

Kawamura K, Shibata T, Saget O et al (1999) A new family of growth factors produced by the fat body and active on Drosophila imaginal disc cells. Development 126:211–219

Kazakova MH, Sarafian VS (2009) YKL-40–a novel biomarker in clinical practice? Folia Med (Plovdiv) 51:5–14

Kjaergaard AD, Bojesen SE, Johansen JS, Nordestgaard BG (2010) Elevated plasma YKL-40 levels and ischemic stroke in the general population. Ann Neurol 68:672–680

Korthagen NM, van Moorsel CH, Barlo NP et al (2011) Serum and BALF YKL-40 levels are predictors of survival in idiopathic pulmonary fibrosis. Respir Med 105:106–113

Krause SW, Rehli M, Kreutz M et al (1996) Differential screening identifies genetic markers of monocyte to macrophage maturation. J Leukoc Biol 60:540–545

Kronborg G, Østergaard C, Weis N et al (2002) Serum level of YKL-40 is elevated in patients with Streptococcus pneumoniae bacteremia and is associated to the outcome of the disease. Scand J Infect Dis 34:323–326

Kucur M, Isman FK, Balci C et al (2008) Serum YKL-40 levels and chitotriosidase activity as potential biomarkers in primary prostate cancer and benign prostatic hyperplasia. Urol Oncol 26:47–52

Kzhyshkowska J, Mamidi S, Gratchev A et al (2006) Novel stabilin-1 interacting chitinase-like protein (SI-CLP) is up-regulated in alternatively activated macrophages and secreted via lysosomal pathway. Blood 107:3221–3228

Lee E, Jin M, Quan Z et al (2008) Effect of dexamethasone on STAT6-dependent Ym1/2 expression in vivo and in vitro. Biol Pharm Bull 31:1663–1666

Lee CG, Hartl D, Lee GR et al (2009) Role of breast regression protein 39 (BRP-39)/chitinase 3-like-1 in Th2 and IL-13-induced tissue responses and apoptosis. J Exp Med 206:1149–1166

Lee CG, Da Silva C, Dela Cruz CS et al (2011) Role of Chitin, Chitinase/Chitinase-Like Proteins in Inflammation, Tissue Remodeling, and Injury. Annu Rev Physiol 73:479–501

Li H, Greene LH (2010) Sequence and structural analysis of the chitinase insertion domain reveals two conserved motifs involved in chitin-binding. PLoS One 5:e8654

Ling H, Recklies AD (2004) The chitinase 3-like protein human cartilage glycoprotein 39 inhibits cellular responses to the inflammatory cytokines interleukin-1 and tumour necrosis factor-α. Biochem J 380:651–659

Malinda KM, Ponce L, Kleinman HK et al (1999) Gp38 k, a protein synthesized by vascular smooth muscle cells, stimulates directional migration of human umbilical vein endothelial cells. Exp Cell Res 250:168–173

Mathiasen AB, Henningsen KM, Harutyunyan MJ et al (2010) YKL-40: a new biomarker in cardiovascular disease? Biomark Med 4:591–600

Matsuura H, Hartl D, Kang MJ et al (2011) Role of Breast Regression Protein (BRP)-39 in the Pathogenesis of Cigarette Smoke-Induced Inflammation and Emphysema. Am J Respir Cell Mol Biol 44:777–786

Mitsuhashi A, Matsui H, Usui H et al (2009) Serum YKL-40 as a marker for cervical adenocarcinoma. Ann Oncol 20:71–77

Mizoguchi E (2006) Chitinase 3-like-1 exacerbates intestinal inflammation by enhancing bacterial adhesion and invasion in colonic epithelial cells. Gastroenterology 130:398–411

Mizoguchi E, Mizoguchi A (2007) Is the sugar always sweet in intestinal inflammation? Immunol Res 37:47–60

Mohanty AK, Singh G, Paramasivam M et al (2003) Crystal structure of a novel regulatory 40 kDa mammary gland protein (MGP-40) secreted during involution. J Biol Chem 278:14451–14460

Monzingo AF, Marcotte EM, Hart PJ et al (1996) Chitinases, chitosanases, and lysozymes can be divided into procaryotic and eucaryotic families sharing a conserved core. Nat Struct Biol 3:133–140

Morrison BW, Leder P (1994) Neu and ras initiate murine mammary tumors that share genetic markers generally absent in c-myc and int-2-initiated tumors. Oncogene 9:3417–3426

Mygind ND, Harutyunyan MJ, Mathiasen AB et al (2011) The influence of statin treatment on the inflammatory biomarkers YKL-40 and HsCRP in patients with stable coronary artery disease. Inflamm Res 60:281–287

Mylin AK, Rasmussen T, Johansen JS et al (2006) Serum YKL-40 concentrations in newly diagnosed multiple myeloma patients and YKL-40 expression in malignant plasma cells. Eur J Haematol 77:416–424

Mylin AK, Andersen NF, Johansen JS et al (2009) Serum YKL-40 and bone marrow angiogenesis in multiple myeloma. Int J Cancer 124:1492–1494

Nielsen AR, Erikstrup C, Johansen JS et al (2008) Plasma YKL-40: a BMI-independent marker of type 2 diabetes. Diabetes 57:3078–3082

Nigro JM, Misra A, Zhang L et al (2005) Integrated array-comparative genomic hybridization and expression array profiles identify clinically relevant molecular subtypes of glioblastoma. Cancer Res 65:1678–1686

Nishikawa KC, Millis AJT (2003) Gp38 k (CHI3L1) is a novel adhesion and migration factor for vascular cells. Exp Cell Res 287:79–87

Nøjgaard C, Høst NB, Christensen IJ et al (2008) Serum levels of YKL-40 increases in patients with acute myocardial infarction. Coron Artery Dis 19:257–263

Nutt CL, Betensky RA, Brower MA et al (2005) YKL-40 is a differential diagnostic marker for histologic subtypes of high-grade gliomas. Clin Cancer Res 11:2258–2264

Ober C, Chupp GL (2009) The chitinase and chitinase-like proteins: a review of genetic and functional studies in asthma and immune-mediated diseases. Curr Opin Allergy Clin Immunol 9:401–408

Ober C, Tan Z, Sun Y, Possick JD et al (2008) Effect of variation in CHI3L1 on serum YKL-40 level, risk of asthma, and lung function. N Engl J Med 358:1682–1691

Pearson G, Robinson F, Beers Gibson T et al (2001) Mitogen-activated protein (MAP) kinase pathways: regulation and physiological functions. Endocr Rev 22:153–183

Peng C, Peng J, Jiang L et al (2010) YKL-40 protein levels and clinical outcome of human endometrial cancer. J Int Med Res 38:1448–1457

Punzi L, Oliviero F, Plebani M (2005) New biochemical insights into the pathogenesis of osteoarthritis and the role of laboratory investigations in clinical assessment. Crit Rev Clin Lab Sci 42:279–309

Qin W, Zhu W, Schlatter L et al (2007) Increased expression of the inflammatory protein YKL-40 in precancers of the breast. Int J Cancer 121:1536–1542

Rathcke CN, Vestergaard H (2009) YKL-40–an emerging biomarker in cardiovascular disease and diabetes. Cardiovasc Diabetol 8:61

Rathcke CN, Johansen JS, Vestergaard H (2006) YKL-40, a biomarker of inflammation, is elevated in patients with type 2 diabetes and is related to insulin resistance. Inflamm Res 55:53–59

Rathcke CN, Persson F, Tarnow L et al (2009) YKL-40, a marker of inflammation and endothelial dysfunction, is elevated in patients with type 1 diabetes and increases with levels of albuminuria. Diabetes Care 32:323–328

Recklies AD, White C, Ling H (2002) The chitinase 3-like protein human cartilage glycoprotein 39 (HC-gp39) stimulates proliferation of human connective-tissue cells and activates both extracellular signal-regulated kinase- and protein kinase B-mediated signalling pathways. Biochem J 365:119–126

Rehli M, Krause SW, Andreesen R (1997) Molecular characterization of the gene for human cartilage gp-39 (CHI3L1), a member of the chitinase protein family and marker for late stages of macrophage differentiation. Genomics 43:221–225

Rehli M, Niller HH, Ammon C et al (2003) Transcriptional regulation of CHI3L1, a marker gene for late stages of macrophage differentiation. J Biol Chem 278:44058–44067

Renkema GH, Boot RG, Au FL et al (1998) Chitotriosidase, a chitinase, and the 39-kDa human cartilage glycoprotein, a chitin-binding lectin, are homologues of family 18 glycosyl hydrolases secreted by human macrophages. Eur J Biochem 251:504–509

Roslind A, Johansen JS, Junker N et al (2007) YKL-40 expression in benign and malignant lesions of the breast: a methodologic study. Appl Immunohistochem Mol Morphol 15:371–381

Roslind A, Johansen JS, Christensen IJ et al (2008) High serum levels of YKL-40 in patients with squamous cell carcinoma of the head and neck are associated with short survival. Int J Cancer 122:857–863

Saidi A, Javerzat S, Bellahcene A et al (2008) Experimental anti-angiogenesis causes upregulation of genes associated with poor survival in glioblastoma. Int J Cancer 122:2187–2198

Schmidt H, Johansen JS, Gehl J et al (2006) Elevated serum level of YKL-40 is an independent prognostic factor for poor survival in patients with metastatic melanoma. Cancer 106:1130–1139

Shuhui L, Mok YK, Wong WS (2009) Role of mammalian chitinases in asthma. Int Arch Allergy Immunol 149:369–377

Sjögren H, Meis-Kindblom JM, Orndal C et al (2003) Studies on the molecular pathogenesis of extraskeletal myxoid chondrosarcoma-cytogenetic, molecular genetic, and cDNA microarray analyses. Am J Pathol 162:781–792

Sohn MH, Lee JH, Kim KW et al (2009) Genetic variation in the promoter region of chitinase 3-like 1 is associated with atopy. Am J Respir Crit Care Med 179:449–456

Sun YJ, Chang NC, Hung SI et al (2001) The crystal structure of a novel mammalian lectin, Ym1, suggests a saccharide binding site. J Biol Chem 276:17507–17514

Suzuki K, Taiyoji M, Sugawara N et al (1999) The third chitinase gene (chiC) of Serratia marcescens 2170 and the relationship of its product to other bacterial chitinases. Biochem J 343:587–596

Suzuki M, Fujimoto W, Goto M et al (2002) Cellular expression of gut chitinase mRNA in the gastrointestinal tract of mice and chickens. J Histochem Cytochem 50:1081–1089

Svane IM, Pedersen AE et al (2007) Vaccination with p53 peptide-pulsed dendritic cells is associated with disease stabilization in patients with p53 expressing advanced breast cancer; monitoring of serum YKL-40 and IL-6 as response biomarkers. Cancer Immunol Immunother 56:1485–1499

Tanwar MK, Gilbert MR, Holland EC (2002) Gene expression microarray analysis reveals YKL-40 to be a potential serum marker for malignant character in human glioma. Cancer Res 62:4364–4368

Terwisscha van Scheltinga AC, Hennig M, Dijkstra BW (1996) The 1.8 Å resolution structure of hevamine, a plant chitinase/lysozyme, and analysis of the conserved sequence and structure motifs of glycosyl hydrolase family 18. J Mol Biol 262:243–257

Tsai ML, Liaw SH, Chang NC (2004) The crystal structure of Ym1 at 1.31 A resolution. J Struct Biol 148:290–296

Tsuruha J, Masuko-Hongo K, Kato T et al (2002) Autoimmunity against YKL-39, a human cartilage derived protein, in patients with osteoarthritis. J Rheumatol 29:1459–1466

Ujita M, Sakai K, Hamazaki K et al (2003) Carbohydrate binding specificity of the recombinant chitin-binding domain of human macrophage chitinase. Biosci Biotechnol Biochem 67: 2402–2407

Vaaje-Kolstad G, Vasella A, Peter MG, Netter C, Houston DR et al (2004) Interactions of a family 18 chitinase with the designed inhibitor Hm508 and its degradation product, chitobiono-δ-lactone. J Biol Chem 279:3612–3619

van Aalten DM, Komander D, Synstad B et al (2001) Structural insights into the catalytic mechanism of a family 18 exochitinase. Proc Natl Acad Sci USA 98:8979–8984

van Dekken H, Wink JC, Vissers KJ et al (2007) Wnt pathway-related gene expression during malignant progression in ulcerative colitis. Acta Histochem 109:266–272

van Eijk M, van Roomen CP, Renkema GH et al (2005) Characterization of human phagocyte-derived chitotriosidase, a component of innate immunity. Int Immunol 17:1505–1512

Varela PF, Llera AS, Mariuzza RA, Tormo J (2002) Crystal structure of imaginal disc growth factor-2. A member of a new family of growth-promoting glycoproteins from Drosophila melanogaster. J Biol Chem 277:13229–13236

Vind I, Johansen JS, Price PA, Munkholm P (2003) Serum YKL-40, a potential new marker of disease activity in patients with inflammatory bowel disease. Scand J Gastroenterol 38:599–605

Volck B, Price PA, Johansen JS et al (1998) YKL-40, a mammalian member of the chitinase family, is a matrix protein of specific granules in human neutrophils. Proc Assoc Am Physicians 110:351–360

Wang Y, Ripa SR, Johansen JS et al (2008) YKL-40 a new biomarker in patients with acute coronary syndrome or stable coronary artery disease. Scand Cardiovasc J 42:295–302

Ward JM, Yoon M, Anver MR et al (2001) Hyalinosis and Ym1/Ym2 gene expression in the stomach and respiratory tract of 129 S4/SvJae and wild-type and CYP1A2-null B6, 129 mice. Am J Pathol 158:323–332

Xavier RJ, Podolsky DK (2007) Unravelling the pathogenesis of inflammatory bowel disease. Nature 448:427–434

Yamac D, Ozturk B, Coskun U (2008) Serum YKL-40 levels as a prognostic factor in patients with locally advanced breast cancer. Adv Ther 25:801–809

Zhang W, Murao K, Zhang X (2010) Resveratrol represses YKL-40 expression in human glioma U87 cells. BMC Cancer 10:593

Zhao X, Tang R, Gao B et al (2007) Functional variants in the promoter region of Chitinase 3-like 1 (CHI3L1) and susceptibility to schizophrenia. Am J Hum Genet 80:12–18

Zhu Q, Deng Y, Vanka P et al (2004a) Computational identification of novel chitinase-like proteins in the Drosophila melanogaster genome. Bioinformatics 20:161–169

Zhu Z, Zheng T, Homer RJ et al (2004b) Acidic mammalian chitinase in asthmatic Th2 inflammation and IL-13 pathway activation. Science 304:1678–1682

Zou L, He X, Zhang JW (2010) The efficacy of YKL-40 and CA125 as biomarkers for epithelial ovarian cancer. Braz J Med Biol Res 43: 1232–1238

Z-u H, Dalal P, Aronson NN Jr, Madura JD (2007) Family 18 Chitolectins: Comparison of MGP40 and HUMGP39. Biochem Biophys Res Commun 359:221–226

Novel Groups of Fuco-Lectins and Intlectins 20

Rajesh K. Gupta and G.S. Gupta

20.1 F-type Lectins (Fuco-Lectins)

A novel lectin family "F-type" constituted by a large number of proteins exhibit greater multiples of F-type motif, either tandemly arrayed or in mosaic combinations with other domains, including a putative transmembrane receptor that suggests an extensive functional diversification of this lectin family. F-type domains are found in proteins from a range of organisms from bacteria to vertebrates, but exhibit patchy distribution across different phylogenetic taxa suggesting that F-type lectin genes have been selectively lost even between closely related lineages, and making it difficult to trace the ancestry of the F-type domain. F-type lectins were first characterized in eels, which are teleost fish. Teleost fish species commonly have a number of F-type lectins (containing multiple tandem F-type domains) resulting from gene duplication events that have occurred independently in different teleost fish lineages. The F-type domain has clearly gained functional value in fish, whereas the fate of F-type domains in higher vertebrates is not clear, rather it has become defunct. While the functions of many F-type lectins have not yet been characterized, a pathogen-recognition role has been established for F-type lectin-like proteins in both invertebrates (e.g. the horseshoe crab *Tachypleus tridentatus*) and vertebrates (e.g. the Japanese eel *Anguilla japonica*). The F-type lectin fold from *Anguilla anguilla* agglutinin (AAA) (Bianchet et al. 2002) is widely distributed in proteins from horseshoe crabs to amphibians, with some representatives proposed to play a role in immunity (Honda et al. 2000; Saito et al. 1997). Although AAA and the rat MBL specifically recognize terminal L-fucose, they do so through differing binding interactions (Ng et al. 1996), which illustrate convergent functional properties among unrelated lectin families. Interestingly, the F-type fold is also shared with several proteins of distinct functional properties, including C-terminal domain of blood coagulation factors V and VIII, C-terminal domain of bacterial sialidases, N-terminal domain of fungal galactose oxidase, a human ubiquitin ligase subunit APC10/DOC1, a domain of single-strand DNA repair protein XRCC1, the b1 domain of neuropilin, and yeast allantoicase (Odom and Vasta 2006) pointing towards an ancient origin for F-type fold. The structural analogy of these seemingly unrelated proteins is indicative of the archaic origin of this lectin fold. Proteins from fish and amphibians with a range of domain organizations contain F-type domains, most of which are likely to bind fucose based on the conservation of key residues. The single-domain F-type lectins found in eels, which represent an early branch of teleost fish, are generally not found in modern teleosts, having been replaced by tandem two-domain proteins in species including striped bass (*Morone saxatilis*), zebrafish (*Danio rerio*), pufferfish (*Tetraodon nigroviridans* and *Fugu rubripes*) and stickleback (*Gasterosteus aculeatus*). In two-domain F-type lectins, both domains feature the fucose and Ca^{2+}-binding residues, but the N-terminal domain lacks one inter-strand disulfide bond, and the C-terminal domain lacks the adjacent cysteines in loop 4, which may affect sugar binding. In four-domain tandem F-type lectins, which are exclusive to trout species (e.g. *Oncorhynchus mykiss*), the first and second domains lack the adjacent cysteines, and the second domain has strikingly lost all three residues from His/Arg/Arg triad, suggesting that fucose binding has been lost in this domain.

20.1.1 F-type Lectins in Mammalian Vertebrates

20.1.1.1 Characterization

The fate of F-type domains in higher vertebrates is not clear. Two genes encoding three-domain F-type proteins are predicted in the genome of the opossum (*Monodelphis domestica*), an early-branching mammal. The genes are present in tandem within a genomic region that is absent in placental mammals and in birds, but which in *Xenopus tropicalis* contains a gene encoding a three-domain F-type lectin ~40 % identical to the opossum sequences. In humans

G.S. Gupta et al., *Animal Lectins: Form, Function and Clinical Applications*,
DOI 10.1007/978-3-7091-1065-2_20, © Springer-Verlag Wien 2012

this region would lie within chromosome 12, between the conserved genes *Wbp11* and *Foxj2*. Some remnants of F-type lectin genes are evident in this region of human genome. Human macrophages express a membrane lectin, or sugar-specific receptor, which specifically mediates the binding and endocytosis of mannose- and fucose-terminated glycoproteins and is involved in the phagocytosis of pathogens (Leu et al. 1985). A similar lectin activity was sought on cultured human DC (Avraméas et al. 1996).

A lectin with a high affinity for fucose residues has been purified from rat liver. The rat hepatic fucose binding protein (FBP) contains two polypeptide subunits with M_r of 88 kDa and 77 kDa. Peptide maps of the subunits, however, showed that they are structurally very similar but distinct from other rat hepatic lectins (Lehrman et al. 1986). The hepatic FBP has a high affinity for Fuc-BSA and galactosyl-BSA but a low affinity for N-acetylglucosaminyl-BSA. In contrast, the mannose/N-acetylglucosamine lectin binds N-acetylglucosaminyl-BSA and Fuc-BSA but not galactosyl-BSA (Lehrman and Hill 1986). The interaction between a rat alveolar macrophage lectin (M_r of 180 kDa) and its ligands is dependent on Ca^{2+} over a optimal pH range. The apparent K_D for fucosyl bovine serum albumin is 20.4×10^{-10} M. D-Mannose, L-fucose, and N-acetyl-D-glucosamine were the most effective inhibitors, and D-galactose was the least. The lectin isolated from alveolar macrophages is widely distributed in other rat tissues. Hepatocytes are devoid of this lectin, but hepatic Kupffer cells and endothelial cells contain significant amounts (Haltiwanger and Hill 1986). Of all rat tissues examined, only liver contained the fucose-binding lectin, whereas both liver and blood serum contained the mannose/N-acetylglucosamine lectin. The fucose-binding lectin was neither responsible for the uptake nor more than one lectin was acting. With the identification of another lectin (Mr = 180 kDa) only two lectins appeared to be involved. The hepatic mannose/N-acetylglucosamine lectin had a higher affinity for L-fucosyl-bovine serum albumin than the majority of the lectin in hepatocytes. This lectin, called the high affinity form, showed K_D of 2.3×10^9 for L-fucosyl-bovine serum albumin compared to 3.5×10^8 for the normal form. The two forms, however, had identical molecular weights (32 kDa) under reducing and nonreducing conditions and produced identical peptide maps after protease digestion (Haltiwanger et al. 1986).

20.1.1.2 Fucose Binding Proteins/Lectins in Fertilization

Bolwell et al. (1979) suggested that fertilization in *Fucus serratus* is based on an association between fucosyl- and mannosyl-containing ligands on the egg surface and specific carbohydrate-binding receptors on the sperm surface like *Fucus serratus,* the fucose-binding proteins (FBP) have been identified in mammamlian spermatozoa. In boar spermatozoa, during ionophore induced early stage acrosome reaction (AR), the FBP is first localized at the border between equatorial segment and anterior acrosome. With the propagation of AR the FBP was dramatically expressed and visible over the entire surface of acrosome and equatorial segment. Results suggested that FBP is responsible for the specific binding of the ghost-sperm unit to the zona pellucida (Friess et al. 1987). Oviductal sperm in cattle (Bos Taurus) revealed a prominent FBP of 16.5 kDa, which was labeled over live uncapacitated sperm acrosome, while capacitated sperm did not. The presence of a Le^a-binding protein with an apparent mass of 16.5 kDA appeared to originate from seminal plasma (Ignotz et al. 2001).

20.2 F-type Lectins in Fish and Amphibians

20.2.1 Anguilla Anguilla Agglutinin (AAA)

The fucose-binding lectins are secretory proteins and exhibit only weak similarities to frog pentraxin, horseshoe crab tachylectin-4, and fly fw protein. There are at least seven closely related members whose messages are abundantly expressed in liver and in significant levels in gill and intestine. The serum fucolectins are derived from liver. The message levels were increased by lipopolysaccharide, suggesting a role for fucolectins in host defense. Eel fucolectins have a SDS-resistant tetrameric structure consisting of two disulfide-linked dimers (Honda et al. 2000). Eel serum lectins have been useful as anti-H hemagglutinins and also in lectin histochemistry as fucose-binding lectins (fucolectins). This domain of unknown function has been identified in Japanese eel (Anguilla japonica) fucolectins and at least one frog pentraxin. The AAA acts as a defensive agent, recognizes blood group fucosylated oligosaccharides including A, B, H and Lewis B-type antigens, but does not recognize Lewis A antigen. AAA has low affinity for monovalent haptens.

20.2.1.1 AAA: A Homotrimer with Three Fucose Binding Sites

AAA is a non-covalent homotrimer in which all three fucose binding sites are oriented in the same direction. This arrangement resembles that of the C-type CRDs in mannose binding protein and may confer high affinity for pathogen surfaces displaying repetitive arrays of oligosaccharide ligands. The AAA F-type CRD has a β-barrel structure, with one three-stranded β-sheet and one five-stranded sheet, connected by two disulfide bonds. One end of the barrel features five loop regions (here termed loops 1–5) that form a ring enclosing the site of fucose binding, which is a positively-charged hollow (Fig. 20.1). A Ca^{2+} ion is bound within a sub-domain that largely lacks regular

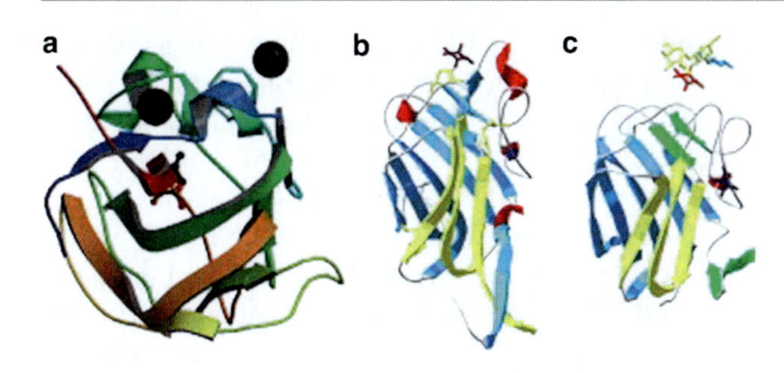

Fig. 20.1 (**a**) Structure of Anguilla anguilla agglutinin asymmetric unit (PDB ID 1 K12); (**b**) Fucose Binding lectin with bound fucose: β-strands in three-stranded sheet are colored *yellow* and those in the five-stranded sheet are colored *blue*. 310 helices are colored *red*. The fucose ligand is shown in *purple* and the Ca^{2+} ion in *blue* (PDB ID: 1 K12); (**c**) Structure of F-type domain 1 from *S. pnemoniae* SP2159 with bound blood group A tetrasaccharide. B-strands in the three-stranded sheet are colored *yellow*, those in the five-stranded sheet are colored *blue*, and those in other substructures are colored *green*. The 310 helix is colored red. Ca^{2+} is shown in *blue*. The ligand is colored by monosaccharide: *yellow*, Gal/GalNac; *blue*, Glc; *red*, Fuc (PDB ID: 2J1U)

secondary structure, and stabilization of tertiary or quaternary structure by Ca^{2+} may enhance sugar binding activity in AAA and other F-type lectins. Within the positively-charged hollow, side chains from a triad of residues characteristic of fucose binding make hydrogen bonds to the 3-OH, 4-OH and ring oxygen of fucose. In AAA these residues—a histidine in loop 3 and two arginine residues in loop 4—are present in the fucose binding sequence motif **H**. . .**R**GDCCGE**R**, which is found in other F-type domains in more general form **H**. . .**R**XDXXXX(**R/K**), where the gap between **H** and **R** is around 26 residues. Carbon atoms C1 and C2 of the fucose ring make van der Waals contacts with the adjacent disulfide-bonded cysteine residues in loop 4 (**H**. . .**R**GDCCGE**R**), and the methyl group (C6) is accommodated in a hydrophobic pocket formed by side chains from loops 1 and 2. Other monosaccharide ligands (e.g. 3-O-methyl-D-galactose, 3-O-methyl-D-fucose) are bound through the same set of interactions. Oligosaccharide ligands, which include the blood group antigens H and Le^a, interact with AAA primarily through the terminal fucose, but non-terminal saccharide residues also interact with a subset of the residues involved in fucose binding as well as additional residues from loop regions 1, 2 and 4. In the seven fucolectins from the Japanese eel (*Anguilla japonica*), various combinations of residue substitutions in loops 1 and 2 open up the methyl binding pocket and/or allow for polar interactions, which may adjust the oligosaccharide binding specificities of these lectins and tune them to recognize different pathogens. Non-eel F-type domains have a shorter loop 1, which may remove some interaction with oligosaccharide ligands and broaden oligosaccharide specificity.

20.2.1.2 Glycans Interacting AAA

AAA is suggested to be associated with innate immunity by recognizing disease-associated cell surface glycans such as bacterial LPSs, and has been widely used as a reagent in hematology and glycobiology. Among the glycans tested, AAA reacted well with nearly all human blood group A_h (GalNAcα1 → 3[L Fucα1 → 2]Gal), B_h (Galα1 → 3[L Fucα1 → 2]Gal), (H LFucα1 → 2Gal) and Le^b (Fucα1 → 2 Galβ1 → 3[Fucα1 → 4]GlcNAc) active glycoproteins (gps), but not with blood group Le^a (Galβ1 → 3[Fucα1 → 4] GlcNAc) substances, suggesting that residues and optimal density of α1-2 linked LFuc to Gal at the non-reducing end of glycoprotein ligands are essential for lectin-carbohydrate interactions. Blood group precursors, Galβ1-3GalNAc (T), GalNAcα1-Ser/Thr (Tn) containing glycoproteins and *N*-linked plasma gps, gave only negligible affinity. Among the mammalian glycotopes tested, A_h, B_h and H determinants were the best, being about 5 to 6.7 times more active than LFuc, but were weaker than *p*-nitrophenylαFuc indicating that hydrophobic environment surrounding the LFuc moiety enhance the reactivity. The hierarchy of potency of oligo- and mono-saccharides can be ranked as follows: *p*-nitrophenyl-αFuc > A_h, B_h and H > LFuc > LFucα1 → 2 Galβ1 → 4Glc (2′-FL) and Galβ1 → 4[L Fucα1 → 3]Glc (3′-FL), while LNDFH I (Le^b hexa-), Le^a, Le^x (Galβ1 → 4 [Fucα1 → 3]GlcNAc), and LDFT (gluco-analogue of Le^y) were inactive. From the present observations, it can be concluded that the combining site of AAA should be a small cavity-type capable of recognizing mainly H/crypto H and of binding to specific polyvalent ABH and Le^b glycotopes (Wu et al. 2004).

20.2.1.3 Comparison of AAA with MBL and UEA-1

MBL from eel serum has a M_r of ~246 kDa and is composed of identical subunits of ~24 kDa, two of each were always covalently linked. AAA, a FBL from eel serum, had a M_r of about 121 kDa and consisted four subunits of 30 kDa, which was made of two identical subunits of 15 kDa. Upon isoelectric focusing MBL displays four bands ranging from pH 4.8 to 5.2. FBL shows 17–20 bands between pH 5.5 and 6.2. Hemagglutination activity of MBL was inhibited only by mannan, whereas FBL activity was inhibited by several glycosubstances. MBL and FBL activity was constant between pH 4–5 and 10. Temperatures above 55º C totally destroyed MBL activity whereas FBL activity remained constant up to 75 º C (Gercken and Renwrantz 1994).

The binding site of AAA has been compared with related fucose-specific lectin from *Ulex europaeus* (UEA-I). Both AAA and lectin from *Ulex europaeus* (UEA-I) recognize Fuc α1-2Gal β-HSA. In addition, AAA cross-reacts with neoglycolipids bearing lacto-N-fucopentaose (LNFP) I [H type 1] and II [Le^a], and lactodifucotetraose (LDFT) as glycans. UEA-I, on the other hand, binds to a LDFT-derived neoglycolipid but not to the other neoglycolipids tested. According to these results, AAA reacts with fucosylated type 1 chain antigens, whereas UEA-I binds only to the α1-2-fucosylated LDFT-derived neoglycolipid. AAA showed a broad reaction in the superficial pyloric mucosa from secretors and non-secretors, but AAA reactivity was more pronounced in Le(a$^+$b$^-$) individuals. On the other hand, UEA-I stained the superficial pyloric mucosa only from secretor individuals. Both reacted with most human carcinomas of different origin (Baldus et al. 1996).

20.2.2 FBP from European Seabass

A Binary Tandem Domain F-Lectin from Striped Bass (Morone Saxatilis): The FBP32 is a recently-characterized two-domain F-type lectin from striped bass, which is expressed abundantly in liver, and to a minor extent in other tissues. FBP32 is present in serum and up-regulated by inflammatory challenge. Gene duplication has produced at least one FBP32 paralogue (the more widely expressed FBP32II) in striped bass, while multiple two-domain F-type lectins are present in other fish species as a result of independent gene duplications. FBP32 binds specifically, in a Ca^{2+}-independent manner, to terminal fucose, but does not share the specificity of AAA for the H and Le^a antigens. FBP32 is monomeric and the presence of two F-type domains within one polypeptide may serve to increase ligand binding affinity, as an alternative mechanism to the oligomerization seen in AAA. Odom and Vasta (2006) described the molecular characterization of a unique L-Fuc-binding lectin from serum of striped bass (*Morone saxatilis*) that not only shares the CRD motif of AAA but possesses two similar yet distinct CRDs in the polypeptide subunit. In contrast with the N-terminal CRD, which presents the highest similarity to the AAA CRD, the C-terminus-CRD presents unique features. Among F-type tandem lectins, MS-FBP32 and other tandem binary homologues appear unique in that although their N-terminal domain shows close similarity to the fucose recognition domain of the eel agglutinin, their C-terminal domain exhibits changes that potentially could confer a distinct specificity for fucosylated ligands. In contrast with the amniotes, in which F-type lectins appear conspicuously absent, the widespread gene duplication in teleost fish suggests these F-type lectins acquired increasing evolutionary value within this taxon.

The crystal structure of the complex of MS-FBP32 with L-fucose shows a cylindrical 81-A-long and 60-A-wide trimer divided into two globular halves: one containing N-terminal CRDs (N-CRDs) and the other containing C-terminal CRDs (C-CRDs). The resulting binding surfaces at the opposite ends of the cylindrical trimer have the potential to cross-link cell surface or humoral carbohydrate ligands. The N-CRDs and C-CRDs of MS-FBP32 exhibit significant structural differences, suggesting that they recognize different glycans. Analysis of the carbohydrate binding sites provides the structural basis for the observed specificity of MS-FBP32 for simple carbohydrates and suggests that the N-CRD recognizes more complex fucosylated oligosaccharides and with a relatively higher avidity than the C-CRD. Modeling of MS-FBP32 complexed with fucosylated glycans that are widely distributed in nature suggests that the binary tandem CRD F-type lectins functions as opsonins by cross-linking "non-self" carbohydrate ligands and "self" carbohydrate ligands, such as sugar structures displayed by microbial pathogens and glycans on the surface of phagocytic cells from the host (Bianchet et al. 2010) (Fig. 20.2).

A lectin specific for fucose and galactose from serum of *Dicentrarchus labrax* showed hemagglutinating activity (HA) against rabbit erythrocytes. The HA was calcium-independent and comprised of two components, but only 34 kDa component (DLL2) showed activity against rabbit erythrocytes. The HA is a dimeric structure stabilized by disulfide bonds. The Ca^{2+}-independent fucose-binding specificity, a significant amino acid sequence homology of the N-terminal end, and cross-reaction of eel fucolectin with antibodies to DLL2 suggested that this lectin belongs to fucolectin family (Cammarata et al. 2001).

20.2.3 Other F-type Lectins in Fish

A F-type lectin (DlFBL) has been isolated from sea bass (Dicentrarchus labrax) serum. DlFBL exhibits two tandemly arranged CRDs that display F-type sequence motif. DlFBL is specifically expressed and localized in hepatocytes and intestinal cells. Exposure of formalin-killed *E. coli* to DlFBL enhanced their phagocytosis by D. labrax peritoneal macrophages relative to the unexposed controls, suggesting that DlFBL may function as an opsonin in plasma and intestinal mucus (Salerno et al. 2009).

A fucose-binding lectin, designated SauFBP32, was purified from the serum of the *gilt head bream Sparus* with subunit M_r of 35 and 30 kDa under reducing and non-reducing conditions, respectively. The native lectin is a monomer under the selected experimental conditions and agglutinated rabbit erythrocytes (Cammarata et al. 2007). A 23 kDa L-fucose-binding lectin from serum of Nile tilapia (*Oreochromis niloticusL.*), designated as TFBP, was isolated by Argayosa and Lee (2009). The fucose-binding proteins were detected in the soluble protein extracts from the gills, gut, head kidneys, liver, serum and spleen using a fucose-binding protein. The fucose-binding lectin from gill

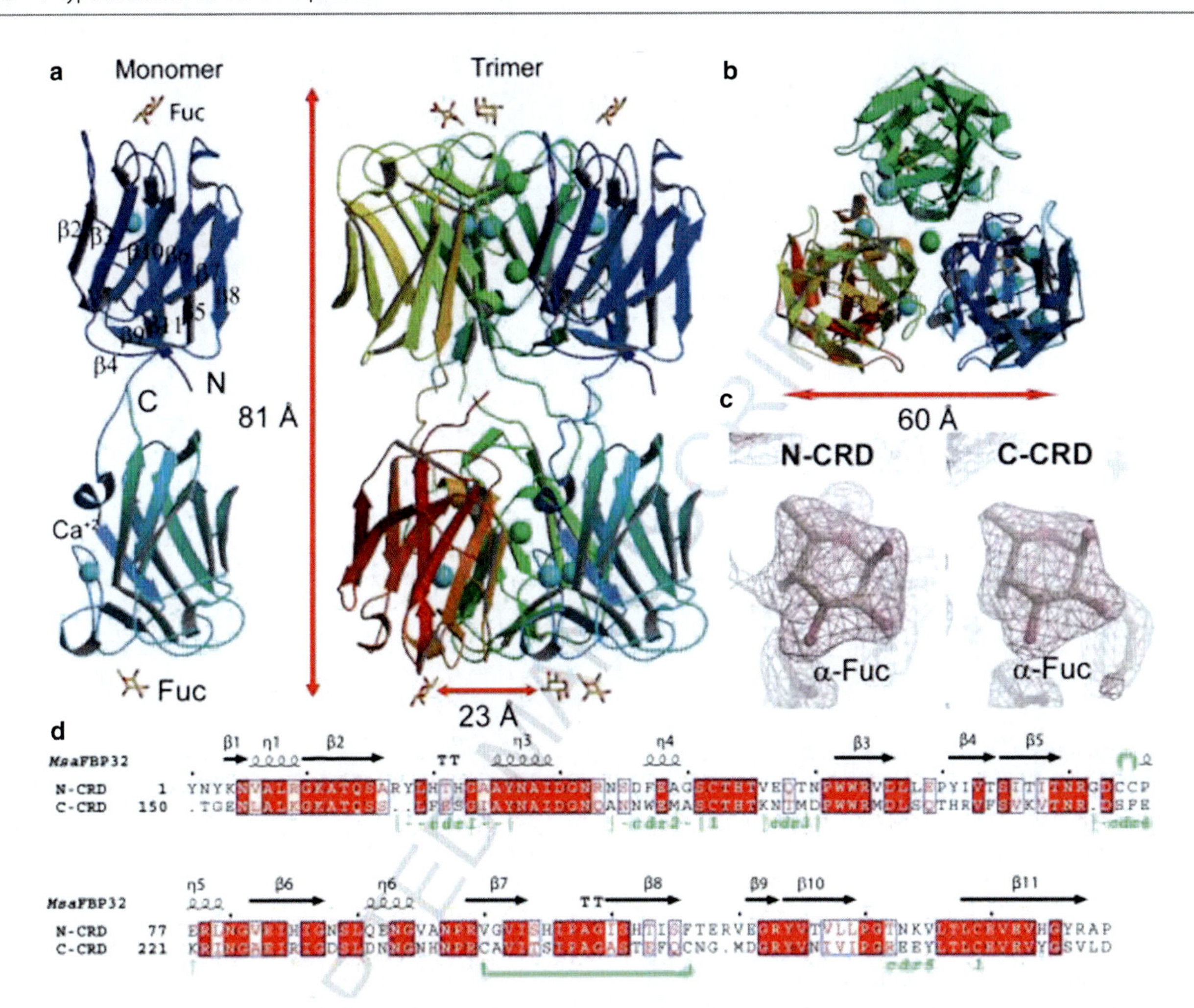

Fig. 20.2 *Structure of* MSFPB32: (**a**) View of MSFPB32 asymmetric unit of the crystal (trimer) and of an isolated monomer. (**b**) top view of the ASU trimer (N-CRD side). Monomers are colored with different hues. L-Fucose atoms are colored *yellow* for carbon atoms, and *red* for oxygen atoms. Ca^{+2} ions are colored *cyan* and Cl^{-} ions are colored *green*. (**c**) 2mFo-DFc sigmaA-weigthed electronic density at the CRDs binding sites. The *magenta* map is contoured at 0.35 e/Å^3. (**d**) Sequence alignment of the two MSFPB32 CRDs resulting in a 48.6 % and a 51 % of identity for the N- and the C-CRD respectively. The first line indicates the observed secondary structure elements of each CRD. Coils indicate 3–10 helices, T indicates turns and arrows indicate β-sheets strands. *Red* solid boxes over the sequences indicate identity and boxed *red* characters indicate amino acids with similar polarity. The last line shows the positions of the different CRDs. Disulfide bridges are indicated either by number "1" under the connected cysteines or a green line between them when they are within one CRD (Printed with permission from Bianchet et al. 2010 © Elsevier)

of bighead carp (*Aristichthys nobilis*) designated GANL had a M_r of 37 kDa under reducing conditions. GANL is a homomultimeric glycoprotein with a native molecular mass of 220 kDa and a carbohydrate content of ~13.4 %. The purified lectin only agglutinated rabbit native erythrocytes, and did not require Ca^{2+}. Its activity was inhibited by only fucose. GANL contains a high proportion of Asp, Glu, Leu, Val, and Lys. The 10 residues of N-terminal region were AGEQGGQCSA. GANL agglutinates and inhibits the growth of *Vibrio harveyi* but has no antifungal activity (Pan et al. 2010).

20.2.4 F-Type Lectins in Amphibians

Xenopus species possess proteins containing F-type domains in copies of 1 (*X laevis* X-epilectin), 2 (*X tropicalis* II-FBPL), 3 (*X tropicalis* III-FBPL) or 4 (*X laevis* II-FBPL), as well as 5 F-type domains in combination with a pentraxin domain (*X laevis* PXN-FBPL). *Xenopus* F-type domains sometimes lack fucose-binding His residue and/or a small number of Ca^{2+}-binding residues. A posterior epidermal marker, X-epilectin in *X. laevis* gene encodes for a fucolectin, which specifically binds fucose residues. The expression of this gene is switched on during gastrulation

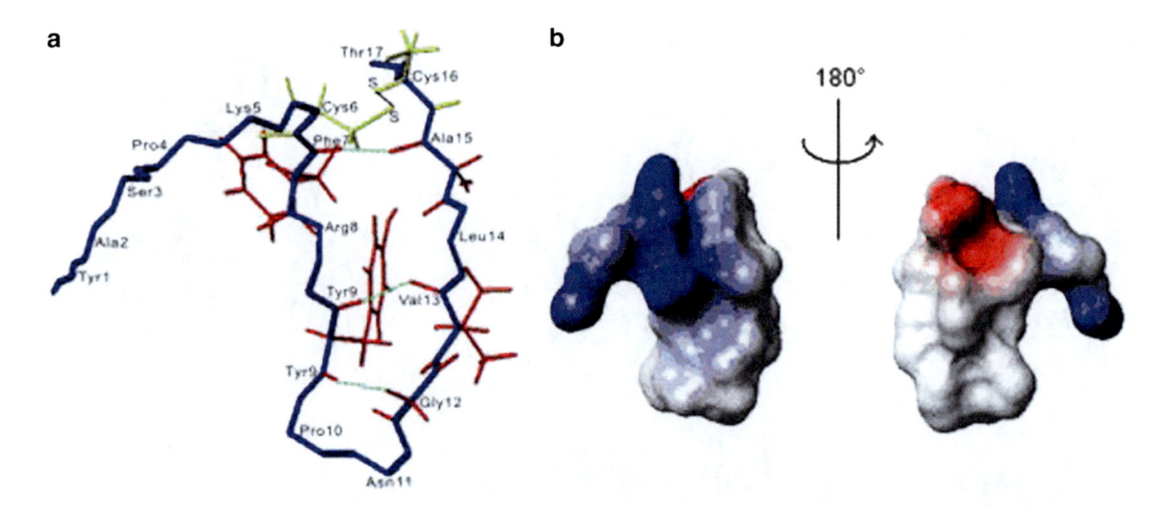

Fig. 20.3 The solution structure of odorranalectin. Backbone, side-chains of residues 4–16 and side-chains of residues 1–3, 4 were shown in *blue*, *red* and *green*, respectively. (**a**) The mean structure calculated from the 20 lowest-energy structures which highlighted three hydrogen bonds (*green broken lines*) and one disulfide bonds (*black solid line*). (**b**) Electrostatic surface of odorranalectin. Positively charged region and negatively charged region were shown in *blue* and *red*, respectively (Adapted from Li et al. 2008)

and up-regulated during neurula stages and found expressed ubiquitously throughout the epidermis. In adult, X-epilectin is mainly expressed in intestinal components, kidney, spinal cord and skin. Change of fate of animal caps into cement gland or dorsal mesoderm induces a down-regulation of X-epilectin expression in explants treated respectively with ammonium chloride and activin A. X-epilectin expression is down-regulated by Noggin and tBR, an effect which is inhibited by BMP4 over-expression, suggesting X-epilectin expression is mediated by BMP signaling pathway (Massé et al. 2004).

A lectin-like peptide named odorranalectin was identified from skin secretions of *Odorrana grahami*. The lectin was composed of 17 aa with a sequence of YASPKCFRY-PNGVLACT. L-fucose could specifically inhibit the haemagglutination induced by this lectin. In mice, odorranalectin mainly conjugated to liver, spleen and lung. The cyclic peptide of odorranalectin in solution NMR adopts a β-turn conformation stabilized by one intramolecular disulfide bond between Cys6-Cys16 and three hydrogen bonds between Phe7-Ala15, Tyr9-Val13, Tyr9-Gly12 (Fig. 20.3). NMR titration and mutant analysis showed that residues K5, C6, F7, C16 and T17 on odorranalectin consist of binding site of L-fucose. The structure of odorranalectin in bound form is more stable than in free form. Smaller peptides which can mimic the function of lectins are promising candidates for drug targeting (Li et al. 2008).

20.3 F-type Lectins in Invertebrates

20.3.1 Tachylectin-4

The F-type domain is found in a range of invertebrate species, often within lineage-specific protein contexts. Sugar binding has been demonstrated in an F-type lectin from the Japanese horseshoe crab *Tachypleus tridentatus*. Like many invertebrates, the horseshoe crabs hemolymph fluid, containing solutes and hemocytes, bathes the internal organs. The horseshoe crab hemocytes are a single type of granular cell, which mounts an innate immune response upon recognition of bacterial LPS. Tachylectin-4 from horseshoe crab hemocyte has Ca^{2+} dependendent hemagglutinating activity against human A-type erythrocytes and was more potent than hemocyte lectin with an affinity to *N*-acetylglucosamine, tachylectin-2. The tachylectin-4 is an oligomeric glycoprotein of 470 kDa, composed of subunits of 30 and 320.5 kDa. The activity was inhibited by L-Fucose, *N*-acetylneuraminic acid and bacterial S-type LPS but not by R-type LPS lacking O-antigen. Colitose (3-deoxy-L-fucose), a unique sugar present in the O-antigen of *E. coli* O111:B4 with structural similarity to L-fucose, is the most probable candidate for a specific ligand of tachylectin-4. The ORF of 1344-bp cDNA coding for tachylectin-4 coded for the mature protein with 232 amino acids. Tachylectin-4 is homologous to the

NH2-terminal domain with unknown functions of *Xenopus laevis* pentraxin 20.

20.3.2 F-type Lectins from *Drosophila melanogaster*

Furrowed and CG9095 from *Drosophila melanogaster* contain single F-type domains within an architecture that also includes a C-type lectin-like domain and a number of complement control repeats. These proteins, which have homologs in other insects (e.g. bee and mosquito) are the only F-type proteins which have transmembrane domains. Signaling by these receptors is influenced by O-fucosylation, but the F-type domains in the receptors do not contain complete His/Arg/Arg triads (the C-type lectin-like domain in these receptors is also unlikely to bind sugar). It has been stated that F-type domains are not present in the nematode worm *Caenorhabditis elegans*, but a highly divergent F-type domain, which is very unlikely to bind fucose, is present in the CG9095 orthologue, C54G4.4. The divergence of F-type domains between CG9095 proteins in different species is suggestive of a non-sugar-binding function of the F-type domain in these receptors.

20.3.3 F -Type Lectins in Sea Urchin

In the sea urchin *Strongylocentrotus purpuratus*, a single F-type domain is present in complex protein architectures, as well as in a simple single-domain fucolectin. The complex proteins include CRL, which is involved in the complement system, and a protein containing a CCP and an EGF domain, both of which have very similar F-type domains, and a protein containing scavenger receptor Cys-rich, Kringle and other domains. F-type domains in the sea urchin are distinctive due to a number of insertions and the absence of fucose-binding His residue and adjacent cysteines. The sea urchin lectin from *Toxopneustes pileolus* is galactose and fucose specific. Incubation of rat peritoneal mast cells with this lectin in presence of CaCl2 inhibited the histamine release induced by GlcNAc-specific *Datura stramonium* agglutinin (DSA). It is suggested that the lectin binds to D^{+}-Gal residues of DSA to interfere with mast cell activation induced by DSA, a glycoprotein with arabinose and Gal residues (Suzuki-Nishimura et al. 2001). Discoidin I and discoidin II are N-acetylgalactosamine (GalNAc)-binding proteins from *Dictyostelium discoideum*. They consist of two domains: an N-terminal discoidin domain and a C-terminal H-type lectin domain. The N-terminal discoidin domain presents a structural similarity to F-type lectins such as eel agglutinin, where an amphiphilic binding pocket suggests possible carbohydrate-binding activity (Mathieu et al. 2010).

20.3.4 Bindin in Invertebrate Sperm

In free-spawning, invertebrate's sperm–egg incompatibility is a barrier to mating between species, and divergence of gamete recognition proteins (GRPs) can result in reproductive isolation. Bindin is a major protein for species-specific recognition between sperm and congenetic egg during fertilization in many free-spawning marine invertebrates. Bindins have identical 24-residue signal peptides and conserved 97-residue N-terminal sequences, and they differ in mass because of the presence of between 1 and 5 tandemly repeated 134-residue fucose-binding lectin (F-lectin) domains. Oyster bindin is a single copy gene, but F-lectin repeat number and sequence are variable within and between individuals. Eight residues in F-lectin fucose-binding groove are subject to positive diversifying selection, indicating a history of adaptive evolution at lectin's active site. There is one intron in middle of each F-lectin repeat, and recombination in this intron creates many combinations of repeat halves. Alternative splicing creates many additional size and sequence variants of the repeat array. Males contain full-length bindin cDNAs of 5 possible sizes, but only one or two protein mass forms exist in each individual. Sequence analysis indicates that positive selection, alternative splicing, and recombination can create thousands of bindin variants within *C. gigas*. The extreme sequence variation in F-lectin sequence of oyster bindin within-species is a novel finding (Springer et al. 2008). The full-length bindin cDNA from oyster *Crassostrea angulata* comprises 1,049 bp with a 771-bp ORF encoding 257 amino acids. The deduced amino acid sequence contained a putative signal peptide of 24 amino acids. The length of the bindin genomic DNA was 8,508 bp containing four exons and three introns. Three haplotypes of F-lectin repeat were detected from seven sequences of F-lectin repeat of six male oysters. Intron-4 may play an important role in recombination. The amount of intraspecific polymorphism in male GRPs may be a consequence of the relative efficiency of local (molecular recognition) and global (electrical, cortical, and physical) polyspermy blocks that operate during fertilization (Moy et al. 2008; Wu et al. 2010).

20.4 F-type Lectins in Plants

The fucose binding proteins extracted from Lotus tetragonolobus seeds consists of two distinct classes of components which correspond to tetrameric glycoproteins of 118–120 kDa with potent temperature dependent hemagglutinating activity and a highly aggregated dimeric component of 58 kDa with macrophage activating properties (Leu et al. 1980). Immobilized fucose-binding lectin Lotus tetragonolobus agglutinin (LTA1), also known as lotus lectin LTA is effective in isolating glycans containing the Le^x antigen and is useful in analyzing specific fucosylation of glycoconjugates (Yan et al. 1997). A well ordered, two-dimensional lattice is formed from fucose-specific isolectin A from Lotus tetragonolobus cross-linked with difucosyllacto-N-neohexaose, an oligosaccharide possessing the Le^x determinant, which is an oncofetal antigen. Using the symmetry and dimensions of the lattice and its appearance in filtered electron micrographs, molecular models were used to determine the orientation of the lectin in the lattice, and to define the range of lectin-oligosaccharide interactions consistent with the structural data (Cheng et al. 1998). Detailed discussion on plant lectins is beyond scope of this book.

20.5 F-type Lectins in Bacteria

The F-type domain is found in a diverse type of bacterial species, in a variety of architectural contexts. Bacterial F-type domains frequently conserve the fucose binding motif, but they do not include disulfide bonds. Mutagenesis experiments have suggested that genes encoding putative glycoside hydrolases are necessary for full virulence of gram positive *S. pneumonia* in a murine lung model of pneumonia. One such gene (SP2159) of *S. pneumoniae* TIGR4 is "fucolectin-related protein" but here called SpGH98. This gene is part of a fucose utilization operon that is conserved among three sequenced strains of *S. pneumoniae*. This protein is a 1038-amino acid protein of which the ~600 amino acid N-terminal region shows amino acid sequence identity with family glycoside hydrolase 98 (GH98). Following GH98 catalytic module is a triplet of modules (Fig. 20.4) which have ~ 50–60 % identity with one another and ~25 % amino acid sequence identity to AAA (Bianchet et al. 2006). This protein contains a C-terminal triplet of fucose binding modules that have significant amino acid sequence identity with the Anguilla anguilla fucolectin. Boraston et al. (2006) dissected the modular structure of SpGH98 by heterologous production of two single modules (SpX-1 and SpX-3), a tandem construct (SpX-20.2) and the triplet (SpX-20.2.3). Functional studies of these fucose binding modules reveal binding to fucosylated oligosaccharides and suggest the importance of multivalent binding. The crystal structures of ligand bound forms of one fucose binding module uncovered the molecular basis of fucose, ABH blood group antigen, and LeY antigen binding. These studies, extended by fluorescence microscopy, showed specific binding to mouse lung tissue

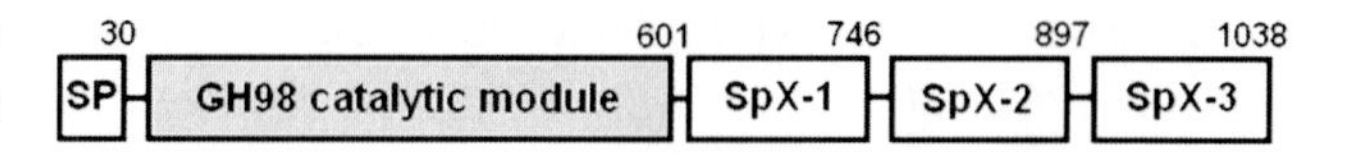

Fig. 20.4 Modular arrangement of SpGH98. The number above the box shows the number of amino acid residue that defines the module boundaries. SP denotes the signal peptide (Boraston et al. 2006)

and defined a new family of carbohydrate binding modules (CBM) now classified as family 47 (Boraston et al. 2006).

Among other Gram-positive bacteria, an F-type domain is found in a protein from *Solibacter usitatus* Ellin6076, combined with a number of bacterial immunoglobulin-like domains and FG-GAP repeats. A lectin is associated with the bacterial cell surface of Rhizobium lupini strain LL13 with a Mr ~19,000. This protein specifically aggregated L-Fuc-BSA-coated microspheres (Wisniewski et al. 1994). Among Gram-negative purple bacteria, an F-type domain follows the structurally similar coagulation factor V/VIII domain in a protein from Saccharophagus degradans 2–40, an organism which degrades a range of polysaccharide substrates, whereas in *Acidiphilium cryptum* JF-5 an F-type domain is present at the C-terminus of the fucolectin tachylectin-4 pentraxin-1 protein, and in Gluconobacter oxydans 621 H F-type domains are present in the proteins GOX0967 and GOX0982. Among bacteria of the planctomycetes phylum, an F-type domain is present at the centre of a large protein that is a periplamic component of a sugar transport system in *Blastopirellula marina* DSM 3645, and a more distantly related domain is present in a probable cytochrome c precursor (NPβ868124) in *Rhodopirellula baltica* SH 20.

PA-IIL is a fucose-binding lectin from *Pseudomonas aeruginosa* that is closely related to the virulence factors of the bacterium and has high affinity for the monosaccharide ligand. Structural studies revealed a new carbohydrate-binding mode with direct involvement of two calcium ions (Mitchell et al. 2002). The crystal structure of the tetrameric PA-IIL in complex with fucose and calcium allowed a proposal which suggested hydrogen-bond network in the binding site. Computational methods indicated that extensive delocalization of charges between the calcium ions, the side chains of the protein-binding site and the carbohydrate ligand is responsible for the high enthalpy of binding and therefore for the unusually high affinity observed for this unique mode of carbohydrate recognition (Mitchell et al. 2005). Mutagenesis of amino acids forming the specificity binding loop allowed identification of one amino acid that is crucial for definition of lectin sugar preference. Altering specificity loop amino acids causes changes in saccharide-binding preferences of lectins derived from PA-IIL, via creation or blocking possible binding interactions (Adam et al. 2007).

The fucose-specific lectin LecB is implicated in tissue binding and biofilm formation by the opportunistic pathogen

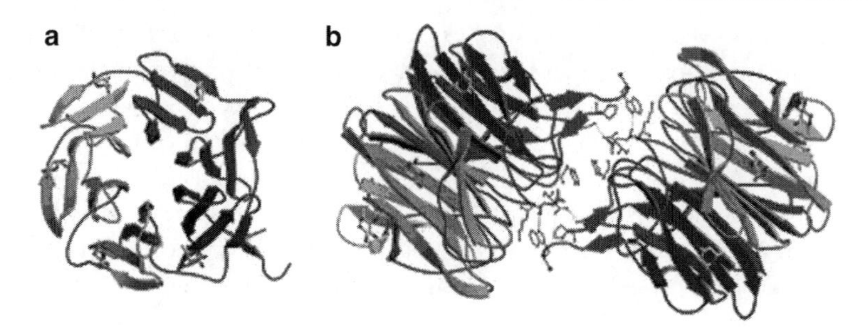

Fig. 20.5 (**a**) Ribbon diagram of monomer A of AAL complexed with fucose shown as sticks. (**b**) Dimer of AAL with stick representation of the amino acids involved in the interaction of monomers (Adapted with permission from Wimmerova et al. 2003 © The American Society for Biochemistry and Molecular Biology)

Pseudomonas aeruginosa, which causes severe respiratory tract infections mainly in immuno-compromised patients or cancer patients undergoing chemotherapy (Kolomiets et al. 2009). The plant pathogen *Ralstonia solanacearum*, which causes lethal wilt in many agricultural crops, produces a potent L-fucose-binding lectin (RSL: subunit Mr 9.9 kDa) exhibiting sugar specificity similar to that of PA-IIL of the human aggressive opportunistic pathogen *Pseudomonas aeruginosa* (Sudakevitz et al. 2002). The properties of the first lectin are related to fungal lectins. The second one, RS-IIL (subunit Mr 120.6 kDa) is a tetrameric lectin with high sequence similarity to fucose-binding lectin PA-IIL of *Pseudomonas aeruginosa*. RS-IIL recognizes fucose but displays much higher affinity to mannose and fructose, which is opposite to the preference spectrum of PA-IIL (Sudakevitz et al. 2004).

20.6 Fuco-Lectins in Fungi

20.6.1 Fuco-Lectin from Aleuria Aurantia (AAL)

A fucose-binding lectin was purified from fruiting bodies of *Aleuria aurantia*, **a** widespread ascomycete fungus. The lectin combines the terminal fucose in the carbohydrate chain. AAL shows sugar-binding specificity for L-fucose. Recombinant form of AAL is a fucose-binding lectin composed of two identical 312 aa subunits (Kochibe and Furukawa 1980). Each subunit contains five binding sites for fucose. One of the binding sites in rAAL had unusually high affinities towards fucose and fucose-containing oligosaccharides with K_D values in the nM range. (Olausson et al. 2008). Isolectin A from *Lotus tetragonolobus* (LTL-A), and AAL were found to be specific for fucose residues. While α-L-Fuc-(1,3)-β-D-GlcNAc and Lewis^x (Le^X) bound to both lectins, sialyl-Lewis^x (sLe^X) bound to AAA only. AAA bound to the ligands more tightly. Binding strength for both lectins decreased from α-L-Fuc-(1,3)-β-D-GlcNAc to Le^X and was lowest for sLe^X. STD NMR experiments suggest that only the L-fucose residues are in intimate contact with the protein (Haselhorst et al. 2001).

AAL: Overall Fold and Organization: The crystal structure of the lectin complexed with fucose revealed that each monomer consists of a six-bladed β-propeller fold and of a small antiparallel two-stranded β-sheet that plays a role in dimerization. Five fucose residues were located in binding pockets between the adjacent propeller blades (Fig. 20.5a) (Paoli 2001). The global shape is a short cylinder, or tore, with an approximate diameter of 45 Å and a height of 35 Å. In the β-propeller fold, each consecutive β-sheet has its first strand (number 1) lining the central cavity of the protein and the last one (number 4) most exposed to solvent at the cylinder surface. Loops connecting the strands within each module are rather short, with the exception of blade III (amino acids 108–162) that displays longer loops. The consecutive blades are connected by long segments that run from the outside of the protein to the central tunnel. AAL has been described as a dimer in solution and also confirmed as a dimer in crystal. The two monomers are very similar, and superimposition of their backbones gives an r.m.s. value of 0.26 Å. A pseudo-2-fold axis of symmetry generates this dimer in the crystal (Fig. 20.5b). The small domain created by the antiparallel association of the N-terminal and C-terminal peptides plays a key role in the dimerization, additional contact being mediated by four loops (those interconnecting blades I and II and blades II and III and the loops between strands β_2-I and β_3-I and between strands β_2-II and β_3-II). Hydrophobic contacts involve the C terminus Trp^{312} from each monomer with Lys^{83} from the other. In addition, tyrosine residue, Tyr^6, interacts via aromatic ring stacking with its counterpart on the other monomer through 2-fold axis. Four main hydrogen bonds are also established between the side chains of Asp^{263} and Ser^{283} and between the Trp^{312} nitrogen side chain and backbone carbonyl back bone of Leu^{59}. Several amino acids at the N and C termini of the peptide chains protrude from the base of the β-propeller cylinder, associating in a small anti-parallel two-stranded β-sheet that forms a separated domain. The inner cavity of AAL has a tunnel shape with a diameter of about 8 Å in its middle part and almost closed off on the N terminus side of the first inner β-strands. This cavity has a strong hydrophobic character, being formed mostly by the conserved alanine residues of the first strands of each propeller blade. The core is filled with a set of about 50 water molecules forming a well ordered hydrogen bond network.

Due to repeats in the amino acid sequence, there are strong similarities between the sites. Oxygen atoms O-3, O-4, and O-5 of fucose are involved in hydrogen bonds with side chains of amino acids conserved in all repeats, whereas O-1 and O-2 interact with a large number of water molecules. The nonpolar face of each fucose residue is stacked against the aromatic ring of a Trp or Tyr amino acid, and the methyl group is located in a highly hydrophobic pocket. Depending on the precise binding site geometry, the alpha- or beta-anomer of the fucose ligand is observed bound in the crystal. Experiments conducted on a series of oligosaccharides confirmed the broad specificity of the lectin, with a slight preference for alphaFuc1-2Gal disaccharide. This multivalent carbohydrate recognition fold is a new prototype of lectins that is proposed to be involved in the host recognition strategy of several pathogenic organisms including not only the *Aspergillus* but also the phytopathogenic bacterium *Ralstonia solanacearum* (Wimmerova et al. 2003).

20.7 Intelectins

20.7.1 Intelectin-1 (Endothelial Lectin HL-1/ Lactoferrin Receptor or *Xenopus* Oocyte Lectin)

The murine gene encoding Intelectin (ITLN) was identified by Komiya et al. (1998) and was renamed Intelectin-1 (ITLN-1), following the isolation of Intelectin-2 by Lee et al. (2001). Murine intelectin-1 was called as HL-1 (Endothelial lectin HL-1) and suggested it to be a human homolog of *Xenopus* oocyte lectin XL-35. The protein has been identified independently as Omentin of adipose tissue. The protein is structurally identical to the intestinal receptor for lactoferrin (Wrackmeyer et al. 2006). The cDNA was cloned by Suzuki et al. (2001) and called LFR [Lactoferrin (LF) receptor (R)].

The mature human intelectin-1 (hITLN-1) is a secretory glycoprotein consisting of 295 amino acids and *N*-linked oligosaccharides, and its basic structural unit is a 120-kDa homotrimer in which 40-kDa polypeptides are bridged by disulfide bonds. The *hITNL* gene was split into 8 exons on chromosome 1q21.3. The hITNL showed high levels of homology with mouse intelectin, *Xenopus laevis* cortical granule lectin/oocyte lectin, lamprey serum lectin, and ascidian galactose-specific lectin. These homologues commonly contained no CRD, which is a characteristic of C-type lectins, although some of them have been reported as Ca^{2+}-dependent lectins. Recombinant hITNL revealed affinities to D-pentoses and a D-galactofuranosyl residue in presence of Ca^{2+}, and recognized bacterial arabinogalactan of *Nocardia* containing D-galactofuranosyl residues. These results suggested that hITNL is a galactofuranose binding lectin that plays a role in the recognition of bacterial- specific components in the host, and is not a member of C-type lectin family (Tsuji et al. 2001).

Galactofuranosyl residues, which are not found on mammalian tissues, are recognized as dominant immunogens (Daffe et al. 1993; Leitao et al. 2003). The hITLN-1 is expected to recognize not only *N. rubra* (Daffe et al. 1993), but also *Aspergillus fumigatus* (Leitao et al. 2003), *Mycobacterium tuberculosis* (Pedersen and Turco 2003), *Streptococcus oralis* (Abeygunawardana et al. 1991), *Leishmania major*, and *Trypanosoma cruzi* (Suzuki et al. 1997), all of which contain galactofuranosyl residues in cell walls. Some observations indicate that hITLN-1 and mouse intelectin-1 (mITLN-1) may play an immunological role against selected microorganisms or foreign antigens (Pemberton et al. 2004a, 2004b; Datta et al. 2005; Kuperman et al. 2005; Wali et al. 2005). Since intelectin homologues are generally oligomeric (Abe et al. 1999; Suzuki et al. 2001; Tsuji et al. 2001; Nagata 2005), this structure may influence their function.

Most mammalian ITLNs have 10 conserved Cys residues but two of these are not found in the N-terminal regions of mouse and rat ITLNs. The recombinant hITLN-1 is a trimer, disulfide-linked through Cys-31 and Cys-48, and N-glycosylated at Asn-163. Despite 84.9 % amino acid identity to hITLN-1, recombinant and intestinal mITLN-1 are unglycosylated 30-kDa monomers. Elution profiles of recombinant hITLN-1, as well as recombinant and intestinal mITLN-1 indicated that the two intelectins have different saccharide-binding specificities. Despite 84.9 % amino acid identity between these two proteins, the oligomeric structure and carbohydrate-recognition specificity of mITLN-1 differ from those of hITLN-1 (Tsuji et al. 2007). In mammals, the galactose-binding C-type lectins, a macrophage lectin or asialoglycoprotein receptors, bind to asialoglycoproteins. However, the binding affinity of hITNL-1 to asialoglycoproteins was weak (Tsuji et al. 2001). This was thought to be because the galactosyl residues of asialoglycoproteins are galactopyranosides although hITNL has affinity to the galactofuranosyl residue.

The hITLN-1 mRNA level increases in airway epithelial cells from individuals with asthma (Kuperman et al. 2005). A high percentage of malignant pleural mesothelioma show up-regulated expression of hITLN-1 (Wali et al. 2005). A marked induction in Intelectin gene expression was observed among human primary mesothelial cells as a consequence of crocidolite asbestos exposure and simian virus 40 infection. However, expression of hITNL-1 is decreased in human airway epithelium of smokers compared to nonsmokers (Carolan et al. 2008). Others have shown that infection with *T. spiralis* (Pemberton et al. 2004a, 2004b) or *Trichuris muris* (Datta et al. 2005) induced mITLN-1 and mITLN-2 mRNA expression. These observations and the galactofuranose binding by hITLN-1 and mITLN-1 suggest that ITLN homologues may have an immunoregulatory role in host defense. For instance, the serum ITLN homologue of ascidian (galactose-specific lectin) functions as an opsonin (Abe et al. 1999). However, Voehringer et al. (2007) indicated no significant modification of immune responses or clearance of microorganisms in mITLN transgenic mice. Although hITLN-1 is a soluble intestinal protein (Tsuji et al. 2001), hITLN-1 is also a glycosylphosphatidylinositol-anchored intestinal lactoferrin receptor (Suzuki et al. 2001), HL-1 of vascular endothelial cells (Lee et al. 2001), or omentin of adipose tissue (Schäffler et al. 2005) and human visceral fat (Tsuji et al. 2007). However, the physiological function of mITLN-1 may differ from that of hITLN-1. It was reported that swine ITLN, which is expected to be oligomeric, is associated with lipid rafts on the enterocyte brush border (Wrackmeyer et al. 2006). It is possible that hITLN-1 is similarly located. Additional investigation will be required to further establish the physiological function of ITLN-1 in both humans and mice.

Human hITLN-1 mRNA is expressed in heart, small intestine, colon, and thymus and exclusively localized in endothelial cells (Lee et al. 2001). Primary cultures of human aortic endothelial cells are positive for HL-1 expression, but several other human cell types are not. The brush border membrane is organized in stable glycolipid-based lipid raft microdomains, and like divalent lectin galectin-4, intelectin was enriched in microvillar "superrafts", i.e., membranes that resist solubilization with Triton X-100 at 37º C. Wrackmeyer et al. (2006) reported that trimeric Intelectin serves as an organizer and stabilizer of the brush border membrane, preventing loss of digestive enzymes to the gut lumen and protecting the glycolipid microdomains from pathogens.

Intelectin is expressed in Paneth and goblet cells of small intestine and serves a protective role in the innate immune response to parasite infection (Komiya et al. 1998; Peebles 2010; Wrackmeyer et al. 2006). Secretory granules of lysozyme-positive Paneth cells in the bottom of crypts as well as goblet cells along crypt-villus axis were intensively labeled with intelectin antibodies, but quantitatively, the major site of intelectin deposition was the enterocyte brush border. Tsuji et al. (2009) found that human intelectin-1 is a serum protein and binds to *Mycobacterium bovis* bacillus Calmette-Guérin (BCG). Human ITLN-1-binding to BCG was inhibited by Ca^{2+}-depletion, galactofuranosyl disaccharide, ribose, or xylose, and was dependent on the trimeric structure of human ITLN-1. Human ITLN-1-transfected cells express ITLN-1 on the cell surface and secrete it in the culture supernatant. Study indicates that intelectin is a host defense lectin that assists phagocytic clearance of microorganisms.

As the *X. laevis* oocyte lectin, a homologue of hITNL, has been shown to participate in the formation of fertilization envelope that blocks sperm entry, hITNL may also participate in fertilization. However, hITNL was also expressed on various tissues other than oocytes, and it has been demonstrated that the other homologues are also present in various tissues (Komiya et al. 1998; Abe et al. 1999). Thus, hITNL and its homologues may not only participate in the formation of fertilization envelope but also have other physiological functions.

The hITLN is plentifully expressed in heart. Since *viridans streptococci* invading blood attack heart and cause subacute infectious endocarditis, and since the surface polysaccharide of *Streptococcus oralis* contains galactofuranosyl residues (Abeygunawardana et al. 1991), hITNL may function in heart as a defense protein against these pathogens (Komiya et al. 1998).

20.7.2 Intelectin-2 (HL-2) and Intelectin-3

Intelectin-2 shows 91 % sequence homology with Intelectin, later named Intlectin-1 (Pemberton et al. 2004). Intelectin-2 also called HL-2 (Endothelial lectin HL-2) was identified as the human homologue of *Xenopus* oocyte lectin XL-35 (Lee et al. 2001). The Intelectin-2 gene is absent from C57BL/10 genome but present in genome of BALB/c mice. The protein is expressed by intestinal goblet cells, and its expression is up-regulated after infection with the nematode parasite *Trichinella spiralis*. BALB/c mice, which express Intelectin-2 are resistant to this nematode, whereas C57BL/10 mice, which do not express Intelectin-2 are susceptible to *Trichinella spiralis* infection. The deduced amino acid sequence of each homologue of HL-1 and HL-2 is about 60 % identical and 80 % similar to that of XL35, and none of these sequences contains the C-type lectin motif, although it is known that XL35 requires calcium for ligand binding.

Sheep Intlectins: Intelectin 2 (sITLN2) from sheep abomasal mucosa shared 76–83 % homology with other mammalian intelectins. Expression of sheep abomasal ITLN2 mRNA was up-regulated on post-challenge of worm-free sheep with *Teladorsagia circumcincta*. Infection with *T. circumcincta* was also associated with increased levels of abomasal transcripts encoding sheep mast cell protease-1, ovine galectin-14 and IL4, which indicated a Th2 type response. The amino acid sequences of sheep sITLN1 and sITLN3 share 86 % and 91 % homology with sheep sITLN2 respectively. While sITLN1 and sITLN3 transcripts are expressed ubiquitously in most of the normal sheep tissues, sITLN2 transcript was restricted to the abomasal mucosa in normal sheep tissues. The three sITLN were absent in unchallenged naïve lambs but induced in abomasal mucosa following challenge of both *Teladorsagia circumcincta* and *Dictyocaulus filaria* natural infection suggesting that intelectins may play an important role in mucosal response to nematode infections in ruminants (French et al. 2009).

Intelectins are capable of binding bacteria via galactofuranose residues and function as intestinal receptors for the antimicrobial glycoprotein lactoferrin (Lf). Lf binds strongly to enterohemorrhagic *E. coli* (EHEC) and the Lf receptor expressed in terminal rectum, the site of predilection of EHEC in cattle. Bovine intelectins (bITLN1 and bITLN2) were expressed in abomasum and rectum, but expression appeared minimal in the jejunum. There was significantly higher expression of bITLN2 in terminal rather than proximal rectum. Thus two bovine ITLNs are expressed along with Lf in the gastrointestinal tract, where they may interact with microbial pathogens (Blease et al. 2009).

20.7.3 Intelectins in Fish

Lin et al. (2009) characterized seven zebrafish intelectins (zINTLs) and made comparative analysis of intelectins from various species. zINTL1-3 are highly expressed in one or several adult tissues. zINTL4-7, however, were expressed at quite low levels both in adults and various development stages. Of the seven zINTLs, zINTL3 was expressed predominantly in the liver and highly up-regulated upon infection, suggesting its important roles in immunity. Based on the characterization of intelectin members in various species it was indicated that intelectin family may be a deuterostome specific gene family; and their expression patterns, quaternary structures and glycosylations vary considerably among various species, though their sequences are highly conserved. Moreover, these varied features have evolved multiple times independently in different species, resulting in species-specific protein structures and expression patterns (Lin et al. 2009). Rainbow trout plasma Intelectin exhibited calcium-dependent binding to N-acetylglucosamine (GlcNAc) and mannose conjugated Toyopearl Amino 650 M matrices. The lectin appeared at ~37 kDa and ~ 72 kDa bands. Similar analysis of plasma revealed a single 72 kDa band under reducing conditions. MALDI-TOF MS demonstrated five, ~37 kDa isoforms (pI 5.3–6.1). A 975 bp cDNA sequence encoded a 325 amino acid secretory protein with homology to human and murine intelectins, which bind bacterial components and were induced during parasitic infections. Rainbow trout IntL was detected ubiquitously in many tissues. Rainbow trout IntL plays a role in innate immune defense against bacterial and chitinous microbial organisms (Russell et al. 2008).

Two types of IntL genes have been identified from catfish. The genomic structure and organization of IntL2 were similar to those of the mammalian species and of zebrafish and grass carp, but orthologies could not be established with mammalian IntL genes. The IntL genes are highly conserved through evolution. Sequence analysis also indicated the presence of the fibrinogen-related domain in the catfish IntL genes. Phylogenetic analysis suggested the presence of at least two prototypes of IntL genes in teleosts, but only one in mammals. The catfish IntL1 gene is widely expressed in various tissues, whereas the channel catfish IntL2 gene was mainly expressed in the liver. While the catfish IntL1 is constitutively expressed, the catfish IntL2 was drastically induced by i.p. injection of *Edwardsiella ictaluri* and/or iron dextran. While IntL1 was expressed in all leukocyte cell lines, no expression of IntL2 was detected in any of the leukocyte cell lines, suggesting that the up-regulated channel catfish IntL2 expression after bacterial infection may be a consequence of the initial immune response, and/or a downstream immune response rather than a part of the primary immune responses (Takano et al. 2008).

20.7.4 Eglectin (XL35) or Frog Oocyte Cortical Granule Lectins

Xenopus Laevis Egg Cortical Granule: *Xenopus laevis* egg cortical granule, calcium-dependent, galactosyl-specific lectin participates in forming the fertilization layer of the egg envelope and functions in establishing a block to polyspermy. This oligomeric lectin is released extracellularly at fertilization and binds to its polyvalent glycoprotein ligand that is cross-linked in the jelly coat layer surrounding the oocyte. The lectin is expressed again at gastrulation and may function as well in cell–cell or cell–matrix adhesion events in the embryo (Outenreath et al. 1988; Lee et al. 1997). A cDNA encoding XL35 was isolated from a *Xenopus* oocyte cDNA library (Lee et al. 1997). The amino acid sequence it encodes did not display the C-type lectin motif, although it does require calcium for binding. The cDNAs encoding two human homologues of the *Xenopus* oocyte lectin, XL35, were isolated from a small intestine cDNA library and termed HL-1 and HL-2. The deduced amino acid sequence of each homologue is about 60 % identical and 80 % similar to that of XL35, and none of these sequences contains the C-type lectin motif, although XL35 requires calcium for ligand binding. HL-1 transcripts are present at relatively high levels in heart, small intestine, colon, thymus, ovary, and testis. HL-2 transcripts, by contrast, are expressed only in small intestine (Lee et al. 2001).

The translated cDNA for the cortical granule lectin had a signal peptide, a structural sequence of 298 amino acids, a molecular weight of 32.7 kDa, contained consensus sequence sites for N-glycosylation and a fibrinogen domain. The lectin cDNA expressed during early stages of oogenesis and 2/3 of the lectin was associated with the extracellular perivitelline space and the egg/embryo fertilization envelope. Lectin mRNA levels were from 100- to 1000-fold greater in ovary than in other adult tissues. The lectin did not have sequence homology to any known lectin families but had 41–88 % amino acid identity with nine translated cDNA sequences from an ascidian, lamprey, frog, mouse, and human. Based on the conserved carbohydrate binding and structural properties, it was named as eglectin (Chang et al. 2004).

Several nucleic acid sequences that predict proteins homologous to XL35 have been reported in frog, human, mouse, lamprey, trout, ascidian worm. These proteins also showed high degrees of amino acid sequence homology to a common fibrinogen-like motif that may involve carbohydrate binding. Several independent studies on these lectins strongly suggest that the lectins are expressed and stored in specialized vesicles that may be released upon infection by pathogens. In addition, some family members have been shown to bind to oligosaccharides from bacterial pathogens. Therefore, this family of lectins likely participates in pathogen surveillance as part of the innate immune system. These were named as homologues of XL35 (Lee et al. 2004).

Xenopus Embryonic Epidermal Lectin (XEEL): The *Xenopus laevis* embryonic epidermal lectin (XEEL) belong to the group of mammalian intelectins, frog oocyte cortical granule lectins, and plasma lectins in lower vertebrates and ascidians. A cDNA from a *Xenopus laevis* embryo library encodes a predicted translation product of 342 amino acids containing a signal sequence for secretion. The predicted protein has 62–70 % amino acid identity with the *Xenopus* oocyte cortical granule lectin (XCGL), the mouse intelectin, the human HL-1/intelectin and HL-2. Onset of gene expression occurs by gastrulation and the transcripts localize in non-ciliated epidermal cells all over the tail bud embryos. The designated XEEL is secreted from the embryonic epidermis (Nagata et al. 2003).

XCL-1: Ishino et al. (2007) isolated a 35-kDa Ca^{2+}-dependent lectin (XCL-1) from adult *Xenopus* serum. Although XCL-1 gene was not induced in the regenerating tails, Ishino et al. (2007) isolated a cDNA for an XCL-1-related protein (XCL-2). In contrast to the XCL-1 gene, XCL-2 gene expression was significantly increased in the regenerating tails, suggesting its role in tail regeneration. Although both XCL-1 and XCL-2 belong to *Xenopus* lectin family (X-lectins), XCL-1 and XCL-2 exhibit distinct developmental gene expression from two other known X-lectin members, both of which are expressed in the embryonic stage, whereas the XCL-1 and XCL-2 genes are predominantly expressed in the adult and middle/late tadpole stages, respectively, suggesting multiple functions of X-lectins. A gene for a humoral C-type lectin family is transiently expressed in the regenerating legs of the American cockroach (Arai et al. 1998). It suggests that the induction of a gene in regenerating organs is conserved among insects and vertebrates.

References

Fuco-Lectins

Adam J, Pokorná M, Sabin C et al (2007) Engineering of PA-IIL lectin from *Pseudomonas aeruginosa* – Unravelling the role of the specificity loop for sugar preference. BMC Struct Biol 7:36

Argayosa AM, Lee YC (2009) Identification of (L)-fucose-binding proteins from the Nile tilapia (Oreochromis niloticus L.) serum. Fish Shellfish Immunol 27:478–485

Avraméas A, McIlroy D, Hosmalin A et al (1996) Expression of a mannose/fucose membrane lectin on human dendritic cells. Eur J Immunol 26:394–400

Baldus SE, Thiele J, Park YO et al (1996) Characterization of the binding specificity of Anguilla anguilla agglutinin (AAA) in comparison to Ulex europaeus agglutinin I (UEA-I). Glycoconj J 13:585–590

Bianchet MA, Odom EW, Vasta GR, Amzel LM (2002) A novel fucose recognition fold involved in innate immunity. Nat Struct Biol 9:628–634

Bianchet MA, Odom EW, Vasta GR, Amzel LM (2010) Structure and specificity of a binary tandem domain F-lectin from striped bass (Morone saxatilis). J Mol Biol 401:239–252

Bolwell GP, Callow JA, Callow ME, Evans LV (1979) Fertilization in brown algae. II. Evidence for lectin-sensitive complementary receptors involved in gamete recognition in Fucus serratus. J Cell Sci 36:19–30

Boraston AB, Wang D, Burke RD (2006) Blood group antigen recognition by a *streptococcus pneumoniae* virulence factor. J Biol Chem 281:35263–35271

Cammarata M, Benenati G, Odom EW et al (2007) Isolation and characterization of a fish F-type lectin from gilt head bream (Sparus aurata) serum. Biochim Biophys Acta 1770:150–155

Cammarata M, Vazzana M, Chinnici C, Parrinello N (2001) A serum fucolectin isolated and characterized from sea bass Dicentrarchus labrax. Biochim Biophys Acta 1528:196–202

Cheng W, Bullitt E, Bhattacharyya L et al (1998) Electron microscopy and x-ray diffraction studies of Lotus tetragonolobus A isolectin cross-linked with a divalent $Lewis^X$ oligosaccharide, an oncofetal antigen. J Biol Chem 273:35016–35022

Friess AE, Toepfer-Petersen E, Nguyen H, Schill WB (1987) Electron microscopic localization of a fucose-binding protein in acrosome reacted boar spermatozoa by the fucosyl-peroxidase-gold method. Histochemistry 86:297–303

Gercken J, Renwrantz L (1994) A new mannan-binding lectin from the serum of the eel (Anguilla anguilla L.isolation, characterization and

comparison with the fucose-specific serum lectin. Comp Biochem Physiol Biochem Mol Biol 108:449–461
Haltiwanger RS, Hill RL (1986) The ligand binding specificity and tissue localization of a rat alveolar macrophage lectin. J Biol Chem 261:15696–15702
Haltiwanger RS, Lehrman MA, Eckhardt AE, Hill RL (1986) The distribution and localization of the fucose-binding lectin in rat tissues and the identification of a high affinity form of the mannose/N-acetylglucosamine-binding lectin in rat liver. J Biol Chem 261:7433–7439
Haselhorst T, Weimar T, Peters T (2001) Molecular recognition of sialyl Lewis(x) and related saccharides by two lectins. J Am Chem Soc 123:10705–10714
Honda S, Kashiwagi M, Miyamoto K et al (2000) Multiplicity, structures, and endocrine and exocrine natures of eel fucose-binding lectins. J Biol Chem 275:33151–33157
Ignotz GG, Lo MC, Perez CL et al (2001) Characterization of a fucose-binding protein from bull sperm and seminal plasma that may be responsible for formation of the oviductal sperm reservoir. Biol Reprod 64:1806–1811
Kochibe N, Furukawa K (1980) Purification and properties of a novel fucose-specific hemagglutinin of Aleuria aurantia. Biochemistry 19:2841–2846
Kolomiets E, Swiderska MA, Kadam RU et al (2009) Glycopeptide dendrimers with high affinity for the fucose-binding lectin LecB from Pseudomonas aeruginosa. Chem Med Chem 4:562–569
Lehrman MA, Haltiwanger RS, Hill RL (1986) The binding of fucose-containing glycoproteins by hepatic lectins. The binding specificity of the rat liver fucose lectin. J Biol Chem 261:7426–7432
Lehrman MA, Hill RL (1986) The binding of fucose-containing glycoproteins by hepatic lectins. Purification of a fucose-binding lectin from rat liver. J Biol Chem 261:7419–7425
Leu RW, Herriott MJ, Worley DS (1985) Characterization of Lotus tetragonolobus fucolectin components for differences in hemagglutinating and macrophage activating activities. Immunobiology 169:250–262
Leu RW, Whitley SB, Herriott J, Huddleston DJ (1980) Lotus tetragonolobus fucolectin as a potential model for "MIF-like" modulation of macrophage function: comparison of the interaction of Lotus fucose binding protein (FBP) and migration inhibitory factor (MIF) with macrophages in the migration inhibition assay. Cell Immunol 52:414–428
Li J, Wu H, Hong J et al (2008) Odorranalectin is a small peptide lectin with potential for drug delivery and targeting. PLoS One 3:e2381
Massé K, Baldwin R, Barnett MW, Jones EA (2004) X-epilectin: a novel epidermal fucolectin regulated by BMP signaling. Int J Dev Biol 48:1119–1129
Mathieu SV, Aragão KS, Imberty A, Varrot A (2010) Discoidin I from Dictyostelium discoideum and Interactions with oligosaccharides: specificity, affinity, crystal structures, and comparison with discoidin II. J Mol Biol 400:540–554
Mitchell E, Houles C, Sudakevitz D et al (2002) Structural basis for selective recognition of oligosaccharides from cystic fibrosis patients by the lectin PA-IIL of Pseudomonas aeruginosa. Nat Struct Biol 9:918–921
Mitchell EP, Sabin C, Snajdrová L et al (2005) High affinity fucose binding of Pseudomonas aeruginosa lectin PA-IIL: 1.0 A resolution crystal structure of the complex combined with thermodynamics and computational chemistry approaches. Proteins 58:735–746
Moy GW, Springer SA, Adams SL et al (2008) Extraordinary intraspecific diversity in oyster sperm bindin. Proc Natl Acad Sci USA 105:1993–1998
Ng KK, Drickamer K, Weis WI (1996) Structural analysis of monosaccharide recognition by rat liver mannose-binding protein. J Biol Chem 271:663–674
Odom EW, Vasta GR (2006) Characterization of a binary tandem domain f-type lectin from striped bass (*Morone saxatilis*). J Biol Chem 281:1698–1713
Olausson J, Tibell L, Jonsson BH et al (2008) Detection of a high affinity binding site in recombinant Aleuria aurantia lectin. Glycoconj J 25:753–762
Outenreath RL, Roberson MM, Barondes SH (1988) Endogenous lectin secretion into the extracellular matrix of early embryos of Xenopus laevis. Dev Biol 125:187–194
Pan S, Tang J, Gu X (2010) Isolation and characterization of a novel fucose-binding lectin from the gill of bighead carp (Aristichthys nobilis). Vet Immunol Immunopathol 133:154–164
Paoli M (2001) Protein folds propelled by diversity. Prog Biophys Mol Biol 76:103–130
Saito T, Hatada M, Iwanaga S et al (1997) A newly identified horseshoe crab lectin with binding specificity to O-antigen of bacterial lipopolysaccharides. J Biol Chem 272:30703–30708
Salerno G, Parisi MG, Parrinello D et al (2009) F-type lectin from the sea bass (Dicentrarchus labrax): purification, cDNA cloning, tissue expression and localization, and opsonic activity. Fish Shellfish Immunol 27:143–153
Schäffler A, Neumeier M, Herfarth H, et al. (2005) Genomic structure of human omentin, a new adipocytokine expressed in omental adipose tissue. Biochim Biophys Acta 1732:96–102
Springer SA, Moy GW, Friend DS et al (2008) Oyster sperm bindin is a combinatorial fucose lectin with remarkable intra-species diversity. Int J Dev Biol 52:759–768
Sudakevitz D, Imberty A, Gilboa-Garber N (2002) Production, properties and specificity of a new bacterial L-fucose- and D-arabinose-binding lectin of the plant aggressive pathogen Ralstonia solanacearum, and its comparison to related plant and microbial lectins. J Biochem 132:353–358
Sudakevitz D, Kostlánová N, Blatman-Jan G et al (2004) A new Ralstonia solanacearum high-affinity mannose-binding lectin RS-IIL structurally resembling the Pseudomonas aeruginosa fucose-specific lectin PA-IIL. Mol Microbiol 52:691–700
Suzuki-Nishimura T, Nakagawa H, Uchida MK (2001) D-galactose-specific sea urchin lectin sugar-specifically inhibited histamine release induced by datura stramonium agglutinin: differences between sugar-specific effects of sea urchin lectin and those of D-galactose- or L-fucose-specific plant lectins. Jpn J Pharmacol 85:443–452
Wimmerova M, Mitchell E, Sanchez JF et al (2003) Crystal structure of fungal lectin: six-bladed beta-propeller fold and novel fucose recognition mode for Aleuria aurantia lectin. J Biol Chem 278:27059–27067
Wisniewski JP, Monsigny M, Delmotte FM (1994) Purification of an α-L-fucoside-binding protein from Rhizobium lupini. Biochimie 76:121–128
Wu AM, Wu JH et al (2004) Lectinochemical studies on the affinity of *Anguilla anguilla* agglutinin for mammalian glycotopes. Life Sci 75:1085–1103
Wu Q, Li L, Zhang G (2011) Crassostrea angulata bindin gene and the divergence of fucose-binding lectin repeats among three species of Crassostrea. Mar Biotechnol (NY) 13:327–335
Yan L, Wilkins PP, Alvarez-Manilla G et al (1997) Immobilized Lotus tetragonolobus agglutinin binds oligosaccharides containing the Le(x) determinant. Glycoconj J 14:45–55

Intlectins/X-Lectins

Abe Y, Tokuda M, Ishimoto R et al (1999) A unique primary structure, cDNA cloning and function of a galactose-specific lectin from ascidian plasma. Eur J Biochem 261:33–39

Abeygunawardana C, Bush CA, Cisar JO (1991) Complete structure of the cell surface polysaccharide of *Streptococcus oralis* C104: a 600-MHz NMR study. Biochemistry 30:8568–8577

Arai et al (1998) A gene for a humoral C-type lectin family is transiently expressed in the regenerating legs of the American cockroach. Insect Biochem Mol Biol 28:987–994

Blease SC, French AT, Knight PA et al (2009) Bovine intelectins: cDNA sequencing and expression in the bovine intestine. Res Vet Sci 86:254–256

Carolan BJ, Harvey B-G, De BP et al (2008) Decreased expression of intelectin 1 in the human airway epithelium of smokers compared to nonsmokers. J Immunol 181:5760–5767

Chang BY, Peavy TR, Wardrip NJ, Hedrick JL (2004) The *Xenopus laevis* cortical granule lectin: cDNA cloning, developmental expression, and identification of the eglectin family of lectins. Comp Biochem Physiol A Mol Integr Physiol 137:115–129

Daffe M, McNeil M, Brennan PJ (1993) Major structural features of the cell wall arabinogalactans of *Mycobacterium*, *Rhodococcus*, and *Nocardia* spp. Carbohydr Res 249:383–398

Datta R, de Schoolmeester ML, Hedeler C et al (2005) Identification of novel genes in intestinal tissue that are regulated after infection with an intestinal nematode parasite. Infect Immun 73:4025–4033

French AT, Knight PA, Smith WD et al (2009) Expression of three intelectins in sheep and response to a Th2 environment. Vet Res 40:53

Ishino T, Kunieda T, Natori S et al (2007) Identification of novel members of the Xenopus Ca2 + – dependent lectin family and analysis of their gene expression during tail regeneration and development. J Biochem 141:479–488

Komiya T, Tanigawa Y, Hirohashi S (1998) Cloning of the novel gene Intelectin, which is expressed in intestinal paneth cells in mice. Biochem Biophys Res Commun 25:759–762

Kuperman DA, Lewis CC, Woodruff PG et al (2005) Dissecting asthma using focused transgenic modeling and functional genomics. J Allergy Clin Immunol 116:305–311

Lee JK, Baum LG et al (2004) The X-lectins: a new family with homology to the Xenopus laevis oocyte lectin XL-35. Glycoconj J 21:443–450

Lee JK, Buckhaults P, Wilkes C et al (1997) Cloning and expression of a *Xenopus laevis* oocyte lectin and characterization of its mRNA levels during early development. Glycobiology 7:367–372

Lee JK, Schnee J, Pang M et al (2001) Human homologs of the *Xenopus* oocyte cortical granule lectin XL35. Glycobiology 11:65–73

Leitao EA, Bittencourt VC, Haido RM et al (2003) β-galactofuranose-containing O-linked oligosaccharides present in the cell wall peptidogalactomannan of *Aspergillus fumigatus* contain immunodominant epitopes. Glycobiology 13:681–692

Lin B, Cao Z, Su P et al (2009) Characterization and comparative analyses of zebrafish intelectins: highly conserved sequences, diversified structures and functions. Fish Shellfish Immunol 26:396–405

Nagata S, Nakanishi M, Nanba R, Fujita N (2003) Developmental expression of XEEL, a novel molecule of the Xenopus oocyte cortical granule lectin family. Dev Genes Evol 213:368–370

Nagata S (2005) Isolation, characterization, and extra-embryonic secretion of the *Xenopus laevis* embryonic epidermal lectin, XEEL. Glycobiology 15:281–290

Pedersen LL, Turco SJ (2003) Galactofuranose metabolism: A potential target for antimicrobial chemotherapy. Cell Mol Life Sci 60:259–266

Peebles RS Jr (2010) The intelectins: a new link between the immune response to parasitic infections and allergic inflammation? Am J Physiol Lung Cell Mol Physiol 298:L288–L289

Pemberton AD, Knight PA, Gamble J et al (2004a) Innate BALB/c enteric epithelial responses to *Trichinella spiralis*: Inducible expression of a novel goblet cell lectin, intelectin-2, and its natural deletion in C57BL/10 mice. J Immunol 173:1894–1901

Pemberton AD, Knight PA, Wright SH, Miller HR (2004b) Proteomic analysis of mouse jejunal epithelium and its response to infection with the intestinal nematode, *Trichinella spiralis*. Proteomics 4:1101–1108

Russell S, Young KM, Smith M et al (2008) Identification, cloning and tissue localization of a rainbow trout (Oncorhynchus mykiss) intelectin-like protein that binds bacteria and chitin. Fish Shellfish Immunol 25:91–105

Suzuki E, Toledo MS, Takahashi HK, Straus AH (1997) A monoclonal antibody directed to terminal residue of beta-galactofuranose of a glycolipid antigen isolated from *Paracoccidioides brasiliensis*: Cross-reactivity with *Leishmania major* and *Trypanosoma cruzi*. Glycobiology 7:463–468

Suzuki YA, Shin K, Lönnerdal B (2001) Molecular cloning and functional expression of a human intestinal lactoferrin receptor. Biochemistry 40:15771–15779

Takano T, Sha Z, Peatman E et al (2008) The two channel catfish intelectin genes exhibit highly differential patterns of tissue expression and regulation after infection with Edwardsiella ictaluri. Dev Comp Immunol 32:693–705

Tsuji S, Uehori J, Matsumoto M et al (2001) Human intelectin is a novel soluble lectin that recognizes galactofuranose in carbohydrate chains of bacterial cell wall. J Biol Chem 276:23456–23463

Tsuji S, Yamashita M, Hoffman DR et al (2009) Capture of heat-killed mycobacterium bovis bacillus Calmette-Guerin by intelectin-1 deposited on cell surfaces. Glycobiology 19:518–526

Tsuji S, Yamashita M, Nishiyama A et al (2007) Differential structure and activity between human and mouse intelectin-1: Human intelectin-1 is a disulfide-linked trimer, whereas mouse homologue is a monomer. Glycobiology 17:1045–1046

Voehringer D, Stanley SA, Cox JS et al (2007) *Nippostrongylus brasiliensis:* identification of intelectin-1 and -2 as Stat6-dependent genes expressed in lung and intestine during infection. Exp Parasitol 116:458–466

Wali A, Morin PJ, Hough CD et al (2005) Identification of intelectin overexpression in malignant pleural mesothelioma by serial analysis of gene expression (SAGE). Lung Cancer 48:19–29

Wrackmeyer U, Hansen GH, Seya T et al (2006) Intelectin: a novel lipid raft-associated protein in the enterocyte brush border. Biochemistry 45:9188–9197

Annexins (Lipocortins)

21

G.S. Gupta

21.1 Annexins

21.1.1 Characteristics of Annexins

The annexins or lipocortins are a multigene family of proteins that bind to acidic phospholipids and biological membranes in a Ca^{2+}-dependent manner (Gerke and Moss 2002; Gerke et al. 2005; Raynal and Pollard 1994; Swairjo and Seaton 1994). Annexins are ubiquitous and characterized by an ability to bind to anionic phospholipids at membrane surfaces in response to elevated Ca^{2+}. Annexins are amphipathic and distinct from soluble and integral membrane proteins, but share features of both (Kojima et al. 1994; Brisson et al. 1991). Annexins have molecular weights ranging between 30 and 40 kDa (only annexin 6 is 66 kDa) and possess striking structural features. The characteristic annexin structural motif is a 70-amino-acid repeat, called the annexin repeat. Four annexin repeats packed into an α-helical disk are contained within the C-terminal polypeptide core (Gerke and Moss 2002). While all annexins share this core region, aminoterminal domains of annexins are diverse in sequence and length (ranging from 11 to 196) on each annexin member. It is this diversity of N-terminal amino-acid sequence that gives the individual annexins their functional differences and biological activities and appears to differentiate the cellular function and location (Gerke and Moss 2002; Gerke et al. 2005; Raynal and Pollard 1994). Cysteine 198 is relatively conserved in annexins, and three of four cysteines (198, 242, and 315) in annexin A4 are conserved in annexin A3. Phospholipids are suggested to bind via hydrophilic head groups to annexins, and the phospholipid-binding region is proposed to be localized on the convex surface side where calcium-binding sites are located in the crystal structure of annexin 5 (Huber et al. 1990). The calcium- and phospholipid-binding sites are located in the carboxy terminal domains. Some of the annexins bind to glycosaminoglycans (GAGs) in a Ca^{2+}- dependent manner. While annexin 2 has specific and high-affinity heparin-binding activity (Kassam et al. 1997), annexin A4 binds to heparin, heparan sulfate and chondroitin sulfate (CS) columns in a Ca^{2+}- dependent manner, annexin 5 to heparin and heparan sulfate columns in a Ca^{2+}-dependent manner and annexin 6 to heparin and heparan sulfate columns in a Ca^{2+}-independent manner and to CS columns in a Ca^{2+}-dependent manner (Ishitsuka et al. 1998) (see Table 21.1). Reports suggest that some annexin species may function as recognition elements for L-α-dipalmitoylphosphatidylethanolamine (PE)-derivatized GAGs under some conditions. The crystal structure of several of the annexins has been reported (Favier-Perron et al. 1996; Luecke et al. 1995; Swairjo et al. 1995). It has been established that the annexins are composed of two distinct sides. The convex side faces the biological membrane and contains the Ca^{2+}- and phospholipid-binding sites. The concave side faces the cytosol and contains the N and C termini. Although the annexins have been studied mostly as calcium-dependent phospholipid-binding proteins mediating membrane-membrane and membrane-cytoskeleton interactions, annexins A4, A5 and A6 bind also to carbohydrate structures suggesting that these annexin possess lectin-like domains.

Annexins bind to phosphatidylserine, phosphatidylethanolamine, and phosphatidylinositol, which are present in the inner leaflet or cytosolic surface of plasma membrane and hardly appear on cell surface, in contrast to phosphatidylcholin and sphingomyelin, which are major components of the outer leaflet of plasma membrane. However, latter are not recognized by annexins. Annexins also bind to organelle membranes such as the Golgi apparatus. This binding can be reversed by the removal of calcium, freeing the annexin from the phospholipid membrane. However, the functional significance of their reversible membrane-binding ability remains unknown in many annexins, although in some it is thought to be important for vesicle aggregation and membrane organization (Liemann and Huber 1997; Rand 2000; Rescher and Gerke 2004; Lim and Pervaiz 2007). Although all annexins share this binding property, there is variation in calcium sensitivity and phospholipid specificity between

G.S. Gupta et al., *Animal Lectins: Form, Function and Clinical Applications*,
DOI 10.1007/978-3-7091-1065-2_21, © Springer-Verlag Wien 2012

Table 21.1 Proteins that interact with vertebrate annexins (Adapted from Moss and Morgan 2004, © Genome Biol. 5: 219; 2004)

Annexin	Interacting proteins
ANXA1	Epithelial growth factor receptor, formyl peptide receptor, selectin, actin, integrin A4
ANXA2	Tissue plasminogen activator, angiostatin, insulin receptors, tenascin C, caveolin 1
ANXA3	None known
ANXA4	Lectins, glycoprotein 2
ANXA5	Collagen type 2, vascular endothelial growth factor receptor2, integrin B5, protein kinase C, cellular modulator of immune recognition (MIR), G-actin, helicase, DNA (cytosine-5-) methyltransferase 1 (DNMT1)
ANXA6	Calcium-responsive heat stable protein-28 (CRHSP-28), ras GTPase activating protein, chondroitin, actin
ANXA7	Sorcin, galectin
ANXA8	None known
ANXA9	None known
ANXA10	None known
ANXA11	Programmed cell death 6 (PDCD6), sorcin
ANXA13	Neural precursor cell expressed, developmentally down-regulated 4 (NEDD4)

individual annexins. For example, within one cell there can be differences in the distribution of annexins, with annexin A1 having an endosomal localisation, A2 to be found in cytosol and A4 being associated with the plasma membrane (Liemann and Huber 1997). Furthermore annexins are exported from cytosol to the outside of cells across the plasma membrane by unknown mechanisms. Although annexins lack hydrophobic signal peptides, secretion and expression on cell surface experienced by some annexins are e6dent in some cell types.

Like galectins, certain members of annexin family can be found both inside and outside cells. In particular, annexins A1, A2, A4, A5, and A11 can be found in the nucleus. This localization is consistent with the findings that annexin A1 possesses unwinding and annealing activities of a helicase and that annexin A2 is associated with a primer recognition complex that enhances the activity of DNA polymerase α. Despite these efforts and accomplishments, however, there is little e6dence or information on an endogenous carbohydrate ligand for these lectins that show nuclear and/or cytoplasmic localization (Wang et al. 2004).

21.1.2 Classification and Nomenclature

Annexins are expressed in a wide range of organisms such as higher plants, slime molds, metazoans, insects, birds, and mammals. Studies of the amino acid sequence of the annexins have established the homology of these proteins. Over 20 types have been found in all eukaryotic kingdoms as well as plants and animals. There are 12 human annexin subfamilies (A1–A11 and A13) that have been found to have various intra- and extracellular roles in a range of cellular processes (Gerke and Moss 2002; Gerke et al. 2005). The annexins are classified into five groups (A–E), and within each of these groups, individual annexins are identified numerically. Annexins in group A are human annexins, with group B referring to animal annexins without human orthologs, group C to fungi and moulds, group D to plants and group E to protists (Liemann and Huber 1997; Rand 2000; Hayes and Moss 2004; Rescher and Gerke 2004; Lim and Pervaiz 2007). At least one of the members can be found expressed in nearly every eukaryotic cell types. Almost all cells produce several kinds of annexins simultaniously and the expression levels are rather high. Why do they need annexins in plenty? It may be helpful in defining the question that the task of annexins is not a unique and annexins have to interact with various ligands both outside and inside of cells to act for multiple roles (Ponnampalam and Rogers 2006).

By definition, an annexin protein must be capable of binding in a Ca^{2+}-dependent manner to negatively charged phospholipids and has to contain a conserved structural element the so-called annexin repeat, a segment of some 70 amino acid residues. Molecular structures obtained for a number of annexins over the past decade helped to extend the similarities to the three-dimensional level. Moreover, they defined a hitherto unknown structural fold, the conserved annexin domain, which is built of four annexin repeats packed into a highly α-helical disk, and which now is considered to be a general membrane binding module. The annexin family has grown steadily in 1990s, and with the turn of the century, more than 160 unique annexin proteins have been discovered in more than 65 different species ranging from fungi and protists to plants and higher vertebrates (Fig. 21.1) (Morgan and Fernandez 1997; Morgan et al. 1999). For a detailed discussion of annexin properties, their structural organization, and intracellular as well as tissue distribution, the interested reader is referred to previous review (Gerke and Moss 2002).

The vertebrate A family includes the 13 annexins that make the family in mammals, but the number of annexins may vary in other classes of vertebrates as genes have been gained and lost. Ancient polyploidization events in bony fish, and more recent genome duplications in pseudotetraploid frogs (Xenopus), have duplicated many of the annexin genes. Thus, annexin A1 has undergone two successive duplications to yield up to four copies in some fish, amphibians and birds. Mammalian ANXA6 is a compound gene, probably derived from the fusion of duplicated ANXA5 and ANXA10 genes in early vertebrate evolution. Annexins A7, A8 and A10 have not yet been detected in fish, although genes similar to annexin A7 have been found in earlier-diverging species such as the sea urchin, the earthworm and

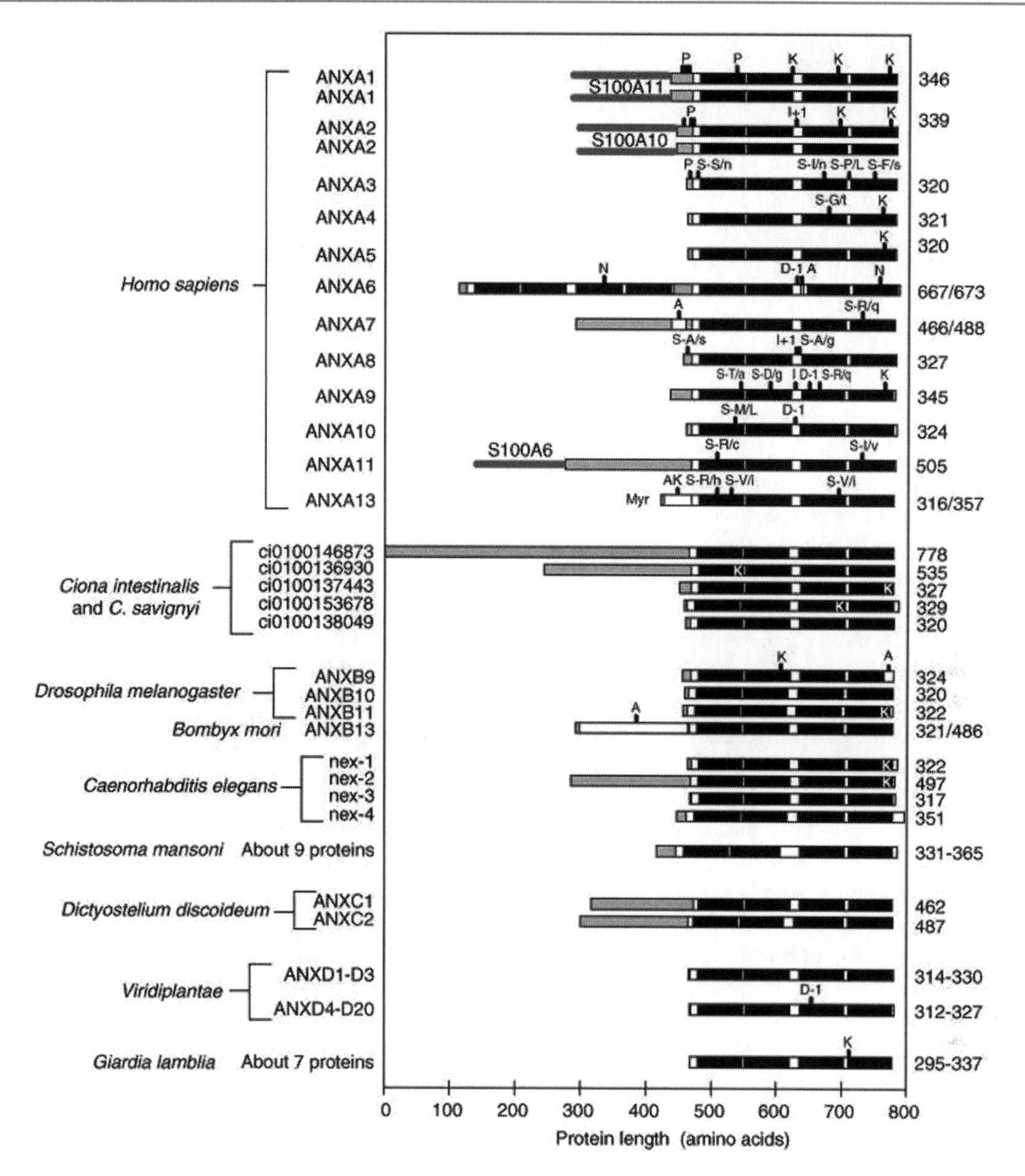

Fig. 21.1 Domain organization of representative annexin proteins: orthologs of the 12 human annexins shown in other vertebrates have the same structures, with strict conservation of the four repeats in the core region (*black*) and variation in length and sequence in the amino-terminal regions (*shaded*). Human ANXA1 and ANXA2 are shown as dimers, with the member of the S100 protein family that they interact with. Domain structures for other model organisms are derived from public data made available by the relevant genome-sequencing projects. Features: *S100Ax* sites for attachment of the indicated member of the S100 family of calcium-binding proteins, *P* known phosphorylation sites, *K*, KGD synapomorphy (a conserved, inherited characteristic of proteins), *I* codon insertions (+x denotes the number of codons inserted), *S-A/b* nonsynonymous coding polymorphisms (SNPs) with the amino acid in the major variant (*A*) and that in the minor variant (*b*), *N* putative nucleotide-binding sites, *D* codon deletions (−x denotes the number of codons deleted), *A* alternatively spliced exons, *Myr* myristoylation. The total length of each protein is indicated on the *right* (Printed from Moss and Morgan, Genome Biol. 5: 219 © 2004)

Hydra. The reasons for the tendency of annexin genes (or their chromosomal regions) to duplicate, their successful preservation, and the extent to which they contribute to vertebrate complexity are as yet unknown. The 12 human annexin genes range in size from 15 kb (*ANXA9*) to 96 kb (*ANXA10*) and are dispersed throughout the genome on chromosomes 1, 2, 4, 5, 8, 9, 10 and 15. Annexin genes from other vertebrates may vary slightly in size and chromosomal linkage, but orthologs are grossly similar in their sequence and splicing patterns (Moss and Morgan 2004).

21.1.3 Annexins in Tissues

Annexins are expressed in a wide range of organisms such as higher plants, slime molds, metazoans, insects, birds, and mammals. The expression level and tissue distribution of annexins span a broad range, from abundant and ubiquitous (annexins A1, A2, A4, A5, A6, A7, A11) to selective (such as annexin A3 in neutrophils and annexin A8 in the placenta and skin) or restrictive (such as annexin A9 in the tongue, annexin A10 in the stomach and annexin A13 in the small intestine). Annexin A1, A2, A3, A4, A5 and A6 are present in cardiac tissue. Annexin A5 is present in both cardiomyocytes and non-myocytal cells of the heart. Annexin A5 mRNA levels were highest in the fibroblast-like cells, followed by the endothelial cells, and a weak signal in the cardiomyocytes (Jans et al. 1995; Matteo and Moravec 2000). Porcine heart expresses annexins A5 and A6 in large amounts, and annexins A3 and A4 in much smaller amount. Annexins A5 and A6 are involved in the regulation of membrane-related processes (Pula et al. 1990). Annexins play a role in the regulation of Ca^{2+} pumps and exchangers on the sarcolemma, and are altered in some cardiac disease states. The plasma membrane of the heart muscle cell and its underlying cytoskeleton are vitally important to the function of the heart. Two major annexin A6 binding proteins were identified as actin and annexin A6 itself. Annexin A6 bound to itself both in the presence and in the absence of calcium ions. Annexin A6 bound preferentially not only to the N terminal fragment (domains I-4, residues 1–352) but also to C-terminal fragments corresponding to domains V + VI and domains VII + VIII (Locate et al. 2008). During embryonic delopment, annexins 1, 2 and 4 have been identified in murine CNS with distinct patterns of temporal and spatial expression. Annexin A4 is the first annexin to be expressed on embryonic day E9.5 while annexin 1A is the last to be expressed (E21.5) (Hamre et al. 1995).

Annexin A1 is expressed by small sensory neurons of rat dorsal root ganglia (DRG), by most neurons of the spinal cord (SC), and by ependymal cells lining the central canal. Annexin A2 is expressed by most sensory neurons of the DRG but is primarily expressed in the SC by glial cells. Annexin A3 is expressed by most sensory neurons, regardless of size, by endothelial cells lining the blood vessels, and by the perineurium. In the SC, annexin A3 is primarily expressed by astrocytes. In the DRG and the SC, annexin A4 is primarily expressed by glial cells and at lower levels by neurons. In the DRG, annexin A5 is expressed in relatively high concentrations in small sensory neurons in contrast to the SC, where it is expressed mainly by ependymal cells and by small-diameter axons located in the superficial laminae of the dorsal horn areas. Annexin A6 is differentially expressed by sensory neurons of the DRG, being more concentrated in small neurons. In the SC, annexin 6 has the most striking distribution and concentrated subjacent to the plasma membrane of motor neurons and their processes. The differential localization pattern of annexins in cells of the SC and DRG could reflect their individual biological roles in Ca^{2+}-signal transduction within the CNS (Naciff et al. 1996). Annexin A3 is implicated in the microglial response to motor nerve injury. It is induced after hypoglossal nerve injury in rat, specifically in activated (axotomy-stimulated) microglia. A3 was the most prominent among annixins expressed in microglia. Results suggested that Annexin 3 may be a Ca^{2+}-dependent mediator between phospholipids and F-actin in microglia stimulated by peripheral nerve injury (Konishi et al. 2006).

21.1.4 Functions of Annexins

Annexins are generally cytosolic proteins, with pools of both a soluble form and a form stably or reversibly associated with components of the cytoskeleton or proteins that mediate interactions between the cell and the extracellular matrix (matricellular proteins). Although the broad themes in terms of cellular function have been uncovered, but the precise role of these proteins is unclear (Gerke and Moss 2002; Hill et al. 2008). Because of the ability of annexins to bind to and "annex" or aggregate membrane surfaces, they appear to participate in Ca^{2+}-regulated membrane dynamics. Thus they have been shown to be involved in exocytosis (Creutz 1992; Gerke and Moss 2002), membrane domain organization and ion channel activity regulation (Hill et al. 2008).

Some annexins such as A1 and A2, have been found in the nucleus (Tomas and Moss 2003; Eberhard et al. 2001) and in certain instances, annexins may be expressed at the cell surface, despite the absence of any secretory signal peptide; for example, annexin A1 translocates from the cytosol to the cell surface following exposure of cells to glucocorticoids (Solito et al. 1994), and annexin A2 is constitutively expressed at the surface of vascular endothelial cells where it functions in the regulation of blood clotting (Brownstein et al. 2001). The A5 and A6 knockout mice have subtler phenotypes and need further investigation (Brachvogel et al. 2003), and two independently derived A7 null mutant mouse strains are either embryonic lethal (Srivastava et al. 1999) or show changes in calcium homeostasis (Herr et al. 2001). The diversity of phenotype in the annexin knockout mice is consistent with these proteins having largely independent functions. Roles for annexins that have been established from studies using cultured cells are not always reflected in phenotypic abnormalities in the corresponding knockout mice, suggesting that functional redundancy may, in some instances, obscure the full range of functions of these multifunctional proteins. In mice that lack an overt phenotype, there is now the opportunity to test molecular theories of annexin function, such as the proposed calcium channel activity of A5.

The annexins have a wide range of biological functions related to their phospholipid/membrane-binding properties (Crompton et al. 1988), such as inhibition of coagulation, interactions with membranous and cytoskeletal elements, vesicular transport and exocytosis and endocytosis. Shifts in subcellular locations (from the cytosol to membrane) are observed on some intracellular annexins, suggesting the active movement of annexins corresponding to dynamic lipid vesicle transport. Since, the expression of some annexins depends on rate of cellular proliferation (Schlaepfer and Haigler 1990), it has been proposed that the cellular level of the annexins might be critical for the regulation of cell growth. However, this hypothesis has been questioned. Results show that cellular expression of annexins plays a general role in cell growth and support the concept that post-transcriptional mechanisms may control levels of annexin 1 and 7 (Raynal et al. 1997). They are anti-inflammatory proteins that inhibit phospholipase A2 activity in vitro by sequestering the substrate phospholipids from phospholipase A2 (Blackwell et al. 1980). Furthermore, the annexins exhibit anti-coagulant activity (Tait et al. 1989), calcium channel activity (Rojas et al. 1990), and cyclic phosphate phosphohydrolase activity (Ross et al. 1990). They function in the membrane fusion process, exocytosis, endocytosis, membrane-cytoskeleton interactions (Gerke et al. 2005) and regulation of calcium-dependent anion current activation (Chan et al. 1994).

Some annexins are capable of calcium-independent binding and several have roles in vesicle aggregation. Annexins A1, A2 and A11 function in cooperation with other calcium-binding proteins to form complexes while annexins A1, A2 and A5 interact with cytoskeletal proteins. Many annexins are involved in exocytic and endocytic pathways and some have roles in ion channel regulation (Gerke and Moss 2002). Extracellularly, annexin A1 has a role in controlling the inflammatory response while annexin A2 is present on the external surface of endothelial cells, where it may act as a receptor for ligands, including plasminogen and tissue plasminogen activator. Extracellular annexin A5 is thought to be involved in the anticoagulation process (Rand 2000; Hayes and Moss 2004) (see also Table 21.1).

21.2 Annexin Family Proteins and Lectin Activity

After being exported outside the cells, some annexins have been shown to function as receptors for extracellular proteins. There exists evidence that shows that annexins interact with glycoconjugates. Annexin A4 was first identified as a lectin binding to carbohydrate moieties of sialoglycoproteins and glycosaminoglycans in the presence of calcium; afterwards annexin 5 and 6 and some other annexins exhibited similar lectin activity. The localization sites of annexin A4 in major expression tissues, the kidney and pancreas, and the result of in vitro binding assay of several glycoconjugates suggest that annexin A4 is involved in the formation of apical sorting (secretory) vesicles due to the interaction with some GPI-anchored glycoproteins and proteoglycans. Exocrine-type neurotrophic activity is found in annexin 5 and involvement of some annexins in cell-adhesion (or inhibition of cell-adhesion) has been found. Future studies should aim at identifying endogenous glycoconjugate ligands recognized by annexins in a variety of cases and investigating the carbohydrate recognition mechanism. While annexin A4 binds glycosaminoglycans (GAGs) in a calcium-dependent manner (Kojima et al. 1996), annexin A5 bound to heparin and heparan sulfate column but not to chondroitin sulfate column. Annexin A5 binds to collagen and annexin A2 binds to tissue plasminogen activator, tenascin and heparin. Moreover, Annexin A6 was adsorbed to heparin and heparan sulfate columns in a calcium-independent manner and to chondroitin sulfate columns in a calcium-dependent manner. Binding of annexins to a wide variety of other proteins is also known (Moss and Morgan 2004) (Table 21.1). Thus, annexins A2, A4, A5, and A6 have different GAG binding properties and function as recognition elements for GAGs in extracellular space (Ishitsuka et al. 1998). Many annexins have posttranslational modifications, such as phosphorylation and myristoylation. Such modifications and surface remodeling of individual members presumably account for much of the subfamily specificity in annexin interactions (Moss and Morgan 2004).

21.3 Annexin A2 (p36)

Annexin A2/(p36) contains three distinct functional regions, the N-terminal region, the C-terminal region, and the core region. The core region of p36 contains Ca^{2+}- and phospholipid-binding sites, whereas C-terminal region contains 14-3-3 homology domain and the plasminogen-binding domain. The N terminus of annexin A2 (p36) contains two important regulatory domains, the L and P domains. The L domain consists of the first 14 residues of the N terminus that contains a high affinity binding site for the p11 protein. The P domain of p36 contains the phosphorylation sites for protein kinase C (Ser^{25}) and pp60src (Tyr23). The N-terminal L and P domains play regulatory roles; activation of the phosphorylation sites of annexin A2 tetramer results in an increase in the $A_{1/2(Ca2+)}$ for chromaffin granule aggregation and F-actin binding, whereas binding of the p11 subunit decreases the $A_{1/2(Ca2+)}$ for these activities. Annexin A2 is a receptor for the alternatively spliced segment of fibronectin type III domains in tenascin-C (Chung and Erickson 1994; Chung et al. 1996; Ling et al.

2004) and for plasminogen and tissue plasminogen activator (Hajjar et al. 1994).

21.3.1 Annexin 2 Tetramer (A2t)

Annexin A2 often co-expresses with another protein, p11 (S100A10), forming an (annexin $A2)_2$-$(S100A10)_2$ heterotetramer (Annexin 2 tetramer (A2t)). Monomeric and tetrameric annexin A2 share many of the same binding properties, e.g. the abilities to bind Ca^{2+}, phospholipid membranes, and certain GAGs (including heparin and heparan sulfate), albeit typically with some modification in affinity. The p11 component alone, despite belonging to the E-F hand family of Ca^{2+}-binding proteins, does not bind Ca^{2+} nor does it bind phospholipids or GAG. The association between annexin A2 and p11 does not require Ca^{2+} and utilizes the concave annexin molecular surface (reviewed in Shao et al. 2006).

Interaction of A2t with Heparin: The heterotetrameric complex formed by the binding of p11 to p36 (A2t) is the predominant form in most cells (Kassam et al. 1997). Interactions of heparin with Ca^{2+}- and A2t have been studied. A2t has been shown to be present at both the cytosolic and extracellular surfaces of the plasma membrane of many cells (Waisman 1995). Extracellular A2t has been proposed to function as a cell adhesion factor (Tressler and Nicolson 1992; Tressler et al. 1993), a receptor for plasminogen and tissue plasminogen activator (Cesarman et al. 1994), and a receptor for tenascin-C (Chung and Erickson 1994; Hajjar et al. 1994; Hubaishy et al. 1995). It is possible that heparin might be involved in the regulation of the interaction of A2t with these ligands. In the absence of Ca^{2+}, heparin induces a more moderate change in the conformation of A2t. Of interest was the heparin-induced increase in the β-sheet from ~21% to 27% and decrease in unordered structure from 22% to 18%. Hubaishy et al. (1995) reported that A2t bound to a heparin affinity column and that the phosphorylation of A2t on tyrosine residues blocked the heparin-binding activity of the protein. The study on interaction of A2t with heparin identified A2t as a specific, high affinity heparin-binding protein. Furthermore, the Ca^{2+}-dependent binding of heparin to A2t causes a dramatic conformational change in the protein. The p36 subunit of A2t contains a Cardin-Weintraub glycosaminoglycan recognition site (Cardin and Weintraub 1989) and that a peptide to this region of A2t binds heparin.

The interaction of A2t with heparin was also shown to be inhibited by tyrosine phosphorylation of A2t (Hubaishy et al. 1995). Since the role that heparin binding plays in the structure or function of A2t is unknown, the current study was aimed at the characterization of the interaction of heparin with A2t. Analysis of the CD spectra of A2t showed that the binding of heparin to A2t resulted in a profound change in the conformation of A2t. We also found that in the absence of Ca^{2+}, a small change in the conformation of A2t occurred upon heparin binding. It was also observed that A2t formed a large complex with heparin. A2t bound heparin with an apparent K_D of 32 ± 6 nM and a stoichiometry of 11 ± 0.9 mol of A2t/mol of heparin. A2t does not bind to disaccharides of heparin, but does bind to 3-kDa heparin that contains ~10 monosaccharides. The binding of ~11 molecules of A2t to a single 17-kDa heparin strand that contains ~50 monosaccharide units suggests that A2t requires ~4–5 monosaccharide units for binding.

Several consensus sequences have been identified among members of the heparin-binding family of proteins. For example, the heparin-binding sequence of the C-terminal region of fibronectin has been identified as WQPPRARI. The p36 subunit of A2t contains a Cardin-Weintraub heparin-binding consensus sequence. Furthermore, a peptide to this region of p36 subunit of A2t (300LKIRSEFKK-KYGKSLYY316) undergoes a conformational change upon heparin binding. These results therefore suggest that residues 300–316 of the p36 subunit of A2t are involved in heparin binding (Kassam et al. 1997).

Fucoidan, a sulfated fucopolysaccharide, mimics the fucosylated glycans of glycoproteins and has been used as a probe for investigating the role of membrane polysaccharides in cell – cell adhesion. A2t is bound to fucoidan with an apparent K_D of 1.24 ± 0.69 nM; the binding of fucoidan to A2t was Ca^{2+}-independent. Furthermore, in the presence but not the absence of Ca^{2+}, the binding of fucoidan to A2t decreased the α-helical content from 32% to 7%. A peptide corresponding to a region of the p36 subunit of A2t, F (306) – S(313), which contains a sequence for heparin binding, was shown to undergo a conformational change upon fucoidan binding. This suggests that heparin and fucoidan bound to this region of A2t. Thus, the binding of A2t to the carbohydrate conjugates of certain membrane glycoproteins have profound effects on the structure and biological activity of A2t (Sandra et al. 2000).

Ca^{2+}-dependent binding of A2t to heparin caused a large decrease in the α-helical content of A2t from ~44% to 31%, a small decrease in the β-sheet content from ~27% to 24%, and an increase in the unordered structure from 20% to 29%. The binding of heparin also decreased the Ca^{2+} concentration required for a half-maximal conformational change in A2t from 360 to 84 μM. Data suggests that in the presence of Ca^{2+}, heparin induces a large conformational change in A2t, resulting in a substantial change in the conformation of the protein. However, heparin can also interact with A2t in the absence of Ca^{2+} and, to a much smaller degree, affect the conformation of the protein (Kassam et al. 1997). A2t appears to be a unique member of the heparin-binding proteins because A2t can discriminate between heparin and heparan sulfate ligands.

21.3.2 Crystal Analysis of Sugar-Annexin 2 Complex

Identification of annexin A2 as a heparin-binding protein (Kassam et al. 1997) raised the possibility that this protein-GAG interaction may participate in the regulation of thrombotic processes. To characterize the heparin-annexin A2 interaction and to determine the basis for its Ca^{2+} dependence, crystallographic studies were carried out on human annexin A2 in complex with heparin-derived oligosaccharides of varying lengths. Binding measurements in solution were also performed using surface plasmon resonance (SPR) (Shao et al. 2006). Crystallographic analysis revealed that the common heparin-binding site is situated at the convex face of domain IV of annexin A2. At this site, annexin A2 binds up to five sugar residues from the nonreducing end of the oligosaccharide. Unlike most heparin-binding consensus patterns, heparin binding at this site does not rely on arrays of basic residues; instead, main-chain and side-chain nitrogen atoms and two calcium ions play important roles in the binding. Especially significant is a novel calcium-binding site that forms upon heparin binding. Two sugar residues of the heparin derivatives provide oxygen ligands for this calcium ion. Comparison of all four structures shows that heparin binding does not elicit a significant conformational change in annexin A2. The combined data with surface plasmon resonance measurements provide a clear basis for the calcium dependence of heparin binding to annexin A2 (Shao et al. 2006).

21.3.3 Functions of Annexin A2

Dual Action of Annexin 2 Tetramer and Arachidonic Acid: Annexin 2 has been implicated in membrane fusion during the exocytosis of lamellar bodies from alveolar epithelial type II cells. Immunodepletion of Annexin 2 from type II cell cytosol reduced its fusion activity. The A2t induced the fusion of lamellar bodies with the plasma membrane in a dose-dependent manner. This fusion is Ca^{2+}-dependent and is highly specific to A2t because other annexins (1 and 2 monomer, 3, 4, 5, and 6) were unable to induce the fusion. The fusion between lamellar bodies with the plasma membrane is driven by the synergistic action of A2t and arachidonic acid (Chattopadhyay et al. 2003).

In an earlier report, though annexins 1–4 mediated liposome aggregation in the presence of 1 mM Ca^{2+}, only A2 tetramer had aggregation activity at 10 μM Ca^{2+}, where as A5 and A6 had negligible aggregation activity at Ca^{2+} concentrations up to 1 mM Ca^{2+}. Of six purified annexins (A1–A6) tested for their ability to reconstitute secretion from permeabilized cells, only Annexin A2 was effective. Annexin A2 was not involved in the exocytosis of lamellar bodies (Liu et al. 1996; Blanchard et al. 1996). The neuroblastoma cells are known to express annexins and confirmed in cell membrane. Following stimulated release of noradrenaline by K^+ depolarisation or by treatment with the ionophore A23187, results favoured for a general role in calcium signaling at discrete intracellular locations by annexins 2 and 5. The results did not support the specific involvement for Annexin 2 in membrane fusion at sites of vesicle exocytosis (Blanchard et al. 1996).

Antiphospholipid (aPL) antibodies recognize receptor-bound β2 glycoprotein I (β2GPI) on target cells, and induce an intracellular signaling and a procoagulant/proinflammatory phenotype that leads to thrombosis. Evidence indicates that annexin A2 binds β2GPI on target cells. Romay-Penabad et al. (2009) studied the effects of human aPL antibodies in A2-deficient ($A2^{-/-}$) mice. After IgG-APS (antiphospholipid syndrome) or 4C5 injections and vascular injury, mean thrombus size was significantly smaller and tissue factor activity was significantly less in $A2^{-/-}$ mice compared with $A2^{+/+}$ mice. The expression of VCAM-1 induced by IgG-APS or 4C5 in explanted $A2^{-/-}$ aorta was also significantly reduced compared with $A2^{+/+}$ mice. It was suggested that annexin A2 mediates the pathogenic effects of aPL antibodies in vivo and in vitro APS.

Soluble A2t activates human monocyte-derived macrophages (MDM), resulting in secretion of inflammatory mediators and enhanced phagocytosis. The modulation of macrophage function by A2t is mediated through TLR4, suggesting an important role for this stress-sensitive protein in the detection of danger to the host, whether from injury or invasion (Swisher et al. 2010). Up-regulation of annexin A2 in differentiated retinal pigment epithelial (RPE) cells may be required for development of phagocytic capability. Law et al. (2009) showed that annexin A2 is highly enriched on newly formed phagosomes in RPE cells and that siRNA-mediated depletion of annexin A2 results in impairment of photoreceptor outer segments (POS) internalization. Findings provide direct evidence that annexin A2 is necessary for the normal circadian phagocytosis of POS by RPE cells.

21.4 Annexin A4 (p33/41)

21.4.1 General Characteristics

Annexin A4 (also called endonexin, protein II, chromobindin 4, placental anticoagulant protein II, and PP4-X) is one of a family of proteins that interact with phospholipids in the presence of calcium. Annexin A4 is a 36-kDa protein that can aggregate on the inner leaflet of cellular membranes. It has additional unique property of recognizing carbohydrates. Calcium-dependent phospholipid binding activities are common and characteristic properties of the annexin family proteins. Phospholipids are suggested to bind via hydrophilic head groups to annexins, and the phospholipid-binding region is proposed to be localized on the convex surface side where calcium-binding sites are located in the crystal structure of annexin A5. Annexin A4 has been proposed to be involved in exocytosis and in the coagulation process. These functions are related to the ability of the annexins to bind to acidic phospholipids. Annexin A4 strongly binds to either lipid at acidic pH. At neutral pH only weak binding to phosphatidic acid (PA) and no binding to phosphatidylserine (PS) occurs. But addition of Ca^{2+} leads to a strong binding to the lipids also at neutral pH. Binding of annexin A4 induces dehydration of the vesicle surface (Zschörnig et al. 2007). Annexin A4 shares 50–60% sequence homology to annexin 5 and was shown to bind heparin, but the binding of heparin to this protein was inhibited by a variety of carbohydrates including glucose, N-acetylneuraminic acid, heparan sulfate, and chondroitin sulfate (Kojima et al. 1996). In contrast, heparan sulfate or other glycosaminoglycans do not induce a conformational change in A2t (Becker et al. 1990).

Gene trap disruption of the first intron revealed that there were in fact three splice variants of annexin A4 with differing tissue expression (Li et al. 2003). The knockout animal lacked the major transcript, annexin A4a, which has a broad tissue distribution. However, two further transcripts, designated annexin A4b and Annexin A4c, were unaffected by the intron disruption in this region. Annexin A4b was shown to be expressed only in the digestive tract and annexin A4c exhibited a restricted expression pattern within solitary chemosensory cells. In nonstratified epithelia, they extend from the basement membrane to the lumen and appear to perform paracrine/endocrine functions (Li et al. 2003).

Annexin A4 is localized in the apical cytoplasmic region of pancreatic acinar cells where zymogen granules are concentrated. Since it is the major component of the zymogen granule membrane, the glycosylphosphatidylinositol-anchored glycoprotein GP-2 was suggested to play a role in apical sorting and secretion of zymogens. The major carbohydrate structures of porcine GP-2 were trisialo-triantennary and tetrasialo-tetra-antennary complex-type oligosaccharides. Annexin IV interacts with GP-2 in the presence of calcium and it recognizes the terminal sialic acid residues linked through α2-3 linkages to the carbohydrate of GP-2. Thus, GP-2 is an endogenous ligand of annexin IV in the exocrine pancreas (Tsujii-Hayashi et al. 2002).

21.4.2 Tissue Distribution

Annexin A4 is found at high levels in secretory epithelia in the lung, intestine, stomach, trachea, and kidney. It is thought to be a marker for polarized epithelial cells. Although the biological roles for annexin A4 remain largely unclear, it has been implicated in the regulation of calcium-activated epithelial chloride channels (Chan et al. 1994) and shown to have anti-inflammatory properties (Katoh 2000; Gotoh et al. 2005). Annexin A4 has also been shown to play a role in kidney organogenesis in *Xenopus laevis*, where ablation of the gene product results in abnormal development of pronephric tubules (Seville et al. 2002). Hill et al. (2003) showed that Annexin A4 can regulate passive membrane permeability to water and protons and can alter physical properties of the membranes by associating with them (Ponnampalam and Rogers 2006). Hill et al. (2008) demonstrated that the major transcript of annexin A4, annexin A4a, is present in the superficial and transitional epithelium of the bladder, where it is expressed throughout the urothelium. Umbrella cells have large numbers of unique elongated vesicles underlying the apical membrane. These "fusiform vesicles" are thought to play a key role in the bladder's ability to stretch by providing a large amount of membrane available for exocytosis and endocytosis. This allows umbrella cells to increase their apical surface area in response to filling and then to decrease it upon emptying (Hill et al. 2008). Given the biophysical properties and urothelial expression of annexin A4, it was hypothesized that this annexin could be important to the integrity or the regulation of bladder permeability barrier.

Bladder filling has been shown to activate a complex set of mechano-sensitive responses in umbrella cells, including ATP release and purinergic receptor-dependent membrane trafficking (Wang et al. 2005). Hydrostatically induced stretch has been shown to raise intracellular Ca^{2+} in the urothelium; furthermore, blocking Ca^{2+} release from intracellular stores inhibited exocytosis (Wang et al. 2005). Given the known membrane-organizing ability of annexins and sensitivity to $Ca^{2+,}$ annexin A4 might play a specific role in umbrella cell membrane trafficking (reviewed in Hill et al. 2008).

It was demonstrated in wild-type bladders that stretch induces a redistribution of annexin A4 within basal and intermediate cells to the cellular periphery. A genetically modified mouse model in which the protein is not expressed in renal epithelia (Li et al. 2003) revealed no alterations in normal bladder function or morphology in the annexin $A4a^{-/-}$ animals, suggesting that the role of annexin A4 in the bladder does not include barrier function or stretch-regulated intracellular trafficking. $Anx4^{-/-}$ mouse model shows no protein in the urothelium where as wild-type umbrella cells showed uniform cytoplasmic staining and some association with the nuclear membrane. It was indicated that loss of anaxin A4 from the urothelium does not affect barrier function, membrane trafficking, or normal bladder-voiding behavior (Hill et al. 2008).

Microarray studies of human endometrium have shown that Annexin A4 mRNA is significantly up-regulated during the secretory phase of the menstrual cycle compared with the proliferative phase (Kao et al. 2002; Riesewijk et al. 2003; Ponnampalam et al. 2004; Mirkin et al. 2005). Anx4 mRNA is up-regulated during mid-secretory (MS) and late-secretory (LS) phases compared with proliferative phases during the menstrual cycle. Anx4 was localized to glandular and luminal epithelium and was present in high levels throughout the menstrual cycle except during early-secretory (ES) phase, when it was significantly reduced. Results suggest that, in proliferative explants, progesterone significantly increased the Anx4 mRNA and protein after 48 h in culture. Estrogen did not have any significant effects. Anx4 transcription and translation are regulated by progesterone and suggests that Anx4 may be important in regulating ion and water transport across the endometrial epithelium (Ponnampalam and Rogers 2006).

21.4.3 Characterization

Soluble monomers of annexin A4 trimerize in the cytoplasm bind to the membrane and then assemble into higher order structures at the membrane interface creating a crystallization cascade in 2D across a large cross-sectional area of membrane (Zanotti et al. 1998). In vitro cross-linking studies demonstrate that trimer, hexamer and higher aggregates of annexin form in the presence of Ca^{2+} and phospholipid-containing vesicles (Concha et al. 1992). Annexin A4 has a short amino-terminal of 12 amino acids, which is susceptible to phosphorylation by protein kinase C. Phosphorylation by protein kinase C causes the release of the N-terminal of annexin A4 and inhibits annexin A4's ability to aggregate on the membrane (Kaetzel et al. 2001). Annexin A4 has also been found in the cytoplasm (Zimmermann et al. 2004) and nucleus (Raynal et al. 1996), and it can be secreted (Masuda et al. 2004). Intranuclear annexin A4 has been shown to translocate to the cytoplasm because of an increase in intracellular calcium (Mohiti et al. 1995; Barwise and Walker 1996; Raynal et al. 1996), and during Fas-induced cell death (Gerner et al. 2000), however, the functional role of nuclear annexin A4 is currently not well known (reviewed in Ponnampalam and Rogers 2006).

Annexin (A4) possesses four repeat domains with one Ca^{2+} -binding site (CBS) in each domain. A4 binds the Na^+ ion in CBSs. One structure (1.58 A) bound Na^+ ion in CBS I, whereas another structure (1.35 A) bound the Na^+ ion in CBS II and CBS III. The Cα atoms of CBS III largely moved by coordination of Na^+ ion. In the Cα atoms of CBS I, however, little change resulted from Na^+ -coordination. Only the side chain of Glu71 was moved by Na^+ -coordination in CBS I. These results indicate that Annexin A4 binds not only Ca^{2+} but also Na^+ ion in CBS (Butsushita et al. 2009).

21.4.4 Pathophysiology

Annexins have been implicated in tumor progression. There is significant increase in expression in annexins A1, A2, A4 and A11 in primary tumors compared with normal colon. Expression of annexins A2, A4 and A11 was related with increasing tumor stage (Duncan et al. 2008). Annexin A4 is elevated in ovarian CCC tumors and is associated with chemoresistance in cultured ovarian cancer cells. Results demonstrate that Annexin A4 confers chemoresistance in ovarian cancer cells in part by enhancing drug efflux (Kim et al. 2009).

The presence of antiphospholipid (aPL) antibodies increases the risk for recurrent miscarriage (RM). Annexins bind to anionic phospholipids (PLs) preventing clotting on vascular phospholipid surfaces. Plasma levels of annexin 5 are significantly higher at the beginning of pregnancy, at the sixth and eighth week of pregnancy in women with aPL antibodies compared with those without aPL antibodies, where as there were no significant differences in plasma annexin A4 levels between women with and without aPL antibodies. These antibodies could displace annexin from anionic phospholipid surfaces of syncytiotrophoblasts (STBs) and hereby promote coagulation activation (Ulander et al. 2007).

Annexins have well characterized anti-inflammatory properties and Lipoxin A4 (Annexin A4) has been shown to exert protective effects in stomach. Suppression of aspirin-triggered lipoxin synthesis, through co-administration of a selective COX-2 inhibitor, results in a significant exacerbation of gastric injury. The gastroprotective effects of lipoxin A4 appear to be receptor mediated, and attributable to suppress leukocyte adherence to the vascular endothelium and to elevate gastroduodenal blood flow, and mediated via lipoxin-induced nitric oxide generation (Wallace et al. 2005).

21.4.5 Doublet p33/41 Protein

Ca^{2+}-dependent carbohydrate-binding proteins from bovine kidney under nonreducing conditions form doublet protein bands corresponding to 33 kDa (p33) and 41 kDa (p41) where as under reducing conditions, a single protein band (p33) was observed. The p33/41 is a lectin which binds to sialoglycoproteins and glycosaminoglycans in a calcium-dependent manner. Amino acid sequences of p33/41 are highly homologous to those of calcium/phospholipid-binding annexin protein, annexin 4 (endonexin) especially in the consensus sequences. The p33/41 exhibited calcium/phospholipid-binding activity similar to annexin 4 (Kojima et al. 1994). The p33/41 cDNA encodes a protein of 319 amino acids with a molecular mass of 35,769 Da. The deduced amino acid sequence was identical to that of bovine annexin 4 except for one amino acid substitution. The recombinant protein revealed that p41 is a dimer of p33 cross-linked at Cys-198 via a disulfide bond. The recombinant protein bound to columns of heparin and fetuin glycopeptides in a calcium dependent manner and to phospholipid vesicles composed of phosphatidylserine (PS)/phosphatidylcholine (PC), phosphatidylethanolamine (PE)/PC or phosphatidylinositol (PI)/PC. Thus, p33/41 binds two types of ligands via different sites and that phospholipids modulate the carbohydrate binding activity of p33/21. The p33/41 does not contain any of the consensus sequences conserved in the CRDs of any animal lectins. Furthermore, consensus heparin-binding motifs (such as BXBXBXXXXB), identified glycosaminoglycan-binding sequences, or basic amino acid clusters capable of binding to acidic polysaccharides through simple ionic interaction were not observed within the primary sequence of p33/41, that suggested that p33/41 may contain a unique CRD. The p33/41 is highly concentrated in the apical plasma membrane of the epithelial cells in the renal proximal tubules, and integrated into the renal brush border membrane. Heparin does not compete with phospholipids in the binding to p33/21.

Chen et al. (1993) reported three-dimensional models of annexins (1, 2, 3, 5 and 7) constructed by homology modeling using the crystal structure of annexin 5 as a template, and showed that cysteines 198, 242, and 315 in the annexin 3 model do not appear to be exposed on the protein surface. A conformational change to expose cysteine 198 to the outside and to allow it to form an intermolecular disulfide bridge appears to occur upon dimer formation of annexin A4.

21.5 Annexin A5/Annexin V

Annexin A5/Annexin V is a protein of unknown biological function that undergoes Ca^{2+}-dependent binding to phospholipids located on the cytosolic face of the plasma membrane. Annexin 5 has been isolated as placental anticoagulant protein I, inhibitor of blood coagulation, vascular anticoagulant-α, endonexin II, lipocortin 5, placental protein 4, and anchorin CII. The function of the protein is unknown. However, based on in vitro experiments, annexin A5 has been proposed to play a role in the inhibition of blood coagulation by competing for phosphatidylserine binding sites with prothrombin and also to inhibit the activity of phospholipase A1. Annexin 5 is a Ca^{2+}-dependent membrane-binding protein that forms voltage-dependent Ca^{2+} channels in phospholipid bilayers and is structurally and functionally characterized. Data indicate that key amino acid residues act as selectivity filters and voltage sensors, thereby regulating the permeability of the channel pore to ions (Demange et al. 1994).

21.5.1 Gene Encoding Human Annexin A5

Characterization of three genomic clones for human annexin 5 revealed that annexin 5 spans at least 29 kb of the human genome and contains 13 exons ranging in length from 44 to 513 bp and 12 introns from 232 to 8 kb. The absence of a typical tata box and the presence of high G + C content and Sp1-binding sites in its promoter characterize it as a 'housekeeping' gene and account for its broad pattern of expression. Potential binding sites for Cis-regulatory elements identified in 5′-upstream region of annexin 5 are consistent with its known regulation by oncogenic and growth-related stimuli. Annexin 5, like its chick homologue, differs from the genes encoding annexins I, II and III in features of its promoter and in the size of its exons 1, 2 and 3 in ways that may impart individuality to its regulation and function (Fernández et al. 1994). The human gene encoding annexin 5 was localized to 4q26-q28. This localization overlaps but differs slightly from the previous assignment of annexin 5 to 4q28-q32 (Tait et al. 1991).

21.5.2 Interactions of Annexin A5

Annexin A5 is a collagen binding protein of the annexin family associated with plasma membranes of chondrocytes, osteoblasts, and many other cells. As a major constituent of cartilage-derived matrix vesicles it has been shown to bind to native type II and X collagen. In accordance with this observation, annexin A5 is localized in the extracellular matrix of calcifying cartilage in the fetal human growth plate, and that it was restricted to the chondrocyte surface in proliferating and resting cartilage. Furthermore, annexin A5 not only binds to native type II and X collagen, but also to chondrocalcin, the carboxy-terminal extension of type II procollagen, in a calcium-independent manner (Table 21.1). Pepsin digestion of type II collagen results in loss of annexin A5 binding. This confirms a notion that the telopeptide region of type II collagen carries annexin A5 binding sites (Kirsch and Pfaffle 1992; Rahman et al. 1997).

Collagen/Annexin A5 Interactions Regulate Chondrocyte Mineralization: Physiological mineralization in growth plate cartilage is highly regulated and restricted to terminally differentiated chondrocytes. Extracellular matrix components (collagens) of growth plate cartilage are directly involved in regulating the mineralization process. Findings showed that types II and X collagen interact with cell surface-expressed annexin A5. These interactions lead to stimulate annexin A5-mediated Ca^{2+} influx resulting in an increased $[Ca^{2+}]i$, and ultimately increased alkaline phosphatase activity and mineralization of growth plate chondrocytes. Thus, the interactions between collagen and annexin A5 regulate mineralization of growth plate cartilage. Because annexin A5 is up-regulated during pathological mineralization events of articular cartilage, it is possible that these interactions also regulate pathological mineralization (Kim and Kirsch 2008).

Annexin A5 Associates with the IFN-γ Receptor and Regulates IFN-γ Signaling: Many of the biological activities of IFN- γ are mediated through the IFN-γR3-linked Jak-Stat1α pathway. However, regulation of IFN-γ signaling is not fully understood, and not all responses to IFN-γ are Stat1α dependent. Annexin A5 is a putative IFN-γR binding protein. Through an inducible association with the R2 subunit of the IFN-γR, annexin A5 modulates cellular responses to IFN- γ by modulating signaling through the Jak-Stat1 pathway (Leon et al. 2006). Annexin A5 is a specific high-affinity inhibitor of PKC-mediated phosphorylation of annexin 1 and myosin light chain kinase substrates. It appears that inhibition occurred by direct interaction between annexin A5 and PKC (Schlaepfer et al. 1992). Annexin A5 has been shown to interact with kinase insert domain receptor and integrin β5 (Marina et al. 2003; Wen et al. 1999).

21.5.3 Molecular Structure of Annexin A5

Each annexin is composed of two principal domains: the divergent NH2-terminal "head" and the conserved COOH-terminal protein core. The latter harbors the Ca^{2+} and membrane binding sites and is responsible for mediating the canonical membrane binding properties. An annexin core comprises four (in annexin A6 eight) segments of internal and interannexin homology that are easily identified in a linear sequence alignment (Gerke and Moss 2002). It forms a highly α-helical and tightly packed disk with a slight curvature and two principle sides. The more convex side contains novel types of Ca^{2+} binding sites, the so-called type II and type III sites (Weng et al. 1993), and faces the membrane when an annexin is associated peripherally with phospholipids. The more concave side points away from the membrane and thus appears accessible for interactions with the NH2-terminal domain and/or possibly cytoplasmic binding partners (Fig. 21.2).

The first structure for an annexin core was solved by Huber et al. (1990) for annexin A5. More than ten crystal structures for annexin cores have been described showing a remarkable conservation of the overall three-dimensional fold (Huber et al. 1990; Liemann and Lewit-Bentley 1995; Swairjo and Seaton 1994). Human annexin A5 (PP4) has been analysed by crystallography. Two crystal forms of human annexin A5 have been refined at 2.3 A and 2.0 A resolution to R-values of 0.184 and 0.174, respectively, applying very tight stereochemical restraints with deviations from ideal geometry of 0.01 A and 2°. The polypeptide chain of 320 amino acid residues is folded into a planar cyclic arrangement of four repeats. The repeats have similar structures of five α-helical segments wound into a right-handed compact superhelix. Three calcium ion sites in repeats I, II and IV and two lanthanum ion sites in repeat I have been found in the R3 crystals. They are located at the convex face of the molecule opposite the N terminus. Repeat III has a different conformation at this site and no calcium bound. The calcium sites are similar to the phospholipase A2 calcium-binding site, suggesting analogy also in phospholipid interaction. The center of the molecule is formed by a channel of polar charged residues, which also harbors a chain of ordered water molecules conserved in the different crystal forms. Comparison with amino acid sequences of other annexins shows a high degree of similarity between them. Long insertions are found only at the N termini. Most conserved are the residues forming the metal-binding sites and the polar channel. Annexins A5 and A7 form voltage-gated calcium ion channels when bound to membranes in vitro (Huber et al. 1992).

The molecule has dimensions of $64 \times 40 \times 30$ A3 and is folded into four domains of similar structure. Each domain consists of five α-helices wound into a right-handed superhelix yielding a globular structure of ~18 Å diameter. The domains have hydrophobic cores whose amino acid sequences are conserved between the domains and within the annexin family of proteins. The four domains are folded into an almost planar array by tight (hydrophobic) pair-wise packing of domains II and III and I and IV to generate modules (II–III) and (I–IV), respectively. The assembly is symmetric with three parallel approximate diads relating II to III, I to IV and the module (II–III) to (I–IV), respectively. The latter diad marks a channel through the centre of the molecule coated with charged amino acid residues. The protein has structural features of channel forming membrane proteins and a polar surface characteristic of soluble proteins. It is a member of the third class of amphipathic proteins different from soluble and membrane proteins (Huber et al. 1990; Voges et al. 1995).

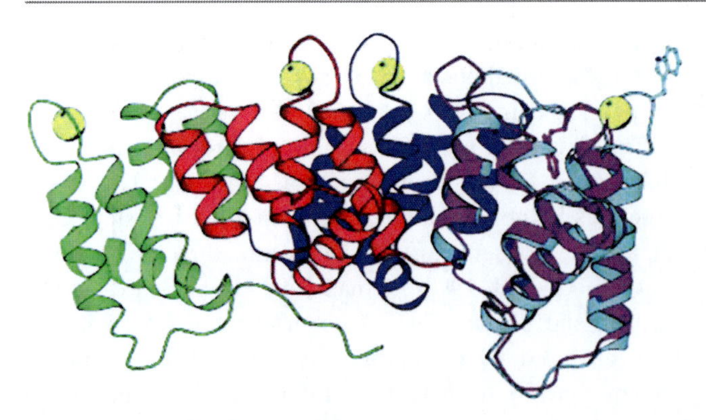

Fig. 21.2 Crystal structure of human annexin A5. The ribbon drawing illustrates the highly α-helical folding of the protein core that forms a slightly curved disk. Different colors were chosen to highlight the four annexin repeats that are given in *green* (repeat I), *blue* (repeat II), *red* (repeat III), and *violet/cyan* (repeat IV). The NH2-terminal domain appears unstructured and extends along the concave side of the molecule (*green*). The high and low Ca^{2+} forms are shown in a superposition revealing the conformational change in repeat III, which leads to an exposure of Trp-187 (*violet* for the low and *cyan* for the high Ca^{2+} form). Bound Ca^{2+} are depicted as *yellow spheres* (Printed with permission from Liemann and Lewit-Bentley 1995 © Elsevier)

Annexin 5 is known to form doublet bands exhibiting an apparent difference of 4 kDa on SDS-PAGE. Unlike annexin A4, the annexin 5 isoforms are not sensitive to reducing agents and occur as a result of only two amino acid substitutions between the isoforms (Learmonth et al. 1992; Bianchi et al. 1992). Emerging evidences suggest that both annexin 1 and annexin 5 can bind to many types of cell surfaces, and the identification of annexins as new receptors of tissue plasminogen activator in human endothelial cells (Hajjar et al. 1994) and of influenza virus in cultured cells (Otto et al. 1994) implies functional expression of annexins on the cell surface. Annexin 5, also identified as a collagen-type II binding protein (anchorin CII), was shown to be secreted and present in plasma, amniotic fluid, and post-culture medium (Kirsch and Pfaffle 1992).

21.5.4 Annexin A5-Mediated Pathogenic Mechanisms

Antiphospholipid syndrome: Annexin A5 forms a shield around negatively-charged phospholipid molecules. Without the shield, there is an increased quantity of phospholipid molecules on cell membranes, speeding up coagulation reactions and causing the blood-clotting characteristic of the antiphospholipid antibody syndrome. Antibodies directed against annexin A5 are the cause of a syndrome called the antiphospholipid syndrome. Annexin A5 binds to phospholipid bilayers, forming two-dimensional crystals that block the phospholipids from availability for coagulation enzyme reactions. Antiphospholipid (aPL) antibodies cause gaps in the ordered crystallization of Annexin A5 which expose phospholipids and thereby accelerate blood coagulation reactions. Recently, hydroxychloroquine, a synthetic antimalarial drug, could reverse this antibody-mediated process. In another translational application, Annexin A5 resistance may identify a subset of aPL syndrome patients for whom this is a mechanism for pregnancy losses and thrombosis. The elucidation of aPL-mediated mechanisms for thrombosis and pregnancy complications may open new paths towards addressing this disorder with targeted treatments and mechanistic assays (Rand et al. 2010).

Annexin A5 in prevention of atherothrombosis in SLE: It is becoming evident that atherosclerosis is an inflammatory disease, which is modulated autoimmunity in animal models. An interesting example of how autoimmune reactions can influence atherosclerosis and consequences thereafter is systemic lupus erythematosus (SLE)-associated cardiovascular disease (CVD). Antithrombotic effect exerted by Annexin A5 is thought to be mediated mainly by forming a mechanical shield over phospholipids (PLs) reducing availability of PLs for coagulation reactions. It may be hypothesized that Annexin A5 can be effective as a treatment to prevent plaque rupture and atherothrombosis not only in SLE, but also in the general population prone to CVD (Cederholm and Frostegård 2007).

21.5.5 A Novel Assay for Apoptosis

Annexin A5 is used as a probe in the annexin A5 affinity assay to detect cells that have expressed phosphatidylserine on the cell surface, a feature found in apoptosis as well as other forms of cell death (Koopman et al. 1994; Wen et al. 1999). Platelets also expose phosphatidylserine on their surface when activated, which serves as binding site for various coagulation factors. In early stages of apoptosis changes occur at the cell surface, which have remained difficult to recognize. One of these plasma membrane alterations is the translocation of phosphatidylserine (PS) from the inner side of the plasma membrane to the outer layer, by which PS becomes exposed at the external surface of the cell. Annexin A5 can be used as a sensitive probe for PS exposure upon the cell membrane. Translocation of PS to the external cell surface is not unique to apoptosis, but occurs also during cell necrosis. The difference between these two forms of cell death is that during the initial stages of apoptosis the cell membrane remains intact, while at the very moment that necrosis occurs the cell membrane looses its integrity and becomes leaky. Therefore the measurement of Annexin A5 binding to the cell surface as indicative for apoptosis has to be performed in conjunction with a dye exclusion test to establish integrity of the cell membrane. In comparison with other tests the Annexin A5 assay is sensitive and easy to perform. The Annexin A5 assay permits measurements of the kinetics of apoptotic death in relation to the cell cycle (Vermes et al. 1995).

21.5.6 Calcium-Induced Relocation of Annexins 4 and 5 in the Human Cells

Cultured human DCs, human osteoblasts and the osteosarcoma cell line MG-63 express annexins A1, A2, A4, A5 and A6. The bulk of these annexins is intracellular express. During endocytosis by DCs, there was a redistribution of annexin A5 which was found to colocalize with vesicles (Larsson et al. 1995). In non-confluent cells, annexin A4and annexin A5 are strongly present throughout the nucleus and are also present in the cytoplasm. On elevation of the intracellular calcium with the ionomycin, the intranuclear pools of annexin A4 and annexin A5 cells showed relocation to the nuclear membrane within seconds. Results support a role for annexins at cellular membranes in response to elevation of cytosolic calcium levels (Mohiti et al. 1995; Raynal et al. 1996). Studies are consistent with the role for annexins in mediating the calcium signal at the plasma membrane and within the nuclei of fibroblasts (Barwise and Walker 1996).

21.6 Annexin A6 (Annexin VI)

21.6.1 Structure

Annexin A6 participates in the formation of a reversible, membrane-cytoskeleton complex in smooth muscle cells through association of protein kinase C (Babiychuk et al. 1999; Schmitz-Peiffer et al. 1998). In contrast to other annexins which have a structural motif of four repeats in the central core region, annexin A6 has eight repeats (Raynal and Pollard 1994). Analysis of the crystal structure of annexin A6 also indicated that it is uniquely organized into two lobes, the N-terminal half (from repeat one to four) and the C-terminal half (from repeat five to eight) of the molecule (Kawasaki et al. 1996; Benz et al. 1996), and each lobe has convex and concave sides and a hydrophilic pore surrounded by the four repeats that might be involved in GAG interactions. Bovine brain annexin A6 is bound to chondroitin sulfate in a Ca^{2+}-dependent manner (Ishitsuka et al. 1998). Furthermore, exon 21 was alternatively spliced, giving rise to two annexin 6 isoforms that differ with respect to a six amino-acid insertion at the starting site of repeat seven (Smith et al. 1994). Results suggest that both lobes of annexin A6, the N- and C-terminal halves of the molecule, are necessary for specific attachment of the annexin-6-expressing cells to CS chains. Calcium-free recombinant human annexin A6 consists of two similar halves closely resembling annexin A1 connected by an alpha-helical segment and arranged perpendicular to each other. The calcium and membrane binding sites assigned by structural homology are therefore not located in the same plane. Analysis of the membrane-bound form of annexin A6 by electron microscopy shows the two halves of the molecule coplanar with the membrane, but oriented differently to the crystal structure, suggesting a flexible arrangement. Ion channel activity has been found for annexin 6 and the half molecules by electrophysiological experiments (Benz et al. 1996).

21.6.2 Functions

Though annexin A6 is a cytoplasmic protein, it functions as a receptor for CS chains. Some reports have described the extracellular expression of annexins on the outer plasma membrane (Kirsch and Pfaffle 1992; Yeatman et al. 1993; Chung and Erickson 1994; Tressler et al. 1994). Since, a significant amount of annexin A6 is exposed on the external cell surface membrane, it is likely that annexin A6 functions as a receptor for CS chains and is involved in the anti-adhesive activity of CS proteoglycans.

During cell-substratum adhesion, cells undergo attachment, spreading and form stress fibers and focal adhesion; these are sequential steps requiring different molecular mechanisms (Sage and Bornstein 1991; Murphy-Ullrich 1995; Murphy-Ullrich 2001). Chondroitin sulfate (CS) proteoglycan PG-M/versican has an inhibitory effect on cell-substratum adhesion. CS chains are needed for this activity, and the immobilization of CS chains on substrata is essential (Yamagata et al. 1989). Takagi et al. (2002) isolated a 68 kDa protein as a candidate receptor for CS chains and identified that protein as annexin 6. Moreover, taking advantage of A431 cells that do not express annexin A6 and transfecting them with exogenous annexin A6, it was demonstrated that annexin 6 is directly involved in the attachment of cells to CS chains and is expressed on cell surfaces. Taken together, it was concluded that annexin A6 is a receptor for CS chains or that it, at least, binds to CS on the cell surface in the presence of Ca^{2+}.

Annexins A2 and A6, which contain KFERQ-like sequences, are degraded more rapidly in response to serum withdrawal, while annexins A5 and A11, without such sequences are degraded at the same rate in the presence and absence of serum. Using isolated lysosomes, only the annexins containing KFERQ-like sequences are degraded by chaperone mediated-autophagy. These results provide evidence for the importance of KFERQ motifs in substrates of chaperone-mediated autophagy (Cuervo et al. 2000).

References

Babiychuk EB, Palstra RJ, Schaller J et al (1999) Annexin VI participates in the formation of a reversible, membrane-cytoskeleton complex in smooth muscle cells. J Biol Chem 274:35191–35195

Barwise JL, Walker JH (1996) Annexins II, IV, V and VI relocate in response to rises in intracellular calcium in human foreskin fibroblasts. J Cell Sci 109:247–255

Becker T, Weber K, Johnsson N (1990) Protein-protein recognition via short amphiphilic helices; a mutational analysis of the binding site of annexin II for p11. EMBO J 9:4207–4213

Benz J, Bergner A, Hofmann A et al (1996) The structure of recombinant human annexin VI in crystals and membrane-bound. J Mol Biol 260:638–643

Bianchi R, Giambanco I, Ceccarelli P et al (1992) Membrane-bound annexin V isoforms (CaBP33 and CaBP37) and annexin VI in bovine tissues behave like integral membrane proteins. FEBS Lett 296:158–162

Blackwell GJ, Carnuccio R, DiRosa M et al (1980) Macrocortin: a polypeptide causing the anti-phospholipase effect of glucocorticoids. Nature 287:147–149

Blanchard S, Barwise JL et al (1996) Annexins in the human neuroblastoma SH-SY5Y: demonstration of relocation of annexins II and V to membranes in response to elevation of intracellular calcium by membrane depolarisation and by the calcium ionophore A23187. J Neurochem 67:805–813

Brachvogel B, Dikschas J, Moch H et al (2003) Annexin A5 is not essential for skeletal development. Mol Cell Biol 23:2907–2913

Brisson A, Mosser G, Huber R (1991) Structure of soluble and membrane-bound human annexin V. J Mol Biol 220:199–203

Brownstein C, Falcone DJ, Jacovina A, Hajjar KA (2001) A mediator of cell surface-specific plasmin generation. Ann NY Acad Sci 947:143–155

Butsushita K, Fukuoka S, Ida K, Arii Y (2009) Crystal structures of sodium-bound annexin A4. Biosci Biotechnol Biochem 73:2274–2280

Cardin AD, Weintraub HJ (1989) Molecular modeling of protein-glycosaminoglycan interactions. Arteriosclerosis 9:21–32

Cederholm A, Frostegård J (2007) Annexin A5 as a novel player in prevention of atherothrombosis in SLE and in the general population. Ann N Y Acad Sci 1108:96–103

Cesarman GM, Guevara CA, Hajjar KA (1994) An endothelial cell receptor for plasminogen/tissue plasminogen activator (t-PA). II. Annexin II-mediated enhancement of t-PA-dependent plasminogen activation. J Biol Chem 269:21198–21203

Chan HC, Kaetzel MA, Gotter AL et al (1994) Annexin IVinhibits calmodulin-dependent protein kinase II-activated chloride conductance. A novel mechanism for ion channel regulation. J Biol Chem 269:32464–32468

Chattopadhyay S, Sun P, Wang P et al (2003) Fusion of lamellar body with plasma membrane is driven by the dual action of annexin II tetramer and arachidonic acid. J Biol Chem 278:39675–39683

Chen JM, Sheldon A, Pincus MR (1993) Structure-function correlations of calcium binding and calcium channel activities based on 3-dimensional models of human annexins I, II, III, V and VII. J Biomol Struct Dyn 10:1067–1089

Chung CY, Erickson HP (1994) Cell surface annexin II is a high affinity receptor for the alternatively spliced segment of tenascin-C. J Cell Biol 126:539–548

Chung CY, Murphy-Ullrich JE, Erickson HP (1996) Mitogenesis, cell migration, and loss of focal adhesions induced by tenascin-C interacting with its cell surface receptor, annexin II. Mol Biol Cell 7:883–892

Concha NO, Head JF, Kaetzel MA et al (1992) Annexin V forms calcium-dependent trimeric units on phospholipid vesicles. FEBS Lett 314:159–162

Creutz CE (1992) The annexins and exocytosis. Science 258:924–931

Crompton MR, Moss SE, Crumpton MJ (1988) Diversity in the lipocortin/calpactin family. Cell 55:1–3

Cuervo AM, Gomes AV, Barnes JA, Dice JF (2000) Selective degradation of annexins by chaperone-mediated autophagy. J Biol Chem 275:33329–33335

Demange P, Voges D, Benz J et al (1994) Annexin V: the key to understanding ion selectivity and voltage regulation? Trends Biochem Sci 19:272–276

Duncan R, Carpenter B, Main LC et al (2008) Characterisation and protein expression profiling of annexins in colorectal cancer. Br J Cancer 98:426–433

Eberhard DA, Karns LR, VandenBerg SR, Creutz CE (2001) Control of the nuclear-cytoplasmic partitioning of annexin II by a nuclear export signal and by p11 binding. J Cell Sci 114:3155–3166

Favier-Perron B, Lewit-Bentley A, Russo-Marie F (1996) The high-resolution crystal structure of human annexin III shows subtle differences with annexin V. Biochemistry 35:1740–1744

Fernández MP, Morgan RO, Fernández MR, Carcedo MT (1994) The gene encoding human annexin V has a TATA-less promoter with a high G + C content. Gene 149:253–260

Gerke V, Creutz CE, Moss SE (2005) Annexins: linking Ca2+ signalling to membrane dynamics. Nat Rev Mol Cell Biol 6:449–461

Gerke V, Moss SE (2002) Annexins: from structure to function. Physiol Rev 82:331–371

Gerner C, Frohwein U, Gotzmann J et al (2000) The Fas-induced apoptosis analyzed by high throughput proteome analysis. J Biol Chem 275:39018–39026

Gotoh M, Takamoto Y, Kurosaka K et al (2005) Annexins I and IV inhibit *Staphylococcus aureus* attachment to human macrophages. Immunol Lett 98:297–302

Hajjar KA, Jacovina AT, Chacko J (1994) An endothelial cell receptor for plasminogen/tissue plasminogen activator. I. Identity with annexin II. J Biol Chem 269:21191–21197

Hamre KM, Chepenik KP, Goldowitz D (1995) The annexins: specific markers of midline structures and sensory neurons in the developing murine central nervous system. J Comp Neurol 352:421–435

Hayes MJ, Moss SE (2004) Annexins and disease. Biochem Biophys Res Commun 322:1166–1170

Herr C, Smyth N, Ullrich S et al (2001) Loss of annexin A7 leads to alterations in frequency-induced shortening of isolated murine cardiomyocytes. Mol Cell Biol 21:4119–4128

Hill WG, Kaetzel MA, Kishore BK et al (2003) Annexin A4 reduces water and proton permeability of model membranes but does not alter aquaporin 2-mediated water transport in isolated endosomes. J Gen Physiol 121:413–425

Hill WG, Meyers S, von Bodungen M et al (2008) Studies on localization and function of annexin A4a within urinary bladder epithelium using a mouse knockout model. Am J Physiol Renal Physiol 294: F919–F927

Hubaishy I, Jones PG, Bjorge J et al (1995) Modulation of annexin II tetramer by tyrosine phosphorylation. Biochemistry 34:14527–14534

Huber R, Berendes R, Burger A et al (1992) Crystal and molecular structure of human annexin V after refinement. Implications for structure, membrane binding and ion channel formation of the annexin family of proteins. J Mol Biol 223:683–704

Huber R, Römisch J, Paques EP (1990) The crystal and molecular structure of human annexin V, an anticoagulant protein that binds to calcium and membranes. EMBO J 9:3867–3874

Ishitsuka R, Kojima K, Utsumi H et al (1998) Glycosaminoglycan binding properties of annexin IV, V, and VI. J Biol Chem 273:9935–9941

Jans SW, van Bilsen M, Reutelingsperger CP et al (1995) Annexin V in the adult rat heart: isolation, localization and quantitation. J Mol Cell Cardiol 27:335–348

Kaetzel MA, Mo YD, Mealy TR et al (2001) Phosphorylation mutants elucidate the mechanism of annexin IV-mediated membrane aggregation. Biochemistry 40:4192–4199

Kao LC, Tulac S, Lobo S et al (2002) Global gene profiling in human endometrium during the window of implantation. Endocrinology 143:2119–2138

Kassam G, Manro A, Braat CE et al (1997) Characterization of the heparin binding properties of annexin II tetramer. J Biol Chem 272:16093–16100

Katoh N (2000) Detection of annexins I and IV in bronchoalveolar lavage fluids from calves inoculated with bovine herpes virus-1. J Vet Med Sci 62:37–41

Kawasaki H, Avila-Sakar A, Creutz CE et al (1996) The crystal structure of annexin VI indicates relative rotation of the two lobes upon membrane binding. Biochim Biophys Acta 1313:277–282

Kim A, Enomoto T, Serada S et al (2009) Enhanced expression of annexin A4 in clear cell carcinoma of the ovary and its association with chemoresistance to carboplatin. Int J Cancer 125:2316–2322

Kim HJ, Kirsch T (2008) Collagen/Annexin V interactions regulate chondrocyte mineralization. J Biol Chem 283:10310–10317

Kirsch T, Pfaffle M (1992) Selective binding of annexin V(annexin V) to type II and X collagen and to chondrocalcin (C-propeptide of type II collagen). Implications for anchoring function between matrix vesicles and matrix proteins. FEBS Lett 310:143–147

Kojima K, Utsumi H, Ogawa H, Matsumoto I (1994) Highly polarized expression of carbohydrate-binding protein p33/41 (annexin IV) on the apical plasma membrane of epithelial cells in renal proximal tubules. FEBS Lett 342:313–318

Kojima K, Yamamoto K, Irimura T et al (1996) Characterization of carbohydrate-binding protein p33/41: relation with annexin IV, molecular basis of the doublet forms (p33 and p41), and modulation of the carbohydrate binding activity by phospholipids. Biol Chem 271:7679–7685

Konishi H, Namikawa K, Kiyama H (2006) Annexin III implicated in the microglial response to motor nerve injury. Glia 53:723–732

Koopman G, Reutelingsperger CP, Kuijten GAM et al (1994) Annexin V for flow cytometric detection of phosphatidylserine expression on B cells undergoing apoptosis. Blood 84:1415–1420

Larsson M, Majeed M, Stendahl O et al (1995) Mobilization of annexin V during the uptake of DNP-albumin by human dendritic cells. APMIS 103:855–861

Law A-L, Ling Q, Hajjar KA et al (2009) Annexin A2 regulates phagocytosis of photoreceptor outer segments in the mouse retina. Mol Biol Cell 20:3896–3904

Learmonth MP, Howell SA, Harris AC et al (1992) Novel isoforms of CaBP 33/37 (annexin V) from mammalian brain: structural and phosphorylation differences that suggest distinct biological roles. Biochim Biophys Acta 1160L:76–83

Leon C, Nandan D, Lopez M, Moeenrezakhanlou A, Reine NE (2006) Annexin V associates with the IFN-γ receptor and regulates IFN-γ Signaling. J Immunol 176:5934–5942

Li B, Dedman JR, Kaetzel MA (2003) Intron disruption of the annexin IV gene reveals novel transcripts. J Biol Chem 278:43276–43283

Liemann S, Huber R (1997) Three-dimensional structure of annexins. Cell Mol Life Sci 53:516–521

Liemann S, Lewit-Bentley A (1995) Annexins: a novel family of calcium- and membrane-binding proteins in search of a function. Structure 3:233–237

Lim LH, Pervaiz S (2007) Annexin 1: the new face of an old molecule. FASEB J 21:968–975

Ling Q, Jacovina AT, Deora A et al (2004) Annexin II regulates fibrin homeostasis and neoangiogenesis in vivo. J Clin Invest 113:38–48

Liu L, Wang M, Fisher AB et al (1996) Involvement of annexin II in exocytosis of lamellar bodies from alveolar epithelial type II cells. Am J Physiol 270:L668–L676

Locate S, Colyer J, Gawler DJ, Walker JH (2008) Annexin A6 at the cardiac myocyte sarcolemma - evidence for self-association and binding to actin. Cell Biol Int 32:1388–1396

Luecke H, Chang BT, Mailliard WS et al (1995) Crystal structure of the annexin XII hexamer and implications for bilayer insertion. Nature 378:512–515

Marina C-V, Wadih A, Renata P (2003) αvβ5 integrin-dependent programmed cell death triggered by a peptide mimic of annexin V. Mol Cell 11:1151–1162

Masuda J, Takayama E, Satoh A et al (2004) Levels of annexin IV and V in the plasma of pregnant and postpartum women. Thromb Haemost 91:1129–1136

Matteo RG, Moravec CS (2000) Immunolocalization of annexins IV, V and VI in the failing and non-failing human heart. Cardiovasc Res 45:961–970

Mirkin S, Arslan M, Churikov D et al (2005) In search of candidate genes critically expressed in the human endometrium during the window of implantation. Hum Reprod 20:2104–2117

Mohiti J, Caswell AM, Walker JH (1995) Calcium-induced relocation of annexins IV and V in the human osteosarcoma cell line MG-63. Mol Membr Biol 12:321–329

Morgan RO, Fernandez MP (1997) Distinct annexin subfamilies in plants and protists diverged prior to animal annexins and from a common ancestor. J Mol Evol 44:178–188

Morgan RO, Jenkins NA, Gilbert DJ et al (1999) Novel human and mouse annexin A10 are linked to the genome duplications during early chordate evolution. Genomics 60:40–46

Moss SE, Morgan RO (2004) The annexins. Genome Biol 5:219

Murphy-Ullrich JE (2001) The de-adhesive activity of matricellular proteins: is intermediate cell adhesion an adaptive state? J Clin Invest 107:785–790

Naciff JM, Kaetzel MA, Behbehani MM, Dedman JR (1996) Differential expression of annexins I-VI in the rat dorsal root ganglia and spinal cord. J Comp Neurol 368:356–370

Otto M, Gunther A, Fan H et al (1994) Identification of annexin 33 kDa in cultured cells as a binding protein of influenza viruses. FEBS Lett 356:125–129

Ponnampalam AP, Rogers PA (2006) Cyclic changes and hormonal regulation of annexin IV mRNA and protein in human endometrium. Mol Hum Reprod 12:661–669

Ponnampalam AP, Weston GC, Trajstman AC et al (2004) Molecular classification of human endometrial cycle stages by transcriptional profiling. Mol Hum Reprod 10:879–893

Pula G, Bianchi R, Ceccarelli P et al (1990) Characterization of mammalian heart annexins with special reference to CaBP33 (annexin V). FEBS Lett 277:53–58

Rahman MM, Iida H, Shibata Y (1997) Expression and localization of annexin V and annexin VI during limb bud formation in the rat fetus. Anat Embryol (Berl) 195:31–39

Rand JH, Wu X-X, Quinn AS, Taatjes DJ (2010) The annexin A5-mediated pathogenic mechanism in the antiphospholipid syndrome: role in pregnancy losses and thrombosis. Lupus 19:460–469

Rand JH (2000) The annexinopathies: a new category of diseases. Biochim Biophys Acta 1498:169–173

Raynal P, Kuijpers G, Rojas E, Pollard HB (1996) A rise in nuclear calcium translocates annexins IV and V to the nuclear envelope. FEBS Lett 392:263–268

Raynal P, Pollard HB, Srivastava M (1997) Cell cycle and post-transcriptional regulation of annexin expression in IMR-90 human fibroblasts. Biochem J 322:365–371

Raynal P, Pollard HB (1994) Annexins: the problem of assessing the biological role for a gene family of multifunctional calcium-

and phospholipid-binding proteins. Biochim Biophys Acta 1197: 63–93
Rescher U, Gerke V (2004) Annexins – unique membrane binding proteins with diverse functions. J Cell Sci 117:2631–2639
Riesewijk A, Martin J, van Os R et al (2003) Gene expression profiling of human endometrial receptivity on days LH + 2 versus LH + 7 by microarray technology. Mol Hum Reprod 9:253–264
Rojas E, Pollard HB, Haigler HT et al (1990) Calcium-activated endonexin II forms calcium channels across acidic phospholipid bilayer membranes. J Biol Chem 265:21207–21215
Romay-Penabad Z, Montiel-Manzano MG, Shilagard T et al (2009) Annexin A2 is involved in antiphospholipid antibody-mediated pathogenic effects in vitro and in vivo. Blood 114:3074–3083
Ross TS, Tait JF, Majerus PW (1990) Identity of inositol 1,2-cyclic phosphate 2-phosphohydrolase with lipocortin III. Science 248 (4955):605–607
Sage EH, Bornstein P (1991) Extracelluar proteins that modulate cell-matrix interactions: SPARC, tenascin, thrombospondin 1. J Biol Chem 266:14831–14834
Sandra L, Fitzpatrick SL, Kassam G et al (2000) Fucoidan-dependent conformational changes in annexin II tetramer. Biochemistry 39:2140–2148
Schlaepfer DD, Haigler HT (1990) Expression of annexins as a function of cellular growth state. J Cell Biol 111:229–238
Schlaepfer DD, Jones J, Haigler HT (1992) Inhibition of protein kinase C by annexin V. Biochemistry 31:1886–1891
Schmitz-Peiffer C, Browne CL, Walker JH et al (1998) Activated protein kinase C α associates with annexin VI from skeletal muscle. Biochem J 330:675–681
Seville RA, Nijjar S, Barnett MW et al (2002) Annexin IV (Xanx-4) has a functional role in the formation of pronephric tubules. Development 129:1693–1704
Shao C, Zhang F, Kemp M et al (2006) Crystallographic analysis of calcium-dependent heparin binding to annexin A2. J Biol Chem 281:31689–31695
Smith PD, Davies A, Crumpton MJ, Moss SE (1994) Structure of the human annexin VI gene. Proc Natl Acad Sci USA 91:2713–2717
Solito E, Nuti S, Parente L (1994) Dexamethasone-induced translocation of lipocortin (annexin) 1 to the cell membrane of U-937 cells. Br J Pharmacol 112:347–348
Srivastava M, Atwater I, Glasman M et al (1999) Defects in inositol 1,4,5-trisphosphate receptor expression, Ca(2+) signaling, and insulin secretion in the anx7(+/-) knockout mouse. Proc Natl Acad Sci USA 96:13783–13788
Swairjo MA, Concha NO, Kaetzel MA et al (1995) Ca2+-bridging mechanism and phospholipid head group recognition in the membrane-binding protein annexin V. Nat Struct Biol 2:968–974
Swairjo MA, Seaton BA (1994) Annexin structure and membrane interactions: a molecular perspective. Annu Rev Biophys Biomol Struct 23:193–213
Swisher JF, Burton N, Bacot SM et al (2010) Annexin A2 tetramer activates human and murine macrophages through TLR4. Blood 115:549–558
Tait JF, Gibson D, Fujikawa K (1989) Phospholipid binding properties of human placental anticoagulant protein-I, a member of the lipocortin family. J Biol Chem 264:7944–7946
Tait JF, Frankenberry DA, Shiang R et al (1991) Chromosomal localization of the human gene for annexin V (placental anticoagulant protein I) to 4q26–q28. Cytogenet Cell Genet 57:187–192
Takagi H, Asano Y, Yamakawa N et al (2002) Annexin 6 is a putative cell surface receptor for chondroitin sulfate chains. J Cell Sci 115:3309–3318
Tomas A, Moss SE (2003) Calcium- and cell cycle-dependent association of annexin 11 with the nuclear envelope. J Biol Chem 278:20210–20216
Tressler RJ, Nicolson GL (1992) Butanol-extractable and detergent-solubilized cell surface components from murine large cell lymphoma cells associated with adhesion to organ microvessel endothelial cells. J Cell Biochem 48:162–171
Tressler RJ, Updyke TV, Yeatman T et al (1993) Extracellular annexin II is associated with divalent cation-dependent tumor cell-endothelial cell adhesion of metastatic RAW117 large-cell lymphoma cells. J Cell Biochem 53:265–276
Tressler RJ, Yeatman T, Nicolson GL (1994) Extracellular annexin VI expression is associated with divalent cation-dependent endothelial cell adhesion of metastatic RAW117 large-cell lymphoma cells. Exp Cell Res 215:395–400
Tsujii-Hayashi Y, Kitahara M, Yamagaki T et al (2002) A potential endogenous ligand of annexin IVin the exocrine pancreas. Carbohydrate structure of GP-2, a glycosylphospha-tidylinositol-anchored glycoprotein of zymogen granule membranes. J Biol Chem 277:47493–47499
Ulander VM, Stefanovic V, Masuda J et al (2007) Plasma levels of annexins IV and V in relation to antiphospholipid antibody status in women with a history of recurrent miscarriage. Thromb Res 120:865–870
Vermes I, Haanen C, Steffens-Nakken H et al (1995) A novel assay for apoptosis. Flow cytometric detection of phosphatidylserine expression on early apoptotic cells using fluorescein labelled annexin V. J Immunol Methods 184:39–46
Voges D, Berendes R, Demange P et al (1995) Structure and function of the ion channel model system annexin V. Adv Enzymol Relat Areas Mol Biol 71:209–239
Waisman DM (1995) Annexin II tetramer: structure and function. Mol Cell Biochem 149/150:301–322
Wallace JL, de Lima OM, Jr FS (2005) Lipoxins in gastric mucosal health and disease. Prostaglandins Leukot Essent Fatty Acids 73:251–255
Wang EC, Lee JM, Ruiz WG et al (2005) ATP and purinergic receptor-dependent membrane traffic in bladder umbrella cells. J Clin Invest 115:2412–2422
Wang JL, Gray RM, Haudek KC et al (2004) Nucleocytoplasmic lectins. Biochim Biophys Acta 1673:75–93
Wen Y, Edelman JL, Kang T, Sachs G (1999) Lipocortin V may function as a signaling protein for vascular endothelial growth factor receptor-2/Flk-1. Biochem Biophys Res Commun 258:713–721
Weng X, Luecke H, Song IS et al (1993) Crystal structure of human annexin I at 2.5 A resolution. Protein Sci 2:448–458
Yamagata M, Suzuki S, Akiyama SK et al (1989) Regulation of cell-substrate adhesion by proteoglycans immobilized on extracellular substrates. J Biol Chem 264:8012–8018
Yeatman TJ, Updyke TV, Kaetzel MA et al (1993) Expression of annexins on the surfaces of non-metastatic and metastatic human and rodent tumor cells. Clin Exp Metastasis 11:37–44
Zanotti G, Malpeli G, Gliubich F et al (1998) Structure of the trigonal crystal form of bovine annexin IV. Biochem J 329:101–106
Zimmermann U, Balabanov S, Giebel J et al (2004) Increased expression and altered location of annexin IV in renal clear cell carcinoma: a possible role in tumor dissemination. Cancer Lett 209:111–118
Zschörnig O, Opitz F, Müller M (2007) Annexin A4 binding to anionic phospholipid vesicles modulated by pH and calcium. Eur Biophys J 36:415–424

Printed by Publishers' Graphics LLC